CELLULAR RADIO AND PERSONAL COMMUNICATIONS
Volume 2:

Advanced Selected Readings

Edited by

Dr. Theodore S. Rappaport
Mobile and Portable Radio Research Group
Bradley Department of Electrical Engineering
Virginia Polytechnic Institute and State University
Blacksburg, Virginia

Printed in the United States of America.

Editor - Theodore S. Rappaport and Barbara Coburn
Typographer - Jill R. Cals

Published by the Institute of Electrical and Electronics Engineers, Inc.
445 Hoes Lane, PO Box 1331, Piscataway, NJ 08855-1331.

Library of Congress Cataloging-in-Publication Data available upon request

http://www.ieee.org/eab/

CONTENTS

CHAPTER 3 MODERN PROPAGATION PREDICTION TECHNIQUES

CHAPTER 4 SPREAD SPECTRUM MULTIPLE ACCESS

CHAPTER 5 MODERN CODING & MODULATION

CHAPTER 6 WIRELESS LINK & SYSTEM SIMULATION

CHAPTER 7 DIGITAL RECEIVERS: IMPLEMENTATION AND DESIGN

CHAPTER 8 ADAPTIVE ARRAYS AND SPATIAL FILTERING

INTRODUCTION

The field of Cellular Radio and Personal Communications is expanding at a rate which is difficult to fathom. While it has taken the landline telephone system nearly 120 years to serve 700 million inhabitants around the globe, over 50 million wireless subscribers have been added in the first 15 years of the wireless revolution. And while the annual wireline subscription growth rate of 3% mirrors global population growth, the number of wireless subscribers doubles every 18 months, and continues to accelerate with time. By 2010, it is likely that over half of the information throughout the world will be transmitted using some form of wireless communications.

International radio bands near 1800 MHZ are now being allocated for world wide personal communications systems (PCS). This promises a major growth in competition and expanded services, as compared to the first generation cellular telephone systems. The demand for trained technical experts in the manufacturing, service, and research sectors of the wireless communications industry is at an all time high. Modern communications techniques that involve digital signal processing, spread spectrum multiple access, advanced coding and modulation methods, and sophisticated simulation and design approaches are just beginning to form the technical foundation for the wireless communications systems of the future.

The papers in CELLULAR RADIO AND PERSONAL COMMUNICATIONS: ADVANCED SELECTED READINGS have been compiled to expand the knowledge of readers who are already familiar with the basic concepts of wireless communications. The text has been organized in eight chapters, each designed to provide information on recent research or design triumphs that will govern the direction of wireless communications over the next two decades.

This collection is based upon the educational and research needs of graduate students, faculty, and sponsor companies of Virginia Tech's Mobile and Portable Radio Research Group (MPRG). Industry interest and technology trends have been identified by the more than 30 corporate and government sponsors of MPRG, and the most often cited wireless areas of interest are represented in this text. Many of these papers are frequently cited in the literature and offer unusually good insight into practical design or research issues for wireless signals and systems. Many of these papers also provide a fundamental basis for emerging products and services.

All papers are being reprinted by permission of the publishers, and their cooperation is gratefully acknowledged. It is also a pleasure to acknowledge the help of Brian Woerner and Jeff Reed, colleagues who assisted with the selection of papers in this text.

PROF. TED S. RAPPAPORT
MOBILE & PORTABLE RADIO RESEARCH GROUP
BRADLEY DEPARTMENT OF ELECTRICAL ENGINEERING
VIRGINIA TECH
BLACKSBURG, VA 24061-0111

Chapter 1

TUTORIALS–WHAT IS PCS?

This chapter provides tutorials on what PCS really means, and how it is being deployed throughout the world.

Reprinted from *IEEE Communications Magazine*, Vol. 33, No. 1, Jan 1995, pp. 28-41.

Overview of Wireless Personal Communications

Cellular radio and cordless telephony have demonstrated the demand for wireless communications and provided a foundation for the development of future wireless telecommunications systems and services.

Jay E. Padgett, Christoph G. Günther, and Takeshi Hattori

This century has seen the development of a public wireline network that allows reliable and affordable communication of voice and low-rate data around the globe. There also is a multiplicity of specialized wired networks optimized for purposes such as the local communication of high-speed data. The goal of wireless communication is to allow the user access to the capabilities of the global network at any time without regard to location or mobility. Cellular and cordless telephony, both of which have gained widespread user acceptance during the last ten years, have begun this process but do not yet allow total wireless communication. Cellular systems currently are limited to voice and low-speed data within areas covered by base stations. The basic cordless telephone provides a wireless counterpart to the standard telephone. The handset typically operates within 50 to 100 m of the user's base station, which connects to the public switched telephone network (PSTN). With the advent of digital cordless telephony, cordless "systems" with enhanced functionality (CT2, DECT, PHS) have been developed that can support higher data rates and more sophisticated applications such as wireless private branch exchanges (PBXs) and public-access Telepoint systems.

This article presents an overview of the current state of wireless communications, including relevant ongoing activities in technology development, standards, and spectrum allocation. The next three sections discuss cellular radio, cordless telephony, and wireless data systems, respectively. Following that, ongoing and planned future developments are summarized. The presentation here is oriented toward broad coverage rather than technical depth. However, brief discussions of the air interfaces for existing digital cellular and digital cordless systems are provided, because the air interface bears heavily on system capacity and the environments in which the system can be used, as well as on the cost and complexity of the equipment. More detailed discussions on topics of interest to the reader, including network aspects, can be found in other papers in this issue as well as in the references provided here.

JAY E. PADGETT is a DMTS at AT&T Bell Laboratories and is chair of the TIA Mobile & Personal Communications Consumer Radio Section.

CHRISTOPH G. GÜNTHER is head of the Cellular System Group at Ascom Tech.

TAKESHI HATTORI is executive manager of the Personal Communication Systems Laboratory of the NTT Wireless Systems Laboratories.

A guide to wireless acronyms and abbreviations is on page 40.

Cellular

Cellular radio can be regarded as the earliest form of wireless "personal communications." It allows the subscriber to place and receive telephone calls over the wireline telephone network wherever cellular coverage is provided; "roaming" capabilities extend service to users traveling outside their "home" service areas. Cellular system design was pioneered during the '70s by Bell Laboratories in the United States, and the initial realization was known as AMPS, for Advanced Mobile Phone Service. As detailed below, systems similar to AMPS were soon deployed internationally. All of these "first-generation" cellular systems use analog frequency modulation (FM) for speech transmission and frequency shift keying (FSK) for signaling. Individual calls use different frequencies. This way of sharing spectrum is called frequency division multiple Access (FDMA).

The distinguishing feature of cellular systems compared to previous mobile radio systems is the use of many base stations with relatively small coverage radii (on the order of 10 km or less vs. 50 to 100 km for earlier mobile systems). Each frequency is used simultaneously by multiple base-mobile pairs. This "frequency reuse" allows a much higher subscriber density per MHz of spectrum than previous systems. System capacity can be further increased by reducing the cell size (the coverage area of a single base station), down to radii as small as 0.5 km. In addition to supporting much higher subscriber densities than previous systems, this approach makes possible the use of small, battery-powered portable handsets with lower RF (radio frequency) transmit power than the large vehicular mobile units used in earlier systems. In cellular systems, continuous coverage is achieved by executing a "handoff" (the seamless transfer of the call from one base station to another) as the mobile unit crosses cell "boundaries." This requires the mobile to change frequencies under control of the cellular network. For a more detailed overview of cellular radio, see [1].

Standard	Mobile TX/Base TX (MHz)	Channel spacing (kHz)	Number of channels	Region
AMPS	824-849/869-894	30	832	The Americas[1]
TACS	890-915/935-960[2]	25	1000	Europe
ETACS	872-905/917-950	25	1240	United Kingdom
NMT 450	453-457.5/463-467.5	25	180	Europe
NMT 900	890-915/935-960[2]	12.5[3]	1999	Europe[1]
C-450	450-455.74/460-465.74	10[3]	573	Germany, Portugal
RTMS	450-455/460-465	25	200	Italy
Radiocom 2000[4]	192.5-199.5/200.5-207.5 215.5-233.5/207.5-215.5 165.2-168.4/169.8-173 414.8-418/424.8-428	12.5	560 640 256 256	France
NTT	925-940/870-885[5] 915-918.5/860-863.5[6] 922-925/867-870[6]	25/6.25[3] 6.25[3] 6.25[3]	600/2400 560 480	Japan
JTACS/NTACS	915-925/860-870[7] 898-901/843-846[6, 7] 918.5-922/863.5-867[6]	25/12.5[3] 25/12.5[3] 12.5[3]	400/800 120/240 280	Japan

Notes:

[1] AMPS is also used in Australia. AMPS, TACS, and NMT are all used in parts of Africa and Southeast Asia.

[2] The bands 890-915/935-960 MHz were subsequently allocated to GSM in Europe.

[3] Frequency interleaving using overlapping or "interstitial" channels: the channel spacing is half the nominal channel bandwidth.

[4] The latter two band pairs are used to provide coverage throughout the country, and the former ones are used to support high traffic densities in areas like Paris and other major cities.

[5] NTT DoCoMo, nationwide.

[6] IDO, in the Kanto-Tokaido areas.

[7] DDI, outside the Kanto-Tokaido areas.

■ **Table 1.** *Summary of analog cellular systems.*

There are more than 30 million cellular subscribers worldwide.

Analog Cellular Systems

In the United States, a total of 50 MHz in the bands 824-849 MHz and 869-894 MHz is allocated to cellular mobile radio. In a given geographical licensing region, each of two carriers (service providers) controls 25 MHz. The "A" and "B" bands are allocated to "non-wireline" and "wireline" carriers, respectively.[1] Under the AMPS standard, this spectrum is divided into 832 frequency channels, each 30 kHz wide. Frequency modulation (8 kHz deviation) is used for speech, and the signaling channels use binary FSK with a 10-kb/s rate. To meet cochannel interference objectives,[2] the typical frequency reuse plan employs either a 12-group frequency cluster with omnidirectional antennas or a 7-group cluster with three sectors per cell. AMPS cellular service has been available to the public since 1983, and there currently are roughly 20 million subscribers in the United States. AMPS is also used in Canada, Central and South America, and Australia.

In Europe, several first-generation systems similar to AMPS have been deployed, including: Total Access Communications System (TACS) in the United Kingdom, Italy, Spain, Austria and Ireland; Nordic Mobile Telephone (NMT) in many countries; C-450 in Germany and Portugal; Radiocom 2000 in France; and Radio Telephone Mobile System (RTMS) in Italy. Those systems use frequency modulation for speech, FSK for signaling, and channel spacings of: 25 kHz (TACS, NMT-450, RTMS); 10 kHz (C-450); and 12.5 kHz (NMT-900, Radiocom 2000).[3] Handover decisions are usually based on the power received at the base stations surrounding the mobile. C-450 is an exception, and uses round trip delay measurements. The total number of subscribers to the analog systems is around 8 million (3.7 million for TACS, 2.9 million for NMT, 0.9 million for C-450, and smaller amounts for the remaining systems). A few additional details are found in Table 1.

In Japan, a total of 56 MHz is allocated for analog cellular systems (860-885/915-940 MHz and 843-846/898-901 MHz). The first analog cellular system, the Nippon Telephone and Telegraph (NTT) system, began operation in the Tokyo metropolitan area in 1979. The frequencies were 925-940 MHz (mobile transmit) paired with 870-885 MHz (base transmit), and the channel spacing was 25 kHz, giving a total of 600 duplex channels. The control channel signaling rate was 300 b/s. In 1988, a high capacity system

[1] A wireline carrier is a fully-separated subsidiary of a Local Exchange Carrier (LEC) in the area where the LEC is a monopoly provider of local wired telephone service.

[2] The objective for the carrier-to-interference ratio, averaged over the multipath fading, is 17 dB for at least 90 percent of the covered area [1].

[3] NMT-900 and C-450 use frequency interleaving, whereby the "channel separation" is half the separation between adjacent non-overlapping channels (the nominal channel "bandwidth"). The channel spacings given here and in Table 1 refer to the separation between the "interleaved" center frequencies. Overlapping channels cannot be used at the same base station.

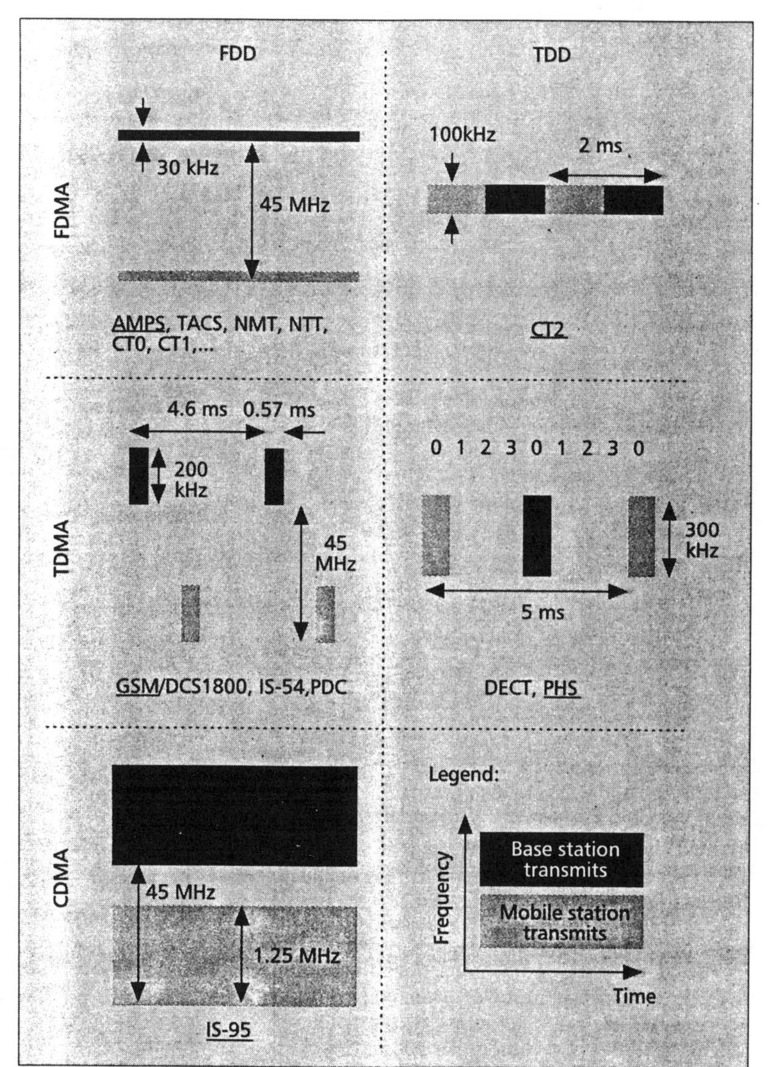

■ **Figure 1.** *Illustration of multiple access and duplexing techniques. The parameters given refer to the systems that are underlined. The widths of the frequency bands and time slots are artificially scaled for illustration purposes.*

was introduced in the same band, with an increased control channel signaling rate of 2.4 kb/s and a reduced channel spacing of 12.5 kHz. The number of channels was further increased by frequency interleaving (channel center frequencies 6.25 kHz apart), resulting in an overall fourfold increase in the number of channels to 2400. In addition, an associated control channel (below the voice band) with a rate of 100 b/s was introduced, which permits signaling without interrupting voice communication. To improve transmission quality and frequency reuse, reception diversity was incorporated into both the base and mobile stations. The mobile terminal of the high capacity system is dual-mode, and can also access the initial (25 kHz) system. The cellular operator NTT DoCoMo currently provides nationwide coverage in the bands 870-885/925-940 MHz.

In 1987, cellular radio was deregulated in Japan and two new operators were introduced. One of these is named "IDO" and began operation of the NTT high capacity system in December 1988, covering the Kanto-Tokaido areas in the bands 860-863.5/915-918.5 MHz. The third cellular carrier is the DDI Cellular Group, which provides coverage outside the metropolitan areas using the JTACS/NTACS systems (based on the European TACS system) in the bands 860-870/915-925 MHz and 843-846/898-901 MHz. Following DDI, IDO has also introduced NTACS in the bands 843-846/898-901 MHz and 863.5-867/918.5-922 MHz. IDO and DDI have formed a partnership to provide nationwide service by introducing roaming capabilities between the two systems. There are currently about 2.6 million subscribers to analog cellular systems in Japan.

Digital Cellular Systems

The development of low-rate digital speech coding techniques and the continuous increase in the device density of integrated circuits (i.e., transistors per unit area), have made completely digital second-generation systems viable. Digitization allows the use of time division multiple access (TDMA) and code division multiple access (CDMA) as alternatives to FDMA. With TDMA, the usage of each radio channel is partitioned into multiple timeslots, and each user is assigned a specific frequency/timeslot combination. Thus, only a single mobile in a given cell is using a given frequency at any particular time. With CDMA (which uses direct sequence spreading), a frequency channel is used simultaneously by multiple mobiles in a given cell, and the signals are distinguished by spreading them with different codes. One obvious advantage of both TDMA and CDMA is the sharing of radio hardware in the base station among multiple users. Figure 1 illustrates the FDMA, TDMA, and CDMA concepts.[4]

Digital systems can support more users per base station per MHz of spectrum, allowing wireless system operators to provide service in high-density areas more economically. The use of TDMA or CDMA digital architectures also offers additional advantages, including:

- A more natural integration with the evolving digital wireline network.
- Flexibility for mixed voice/data communication and the support of new services.
- A potential for further capacity increases as reduced rate speech coders are introduced.
- Reduced RF transmit power (increasing battery life in handsets).
- Encryption for communication privacy.
- Reduced system complexity (mobile-assisted handoffs, fewer radio transceivers).

Second-generation cellular systems based on digital transmission are currently being deployed. These systems are discussed in some detail in this section. Figure 2 shows the associated frequency allocations.

The Pan-European GSM System and DCS 1800 — The large number of different analog systems used in Europe did not represent an ideal situation from a subscriber point of view. This, together with the need to accommodate an increasing number of users and to establish compatibility with the evolution of the fixed network towards digital systems, led theConférence Européenne des Postes et Télécommunications (CEPT) to establish a "Groupe Spécial Mobile" in 1982. The work of that group became the GSM system (now "Global System for Mobile communications").

[4] *The digital cordless systems CT2, DECT, and PHS noted in Fig. 1 are discussed in the section to follow on cordless telephony.*

The new system was primarily expected to provide better quality, pan-European roaming, and the transmission of data for fax, e-mail, files, etc. A new design also offered the opportunity to specify a system for lower-cost implementations and the potential for increased spectral efficiency. Finally, a high degree of flexibility and openness to future improvements were recognized as important and taken into account [2].

The decision to use a digital approach was rather obvious because of the advantages already discussed. TDMA was chosen with eight timeslots per radio channel. Each user transmits periodically in every eighth slot (of duration 0.57 ms) and receives in a corresponding slot (Fig. 1). With such an approach, a base station only needs one (faster) transceiver for eight channels. In addition, transmit/receive slot staggering allows a relaxation of the duplex filter requirements for the mobile. The intermittent activity of the mobile transceiver also provides the opportunity (between the transmit and receive bursts at the mobile) to measure the strength of the signals from surrounding base stations. These measurements are reported to the serving base station and used for handover decisions. Note that contrary to the traditional FDMA systems, no additional hardware is needed for finding candidate base stations.

The time-compression of the user data (22.8 kb/s including error correction coding), by a factor of around eight inherent in the TDMA format, implies a bandwidth expansion of the signal by a corresponding factor. This has consequences on the fading of the received signal. The presence of reflectors, like mountains, hills, buildings, and others, leads to a multitude of echoes. In a narrowband system, the resulting signal paths cannot be resolved in time. With a bandwidth of 200 kHz, on the other hand, some degree of resolution becomes possible. In GSM, the staggering of the paths is estimated using a fixed mid-amble (training sequence in the middle of the slot). The intersymbol interference is then resolved with a Viterbi equalizer, for example. The multipath interference is a form of diversity, which, depending on the environment, can significantly reduce fading. With fast-moving mobiles, residual errors are corrected by the rate 1/2 convolutional coding and convolutional interleaving over eight bursts (for class 1 bits).

In some environments, like cities (delays around 1 to 2 μs), 200 kHz of bandwidth is no longer sufficient to resolve the multipath, and slow mobiles can experience long error bursts. This situation can be greatly improved by changing the frequency from slot to slot (frequency hopping). With different hopping patterns used at different base stations, interference diversity can be realized. This is particularly attractive with discontinuous transmission (speech pauses are not transmitted).

In its present version, GSM supports full-rate (22.8 kb/s, eight slots/frame) and half-rate (11.4 kb/s, 16 slots/frame) operation. Speech coders have been specified for both rates, and improved full-rate coders for "radio local loop" (RLL) applications can be accommodated. On the data side, various synchronous and asynchronous services at 9.6, 4.8, and 2.4 kbps are specified for both full-rate and half-rate operation. In particular, these data services interface to audio modems (e.g., V.22bis or V.32) and ISDN. Additionally, a connectionless packet service is in preparation, with an emphasis on interworking with X.25 and Internet (see the section on wireless data). Group 3 fax is also supported by the standard.

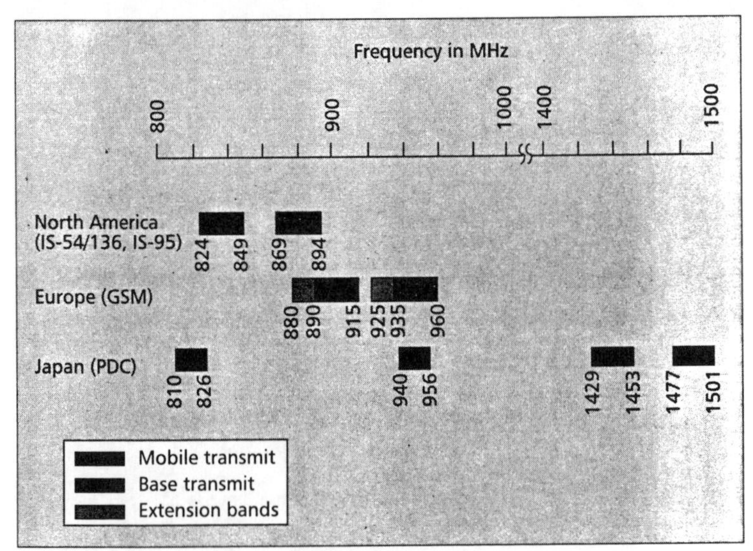

■ **Figure 2.** *Digital cellular frequency allocations.*

Originally, GSM was intended to be operated only in bands around 900 MHz. In early 1989, the UK Department of Trade and Industry started an initiative which finally led to the assignment of 150 MHz near 1.8 GHz for Personal Communications Networks (PCN) in Europe, and to the choice of GSM as a standard for that application. This system is called DCS1800, for Digital Cellular System 1800. Its definition meant translating the specifications to the new band and modifying some parts for accommodating overlays of micro- and macrocells. Cellular and PCN are certainly the most prominent applications, but GSM is currently also extended to include "group calls" and "push to talk" for private mobile radio (PMR) applications.

In the cellular area, GSM has experienced tremendous growth since the start of deployment in 1993. As of November 1994, there were two million subscribers throughout Europe, and the system had been adopted by many non-European countries (a total of 26 European and 26 non-European countries). Operators in the United Kingdom and Germany have started DCS1800 networks recently. Coverage is not country-wide at the moment, but ultimately a similar success is expected as for GSM.

For an overview of GSM, see [3], and for an in-depth description, see [4] or the standard [5]. For an overview of DCS 1800, see [6]. Figure 3 shows the frequency allocations for DCS1800, as well as allocations for wireless communications near 2 GHz in the United States and Japan.

IS-54 in North America — To meet the growing need to increase cellular capacity in high-density areas, the Electronic Industries Association (EIA) and the Telecommunications Industry Association (TIA) adopted the IS-54 standard based on TDMA [7]. IS-54 retains the 30-kHz channel spacing of AMPS to facilitate evolution from analog to digital systems. Each frequency channel

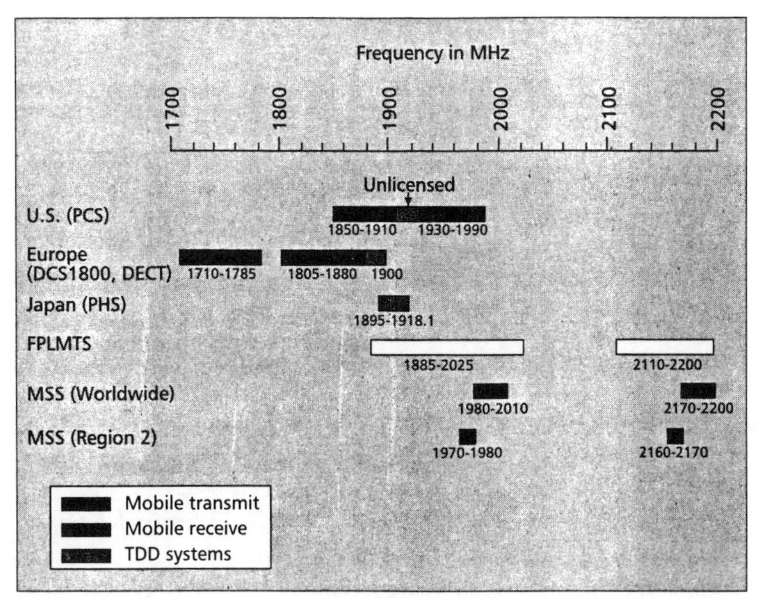

■ **Figure 3.** *Regional and international wireless frequency allocations near 2 GHz.*

provides a raw RF bit rate of 48.6 kb/s, achieved using π/4 DQPSK[5] at a 24.3-kbaud channel rate. This capacity is divided among six timeslots, two of which are assigned to each user in the current implementation, which uses a 7.95-kb/s vector sum excited linear prediction (VSELP) speech coder (13 kb/s with error protection). Thus, each 30-kHz frequency pair can serve three users simultaneously, and with the same reuse pattern, IS-54 provides triple the capacity (user channels per cell) of AMPS.[6] When half-rate coders are introduced, each 30-kHz frequency channel will be able to accommodate six user channels, giving another doubling of capacity. The IS-54 standard provides for an adaptive equalizer to mitigate the intersymbol interference caused by large delay spreads, but due to the relatively low channel rate (24.3 kbaud), the equalizer will be unnecessary in many situations.[7]

Since systems using the IS-54 standard must operate in the same spectrum used by the existing AMPS systems, the IS-54 standard is "dual mode," meaning that it provides for both analog (AMPS) and digital operation. This is necessary to accommodate "roaming" subscribers, given the large embedded base of AMPS equipment. Although service providers have already begun to deploy IS-54 equipment in major metropolitan areas in the United States, the conversion to digital will be slower in less dense areas. Consequently, there will be a mix of analog and digital terminals as well as base station equipment for a considerable period of time.

While the initial version of the IS-54 standard used the AMPS control channel specification (10 kb/s Manchester-encoded FSK), IS-136 (formerly IS-54 rev. C) includes a digital control channel (DCC) which uses the 48.6-kb/s modem. In addition to the increased signaling rate, the DCC offers such capabilities as point-to-point short messaging, broadcast messaging, group addressing, private user groups, hierarchical cell structures, and slotted paging channels to support a "sleep" mode in the terminal (to conserve battery power). The companion EIA/TIA standard IS-41 provides network signaling protocols to support inter-system handoff, roaming, and delivery of network-based features and services [8]. IS-54 equipment has already been deployed and is operational in a majority of the top ten cellular markets in the United States, and the rate of customer adoption is increasing.

[5] *Differential quadrature phase-shift keying (DQPSK) with a π/4 radian phase shift between successive symbols, to reduce amplitude fluctuations in the signal envelope.*

[6] *The Erlang capacity (offered load at a given blocking rate) actually is increased more than threefold because of improved "trunking efficiency" due to the larger number of available channels.*

[7] *As a rule-of-thumb, an equalizer is needed when the rms delay spread is on the order of 20 percent or more of the symbol duration (for binary and quaternary modulation). The symbol duration is about 40 µs for IS-54. Delay spreads for cellular environments usually are less than 8 µs, except for hilly and mountainous areas.*

[8] *Psi-CELP is pitch synchronous innovation code excited linear prediction, which is a newly established half-rate codec in Japan.*

Personal Digital Cellular in Japan — In Japan, there are two different types of analog cellular systems: (NTT and JTACS), which developed from different backgrounds. From the perspective of the user, a single air interface was desirable for providing roaming capability among different mobile networks. A development study for digital cellular systems with a common air interface was initiated in 1989 under the auspices of the Ministry of Posts and Telecommunications (MPT). The new digital system was established in 1991 and named Personal Digital Cellular (PDC).

The PDC system is also based on TDMA, with three slots multiplexed onto each carrier, similar to IS-54. The channel spacing is 25 kHz with interleaving to facilitate migration from analog to digital. The RF signaling rate is 42 kb/s, and the modulation is π/4 DQPSK. A key feature of PDC is mobile-assisted handoff, which facilitates the use of small cells for efficient frequency usage. The full-rate VSELP speech codec operates at 6.7 kb/s (11.2 kb/s with error correction). Currently, the full-rate codec is available; however, a 5.6-kb/s psi-CELP[8] half-rate codec has already been standardized and will soon be introduced, resulting in a doubling of capacity.

A total of 80 MHz is allocated to PDC; the frequency bands are 810-826 MHz paired with 940-956 MHz and 1429-1453 MHz paired with 1477-1501 MHz. With antenna space diversity, the required C/I is reduced, giving a reuse factor of 4. Group 3 FAX (2.4 kb/s) as well as 4.8-kb/s modem transmission with MNP class 4 are supported using an adaptor to provide the required transmission quality. There are five PDC operators, and are currently 250,000 subscribers to PDC. It is gaining popularity due to high quality, high security, a longer handset battery life, etc. See [9] or the standard [10] for more details.

IS-95 in North America — The EIA/TIA IS-95 standard is based on the CDMA system originally described in [11]. With IS-95, many users share a common channel for transmission. The basic user channel rate is 9.6 kb/s. This is spread to a channel chip rate of 1.2288 Mchip/s (a total spreading factor of 128) using a combination of techniques. The spreading process is different on the forward (base-to-mobile) and reverse links. On the forward link, the user data stream is encoded using a rate 1/2 convolutional code, interleaved, and spread by one of 64 orthogonal spreading sequences (Walsh functions). Each mobile in a given cell is assigned a different spreading sequence, providing perfect separation among the signals from different users, at least for a single-path channel. To reduce interference between mobiles that use the same spreading sequence in different cells and to provide the desired wideband spectral characteristics (not all of the Walsh functions yield a wideband power spectrum), all signals in a particular cell are scrambled using a pseudo-random sequence of length 2^{15} chips.

Orthogonality among users within a cell is preserved because their signals are scrambled identically. A pilot channel (code) is provided on the forward link for channel estimation. This allows coherent detection. The pilot channel is transmitted at higher power than the user channels.

On the reverse link, a different spreading strategy is used because each received signal arrives at the base via a different propagation channel. The user data stream is first convolutionally encoded at rate 1/3. After interleaving, each block of six encoded symbols is mapped to one of the 64 orthogonal Walsh functions (i.e., 64-ary orthogonal signaling). A final fourfold spreading, giving a rate of 1.2288 Mchip/s, is achieved by spreading the resulting 307.2-kchip/s stream by user-specific and base-specific codes of periods 2^{42}-1 and 2^{15} chips, respectively. The rate 1/3 coding and the mapping onto Walsh functions result in a greater tolerance for interference than would be realized from traditional spreading (i.e., a repetition code). This added robustness is important on the reverse link, due to the non-coherent detection and the in-cell interference (no orthogonality among in-cell users).

Another essential element of the reverse link is tight control of the mobile's transmit power, to avoid the "near-far" problem that arises from the different fading, shadowing, and path loss situations experienced by the different signals. A combination of open-loop and fast closed-loop control is used, with the commands for the closed-loop control being transmitted at a rate of 800 b/s. These bits are stolen from the speech frames.

At both the base station and the mobile, RAKE receivers are used to resolve and combine multipath components, significantly reducing the fading amplitude. This receiver architecture also is used to provide base station diversity during "soft" handoffs, whereby a mobile making the transition between cells maintains links with both base stations during the transition. The mobile's receiver combines the signals from the two base stations in the same manner as it would combine signals associated with different multipath components.

The IS-95 CDMA approach offers a number of benefits, including increased capacity, elimination of the need for planning frequency assignments to cells, and flexibility for accommodating different transmission rates (a higher-quality 13 kb/s speech coder, for example). Moreover, the variable-rate speech coding, power control, reduced fade margin, and forward error correction (FEC) all contribute to the reduction of the required RF transmit power.

Like IS-54, IS-95 is compatible with the IS-41 signaling protocol, and is a dual-mode standard designed for the existing North American cellular bands; IS-95 terminals can operate either in the CDMA mode or the AMPS mode. Deployment of IS-95 systems in the Los Angeles, California area is expected this year. For a more detailed overview of IS-95, see [11, 12] and for a focus on the reverse link, see [13].

Cordless Telephony

First-Generation Analog Cordless

Since 1984, analog cordless telephones in the United States have operated on ten frequency pairs in the bands 46.6-47.0 MHz (base transmit) and 49.6-50.0 MHz (handset transmit). Prior to 1984, five of the 49 MHz frequencies were paired with five frequencies near 1.6 MHz, an arrangement that proved less than satisfactory due to the imbalance in the performance of the two links and the limited number of channels.

The allowed emission bandwidth is 20 kHz, and the effective radiated power (ERP) is very low, roughly 20 μW (compared to 10 mW for most other cordless telephone systems). Analog FM is used for the voice signal, and the U.S. Federal Communications Commission (FCC) rules require digital coding of the signaling functions for security. There are an estimated 60 million 46/49 MHz cordless telephones in use in the United States, and total sales are roughly 15 million units per year. Despite the recent availability of higher-power digital cordless telephones operating in the 915 MHz Industrial, Scientific, and Medical (ISM) band (discussed below), the popularity of 49 MHz analog cordless telephones is expected to continue for a considerable time due to their low cost (U.S. $50 to $100 is typical for a basic unit).

Because of the large embedded base of these devices, the existing ten channel pairs have become inadequate, particularly in high-density areas. In August 1992, the TIA petitioned the FCC to make 15 additional frequency pairs near 44 MHz (base transmit) and 49 MHz (handset transmit) available for cordless telephones. In August 1993, the FCC adopted a Notice of Proposed Rule Making (NPRM) in response to TIA's petition, proposing specific provisions to be added to the FCC Rules [14]. A final ruling on the new frequencies by the FCC is expected this year.

The first cordless telephones were imported to Europe from the Far East and the United States. In most of Europe, such equipment was illegal but was sold in large quantities "for export only." In the United Kingdom, a standard very similar to the one originally used in the United States was introduced (MPT 1322) to offer an alternative to the illegal imports [15]. This standard (sometimes referred to as "CT0"), allowed for eight channel pairs near 1.7 MHz (base unit transmit) and 47.5 MHz (handset transmit), and most units could access only one or two channel pairs. A similar standardization approach was adopted in France. There are an estimated 4.7 million illegal units in operation; 1994 sales of legal and illegal units are estimated at 2.4 and 1 million, respectively.

In the rest of Europe, the reaction to the demand for cordless communication was to develop the analog cordless standard known as CEPT/CT1 [16]. This cordless standard provides for forty 25-kHz duplex channel pairs in the bands 914-915/959-960 MHz (80 pairs are allocated for CT1+ in the bands 885-887/930-932 MHz, which do not overlap the GSM allocation) and for a form of Dynamic Channel Assignment (DCA), whereby one of the 40 (or 80) duplex frequency pairs is selected at the beginning of each call. With this large number of channels and DCA, the blocking probability is low, even in densely populated areas. CT1 historically is a coexistence standard rather than an interoperability standard, which has the consequence that equipment from different manufacturers is typically incompatible. This situation also has advantages, however; it provides opportunities for new features such as scrambling to provide speech privacy, and cordless PBX applications. Such enhancements

There are an estimated 60 million 46/49 MHz cordless telephones in use in the United States, and total sales are roughly 15 million units per year.

CT2 is an FDMA/TDD air interface optimized for low-cost implementation of digital cordless and Telepoint systems.

would have been impossible to implement under a tightly-specified standard. Total annual sales for CT1/CT1+ are estimated at 2.2 million units and are expected to increase to 2.5 million and 2.7 million in 1995 and 1996, respectively. The current embedded base is estimated at 5.4 million units.

In Japan, there are 89 duplex channels near 254 MHz (handset transmit) and 380 MHz (base transmit) allocated to analog cordless telephones using FM [17]. The channel spacing is 12.5 kHz and the allowed transmit power is 10 mW. Unlike CEPT/CT1, there are two dedicated control channels, to facilitate fast connections and conserve battery life, at the price of somewhat reduced robustness against interference. Typical battery life in the "standby" mode between charges is 150 to 190 hours. The total market penetration is roughly 20 million units, with retail sales of 3 to 4 million units annually.

In addition, very low-power (500 μV/m at 3 meters, or roughly 50 nW ERP) cordless telephones are allowed in Japan below 350 MHz and above 1 GHz. However, the demand for these devices is fairly small due to their very limited operating range (on the order of 10 m).

Digital Cordless

The CT2 Common Air Interface — The deficiencies of the analog MPT1322 cordless systems in the United Kingdom (most notably the limited number of channels and resulting high blocking probability) stimulated the development of an alternative that later became the CT2/Common Air Interface (CAI). The most salient features of that standard are the digital transmission format and the use of time division duplexing (TDD). With CT2, speech is first digitized using a 32 kb/s adaptive differential pulse code modulation (ADPCM) encoder (CCITT G.721). The time-compressed digitized speech and control data are modulated onto a carrier at a rate of 72 kb/s using Gaussian filtered FSK (GFSK) and are transmitted in 2-ms frames. Each frame includes one base-to-handset and one handset-to-base burst (Fig. 1). This mode of duplexing (TDD) has significant advantages: the expensive duplex filter required with frequency duplexing is replaced by a switch; both transmission and reception antenna diversity can be used (due to propagation path reciprocity); and rather arbitrary frequency bands can be allocated (frequency-duplexed systems need symmetric pairs of bands separated by several MHz).[9]

The CT2 spectrum allocation consists of 40 FDMA channels with 100-kHz spacing in the range 864-868 MHz. The 100-kHz spacing reduces the bandwidth efficiency to half its value for CT1 (50 kHz per duplex channel). The maximum transmit power is 10 mW, and two-level power control helps to prevent the desensitization of base station receivers, and as a by-product, contributes to improved frequency reuse. A call re-establishment procedure on another carrier after three seconds of handshake failure gives robustness against the usual impairments of radio channels. The control data bits are protected against errors.

CT2 also supports the transmission of data, up to 2.4 kb/s through the speech codec (4.8 kb/s with an increased error rate), and higher rates by accessing the 32 kb/s bearer channel directly.

CT2 was promoted as a Telepoint standard. Telepoint networks use cordless base stations to provide wireless pay phone services. Incoming calls are not supported with the basic service. They require additional functionality such as manual registration and paging. User acceptance of CT2-Telepoint varies widely over the continents. No breakthrough was achieved in Europe; in the United Kingdom, for example, the last operator terminated service in 1993. In Hong Kong, however, there are 150 000 subscribers to CT2 Telepoint service, and systems are being deployed in other parts of Southeast Asia, including China, Malaysia, Singapore, and Thailand. European sales of CT2 handsets for 1994 are estimated to be 0.4 million, with projected increases to 0.5 and 0.6 million in 1995 and 1996, respectively. The current embedded base is estimated at 0.6 million units.

A Canadian enhancement of the CT2 CAI, called CT2+, is designed to provide some of the missing mobility management functions. For that purpose, 5 of the 40 carriers are reserved for signaling. Each carrier provides 12 Common Signaling Channels (CSC) using TDMA. These channels support location registration, location updating, and paging, thereby enabling the Telepoint subscriber to receive calls. CT2+ operates in the frequency band 944-948 MHz.

In summary, the family of CT2 standards is an attractive option for cordless and Telepoint systems optimized with respect to cost. A more detailed description of CT2 can be found in [18].

DECT — Digital European Cordless Telecommunications — DECT is designed as a flexible interface to provide cost-effective communication services to high user densities in picocells, even with colocated systems that are not coordinated. The standard is intended for applications such as domestic cordless telephony, Telepoint, cordless PBXs, and RLL. It supports multiple bearer channels for speech and data transmission (which can be set up and released during a call), handover, location registration, and paging. Functionally, DECT is closer to a cellular system than to a classical cordless telephone. However, the interface to the PSTN or ISDN network remains the same as for a PBX or corded telephone.

DECT uses TDMA and TDD, with 12 slots per carrier in each direction (Fig. 1). A DECT base station therefore can support multiple handsets simultaneously with a single transceiver. Furthermore, it can allocate several slots to a single call to provide higher data rates. In addition to the advantages of TDD discussed above, the flexible allocation of slots to one or the other direction allows DECT to adjust for traffic asymmetries such as can occur during data base retrieval. To meet the expected traffic density, 10 carriers in the band 1880-1900 MHz are allocated in the initial implementation. The basic speech teleservice uses the same 32 kb/s ADPCM coding as CT2.

A key element of DECT is its "interference confinement" and "interference avoidance" strategy. The former involves the concentration of interference to a small time-frequency element even at the price of a reduced robustness, and the latter implies the avoidance of time-frequency slots with a significant level of interference by a handover to another slot at the same or another base station. This is a powerful approach for the uncoordinated

[9] TDD is not used more widely (e.g., in cellular systems) because it has disadvantages that are significant for some applications. TDD requires systemwide frame synchronization to avoid base-to-base and mobile-to-mobile interference, doubles the RF symbol rate (or halves the utilization of the front-end radio hardware in the base station), and requires additional guard time (which imposes substantial overhead for large cells).

	CT2	CT2+	DECT	PHS	PACS
Region	Europe	Canada	Europe	Japan	United States
Duplexing	TDD		TDD	TDD	FDD
Frequency band (MHz)	864-868	944-948	1880-1900	1895-1918	1850-1910/1930-1990*
Carrier spacing (kHz)	100		1728	300	300/300
Number of carriers	40		10	77	16 pairs/10 MHz
Bearer channels/carrier	1		12	4	8/pair
Channel bit rate (kb/s)	72		1152	384	384
Modulation	GFSK		GFSK	π/4 DQPSK	π/4 QPSK
Speech Coding	32 kb/s		32 kb/s	32 kb/s	32 kb/s
Average handset TX power (mW)	5		10	10	25
Peak handset TX power (mW)	10		250	80	200
Frame duration (millisec)	2		10	5	2.5

*General allocation to PCS (see Figs. 3 and 4); licensees may use PACS.

■ **Table 2.** *Summary of digital cordless air interface parameters.*

operation of base stations, since the interference caused by a foreign mobile station approaching a base station may reach a level at which no practical amount of baseband processing can recover the desired signal. In a cellular system, a handover to the closer base station takes place under such circumstances. With DECT, this is only possible if the foreign mobile station has the required access rights.

DECT is designed for low cost, flexibility, and operation in an uncoordinated environment. Among other things, this means that the base stations need not be synchronized. Various mechanisms, including fields for synchronization and parity, allow the detection of sliding collisions and the initiation of corrective actions such as handover. The information about access rights, base station capabilities, paging messages, etc., is multiplexed onto the control channel of each active transmission to optimize the utilization of the base station transmitter and to obtain robustness; a dummy bearer is used when there is no active call. The problem of inadequate frequency-switching speed in the base station synthesizer is taken into account by the provision of a list of "blind" slots. Due to the high signaling rate (1152 kb/s) and the correspondingly large bandwidth, either equalization or antenna diversity typically is needed for using DECT in the more dispersive microcells (Telepoint or RLL applications). This issue is currently under study, and field tests are being performed. Finally, a multiframe structure, which allows for a sleep mode of the handset, contributes to the conservation of battery life. Clearly, this short summary of DECT characteristics is only a representative selection to convey the spirit of the design.

For flexibility and broad applicability, DECT closely follows the Open Systems Interconnect (OSI) reference model, and it allows for a series of escape routes to proprietary additions and alternatives, which in particular include the possibility of specifying alternative Medium Access Control (MAC), Data Link Control (DLC) and network-layer protocols. This possibility is key to the future evolution.

In Europe, the first DECT systems are currently being shipped, primarily for business and domestic applications. DECT unit sales for 1994 are estimated at about 0.1 million. DECT and possible variants of it are the potential basis for low cost, picocell based systems in the near future.

In summary, DECT is a flexible standard for providing a wide range of services in small cells. An overview of basic design considerations is given in [19]. Detailed specifications are included in [20].

Personal Handyphone System — A study for the next generation portable telephone systems in Japan was initiated in January 1989 under the auspices of MPT. Following the report, the concept of the Personal Handyphone System (PHS, formerly PHP) was launched, on the basis of a digital cordless telephone and a digital network. The system objectives of PHS are to provide for not only home and office use, but also for public access capability [21]. The air interface protocol for PHS was determined by the Research & Development Center for Radio Systems (RCR), and the network interface was determined by the Telecommunications Technical Committee (TTC) [22].

Like DECT, the PHS standard uses TDMA and TDD, but each frequency carries 4 duplex bearer channels rather than 12. The RF channel rate of 384 kb/s was chosen based on the tradeoff between maximizing the multiplexing number and minimizing the effects of frequency-selective fading, particularly in outdoor environments (250-ns delay spread).

The PHS allocation consists of 77 channels, 300 kHz in width, in the band 1895-1918.1 MHz. The band 1906.1-1918.1 MHz (40 frequencies) is designated for public systems, and the band 1895-1906.1 MHz (37 frequencies) is used for home/office applications. The channel is autonomously selected by measuring the field strength and selecting a channel on which it is below a prescribed level, i.e., fully dynamic channel assignment is used. The modulation is π/4 DQPSK, and the average

DECT is a flexible standard for providing a wide range of services in small cells. Like DECT, PHS uses TDMA and TDD, but has four duplex bearer channels per frequency rather than twelve.

D*irect sequence and frequency hopping digital cordless telephones operate in the 902-928 MHz band in the United States.*

transmit power per direction is 10 mW (80 mW peak power) for the handset and no greater than 500 mW (4W peak) for the cell site. The frame duration is 5 ms. Like DECT and CT2, PHS uses 32 kb/s ADPCM speech coding, and error detection in the form of a cyclic redundancy check (CRC) is provided, but there is no error correction. Unlike DECT, however, PHS provides dedicated control channels. The typical talk and standby time are 5 hours and 150 hours, respectively. PHS supports handoffs (as an option), although it is confined to walking speed. Taking advantage of the channel reciprocity inherent with TDD and low-speed mobility, transmission diversity is provided on the forward link. Reception diversity at the base station can be used on the reverse link. Moreover, muting of the audio speech signal improves voice quality on both links in fading channels. As for data transmission capability, PHS currently can support G3 fax at 4.2 to 7.8 kb/s and full-duplex modem transmission at 2.4 to 9.6 kb/s through the speech codec. A new standard will be established to support 32- or 64-kb/s data by direct access to one or two bearer channels, respectively.

The home/office application for PHS has already been introduced in Japan, and the public application will be introduced this year. For purposes of providing service, Japan is divided into 11 regions, with at most three operators per region licensed to offer public access systems using the 12 MHz of available spectrum. The potential subscriber base for PHS is estimated to be 5.5 million in 1998 and 39 million in 2010. PHS also is expected to be introduced soon to provide cost-effective local wireless access (i.e., RLL). See [23] for a discussion of PHS network aspects and anticipated services.

WACS and PACS — In the United States, Bell Communications Research (Bellcore) developed an air interface for Wireless Access Communications Systems (WACS) [24]. This interface is intended to provide wireless connectivity to the local exchange carrier (LEC), and is designed with low-speed portable applications and small-cell systems in mind. Base stations are envisioned as shoebox-sized enclosures mounted on telephone poles, separated by about 600 m. The WACS air interface is similar to the digital cordless interfaces, with two notable exceptions: frequency-division duplexing (FDD) is used rather than TDD, and greater effort has been made to optimize the link budget and frequency reuse.

In the original design, each frequency carried ten user timeslots. Speech coding was 32 kb/s, with a superframe structure to allow for lower rate speech codecs. The frame duration was 2 milliseconds to minimize the delay added to the speech path. The modulation was QPSK (Quaternary Phase-Shift Keying) with coherent detection, which provides substantially better performance than the discriminator-based receivers used in most digital cordless systems. Two-branch polarization diversity at both the handset and base with feedback gives an advantage approaching that of four-branch reception diversity. Like the digital cordless air interfaces, the WACS design provides for error detection but not for FEC or adaptive equalization. Potential applications envisioned for WACS include RLL, portable public service, and wireless PBX.

As part of the standards process in the United States related to the recently-allocated spectrum near 2 GHz for Personal Communications Services (PCS) discussed below, attributes of WACS and PHS have been combined to create an industry standard proposal for Personal Access Communications Services (PACS). PACS is intended as a "low-tier" air interface for the licensed portion of the new 2-GHz spectrum. PACS retains many of the attributes of the original WACS design. The main changes include a reduction of the number of timeslots from ten to eight, and a corresponding reduction in the channel bit rate and bandwidth, as well as a slight increase in the frame duration. The modulation has been changed to $\pi/4$ QPSK, and coherent detection is used.

Table 2 shows the key physical layer parameters for CT2/CT2+, DECT, PHS, and PACS.

Digital Cordless in the North American ISM Bands — There are a number of frequency bands reserved for Industrial, Scientific, and Medical (ISM) devices. These devices generally use RF energy for heating rather than communication, and include applications such as microwave ovens, RF welders, and plywood heaters. In the United States and Canada, these bands include 902-928 MHz, 2400-2483.5 MHz, and 5725-5850 MHz. Unlicensed devices such as cordless telephones that use either frequency hopping or direct sequence spreading[10] are allowed to operate in these ISM bands with up to 1 watt of transmit power. Devices not using frequency hopping or direct sequence spreading are limited to a field strength of 50 mV/m at 3 meters, or about 0.5 mW ERP.

There currently are digital cordless telephones using both frequency hopping and direct sequence spreading, operating in the 902-928 MHz band. There are no detailed standards specifically governing ISM band cordless telephones. Manufacturers therefore have considerable design freedom for innovation and application of advances in technology without being constrained by channelization plans and operational restrictions based on older technologies. However, the interference environment tends to be undisciplined and difficult to predict. This situation places a premium on designs that are sufficiently robust and flexible to operate in the presence of unpredictable interference or to avoid it.

Digital Cordless Compared to Digital Cellular

From the foregoing summaries of the various digital cordless air interfaces, it is clear that while there are significant differences among them, they have a number of characteristics in common which distinguish them from the digital cellular technologies discussed earlier. In general, the digital cordless systems are optimized for low-complexity equipment and high-quality speech in a quasi-static environment (with respect to user mobility). Conversely, the digital cellular air interfaces are geared toward maximizing bandwidth efficiency and frequency reuse in a macrocellular, high-speed fading environment. This is achieved at the price of increased complexity in the terminal and base station. As summarized in Table 3, the physical layer parameters for digital cordless and digital cellular technologies reflect these respective design objectives.

[10] Frequency hopping systems are required to use pseudo- random hopping patterns of at least 50 frequencies in the 902-928 MHz band, and at least 75 frequencies in the other bands. Direct sequence systems must have a processing gain of at least 10 dB and a 6 dB bandwidth of at least 500 kHz.

Wireless Data

Wireless data systems are designed for packet-switched ("asynchronous") rather than circuit-switched ("isochronous") operation. Operators of wide-area messaging systems use licensed spectrum, and sell service to customers. Conversely, wireless local area networks (LANs) are usually privately owned and operated, and provide high-rate data communication over a small area. With the exception of the Altair system, wireless LANs are unlicensed and typically operate in the ISM bands (except for the infrared systems). A brief summary of wireless data systems is provided here; see [25] for a detailed discussion and an extensive reference list.

Wide Area Data Service

Advanced Radio Data Information Service (ARDIS) and RAM Mobile Data (RMD) offer wireless packet data messaging service over their dedicated networks using the specialized mobile radio (SMR) frequencies near 800/900 MHz. Both are available to 90 percent of the urban business population in the United States, and have a combined total of roughly 52,000 subscribers. ARDIS offers service in over 400 metropolitan areas. The data rate is 4.8 kb/s, with upgrades underway to 19.2 kb/s in some areas. RMD offers service over its Mobitex network, providing coverage in 216 metropolitan areas, with 10 to 30 duplex channels available in each area. The data rate is 8 kb/s. The Mobitex architecture was originally developed by Telia, the Swedish national operator. To encourage the development of multiple equipment sources, Mobitex software and hardware specifications are made available without any license or fee. The specifications are published by the Mobitex Operators' Association (MOA). As a result, there are a number of terminal suppliers. Mobitex networks are operational in 10 countries besides the United States.

Cellular Digital Packet Data (CDPD) does not require a specialized network but rather uses the existing analog cellular network. CDPD takes advantage of the idle time on the analog AMPS channels to transmit packet data at a rate of 19.2 kb/s. It is designed to operate as a "transparent" overlay on the AMPS system.

The General Packet Radio Service (GPRS) standard is being developed to provide packet data service over the GSM infrastructure. Two alternative approaches are being considered: 1) the allocation of specific GSM channels for packet transmission, which are shared by all active packet subscribers; and 2) the fast establishment of a GSM traffic channel on any radio resource available. The high grade service aims at a packet error rate of 10^{-4} and a delay less than 1 s. Particular attention is given to the interworking with Public Switched Packet Data Networks and Internet.

There also are wireless data services emerging in the United States that operate on an unlicensed basis in ISM spectrum. The recently-announced Ricochet™ wireless packet data system uses a microcell architecture with small, inexpensive, easily-installed base stations mounted on the tops of lamp posts, utility poles, and buildings. The customer accesses the network via a wireless modem that connects to the serial port of a laptop or notebook computer.

	Digtial Cordless	Digital Cellular
Characteristics		
Cell size	small (50 to 500 m)	large (0.5 to 30 km)
Antenna elevation	low (15m or less)	high (15m or more)
Mobility speed	low (6 kph or less)	high (up to 250 kph)
Coverage	zonal	wide-area continuous
Handset complexity	low	moderate
Base complexity	low	high
Spectrum access	shared	exclusive
Design attributes		
Handset TX power (average)	5 to 10 mW	100 to 600 mW
Duplexing	TDD*	FDD
Speech coding	32 kb/s ADPCM	8 to 13 kb/s vocoder
Error control	CRC	FEC/interleaving
Detection	discrim/differential*	coherent/differential
Multipath mitigation	antenna diversity (opt.)	diversity/equalizer/Rake

*PACS uses frequency duplexing and coherent detection.

■ **Table 3.** *General comparison of digital cordless and digital cellular air interfaces.*

Wireless Local Area Networks (LANs)

Wireless LANs are targeted primarily for high data rates (generally ≥ 1 Mb/s) and in-building applications, and may be preferable to their wired counterparts for situations in which wiring is difficult or impractical, or some degree of mobility is needed. There currently are a number of products available that operate on an unlicensed basis in the ISM bands, such as FreePort™ and WaveLAN.® FreePort provides a wireless Ethernet® (IEEE 802.3) hub and operates in the 2400-2483.5 MHz (hub receive) and 5725-5850 MHz (hub transmit) ISM bands, using direct sequence spreading. WaveLAN [26] provides peer-to-peer communication in the 902-928 MHz band in the United States and in the 2.4-2.48 GHz band in most other countries (it is available in 39 countries outside the United States). WaveLAN uses direct sequence spreading with a Carrier Sense Multiple Access/Collision Avoidance (CSMA/CA) protocol. Altair uses the Ethernet protocol and operates in the terrestrial microwave spectrum near 18 GHz; a site-specific U.S. FCC license is required.

Standards are being developed for wireless LANs, under IEEE 802.11 in the United States and ETSI/RES10 in Europe (known as HIPERLAN, for High Performance Radio LAN). There are a number of similarities between the IEEE 802.11 and the HIPERLAN work. Both standards are intended for rates exceeding 1Mb/s, and will support architectures with an infrastructure as well as "ad hoc" architectures, whereby terminals communicate directly with each other (peer-to-peer) without the mediation of a fixed base station. Point-to-point, point-to-multipoint, and broadcast services will be available. While asynchronous packet transmission will be

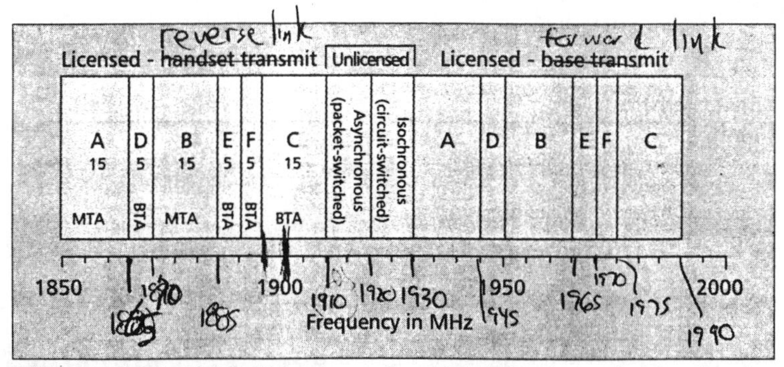

■ **Figure 4.** *The U.S. PCS frequency plan. MTA is "major trading area" and BTA is "basic trading area."*

the dominant mode, distributed time-bounded services (DTBS) will also be supported. Finally, since it is anticipated that many terminals will be battery-powered, the standards will incorporate a sleep mode for power management.

There also are some differences between IEEE 802.11 and HIPERLAN. The initial focus of IEEE 802.11 is the development of a single standard for the MAC layer,[11] and multiple physical-layer standards compatible with the single MAC. Initial work on physical-layer standards is focused on both direct sequence and frequency hopping for the 2.4-GHz ISM band. Other physical layer standards, including one for the 1.9 GHz unlicensed PCS band (discussed below), are anticipated. In addition, a standard for a baseband infrared (IR) physical layer is under development.

HIPERLAN is focusing on higher data rates than IEEE 802.11. This is made possible by the allocation of large dedicated bands: 5150-5300 MHz plus another 200 MHz near 17 GHz. Furthermore, the HIPERLAN standard is planned to include provisions for a flexible forwarding mechanism for ad-hoc (non-infrastructure) networks, to extend the effective range of the terminals. The nodes are subdivided into forwarding and non-forwarding nodes, and a self-adapting wireless LAN is created. Finally, HIPERLAN is placing more emphasis on DTBS.

In Japan, two types of wireless LANs have been standardized. One is for medium rates in the range of 256 kb/s to 2 Mb/s using spread spectrum in the 2.4 GHz ISM band. The other is for high rates (≥ 10 Mb/s) using Quadrature Amplitude Modulation (QAM), QPSK, or 4-level FSK, and operating near 18 GHz.

Ongoing Work and the Future

The future of wireless personal communications offers many possibilities. Continuous improvements in microelectronics technology and radio link techniques coupled with advances in network signaling and control capabilities will support increasingly sophisticated features and services. Part of the challenge in planning future wireless systems is to determine the services that they will be required to support. This has been the major thrust of TG8/1 of the ITU-R (the International Telecommunication Union-Radiocommunication sector, formerly CCIR), which is defining Future Public Land Mobile Telecommunication Systems (FPLMTS). Spectrum was allocated on an international basis to FPLMTS at the 1992 World Administrative Radio Conference (WARC '92), as shown in Fig. 3. For more detail on the WARC '92 allocations, refer to reference [27].

Universal Mobile Telecommunications Systems (UMTS) in Europe

In Europe, the long term goal is a Universal Mobile Telecommunications System (UMTS), which unifies the worlds of cellular, cordless, RLL, low-end wireless LAN, private mobile radio (PMR), and paging. The idea is to provide the same type of services everywhere, with the only limitation being that the available data rate may depend on the location (environment) and the load of the system. The scope is a multi-operator system with mixed cell architectures and multimedia capabilities. These requirements are difficult to meet. They have implications on both the radio interface and the protocol structure.

GSM and DCS1800 are actively being pushed towards a UMTS and get rather close to it. Radio local loop service to the home and business, as well as PBX functionality in the public switches, could take over cordless functions if service is provided at an adequate price. A major unsolved issue is the provision of higher data rates. Under rather optimistic conditions, an alternative modulation scheme and the use of asymmetric slot allocations could boost the rate perhaps up to 64 kb/s, but certainly not to 2 Mb/s. Spectrum efficiency and costs for the low end are issues also.

DECT could grow beyond its present application areas of cordless phone, PBX and Telepoint. Studies and field trials currently are being performed for local loop replacement. The extension of the range by more advanced antenna systems and some form of equalization, as well as repeaters, are important issues in this context. DECT could support the low-cost mass-market in densely populated areas, leaving wide area coverage to cellular systems (partial UMTS). Dynamic channel allocation and the high degree of flexibility in providing new services, by combining channels to achieve higher data rates (without modification of the standard perhaps up to 200 kb/s/transceiver) are other assets. Finally, the use of a DECT air interface in conjunction with the GSM infrastructure also is being studied.

The projects Advanced TDMA (ATDMA) and Code Division Testbed (CODIT) from the R & D in Advanced Communications Technologies in Europe (RACE) program follow more revolutionary approaches to UMTS. The former project uses TDMA, and the latter one, CDMA as the principal access technique. Both projects seek higher data rates. In the testbeds, rates up to 64 and 128 kb/s are being considered, and the concept work includes burst rates up to 2 Mb/s. Low delay connectionless packet transmissions and access to the Internet also are being addressed. A type of multimedia capability with the parallel transmission of speech and fax or a concurrent retrieval of data from a data base is possible in CODIT, for example. Many aspects of the TDMA and CDMA approaches have been introduced in the previous sections. A few properties that have received particular attention in view of the UMTS objectives are as follows. With CDMA, a bandwidth that is in excess of the optimum bandwidth for a given environment can be tolerated

---

[11] The MAC layer manages the access to the radio channel(s). In terms of the OSI reference model, the MAC is the lower part of the data link layer, directly above the physical layer. The upper part of the data link layer is the logical link control (LLC) layer.

and cells of widely varying size can be accommodated with almost no penalty on spectrum efficiency. With TDMA, different operators (smaller bandwidth) and uncoordinated environments (see DECT) can be supported more easily. Both projects have tried to solve some of the inconveniences of the respective approaches. In ATDMA, a type of soft performance degradation is achieved by trading the rate of the speech coding against the rate of the error correction coding. In order to take advantage of speech activity, the combination of talk-spurt detection and a Packet Reservation Multiple Access (PRMA) scheme has been studied. In CODIT, interfrequency handover is included to switch from embedded microcells to overlaying macrocells or from a cellular environment to an uncoordinated one. In uncoordinated environments, interference from neighbor systems is detected by the mobile station. The presence of the latter interference, which is an indication of a close base station that could be blocked, triggers a handover to another frequency or to a cellular operator. As is to be expected, both projects try to develop the specific strengths of their own approach, but try also to find mechanisms for including the advantages of the competing approach. The soft degradation introduced in TDMA or the interfrequency handover introduced in CDMA, in order to support interoperability between collocated operators, are examples. Both approaches look promising, with some advantage for CDMA in cellular environments and for TDMA in uncoordinated environments.

Additional concepts are considered outside ETSI and RACE also. "Page and answer" is an idea being investigated at Ascom, for example. In its simplest form, an incoming call is put on hold at a server, which issues a page. The called party responds on the most suitable terminal (wired, cordless, or cellular) and answers using a wrist pager sending dual-tone multifrequency (DTMF) tones to connect (toll-free) and authenticate to the server. The server then completes the connection. More advanced implementations are also being conceived. Advantages of such approaches are that they save investment, do not require heavy handsets with rather limited operation times, avoid unnecessary restrictions on data rates (in the office, at home), and maintain the traditional charging of the calling side.

In summary, the mainstream in Europe is currently exploiting the potential of GSM and DECT, including further evolutions and the interworking of these standards. Other standards are being finalized, for wireless LAN (HIPERLAN) and PMR (TETRA, for Trans European Trunked Radio System), for example. Concurrently, the definition and specification of a UMTS is being attempted within ETSI and supported by RACE. The design of ISM band and other non-standardized systems also is being pursued but with a lesser emphasis.

PCS in the Emerging Technologies Bands in the United States

The path being taken toward future wireless systems in the United States is very different from that in Europe. In support of a free-market approach, the FCC has recently allocated 140 MHz of spectrum near 2 GHz to PCS. Figure 4 shows the exact frequencies. Note that the A and B blocks are designated for major trading areas (MTAs), while blocks C through F are to be licensed on the basis of BTAs (basic trading areas). There are 51 MTAs and 492 BTAs in the United States (which suggests a rough comparison of MTAs to states and BTAs to counties). In addition to the allocations for licensed terrestrial systems, a total of 20 MHz was allocated for unlicensed applications (and more is expected in the near future). Of this, 10 MHz is designated for "isochronous" (i.e., circuit-switched) applications such as voice, and 10 MHz for "asynchronous" applications like wireless packet data. Equipment operating in the 2-GHz unlicensed band will be required to comply with a "spectral etiquette" incorporated into Part 15 of the FCC rules. This etiquette was developed by the technical subcommittee WINTech of the industry body WINForum, and is intended to promote harmonious coexistence of diverse systems in the unlicensed band while allowing designers considerable flexibility with respect to system architecture (modulation, coding, signaling protocols, frame structure, etc.). Two of the essential ingredients in the etiquette are 1) a "listen before talk" (LBT) requirement, intended to prevent a transmitter from interrupting communication already in progress on a frequency; and 2) a transmit power limit that varies as the square root of the signal bandwidth, intended to put wideband and narrowband systems on a relatively even footing with respect to interference and use of the spectrum.

The FCC is awarding licenses in the PCS spectrum via "competitive bidding" (i.e., auctions). The winner of each license is free to use any desired air interface and system architecture, provided it complies with the FCC rules governing transmit power levels, etc. Hence, there are no predetermined standards for systems that will be operating in the 2-GHz PCS spectrum. This has led the TIA and Committee T1 of the Alliance for Telecommunications Industry Solutions (ATIS, formerly the Exchange Carrier Standards Association) to form a Joint Technical Committee (JTC) to review potential PCS standards submitted by contributors, and to make recommendations. The JTC has recognized that PCS standards fall naturally into two categories: "high tier," supporting macrocells and high-speed mobility; and "low tier," optimized for low power and low complexity. These two tiers essentially correspond to the "digital cellular" and "digital cordless" categories, respectively, discussed earlier. At this point, the JTC has narrowed the initial list of 16 possible standards down to seven. Five of these are variations of existing air interfaces: PACS (low-tier) and proposals based on GSM, IS-54, and IS-95 (high-tier) as well as DECT (low-tier). The other two are based on a hybrid TDMA/CDMA approach, and wideband CDMA (W-CDMA), respectively. In addition, the TIA has initiated an activity under its TR41 technical committee to develop standards for wireless user premises equipment (WUPE) operating in the unlicensed PCS band.

The 2-GHz PCS spectrum currently is occupied by operating point-to-point microwave radio systems. For the most part, these incumbent transmission links must be relocated to other frequencies (e.g., 6 GHz) or converted to optical fiber facilities before PCS systems can be deployed.[12] The PCS licensee is required to compensate the incumbent for the cost of relocating. For the unli-

The mainstream in Europe is currently exploiting the potential of GSM and DECT, including further evolutions and the interworking of these standards.

[12] *In some cases, it may be possible to frequency-coordinate between the PCS operator and an incumbent to avoid interference.*

censed band, financing the cost of relocating the incumbents is less straightforward, since an equitable mechanism must be developed for sharing the cost of relocation among the purveyors of unlicensed equipment. In addition, during the transition period, the usage of specific frequencies in specific areas must be limited to those which have been cleared. The industry forum UTAM, Inc. (for Unlicensed PCS Ad Hoc Committee for 2 GHz Microwave Transition And Management) was formed to deal with these issues.

FPLMTS Work in Japan

Japan has estaablished a new standardization committee and related working groups in connection with FPLMTS work. CDMA as well as advanced TDMA technologies are being studied, including progagation and access tests, to contribute to ITU-R standardization activities.

Global:Mobile Satellite Services

There are some situations in which providing radio coverage with cellular-like terrestrial wireless networks is either not economically viable (such as in remote, sparsely-populated areas), or physically impractical (such as over large bodies of water). In these cases, mobile satellite services (MSS) could fill the gap, allowing complete global coverage. Spectrum has been designated by the ITU for MSS (as shown in Fig. 3), and there are many MSS systems in various stages of concept, design, and operation. Some support only data services while others accommodate voice as well. Some are designed for special purposes and/or private user groups while others are intended for general (public) use and interconnection to the PSTN. The latter could support universal wireless communications.

One way to broadly categorize MSS systems is according to the orbital altitude of the satellites: geostationary satellites (GEOS), at an altitude of 35,786 km; low earth orbit satellites (LEOS), at altitudes on the order of 1,000 km; medium earth orbit satellites (MEOS), at altitudes on the order of 10,000 km; and highly elliptical orbit satellites (HEOS), with widely varying altitudes. GEOS systems for public use include INMARSAT-M, MSAT, ACTS, MOBILESAT, and NSTAR. LEOS systems include Iridium [28] (66 satellites at roughly 770 km), Globalstar (48 satellites at 1400 km), and Teledesic (840 satellites at 700 km). Odyssey is a MEOS proposal with 12 satellites at about 10,600 km, and the ELMSAT proposal specifies a HEOS approach with two or three satellites.

Guide to Wireless Acronyms and Abbreviations

ADPCM	Adaptive differential pulse code modulation
AMPS	Advanced Mobile Phone Service
CDMA	Code division multiple access
CDPD	Cellular Digital Packet Data (U.S.)
CEPT	Conférence Européenne des Postes et Télécommunications
CRC	Cyclic redundancy check
CT	Cordless Telephone (interim ETSI standards, e.g., CT1)
DCA	Dynamic channel assignment
DCS1800	Digital Cellular System 1800 (Europe)
DECT	Digital European Cordless Telecommunications
DLC	Data link control (layer)
DQPSK	Differential quaternary phase shift keying
EIA	Electronic Industries Association (U.S.)
ERP	Effective radiated power
ETSI	European Telecommunications Standards Institute
FCC	Federal Communications Commission (U.S.)
FDD	Frequency division duplexing
FDMA	Frequency division multiple access
FEC	Forward error correction (channel coding; e.g., convolutional coding)
FM	Frequency modulation
FPLMTS	Future Public Land Mobile Telecommunication Systems
FSK	Frequency shift keying
GFSK	Gaussian filtered FSK
GMSK	Gaussian minimum shift keying
GPRS	General Packet Radio Service (GSM connectionless packet service, ETSI, Europe)
GSM	Groupe Spécial Mobile (originally) currently Global System for Mobile Communication (ETSI, Europe)
HIPERLAN	High Performance Radio Local Area Network (ETSI, Europe)
INMARSAT	International Maritime Satellite Organization
IR	Infrared
IS-41	Interim Standard 41 (TIA/EIA cellular network signaling standard, U.S.)
IS-54	Interim Standard 54 (TIA/EIA TDMA cellular standard, U.S.)
IS-95	Interim Standard 95 (TIA/EIA CDMA cellular standard, U.S.)
ISDN	Integrated Services Digital Network
ISM	Industrial, Scientific, and Medical (bands, devices)
ITU-R	International Telecommunication Union-radio communication sector
LEOS	Low earth orbit satellite
MAC	Medium access control (layer)
MNP	Microcom Networking Protocol
MPT	Ministry of Posts and Telecommunications (Japan)
MSS	Mobile satellite services (or systems)
NMT	Nordic Mobile Telephone (Europe)
PACS	Personal Access Communications Services
PBX	Private branch exchange
PCN	Personal Communications Network (Europe)
PCS	Personal Communications Services (U.S.)
PDC	Personal Digital Cellular (Japan)
PHS	Personal Handyphone System (Japan, formerly PHP)
PMR	Private mobile radio
PRMA	Packet Reservation Multiple Access
PSTN	Public Switched Telephone Network
QPSK	Quaternary Phase Shift Keying
RACE	R & D in Advanced Communications Technologies in Europe
RCR	Research & development Center for Radio systems (Japan)
RF	Radio frequency
RLL	Radio local loop
RTMS	Radio Telephone Mobile System (Italy)
RX	Receive or reception
SMR	Specialized mobile radio
TACS	Total Access Communication System (Europe)
TDD	Time division duplexing
TDMA	Time division multiple access
TETRA	Trans European Trunked Radio System
TIA	Telecommunications Industry Association (U.S.)
TTC	Telecommunication Technical Committee (Japan)
TX	Transmit or transmission
UMTS	Universal Mobile Telecommunications System
VSELP	Vector sum excited linear prediction (speech coding)
WACS	Wireless Access Communications Systems
WARC	World Administrative Radio Conference

A major advantage of using GEOS is that continuous global coverage up to 75° latitude can be provided with only 3 satellites. Drawbacks include the 240-to-270-ms round trip propagation delay and the high RF power required. Although the LEOS approach minimizes the required transmit power (making small portable handsets viable) as well as the delay, there are also disadvantages. Due to the large number of satellites required to provide global coverage and their limited lifetime (five to ten years due to orbital decay), replacement satellites will have to be launched frequently. Moreover, the rapid movement of the LEOS "cells" relative to the earth (about 7.4 km/s for Iridium) requires frequent handoffs.

A detailed discussion of mobile satellite communications is beyond the scope of this article (see [29] and [30] for details), but it should be noted that there are some economic viability issues. The network operating and maintenance costs must be balanced against the willingness-to-pay of a sufficiently large market to support the service. Despite the technical and economic hurdles, however, MSS may become an important component in the future wireless global network.

Conclusion

Wireless personal communications is in the process of revolutionizing telecommunications services and the way in which people use them. Overall growth in the cordless and cellular markets during recent years has exceeded expectations. There is widespread anticipation that customer demand for wireless telecommunications will continue to expand for the foreseeable future. This is reflected by the high level of engineering activity and standards development worldwide. There are many different views on what the future will bring in terms of wireless capabilities and designs. However, one thing is clear: wireless personal communications has achieved "mainstream" status and will be a major force in driving the development of telecommunications systems and services.

References

[1] V. H. McDonald, "The Cellular Concept,"*BSTJ*, vol. 58, no. 1, part 3, pp. 15-41, Jan. 1979.

[2] M. Mouly and M.-B. Pautet, "The Evolution of GSM," in Mobile Communications - Advanced Systems and Components, Proc. 1994 Int. Zurich Sem. on Dig. Commun., (Springer-Verlag, LNCS, vol. 783).

[3] M. R. L. Hodges, "The GSM Radio Interface," *Br. Telecom Tech. J.*, vol. 8, pp. 31-43, Jan. 1990.

[4] M. Mouly and M.-B. Pautet, *The GSM System for Mobile Communications*, (Paris, 1992, ISBN 2-9507190-0-7).

[5] ETSI GSM Specifications, Series 01-12.

[6] P. A. Ramsdale, "Personal Communications in the UK — Implementation of PCN Using DCS 1800," *Internat. J. Wireless Info. Networks*, pp. 29-36, vol. 1, no. 1, Jan. 1994.

[7] TIA/EIA IS-54, "Cellular System Dual-Mode Mobile Station-Base Station Compatibility Standard," Telecommunications Industry Association, April, 1992

[8] TIA/EIA IS-41, "Cellular Radio Telecommunications Intersystem Operations," Telecommunications Industry Association, December, 1991

[9] K. Kinoshita, M. Kuramoto, and N. Nakajima, "Development of a TDMA Digital Cellular System Based on Japanese Standard," Proc. 41st IEEE Veh. Technol. Conf., pp. 642-645, 1991.

[10] Research & Development Center for Radio Systems (RCR), "Digital Cellular Telecommunication Systems," RCR STD-27, Apr. 1991.

[11] A. Salmasi and K. S. Gilhousen, "On the System Design Aspects of Code Division Multiple Access (CDMA) Applied to Digital Cellular and Personal Communications Networks," Proc. 41st IEEE Veh. Technol. Conf., pp. 57-62, 1991.

[12] TIA/EIA IS-95, "Mobile Station-Base Station Compatibility Standard for Dual-Mode Wideband Spread-Spectrum Cellular Systems," Telecommunications Industry Association, July, 1993.

[13] R. Padovani, "Reverse Link Performance of IS-95 Based Cellular Systems," *IEEE Pers. Commun.*, vol. 1, no. 3, pp. 28-34, 3Q 1994.

[14] Federal Communications Commission, Notice of Proposed Rule Making in the matter of Amendment of Parts 15 and 90 of the Commission's Rules to Provide Additional Frequencies for Cordless Telephones, ET Docket 93-235, adopted Aug. 20, 1993.

[15] R. S. Swain, "Cordless Communication in the UK," *British Telecom Technol. J.*, vol. 3, no. 2, pp. 32-38, April 1985.

[16] ETSI, "Radio Equipment and Systems - Technical characteristics, test conditions and methods of measurement for radio aspects of cordless telephones CT1."

[17] T. Hattori, A. Sasaki, and K. Momma, "Emerging Technology and Service Enhancement for Cordless Telephone Systems," *IEEE Commun. Mag.*, vol. 26, no. 1, pp. 53-58, Jan. 1988.

[18] R. Steedman, "The Common Air Interface MPT 1375," in *Cordless Telecommunications in Europe*, W. H. W. Tuttlebee, ed., (Springer Verlag, 1990).

[19] D. Åkerberg, "Novel Radio Access Principles Useful for Third Generation Mobile Radio Systems," Proc. IEEE PIMRC '92 (1992 IEEE Symposium on Personal, Indoor, and Mobile Radio Communications), Boston MA, Oct. 19-21, 1992.

[20] ETSI, Digital European Cordless Telecommunications - Common Interface, Radio Equipment and Systems, Valbonne, France.

[21] T. Hattori *et al.*, "Personal Communication - Concept and Architecture," Proc. 1990 IEEE Internat'l Commun. Conf. (ICC '90), pp. 1351-1357, Apr. 16-19, 1990.

[22] Research & Development Center for Radio Systems (RCR), "Personal Handy Phone Systems," RCR STD-28, Dec. 1993.

[23] K. Ogawa, K. Kohiyama, and T. Kobayashi, "Toward the Personal Communication Era - A Proposal of the 'Radio Access' Concept from Japan," *Internat. J. Wireless Info. Networks*, vol. 1, no. 1, pp. 17-27, Jan. 1994.

[24] D. C. Cox, "Wireless Network Access for Personal Communications," *IEEE Commun. Mag.*, vol. 30, no. 12, pp. 96-115, Dec. 1992.

[25] K. Pahlavan and A. H. Levesque, "Wireless Data Communications," *Proc. IEEE*, vol. 82, no. 9, pp. 1398 - 1430, Sep. 1994.

[26] B. Tuch, "Development of WaveLAN, an ISM Band Wireless LAN," *AT&T Tech. J.*, vol. 72, no. 4, pp. 27-37, July/Aug. 1993.

[27] C. M. Rush, "How WARC '92 Will Affect Mobile Services," *IEEE Commun. Mag.*, vol. 30, no. 10, pp. 90-96, Oct. 1992.

[28] J. L. Grubb, "The Traveler's Dream Come True," *IEEE Commun. Mag.*, vol. 29, no. 11, pp. 48-51, Nov., 1991.

[29] W. W. Wu *et al.*, "Mobile Satellite Communications," *Proc. IEEE*, vol. 82, no. 9, pp. 1431-1448, Sep. 1994.

[30] F. Abrishamkar and E. Biglieri, "An Overview of Wireless Communications," Proc. 1994 IEEE Military Commun. Conf., (MILCOM '94), pp. 900-905, October 2-5, 1994.

Biographies

JAY E. PADGETT [SM '90] received Bachelor's and Master's degrees from the University of Virginia in 1976 and 1977, respectively, and a Ph.D. from Polytechnic University (Brooklyn, New York) in 1987, all in electrical engineering. He has been with AT&T Bell Laboratories in New Jersey since 1977, and became a distinguished member of technical staff in 1986. His activities have included systems engineering for analog and digital cordless telephones, microwave terrestrial radio links, and personal communications systems, focusing on mathematical modeling and analysis of performance and capacity. He is chair of the TIA Mobile & Personal Communications Consumer Radio Section, and a member of Eta Kappa Nu and Tau Beta Pi.

CHRISTOPH G. GÜNTHER [M '87] received his diploma and doctor degrees in theoretical physics at the Swiss Federal Institute of Technology (ETH) in Zurich in 1979 and 1984, respectively. In 1984, he joined Brown Boveri Corporate Research and worked in cryptography, mainly in the development and analysis of stream cipher algorithms, homophonic coding, and key-exchange protocols. In 1989, he started working in spread spectrum transmissions and joined Ascom Tech, Mägenwil, Switzerland, in January 1990. Since 1991, he has been the Head of the Cellular System Group of that company, which is the research organization of Ascom. He is working on modulation, synchronization, multi-path receivers, packet transmission, and various aspects of personal communication systems. His interests also include information and coding theory.

TAKESHI HATTORI [M '71] received B.S., M.S., and Ph.D. degrees from the University of Tokyo in 1969, 1971, and 1974, respectively. He joined the Electrical Communication Laboratory, NTT, Japan in 1974. From 1974 to 1986, he was engaged in research on 800-MHz land mobile telephone systems, high-capacity mobile communication systems, and high-speed paging systems and technologies. From 1984 to 1986, he was senior manager at the ECL Research and Development Headquarters and worked on systems engineering and research planning. From 1986 to 1987, he was head of the Mobile Communication Applications Section in the Radio Communications Networks Laboratory, and responsible for the development of new cordless telephone and maritime telephone systems. From 1987 to 1989, he was Research Group Leader of the Radio Communication Systems Laboratory, and was involved in research on high-speed digital mobile radio transmission technologies. From 1989 to 1991, he was Project Team-3 Leader for R & D of personal communications systems. From 1991 to 1992, he was executive manager of the Research Planning Department in the NTT Radio Communication Systems Laboratories. He is currently executive manager of the Personal Communication Systems Laboratory in NTT Wireless Systems Laboratories, Kanagawa, Japan, and is responsible for R & D of wireless personal communication systems. He was awarded the IEEE Vehicular Technology Society Paper of the Year in 1981. He is a member of the IEEE Communications, Vehicular Technology, and Computer Societies, and of the Institute of Electronics and Communications Engineers (IEICE) of Japan.

Wireless personal communications has achieved mainstream status and will be a major force in driving the development of telecommunications systems and services.

Reprinted from *IEEE Personal Communications*, Vol. 1, No. 2, Second Quarter 1994, pp. 6-11.

A speculative commentary on the future

The Evolution of Personal Communications

RAYMOND STEELE

The term *personal communications* is not well defined, as are the related terms personal communications networks (PCN) and personal communications systems (PCS). PCN and PCS appear to synonymous, with the Europeans preferring the former and the Americans the latter nomenclature. However, usage of PCN/PCS varies widely. To some it is public communications where the communicator is personally owned, while others would argue that PCN does not yet exist, and the first one will be the global mobile network planned for beginning of the next century.

We are concerned here with examining how we might evolve from the current chaotic scene in mobile radio communications to the global network known in Europe as the universal mobile telecommunication system (UMTS), and by the ITU as the future public land mobile telecommunications system (FPLMTS). A pre-requisite for our discussion is a brief statement of the current scene followed by list of the cardinal goals of UMTS. From this framework we may speculate on evolutionary routes.

The Current Scene

The mobile communications scene of today is an international bazaar, with a wide variety of equipment on offer to provide different types of networks. In recent times many national telecommunications companies are, or have been, deregulated, only to be partially re-regulated to give newly formed competing companies a chance to establish themselves. Some mobile companies are closed user groups, able to exist without requiring the services of the large public switch telephone networks (PSTNs) or integrated service digital networks (ISDNs), while many mobile networks are forced to use the PSTN/ISDN and pay charges which inevitably are passed on to the mobile subscriber. The deregulating process may yield financial benefits to the consumer due to the competitive pressure on price tariffs, but they also introduce technical solutions that are more expensive, sub-optimum, and may even delay the realization of the universal mobile telecommunications system (UMTS).

The current mobile scene is composed of many disconnected systems. Table 1 shows a compilation of some of them [1-5]. There are first generation analog cellular, second generation digital cellular, analog and digital cordless, long-standing analog private mobile radio (PMR) and the latest digital PMR systems, a range of wireless LANs, existing satellite networks for land, sea and air mobiles, a group of proposed digital mobile satellite networks, entrenched and new paging systems, terrestrial flight telecommunication systems (TFTS), and so on. Space does not allow us to describe these systems. We may note that the analog cellular systems are conceptually similar, although incompatible, and based on FDMA, while the digital cellular employ either TDMA or CDMA. Cellular systems use frequency division duplex (FDD). Following the analog cordless telecommunications (CT) the digital CT2 is FDMA using time division duplex (TDD), while the digital European cordless telecommunications (DECT) system uses time division multiple access (TDMA) with TDD. Wireless LANs will be a complement to, and not a replacement for, fixed LANs.

Aims of UMTS/FPLMTS

Although UMTS and FPLMTS are sponsored by different organizations their goals are fundamentally the same. UMTS is powered by the European RACE program, and a sub-committee of the European Telecommunications Standards Institute (ETSI), called SMG5, is tasked with defining the UMTS standards. The primary aims of UMTS is global coverage for speech and low-to-medium bit rate services, with the provision of high bit rate services over limited coverage areas. To achieve these aims the air interfaces need to be defined; functions, signaling systems and architecture to support the fixed part of the UMTS network is required; integration with the PSTN/ISDN in terms of the intelligent network (IN) and universal personal telecommunications (UPT) needs quantifying; the ability to be ISDN-compatible; to be technologically independent; to have end-to-end encryption; to accommodate a range of cells from microcells to satellite cells, and so forth. The spectrum allocated is 230 MHz in the 1.885 to 2.200 GHz band. The services may be sub-divided into dialogue (eg., speech, fax, low resolution video telephony), messaging (eg., paging voice and e-mail), informational retrieval (eg., voice, music, data). The bit rates for these services vary from 1 kb/s to 2 Mb/s. UMTS/ FPLMTS is scheduled to come into operation at the beginning of the 21st century.

The Virtues of Evolution

Many of the digital systems in Table 1 have yet to be installed in any significant numbers. They are all stand-alone systems and only a few are moderately compatible. They range from satellite, PMR, and cellular to CT. The global PCN will embrace all these types of networks. However, will this global PCN evolve from existing second generation mobile networks, and if so, from which ones? Slow evolution has the virtue of allowing each new service to be commercially established before the next one is introduced. Evolution would also facilitate the steady improvement to the existing networks, avoiding the massive investment required to deploy a new network over a relatively short time scale. Thus evolution would allow operators to maintain or increase profitability during the evolutionary phase, while providing more services.

Evolution does occur naturally and continuously at the local level as equipment suppliers improve their products, operators enhance their managerial skills, and service providers add new services. The evolution we are concerned with here is the evolution imposed by international standards bodies. These bodies are often technologically rather than market driven. They concern themselves with elegant and commercially viable engineering improvements to the current state-of-the-art. They are often unaware or may appear indifferent to the aspirations of the public. Like it or not the global PCN will arise from technocrats.

Analog cellular	Nordic mobile telephone Advanced mobile phone service Total access communications system NETZ C, D Nippon advanced mobile telephone system
Digital cellular	GSM Digital communication system at 1800 MHz IS-54 (D-AMPS) IS-95 (Qualcomm CDMA) Japanese digital cellular Terrestrial flight telecommunication system
Cordless communications (CT)	Analog domestic cordless telecommunications (CT) Improved CTI Common air interface CT 2 system Digital CT at 900 MHz Digital European CT
Wireless LAN	European Telecommunication Standards Institute categories 1, 2, 3 ALTAIR (proprietary wireless LAN)
Private mobile radio (PMR)	Analog 12.5 kHz/FM Trans-European trunked radio Digital short range radio
Mobile satellite	Aeronautical in-flight Land: EUTELTRACS, PRODAT Maritime: Inmarsat A, B, C Proposed: IRIDIUM, GLOBALSTAR, ODYSSEY CONSTELLATION ELLIPSAT
Paging	Many entrenched systems European radio message system

■ **Table 1.** *Some existing or soon to be deployed mobile radio systems.*

Factors Affecting Evolution to UMTS

The public perceive mobile phones as a luxury item, that calls are relatively expensive and often of poorer quality than those of the fixed network, and that there may be biological hazards. Persuading people to buy more complex products will require technological improvements, a decrease in user equipment costs, lower tariffs, and creative marketing of new services. Then there are a number of critical technologies. These include improvements in batteries and integrated circuit devices, introduction of flat screens and voice recognition and speaker verification systems, end-to-end encryption, IN to support UMTS, introduction of multicellular, multimode interfaces, and widescale deployment of special purpose optical supporting networks. It is also imperative that the second generation networks are a financial success. Without this success UMTS is likely to be significantly retarded. Some existing networks are of a brickwall nature, being incapable of significant development. We note that development is synonymous with evolution in this context. Others are capable of much development, and possibly the network that will evolve the most is the global system of mobile telecommunications (GSM), as developments are planned for both its radio interface and its network. Evolutionary aspects of existing networks may have considerable influence on UMTS. However, if these existing networks have fundamental weaknesses UMTS should abandon them at the outset.

Evolution of GSM

The most complex of all the second generation systems is GSM, having an advanced radio system and supporting fixed network. GSM supports eight users in a TDMA format on each radio carrier. The TDMA carriers are FDMA multiplexed. GSM has many advanced features to combat fast fading, such as FEC coding, bit interleaving, channel sounding and equalization, power control, and voice activity detection. Currently we have GSM, phase 1, and because of the introduction of DCS 1800, this is now referred to as GSM 900. DCS 1800 is itself an evolution of GSM, where the operation band is at 1800 rather than 900 MHz. There are a number of marginal differences between GSM 900 and DCS 1800, the most important of which is that the transmitted power levels of DCS 1800 are lower to promote the deployment of small cells. Phase 2 GSM 900 will have lower bit rate speech codecs and with their new FEC codecs will enable 16 mobile channels per carrier. This doubling of users per carrier will not change the bandwidth occupancy of each carrier, but it will significantly increase system capacity. Phase 2 will also add data services, fax, and supplementary services. The specifications of both GSM 900 and DCS 1800 will merge to allow both systems to provide coverage in microcells and in large cells. Macrocells will overlay microcellular clusters. It is envisaged that slow moving mobiles will use microcells, faster mobiles will use the macrocells. The mobile will be able to identify itself as fast moving and do fewer cell handovers and location updates. The handover algorithms will be modified to allow for the small distances between a mobile and its microcellular BS by simplifying the mobile's timing advance procedure.

The supplementary services of call-forwarding and call-barring will be extended to conference calls, call-waiting indication, mobile PABX. GSM 900 will support improved short messaging, computer data, and video phones. Vehicle speeds of up to 500 km/hr are anticipated and the system will be adapted to cope with the high Doppler speeds that will occur. We note that there is no intention to have a GSM phase 3; instead all changes after phase 2 will be referred to as phase 2+, a term that implies continual refinement or evolution.

The Influence of GSM on Other Systems

Representatives of equipment manufacturing companies pollinate standard bodies. Having influenced the design of GSM 900 and produced the equipment to deploy the system, they modified GSM 900 to formulated DCS 1800. Having a grasp on TDMA, they influenced IS-54 (which became D-AMPS). Turning their attention to cordless telecommunications (CTs) for office and telepoint applications resulted in ETSI standardising the digital European CT (DECT) which has enhanced TDMA features. Each TDMA carrier supports 12 channels but instead of using FDD it employs TDD. A number of slots rather than a single slot per carrier can be assigned to wideband users, and it can also configure its radio interface so that most of the uplink and downlink slots can be used for transmission of data in one direction. Dynamic channel allocation (DCA) can also be used. Here clusters of BSs are given the same set of frequencies and time slots and under network control can assign them in such a way that the capacity is increased compared to earlier fixed channel assignment methods.

The multiple access method has a significant influence on the spectral efficiency, and there is much research in progress to enhance the different multiple access techniques.

Evolution of Multiple Access Methods

While the capacity of the fixed component of the mobile network is not a fundamental problem, maximizing the capacity of the radio part of the network has not yet been attained. All mobile radio systems are interference-limited, as the precious allocated radio spectrum must be continually reused. Spectral efficiency of mobile radio systems is defined in terms of Erlangs/MHz/km^2, and systems that can cope with cochannel interference with few or single cell clusters will have a higher spectral efficiency than systems that have a large number of cells per cluster. The multiple access method does have a significant influence on the spectral efficiency, and there is much research in progress to enhance the different multiple access techniques.

A discourse on evolutionary issues in PCN must therefore be concerned with the evolution of multiple access methods. We need to identify the basic features of an access method when it is obfuscated by numerous enhancements. Then we must ask ourselves if these basic features will seriously limit the performance, no matter how inventive we are. Research should then be directed to those multiple access methods that innately have good performances.

It will be recalled that all first generation mobile systems use frequency division multiple access (FDMA). Using large cells and therefore high transmission powers, the BS equipment is bulky, and one transceiver is required per channel. These features persuaded second generation designers to avoid many of these problems by opting for time division multiple access (TDMA), where a new set of difficulties relating to signal processing could be handled. We note that two of the contending systems for GSM did employ hybrid multiple access methods that involved TDMA with spectral spreading [6, 7]. Following the specification of the TDMA cellular radio interfaces, code division multiple access (CDMA) interfaces were designed. Presented with three basic multiple access options we need to identify if there is a role for all of these multiple access methods in UMTS, or is there a clear long term winner. Let us now examine the evolutionary steps that have occurred, and speculate as to what might happen.

TDMA Evolution

In the section on GSM's influence on other systems, the evolution of TDMA that has occurred in DECT was mentioned. Another embellishment is packet reservation multiple access (PRMA), which comes to mobile radio from earlier systems concerned with packet transmissions, such as those involving ALHOA techniques. Basically PRMA enables more users to be supported than there are time slots [8]. It achieves this by exploiting the statistics of an ensemble of users. For example, in voice communications users who are not currently speaking vacate their channels for those who are. Essentially PRMA exploits the condition that at any given instant a substantial number of users are not talking. As there are more users than channels there will be times when not enough channels are available when the quality of some channels becomes severely degraded.

In fixed channel assignment (FCA) systems the graph of signal-to-interference ratio (SIR) versus the number of cells per cluster M is a slowly increasing monotonic function with M. This means that to improve the SIR to meet radio link requirements may necessitate a significant increase in M that results in a substantial decrease in capacity. It is because of the shape of this curve that the theoretical minimum cluster size in both TDMA (and in FDMA) is not realized in practice. Cluster size has no meaning in DCA, as all the BSs can have the complete set of TDMA carriers. By not requiring frequency planning, DCA may have an important role to play in UMTS where three-dimensional microcells having very irregular volumes will be deployed. DCA will require accurate and rapid power control together with fast handover algorithms to enable the mobile to escape from high cochannel conditions. New protocols will be required, as well as distributed network intelligence to support spectrally efficient DCA.

The evolution of TDMA will continue and we may speculate on its inherent limitations. TDMA implies that users' transmissions are bursty rather than continuous, and that a time frame structure prevails. Consequently the transmission rate is much higher than the source date rate, so that in the future when users' data rates become very high the burst rate may become unacceptably excessive. Further, and perhaps ultimately the most significant limitation, is that the frame structure means that TDMA signals may cause serious interference with non-radio systems.

FDMA Evolution

The first generation mobile radio systems use FDMA. The development of these systems continues and it is interesting to note that systems like TACS have fax and data services, whereas GSM 900, Phase 1, does not. The innate weakness of these analog systems is their inability to provide affordable security. But what of the evolution of FDMA. By using digital FDMA the security can become the same as for TDMA. In FDMA users are assigned a specific frequency band for the duration of their call. One development of FDMA is to arrange for a user's data to be simultaneously carried on a number of different frequency channels. This can combat frequency selective fading, and is particularly appropriate for wideband signals. Frequency hopping of FDMA channels may be used, where a mobile's transmissions is continuous, but in a different frequency band during each hop. PRMA may also be applied to FDMA where more than one user may share an FDMA channel according to activity of the users' data sources.

As previously mentioned, it was equipment factors that prompted the move from FDMA to TDMA. However, most of the teletraffic carried in UMTS will be from microcells in streets and buildings. As this can be accomplished using low power levels, and due to the likely advancements in technology, FDMA equipment can be small and lightweight. Another innate virtue of FDMA is that the data rate is essentially the same as the multiplexed rate. Consequently for the highest bearer rates in UMTS, FDMA can be implemented without the signal processing complexity required in TDMA and CDMA systems. We also note that FDMA is an enabling multiple access method, as both TDMA and CDMA multiplex their carriers using FDMA.

CDMA Evolution

The roots of CDMA are in military communications, as it was conceived for covert communications with a relatively strong immunity to enemy jamming. By representing each data symbol by a code whose elements are called chips, the CDMA signal spans a wide band of frequencies and is a noise-like time signal. An important feature of CDMA is that the jamming noise is decreased by the processing gain, the ratio of the symbol duration to the chip duration.

CDMA was not seriously considered for cellular communications because it was perceived to have three fundamental weaknesses. These were the necessity to give each user a unique code, the ability to synchronize the network at the chip time, and the tight power control to ensure that the received power from all the mobiles at the base station was the same to within a small error. These weaknesses were all overcome by Qualcomm. Their system is now the second U.S. digital cellular standard, IS-95.

Qualcomm's activities in CDMA have had a profound catalytic effect as witnessed by the current world wide explosion of research in CDMA and the new CDMA systems that are being offered. We may expect to see variable rate CDMA, with high chip rates being used in microcells and lower chip rates in larger cells. Interference cancellation techniques will be enhanced to significantly increase capacity, macrodiversity will be deployed, co-existence with other CDMA services using the same frequency band will be provided, and so forth. CDMA does not seem to have the serious innate deficiencies of TDMA, except that if services require very high bit rates, the demands on the technology to generate much higher chip rates may be prohibitive. The spectral efficiency of CDMA versus an efficient DCA system needs to be resolved. CDMA would get my vote with its innate robustness to interference.

Network Evolution

At the present time mobile communication networks are minuscule compared to PSTN/ISDN networks that handle vast volumes of traffic, with their trunk transmission rates many orders greater than mobile rates. PSTN/ISDN operators have their own evolutionary agenda. It involves establishing an intelligent network (IN) in which functions, such as call diversion, call holding, multimedia facilities, etc., that are often limited to PBXs, will be stored in a central location accessible from the PSTN/ISDN. A flexible hardware architecture enables the stored functions to be realised via separate software modules. New features and new services only require new software modules to be introduced.

The IN conceptual model has four planes of abstraction. The service plane describes services and service features from a users perspective. The global plane has serviceindependent building blocks (SIBs), which are units of service functionality. SIBs when combined with service logic in the global plane are able to realize services and service features on the service plane. Below the global plane is the distributed function plane. This plane has units of network functionality called functional entities (FEs) and the information flow between FEs is called a relationship. A sequence of FE actions (FEA) and the resulting relationships will realise a SIB on the global plane. The bottom plane is the physical plane. The potential physical systems, called physical entities (PEs) and their interfaces describe different physical architectures that can be realised. FEs on the distributed plane are physically realised on the physical plane, while relationships on the distributed plane are mapped on the physical plane as physical interfaces [9].

IN is evolving from the existing PSTN/ISDN. Each set of IN capabilities is called a capability set (CS) and each IN CS is to be evolutionary toward the goal of service creation, management, interaction and processing; and network management and interworking. Only IN CS-l currently exists.

The roots of CDMA are in miliary communications as it was conceived as for covert communications with a relatively strong immunity to enemy jamming.

The services will include freephone, UPT, virtual private network (VPN), and IN CS-l will support flexible routing, user interaction and charging. IN CS-l is concerned with actions at the initiation and termination of calls, not actions during calls. There are 14 SIBs.

Now let us turn our attention to IN for mobile networks. All fixed network IN features are required by mobiles, plus IN functions relating to subscriber service mobility and call management. Current mobile networks have these features and could develop their own INs. With many different service providers and mobile network operators, plus the numerous national PSTN/ISDN networks, we could have a myriad of INs needing to be themselves networked. This scenario is at variance with the PSTN/ISDN IN, which has a single centralized location for all software and data bases. Mobile communications in microcellular environments will require large information flows of location updating, which implies distributed data bases. Along with distributed data bases is distributed control due to the numerous handovers during a call.

Some mobile networks will grow and offer all the services of the current PSTN/ISDN, while some PSTN/ISDN will develop wireless local loop access. Each may have its own IN with gateways between INs. The evolution of IN in the fixed network has a momentum of its own. Nevertheless, given the size of the problems the IN deployment may be slow, and UMTS is likely to develop its own independent IN if it is to be deployed according to its time scales.

IS-95 does have a greater capacity than GSM, but unlike GSM it is only designed to the base station controller (BSC) level in GSM architecture. However, a full network design of IS-95 is imminent as operators are in the process deploying this system.

Evolutionary Routes

Route 1

GSM, Phase 2+ will continue to develop and will be doing so by the turn of the century when deployment of UMTS is scheduled to start. By then GSM may be widespread with an effective IN in place. Its radio interface will have been honed, and microcells and many advance features and services required by UMTS will exist in GSM. As a consequence, GSM may have prevented DECT and CT2 from developing, as GSM will operate in CT environments and, equipped with FEC codecs and equalizers, it will not have difficulties with dispersive channels. GSM will be a combined cellular and CT network. An alternative is that GSM operators will use DECT equipment coverage in both CT and cellular environments. Further, GSM may develop to accommodate closed user groups and seriously compete with second generation digital private mobile ratio (PMR).

If GSM develops in this way with mobiles able to log on to either GSM 900 or its sister DCS 1800, then how much more will GSM have to develop to satisfy the goals of UMTS. Firstly, it must become a global standard. This requires all countries to adopt it and in particular for the USA to do so. While it is unlikely that GSM 900 will be deployed in the United States, DCS 1800 may be introduced as a PCS network. Armed with a dual GSM 900/DCS 1800 portable, a user would then be able to communicate from anywhere in the world. While GSM 900 Phase 2+ will be able to support many UMTS services, it cannot provide a significant number of them, particularly those requiring high bit rates, such as 2 mb/s.

We conclude by considering that GSM will be deployed over most of the world, will have many UMTS features, but that cannot evolve into the UMTS standard because of its relatively low TDMA rate.

Route 2

UMTS is driven in Europe by ETSI committee SMG5. Technical information is provided by the RACE II program dealing with mobile communications. Other inputs come from academic and industrial organizations. Some members of SMG5 sit on the ITU committee designing FPLMTS, along with other international representatives. The result is that the SMG5 and ITU committees have essentially common goals. That is why the term UMTS has also been used to mean, in the final outcome, FPLMTS.

IS-95 does have a greater capacity than GSM, but unlike GSM it is only designed to the base station controller (BSC) level in GSM architecture. However, a full network design of IS-95 is imminent as operators are in the process deploying this system. If IS-95 significantly evolves (Qualcomm has an evolved CDMA at 1.7 GHz with higher bit rate services compared to IS-95), and the GSM fixed network meets UMTS criteria, then an up-graded IS-95 radio interface operating with the GSM, Phase 2+ fixed network might be able to provide more UMTS services than a GSM Phase 2+ system. However, this arrangement would not satisfy the goals of UMTS. It is important to note that evolved GSM network with an evolved IS-95 radio interface would provide strong competition to a fledgling UMTS network, and may slow the introduction of an effective UMTS service.

Route 3

This route is different because it does not adhere to the simple notion that one type of radio interface will suffice. The objectives for UMTS are relatively modest, except in one aspect, i.e., the global dimension. If we consider that UMTS should be an evolving network, then we need to design into UMTS the flexibility to continually introduce more services and high data rate links. We need to think in terms of a plethora of cell types to cover our global environment: picocells (few meters), nanocells (up to 10 m), nodalcells (10 to 200 m), microcells (10 to 400 m for pedestrians, 300 m to 2 km for vehicles), minicells (500 m to 3 km), macrocells (up to 5 km), large cells (5 to > 35 km), megacells (20 to 100 km) and satellite cells (> 500 km). Currently cellular radio employs minicells and above, while CT uses microcells and below.

UMTS will have the complete range of cells. Cochannel interference in macrocells and in street microcells depends on the size of buildings, whereas in CTs it is dependent on office construction, furniture and people. It is interference that effects cluster size and forces TDMA and FDMA to work with cluster sizes that exceed unity. CDMA can operate with single cell clusters. However, in city streets the buildings may significantly shield the cochannel propagation and TDMA with DCA may be a viable option. For nodalcells or megacells, we advocate FDMA due to the high data rates. Suffice to say that the optimum multiple access method is dependent on the type of cell and service, and that a rigid radio interface would be suboptimum in some situations. This leads to the notion of a radio interface where the network selects the type of multiple access to enhance performance. From this concept we propose that other radio link elements, such as source codec, FEC codec, and modulation be modified under system control. The UMTS terminal is now seen to have an adaptive structure. It will adapt to different channel conditions, teletraffic loading, and required service. We call this terminal an intelligent multimode terminal (IMT) [10].

The IMT is able to operate in a multicellular environment, although some IMTs may be designed to operate in a limited range of cell types, e.g., megacells and satellite cells. Base stations will emit radio beacons and the IMT may receive a number of these beacons, each uniquely coded. The IMT selects the beacon associated with the smallest cell as the smaller the cell the wider the range of services, particularly high bit rate services, it can support. Each type of cell, e.g., a microcell, has a specific default set of radio-link sub-systems for logging on to the base station. Authentication and registration ensues.

Each sub-system in the IMT is programmable under software control from its management entity (ME). When the user dials up a service the ME using the default sub-systems arranges for the IMT to establish a call for the required service. Once the service has been granted, and the call connected, the BS informs the ME in the IMT of the type of multiple access, modulation, FEC coding that is required. The BS makes this decision in conjunction with network control, which knows what the other adjacent cells are doing in terms of local area teletraffic levels and estimated cochannel levels. As the call proceeds, the IMT receiver determines the quality of the downlink radio channel, while the BS measures the quality of the uplink channel. The channel quality information, together with information related to local teletraffic loading and cochannel levels at the BS, are used to fine tune the radio link subsystems.

Evolution from second generation to this global high capacity multicellular network with IN, UPT, fluid radio interfaces, and a wide range of services can be done in stages. The generic forms of second generation mobile systems such as cellular, CT, and PMR are identified, and a set of stage 1 UMTS systems is conceived. For CTs there is m-UMTS; for cellular, M-UMTS; and for PMR, P-UMTS. Each ( )-UMTS can emulate the second generation network with which it is associated, but more, it can provide enhanced services that are expected of a UMTS working in that specific environment. It is observed that ( )-UMTS terminal has a flexible radio interface, but not the IMT mentioned above. This method provides evolution to Stage I with backward compatibility to its second generation system.

In Stage II m-UMTS and M-UMTS merge to give m/M-UMTS. This evolution merges CT and cellular radio LANs in offices, while supporting a wide range of services. If cellular and CT networks have already merged by the time of Stage I, then m/M would be introduced without recourse to the independent stages m and M. Once the satellite and megacellular systems are in place, S-UMTS is formed. Now there are three intermediate stage networks: m/M- UMTS, P-UMTS and S-UMTS, all with the same architecture and compatible protocols. These networks are then forged together to yield UMTS.

In Stage III the adaptation of the UMTS network yields the full IMT. The network becomes an adaptive one with flexible architecture, distributive data bases, personalized mobility profiles, rapid introduction of new services, and the ability to interwork with other networks.

If we consider that UMTS should be an evolving network, then we need to design into UMTS the flexibility to continually introduce more services and high data rate links.

Acknowledgements

This article is itself an evolution. Starting from a report for the European Commission, to whom much thanks are given, it continued from the keynote lectures at Wireless '93 in Calgary and the International Symposium of Personal Communications, in Nanjing, China, October 1993, to this present discourse.

References

[1] R. Steele, Mobile Radio Communications, (London: Pentech Press, 1992).
[2] M. Mouly and M-B Pautet, The GSM System for Mobile Communications, Published by authors, 1992.
[3] W. Tuttlebee, Cordless Communications in Europe, (London: Springer-Verlag, 1990).
[4] Proc. Fifth Nordic Seminar on Digital mobile radio communications, DMRV, Helsinki, Dec., 1992.
[5] PCS: The Second Generation, special issue of *IEEE Communications Magazine*, vol. 30, no. 12, Dec. 1992.
[6] W-H. Lam and R. Steele, 'The Error Performance of CD900-like Cellular Mobile Radio Systems,' *IEEE Trans. on Veh. Tech.*, vol. 40, no. 4, Nov. 1991, pp. 671-685.
[7] D. G. Appleby, *et al.*, 'The Proposed Multiple Access Methods for the Pan-European Mobile Radio System,' IEE Colloquium, London, Digest No. 1986/95, Oct. 1986, pp. 1/1-1/26.
[8] D. J. Goodman and S.X. Wei, 'Efficiency of Packet Reservation Multiple Access,' *IEEE Trans. on Veh. Tech.*, vol. 40, no. 1, Feb. 1991, pp. 170-176.
[9] J. M. Duran and J. Visser, "International Standards for Intelligent Networks," *IEEE Commun. Mag.*, vol. 1, no. 2, Feb. 1992, pp 27-32.
[10] R. Steele and J. E. B. Williams, "Third generation PCN and the intelligent multimode mobile portable," *IEE Elec. & Commun. Eng. J.*, vol. 5, no. 3, June 1993, pp. 147-156.

Biography

RAYMOND STEELE [SM '80] received a B.Sc. in electrical engineering from Durham University, England, in 1959 and Ph.D. and D.Sc. degrees from Loughborough University of Technology, England, in 1975 and 1983, respectively. Before attaining his B.Sc., he was an indentured aprentice radio engineer. After R & D posts with E.K. Cole, Cossor Radar and Electronics, and Marconi, he joined the lecturing staff at the Royal Naval College. He moved to Loughborough University in 1968 where he lectured and directed a research group in digital encoding of speech and picture signals. During the summers of 1975, 1977, and 1978 he was a consultant to the Acoustics Research Department at Bell Laboratories in the United States, and in 1979 he joined the company's Communications Methods Research Department, Crawford Hill Laboratory. He returned to England in 1983 to become professor of communications in the Department of Electronics and Computer Science at the University of Southampton, a post he retains. From 1983 to 1986, he was a non-executive director of Plessey Research alnd Technology and from 1983 to 1989, a consultant to British Telecom Research Laboratories. In 1986, he formed Multiple Access Communications Ltd., a company concerned with digital mobile radio systems. He is a senior technical editor of *IEEE Communications Magazine*, a Fellow of the Royal Academy of Engineering, and a Fellow of the IEE in the United Kingdom.

Reprinted from *IEEE Personal Communications*, Vol. 2, No. 2, Apr. 1995, pp. 20-35.

An evolution toward three large groups of applications and services

Wireless Personal Communications: What Is It?

DONALD C. COX

Wireless Personal Communications has captured the attention of the media, and with it, the imagination of the public. Hardly a week goes by without one seeing an article on the subject appearing in a popular U.S. newspaper or magazine. Articles ranging from a short paragraph to many pages regularly appear in local newspapers, as well as in nationwide print media, e.g., *The Wall Street Journal*, *The New York Times*, *Business Week*, and *U.S. News and World Report*. Countless marketing surveys continue to project enormous demand, often projecting that at least half of the households, or half of the people, want wireless personal communications. Trade magazines, newsletters, conferences, and seminars on the subject by many different names have become too numerous to keep track of, and technical journals, magazines, conferences and symposia continue to proliferate and to have ever increasing attendance and numbers of papers presented. It is clear that wireless personal communications is, by any measure, the fastest growing segment of telecommunications.

However, if you look carefully at the seemingly endless discussions of the topic, you cannot help but note that they are often describing different "things", i.e., different versions of wireless personal communications [1, 2]. Some discuss pagers, or messaging, or data systems, or access to the National Information Infrastructure, while others emphasize cellular radio, or cordless telephones, or dense systems of satellites. Many make reference to popular fiction entities like Dick Tracy, Maxwell Smart, or *Star Trek*.

Thus, it appears that almost everyone wants Wireless Personal Communications, but, *What Is It*?!! There are many different ways to segment the complex topic into different communications applications, modes, functions, extent of coverage, or mobility [1, 2]. The complexity of the issues has resulted in considerable confusion in the industry, as evidenced by the many different wireless systems, technologies, and services being offered, planned, or proposed. Many different industry groups and regulatory entities are becoming involved. The confusion is a natural consequence of the massive dislocations that are occurring, and will continue to occur, as we progress along this large change in the paradigm of the way we communicate. Among the different changes that are occurring in our communications paradigm, perhaps the major ingredient is the change from wired fixed place-to-place communications to wireless mobile person-to-person communications. Within this major change are also many other changes, e.g., an increase in the significance of data and message communications, a perception of possible changes in video applications, and changes in the regulatory and political climates.

This article attempts to identify different issues and to put many of the activities in wireless into a framework that can provide perspective on what is driving them, and perhaps even yield some indication of where they appear to be going in the future. However, like any attempt to categorize many complex interrelated issues, there are some that don't quite fit into neat categories, so there will remain some dangling loose ends. Like any major paradigm shift, there will continue to be considerable confusion as many entities attempt to interpret the different needs and expectations associated with the new paradigm.

Background and Issues

Mobility and Freedom from Tethers

Perhaps the clearest ingredients in all of the wireless personal communications activity are the desire for mobility in communications, and the companion desire to be free from tethers, i.e., from physical connections to communications networks. These desires are clear from the very rapid growth of mobile technologies that provide primarily two-way voice services, even though economical wireline voice services are readily available. For example, cellular mobile radio has experienced rapid growth. Growth rates have been between 35 and 60 percent per year in the United States for a decade, with the total number of subscribers reaching 20 million by year-end 1994. The often neglected wireless companions to cellular radio, i.e., cordless telephones, have experienced even more rapid, but harder to quantify, growth with sales rates often exceeding 10 million sets a year in the United States, and with an estimated usage significantly exceeding 50 million in 1994. Telephones in airliners, have also become commonplace. Similar, or even greater, growth in these wireless technologies has been experienced throughout the world.

Paging and associated messaging, while not providing two-way voice, do provide a form of tetherless mobile communications to many subscribers worldwide. These services have also experienced significant growth. There is even a glimmer of a market in the many different specialized wireless data applications evident in the many wireless local area network (WLAN) products on the market, the several wide area data services being offered, and the specialized satellite-based message services being provided to trucks on highways.

 1070-9916/95/$04.00 © 1995

The topics discussed in the previous two paragraphs indicate a dominant issue separating the different evolutions of wireless personal communications. That issue is the voice versus data communications issue that permeates all of communications today; this division also is very evident in fixed networks. The packet-oriented computer communications community and the circuit-oriented voice telecommunications (telephone) community hardly talk to each other, and often speak different languages in addressing similar issues. Although they often converge to similar overall solutions at large scales (e.g., hierarchical routing with exceptions for embedded high usage routes), the small scale initial solutions are frequently quite different. Asynchronous Transfer Mode (ATM)-based networks are an attempt to integrate, at least partially, the needs of both the packet-data and circuit-oriented communities.

Superimposed on the voice-data issue is an issue of competing modes of communications that exist in both fixed and mobile forms. These different modes include:

Messaging, where the communication is not real time, but is by way of message transmission, storage, and retrieval. This mode is represented by voice mail, electronic facsimile (fax), and electronic mail (e-mail), the latter of which appears to be a modern automated version of an evolution that includes telegraph and telex. Radio paging systems often provide limited one-way messaging, ranging from transmitting only the number of a calling party, to longer alpha-numeric text messages.

Real-time two-way communications, represented by the telephone, cellular mobile radio telephone, and interactive text (and graphics) exchange over data networks. Two-way video phone always captures significant attention and fits into this mode; however, its benefit/cost ratio has yet to exceed a value that customers are willing to pay.

Paging, i.e., broadcast with no return channel, alerts a paged party that someone wants to communicate with him/her. Paging is like the ringer on a telephone, without having the capability for completing the communications.

Agents, new high level software applications or entities being incorporated into some computer networks. When launched into a data network, an "agent" is aimed at finding information by some title or characteristic, and returning the information to the point from which the agent was launched.

There are still other ways in which wireless communications have been segmented in attempts to optimize a technology to satisfy the needs of some particular group. Examples include:

- User location, that can be differentiated by indoors or outdoors, or on an airplane or a train.
- Degree of mobility, that can be differentiated either by speed, e.g., vehicular, pedestrian, or stationary, or by size of area throughout which communications are provided.

At this point one should again ask: "Wireless Personal Communications — *What Is It*?!!" The evidence suggests that what is being sought by users, and produced by providers, can be categorized according to the following two main characteristics.

Communications Portability and Mobility on many different scales:
- Within a house or building (cordless telephone, wireless local area networks (WLANs)).
- Within a campus, a town, or a city (cellular radio, WLANs, wide area wireless data, radio paging, extended cordless telephone).
- Throughout a state or region (cellular radio, wide area wireless data, radio paging, satellite-based wireless).
- Throughout a large country or continent (cellular radio, paging, satellite-based wireless).
- Throughout the world?!!

Communications by many different modes for many different applications:
- Two-way voice.
- Data.
- Messaging.
- Video?

Thus, it is clear why wireless personal communications today is not one technology, not one system, and not one service, but encompasses many technologies, systems and services optimized for different applications.

Evolution of Technologies, Systems, and Services

Technologies and systems [1-7] that are currently providing, or are proposed to provide, wireless communications services can be grouped into about seven relatively distinct groups, although there may be some disagreement on the group definitions, and in what group some particular technology or system belongs. All of the technologies and systems are evolving as technology advances and perceived needs change. Some trends are becoming evident in the evolutions. In this section, different groups and evolutionary trends are explored along with factors that influence the characteristics of members of the groups. The grouping is generally with respect to scale of mobility and communications applications or modes.

Perhaps the clearest ingredients in all of the wireless personal communications activity are the desire for mobility in communications, and the companion desire to be free from tethers, i.e., from physical connections to communications networks.

Cordless Telephones

Cordless telephones [1-3] generally can be categorized as providing low mobility, low-power, two-way tetherless voice communications, with low mobility applying both to the range and the user's speed. Cordless telephones using analog radio technologies appeared in the late 1970s, and have experienced spectacular growth. They have evolved to digital radio technologies in the forms of second-generation cordless telephone (CT-2), and Digital European Cordless Telephone (DECT) standards in Europe, and several different Industrial Scientific Medical (ISM) band technologies in the United States.[1]

[1] These ISM technologies either use spread spectrum techniques (direct sequence or frequency hopping), or very low transmitter power ($\lesssim$ 1 mw) as required by the ISM band regulations.

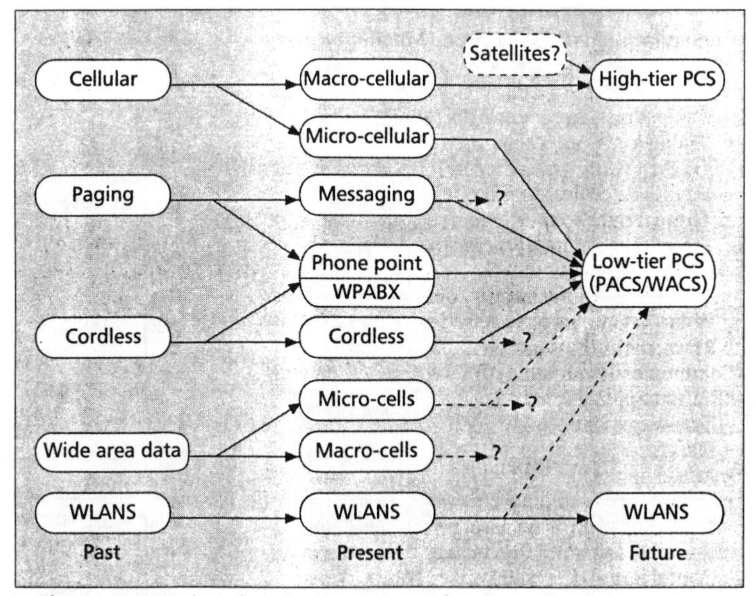

■ **Figure 1.** *Digital wireless access systems evolution.*

Cordless telephones were originally aimed at providing economical, tetherless voice communications inside residences, i.e., at using a short wireless link to replace the cord between a telephone base unit and its handset. The most significant considerations in design compromises made for these technologies are to minimize total cost, while maximizing the "talk time" away from the battery charger. For digital cordless phones intended to be carried away from home in a pocket, e.g., CT-2 or DECT, handset weight and size are also major factors. These considerations drive designs toward minimizing complexity, and minimizing the power used for signal processing and for transmitting.

Cordless telephones compete with wireline telephones. Therefore, high circuit quality has become a requirement. Early cordless sets had marginal quality. They were purchased by the millions, and discarded by the millions, until manufacturers produced higher-quality sets. Cordless telephones sales then exploded. Their usage has become commonplace, approaching, and perhaps exceeding, usage of "corded" telephones.

The compromises accepted in cordless telephone design in order to meet the cost, weight, and talk-time objectives are:

- Few users per MHz.
- Few users per base unit (many link together a particular handset and base unit).
- Large number of base units per unit area; one or more base units per wireline access line (in high-rise apartment buildings the density of base units is very large).
- Short transmission range.

There is no added network complexity since a base unit looks to a telephone network like a wireline telephone. These issues are also discussed in [1, 2].

Digital cordless telephones in Europe have been evolving for a few years to extend their domain of use beyond the limits of inside residences. Cordless telephone, second generation, (CT-2) has evolved to provide telepoint or phone-point services. Base units are located in places where people congregate, e.g., along city streets and in shopping malls, train stations, etc. Handsets registered with the phone-point provider can place calls when within range of a telepoint. CT-2 does not provide capability for transferring (handing off) active wireless calls from one phone point to another if a user moves out of range of the one to which the call was initiated. A CT-2+ technology, evolved from CT-2 and providing limited handoff capability, is being deployed in Canada. Phone-point service was introduced in the United Kingdom twice, but failed to attract enough customers to become a viable service. However, in Singapore and Hong Kong, CT-2 phone-point has grown rapidly, reaching over 150,000 subscribers in Hong Kong [8] in mid-1994. The reasons for the success in some places and failure in others are still being debated, but it is clear that the compactness of the Hong Kong and Singapore populations make the service more widely available, using fewer base stations than in more spreadout cities. Complaints of CT-2 phone-point users in trials have been that the radio coverage was not complete enough, and/or they could not tell whether there was coverage at a particular place, and the lack of handoff was inconvenient. In order to provide the "alerting" or "ringing" function for phone-point service, conventional radio pagers have been built into some CT-2 handsets. (The telephone network to which a CT-2 phone point is attached has no way of knowing from which base units to send a ringing message, even though the CT-2 handsets can be "rung" from a home base unit).

Another European evolution of cordless telephones is Digital European Cordless Telephone (DECT) which was optimized for use inside buildings. Base units are attached through a controller to private branch exchanges (PBXs), key telephone systems, or phone company CENTREX telephone lines. DECT controllers can hand off active calls from one base unit to another as users move, and can "page" or "ring" handsets as a user walks through areas covered by different base units.

These cordless telephone evolutions to more widespread usage outside and inside with telepoints, and to usage inside large buildings are illustrated in Fig. 1, along with the integration of paging into handsets to provide alerting for phone-point services. They represent the first attempts to increase the service area of mobility for low-power cordless telephones.

Some of the characteristics of the digital cordless telephone technologies, CT-2 and DECT, are listed in Table 1. Additional information can be found in References [2, 3]. Even though there are significant differences between these technologies, e.g., multiple access technology (FDMA or TDMA/FDMA), and channel bit rate, there are many similarities that are fundamental to the design objectives discussed earlier, and to a user's perception of them. These similarities and their implications are as follows.

32 kb/s adaptive differential pulse code modulation (ADPCM) digital speech encoding: this is a low complexity (low signal processing power) speech encoding process that provides wireline speech quality and is an international standard.

Average transmitter power ≤ 10 milliwatts: this permits many hours of talk time with small, low-cost, lightweight batteries, but provides limited radio range.

	High Power Systems				Low Power Systems			
	Digital Cellular (High Tier PCS)				Low Tier PCS		Digital Cordless	
System	IS-54	IS-95 (DS)	GSM	DCS-1800	WACS/PACS	Handi-Phone	DECT	CT-2
Multiple access	TDMA/FDMA	CDMA/FDMA	TDMA/FDMA	TDMA/FDMA	TDMA/FDMA	TDMA/FDMA	TDMA/FDMA	FDMA
Freq. band (MHz) Uplink (MHz) Downlink (MHz)	 869-894 824-849 (USA)	 869-894 824-849 (USA)	 935-960 890-915 (Eur.)	 1710-1785 1805-1880 (UK)	 Emerg. Tech.* (USA)	1895- 1907 (Japan)	1880- 1900 (Eur.)	864-868 (Eur. and Asia)
RF ch. spacing Downlink (KHz) Uplink (KHz)	 30 30	 1250 1250	 200 200	 200 200	 300 300	300	1728	100
Modulation	π/4 DQPSK	BPSK/QPSK	GMSK	GMSK	π/4 QPSK	π/4 DQPSK	GFSK	GFSK
Portable txmit Power, max./avg.	600 mW/ 200 mW	600 mW	1 W/ 125 mW	1 W/ 125 mW	200 mW/ 25 mW	80 mW/ 10 mW	250 mW/ 10 mW	10 mW/ 5 mW
Speech coding	VSELP	QCELP	RPE-LTP	RPE-LTP	ADPCM	ADPCM	ADPCM	ADPCM
Speech rate (kb/s)	7.95	8 (var.)	13	13	32/16/8	32	32	32
Speech ch./RF ch.	3	-	8	8	8/16/32	4	12	1
Ch. bit rate (kb/s) Uplink (kb/s) Downlink (kb/s)	 48.6 48.6		 270.833 270.833	 270.833 270.833	 384 384	384	1152	72
Ch. coding	1/2 rate conv.	1/2 rate fwd 1/3 rate rev.	1/2 rate conv.	1/2 rate conv.	CRC	CRC	CRC (control)	None
Frame (ms)	40	20	4.615	4.615	2.5	5	10	2

* Spectrum is 1.85 to 2.2 GHz allocated by the FCC for emerging technologies; DS is direct sequence.

■ **Table 1.** *Wireless PCS technologies.*

Low-complexity radio signal processing: there is no forward error correction and no complex multipath mitigation (i.e., no equalization or spread spectrum).

Low transmission delay, e.g., < 50 ms, and for CT-2 < 10 ms round trip: this is a speech-quality and network-complexity issue. A maximum of 10 ms should be allowed, taking into account additional inevitable delay in long-distance networks. Echo cancellation is generally required for delays > 10 ms.

Simple frequency-shift modulation and non-coherent detection: while still being low in complexity, the slightly more complex 4QAM modulation with coherent detection provides significantly more spectrum efficiency, range and interference immunity.

Dynamic channel allocation: While this technique has potential for improved system capacity, the cordless-telephone implementations do not take full advantage of this feature for handoff, and thus cannot reap the full benefit for moving users [9, 10].

Time division duplex (TDD): this technique permits the use of a single contiguous frequency band, and implementation of diversity from one end of a radio link. However, unless all base station transmissions are synchronized in time, it can incur severe cochannel interference penalties in outside environments [9, 11]. Of course, for cordless telephones used inside with base stations not having a propagation advantage, this is not a problem. Also, for small indoor PBX networks, synchronization of base station transmission is easier than is synchronization throughout a widespread outdoor network, which can have many adjacent base stations connected to different geographic locations for central control and switching.

Cellular Mobile Radio Systems

Cellular mobile radio systems are becoming known in the United States as high-tier Personal Communications Service (PCS), particularly when implemented in the new 1.9 GHz PCS bands [12]. These systems generally can be categorized as providing high-mobility, wide-ranging, two-way tetherless voice communications. In these systems, high mobility refers to vehicular speeds, and also to widespread regional to nationwide coverage [1, 2, 7]. Mobile radio has been evolving for over 50 years. Cellular radio integrates wireless access with large-scale networks having sophisticated intelligence to manage mobility of users.

Cellular radio was designed to provide voice service to wide-ranging vehicles on streets and highways [1-3, 13], and generally uses transmitter power on the order of 100 times that of cordless telephones (≈ 2 watts for cellular). Thus, cellular systems can only provide reduced service to handheld sets that are disadvantaged by using somewhat lower transmitter power (< 0.5 watts) and less efficient antennas than vehicular sets. Handheld sets used inside buildings have the further disadvantage of attenuation through walls that is not taken into account in system design.

Cellular radio or high-tier PCS has experienced large growth as noted earlier. In spite of the limitations on usage of handheld sets noted above, handheld cellular sets have become very popular, with their sales becoming comparable to the sales of vehicular sets. Frequent complaints from handheld cellular users are that batteries are too large and heavy, and both talk time and standby time are inadequate.

Cellular radio at 800 MHz has evolved to digital radio technologies [1-3] in the forms of the deployed systems standards:

- Global Standard for Mobile (GSM) in Europe.
- Japanese or Personal Digital Cellular (JDC or PDC) in Japan.
- U.S. TDMA digital cellular known as USDC or IS-54.

and in the form of the code division multiple access (CDMA) standard, IS-95, which is under development, but not yet deployed.

The most significant consideration in the design compromises made for the U.S. digital cellular or high-tier PCS systems was the high cost of cell sites (base stations). A figure often quoted is U.S. $1 million for a cell site. This consideration drove digital system designs to:

- Maximize users per MHz.
- Maximize the users per cell site.

Because of the need to cover highways running through low population-density regions between cities, the relatively high transmitter power requirement was retained to provide maximum range from high antenna locations.

Compromises that were accepted while maximizing the above parameters are:

- High transmitter power consumption.
- High user-set complexity, and thus high signal-processing power consumption.
- Low circuit quality.

The use of microcell base stations provides large increases in overall system capacity, while also reducing the cost per available radio channel, and the battery drain on portable subscriber equipment.

- High network complexity, e.g., the new IS-95 technology will require complex new switching and control equipment in the network, as well as high-complexity wireless-access technology.

Cellular radio or high-tier PCS has also been evolving for a few years in a different direction, toward very small coverage areas or microcells. This evolution provides increased capacity in areas having high user density, as well as improved coverage of shadowed areas. Some microcell base stations are being installed inside, in conference center lobbies and similar places of high user concentrations. Of course, microcells also permit lower transmitter power that conserves battery power when power control is implemented, and base stations inside buildings circumvent the outside wall attenuation. Low complexity microcell base stations also are considerably less expensive than conventional cell sites, perhaps two orders of magnitude less expensive. Thus, the use of microcell base stations provides large increases in overall system capacity, while also reducing the cost per available radio channel, and the battery drain on portable subscriber equipment. This microcell evolution, illustrated in Fig. 1, moves handheld cellular sets in a direction similar to that of the expanded-coverage evolution of cordless telephones to phone points and wireless PBX.

Some of the characteristics of digital-cellular or high-tier PCS technologies are listed in Table 1 for IS-54, IS-95, and GSM at 900 MHz, and DCS-1800, which is GSM at 1800 MHz. Additional information can be found in [1-3]. The JDC or PDC technology, not listed, is similar to IS-54. As with the digital cordless technologies, there are significant differences among these cellular technologies, e.g., modulation type, multiple access technology, and channel bit rate. However, there are also many similarities that are fundamental to the design objectives discussed earlier. These similarities and their implications are as follows.

Low bit-rate speech coding; ≤ 13 kb/s with some ≤ 8 kb/s: low bit-rate speech coding obviously increases the number of users per MHz and per cell site. However, it also significantly reduces speech quality [1], and does not permit the tandemming of speech encoding while traversing a network. That is, when low bit rate speech is transcoded to a different encoding format, e.g., to 64 kb/s as is used in many networks, or from an IS-54 phone on one end to a GSM or IS-95 phone on the other end, the speech quality deteriorates precipitously. While this may not be a serious issue for a vehicular mobile user who has no choice other than not to communicate at all, it is likely to be a serious issue in an environment where a wireline telephone is available as an alternative. It is also less serious when there are few mobile-to-mobile calls through the network, but, as wireless usage increases, and digital mobile-to-mobile calls become commonplace, the marginal transcoded speech quality is likely to become a serious issue.

Some implementations make use of speech inactivity: this further increases the number of users per cell site, i.e., the cell-site, capacity. However, it also further reduces speech quality [1] because of the difficulty of detecting the onset of speech. This problem is even worse in an acoustically noisy environment like an automobile.

High transmission delay; ≈ 200 ms round trip: this is another important circuit-quality issue. Such large delay is about the same as one-way transmission through a synchronous-orbit communications satellite. A voice circuit with digital cellular technology on both ends will experience the delay of a full satellite circuit. It should be recalled that one reason long-distance circuits have been removed from satellites and put onto fiber-optic cable is because customers find the delay to be objectionable. This delay in digital cellular technology results from both computation for speech bit-rate reduction, and from complex signal processing, e.g., bit interleaving, error correction decoding, and multipath mitigation (equalization or spread spectrum (CDMA)).

High-complexity signal processing, both for speech encoding and for demodulation: signal processing has been allowed to grow without bound, and is about a factor of 10 greater than that used in the low-complexity digital cordless telephones [1]. Since several watts are required from a battery to produce the high transmitter power in a cellular or high-tier PCS set, signal-processing power is not as significant as it is in the low-power cordless telephones.

Fixed channel allocation: the difficulties associated with implementing capacity-increasing dynamic channel allocation to work with handoff [9, 10] have impeded its adoption in systems requiring reliable and frequent handoff.

Frequency division duplex (FDD): cellular systems have already been allocated paired-frequency bands suitable for FDD. Thus, the network or system complexity required for providing synchronized transmissions [9, 11] from all cell sites for TDD has not been embraced in these digital cellular systems. Note that TDD has not been employed in IS-95 even though such synchronization is required for other reasons.

Mobile/portable set power control: the benefits of increased capacity from lower overall co-channel interference, and reduced battery drain have been sought by incorporating power control in the digital cellular technologies.

Wide Area Wireless Data Systems

Existing wide area data systems generally can be categorized as providing high mobility, wide-ranging, low-data-rate digital data communications to both vehicles and pedestrians [1, 2]. These systems have not experienced the rapid growth that the two-way voice technologies have, even though they have been deployed in many cities for a few years and have established a base of customers in several countries. Examples of these packet data systems are shown in Table 2.

	CDPD	RAM Mobile (Mobitex)	ARDIS (KDT)	Metricom (MDN)
Data rate	19.2 KB/s	8 Kb/s [19.2 Kb/s]	4.8 Kb/s [19.2 Kb/s]	~ 76 Kb/s
Modulation	GMSK BT = 0.5	GMSK	GMSK	GMSK
Frequency	~ 800 MHz	~ 900 MHz	~ 800 MHz	~ 915 MHz
Chan. spacing	30 KHz	12.5 KHz	25 KHz	160 KHz
Status	1994 service	Full service	Full service	In service
Access means	Unused AMPS channels	Slotted Aloha CSMA		FH SS (ISM)
Transmit power			40 watt	1 watt

Note: data in square brackets [ ] indicates proposed.
CDPD: Cellular Digital Packet Data
MDN: Microcellular Data Network
ARDIS: Advanced Radio Data Information Service

Table 2. *Wide area wireless packet data systems.*

The earliest and best known of these systems in the United States are the ARDIS network developed and run by Motorola, and the RAM mobile data network based on Ericsson Mobitex Technology. These technologies were designed to make use of standard, two-way voice, land mobile-radio channels, with 12.5 KHz or 25 kHz channel spacing. In the United States these are specialized mobile radio services (SMRS) allocations around 450 MHz and 900 MHz. Initially, the data rates were low: 4.8 kb/s for ARDIS and 8 kb/s for RAM. The systems use high transmitter power (several tens of watts) to cover large regions from a few base stations having high antennas. The relatively low data capacity of a relatively expensive base station has resulted in economics that have not favored rapid growth.

The wide area mobile data systems also are evolving in several different directions in an attempt to improve base station capacity, economics, and the attractiveness of the service. The technologies used in both the ARDIS and RAM networks are evolving to higher channel bit rates of 19.2 kb/s.

The cellular carriers and several manufacturers in the United States are developing and deploying a new wide area packet data network as an overlay to the cellular radio networks. This Cellular Digital Packet Data (CDPD) technology shares the 30 kHz spaced 800 MHz voice channels used by the analog FM Advanced Mobile Phone Service (AMPS) systems. Data rate is 19.2 kb/s. The CDPD base station equipment also shares cell sites with the voice cellular radio system. The aim is to reduce the cost of providing packet data service by sharing the costs of base stations with the better-established and higher cell-site capacity cellular systems. This is a strategy similar to that used by nationwide fixed wireline packet-data-networks that could not provide an economically viable data service if they did not share costs by leasing a small amount of the capacity of the interexchange networks that are paid for largely by voice traffic.

Another evolutionary path in wide area wireless packet data networks is toward smaller coverage areas or microcells. This evolutionary path also is indicated on Fig. 1. The microcell data networks are aimed at stationary or low-speed users. The design compromises are aimed at reducing service costs by making very small and inexpensive base stations that can be attached to utility poles, the sides of buildings, and inside buildings, and can be widely distributed throughout a region. Base-station-to-base-station wireless links are used to reduce the cost of the interconnecting data network. In one network this decreases the overall capacity to serve users, since it uses the same radio channels that are used to provide service. Capacity is expected to be made up by increasing the number of base stations that have connections to a fixed-distribution network as service demand increases. Another such network uses other dedicated radio channels to interconnect base stations. In the high-capacity limit, these networks will look more like a conventional cellular network architecture, with closely spaced, small, inexpensive base stations, i.e., microcells, connected to a fixed infrastructure. Specialized wireless data networks have been built to provide metering and control of electric power distribution, e.g., Celldata, and Metricom in California.

A large microcell network of small inexpensive base stations has been installed in the lower San Francisco Bay Area by Metricom, and public packet-data service was offered during early 1994. Most of the small (shoe-box-size) base stations are mounted on street light poles. Reliable data rates are about 75 kb/s. The technology is based on slow frequency-hopped spread spectrum in the 902-928 MHz U.S. Industrial Scientific Medical (ISM) band. Transmitter power is 1 watt maximum, and power control is used to minimize interference and maximize battery life time.

High-Speed Wireless Local-Area Networks (WLANs)

Wireless local-area data networks (WLANs) can be categorized as providing low-mobility high-speed data communications within a confined region, e.g., a campus or a large building. Coverage range from a wireless data terminal is short, tens to hundreds of feet, like cordless telephones. Coverage is limited to within a room or to several rooms in a building. WLANs have been evolving for a few years, but overall, the situation is chaotic, with many different products being offered by many different vendors [1, 6]. There is no stable definition of the needs or design objectives for WLANs, with data rates ranging from hundreds of kb/s to more than 10 MB/s, and with several products providing one or two MB/s wireless link rates. The best description of the WLAN evolutionary process is: "having severe birth pains." An IEEE standards committee, 802.11, has been attempting to put some order into this topic, but their success has been

Product Company Location	Freq. (MHz)	Link rate	User rate	Protocol(s)	Access	No. of chan. or spread factor	Mod./coding	Power	Network topology
Altair Plus II Motorola Arlington Hts., IL	18-19 GHz	15 Mb/s	5.7 Mb/s	Ethernet			4-level FSK	25 mW peak	Eight devices/ radio; radio to base to Ethernet
WaveLAN NCR/AT&T Dayton, OH	902-928	2 Mb/s	1.6 Mb/s	Ethernet-like	DS SS		DQPSK	250 mW	Peer-to-peer
AirLAN Solectek San Diego, CA	902-928		2 Mb/s	Ethernet	DS SS		DQPSK	250 mW	PCMCIA w/ant; radio to hub
Freeport Windata Inc. Northboro, MA	902-928	16 Mb/s	5.7 Mb/s	Ethernet	DS SS	32 chips/bit	16 PSK trellis coding	650 mW	Hub
Intersect Persoft Inc. Madison, WI	902-928		2 Mb/s	Ethernet, token-ring	DS SS		DQPSK	250 mW	Hub
LAWN O'Neill Comm. Horsham, PA	902-928		38.4 kb/s	AX.25	SS	20 users/chan.; max. 4 chan.		20 mW	Peer-to-peer
WiLAN Wi-LAN Inc. Calgary, Alberta	902-928	20 Mb/s	1.5 Mb/s/ chan.	Ethernet, token ring	CDMA/ TDMA	3 chan. 10-15 links each	"unconventional"	30 mW	Peer-to-peer
RadioPort ALPS Electric USA	902-928		242 kb/s	Ethernet	SS	?/3 channels		100 mW	Peer-to-peer
ArLAN 600 Telesys. SLW Don Mills, Ont.	902-928; 2.4 GHz		1.35 Mb/s	Ethernet	SS			1 W max	PCs with ant.; radio to hub
RadioLink Cal. Microwave Sunnyvale, CA	902-928; 2.4 GHz	250 kb/s	64 kb/s		FH SS	250 ms/hop 500 kHz space			Hub
Range LAN Proxim, Inc. Mountain View, CA	902-928		242 kb/s	Ethernet, token ring	DS SS	3 chan.		100 mW	
RangeLAN2 Proxim, Inc. Mountainview, CA	2.4 GHz	1.6 Mb/s	50 kb/s max.	Ethernet, token ring	FH SS	10 chan. @ 5 kb/s; 15 sub-ch. each		100 mW	Peer-to-peer bridge
Netwave Xircom Calabasas, CA	2.4 GHz	1 Mb/s/ adaptor		Ethernet, token ring	FH SS	82 1-MHz chan. or "hops"			Hub
Freelink Cabletron Sys. Rochester, NH	2.4 and 5.8 GHz		5.7 Mb/s	Ethernet	DS SS	32 chips/bit	16 PSK trellis coding	100 mW	Hub

■ **Table 3.** *Partial list of WLAn products.*

somewhat limited. A partial list of some advertised products is given in Table 3. Users of WLANs are not nearly as numerous as the users of more voice-oriented wireless systems. Part of the difficulty stems from these systems being driven by the computer industry that views the wireless system as just another plug-in interface card, without giving sufficient consideration to the vagaries and needs of a reliable radio system.

There are two overall network architectures pursued by WLAN designers. One is a centrally coordinated and controlled network that resembles other wireless systems. There are base stations in these networks that exercise overall control over channel access [14].

The other type of network architecture is the self organizing and distributed controlled network where every terminal has the same function as every other terminal, and networks are formed ad-hoc by communications exchanges among terminals. Such ad-hoc networks are more like citizen band (CB) radio networks, with similar expected limitations if they were ever to become very widespread. Nearly all WLANs in the United States have attempted to use one of the ISM frequency bands for unlicensed operation under part 15 of the FCC rules. These bands are 902 to 928 MHz, 2400 to 2483.5 MHz, and 5725 to 5850 MHz, and they require users to accept interference from any interfering source that may also be using the frequency. The use of ISM bands has further handicapped WLAN development because of the requirement for use of either frequency hopping or direct sequence spread spectrum as an access technology, if transmitter power is to be adequate to cover more than a few feet. One exception to the ISM band implementations is the Motorola ALTAIR, which operates in a licensed band at 18 GHz. The technical and economic challenges of operation at 18 GHz have hampered the adoption of this 10 to 15 MB/s technology. The frequency-spectrum constraints have been improved in the United States with the recent FCC allocation of spectrum from 1910 to 1930 MHz for unlicensed "data PCS" applications. Use of this new spectrum requires implementation of an access "etiquette" incorporating "Listen before Transmit" in an attempt to provide some coordination of an otherwise potentially chaotic, uncontrolled environment [15]. Also, since spread spectrum is not a requirement, access technologies and multipath mitigation techniques more compatible with the needs of packet data transmission [6], e.g., multipath equalization or multicarrier transmission can be incorporated into new WLAN designs.

Three other widely different WLAN activities also need mentioning. One is a large European Telecommunications Standards Institute(ETSI) activity to produce a standard for High Performance Radio Local Area Network (HIPERLAN), a 20 MB/s WLAN technology to operate near 5 GHz. Other activities are large, U.S. Advance Research Projects Agency (ARPA)-sponsored, WLAN research projects at the Universities of California at Berkeley (UCB), and at Los Angeles (UCLA). The UCB Infopad project is based on a coordinated network architecture with fixed coordinating nodes and direct-sequence spread spectrum (CDMA), whereas, the UCLA project is aimed at peer-to-peer networks and uses frequency hopping. Both ARPA sponsored projects are concentrated on the 900 MHz ISM band.

As computers shrink in size from desktop, to laptop, to palmtop, mobility in data network access is becoming more important to the user. This fact, coupled with the availability of more usable frequency spectrum, and perhaps some progress on standards, may speed the evolution and adoption of wireless mobile access to WLANs. From the large number of companies making products, it is obvious that many believe in the future of this market.

As computers shrink in size from desktop, to laptop, to palmtop, mobility in data network access is becoming more important to the user. This fact, coupled with the availability of more usable frequency spectrum, and perhaps some progress on standards, may speed the evolution and adoption of wireless mobile access to WLANs.

Paging/Messaging Systems

Radio paging began many years ago as a "one bit" messaging system. The one bit was "some one wants to communicate with you." More generally, paging can be categorized as one-way messaging over wide areas. The one-way radio link is optimized to take advantage of the asymmetry. High transmitter power (hundreds of watts to kilowatts), and high antennas at the fixed base stations permit low complexity, very-low-power-consumption, pocket paging receivers that provide long usage time from small batteries. This combination provides the large radio-link margins needed to penetrate walls of buildings without burdening the user set battery. Paging has experienced steady rapid growth for many years and serves about 15 million subscribers in the United States

Paging also has evolved in several different directions. It has changed from analog tone coding for user identification to digitally encoded messages. It has evolved from the one-bit message, "someone wants you," to multibit messages from, first, the calling party's telephone number to, now, short e-mail text messages. This evolution is noted in Fig. 1.

The region over which a page is transmitted has also increased from a) local, around one transmitting antenna; to b) regional, from multiple widely-dispersed antennas; to c) nationwide, from large networks of interconnected paging transmitters. The integration of paging with CT-2 user sets for phone-point call alerting was noted previously.

Another "evolutionary" paging route sometimes proposed is "two-way" paging. However, this is an ambiguous and unrealizable concept, since the requirement for two-way communications destroys the asymmetrical link advantage so well exploited by paging. "Two-way" paging puts a transmitter in the user's set, and brings along with it all the design compromises that must be faced in such a two-way radio system. Thus, the word "paging" is not appropriate to describe a system that provides two-way communications.

Satellite-Based Mobile Systems

Satellite-based systems are the epitome of wide-area-coverage, expensive, base station systems. They generally can be categorized as providing two-way (or one-way) limited quality voice, and/or very limited data or messaging, to very wide-ranging vehicles (or fixed locations). These systems can provide very widespread, often global, coverage, e.g., to ships at sea by INMARSAT. There are a few messaging systems in operation, e.g., to trucks on highways in the United States by Qualcomm's Omnitracs system.

It remains to be seen whether there will be enough users with enough money in low population density regions of the world to make satellite mobile systems economically viable.

A few large scale mobile satellite systems have been proposed and are being pursued; perhaps the best known is Motorola's Iridium, and others include Odyssey, Globalstar, and Teledesic. The strength of satellite systems is their ability to provide large regional or global coverage to users outside buildings. However, it is very difficult to provide adequate link margin to cover inside buildings, or even to cover locations shadowed by buildings, trees or mountains. A satellite system's weakness is also its large coverage area. It is very difficult to provide from earth orbit the small coverage cells that are necessary for providing high overall systems capacity from frequency reuse. This fact, coupled with the high cost of the orbital base stations, results in low capacity along with the wide overall coverage, but also in expensive service. Thus, satellite systems are not likely to compete favorably with terrestrial systems in populated areas, or even along well traveled highways. They can complement terrestrial cellular or PCS systems in low population density areas. It remains to be seen whether there will be enough users with enough money in low population density regions of the world to make satellite mobile systems economically viable.

Proposed satellite systems range from a) low-earth-orbit (LEOS) systems, having tens to hundreds of satellites, through b) intermediate or medium height systems (MEOS?), to c) geostationary or geosynchronous orbit systems (GEOS), having fewer than ten satellites. LEOS require more, but less expensive, satellites to cover the earth, but they can more easily produce smaller coverage areas, and thus provide higher capacity within a given spectrum allocation. Also, their transmission delay is significantly less (perhaps two orders of magnitude!), providing higher-quality voice links as discussed previously. On the other hand, GEOs require only a few, somewhat more expensive, satellites (perhaps only three), and are likely to provide lower capacity within a given spectrum allocation, and suffer severe transmission-delay impairment on the order of 0.5 seconds. Of course, MEOS fall in-between these extremes. The possible evolution of satellite systems to complement high tier PCS is indicated in Fig. 1.

Evolution Toward the Future and To Low-Tier Personal Communications Services

After looking at the evolution of several wireless technologies and systems in the previous sections, it appears appropriate to ask again: "Wireless Personal Communications — What Is It?" All of the technologies in the previous sections claim to provide wireless personal communications, and all do to some extent. However, all have significant limitations and all are evolving in attempts to overcome the limitations. It seems appropriate to ask, what are the likely endpoints? Perhaps some hint of the endpoints can be found by exploring what users see as limitations of existing technologies and systems, and by looking at the evolutionary trends.

In order to do so, we summarize some important clues from the previous sections, and project them, along with some U.S. standards activity, toward the future.

Digital Cordless Telephones

- Strengths: good circuit quality; long talk time; small lightweight battery; low-cost sets and service.
- Limitations: limited range; limited usage regions.
- Evolutionary trends: phone-points in public places; wireless PBX in business.
- Remaining limitations and issues: limited usage regions and coverage holes; limited or no hand-off; limited range.

Digital Cellular Pocket Handsets

- Strength: widespread service availability.
- Limitations: limited talk time; large heavy batteries; high-cost sets and service; marginal circuit quality; holes in coverage and poor in-building coverage; limited data capabilities; complex technologies.
- Evolutionary trends: microcells to increase capacity and in building coverage, and to reduce battery drain; satellite systems to extend coverage.
- Remaining limitations and issues: limited talk time and large battery; marginal circuit quality; complex technologies.

Wide Area Data

- Strength: digital messages.
- Limitations: no voice; limited data rate; high cost.
- Evolutionary trends: microcells to increase capacity and reduce cost; share facilities with voice systems to reduce cost.
- Remaining limitations and issues: no voice; limited capacity.

Wireless Local Area Networks (WLANs)

- Strength: high data rate.
- Limitations: insufficient capacity for voice; limited coverage; no standards; chaos.
- Evolutionary trends: hard to discern from all the churning.

Paging/messaging

- Strengths: widespread coverage; long battery life; small lightweight sets and batteries; economical.
- Limitations: one-way message only; limited capacity.
- Evolutionary desire: two-way messaging and/or voice; capacity.
- Limitations and issues: two-way link cannot exploit the advantages of one-way link asymmetry.

There is a strong trajectory evident in these systems and technologies, aimed at providing the following features.

High Quality Voice and Data

- To small, lightweight, pocket carried communicators.
- Having small lightweight batteries.
- Having long talk time, and long standby battery life.
- Providing service over large coverage regions.
- For pedestrians in populated areas (but not requiring high population density).
- Including low to moderate speed mobility with handoff.

Economical Service

- Low subscriber-set cost.
- Low network-service cost.

Privacy and Security of Communications

- Encrypted radio links.

This trajectory is evident in all of the evolving technologies, but can only be partially satisfied by any of the existing and evolving systems and technologies! Trajectories from all of the evolving technologies and systems are illustrated in Fig. 1 as being aimed at low-tier personal communications systems or services, i.e., low-tier PCS.

Taking characteristics from cordless, cellular, wide area data and, at least moderate-rate, WLANs, suggests the following attributes for this low-tier PCS:

• 32 kb/s ADPCM speech encoding in the near future to take advantage of the low complexity and low power consumption, and to provide low-delay high-quality speech.

• Flexible radio link architecture that will support multiple data rates from several kb/s to several hundred kb/s. This is needed to permit evolution in the future to lower bit rate speech as technology improvements permit high quality without excessive power consumption or transmission delay, and to provide multiple data rates for data transmission and messaging.

• Low transmitter power ($\leq$ 25 mW average) with adaptive power control to maximize talk time and data transmission time. This incurs short radio range which requires many base stations to cover a large region. Thus, base stations must be small and inexpensive, like cordless telephone phone points or the Metricom wireless data base stations.

• Low complexity signal processing to minimize power consumption. Complexity one-tenth that of digital cellular or high-tier PCS technologies is required [1]. With only several tens of milliwatts (or less under power control) required for transmitter power, signal processing power becomes significant.

• Low co-channel interference and high coverage area design criteria. In order to provide high-quality service over a large region, at least 99 percent of any covered area must receive good or better coverage, and be below acceptable co channel interference limits. This implies less than 1 percent of a region will receive marginal service. This is an order-of-magnitude higher service requirement than the ten percent of a region permitted to receive marginal service in vehicular cellular system (high-tier PCS) design criteria.

• Four-level phase modulation with coherent detection to maximize radio link performance and capacity with low complexity.

• Frequency division duplexing to relax the requirement for synchronizing base station transmissions over a large region.

Such technologies and systems have been designed, prototyped, and laboratory-and field-tested and evaluated for several years [1, 2, 7, 16-23]. The viewpoint expressed here is consistent with the progress in the Joint Technical Committee (JTC) of the U.S. standards bodies, Telecommunications Industry Association (TIA) and Committee T1 of the Alliance for Telecommunications Industry Solutions (ATIS). Many technologies and systems were submitted to the JTC for consideration for wireless PCS in the new 1.9 GHz frequency bands for use in the United States [12] Essentially all of the technologies and systems listed in Table 1, and some others, were submitted in late 1993. It was evident that there were at least two, and perhaps three distinct different classes of submissions. No systems optimized for packet data were submitted, but some of the technologies are optimized for voice.

One class of submissions was the group labeled High Power Systems, Digital Cellular (High-Tier PCS) in Table 1. These are the technologies discussed previously in this article. They are highly optimized for low bit-rate voice, and therefore have somewhat limited capability for serving packet-data applications. Since it is clear that wireless services to wide ranging high speed mobiles will continue to be needed, and that the technology described above for low-tier PCS may not be optimum for such services, Fig. 1 shows a continuing evolution and need in the future for high-tier PCS systems that are the equivalent of today's cellular radio. There are more than 100 million vehicles in the United States alone. In the future, most, if not all, of these will be equipped with high-tier cellular mobile phones. Therefore, there will be a continuing and rapidly expanding market for high-tier systems.

It is not clear what the future roles are for paging/messaging, cordless-telephone appliances, or wide area packet-data networks in an environment having widespread contiguous coverage by low-tier and high-tier PCS.

Another class of submissions to the JTC [12] included the Japanese Personal Handiphone System (PHS), and a technology and system originally developed at Bellcore, but carried forward to prototypes, and submitted to the JTC, by Motorola and Hughes Network Systems. This system was known as Wireless Access Communications System (WACS).[2] These two submissions were so similar in their design objectives and system characteristics that, with the agreement of the delegations from Japan and the United States, the PHS and WACS submissions were combined under a new name, Personal Access Communication Systems (PACS), that was to incorporate the best features of both. This advanced, low-power wireless access system, PACS, was to be know as low-tier PCS. Both WACS/PACS and Handiphone (PHS) are shown in Table 1 as Low-Tier PCS and represent the evolution to low-tier PCS, on Fig. 1. The WACS/PACS/

[2] *WACS was known previously as Universal Digital Portable Communications (UDPC).*

Parameter	Cellular (high tier)	Low tier PCS	Capacity factor
Speech coding	8 kb/s (MOS 3.4) No tandem coding	32 kb/s (MOS 4.1) 3 or 4 tandem	x 4
Speech activity	Yes (MOS 3.2)	No (MOS 4.1)	x 2.5
Percentage of good areas	90%	99%	x 2
Propagation σ	8 dB	10 dB	x 1.5
Total: trading quality for capacity			x 30

■ **Table 4.** *A comparison of cellular (IS-54/IS-95) and low tier PCS (WACS/PACS). Capacity comparisons made without regard to quality factors, complexity, and cost per base station are not meaningful.*

UDPC system and technology are discussed in [1, 2, 16-23].

In the JTC, submissions for PCS of DECT and CT-2 and their variations were also lumped under the class of low-tier PCS, even though these advanced digital cordless telephone technologies were somewhat more limited in their ability to serve all of the low-tier PCS needs. They are included under Digital Cordless technologies in Table 1. Other technologies and systems were also submitted to the JTC for high-tier and low-tier applications, but they have not received widespread industry support.

One wireless access application discussed earlier that is not addressed by either high-tier or low-tier PCS is the high-speed WLAN application. Specialized high-speed WLANs also are likely to find a place in the future. Therefore, their evolution is also continued in Fig. 1. The figure also recognizes that widespread low-tier PCS can support data at several hundred kb/s, and thus can satisfy many of the needs of WLAN users.

It is not clear what the future roles are for paging/messaging, cordless telephone appliances, or wide area packet-data networks in an environment with widespread contiguous coverage by low-tier and high-tier PCS. Thus, their extensions into the future are indicated with a (?) on Fig. 1.

Those who may object to the separation of Wireless PCS into high tier and low tier, should review this section again, and note that we have two tiers of PCS now. On the voice side there is Cellular Radio, i.e., high-tier PCS, and cordless telephone, i.e., an early form of low-tier PCS. On the data side there is wide area data, i.e., high-tier data PCS, and WLANs, i.e., perhaps a form of low-tier data PCS. In their evolutions, these all have the trajectories discussed and shown on Fig. 1 that point surely toward low-tier PCS. It is this low-tier PCS that marketing studies continue to project is wanted by more than half the U.S. households or by half of the people, a potential market of over 100 million subscribers in the United States alone. Similar projections have been made worldwide.

Quality, Capacity, and Economic Issues

Although the several trajectories toward low-tier PCS discussed in the previous section are clear, it does not fit the existing wireless communications paradigms. Thus, low-tier PCS has attracted less attention than the systems and technologies that are compatible with the existing paradigms. Some examples are cited in the following paragraphs.

The need for intense interaction with an intelligent network infrastructure in order to manage mobility is not compatible with the cordless telephone appliance paradigm. In that paradigm, independence of network intelligence, and base units that mimic wireline telephones, are paramount.

Wireless data systems often do not admit to the dominance of wireless voice communications, and, thus, do not take advantage of the economics of sharing network infrastructure and base station equipment. Also, wireless voice systems often do not recognize the importance of data and messaging, and, thus, only add them in as "bandaids" to systems.

The need for a dense collection of many low-complexity low-cost low-tier PCS base stations interconnected with inexpensive fixed-network facilities (copper or fiber based) does not fit the cellular high-tier paradigm that expects sparsely distributed $1 million cell sites. Also, the need for high transmission quality to compete with wireline telephones is not compatible with the drive toward maximizing users-per-cell-site and per MHz to minimize the number of expensive cell sites. These concerns, of course, ignore the hallmark of frequency-reusing cellular systems. That hallmark is the production of almost unlimited overall system capacity by reducing the separation between base stations.

This list could be extended, but the above examples are sufficient, along with the earlier sections of the paper, to indicate the many complex interactions among circuit quality, spectrum utilization, complexity (circuit and network), system capacity, and economics that are involved in the design compromises for a large, high-capacity wireless-access system. Unfortunately, the tendency has been to ignore many of the issues, and focus on only one, e.g., the focus on cell site capacity that drove the development of digital-cellular high-tier systems in the United States. Interactions among circuit quality, complexity, capacity and economics are considered in the following sections.

Capacity, Quality, and Complexity

Although "capacity" comparisons frequently are made without regard to circuit quality, complexity, or cost per base station, such comparisons are not meaningful. An example in Table 4 compares capacity factors for U.S. cellular or high-tier PCS technologies with the low-tier PCS technology, PACS/WACS. The Mean Opinion Scores (MOS) (noted in Table 4) for speech coding are discussed in reference [1]. Detection of speech activity and turning off the transmitter during times of no activity is implemented in IS-95. Its impact on MOS also is noted in reference [1]. A similar technique has been proposed as E-TDMA for use with IS-54, and is discussed with respect to TDMA systems in reference [1]. Note that the use of low bit-rate speech coding combined with speech activity degrades the high-tier system's quality by nearly one full MOS point on the 5-point MOS scale when compared to 32 kb/s ADPCM. Tandem encoding is discussed in the previous section. These speech quality-degrading factors alone provide a base station capacity increasing factor of x 4 x 2.5 = x 10 over the high speech-quality low-tier system! Speech coding, of course, directly affects base station capacity and thus overall system capacity by its effect on the number of speech channels that can fit into a given bandwidth.

The allowance of extra system margin to provide coverage of 99 percent of an area for low-tier PCS *versus* 90 percent coverage for high-tier is discussed in the previous section and [1]. This additional quality factor costs a capacity factor of x 2. The last item in Table 4 does not change the actual system, but only changes the way that frequency reuse is calculated. The additional 2-dB margin in standard deviation, σ, allowed for coverage into houses and small buildings for low-tier PCS, costs yet another factor of x 1.5 in calculation only. Frequency reuse factors affect the number of sets of frequencies required, and thus the bandwidth available for use at each base station. Thus, these factors also affect the base station capacity and the overall system capacity.

For the example in Table 4, significant speech and coverage quality has been traded for a factor of x 30 in base station capacity!! While base station capacity affects overall system capacity directly, it should be remembered that overall system capacity can be increased arbitrarily by decreasing the spacing between base stations. Thus, if the PACS low-tier PCS technology were to start with a base station capacity of x 0.5 of AMPS cellular[3] (a much lower figure than the x 0.8 sometimes quoted [12]), and then were degraded in quality as described above to yield the x 30 capacity factor, it would have a resulting capacity of x 15 of AMPS! Thus, it is obvious that making such a base station capacity comparison without including quality is not meaningful.

System Capacity/Coverage Area Size/Economics

Example 1

Assume channels/MHz are the same for cellular and PCS
Cell site: spacing = 20,000 ft cost = $1 M
PCS port: spacing = 1,000 ft
PCS system capacity is $(20000/1000)^2$ = 400 x cellular capacity

Then, for the system costs to be the same
Port cost = ($1M/400) = $2,500, a reasonable figure

If, cell site and port each have 180 channels
Cellular cost/circuit = $1M/180 = $5,555/circuit
PCS cost/circuit = $2500/180 = $14/circuit

Example 2

Assume equal cellular and PCS system capacity
Cell site: spacing = 20,000 ft
PCS port: spacing = 1,000 ft
If, a cell site has 180 channels

Then, for equal system capacity, a PCS port needs 180/400 < 1 channel/port!!

Example 3

A Quality/cost trade
Cell site: Spacing = 20,000 ft cost = $1 M channels = 180
PCS port: Spacing = 1,000 ft cost = $2,500

Cellular to PCS: base station spacing capacity factor = x 400
PCS to Cellular "quality" reduction factors:

32 kb/s to 8 kb/s speech	x 4
Voice activity (buying)	x 2
99% to 90% good areas	x 2
Both in same environment (same σ)	x 1
capacity factor traded	x 16

180 ch/16 = 11.25 channels/port then, $2500/11.25 ≈ $222/circuit
and remaining is x 400/16 = x 25 system capacity of PCS over cellular

Economics, System Capacity, and Coverage Area Size

Claims are sometimes made that low-tier PCS cannot be provided economically, even though IT is what the user wants. These claims are often made based on economic estimates from the "cellular paradigm." These include:

- Very low estimates of market penetration, much less than cordless telephones, and often even less than cellular.
- High estimates of base station costs more appropriate to high-complexity high-cost cellular technology than to low-complexity low-cost low-tier technology.

Such economic estimates are often done by making "absolute" economic calculations based on very uncertain input data. The resulting estimates for low-tier and high-tier are often closer together than the large uncertainties in the input data. A perhaps more realistic approach for comparing such systems is to vary only one or two parameters while holding all others fixed, and then look at relative economics between high-tier and low-tier systems. This is the approach used in the following examples.

Example 1 – In the first example (see textbox), the number of channels per MHz is held constant for cellular and for low-tier PCS. Only the spacing is varied between base stations, e.g., cell sites for cellular and radio ports for low-tier PCS, to account for the differences in transmitter power, antenna height, etc. In this example, overall system capacity varies directly as the square of base station spacing, but base station capacity is the same for both cellular and low-tier PCS. For the typical values in the example, the resulting low-tier system capacity is x 400 greater, only because of the closer base station spacing. If the two systems were to cost the same, the equivalent low-tier PCS base stations would have to cost less than $2,500.

This cost is well within the range of estimates for such base stations, including equivalent infrastructure. These low-tier PCS base stations are of comparable or lower complexity than cellular vehicular subscriber sets, and large-scale manufacture will be needed to produce the millions that will be required. Also, land, building, antenna tower and legal fees for zoning approval, or rental of expensive space on top of commercial buildings, represent large expenses for cellular cell sites. Low-tier PCS base stations that are mounted on utility poles and sides of buildings will not incur such large additional expenses. Therefore, costs of the order of magnitude indicated above seem reasonable in large quantities. Note that, with these estimates, the per-wireless-circuit cost of the low-tier PCS circuits would be only $14/circuit compared to $5,555/circuit for the high-tier circuits. Even if there were a factor of 10 error in cost estimates, or a reduction of channels per radio port of a factor of 10, the per-circuit cost of low-tier PCS would

[3] *Note that the x 0.5 factor is an arbitrary factor taken for illustrating this example. The so called x AMPS factors are only with regard to base station capacity, although they are often misused as system capacity.*

still be only $140/circuit, which is still much less than the per-circuit cost of high-tier.

Example 2 – In the second example (see textbox), the overall system capacity is held constant, and the number of channels/port, i.e., channels/(base station) is varied. In this example, less than 1/2 channel/port is needed, again indicating the tremendous capacity that can be produced with close-spaced low-complexity base stations.

Example 3 – Since the first two examples are somewhat extreme, the third example (see textbox), uses a more moderate, intermediate approach. In this example, some of the cellular high-tier channels/(base station) are traded to yield higher quality low-tier PCS as in the previous subsection. This reduces the channels/port to 11 +, with an accompanying increase in cost/circuit up to $222/circuit, which is still much less than the $5,555/circuit for the high-tier system. Note, also, that the low-tier system still has x 25 the capacity of the high-tier system!

Low-tier base station (PORT) cost would have to exceed $62,500 for the low-tier per-circuit cost to exceed that of the high-tier cellular system. Such a high port cost far exceeds any existing realistic estimate of low-tier system costs.

It can be seen from these examples, and particularly Example 3, that the circuit economics of low-tier PCS are significantly better than for high-tier PCS, IF the user demand and density is sufficient to make use of the large system capacity. Considering the high penetration of cordless telephones, the rapid growth of cellular handsets, and the enormous market projections for "wireless PCS" noted earlier in this paper, filling such high capacity in the future would appear to be certain. The major problem is providing rapidly the widespread coverage (buildout) required by the FCC in the United States. If this unrealistic regulatory demand can be overcome, low-tier wireless PCS promises to provide *the* wireless personal communications that everyone wants.

With the continuing problems and delays in initial deployments, there is increasing concern throughout the industry as to whether CDMA is a viable technology for high capacity cellular applications.

Other Issues

Several issues in addition to those addressed in the pervious two sections continue to be raised with respect to low-tier PCS. These are treated in this section.

Improvement of Batteries

Frequently, the suggestion is made that battery technology will improve so that high-power handsets will be able to provide the desired five or six hours of talk time in addition to 10 or 12 hours of standby time, and still weigh less than half of the weight of today's smallest cellular handset batteries. This "hope" does not take into account the maturity of battery technology, and the long history (many decades) of concerted attempts to improve it. Increases in battery capacity have come in small increments, a few percent, and very slowly over many years, and the shortfall is well over a factor of 10. In contrast, integrated electronics and radio frequency devices needed for low-power low-tier PCS continue to improve and to decrease in cost by factors of greater than 2 in time spans on the order of a year or so. It also should be noted that, as the energy density of a battery is increased, the energy release rate per volume must also increase in order to supply the same amount of power. If energy storage density and release rate are increased significantly, the difference between a battery and a bomb become indistinguishable! The likelihood of a x 10 improvement in battery capacity appears to be essentially zero. If even a modest improvement in battery capacity were possible, many people would be driving electric vehicles.

New Technology

New technology, e.g., spread spectrum or CDMA, is sometimes offered as a solution to both the high-tier cell site capacity and transmitter power issues. However, as these new technologies are pursued vigorously, it becomes increasingly evident that the early projections were considerably over-optimistic, that the base station capacity will be about the same as other technologies [1], and that the high complexity will result in more, not less, power consumption.

With the continuing problems and delays in initial deployments, there is increasing concern throughout the industry as to whether CDMA is a viable technology for high capacity cellular applications. With the passage of time, it is becoming more obvious that Viterbi was correct in his 1985 paper in which he questioned the use of spread spectrum for commercial communications [33].

Thus, it is clear that new high-complexity high-tier technology will not be a substitute for low-complexity, low-power low-tier PCS.

People Only Want One Handset

This issue is often raised in support of high-tier cellular handsets over low-tier handsets. While the statement is likely true, the assumption that *the* handset must work with high-tier cellular is not. Such a statement follows from the current large usage of cellular handsets; but such usage results because that is the only form of widespread wireless service currently available, not because it is what people want. The statement assumes inadequate coverage of a region by low-tier PCS, and that low-tier handsets will not work in vehicles. The only way that high-tier handsets could serve the desires of people discussed earlier would be for an unlikely "breakthrough" in battery technology to occur [7]. However, a low-tier system can cover economically any large region having some people in it. (It will not cover rural or isolated areas — but, by definition, there is essentially no one there to want communications anyway).

Low-tier handsets will work in vehicles on village and city streets at speeds up to 30 or 40 miles per hour, and the required handoffs make use of computer technology that is rapidly becoming inexpensive. Highways between populated areas, and also streets within them, will need to be covered by high-tier cellular PCS, but, users are likely to use vehicular sets in these cellular systems. Frequently the vehicular mobile user will want a different communications device anyway, e.g., a hands-free phone. The use of hands-free phones in vehicles is becoming a legal requirement in some places now, and is likely to become a requirement in many more places in the future. Thus, handsets may not be legally usable in vehicles anyway. With widespread deployment of low-tier PCS systems, *the one* handset of choice will be the low-power low-tier PCS pocket handset or voice/data communicator.

There are approaches for integrating low-tier pocket phones or pocket communicators with high-tier vehicular cellular mobile telephones. The user's identity could be contained either in memory in the low-tier set, or in a small smart card inserted into the set, as is a feature of the European GSM system. When entering an automobile, the small low-tier communicator or card could be inserted into a receptacle in a high-tier vehicular cellular set installed in the automobile.[4] The user's identity would then be transferred to the mobile set. The mobile set could then initiate a data exchange with the high-tier system, indicating that the user could now receive calls at that mobile set. This information about the user's location would then be exchanged between the network intelligence so that calls to the user could be correctly routed.[5] In this approach the radio sets are optimized for their specific environments, high-power high-tier vehicular or low-power low-tier pedestrian, as discussed earlier, and the network access and call routing is coordinated by the interworking of network intelligence. This approach does not compromise the design of either radio set or radio system. It places the burden on network intelligence technology that benefits from the large and rapid advances in computer technology.

The approach of using different communications devices for pedestrians than for vehicles is consistent with what has actually happened in other applications of technology in similarly different environments. For example, consider the case of audio cassette tape players. Pedestrians often carry and listen to small portable tape players with lightweight headsets (e.g., a Walkman[6]). When one of these people enters an automobile, he or she often removes the tape from the Walkman and inserts it into a tape player installed in the automobile. The automobile player has speakers that fill the car with sound. The Walkman is optimized for a pedestrian, whereas the vehicular-mounted player is optimized for an automobile. Both use the same tape, but they have separate tape heads, tape transports, audio preamps, etc. They do not attempt to share electronics. In this example, the tape cassette is the information-carrying entity similar to the user identification in the personal communications example discussed earlier. The main points are that the information is shared among different devices, but the devices are optimized for their environments and do not share electronics.

Similarly, a high-tier vehicular-cellular set does not need to share oscillators, synthesizers, signal processing, or even frequency bands or protocols with a low-power low-tier pocket-size communicator. Only the information identifying the user and where he or she can be reached needs to be shared among the intelligence elements, e.g., routing logic, databases, and common channel signaling [1, 22] of the infrastructure networks. This information exchange between network intelligence functions can be standardized and coordinated among infrastructure subnetworks owned and operated by different business entities (e.g., vehicular cellular mobile radio networks, and intelligent low-tier PCS networks). Such standardization and coordination are the same as are required today to pass intelligence among local exchange networks and interexchange carrier networks.

The approach of using different communications devices for pedestrians than for vehicles is consistent with what has actually happened in other applications of technology in similarly different environments.

Other Environments – Low-tier personal communications can be provided to occupants of airplanes, trains, and buses by installing compatible low-tier radio access ports inside these vehicles. The ports can be connected to high-power high-tier vehicular cellular mobile sets or to special air-ground or satellite-based mobile communications sets. Intelligence between the internal ports and mobile sets could interact with cellular mobile, air-ground, or satellite networks in one direction, using protocols and spectrum allocated for that purpose, and with low-tier personal communicators in the other direction to exchange user identification and route calls to and from users inside these large vehicles. Radio isolation between the low-power units inside the large metal vehicles and low-power systems outside the vehicles can be ensured by using windows that are opaque to the radio frequencies. Such an approach also has been considered for automobiles (i.e., a radio port for low-tier personal communications connected to a cellular mobile set in a vehicle so that the low-tier personal communicator can access a high-tier cellular network. This could be done in the United States using unlicensed PCS frequencies within the vehicle.)

High-Tier to Low-Tier or Low-Tier to High-Tier Dual Mode

Industry and the FCC in the United States appear willing to embrace multi-mode handsets for operating in very different high-tier cellular systems, e.g., analog FM AMPS, TDMA IS-54, and CDMA IS-95. Such sets incur significant penalties for dual mode operation with dissimilar air interface standards, and, of course, incur the high-tier complexity penalties.

It has been suggested that multi-mode high-tier and low-tier handsets could be built around one air-interface standard, for example, TDMA IS-54 or GSM. When closely spaced low-power base stations were available, the handset could "turn off" unneeded power-consuming circuitry, e.g., the multipath equalizer. The problem with this approach is that the handset is still encumbered with power-consuming and quality-reducing signal processing inherent in the high-tier technology,

[4] *Inserting the small personal communicator in the vehicular set would also facilitate charging the personal communicator's battery.*

[5] *This is a feature proposed for FPLMTS in CCIR Rec. 687.*

[6] *Walkman is a registered trademark of Sony Corporation.*

e.g., error correction decoding, and low-bit-rate speech encoding and decoding.

An alternative "dual-mode" low-tier, high-tier system based on a common air-interface standard can be configured around the low-tier PACS/WACS system, if such a dual-mode system is deemed desirable in spite of the discussion in this article. The range of PACS can readily be extended by increasing transmitter power and/or the height and gain of base station antennas. With increased range, the multipath delay-spread will be more severe in some locations [24-26]. Two different solutions to the increased delay-spread can be employed, one for the downlink and another for the uplink. The

The signaling, control processing, and data base interactions required for wireless access PCS are considerably greater than those required for fixed place-to-place networks, but that fact must be accepted when considering such networks.

PACS radio-link architecture has a specified bit sequence, i.e., a unique word, between each data word on the TDM downlink [16, 17]. This unique word can be used as a training sequence for setting the tap weights of a conventional equalizer added to subscriber sets for use in a "high-tier" PACS mode. Since received data can be stored digitally [27, 28], tap weights can be trimmed, if necessary, by additional "passes" through an adaptive equalizer algorithm, e.g., a decision feedback equalizer algorithm.

The PACS TDMA uplink has no "unique word." However, the "high-tier" uplink will terminate on a base station that can support greater complexity, but still be no more complex than the high-tier cellular technologies. Research at Stanford University has indicated that blind equalization, using constant-modulus algorithms (CMA) [29, 30], can be effective for equalizing the PACS uplink. Techniques have been developed for converging the CMA equalizer on the short TDMA data burst.

Advantages of building a dual-mode high-tier, low-tier PCS system around the low-tier PACS air-interface standard are that:

- The interface can still support small low-complexity, low-power, high-speech-quality low-tier handsets.
- Both data and voice can be supported in a PACS personal communicator.
- In high-tier low-tier dual mode PACS sets, circuits used for low-tier operation will also be used for high-tier operation, with additional circuits being activated only for high-tier operation.
- The flexibility built into the PACS radio link to handle different data rates from 8 kb/s to several hundred kb/s will be available to both modes of operation.

Infrastructure Networks

It is beyond the scope of this article to consider the details of PCS network infrastructures. However, there are perhaps as many network issues as there are wireless access issues discussed herein [22, 23, 31, 32]. With the possible exception of the self-organizing WLANS, wireless PCS technologies serve as access technologies to large integrated intelligent fixed communications infrastructure networks.

These infrastructure networks must incorporate intelligence i.e., data-base storage, signaling, processing and protocols, to handle both small-scale mobility, i.e., handoff from base station to base station as users move, and large-scale mobility, i.e., providing service to users who roam over large distances, and perhaps from one network to another. The fixed infrastructure networks also must provide the interconnection among base stations and other network entities, e.g., switches, data bases, and control processors. Of course, existing cellular mobile networks now contain or are incorporating these infrastructure network capabilities. However, existing cellular networks are small compared to the expected size of future high-tier and low-tier PCS networks, e.g., 20 million cellular users in the United States compared with perhaps 100 million users or more each in the future for high-tier and low-tier PCS.

Several other existing networks have some of the capabilities needed to serve as access networks for PCS. Existing networks that could provide fixed base station interconnection include:

- Local exchange networks that could provide interconnection using copper or glass-fiber distribution facilities.
- Cable TV networks that could provide interconnection using new glass-fiber and coaxial-cable distribution facilities.
- Metropolitan fiber digital networks that could provide interconnection in some cities in which they are being deployed.

Networks that contain intelligence, e.g., databases, control processors, and signaling that is suitable, or could be readily adapted, to support PCS access include:

- Local exchange networks that are equipped with signaling system 7 common channel signaling (SS7 CCS), data bases and digital control processors.
- Interexchange networks that are similarly equipped.

Data networks, e.g., the Internet, could perhaps be adapted to provide the needed intelligence for wireless data access, but it does not have the capacity needed to support large voice/data wireless low-tier PCS access.

Many entities and standards bodies worldwide are working on the access network aspects of wireless PCS. The signaling, control processing, and data base interactions required for wireless access PCS are considerably greater than those required for fixed place-to-place networks, but that fact must be accepted when considering such networks.

Low-tier PCS, when viewed from a cellular high-tier paradigm, requires much greater fixed interconnection for the much closer spaced base stations. However, when viewed from a cordless telephone paradigm of a base unit for every handset, and perhaps several base units per wireline, the requirement is much less fixed interconnection because of the concentration of users and trunking that occurs at the multi-user base stations. One should remember that there are economical fixed wireline connections to almost all houses and business offices in the United States now. If wireless access displaces some of the wireline connections, as expected, the overall need for fixed interconnection could decrease!

Conclusion

Wireless personal communications embraces about seven relatively distinct groups of tetherless voice and data applications or services having different degrees of mobility for operation in different environments. Many different technologies and systems are evolving to provide the different perceived needs of different groups. Different design compromises are evident in the different technologies and systems. The evidence suggests that the evolutionary trajectories are aimed toward at least three large groups of applications or services, namely, high-tier PCS (current cellular radio), high-speed wireless local-area networks (WLANS), and low-tier PCS (an evolution from several of the current groups). It is not clear to what extent several groups, e.g., cordless telephones, paging, and wide area data, will remain after some merging with the three large groups. Major considerations that separate current cellular technologies from evolving low-tier low-power PCS technologies are speech quality, complexity, flexibility of radio-link architecture, economics for serving high-user-density or low-user-density areas, and power consumption in pocket carried handsets or communicators. High-tier technologies make use of large complex expensive cell sites and have attempted to increase capacity and reduce circuit costs by increasing the capacity of the expensive cell sites. Low-tier technologies increase capacity by reducing the spacing between base stations, and achieve low circuit cost by using low-complexity low-cost base stations. The differences between these approaches result in significantly different compromises in circuit quality and power consumption in pocket sized handsets or communicators. These kinds of differences also can be seen in evolving wireless systems optimized for data. Advantages of the low-tier PACS/WACS technology are reviewed in the article, along with techniques for using that technology in high-tier PCS systems.

Wireless personal communications embraces about seven relatively distinct groups of tetherless voice and data applications or services having different degrees of mobility for operation in different environments. Many different technologies and systems are evolving to provide the different perceived needs of different groups.

References

[1] D. C. Cox, "Wireless Network Access for Personal Communications," *IEEE Commun. Mag.*, Dec. 1992, pp. 96-115.
[2] J. E. Padgett, T. Hattori and C. Gunther, "Overview of Wireless Personal Communications," *IEEE Commun. Mag.*, Jan. 1995, pp. 28-41.
[3] D. J. Goodman, "Trends in Cellular and Cordless Communications," *IEEE Commun. Mag.*, June 1991, pp. 31-40.
[4] R. Steele, "Deploying Personal Communications Networks," *IEEE Commun. Mag.*, Sept. 1990, pp. 12-15.
[5] *IEEE Commun. Mag.*, Special Issue on "Wireless Personal Communications," Jan. 1995.
[6] R. Schneideman, "Spread Spectrum Gains Wireless Applications," *Microwaves and RF*, May 1992, pp. 31-42.
[7] D. C. Cox, "Personal Communications-A Viewpoint," *IEEE Commun. Mag.*, Nov. 1990, pp. 8-20.
[8] A. Wong, "Regulating Public Wireless Networks," Workshop on Lightwave, Wireless and Networking Technologies, the Chinese University of Hong Kong, Hong Kong, Aug. 24, 1994.
[9] J. C-I, Chuang, "Performance Issues and Algorithms for Dynamic Channel Assignment," *IEEE JSAC*, Aug. 1993.
[10] J. C-I Chuang, N.R. Sollenberger, and D.C. Cox, "A Pilot Based Dynamic Channel Assignment Scheme for Wireless Access TDMA/FDMA Systems," Proc. IEEE ICUPC'93, Ottawa, Canada, Oct. 12-15, 1993, pp. 706-712, and *Int'l J. of Wireless Info. Networks*, vol. 1, no. 1, Jan. 1994, pp. 37-48,.
[11] J. C-I, Chuang, "Performance Limitations of TDD Wireless Personal Communications with Asynchronous Radio Ports," *Electronic Letters*, vol. 28, March 12, 1992, pp. 532-533.
[12] C. I. Cook, "Development of Air Interface Standards for PCS," *IEEE Personal Commun.*, Fourth Quarter, 1994, pp. 30-34.
[13] Bell System Technical Journal (BSTJ), Special Issue on Advanced Mobile Phone Service (AMPS), vol. 58, Jan. 1979.
[14] R. H. Katz, "Adaptation and Mobility in Wireless Information Systems," *IEEE Personal Commun.*, First Quarter, 1994, pp. 6-17.
[15] D. G. Steer, "Coexistence and Access Etiquette in the United States Unlicensed PCS Band," *IEEE Personal Communications Magazine*, Forth Quarter, 1994, pp. 36-43.
[16] D. C. Cox, "A Radio System Proposal for Widespread Low-power Tetherless Communications," *IEEE Trans. on Commun.*, Feb. 1991, pp. 324-335.
[17] Bellcore Technical Advisories, "Generic Framework Criteria for Universal Digital Personal Communications Systems (PCS)," FA-TSY-001013, Issue 1, March 1990 and FA-NWT-001013, Issue 2, Dec. 1990, and Technical Reference, "Generic Criteria for Version 0.1 Wireless Access Communications Systems (WACS), Issue 1, Oct. 1993 and Revision 1, June 1994.
[18] D.C. Cox, "Universal Portable Radio Communications," *IEEE Trans. on Veh. Tech.*, Aug. 1985, pp. 117-121.
[19] D. C. Cox, H. W. Arnold, and P.T. Porter, "Universal Digital Portable Communications-A System Perspective," *IEEE JSAC*, Vol. JSAC-5 pp. 764-773, June 1987.
[20] D. C. Cox, "Research Toward a Wireless Digital loop," *Bellcore Exchange*, vol. 2, Nov./Dec. 1986, pp. 2-7.
[21] D. C. Cox, "Universal Digital Portable Radio Communications," Proc. IEEE, vol. 75, April 1987, pp. 436-477.
[22] D. C. Cox, "Portable Digital Radio Communications-An Approach to Tetherless Access," *IEEE Commun. Mag.*, July 1989, pp. 30-40.
[23] D. C. Cox, W. G. Gifford, and H. Sherry, "Low-Power Digital Radio as a Ubiquitous Subscriber Loop," *IEEE Commun. Mag.*, March 1991, pp. 92-95.
[24] D. M. J. Devasirvatham, "Radio Propagation Studies in a Small City for Universal Portable Communications," Conference Record, IEEE VTC'88, Philadelphia, PA, June 15-17, 1988, pp. 100-104.
[25] D. C. Cox, "Delay-Doppler Characteristics of Multipath Propagation at 910 MHz in a Suburban Mobile Radio Environment," *IEEE Trans. on Antennas and Propagation*, Sep. 1972, pp. 625-635.
[26] D. C. Cox, "Multipath Delay Spread and Path Loss Correlation for 910 MHz Urban Mobile Radio Propagation," *IEEE Trans. on Vehicular Technology*, Nov. 1977, pp. 340-344.
[27] N. R. Sollenberger and J. C-I. Chuang, "Low-Overhead Symbol Timing and Carrier Recovery for TDMA Portable Radio Systems," *IEEE Trans. on Commun.* Oct. 1990, pp. 1886-1892,.
[28] N.R. Sollenberger, "An Experimental VLSI Implementation of Low-Overhead Symbol Timing and Frequency Offset Estimation for TDMA Portable Radio Applications," IEEE GLOBECOM '90, San Diego, CA, Dec. 2-5, 1990, pp. 1701-1711.
[29] J. R. Treichler and B. G. Agee, "A New Approach to Multipath Correction of Constant Modulus Signals," *IEEE Trans. on Acoustics, Speech and Signal Processing*, April 1983, pp. 459-472.
[30] Y. Sato, "A Method of Self-Recovering Equalization for Multilevel Amplitude Modulation Systems," *IEEE Trans. on Commun.*, June 1975, pp. 679-682.
[31] B. Jabbari *et al.*, "Network Issues for Wireless Personal Communications," *IEEE Commun. Mag.*, Jan. 1995, pp. 88-98.
[32] M. Zaid, "Personal Mobility in PCS." *IEEE Personal Commun.*, Fourth Quarter, 1994, pp. 12-16.
[33] A. J. Viterbi, "When Not to Spread Spectrum-A Sequel," *IEEE Commun. Mag.*, April 1985, pp. 12-17.

Biography

DONALD C. COX [F '79] did research at Bell Laboratories from 1968 to 1973 on advanced mobile radio systems that still provides basic input to the design of digital cellular, cordless, and PCS systems. From 1978 to 1993 he led and was actively involved in pioneering wireless research, first at Bell Labs and then at Bellcore, that started and fueled the current explosion in wireless personal communications. He was instrumental in evolving this research into the WACS/PACS specification being standardized by the U.S. TIA/T1 JTC. For this pioneering work, he received the IEEE 1993 Alexander Graham Bell Medal and Bellcore Fellow Award, and was elected into the National Academy of Engineering. He received the IEEE Morris E. Leeds award in 1985 and the Prize Guglielmo Marconi from Italy in 1983 and is a Fellow of the AAAS and the RCA. He holds 12 patents, has authored or coauthored more than 75 journal papers, including three that won prizes, was coauthor of a book, *Microwave Mobile Communications*, and has been guest editor of special issues on Wireless Communications in IEEE journals. He managed Radio Research at Bellcore from 1984 to 1993, and is currently the Harald Trap Friis Professor of Engineering and director of the Telecommunications Center at Stanford University, Stanford, California. He received B.S. and M.S. degrees in electrical engineering and an Honorary Dr. of Science from the University of Nebraska, and a Ph.D. in electrical engineering from Stanford.

Reprinted from *IEEE Communications Magazine*, Vol. 33, No. 3, Mar. 1995, pp. 88-95.

Trends in Local Wireless Networks

The authors' vision of the future, in which a ubiquitous local wireless computing environment leads to a fusion of communications and computation, must overcome significant technical obstacles before becoming a reality.

Kaveh Pahlavan, Thomas H. Probert, and Mitchell E. Chase

In June 1985 the lead author of this article published an article entitled "Wireless Office Information Networks" in this magazine [1]. The article examined spread spectrum, standard radio and infrared (IR) technologies for intra-office wireless networking. This article was published in a timely manner. In May of that same year the FCC released the ISM (industrial, scientific, and medical) bands for spread spectrum local communications. Although ISM bands are not restricted to any specific application, wireless local area networks (LANs) were one of the most prominent applications that were envisioned by the rule makers in the FCC [2, 3]. Since 1985, many small startup companies, as well as small groups in larger companies, have started to develop wireless LANs.

In this article we provide a sequel to the previously mentioned article by providing an overview of the past and present of the wireless LAN industry, as well as a perspective of the future directions that encompass a vision for a ubiquitous local wireless computing environment that leads to a fusion of communications and computation.

Historical Trends

In the late 1970s IBM Laboratories in Ruschlikon, Switzerland published the results of their experimental work on the design of a wireless indoor network using diffused IR technology that was envisioned to be used on manufacturing floors [4]. Around the same time, another experiment at Hewlett-Packard (HP) laboratories [5] examined the use of direct sequence spread spectrum technology for wireless inter-terminal communications. The data rates experimented by both methods were around 100 Kb/s [1] and none of the projects were turned into a commercial product. The diffused IR design could not provide a reliable link to meet the project goal of 1 Mb/s, and the spread spectrum project had to wait for the approval of a commercial band. However, these works initiated further research in high-speed wireless indoor communications. HP laboratories and others developed directed-beam IR networks [6, 7]; Motorola's Codex worked on a wireless LAN using ordinary radio modems at 1.7 GHz and petitioned the FCC for that band [2]; and GTE Laboratories worked on a fiber optic high-speed LAN with wireless drops [8], none of which turned into a commercial product.

Motivated by HP's petition, in 1981 the FCC began exploring the feasibility of regulating a band for the commercial application of spread spectrum technology that led to the adoption of the ISM bands released in May 1985 [2, 3]. Development of commercial wireless LAN products entered a more serious phase after the announcement of the ISM bands. Motorola's petition for a 1.7 GHz band was not granted. However, it revealed very encouraging market evaluations for wireless LAN products. Later on, Motorola was able to secure licensed bands at 18 to 19 GHz for wireless LAN applications that pointed at the availability of bands at higher frequencies for wireless local communications. The release of the ISM bands, results of the market evaluations, and timely publication of some papers [1, 5, 9, 10] prompted a significant interest in the industry for the design of numerous wireless LAN products, mostly operating in the ISM bands. By 1990, wireless LAN products using direct sequence spread spectrum (DSSS) in the ISM bands [11], licensed radio at 18-19 GHz [12], and IR technology appeared in the market. These products were the first that could be called wireless LANs because they were operating at high speeds[1] and they could communicate with at least one standard LAN software product, such as Novell. In that year, the IEEE 802.11 committee was formed as an independent standards group within IEEE 802 to follow up the work of IEEE 802.4L, which had started earlier as a branch of 802.4. In May 1991 the first IEEE wireless LAN workshop dealt with this evolving technology [13]. In 1992, WINFORUM, an alliance among the major computer and communication companies to obtain bands from the FCC for the so called data-PCS, was initiated by Apple Computer in the United States, and the HIPERLAN standards activities in the European Community (EC) were initiated under ETSI direction. In 1993 the EC announced bands at 5.2 and 17.1 GHz for HIPERLAN, and the FCC announced plans for the release of unlicensed PCS bands that can be used for wireless local data communications. Both IEEE 802.11 and HIPERLAN are expecting to complete their standards in 1995. WINFORUM is working with the FCC to develop a "Spectrum Etiquette" for the PCS bands [14].

KAVEH PAHLAVAN is the director of the Center for Wireless Information Network Studies at the Worcester Polytechnic Intstitute.

THOMAS H. PROBERT is the president of the Enterprise Computing Institute, Inc.

MITCHELL E. CHASE is a research scientist with the Enterprise Computing Insitute, Inc.

[1] *1 Mb/s is considered by IEEE 802 LAN standards as the lowest data rate for a LAN.*

Trends in Applications

During the development of the first-generation wireless LANs at the end of the last decade, it was thought that the savings from the installation and relocation costs of wired LANs justified the additional cost of wireless equipment. It was expected that a market close to a billion dollars would evolve around workstations adopting wireless LANs. In reality, old buildings were already wired and the cost of wiring in new construction was so low that it would be done during construction. Meanwhile, the twisted pair LAN technology dominated the more expensive coaxial cable LANs, which reduced the installation costs significantly. As a result, the first-generation wireless LANs did not closely meet the predicted market.

The first generation of wireless LANs were designed to operate with workstations and had an electronic power consumption of approximately 20 Watts. These devices are not suitable for battery operated portable computers. The next-generation wireless LANs are developing around the lap-top, palm-top, and pen-pad computers in a PCMCIA card operating with small batteries. Wireless connection is the natural medium for the personal portable computing devices that are growing in popularity. In this environment researchers are thinking of new concepts such as ad hoc networking, nomadic access and mobile computing that leads to the fusion of the computer and communications in a ubiquitous computing environment.

Today, first-generation wireless LANs are marketed mostly as LAN extension. LAN extension refers to applications in which the coverage of a wireline LAN is extended to areas with wiring difficulties. Examples of buildings with wiring difficulties are buildings with large open areas such as manufacturing floors, stock exchange halls, or warehouses; historical buildings where drilling holes for wiring is prohibited: and small offices such as a branch of a real estate agency where maintenance of wireline LANs is not economically attractive. Recently, wireless LAN manufacturers were also marketing their products aggressively for connecting LANs located in two different buildings. Here, the wireless LAN is a LAN interconnect device that is easy and inexpensive to install.

Figure 1 shows wireless LANs used for LAN extension, cross-building interconnect, and nomadic access. A nomadic access provides a wireless connection between a hub and a personal portable computer. This access is particularly useful for a person returning to the office from traveling who needs to transfer large data files between his personal portable computer and the backbone information network. Another situation related to the same users is ad hoc networking in which a group of portable users, for example in a classroom or a meeting, intend to set up a network among themselves in an unpredicted situation. Figure 2 shows an ad-hoc network among several lap-tops. Figure 3 depicts a ubiquitous computing environment in which a large distributed information and computing base is available to the pen-pad users in a local area through a high-speed wireless link. Typical applications include a wireless campus or a wireless battlefield in which a user (a student or a soldier) can move within a local area with continuous access to the backbone distributed computational facilities.

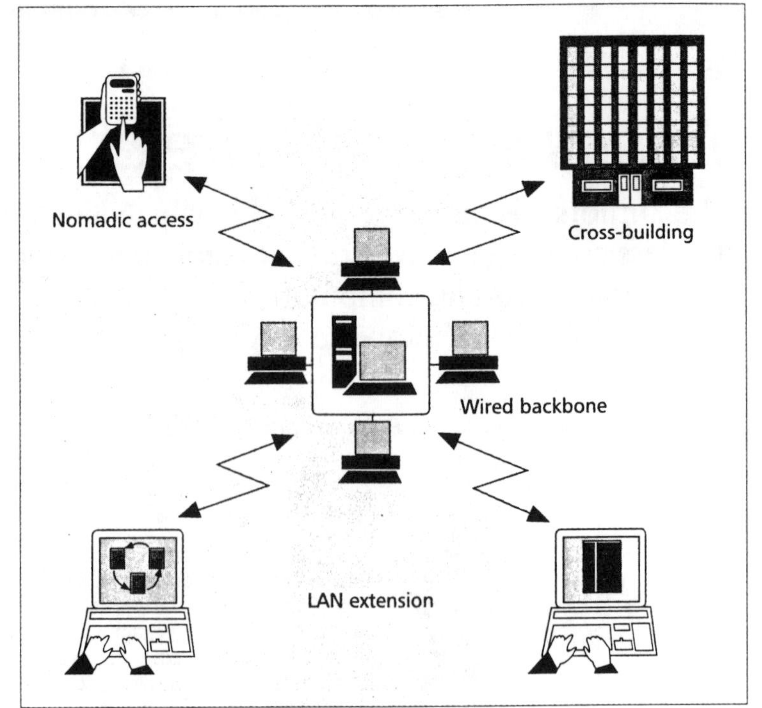

■ **Figure 1.** *Applications for wireless LANs.*

■ **Figure 2.** *Ad hoc networking.*

Trends in Frequency Administration

A major difference between a wireline and a radio service is that the transmission medium for radio communications is regulated and needs frequency administration. All new services and products must go through an exploratory phase to examine various technologies against the available market. In a wireline environment, innovative ideas and new technologies can be examined against the market immediately. To examine the market for an innovative idea or a new concept using radio technology, one first needs to convince the frequency administration organization. On one hand, wireless LANs need a relatively large bandwidth for high-speed transmission, and the current market needs more time to grow to a reasonable size. On the other hand, it is a new technology that is expected

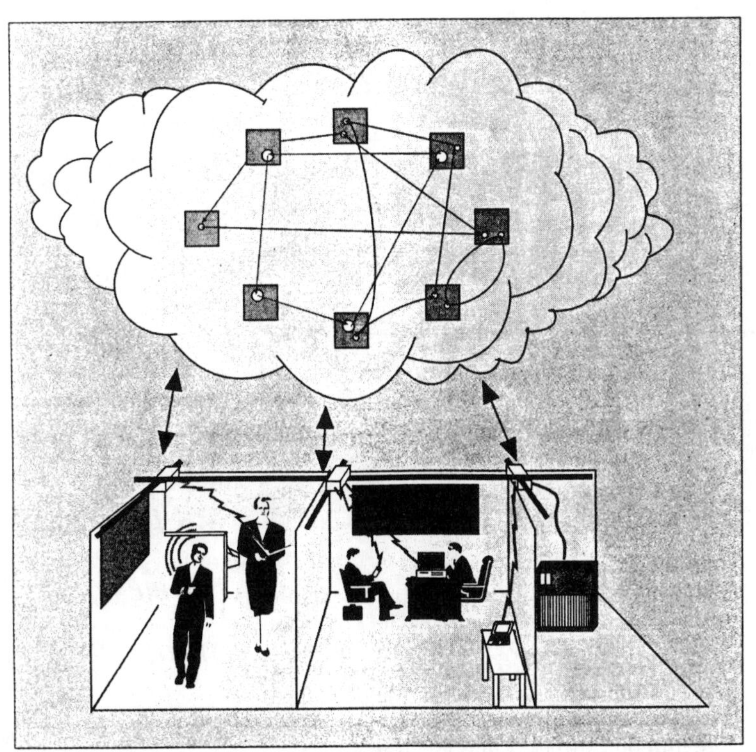

■ **Figure 3.** *Fusion of computers and communications into a computational cloud.*

to bring revolutionary applications to the market. This situation leaves the frequency administration agencies in a difficult position to justify the assignment of adequate bands to this application while other traditional voice-oriented services may use the same bands to stimulate a larger market. Two steps are taken to resolve this situation: one is to resort to higher frequencies where larger bandwidths are available; the other is to release unlicensed bands. Raising the frequency of operation increases the size and power consumption and limits the coverage of the signal in indoor areas. The size and power consumption will reduce in time as the electronic design technologies advance. Limitations on coverage is not important in many applications and can be solved easily by reducing the size of the cells. In fact, in some applications confinement of the signal to a small area is a desirable feature for wireless LANs [15].

An unlicensed band with minimum regulation provides a ground for exploration of different wireless technologies against the market in a multivendor environment. This approach particularly suits listen-before-talk random access methods commonly used in wireless LANs. Today, all activities around wireless LANs are focused on unlicensed operation. The major criticism of the unlicensed bands is that they provide a self-defeating purpose. In an unlicensed band a successful market generates high utilization by devices from different manufacturers possibly using different technologies. In the contention access methods used in data communication applications, high utilization causes instability in the system [15]. Any sort of traffic control to stabilize the system requires a hub base network which is an extremely challenging task in a multivendor environment. This situation pushes the trend toward integration with other cellular networks.

Trends in Spread Spectrum

Spread spectrum wireless LANs have been more successful in the market than any other wireless LAN technology, and most firms involved in wireless LANs are using this technology. These spread spectrum wireless LANs are developed in the ISM bands that were the first unlicensed bands in which a high-speed wireless LAN could be implemented. These bands allow unlicensed transmissions of up to 1 W of spread spectrum signal with a minimal spreading factor specified for direct sequence and frequency hopping. Operating in these bands is secondary to other users that already existed in these bands, such as microwave ovens at 2.4 GHz or amateur radios at 900 MHz. These bands can be used for any application, but wireless LANs were one of the primary applications considered in the rule making [2, 3].

At the time of rule making it was thought that spread spectrum technology had two interesting features for wireless LAN applications: It allowed different systems in a multivendor environment to operate simultaneously using CDMA, and its anti-multipath nature provided a reliable transmission at high data rates. It was thought that other access methods such as TDMA or FDMA require explicit cooperation among those sharing the channel. This is not practical in a multivendor environment. Therefore, CDMA was thought to be the only choice [2, 3].

Shortly after the announcement of the ISM bands, the efficiency of CDMA for wireless PBX operation was shown [10]. However, it was pointed out that to avoid the near-far problem, power control is essential for proper operation in a CDMA environment [10, 16]. Power control, however, requires communication among all terminals and a central unit and it cannot be implemented in a multivendor environment unless all manufacturers agree to certain principles that were not envisioned in the rule making for ISM bands. In a single-vendor environment, it was noticed that wireless LAN users need the entire capacity of the channel for quick transmission of data bursts generated by each terminal, and CDMA is not suitable for this purpose [11]. As a result, the first wireless LANs in ISM bands that appeared in the market used spread spectrum without CDMA.

Using spread spectrum without CDMA, it can be argued that the maximum supportable data rate (bandwidth) is sacrificed to gain transmission reliability. But data rate is the most important technical feature appealing to a wireless LAN user. The transmission reliability of spread spectrum is due to the anti-multipath and anti-interference nature of this technology. Other techniques such, as decision-feedback equalization (DFE), multicarrier transmission, and sectored antennas [15], are also anti-multipath, but do not sacrifice bandwidth. These could be adopted as well. So far as the resistance to interference is considered, the low spreading factors used in the ISM band wireless LANs are insufficient to provide very much resistance. Another related important issue is the reliability of the delivered packets of data. In a wireless LAN, the delivery is checked with feedback acknowledgment in the communication software; this reduces the need for an extremely reliable transmission media. Due to the channel fading

Technique	Optical		RF		
	DF/IR	DB/IR	RF	DSSS	FHSS
Data rate (Mb/s)	1-4	10	5-10	2-20	1-3
Mobility	Stationary/ mobile	Stationary with LOS	Stationary/mobile		Mobile
Range (ft.)	50-200	80	40-130	100-800	100-300
Detectability	Negligible		Some	Little	Little
Wavelength/frequency	λ=800 - 900 nm		18 GHz or ISM	ISM bands	
Modulation technique	OOK		FS/QPSK	QPSK	GFSK
Radiated power	-		25 mW	<1W	
Access method	CSMA	Token Ring, CSMA	Reservation ALOHA, CSMA	CSMA	

■ **Table 1.** *Comparison of wireless LAN technologies.*

Spread spectrum technology will play an important role in the future multimedia wireless communication industry.

condition or interference, lost packets are simply retransmitted. Therefore, the reliability of transmission is not as necessary as it is in the case of real-time voice transmission in which there is no acknowledgment mechanism.

From the above discussion we conclude that the most practical reason for using spread spectrum in the ISM bands for multivendor wireless LAN applications is the availability of this band to host a high-speed, unlicensed data link. The main problem with this technique is the reduction in the maximum supportable data rate for a given bandwidth. To compensate for the loss in the maximum data rate in the ISM bands, some companies have adopted multiamplitude and multiphase modulation. Other companies have adopted single channel CDMA, in which each transmitter uses several orthogonal codes simultaneously in the same channels. Since all the codes are modulated over the same carrier, the received power for each code is the same. This is equivalent to having perfect power control in place. Using this method, some companies have been able to achieve data rates on the order of 20 Mb/s in an ISM band. The problem with this approach is the complexity of the design of the receiver and the transmitter.

The fact that without power control and in a multivendor environment spread spectrum is not appealing for wireless LAN applications does not imply that spread spectrum is not suitable for wireless data communications. Indeed, in a multimedia environment where various information sources have different requirements, CDMA is a promising technique. However, an efficient communication in that environment requires power control, which restricts the independence of the designs in a multivendor environment. Although the current trends in spread spectrum wireless LANs in the ISM bands have their own problems, spread spectrum technology will play an important role in the future multimedia wireless communication industry.

Trends in Products

Wireless LANs are designed for a small number of users, usually operating in indoor areas. The range of coverage is small, which leaves many options open for the transmission technology. Technologies used in the existing wireless LAN products are divided into five categories: diffused IR (DFIR), directed beam IR (DBIR), standard radio (RF), direct sequence spread spectrum (DSSS), and frequency hopping spread spectrum (FHSS). In each category several products with different specifications are available in the market. Table 1 shows a comparison among various features of the existing products in each of these categories. These technologies have evolved around the availability of the channel and the suitability of the transmission technique to provide a high data rate link in the wireless media. The IR products are designed to operate in the optical frequencies that are not regulated by the FCC. The spread spectrum LANs are designed to operate in the ISM bands. The RF products are either implemented in the 18-19 GHz licensed bands or in the ISM bands using very low power. The ISM bands allow non-spread spectrum transmission devices if the power is very low.

Data rate is an essential ingredient of local communication networks and the most important aspect for marketing and sale of the product. The higher the data rate, the more likely the impact of the product in the market. Another important feature affecting the market for wireless LANs is the mobility of the terminal, which is a function of the power consumption and the size of the product, except for DBIR, in which the terminal must stay stationary to keep the radiation pattern of the device effective. If the technology can be implemented with small batteries and light weight, it would be suitable for mobile applications and can be used for personal portable computers. Power consumption is a function of the electronic implementation of the device, and some technologies can be implemented either with or without battery operation for mobile and stationary applications, respectively. Most wireless LANs use a version of CSMA, and some use reservation slotted ALOHA or token ring.

Current Trends in Standards

Although none of the standards for wireless LANs are completed, there are numerous wireless LAN products on the market. Wireless LANs are stand-alone products that can be man-

The future of this industry is toward interoperability; however, a discovery period, in which the technology and the market settles itself, is being planned.

ufactured without a widely accepted standard. As the penetration of wireless LANs in the market grows and the standards are completed, this situation will change.

Currently, all standard activities for wireless LANs use unlicensed bands, and there are two approaches to regulate an unlicensed band. One approach is to develop a standard that will allow different vendors to communicate with one another using a set of interoperable rules. This approach is taken by IEEE 802.11 and ETSI's RES10, HIPERLAN. The second approach is to provide a minimum set of rules or "Spectrum Etiquette" [14] that allow terminals designed by different vendors to use a fair share of the available channel frequency-time resources and coexist in the same band. The second approach does not preclude the first approach and it is pursued by the WINFORUM. In a coexisting environment a vendor can interoperate with another vendor by using the same protocol and transmission scheme. The future of this industry is toward interoperability. However, a discovery period in which the technology and the market settles itself is being planned. An analogy existed in the development of voiceband modems: at the beginning there was no standard, and as the industry evolved CCITT adopted successful modems as the standard [17].

The three major activities related to wireless LANs are IEEE 802.11, HIPERLAN, and WINFORUM. IEEE 802.11 is developing a standard for DSSS, FHSS, and DFIR technologies and uses the ISM bands as the radio channel. The HIPERLAN standard is concerned with the recently released 5.2 and 17.1 GHz bands in the EC. It is expected that HIPERLAN will adopt non-spread spectrum modulation techniques. To achieve high data rates in multipath fading, HIPERLAN may resort to techniques such as adaptive equalization [18-20], sectored antenna [21] or multicarrier modulation [22-24]. The WINFORUM goal is to obtain parts of the PCS band for unlicensed data and voice applications and develop a "spectrum etiquette" for them.

IEEE 802.11 and ISM Bands

IEEE 802.11 focuses on the physical and media access protocol (MAC) layers for peer-to-peer and centralized topologies accommodating DFIR, DSSS and FHSS. Both spread spectrum systems operate in the 2400-2483.5 MHz ISM band. This band is selected over the 902-928 MHz and 5725-5850 MHz ISM bands because it is widely available in most leading countries. Figure 4 shows the map of four major geographic areas with their position toward the ISM bands at 2.4-2.5 GHz. In this band, more than 80 MHz of bandwidth is available that is suited to high-speed data communication. The implementation in this band is also more cost effective as compared with implementation in frequencies that are a few GHz higher. IEEE 802.11 supports DSSS with BPSK and QPSK modulation for data rates of 1 and 2 Mb/s, respectively; FHSS with GFSK modulation and two hopping patterns with data rates of 1 and 2 Mb/s; and DFIR with OOK modulation with a data rate of 1 Mb/s. For DSSS the band is divided into five overlapping 26 MHz sub-bands centered at 2412, 2442, 2470, 2427 and 2457 MHz, with the last two overlapping the first three. This setup provides five orders of frequency selectivity for the user, which is very effective in improving the transmission reliability in the presence of interference or severe frequency selective multipath fading. For FHSS, the channel is divided into 79 subbands, each with 1 MHz bandwidth, and three patterns of 22 hops are left as options for the users. A minimum hop rate of 2.5 hops/s is assigned to provide an opportunity for slow frequency hopping in which each packet can be sent in one hop and, if it is destroyed, the following packet can be sent from another hop for which the channel condition would be different. This approach will provide a very effective time-frequency diversity that takes advantage of a retransmission scheme to provide a robust transmission. The standard supports CSMA and intends to provide interoperability among all users.

WINFORUM and PCS Bands

An important issue that was not addressed in the ISM bands was the enforcement of a time limit for the air time of a terminal. Suppose that a wireless LAN and another device operate in the ISM bands near one another. Further assume that the other device serves an application that constantly radiates a signal in the band. The wireless LAN is a data communication device and it communicates with bursts of information. In this situation, all information bursts generated from the wireless LAN will suffer from the interference caused by the other device, while the wireless LAN only produces bursts of interference to the device. This is not fair to the wireless LAN because it is not getting a fair share of the available resources. A fair method to allow several devices to operate in the same band is to restrict their frequency-time share of the channel according to their transmission power. This issue is addressed by WINFORUM.

WINFORUM was initiated by Apple Computer to form an alliance in the industry to obtain frequency bands for the so called data-PCS. Today, WINFORUM has approximately 40 members from leading information technology companies, and its objective is to obtain and effectively employ radio spectrum for unlicensed user-provided voice and data personal communication services referred to as User-PCS. Currently, WINFORUM intends to foster technical advances in applications such as wireless LANs and wireless PBX services. Originally, WINFORUM was looking for a 40 MHz unlicensed band for the data-PCS, and the FCC has shown some indications to provide them with a 20 MHz band divided into two 10 MHz separate sub-bands for voice and data applications. The technical innovation initiated by the WINFORUM is the development of the so-called "spectrum etiquette" that is a means to provide fair access to an unlicensed band to widely different applications and devices [14]. The etiquette does not intend to preclude any common air interface standards or access technologies. Spectrum etiquette demands listen-before-talk (LBT), which means a device may not transmit if the spectrum it will occupy is already in use within its range. The power is limited to keep the range short. That allows operation in densely populated office areas. The power and connection time is related to the occupied bandwidth to equalize the

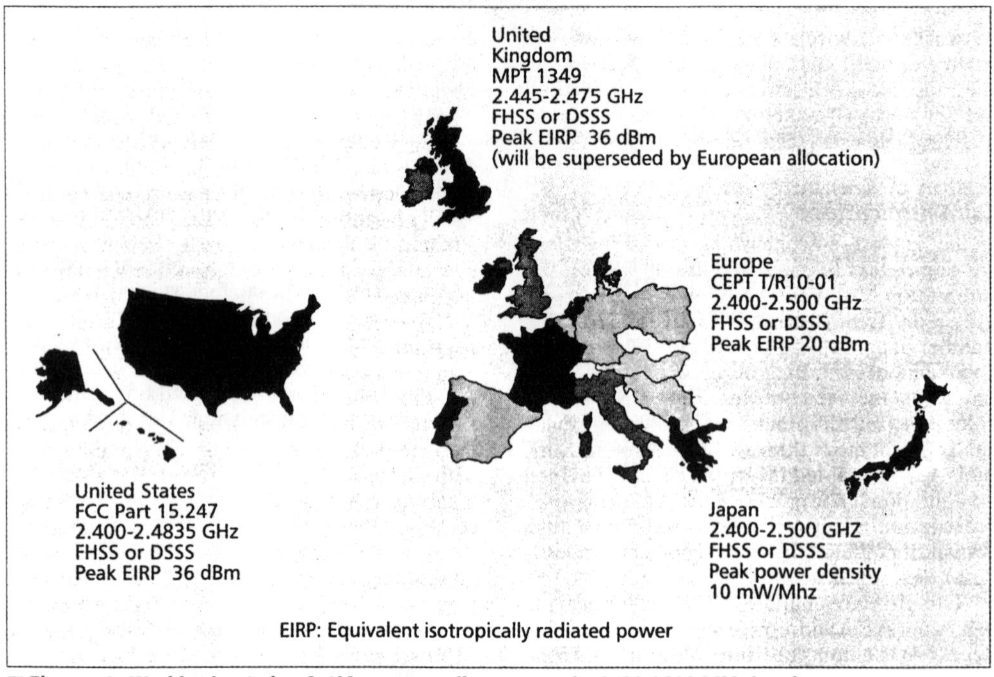

■ **Figure 4.** *Worldwide wireless LAN spectrum allocation in the 2400-2500 MHz band.*

interference and provide a fair access to frequency-time resources. In May 1993 WINFORUM filed its spectrum etiquette with the FCC.

In the view of WINFORUM, there are two classes of information type generated from the asynchronous and the isochronous transmissions. The asynchronous transmission, typified by wireless-LAN-like applications is bursty, begins transmission within milliseconds, uses short bursts that contain large amounts of data, and releases the link quickly. The isochronous transmission, typified by voice services such as wireless PBX, exhibits long holding times, periodic transmissions, and flexible link access times that may be extended up to a second. The asynchronous sub-bands may range from 50 KHz to 10 MHz, while the isochronous sub-bands may be divided into 1.25 MHz segments. The two types are technically contrasting and cannot share the same spectrum.

HIPERLAN and 5.2 and 17.1 GHz Bands

Europeans are approaching wireless LAN development from a different angle. They intend to establish a standard first and then develop the products based on the standard. ETSI has asked the Sub-Technical Committee RES10 to develop a standard for High Performance Radio Local Area Networks (HIPERLAN). The committee has secured two bands at 5.12-5.30 GHz and 17.1-17.3 GHz for the development of the HIPERLAN to operate at a minimum useful bit rate of 20 Mb/s for point-to-point data communications with a range of 50 m. They expect that at this rate and range they can provide 500-1000 Mb/s for a standard building floor of approximately 100 meters square that is comparable with FDDI [25]. RES10 is chartered to define a radio transmission technique that includes type of modulation, coding, and channel access, as well as the specific protocols. The first meeting of the HIPERLAN took placed in December 1991. They secured the bands in 1992-93 and they expect to receive ETC approval this year.

Trends in Research

Research today is directed toward enabling the average person to access vast compute power and an enormous amount of information through a "cloud" comprised of a synergistic fusion of communications and distributed computers. The student, nurse, business person, or firefighter dealing with a hazardous materials incident will be able to tap into an enormous quantity of information and computing power assembled to benefit all.

From the Past to the Present

In the view of many, computers and communications are perceived as complementary technologies: related, but different, and not integrated. This is borne out by the focus of research during the past 20 years.

An earlier direction of U.S. government spending for computer research was directed toward supercomputers applied to complex problems, the so-called "grand challenges." The object was to apply massive computing power to solve complex tasks. Wireless networks were not a factor in these programs. Over the past decade, the supercomputer market has been eroded by the emergence of cost effective high-performance desktop workstations and personal computers [26]. This is partially responsible for an expansion in research direction to the "national challenges," which includes wireless networks.

Yielding to the demands of businesses, healthcare, digital libraries, and other interests and applications, the U.S. government has broadened its focus from the "grand challenges" to the "national challenges." The expanded goal now includes funding to provide computational power to the average person in a meaningful way through the National Information Infrastructure (NII) [27]. Support for the range of applications envisioned for the NII requires a fusion of computers and

Research is directed toward enabling the average person to access an enormous amount of information through a fusion of communications and distributed computers.

In an increasing number of applications, the network is providing computation resources through the communications pathway.

networks with wireless access. The network can no longer be thought of as simply a pathway for the transmission of files from one computer to another. The network must meld with the computer, enabling a new paradigm referred to as a "cloud."

Fusion of Computer and Communications

In the simplest sense, networks provide the ability for computers to communicate. However, the integration of computers and networks can provide more than file transfers. In an increasing number of applications, the network is providing computation resources through the communications pathway. One need only look at the burgeoning Internet to see this trend. This coupling has made the Internet more than a vehicle for electronic mail. It has enabled information and provided computational power independent of geographical location. This trend promises to fuel a huge demand for widely distributed computers seamlessly networked together.

The ultimate challenge is to exploit wasted computing cycles and provide location-independent access to the computational resources. If the number of personal computers is 50 million and each is capable of 2 MIPS, then the combined computing power is 10^{14} MIPS with, perhaps, 95 percent of that idle. When one considers workstations, minicomputers and mainframes, the combined computing power is immense. The difficulty is how to put this incredible resource to productive work and provide ubiquitous local wireless access to the cloud anytime and from anywhere.

Impact of the Cloud

To illustrate how the fusion of computers and wireless LANs will impact the future, we present three applications under development or envisioned: a military application, the military concept applied to the office or hospital, and a community emergency services response.

Digitalization of the Battlefield — Command, control, communications, and intelligence (C^3I) are a difficult task in the hostile environment of a battlefield. The field commander needs to know the location and situation of the troops under his command, and have all available intelligence as well, in order to make the appropriate decisions. The lives of the troops and the success of the mission depend on the commander and his C^3I. If the commander along with all support and communications is at one site, the battalion is vulnerable. To reduce the battalion's exposure and assure the integrity of the chain of command, the integrated wireless digital battlefield has been proposed.

In the digitized battlefield, each soldier carries a backpack computer connected through a wireless LAN to all others in the battalion. Through the network, each soldier's physical condition and location can be monitored through the use of a personal network [28], and instructions can be given accurately and without delay. Visual data can be relayed to the commander. More importantly, the backpack computers form a distributed, fault-tolerant computer, survivable after the loss of one or more nodes. The wireless LAN enables the unencumbered movement of troops and machines, as well as providing a communication infrastructure. The commander, or his chain of command, can monitor the health of the troops, monitor movements, gather intelligence, communicate with neighboring commanders or superiors, and give orders.

The wireless digital battlefield provides a significant strategic advantage. The enemy is no longer able to destroy the effectiveness of a battalion by striking a central command site. The C^3I function is more difficult to compromise. Command is spread over a larger area, with each soldier a part of the distributed C^3I structure.

Jacquard — The Jacquard project[2] [29] fuses computers and communications into a scalable, real-time, distributed, fault-tolerant architecture. Jacquard provides for both shared-memory and message-passing distributed-processor paradigms with wireless access. It is essentially a scalable local *cloud* connectable to a global *cloud*. Processes on the system can be run on processors with available compute cycles, thus putting wasted resources to use. If a machine fails, the process can be automatically restarted on another machine under control of the operating system. The user need not be aware of the intricacies of this computational backbone, only that his mission-critical process will run on the available cycles, thus increasing resource utilization. Consider the battleship captain needing to compute missile trajectories to launch a counter-attack after suffering damage to the weapons control. The Jacquard architecture will still allow the ship to compute and launch a retaliatory strike by utilizing the remaining processors distributed throughout the ship. The captain keeps informed of the situation through his personal digital assistant (PDA), connected to the ship's Jacquard framework, regardless of where he roams on the ship.

In the office or campus environment, an individual is able to access the Jacquard *cloud* through a PDA with ubiquitous wireless communications. The PDA is of limited processing capability to conserve battery power, and has a user-friendly graphical interface [30, 31]. The businessperson, student, or engineer can access vast quantities of data, run simulations, utilize decision support software, and teleconference through their PDA. Beyond ad hoc networks, the PDA provides mobility to the user by maintaining connectivity throughout the local area. Discussions in a conference room, hallway, or associate's office can be more productive by testing hypothesis, running simulations, searching databases, or calculating projections quickly, utilizing the heretofore wasted compute cycles of traditional computers through wireless LANs.

Emergency Service Response — The local fire department is called to respond to a hazardous material spill on a busy highway in the community. Speed is critical to mitigate the situation and to insure citizen safety and protect property. Upon arrival at a safe distance from the spill, the highly-trained response team dons protective clothing and attaches a PDA with wireless LAN. The incident commander and his assistants monitor the vital signs of the response team and observe the situation through video, audio, and data transmitted over the wireless LANs. As the team leader nears the spill, he observes the placard on

[2] *Funded by ARPA under contract #DABT63-91-C-0016.*

the trunk and on the leaking drums.

The chemical identification is entered into the PDA and, through the *cloud*, identification, health and fire hazards, recommendation for evacuation, and containment and neutralization procedures are relayed to the team. Additionally, the mixture of different chemicals is relayed to a network of computers that quickly analyzes the resultant chemical compound, which may not be contained in any reference material, and suggests a course of action.

This represents a vast improvement over the current method of limited voice-only communications. Similar technologies can be effectively applied to emergency medical services, law enforcement, highway workers, and so on.

Technical Challenges

This vision of the future, the integration of computers and communications into one entity or *cloud*, must overcome significant technical obstacles before becoming a reality. The communications technologist is faced with providing a ubiquitous wireless LAN for connectivity anywhere, anytime. The computer technologist is faced with the problem of integrating the computers into a cohesive distributed network through the operating system and software applications.

The communications problems to be overcome include limited bandwidth, latency, dropout, limited available power, and secure communications. In distributed computing applications, latency is one of the biggest problems.

The computer component of the *cloud* has its hurdles to overcome. The speed of the processor is not the limiting factor. Density, power requirement, performance, and cost of memory are limiting factors. Display technology is another area where improvement must be made. Resolution, power, size, and cost are areas for improvement in PDA display technology.

The system issue of resource management must be considered and may be intractable. Imagine allocating resources on 10 different computers or processors to run 48 different applications or processes. There are on the order of 10^{48} different allocation assignments or possibilities, each with its own system-wide performance implications. The complexities of efficient, fair, and equitable allocation will be a challenge for the future.

The fusion of computers and communications into a unified entity may appear a subtle distinction from the computers and networks popular today. However, the impact of the fusion along with wireless LANs will be profound. It will affect everything from the way we do business to the way we live. It will be the harbinger of an information wave.

References

[1] K. Pahlavan, "Wireless office information networks," *IEEE Commun. Mag.*, vol. 23, no. 6, June 1985, pp. 19-27.

[2] M. J. Marcus, "Regulatory policy considerations for radio local area networks," IEEE *Commun. Mag.*, vol. 25, no. 7, July 1987, pp. 95-99.

[3] M. J. Marcus, "Regulatory policy considerations for radio local area networks," Proc. IEEE Workshop on Wireless LANs, Worcester, MA, May 1991, pp. 42-48.

[4] F. R. Gfeller and U. Bapst, "Wireless in-house data communication via diffuse infrared radiation," *Proc. IEEE*, vol. 67, Nov. 1979, pp. 1474-1486.

[5] P. Ferert, "Application of spread spectrum radio to wireless terminal communications," Proc. IEEE NTC'80, Houston, TX, 1980, pp. 244-248.

[6] C. S. Yen and R. D. Crawford, "The use of directed beams in wireless computer communications," Proc. IEEE GLOBECOM '85, Dec. 1985, pp. 1181-1184.

[7] Y. Nakata *et al.*, "In-house wireless communication systems using infrared radiations," Proc. Int. Conf. Comp. Commun., Sydney Australia, 1984, pp. 333-338.

[8] R. Mednick, "Office information network: an integrated LAN," Proc. IEEE GLOBECOM '85, New Orleans, LA, 1985, pp. 15.2.1-5.

[9] T. S. Chu and M. J. Gans, "High speed infrared local wireless communication," *IEEE Commun. Mag.*, vol. 25, no. 7, July 1987, pp. 4-10.

[10] B. Ramamurthi and M. Kavehrad, "Direct-sequence spread spectrum with DPSK modulation and diversity for indoor wireless communications," *IEEE Trans. Commun.*, vol. COM- 35, Feb. 1987, pp. 224-236.

[11] B. Tuch "An ISM band spread spectrum local area network: WaveLAN," Proc. IEEE Workshop on Wireless LANs, Worcester, MA, May 1991, pp. 103-111.

[12] T. Freeburg, "A new technology for high speed wireless local area networks," Proc. IEEE Workshop on Wireless LANs, Worcester, MA, May 1991, pp. 127-139.

[13] IEEE Workshop on Wireless LANs, Worcester Polytechnic Institute, Worcester, MA, May 1991.

[14] D. G. Steer, "Coexistence and access etiquette in the United States unlicensed PCS band," *IEEE Personal Commun.*, vol. 1, no. 4, 4Q 1994.

[15] K. Pahlavan and A. H. Levesque, "Wireless data communications," *Proc. IEEE*, vol. 82, no. 9, Sept. 1994, pp. 1398-1430.

[16] K. Pahlavan, "Spread spectrum for wireless local networks," Proc. IEEE PCCC, Feb. 1987.

[17] K. Pahlavan and J. L. Holsinger, "Voice-band data communications, a historical review: 1919-1988," Invited paper, *IEEE Commun. Mag.*, Jan. 1988

[18] K. Pahlavan, S. Howard, and T. Sexton, "Adaptive equalization of indoor radio channel," *IEEE Trans. Commun.*, vol. 41, Jan. 1993, pp. 164-170.

[19] S. W. Wales, "Modulation and equalization techniques for HIPERLAN," Proc. of IEEE PIMRC'94, Sept. 1994, pp. 959-963.

[20] A. Nix *et al.*, "Modulation and equalization considerations for high performance radio LANs (HIPERLAN)," Proc. of IEEE PIMRC'94, Sept. 1994, pp. 964-968.

[21] G. Yang and K. Pahlavan, "Sector antenna and DFE modems for high speed indoor radio communications," *IEEE Trans. on Vehic. Technol.*, Nov. 1994.

[22] G. Yang and K. Pahlavan, "Performance analysis of multicarrier systems in office environment using 3D ray tracing," Proc. IEEE GLOBECOM, Dec. 1994.

[23] M. Aldinger, "Multicarrier COFDM scheme in high bit rate radio local area net- works," Proc. of IEEE PIMRC'94, Sept. 1994, pp. 969-973.

[24] G. Yang and K. Pahlavan, "Performance analysis of multicarrier modems in an office environment using 3D ray tracing," Jan. 1994. (In review.)

[25] B. Bourin, "HIPERLAN markets and applications standardization issues," Proc. of IEEE PIMRC'94, Sept. 1994, pp. 863-868.

[26] "The reinvention of supercomputing: Silicon Graphics and the changing technology," International Data Corporation White Paper, 1994.

[27] "High performance computing and communications: Technology for the National Information Infrastructure," Committee on Information and Communication, National Science and Technology Council, 1995.

[28] P. P. Carvey, "BodyLAN: A wireless body local area network," Wireless, Adaptive and Mobile Information Systems Program Summary, Advanced Research Projects Agency, 1994.

[29] R. P. LaRowe Jr., *et. al.*, "Jacquard: An architecture for shared memory networks of workstations," Enterprise Computing Institute Technical Report, Nov. 1994 (submitted for publication).

[30] A. Smailagic and D. P. Siewiorek, "The CMU mobile computers: A new generation of computer systems," Proc. IEEE COMPCON 94, Feb. 1994.

[31] S. Sheng, A. Chandrakasan, and R. W. Brodersen, "A portable multimedia terminal," *IEEE Commun. Mag.*, vol. 30, no. 12, Dec. 1992, pp. 64-75.

Biographies

KAVEH PAHLAVAN is the Westin Hadden Professor of Electrical and Computer Engineering and the director of the Center for Wireless Information Network Studies at the Worcester Polytechnic Institute, Worcester, Massachusetts. His recent research has been focused on indoor radio propagation modeling and analysis of the multiple access and transmission methods for wireless local networks. His previous research background is on modulation, coding and adaptive signal processing for digital communication over voice-band and fading multipath radio channels. He is a senior member of the IEEE Communication Society.

THOMAS H. PROBERT is the founding President of the Enterprise Computing Institute, Inc. (ECI), a private not-for-profit organization conducting a program of research, development, and education in advanced computer and communications technology. He has been a principal investigator on a number of ARPA-funded research projects. In addition, he has held academic appointments at numerous universities and colleges. He holds M.S. and Ph.D. degrees in computer and information science from the University of Massachusetts at Amherst.

MITCHELL E. CHASE is a research scientist with the Enterprise Computing Institute Inc. (ECI), Hopkinton, Massachusetts. Prior to joining ECI, he was involved with communication systems and network simulation at the Center for High Performance Computing and Comdisco Systems. He received a B.E. in electrical engineering from City College of New York, an M.S. in biomedical engineering from Iowa State University, an M.B.A. from Northeastern University, and a Ph.D. in electrical and computer engineering from Worcester Polytechnic Institute.

The system issue of resource management may be intractable. Imagine allocating resources on 10 different computers or processors to run 48 different applications or processes.

Chapter 2

WIRELESS COMMUNICATION STANDARDS & SYSTEM DESIGN

This chapter covers many of the emerging cellular, PCS, data, and mobile satellite standards, and includes an in-depth treatment of Cellular Digital Packet Data (CDPD) and Global System Mobile (GSM) standards.

CELLULAR DIGITAL PACKET DATA (CDPD)

AN OVERVIEW

GTE Personal Communications Services
Mobile Data Group

Cellular Digital Packet Data: An Overview

Release 1.0
June 9, 1994

Author

This document was written and compiled by Kevin Puk.

Acknowledgments

Thanks to Bob Brenner and Steve Isaacson for providing information and clarification to material covered in this document.

Table of Contents

List of Figures

1.0 INTRODUCTION

The latest trend in telecommunications is toward freeing users from the constraints of their wired environment. This first started with the cordless phone but range was rather limited. With advancements in radio engineering and radio frequency (RF) spectrum availability, a wireless telecommunications industry has been born which allows customers to take their phone wherever they go.

This mobility is eagerly accepted and an increase in mobility is desired. Not only do customers want to be able to send and receive voice in a mobile manner, they also want to send and receive data and video. The infrastructure for wireless (or cellular) voice communications has already been established. Now user mobility is being extended to include data services.

The purpose of this document is to introduce one particular wireless data technology, Cellular Digital Packet Data (CDPD). Basic background material will first be presented on cellular voice/data and networking. This will be followed by a discussion of CDPD interfaces, components, and network services.

1.1 Cellular Voice

Currently, the widest implementation of cellular technology in the US is the Advanced Mobile Phone System (AMPS), more commonly known as cellular. Three basic building blocks compose the AMPS system. These blocks are Mobile Telephone Switching Offices (MTSOs), cell sites, and cellular (mobile) terminals.

A mobile terminal is any cellular device that can send and receive voice signals, such as a portable cellular phone. A cell site provides an interface between the mobile terminals and the MTSO. It has an antenna for receiving cellular signals and a transmitter for sending cellular signals. The MTSO coordinates many cell sites and mobile terminals, acting as a "cellular switch" for cellular phone calls. Also, the MTSO provides an interface with local telephone companies and keeps track of billing information.

Figure 1-1 illustrates how the mobile terminal (car phone), cell site, and MTSO work together to complete a call.

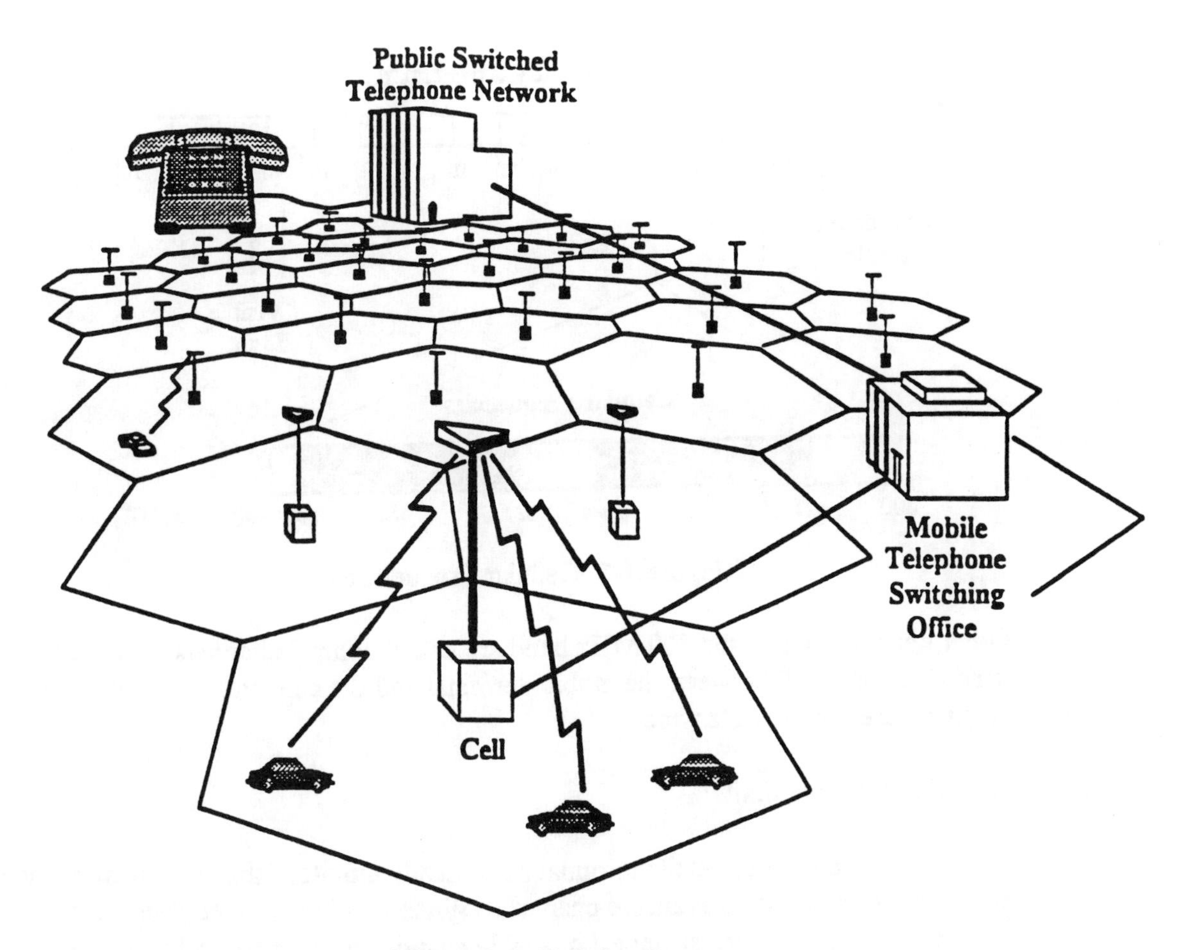

Figure 1-1 Cellular Coverage

1.1.1 Cellular Frequencies

The Federal Communications Commission (FCC) allocated 50 megahertz (MHz is 10^6 Hertz) of the frequency spectrum to be used for cellular communications. Figure 1-2 illustrates where this 50 MHz band is located within the frequency spectrum.

This 50 MHz is broken up into two 25 MHz sections: one for forward communications (cell to mobile terminal) and one for reverse communications (mobile terminal to cell). The forward frequencies are 869 MHz to 894 MHz and the reverse frequencies are 824 MHz to 849 MHz. The 20 MHz gap in the cellular band helps minimize interference between forward and reverse communications.

For cellular voice communications, forward and reverse connections are required. Each connection uses 30 kilohertz (kHz) of the cellular frequency band which is called a channel. Therefore, for one cellular phone call, 2 channels (a channel pair) are required for communications. Within the cellular domain there are 832 channel pairs available.

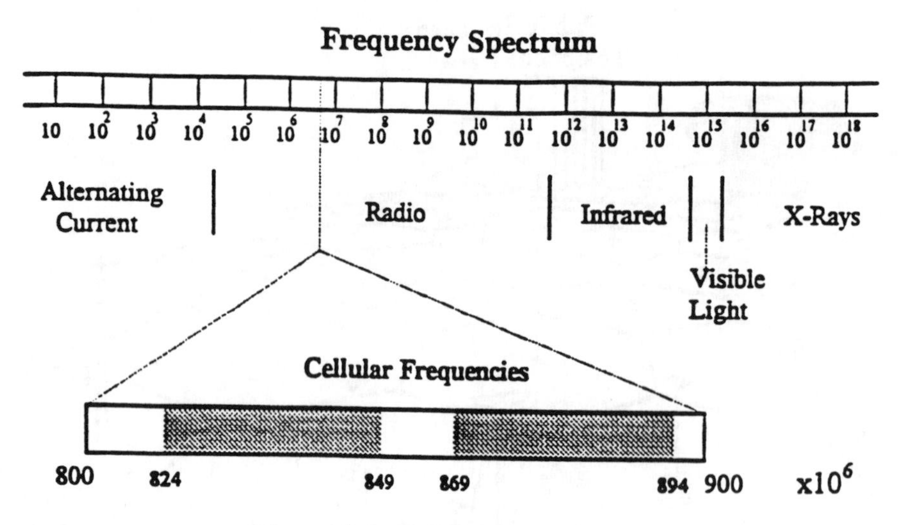

Figure 1-2 Cellular Frequencies

Other channels within the 800 MHz band are called control channels which are used to setup a call between the mobile terminal and the cell site. Each of these channels are also 30 kHz wide.

1.1.2 A Cellular Phone Call

When powered on, the mobile terminal periodically monitors the cellular frequency spectrum for a cell site's available channels. When a call is initiated, the mobile terminal "locks" onto the strongest available channel in order to exchange data with the cell site. This data includes the called-party telephone number, the calling-party telephone number, and the calling-party electronic serial number (unique number that identifies the mobile terminal). The exchange of this data is done on what is called a control channel because no voice communication is conducted on this channel. Then, after the mobile terminal is authenticated as a valid user and the connection is established, the cell instructs the mobile terminal to change frequency to a particular voice channel for communication, connecting both parties. The time spent for the original communication between the mobile terminal and the cell site (on the control channel) is the call setup time.

During communication, there is a possibility that the mobile terminal will move outside the current cell's coverage area and need to be transferred to another cell site. When this happens, the MTSO is responsible for maintaining the connection by initiating a hand-off (an instruction for the mobile terminal to communicate with a new cell site). The MTSO determines which surrounding cell site would provide the strongest available signal for the mobile terminal and then sets up the communication between that cell and the mobile terminal. Figure 1-3 shows the decision flow for a hand-off.

As the user moves further away from the cell the cellular signal strength weakens and the voice communication quality drops (i.e. the user starts to hear static). As

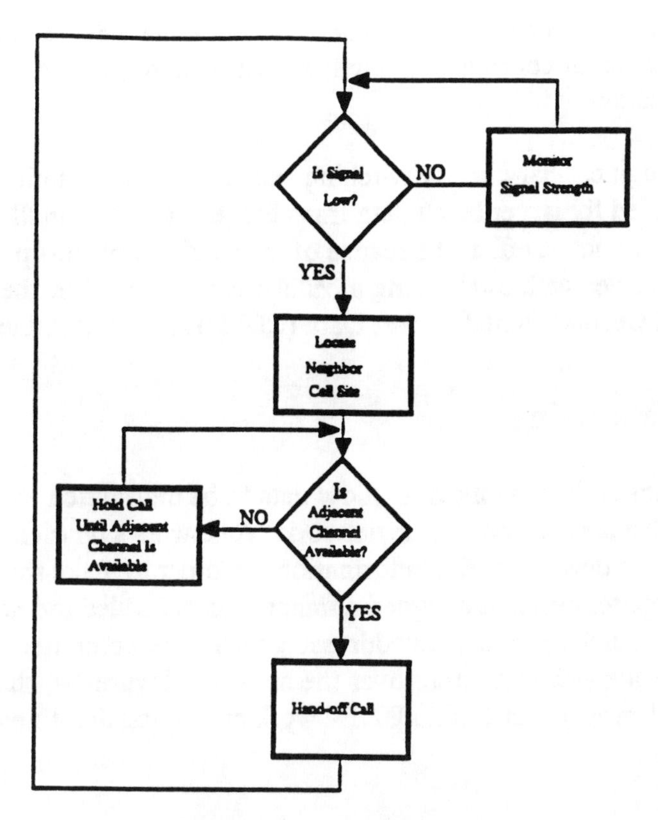

Figure 1-3 Hand-Off Flow

the MTSO searches for an available channel in the neighbor cell, it is possible no channels are available. As Figure 1-2 indicates, the MTSO will hold the call until a channel becomes available. However, if the user moves far enough away before a channel is freed then communication will be cut c... i.e. the call is dropped).

1.2 Cellular Data (Circuit-Switching)

Circuit-switched technology is one way of transmitting data over the cellular network using the same cellular frequencies (same channels) as voice. Circuit-switching provides a constant connection between originating and terminating points of communication for an entire data transmission. Data can be sent using circuit-switched technology by attaching a wireless modem to a mobile computing device (e.g. PC laptop). A wireless modem translates the digital data into an analog signal and transmits the data over the cellular network. On the receiving end of the call, only a compatible modem will be able to "hear" the transmitted data.

The wireless modem allows the user to establish a connection ("lock" a cellular channel) with a cell site in exactly the same manner as making a cellular phone call (mentioned above). Once the channel is "locked" by the user, data can be transmitted over the cellular network. Similar to a voice transmission, the MTSO is responsible for coordinating any hand-offs that may occur by the user moving to a new location.

Even though cellular circuit-switching can be used to transmit any type of data, it is best suited for larger batch data transfers (e.g. a fax). Small short bursty data transfers are not as efficient because of the overhead of set-up and tear down delays involved with establishing a cellular connection. For these types of data transfers, Cellular Digital Packet Data (CDPD) is more suitable.

1.3 Data Networking

The purpose of a network is to allow data to be transmitted to and received from devices that are attached to the network. A network's physical makeup is a group of computer devices sharing information. In order to make this sharing possible, each computer device is assigned a unique number called the network address (i.e. telephone number). Using an addressing technique, computer devices can directly communicate with each other over the network. Figure 1-4 shows devices A, B (e.g. PC laptops), and C (e.g. UNIX workstation) talking to each other via the network.

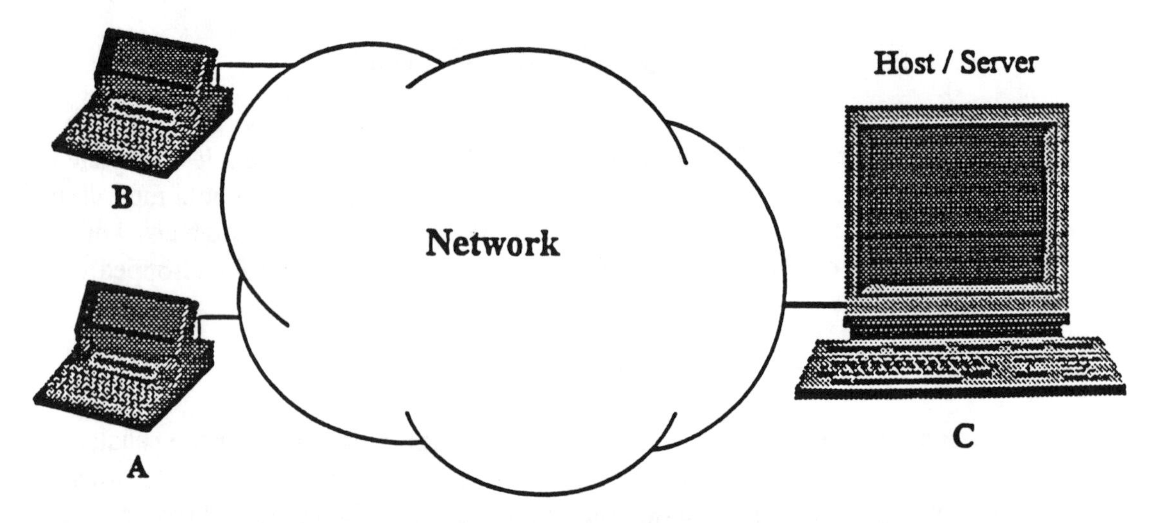

Figure 1-4 Generic Network

Every network has a defined set of protocols which establishes the rules and formats for communications within that network. Therefore, any applications or products developed within a particular network will be compatible with each other. For example, if a word processing application (i.e. Microsoft Word) is created with the above network's protocols (Figure 1-4) then both device A and device B will be able to use that application whether it is running off a local disk or the remote host (device C). However, in order to use data networking to help solve incompatibility problems between different types of computing devices, translation or conversion equipment, such as gateways, may have to be used.

1.4 What is CDPD?

Cellular Digital Packet Data (CDPD) is a technique for transmitting small chunks of data over the cellular network (using cellular channels) in a reliable manner. CDPD allows users to send and receive data from anywhere within the cellular domain, at anytime, quickly and efficiently.

Sending data through the CDPD network is analogous to mailing an envelope through a Postal system (Figure 1-5). First, a letter is placed into an envelope which contains the mailing address and return address. After the envelope is placed in the mail box, it goes to the local Post Office. From there it is routed to the destination Post Office. Then, it is routed again to a distribution center where, based on the mailing address, it is given to the appropriate deliverer. At the distribution center the accounting is done which includes recording the number of envelopes sent to a particular user. Then, the deliverer relays the envelope from the distribution center to its final destination, the mobile user.

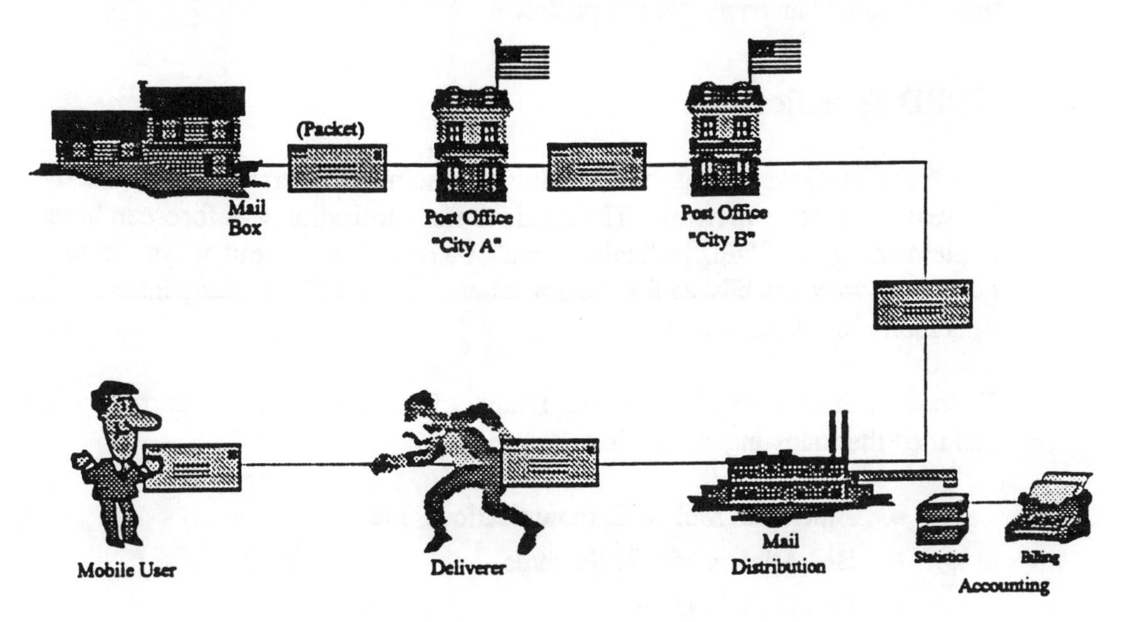

Figure 1-5 Post Office Analogy

After presenting more detail on CDPD, we will refer back to this analogy and relate it with the CDPD components (see page 26).

The CDPD Network, in its most basic form, is just an extension of common networking techniques, which allows connected devices to communicate with each other over the cellular infrastructure (Figure 1-6).

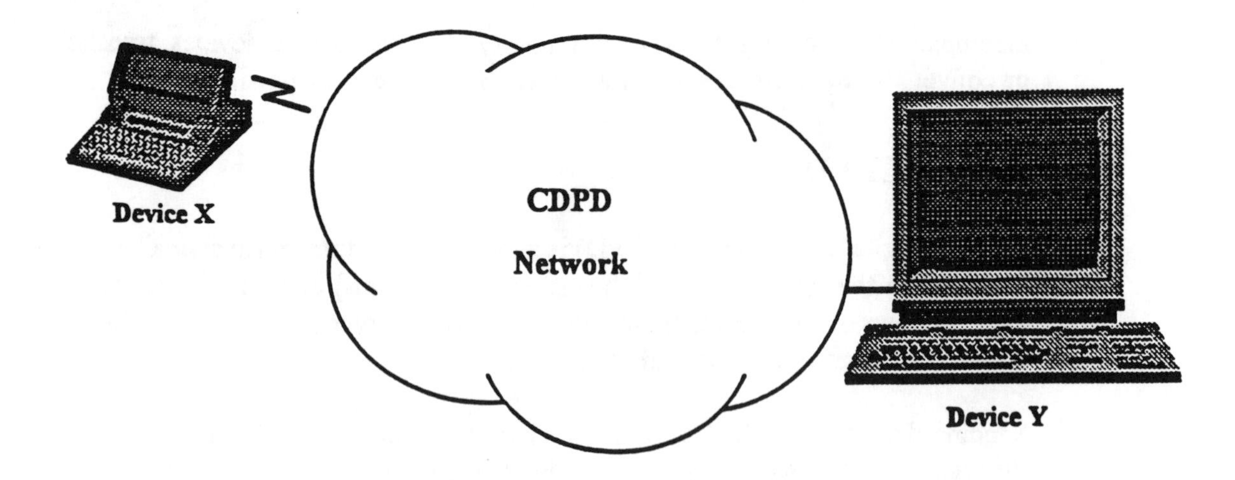

Figure 1-6 CDPD Network

The chunks of data transmitted through the CDPD Network are officially called Network Protocol Data Units (NPDUs), but are more commonly known as packets. Sending a data block broken up into packets is similar to mailing a 100 page document one page at a time. The whole document is a data block and each page, including an envelope, is a packet.

1.5 CDPD Specifications

In 1992, CDPD was recognized as a business solution for transmitting data over the existing cellular network. The need for standardization was foreseen because implementing the CDPD technology could be done in numerous ways. If no guidelines were established for the implementation of CDPD, many inter-operability problems would arise.

Therefore, the leaders of the cellular industry formed a CDPD consortium which included the following companies:

- Ameritech Mobile Communications, Inc.
- Bell Atlantic Mobile Systems
- Contel Cellular, Inc.
- GTE Mobile Communications, Inc.
- McCaw Cellular Communications, Inc.
- NYNEX Mobile Communications, Inc.
- Pactel Cellular
- Southwestern Bell Mobile Systems

The CDPD consortium published a document in July 1993 called *Cellular Digital Packet Data System Specifications Release 1.0* which details the agreed upon guidelines. The CDPD consortium has been continuously collecting data and revising the specifications; Release 2.0 will be published at the end of 1994.

1.6 CDPD Infrastructure

Even though the cellular infrastructure is extensive with nation wide coverage, it was designed specifically to transmit analog voice signals. Therefore, in order to provide a national seamless digital data overlay of the cellular network, a new technology needs to be incorporated. CDPD takes advantage of the current cellular design by using the idle times between cellular voice transmissions (i.e. cellular phone calls) to maximize the use of the cellular network. However, CDPD can also use dedicated channels to transmit data if needed.

An important aspect of CDPD is that it utilizes the existing cellular equipment, saving a significant amount of capital expenditure. The technology required to make CDPD a reality is simply an overlay to the established cellular infrastructure. CDPD uses the same cellular frequencies and has the same coverage as cellular voice.

There are other non-cellular packet data methods for transmitting data (such as Ardis and RAM). However, because the cellular infrastructure has already been installed, CDPD is more attractive and cost effective. In addition, CDPD technology will allow mobile subscribers to use existing wire-line network applications and services because it uses established network protocols (e.g. TCP/IP).

As mentioned earlier, data can also be sent with circuit-switched technology over the current cellular infrastructure. Let us now compare circuit-switching to CDPD (packet-switching).

CDPD vs. Circuit-Switching

One basic difference between CDPD and circuit-switching is that CDPD is connectionless oriented and sends each packet intermittently while circuit-switching sends the data over a continuous connection. Referring back to our mail example, sending all 100 pages in the same box is like using circuit-switching, while sending each page in a separate envelope is like CDPD.

With CDPD technology, the interconnection of all Service Providers (local carriers) allows seamless packet data services to be available to every CDPD customer (seamless mobility). However with circuit-switched, customer mobility is usually restricted to the cellular switch (MTSO) which means that when moving to a new service area, the user becomes disconnected from the cellular network and must reestablish a connection for continued communications.

Circuit-switching occupies the whole channel full time, hence it is more suitable for very large data transmissions (i.e. greater than 1 Megabyte). CDPD allows the sharing of a cellular channel among many users and among data and voice applications, hence it is more efficient and cost effective for small data transmissions (i.e. less than 1 Megabyte).

Another difference is that CDPD uses less power than circuit-switching. When a device uses circuit-switching to send data, it needs to ramp up power and stay at a high level until all of the data is sent successfully. CDPD sends data packets in short bursts, so power only needs to ramp up for short intervals. The result is less average power used (see Figure 1-7).

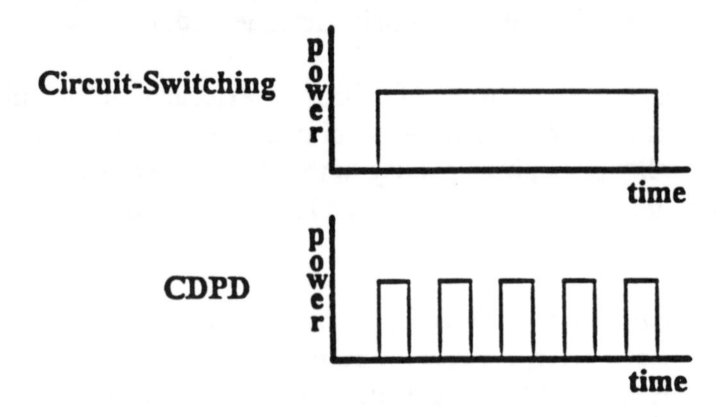

Figure 1-7 Power Usage

Also, CDPD has a "sleep" mode which allows the device to conserve power, when not currently being used, without having to log off the network. Therefore, a mobile computing device will have a longer battery life using CDPD technology than using circuit-switched.

Because circuit-switching data is the same as a voice call (the cellular network cannot distinguish between circuit-switched data and circuit-switched voice), transmission set-up and tear down delays are incurred. However, CDPD never establishes a direct connection, resulting in the avoidance of these delays.

Furthermore, circuit-switched data is unsecured and susceptible to casual eavesdropping by a third party, just as a cellular voice call is. However, CDPD uses an encryption technique to prevent an outside source form receiving the transmitted data (note that this excludes multicast and broadcast type data transfers).

Some of the differences between CDPD and circuit-switching technologies are summarized in Table 1-1.

Table 1-1 CDPD vs. Circuit Switching

CDPD	Circuit-Switching
Efficient for short to large burst transmissions	Efficient for very large transmissions
No call set-up or take down delays	Call set-up and take down required
Broadcast capabilities	Point-to-point connection
One log-on at power-up	Log-on (call) for every transmission
Power ramping for short bursts	Power ramping for entire connection
Airlink security	Airlink unsecured

1.7 CDPD Applications

CDPD is an integrated technology that can be used in a large variety of applications. An important advantage of CDPD is that it is designed to accommodate new technologies and maximize utilization of the existing cellular infrastructure. CDPD based solutions can be used in the following wide-ranging application types:

- Transaction applications
- Batch applications
- Broadcast/Multicast applications

Each of these are described below accompanied with some examples.

1.7.1 Transaction Applications

Transaction applications provide communication between two network users. A transaction normally consists of two messages (inquiry and response): one to request an operation and another to signal the success or failure of that operation. Figure 1-8 illustrates a transaction.

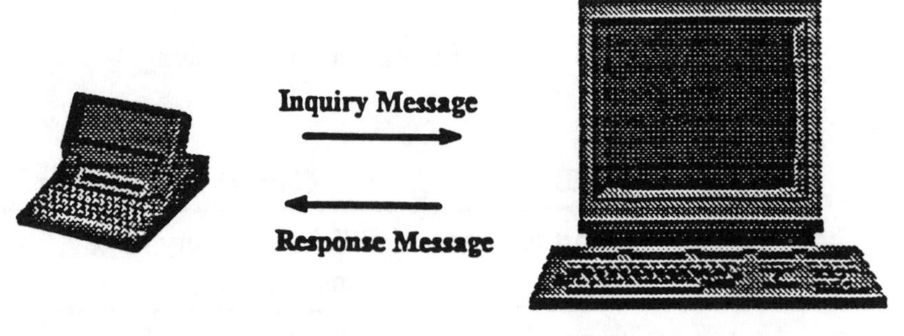

Figure 1-8 A Transaction

Credit Card Verification

A credit card number may be read in at the point of sale and a inquiry is sent to authorize the purchase. The inquiry message contains parameters to identify the business, credit card numbers, and the amount of purchase. The responding message could grant or deny authorization.

Point-of-Sale

Messages can be exchanged to query individual prices of items as well as to report the transaction total. The point-of-sale transactions may include transactions for credit card verification, inventory control, and other services. Imagine how hand-held, portable "registers" could eliminate some check-out counters, saving floor space and putting sales people on the floor with customers.

Taxi, Truck, and Job Dispatch

A message can be sent to a specific mobile computing device to direct it toward a destination for a pickup or next job. The response acknowledges receipt of the dispatch message.

Package Pickup, Delivery, and Tracking

A message can be sent to a specific mobile computing device to direct it to pickup or deliver a package. The response message confirms the receipt of the dispatch message. When a package is delivered or picked up, the package tracking number and the disposition of the package may be registered with a package tracking system.

Fleet Management

Messages can be exchanged to dispatch a fleet of vehicles and to track their current locations.

Inventory Control

The amount of individual items sold can be reported by employees, point-of-sale terminals, or vending machines to inventory control applications. Movement of material within a warehouse can be reported from terminals, or forklifts, or other material-handling equipment.

Emergency Services

A message can be sent to report a fire or burglary break-in to a public emergency (911) dispatch service. The request can contain additional information, such as the type of alarm, street address, additional telephone number, *etc.* Such an alarm could be silent, and not as susceptible to tampering as wired services. A similar service can be provided for private security and emergency services to monitor homes, report alarms or for personal distress signaling.

Mobile Computing Device or Vehicle Theft Recovery

When a mobile computing device is reported stolen, it may be able to receive a query and automatically report its present location. However, CDPD Location Services can only locate the particular sector/cell the device is in. Therefore, a global positioning system (GPS) is required in order to pinpoint the exact location of the device.

Telephone Activity Paging

This feature provides pager services, indicating activity on the cellular or land line telephone network. The subscriber is notified of any voice call attempts to its number. The treatment given to the call is included in the message to assist the subscriber in deciding how to dispose of the message.

Bi-directional Paging

With a bi-directional pager, a network user is able to receive a small alphanumeric message and notify the sender that the page was received. The recipient may also respond directly to a message.

Display of Calling Party Name or Number

A cellular subscriber may elect to have the calling party identity displayed as a name or telephone number. For the name service, the network requests a reverse directory search for the name associated with the calling party number. This way, a subscriber can selectively answer his or her calls, resulting in saved money and time.

Message Service

Calls destined to a subscriber can be routed to a message service. The message service converts a voice message into a text message that would be delivered to the corresponding subscriber.

Notice of Voice Mail

When a caller leaves a message at a voice mail system, a message is sent to the corresponding subscriber. This notification message could include the day, date, time message was left, calling party number, calling party name and length of message.

Notice of Electronic Mail

When a message is delivered to an electronic mail server, the server could generate a message notifying the subscriber of the receipt of a message.

Delivery of Electronic Mail

Electronic mail may be delivered directly to the mobile computing device. Users may be in contact with their most important information sources on a real time basis.

Telemetry (Usually Stationary)

Telemetry is used to periodically send information from wireless data acquisition equipment to a collector or user of the data. Such data acquired may be one of the following:

- Approximate subscriber location
- Engine or vehicle statistics (mileage, speed, oil pressure, temperature, vacuum, voltage, RPM, *etc.*)
- Burglar or fire alarm reporting
- Vending machine status (temperature, sales, unauthorized opening, tilt, location, velocity, *etc.*)
- Weather statistics (ambient temperature, barometric pressure, sunlight intensity, rainfall rate, snowfall, wind speed, wind direction, visibility, *etc.*)
- Stream statistics (flow rate, level, pump status, gate status, *etc.*)
- Vehicular traffic statistics (traffic density, traffic speed, queue length)
- Field measurements or other user-specified data collection, such as meter reading / monitoring for gas, electric, and water meters.

Information Retrieval Services

An information retrieval service is a transaction-based service that is more casual than the other transaction services. A subscriber would send a query to an information service, which will respond with the requested information. Such services could include:

- Restaurant reservation service
- Weather report and forecast for a particular place
- Traffic report and advisory for a particular highway, location, or area
- Directory assistance (electronic white pages)
- Local service assistance (electronic yellow pages)
- Driving instructions and location information

1.7.2 Batch Applications

Batch type applications provide subscribers a means to transmit data in one transaction. A batch transfer allows buffered, or saved, data to be transmitted entirely at a convenient time for the mobile users. Some example of batch type applications are as follows.

File Transfers

Complete files can be copied from an external system to a mobile subscriber's system, or vice versa, using a file transfer application (e.g. FTP).

Statistical Information Transfers

Telemetry devices recording less critical information can transmit statistical data during non-peak evening hours in order to save money on transaction costs. An example of this is a local utility power company may have a central unit located at an apartment complex which is connected to every apartment's utility meter. During the day, the central unit collects the power usage from each unit and at midnight it transmits the accounting information back to the company for billing.

1.7.3 Broadcast/Multicast Applications

Broadcasting

Broadcast messages are of interest to a wide audience. The messages can be delivered to a specific geographical location (a single sector/cell or to a group of sectors and or cells). Figure 1-9 below illustrates broadcasting a message to two surrounding cells.

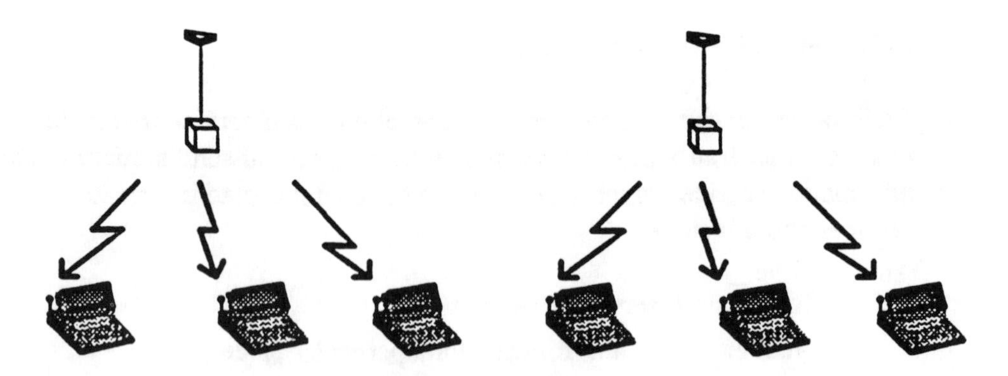

Figure 1-9 Broadcasting

Although broadcast messages are not acknowledged, broadcast messages may be sent several times over a period of time to increase the probability of reception. Also, because of point-to-multipoint service, CDPD does not encrypt any broadcast or multicast data packets.

General information Services

Any information that can be distributed over broadcast radio can be delivered using CDPD broadcast applications. Such items include:

- International news
- National news
- State news
- Local news
- Local date and time
- Prompts advertising to encourage access of information retrieval services for more information.

Weather Advisory

A weather advisory notice would be issued when the weather becomes a threat to public safety, such as storm warnings, small craft advisories, and hurricane watches.

Traffic Advisory

A traffic advisory notice will be issued to announce road closures, detours, *etc.* It can also be used to report localized accidents and congestion, and offer alternative routes.

Advertising

The cell broadcast service may be used for advertising. The advertising distribution can be controlled so that the message is delivered to its intended audience.

Multicasting

The Multicast Service is a value-added function that allows delivery of the same message to a particular group of subscribers. The CDPD Multicast Service is a point-to-multipoint, one-way unacknowledged message transfer service in which data packets are not encrypted. The Multicast Service sends the message to all members of a multicast group regardless of their current location. Figure 1-10 below illustrates a multicast to a group of three users.

Figure 1-10 Multicasting

Information Service Subscriptions

An information service provider collects information and sells it on a subscription or inquiry basis to subscribers. Subscription information services will deliver messages to individual subscribers on a periodic basis or as the desired information becomes available. Examples of information include stock quotes, business news, general news, news related to specific criteria, want ads for a particular classification, recreational information, road conditions, *etc.*

Private Bulletin Boards

A company with a mobile work force may use the multicast messages to communicate with many of its employees or users to disseminate service bulletins, company functions, product updates, *etc.*

2.0 CDPD SYSTEMS OVERVIEW AND ARCHITECTURE

Cellular Digital Packet Data (CDPD) technology will provide extensive coverage of high speed and high capacity data services to mobile users. In this section, basic concepts and components of the CDPD System will be discussed.

2.1 CDPD Specifics

With the CDPD technology, both voice and data will be able to be transmitted over the cellular channels. In order to integrate voice and data traffic on the cellular system, the CDPD network implements a technique called channel hopping. In this section we will illustrate a channel hopping example and examine how a data block is broken down into packets.

2.1.1 Channel Hopping

To help explain channel hopping, we will present a simple scenario. Imagine a cell which has three cellular channels (A, B, C) allocated to both CDPD and voice usage. Also imagine four mobile terminals, 1 mobile data unit and 3 cellular phones (X, Y, & Z). In the CDPD environment the mobile data unit is called a mobile end system (M-ES). All units are within the cell's boundary where the Received Signal Strength Indicator (RSSI) is strong enough for voice and data communications (Figure 2-1).

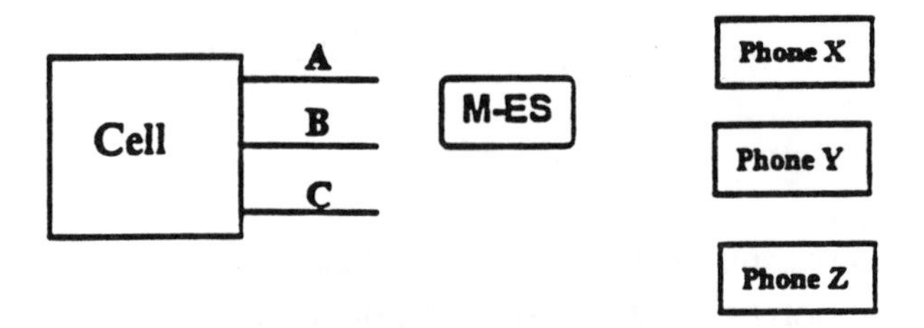

Figure 2-1 Hop Scenario Layout

When a data transmission is desired, the M-ES monitors the RSSI and availability of the cell's channels (A, B, C). Once a channel is located, the M-ES establishes a data link with that cell. In our example, the M-ES selects channel C (Figure 2-2).

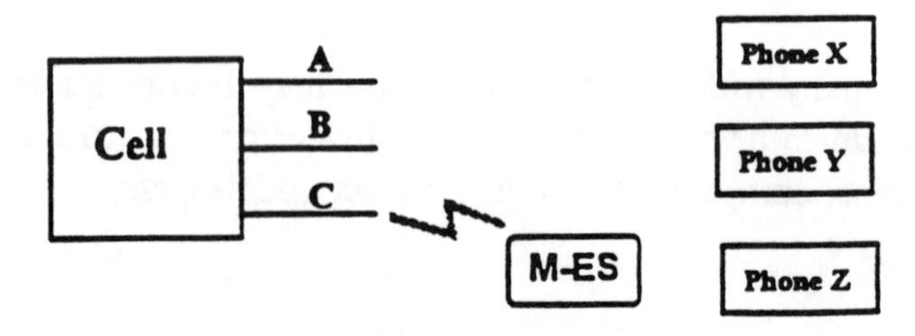

Figure 2-2 Hop Scenario Step 1

Establishing the data link with the cell includes having the M-ES verified as a valid subscriber by the CDPD Network.

As long as the cellular channel is not needed for voice communications, the M-ES can continue to transmit data packet bursts on it. However, the M-ES is limited to a preset burst time in order to allow other surrounding M-ESs to send data. Also, the CDPD serving entity in the cell, called the Mobile Data Base Station (MDBS, described in Section 2.4) constantly "sniffs" the RF to check for potential voice communication on the channel. In Figure 2-3, phone X uses channel B but since channel C is not needed for voice communications, or by other M-ESs, the M-ES can continue to transmit on it.

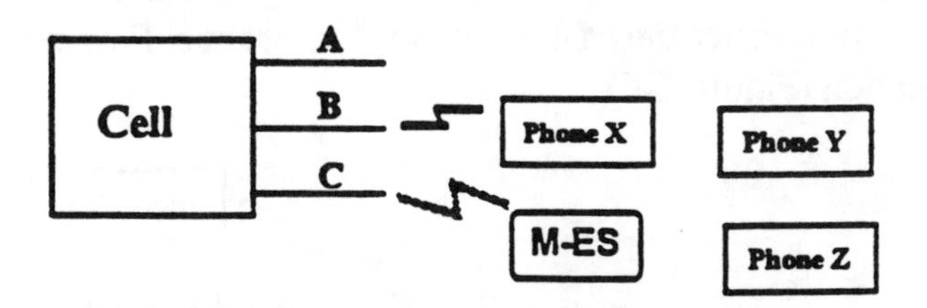

Figure 2-3 Hop Scenario Step 2

Now, phone Z wants to setup voice communication. From our example, the strongest channel near phone Z is C. Even though the M-ES is using that channel, voice always has priority and the M-ES will, within 2 milliseconds as defined in the CDPD Specifications, have to power down and give channel C up. The MDBS is responsible for initiating a channel "hop" (switching to another cellular channel) when the presence of voice forces the M-ES to give up the channel. In this scenario, the MDBS instructs the M-ES to "hop" to channel A for continued data transmission (Figure 2-4).

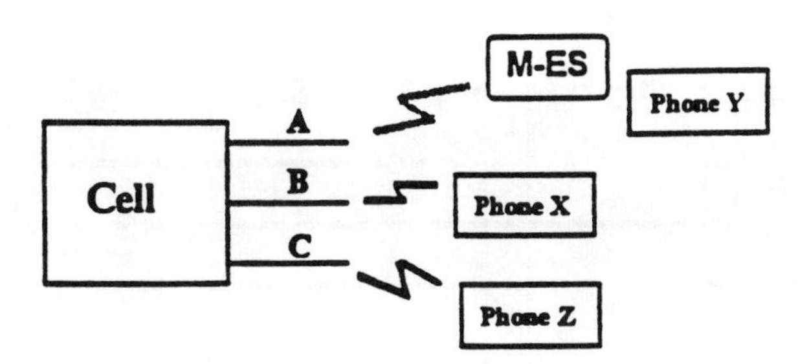

Figure 2-4 Scenario Step 3

Now, phone Y wants to establish voice communication. Again, because voice has priority, phone Y will connect with channel A and the M-ES will have to find another channel. However, all of the channels on the cell are busy so the M-ES cannot "hop" to another channel. Therefore, the M-ES has to either wait for one of cell's channels to become free (i.e. one of the phones ends voice communication or moves out of the cell's boundary), or move to another cell location. Until a channel becomes free, the data flow between the cell and the M-ES is discontinued (Figure 2-5).

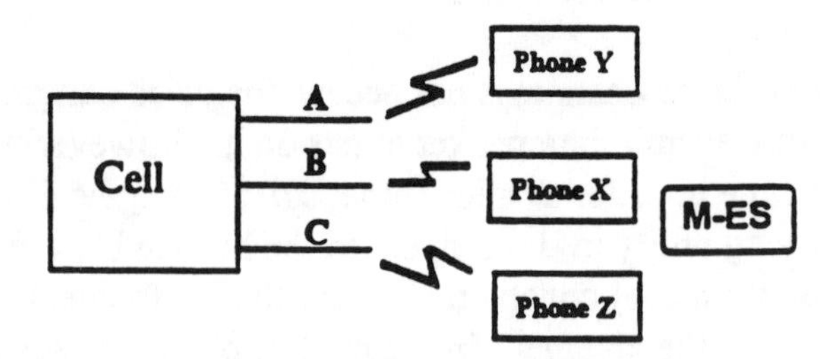

Figure 2-5 Scenario Step 4

Suppose that phone X leaves the cell's boundary and channel B is freed up. Now, the MDBS can instruct the M-ES to "hop" to channel B and continue data communication (Figure 2-6).

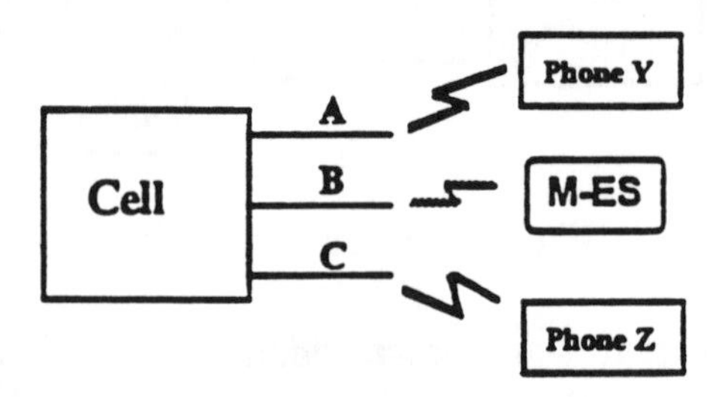

Figure 2-6 Scenario Step 5

An important thing to remember is these channel "hops" are completely transparent to the mobile data user. To the user, only one data channel stream was used to complete the entire transmission. In Figure 2-7, the data transmission for the above sequence of events is summarized.

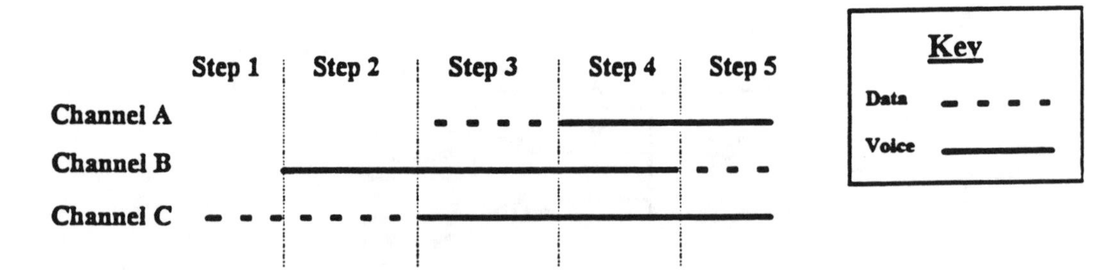

Figure 2-7 Scenario Summary

2.1.2 Breaking Up The Data

Before CDPD technology and its services can be used, some kind of data needs to be created (i.e. the 100 page document). Once this data exists in digital form, it can be sent and received over the CDPD Network. The first step is transforming the digital data into packets.

Data can just be thought of as a long series of zeros and ones which are lumped together in one big group. Figure 2-8 shows a simplified version of how the original data is broken down into packets paralleled with the example of dividing the document into separate envelopes.

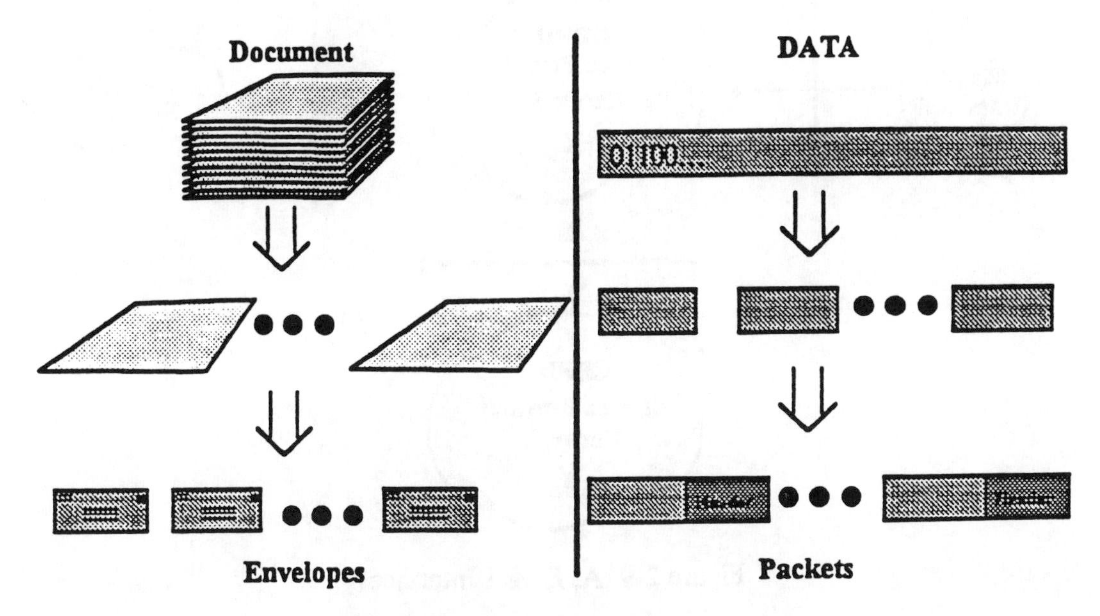

Figure 2-8 Envelope / Packet Analogy

First, the data is divided into separate sections (the document is divided into 100 separate pages). Then a header is added to each section (each page is placed into a separate addressed envelope). The header (envelope) contains important information including where the packet is from (return address) and where the packet is going (destination address). Packets are sent through the CDPD Network individually and each may take a different path to the final destination. However, if any of the packets are received out of order then the receiving device will reassemble the packets in the original sequence.

CDPD packets (NPDUs) will be able to hold up to 2048 bytes of data. Therefore, if a user has a packet size larger than 2048 then it will be broken down into 2048 byte segments by the CDPD Network.

2.3 CDPD Interfaces

Different systems of any network are able to communicate with each through an interface. Three distinct interfaces are defined for a CDPD Network. These are:

- Airlink (A) Interface
- External (E) Interface

- Inter-Service Provider (I) Interface

Figure 2-3 shows the three interfaces (A, E, I) within the CDPD Infrastructure.

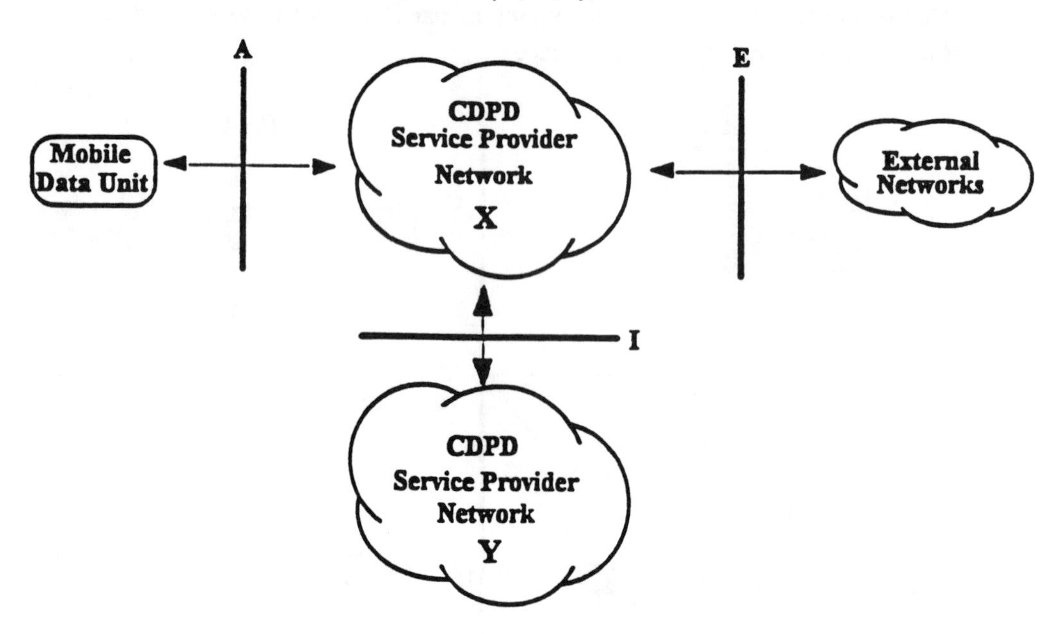

Figure 2-9 A, E, & I Interfaces

2.2.1 Airlink (A) Interface

The Airlink (A) Interface is the only wireless portion of the CDPD Network. This is where data communications between CDPD mobile end systems (M-ESs) and the serving portions of the CDPD Network occur. As with cellular phones, these communications are done over a cellular channel pair; a forward channel and a reverse channel. The A Interface encrypts data sent other these channels for non broadcast/multicast type data transfers, in order to prevent casual eavesdropping. Also, the connection oriented A Interface provides data flow control and error correction.

2.2.2 External (E) Interface

The External (E) Interface provides wired communications between the CDPD networks and external networks. These might include existing networks such as the Internet, OSInet, or other (e.g., private or VAN) external networks having compatibility with the CDPD network.

2.2.3 Inter-Service Provider Network (I) Interface

The Inter-Service Provider (I) Interface allows different CDPD carriers to communicate and exchange information with each other. Therefore, CDPD subscribers in New York can communicate with CDPD subscribers in San

Francisco regardless of who is providing the local service. Also, by providing data encapsulation, the I Interface allows data packets to be forwarded from an M-ES's home location to its current location.

The I Interface supports CDPD Network services across all geographical areas where CDPD is available. These services include M-ES authentication, network management and remote activation. An important characteristic of the I Interface is that it is not visible outside the CDPD Network.

2.4 CDPD Network Entities

In order to utilize the current cellular infrastructure, CDPD Network entities need to be added. Figure 2-10 below illustrates how the CDPD elements interface with each other.

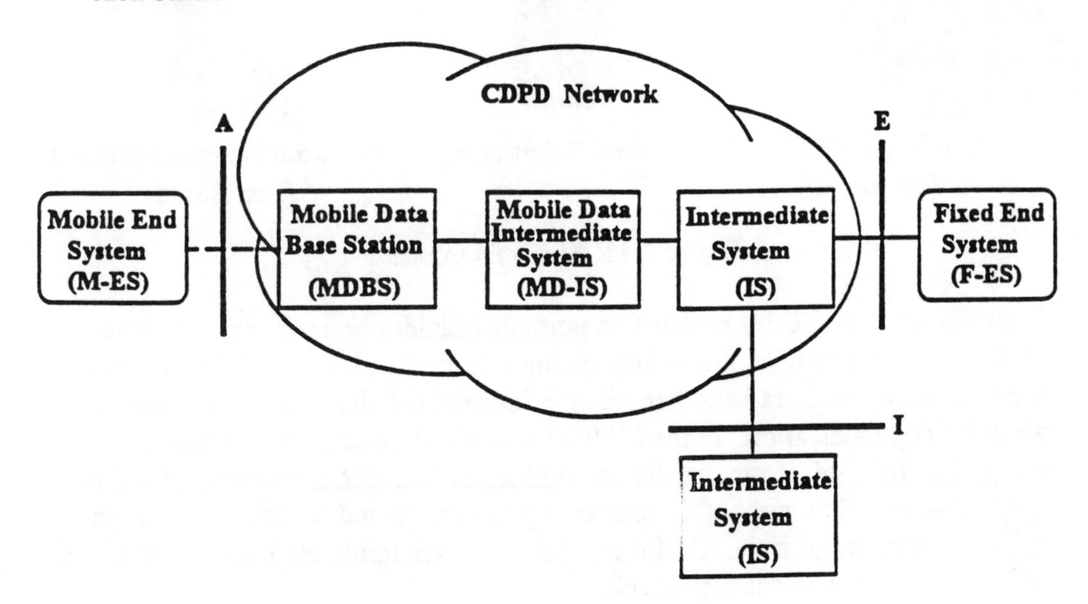

Figure 2-10 CDPD Network Entities

There are two basic classes of CDPD Network entities which are End Systems (ESs), and Intermediate Systems (ISs). The ESs are the CDPD Network hosts and the ISs are the CDPD Network routers.

2.4.1 End Systems

End Systems can communicate with each other via the CDPD Network. An End System is like a telephone (moveable or stationary), except in the CDPD sense it is used to send and receive digitized data. There are two basic types of End Systems:

- Mobile End System (M-ES), and

- Fixed End System (F-ES)

2.4.1.1 Mobile End System (M-ES)

The M-ES can be any mobile computing device (Figure 2-11) which has a CDPD modem installed or attached to it via cable. Also, an M-ES's physical location can change as the user desires without losing CDPD connectivity. The CDPD Network stores the last known M-ES location and routes the appropriate data packets to and from it accordingly.

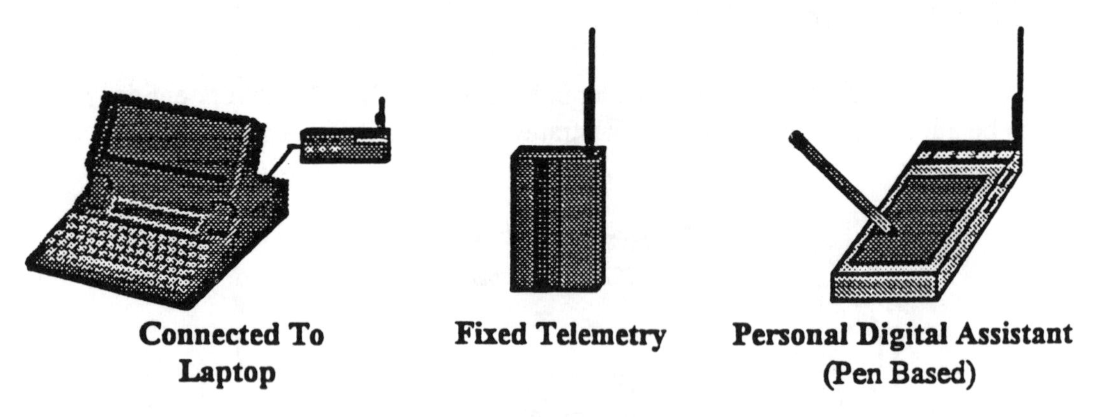

Figure 2-11 M-ES Examples

Even though the M-ES is similar to a portable cellular phone, it is much more intelligent. For example, the cellular phone is instructed by the MTSO to initiate a hand-off from one cell to another when the current cellular signal becomes too weak for communication. In the CDPD Network, the decision to initiate a transfer, or hand-off, from one cell to another cell is under the control of the M-ES itself. The M-ES monitors the received signal strength indicator (RSSI) of the cellular channels and if the RSSI drops below a predetermined level then the M-ES will transfer to a new channel or cell.

A major objective of the CDPD Network is to provide compatibility with existing networks and applications. Because of this goal, the M-ES was designed to appear like any other End System on the CDPD Network in order to utilize current applications, services, and systems.

M-ESs communicate with serving portions of the CDPD Network through the A-Interface.

2.4.1.1 Fixed End System (F-ES)

F-ESs can be one of many stationary computing devices, such as a UNIX workstation or a host computer. F-ES configuration and applications do not have to be modified for the CDPD Network and an F-ES can communicate with M-ESs in the same way as it does with other F-ESs.

Even though F-ESs are located in external networks, they can still communicate with the CDPD Network through the E-Interface.

2.4.2 Intermediate Systems

In order to modify the existing cellular infrastructure by incorporating the CDPD technology, some new internal entities, or components, need to be added. These include the following:

- **Mobile Data Intermediate System (MD-IS)**
- **Mobile Data Base Station (MDBS), and**
- **Intermediate System (IS)**

As mentioned earlier the CDPD architecture was designed to be integrated with the current existing cellular infrastructure. Therefore, the added entities overlay the cellular voice entities. Figure 2-12 illustrates the CDPD overlay.

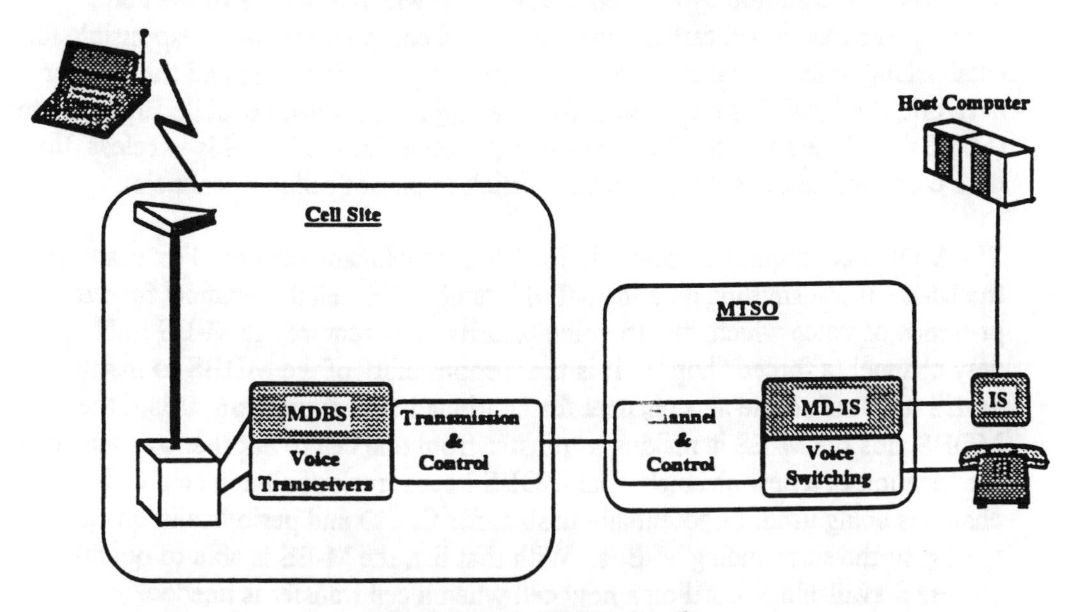

Figure 2-12 CDPD Overlay

Typically, MDBSs are located within the cell sites, while the MD-ISs are in the Mobile Telephone Switching Offices (MTSOs).

2.4.2.1 Mobile Data Intermediate System (MD-IS)

The MD-IS is a stationary network component which has similar responsibilities as the MTSO components mentioned earlier. Just as the MTSO components were responsible for tracking a portable cellular phone's location and the routing of calls, the MD-IS is responsible for keeping track of the M-ES's location and routing data packets to and from the CDPD Network and the M-ES appropriately.

The MD-IS can be considered the "brain" of the CDPD Network because of its functionality. It is responsible for making sure that an M-ES is valid to log on the CDPD Network. Also, it stores information including the M-ES's last known location (the last known location is just the last location the M-ES was used), traffic statistics, and billing information.

MD-ISs are the only network-relay systems that have any knowledge of mobility and operate a CDPD-specific protocol (Mobile Network Location Protocol) to exchange location information with each other. This exchanging of information allows an M-ES to go anywhere in the cellular domain without losing CDPD connectivity (seamless mobility).

2.4.2.2 Mobile Data Base Station (MDBS)

The MDBS is a stationary network component which is similar to the cell components mentioned earlier. Just as the cell components were responsible for establishing voice communications between the portable phone and the cellular network, the MDBS is responsible for relaying data between Mobile End Systems (M-ESs) and the MD-IS. Because the connection for the M-ES is wireless, the data communication is done over the Airlink (a pair of cellular channels).

The MDBS is primarily responsible for RF channel management. For example, as the M-ES is transmitting data the MDBS “sniffs” the cellular channel for the presence of voice which, due to voice priority, may require the M-ES to "hop" to a new channel (a forced “hop”). It is the responsibility of the MDBS to instruct the M-ES to "hop" to the new channel for continued communication. Also, the MDBS aids the M-ES in making a transfer from one cell to another by assisting in the location of a new channel. The MDBS keeps track of all adjacent cells' channels being used, or potentially usable, for CDPD and periodically broadcasts the list to the surrounding M-ESs. With that list, the M-ES is able to quickly choose a available channel on a new cell when a cell transfer is needed.

2.4.2.3 Intermediate System (IS)

ISs are off-the-shelf routers that are CDPD compatible with a primary responsibility of relaying data packets. An important characteristic of an IS's routing capabilities is that it can receive data packets on one interface and forward them out over another interface (similar to a gateway).

Also, multiple ISs are usually configured so that there is more than one potential path between any two ISs.

Post Office Analogy Re-Visited

Earlier we made an analogy with CDPD to the postal system. Now that the CDPD components have been presented, we can refer back to that analogy and examine it in more depth. Figure 2-13 is the same as the figure shown earlier except that a relation to the CDPD systems has been added.

Keep in mind that communication between the M-ES and F-ES can be in both directions, just as a person can send and receive mail. In Figure 2-13, the F-ES is sending data packets (envelopes) to a M-ES. Once the packet leaves the F-ES, it is sent through some number of routers (the number will vary depending on the data path). From the routers, or ISs, the data is sent to the home MD-IS. The home MD-IS is responsible for locating the M-ES and forwarding its packets appropriately. Also, the serving MD-IS keeps accounting information which can be used for statistical analysis or customer billing. The MDBS (cell which the M-ES is located) then relays the data from the MD-IS to the designated M-ES.

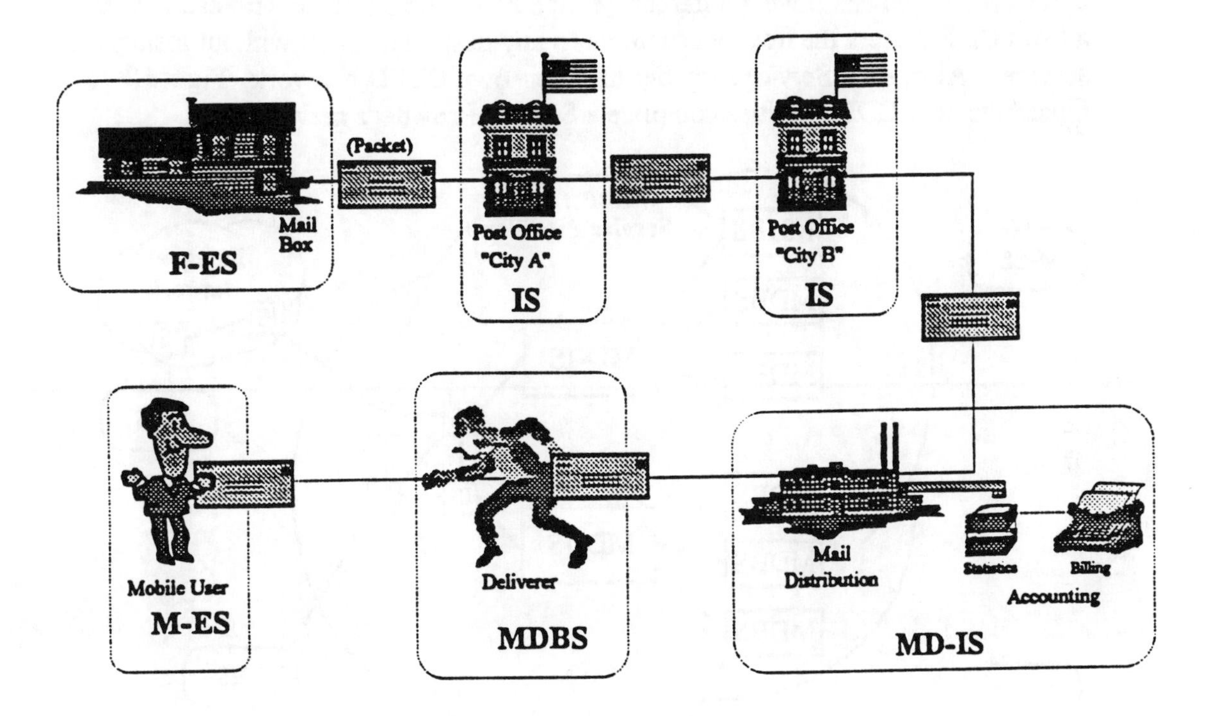

Figure 2-13 Post Office Analogy and CDPD

2.5 Service Providers

The CDPD domain will include many geographical locations with a large variety of companies offering CDPD services. These companies are called the CDPD Service Providers, which are the local CDPD carriers.

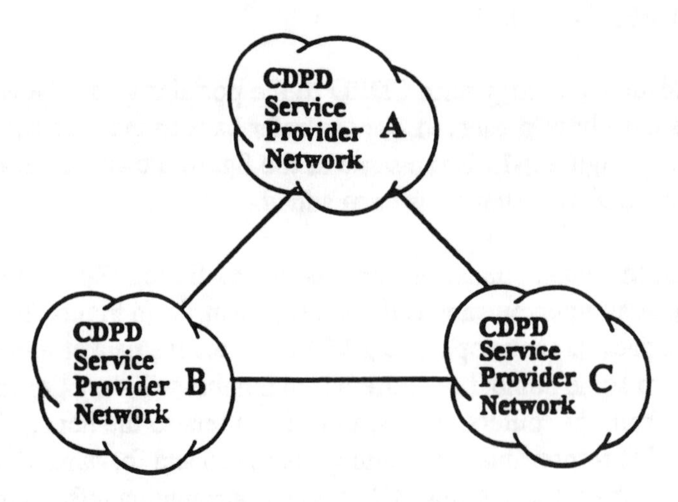

Figure 2-14 Service Providers

Figure 2-14 illustrates how different CDPD Service Provider Networks are connected together. Having different Service Providers interface with each other allows CDPD users the freedom to move to any desired location without losing service. Also, each Service Provider has their own CDPD Network. Figure 2-15 illustrates how CDPD entities comprise a Service Provider's network.

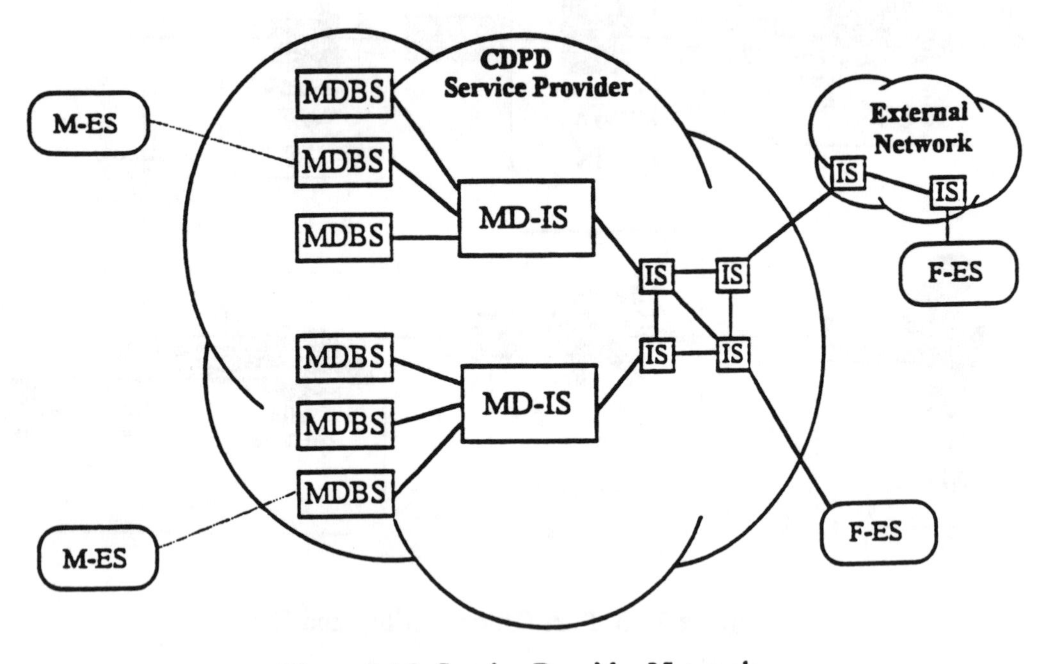

Figure 2-15 Service Provider Network

In order for a CDPD Service Provider to interface with all other CDPD Service Providers, it needs to be connected to the CDPD Backbone Network System (BNS).

2.6 Backbone Network System (BNS)

The BNS supplies the inter-city and inter-carrier high speed wired data transport service. Through the BNS, mobile subscribers with different CDPD Service Providers can communicate with each other (see Figure 2-16). Also, if a subscriber moves outside the local Service Provider Network, uninterrupted CDPD service will continue to be available.

The BNS transport service uses the most cost effective technology that is available at the time of data transmission. This will guarantee the most economical CDPD communication for the mobile user.

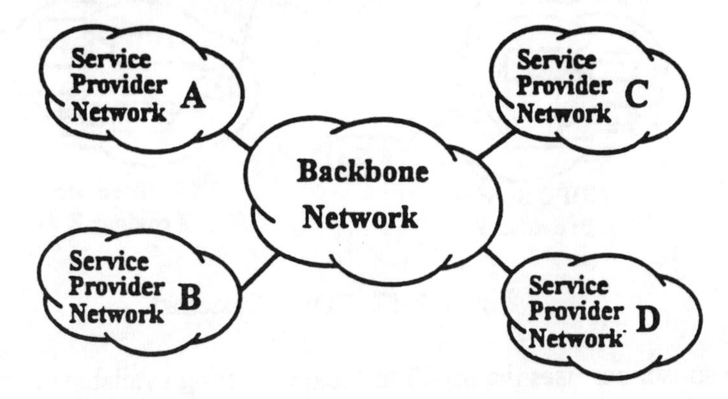

Figure 2-16 Backbone Network

The transport interconnection of the major metropolitan serving areas (MSAs) is completely transparent to end-users (seamless mobility). Therefore, unlike the cellular phone system, no roamer access numbers will be needed in order to maintain CDPD connectivity. The user will be able to send and receive data in exactly the same manner whether he or she is located in the home area or 2000 miles away.

2.7 CDPD Systems

CDPD Service Providers each will provide CDPD services to their MSA. With the help of the BNS, users will be able to maintain CDPD connectivity no matter where they go.

Many different End Systems will be connected to a particular Service Provider, and many different Service Providers will be connected to the Backbone Network System. Figure 2-17 illustrates an example of how the CDPD systems are used to connect a mobile subscriber to his or her corporate network. This connection could be over a couple of miles or 3000 miles. We will trace through the interfaces, end systems, and intermediate systems used in this connection. In the diagram, one possible data path is highlighted.

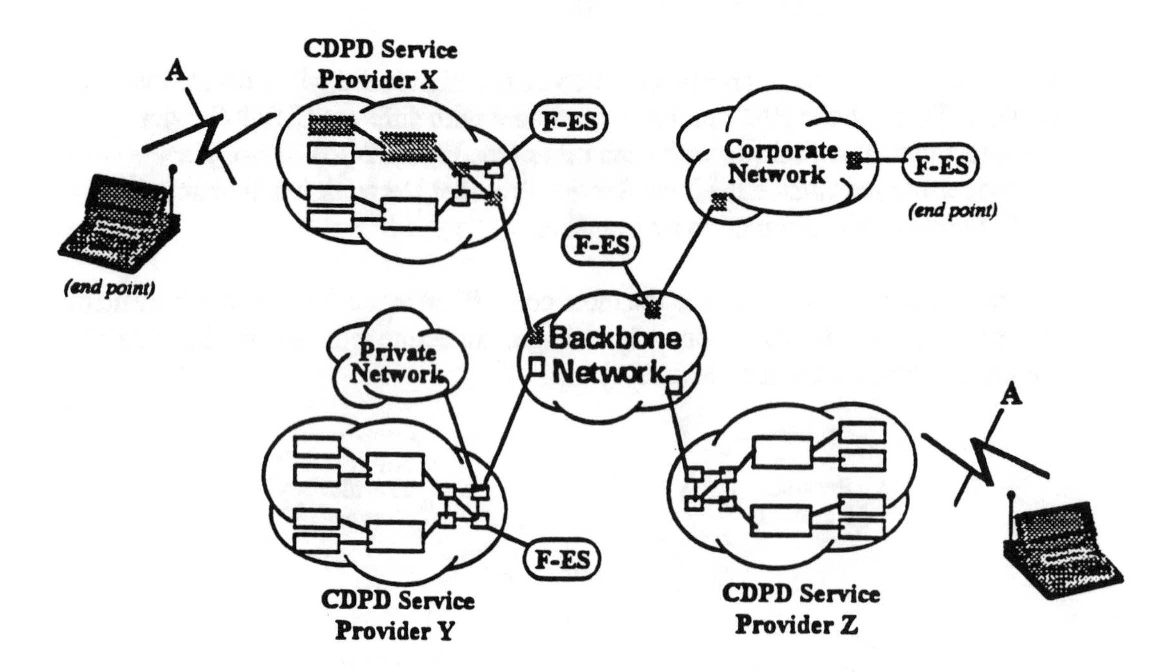

Figure 2-17 CDPD Systems

First the subscriber uses the M-ES to locate a strong available channel for data transmission. In Figure 2-17 the M-ES connects to the MDBS in CDPD Service Provider X's network (note: the MDBS could change due to either channel hoping or user movement). Once the data link is established with the serving MD-IS, the mobile user can log on in a normal manner and begin to transmit data packets over the Airlink Interface. After Service Provider X's MDBS receives the data, it is relayed to the MD-IS, which will act as a CDPD "switch". From there the data is routed to a group of ISs, which are just off the shelf routers. The ISs will, based on the information in the packet headers, route the data to the Backbone Network. Note that different packets may follow different paths. After the data travels through the Backbone Network, it is then sent to the corporate network. From there, the data will be routed to the correct end system.

If any data is returned from the corporate network, it is always sent to the M-ES's home MD-IS and then forwarded to its serving MD-IS.

3.0 MOBILITY MANAGEMENT

One of the unique attributes of the CDPD Network is M-ESs change their physical location at will. For example, a salesperson may use the CDPD network in Los Angeles on one day and access the network again in New York on the next day. Standard networks are not equipped to handle this movement. In a standard network, moving a device to a different geographical location requires reconfiguring the network software.

The CDPD network is designed to work with standard off-the-shelf TCP/IP products, therefore it has to manage mobility yet keep this movement transparent to those standard products. To achieve this, the architects of CDPD created a protocol sub-layer named Mobile Network Location Protocol (MNLP). The MNLP sub-layer manages the movement of an M-ES from one location to another.

So how does the CDPD network know where to send a message if the destination is changing from day to day? The method is not too different from the forwarding of letters in the Post Office system. For example assume someone has just moved to a new city and needs to receive his or her mail at the new address. Many people will be unaware of this move and continue to send letters to the old address. If the Post Office has a Change Of Address form however, it knows where to forward the mail. Letters arrive at the old Post Office with the old address on the envelope. The Post Office sees the old address and sticks on a yellow sticker indicating the new address and forwards the letter to the new Post Office. Upon arrival in the new Post Office, the letter is loaded on the postal vehicle for delivery to the new address. Continuing the analogy, letters sent in the other direction go directly to the original sender.

In the CDPD Network, the M-ES movement can cause a cell transfer. The decision to initiate a transfer, or hand-off, from one cell to another cell is under the control of the M-ES. There are two basic types of cell transfers. These are:

- Intra-Domain Network (same MD-IS), and
- Inter-Domain Network (different MD-IS)

3.1 Intra-Domain Cell Transfer

Most of the transfers due to the M-ES movement occur at the intra-area (i.e. from one MDBS to another MDBS controlled by the same MD-IS) level. If we refer to the above postal analogy, this type of transfer is similar to a person moving to a new house which is still within the same "home" Post Office service area.

Therefore, when the Post Office forwards the letter, the envelope will just go on a different delivery truck and not to a new Post Office.

In CDPD terms, after the M-ES locates a free channel on the new MDBS, the M-ES ends the connection with the current MDBS and establishes connection with the new MDBS. Then the serving MD-IS is notified of the transfer by the M-ES (i.e. Change Of Address Form) and the packets are then sent to the new MDBS (i.e. new delivery truck).

3.2 Inter-Domain Cell Transfer

A M-ES cell transfer could be to a MDBS controlled by a different MD-IS which would require a transfer at the inter-area level. This type of transfer was just illustrated in above Post Office example (**Section 3.0**) where the person moved to a new house which was in a different Post Office service area, requiring the mail to be forward to the new Post Office.

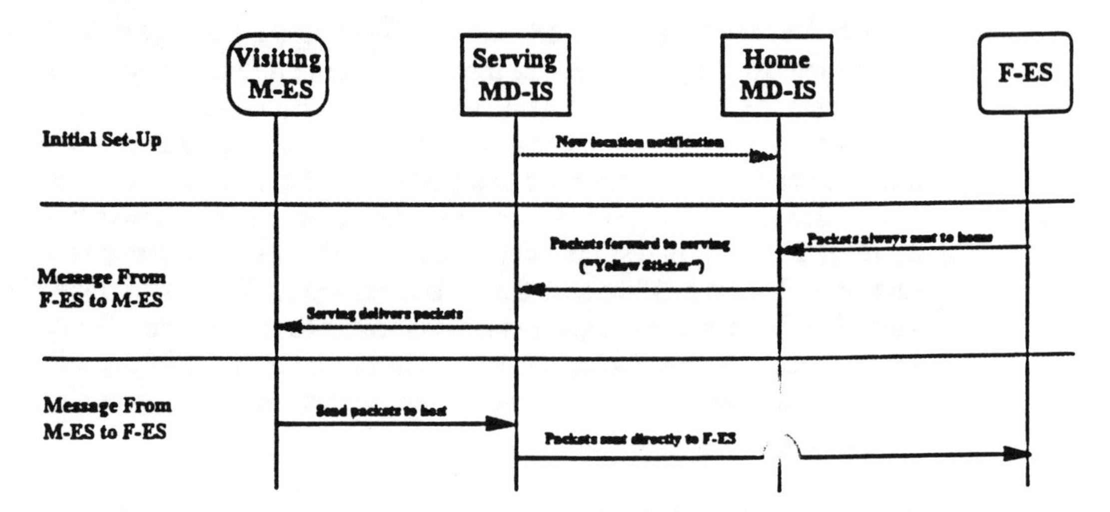

Figure 3-1 Inter-Area Transfer

With the help of Figure 3-1, we can look at how an F-ES transmits data to an M-ES that has moved to a new location. The visiting M-ES registers with the serving MD-IS in the new city. The serving MD-IS notes the "visiting" status and notifies the home MD-IS of the new location through the CDPD network. Also, the home MD-IS is responsible for authenticating the M-ES. The F-ES always sends messages to the home MD-IS, which forwards them to the serving MD-IS for delivery to the visiting M-ES. Upon receipt, the visiting M-ES replies to the F-ES by sending a message to the serving MD-IS. The serving MD-IS then routes the message directly to the F-ES rather than sending it to the home MD-IS because the home MD-IS does not need to be aware of any traffic going from the visiting M-ES to the F-ES.

Remember that the cell transfer process, whether it is intra-area or inter-area, is completely transparent to the users of the CDPD Network (M-ES and F-ES). As far as the F-ES is concerned, the M-ES has not moved and expects the M-ES to receive the message at its home location. As far as the M-ES is concerned, it has received the message directly from the F-ES and is not aware of the location change. It is this transparency to the standard products and applications that make this movement of an M-ES from location to location possible.

4.0 DATA OVER THE AIRLINK

The only wireless portion of the CDPD Network is the "last mile", which is just the communications between the Mobile End System (M-ES) and Mobile Data Base Station (MDBS) through the Airlink (A) Interface. However, the MDBS acts solely as a relay between the M-ES and the Mobile Data Intermediate System (MD-IS) using the CDPD Mobile Data Link Protocol (MDLP). Data packets can be sent between M-ESs and the MDBS, in either the forward or reverse direction, over the Airlink with a data rate of 19.2 kilobits per second (kbps).

The Airlink, in its physical form, consists of the two cellular channels: one for forward communication (MDBS to M-ES) and one for reverse communication (M-ES to MDBS). Figure 4-1 below illustrates how the surrounding M-ESs share the RF channels when communicating with the local MDBS.

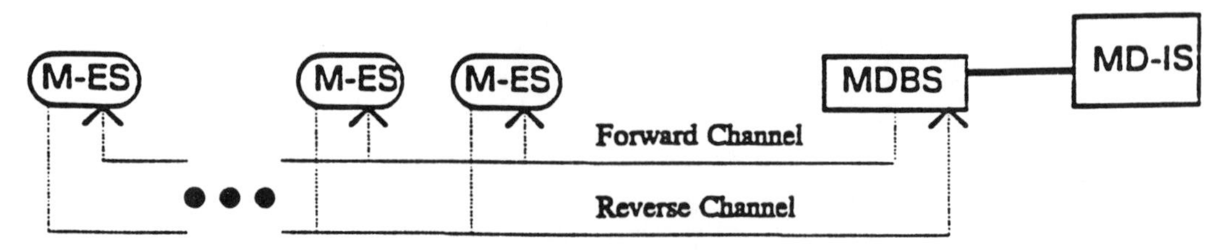

Figure 4-1 The Airlink

In order to send data over these channels, the data packets first need to be broken down into frames and then blocks, which then can be transmitted over the cellular channels by using frequency modulation.

4.1 Data Packet Transformation

Before an M-ES can transmit data packets across the reverse channel of the Airlink, it needs to transform the packets into a bit stream (a series of zeros and ones) because only one bit can be sent across the cellular channel at a time. This transformation of the data packet (on reverse channel) is illustrated in Figure 4-2. Applications running on the M-ES will store the data packets in the network layer (e.g. IP) packet format which includes a header and the user data. The maximum size of the packet (NPDU) is 2048 bytes. The transformation takes this packet and creates the bit stream by going through the following steps:

1. The packet header is compressed (for TCP/IP packets only).
2. The packet is segmented and each segment is provided with a segment header. Each segment may hold up to 128 bytes of user data.
3. Each point-to-point segment (i.e. non broadcast/multicast) is encrypted for Airlink security to protect against casual eavesdropping.

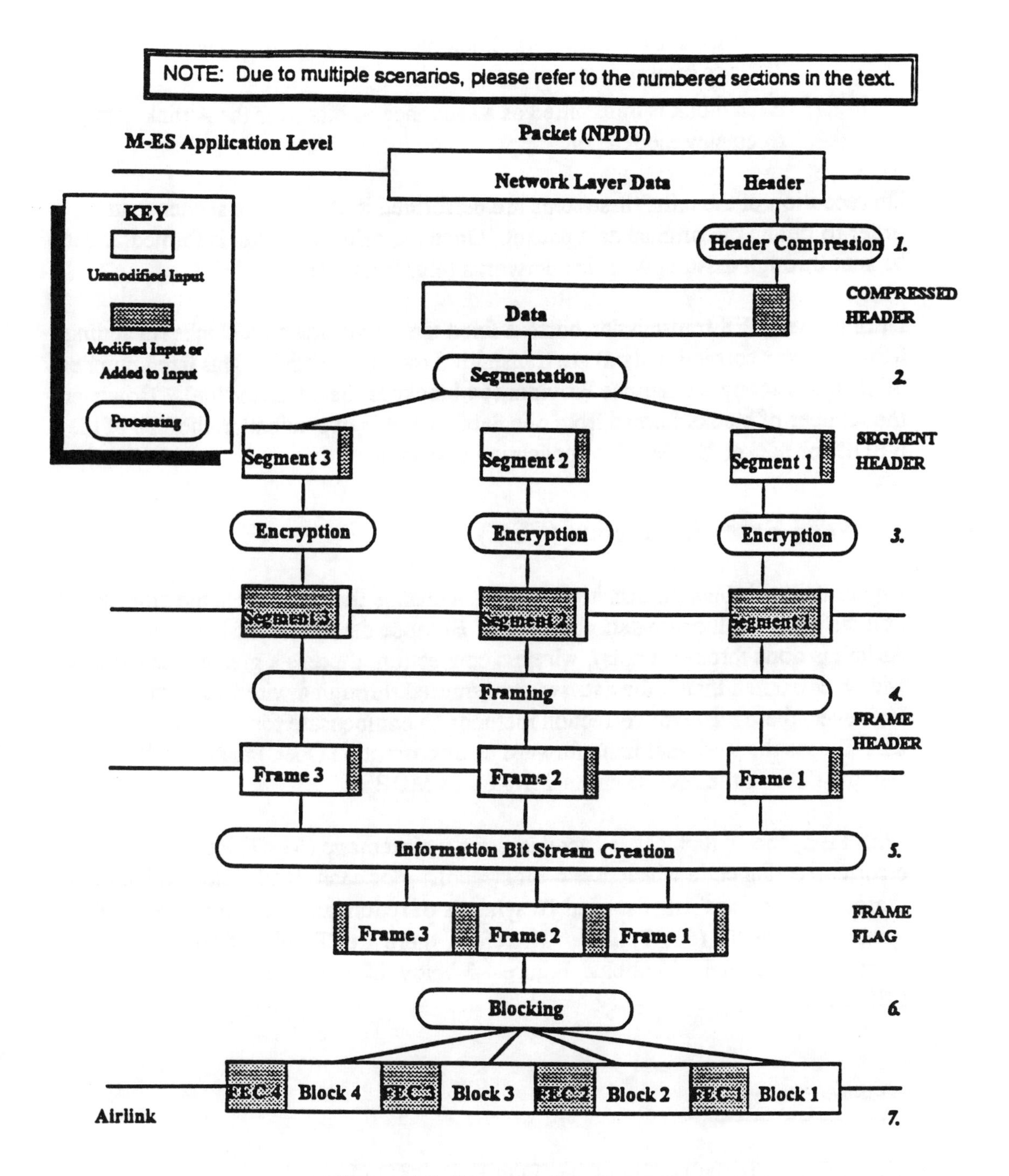

Figure 4-2 Data Packet Transformation

4. Each segment is encapsulated within a frame with a frame header. Each frame may contain up to 130 bytes of user data.
5. The sequence of frames are transformed into an information bit stream by the insertion of frame flags to delimit frames from each other.
6. The information bit stream is blocked into consecutive sets of bits and the blocks are subjected to Reed-Solomon forward error correction

(described in next section). Each block has a fixed size which holds 282 bits of user data.

7. Each block is transmitted as a sequence of bits over the Airlink using frequency modulation.

On reception of the data, these steps are performed in the reverse sequence in order to obtain the original data packet. Once the original packet is formed, it can be sent through existing wire-line networks (e.g. Internet).

During one M-ES transmission burst, a fixed number of the data blocks (including forward error correction data) are transmitted over the Airlink. This fixed number of blocks is set by the Service Provider (64 blocks is the default value). However, the number of blocks formed from one data packet is dependent on the size of the NPDU, or packet, the M-ES's application is sending.

4.2 Forward Error Correction (FEC)

Any time data is being transmitted, there is a chance that noise or some other form of interference will cause data corruption. Because data transmission over the Airlink is done through a noisy, wireless connection, there is a greater chance data will be corrupted than if the data was transmitted through a wired connection. However, there are error correction methods to compensate for this corruption. CDPD uses the Reed-Solomon forward error correction (FEC) method when sending data on the reverse channel (M-ES to MDBS).

With FEC, each data block is encoded using a systematic (63,47) Reed-Solomon error correcting code which uses 6-bit symbols. For each Reed-Solomon block, there 47 symbols of user data and 16 symbols of redundant data for a total of 63 symbols, hence the (63,47) name. Therefore, there are 282 bits of user data encoded into a total of 378 bits. Figure 4-3 below illustrates one Reed-Solomon FEC data block.

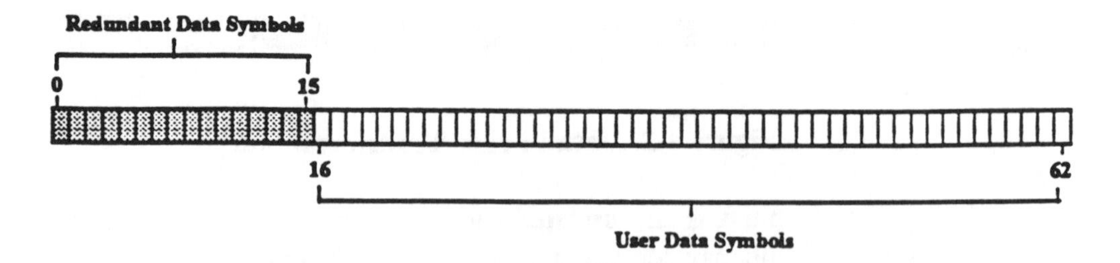

Figure 4-3 Data Block Bits

By transmitting the extra data bits, any data corrupted during transmission can then be corrected by the MDBS (or the M-ES if data is being sent in the forward direction). The benefit of using FEC is that in the likely case that data is corrupted over the noisy Airlink environment, then the very inefficient re-transmission process can be avoided. Therefore, FEC improves the CDPD data transmission performance over the Airlink Interface.

Also, there are 7 continuity bits added to the Reed-Solomon block (not shown in Figure 4-3) in order to signal the MDBS that either the transmission is complete ("0000000") or that there is at least one more block to be transmitted ("1111111"). Therefore, the total fixed block size transmitted over the Airlink is 385 bits.

4.3 Frequency Modulation

Frequency modulation of a data signal is just altering the signal's frequency for data transmission over another frequency. In the case of CDPD, the data bit stream is modulated in order to transmit data over one of the 800 MHz cellular channels.

The CDPD Infrastructure uses the Gaussian Minimum Shift Keying (GMSK) method for modulating the frequency over the Airlink. It is important to know that GMSK was chosen because it works well for a "noisy" environment.

4.4 Mobile Data Link Protocol (MDLP)

The Mobile Data Link Protocol (MDLP) is used to provide a logical link between Mobile End Systems (M-ESs) and Mobile Data Intermediate Systems (MD-ISs). Even though the Airlink connection is between M-ESs and Mobile Data Base Stations (MDBSs), the MDBS uses the MDLP to act as a relay between the M-ES and MD-IS. The relay connection between the M-ES and the MD-IS is known as a data link. With the use of the MDLP layer, the MDBS can quickly and efficiently transmit the data between the M-ES and the MD-IS.

5.0 CDPD NETWORK SERVICES

From the M-ES and F-ES perspectives, the CDPD Network is simply a wireless mobile extension of traditional networks. By using a CDPD Service Provider Network's service, a subscriber is able to seamlessly access needed data applications.

Services provided through the CDPD Network may be classified into the following three broad categories:

- CDPD Network Services
- CDPD Network Application Services
- CDPD Network Support Services

5.1 CDPD Network Services

Network services are data transfer services that allow the subscribers to communicate with applications. The unique value of the CDPD Network Service lies in the fact that one or both ends of the data communications may be a Mobile End System (M-ES).

To support the mobile subscriber's use of application services on existing data networks, the CDPD Network provides the following services:

- Connectionless Network Services (CLNS)
- Mobility Management Services
- Effective Interconnection Service to Existing Networks

Connectionless Network Services (CLNS)

CLNS is sometimes referred to as a *datagram service*. In CLNS, the network routes each packet individually based on the destination address carried in the packet and knowledge of current network topology.

Mobility Management Services

Mobility Management service is a distinguishing feature of the CDPD Network service. Data communications to the subscriber is managed over a wide geographic area, whether in the "home" CDPD Service Provider area or traveling outside of their service area.

Effective Interconnection Service to Existing Networks

Each IP/CLNP based CDPD Service Provider Network appears to the existing network as an extension of the network, providing transparent data networking service to the mobile subscriber.

5.2 CDPD Network Application Services

CDPD is a high speed (19.2 Kilobits per second over Airlink) wireless packet data system that can utilize the existing circuit-switched based cellular environment. Because CDPD uses existing technologies and network protocols, data received from the wireless side of the network can then be sent through existing wired networks such as the Public Switched Telephone Network (PSTN), or other public and private data networks.

A CDPD Service Provider Network, can provide specific value-added network application services through the CDPD Network, such as electronic mail services, directory services (e.g., white pages), and virtual terminal services. Many existing application services may be immediately available through the CDPD Network from CDPD Service Providers and from application providers external to the CDPD Network.

CDPD Service Provider Network Applications

A variety of network application services may be provided by CDPD Service Providers specifically for use by CDPD subscribers. Examples of such applications include Broadcast Services and Multicast Services already described. Others include:

Directory Services

The CDPD Directory Services provide access to a collection of information about subjects of interest to CDPD Network users that are found in a Directory Information Base and can be accessed through communications with a Directory Server.

Message Handing Services

The Message Handling Service could provide store-and-forward messaging capability that includes automated storage of messages on failure of normal delivery.

Domain Name Service

The Domain Name Service will provide translation between host names and addresses in the CDPD Network. This service will allow users and applications to use names formed from text, rather than addresses formed from numbers. Thus, making the CDPD System more user friendly.

Subscriber Location Service

The Subscriber Location Service will report the location of an M-ES. However, this is limited to the particular cell's location the M-ES is linked to. This service may be of value to delivery services and fleet vehicle owners in order to track the location of their vehicles. However, this information is available to only authorized customers for their authorized subscribers.

CDPD-Specific Network Applications

A variety of CDPD specific network application services may be provided by service providers external to the network. These services would provide value-added information and applications to wireless users.

5.3 CDPD Network Support Services

Network support services perform functions that are necessary for Service Provider network management, usage accounting, network security, *etc.*, and typically are transparent to the subscriber.

The CDPD Network is a logical, commercial public mobile data communications network. As such, functions must be performed that assist the CDPD Service Provider's ability to manage the network, to collect usage information that allows billing, to protect the Service Provider network from unauthorized usage, and monitor the network. Even though such services are critical to the success of the network, CDPD subscribers will be unaware that these services exist.

Network Entity Identifier (NEI) Authentication Service

The NEI Authentication service will provide identification and verification of M-ESs using the network. Every M-ES will have one or more NEIs (similar to a network address) in order to be recognized by the CDPD Network. The home MD-IS is responsible for authenticating an M-ES's NEI(s).

Data Confidentiality Services

The CDPD Network will provide basic data confidentiality over the Airlink that protects the subscriber's data from casual eavesdropping. A CDPD Service Provider may provide additional value-added services that ensure higher levels of data security beyond the Airlink.

Access Control Service

The access control service will provide the prevention of unauthorized use of a resource. This service will control the access to the data networks to alleviate the problem of unauthorized users entering and corrupting the CDPD system.

Accounting Service

The accounting service collects necessary data in order to calculate user charges. The data collected includes packet count, packet size, source and destination addresses, registrations, de-registrations, service provider, cell ID, and time stamp. This service will provide near real-time information of the network usage so that CDPD Service Providers can manage their networks most effectively. Usage accounting data also allows Service Providers to plan system configurations on the basis of actual usage. Accounting information can be exchanged by Service Providers through the I Interface using the message transfer service.

Network Management Services

Network Management services will allow CDPD Service Providers to operate the network with up to the second information to detect network errors, to exercise controls to correct errors, and to configure the network for optimal operation. Also, this allows for the proactive exchange of network status between different Service Providers.

Network Information Center - Administration Services

Because the CDPD Network is designed as an interconnected set of CDPD Service Provider networks, effective procedures must coordinate shared resources (such as address assignment and inter-domain cell hand-off) and shared information (such as accounting data for mobile subscribers). Also, CDPD Service Providers and IXCs providing IP/CLNP backbone services are assigned unique identifiers. This requires the establishment of a CDPD Network Information Center (NIC) for all Service Providers and IXC backbone providers.

6.0 Summary

The Cellular Digital Packet Data (CDPD) technology was designed to fulfill the growing need of users to send and receive mobile data. CDPD allows user to "cut the cord" of their restraining wired environment, allowing people access to valuable information from anywhere at any time seamlessly.

CDPD is simply an extension of existing data networks. Services on those networks may be easily integrated with the CDPD Network without imposing special technical requirements on network service providers who wish to extend their services to CDPD subscribers.

By designing these systems to overlay the existing cellular systems, CDPD will quickly provide economical wireless data communications to more users, covering a larger area, and at a faster transmission rate than any other wireless data solution.

Cellular communications, and telecommunications as a whole, have witnessed many technological improvements. In a few years, Advance Mobile Telephone System (AMPS) may be replaced with a higher capacity digital cellular technology. However, CDPD was designed to adapt to such changes in order to re-use existing, and future, applications and services. Therefore, not only will CDPD provide mobile data solutions now, but in the future as well.

7.0 REFERENCES

1. *Cellular Digital Packet Data System Specification* Release 1.0 July 19, 1993.

2. Dewire, Dawna Travis, *Client/Server Computing*, New York: McGraw-Hill, Inc., 1993.

3. Lee, William C. Y., *Mobile Cellular Telecommunications Systems*, New York: McGraw-Hill Book Company, 1989.

4. Losee, Michael, *The Cellular Telephone Installation Handbook*, Mendocino CA: Quantum Publishing, Inc., 1989.

5. Minoli, Daniel, *Telecommunications Technology Handbook*, Boston: Artech House, Inc., 1991.

6. Tanenbaum, Andrew S., *Computer Networks*, Englewood Cliffs, N.J.: Prentice-Hall, Inc., 1990.

APPENDIX A GLOSSARY

A-Interface The Airlink interface between the M-ES and the serving portions of the CDPD Network (MDBS, MD-IS). The only wireless portion of the CDPD Network.

Airlink A pair of RF cellular channels.

AMPS Advanced Mobile Phone System - currently the most widely used radio base communications technology. More commonly known as cellular.

BNS Backbone Network System - the transport service which allows different Service Providers to interface with each other and other private networks (F-ESs). Helps provide seamless roaming.

CDPD Cellular Digital Packet Data - packet based, high speed seamless wireless data technology that overlays the existing cellular network.

CLNP Connectionless Network Protocol

CLNS Connectionless Network Services

E-Interface The external interface between the CDPD Network and private networks (F-ESs).

ES End System - hosts or customers and third party networks of the CDPD Network.

FCC Federal Communications Commission

FEC Forward Error Correction - CDPD uses Reed-Solomon forward error correction method to minimize data corruption due to Airlink interference.

GPS Global Positioning System - needed to pinpoint an M-ES's exact location.

Hz Hertz - a measure of frequency in cycles per second. One million hertz is represented by MHz. The cellular frequencies are in the 800 MHz range.

I-Interface The inter-Service Provider interface which allows different CDPD carriers to interface with each other in order to provide seamless roaming.

IP	Internet Protocol
IS	Intermediate System - an off the shelf router which routes the CDPD data packets appropriately. This is also used to identify CDPD internal systems (MDBS, MD-IS, an IS router) together.
IXC	Inter Exchange Carrier - CDPD IP/CLNP backbone service provider.
MD-IS	Mobile Data Intermediate System - stationary unit of the CDPD system which is responsible for relaying the CDPD data packets to and from the network. Also responsible for storing information such as the last known M-ES location and accounting information for billing.
MDBS	Mobile Data Base Station - stationary unit of the CDPD system which is responsible for RF management and communication with the M-ES over the Airlink.
MNLP	Mobile Network Location Protocol - procedures used by the MD-ISs of the CDPD network in order to manage a M-ES's mobility.
MSA	Metropolitan Serving Area - geographical area where there is cellular coverage.
MTSO	Mobile Telephone Switching Office - cellular system responsible for maintaining connectivity between a cellular phone and the cellular network.
NEI	Network Entity Identifier - this is used to identify and validate an M-ES in the CDPD Network. One or more IP or CLNP (NSAP) addresses are assigned to each M-ES.
NPDU	Network Protocol Data Unit - IP or CLNP packet.
NSAP	Network Service Access Point - CLNP address.
NSDU	Network Service Data Unit - Data portion of an IP or CLNP packet.
NIC	Network Information Center
OSInet	Open Systems Interface Network

PDA Personal Digital Assistant

PSTN Public Switched Telephone Network - the currently used wired network for telecommunications.

RF Radio Frequency - CDPD uses the same frequencies as cellular voice which are in the 800 Mhz band.

RSSI Received Signal Strength Indicator - the measured power of the received signal a unit is receiving.

VAN Value-Added Networks

Reprinted from *IEEE Personal Communications*, Vol. 2, No. 5, Oct. 1995, pp. 9-19.

Technical aspects of the GSM system and its derivatives

Current Evolution of the GSM Systems

MICHEL MOULY AND MARIE-BERNADETTE PAUTET

The central topic of this article is the past, present and future evolution of the technical aspects of the GSM system and its derivatives such as DCS1800 and DCS 1900, as viewed from the European standardization bodies. An important part of the article outlines the current evolution, trying to identify the major trends that are likely to show where the future lies. While the past can be reported rather objectively, this is not the case for the present or the future. In a number of places, working assumptions are presented which are not fully stabilized; they may or may not represent the solution finally adopted by the engineers. This article is thus somewhat dated; to be precise, it reflects the state of the work as of summer 1995. Similarly, the stress on this evolving trend reflects the authors choice, as independent consultants, and does not necessarily reflect the ideas of other involved parties, nor even the general feeling in the GSM community.

GSM: A System Designed for Evolution

The 1980s will certainly be remembered as a period of important technological evolution, in particular in the area of telecommunications. Mainly because of the extraordinary progress of digital electronic circuit integration, the potential complexity of low cost, mass produced, man-designed machines increased considerably. GSM, which was designed during this period, is an excellent example of this trend. Though numerous were the attacks vituperating the complexity of such or such an approach, we see nowadays that electronics kept their promises, and more: low-cost handheld GSM mobile stations arrived sooner than was expected many years ago, to the greatest benefit of the system and of those manufacturers and operators who placed their hope in its success.

While the trend of technological progress antedates the 1980s, and will certainly go on, this period will also be remembered for a qualitative evolution. Electronics technology provides just the substrate from which systems are built. In a wide sense what has to be added, the *software* (including, in particular, custom designed hardware), is now by far the most important part. In addition, the design of this part does not show the increase in productivity that characterizes hardware. Sometime in the 1980s this divergence had reached a point where the material parts of the system could become obsolete sooner than the return of investment of the "soft" part design. This qualitative change is very important to system designers: now systems have to be thought of as evolutionary, or they will be rapidly taken over by newer systems, which better utilize state-of-the-art capabilities. A main quality of evolution is its capability to incorporate not only expected technological advances, but also the unexpected ones. Sheer increases of computational power translate sooner or later into new solutions to existing problems, new services to users, and new ways to improve the quality and value of a system. Designing an evolving system is certainly a most interesting challenge for system designers.

GSM was outlined at a time when these considerations were not given a pre-eminent importance. Little by little the extent of the issue dawned in the mind of the design team. The result is that the system, although not designed from the start to be capable of evolving smoothly, includes many features facilitating such an evolution, even in its earliest release. These include obvious cases, purposefully introduced for a specific and expected evolution, (e.g., the half-rate channel), as well as more indirect features, paving the way for unexpected change requests (e.g., a sound material and logical architecture).

Evolutions Foreseen from the Start

Historically, some precise evolving points were anticipated very early on. The chief example is the introduction of the half-rate speech encoding scheme. Speech transmission thoroughly influences the whole transmission system in GSM (as in all telecommunication systems for which speech transport is the main service). The choice of a speech encoding rate had to be made early, not to delay the design of the transmission aspects. When the choice was made for the introductory release of GSM ("GSM phase 1"), i.e., the choice of a rate around 16 kb/s made formally in 1987, it was already clear that speech algorithm technology had not yet yielded the ultimate schemes. This, combined with the expected decrease of cost and consumption of computing power, led to the decision at the same time to design the system from the start to also support speech compressed at an halved rate, around 8 kb/s. This meant dual-speed transmission devices (in fact, multi-speed, since eighth-rate channels also exist), and more generally the flexibility to choose transmission channels types on a call-by-call basis, in order to cope with mobile sets of different generations. This early choice had a profound impact on the design of the radio channel managing protocols. It is unlikely that a simple result would have been obtained would the early design be centered on a single user channel type.

Past Evolutions

In the years since 1987, several punctual requests for evolution were made and it is interesting to see how these requests were met. A main example is the flexibility of the frequency band used for radio transmission, a common problem for many cellular systems. The possibility to extend the 900 MHz band, with mobile stations having different bandwidth capabilities able to cohabit in the system, has been introduced without difficulties. A more thorough case was the design of DCS1800, the 1800 MHz version of the GSM system. The adaptation of the standard to a completely distinct radio bandwidth was not expected from the start. Nevertheless, this adaptation required only minor adaptations to the signaling scheme on one hand and on the other hand, the radio transmission techniques proved versatile enough to require no modification except the revising of the performance requirements. More recently, the specifications have been adapted in the United States for the 1900 MHz band, with no further difficulties.

Sheer increases in computational power translate sooner or later into new solutions to existing problems, new services to the users, and new ways to improve the quality for value of a system. Designing an evolutive system is certainly a most interesting challenge for system designers.

Another example worth mentioning is the possibility to introduce mobile stations supporting a ciphering algorithm that is different from the single one specified originally. There again the architectural concepts, as well as the concerned signaling protocols, proved flexible and adaptable enough to introduce this feature without difficulty.

Phase 2 and the Upward Compatibility Issue

The moment when the true problems of evolution arose was when the concept of a Phase 2, as a system version distinct from the Phase 1 with which the first systems would be launched, came into existence. As often occurs in standardization committees, the decision cannot be dated with precision. However, in 1990, the first papers raising the "upward compatibility" issue were put on the table. This issue has been the main source of difficulties for the work on the Phase 2 standard, and lies at the core of the evolution problem. The original concept of Phase 1 and Phase 2 is quite different from the one which eventually crystallized. Currently, Phase 1 refers to a self-contained version of the standard, supporting only a subset of the services that were thought to be in GSM. Phase 1 is the basis for the first launch of present commercially operated GSM systems; Phase 2 is the full-fledged version of the standard. Superficially, Phase 2 differs from Phase 1 only in a number of added services, mainly in the area of the "supplementary services". When examined more closely, it appears that the major signaling protocols, namely the application protocols between switches and data bases on one hand (the MAP), and the protocol between the mobile stations and the infrastructure on the other hand, have been modified in many points. The reason behind these changes is that the elaboration of the Phase 2 standard, with the introduction of the left-aside services as its main goal, was also the occasion to face the issue of upward compatibility. Ultimately, upward compatibility seems to indicate that old-generation pieces of equipment need not be modified to be able to interwork with newer, functionally richer, machines. This makes sense with user equipment, since it is economically unthinkable to take back and modify all the mobile stations already sold. It also makes sense for infrastructure equipment with the roaming feature: it would be unacceptable for one operator to be able to modify its equipment only if the others also do the corresponding modifications at the same time.

And there lies the core of the evolutivity challenge: the system must be designed so that new functionalities and new ideas, can be introduced without jeopardizing the functioning of already in-use equipment. The work on Phase 2 has shown that because of a number of small things the Phase 1 standard is not totally fulfilling this goal. The upward compatible introduction of many seemingly innocuous features proved not to be straightforward. Diverse solutions have been retained, from adopting a more complicated but upward compatible scheme in Phase 2 for some features to enforcing an upgrade of the Phase 1 infrastructure equipment before they can interwork with Phase 2 mobile stations. These practical exercises taught the designers that upward compatibility is not a simple thing, and a lot of effort was involved in modifying the standard so that evolutions from Phase 2 onward would be easier to introduce. Some mechanisms have been introduced or improved for that purpose, such as the version level in the classmark that enables a mobile station to indicate which kind of protocol it uses. In addition, many (hopefully, all) standardization points — the ambiguity of which could complicate the later introduction of something else — have been clarified. For instance, the reaction of mobile stations and infrastructure machines to unknown messages or parts of messages from the other side has been thoroughly specified, so that what happens when new items are introduced can be predicted. Another example involves the mechanisms enabling MSCs, HLRs, and mobile stations to accept new, yet unforeseen, supplementary services. The Phase 2 GSM is then ready for further, piece per piece, evolutions: this piece-wise introduction of a number of new functionalities is what is called the Phase 2+ of the standard. The dates of standardization of the different phases are recalled in Fig. 1.

Architecture and Capacity to Evolve

The cleaning up of GSM signaling standards which took place during the elaboration of the Phase 2 version was really only that — a cleaning up. The fact that an upward-compatible evolution is at all possible with only a cleaning up has to be put to the credit of the original *architectural design*, based on a number of concepts propitious to flexibility and evolution capability. These concepts were not invented *ex nihilo*. Protocol and software design techniques have been progressing quite a lot since the 1970s, and the concepts of interfaces, protocols, and protocol stacks were well identified, in particular with the focus of the Open System Interconnection (OSI) principles. OSI was originally developed for data transmission between computers, but was incorporated readily in all telecommunications domains, and in particular in the Signaling System No. 7, in the ISDN access protocols and in the Telecommunications Management Network (TMN) protocols, which constitute the firm foundations on which the GSM signaling standards are based.

The architecture of GSM can be seen from different directions: splitting in different machines, such as MSCs or BTSs; splitting between functional domains, such as Radio Resource Management or Operation and Maintenance; and splitting in procedures, such as handovers or access procedures. In all cases, architectural choices, following the spirit of OSI, were led by the search for maximum independence between parts. Such an approach simplifies design, test, and implementation, following the long established principle of "divide to conquer." It also favors evolvability, since independence means the possibility to modify one part with minimum impact on the others.

Many examples can be presented that illustrate how architectural choices ease evolution. For instance, the distinction between services and transmission capabilities, which in fact is a source of simplification by itself in a telecommunication system that supports a wide variety of services, enables an easy introduction of new services. The management of transmission capabilities is versatile enough from the start to cover readily the needs of many, not yet introduced, end-to-end services. Another example in the physical architecture area is the split between BSS and NSS, which enables to modify radio related aspects with minimum impacts on the switches and data bases. The same principles apply to the operation and maintenance area, where the TMN approach has been followed. A final example that illustrates the time axis is the division of the protocols in many, rather independent, procedures, which enables their execution in various orders and provides a flexibility that opens possibilities of finding new compromises, for example, between time performances and functionalities.

Open Specifications

A final aspect of GSM evolvability worth noting is, paradoxically, the fact that some points are not specified in the standard. In a number of cases the specifications have been left to operators and manufacturers. This was done whenever possible, i.e., when interworking between machines of different makes was not at stake. In essence these are functions in the system which can be modified without impacting the rest of the system, and are then ready areas for evolution. This is exemplified by present evolution trends.

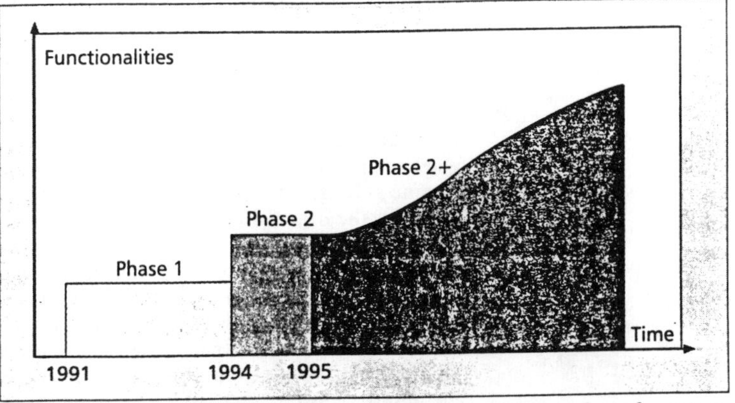

■ **Figure 1.** *The GSM phases. Phase 1 was the first adequate release for system launch, Phase 2 is the evolutive platform on which Phase 2+ items will be added gradually on demand.*

The Present Evolution of GSM

At the date of writing, Phase 2 is completed. The next step is under way, and is called the Phase 2+. The notation has been chosen to stress that it is not a new phase, such as the abrupt modification between Phase 1 and Phase 2, but a continuous program of enhancements of the Phase 2 platform.

The Phase 2+ work is organized as a set of independent items, so that each of them could be introduced with little or no influence on the others. This allows the study for each item to follow its own rhythm, and to adapt priorities to the market needs as analyzed by the operators. New items can be introduced, making the method suited to a long term, possibly open-ended, evolution.

To date, more than 80 items are listed in the SMG committee, covering most aspects from radio transmission to call management. Overall, the program is already quite ambitious and could easily lead to fundamental modifications of the whole system. It should be noted that the accumulation of such fundamental modifications, and of others that may be asked for in the next years, can change the system so as to make it only remotely related to the original. The real challenge of the Phase 2+ program is to introduce gradually important changes whilst keeping upward compatibility.

A complete description and analysis of the program requires much more elaboration than this article can devote. The following sections present a selection of evolving points, to provide the gist of ongoing activity, and an idea of the important evolving trends of the system.

New Basic Services

Some work items aim at modifying or introducing basic services.

Speech – Speech remains, and is very likely to remain, the prominent service of public cellular systems, which aim first to provide ubiquitous interpersonal communications. The optimization of spectral efficiency reflects the search for a good compromise between the cost of the system on one side, and the quality of service to the users on the other side. The speech encoding algorithm is a key part of this compromise, and it is not astonishing that evolution in this domain is looked for. By design, the GSM system can accept different speech encoding algorithms. Mechanisms exist in the signaling for the necessary negotiations between mobile stations and the infrastructure.

The introduction of the half-rate coder has already been presented. The point at stake was more the increase of spectral efficiency (and hence the decrease of the system costs for a given capacity) than the improvement of the speech quality. The intrinsic quality of the GSM half-rate speech coding algorithm is comparable to the GSM full-rate algorithm in most conditions. However, recent analyses of the half-rate vocoders have shown some weaknesses in the tandeming case, i.e., in the case of a communication between two GSM users, and hence with two speech encoding/decoding processes in cascade. A work item aims at mitigating this shortcoming by transporting GSM-encoded speech from end to end.

The Phase 2+ work is organized as a set of independent items, i.e., the goal is that each of them could be introduced with little or no influence on the others. This allows the study for each item to follow its own rhythm, and to adapt priorities to the market needs as analyzed by the operators.

Other ideas for evolution of the speech service have been put on the table. The foremost is the future introduction of a new full rate coder or coders, at a rate a bit lower than 16 kb/s (i.e., able to operate on the GSM full rate channel) and offering wireline quality telephony, Enhanced Full Rate (EFR) speech. Because of the continuous improvement of voice coding techniques, it is now possible to design vocoders complying with these constraints and yet yielding a quality of speech transmission fully competitive with wireline. This is particularly important when cellular is assessed as an alternative by users of wireline telephone systems. This is the case as soon as cellular users represent a very substantial part of the population, or in countries where it may be more economical to provide telephony by cellular means rather than by wireline. Another aspect is the place of GSM services in relation with those provided by quasi-cellular systems derived from cordless telephony, such as those based on CT2 or DECT.

The U.S. adaptation of GSM for the 1900 MHz band provides from the start multiple speech coder support (this is present in an embryonic form in GSM900 and DCS1800) and the specification of an enhanced full rate speech coder. Current tests yet to be completed tend to show that this vocoder meets the requirements as set by the European committees, with some added advantages such as the reuse of the present channel coding scheme, which translates in lower costs for introduction in a deployed network. The main debate at the date of writing in the European committees is whether to adopt the U.S. EFR here and now, or to launch a competition between a number of candidates. The authors think that the overall GSM project can only gain from an immediate adoption of the U.S. EFR. The definite conclusion of the debate will wait for autumn 1995.

The EFR coder will not be the end of the story. The U.S. EFR in particular is somewhat conservative, since it does not include truly evolutionary ideas, such as adaptative trade-off between channel and source coding. However, the intrisic quality (i.e., in perfect conditions) seems high enough to imagine that future progress will not aim at improving this intrinsic quality, but rather be targeted at improving the quality in difficult propagation conditions, in order to gain in range or spectral efficiency and/or decreasing the spectrum consumption, e.g., introducing an Enhanced Half-Rate coder.

Group Call and Related Services – In 1992, UIC (Union Internationale des Chemins de fer), the international railway organization, chose the GSM technology for their future standardized radio cellular system for railway companies' usage. The onset of trains which cannot change engines when crossing borders, such as the French TGV, makes it important to have standardized means of communications between trains and ground-based staff.

Several features not yet present in GSM are needed to fulfill the UIC requirements. Some of them, such as the possibility to support high speed, were already identified as study points in SMG. The area where UIC requirements are new are services usually associated with Private Mobile Radio. The most interesting one is the support of group calls, i.e., calls involving several users, on a push-to-talk basis, and sharing, for the sake of efficiency, radio channels.

In a group call, there is at a given time a single speaker who is heard by all the other participants of the group in the service area. The service area includes a number of cells. As defined, a given group call is supported in each cell by a single traffic channel, comparable to the one used in a point-to-point speech call. From network to mobile ("downlink"), the speech of the talker is broadcast in each cell on a group channel. The right to talk from mobile to network ("uplink") is managed by signaling exchanges, while the infrastructure guarantees that only a single talker is allowed in the uplink direction.

The proposed technical solution makes use of elementary mechanisms already present in the system and supported by the Mobile Stations.

At the date of writing, the study is well advanced, in the phase of the detailed signaling specifications, and completion of the standardization work may happen before the end of 1995.

Packet Radio – The GSM standard already offers the possibility to support packet services, for instance, by accessing X.25 Public Switched Packet Data Networks. However, for the moment the radio segment uses a dedicated circuit access, which is adapted to users with an important data rate. In present GSM there exists a true packet switched service, the Short Message Service, which is adapted to infrequent transmission. There is a place in-between for a service adapted to traffic per user, consisting of frequent but noncontinuous transmission of small bursts of data. The need for such a service has been identified by public operators, as well as for the application to trains. This is the aim of the General Packet Radio Service under study in Phase 2+ (GPRS).

These services, as defined so far and viewed by users, are aimed at covering many different things, including, for instance, access to Internet and automatic toll-ticketing on highways. From a technical point of view, GPRS raises several, rather independent, issues. One is at the application level, i.e., the service as seen by the user, independent of the underlying implementation. A second issue is the interworking between the GSM system (as operated by a cellular operator) and external data networks. A third issue is the architecture of GSM with regards to GPRS, and a fourth is the radio interface.

On the two first points, the work focuses on Internet, its TCP/IP, and on standard OSI protocols such as X.25 or "Connectionless Network Protocol" (CLNP). Most of the other applications will quite probably not be specified in the public standard.

The architecture of GSM with regard to GPRS is a main topic of debate at the time of writing. The current basic approach is to have the implementation of "GPR" Services rather independent from that of the already specified services. The common part is mainly the radio subsystem. However, many basic functions are common to GPRS and to the rest of GSM, such as mobility management, subscriber management, or the management of simultaneous services, including the warning of new calls of whatever service to Mobile Stations involved in one service. The committees are currently working on a functional architecture, and it is likely that the actual implementations will in many cases see the GSM machines in charge of non-GPRS services dealing also with GPRS-specific functions. An important (possible) departure of GPRS compared to the existing architecture is the routing of all packets toward a given subscriber through a single point in the network, independent of the user's location. This is not the case for circuit calls or short messages in GSM, where different GMSCs (depending on the calling party's location) can route the call directly to the current VMSC, which is an important factor for routing optimality. The rationale of the choice for GPRS is to avoid interrogating the HLR for each packet. The single point is functionally close to the HLR itself.

The orientation of the work on the radio interface is twofold. The GPRS study point spawned two work items, GPRS proper and "Packet Data on Signalling." The latter corresponds to the use of existing signaling transport means on the radio interface to carry user packets. At the same time, this can be considered a simplified approach to GPRS or an enhancement of the Short Message Service already defined in GSM. GPRS proper aims at designing a transmission scheme on the radio path where channels dedicated to GPRS are shared between many Mobile Stations. The radio subsystem would be in charge of orchestrating the uplink direction on these channels, attributing the right to send in a fair way to the different Mobile Stations waiting for transmission.

The work on GPRS is progressing rapidly; the official aim is to complete the work by the end of 1995.

High Rate Data – The other important evolution of the basic services is the move toward higher rates for data services. The Phase 1 and 2 GSM is limited to a user rate of 9600 b/s (in fact 12,000 b/s, though all communications with external networks do not extend beyond 9600 b/s plus user interface control signals). This is considered insufficient for a true success of data services. (Unfortunately it is unclear if any data rate will ever be sufficient, in view of the ever increasing size of files used for any application...) Even without changing the transmission bandwidth of GSM of around 200 kHz, there is a definite possibility of providing data channels of up to 100 kb/s with minimal impact on the infrastructure and not too much impact of the radio frequency part of the Mobile Stations. This is the topic of the "High-Speed Circuit Switched Data" work item. In practice, the current working assumption is to combine several 9.6 kb/s communications to provide one high-rate communication. Impact on the Base Station can be almost nil, to limit the introduction costs. The impact on the Mobile Station is unavoidably substantial, since the demodulation and channel decoding requirements increase in proportion with the user data rate. This simply means that new types of Mobile Stations will have to be developed for this service, to be purchased by those users desiring these services.

Major issues are in fact not the different services per se, but how to introduce these new services taking into account somewhat contradictory aspects such as roaming and mobility, upward compatibility with existing Mobile Stations, compatibility with the wireline network when this applies and ease of implementation.

Supplementary Services

In the distant past, telecommunications services were limited to sheer transmission and the means to set up and release calls. In recent decades, interest in additional facilities, as well as the technological possibility to provide them, has increased. Such facilities aim at offering users some level of control on how calls are set up (e.g., call forwarding or call barring), or at providing better management of calls by the users (e.g., advice of charge, call waiting, call hold, etc.). The intrusiveness of the telephone is already noticeable, and can only increase with pocket cellular phones. This makes facilities a very important part of a wide-market public cellular service.

GSM has already incorporated most of the facilities offered by ISDN, for instance. However, new facilities are being introduced in wireline systems, and new ones can be imagined for cellular systems. Already, many evolution points are identified in this area, like the Completion of Calls, the support of private numbering plans, Call Deflection, Mobile Access Hunting, etc. In a related area, study points are open for adapting GSM continuously to the evolution of wireline networks. Major concepts in this domain are

An obviously important trend of GSM is its success worldwide. The original standard, GSM900, has been adopted in many countries, including the major part of Europe, North Africa and the Middle East; many countries in East Asia and Australasia, as well as countries in sub-Saharan Africa.

Universal Personal Telecommunications (UPT) and the Intelligent Network (IN) technologies; their implications on GSM will be discussed further in this article.

In fact, a major issue is not the different services per se, but how to introduce these new services taking into account somewhat contradictory aspects such as roaming and mobility, upward compatibility with existing Mobile Stations, compatibility with the wireline network when this applies, and ease of implementation. Mobility can be a source of extra complexity, as the example of CCBS shows (see the following section). Some of the other aspects (roaming and upward compatibility in particular) are addressed in SMG committees on a general basis.

In fact, a large part of the Phase 2+ work program is concerned with Supplementary Services, many of which are quite interesting. Only one is presented in detail in this article, as an example. Others include features such as call deflection, explicit call transfer, short message forwarding, malicious call identification, mobile access hunting, multiple subscriber profile, and support of private numbering plan.

Completion of Calls to Busy Subscriber – This supplementary service (CCBS) consists in automatically reestablishing a call which did not succeed because the called party was already engaged in another call. The interest of the feature comes from the automatic handling by the network and from the possibility to attempt the reestablishment at the earliest moment, i.e., when the called party leaves the previous call.

CCBS is a well defined service in the ISDN environment. It could be thought that its inclusion in GSM is a simple matter, consisting in copying the ISDN specifications. However, mobility makes things more difficult. In a non-mobile environment, two switches are involved: one in charge of the calling party and one in charge of the called party. The first one manages the queue of the calls asked by the calling party to be reestablished when possible, and the second a queue of call requests (typically from different calling parties) waiting for the called party to be free. In a mobile environment, more switches can be involved since each of the two users can move from a switch to another between the moment of the failed call attempt and the reestablishment. In GSM not only the serving switch can change because of movement but also the serving network. Implementation of CCBS must then cater for either a centralized management of the queues, or the transfer of the current CCBS data from switch to switch (possibly in different networks), following user movements. The working assumption is that of a centralized management: the CCBS queues for a given subscriber are copied in the corresponding HLR in addition to the storing in the serving switch. When a subscriber registers in a new switch, the queues are transferred by the HLR to the new switch.

This examplifies the changing role of the GSM HLR. Originally it was, as its name indicates, a repository for the subscribers location data. Rapidly it was augmented by the role of a database where to find the technical subscription data, i.e., the subscription related data needed to handle subscriber communications. In Phase 1 and Phase 2 already the HLR is more than a simple database: it is an active part in call forwarding, in order to reroute calls from a point as close as possible to the calling party. With CCBS, and a number of other Phase 2+ work items, the HLR is increasingly playing the role of a Service Control Point, orchestrating the services provided to the subscribers it manages, whether served by the Home network or roaming in another network.

The Globalization of GSM

An obviously important trend of GSM is its success worldwide. The original standard, GSM900, has been adopted in many countries, including the major part of Europe, North Africa and the Middle East; many countries in East Asia and Australasia, as well as countries in sub-Saharan Africa. In most cases, roaming agreements exist, making it possible for users to travel in many places in the world, enjoying continuity of their telecommunications services with a single number and a single bill.

The adaptation at 1800 MHz, DCS1800, is also spreading outside Europe to East Asia and some South American countries. The U.S. adaptation at 1900 MHz, DCS1900, seems likely to cover a substantial part of the United States. These systems will also enjoy a form of roaming (referred to as SIM-roaming) between them and with all other GSM-based systems, providing the Subscriber Identity Module (SIM) standard as well as internetwork protocols are kept compatible: a subscriber of any of these systems could maintain access to telecommunications services (including receiving calls and being billed at home) by using his SIM card in a handset suitable to the network from which coverage is provided. The concept of SIM-roaming is illustrated in Fig. 2.

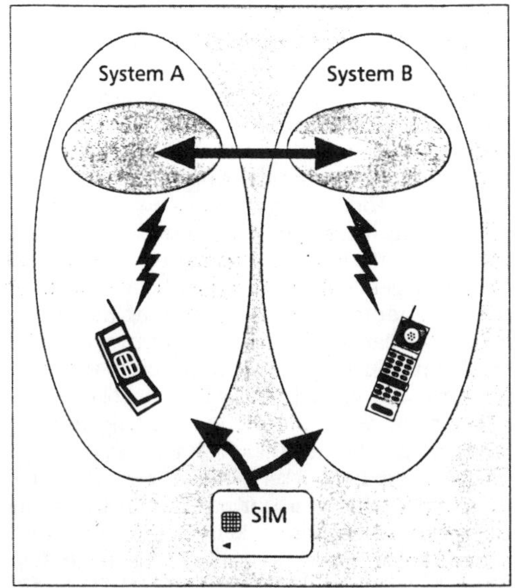

■ **Figure 2.** *SIM-roaming. Two systems with totally different radio access means may still offer a form of roaming to subscribers without dual-system mobiles, if both the SIM to mobile interface and the inter-network interface are standardized.*

This globalization is putting GSM and its derivatives as the leading cellular digital system. The gathered momentum amplifies the trend, but some negative forces may tamper with it. The most important difficulty is the fact that the technical evolution is done by regional committees in Europe and the United States, with some loose and indirect coordination done by active parties, in particular, the manufacturers. On one side, this is a portent of divergencies that will partition the market, and on the other side this excludes a substantial portion of the GSM networks. A better approach would involve cooperation between the two sides of the Atlantic, serving henceforth as a focus for a true global approach to the technical evolution of GSM. A non-negligible part of the GSM community shares these views; the future will tell whether common sense will prevail over hidebound thinking.

Extended Roaming and the Relation with Other Systems

Because GSM will provide wide seamless coverage, in Europe in particular, with a corresponding widespread infrastructure, many ideas have been put on the table regarding use of part of the GSM system to support other systems. The architectural approach of GSM, with clear-cut separation of functionalities, is well adapted to the borrowing of parts. GSM provides powerful tools for roaming, tools that are totally independent from any access technique, radio or otherwise, and this can be utilized in different contexts.

A first application will not take us very far. For historical reasons, GSM900 and DCS1800 has been designed without intersystem roaming being a requirement. This was corrected with two Phase 2+ work items which are finalized at the date of writing. A first easy step was to ensure the compatibility of the SIMs, so that SIM-roaming can be provided between GSM900 and DCS1800. The second step was to introduce the technical modifications enabling an operator to run a network at both 900 MHz and 1800 MHz, including the support of dual-system Mobile Stations. This includes the possibility of automatic choice of the band by the Mobile Station in idle mode, and the possibility to handover calls between the two bands. All things considered, the modifications are quite limited, thanks to the openness of the standard and to some precautions taken long ago. It is a bit annoying to note that such multiple band operation and Mobile Stations will be more difficult to introduce with DCS1900 because of a very little technical point, the frequency numbering, which overlapps that of DCS1800.

As an another example, outside the realm of GSM and its derivatives, we can cite the current efforts to combine GSM and DECT, either by providing DECT access to GSM, or by using part of the GSM infrastructure to support DECT traffic. From the point of view of GSM operators, this feature enables them to offer DECT services with minimal infrastructure investment. For operators interested in DECT alone, this provides ready equipment to support full mobility.

A last example is the interworking of GSM with satellite services. This is an identified Phase 2+ item, but with no inputs in the committees for the moment. Many studies concerning interworking between cellular and satellite are nevertheless under way, with the idea that these services complement each other in high and low density areas, respectively. Some obvious limitations of terrestrial cellular systems can be overcome with satellites: it suffices to cite the oceans. On the other hand, capacity is severely limited with satellites compared with on-the-ground coverage, and satellite-to-handheld communications are viable only with line-of-sight propagation, excluding cities and the inside of buildings. The complementarity is sufficiently clear that the different satellite projects such as Globalstar, Iridium, Inmarsat-P or TRW's Odyssey are looking to some interworking with GSM (among other cellular systems). This can take the form of roaming: GSM provides all needed tools for this.

The architectural approach of GSM, with clear-cut separation of functionalities, is well adapted to the borrowing of parts. GSM provides powerful tools for roaming, tools that are totally independent from any access technique, radio or otherwise, and this can be made utilized in different contexts.

Easing the Introduction of New Services

The approach to the development of the GSM standard leads to a difficult balance between cooperation and competition. The more a feature is standardized, the lesser the research and development costs, since they are shared by the whole community. But also, the more a feature is standardized, the lesser the place for competition, among operators in particular. In addition, the absence of a standard for a given service often means that the service is not provided when the users are away from their home networks, which undermines the concept of roaming, a strong point of GSM. Finally, the standardization process is often too long. Each time a new service is conceptualized by an operator, the dilemma arises between independent development and standardization.

The way out of this dilemma is twofold: first, to standardize mechanisms to introduce new services rather than the services themselves, so as to streamline service standardization; and second, to restrict the nodes having to implement the specificities of a given service to a few, and to the extent possible to those under control of the home network. The introduction of generic tools and mechanisms for that purpose is the aim of several work items, presented in the next section.

The dominant GSM approach for SOSS gives the leading role in service provision to the home network, i.e., the one with which the subscriber has a contract. The role of the access network is reduced to ancillary functions.

The Question of a Humpbacked Animal – The foremost of the work items studying generic service support means is called "Support of Operator Specific Services" (SOSS), nicknamed CAMEL, and being the main road to combining GSM and IN techniques. As an anecdote, the nickname CAMEL is self-derogatory, coming from the joke "What is a camel? A horse designed by a committee!" This simply shows the weird humor of some committees and the fact that they are aware that the matter is not a simple one.

There is already a lot of kinship between IN and GSM. Though neither IN terminology nor the specific IN protocols like INAP are used in the GSM standard, the roles of GSM functional entities such as MSC, GMSC, or HLR, and the use of SS7 and specific application protocols are in line with IN philosophy. IN supports features such as the Service Independent Building Blocks that are not present for the moment in GSM, and which certainly bears some relationship with the issue of how to introduce easily new services. The undergoing study within the scope of the SOSS work item sees the full usage of IN in GSM as a long term goal, while looking for a first step introduced rapidly and fulfilling a substantial part of the identified needs.

The key point (using IN terminology) is to split service implementation between, on one side, generic functions in the (visited) MSC and the GSMC, acting as Service Switching Points (SSPs) and, on the other side, Service Control Points (SCPs) under control of the home network. The HLR is an obvious candidate for fulfilling such an SCP function (it does this already), but provision is made for independent SCPs. An example can be specific SCPs providing short number translation for a set of Virtual Private Network users. Figure 3 represents a mapping of the GSM architecture using the IN concepts, with its evolution to cover the wider Phase 2+ service range and further separate the role of the home controlling network vs. the serving network.

In practice, a set of triggering points are defined on the basis of a standardized call model (or more generally a subscriber behavior model, since in addition to calls it will include topics such as location updatings). In the subscriber profile provided by the HLR, hence by the home network, the Visited MSC will find a list of associations between on one hand a trigger point and on the other hand an SCP address (typically in the home network). When a trigger point is reached as the result of interactions with the subscriber, a message is sent to the SCP, which can analyze the event and as a result dictate some behavior to the Visited MSC. With a suitable list of trigger points, and the support of a suitable array of different actions in the VMSC (including call forwarding, call release, sending stimuli to the user, etc.), a vast panoply of services can be built by operators for their own subscribers without requiring specific service logic to be implemented in the visited networks.

The scenario above is the one dominant in the debate so far, but not by consensus. The incorporation of such a feature has many and far-reaching consequences for GSM, and in a way on all telecommunications. It has more than a flavor of IN, but it is not IN. The stress in GSM is put on the roaming issue, which can be seen as the utmost in mobility, something which is absent to IN as it was studied until recently. Traditionally, the most open IN interface is the one between Service Data Points (which hold subscription related data in particular) and Service Control Points. The philosophy is that a telecommunications service is provided by the serving network, modified if need be by the subscription parameters. The dominant GSM approach for SOSS gives the leading role in service provision to the home network, i.e., the one with which the subscriber has a contract. The role of the access network is reduced to ancillary functions. This translates for a given network in more control on the services provided to its subscribers, and less control for the users to which it provides access directly to telecommunications services. This has an important impact on the competition among operators and opens the door for a different approach to the role of operators, with more separation between the role of *access network* and *service provision* to subscribers.

A number of other items in the Phase 2+ work program aim at providing generic means which can be used as elementary bricks to build services. One, barely starting at the date of writing, is the design of a generic protocol to carry exchanges between the user and the service control points, so that new services can be provided to Mobile Stations not specifically programmed for them. Other items in this category are related to two powerful GSM features, the SIM and the Short Message Service, and will be studied on their own in the following sections.

The Short Message Service – Historically, the Short Message Service was introduced as a kind of paging service. While not competing with standalone paging services in terms of high penetration or coverage (at least at the beginning), the bidirectional capability of a cellular handset allows acknowledgment, and intelligent reemission of messages whose transmission failed: the infrastructure knows when the Mobile Station comes back to activity (or coverage) and re-sends then the waiting messages.

Nowadays, the Short Message Service plays a much more significant role, and its importance increases. Beside the basic paging usage, the SMS is often used as a true message service, e.g., to broadcast news to members of a company. But the more interesting phenomenon is the use of the SMS as an additional signaling means. The automatic intelligent re-sending makes it a wonderful tool to indicate to the user other waiting events, such as voice mail messages, e-mail messages, faxes, etc. This feature was introduced independently from standardization by some operators, and now a Phase 2+ work item aims at federating its implementation. More generally, a number of Phase 2+ work items address the SMS, to make it even more versatile.

Another aspect of the use of the SMS as a general signaling means makes use serendipitiously of a feature introduced simply for the comfort of the user. In GSM Phase 1 and 2, the Mobile Station is required to store incoming Short Messages on the SIM. Originally, this was meant only as a convenient storage for the user. Now, this is more and more used as a protocol between the SIM and the infrastructure (in particular the Home network of a subscriber), enabling commands to be given to the SIM, or data to be downloaded onto it. Information from the message header, or within the message itself, is used to distinguish the different meanings of the messages. This is done outside the standard for the moment, since the visited network needs not be aware of the message semantics. However, the lack of standardization is annoying for SIM manufacturers, and risky for the future. As the next section will show, the SIM can be a powerful tool for introduction of operator specific services, and the SMS will play a significant role.

The Increasing Role of the SIM – The Subscriber Identity Module appears as a key feature of the evolution of GSM. As for the SMS, its role now evolves beyond the original intent. The main characteristic of the SIM which explains its importance is the fact that it is provided by the home operator, and under full control of this operator. An original usage of the SIM, compatible with existing Mobile Stations, can well be used to support differentiating services. So far, the interface between the SIM and the rest of the system cramps the engineers somewhat. We have seen that the storage of the SMS opened a useful communication channel, which can be used to modify the contents of the SIM fields. An example of an already existing application is the use of the short number directory of the SIM for storing and making easily usable operator-specific numbers such as one for voice mail. This can be initialized during the personalization

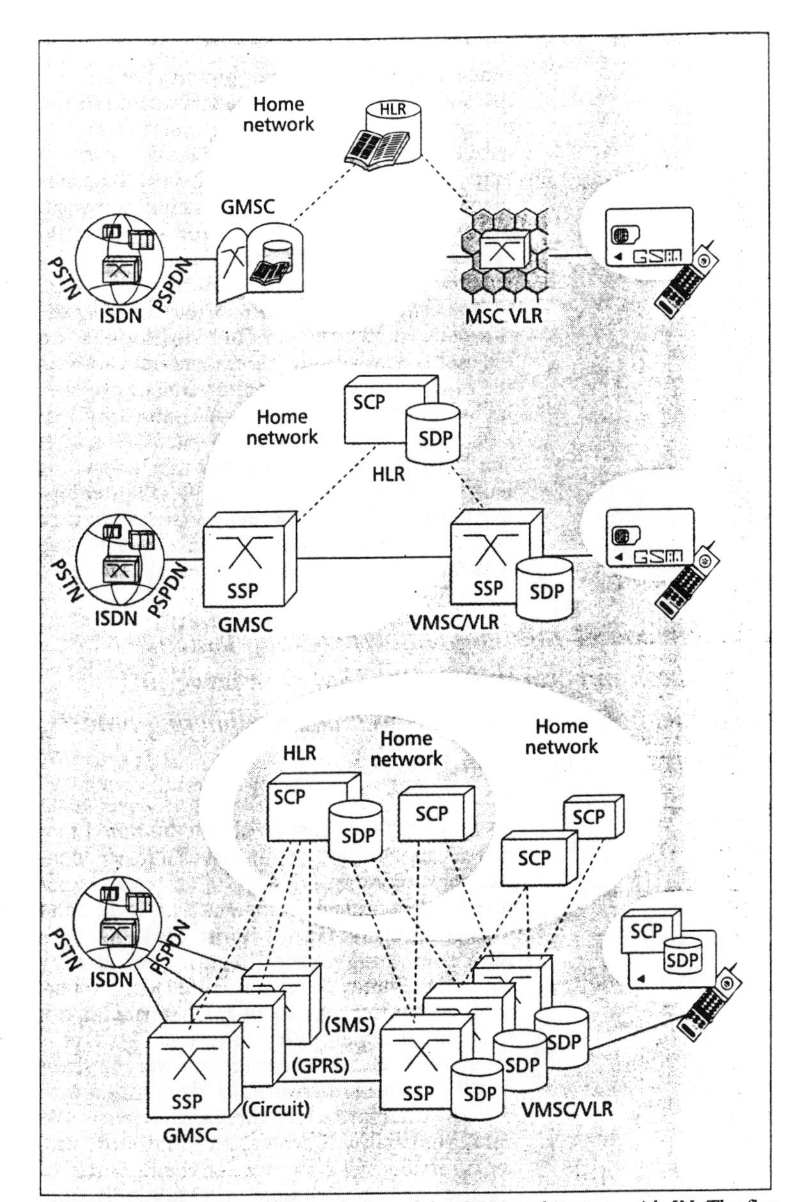

■ **Figure 3.** *One view of the evolution of the GSM architecture with IN. The first box represents the standard GSM architecture, as from phase 1, the second box represents the mapping for this architecture onto IN concepts, the third box represents a potential evolution within phase 2+. The dashed lines represent packet-type exchanges (including signaling), and the continuous lines represent circuit-type exchanges.*

process, before the SIM is given to the user, and may be updated later, if need be, thanks to the SMS channel. This feature was introduced independently by some operators and is now the topic of one work item.

As already mentioned, there would be many more such applications, would the SIM be more open to the outside world. The evolution in that direction is the topic of the "SIM toolkit" work item. A way of presenting this work item is to say that it is the analog of the SOSS feature, but on the Mobile Station side. For the moment, the official description of the "toolkit" is to include in the SIM-MS protocol means for the SIM to order some actions to the Mobile Station, such as sending a Short Message. Part of the requirements also cover the need for the SIM to be made aware of some events, such as a call request. A possible development (debates on these ideas have barely started) is to define trigger points in an extended call model, and to have the SIM indicate at which points it wants to be notified. This is simply using the IN technique on the user side. The possibilities for provision of operator-specific services can be tremendous with the combination of IN techniques on the infrastructure side, which gives the home network some control of the services over the visited network, and the "SIM toolkit" on the user side, which gives the home operator (via the SIM) some control of the services over the Mobile Station. In a way, this is an application of a sound architectural principle, which is to design the implementation of a feature so that it impacts a minimum number of pieces of equipment. Many services are really between the user and his subscription operator, and moving the implementation of such services at the two ends makes only good sense. It should be noted that there is an atavistic trend in classical telecommunications to put the stress on the network rather than on the terminal equipment. This is different with computers. GSM Mobile Stations, with their computing power, displays, and already imposing quantity of software, are closer to personal computers than to an ISDN terminal, not to speak of a plain telephone set. This fact is taken into account to some extent, yet the stress is still put on the network, as exemplified with the current approach to the CCBS service. We think that this will evolve, and that more and more services will draw more on the resources of the user premises equipment (Mobile Station, SIM, terminals) and less on those of the infrastructure.

Fraud prevention and confidentiality have always been important issues in GSM. The powerful authentication mechanism, as well as the ciphering process, are the result of these concerns.

Radio Aspects

The examination of the Phase 2+ work program shows that radio coverage is the poor parent of the lot. A pretentious explanation would be that GSM is so good than evolution is not needed. More seriously, things move in this domain also, but not in the part which is standardized. The standard leaves a lot of freedom to manufacturers and operators on coverage techniques.

In basic transmission aspects, progress concerns adaptation to specific cases. The more visible case is the issue of high speed. The standard was devised originally with a maximum speed of 250 km/h (around 150 miles/h). Since the main limitation comes from Doppler and phase shift effects, this translates in half the speed when the operation band is around 2 GHz. The adaptation to the trains, as asked by UIC, included a study item for the speed, since some European trains have a commercial speed close to 400 km/h. The study has shown first that the degradation at high speed was low, that there was no intrisic limitations due to the technical choice of GSM, and that an adapted demodulation scheme could circumvent the detrimental effects of speed.

Another work item shedding some light on the current evolution of coverage technique is the one related to the enhancement of cell selection to take into account "multilayer" coverage, in which multiple cells of different properties cover some area. A typical example is the coexistence of microcell, high density, urban coverage with larger umbrella cells. More generally, small but continuous progress is made in the cell and network selection algorithm as specified by the standard, to take into account ever more complex cell planning techniques.

In fact, the domain where radio coverage makes progress important for the future of the system does not appear at all in the Phase 2+ program, because its topic is part of what has been left open in the standard. There are many such points in relation to radio transmission, such as the demodulation and the decoding process, the antennas to be used, or features such as antenna diversity. This enables manufacturers to introduce evolution toward higher spectral efficiency, still the rule against which many people judge and compare cellular systems. An example is the study of antenna arrays and Space Multiple Access, which indeed can be introduced easily in GSM since they are independent from what is specified. Another case is the handover algorithm. While the measurements (on which the decision to handover a given call is taken) are quite constrained by the standard, this is not true for the decision algorithm itself. There also, progress is made and introduced in the system easily.

Fraud Prevention

Fraud prevention and confidentiality have always been important issues in GSM. The powerful authentication mechanism, as well as the ciphering process, are the result of these concerns. It is not appropriate to detail all the frauds ill-intentioned people can imagine, and only one Work Item is cited here, because of its architectural importance. The so-called "Subscriber Supervision System" aims at making the home operator aware of all ongoing communications for part or all of its subscribers. For telecommunication engineers not acquainted with mobility, it may be astonishing that this is not already the case! For charging purposes, the home network is indeed aware of all communications of its subscribers, but not in real time. With roaming, communications can be established entirely autonomously by visited networks and its is only after the completion of the call that the home network is notified.

The major difficulty of the work item comes from mobility once again. In contrast to what happens in classical wireline networks, it is not only possible, but quite usual that several ongoing calls exist at a point in time for a given subscriber, and that they have no switch in common. This comes from features such as call forwarding, and the movements of the users between the setup of two calls. As a consequence, the only way to supervise the traffic of one subscriber is to do it centrally, typically as a function of the HLR. Once again, the role of the HLR evolves beyond that of a simple data repository. The kinship of this approach with that of the SOSS feature is clear, and this comforts the whole picture of subscription networks as being more and more in control of the services provided to their users, with the switches of the visited networks being more and more confined to the provision of a telecommunications access.

GSM and its derivatives are now fully evolutive systems, a platform which can accept improvements, new features, or modifications needed to adapt it to something which was not foreseen from the start.

Conclusion

GSM and its derivatives are now fully evolving systems — a platform which can accept improvements, new features, or modifications needed to adapt it to something which was not foreseen from the start. This is important for companies operating a GSM network, which can expect that there investment will be a sufficiently long-term one, but also to conquer new markets. The choice of UIC is an example. In another direction, some non-European markets are conquered with the unadulterated standard, but some other markets, like the United States, requires adaptations. The success of GSM in these cases will depend on its flexibility and its adaptiveness to new constraints.

The Phase 2+ Work Program, which is more and more ambitious, shows the present very dynamic evolution of GSM. It is worth repeating that most of the program is market-driven, most points being introduced at the request of operators. This dynamism, together with the openness of the system, may, as we hope to have shown in this article, have the paradoxical result of changing profoundly the GSM system. This is just an avatar of the well known paradox of evolution: if something evolves, it becomes different! There is no true core of GSM to be preserved at all cost. What has to be kept, because of the economical impact, is the operational continuity. Therefore changes must be gradual, and several generations of equipment must cohabit. This does not prevent in-depth modifications, provided too old equipment can be put aside after some time.

This statement unerringly leads to the question of the long-term future of GSM and its relationship with "third-generation systems" such as FPLMTS. A few years ago, the scheme of a stable GSM system being replaced after some years of "good and faithful services" (as we say in French) by a brand new technology was in the mind of many. The new guise of GSM as an evolving platform makes this future less convincing. Already, the differences in term of services provided to the users between GSM plus all Phase 2+ ideas and FPLMTS as presently described are confined to a few points, such as wideband services. In-depth changes of GSM to meet even these needs cannot be excluded. It is difficult to predict what will happen, but one thing is certain, and that is that GSM and its derivatives DCS1800 and DCS1900 have a big potential for evolution and can have a preeminent role in the evolution of telecommunications as a whole. This will depend on whether all the involved parties, in Europe, but also in the United States, make the best coordinated usage out of this evolution potential.

References

[1] "Phase 2+ work program," internal document of ETSI Technical Committee SMG.

[2] P. Dupuis, "A European view on the Transition Path towards Advanced Mobile Systems," *IEEE Personal Commun. Mag.*, vol. 2, no. 1, Feb. 1995.

[3] M. Mouly and M. B. Pautet, "The GSM System for Mobile communications," 1992, published by the authors, fax +33-1-69310338.

Biographies

MICHEL MOULY was graduated from Ecole Polytechnique, and Ecole National Supérieure des Télécommunications. He worked first for France Télécom then for Matra Communication (whose GSM activities are now part of NorTel) before becoming an independent consultant. He has participated in the design of GSM continuously since its origin, and is now chair (under contract with NorTel) of SMG3, the committee in charge of architecture and general signaling protocols for the GSM system; he also chairs the SMG3 subgroup dealing with air interface signaling protocols and the one dealing with architecture.

MARIE-BERNADETTE PAUTET was graduated from Ecole Polytechnique and Ecole National Supérieure des Télécommunications. She worked at France Télécom before becoming an independent consultant; she was part of the GSM Permanent Nucleus at the more hectic moments of the initial GSM design, and chaired for some time the GSM sub-committee in charge of radio interface signaling protocols. Before leaving France Télécom, she was in charge of its system development team.

Reprinted by permission from Hewlett-Packard, Sept. 1995.

GSM Basics, An Introduction

Bob Garner &
Ian Reading

Hewlett-Packard Ltd.
Queensferry Microwave Division
South Queensferry
West Lothian EH30 9TG
Scotland

Symposium Paper

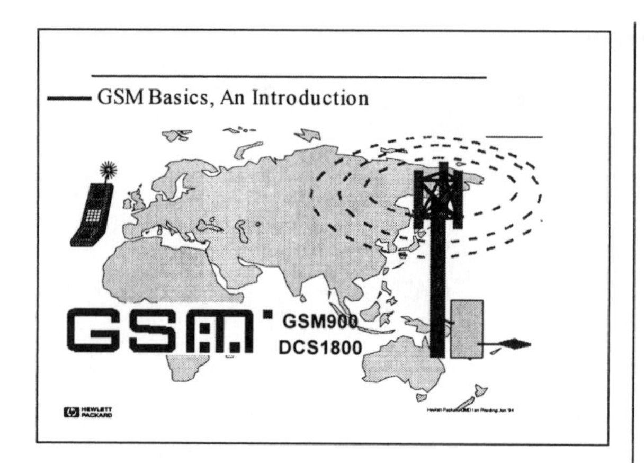

The Global System for Mobile communications (GSM) is a huge, rapidly expanding and successful technology. Less than three years ago, there were a few 10's of companies working on GSM. Each of these companies had a few GSM experts who brought knowledge back from the European Telecommunications Standards Institute (ETSI) committees designing the GSM specification. Now there are 100's of companies working on GSM and 1000's of GSM experts. GSM is no longer state-of-the-art. It is everyday-technology, as likely to be understood by the service technician as the ETSI committee member.

GSM is quickly moving out of Europe and is becoming a world standard. HP has become expert in GSM through our involvement in Europe. With excellent internal communications, HP is in an excellent position to help our customers, in other regions of the world, benefit from our GSM knowledge.

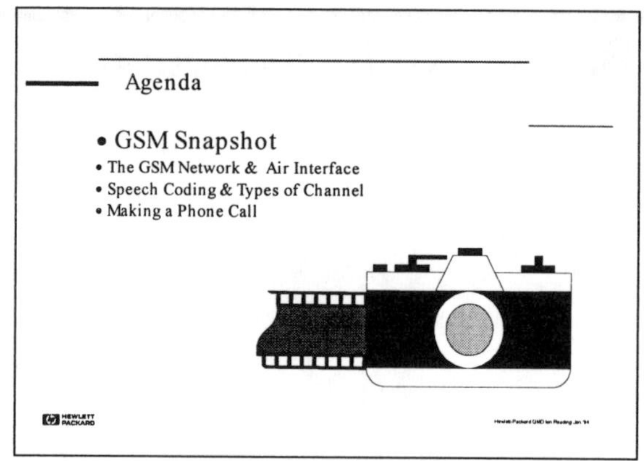

I. GSM Snapshot

GLOBAL System for Mobiles

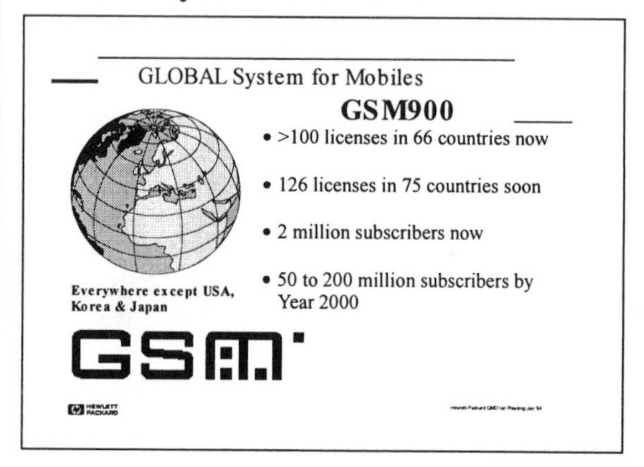

GSM is truly becoming the GLOBAL System for Mobile Communications. It's been clear for a long time that GSM would be used across Europe. Now, many countries around the world, who have been delaying their decision, have selected GSM. GSM has become a Pan Asian standard and is going to be used in much of South America.

Personal Communications Networks

PCN started in the UK with Mercury One-to-One and Hutchison Microtel (Orange) offering the first two networks to use DCS1800. Mercury's network has coverage in the South East of England. Microtel turned on in 1994 with many UK population centres and transport routes covered. Mercury have been overwhelmed by the success of their system. With free calls off-peak, and competitive business tariffs, Mercury are not just competing with existing cellular operators, they plan to take business from British Telecom's wired network as well.

Germany's E net followed the UK PCNs. DCS1800 is becoming more widespread with systems in Thailand being followed by France and Australia.

Even the USA, which has shunned GSM is about to use the GSM based DCS1900 for it's PCS system. In the USA, GSM will share the allocated bands with other systems based on CDMA and DECT.

How is GSM different from CT2 and DECT?

GSM900 and DCS1800 are cellular systems, DECT and CT2 are cordless systems. GSM like AMPS and TACS allows users to make and receive calls over a wide geographic area. The system uses a register to log the position of all mobiles, allowing calls to be routed to the correct base station. DECT and CT2, like other cordless systems do not include this tracking capability. They operate in much the same way as a conventional domestic cordless phone (CT0 or CT1). Calls can be received when the mobile is within range of it's home base station, but not at other locations.

How are GSM900, DCS1800 and DCS1900 Different?

GSM900 is the original GSM system, using frequencies in the 900 MHz band and designed for wide area cellular operation. Mobiles with output powers from 1 to 8W are typical. DCS1800 is an adaptation of GSM900. The term GSM can be used collectively to describe the GSM900 and DCS1800 standards. Creating DCS1800 involved widening the bands assigned to GSM and moving them up to 1.8 GHz. The DCS1800 standard was created to allow PCN (Personal Communications Networks) to form, increasing competition in the cellular communications industry, particularly in the UK. To avoid confusion, the channel numbers (ARFCN) used for DCS run from 512 to 885. GSM900 channels run from 1 to 124. With wider frequency allocation, leading to more channels, DCS1800 is able to cope with higher user densities. DCS1800 mobiles are also designed for lower output powers (up to 1W), so cell sizes have to be smaller, meaning even higher densities. In all other respects, GSM900 and DCS1800 are the same. The GSM phase 2 specifications (a revised and re-written standard due for release in 1994) brings the two systems even closer. GSM900 gets additional bandwidth and channels, called E-GSM (Extended band GSM) and lower power control levels for mobiles, allowing micro-cell operation. These two features allow increased user densities in GSM systems. While not expected in phase 2, DCS1800 will soon get higher power 4W mobiles. This will allow larger cells, reducing the cost of covering rural areas.

PCS1900 is not completely defined. In the USA, bands have been released around 2 GHz for a PCS (Personal Communications System). Unlike Europe and the Far East, the PCS licence holders will not be forced to use any particular radio technology. CDMA has received a great deal of attention, and it's likely some of the operators will use this standard. The ready availability of GSM equipment and expertise makes it extremely likely that some of the operators will use GSM at 1.9 GHz. This version of GSM is variously called DCS1900 or PCS1900. Other systems are also on trial in the US, including DECT. Many of the large GSM manufacturers are backing PCS1900 including Nokia,

Ericsson, Matra, AEG, Northern Telecom. MCI and other US operators are trying the systems. In technical terms it seems likely that PCS1900 will be identical to DCS1800 except for frequency allocation.

Some GSM History

Before we go into how the GSM system actually operates, let's take a look at the past and see how we got where we are today. In 1981 analogue cellular was introduced and at about the same time there was a joint Franco-German study looking at digital cellular technology and the possibility of making a pan-european system. In 1982 a special working committee, Groupe Spécial Mobile (GSM), was formed within the CEPT to look at and continue the Franco-German study. In 1986 the working committee was taken a step further by the establishment of a permanent nucleus of people to continue the work and create standards for a digital system of the future. About a year later, the memorandum of understanding, or MoU, as it is referred to, was signed by over 18 countries. It stated that they would participate in the GSM system and get it into operation by 1991. In 1989 GSM was moved into the ETSI (European Telecommunications Standards Institute) organisation.

Once under the control of ETSI, the GSM system had it's name changed to Global System for Mobile communications. The committees working on the system changed from GSM to SMG (Special Mobile Group). These changes avoided confusion between the system name (GSM), and the people working on the specification (SMG). It also brought the naming in line with the official working language of ETSI (English).

In 1990 the GSM specification developed an offshoot - DCS1800. The Original DCS1800 specifications were developed simply as edited versions of the GSM900 documents.

Interest in GSM quickly spread outside Europe. Australia was the first non-European country to join the MoU in 1992. Since then, many other Asian countries have adopted GSM. There's now a Pan-Asian MoU, investigating international roaming agreements.

The Next step for GSM will be the phase 2. Phase 2 is a completely revised set of GSM specifications, due for release in 1994. GSM900 and DCS1800 documents are merged, a number of new features are added to the system, along with many minor adjustments.

II. The GSM Network & Air Interface

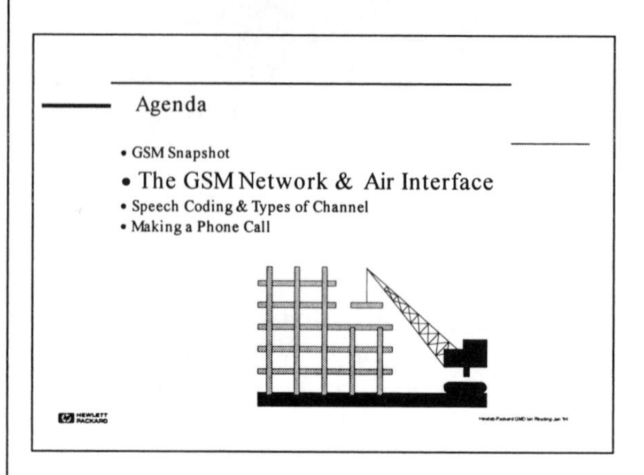

GSM Networks

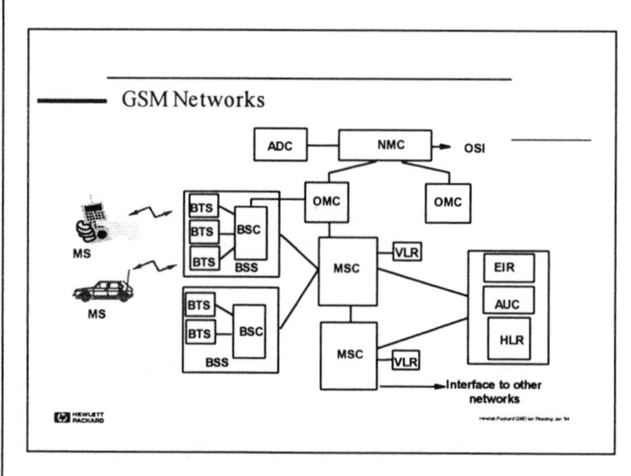

This is the GSM system. The Mobile Stations (MS), both hand held (or portables) and traditional mobiles in a car, talk to the Base Station System (BSS) over the RF air interface. The Base Station System (BSS) consists of a Base Transceiver Station (BTS), and a Base Station Controller (BSC). It's typical for several BTS to be located at the same site, producing 2 to 4 sectored cells around a common antenna tower. BSC's are often connected to BTS via microwave links. The BSC to BTS link is called the Abis interface. Typically 20 to 30 BTS will be controlled by one BSC. A number of BSSs would then report back to the Mobile Switching Centre (MSC) which controls the traffic among a number of different cells. Each Mobile Switching Centre (MSC) will have a Visitors Location Register (VLR) in which mobiles that are out of their home cell will be listed so that the network will know where to find them. The MSC will also be connected to the Home Location Register (HLR), the Authentication Centre (AUC), and the Equipment Identity Register (EIR) so the system can verify that users and equipment are legal subscribers. This helps avoid the use of stolen or fraud mobiles. There are also facilities within the system for Operations and Maintenance (OMC) and Network Management (NMC) organisations. The Mobile Switching Centre (MSC) also has the interface to other networks such as Private Land Mobile Networks (PLMN) and Public Switched Telephone Networks (PSTN) and ISDN networks.

Testing GSM

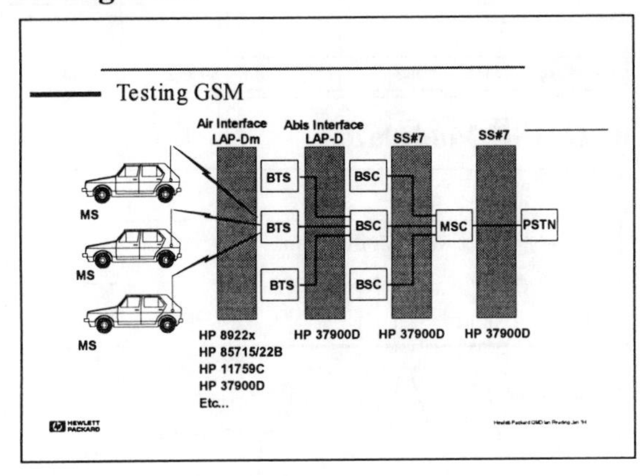

Taking a simpler view of the GSM system, we can see the base transceiver station , base station controller, mobile switching centre, and public switch telephone network are tied together with hard lines (optical fibre or microwave links). The link between the mobiles and the base station is the air interface. Hewlett-Packard has many measurement solutions, designed to test most areas of the GSM system.

A GSM Cell

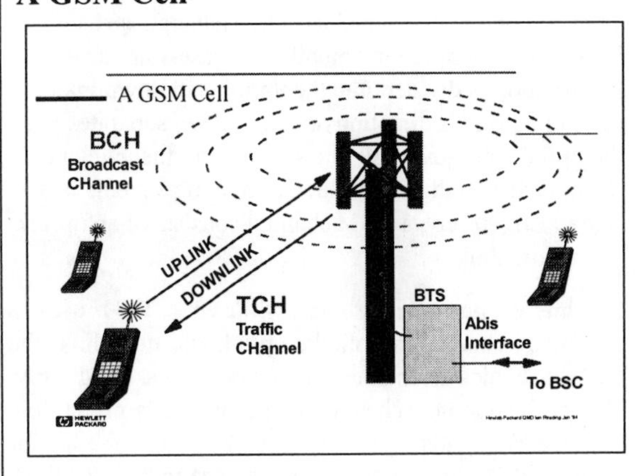

This is a close-up view of a typical GSM cell. Cells can be up to 35km radius for GSM900 and 2km for DCS1800 (because of lower power DCS1800 mobiles). The most obvious part of the GSM cell is the base station and it's antenna tower. It's common for several cells to be sectored around a common antenna tower. The tower will have several directional antennas, each covering a particular area. This co-location of several BTS is sometimes called a cell-site, or just a base station. The BTS are connected to their BSC by the Abis interface. This sometimes a cable, or an optical fibre. DCS1800 networks often use a microwave link for the Abis interface.

Each BTS will be fitted with a number of TX/RX pairs or transceiver modules. The number will determine how many frequency channels can be used in the cell, and depends on the expected number of users.

All BTS produce a BCH (Broadcast CHannel). The BCH is like a lighthouse or beacon. It's on all the time and allows mobile to find the GSM network. The BCH signal strength is also used by the network for a variety of user functions. It's a useful way of telling which is the closest BTS to the mobile. It also has information coded onto it, such as the identity of the network (e.g. Mannesmann, Detecon, or Optus), paging messages for any mobiles needing to accept a phone call, and a variety of other information. The BCH is received by all mobile's in the cell, whether they are on a call, or not.

The frequency channel used by the BCH is different in each cell. Channels can be re-used by distant cells, where the risk of interference is low.

Mobile's on a call use a TCH (Traffic CHannel). The TCH is a two way channel used to exchange speech information between the mobile and base-station. Information is divided into the uplink and downlink, depending on it's direction of flow. GSM separates out the uplink and downlink into separate frequency bands. Within each band, the channel numbering scheme is the same. Effectively, a GSM channel consists of an uplink and a downlink.

It's interesting to note that while the TCH uses a frequency channel in both the uplink and downlink, the BCH occupies a channel in the downlink band only. The corresponding channel in the uplink is effectively left clear. This can be used by the mobile for unscheduled or random access channels (RACH). When the mobile wants to grab the attention of the base station (perhaps to make a call), it can ask for attention by using this clear frequency channel to send a RACH. Since more than one mobile may want to grab attention at the same time, colliding RACHs are possible, and mobiles may need to make repeated attempt to get heard.

GSM Air Interface

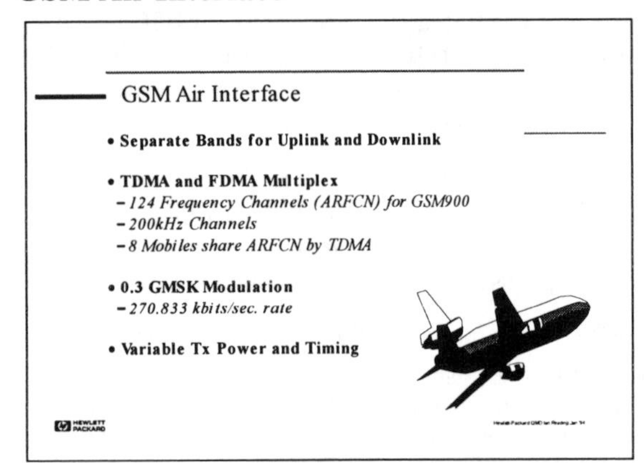

Here are some more specifics about the air interface.

	Phase 1 GSM900	Phase 2 GSM900	Phase 1 DCS1800	Phase 2 DCS1800
Uplink	890 to 915MHz	880 to 915MHz	1710 to 1785MHz	1710 to 1785MHz
Downlink	935 to 960MHz	925 to 960MHz	1805 to 1880MHz	1805 to 1880MHz
ARFCN range	1 to 124	0 to 124 and 975 to 1023	512 to 885	512 to 885
TX/RX Spacing (Freq.)	45MHz	45MHz	95MHz	95MHz
TX/RX Spacing (Time)	3 Timeslots	3 Timeslots	3 Timeslots	3 Timeslots
Modulation Data Rate	270.833kbit/s	270.833kbit/s	270.833kbit/s	270.833kbit/s
Frame Period	4.615ms	4.615ms	4.615ms	4.615ms
Timeslot Perios	576.9μs	576.9μs	576.9μs	576.9μs
Bit Period	3.692μs	3.692μs	3.692μs	3.692μs
Modulation	0.3GMSK	0.3GMSK	0.3GMSK	0.3GMSK
Channel Spacing	200kHz	200kHz	200kHz	200kHz
TDMA Mux	8	8	8	8
MS Max Power	20W (8W is max in use)	20W (none expected above 8W)	1W	1W
MS Min Power	13dBm	5dBm	4dBm	0dBm
MS Power Control Steps	0 to 15	0 to 19	0 to 13	0 to 15
Voice Coder Bit Rate	13kbit/s	13kbit/s	13kbit/s	13kbit/s

0.3GMSK Modulation

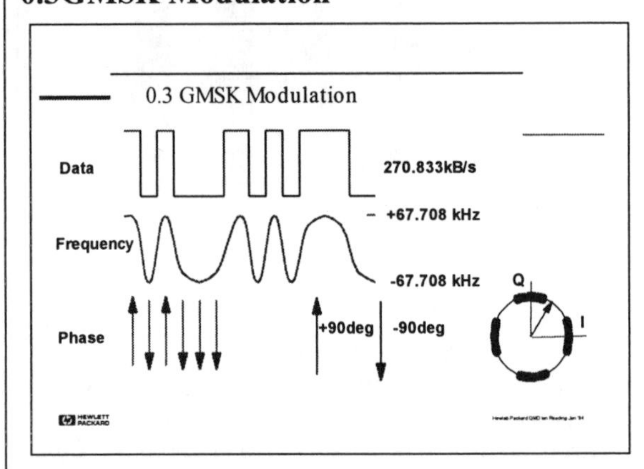

GSM uses a digital modulation format called 0.3GMSK (Gaussian Minimum Shift Keying). The 0.3 describes the bandwidth of the Gaussian filter with relation to the bit rate.

GMSK is a special type of digital FM modulation. One's and zero's are represented by shifting the RF carrier by plus or minus 67.708kHz. Modulation techniques which use two frequencies to represent one and zero are denoted FSK (Frequency Shift Keying). In the case of GSM, the data rate of 270.833kbit/sec is chosen to be exactly four times the RF frequency shift. This has the effect of minimising the modulation spectrum and improving channel efficiency. FSK modulation, where the bit rate is exactly four times the frequency shift is called MSK (Minimum Shift Keying). The modulation spectrum is further reduced by applying a Gaussian pre-modulation filter. This slows down the rapid frequency transitions which would otherwise spread energy into adjacent channels.

0.3GMSK is not phase modulation. Information is not conveyed by absolute phase states, as in QPSK, for example. It's the frequency shift, or change of phase state which conveys information. It is sometime useful though, to try to visualise GMSK on an I/Q diagram. Without the Gaussian filter, if a constant stream of 1's is being transmitted, MSK will effectively stay 67.708kHz above the carrier centre frequency. If the carrier centre frequency is taken as a stationary phase reference, the +67.708kHz signal will cause a steady increase of phase. The phase will role +360 degrees at a rate of 67,708 revolutions per second. In one bit period (1/270.833kHz) the phase will get a quarter of the way round the I/Q diagram, or 90 degrees. One's are seen as a phase increase of 90 degrees. Two one's causes a phase increase of 180 degrees, three one's 270 degrees, and so on. Zero's cause the same phase change in the opposite direction. Adding the Gaussian filter does not affect this average 90 degree transition for one's and zero's. Because the Bit rate and frequency shift are tied together by a factor of 4, filtering can not affect the average phase relationships. The filtering does slow down the rate of change of phase velocity (the acceleration of the phase). When Gaussian filtering is applied, the phase makes slower direction changes, but may reach higher peak velocities to catch up again. Without Gaussian filtering, the phase makes instantaneous direction changes, but moves at a constant velocity.

The exact phase trajectory is very tightly controlled. GSM radios need to use digital filters and I/Q or digital FM modulators to accurately generate the correct trajectory. The GSM specifications allow no more than 5 degrees rms and 20 degrees peak deviation from the ideal trajectory.

TDMA and FDMA

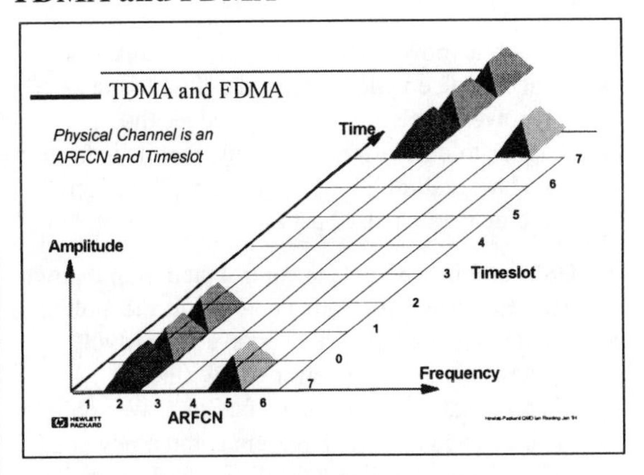

GSM uses TDMA (Time Division Multiple Access) and FDMA (Frequency Division Multiple Access). The frequencies available are divided into two bands. The uplink is for mobile transmission, while the downlink is for base station transmission. The slide shows part of one of these bands. Each band is divided into 200kHz slots called ARFCN (Absolute Radio Frequency Channel Number). As well as slicing up frequency, we also slice up time. Each ARFCN is shared between 8 mobiles, each using it in turn. Each mobile uses the ARFCN for one TS (Timeslot) and then waits for its turn to come round again. Mobiles get the use of the ARFCN once per TDMA frame.

The slide illustrates 4 TCH (Traffic CHannels). Each one of the TCH uses a particular ARFCN and Timeslot. Three of the TCH are on the same ARFCN, using different timeslots. The fourth TCH is on a different ARFCN.

The combination of a TS number and ARFCN is called a physical channel.

There's not much space between timeslots and ARFCN's. It's important for the mobile or base-station to transmit their TDMA bursts at exactly the right time and with exactly the right frequency and amplitude. Too early or too late and a burst may collide with an adjacent burst. Poorly controlled modulation spectrum or spurious will cause interference with adjacent ARFCN.

Changing Power

As the mobile moves around the cell, it's transmitter power needs to be varied. When it's close to the base station, power levels are set low to reduce the interference to other users. When the mobile is further from the base station, it's power level needs to increase to overcome the increased path loss.

All GSM mobiles are able to control their output power in 2dB steps. The base station commands the mobile to a particular MS Tx Level (Power level). GSM900 mobile have a maximum power of 8W (the specifications allow 20W, but so far, no 20W mobiles exist). DCS1800 mobiles have a maximum power of 1W. Consequently DCS1800 cells need to be smaller.

Timing Advance

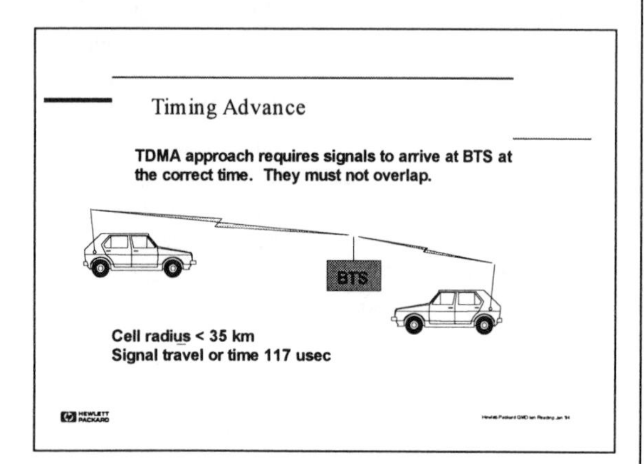

Timing advance is required in GSM because it uses TDMA with cells up to 35 km radius. Since a radio signal take a finite period of time to travel from the mobile to the base-station, there must be some way to make sure the signal arrives at the base-station at the correct time.

Without timing advance, the transmitted burst from a user at the edge of a cell would arrive late and overlap (and corrupt) the signal from a user right next to the base station (unless a guard time, between timeslots, greater than the longest signal travel time was used). By advancing the timing of the mobiles, their transmissions arrive at the base station at the correct time. As a mobile (MS) moves, the Base Station (BTS) will signal the MS to reduce its timing advance as it gets closer to the centre of the cell, and increase its timing advance as it away from the centre of the cell.

Mobile's in idle mode (not on a call, but still camped to the network) receive and decode the BCH (Broadcast CHannel) from the base station. One element of the BCH, the SCH (Synchronisation CHannel) allows the mobile to adjust it's internal timing. When the mobile is receiving the SCH, it doesn't know how far it is from the base station. A distance of 30km will cause the mobile to set it's internal timing 100µs behind the base-station. When the mobile sends it's first RACH burst, it will leave 100µs late, after a 100µs transit delay, it will arrive 200µs late, colliding with the bursts from mobile's closer to the base station. For this reason, the RACH, and other types of access burst are shorter than normal. The mobile only sends normal length bursts once it's received timing advance information from the base-station. The mobile in our example would need to advance it's timing by 200µs.

We'll see later how the base station commands the mobile to change it's timing advance or transmitter power using the SACCH (Slow Associated Control CHannel)

GSM TDMA Power Burst

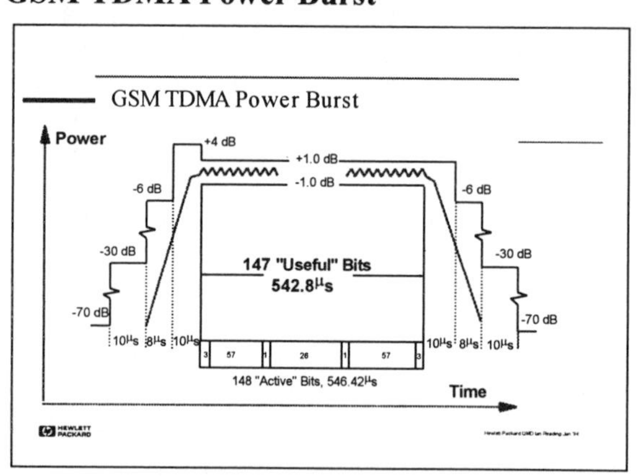

Since GSM is a TDMA system and there are 8 users on a frequency pair, each user must only turn his transmitter on at the allowed time and, have his transmitter off in time so that he does not interfere with other users in the adjacent timeslots. Because of this need, GSM has specified an amplitude envelope for the RF burst of the timeslots. There's also a demanding flatness specification over the active part of the useful bits in the timeslot. The amplitude envelope has greater than 70dB of dynamic range yet needs to measure less than +/-1dB flatness over the active part of the timeslot. All of this is happening over the 577μs period of a timeslot.

Frames and Multiframes

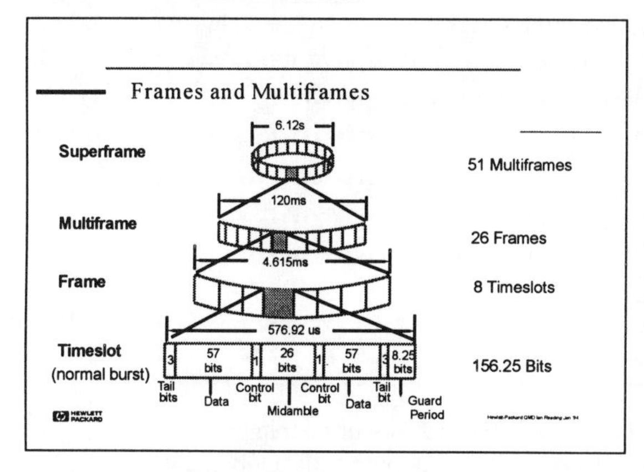

The GSM system is a time division multiplex system. The smallest unit within that system is the individual data bits. Each date bit is 3.69μs long. A timeslot, the amount of time each mobile has to transmit or receive information, has a time period equal to 156.25 of these data bits. Since there are 8 users on each frequency, there are 8 timeslots per frame. The pattern repeats giving the users another timeslot each frame. The frame period is 4.615 ms. Frames are grouped into larger structures called multiframes. There are two sizes of multiframes, 26 frame multiframes and 51 frame multiframes. TCH use 26 frame multiframes, while BCH use pairs of 51 frame multiframes stacked end-to-end to make a 102 frame sequence. A superframe consists 51 or 26 multiframes and a hyperframe is made up of superframes

These multiframe structures are necessary to allow the partitioning of physical channels (an ARFCN and a timeslot) into logical channels. A logical channel is simply an end-to-end conduit for information. In later slides, we will see how the TCH is mainly used for carrying speech data. Once per multiframe, one of the TCH's physical channel timeslots is used to carry control information. This logical control channel which shares the same physical channel as the TCH is called an SACCH. There are long repeat patterns on the BCH too. Times are set aside for different types of logical channels to coexist on the same physical channel.

The midamble or training sequence in the centre of the burst is a known pattern. It allows the equaliser in the mobile or base station to analyse the RF path characteristics before decoding the other useful data. Midambles have a few allowed patterns or colour codes. On either side of the midamble there are control bits called steeling flags. Sometimes the TCH has to be interrupted with urgent control information in an FACCH (Fast Associated Control CHannel). The FACCH is used to tell the MS to change ARFCN or TS, for example, and results in some lost TCH data. The steeling flags allow the TCH and FACCH to be distinguished. The remainder of the burst carries data (speech for example) and tail/guard bits to fill the gaps between bursts.

It's easy to get confused about the number of bits in a timeslot, are there 148 bits in a timeslot or 147 bits in a timeslot? There are 148 ACTIVE bits in a timeslot, consisting of the mid-amble, the control bits, the data and the tail-bits. There are 147 USEFUL bits from the middle of the first bit to the middle of the last. Effectively 1/2 a bit off each end is lost.

Downlink and Uplink

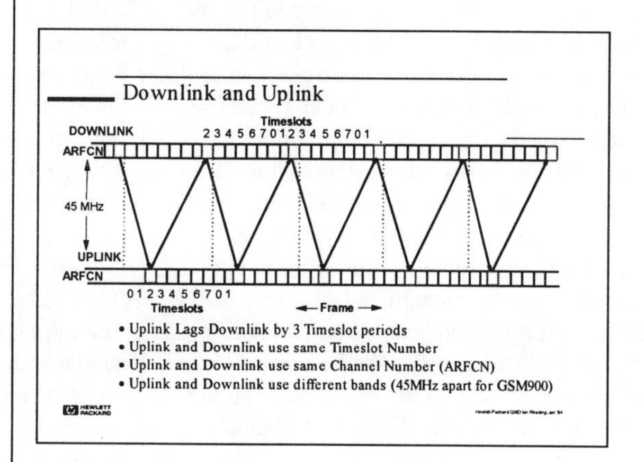

To see how information is transmitted let's look at an example. We have been assigned timeslot 2 and we're in a traffic mode, receiving and transmitting information to the base station. The downlink, on which we receive information, will be in the frequency range of 935 to 960MHz. The uplink, the frequency which the mobile will transmit information to the base station, will be in the frequency range of 890 to 915MHz. The uplink and the downlink make up a frequency pair, which for GSM900, is always separated by 45MHz. We can see that the timeslots are offset by 3 between the downlink and the uplink. We receive information in timeslot two in the downlink we have two timeslots in which to switch to the uplink frequency and be ready to transmit information. Then, we have to get ready to receive our next time slot of information in the next frame.

Measuring Adjacent Cell BCH Power

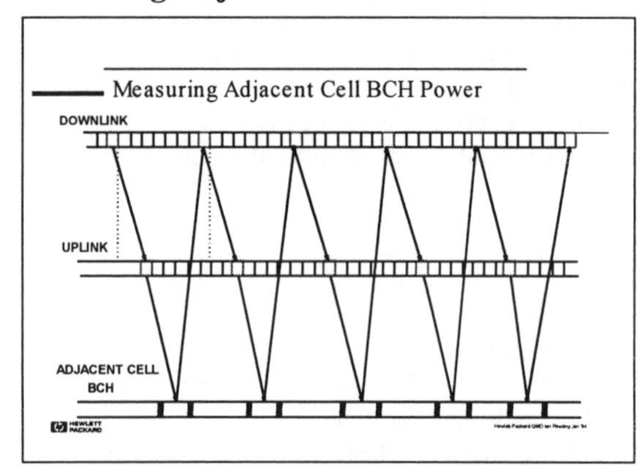

Besides receiving and transmitting information, the mobile must switch frequency and get ready to receive and measure the level of the adjacent cell's broadcast channels. It then reports this (RXLev) information to its own base station in order to establish when a handover is appropriate between cells. Again, information is received on timeslot 2, we switch 45MHz to transmit information and then, need to switch back 45MHz +/- a few MHz to monitor and measure the level of the adjacent cell's broadcast channels. This information will be reported back to the base station at least every 30 seconds so that the base station can determine the appropriate time to do a handoff. The RxLev information is reported back to the base-station on the uplink SACCH (Slow Associated Control CHannel).

The mobile uses a list of ARFCN in the BA (Base Allocation) table to know which BCH frequencies to go out and measure. The BA table is coded onto the BCH, and also the downlink SACCH.

This is the primary (or non-hopped) mode of operation in the GSM system. If there is an area which has bad multipath, such as urban areas with lots of reflections from buildings, the cell may need to be defined as a hopping cell.

Hopping Traffic Channel

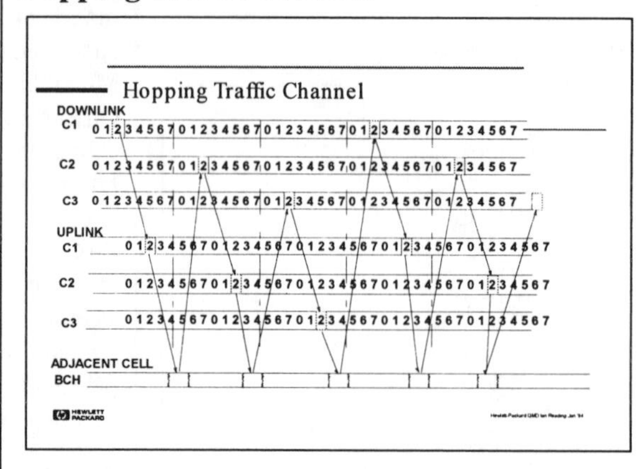

All mobiles must have the capability of hopping. However, not all cells will be hopping cells. Only those cells which have bad multipath problems will be defined as hopping cells. In this example, there are three frequencies pairs to hop among. The mobile still needs to go out and measure the adjacent cells' broadcast channel (BCH). In the first frame, the mobile receives information on channel 1 downlink, then switches to the uplink for channel 1 (45MHz away), transmits it's information, and finally monitors one of the adjacent cells to measure its level. The mobile must move to the downlink for channel 2 and receive information in timeslot 2, switch 45MHz, and transmit on the uplink for channel 2. Then it monitors another cell's broadcast channel and measures its level. This continues through the sequence of frequencies that have been assigned to the cell. The hopping sequence is defined by the CA (Cell Allocation) and MA (Mobile Allocation) tables. The CA table is a master list of all the hop frequencies available in a particular cell. It's sent to the mobile on the BCH and also the downlink SACCH. The MA table is an index into the CA table, and gives a hopping sequence for a particular mobile. The MA table is sent to the mobile as part of the handover or channels assignment process.

III. Speech Coding & Types of Channel

Speech Coder

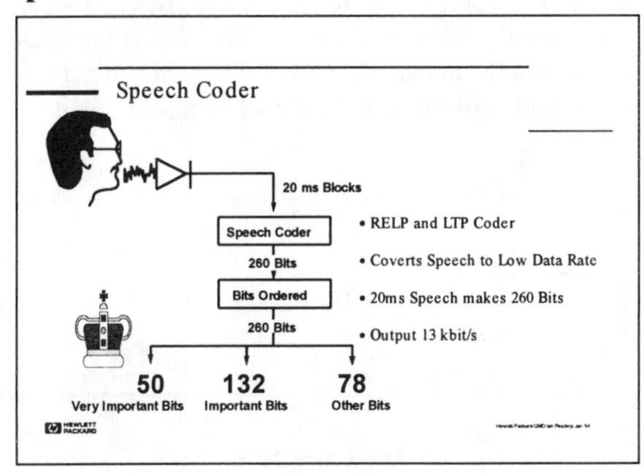

Most modern digital communications systems use some sort of voice compression. GSM is no exception. It uses a voice coder to model the tone and noise generation in the human throat and the acoustic filtering of the mouth and tongue. These characteristics are used to produce coefficients which are sent via the TCH.

The speech coder is based on a residually excited linear predictive coder (RELP), this is enhanced by including a long term predictor (LTP). The LTP improves speech quality by removing the structure from vowel sounds prior to coding the residual data. The coder outputs 260 bits for each 20 ms block of speech. This yields a 13kbit per second rate. Output bits are ordered, depending on their importance, into groups of 182 and 78bits. The most important 182 bits get further subdivided, with the 50 very important bits being separated out

The data rate of 13kbit/sec is considerably lower than for direct speech digitising as in PCM. In the future, more advanced voice coders will cut this to 6.5kbit/sec (half rate coding)

Error Correction

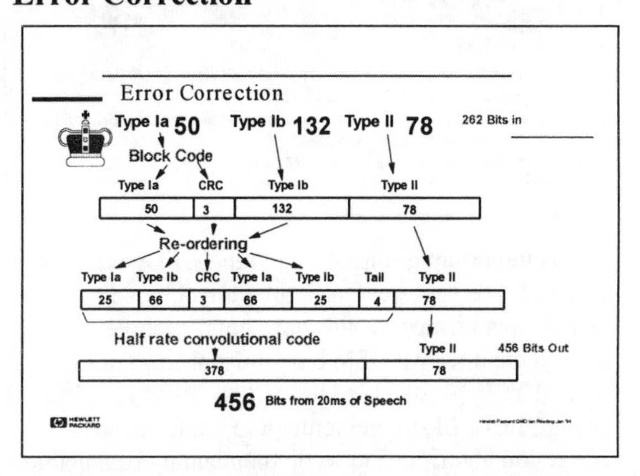

The nature of the GSM air interface means that some bit errors will be introduced. The bits are handled in such a way that errors are more likely where they matter least. The sound quality is affected more by the most significant coefficient bits than the least significant. The least important or type II bits have no error correction or detection. The premier type Ia bits have error detection CRC bits added. Both type Ia and the medium importance type Ib have convolutional error correction bits added.

It's sometimes interesting to think of GSM bits as aircraft passengers! There are three classes, Ia, Ib and II. The most important bits get first-class treatment, they get surrounded by lot's of error correction, and in the case of Ia bits, error detection as well. These extra bits take up space in the TCH bursts. The second class, type II bits, take up the least space on the TCH, just like first and second class passengers on an aeroplane.

We will see in the next slide how the final 456bits are sent over the TCH. To minimise the effects of a whole lost frame, the bits are re-ordered before convolutional error correction coding

Diagonal Interleaving

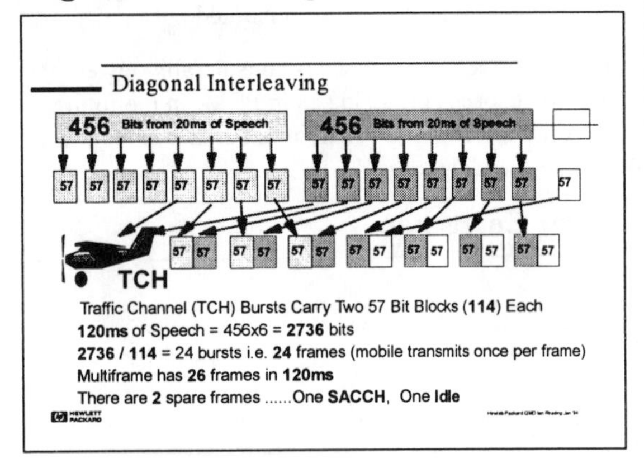

Just as important groups of individuals, like a company board of directors, generally don't travel together (in case the plane crashes and wipes out the whole management team), GSM bits spread themselves over several TCH bursts. If a burst is lost due to interference, enough bits will still get through to allow the error correction algorithms to work, maintaining reasonable speech quality. The 456 bits of speech data are sliced up into 8 blocks of 57. Each TCH frame carries two 57 bit blocks of data from two different 20ms 456bit speech segments.

From the arithmetic on the slide, notice that in the period taken up by 1 frame (120ms), six 20ms blocks of speech are processed by the speech coder. Each of these blocks results in 456 bits. A 120ms segment of speech will produce 2736 bits. Each TCH burst has a pair of 57 bit data sections on either side of the midamble. Effectively, each TCH burst carries 114 bits. It takes 24 of these TCH bursts to ship the 2736 bits from 120ms of speech. In an earlier slide, we saw how the TCH frame structures has 26 frames in a multiframe, lasting 120ms. Since the mobile or base station transmits one burst per frame, there are two more bursts available in 120ms than are actually needed to transmit the voice data. One of these spare bursts is used for an SACCH, the other is an idle burst.

Multiframe

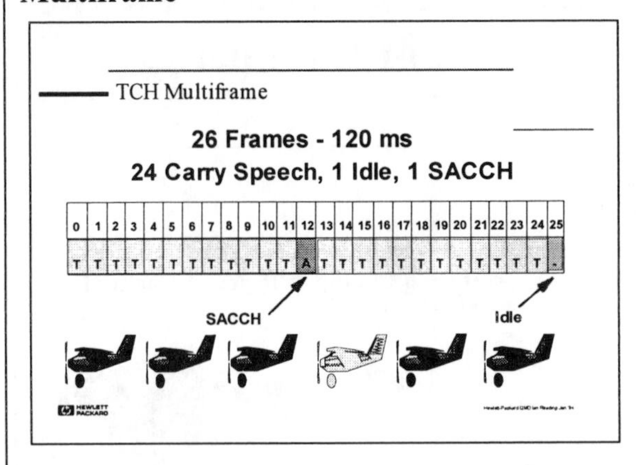

This is how the SACCH and Idle bursts fit in with the other TCH frames. The Idle burst is used by the mobile to make more detailed measurements on the adjacent cell BCH. It stays tuned to the adjacent BCH ARFCN long enough to decode the midamble. The colour code, encoded in the midamble, allows the mobile to get a positive identification of the signal being measured.

SACCH

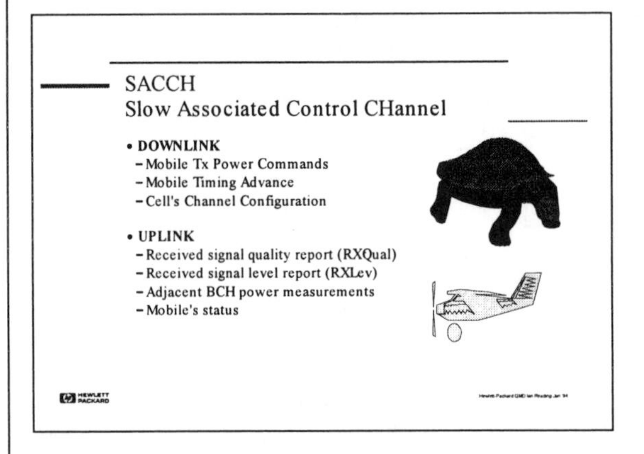

One of the two spare frames every 12 TCH frames is used for the SACCH (Slow Associated Control Channel). On the down-link, the SACCH is used to send slowly but regularly changing control information to the mobile. Examples are instructing the mobile to change its transmitter power (MS TX Lev) and burst timing advance (to compensate for RF transit time) as it moves around the cell. It also carries the BA and CA tables.

The up-link SACCH carries information about received signal strength (RXLev) and quality (RXQual) of the TCH and the adjacent cell BCH measurement results (also RXLev).

FACCH

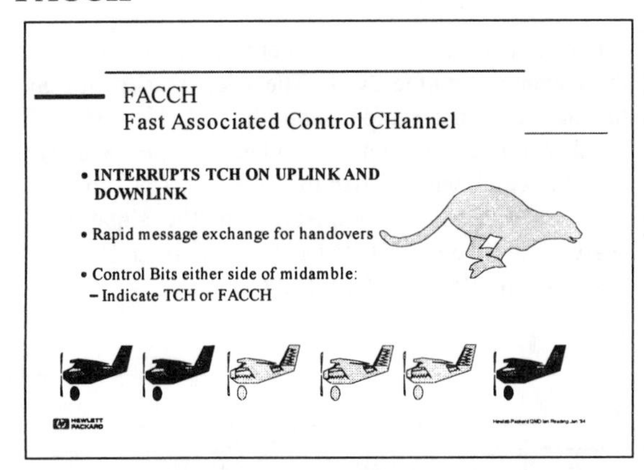

When the SACCH reports coming back to the base station indicate that another cell would offer the mobile better signal quality, a handover is necessary. The SACCH just doesn't have the bandwidth to transfer all the information associated with a handover (like the new ARFCN and timeslot, or the MA table). For a short period of time, the TCH is replaced by an FACCH. The FACCH uses consecutive bursts, so has a much higher data rate that the SACCH, which uses only one burst in 26. The frame stealing flags (the control bits on either side of the midamble) are set to indicate that the data being sent is an FACCH, not the TCH. In other respects, the FACCH looks just the same as the TCH. It uses the same physical channel (ARFCN and timeslot). When the FACCH steals bursts from the TCH, speech data is lost. It's often possible to hear a small speech drop-out when handovers take place.

RACH

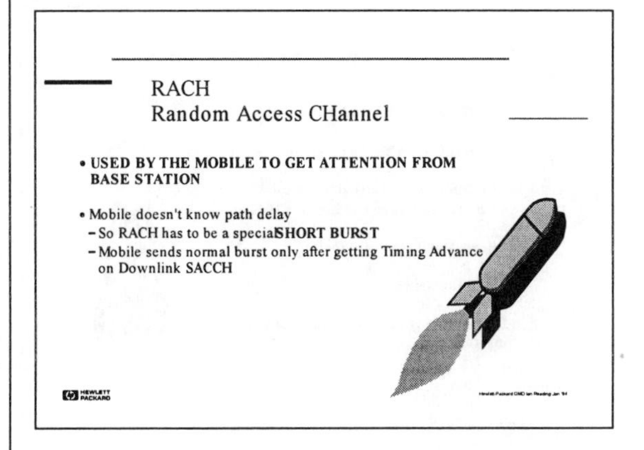

When the mobile has become synchronised to the frequency and frame timing of the cell, and looked at the other information on the BCH it is ready to make and receive calls. Once the mobile is in this state it is 'camped' to the base station. If the mobile is near the base station their timing will be closely aligned. If the mobile is on the edge of the cell, maybe 30km from the base station, the SCH will have a propagation delay of 100µs. The mobile's timing will be 100µs in error. When the mobile sends out a RACH, to start a call, the RACH is transmitted 100µs late, with another 100µs transit time to the base station, it arrives 200µs late. To avoid collisions with bursts in adjacent TS, RACH busts are shorter than normal.

The RACH is not the only type of short access burst. When a mobile is handed over to another cell, there will be a short period of time before it receives timing advance information on the downlink SACCH from the new cell. During this period, there's a risk of the mobiles bursts colliding with bursts in the new cell. Until it gets timing advance information from the new cell it sends short access bursts.

BCH

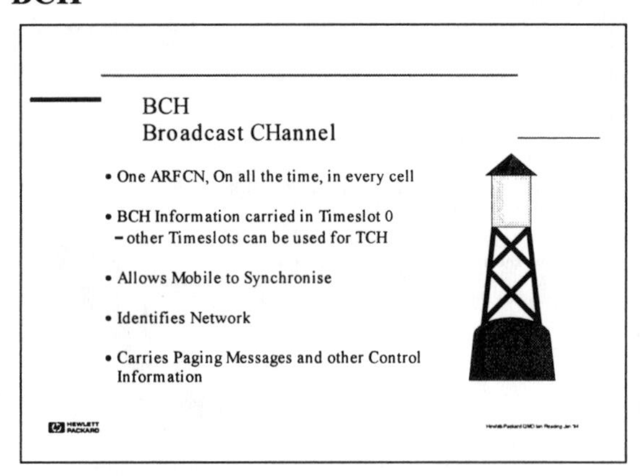

The concept of a BCH is very simple, but the details can get a little complicated. In simple terms, the BCH acts like a beacon, or lighthouse. It's on all the time and is the first thing the mobile looks for when it's trying to find service. The BCH ARFCN has to be active in all timeslots to allow mobiles synchronised to other cells to measure it's power. The useful BCH information is always carried in timeslot 0. The other timeslots are filled with dummy bursts, or are available for TCH. There are a number of interesting parts to the BCH:

- The FCH (Frequency correction CHannel) uses a special burst which repeats on the BCH, it has a special fixed bit sequence to allow the mobile to tune it's internal frequency reference when it first turns on.
- The SCH (Synchronisation CHannel) has a burst with extended midamble. It's used by the mobile after the FCH to adjust it's internal timing and get synchronised to the multiframe sequence.
- The BCCH (Broadcast Control CHannel) has information encoded on it which identifies the network. It also carries lists of the channels in use in the cell (BA and CA tables)
- The CCCH (Common Control CHannel) is like a message board. Just like the FCH, SCH and BCCH, it can be received by any mobile. Sub-channels like PCH (Paging CHannel) are posted on the CCCH. When the mobile sees its number on the PCH it recognises that it should respond by requesting service with a RACH.
- Another CCCH sub-channel is the AGCH (Access Grant CHannel). Once a mobile has sent a RACH, the base station responds by putting an AGCH on the CCCH, bearing the mobiles random number (read from the RACH). The AGCH instructs the mobile to go to an SDCCH or TCH.

There are a variety of different configurations for all these channels on the BCH. The selection depends on the number of users expected in the cell. If a large number of users are expected, a large CCCH capacity is needed, which when added to the SCH, FCH and BCCH, fills the BCH completely. In other situations, spare capacity on the BCH can be used for an SDCCH (Stand-alone Dedicated Control CHannel).

SDCCH

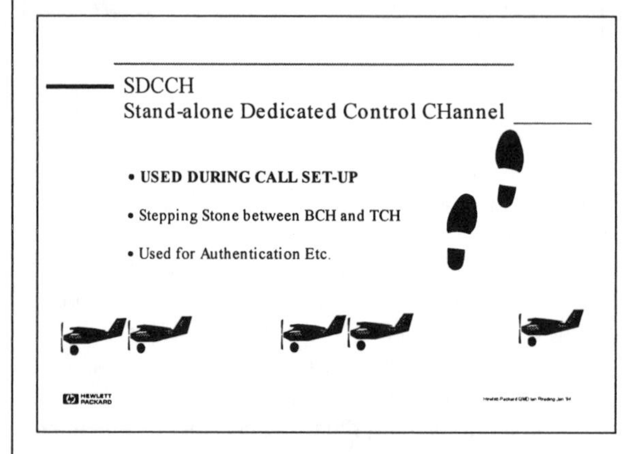

The SDCCH is sometimes configured as a logical channel on the BCH, and sometimes on it's own physical channel. The SDCCH has a different multiframe structure to the TCH. SDCCH bursts repeat less frequently than once per frame. For this reason, more than 8 SDCCH can share a physical channel. As a consequence, the data rate on the SDCCH is lower than on the TCH.

The SDCCH is used like a stepping stone. During the call set-up process, there can be a lot of time between the mobile sending a RACH and getting service, to the start of conversation. Time is taken up while the phone is ringing and waiting to be answered. During this period, there's a need to exchange control information between the mobile and base station. Alerting messages are sent, and authentication takes place, but there's no need to send speech information. The SDCCH, by using less of the cells resource of physical channels, improves efficiency, and provides a useful holding channel for the mobile until speech data needs to be exchanged. Just like the TCH, the SDCCH has an SACCH associated with it.

SIM

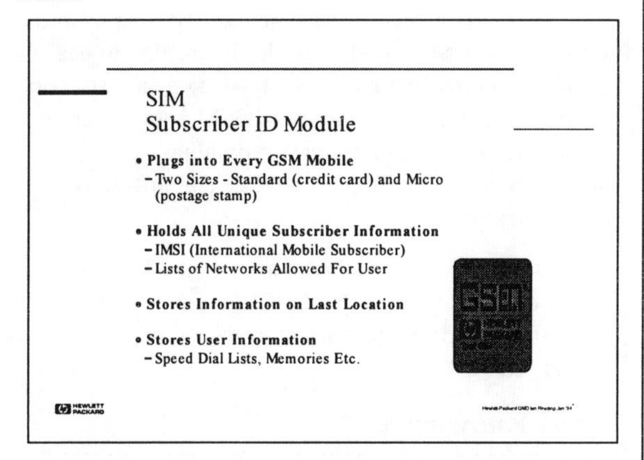

The SIM card comes in two sizes: standard (credit card size) and micro (postage stamp size). SIMs (subscriber Identification Modules) plug into the GSM mobile. The SIM holds all the information related to a subscriber. For example:

- His unique subscriber number or IMSI (International Mobile Subscriber Identification)
- The networks and countries where he is entitled to service (MCC and MNC)
- Any other user specific information like speed dial numbers and memories

Without a SIM installed, all GSM mobiles are identical. It's the SIM card which gives a mobile it's identity. If a user (Fred) takes his SIM on a business trip and plugs it into the GSM mobile fitted to his rental car, the car's phone takes on the SIM's identity. Fred's network access rights, his speed-dial memories and any other saved features, are transferred to the rental car phone. The really nice feature of SIM's is that they also carry your phone number. If Fred's office want to call him, they simply dial his normal mobile number. The network knows the location of the phone with Fred's SIM in it and so routes the call directly to the rental car.

For test purposes, there are special Test-SIMs. Test SIMs allow mobiles to enter a special loop-back mode for receiver BER test.

IV. Making a Phone Call

Mobile Turn-On

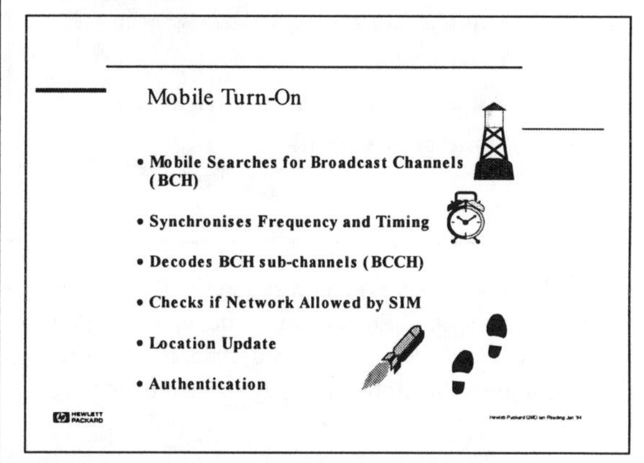

When a mobile first turns on, it searches all 124 channels in the downlink for signals. It will then order the channels by received signal strengths and check to determine if the channel was a BCH (Broadcast CHannel). Once the MS finds a BCH, it adjusts internal frequency and timing from the FCH and SCH, then checks to determine if the BCH is from its PLMN (Public Land Mobile Network). This involves comparing the allowed network and country codes stored on the SIM card with the information encoded on the BCCH. The mobile repeats this cycle until a good broadcast channel is found. If the mobile recognises that it's in a different cell from the last time it was used, it needs to tell the network where it is. The network has to keep track of where every mobile is so that it can route calls to the correct cell for any particular mobile. This process of telling the network "here I am" is called a location update. The mobile sends a RACH, gets assigned to an SDCCH, exchanges control information, then ends the call. The user will typically not be aware that this process is taking place.

Some networks have IMSI attach enabled. This forces the mobile to do a location update every time it turns on, even if it has not moved to different location.

Mobile Call Origination

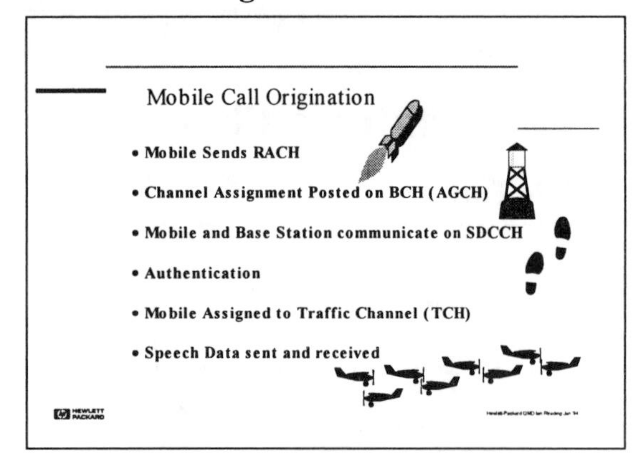

Once the mobile has synchronised to the BCH, determined that it's allowed to use the network (PLMN) and if necessary done a location update, it's camped. Once camped, the mobile is ready to send or receive calls.

When a user dials a number, and presses the send button on the mobile, call origination takes place. The mobile transmits a short RACH burst on the uplink, using the same ARFCN as the BCH is using on the downlink. The base station responds to the RACH by posting an AGCH (Access Grant CHannel) on the CCCH. These are logical channels on the BCH physical channel. The mobile listens on the BCH for the AGCH, when it receives it and decodes the instructions, it re-tunes to another ARFCN and/or timeslot and begins a two-way dialogue with the base station on an SDCCH. One of the first things that the mobile will receive is the SACCH associated with the SDCCH. Once it receives the SACCH, it will get timing advance and transmitter power information from the base station. The base station will have calculated the correct timing advance from the arrival time of the RACH. Once the mobile gets timing advance information, it can send normal length bursts. The SDCCH is used to send messages back and forth, taking care of alerting (making the mobile ring) and authentication (verifying that this mobile is allowed to use the network). After a short period of time (1 to 2 seconds), the mobile is commanded over the SDCCH to re-tune to a TCH. Once on the TCH, speech data is transferred on the uplink and downlink.

The process for base station originated calls is very similar. The base-station posts a PCH (paging CHannel) on the CCCH part of the BCH. When the mobile receives the PCH, it responds by sending a RACH. The remainder of the process is identical to the mobile originated case.

If you can find a way to translate the GSM bursts into audio tones (AM demodulate), it's interesting to hear the difference between the channel types as a call is set up. A good way to do this is to use a GSM phone near an old TV set or a conventional wired phone. The interference created in these devices amounts to AM demodulation.

The RACH burst can be heard as a single 'Tick' sound. It's quickly followed by the SDCCH 'Tat, Tat-tat-tat, tat-tat-tat ...'. After a few seconds, the TCH is connected 'Buzzzzzzzzz'

GSM Knowledge

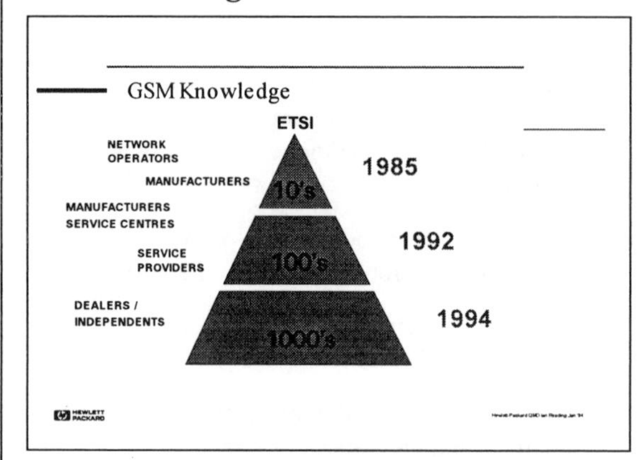

Like any technical subject, GSM can be confusing at first. You may need to read through the hand-out several times to understand the system.

In 1985, there were just a few GSM experts, working on the ETSI committees. Today there are thousands of people in the GSM business. GSM is no longer state-of-the-art, it's everyday-technology. GSM understanding is as important for the service technician or manager as the ETSI committee member.

Further Reading

GSM Measurement Basics
Hewlett-Packard Symposium Paper

Selecting GSM Measurements for Your Application
Hewlett-Packard Symposium Paper

Repairing GSM Mobiles
Hewlett-Packard Symposium Paper

GSM Mobile Service
Hewlett-Packard Symposium Paper

GSM900 & DCS1800 Base Station Installation & Maintenance
Hewlett-Packard Symposium Paper

Key Test Concerns in GSM Mobile Phone Manufacture
Hewlett-Packard Symposium Paper

Test and measurement solutions for wireless communications
Hewlett-Packard Literature No. 5091-7273E

GSM Mobile Service Solutions
Hewlett-Packard Literature No. 5963-0037E

GSM Mobile Manufacturing Solutions
Hewlett-Packard Literature No. 5962-0197E

GSM 11.10 Specification &
GSM 05.05 Specification
European Telecommunications Standards Institute

The GSM System for Mobile Communications
M. Mouly and M. B. Pautet
ISBN 2-9507190-0-7
Order direct from M. Mouly
Phone: +33 1 69 31 03 18
Fax: +33 1 69 31 03 38

Appendix

More GSM Topics

Logical and Physical Channels

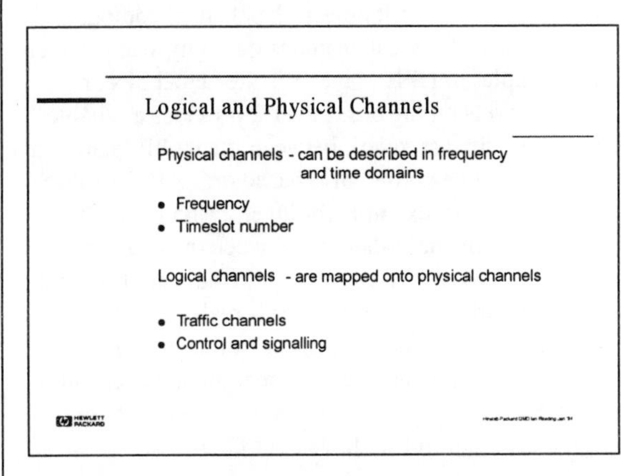

What's the difference between logical and physical channels? Physical channels can be described in terms of the frequency domain and time domain. They are the actual frequencies and/or the timeslot the MS or BS are transmitting or receiving on. The logical channels are mapped onto these physical channels. At any particular instant a frequency/timeslot may be either a traffic channel or some control or signalling channel. A logical channel describes the function of a physical channel is at that point in time.

Signalling Layers 1, 2, 3

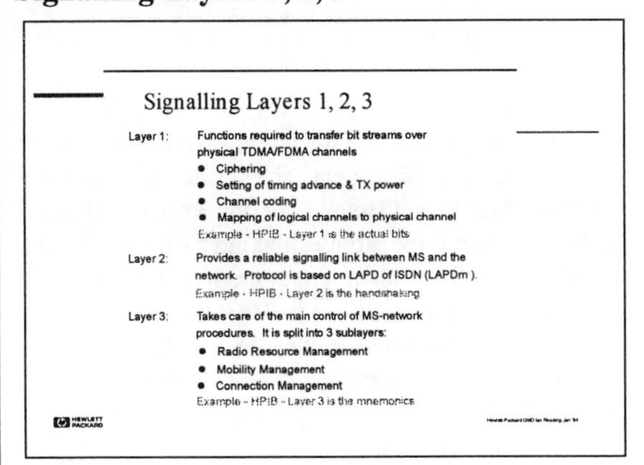

Some other terms you will hear thrown about in the GSM world are the different layers of signalling (layer 1, layer 2, layer 3). These layers are from the OSI (Open

Systems Interconnection) seven layer model. The GSM system uses the first 3 layers from the OSI model. Layer 1 can be thought of as the functions required to transfer bit streams over physical TDMA and FDMA channels. This includes things like ciphering, the setting of timing advance, and transmit power, the channel coding, and the mapping of logical channels onto physical channels. An example: In HPIB, layer 1 is the actual bits or voltage levels on the bus. Layer 2 provides a reliable signalling link (protocol) between the mobile station and the network. The protocol is based on LAP-D of ISDN or LAP-Dm. An example for layer 2 (in the HPIB example) is the handshaking between the listener and the talker. Layer 3 takes care of the main control on the MS - network procedures and it is really split into 3 sub-layers. These include radio resource management, mobility management, and connection management. Our HPIB example of layer 3 is the mnemonics or higher level control we have over the bus.

Location Update

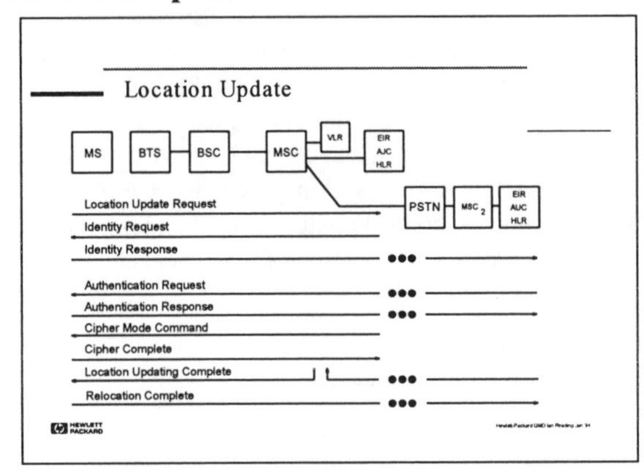

The process begins when the mobile sends a location update request to the system. The system asks for the mobile's identity and Home Location Register (HLR), and authenticates the mobile, making sure it is a legal mobile. The BS will then turn ciphering on or off. The new MSC lists the mobile in its Visitors Location Register (VLR) and notifies the MS's HLR of the fact that the MS is in the new location and will be serviced by the new MSC.

Mobile Call Origination

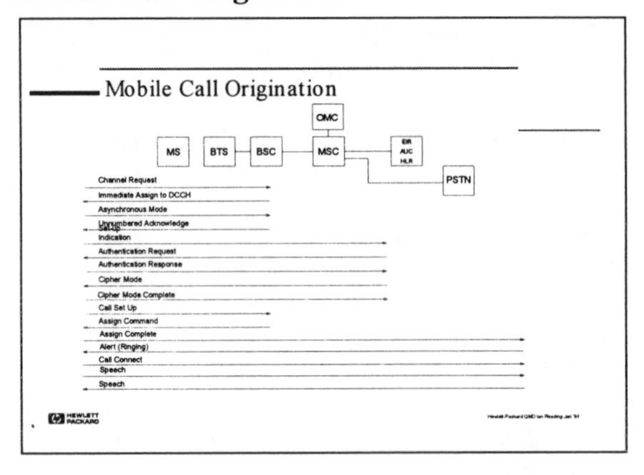

To originate a call, the mobile sends out a channel request to the base station. The mobile is immediately assigned to a SDCCH (Stand-alone Dedicated Control Channel, or sometimes just DCCH) and responds to the authentication request to ensure that it is legal. Again, the Cipher mode will turn off or on ciphering and when completed the MS will then do a call set-up. The system will assign the MS to a frequency and timeslot. Once the connection is made, the MS is in a traffic mode and the information goes back and forth as we saw earlier. (receives, transmits, measures the adjacent BCH and then repeats for the next frame).

Midamble and Training Bits

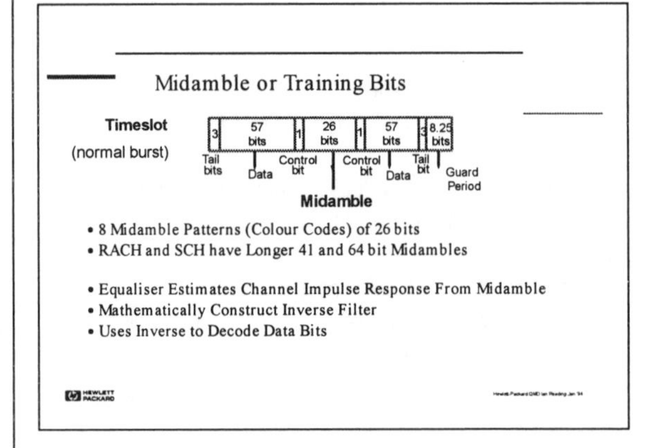

In the timeslot structures we saw, there was 26 bits referred to as either a mid-amble or a training sequence. For a normal burst this mid-amble will consist of 8 base station colour codes and these are numbered 0 through 7. They are 26 bits long. Another mid-amble or training sequence is used in the random access channel (RACH) and it is 41 bits long. There is also a 64 bit long sequence that is used on the SCH or synchronisation channel. Mid-ambles are placed in the centre of the burst to minimise the time difference from to any bit in the burst. The mid-amble has a number of different uses, the most important is equalisation to improve bit error rate. The mobile knows the mid-amble it should be receiving (part of the information the MS gets when assigned to a BS). This is a pre-defined sequence is 26 bits in the case of a traffic channel. It receives the mid-amble and compares the it to what it should have been. From the difference it can estimate the impulse response of the transmission path at that instant in time. Once it knows the impulse response it can mathematically calculate an inverse filter, it can apply this filter to the data bits on each side of the mid-amble and clean them up, reducing the chance of detecting a bit wrong. This is referred to as equalisation or the equaliser within the radio. Equaliser mechanisms are a closely guarded design feature of most mobiles. It's a key area of competition between mobile manufacturers.

Multipath and Equalisation

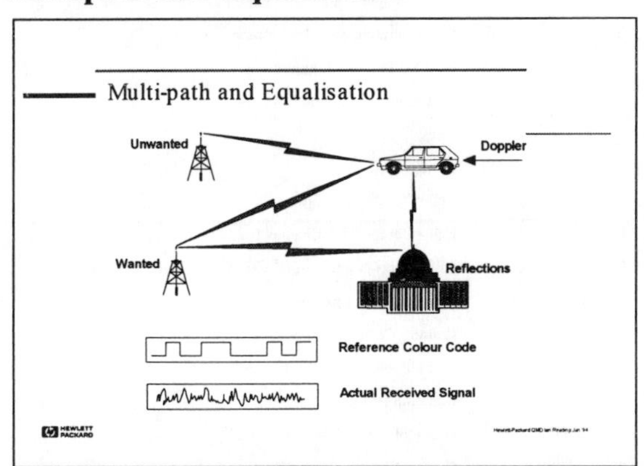

This is a simple example of a base station, a mobile, a direct transmission path, one large reflection from a mountain or a building and some slight frequency shift due to the movement of the mobile. The mobile knows it should be receiving a particular colour code for the base station. By calculating the characteristics of the RF path from the disturbance it caused to the midamble, the mobile's equaliser can more effectively reconstruct the other parts of the burst, reducing the chance of detecting a bit wrong.

DTX and DRX

DTX and DRX

DISCONTINUOUS RECEPTION (DRX)

- Idle mode to save battery power
- MS's divided into paging groups based on IMSI
  - Paging requests transmitted by network at predefined intervals

DISCONTINUOUS TRANSMISSION (DTX)

- Saves battery power
- Inserts comfort noise

Discontinuous Reception or DRx and Discontinuous Transmission or DTx are modes used by the mobile to save battery power. Mobiles are divided into paging groups (depending on their subscriber identity number). Because paging groups are only paged or called at pre-defined times, the mobile only needs to listen to see if the network has any messages or calls for it at these times. In DRx the mobile "goes to sleep" (conserving battery power), wakes up when it is supposed to listen for pages (dependent on its paging group) and then go back to sleep.

Discontinuous Transmission occurs if the user is just listening and not talking. In order to conserve battery power, the radio will not transmit a burst (transmitting is the biggest power drain) until there is information to be sent. When DTx occurs the system will insert "comfort noise" so that the caller on the other end will know that a link is still established.

Encryption

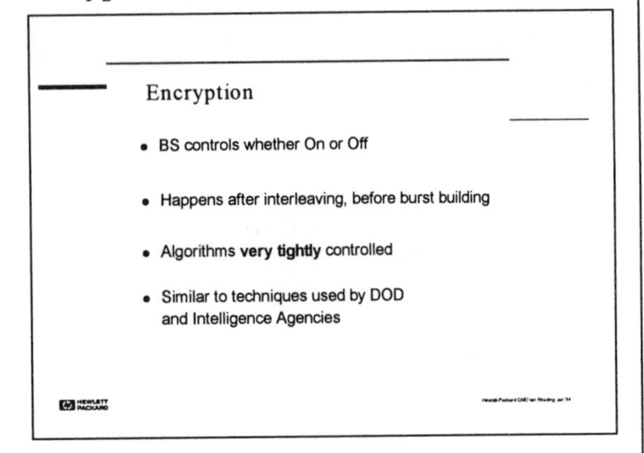

One of the key features of the GSM system is security. This comes about because of the use of encryption or ciphering. The base station controls whether ciphering is on or off. The encryption of the data occurs after the data is interleaved and formed into the eight data blocks. (before the actual bursts are built). The encryption algorithms are very tightly controlled. They are very similar to the techniques used by many of the top intelligence agencies around the world. The security of this is enhanced with the fact that the encryption algorithms change from call to call (even if it is deciphered for one call, the encryption on the next call will be different).

Abbreviations and Acronyms

A	See MS-PWR-CLASS
AB	Access Burst
AC	Administration Centre
ACCH	Associated Control Channel
ACM	Address Complete Message
ACS	Adjacent Channel/Carrier Suppression
ACU	Antenna Combinary Unit
AEF	Additional Elementary Functions
AFC	Automatic Frequency Control
AGC	Automatic Gain Control
AGCH	Access Grant Channel
ARFCN	Absolute Radio Frequency Channel Number
ARQ	Automatic Request for Retransmission
AUC	Authentication Centre
AUT	Authentication
BCC	Base station Colour Code
BCCH	Broadcast Control Channel
BCD	Binary Coded Decimal
BCF	Base Control Function
BCH	Broadcast Channel
BER	Bit Error Rate
BFI	Bad Frame Indication
Bm	Traffic channel for full rate voice coder
BN	Bit Number
BNHO	Barring all outgoing calls except those to Home PLMN
BS	Base Station
BS-AG-BLKS-RES	The number of blocks on each common control channel reserved for access grant messages
BS-BCCH-SDCCH-COMB	Combination of dedicated and associated control channels on the same physical channel
BSC	Base station Controller
BS-CC-CHANS	The number of basic physical channels supporting common control channels
BSCU	Base Station Controller Unit
BS-FREQ-NCELL	Frequency of the RF carrier on which the BCCH of an adjacent cell is transmitted
BSI	Base Station Interface
BSIC	Base Station Identity Code
BSIC-NCELL	BSIC of an adjacent cell
BS-PA-MFRMS	The number of multiframes between two transmissions of the same paging message to MSs of the same paging group
BSS	Base Station System
BSSAP	Base Station Application Part
BSSMAP	BSS Management Application Part
BSSOMAP	BSS Operation and Maintenance Application Part
BSSTE	Base Station Test Equipment
BTS	Bast Transceiver Station
C	Conditional
CA	Cell Allocation
CA-CN	Cell Allocation RF Channel Number
CA-NO	RF Channel Number of BCCH in a Particular Cell Allocation
CBCH	Cell Broadcast Channel
CC	Country Code
CCBS	Completion of Calls to Busy Subscribers
CCCH	Common control channel
CCCH GROUP	Group of MS in Idle Mode
CCPE	Control Channel Protocol Entity
CELL-BAR-ACCESS	Cell Access Barred
CELL-RESELECT-HYSTERESIS	RXLEV Hysteresis required for Cell reselection
CI	Cell Identify
CM	Connection Management
CMD	Command
CNIP	Calling Number Identification Presentation
CNIR	Calling Number Identification Restriction
COM	Complete
CONI	Connect Number Identify
CONN	Connect

CONP	Connect Number Identification Presentation
CRC	Cyclic Redundancy Check
CRE	Call RE-establishment procedure
CSPDN	Circuit Switched Public Data Networks
CU	Central Unit of a MS
CT	Channel Tester
CUG	Closed User Group
C/I	Carrier to Interference Ratio
D	Downlink
DB	Dummy Burst
DCF	Data Communications Function
DCCH	Dedicated Control Channel
DCN	Data Communication Network
DCPE	Data Connection Physical Endpoint
DCS	Digital Communication System
DET	Detach
DISC	DISConnect
DL	Data Link (layer)
DLCI	Data Link Connection Identifier
DLD	Data Link Discriminator
Dm	Control Channel (ISDN terminology applied to mobile service)
DMR	Digital Mobile Radio
DP	Dialled Pulse
DRM	Discontinuous Reception Mechanisms
DTAP	Direct Transfer Application Part
DTE	Data Terminal Equipment
DTMF	Dual Tone Multi-Frequency (signalling)
DRX	Discontinuous Reception
DTX	Discontinuous Transmission Mechanism
EA	External Alarms
Ec/No	Ratio of energy per modulating bit to the noise spectral density
EIR	Equipment Identify Register
ERR	ERRor
FA	Full Association
FB	Frequency correction Burst
FACCH	Fast ACCH
FACCH/F	Full-rate FACCH
FACCH/H	Half-rate FACCH
FCH	Frequency Correction Channel
FDMA	Frequency Division Multiple Access
FEC	Forward Error Correction
FER	Frame Erasure Rate
FFS	No Further Study
FN	Frame Number
FN-MAX	Maximum TDMA Frame Number
FS	Further Study

GB	Guard Bits
GMSC	Gateway Mobile Services Switching Centre
GMSK	Gaussian Minimum Shift Keying
GSA	GSM System Area
GSM	Global Syatem for Mobile communications
GSM PLMN	GSM Public Land Mobile Network
HANDO	Handover
HDLC	High Level Data Link Control
HLR	Home Location Register
HMSC	Home Mobile-services Switching Centre
HO-MARGIN	SDL Message name for Handover Margin
HPLMN	Home PLMN
HPU	Hand Portable Unit
HSN	Hop Sequence Number
I	Information (frames)
IAM	Initial Address Message
ICB	Incoming Calls Barred
ID	Identification
IDN	Integrated Digital Network
IE	Signalling Information Element
IF	Intermediate Frequency
IMEI	International Mobile station Equipment Identity
IMSI	International Mobile Subscriber Identity
INU	Interworking Unit
ISDN	Integrated Services Digital Network
IWF	Inter Working Function
I/Q	In-phase and Quadrature
K	Constraint Length of the Convolutional Code
Kc	Cipher Key
Ki	Key used to calculate SRES
Kl	Location Key
Ks	Session Key
LAC	Location Area Code
LAI	Location Area Identify
LAN	Local Area Network
LAP-Dm	Link Access Protocol on Dm Channel
L2R	Layer 2 Relay
LCN	Local Communication Network
LE	Local Exchange
Lm	Traffic channel with capacity lower than Bm.
LPC	Linear Prediction Coding (Voice Codec)
LR	Location Register
M	Mandatory
MA	Mobile Allocation
MACN	Mobile Allocation Channel Number

MAF	Mobile Additional Function
MAIO	Mobile Allocation Index Offset
MAP	Mobile Application Part
MCC	Mobile Country Code
MCI	Malicious Call Identification
MD	Mediation Device
MDL	(mobile) Management (entity) - Data Link (layer)
ME	Maintenance Entity
MEF	Maintenance Entity Function
MIC	Mobile Interface Controller
MM	Man Machine
MME	Mobile Management Entity
MMI	Man Machine Interface
MNC	Mobile Network Code
MPH	(mobile) Management (entity) - PHysical (layer) [primitive]
MS	Mobile Station
MSC	Mobile-services Switching Centre
MSCU	Mobile Station Control Unit
MS ISDN	Mobile Station ISND Number
MSL	Main Signalling Link
MSRN	Mobile Station Roaming Number
MS-RANGE-MAX	Mobile Station Range Maximum
MS-RXLEV-L	Lower Receive Level
MS-TXPWR-CONF	MS Transmitted RF Power Confirmation
MS-TXPWR-MAX-CCH	Maximum Allowed Transmitted RF Power for MSs to Access the System
MS-TXPWR-REQUEST	MS Transmitted RF Power Request. Parameter sent by the BS that commands the required MS RF Power Level.
MT	Message Transfer Part
MT	Mobile Termination
MTP	Message Transfer Part
MUMS	Multi User Mobile Station
NB	Normal Burst
NBIN	A parameter in the hopping sequence
NCELL	Neighbouring (adjacent) Cell
NDC	National Destination Code
NE	Network Element
NEF	Network Element Function
NER	Normal Error Rates
NF	Network Function
NM	Network Management
NMC	Network Management Centre
NMSI	National Mobile Station identification number
NMT	Nordic Mobile Telephone
NSAP	Network Service Access Point
NT	Network Termination
N/W	Network
O	Optional
OACSU	Off-Air-Call-Set-Up
OD	Optional for operators to implement for their aim
O&M	Operations & Maintenance
OCB	Outgoing Calls Barred
OMC	Operations & Maintenance Centre
OS	Operating System
OSI	Open System Interconnection
OSI RM	OSI Reference Model
PAD	Packet Assembly/Disassembly facility
PCH	Paging Channel
PD	Public Data
PCS	Personal Communications System
PDN	Public Data Networks
PH	Physical (layer)
PI	Presentation Indicator
PIN	Personal Identification Number
PLMN	Public Land Mobile Network
PLMN-PERMITTED	PLMN Permitted for handover purposes
PPE	Primitive Procedure Entity
PRBS	Pseudo Random Binary Sequence
Ps	Location Probability
PSPDN	Public Switched Public Data Network
PSTN	Public Switched Telephone Network
PTO	Public Telecommunications Operators
QA	Q (Interface) - Adapter
QAF	Q - Adapter Function
QOS	Quality of Service
R	Value of Reduction of the MS Transmitted RF Power relative to the maximum allowed output power of the highest power class of MS (A)
RA	Random Mode Request information field
RAB	Random Access Burst
RACH	Random Access Channel
RADIO-LINK-TIMEOUT	The timeout period for radio link failure
RAND	RANDom Number (authentication)
RBER	Residual Bit Error Rate (BER after errored frames removed)
REC	RECommendation
REL	RELease
REQ	REQuest
RES	RESponse (authentication)
RF	Radio Frequency
RFC	Radio Frequency Channel
RFCH	Radio Frequency Channel
RFN	Reduced TDMA Frame Number
RLP	Radio Link Protocol
RNTABLE	Table of 128 integers in the hopping sequence
RPE	Regular Pulse Excitation (Voice Codec)

RR	Radio Resource
RSE	Radio System Entity
RX	Receiver
RXLEV	Received Signal Level
RXLEV-MIN	The minimum received signal level at a MS from an adjacent cell for handover into that cell to be permitted
RXLEV-ACCESS-MIN	The minimum received signal level at a MS for access to a cell
RXLEV-NCELL	Received signal level of neighbouring (adjacent) cell
RXLEV-NCELL-[1-N]	The received signal level in adjacent cell
RXLEV-SERVING CELL	The received signal level in the serving cell
RXQUAL	Received Signal Quality
RXQUAL-SERVING-CELL	Received signal quality of serving cell
SABM	Set Asynchronous Balanced Mode
SACCH	Slow Associated Control Channel
SACCH/C4	Slow, SDDCCH/4 Associated, Control Channel
SACCH/C8	Slow, SDDCCH/8 Associated, Control Channel
DACCH/T	Slow, TCH-Associated, Control Channel
SACCH/TF	Slow, TCH/F-Associated, Control Channel
SACCH/TH	Slow, TCH/H-Associated, Control Channel
SAP	Service Access Points
SAPI	Service Access Point Indicator
SB	Synchronisation Burst
SCCP	Signalling Connection Control Part
SCH	Synchronisation Channel
SCN	Sub-Channel Number
SDCCH	Stand alone Dedicated Control CHannel
SDCCH/4	Stand alone Dedicated Control CHannel/4
SDCCH/8	Stand alone Dedicated Control CHannel/8
SDL	Specification Description Language
SE	Support Entity
SEF	Support Entity Function
SEG	Security Experts Group
SFH	Slow Frequency Hop
SI	Service Interworking
SID	Silence Descriptor
SIM	Subscriber Identification Module
SLTM	Signalling Link Test Message
SMG	Special Mobile Group
SMS	Short Message Service Support
SMSCB	Short Message Service Cell Broadcast
SN	Subscriber Number
SP	Signalling Point
SRES	Signal RESponse (authentication)
SS	Supplementary Resource Support
STP	Signalling Transfer Point
S/W	Software

TA	Terminal Adapter
TAC	Type Approval Code
TACS	Total Access Communication System
TAF	Terminal Adaptation Function
TB	Tail Bits
TC	Transaction Capabilities
TCAP	Transaction Capabilities Application Part
TCH	Traffic CHannel
TCH/F	A Full rate TCH
TCH/H	A Half rate TCH
TCH/FS	A Full rate Speech TCH
TCH/HS	A Half rate Speech TCH
TCH/F2.4	A Full rate data TCH (<2.4kbit/s)
TCH/F4.8	A Full rate data TCH (4.8kbit/s)
TCH/F9.6	A Full rate data TCH (9.6kbit/s)
TCH/H4.8	A Half rate data TCH (4.8kbit/s)
TCI	Transceiver Control Interface
TDMA	Time Division Multiple Access
TE	Terminal Equipment
TFA	Transfer Allowed
TFP	Transfer Prohibited
TMN	Telecommunication Management Network
TMSI	Temporary Mobile Subscriber Identity
TN	Timeslot Number
TPS	Three Part Service
TRX	Transceiver
TS	Time Slot
TSC	Training Sequence Code
TSDI	Transceiver Speech & Data Interface
TX	Transmitter
TXPWR	TX power level in the MS-TXPWR-REQUEST and MS-TXPWR-CONF parameters
U	Uplink
UA	Unnumbered Acknowledge
UI	Unnumbered Information (Frame)
UPD	UP to Date
VAD	Voice Activity Detection
VLR	Visitor Location Register
VPLMN	Visited PLMN
WS	Work Station

Reprinted from *IEEE Personal Communications*, Vol. 1, No. 3, Third Quarter, 1994, pp. 28-34.

The application of spread spectrum to PCS has become a reality

Reverse Link Performance of IS-95 Based Cellular Systems

ROBERTO PADOVANI

IS-95 "Mobile Station-Base Station Compatibility Standard for Dual-Mode Wideband Spread Spectrum Cellular System" is a digital cellular standard endorsed by the U.S. Telecommunications Industry Association/Electronic Industries Association (TIA/EIA) based on CDMA technology [1]. The first issue of *IEEE Personal Communications* presented a tutorial review of random-waveform access techniques such as those implemented in IS-95 based systems [2]. This article presents a tutorial review of the reverse link (subscriber to base station link) characteristics and its performance in terms of both coverage and capacity.

Reverse Link Waveform

The reverse link waveform generation is shown in Fig. 1. The speech coder[1] generates variable length packets according to the speech activity at a rate of one packet every 20 ms. The length of the packets in bits and the corresponding data rates are shown in Table 1.

The protocol supports a mix of services within the same high data rate packet. For example, the 171 information bits can be used to support simultaneous transmission of speech and data, speech and signaling, or data and signaling. A specific example is provided by the transmission of in-traffic signaling data, whereby in addition to the common technique of "blank-and-burst" where a speech frame is replaced by signaling information, a "dim-and-burst" approach can be used in which speech (or data) and signaling information share the same high data rate packet.

The packets formatted as described in Table 1 are then convolutionally encoded by a powerful rate 1/3 constraint length $K = 9$ code (thus the eight code tail bits of Table 1) with code

[1] *This discussion, which is limited to the physical layer aspects of the link, is also valid for data applications.*

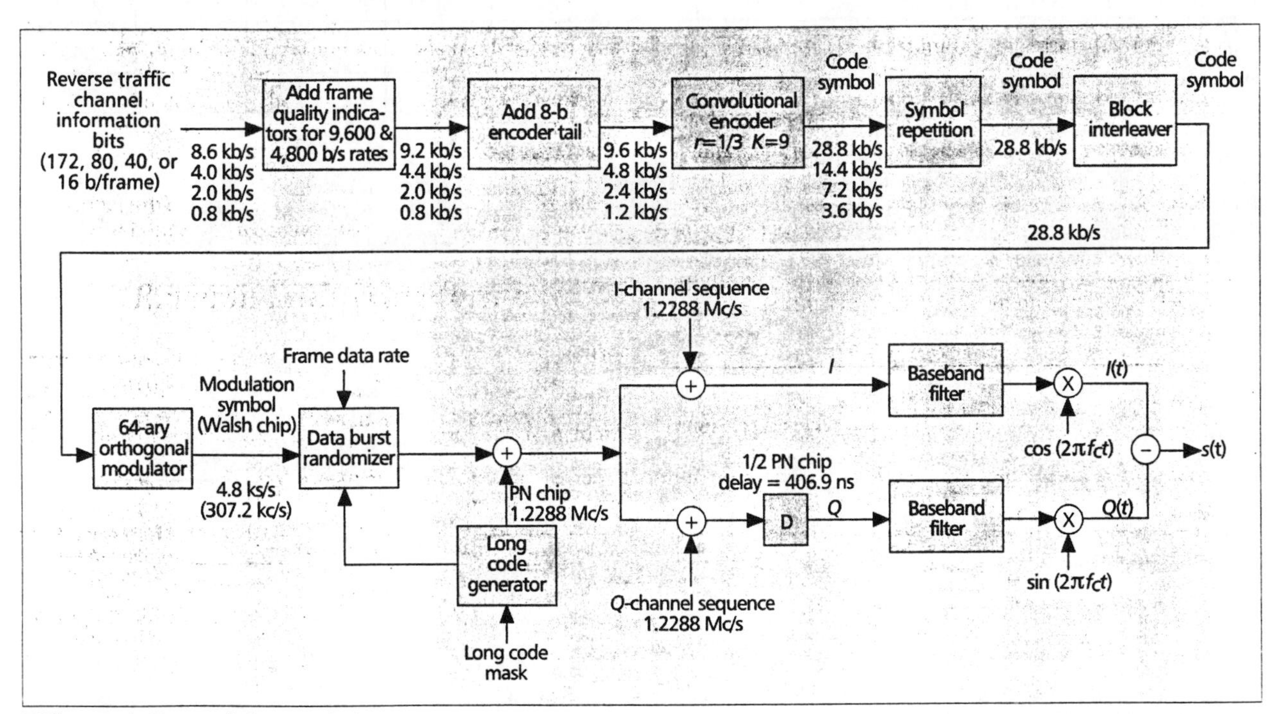

■ **Figure 1.** *Reverse link transmission.*

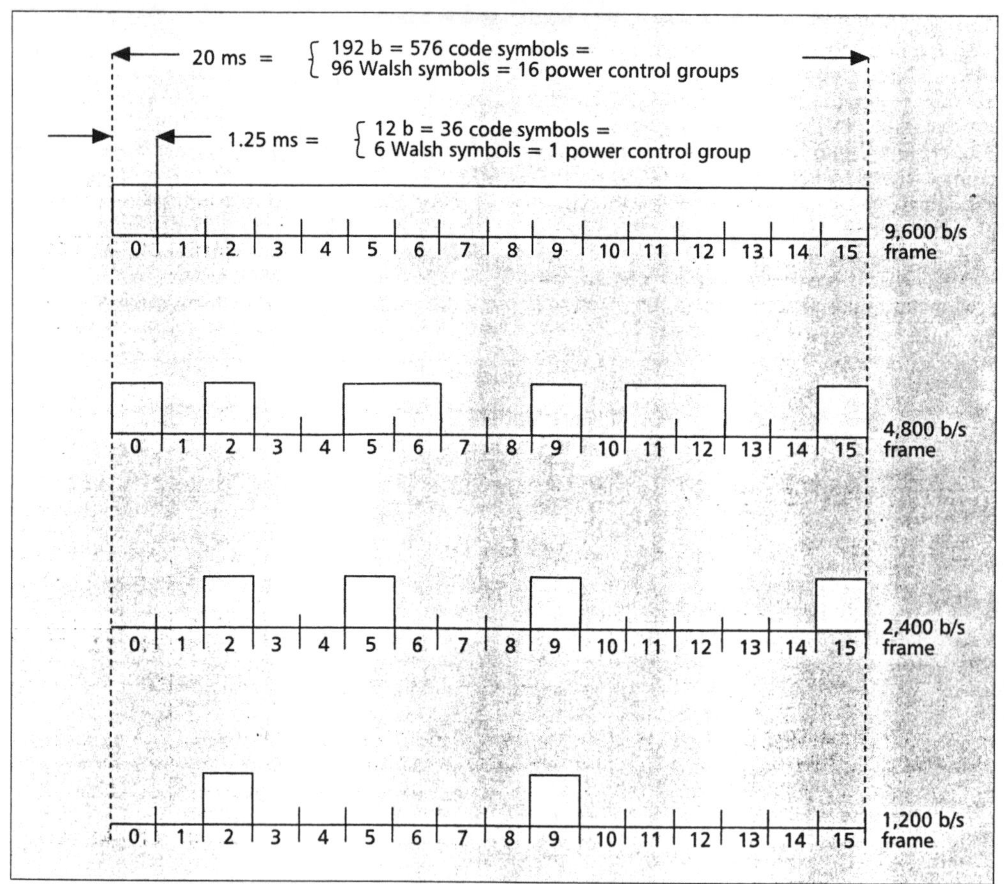

■ **Figure 2.** *Example of reverse link transmission for different data rates.*

generators 557, 663, and 711 (octal) and free distance $d_{free} = 18$. Thus all information bits are equally protected by the error correcting code. In addition, the two high rate packets are also protected by a 12-b and 8-b Cyclic Redundancy Code (CRC) for error detection, as shown in Table 1.

After convolutional encoding, the code symbols are interleaved by a block interleaver with span equal to one frame or 20 ms and modulated by a 64-ary orthogonal modulator. In other words, two information bits which after coding generate six code symbols, select after interleaving one of 64 orthogonal Walsh functions to be transmitted. Therefore, the orthogonal symbols have duration equal to $T = 208.3\,\mu s$ allowing the receiver to take advantage of the partial coherence of the channel.[2]

The final signal processing elements perform the direct-sequence spreading functions. First, the modulation symbols are spread by a subscriber unique PN sequence, i.e., the subscriber address (generated by a maximum length code of period 2^{42}-1) at a rate of 1,228,800 chips per second, i.e., 256 chips per orthogonal modulation symbol. Furthermore, the waveform is spread by a pair of PN codes (maximum length codes of period 2^{15}-1), common to all subscribers, in an OQPSK arrangement. The final waveform is then tightly filtered to generate a spectrum with 1.2288 MHz double-sided 3 dB bandwidth.

The actual transmission is then gated on-off pseudo-randomly[3] at 1.25 ms intervals, as shown in Fig. 2. This effectively reduces the transmit duty cycle from 100 percent for 9600 b/s transmission to 50 percent for 4800 b/s, 25 percent for 2400 b/s, and 12.5 percent for 1200 b/s. The overall result is the reduction of self-interference by a factor directly proportional to the average voice activity of the users, e.g., 40 percent voice activity corresponds to 4.0 dB interference reduction equivalent to a 2.5 time increase in the number of users.

Information [bits]	Signaling [bits]	CRC [bits]	Code tail [bits]	Total [bits]	Bit rate [b/s]
171	1	12	8	192	9,600
80	0	8	8	96	4,800
40	0	0	8	48	2,400
16	0	0	8	24	1,200

■ **Table 1.** *Packet structure.*

Two receiver structures are shown in Fig. 3. For early development and IS-95 field test validation the structure shown in Fig. 3a has been used. This structure implements a four-way RAKE receiver to demodulate the four strongest multipath components received on the two diversity antennas, as shown in Fig. 4. In this configuration a set of four ASIC's is employed each one implementing a complete reverse link demodulation path. The decision output from each of the active demodulators is then fed to an external microprocessor. The microprocessor combines the individual demodulator decisions, weighing each one by the relative strength of the respective multipath component, and generates a single stream of soft-decision inputs to the Viterbi decoder.

[2] *Coherence over the period T is guaranteed for vehicular applications in both cellular and PCS frequencies.*

[3] *This is the function performed by the data burst randomizer block in Fig 1.*

It is quite obvious that the above multipath diversity combining is sub-optimal since an independent decision on the transmitted orthogonal symbol is being made by each individual demodulator. The second demodulator architecture implemented in a new generation ASIC is shown in Fig. 3b. With this architecture, the multipath diversity receiver outputs are optimally combined by first combining the matched filter output energies. Note that this can be easily accomplished in this architecture since all four demodulators reside in the same device.

The second receiver architecture provides substantial performance improvements coupled with a much higher level of integration. Fig. 5 shows a comparison of the performance achieved in a Rayleigh channel with one, two, and four paths combining. Table 2 compares simulation results of the E_b/N_0 required for a 1 percent FER in a Rayleigh channel and two paths combining versus vehicle speed for the two architectures.

The final step in recovering the information

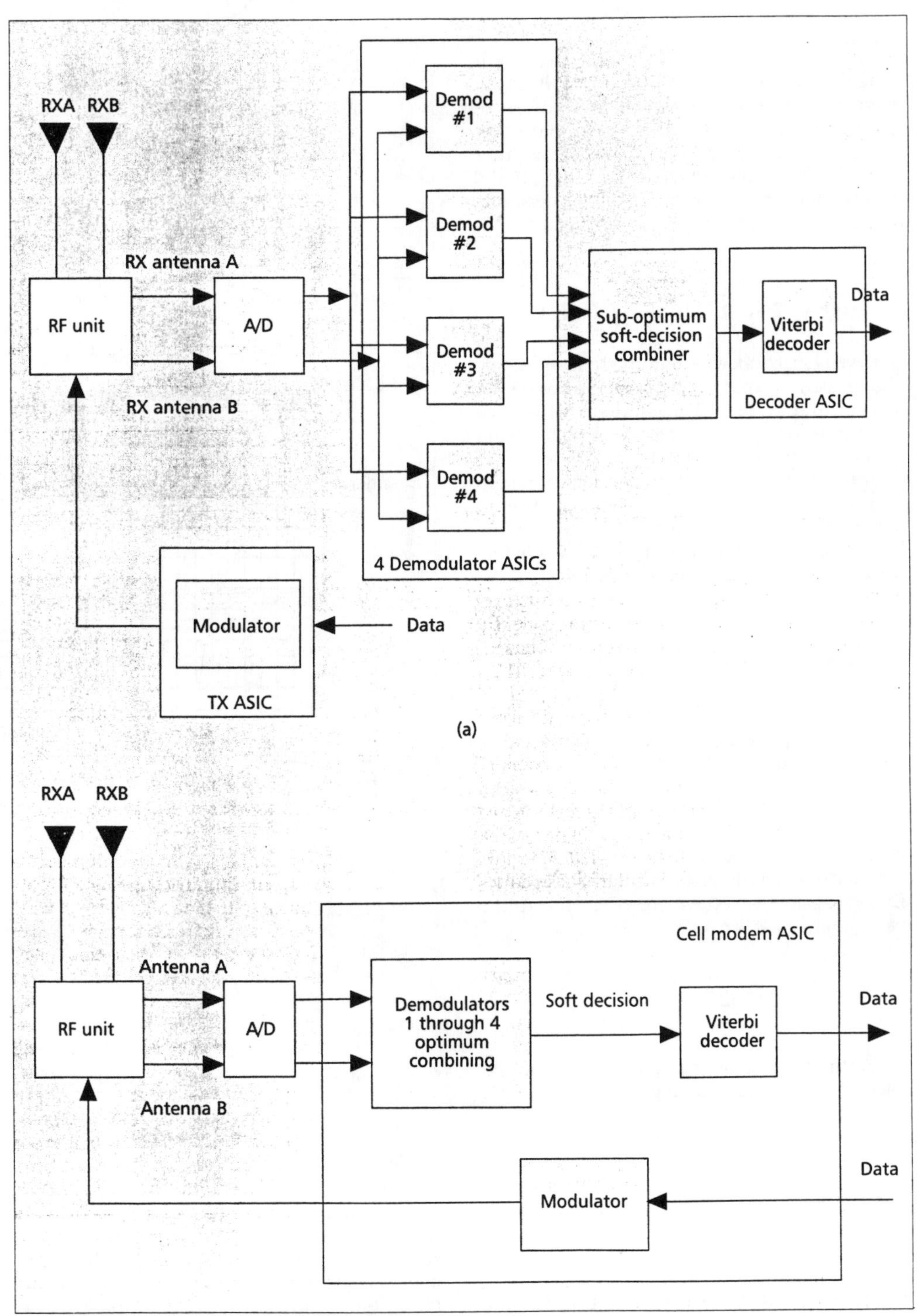

■ **Figure 3.** *a) Sub-optimal multipath diversity receiver architecture; b) optimal multipath diversity receiver architecture.*

	Eb/No [dB] architecture (a)	Eb/No [dB] architecture (b)
AWGN	4.0	2.6
8 [Km/h]	5.7	3.8
30 [Km/h]	7.9	5.5
100 [Km/h]	8.0	5.8

■ **Table 2.** *Comparison of Eb/No requirements in AWGN and Rayleigh fading for two-way combining and FER = 1 percent.*

consists of determining which of the four available packet types was actually transmitted. In order to accomplish this without any overhead penalty the received data is decoded by the Viterbi decoder four times once for each of the four hypothesis. After the multiple decoding, several metrics, such as CRC pass/fail and metrics obtained from the decoding process, are compared to select one final decoded packet.

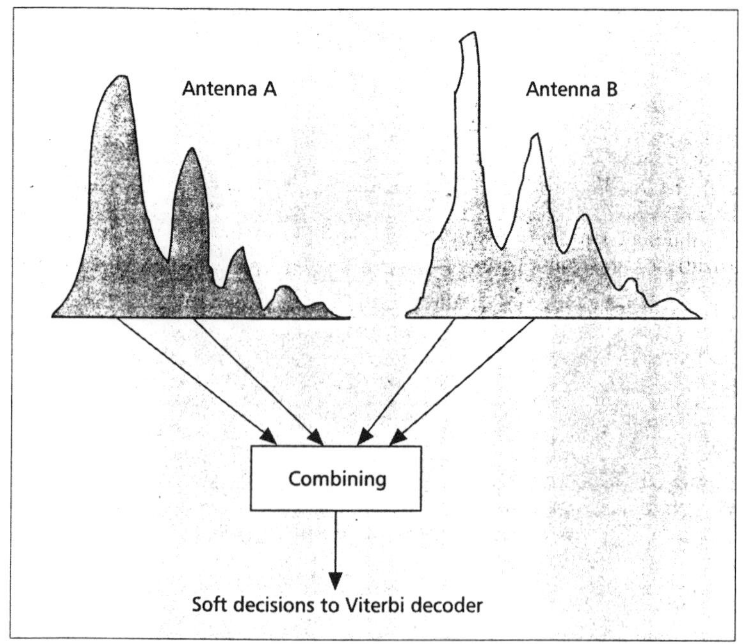

■ **Figure 4.** *Example of multipath and antenna diversity combining.*

Reverse Link Capacity

Several analyses have been carried out concerning the performance and the capacity of a CDMA cellular system [3-6]. In this section, we attempt to reconcile the results obtained from field tests with such analysis.

In the reverse link, one of the fundamental parameters to be analyzed and measured in determining the capacity is the total power received at the base station antennas. With M active users in one isolated sector the total received power C can be expressed as

$$C = N_o W + \sum_{i=1}^{M} \upsilon_i P_i \,, \tag{1}$$

where N_oW represents the background noise power in the bandwidth W, υ_i represents voice activity of the ith user, and P_i is the received power of the ith user. For IS-95, W=1.2288 MHz and $E\{\upsilon\}$ is taken to be equal to 0.4 (during the field tests the precise average voice activity of 40 percent was achieved by setting the mobile stations in test mode). By expressing the receive power relative to the background noise, we obtain

$$Z = \frac{C}{N_o W} = 1 + \sum_{i=1}^{M} \upsilon_i \frac{P_i}{N_o W} \tag{2}$$

Furthermore, the signal-to-noise plus interference ratio for a given user is given by,

$$\frac{E_{bi}}{N_o + I_o} = \frac{W}{R} \cdot \frac{\dfrac{P_i}{N_o W}}{1 + \dfrac{1}{W N_o} \displaystyle\sum_{j=1}^{M-1} \upsilon_i P_i} \,, \tag{3}$$

where R = 9,600 b/s. Combining Eqs. (2) and (3) and approximating M-1 with M in (3), we obtain

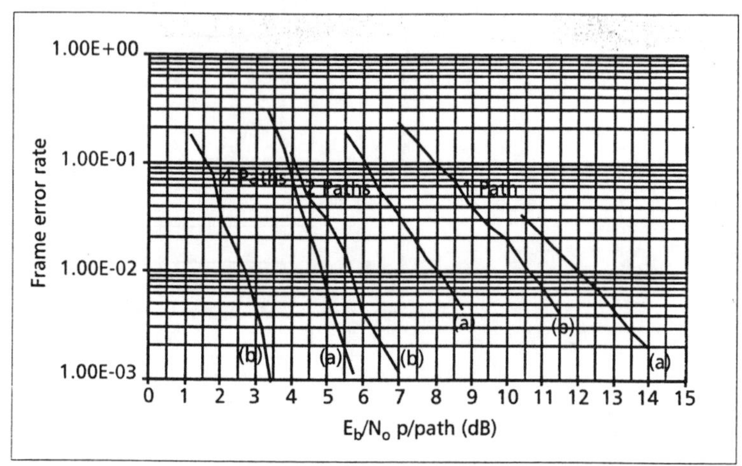

■ **Figure 5.** *Comparison of FER vs.* E_b/N_o *for the two architectures (a) and (b). Vehicle speed of 100 Km/h at a carrier frequency equal to 850 MHz. Results are shown for 1, 2, and 4 independent Rayleigh paths.*

$$Z \cong \frac{1}{1 - \dfrac{R}{W} \displaystyle\sum_{i=1}^{M} \upsilon_i \frac{E_{bi}}{N_o + I_o}} = \frac{1}{1 - X} \tag{4}$$

where

$$X = \frac{R}{W} \sum_{i=1}^{M} \upsilon_i \frac{E_{bi}}{N_o + I_o} \tag{5}$$

The signal-to-interference ratio is closely approximated by a lognormal distribution which has a mean εdB and a standard deviation σdB [2]. The voice activity υ is a quaternary random variable with mean $E\{\upsilon\}$. By central limit arguments, the variable X approaches a normal distribution.[4] Therefore, with $\beta = ln(10)/10$, we obtain

[4] *It should be clear that the approximation of Eq. (4) holds only if the probability of X > 1 is small.*

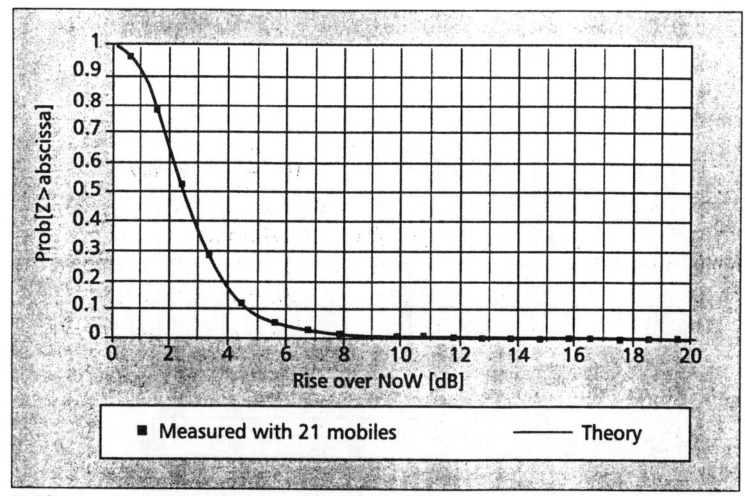

■ **Figure 6.** *Complementary cdf of the cell receiver power rise over background noise Z. Theoretical and measured results with 21 mobiles.*

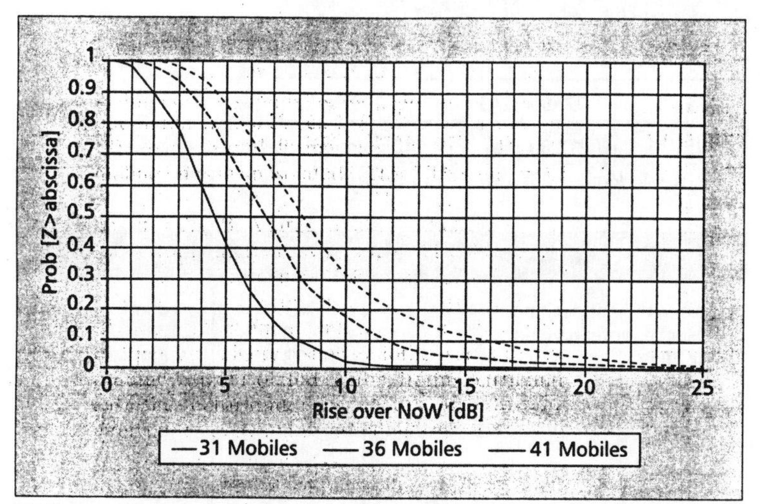

■ **Figure 7.** *Complementary cdf of the cell receiver power rise over background noise Z. Measure with 31, 36, and 46 mobiles.*

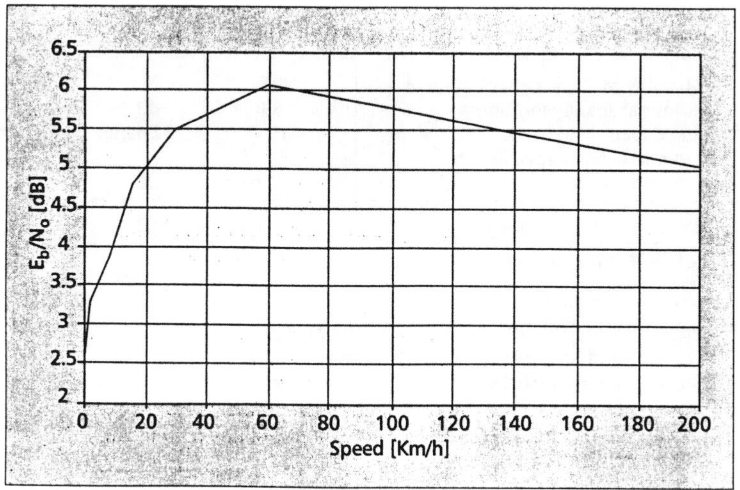

■ **Figure 8.** *Typical* E_b/N_o *performance vs. vehicle speed for 850 MHz links to achieve a FER = 1 percent. Rayleigh channel with two independent path (speed = 0 corresponds to the AWGN channel with two paths.*

$$E\{X\} = \frac{R}{W} M\, E\{v\}\, e^{(\beta\sigma)^2/2+\beta\varepsilon}, \tag{6}$$

$$\mathrm{Var}\{X\} = \left(\frac{R}{W}\right)^2 M e^{(\beta\sigma)^2+2\beta\varepsilon} \bullet \left[E\{v^2\}\, e^{(\beta\sigma)^2} - E\{v\}^2\right]. \tag{7}$$

Expressing Z in dB, i.e., $Z = -10 \log_{10}(1 - X)$, we can easily derive the distribution and density functions of the rise in dB over background noise, namely

$$P_z(z) = \frac{1}{\sqrt{2\pi}} \int_{-\infty}^{\frac{1-e^{-\beta z}-E(x)}{\sqrt{\mathrm{Var}(x)}}} e^{-y^2/2} dy \tag{8}$$

$$p_z(z) = \frac{1}{\sqrt{2\pi \mathrm{Var}(x)}} e^{-\frac{[1-e^{-\beta z}-E(x)]^2}{2\mathrm{Var}(x)}} \beta e^{-\beta z} \tag{9}$$

Returning now to the field test results, Fig. 6 compares the complementary cumulative distribution function of Z, i.e., $1 - P_Z(z)$, calculated from (8) and measured during a test involving $M = 21$ mobile units in an isolated sector.[5] The numerical values used for Eq. (8) are $M = 21$, $\varepsilon = 7.9$dB, and $\sigma = 2.4$dB. A large number of field tests performed in a variety of environments have shown similar performance to that of Fig. 6 with signal-to-noise requirements varying from $\varepsilon = 5$ dB to $\varepsilon = 8.5$ dB needed to maintain a frame error rate (FER) of 1 percent. Figure 7 shows another set of results obtained in an isolated sector. In this particular case the sector under test covers an eight-lane interstate freeway and all the mobiles involved in the test are placed on this freeway.

A distinct improvement in the signal-to-noise ratio requirement is obtained from low mobility users, e.g., pedestrian or in-building users, which are not experiencing the faster fading induced by vehicular motion. This is easily seen from Fig. 8 which shows the signal-to-noise required to achieve a 1 percent FER as a function of vehicle speed with one Rayleigh path per receive antenna. The shape of the curve shown in Fig. 8 is explained by the fact that at relatively low speeds power control is very effective in counteracting the slow fades whereas at higher speeds, where power control is not as effective in counteracting the fast fading, the effects of interleaving become increasingly beneficial.

Since each user is very accurately power controlled to the minimum signal-to-noise value necessary to achieve a given FER, low mobility users produce approximately one half the interference of high mobility users (typical signal-to-noise requirements for low mobility users is 4 to 5 dB). This has the obvious beneficial effect of increasing the capacity of the reverse link when the user population is a mix of high and low mobility users.

[5] *This particular test was conducted in a densely populated residential area in San Diego, California. The base station facilities and antennas subsystems were shared with those of the existing analog system.*

A complete analysis of the reverse link capacity must include the effects of other-cells interference and a model for the traffic load. These analysis have been carried out in detail in [4] and [6]. In the following we present a summary of the main result. The derivation uses the following parameters:

Median $E_b/(N_0+I_0)$: $\varepsilon = 7dB$ — This assumption combines the values measured in the field tests for high mobility users with the improvements achieved by the new receiver architecture described in the previous section.

$E_b/(N_0+I_0)$ standard deviation: $\sigma = 2.5dB$ — This value, induced by the closed loop power control, has been consistently measured in the field tests for high mobility users. Smaller values (σ = 1.5 dB) have been consistently measured in field tests for low mobility users.

Average voice activity: $E\{\upsilon\} = 0.4$, $Var\{\upsilon\} = 0.15$ — This is an estimate that should be refined as large commercial deployments are carried out and large population samples can be measured.

Other-cell interference fraction: $f = 0.55$ — This is the fraction of other-cell interference with respect to in-cell interference generated in an equally loaded network. This assumes a 4th power propagation law with 8 dB lognormal shadowing. Higher propagation exponents will reduce the factor f and lower exponents will increase it [6].

Traffic model — Poisson arrival rate of calls with parameter λ [calls/s] and exponential service time with parameter $1/\mu$ [s/calls], namely

$$Pr(k \text{ active users } / \text{ sector}) = \frac{(\lambda/\mu)^k}{k!} e^{-\lambda/\mu}$$

$$E\{k\} = \frac{\lambda}{\mu} \qquad \text{Var}\{k\} = \frac{\lambda}{\mu}$$

Given all the above assumptions we can now calculate the distribution of Z, i.e., the rise over background noise Z can now be expressed as

$$Z = \frac{C}{N_o W} = 1 + \sum_{i=1}^{k} \upsilon_i \frac{P_i}{N_o W} + \sum_{j}^{\text{other cells}} \sum_{i=1}^{k} \upsilon_i^{(j)} \frac{P_i^{(j)}}{N_o W} \tag{10}$$

The distribution of Z is then given by Eq. (10) where now the mean and variance of X are given by

$$E\{X\} = \frac{R}{W} \bullet \frac{\lambda}{\mu} \bullet E\{\upsilon\} \bullet e^{(\beta\sigma)^2/2+\beta\varepsilon} \bullet (1+f) \tag{11}$$

$$\text{Var}\{X\} = \left(\frac{R}{W}\right)^2 \bullet \frac{\lambda}{\mu} \bullet E\{\upsilon^2\} \bullet e^{2(\beta\sigma)^2+2\beta\varepsilon} \bullet (1+f) \tag{12}$$

From Eqs. (10), (12), and (13) it is straightforward to calculate the offered load in Erlangs for a given blocking probability. The blocking probability is defined as the probability of Z exceeding a given value z in dB [4]. Figure 9 shows the offered load per sector versus z for 1 percent and 2 percent blocking probabilities.

Operationally, a value of z = 10 dB and 2 percent blocking probability are a good compromise between offered load and coverage. As seen in Fig. 9, this corresponds to 19 Erlangs/sector or approximately 27 voice channels per sector. Notice that this result applies to a single CDMA frequency assign-

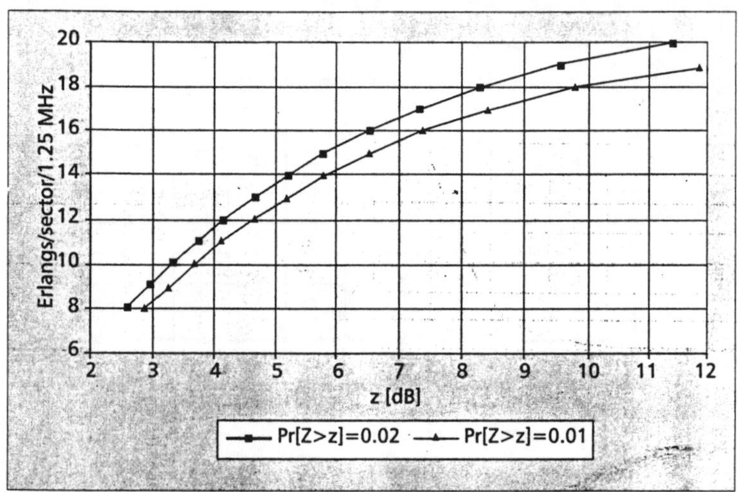

■ **Figure 9.** *Erlangs/sector/1.25 MHz with 1 percent and 2 percent blocking probabilities.*

System	Radio Capacity/ Sector	Erlang Capacity/ Sector
AMPS	19	12.3
IS-95	27•9 = 243	229
IS-95/AMPS	12.8 times	18.6 times

■ **Table 3.** *Reverse link capacity summary.*

Parameter	Value	Units
Power amplifier peak power	23.0	dBm
Subscriber antenna gain	0.0	dBi
Peak EIRP	23.0	dBm
Maximum isotropic path loss Single user — no shadowing	–155.2	dB
Base station antenna gain	12.0	dBi
Base station losses	–2.0	dB
Received signal strength	–122.2	dBm
Receiver noise figure	5.0	dB
Receiver noise density	–169.0	dBm/Hz
Data rate (9,600 b/s)	39.8	dB-Hz
Received Eb/No	7.0	dB
Probability of service at cell edge: Ps	90	%
LogNormal shadowing sigma	8.0	dB
Offered load	19	Erlangs/sector
Margin to achieve specified Ps with soft-handoff	7.7	dB
Maximum isotropic path loss	–147.5	dB

■ **Table 4.** *Link budget.*

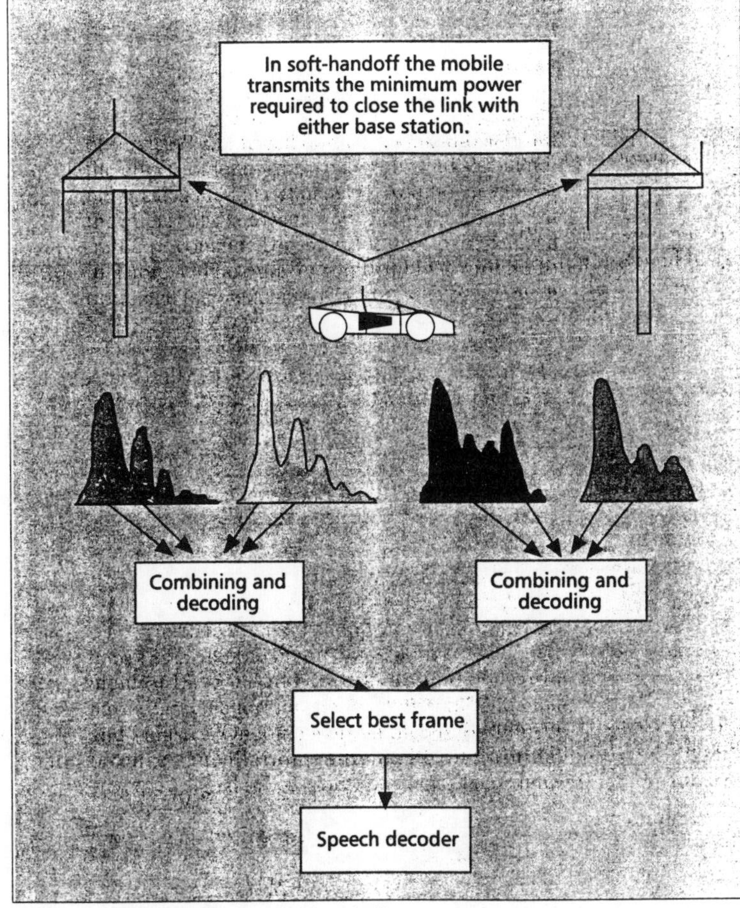

■ **Figure 10.** *Reverse link soft handoff.*

ment, i.e., one 1.25 MHz block. The US cellular spectrum allocation when completely converted to CDMA will utilize nine CDMA frequency blocks in a FDM arrangement.[6]

Finally Table 3 summarizes the capacity results and compares them to those of the analog AMPS system for the same 2 percent blocking probability.

[6] *The U.S. spectrum allocation for mobile station to base station transmission consists of 12.5 MHz for each of the two carriers servicing one market. The 12.5 MHz are not contiguous for either of the two carriers. The wireline carrier will deploy seven CDMA carriers in a block of 10 MHz and 2 more carriers in a block of 2.5 MHz. The non-wireline carrier will deploy eight CDMA carriers in a block of 11 MHz and one more carrier in a block of 1.5 MHz.*

Reverse Link Coverage

An additional advantage of the efficient waveform design implemented in the IS-95 standard is found in the increased coverage as compared to other digital cellular standards. In this section a simple link budget for an IS-95 based system is presented. The link budget is derived for Class III portable subscriber units with a minimum output power requirement of 200 mWatts [1].

Figure 10 shows the complete reverse link processing for a mobile station in soft handoff. In order to close the link, a mobile at the edge of coverage and in soft handoff with two (or multiple) base stations needs to transmit the minimum power required to achieve the desired SNR to either of the two (or multiple) base stations.

The margin required to achieve a given probability of service P_S at the cell border, taking into account the effects of lognormal shadowing and soft-handoff, is calculated in detail in [7]. Additionally, in [8] the effects of shadowing, soft-handoff, and offered traffic are combined to obtain the final margin shown in Table 4. The margin required by a mobile to overcome independent lognormal shadowing (with 8 dB sigma) between two base stations equally loaded at a level of 19 Erlangs is equal to 7.7 dB for a probability of service at the cell border equal to 90 percent. With the above assumptions, the maximum isotropic path loss that a portable unit can sustain equals 147.5 dB.

Conclusion

The application of spread spectrum to personal communication services has become a reality in just a few years. The approval of interim standard IS-95 and the imminent commercial deployment of IS-95 based networks around the world have sparked enormous interest in both academia and industry. Ongoing work in the areas of interference cancellation, antenna beam forming, advanced signal processing, and components integration will certainly improve on the performance reported in this tutorial presentation.

References

[1] TIA/EIA/IS-95 "Mobile Station-Base Station Compatibility Standard for Dual-Mode Wideband Spread Spectrum Cellular System," Telecommunication Industry Association, July 1993.
[2] A. J. Viterbi, "The Orthogonal-Random Waveform Dichotomy for Digital Mobile Personal Communications," *IEEE Personal Commun.*, vol. 1, no. 1, pp. 18-24, First Qrtr. 1994.
[3] K. S. Gilhousen *et al.*, "On the Capacity of a Cellular CDMA System," *IEEE Trans. Veh. Technol.*, vol. 40, pp. 303-311, May 1991.
[4] A. M. Viterbi and A. J. Viterbi, "Erlang Capacity of a Power Controlled CDMA System," *IEEE Jour. on Sel. Areas of Commun.*, vol. 11, pp. 892–890, Aug 1993.
[5] W. C. Y. Lee, "Overview of Cellular CDMA," *IEEE Trans. Veh. Technol.*, vol. 40, pp. 291-301, May 1992.
[6] A. J. Viterbi, A. M. Viterbi, and E. Zehavi, "Other-Cell Interference in Cellular Power-Controlled CDMA," *IEEE Trans. on Commun.*, vol. 42, no. 4, pp.1501-1504, April 1994.
[7] A. J. Viterbi *et al.*, "Soft Handoff Extends CDMA Cell Coverage and Increases Reverse Link Capacity," *IEEE Jour. on Sel. Areas of Commun.*, to be published.
[8] R. Vijayan, R. Padovani, and E. Zehavi, "The Effects of Lognormal Shadowing and Traffic Load on CDMA Cell Coverage," submitted for publications to *IEEE Trans. on Commun.*

Biography

Roberto Padovani received a Laurea degree from the University of Padova, Italy and M.S. and Ph.D. degrees from the University of Massachusetts, Amherst in 1978, 1983, and 1985, respectively, all in electrical and computer engineering. In 1984 he joined M/A-COM Linkabit, San Diego, where he was involved in the design and development of satellite communciation systems, secure video systems, and error-correcting coding equipment. In 1986, he joined QUALCOMM, Incorporated, San Diego, California, and is now vice president of system engineering in the Engineering Department. He has been involved in the design, development, and test of the CDMA cellular system which led to EIA/TIA IS-95 standard.

Reprinted from *IEEE Transactions on Communication*, Vol. 41, No. 4, Apr. 1993, pp. 559-569.

Performance of Power-Controlled Wideband Terrestrial Digital Communication

Andrew J. Viterbi, *Fellow, IEEE,* Audrey M. Viterbi, *Member, IEEE,* and Ephraim Zehavi, *Member, IEEE*

***Abstract*—Performance of a wideband multipath-fading terrestrial digital coded communication system is treated. The analysis has applications to a cellular system employing direct sequence spread spectrum CDMA with M-ary orthogonal modulation on the many-to-one reverse (user-to-base station) link. For these links, power control of each multiple access user by the cell base station is a critically important feature. This is implemented by measuring the power received at the base station for each user and sending a command to either raise or lower reverse link transmitted power by a fixed amount. Assuming perfect interleaving, the effect of the power-control accuracy on the system performance is assessed.**

I. Introduction

In terrestrial wireless transmission for cellular, mobile, and personal communication services, the channel is subject to time varying carrier amplitude and phase. While, in narrowband channels, the multipath propagation causes carrier signal cancellation (by equal-delay opposite phase paths) and consequent deep fades, wideband signals suffer from much shallower fading with multiple paths appearing as interference often separated in time. With spread spectrum code division multiple access (CDMA) techniques, the multiple paths can be demodulated individually by a "RAKE-type" receiver and combined prior to a decision, thus minimizing interference and mitigating fading further. Finally, in systems employing multiple base stations or cell sites, generally referred to as cellular systems, power control, which is desirable or required to control other-user interference, also serves to mitigate shadowing when the control is sufficiently rapid.

This paper deals with the many-to-one reverse links from multiple access users to a cellular base station. Assuming the classical multipath-fading model, for which there is ample experimental evidence for wideband signals [1], [2], and a conventional direct-sequence spread-spectrum waveform, error performance is determined for a convolutionally coded M-ary orthogonal modulation with noncoherent envelope-detector matched filter demodulation. Effects of shadowing, typically modeled as log-normally distributed multiplicative interference, is mitigated through the use of power control whose performance is also analyzed.

Paper approved by the Past Editor-in-Chief of the IEEE Communications Society. Manuscript received May 6, 1991; revised December 11, 1991.

A. J. Viterbi and A. M. Viterbi are with Qualcomm Inc., San Diego, CA 92121-1617

E. Zehavi was with Technion-Israel Institute of Technology, Department of Electrical Engineering, Technion City, Haifa 32000, Israel. He is currently with Quelcomm, Incorporated.

IEEE Log Number 9209478.

However, the main purpose of power control is to maintain all users' signal energy received at the base station nearly equal in the spread spectrum which is shared in common. Since each user's signal appears as interference to all other users, the total capacity of the system [3] depends on tight power control. Another benefit is that each user transmits only as much energy as is required to maintain a given level of error performance; hence, its overall transmitted energy is kept at a minimum, thus prolonging battery life in portable transmitters.

II. Characterization of Spread Spectrum CDMA Communication System and Multipath- Fading Channel

The end-to-end communication system block diagram (as shown in Fig. 1) can be subdivided into five components:

a) the encoder-interleaver-waveform generator;
b) the spreading processors, D/A converters (impulse modulation), shaping filters, upconverter, and power amplifier;
c) the multipath channel;
d) the downconverter, matched filter, despreader, A/D converter (sampler); and
e) The multiple demodulator-deinterleaver-soft decision decoder.

The spread spectrum modulator (b) is as described in [4]. The shaping filters are typically finite impulse response (FIR) digital filters to which the receiver filters are matched (mirror image impulse response and conjugate transfer function). Their purpose is to contain the transmitted energy in the allocated wideband spectrum of bandwidth W, which is the inverse of T_c, the spreading sequence switching time, generally called the chip time. The power amplifier has its output power controlled digitally as described in a later section.

To streamline the diagram, all two-component (I and Q) signals are represented as complex quantities and the corresponding (two-component) branches are shown as double lines. Thus, the quadrature spreading by multiplication by the independent pseudorandom (PN) sequences, the two-branch baseband filters $H(f)$, and the upconverting carrier multipliers are all treated in this way.

The classical model for multipath is a delay line, with delays corresponding to discernable paths each scaled by a complex random variable with Rayleigh distributed amplitude and uniformly distributed phase. The incremental delays $\tau_k - \tau_{k-1}$ must be larger than T_c, the inverse of the spread spectrum bandwidth W, in order for the multipath components to be distinguishable (those that are not distinguishable combine

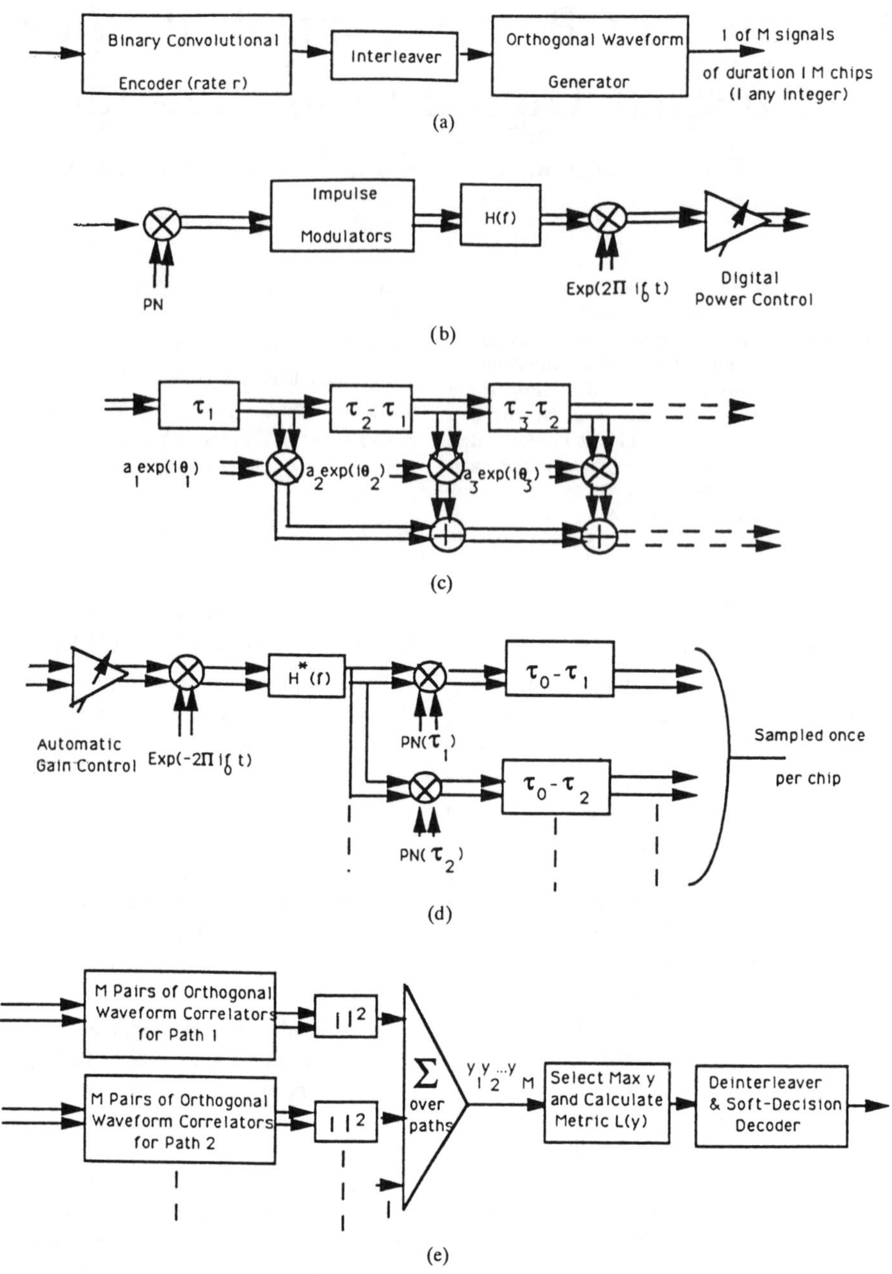

Fig. 1. Overall link system diagram. (a) Encoder–interleaver–waveform generator. (b) Spectrum spreaders–shaping filters–upconverter. (c) Multipath channel model. (d) Downconverter–matched filter–despreaders. (e) Envelope correlators–metric calculator–deinterleaver–decoder.

randomly, thereby giving rise to the Rayleigh distributed amplitudes and uniform phases of the scale factors). Thus, the overall complex transfer function of the m-component multipath channel is

$$\sum_{k=1}^{m} a_k \exp(i\Theta - 2\Pi i f_0 \tau_k).$$

It is assumed that the distinguishable individual path delays can be measured. Of course, in a mobile or otherwise changing propagation environment, these delays must be tracked; also, some will disappear after time while new ones appear. However, these will vary slowly compared to the bit rate, and hence can be accurately estimated and tracked. Amplitudes and phases, on the other hand, will vary more rapidly and are not estimated, but rather are taken to be independent, identically distributed variables constant at least over each transmitted orthogonal waveform.

The optimum receiver for such a channel [5] is as shown in Fig. 1(d). After downconversion and matched filtering by $H^*(f)$, the received signal is despread independently for each multipath component by multiplying by the quadrature spreading sequences delayed by an amount equal to the delay of that multipath component. Thus, the demodulator is effectively replicated m-fold, once for each significant multipath delay. To line up the multipath components after despreading, each must be delayed by a complementary amount $\tau_0 - \tau_k$, where

τ_0 is greater than any of the path delays. This is generally called a "RAKE" receiver [6]. Sampling and A/D conversion can be performed after the despreading or before, provided in the latter case that the received signal is oversampled to the accuracy of the delay measurements.

The form of the demodulator-decoder [Fig. 1(e)] depends on the choice of coder-modulator[1] [Fig. 1(a)]. Since amplitudes and phases can be assumed to remain essentially constant over a few bit times, this suggests the use of M-ary orthogonal waveforms whose time duration is no greater than the duration over which the fading amplitudes and phases remain virtually unchanged. Using binary waveforms only, such M-ary orthogonal waveforms can be generated using Hadamard or Walsh functions [4] whose duration is M chip times of the PN sequence, or any multiple thereof. Thus, if the input bit rate is R, and the code rate of the preceding encoder is r, using M chip times per orthogonal waveform implies that the bandwidth would be $W = 1/T_c = RM/(\log_2 M)r$. The bandwidth can be expanded by any multiple of this quantity by making each symbol of the Hadamard–Walsh function be I chips long. Thus, the ratio of bandwidth-to-bit rate, usually called the processing gain, is

$$W/R = IM/(\log_2 M)r, \qquad \text{where } I \text{ is any integer}.$$

As a specific example, with $M = 64$ orthogonal waveforms and $r = 1/3$, W/R can be made any multiple of 32; with $I = 4$, the processing gain is 128 or 21 dB.

III. Signal Statistics, Metric Calculation, and Soft-Decision Decoder Performance

The output of each envelope-detector correlator pair in Fig. 1(e) is a nonnegative random variable z with probability density function

$$P_C^{(z)} = e^{-z/(S+1)},$$

if the correlator corresponds to the correct signal sent

$$P_I^{(z)} = e^{-z}, \qquad \text{if the correlator corresponds to one of the } M-1 \text{ other (incorrect) signals}$$

where it is assumed that automatic gain control (AGC) has normalized the noise variance to unity, and S is the normalized mean received energy per path. Thus, if $\overline{E}$ is the total received energy per orthogonal waveform summed over all m equal average energy paths, and N_0 is the additive noise density, including other-user spread signals [3], [4], then

$$S = (\overline{E}/N_0)/m.$$

Assuming, as we have, that all m paths are mutually independent, the sum of all m paths for the correct signal correlator $y = \sum_{k=1}^{m} z_k$ has probability density which is the m-fold convolution of that for each path, $p_C(z)$,

$$f_C^{(m)}(y) = \frac{y^{m-1}e^{-y/(S+1)}}{(m-1)!\,(S+1)^m}. \tag{1a}$$

For each of the incorrect signal correlators, it is the m-fold convolution of $p_I(z)$

$$f_I^m(y) = \frac{y^{m-1}e^{-y}}{(m-1)!}. \tag{1b}$$

For the sake of comparison, we also consider the case of only one *unfaded* path, for which it is well known that, for fixed signal energy E,

$$f_C^U(y) = e^{-(y+E/N_0)} I_0\left(2\sqrt{(E/N_0)y}\right) \tag{2a}$$

$$f_I^U(y) = e^{-y}. \tag{2b}$$

The probability that an error is made by the maximum likelihood detector, which results when one of the $M-1$ random variables y_j corresponding to an incorrect signal exceeds that for the correct signal, as derived from (1) or (2), is well known [7], [8]. However, the soft decision decoder operates not only on the decision of which y_j is maximum, but also on their relative magnitudes. To reduce complexity (of both implementation and analysis), we consider only the magnitude of the maximum correlator output

$$y = \text{Max } y_j.$$

The input to the soft decision decoder is this value along with the index of y, which is a binary sequence of length $\log_2 M$ representing the (hard) decision symbols. Note that upon deinterleaving, the value of y (soft decision) must be attached to each (deinterleaved hard decision) symbol of the index, which shall be denoted x.

The optimum choice of binary metrics, based only on the value of y and any one of the binary symbols x to which it pertains, is obtained from the two joint likelihood functions of y and x, given that it did and did not correspond to what was sent. These are, respectively,

$$p(y, x|x) = f_C(y)F_I(y)^{M-1} + (M/2-1)f_I(y)F_C(y)F_I^{M-2}(y), \tag{3a}$$

$$p(y, x|\overline{x}) = (M/2)f_I(y)F_C(y)F_I^{M-2}(y) \tag{3b}$$

where $x = 0$ or 1 and $\overline{x}$ is its complement,[2] and $f_C(y)$ and $f_I(y)$ are given by (1) or (2). $F_C(y)$ and $F_I(y)$ are the corresponding distribution functions, their indefinite integrals. Note that $f_C(y)$ and $F_C(y)$ depend also on $\overline{E}$ and E in the faded and unfaded cases, respectively

The expression (3b) follows from the fact that the density function of the largest of $(M-1)$ incorrect signal measured

[1] Use of a binary convolutional encoder with interleaving prior to M-ary orthogonal waveform selection is superior to encoding the M-ary orthogonal waveforms directly without binary symbol interleaving, as demonstrated in Appendix II.

[2] Alternately, if we take x to be $+1$ or -1 (or any multiple thereof) and $\overline{x}$ its negative, then the joint densities could be denoted as the four one-dimensional densities $p(\pm y|x = \pm 1)$, where $y > 0$.

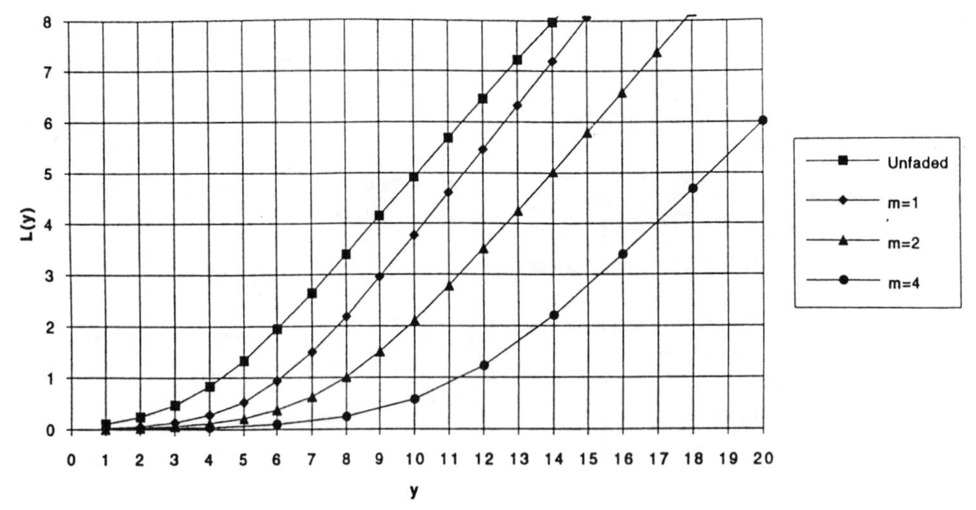

Fig. 2. Log-likelihood ratio ($E/N_o = 10$ dB; M = 64).

energies, when it is greater than the correct signal measured energy, is $(M-1)f_I(y)F_I^{M-2}(y)F_C(y)$, and conditioned on this, a symbol error ($\overline{x}$ mistaken for x) occurs with probability $M/[2(M-1)]$.

The expression (3a) follows from the fact that a symbol can be correct in either of two disjoint events: if the correct decision is made, in which case the correct signal energy is greatest with density $f_C(y)F_I(y)^{M-1}$; or if the incorrect decision is made (with density as given above) but the binary symbol is nonetheless correct, which occurs with probability $(M/2-1)/(M-1)$.

Given the observable y with density function given by (3a) and (3b), the optimum metric is the log-likelihood function, formed from the ratio of (3a) and (3b).

$$\exp[L(y)] \triangleq \frac{p(y,x|x)}{p(y,x|\overline{x})} = \left(1-\frac{2}{M}\right) + \frac{2}{M}\,\frac{f_C(y)F_I(y)}{f_I(y)F_C(y)}. \tag{4}$$

$L(y)$ is plotted in Fig. 2 for $M = 64$ and $E/N_0 = 10$ dB for the unfaded signal case, and for the m-path fading case for $m = 1, 2$ and 4 with $\overline{E}/N_0 = 10$ dB.

Using this ideal soft metric, which requires knowledge of E in the unfaded case and $S = (\overline{E}/N_0)/m$ in the faded cases, the performance parameter Z for a soft decision decoder (see Appendix I) then becomes

$$\begin{aligned} Z &= 2\int_0^\infty \sqrt{p(y,x|x)p(y,x|\overline{x})}\,dy \\ &= 2\int_0^\infty p(y,x|\overline{x})\sqrt{\exp[L(y)]}\,dy \\ &= M\int_0^\infty f_I(y)F_C(y)F_I^{M-2}(y) \\ &\quad \cdot \left[\left(1-\frac{2}{M}\right)+\frac{2}{M}\,\frac{f_C(y)F_I(y)}{f_I(y)F_C(y)}\right]^{1/2} dy. \end{aligned} \tag{5}$$

Using (1) or (2) and (5), for a range of E/N_0 and $\overline{E}/N_0 = mS, Z$ was integrated numerically for the unfaded case and for multipath-fading with $m = 1, 2$, and 4.

From Z, according to Appendix I, we obtain the following parametric relationship between required $\overline{E}/N_0$ and code rate r for any $\alpha = r_0/r > 1$.

$$\frac{1/r}{\alpha} = 1/r_0 = \frac{1}{1-\log_2(1+Z)} \tag{6}$$

$$\begin{aligned} \frac{\overline{E}_b/N_0}{\alpha} &= \frac{\overline{E}/N_0}{\alpha r \log_2 M} = \frac{\overline{E}/N_0}{r_0 \log_2 M} \\ &= \frac{\overline{E}/N_0}{[1-\log_2(1+Z)]\log_2 M}, \\ &\text{where } \overline{E}/N_0 = E/N_0, \text{unfaded} \\ &\text{and } \overline{E}/N_0 = mS, \text{ multipath-fading}. \end{aligned} \tag{7}$$

This relation is plotted as solid curves in Fig. 3 for the unfaded cases and for fading with $m = 1, 2$, and 4 multipath components, and for $M = 64$ orthogonal waveforms.

Since α is established from the error probability requirement ($\alpha = 0.6$ dB and 0.8 dB for the rate 1/2 and rate 1/3 codes, respectively[3]), required $\overline{E}_b/N_0$ can be obtained from Fig. 3 by backing off α (in dB) from $1/r$ (in dB), and, from this finding the corresponding $(\overline{E}_b/N_0)/\alpha$ (in dB) and finally adding α (in dB). With the α values noted, this represents E_b/N_0 requirements of 4.9 dB and 5.2 dB for the unfaded case at rate 1/2 and 1/3, respectively. For the three faded cases at rate 1/2, $\overline{E}_b/N_0$ requirements range from 7.0 dB to 7.9 dB; at rate 1/3, $\overline{E}_b/N_0$ required ranges from 7.1 dB to 7.5 dB. Hence, rate 1/3 is the better choice globally.

IV. Performance Bounds for Integer Metrics

While the optimum metric is the log-likelihood function $L(y)$ given in (4) and shown in Fig. 2 for various fading and unfaded channels, practical implementation requires quantization of the maximum energy y. Scaling any $L(y)$ by an arbitrary amount does not affect decoder performance in any way, although quantization obviously does. A reasonable approximation based on an 8-level quantizer, which uses integer

[3] For error probabilities of approximately 0.01 for 200-bit frames.

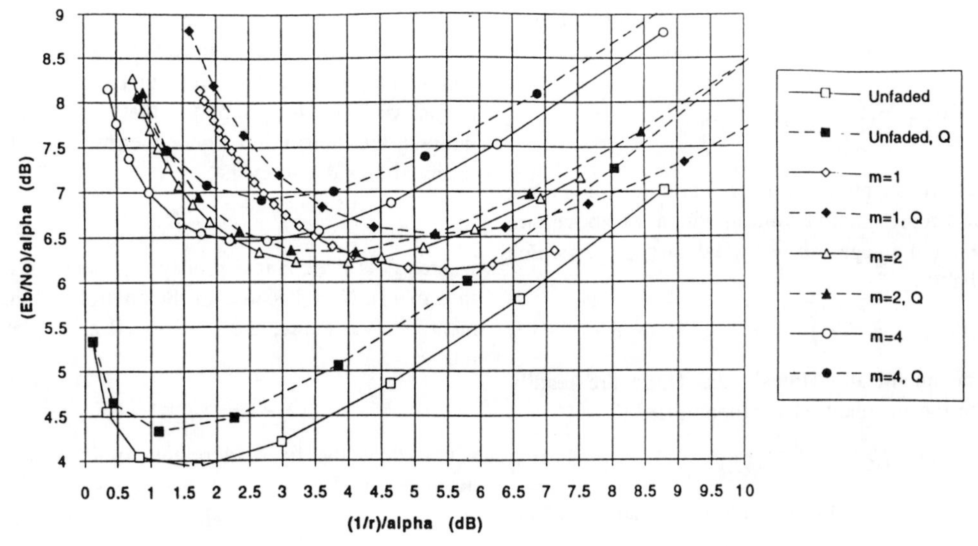

Fig. 3. Performance bound for m-path fading propagation ($M = 64$).

representative values, utilizes the quantization thresholds at values of y : $\theta_0 = 0$, $\theta_1 = 7$, $\theta_2 = 9$, $\theta_3 = 11$, $\theta_4 = 12$, $\theta_5 = 13.5$, $\theta_6 = 15$, $\theta_7 = 16.5$, $\theta_8 = \infty$. This best approximates the metric $L(y)$ for the two-path Rayleigh fading case but, as we shall see, performance is in no case degraded by more than 0.5 dB.

Quantization converts the continuous binary-input channel described by $p(y, x|x)$ and $p(y, x|\overline{x})$, into a binary-input discrete symmetric-output channel with transition probabilities

$$P_{+k} = \int_{\theta_{k-1}}^{\theta_k} p(y, x|x)\, dy, \qquad k = 1, 2, \cdots, 8, \tag{8a}$$

$$P_{-k} = \int_{\theta_{k-1}}^{\theta_k} p(y, x|\overline{x})\, dy, \qquad k = 1, 2, \cdots, 8. \tag{8b}$$

Note that, in this formulation, hard decisions correspond to a single level ($k = 1$) with plus and minus unity metric and $\theta_0 = 0$, $\theta_1 = \infty$.

Substituting (3a) and (3b) into (8a) and (8b) and integrating by parts yields

$$P_{+k} = \frac{M/2 - 1}{M - 1} F_C(y) F_I^{M-1}(y) |_{\theta_{k-1}}^{\theta_k} + \frac{M/2}{M - 1} \int_{\theta_{k-1}}^{\theta_k} f_C(y) F_I^{M-1}(y)\, dy \tag{9a}$$

$$P_{-k} = \frac{M/2}{M - 1} F_C(y) F_I^{M-1}(y) |_{\theta_{k-1}}^{\theta_k} - \frac{M/2}{M - 1} \int_{\theta_{k-1}}^{\theta_k} f_C(y) F_I^{M-1}(y)\, dy. \tag{9b}$$

Coded performance with an ideal interleaver can now be evaluated based on the Chernoff bound for this quantized channel.

$$Z = \operatorname*{Min}_{s>0} E(e^{-s\hat{y}})$$

where $\hat{y}$ is the quantized energy metric which takes on integer value $\pm(k-1)$, where $k = 1, 2, \cdots, 8$. Thus, letting $w = e^{-s}$ $(w < 1)$,

$$Z = \operatorname*{Min}_{w<1} E(w^{\hat{y}}) = \operatorname*{Min}_{w<1} \sum_{k=1}^{8} \left(P_k w^{k-1} + P_{-k} w^{-(k-1)}\right) \tag{10}$$

for the soft quantizer[4] considered here.

As in the ideal unquantized case, $\overline{E}_b/N_0$ and r are related by (6) and (7) through Z, given by (10) in this case. The results are plotted as dotted lines in Fig. 3, where they are seen to degrade performance for the unquantized cases by never more than 0.5 dB.

V. Power Control With Ideal Interleaving and Lognormal Shadowing

Given the integer metric demodulator described in the last section, a natural method of providing power control for a channel, in which the average energy varies slowly, is to take the quantized maximum M-ary signal demodulator energies (as before normalized by AGC so that the average noise-only energy is unity), sum them over N signal periods, and compare these with a threshold. If the threshold is exceeded, a command is sent via a (low data rate) command channel from the cell base station to lower the energy a given amount in decibels and otherwise to raise it by the same amount. It is assumed that the command to raise or lower is sent uncoded so as to minimize its delay, and hence is subject to a higher error rate.

Thus, assuming a given average energy-to-noise level $\overline{E}$ and summing N successive quantized normalized maximum energies

$$Y = \sum_{n=1}^{N} \hat{y}(n)$$

[4] Note that for hard decisions (two levels with metrics ± 1) $Z = 2\sqrt{P_1 P_{-1}}$ with $\Theta_0 = 0$ and $\Theta_1 = \infty$.

and comparing with the threshold ϕ, results in a probability of sending a command to lower energy equal to

$$P_d = \Pr\left(\sum_{n=1}^{N} \hat{y}(n) > \phi\right). \tag{11}$$

Each term of the sum is a discrete random variable which takes on the quantizer's representative values, which, as above, are taken to be the eight integers $k - 1 = 0$ through 7, which occur with probability

$$Q_k = \Pr(\theta_{k-1} \leq \hat{y} < \theta_k) \qquad k = 1, 2, \cdots, 8 \tag{12}$$

where θ_k are the quantization thresholds. These are readily determined to be the unconditioned probabilities

$$Q_k = F_C(y) F_I^{M-1}(y)\Big|_{\theta_{k-1}}^{\theta_k} \tag{13}$$

which are thus related to the transition (conditional) probabilities of (9) by

$$Q_k = P_{+k} + P_{-k}.$$

Now letting

$$Q(w) \triangleq \sum_{k=1}^{8} Q_k w^{k-1} \tag{14}$$

be the moment generating function of each $\hat{y}(n)$, the moment generating function of the sum $Y = \sum_{n=1}^{N} \hat{y}(n)$ is

$$[Q(w)]^N = Q_1^N w^0 + \cdots + Q_8^N w^{7N}.$$

Finally, to compute (11), we must sum all the coefficients of this polynomial for terms whose powers have integer value greater than ϕ. We denote this as

$$P_d = \left\{[Q(w)]^N\right\}_{\phi+}. \tag{15}$$

Clearly, the probability of sending a command to increase power is

$$P_u = 1 - P_d. \tag{16}$$

Finally, as noted, errors can occur in the command link with probability γ. Since this is independent of the above expressions, the overall probability that the command to increase power is received by the transmitter (correctly or incorrectly) is

$$P_U = (1-\gamma)P_u + \gamma P_d = (1-\gamma) - (1-2\gamma)P_d, \tag{17}$$

while the command to decrease power is received by the transmitter with probability

$$P_D = (1-\gamma)P_d + \gamma P_u = \gamma + (1-2\gamma)P_d. \tag{18}$$

Thus, P_U and P_D can be determined from (13), (14), and (15). Throughout the following, the erroneous command probability γ will be taken equal to 0.05.

As assumed above, power control commands are determined on a measurement interval comprising N contiguous M-ary transmissions, over which the average received signal energy is taken to be constant. While it would be desirable to take action immediately, the processing delay (plus the generally small propagation delay) introduces a lag (or latency) in the control loop of one measurement interval (N M-ary symbol transmissions). Now suppose that at the rth measurement interval, the transmitted signal energy is T_r while the average received signal energy is

$$\overline{E}_r = T_r - L_r \tag{19}$$

where L_r is the channel propagation loss, all quantities being in decibels. Now because of the control and the one measurement interval delay, the transmitted energy at the $(r+1)$th interval is

$$T_{r+1} = T_r + C(\overline{E}_{r-1})\Delta \tag{20}$$

where Δ is the fixed increment of increase or decrease in decibels and

$$C(\overline{E}_{r-1}) = \begin{cases} +1 & \text{if an up command was received, with probability } P_U \\ -1 & \text{if a down command was received, with probability } P_D \end{cases} \tag{21}$$

where P_U and P_D are given by (17) and (18), which in turn depend on $\overline{E}$ through (13)–(15).

Further combining (19), (20), and (21) yields

$$\overline{E}_{r+1} = \overline{E}_r + C(\overline{E}_{r-1})\Delta - (L_{r+1} - L_r). \tag{22}$$

We assume that the propagation loss, including distance and fading induced losses, exhibits independent random increments (as a Brownian motion). This would lead to unbounded variance if the control were not present. On the other hand, for a mobile user who may travel rapidly over a variety of terrains and is subject to blockages, the independent increment model is justifiable. In any case, with the control present, the received energy's variance is always finite.

The nonlinear difference equation (22) is reminiscent of similar differential equations for continuous control systems utilizing "bang-bang" control for which closed-form solutions for the probability densities are known. However, with discrete time and the closed-loop delay (of two intervals), standard analyses do not apply. We have instead resorted to simulation with the independent-increment driving function $(L_{r+1} - L_r)$ taken to be Gaussian (in decibels) with standard deviation σ, in accordance with the log-normal shadowing assumption. Note that the probabilities of up and down commands depend on the instantaneous $\overline{E}$ through (13) where F_C is a function of $\overline{E}$ as is evident from (1a) and the preceding definition of S, or (2a).

The results are shown in Fig. 4 for multipath-fading with $m = 1, 2$, and 4, as well as for the unfaded case. In each case, $N = 6$ and the standard deviation of the independent increments $L_r - L_{r-1}$ is taken to be 0.5 dB as is the energy increment Δ.

Also, in each instance, the probability that an up–down command is received incorrectly is taken to be $\gamma = 0.05$. The threshold ϕ in (11) and (15) is set so as to achieve a mean $\overline{E}/N_0 = 10$ dB, since by varying ϕ we may vary the mean of the distribution at will. It is noteworthy that the probability density of $\overline{E}/N_0$ does not differ much among the four cases.

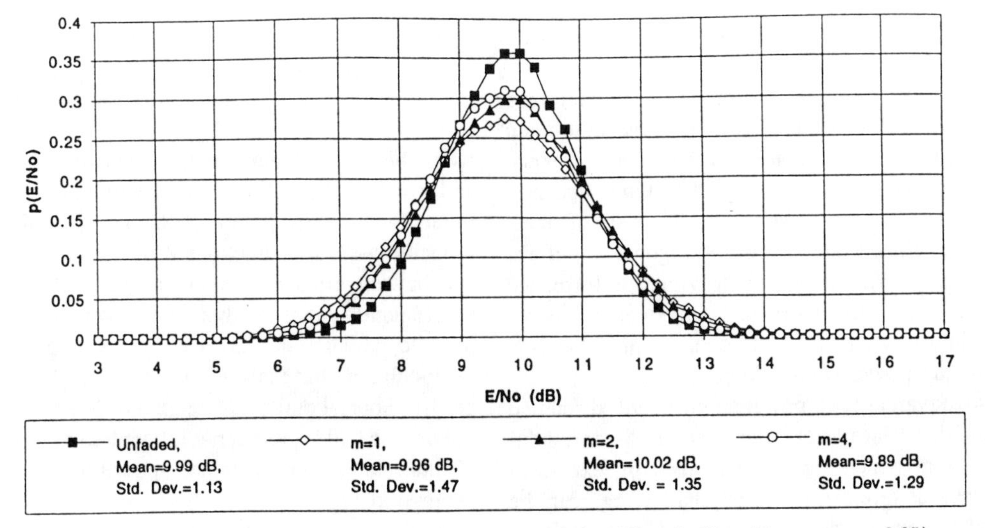

Fig. 4. Steady-state E_b/N_0 probability densities from simulation ($N = 6$, $M = 64$, gamma $= 0.05$).

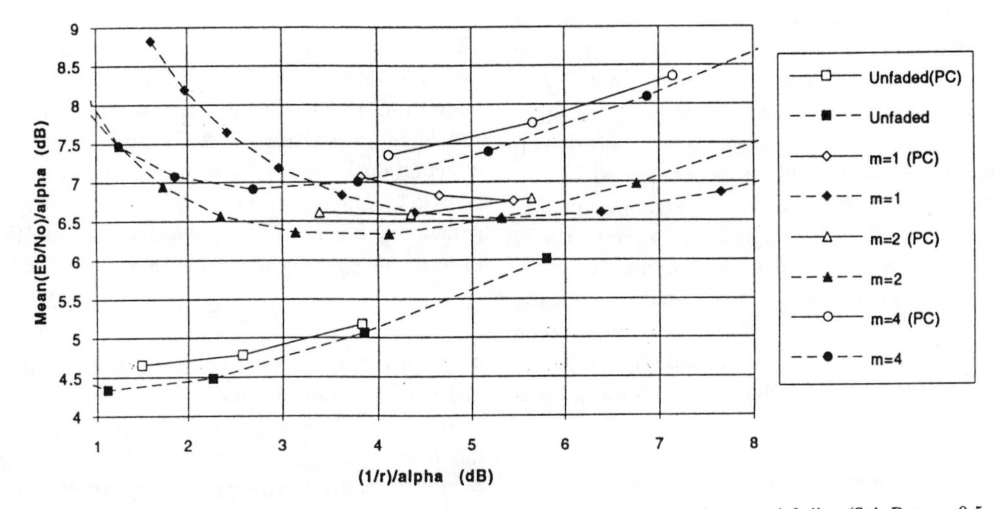

Fig. 5. Performance with integer metrics and power control in independent-increment lognormal fading (Std. Dev. $= 0.5$ dB).

It might be argued that the fading cases apply to very fast fading since the measurements are made on N successive (not interleaved) orthogonal signals and are assumed to be independent. On the other hand, the unfaded case could also be taken to be that for very slow fading where the energy remains constant over the N successive signal periods of the measurement, provided the time constant for the power control loop is much shorter than the bandwidth of the fading process. In the following, we shall take the power-controlled E_b/N_0 distribution in multipath to be that for fast fading in each case, which is a slightly pessimistic assumption since the corresponding standard deviation is somewhat larger than for the fixed energy case.

VI. Coded Error Performance with Ideal Interleaving and Power Control—Conclusions

Performance is determined by the relationship between $\overline{E}/N_0$ and r, through (6) and (7), which are both functions of Z. With integer metrics, Z is given by (10); but with the added effect of power control, each transition probability P_k, which depends on $\overline{E}$, must be replaced by its average.

$$\int_0^{\infty} P_k(\overline{E}/N_0)p(\overline{E}/N_0)d(\overline{E}/N_0), \qquad k = \pm 1, \cdots, 8 \tag{23}$$

where $P_k(\overline{E}/N_0)$ is given by (9a) and (9b) with $f_C(y)$ and $F_C(y)$ being functions of $\overline{E}/N_0$, and $p(\,)$ is the density function of the power-controlled $\overline{E}/N_0$, determined by simulation as shown in Fig. 4 for one choice of power control threshold ϕ.

Fig. 5 shows (as solid lines) the result of this calculation for the power-controlled case for three settings of power control thresholds ϕ for each of the faded and unfaded cases. Also shown as dotted lines are the corresponding values for fixed $\overline{E}_b/N_0$, as taken from Fig. 3. From this it is seen that power control degrades performance by less than 0.2 dB in each case.

It is also apparent from Fig. 5 that, $M = 64$, choosing a rate $1/3$ code with a backoff factor $\alpha = 0.8$ dB, as suggested in Appendix I to achieve $P_f < 0.01$, the resulting abscissa value for $(1/r)/\alpha$ is 4 dB. Then adding α to the corresponding ordinate values leads to mean $E_b/N_0 = 6.1$ dB for the unfaded case and mean $\overline{E}_b/N_0 = 7.4$, 7.8, and 8.1 dB for $m =$ 1-,2-, 4-path multipath.

Comparing with performance for optimum unquantized metrics (solid lines of Fig. 3), it is seen that the combined effect of quantization and power control degraded performance by no more than 0.6 dB. It should be emphasized in conclusion that all performance estimates were for the case of ideal interleaving between the encoder and the orthogonal waveform generator [Fig. 1(a)]. With finite interleaving, performance will generally degrade, particularly in fading which is slow relative to the interleaving span, but more rapid than the response time of the power control loop.

An additional advantage of power control not discussed in this paper is that since each user's power is individually controlled, users which are disadvantaged by excessive multipath, as well as greater range or shadowing, can be given additional power to achieve a higher E_b/N_0 ratio in excess of just what is required to equalize the received power for each user. Thus, power control affords the possibility of achieving a desired level of error probability for all users simply by varying E_b/N_0 requirements. This increased E_b/N_0 for selected users produces increased interference for other users, which ultimately limits the capacity of a given cell in terms of number of supportable users, as treated in [3]. However, when the total number of users is large, the number of disadvantaged users requiring higher E_b/N_0 ratios will to some extent be offset by the number enjoying favorable propagation and hence lower E_b/N_0 ratios. Thus, controlling power to achieve uniform error rates will on the average not severely impact average E_b/N_0 and, consequently, the interference suffered by the collection of users which governs the overall user capacity.

APPENDIX I
GENERIC AND SPECIFIC PERFORMANCE OF BINARY CONVOLUTIONAL CODES ON MEMORYLESS CHANNELS

No digital communication system is complete without forward error control (FEC) coding. With wideband spread spectrum modulation, the coding redundancy is already present and consequently imposes no limitation on code rate. Moreover, FEC coding performance analysis is central to the design of the overall system; hence, a brief review of coding fundamentals is in order.

For any symmetric memoryless channel, characterized by interference and fading which is independent from symbol to symbol, there is a large number of binary convolutional codes with constraint length K and rate r bits/symbol for which the L-bit frame[5] error probability is upper bounded [7] by

$$P_f < \frac{L2^{-Kr_0/r}}{2\left[1 - 2^{-(r_0/r-1)}\right]}, \qquad r < r_0 \tag{A.1}$$

where the parameter r_0 is a function solely of the memoryless channel statistics given by

$$r_0 = 1 - \log_2(1 + Z) \tag{A.2}$$

where

$$Z = \int_{-\infty}^{\infty} \sqrt{p(y|x)p(y|\overline{x})}\, dy \tag{A.3}$$

with y being the channel (soft decision) output random variable and $x = 0$ or 1 and $\overline{x}$ its complement, being the binary channel inputs, with corresponding channel (conditional) transition probability density functions $p(y|x)$ and $p(y|\overline{x})$. Any channel can be converted into a memoryless channel by providing a sufficiently large symbol interleaver after the encoder and prior to modulation, and a corresponding deinterleaver after demodulation but before the decoder. Unfortunately, these generic upper bounds are relatively loose.

For a specific good convolutional code of constraint length K and rate r, a much tighter upper bound is provided by the expression [7]

$$P_f < \frac{L}{2}T(Z) \tag{A.4}$$

where $T(Z)$ is the code generating function and Z is given by (A.3). Based on the convergence region of the generic bound (A.1) and other theoretical considerations, r_0 is often considered a practical limit on the code rate r of the binary convolutional code. Hence, for a specific code of rate r and constraint length K whose P_f is bounded in terms of its generating function $T(Z)$, we may express the bound (A.4) in terms of r_0/r by using the inverse of (A.2),

$$Z = 2^{1-r_0} - 1.$$

Thus, in Fig. 6, for the two codes of constraint length $K = 9$ and rates 1/2 and 1/3, whose shift register tap generators are, respectively, (753, 561) and (557, 663, 711) in octal notation, the frame error probability bound is shown as function of $\alpha = r_0/r > 1$ (in decibels). It appears that to achieve frame error rates below 10^{-2}, which is the acceptable level for vocoded voice traffic, $\alpha = 0.6$ dB for the $r = 1/2$ code, and $\alpha = 0.8$ dB for the $r = 1/3$ code[6].

Tighter bounds yet can be obtained, but these are more complex functions of all the memoryless channel transition probabilities and not just the r_0 parameter.

APPENDIX II
PERFORMANCE OF THE OPTIMUM M-ARY CONVOLUTIONAL CODES EMPLOYING INTERLEAVED-ORTHOGONAL SIGNALS

Consider a convolutional code which directly selects an M-ary orthogonal signal. Let b bits be shifted into the register to select n successive M-ary signals, where both b and n are integers. Hence, the code has rate $R = b/n$ bits/orthogonal signal and has 2^b branches emanating from each trellis node, with n orthogonal signals per branch. The generic L-bit frame error probability upper bound for a constraint length of bK bits is given by

$$P_f < \frac{L}{2}\left(\frac{2^b - 1}{b}\right)\frac{2^{-bKR_0/R}}{1 - 2^{-b[(R_0/R)-1]}} \tag{B.1}$$

[5] The frame or packet is assumed to contain $L - (K - 1)$ bits followed by $K - 1$ "flush tail" zeros.

[6] Even though it appears from this that the rate 1/2 code is superior to the rate 1/3 code for the same α, performance depends on the E_b/N_0 which is related to the r_0 parameter, as described above and shown in Fig. 3.

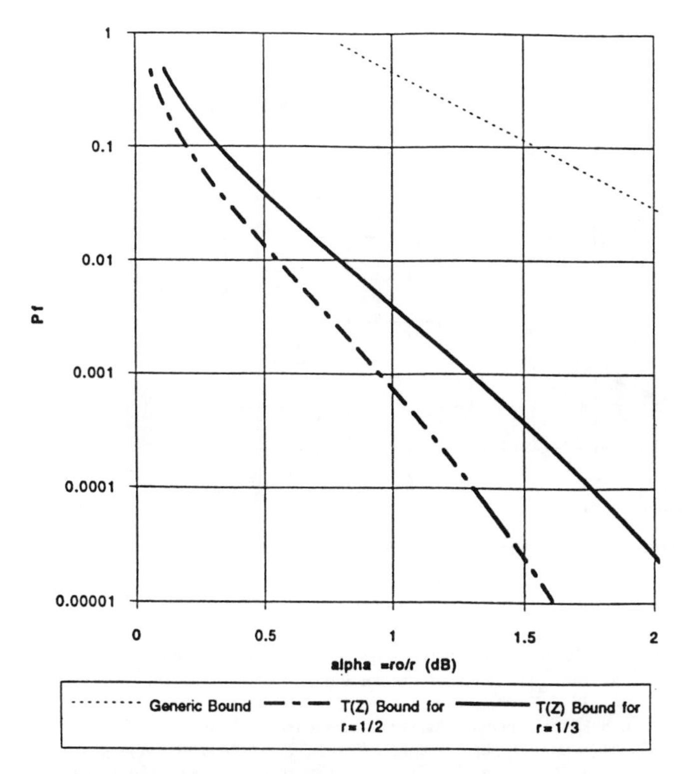

Fig. 6. Frame error probability bounds for $K = 9$, convolutional codes.

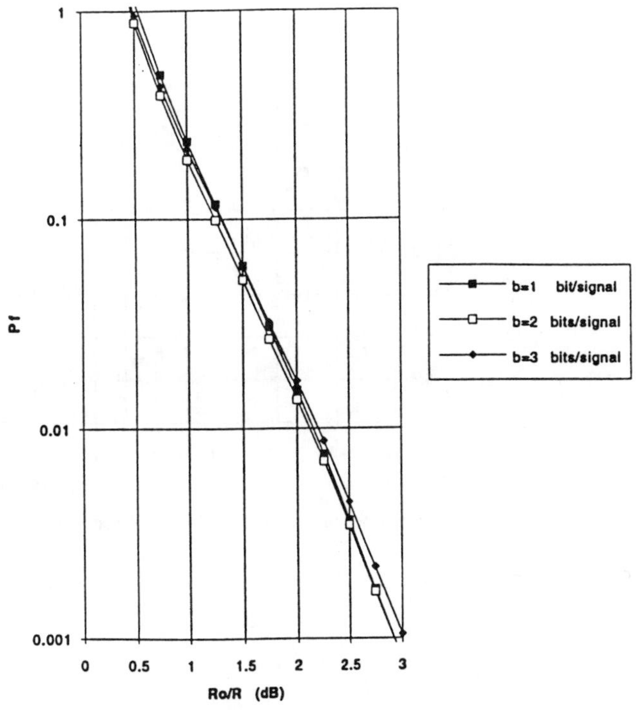

Fig. 7. Frame error rate as a function of Ro/R for $K = 9$ orthogonal convolutional codes.

where

$$R_0 = -\log_2\left(\frac{1}{M} + \frac{M-1}{M}Z\right) \tag{B.2}$$

and Z is the Chernoff (Bhattacharyya) bound for orthogonal signals. Note that here both R and R_0 can be greater than 1, while for $M = 2$, (B.2) reduces to (A.2).

However, if $M \geq 2^{bK}$, the generic bound becomes the specific bound for the orthogonal convolutional code, since in this case each of the 2^{bK} shift register tap combinations of the encoder can select a different orthogonal signal [4]. In this case, the bound (B.1) holds, but with[7]

$$R_0 = -\log_2 Z \tag{B.3}$$

or, equivalently, from (B.1) and (B.3)

$$P_f < \frac{L}{2}\left(\frac{2^b-1}{b}\right)\frac{Z^{nK}}{1-2^bZ^n}.$$

Fig. 7 is a plot of the frame error rate bound (B.1) as a function of

$$\alpha = R_0/R \qquad \text{in decibels} \tag{B.4}$$

for $bK = 9$ and[8] $b = 1, 2$, and 3. It is noteworthy that the results are practically insensitive to the choice of b, but that to achieve $P_f < 10^{-2}$ requires $\alpha \approx 2$ dB.

[7]Obviously, this is also the limit of (B.2) as $M \to \infty$. In fact, (B.1) and (B.3) are closely approximated even for M as small as $2^{bK/2}$.

[8]While only $b = 1$ and $b = 3$ are consistent with $bK = 9$, we also include $b = 2$ in this comparison for completeness. A noninteger value of K could actually be implemented by puncturing a higher rate code.

To compare with the binary interleaved case, it is necessary to compute Z for orthogonal signals (which assumes the use of optimum likelihood function metrics),

$$Z = \iint \sqrt{p_a(y_a, y_b)p_b(y_a, y_b)}\, dy_a\, dy_b$$

where y_a and y_b are the envelope-matched filter outputs for any two branches that are compared by the decoder, with $p_a(y_a, y_b)$ being the joint probability of the two outputs when y_a corresponds to the correct decision, and conversely for $p_b(y_a, y_b)$. Thus, for the m-component multipath fading channel[9], since

$$p_a(y_a, y_b) = \frac{y_a^{m-1}}{(m-1)!}\frac{e^{-y_a/(1+S)}}{(1+S)^m} \cdot \frac{y_b^{m-1}}{(m-1)!}e^{-y_b}$$

and $p_b(y_a, y_b)$ is the same but with y_a and y_b interchanged,

$$\begin{aligned} Z &= \iint \frac{y_a^{m-1}}{(m-1)!}\frac{e^{-[y_a/(1+S)+y_a]/2}}{(1+S)^{m/2}}\frac{y_b^{m-1}}{(m-1)!} \\ &\quad \cdot \frac{e^{-[y_b/(1+S)+y_b]/2}}{(1+S)^{m/2}}\, dy_a\, dy_b \\ &= \frac{1}{(1+S)^m}\left[\int \frac{y^{m-1}}{(m-1)!}\exp\left[\frac{-y(1+S/2)}{1+S}\right]dy\right]^2 \\ &= \left[\frac{1+S}{(1+S/2)^2}\right]^m. \end{aligned} \tag{B.5}$$

[9]See also [8] for the derivation of (B.5).

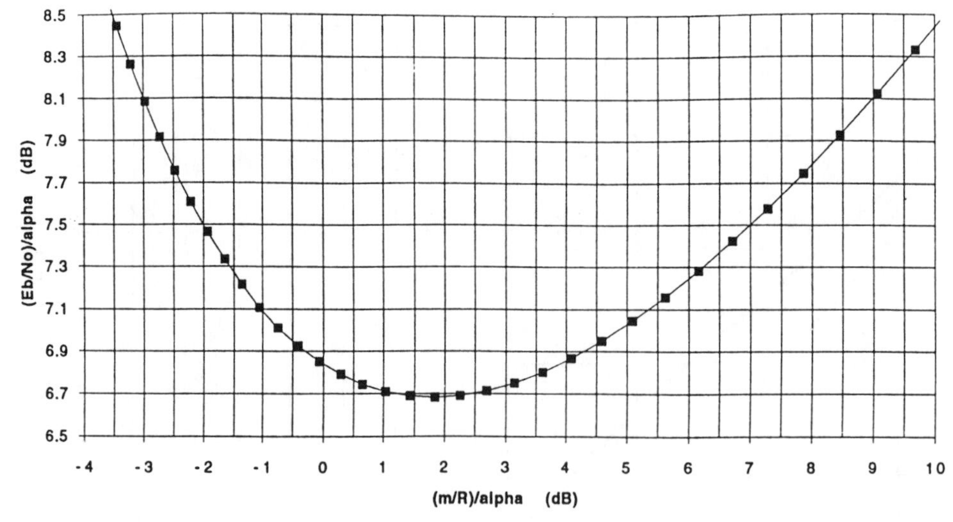

Fig. 8. Performance of optimum convolutional codes for interleaved orthogonal signals on M-path multipath fading channel.

Hence, from (B.3), we have that for orthogonal convolutional codes,

$$R_0 = -m\log_2\left[\frac{1+S}{(1+S/2)^2}\right]. \tag{B.6}$$

Hence, using (B.4), we have

$$\frac{m/R}{\alpha} = \frac{m}{R_0} = -\frac{1}{\log_2\left[\frac{1+S}{(1+S/2)^2}\right]} \tag{B.7}$$

and, using the definition of S,

$$\frac{\overline{E}_b/N_0}{\alpha} = \frac{Sm/R}{\alpha} = \frac{-S}{\log_2\left[\frac{1+S}{(1+S/2)^2}\right]}. \tag{B.8}$$

From the parametric equations (B.7) and (B.8), $\overline{E}_b/N_0$ is obtained as a function of $(m/R/(\alpha))$ and plotted in Fig. 8.

From this it seems that it is possible to nearly optimize for both $m = 2$ and $m = 4$ multipath by choosing $b = n = R = 1$ bit /signal = 1 bit/branch. To achieve $P_f = 10^{-2}$, it is seen from Fig. 7 that $\alpha = 2$ dB. Then, for $m = 4$, $(m/R)/\alpha =$ 4 dB, while for $m = 2$, $(m/R)/\alpha = 1$ dB, which implies from Fig. 8 that $(\overline{E}_b/N_0)/\alpha = 6.85$ dB and 6.7 dB, respectively and, hence, with ideal metrics,

$$\overline{E}_b/N_0 = 8.85 \text{ dB} \qquad \text{for } m = 4$$

and

$$\overline{E}_b/N_0 = 8.7 \text{ dB} \qquad \text{for } m = 2.$$

These appear to be more than 1 dB greater than the values of $\overline{E}_b/N_0$ required for a binary code with interleaving as shown in Fig. 3. Thus, even though for a given K, the code is optimum for the signal set, interleaving is performed only to the level of orthogonal signals, while with an interleaved binary code, interleaving is performed on binary symbols of which there are $\log_2 M$ as many. This implies that the greater diversity provided by ideal interleaving at the binary symbol level improves performance significantly over ideal interleaving only of the orthogonal signals.

REFERENCES

[1] G. L. Turin, "Introduction to spread-spectrum antimultipath techniques and their applications to urban digital radio," *Proc. IEEE,* vol. 68, pp. 328–353, Mar. 1980.

[2] J. Shapira, "Channel characteristics for the land cellular radio, and its system implications," submitted to *IEEE Trans. Antennas Propagat.*

[3] K. S. Gilhousen *et al.,* "On the capacity of a cellular CDMA system," *IEEE Trans. Vehicular Technol.,* vol. VT-40, pp. 303–312, May 1991.

[4] A. J. Viterbi, "Very low rate convolutional codes for maximum theoretical performance of spread-spectrum multiple-access channels," *IEEE J. Select. Areas Commun.,* vol. 8, pp. 641–649. May 1990.

[5] C. W. Helstrom, *Statistical Theory of Signal Detection,* 2nd ed. Oxford, UK: Pergamon, 1968.

[6] R. Price and P. E. Green, Jr., "A communication technique for multipath channels," *Proc. I.R.E.,* vol. 46, pp. 555–570, Mar. 1958.

[7] A. J. Viterbi, and J. K. Omura, *Principles of Digital Communication and Coding.* New York:McGraw-Hill, 1979.

[8] J. M. Wozencraft and I. M. Jacobs, *Principles of Communication Engineering.* New York:Wiley, 1965, Ch. 7.

Andrew J. Viterbi (S'54–M'58–SM'63–F'73) received the S.B. and S.M. degrees in electrical engineering from the Massachusetts Institute of Technology, Cambridge, in 1957, and the Ph.D. degree in electrical engineering from the University of Southern California, Los Angeles, in 1962.

He has devoted approximately equal segments of his career to academic research, industrial development, and entrepreneurial activities. In 1985 he was a cofounder and became Vice Chairman and Chief Technical Officer of QUALCOMM, Incorporated, a company specializing in satellite and terrestrial mobile communications. In 1968 he co-founded LINKABIT Corporation. He was Executive Vice President of LINKABIT from 1974 to 1982. In 1982 he took over as President of M/A-COM LINKABIT, Inc., and from 1984 to 1985 served as Chief Scientist and Senior Vice President of M/A-COM, Inc. Previously, from 1963 to 1973, he was a Professor at the UCLA School of Engineering and Applied Science, and from 1957 to 1962 he was a Research Engineer at C.I.T. Jet Propulsion Laboratory. He holds a part-time appointment as Professor of Electrical and Computer Engineering at the University of California, San Diego.

Dr. Viterbi has received numerous professional society awards and international recognition. These include three paper awards, culminating in the 1968 IEEE Information Theory Group Outstanding Paper Award, and four major society awards: the 1975 Christopher Columbus International Award (from the Italian National Research Council sponsored by the City of Genoa); the 1984 Alexander Graham Bell Medal (from IEEE sponsored by AT&T) "for exceptional contributions to the advancement of telecommunications," the 1990 Marconi International Fellowship Award and the 1992 NEC C&C Foundation Award (jointly). In 1991 he was awarded an Honorary Doctor of Engineering degree by the University of Waterloo, Ont., Canada. He is a member of the National Academy of Engineering.

Audrey M. Viterbi (M'84) received the B.S. degree in electrical engineering and computer science and the B.A. degree in mathematics from the University of California, San Diego, in 1979. She received the M.S. and Ph.D. degrees in electrical Engineering from the University of California, Berkeley, in 1981 and 1985, repectively.

She was an Assistant Professor of Electrical and Computer Engineering at the University of California, Irvine, from 1985 to 1990. Since 1990 she has been at Qualcomm, Inc., San Diego, where she is currently a staff engineer involved in the development of the CDMA cellular telephone system. Her professional interests are in the area of modeling and performance evaluation of computer-communication networks and communications systems.

Ephraim Zehavi (M'77) received the B.Sc. and M.Sc. degrees in electrical engineering from the Technion–Israel Institute of Technology, Haifa, Israel, in 1977 and 1981, respectively, and the Ph.D. degree in electrical engineering from the University of Massachusetts, Amherst, in 1986.

From 1977 to 1983 he was an R&D Engineer and Group Leader in the Department of Communication, Rafael, Armament Development Authority Haifa, Israel. From 1983 to 1985 he was a Research Assistant in the Department of Electrical and Computer Engineering University of Massachusetts, Amherst. In 1985 he joined Qualcomm, Inc., San Diego, CA, as a Senior Engineer, where he was involved in the design and development of satellite communication systems, and VLSI design of Viterbi decoder chips. From 1988 to 1992 he was on the Faculty of Department of Electrical Engineering, Technion–Israel Institute of Technology, Haifa, Israel. He is currently a Principal Engineer at Qualcomm, Inc., working in the areas of satellite communication, digital cellular telephone systems, and combined modulation and coding.

Reprinted from *IEEE Personal Communications*, Vol. 1, No. 2, Second Quarter, 1994, pp. 26-35.

True vehicular/portable mobility with low cost and low complexity

Slow Frequency-Hop TDMA/CDMA for Macrocellular Personal Communications

PHILLIP D. RASKY, GREG M. CHIASSON, DAVID E. BORTH AND ROGER L. PETERSON

Over the last decade, and in particular, over the last several years, there has been a proliferation of proposals for personal communication systems (PCS). These proposals have been prompted by the worldwide frequency allocation for Future Public Land Mobile Telephone Services (FPLMTS) as well as the recent (September 1993) Federal Communications Commission (FCC) allocation of frequencies for PCS in the 1.85-1.97 GHz and 2.13-2.20 GHz bands within the United States. The proposed systems include Bellcore's UDPC proposal [1], CT-2 [2], DECT [3], and the Japan Personal Handy Phone system [4]. All of these proposals can be characterized as being low-power pedestrian systems designed to operate in small service areas and are expected to be available for both low initial and continuing (service) costs. Several of the systems listed are capable of handoff between service areas, or cells; however, in each case, the cell radius extends to at most 1,000 to 2,000 feet, usually due to transmit power and antenna height limitations. Providing continuous coverage in a large metropolitan area would not appear to be economically feasible with such small cells, except in densely populated regions. Furthermore, providing vehicular mobility would seem to be impractical in such systems due to the extreme rate of handoffs.

In order to provide a full range of personal communications services with wide-area coverage and high mobility, it is desirable to define a macrocellular system to complement microcellular PCS. At first glance, merely increasing the antenna height or the transmitter power levels would seem to permit air interfaces designed for microcellular deployment to be suitable for the vehicular environment. However, a high-mobility macrocellular communication system must simultaneously compensate for the effects of delay-spread channels characteristic of higher base antenna heights and the reduction in overall system capacity relative to microcellular PCS.

In this article, a system that overcomes these problems is described. This system was designed to permit true vehicular/portable mobility with larger cells than so-called "conventional" PCS systems would allow while remaining true to the fundamental PCS tenets of low cost and low complexity. Furthermore, the system provides a flexible backbone to support emerging telecommunications services through the exploitation of the concept of "bandwidth-on-demand."

System Concepts

Simulation Study

In order to determine the best method of meeting the goals established for high mobility/wide area PCS, a simulation study was conducted that examined different combinations of options for a mobile communication system with a 500 kb/s gross bit rate on a time-division multiple-access (TDMA) carrier providing up to 10 voice channels or, through the concatenation of multiple time slots, data services with net data rates up to 144 kb/s. The options examined included possible combinations of: channel coding and interleaving; diversity combining; slow frequency hopping (SFH); and channel equalization. The goal of this simulation study was to determine the combination of these options which would yield the lowest carrier-to-interference (C/I) protection ratio requirement, which, in turn, translates into the highest system capacity.

For the purposes of the comparisons made in this section, a linear receiver was employed, and the selection of the channel coding technique was limited to rate-1/2 convolutional codes with soft-decision decoding and interleaving spanning 40 ms to limit the voice channel delay. In each case, the modulation technique was QPSK with burst-coherent demodulation and carrier recovery from a preamble inserted in each TDMA slot. In addition, a source bit rate of 32 kb/s to support industry-standard ADPCM was assumed. This requires two out of the total of 10 time slots per TDMA frame. In the receiver processing, two branch maximal ratio diversity combining was employed in the systems incorporating diversity, and a 16-state maximum likelihood sequence estimation (MLSE) equalizer was used for the equalized results. Interference was modeled as a faded pseudo-random data source with modulation identical to that of the desired signal. Base stations were assumed to be synchronized to the extent that corresponding TDMA slots of adjacent sites were aligned.

The performance of the candidate systems has been evaluated on several slowly fading (5 mph) channels that characterize the range of conditions expected in a macrocellular PCS environment. A non-frequency selective Rayleigh-fading channel and three delay-spread channels were considered. For the case of frequency-hopped systems, it was assumed that the system bandwidth was much greater than the coherence bandwidth, therefore resulting in independent fading from hop-to-hop [5]. The first delay-spread channel was based

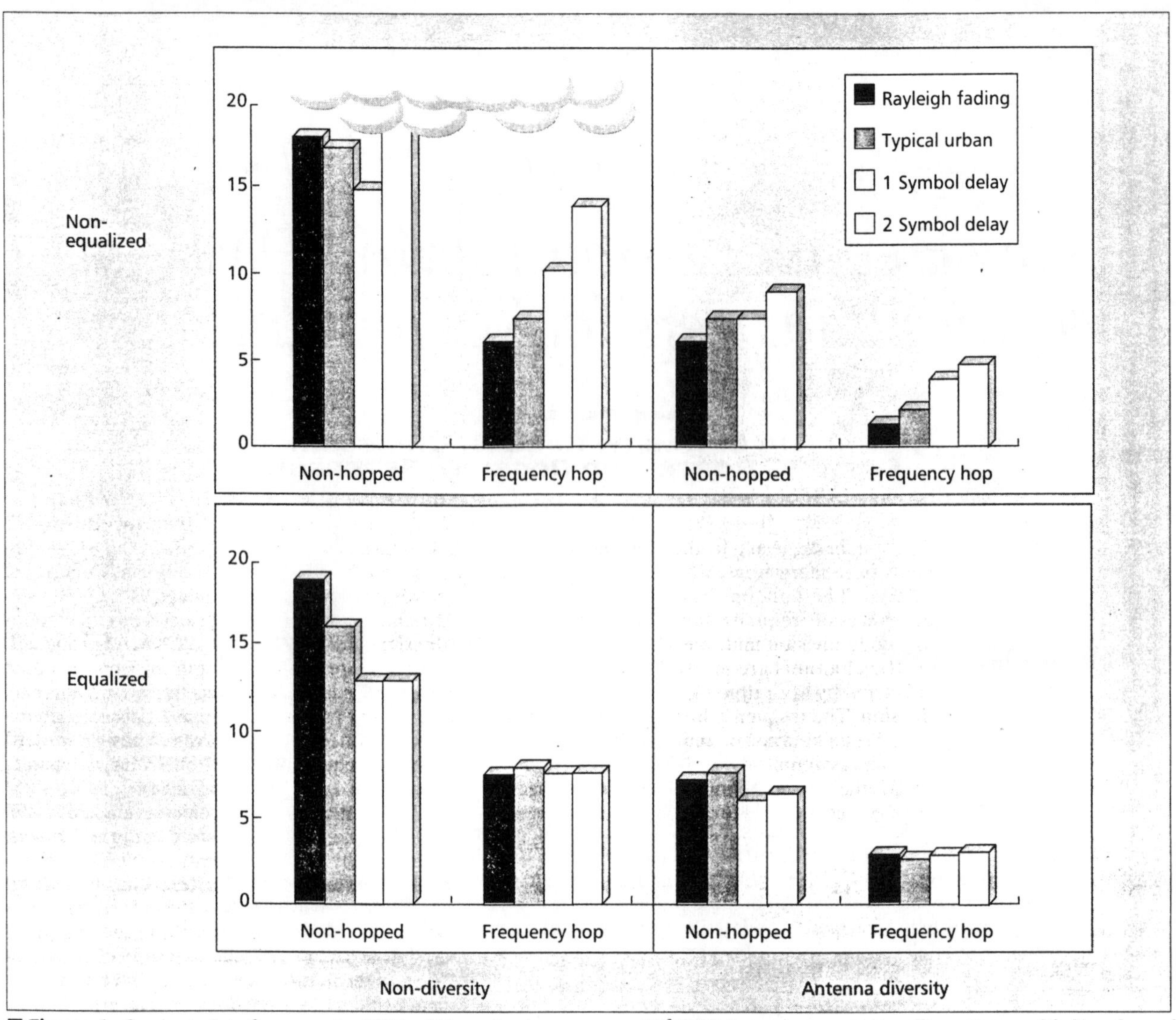

■ **Figure 1.** *Carrier-to-interference ratio (C/I in dB) required to support a 10^{-3} BER for various system configurations and fading channel conditions. Note that without hopping or diversity, a 10^{-3} BER cannot be achieved on severe multipath channels. (Metaphorically, the column reaches into the clouds.)*

upon the COST 207 typical urban profile [6] used in the validation testing of the Pan-European digital cellular system, GSM. The second and third delay-spread channel models were two-ray, equal-gain models with one and two symbols of differential delay — 4 and 8 μs, respectively. For each of the delay-spread channel models, all of the paths were assumed to undergo independent Rayleigh fading with a power spectrum determined by the classical (Jakes) model for Rayleigh fading in the land mobile radio environment.

The bar graphs in Fig. 1 summarize the results of this study by showing the C/I ratio required to achieve a decoded bit error rate (BER) of 10^{-3} for various combinations of the options. The results presented were obtained using a constraint length six convolutional code. The performance of a constraint length nine code was also evaluated; however, the increased constraint length generally resulted in a performance gain of less than 1 dB while adding significantly more complexity to the system. In reviewing the results, it is apparent that the combination of channel coding, antenna diversity combining, and slow frequency hopping best meets the defined goals. The combination of diversity and forward error correction coding allows robust reception on the harsh mobile radio channel, particularly in multipath conditions. Frequency hopping provides a form of frequency diversity to the system by effectively exploiting the frequency selectivity over the entire system bandwidth, thereby decorrelating the fading process from hop-to-hop over the span of the interleaver or equivalently reducing the duration of long fades [7]. This results in good performance at low speeds and, in general, performance that is independent of speed.

It is interesting to note from the results presented in Fig. 1 that given the two system attributes of diversity and frequency hopping, the system does not require an equalizer in multipath of up to several microseconds of RMS delay spread. The equalizer improves performance only slightly on the more severe channels and actually degrades performance on the Rayleigh fading channel and typical urban profile (approximately 1 μs RMS delay spread). This degradation can be attributed to equalization errors resulting from erroneous channel estimation at very low signal-to-noise ratios.

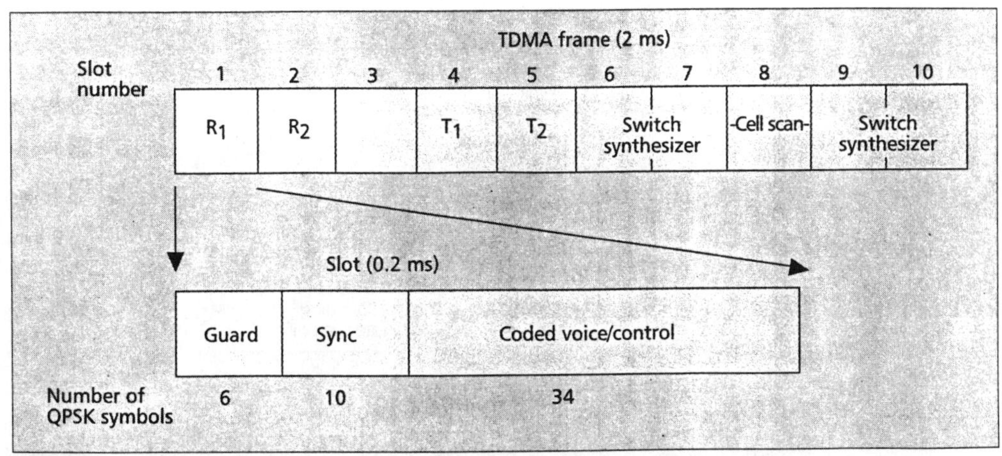

■ **Figure 2.** *TDMA frame and slot structure for hybrid SFH/TDMA PCS.*

Multiple-Access Structure

The system suggested by the previous simulation study is, in many ways, a hybrid multiple-access system. The concept borrows from the best attributes of frequency-division, time-division, and code-division multiple access. At a cell site, traffic channels are multiplexed onto the RF spectrum by both time-division and frequency-division. The frequency hopping of each TDMA carrier results in a new permutation of the frequency-division assignments each TDMA frame. As a result, the frequency-hop patterns are orthogonal within a cell; therefore, there is no intracell interference in the SFH system. Since the orthogonality is based upon frequency division, it is maintained on both the uplink and the downlink and in the presence of multipath channel conditions and propagation delays. This is a significant advantage over direct-sequence code-division multiple access (DS/CDMA) proposals. Furthermore, as inbound transmissions destined to the same base station receiver are orthogonal, no near-far problem exists due to this access technique. Consequently, transmit power control does not require an exceptionally large dynamic range.

Frequency band:	Inbound: 1,899 to 1,929 MHz Outbound: 1,949 to 1,979 MHz
Channel spacing	400 kHz
Channel bit rate	500 kb/s
Frequency hopping rate	500 hops/s
Time slots/frame	10
Time slot duration	0.2 ms
Frame duration	2.0 ms
Gross data rate/slot	34 kb/s
Duplex method	FDD
Intracell multiplexing	TDM/FDM (orthogonal SFH)
Intercell multiplexing	SFH/CDMA
Modulation	QPSK
Pulse shaping	Raised cosine, $\alpha = 0.5$
Receiver gain control	Hard-limiting
Antenna diversity	Pseudo max-ratio
Demodulation	Burst coherent
Transmit power	800 mW
Channel coding: Type Interleaving span	 Rate 1/2, $K = 6$ convolutional, soft decision 40 ms
Speech coder	32 kb/s ADPCM

■ **Table 1.** *SFH/TDMA PCS system specifications.*

Between cells, traffic channels are separated by SFH code division. By virtue of the assignment of minimally correlated frequency-hopping patterns, a given source of interference can affect the system only a small portion of the time. (In [8] and [9] this is referred to as "interferer diversity.") Such an interference diversity effect arises because the interference experienced by any user is not subject to worst-case interference from a single, statistically dominant interferer, but rather from the aggregate of all the multiple-access users. This effect also permits the system capacity or the link reliability to be increased when voice inactivity is thoughtfully exploited in the air interface design.

Furthermore, in SFH/CDMA, the intercell co-channel interference is sampled rather than summed. Over the span of the interleaved FEC code, individual samples of extreme interference may be essentially erased prior to decoding. Hence, there is no danger of a single rogue user disrupting an entire cell site. Taken together, these benefits allow the use of single-cell reuse patterns with high loading factors thereby enabling the macrocellular system to meet the capacity requirements demanded of a PCS.

In much the same way that the system benefits from the combination of access techniques, the system benefits from the features of both wideband and narrowband channelization. As mentioned previously, the communications link is made more robust by the frequency diversity derived from hopping over the entire system bandwidth. However, the relatively narrow RF carrier channelization (400 kHz) permits considerable flexibility in avoiding existing services, such as point-to-point microwave, sharing the same spectrum allocation.

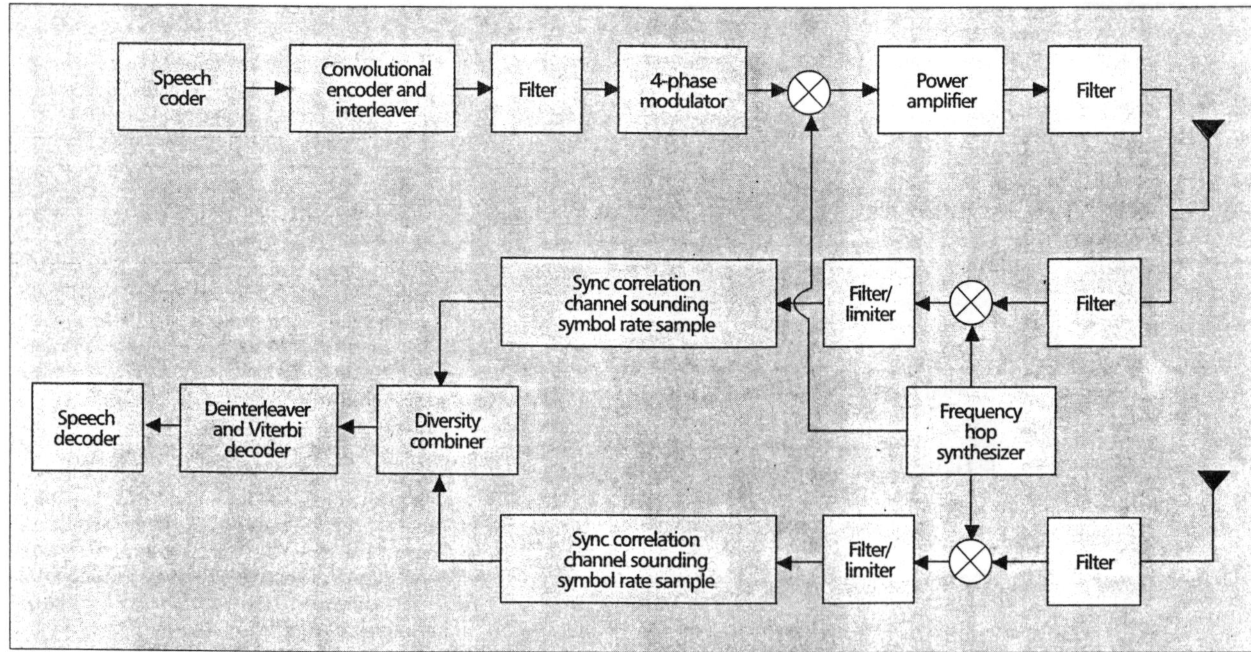

■ **Figure 3.** *SFH/TDMA PCS transceiver high-level block diagram.*

In order to meet the service and economic requirements demanded of a PCS, a TDMA structure is also employed. Figure 2 depicts the TDMA structure of the SFH PCS. The 0.2 ms slot structure is shown along with the 2-ms TDMA frame. The basic rate of this channelization is 16 kb/s per time slot. In the figure, two slots are allocated on each of the downlink and uplink to support a duplex 32 kb/s service such as ADPCM. This structure reduces the cost of the system in several ways. By time-division multiplexing onto the hopping RF carrier, the overall RF transceiver count at the base station is reduced by a factor of 10 for the basic rate service. Furthermore, for services below 64 kb/s, the transmit and receive slots may be offset in time so as not to overlap. By eliminating the need to simultaneously transmit and receive, the cost of a basic subscriber unit may be reduced due to the removal of the diplexer/filter. Finally, the TDMA structure permits a much simpler frequency hop synthesizer in the subscriber PCS unit, since the off-time of a few slots (200 to 400 μs) between active slots may be used to switch the synthesizer. The mapping of the 32 kb/s traffic channel onto the TDMA carrier, shown in Fig. 2, permits two full-time slots to switch and settle the hopping synthesizer while still permitting the subscriber unit to scan the adjacent cell sites for the purpose of aiding the handoff process.

The greatest advantage of the TDMA carrier is the ability to assign traffic channels that are multiples of the basic rate via multiple slot assignments so that high data rate channels can be supported. By the assignment of the first nine time slots in a frame to an individual duplex link, leaving one slot to hop the synthesizer, a high-end subscriber unit equipped with a diplexer can access a 144-kb/s circuit switched service. The ability to support high data rate traffic is considered essential for wireless personal communications to thrive when considering the increasingly broadband nature of the land-line PSTN and CATV networks.

Experimental Study

Prototype Transceiver Description

In order to validate the concepts and level of system performance presented in the previous sections and to demonstrate that these attributes can be achieved at a reasonable level of complexity, a prototype system has been developed. This system is based on the design considerations discussed in the section on system concepts and incorporates slow frequency hopping and diversity reception. The prototype PCS transceiver is depicted in high-level block diagram form in Fig. 3, and the major system specifications appear in Table 1. As indicated, the transceiver may be logically separated into transmitter and receiver functions. The transmitter consists of a speech coder followed by a channel coder and interleaver. The encoded data is four-phase modulated, and frequency hop transmission is employed.

The receiver portion of the transceiver is designed to efficiently detect and decode the transmitted signal. In view of the fading and multipath channels that are prevalent in mobile communications, a diversity receiver is employed to improve performance. At the receiver, an RF front end tracks and downconverts the hopping signal for each branch. Within this conversion process the signal is hard-limited, which results in significant cost and complexity savings. Following the front end, the signals are digitized and combined in a method approximating max-ratio diversity. The combined signal is deinterleaved, and soft decision Viterbi decoding is performed to estimate the transmitted data that is output to the speech decoder.

The mobile and base station hardware are assembled in 19" racks with modules that are 10" high and at most 12" deep. The transceiver hardware is suitably constructed for vehicular field testing. The base and mobile units can transmit a peak power of 800 mW. The system is fully capable of establishing both mobile-originated and mobile-terminated tele-

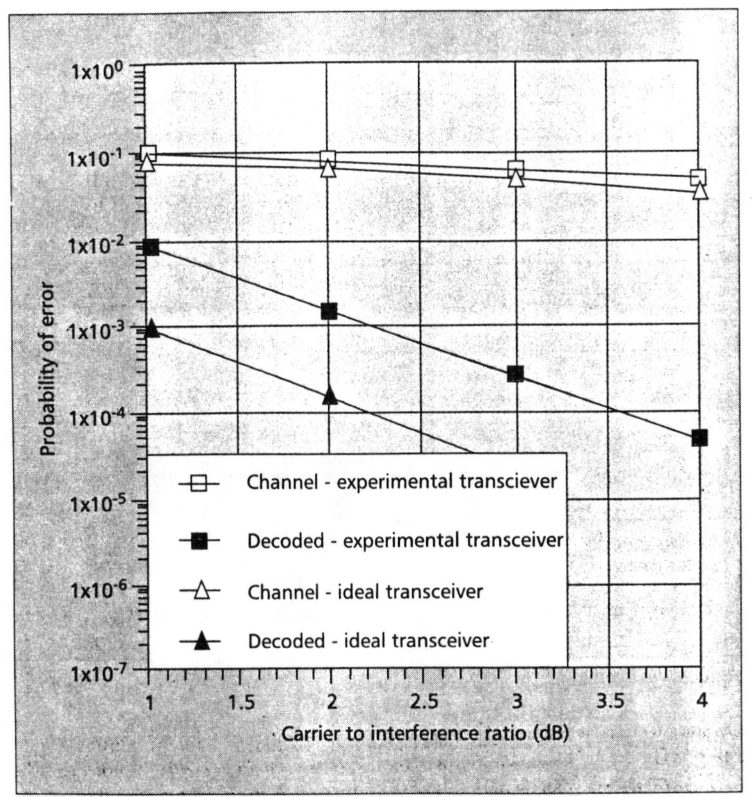

■ **Figure 4.** *Comparison of the BER performance of a low complexity, hard-limited PCS transceiver vs. an idealized linear transceiver.*

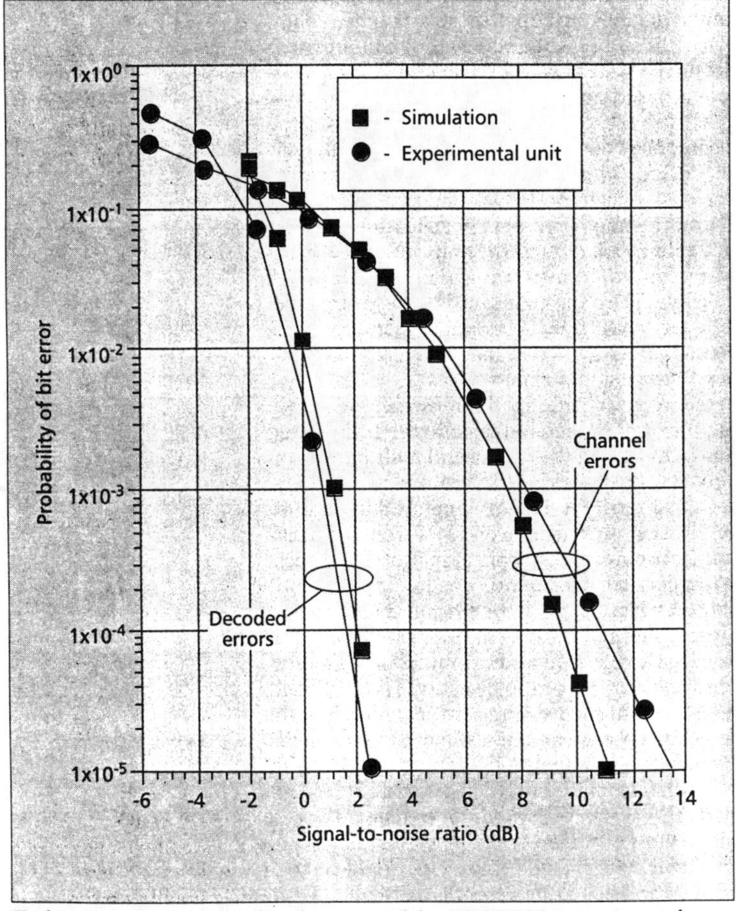

■ **Figure 4.** *Static channel performance of the SFH/TDMA experimental prototype as compared to simulation.*

phone calls with an interface to the PSTN through an echo-canceled conventional analog line. Considerable special purpose testing software has been integrated into the units to allow both objective and subjective quality assessments of the full-duplex link. The experimental hardware is capable of logging exact error rate statistics and estimates of received signal strength, carrier-to-interference ratio, carrier frequency offset, fading envelope standard deviation and correlation, diversity branch correlation, and RMS delay spread. More detailed statistics are available at test points throughout the transceiver. An expanded description of the structure and function of the experimental transceiver can be found in [10].

In keeping with the fundamental PCS requirement of low cost communications, great efforts have been taken to minimize the complexity of the system. In particular, the elimination of an equalizer presented a significant opportunity to reduce the complexity of the slow frequency-hopping receiver. Typically, an adaptive equalizer in the mobile radio environment requires a linear, wide dynamic range front end. With this restriction removed, it was possible to employ a hard-limiting receiver. With the hard-limiting receiver, the personal communicator can synchronize to the system much faster as receiver gain is removed as a variable in the synchronization process. Furthermore, the elimination of the signal processing complexity of the adaptive equalizer, automatic gain control (AGC), and wide-dynamic range A/D conversion —all associated with linear receivers—significantly reduces the complexity impact of the dual receiver branches.

Although great efforts were made to develop a low cost, low complexity system, the required trade-offs were carefully balanced to have a minimal impact on performance. Figure 4 compares the performance of a linear, infinite precision receiver with that of the hard-limited, fixed point receiver described in this article. For these results, the performance of both receivers was obtained using the detailed link simulation. As can be seen, at a 10^{-3} decoded bit error rate (BER), the loss is approximately 1.2 dB. This small degradation is significant, as low-complexity, hard-limited receivers are not usually associated with soft-decision decoding and diversity combining. Thus, based on simulation results, the goal of designing a high performance receiver for PCS has been achieved.

Experimental Results

In order to verify that this level of performance is actually achievable in practice, the hardware prototype was benchmarked against the detailed link simulation. The initial testing was performed in the laboratory using an automated test setup. An example of hardware versus simulation performance is depicted in Fig. 5, which illustrates system performance on an AWGN channel. Both the channel and decoded BER show excellent agreement between the actual and estimated results.

Once the hardware performance was verified in a controlled setting, field testing began from a cell site located at the Motorola Center in Schaumburg, Illinois. This testing served two purposes: to validate the hardware performance in a realistic environment and to characterize the actual SFH propagation environment. Figure 6 depicts a comparison of the hardware performance in a dense suburban environment (metropolitan Chicago) with that of the link simulation using a typical urban channel model. This channel model has an RMS delay spread of 1 μs and is thought to approximate the dense suburban propagation environment. Additionally, within this model the Rayleigh fading process was assumed to be independent between hops. The hardware results, shown as the scatter data, were obtained from short term performance measurements taken while driving along a fixed route. As the figure indicates, there was good agreement between the simulation results and those measured with the hardware. The results are quite encouraging with the system yielding acceptable performance in a range corresponding to a 3 to 4 dB signal-to-noise ratio.

In stating the channel modeling assumptions employed by the link simulation for the above comparison, a key parameter for frequency hopped systems, the frequency selectivity of channel across the hopping bandwidth, was brought to light. Earlier results have relied on the relationship between the coherence bandwidth and the hopping bandwidth [5] to justify the assumption that the Rayleigh fading process was independent between hops. However, the degree to which this assumption is valid in the typical PCS propagation environment has not been established, and in light of the substantial differences in system performance that were obtained earlier with non-hopped and independently-faded hopping models (i.e., the two possible extremes for modeling the effects of frequency hopping), the correct choice of frequency-hopping model is imperative.

In order to ascertain the appropriate model for a frequency-hop PCS, a method of observing the variation in the signal characteristic across the hopping spectrum was developed. The hardware link system was modified to display the received signal strength value for each dwell interval across the span of the hopping pattern. By programming the hopping pattern to increment by a single RF carrier per hop, the hardware was used as a wideband, real-time spectrum analyzer/channel sounder. Figure 7 displays a typical snapshot, a captured oscilloscope trace, of these values across 24 frequency channels spanning a 10 MHz hopping bandwidth covering 1,960 to 1,970 MHz. As the figure indicates, substantial frequency selectivity exists (variations in excess of 20 dB over the hopping bandwidth are apparent), which indicates that the independent fading model is a reasonable choice for performance analysis. It should be noted that the results contained in Fig. 7 were taken with the mobile stationary, and were therefore relatively static. This highlights an advantage of the frequency-hopped system, which is able to sample a number of channels, over a nonhopped system, which, if assigned a poor channel, would experience a prolonged period of unsatisfactory performance.

Figure 8 displays another snapshot of the signal levels over the hopping bandwidth. In this case, a strong direct ray has combined with specular paths to create a standing wave pattern. As indicated, nulls in excess of 2 MHz can occur over the system bandwidth. Therefore, even relatively wideband systems may experience sustained, deep fading in the PCS environment. However, by averaging the effects of these severely faded areas with those of the stronger regions over the span of an interleaved, forward error correction code, the frequency-hopped system can deliver satisfactory performance. It should be stressed that the signal distributions depicted in Fig.s 7 and 8 were readily obtained in a typical suburban area and are characteristic of the range of conditions expected in the expanded PCS environment.

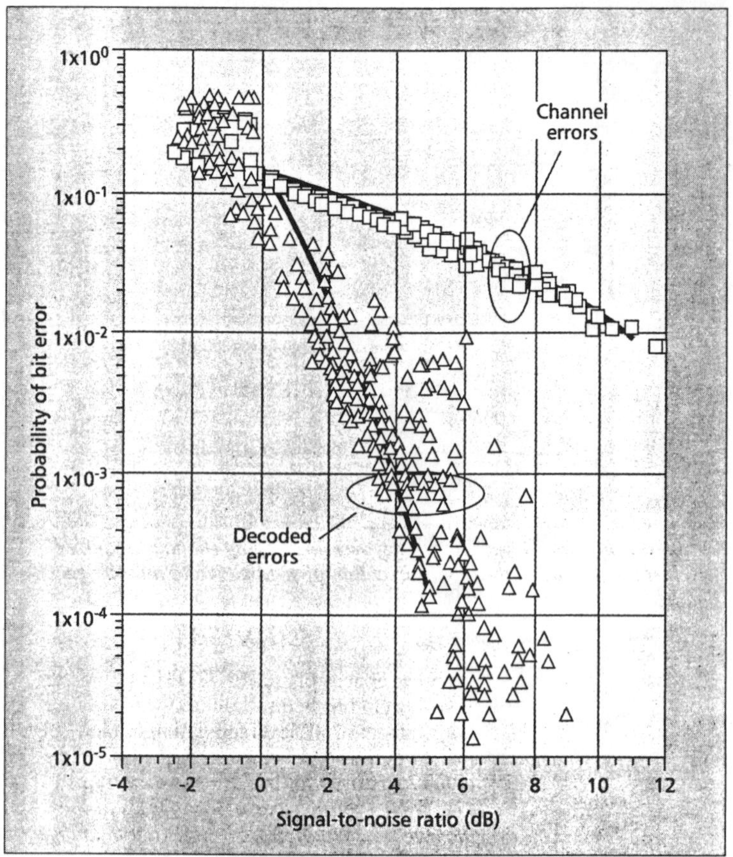

■ Figure 6. *Field performance of the SFH/TDMA experimental prototype as compared to simulation of the GSM typical urban profile.*

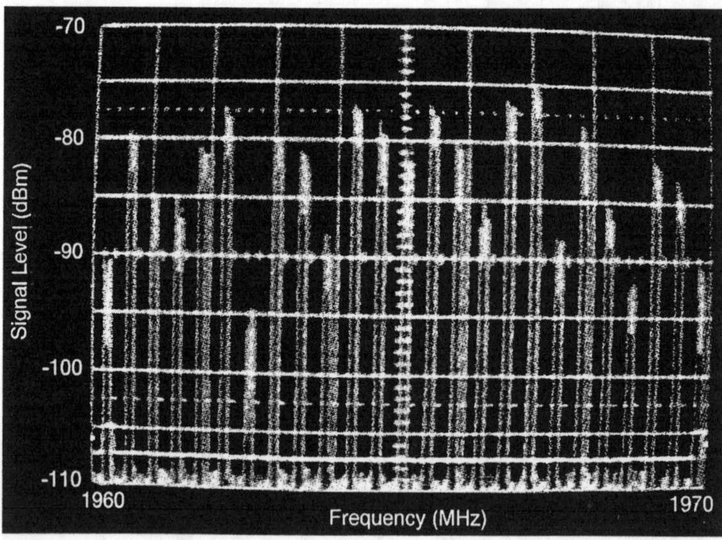

■ Figure 7. *Frequency selective fading over 24 hopping channels spanning 10MHz. The mobile unit is stationary, and there is no direct line-of-sight path.*

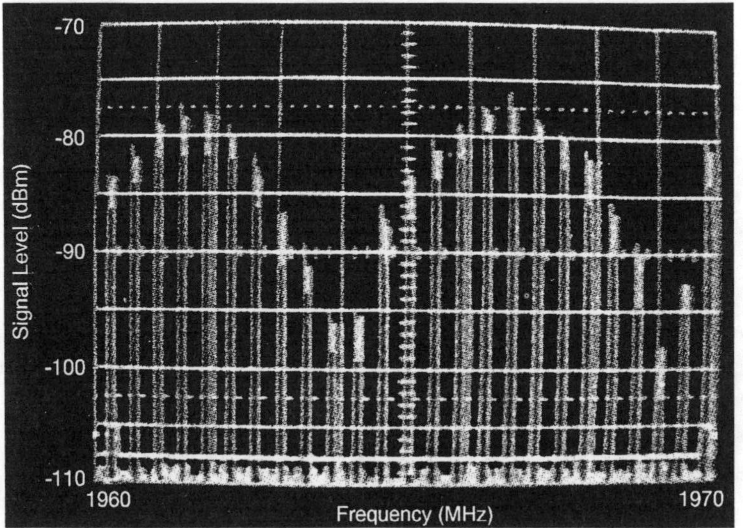

■ **Figure 8.** *Frequency selective fading over 24 hopping channels spanning 10 MHz. In this data, there is a direct line-of-sight path and multiple specular paths.*

Capacity

The capacity of the SFH/CDMA PCS described herein has been evaluated via extensive system- and link-level simulation studies. The following definition of capacity was adopted: *System capacity is the number of continuous basic-rate (16-kb/s) channels per-cell per-MHz of spectrum that the system can support while maintaining a bit error probability of 10^{-3} or better in 95 percent or more of all locations*. Voice activity is intentionally *not* accounted for in this definition so that capacity may apply to non-voice (i.e., data) users. For the purposes of the simulation study, a cell is divided into three 120° sectors.

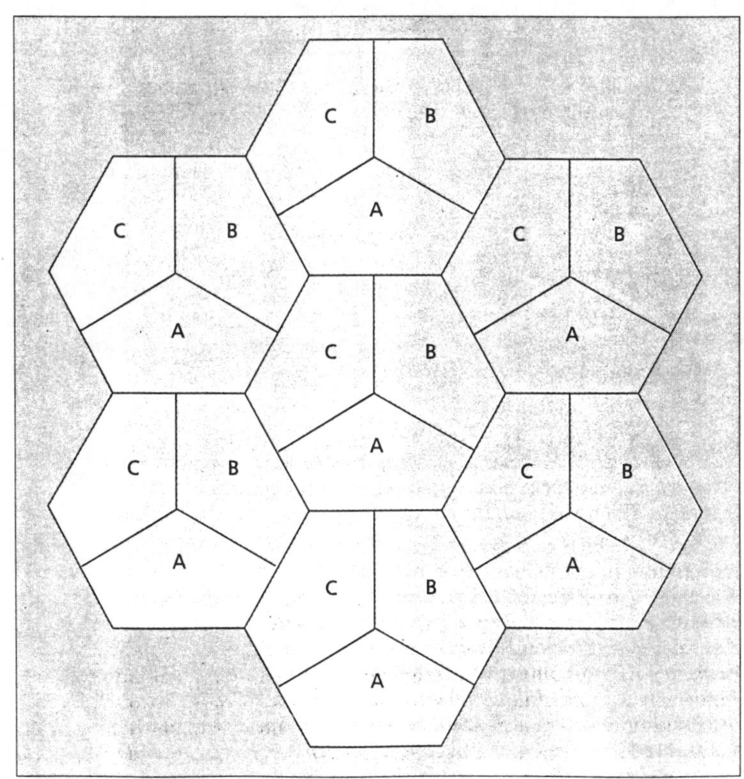

■ **Figure 9.** *Single cell three-sector frequency plan.*

The following assumptions were made in developing the simulation:

- The propagation model includes path loss with path-loss exponent of 3.6, log-normal shadow fading with standard deviation 8 dB, and typical urban multipath Rayleigh fading. For the numerical results, the velocity of the user was assumed to be 5 mph and the appropriate fading spectrum was used. The 120° sector antenna is modeled as providing only 15 dB of attenuation outside the sector due to scattering in land-mobile radio propagation.
- Log-normal shadowing affects the uplink and downlink in different manners. For both the uplink and the downlink, the power received from an interfering transmitter is affected by two transmission paths and the associated log-normal fading. The first of these is the path between the interfering mobile and its serving sector-cell and is independent for each interfering mobile. The loss on this path affects interferer transmit power due to power control. The second path is between the interfering transmitter and the test receiver. For the downlink, the interfering transmitter is the base station in the sector-cell serving the interfering mobile and the transmission path is from that base station to the test mobile. This path and the associated log-normal fading is the same for all mobiles in the interfering sector-cell. For the uplink, the interfering transmitter is the interfering mobile's transmitter, and the transmission path is from that mobile to the test mobile's base station; this path and the associated log-normal fading is different for each interfering mobile.
- The system is modeled as a plane tessellated by uniform hexagons. The propagation model and user density are uniform over the entire system.
- A single-cell, three-sector "frequency plan" is used. The carrier frequency allocation is partitioned into three equal subsets denoted A, B, and C. These subsets are assigned to sectors as illustrated in Fig. 9. Within each sector a signal frequency hops over the carriers in that sector's carrier subset. As mobiles request service in a sector, a hopping channel from that sector's carrier subset is assigned until all available channels are exhausted. System loading is characterized by the parameter L; setting L = 1.0 indicates that every channel of a sector's carrier subset is assigned to a mobile. The "frequency plan" permits L > 1.0 by borrowing carrier frequencies from adjacent sectors only when all channels from the sector's assigned subset are used. For the numerical results 12, 24, or 75 RF carriers are used.
- Power control eliminates received power level variations due to path loss and log-normal shadow fading but makes no attempt to remove Rayleigh fading. Error in power control is modeled by a log-normal random variable with a standard deviation of 1 dB. The link level simulation models both co-channel interference and receiver front-end thermal noise. The target received signal power is calibrated in terms of received E_b/N_0. For the uplink, it is desirable to minimize mobile transmitted power; thus the target E_b/N_0 is +13.0 dB. For the downlink, base station transmit power may be set arbitrarily without significant economic impact; thus the target E_b/N_0 is +40.0 dB making thermal noise negligible.

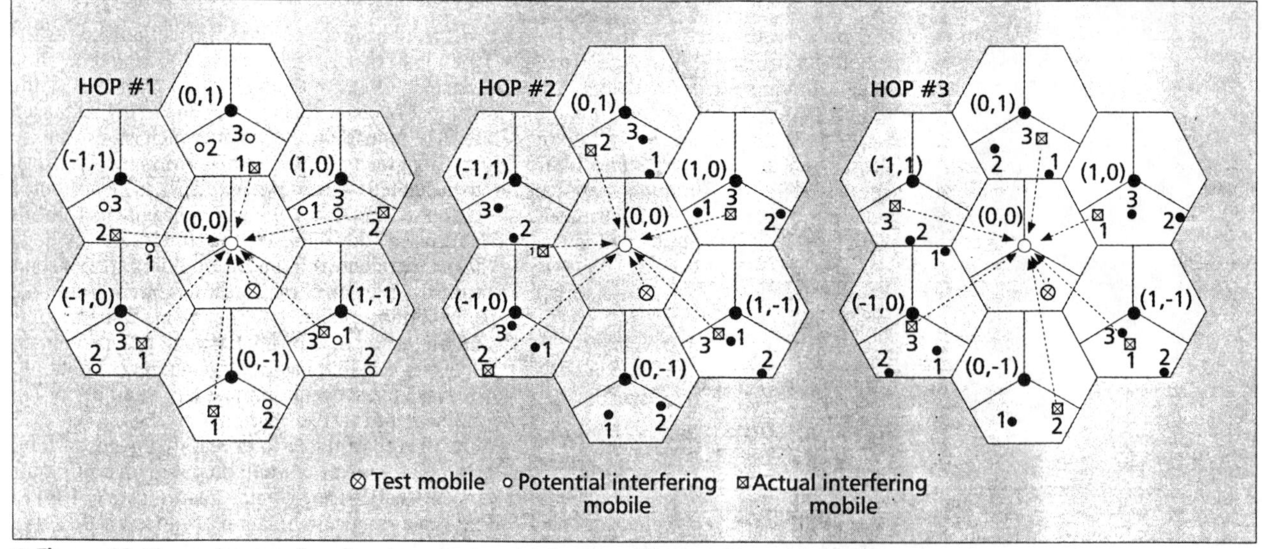

■ **Figure 10.** *Hop-to-hop interferer location selection: (a) first hop; (b) second hop; (c) third hop.*

◆ Handoff is ideal and is based upon perfect knowledge of the sum of path loss and log-normal path loss between the base station and the mobile. Mobiles are served by the sector whose path loss plus shadow fading loss is minimum.

Signal quality, i.e., bit error probability, is a function of the transmit power and locations of all mobiles within the system and the amount of co-channel interference diversity that can occur, which is a function of the number of RF carriers in the frequency allocation. Signal quality estimates were made using a very detailed link-level Monte-Carlo simulation. This simulation models the analog signal processing of the receiver using standard sampling techniques. All digital signal processing within the transceiver is duplicated *exactly* in the simulation. Thus the simulation models degradation due to finite word lengths, synchronization errors, and diversity combining gain errors precisely. The link simulation approximates co-channel interference from multiple interfering mobiles with a single co-channel interference source with power equal to the sum of the powers of the actual interferers.

The system-level simulation supports the link-level simulation by calculating the carrier-to-interference ratio (C/I) for each frequency-hop dwell. Figures 10a through 10c illustrate the uplink interference sources for three consecutive frequency hops for a system with three frequency-hop carriers per sector. Sector-cells are denoted by the cell coordinates and the sector identifier. The test mobile is in sector-cell (0,0,A). Each interfering sector-cell has three potential interfering mobiles since there are only three frequency-hop carriers available. The interfering and test mobile locations, and therefore their transmit powers, are fixed for these three dwells and all subsequent dwells of the link-level quality estimate. For each dwell, a specific interfering mobile is selected randomly from each sector. The sequence of interfering mobiles from (1,-1,A) is 3,3,1, illustrating that the random mobile selection can indeed cause the same mobile to interfere on consecutive dwells. Only two potential interfering mobiles are shown in (0,-1,A), and no interference is generated in this sector-cell during dwell 2. This indicates that $L < 1.0$, and not all channels are active in all sectors for a given random draw of mobile locations.

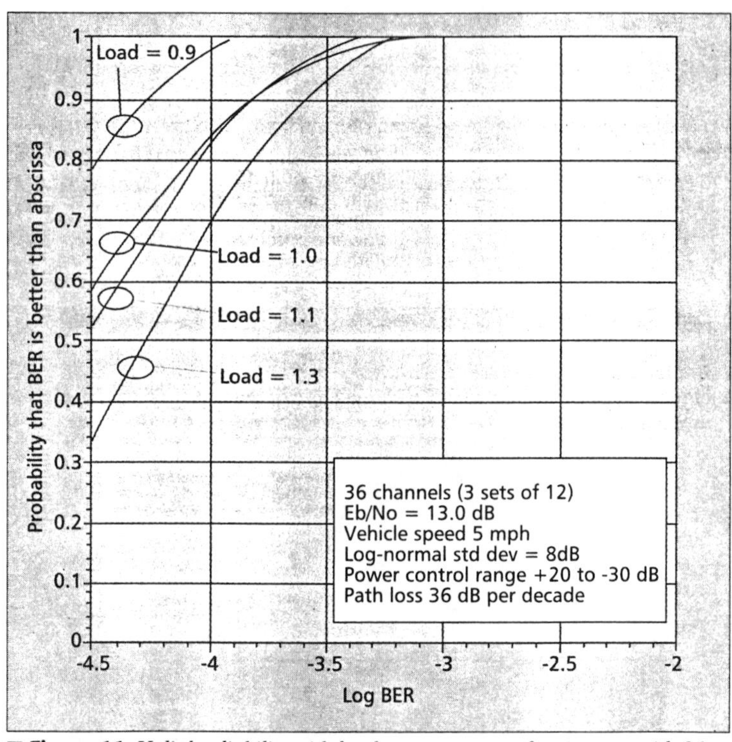

■ **Figure 11.** *Uplink reliability with load as a parameter for a system with 36 available RF carriers.*

Location reliability is the probability that signaling quality will be better than a selected threshold over all simulated mobile locations. Reliability is calculated by measuring signal quality a large number of times and calculating the fraction of the quality estimates for which bit error probability is better than the required 10^{-3} threshold. A new, independent set of mobile locations are randomly selected for each quality estimate. The output of the combined system- and link-level simulations is location reliability as a function of the quality threshold for a particular set of system parameters.

The capacity of the system was estimated by creating and interpreting reliability curves for various system parameters. Figures 11 and 12 illustrate uplink and downlink reliability as a function of bit error probability threshold for system loads between 0.9 and 1.3 for the system parameters given. For a bit error rate threshold of 10^{-3} and reliability of 95 percent, the allowable downlink loading is slightly greater than 1.1. Figures 11 and 12 are based upon the availability of 36 frequency-hop carriers corresponding to 2 x 15 MHz of spectrum and a carrier spacing of 400 kHz. Figures 13 and 14 illustrate uplink and downlink reliability results for a load of 1.1 for 12, 24, and 75 frequency-hop carriers. Observe that reliability improves as the number of channels increases due to the increased level of interference diversity which is possible.

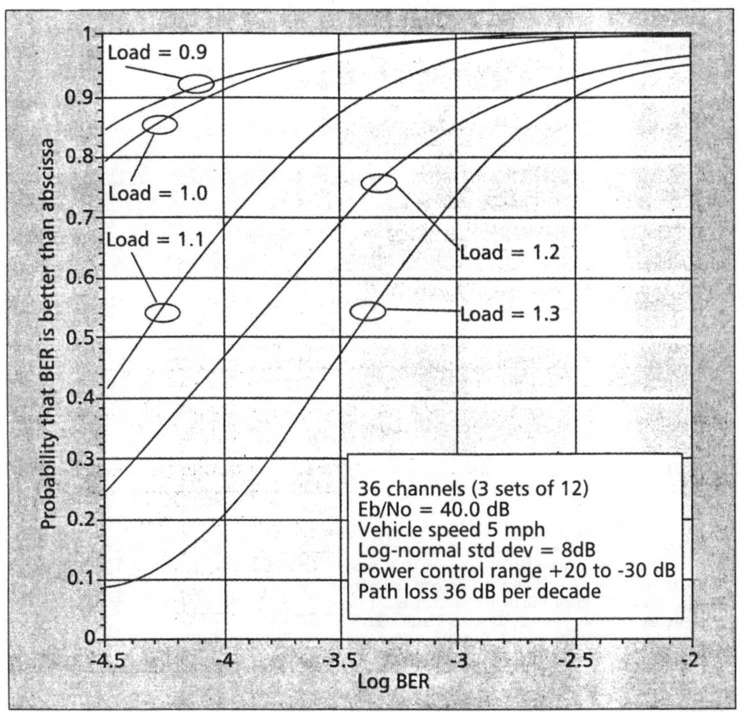

■ **Figure 12.** *Downlink reliability with load as a parameter for a system with 36 available RF carriers.*

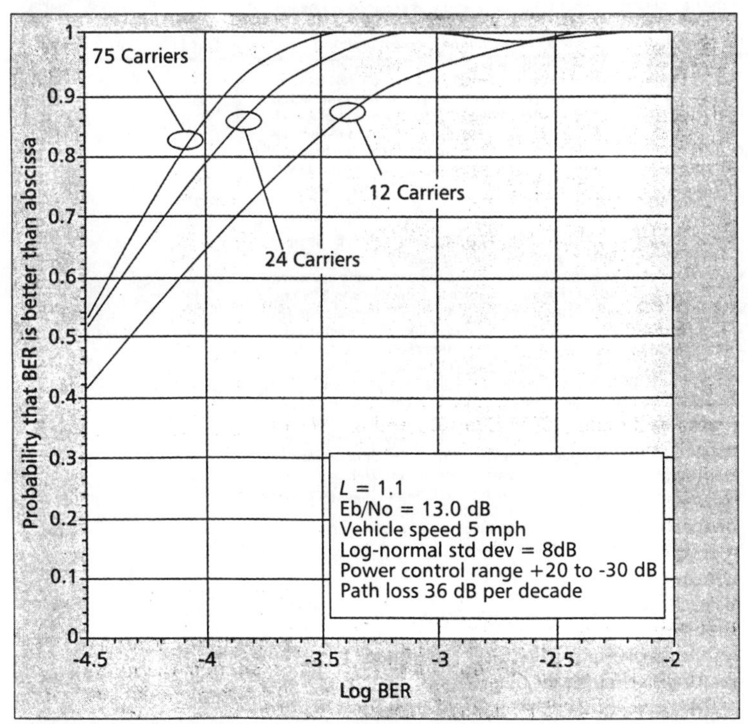

■ **Figure 13.** *Uplink reliability with number of RF carriers as a parameter and* L = *1.1.*

From the figures, system performance is shown to be limited by the downlink. This can be attributed to the reduced level of interference diversity available on the downlink as the path loss and shadowing component from interfering sites are constant from dwell-to-dwell during the observation interval. Furthermore, relative to the uplink, there are few interferers on the downlink (two to three dominant cell sites), so that law-of-large numbers effects do not take place even for a very large number of RF carriers and traffic channels. In the duplex PCS link, it is reasonable to expect that a closed-loop downlink power control mechanism could balance C/I, BER, or frame-erasure rate (FER) in links across a cell. Such an advance would serve to better balance the capacity of the uplink and downlink. Modeling and analysis of such a strategy is a topic for future investigation.

The final desired result is the system capacity in terms of the number of simultaneous 16 kb/s channels per-cell per-MHz, which can be supported at the desired signal quality and reliability. A 15-MHz spectrum allocation will support 37 frequency-hop carriers spaced by 400 kHz. Each carrier supports ten 16 kb/s continuous users. Based on the downlink results, the maximum supportable load is $L = 1.1$ at 95 percent reliability and a bit error probability threshold of 10^{-3}. The system capacity is therefore 27 continuous channels per-cell per-MHz. This capacity would be increased by taking voice activity into account. This level of performance is quite competitive with other state-of-the art proposals for wireless personal communications systems.

Conclusions

A slow frequency-hop time-division/code-division multiple access system for macrocellular personal communication systems has been proposed. With the inclusion of diversity reception in the PCS unit along with the frequency diversity attributes inherent to the frequency-hopping access technique, the design results in a highly reliable, high-capacity macrocellular communication system, even in the presence of multipath fading. Furthermore, this can be achieved without excessive complexity.

It has also been shown that through this hybrid access technique, single-cell frequency reuse is realizable. This is most easily achieved by a frequency mapping assigning orthogonal frequency sets to adjacent sectors in a three-sector cell. Though this is not a requirement of the system, it seems only reasonable that as a logical extension to orthogonal transmissions within a sector, the sources of greatest potential interference be coordinated so as to exploit the propagation loss in wireless communications as an advantage. Such coordination does not require any frequency planning per se, but the ability to map the frequency set onto consistently positioned sector antennas is necessary.

The system concept is further enhanced by the attributes of a time-division multiple access carrier. Through the hybrid of TDMA and CDMA, bandwidth-on-demand can be delivered to support emerging telecommunications services for data, video, and high quality audio transmission. If PCS is to thrive in the marketplace, it must distinguish itself on the basis of high quality voice channels and services not provided by present-day wireless communication services.

Acknowledgments

The authors gratefully acknowledge the efforts of Mark Cudak, Jim Kepler, Steve Kuffner, and Nick Tolli for the development of the slow frequency-hop research prototype and Fuyun Ling for many analytical insights into the operation of the modulation system. All of these individuals are with the Communication Systems Research Laboratory at Motorola's Corporate Systems Research Laboratories.

References

[1] D. C. Cox, "A Radio System Proposal for Widespread Low-Power Tetherless Communications," *IEEE Trans. on Commun.*, vol. 39, Feb. 1991, pp. 324-335.

[2] J. G. Gardiner, "Second Generation Cordless (CT-2) Telephony in the UK: Telepoint Services and the Common Air-Interface," *Electronics and Commun. Eng. J.*, vol. 2, no. 2, April 1990, pp. 71-78.

[3] D. J. Goodman, "Second Generation Wireless Information Networks," *IEEE Trans. on Vehicular Tech.*, vol. 40, May 1991, pp. 366-374.

[4] K. Tanaka, M. Hirono, and I. Horikawa, "Signaling Architecture for Microcell Communication Systems," *Proceedings of the 41st IEEE Vehicular Technology Conference*, May 1991, pp. 240-244.

[5] W. C. Jakes (ed.), *Microwave Mobile Communications*, (New York: John Wiley & Sons, 1974).

[6] COST 207 WG1, "Proposal on Channel Transfer Functions to be Used in GSM Tests Late 1986," COST 207 TD (86)51 Rev. 3, Paris, Sept. 29-30, 1986.

[7] M. Mizuno, "Randomization Effect of Errors by Means of Frequency-Hopping Techniques in a Fading Channel," *IEEE Trans. on Commun.* vol. COM-30, May 1982, pp. 1052-1056.

[8] D. Verhulst, M. Mouly, and J. Szpirglas, "Slow Frequency Hopping Multiple Access for Digital Cellular Radiotelephone," *IEEE Trans. on Vehicular Tech.*, vol. VT-33, Aug. 1984, pp. 179-190.

[9] J. L. Dornstetter and D. Verhulst, "Cellular Efficiency with Slow Frequency Hopping: Analysis of the Digital SFH900 Mobile System," *IEEE JSAC*, vol. SAC-5, June 1987, pp. 835-848.

[10] P. D. Rasky, G. M. Chiasson, and D. E. Borth, "An Experimental Slow Frequency-Hopped Personal Communication System for the Proposed U.S. 1850-1990 MHz Band," ICUPC '93, Oct. 12-15, 1993, pp. 931-935, Ottawa, Canada.

[11] P. D. Rasky, G. M. Chiasson, and D. E. Borth, "Hybrid Slow Frequency-Hop/CDMA-TDMA as a Solution for High-Mobility, Wide Area Personal Communications," Fourth WINLAB Workshop on Third Generation Wireless Information Networks, Oct. 19-20, 1993, pp. 199-215, East Brunswick, NJ.

Biographies

PHILLIP D. RASKY [M '83] received B.S. and M.S. degrees, both in electrical engineering, from the University of Illinois at Urbana-Champaign in 1982 and 1984, respectively. In 1983, he was a research assistant in the Coordinated Science Laboratory at the University of Illinois, where he was engaged in research in multi-user communications systems. In 1984, he joined the Systems Research Laboratory, Chicago Corporate Research and Development Center of Motorola, Inc., where he conducted research in digital land-mobile radio systems, which included combined speech and channel coding for that application. From 1987 to 1990, he was instrumental in the development of signal processing algorithms and architectures for Motorola's GSM cellular radiotelephone system. Presently, as a principal staff engineer in the Communication Systems Research Laboratory, Motorola, Inc., Schaumberg, Illinois, he leads a project studying the application of CDMA techniques to personal communications and cellular radiotelephones. He holds 10 U.S. patents.

GREG CHIASSON [M '89] received B.S. and M.S. degrees in electrical engineering from the University of Illinois in 1989 and 1990, respectively. In 1989, he was a research assistant in the Coordinated Science Laboratory at the University of Illinois, involved in research on the application of concatenated coding to frequency hop, packet radio communication systems. In 1990, he joined the Systems Research Laboratory, Corporate Research and Development Center of Motorola where he has engaged in the analysis, simulation, and exploratory design of spread-spectrum cellular and personal communication systems. He is currently a senior research engineer in the Communication Systems Research Laboratory, Motorola, Inc., Schaumberg, Illinois. He has two issued and five pending U.S. patents. He has been an Office of Naval Research Graduate Fellow and a University of Illinois Fellow. He is also a member of Eta Kappa Nu and Tau Beta Pi.

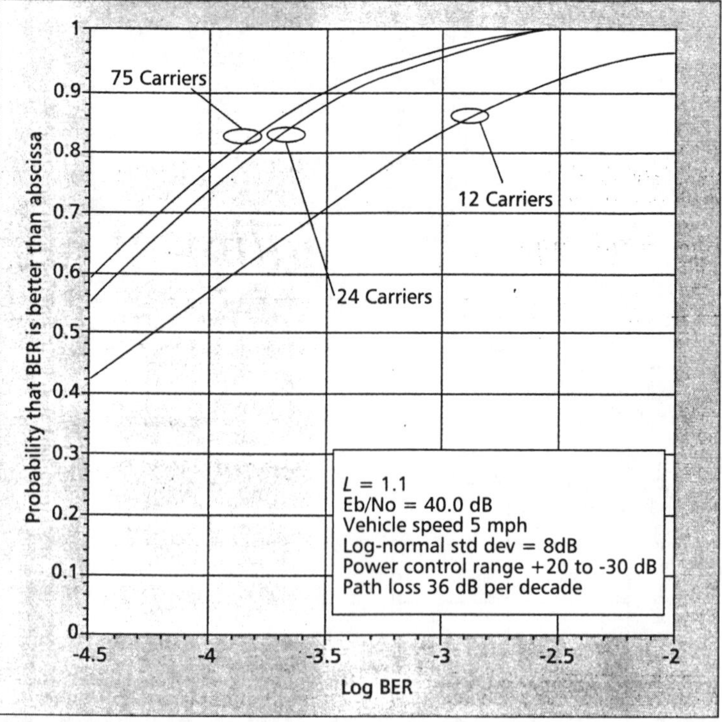

Figure 14. *Downlink reliability with number of RF carriers as a parameter and* L = *1.1.*

DAVID E. BORTH [SM '86] received B.S., M.S., and Ph.D. degrees, all in electrical engineering, from the University of Illinois at Urbana-Champaign in 1974, 1975, and 1979, respectively. From 1975 to 1977, he was a member of the technical staff of the Systems Division of Watkins-Johnson Company. In 1977, he returned to the University of Illinois as a research assistant in the Coordinated Science Laboratory, and was engaged in research on the performance of spread spectrum communication systems over fading channels. From 1979 to 1980, he was an assistant professor in the School of Electrical Engineering, Georgia Institute of Technology. In 1980 he joined the Systems Research Laboratory, Corporate Research and Development Center of Motorola Inc., where he has conducted research in digital modulation techniques, digital processing of speech signals, adaptive digital signal processing methods applied to communication systems, and personal communication systems. He holds 22 U.S. patents. He is currently a senior member of the technical staff and is the manager of the Communication Systems Research Laboratory, Motorola, Inc., Schaumberg, Illinois.

ROGER PETERSON [M '71] received B.S.E.E., M.S.E.E., and Ph.D. degrees from the Illinois Institute of Technology. From 1968 through 1978 he worked for the IIT Research Institute in Chicago performing advanced development for radar tracking subsystems. From 1978 to 1980 he worked for Bell Laboratories in communications system design for cellular telephony and paging systems. With Motorola since 1980, he currently works in the Communication Systems Research Laboratory, Motorola, Inc., Schaumburg, Illinois, where he performs system studies for personal communications. Between 1980 and 1992 he worked for Motorola's Government Electronics Group as lead communications system engineer for a number of spread-spectrum radios using both direct-sequence and hybrid direct-sequence-frequency-hop technology. He is past chair of the Phoenix Chapter of the IEEE Communications Society and is currently an associate editor for the *IEEE Transactions on Communications*. He is co-author of the textbooks *Digital Communications and Spread Spectrum Systems* (1985), *Introduction to Digital Communications* (1992) and the forthcoming *Introduction to Spread Spectrum Communication Systems* (1995).

The Globalstar System

The Globalstar system will employ active phased array antennas and energy management in combination with CDMA technology to achieve greater power efficiency than ever achieved for comparable satellite communications.

Edward Hirshfield
Globalstar
San Jose, California

Active phased array satellite antennas incorporate multibeam functionality in very small space. Such phased array antennas depend upon the use of highly efficient and uniform monolithic microwave integrated circuit (MMIC) amplifiers that were not previously available. MMIC amplifiers coupled directly to each radiating element result in very high G/T system figure of merit (effective gain/system noise temperature) as well as high EIRP (effective isotropic radiated power), given the actual size of the antenna apertures and power capacity and noise figures of the amplifiers. As a result, both the RF power efficiency in the transmit direction and the efficiency from input flux density to C/N (carrier to noise ratio) in the receive phased arrays is superior to performance previously attained for satellite communications.

The Qualcomm CDMA waveform employed in the User Terminals (handsets) and Gateways to the terrestrial networks (PSTNs), spreads the RF energy evenly over the allotted spectrum, as seen by the user antennas, optimizing power handling (signal fidelity) in the MMIC amplifiers.

This CDMA implementation, originally developed for terrestrial cellular and PCS, works particularly well in a system that employs a constellation of low earth orbiting (LEO) satellites.

System Description

The system will enable satellite based extension of terrestrial cellular systems worldwide, except for polar regions. A functional overview of the system is shown in Figure 1.

Calls may originate from either one of the system's subscribers or from any public or private network. For example, I will describe a call initiated by a system hand held unit. When the phone is turned on, it automatically registers with the Gateway in view of the satellites overhead. If there is more than one Gateway in range of the subscriber, it will register with the one with whom the subscriber has a direct subscription. If the user is out of range of his home Gateway, a roaming protocol will be implemented similar to that used in terrestrial cellular telephone networks and the user will be logged into the visitor location register at the accessible Gateway. The process is very similar to the operation of cellular telephone systems in current use.

Figure 1. Globalstar System overview.

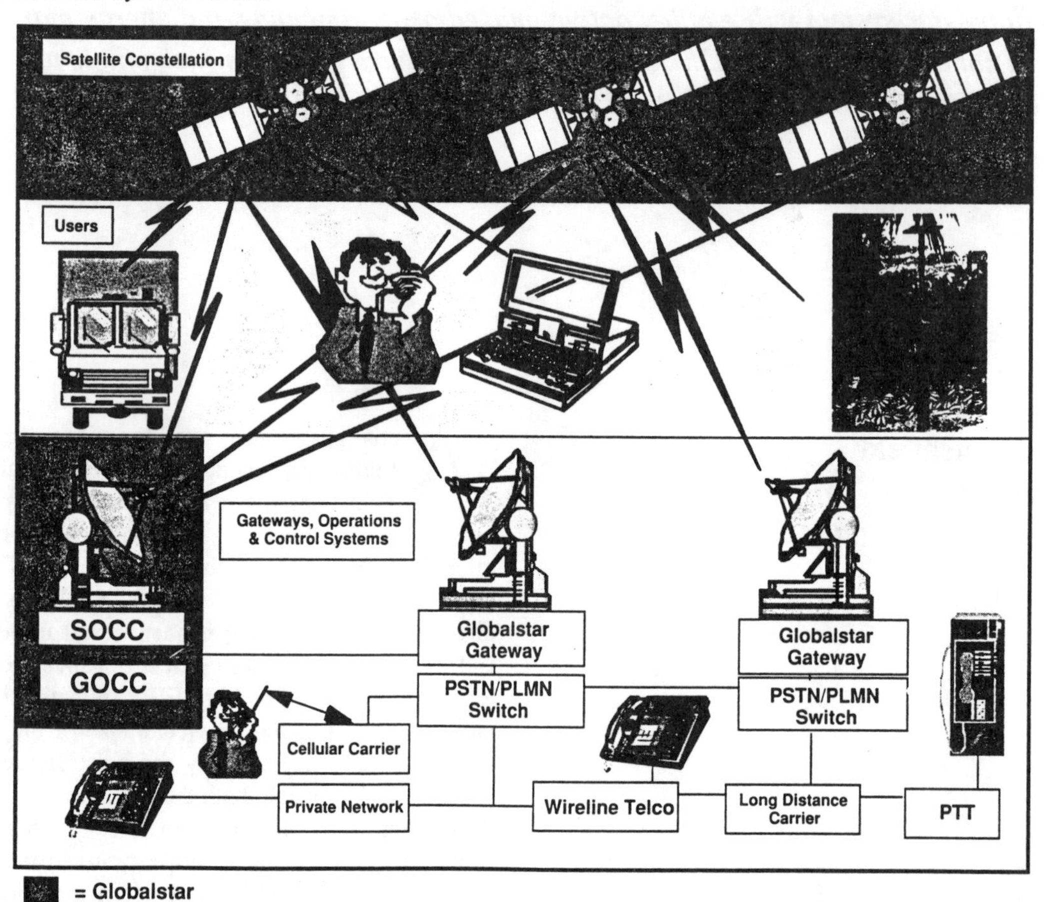

A constellation of 48 satellites in 8 planes (6 satellites per plane) inclined at 52° relative to the equator will be deployed. The first group of satellites will be launched in mid 1997; interim service will begin in mid 1998 and full service will be initiated early in 1999. Figure 2 shows how the system is configured. A hand-held user terminal is used for illustration, but this could equally be a fixed phone in a booth or a phone integrated into an automobile or other vehicle.

The primary difference compared to terrestrial cellular systems is that the radio signals are relayed through satellites to the base stations. In this system, base stations are called Gateways. The operating frequencies are at 1.6 GHz and 2.5 GHz instead of the 800/900 MHz range used by terrestrial systems. The frequency plan is shown in Figure 3. Satellites at a height of 1410 KM provide coverage areas as great as 5,000 KM in diameter, compared to the 20 KM range accommodated by the 20 meter high towers of terrestrial systems.

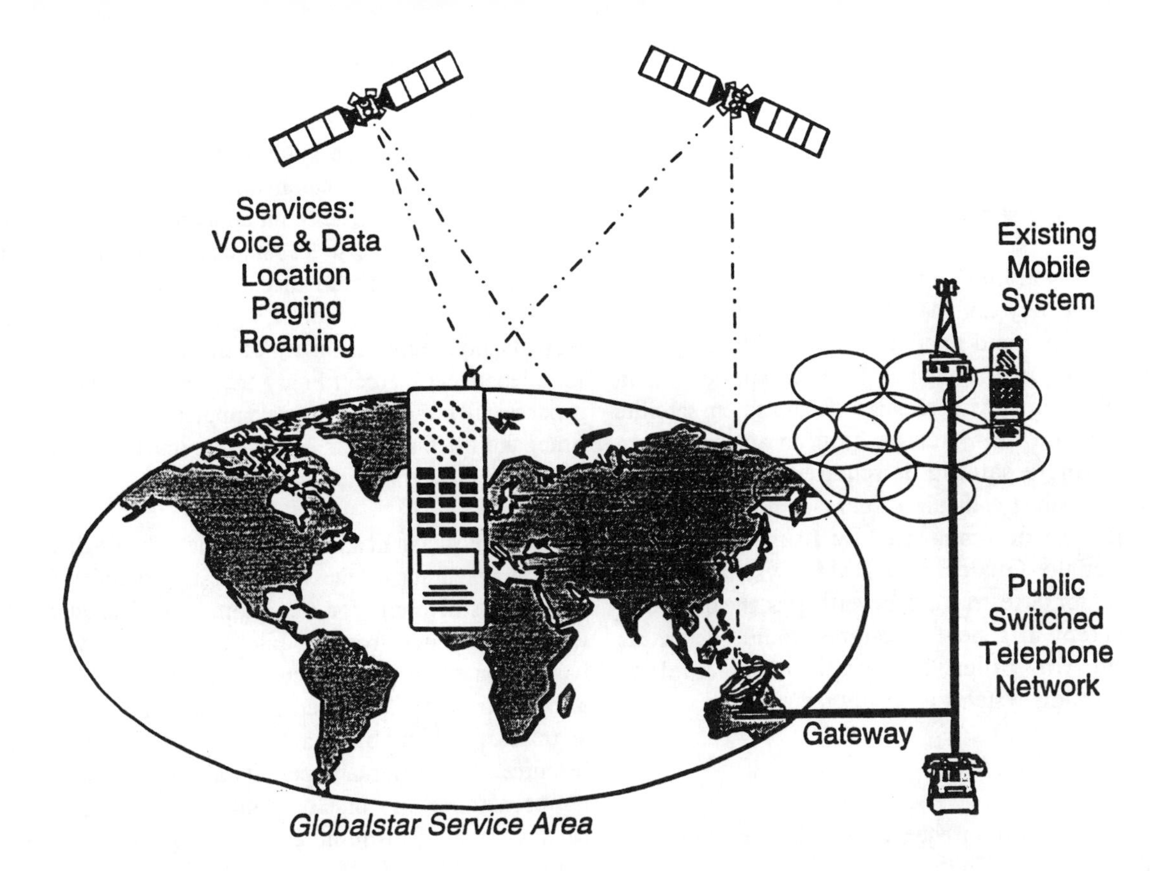

Figure 2. System configuration.

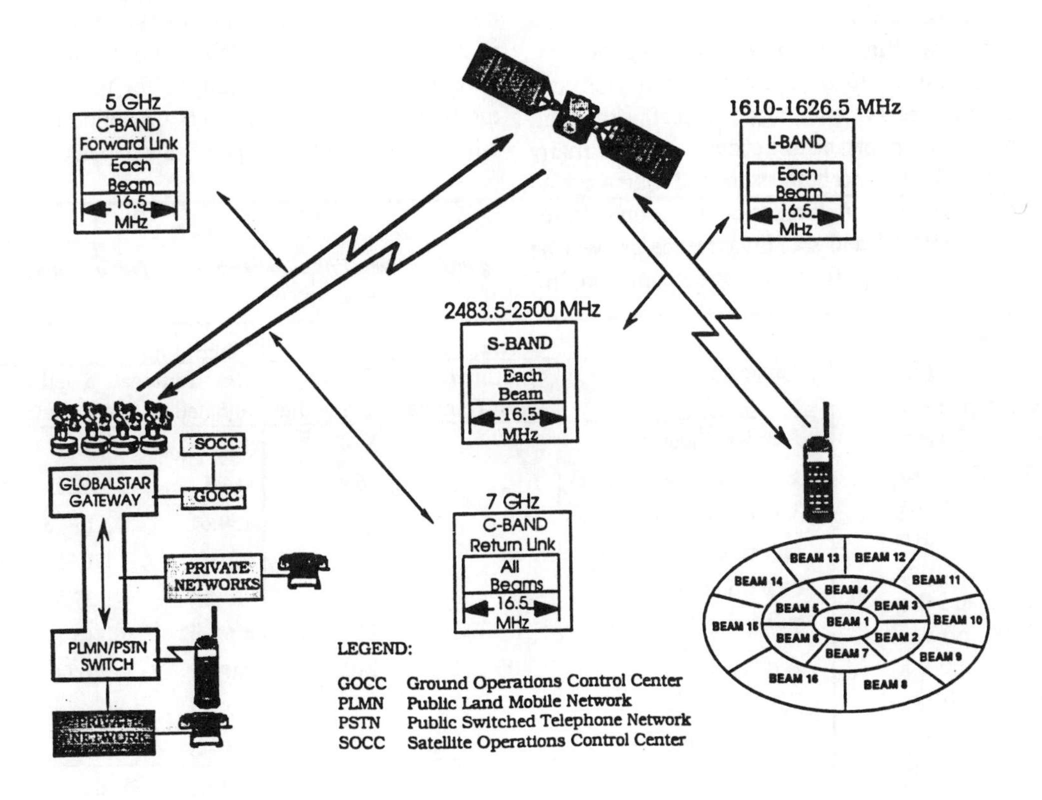

Figure 3. System frequency plan.

The satellites utilize simple frequency translating repeaters. The received signals from the users in the 1.6 GHz range are converted to the 7 GHz range for retransmission to the Gateways. Conversely, signals from the Gateways at about 5 GHz are converted to signals at about 2.5 GHz for retransmission to the users.

Globalstar satellites orbit the earth in 114 minutes. Accordingly they are in view of a user for up to 14 minutes. This rapid relative motion results in higher Doppler shift in the signals than is experienced in terrestrial systems. Also, seamless hand-off from satellite to satellite and from antenna beam to antenna beam within a single satellite is required to provide for continuous phone calls placed at random times. These differences are accommodated by a slightly modified version of the Qualcomm CDMA system already developed for terrestrial use; nevertheless, the degree of risk inherent in a newly designed radio system as proposed for other low earth orbit satellite communication systems is greatly reduced

Communication Links

There are four radio frequency links between the users and the Gateways. These range in frequency between 1.6 and 7 GHz; each having different characteristics. The link parameters are a function of look angle between the ground and the satellites as well as where the users and Gateways fall within the satellite antenna beams. Propagation loss depends upon atmospherics and whether there are trees or buildings in direct line of sight. In addition, it is important to account for how many satellites are available to both users and Gateways to enable path diversity. A further complicating factor depends upon mutual and self interference as well as interference resulting from the spectrum sharing expected in the assigned frequency bands. Finally, geographic distribution of users affects the density of use of frequencies and codes, in turn affecting mutual interference and the amount of power requisite to close a link (i.e. establish communication between the user and the system).

System capacity depends upon all of the above factors as well as regulatory limits on PFD (power-flux density) and interference with other systems in adjoining frequency bands. The system must deal with these employing a worldwide communication network control system, allocating capacity on an as needed basis. For this reason no simple set of link equations can be used to assess system performance. Most elements in the customary link budget are parametric and the relationships between many of these parameters are interdependent. Accordingly the complex allocation process will employ intensive computations in near real-time, and the link parameters will be under continuous management within the system.

Taking into account all of the possible link impairments such as noted above, the system design includes EIRP and G/T sufficient for a margin pool. Resource is temporally drawn from the pool on a frame-by-frame basis using real-time power control for each link. CDMA allows power to be moved from one link to another seamlessly to enable margin to be a system-wide resource. This advantage permits a lower average link power margin than would be required were it necessary to assign a minimum, excessive power to each communication link.

Although the link parameters vary for each link in a dynamic manner, a representative simplified link budget is shown in Table 1. Note that all of the aforementioned parametric factors are grouped for this illustration on the line called *Other Factors*, a detailed description of which is beyond the scope of this paper.

Table 1. Nominal link parameters for the four operating frequencies.

Parameter	Forward Link		Return Link		Units
	Satellite to User Terminal	Gateway to Satellite	User Terminal to Satellite	Satellite to Gateway	
Nominal Frequency	2.5	5	1.6	7	GHz
EIRP/User (nadir, clear)	1.1	36.4	-14.3	-33.3	dBW
Space Loss (nadir, clear)	163.4	169.7	159.6	172.3	dB
Other Factors	1.40	7.90	2.10	1.60	dB
G/T (hand held)	-26	-29.6	-14.25	27.5	dB/K
Eb/No*	5.1	24	4.5	15.1	dB
Average Data Rate	2400	2400	2400	2400	bps
Diversity Benefit			2.2		dB
Composite Eb/No	5		6.3		dB

* Satellite to U.T. & Satellite to Gateway link Eb/No are combined results of both up and down paths.

The Ground Segment

The ground segment consists of Gateways employing up to four bi-directional C-band feederlink 5 Meter diameter antennas. These connect signals to/from the satellites and to/from the terrestrial telephone infrastructure for each coverage region. A coverage region may be a portion of a country, an entire country or a collection of countries sharing a Gateway.

User terminals include hand sets similar to terrestrial cellular telephones. These operate in triple or dual mode for interoperation with existing AMPS (amplitude modulated phone system), GSM or other cellular networks. This system typically will be selected automatically when terrestrial service is unavailable. There also will be fixed service from phone booths and mobile service from automobiles and other vehicles. Handsets will operate with power output that ranges from 150 mW to more than a watt for appropriate applications.

Frequency Re-Use

In order to achieve cost-effective capacity in the 16.5 MHz bandwidth that has been allocated, the field of view of each satellite is divided into 16 beams. CDMA allows 16 times reuse of the frequency ranges. This requires 16 frequency translating converters in both the forward and reverse direction on each satellite. A block diagram of the satellite payload is shown in Figure 4. This includes a 16 beam receive antenna centered at 1618 MHz and a 16 beam transmit antenna centered at 2492 MHz. The 5 and 7 GHz antennas employ a single beam and can operate with lower gain because they are sending and receiving signals from 5.5 Meter diameter high gain Gateway antennas on the ground. The 16 beams of signals received and transmitted from the users are frequency division multiplexed into single channels with 160 MHz bandwidth each for communication with the Gateways, most of which are equipped to track up to 4 separate satellites simultaneously.

Active phased array satellite antennas

Active phased array satellite antennas have been developed for this system which have dedicated low noise amplifiers (LNAs) for the receive and dedicated high power amplifiers (HPAs) for every antenna element. The 2500 MHz HPAs on the satellite are Monolithic Microwave Integrated Circuit (MMIC) solid state power amplifiers. Phased array antennas using passive phase shifters for beam steering have not found wide application in communication systems in the past because there has been too much signal loss between the elements of the antenna and the associated amplifiers.

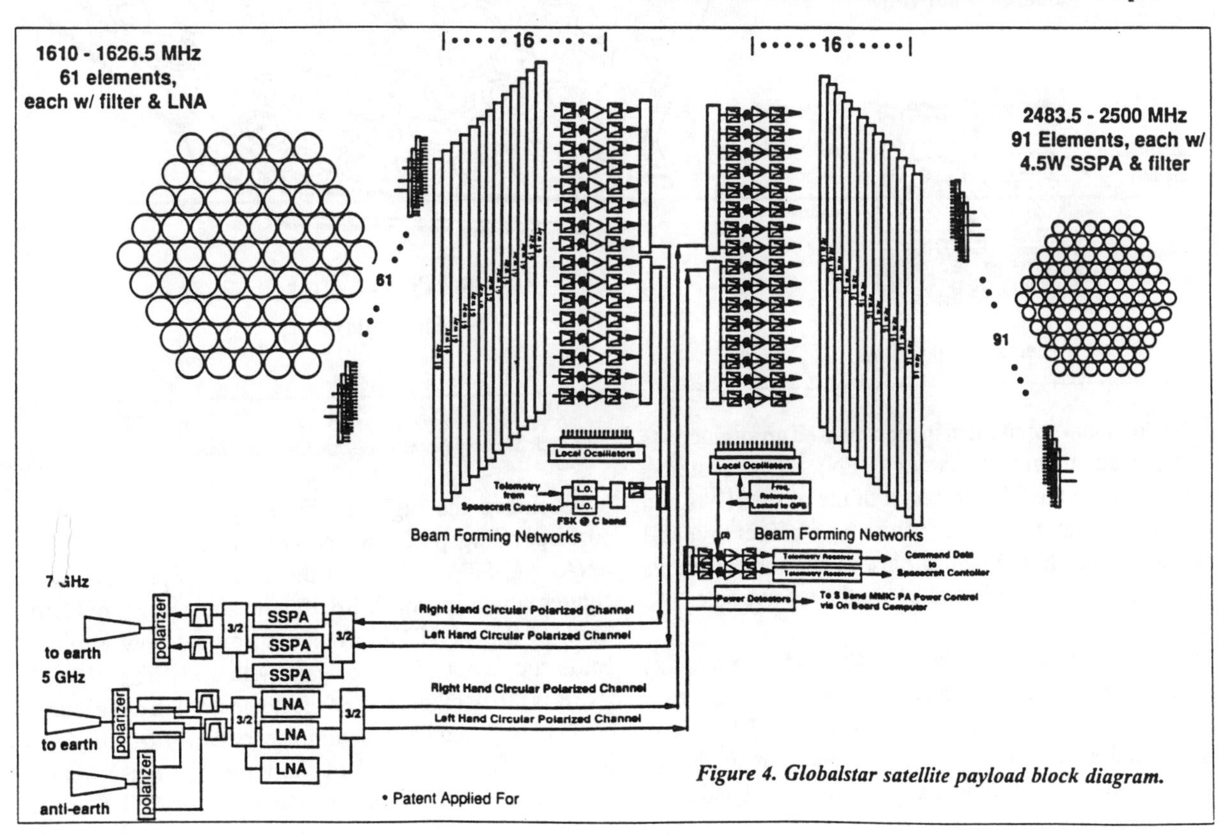

Figure 4. Globalstar satellite payload block diagram.

Phased arrays have been used in radar systems or purely pulse modulated TDMA (time division multiple access) communication systems wherein signal amplitude linearity was not important.

However for this system both efficiency and linearity are very important because they are critical factors in system capacity. Ultimately they are a substantial portion of the basis for establishing the charge for each minute of call time. For these reasons, active arrays are used.

The Transmit antenna is shown in Figure 5. The receive antenna is almost identical to the transmit antenna except that LNAs are substituted for HPAs. Note that the only functional components between these amplifiers and the antenna radiators are filters, transmission media and hybrids which act as polarizers. The net loss due to these passive components is held to less than 2 dB.

temperature stabilized Teflon™ loaded dielectric material. A typical BFN combines or divides the energy (respectively for receive or transmit) from/to each element into 16 equal parts to feed 16 beam generators. Each beam generator establishes the relative amplitude and phase weights for every antenna radiating element. The value of these weights, in turn, establishes the direction of the beam, its gain and shape. A beam generator is comprised of two of the 32 printed circuit boards cited above. Interconnection between these boards is by plated through holes. Interconnection between the 16 beam generators is by intra-layer couplers, consequently plated holes are not required through more than 2 layers. A picture of a beam generator is shown in Figure 6. The 37 to 40 dB of loss referred to above includes the 16-way power dividers/ combiners and the 61 or 91-way power dividers/ combiners for each element as well as the structures for establishing phase and amplitude weights.

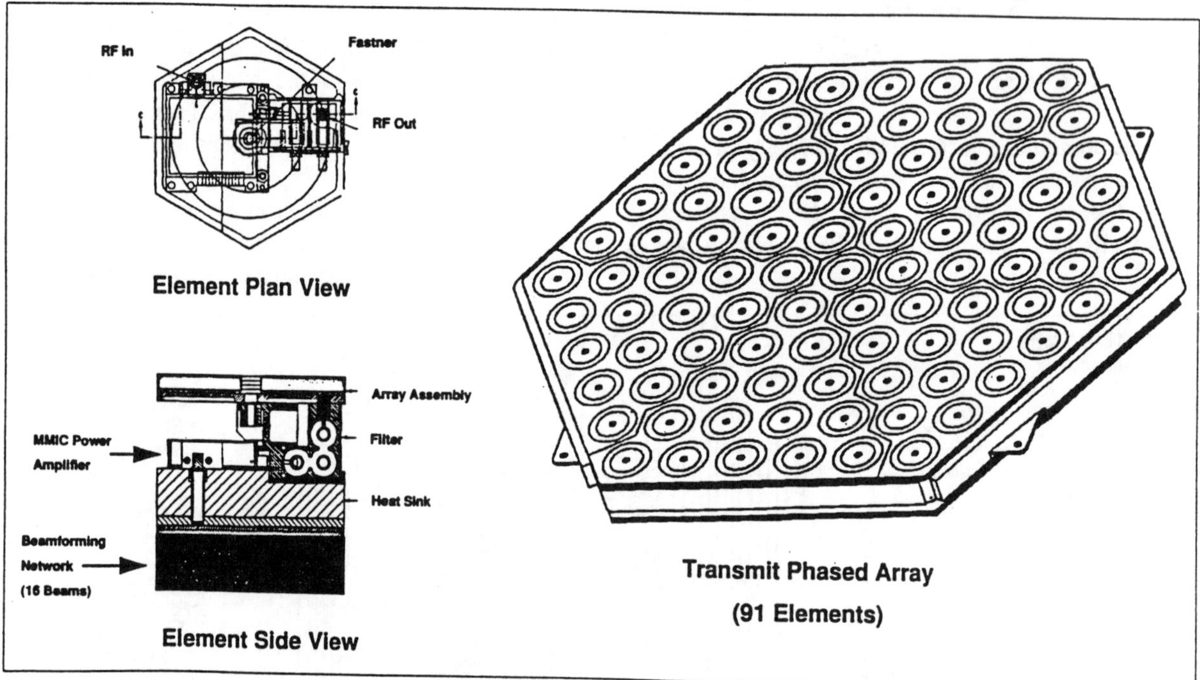

Figure 5. Sixteen beam user transmit antenna.

The dominant component in most phased array antennas is the beam forming network (BFN). This system's antennas have the BFN in-board of the of the HPAs and LNAs so that their loss does not affect overall performance. These antennas have 91 radiating elements on the transmitting side and 61 on the receive side.

BFN loss is approximately 37 to 40 dB (for Rx and Tx) because they are implemented in stripline. There are no coaxial cables or wave guides within the BFN. It is comprised of a 32 layer bonded printed circuit board with 64 surfaces of etched copper circuitry bonded to

To achieve the highest efficiency and performance, corresponding performance is required in the MMIC LNAs and HPAs. Each of the 61 LNAs provide noise figures which approach 1.6 dB with about 40 dB of Gain. Each of the 91 HPAs provide an output power of 4.2 watts and about 50 dB of gain with an NPR of 17 dB. (NPR is the decibel ratio of the noise level in a measured channel with the base band noise loaded—to—the level in that channel with all of the baseband noise loaded except the measured channel. It is a measure of linearity).

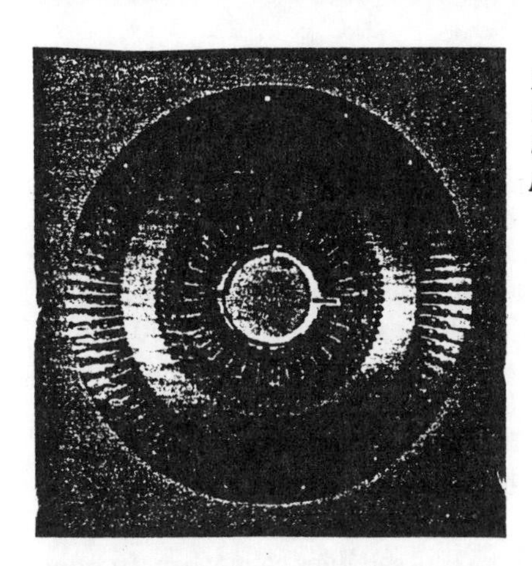

Figure 6. One of the 16 beam generators in the user transmit phased array.

These HPAs provide a power added efficiency of 33% (including integrated isolator). A further feature of the HPAs is that every HPA serves all 16 beams simultaneously. As the satellites pass over subscriber population centers and power is required to pass from beam to beam, it does so without changing the demand on individual amplifiers. Every beamforming generator in the transmit antenna has an amplitude coefficient of unity, so every amplifier experiences the same power at the same time. This virtually eliminates a temperature or signal level gradient across the array that would otherwise result in different amplitude or phase transfer characteristics and attendant antenna performance degradation.

This phased array approach also avoids separate amplifiers for each antenna beam and thus the requirement to handle maximum transmit power in every beam simultaneously. In a practical situation some beams are required to handle close to the maximum power and some considerably less. By using the same amplifiers for all beams simultaneously, the power moves seamlessly from beam to beam, based upon demand; and no energy is wasted in bias currents to unused amplifiers. In another proposed system, for example, this same functionality is achieved by four 8-way multiport amplifiers to power their 32 beams. Multiport amplifiers were considered for this system but rejected because the hybrid network required between the internal solid state power amplifiers (SSPAs) in the multiport amplifiers and the antenna feeds might have 3 dB or greater additional loss. By contrast, the system antennas combine the outputs of their multiple MMIC amplifiers in free space, therefore losslessly. The advantage of this feature can be appreciated by recognizing that with this system the S-Band satellite transmitter requires over 1000 Watts of prime power at maximum output, whereas other satellite communication systems not so configured require double this power for the production of a comparable EIRP.

Consideration of bias power is an important factor in the HPAs. Typically, class A operation would be required in these amplifiers to provide a NPR of 17 dB. In this case, pseudomorphic high electron mobility power transistors (PHEMPT) devices integrated with MMIC drivers developed for this system by the Raytheon Corporation maintain this NPR while operating in a class A-B mode. Furthermore, they operate with virtually the same gain and other transfer characteristics from a supply voltage range of 2 to 8 volts. The power added efficiency across this range varies by about 10%, but the power capacity varies by 16 dB. This means that the power consumption of the transmit antenna/HPA(s) is adjustable from 130 to 1350 Watts with little sacrifice of efficiency. A picture of a MMIC LNA and HPA with associated bandpass filters is shown in Figures 7a and 7b. A summary of the performance characteristics of the MMIC amplifiers is given in Table 2.

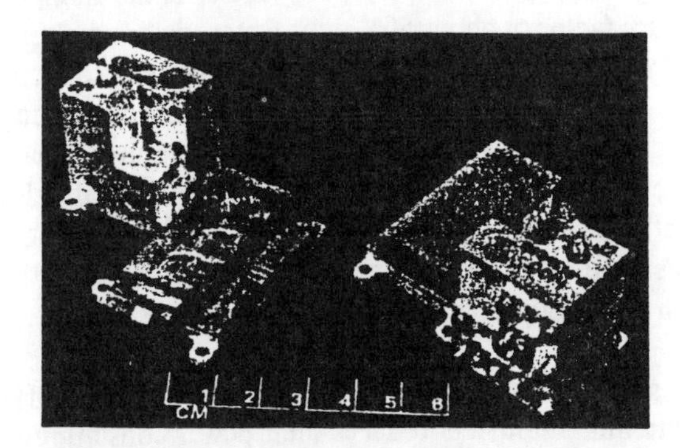

Figure 7a. Low noise amplifier with Rx filter.

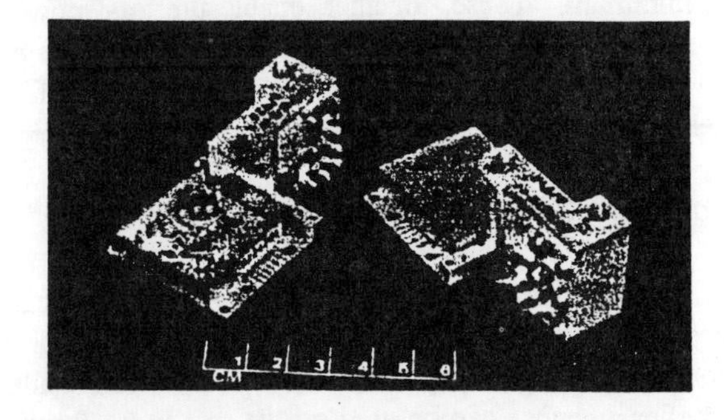

Figure 7b. High power amplifier with Tx filter.

In the satellites, the power flux density from the Gateways is measured on the 5 GHz uplink and used to control the power capacity in the 2.5 GHz HPAs. This is done by controlling special dc/dc converters developed for this system which maintain efficiency of 82 to 92% throughout the 2 to 8 volt range, producing more than 180 amps in their highest power state. The converters are unique by virtue of maintaining such high efficiency at output voltages as low as 2 volts.

	LNAs (61 / spacecraft)	HPAs (91 /spacecraft)
Frequency Range	1610 - 1626.5 MHz	2483.5 - 2500 MHz
Noise Figure	1.6 dB max.	n.a.
Power Output	4.5 W @ NPR = 17dB	n.a.
Gain	40 dB min.	50 dB min.
Efficiency	33 % min.	n.a.
Pwr Capacity Range	16 dB	n.a.
Phase Mismatch	6º max. (including filter)	6ºmax. (including filter)
Amplitude Mismatch	0.5 dB (including filter)	0.5 dB (including filter)

Table 2. MMIC Antenna Amplifier Performance

This is accomplished through the use of *buck converters* wherein the output of a chopper is pulsewidth modulated to vary the output voltage. The average value of the output voltage is proportional to the duty factor of the chopper. A duty factor of about 50% duty factor yields an output of 8 volts, 12.5% duty factor yields 2 volts, and so forth. The resultant output power is then smoothed and integrated by a low loss filter to yield the required steady dc level to bias the amplifiers. The combination of pulse width modulation and a power FET with very low resistance in the on state (a few milliohms) results in the high efficiency over the wide output voltage range.

A similar power control implemented for the 7 GHz feederlink amplifiers reduces their power consumption from 280 Watts at maximum traffic to less than 90 Watts at minimum. These, together, enable the total power consumption of each satellite to vary from 600 Watts in the quiescent state to more than 2000 Watts as traffic demand peaks.

Low Earth Orbiting (LEO) satellites

LEO satellites require less G/T and EIRP than those at greater altitude because of their lower altitude and correspondingly lower propagation path loss. As a result, smaller antennas can be used on the satellites. Lower altitude also results in less propagation delay, thereby obviating the annoying pauses that are inherent in GEO (geosynchronous earth orbits, about 19,000 miles above the earth's surface) or digital MEO (medium earth orbit) systems that use time consuming digital signal processing (DSP) algorithms for vocoding. Low propagation delay can be enjoyed by this system's users because the time saved is not consumed in on-board processing, as is the case for a competing system proposal. This system's satellites employ translating repeaters, which introduce imperceptible delay.

Figure 8 shows a typical system profile of transmit power required as a function of time throughout a 24 hour period. Note that the power peaks are rather narrow and that the average power generated by the Solid-State-Power Amplifiers (SSPAs), while peaking at about 350 watts, average only about 80 watts. This occurs because the satellites move completely around the earth in 114 minutes, passing over subscriber populations rapidly.

By contrast, MEO or GEO satellites cannot benefit from this averaging because they move so slowly that power consumption changes in approximately hourly increments, including the "busy hour" rather than over the shorter periods displayed in Figure 8.

CDMA Waveform

Efficiency in the Qualcomm CDMA architecture begins with voice codecs. They digitize voice at a rate that is dependent upon whether a user is speaking or listening and the rate required to encode the phonemes that are being used when speaking. The Qualcomm codec used for this system operates at up to 9.6 Kbits/sec and at several intermediate rates, depending on user voice characteristics. The codec idles at a very low rate when a user is not speaking. The average rate, including pauses and intermediate rates, is only about 2.4 Kbits/sec. This average rate is used for calculation of link margins and it is valid because of the law of large numbers. Rate and transmit power is varied on a frame by frame basis using link quality feedback as a criterion. The power control link is closed on a quality roughly equivalent to current wireline telephone, close to a Mean Opinion Score (MOS) rating of 3.5.

The second factor that contributes to efficiency is the use of interleaving and forward-error-correction coding (FEC). Viterbi maximum likelihood decoding with FEC at 1/2 rate, constraint length 9 is employed. This allows the transmitted power to be decreased by about 5.2 dB for an equivalent error rate commensurate with the transmission of uncoded data. Half rate FEC means that two symbols are transmitted for every data bit. Constraint length 9 means that each FEC code symbol is propagated across 9 data bits. Interleaving decreases the vulnerability of the link to burst errors.

Reprinted from *IEEE Communications Magazine*, Vol. 29, No. 11, Nov. 1991, pp. 48-51.

The Traveller's Dream Come True

The Iridium^{sm} satellite-based personal communications system will bring cellular telephone service to virtually every point on the globe.

Jerry L. Grubb

Even before the Dick Tracy comic strip introduced the amazing wrist radio, there was a desire in all of us to be able to communicate from and to anywhere. What a comfortable feeling it would be to have a portable telephone to call the kids or the office while hiking in the Andes, sailing in the Caribbean, or driving in the deserts of California or Mexico, when well out of range of a terrestrial cellular system. What a benefit to travelers and residents of small villages in remote areas without telephone infrastructures to have the use of a solar-powered phone booth, connected to the world through a wireless network.

The Iridium system satisfies this age-old desire in a manner similar to terrestrial cellular systems; but with cell-forming antennas and radio relays located on satellites rather than on the ground. It almost resembles a terrestrial cellular system upside down. By using a constellation of 77 satellites, at least one is in radio Line-Of-Sight (LOS) at all times from every point on the earth; thus, you are never out of range of a "repeater," as is often the case with terrestrial cellular service.

The global economic and industrial process providing the momentum for the Iridium system is man's apparently unquenchable desire for mobile and portable communications. The demand for terrestrial cellular telephone service has far outpaced the marketeers' projections, and the usage patterns are tied quite closely to local demographic considerations. The number of cellular telephones now exceeds eight million. In early-1980s projections, seven million were anticipated to be achieved by the year 2000. More recent projections are as high as 100 million worldwide by the year 2000. The Iridium system, however, does not replace or substitute for cellular telephone service. Rather, it extends radio-telephone coverage to the entire world.

The Iridium system, by its very nature, is a lower-density higher-priced service than terrestrial cellular. Where an area is covered with a terrestrial-based cellular system, Iridium is a backup or emergency service. In areas of the world where no mobile service is readily available, Iridium is the mobile system. In areas of the world where mobile service is only provided with geostationary satellites, Iridium provides more channels, shorter delays, smaller lighter-weight radio-telephone equipment, and worldwide networking. Also, in areas of the world where there is no telephone service, Iridium can provide telephone service.

The system is an amalgamation of technologies that were creatively interwoven by a small team of engineers with dissimilar backgrounds inside a company with diverse areas of expertise. The key technologies include wireless communications in two realms: space communications systems and cellular telephone systems. Important supporting technologies include small satellites, phased-array antenna systems, functionally dense radiation-tolerant semiconductors, advanced baseband processing architectures and distributed network architectures. But why call it Iridium? A Motorola engineer noticed that the 77-satellite constellation circling the globe is analogous to Bohr's atom for the element "iridium" with 77 electrons. In an instant, the system had a name.

Many services will be provided to a wide variety of customers, including voice, paging, and messaging services, Radiodetermination Satellite Service (RDSS), and facsimile and data services. An international business person with a portable unit in his coat pocket can have easy access to the home office, and the head of a large multinational corporation can quickly call any of his general managers, whether they are at home or traveling, on land, at sea, or in the air, anywhere in the world. The mountain climber, hunter, skier, or recreational sailor can electronically determine his geographical location and communicate with home or his brokerage business. Third-world countries without a telephone infrastructure can have subsidized solar-powered centrally located telephone "booths" in every village. Land and sea mining operations can have continuous worldwide service. Areas experiencing natural disasters can maintain a reliable communications link to the rest of the world.

Iridium is still in development; and though a com-

Jerry Grubb is a Senior Systems/Communications Engineer at Motorola Satellite Communications, Inc., Chandler, Arizona.

Iridium^{sm} is a trademark and service mark of Motorola, Inc.

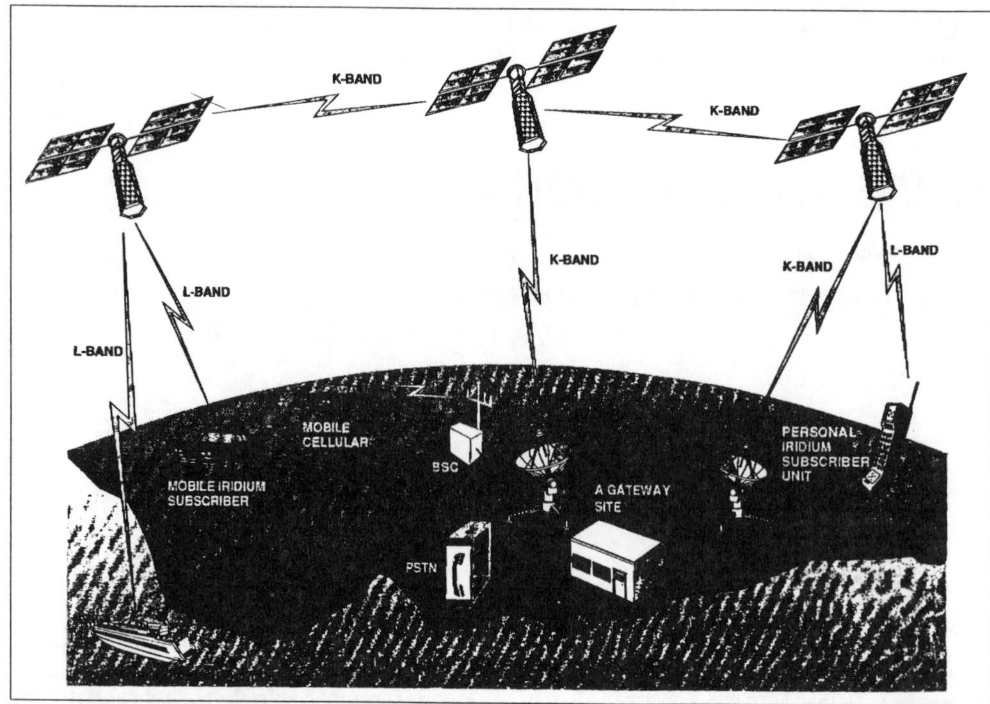

Figure 1. LEO satellite network

plete integrated baseline design exists, tradeoff analyses continue, and the final design for the satellites will not be frozen until early 1992. The system is to become operational by 1997.

Some of the primary technical parameters of Iridium are described here, but quite obviously, Iridium's foremost challenges are not in the technology; rather, the regulatory and licensing aspects of a truly worldwide portable radio-telephone service are clearly the dominant issues.

System Overview

The backbone of the Iridium system is the constellation of 77 small smart satellites in polar Low-Earth-Orbit (LEO) at an altitude of 420 nautical miles (778 km). This relatively low altitude assures that a user will never have to communicate further than about 1,250 nautical miles (2,315 km) to a satellite. This allows the use of lightweight low-power radio-telephones with small low-profile antennas compared to those required to utilize geostationary satellites at ranges of more than 19,000 nautical miles (35,188 km). The constellation comprises seven planes of 11 satellites, each plane equally spaced in longitude. The orbital period of each satellite is approximately 100 minutes.

The satellites are interconnected via microwave crosslinks, thus forming a global network in space, with linkage to the ground via gateways. Each gateway interfaces with the local Public Switched Telephone Network (PSTN), providing local area user/customer record keeping and interconnection caoability between Iridium and non-Iridium users. Networking in space has advantages over ground networking in that it can provide full global coverage over land and sea with relatively few gateways, and it provides direct satellite-to-user service in remote areas without requiring that a gateway be operating within the staellite's LOS. The decision to locate a gateway in any given area can be made by local governments and/or servers solely out of a desire to offer local service, rather than to satisfy a functional need by the system for the gateway to be a node in the network to assure global coverage. Satellite TT&C and overall system control will be accompanied by one or more centrally located System Control Facilities utilizing earth terminals and microwave links.

The radio links between users and satellites will operate somewhere in the 1,600 MHz to 1,700 MHz region of L-Band, while the gateway feeder links and satellite crosslinks will operate in small portions of the 18 GHz to 30 GHz region of K-Band[1]. L-Band is best suited for direct satellite-to-portable-user links because it offers a good mix of relatively short wavelengths and low propagation losses, and it includes the upper frequency limits of economical commercial hardware using available technologies.

Each satellite employs an L-Band phased array antenna complex that forms 37 interlaced spot beams (cells). Each beam operates independently. Each carrier of each beam has its downlink transmit power electronically controlled to maintain link quality for each user, allowing for variations in range and blockage/foliage.

The cell pattern is fixed relative to the space vehicle, but rapidly moving on the earth's surface. Handoffs occur from cell to cell as a subscriber unit is operated, similar to today's cellular telephones; but unlike those cellular telephones, the Iridium cells move through the users rather than the users moving through the cells. Handoffs occur with the same frequency (about one per minute) as with cellular telephones, but the handoffs in Iridium involve fewer handoff options and better information with which to choose between the options than today's cellular telephone systems; in that sense, Iridium is less complex. However, since Iridium's cell patterns overlap little in the low latitudes and very much in the high latitudes, Iridium has two problems to deal with that cellular telephones do not have.

The backbone of the Iridium system is the constellation of 77 small smart satellites in polar Low-Earth-Orbit at an altitude of 420 nautical miles.

[1]Pending FCC and WARC approval.

Year	2001		2006	
Market Segment	USA	Total	USA	Total
Maritime	7,000	35,000	8,000	45,000
High Income	180,000	448,000	270,000	800,000
Business Travel	210,000	820,000	335,000	1,515,000
Aeronautical	13,000	22,000	16,000	34,000
Industrial	50,000	140,000	70,000	230,000
Government	36,000	292,000	50,000	450,000
Rural	0	67,000	0	150,000
Total	496,000	1,824,000	749,000	3,224,000

Table 1. Iridium demand projections (number of subscribers)

First, cells must be turned on/off depending upon their position in the orbit. For example, an individual cell that is turned on when over the equator may be turned off when at a higher latitude due to excessive overlap of cells when approaching the poles. Of course, if operation were prohibited in some part of the world, more would be turned off. Second, the channel reuse pattern must be reset at different times during an orbital period. (A modified seven-cell reuse pattern is used.) The seven-cell reuse pattern was selected in a tradeoff analysis that considered the mainlobe and sidelobe interferences realized in alternative antenna designs. This reuse pattern offers part of the spectral efficiency realized with Iridium.

The user baseband format is fully digital, utilizing high-quality vocoded voice messages at 4,800 b/s with Forward Error Correction (FEC) coding. The vocoding technique used is called Vector Sum Excited Linear Prediction (VSELP). The VSELP speech coder utilizes a codebook with a structure that allows for a very efficient search procedure. Testing has shown that voice digitized with this vocoder sounds equivalent to high-quality analog voice used in today's land mobile radio equipment.

The user radio links employ Quarternary Phase-Shift Keying (QPSK) with a combination of Time-Division Multiplexing (TDM) and Frequency- Division Multiplexing (FDM). In addition, the cellular architecture provides Space-Division Multiplexing (SDM). The combination of both TDM and SDM allows each carrier frequency to be reused globally a great many times(more than five times in the U.S. alone). Techniques such as raised cosine filtering shape and reduce the spectral content of each voice channel to about 2 kHz. The reuse assignment employs a combination of fixed and dynamically assigned channels, so that both the world as a whole and the possibly higher concentration of users in isolated localities can be handled with a balanced efficiency. A single cell, on average, can service approximately 236 simultaneous users, while a limited number of geometrically isolated cells can handle peak loads of two to three times that many with reassignments of reuse functions on a dynamic basis. The worldwide system capacity is approximately 283,000 channels with an occupied bandwidth of 14 MHz, which nets a spectrum utilization of less than 50 Hz per channel.

Though a variety of voice and/or data subscriber units are feasible, the initial development focuses on pagers, personal handheld cellular-like units, and various mobile units that can be installed in an automobile, boat, or aircraft.

The mobile units anticipate the availability of power sources and antenna configurations better suited to wireless communications than the handheld unit, and thus provide less challenging design constraints. The Iridium handheld unit will utilize a small low-profile antenna and will be designed to comply with the same size, weight and battery life constraints as terrestrial cellular units—a nearly impossible task if communicating with geostationary satellites. In addition, radio frequency (rf) radiation levels are compliant with the most restrictive personnel limits recommended by the World Health Organization.

The handheld unit can operate for 24 hours on a single recharge: 23 hours of standby (able to receive a "ring" indicating an incoming call) plus one hour of continuous calling time. The system can be operated with subscriber units with less than 600-mW transmitters (comparable to cellular telephones).

In addition to voice, Iridium customers will find it important to transmit and receive a wide variety of data worldwide. This could be in the form of text or files from a personal portable computer or images from a fax machine. To accommodate this need, the Iridium subscriber equipment will be available with an industry-standard data port, such as RS232, to transport data from an external device. Several levels of data service will be available, from basic 2,400-baud asynchronous circuits without error checking to higher-speed packet transport sold on demand.

Services Offered

The Iridium system is designed to provide global telecommunications service on a continuous basis, with coverage of all points from the North Pole to the South Pole. The system will support many geographic and application markets with a single infrastructure. After the satellite constellation is in service, the Iridium system can provide service to all markets with minimal additional infrastructure costs.

Present plans call for bulk transmission capacity on the Iridium system to be provided to licensed and authorized operators, who in turn will sell service to the public in their authorized area. These operators will collect payments from customers and also serve as an outlet for Iridium subscriber products.

The Iridium system's reliability, quality and performance will be comparable to the emerging digital cellular telephone service. The system is designed to allow worldwide usage with both incoming as well as outgoing calls, regardless of the subscriber's home and current location. Sovereign jurisdictions will, of course, maintain the ability to control calls from or to their sovereign territory.

Five types of basic service are presently planned, each with a separate terminal or a family of terminals:

• Radiodetermination service and two-way messaging: All subscriber units, except some pagers, will contain this capability, which includes automatic location reporting. Units with the Global Positioning Service (GPS), or GLONASS, will also be available, which will provide greatly increased geolocation speed and accuracy.

• Digital voice communications service: The system will provide duplex high-quality 4.8 kb/s voice communications.

• Paging: An alphanumeric pager for instantaneous direct-satellite paging in any region of the world is being designed in a package only slightly larger than today's pagers.

• Facsimile: Two types of mobile facsimile units are planned, a stand-alone facsimile unit and a unit to be used with an Iridium telephone.

• Data: A 2.4 kb/s modem is being developed for use within the Iridium network.

Though designed for these types of services, the Iridium system will evolve as people's imaginations conceive continuously novel uses. The service of the Iridium system is not limited by today's imaginations any more than today's telecommunications are limited by Alexander Graham Bell's inspiration of a century ago.

Market Projections

Since no one can truly estimate the number of subscribers Iridium will generate or the usage patterns of those subscribers, it is difficult to put an exact number on the subscriber units the system can accommodate. Nevertheless, by most accounts the two percent (0.02 Erlang) average user rate experienced by cellular telephones is thought to be high. However, even if Iridium were used at this two percent rate with the baseline point design, several million users in reasonable locations in the temperate latitudes can be serviced. (Virtually all the others, in less likely subscriber areas, can also be serviced.)

As detailed in Table I, nearly 500,000 subscribers are anticipated in the United States alone by the year 2001, with over 1.8 million worldwide. By the year 2006, more than 3.2 million subscribers are expected worldwide. The baseline architecture does allow for growth, and of course the initial constellation's capacity could be adjusted slightly higher or lower to accommodate costs, spectrum allocations, schedule adjustments or other nontechnical considerations.

Innovative design technologies and manufacturing engineering, combined with relatively high quantities, will yield inexpensive satellites. In addition, each satellite will weigh as little as approximately 1,100 pounds (500 kg), including expendables, allowing the use of small inexpensive launch vehicles. These significant economic advantages allow timely satellite replacements and upgrades to assure high availability of state-of-the-art service in a rapidly growing market.

Concluding Remarks

A LEO satellite-based personal communications system was described, as were some of the key system-level considerations that have driven the design. Many specific details were omitted because of the competitive nature of the mobile telecommunications business.

Iridium represents a bold step into the future in terms of portable radio-telephone capability and its worldwide networking capability. It also represents a paradigm shift in satellite communications when compared to geostationary satellite systems with their large expensive equipment and limited coverage areas. Iridium equipment will be small and affordable and will provide worldwide capability 24 hours a day. The age of the Dick Tracy radio has arrived!

Nearly 500,000 subscribers are anticipated in the United States alone by the year 2001.

References

[1] I. Gerson and M. Jasiuk, "Vector Sum Excited Linear Prediction (VSELP)," *IEEE Wksp. on Speech Coding for Telecommun.*, pp. 66-68, Sept. 1989.

Biography

Jerry L. Grubb [M] received a B.S.E.E. from Iowa State Unversity in 1963 and an M.S.E.E. from the University of Nebraska. He is currently a Senior Systems/Communications Engineer at Motorola Satellite Communications, Inc., in Chandler, Arizona, where he has worked on the Iridium Program since late 1989. For more than 25 years, prior to the formation of the Iridium Program, he held numerous technical and project-management positions for government-sponsored programs in the Motorola Government Electronics Group.

Chapter 3

MODERN PROPAGATION PREDICTION TECHNIQUES

This chapter describes modern propagation modeling techniques that will be used to deploy and simulate wireless systems of the future.

Reprinted from *Proceedings of the IEEE*, Vol. 82, No. 9, Sept. 1994, pp. 1333-1359.

UHF Propagation Prediction for Wireless Personal Communications

HENRY L. BERTONI, FELLOW, IEEE, WALTER HONCHARENKO, MEMBER, IEEE, LEANDRO ROCHA MACIEL, AND HOWARD H. XIA

Invited Paper

Propagation characteristics of radio signals in the UHF band place fundamental limits on the design and performance of wireless personal communications systems, such as cellular mobile radio (CMR), wireless LAN's, and personal communication services (PCS). Because the radio link is direct to each subscriber, the prediction of signal characteristics is most important in urban areas where subscriber density is high, and the buildings have a profound influence on the propagation. This paper starts by reviewing the characteristic signal variations observed in CMR systems employing high base station antennas to cover macrocells having radius out to 20 km. Theoretical models incorporating diffraction are shown to explain the observed range dependence and shadow loss statistics. For the low base station antennas envisioned to cover microcells of radius out to 1 km for PCS applications, signal propagation is more strongly dependent on the building environment and on the location of the antennas in relation to the buildings. Various levels of theoretical modeling of this dependence are discussed in conjunction with measurements made in various building environments. Finally, the paper discusses recent advances in site specific prediction for outdoor and indoor propagation.

I. Introduction

The commercial success achieved by the introduction of cellular mobile radio (CMR) telephones has generated strong interest in the development of other wireless communications systems, such as personal communication services (PCS), wireless local area networks (W-LAN), and wireless private branch exchanges (W-PBX). These systems will operate at various frequencies in the UHF band (300 MHz–3 GHz), or even at microwave frequencies.

Manuscript received October 15, 1993; revised May 12, 1994. This work was supported by the New York State Science and Technology Foundation.

H. L. Bertoni is with the Center for Advanced Technology in Telecommunications, Polytechnic University, Brooklyn, NY 11201 USA.

W. Honcharenko and L. R. Maciel were with the Center for Advanced Technology in Telecommunications, Polytechnic University, Brooklyn, NY 11201 USA. They are now with AT&T Bell Laboratories, Whippany, NJ 07981–0903 USA.

H. H. Xia was with Telesis Technologies Laboratory, Walnut Creek, CA. He is now with AirTouch Communications, Walnut Creek, CA 94598 USA.

IEEE Log Number 9403228.

For each proposed system it is necessary to understand the propagation characteristics of the radio channel envisioned for the system.

During the development of CMR, knowledge of the channel was gained almost entirely through measurements closely matched to the anticipated implementation, which employs elevated base station antennas [1]–[8]. Subsequent theoretical studies have uncovered wave processes that are responsible for the observed signal dependence [9]–[14]. Because the theoretical dependence on parameters such as frequency and antenna height is explicit, the theory can be used to extrapolate the extensive CMR measurements to the higher frequency and lower base station antennas envisioned for PCS [15]–[18]. In this way, a well-tested theoretical understanding can enhance the value of a set of measurements by making them transportable to other parameter ranges. In addition, an understanding of the potential wave mechanisms can guide measurement programs by identifying important parameters or geometries. This paper is intended to review theoretical prediction methods that have been used to model propagation in connection with various wireless applications.

The macrocells of CMR make use of base station antennas located well above the surrounding buildings and cover distances R from the base station out to about 20 km. The CMR industry has experience in working with models that give the received power (signal) dependence in the form A/R^n for R in the range 1 km $< R <$ 20 km, where n is typically between 3.5 and 4. When augmented by a statistical description of variations about this simple range dependence, the models have been used to predict coverage within cells and interference between cells. Variations about the range dependence occur over two scale lengths, which are referred to as fast and slow fading [1], [3], [8]. Fast fading has a scale length of one half wavelength, and results from the interference when signals arrive at the subscriber from many directions. It occurs in all situations where the subscriber is located among buildings, vehicles, and other

Ed. Note: The terms "fast" and "slow" fading actually refer to the relationship between the modulation of a signal and the rate of change of the channel. The spatial variations are more accurately referred to as "small-scale" and "large-scale" fading, respectively.

objects that reflect or scatter radio waves, and must be accepted or mitigated with diversity techniques.

When the fast fading is smoothed by averaging over 20 or so wavelengths, the resultant, which is called the sector average, shows a variation over a scale length on the order of 10 m. This slow fading, or shadow loss, is due to variations in the shadowing of the subscriber as he/she moves around in the local urban environment, e.g., past buildings of different height, from mid block into an intersection, or from ground level onto a bridge or elevated highway. Treated as a random variable, the variation can have a standard deviation of as much as 8 dB [4], [6], [19], [20]. Viewed in another way, the slow fading is the error inherent in the simple A/R^n model that results when the effects of environmental features are described in a highly averaged manner through the parameter A.

Proposals for PCS systems have been based on the use of microcells having base station antennas at the height of street lights [15]–[18]. Propagation distances R are envisioned to be 1 km or less, and the coverage area is not expected to be circular. Over such short ranges it is important to distinguish between line-of-sight (LOS) propagation paths, such as down a street, and paths that are shadowed by intervening buildings. At least for the shadowed paths, it is also possible to distinguish between areas having different building types, e.g., the high-rise urban core and residential areas with two-story houses. These distinctions imply that several different theoretical models will be required to accurately predict propagation in the various areas, and that by using area models that reflect the generic building structure, the prediction accuracy can be substantially increased from that tolerated for CMR macrocells. However, CMR-type models may still play a useful role in predicting interference between cells.

At a somewhat smaller distances R, theoretically based site-specific computer programs have been discussed that account for the shape of individual buildings and give essentially deterministic prediction of the sector average signal obtained by averaging the received signal over 20 or so wavelengths to remove the fast fading. For indoor picocells covering all or a portion of a building, computer programs have been proposed to make site-specific predictions of the sector average by making use of the building's floor plan. Validated theoretically based computer programs (models) of this type can be used for system design at specific sites, or as a replacement for measurement campaigns in developing design rules or installation instructions.

At frequencies in the 900-MHz band and above, the electromagnetic wavelength is small compared to the dimensions of buildings, which permits the use of ray-optical methods as a means for describing propagation in and around buildings. Ray optics treats the electromagnetic waves as traveling along localized ray paths. The paths themselves and the associated fields can be traced by sequentially accounting for the local features of the medium as they are encountered by the rays, as discussed in [21]. For CMR and PCS applications, the most important local features encountered by the rays are the ground and planar building elements, such as walls, floors, and roofs.

Disregarding roughness, rays incident on the ground are specularly reflected with amplitude given by the plane wave reflection coefficient. Similarly, rays incident on the central portions of planar building elements are specularly reflected and transmitted with amplitude coefficients found from plane wave analysis. Rays incident on the edges of building elements can be viewed as exciting the edge to act as a secondary line source. The diffracted fields are generated by this secondary source and propagate as a cylindrical wave away from the edge, reaching into regions where the primary incident field is shadowed by the wall.

The foregoing wave mechanisms are central to the prediction of propagation effects in and around buildings, and are discussed in more detail as they are used. In addition, irregular objects such as vehicles, foliage, wall roughness, and furniture scatter fields in all directions in the form of spherical waves. These scattered fields may have a significant influence on the fast fading in the vicinity of the scatterer. However, because the energy is scattered over a sphere, these fields are less important far from the scatterer, as discussed in connection with indoor propagation.

This paper attempts to give a broad review of the theoretical approaches that have been used in order to predict propagation effects in cities and inside buildings. It is intended for those familiar with wireless systems who would like to gain some understanding of the physical mechanisms involved, and for those experienced in electromagnetic propagation and diffraction who lack knowledge of the issues of importance for communications systems. The paper does not attempt to give a comprehensive survey of the measurements that have been reported. However, measurements are discussed in order to evaluate the accuracy of the theoretical predictions.

As a basis for understanding the problem of predicting UHF propagation, we first review the characteristic variations that are observed, as well as discuss the data reduction that is commonly employed to identify properties associated with different scale lengths in space (Section II). Later sections of this paper are devoted to discussions of wide area predictions for propagation in macrocellular systems (Section III), and small area predictions for different building environments anticipated for microcellular systems (Section IV). We also consider site-specific predictions for outdoor PCS systems (Section V) and indoor systems (Section VI). In order to focus on the influence of buildings, for outdoor propagation we restrict the discussion to builtup regions, and for the most part assume that the terrain is flat. Because there has been no theoretical investigation of the influence of foliage on CMR and PCS propagation, we do not discuss this important effect.

II. Observed Signal Characteristics

As a basis for design of the CMR system, extensive measurements of signal characteristics have been made over the macrocellular range 1 km $< R <$ 20 km in various

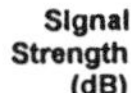

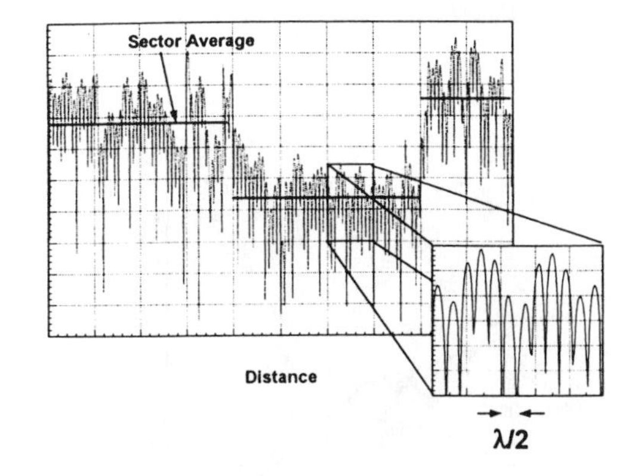

Fig. 1. Variation of the received signal as mobile travels along a street showing both fast fading and differences in the sector average (slow fading).

cities around the world [1]–[8]. Microcell measurements for $R < 1$ km have been made by operators as part of the process of cell splitting for system growth, but they have gone unreported in the literature. However, some microcell measurements intended to explore the potential of low base station antennas for PCS applications have been published, as for example in [22]–[36]. In a similar vein, some of the measurements of UHF propagation inside buildings for wireless local-area networks and private branch exchange (W-LAN and W-PBX) have been published, as for example in [37]–[51].

A. Narrow-Band Signal Characteristics for Macrocells

For the typical urban propagation path, the base station is located above the surrounding buildings while the subscriber is located at street level and is surrounded by buildings so that the base station is not within the line of sight. For CW or narrow-band transmission, the signal received as the subscriber moves along a street is indicated in Fig. 1, and involves fades that are separated by about one half wavelength and can be larger than 20 dB. If measurements of received power are made at points along a line whose length is 20 or more wavelengths, and the measured values are treated as a random variable, then the distribution function of the variable is typically found to be that of a Rayleigh distribution [52]–[54]. This Rayleigh fading, or fast fading, is due to the interference of signals arriving at the mobile from all directions as a result of reflections and scattering by buildings, vehicles, and other obstacles in the vicinity of the subscriber.

Fast fading is similar to the standing-wave pattern set up by waves propagating in opposite directions. However, since it is due to waves arriving with different amplitudes from all directions in the horizontal plane, similar patterns are found when moving horizontally in any direction, and each pattern loses correlation with itself after about one half wavelength [53]. Similar patterns are found for all polarizations of the subscriber antenna, but the patterns for orthogonal polarizations are uncorrelated [55]–[57]. Fading is also observed for a stationary subscriber if the frequency is swept slowly, since the differential phases of the various

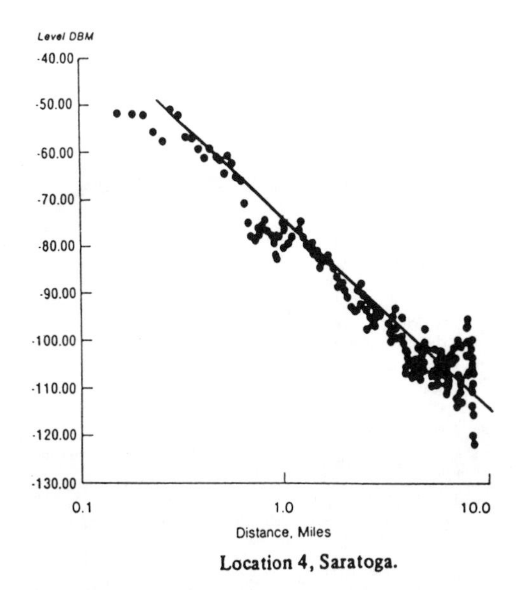

Fig. 2. Measured sector averages signal (dots) plotted versus distance R on a logarithmic scale, and showing clustering about an average dependence of the form A/R^n. Measurements are taken from [6].

multipath components change rapidly with frequency [58]. Fading will even be observed by a stationary subscriber operating at a constant frequency due to motion of scatters, such as people, vehicles, and trees, in the vicinity of the subscriber [8, sec.VI].

When the signal is averaged over a distance of 20 or so wavelengths using either a sliding window or discrete windows, the result is referred to as the sector average. Sector averages found using discrete windows are shown in Fig. 1. Variation of the sector average as the subscriber moves along the street is known as slow fading or shadow fading since the scale over which it takes place is on the order of building widths. Averages for a group of sectors at the same distance from the base station may again be treated as a random variable. When the sector averages are expressed in watts, their variation about the overall average is typically found to have a log-normal distribution.

More commonly, the slow fading statistics are obtained by plotting the sector averages in decibells versus distance R from the base station on a logarithmic scale, as in Fig. 2. A regression or least squares fit, which is a straight line of the type shown in Fig. 2, is then made to the data. The regression fit corresponds to the range dependence of the received power given by A/R^n, where A is an amplitude constant and n is the slope index. The deviations of the sector averages in decibells from the fit line are then treated as a random variable, which is typically found to have a Gaussian or normal distribution, corresponding to a log-normal distribution of the received signal in watts normalized to A/R^n.

Irregularity of the measurement environment greatly affects the degree of slow fading. In the middle of blocks composed of row houses of uniform height, the sector averages show only a few decibell variation. However,

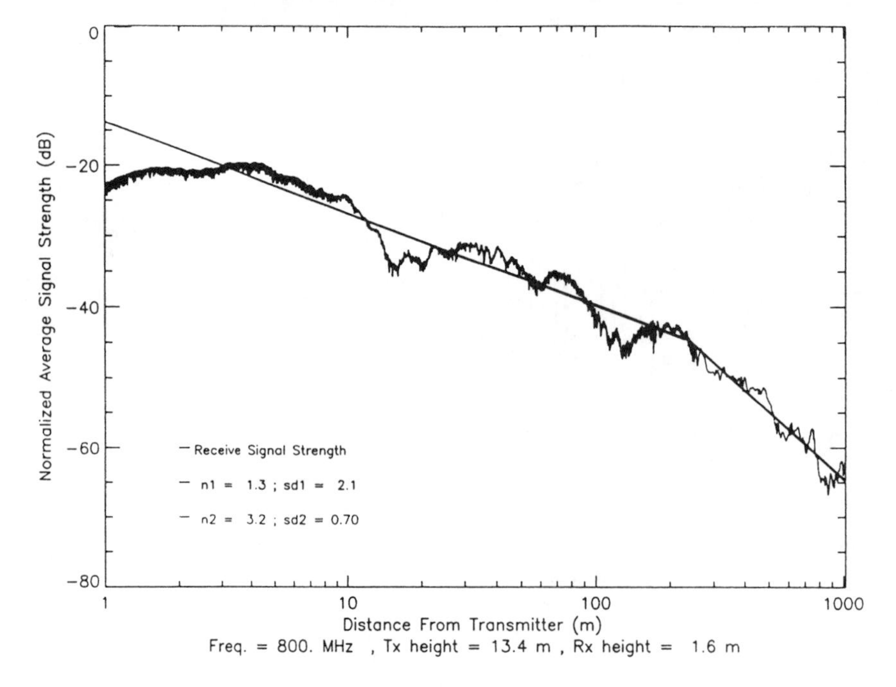

Fig. 3. Signal variation measured on a line-of-sight path in a rural environment—taken from [34].

measurements made over a wide area of the city, including measurements at intersections as well as mid block, give rise to slow fading distributions that have a standard deviation of as much as 8 dB [1], [4], [6], [19], [20], [59].

The log-normal character of the slow fading distribution has been observed in many environments, both outdoor and indoor, and appears to be due to the fact that several random processes act on the signal in sequence [11]. Since multiplication by a sequence of random variables is equivalent to their addition when expressed in decibells, and since the sum of several random variables tends to a Gaussian or normal distribution [60], the signal in decibells tends towards a normal distribution. Examples of processes that act on the signal in sequence for outdoor propagation are discussed in a subsequent section.

The regression fit A/R^n gives the variation of the signal on a macroscopic scale, which is typically taken to cover 1 km $< R <$ 20 km. Close to high base station antennas, n can be less than the free-space index $n = 2$ [8, sec. IV], while further away it is normally between 3.5 and 4. For $R <$ 1 km, the regression fit is not usually accurate, but this deviation is not important for systems designed to operate with cells having radius of several kilometers. Working with the measurements reported by Okumura *et al.* [1], Hata [61] made parametric fits to the data that give the dependence on frequency and base station antenna height of A and n over limited parameter ranges. Hata's results have been widely used for system design.

B. Narrow-Band Signal Characteristics for Microcells

In connection with microcellular PCS systems, several sets of measurements have been made on LOS paths [27], [30], [35], [62], [63], while a few studies of obstructed paths have been reported [23], [24], [31]–[33], [36]. On LOS paths the direct and ground reflected rays dominate over signals reflected from buildings, or scattered from cars and other objects. As a result, the fast fading statistics are expected to be Rician, rather than Rayleigh [8, Appendix II, p. 26].

Although most signals arrive from substantially different directions, and therefore have relative phases that vary rapidly with position of the mobile, the direct and ground reflect rays arrive with only small angular deviation, and hence can have a coherent additive effect, even when the signal is averaged over a 20 λ sector. Taken together, the two ray contributions give large signal variations as they go in and out of phase for distances less than the Fresnel break distance [63], and then partially cancel to give the $1/R^4$ dependence found beyond the break distance, as seen in Fig. 3 for 800-MHz measurements made in a rural environment. In the presence of buildings, rays reflected off the building faces produce additional signal variations [30]. However, the overall signal variation resembles that found in rural environments.

For obstructed paths, the signal must travel down the streets and around corners in high-rise environments, and over the roofs in residential and commercial sections with low buildings. In these cases, the signal is found to exhibit fast fading similar to that for macrocells, so that the sector average is used as a simple meaningful measure of the signal. However, since the base station may be located in the clutter of reflectors and scatters, this method of defining the sector average will have limitations whose nature is similar to that described below for propagation inside buildings.

Because the coverage area for microcells is limited, investigators have been careful to identify different subregions over which to examine the sector average signal. For example, the signal on streets parallel to the LOS street have been considered separately from cross streets, and the signals in intersections have been discussed [23], [36], [64]. Thus rather than aggregating all measurements, as was done with the macrocellular measurements, individual propagation scenarios have been investigated. In later sections, several of these scenarios will be considered in depth.

C. Narrow-Band Signal Characteristics Inside Buildings

When base station and subscriber are located inside a building, many paths exist by which the signal can propagate between them. Since both are located in the clutter, fast fading will be observed when either is moved. If a LOS path exists between them, such as down a hallway, the fast fading statistics are Rician, otherwise they are Rayleigh [44], [45]. Following the tradition established for the macrocells of CMR, a sector average signal is usually defined by averaging the signal measured as one end of the radio link is moved over a path whose length is about 20λ. To fit within a room, this is frequently accomplished by moving the link end in a circle of radius 1 m for 900-MHz signals [65], or on a raster path [47].

For the high base station antennas used for macrocells, the methodology of moving only the subscriber end of the link is appropriate for defining a unique single number describing the propagation conditions of the link. Because the base station is above the clutter of reflectors and scatterers, supposing that the entire base station was moved by a few wavelengths, as suggested in Fig. 4(a), one would expect no change in the average at a given sector. However, in the case of in-building propagation, both the base station and subscriber experience fast fading as a result of the surrounding reflectors and scatterers, as suggested in Fig. 4(b). In this case, the average obtained by moving one end of the link over a 20λ path will change if the other end is moved by a fraction of a wavelength. In one set of measurements, this change was up to 5 dB [66], as shown in Fig. 5.

The significance of the foregoing result can be seen in two ways. First, averaging over both ends of the link is required if a unique number is to be obtained that characterizes propagation from one small region to another. Averaging only over one end bears an uncertainty of $\pm$ 2.5 dB. The second way of interpreting the result is that using diversity at both ends of the link, rather than just one end, can add several decibells to the received signal. A similar effect can be expected for outdoor microcells employing base station antennas near to or below the roof tops, although the magnitude of the effect remains unclear. As yet, no measurements of the effect for microcells appear to have been reported.

When reporting in-building measurements over a single floor, investigators typically plot the signal in decibells versus the direct line distance R between base station and subscriber on a logarithmic scale. For propagation down hallways, the measurements indicate a $1/R^n$ signal dependence with n less than the value 2 for free space [42], [43], [47]. For propagation through rooms and walls, the

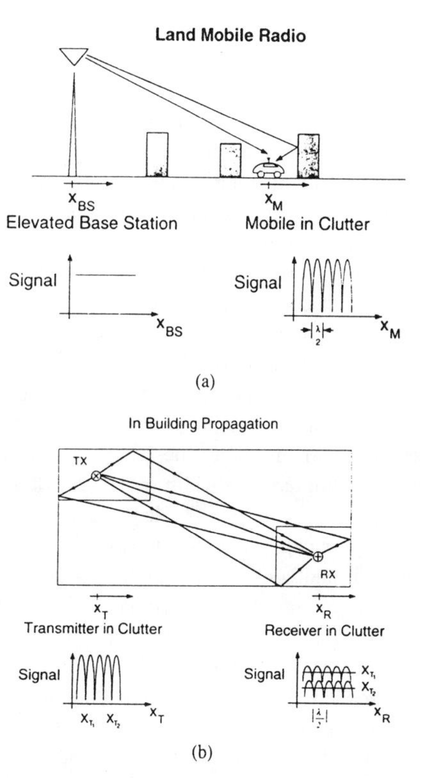

Fig. 4. Fast fading of the received signal as either end of the radio link is moved for the case of: (a) base station elevated above the clutter environment; and (b) both ends located inside the building clutter.

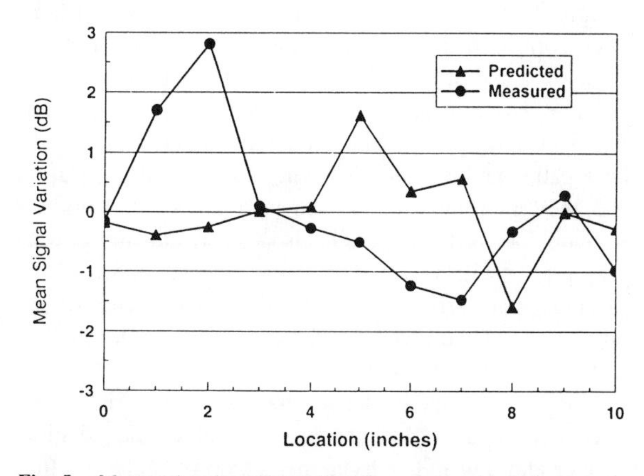

Fig. 5. Measured and predicted dependence on position of one end of an indoor link when the signalis are spatially averaged over the position of the other end of the link—taken from [66].

slope index n is larger than 2, and increases rapidly with R [42], [43], [47], [65], [67]. For propagation between floors, a different dependence is obtained [68]. These variations are discussed in greater detail in subsequent sections.

D. Time Dependence of Pulsed Signals

Measurements of pulse propagation characteristics in UHF channels have been made as a basis for evaluating digital signal transmission. Because of the multipath environment, contributions to the total signal arrive with differential delays ranging from less than an RF period up to hundreds of microseconds in mountainous outdoor environments [24], [26], [28], [29], [34], [37], [38], [40], [44], [69]–[71]. This echo spreads each pulse in time resulting in errors in detecting individual pulses, and in the extreme cases leads to inter-symbol interference when the echo of one pulse overlaps later pulses. The effect of pulse spreading can be mitigated through the use of equalizers that make an estimate of the impulse response of the channel and then constructs an inverse filter to recompress the spread pulse at the receiver. In order to estimate the rate of errors in detecting the bits, and to design equalizers, it is necessary to evaluate the channel response.

Actual telecommunication systems, and the channel sounders used to measure the channel characteristics, operate over a limited bandwidth of 1 to 50 MHz at center frequencies from about 900 MHz to 2.5 GHz, although at least one concept calls for ultra-wideband operation [72]. For such bandwidths the radiated pulses representing each bit of the digital signal contain many RF cycles. As a result it is the finite-bandwidth channel response that is of importance, rather than the infinite-bandwidth impulse response. The limitation to finite bandwidth has lead to particular ways for presenting the results of channel soundings, as discussed below. Channel soundings have be carried out directly in the time domain by radiating and detecting RF pulses [38], or in the frequency domain using a network analyzer.

Because the pulses consist of many RF cycles, it is common to measure the envelope of the received power, which corresponds to averaging the instantaneous received power over one or more RF cycles with a sliding window. The time variation of the power envelope is referred to as the power-delay profile. Since some individual pulses arrive along paths having small differential delay with overlap in time, they interfere with each other in much the same way as found for narrow-band signals. As a result, the power-delay profile will be sensitive to the relative phases of the individual path arrivals. Thus small displacements of the end or ends of the link located in the clutter will result in changes in the power-delay profile.

In order to uniquely define channel characteristics between one small region and another, rather than distinct receiver and transmitter locations, the power-delay profiles are averaged over a number of positions on a several meters long path [40], much as is done to find the sector average signal for narrow-band excitation. Figure 6 shows an example of such an averaged power-delay profile $P(t)$ for propagation inside a large office building [40], where the time origin is taken to coincide with the initial arrival of the signal. The mean excess delay and rms delay spread are then defined in terms of the first and second moments of this average power delay profile [40], and are useful in evaluating the performance of digital systems [73]. As was discussed in the case of indoor propagation, when both ends of the link are located in the clutter, it may be necessary to average over both ends of the link to obtain a unique channel characterization.

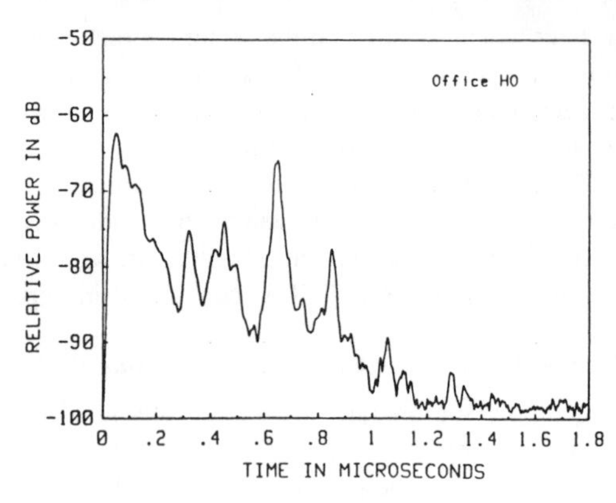

Fig. 6. Spatial average time-delay profile measured in a large office building—taken from [39].

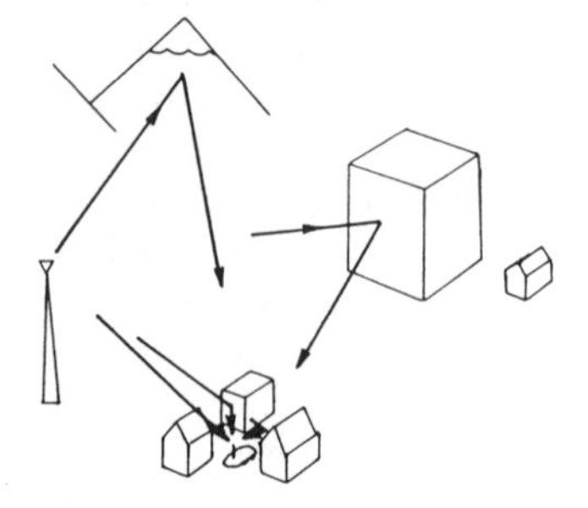

Fig. 7. Scattering by local, remote, and distant obstacles of increasing size leads to a hierarchy of delayed signal arrivals.

A hierarchy of delays for outdoor propagation due to scattering by successively larger and more distant features is suggested in Fig. 7. Local scattering in the vicinity of the subscriber and the base station produces spreading out to several microseconds. Scattering from large buildings can be responsible for arrivals having significant amplitudes and delays out to tens of microseconds, while scattering from mountains can result in measurable arrivals with delays of hundreds of microseconds. Ray-based prediction methods used to evaluate these scattering contributions can also be used to predict the average power-delay profile. For these predictions, the contribution from each individual ray path is assumed to be received with the same undistorted power profile that would be obtained if the antennas were located in free space, except for the time delay associated with the ray-path length. Adding the power profiles of the individual rays, each with its appropriate delay, gives a prediction of the average power profile.

III. Propagation Prediction for Macrocellular Systems

The initial experimental and theoretical modeling of UHF propagation in cities was carried out in connection with the design of macrocellular systems employing high base station antennas and cell radii greater than several kilometers. In most cities, the high rise core occupies a region whose radius is less than about 1 km. Outside of this core, the buildings are of nearly uniform height over large areas with occasional high-rise buildings. Building height decreases

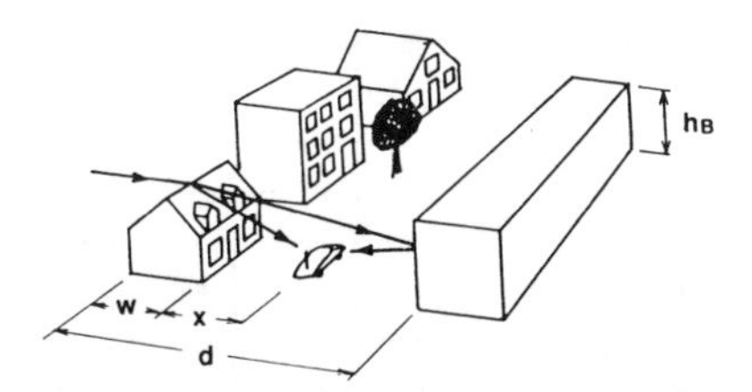

Fig. 8. Diffraction and reflection at buildings next to the mobile bring fields down to street level.

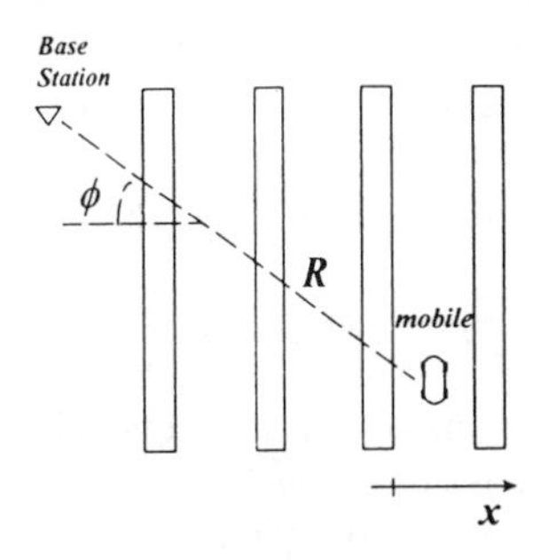

Fig. 9. Top view of a propagation path to the mobile that is oblique to the rows of buildings.

gradually from 4–6 stories near the core to 1–2 stories far away. The most common macrocellular propagation scenario therefore involves propagation from a base station antenna located above the rooftops of surrounding buildings having nearly uniform height down to a subscriber at street level, which will be the focus of this study.

Except along occasional streets aligned with the transmitter, or when there is a large open space in front of the subscriber, the transmitting antenna is not visible from street level. Buildings are organized along streets with gaps between them that are smaller than the building width, or they form continuous rows. Since gaps between buildings are not aligned with the transmitter from row to row, and because of the transmission loss through exterior and interior walls, propagation must take place over the rooftops. Treating the base station as the source, paths by which the rooftop fields reach street level are suggested in Fig. 8, and include diffraction at the last row before the mobile, diffraction at the last row followed by reflection at the next row, or diffraction at the previous row followed by propagation through spaces between buildings in the last row. In order to predict the range dependence of the average signal, the row of buildings is replaced by a rectangular cylinder lying on the ground, as shown on the right in Fig. 8, and all rows are assumed to be of equal height.

We define the sector average path loss S to be the ratio of the sector average received power to the radiated power for isotropic antennas. With this definition, S can be written as the product of the three factors

$$S = P_0 Q^2 P_1. \tag{1}$$

Here P_0 represents free-space path loss, which is the ratio of received to radiated power for isotropic antennas in free space, and is given by

$$P_0 = \left(\frac{\lambda}{4\pi R}\right)^2. \tag{2}$$

The factor Q^2 gives the reduction in the rooftop signal at the row just before the subscriber as a result of propagation past previous rows, and P_1 is the reduction due to the diffraction of the rooftop fields down to street level. The factor P_1 is discussed in Subsection III-A, and Q^2 is discussed in Subsection III-B.

A. Diffraction Down to Street Level

Diffraction of the rooftop fields down to street level was first postulated by Allsbrook and Parsons [4]. Ikegami *et al.* [9], [12] also recognized this mechanism, and compared theoretical predictions with measurements of signal variation due to differences in building height along streets in Tokyo.

We approximate the fields propagating across the rooftops as a plane wave propagating parallel to the ground but at an angle $90° - \phi$ to the row of buildings, ϕ being the angle between the direction of propagation and x, as shown in Fig. 9. The wave incident on the building edge visible to the subscriber, as indicated in Fig. 8, causes the edge to act as a secondary line source that radiates a cylindrical wave propagating away from the edge towards the subscriber. The power density of the cylindrical wave varies inversely as the perpendicular distance r from the edge, and varies with the direction θ from the horizontal [21]. This variation is given by $D^2(\theta)$, where the diffraction coefficient $D(\theta)$ depends on the boundary conditions appropriate to the building elements whose junction constitute the edge [21], [74]. While the diffraction coefficients are different for different boundary conditions, they all approach the same dependence for θ small. We therefore use the simplest diffraction coefficient, which is that for an absorbing wedge [74].

As shown in Fig. 8, one ray reaches the subscriber directly from the edge, and the second reaches the subscriber after reflection by the buildings on the opposite side of the street. Because of the rapid variation of the relative phase of these two ray fields, when the received power (proportional to the magnitude squared of the field) is averaged over a sector, the result is equal to the sum of the individual powers for each ray. With the foregoing simplifications, the signal diffracted down to the subscriber is reduced from the rooftop field by the factor P_1 given by [74], [75]

$$P_1 = \frac{1}{2\pi k \cos\phi}\left[\frac{1}{r_1}D^2(\theta_1) + \frac{1}{r^2}\Gamma^2 D^2(\theta_2)\right] \tag{3}$$

where Γ is the reflection coefficient at the building face, $k = 2\pi/\lambda$, and the diffraction coefficient $D(\theta_i)$ is given by

$$D(\theta_i) = \frac{1}{\theta_i} - \frac{1}{\theta_i + 2\pi}, \qquad i = 1, 2. \tag{4}$$

Let h_B be the height of the building, h_m the height of the subscriber or mobile antenna, and x the horizontal distance (measured perpendicular to the building faces) from the diffracting edge of the building to the subscriber, as shown in Fig. 8. Then the angle θ_1 and distance r_1 of the diffracted ray from the roof to the subscriber are given by

$$\theta_1 = \tan^{-1}\left[\frac{h_B - h_m}{x}\right]$$
$$r_1 = \sqrt{(h_B - h_m)^2 + x^2} \tag{5}$$

where d is the center-to-center spacing of the rows. If the diffracting edge is at the middle of the building, as in Fig. 8, then the angle θ_2 and distance r_2 of the diffracted ray that is also reflected from the building on the opposite side of the street are

$$\theta_2 = \tan^{-1}\left[\frac{h_B - h_m}{2d - 1.5w - x}\right]$$
$$r_2 = \sqrt{(h_B - h_m)^2 + (2d - 1.5w - x)^2} \tag{6}$$

where w is the front-to-back length of the buildings.

It is seen from (3) that P_1 varies inversely with frequency through the wavenumber $k = 2\pi f/c$. As a result, the diffraction loss increases with frequency, which is partially responsible for the frequency dependence of the received power, as discussed in Section IV-B. Expression (4) is not valid for small vertical displacements $h_B - h_m$ that lie within a transition region centered on the shadow boundary $\theta_i = 0$. The half width of the transition region is given by $[x\lambda/\cos\phi]^{1/2}$, where x is the distance along the shadow boundary from the edge. In the case of the reflected ray, the distance from the edge is the unfolded path length $2d - 1.5w - x$. Inside the transition region, $D(\theta)$ is given by the uniform asymptotic theory of diffraction [74], [76], which is required in order to account in a continuous way for the disappearance of the primary illuminating field as the shadow boundary is crossed.

In aggregate, signals reflected from the next row of buildings, giving the second term in (3), together with signals arriving via other paths, have amplitude nearly equal to that of the primary diffracted field, given by the first term in (3). It is this amplitude relation that leads to the deep fast fading that is observed. Thus the expression for P_1 can be further simplified to

$$P_1 = \frac{1}{\pi k \cos\phi}\frac{1}{r_1}D^2(\theta_1) \tag{7}$$

which is twice the first term in (3).

Expression (3) or (7) gives the height gain of the subscriber antenna. It is seen that the gain depends on the height of the buildings relative to the subscriber, and not simply the height of the subscriber antenna above the street. It also depends on the distance x, which is approximately one half the separation d between rows of buildings. In an attempt to define a simple representation for height gain, Okumura *et al.* [1] aggregated all their measurements in Tokyo, therefore averaging out differences in the heights of the buildings surrounding the mobile and the separations between the rows.

In contrast, a set of measurements of mobile height gain were made in a fairly homogeneous building environment

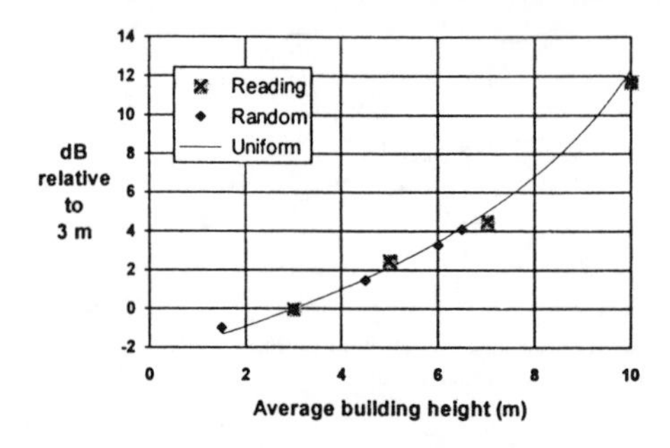

Fig. 10. Measured and computed average mobile antenna height gain at 191.25 MHz for Reading, England. Solid curve computed from (7) and squares are from a numerical simulation accounting for random building heights. Measurements are from [77].

(Reading, England) using signals from a distant TV transmitter broadcasting at 191.25 MHz [77]. A van with an antenna fitted on a telescopic mast measured the received signal at 3-, 5-, 7-, and 10-m heights. This measurement was repeated at many locations, and a cumulative distribution functions for each antenna height were constructed. In Fig. 10 we have indicated with $\times$'s the median signal value, normalized to that for the 3-m-high antenna, for 5-, 7-, and 10-m antenna heights. Using the reported building height of 2–5 stories, we have assumed an average height of 10.5 m and taken the row separation $d \approx 40$ m from maps of the city. For $x = d/2$ we have plotted as a continuous curve the height gain predicted by (7) normalized to the value at $h_m = 3$ m. Simulation results for height gain when the building heights are assumed to be random are also shown, and will be discussed subsequently. The excellent agreement between the measurements and the simple theory is seen as a validation of diffraction as the mechanism by which the rooftop signals reach street level.

B. Range, Frequency, and Base Station Height Dependence

In North American cities the buildings are frequently organized into straight, parallel rows by a rectangular street grid, as suggested in Figs. 8 and 9. Separation between the rows, either back-to-back across the yards or front-to-front across the streets is nearly equal. Thus in predicting the range dependence of the average signal, we assume equal spacing d for the rectangular cylinders representing the rows of buildings, as suggested in Fig. 11. Because the separation d between the rows of buildings is large compared to λ, signals that are multiply diffracted and reflected between rows, such as 3 in Fig. 11, can be neglected since they are substantially smaller than those passing once over the row. In addition, they lose phase coherence due to the irregularities of the buildings so that signals generated by one of these processes between one pair of rows will tend to cancel those generated by the same process at other pairs.

Because of the low glancing angle α in Fig. 11 for CMR, the propagation process is one of multiple forward diffraction past rows of buildings. Since only forward diffraction is involved, the resulting signal will not be sensitive to irregularities in the row spacing, or to lack of parallelism

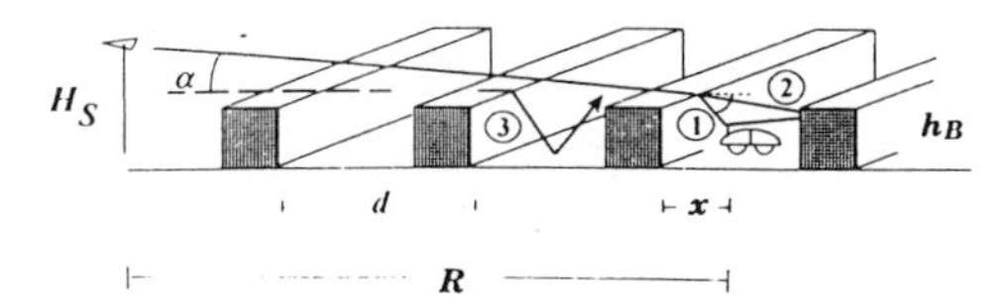

Fig. 11. Diffraction past rows of buildings as a model for computing the range dependence and base station height gain of the signal.

between the rows. Thus results obtained for a city with a rectangular street grid can be applied in cities with buildings in rows along curving streets by using the average row spacing. However, in cities such as Tokyo where the buildings are organized into blocks with very narrow lanes running through the blocks, the propagation process may be somewhat different. This difference in building organization may also explain the somewhat smaller range index n found in Tokyo as compared to North American and European cities.

Row spacing is typically on the order of 50 m, so that to evaluate signals propagating out to 5 km one must consider diffraction past 100 rows. Because of the large number of rows, simplifying approximations in the diffraction analysis are called for. Since forward diffraction through small angles depends only weakly on obstacle shape or the boundary conditions, the rectangular cylinders may be replaced by thin absorbing screens. Also, since ground reflections between the screens is ignored, as discussed above, the screens may be assumed to be semi-infinite.

Diffraction past an array of absorbing screens is a classic electromagnetic problem [78]–[81]. However, solutions applicable to the CMR problem have only recently been obtained [10], [82], [83]. These latter studies evaluated the reduction in the field reaching an edge (rooftop) as a result of passage past the previous edges. The first approach examined the reduction for the case of an incident plane wave [10], and is analogous to the use of the plane wave reflection coefficient to find the transmission between antennas located above a plane earth. This approach is limited to the propagation past many rows from a base station located above the buildings. The second approach considered propagation of a cylindrical wave radiated by a line source [82], [83] parallel to the rows. Diffraction effects in the vertical plane are the same for the line-source fields as they are for point-source fields, while spreading of the point-source field in the horizontal plane introduces an additional $1/R$ factor into the expression for the power density. The second approach is more general in that it is valid for sources above or below the rooftops, although the separation between the source and the plane of the nearest edge is restricted to be equal to d.

Results obtained from the line-source approach confirm those by the plane-wave approach for antennas above the rooftops. It is found that the field reaching the rooftop before the mobile is reduced by the factor Q that depends on the row spacing, frequency and path geometry through

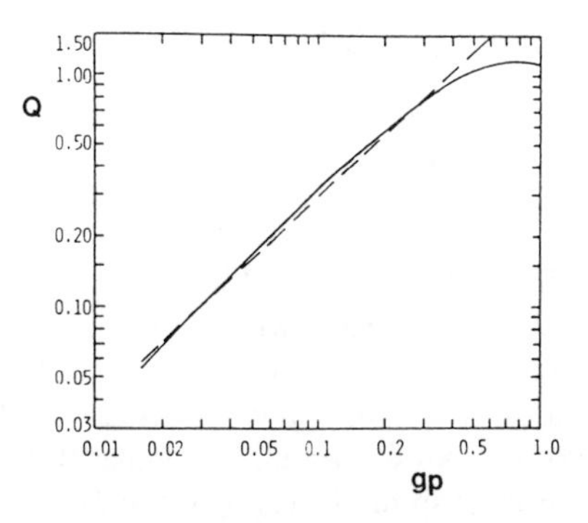

Fig. 12. Dependence on the parameter g_p in (8) of the factor Q giving the reduction of the rooftop fields due to propagation past previous rows of buildings (solid curve). A simple approximation covering the CMR parameter range is shown by the dashed line.

the dimensionless parameter g_p given by

$$g_p = \alpha\sqrt{\frac{d}{\lambda}}\cos\phi \tag{8}$$

where α is in radians. The obliquity factor $\cos\phi$, where ϕ is the angle shown in Fig. 9, is introduced since the wavenumber in the plane perpendicular to the edge is approximately $(2\pi/\lambda)\cos\phi$ for small α [74]. For flat terrain α is given by

$$\alpha = \tan^{-1}\left[\frac{H_S - h_B}{R\cos\phi}\right] \approx \frac{H_S - h_B}{R\cos\phi}. \tag{9}$$

The computed dependence of Q is plotted in Fig. 12 using logarithmic scales. For small g_p the curve is asymptotic to a line having unity slope .

Close to the base station, where g_p approaches unity, the dependence of Q on g_p departs from linearity, approaching unity at small R. The effect of this dependence on the signal has been observed in measurements made in Denmark [84]. Over the range $0.01 < g_p < 1.00$, a polynomial fit to the curve can be used to compute the $Q(g_p)$, and is given by [85]

$$Q(g_p) = 3.502g_p - 3.327g_p^2 + 0.962g_p^3 \tag{10}$$

which exhibits linear dependence for small g_p. In view of (9), very far from the base station (α small) the reduction factor is linearly dependent on α or inversely dependent on R. The received power is proportional to Q^2, and when multiplied by the free-space factor P_0 of (2), gives the range dependence of $1/R^4$. Also, the variation of Q^2 with $H_S - h_B$, which gives the base station height gain, is essentially quadratic for subscribers far from the base station.

For typical values $d = 50$ m, $H_S - h_B = 12$ m, and 1 km $< R <$ 10 km, g_p falls in the range 0.015–0.15 at 900 MHz and 0.021–0.21 at 1800 MHz. An even simpler approximation to $Q(g_p)$ covering this range of g_p is indicated by the dashed line in Fig. 12 having a slope 0.9. This approximation has an error of less than 10% over the range $0.015 < g_p < 0.4$ and in algebraic form is

$$Q \approx 2.35 \left[\alpha \sqrt{\frac{d}{\lambda}} \cos \phi \right]^{0.9} \qquad (11)$$

Since α is inversely proportional to R, the field reduction given by (11) taken together with the free-space path loss give a range index $n = 3.8$, which is close to that reported in North American cities [6]. For comparison with the sector average received power measured in Philadelphia [6], the path loss obtained from (1), (2), (7), and (11) for $\phi = 0$, has been multiplied by the effective radiated power, and is plotted as the straight line in Fig. 2. Excellent agreement with measurements is seen for both the predicted amplitude A and the range index (slope).

Using (1)–(3) and (11), the path loss S is seen to vary inversely with frequency to the 2.1 power. This dependence comes from the near cancellation of the frequency dependence of P_1 and Q^2 for high base station antennas. For the low base station antennas, the frequency dependence of these two terms do not cancel, resulting in a stronger frequency variation of the received signal, as discussed in Section IV.

The dependence on the direction of propagation relative to the street grid is given via the angle ϕ in (3) or (7) and in (8), and reflects the fact that the horizontal distance in the direction of propagation to or between the points of diffraction increases as $1/\cos\phi$. In connection with (7), the angle ϕ over which this variation is valid is limited by the condition $h_B - h_m > (\lambda x/\cos\phi)^{1/2}$. The variation in (7) with ϕ represents an effect due to the last row of buildings and indicates a signal variation of $1/\cos\phi$. Measurements of this effect made in Darmstadt, Germany, with low base station antennas [86] are in general agreement with this variation. The variation of Q with ϕ represents the effect of propagation past many rows of buildings before the subscriber. Unless the city has a strict rectangular grid, it will be difficult to measure this dependence. In addition, measurement of this effect requires separation of mid-block measurements from those at intersections. The authors do not know of measurements that have been made of this effect using elevated antennas.

C. Terrain Effects

The dependence of Q on the angle α suggests a way that some terrain effects can be easily incorporated into the propagation prediction. A hypothetical urban terrain profile is shown in Fig. 13 with three different subscriber locations. At position 1, the value of the angle α_1 is that between the tangent plane to the rooftops and the line to the base station. This angle can be found from the effective base station height H_e above the tangent to the terrain slope at the subscriber location, as shown in Fig. 13, and the horizontal separation R. Terrain slope is then accounted for by using this value of α in determining Q^2. The foregoing approach is equivalent to the method introduced by Lee [87].

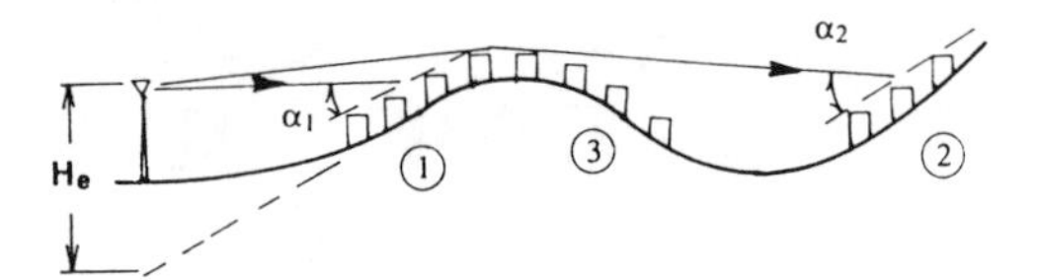

Fig. 13. Terrain effects may be incorporated for locations 1 and 2 by using the local angle α in determining Q and accounting for diffraction by intervening hills.

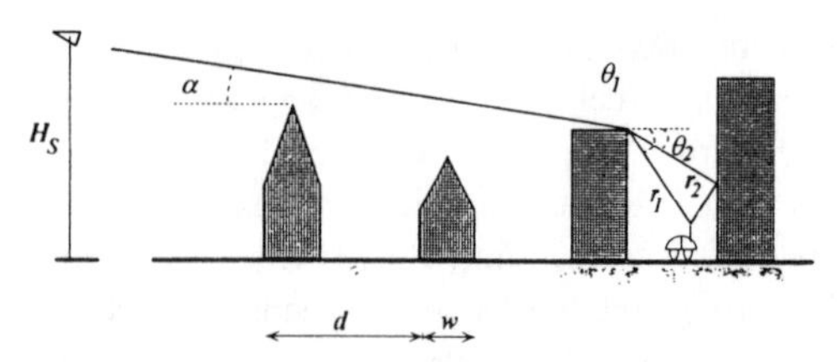

Fig. 14. Variations in building shape, height, and location that, together with construction materials, are used in the diffraction model to predict slow fading statistics.

Propagation paths may have terrain obstruction followed by clear space, such as the path to position 2 in Fig. 13. In this case the free space factor P_0 of (2) is reduced by the appropriate knife-edge diffraction loss [88] computed for the intervening hill. The buildings in the vicinity of the mobile are then accounted for by using the angle α_2 shown at position 2 in Fig. 13 when determining Q^2. In the case when the rooftop fields are obstructed and the field continues to be diffracted from rooftop to rooftop, as at position 3 in Fig. 13, the reduction of the rooftop fields due to previous rows of buildings is made more complex by the curvature of the ground, and remains to be evaluated.

D. Mechanisms Causing Slow Fading

The model described above for computing the range dependence of the average signal, with small modification, can be used to study two effects that produce shadow loss. Differences in building height is one important source of variability of the sector average. Another group of features that causes variability includes differences in roof shape, construction materials, gaps between buildings, as in Fig. 8, and cross streets at the ends of blocks. Some of these sources of variability are indicated in Fig. 14. Foliage is another important source of variability [89], [90]. However, theoretical models of foliage effects on CMR propagation have yet to be developed, and they are not considered here.

Differences in building height occur from row to row along the propagation path, as well as down each row. Numerical methods for studying the effect of both variations simultaneously would be very computer-intensive. Fortunately, the narrowness of the Fresnel zones at CMR and PCS frequencies tends to decouple these effects making it possible to consider them separately. Viewed from above, the Fresnel zone about the primary ray to the subscriber has full width $2(\lambda u)^{1/2}$ close to the subscriber, where $u \ll R$ is the distance measured back along the ray from the subscriber.

At 900 MHz, the width at the first row before the mobile ($u \approx d/2$ with $d \approx 50$ m) is 5.8 m, while at the fifth and tenth rows before the mobile ($u \approx 4.5d$ and $9.5d$) the width is 17.3 and 25.2 m, respectively. Thus the Fresnel zone along the primary ray path encompasses only one or two buildings on either side of the ray, whose heights are averaged by the propagation. However, height variations from row to row are not averaged, and lead to shadow loss as discussed below. Variations of building height along the last row before the subscriber permit diffraction at the vertical building edges, which result in additional contributions to the received signal. This effect can be an important source of fast fading as well, and tends to raise the average signal along an entire block, as discussed briefly at the end of this section.

The signal variations resulting from differences in building height from row to row can be simulated theoretically using direct numerical integration [11], [14], or semi-analytic methods [13], [91]. These simulations show that building height variation alone cannot explain the log-normal distribution observed for the slow fading. For example, if the building heights are chosen randomly according to a uniform distribution, then the distribution of sector averages is more nearly uniform [11]. This discrepancy with measurements is overcome if one also includes variations associated with the final diffraction down to street level.

In Fig. 14 we have indicated sources of variability associated with the final diffraction. The roof shape, which may be peaked or flat, affects the horizontal distance x in (3) from the diffracting edge to the subscriber. The boundary conditions may be conducting, as for houses with aluminum siding or aluminized vapor barrier insulation, rather than absorbing, as for masonry construction. For conducting boundary conditions, the diffraction coefficient in (3) is given by [21], [74]

$$D(\theta) = \frac{1}{2}\left[\csc\left(\frac{\pi+\theta}{2}\right) - \sec\left(\frac{\pi+\theta}{2}\right)\right]. \quad (12)$$

Finally, the building immediately before or immediately after the subscriber may not be present, corresponding to gaps between buildings and intersections. These gaps affect the diffraction path and the values of the terms in (3).

The foregoing sources of variation have been incorporated into simulation calculations for propagation perpendicular to the rows of buildings that also account for the random variation of building heights from row to row [14]. This simulation assumes that half the buildings have peaked roofs and half flat roofs; half require conducting boundary conditions and half absorbing; and gaps or intersections represent 10% of the cases. Each of the three features is selected by an independent random process, which are also independent of the random process for selecting building heights. Buildings are assumed to have height given by a fixed minimum value plus a random variable that is taken either from a uniform distribution or from a Rayleigh distribution giving the same mean building height $\langle h_B \rangle$ as the uniform distribution. A plane wave is assumed incident from above at some small grazing angle α, corresponding to a fixed distance from the base station, and the simulation is run to determine approximately 200 values of the sector average, which are then used to construct cumulative distribution function [11], [14].

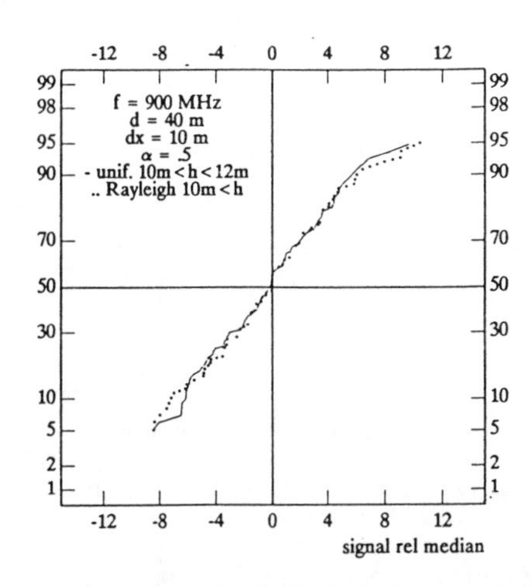

Fig. 15. Cumulative distribution functions of the simulated slow fading distribution assuming: a) uniform distribution of building heights; and b) a Rayleigh distribution.

Cumulative distribution functions of the deviation in decibells from the median signal value obtained for the uniform and Rayleigh distributions of building height are compared in Fig. 15 for the parameters indicated. It is seen that the cumulative distribution functions are nearly the same, and when plotted on a probability scale are nearly linear, at least up to the 90% point. A linear plot on this distorted scale corresponds to a Gaussian or normal distribution of the abscissa variable, which in this case is S in decibells. This in turn corresponds to a distribution of S in watts that is log-normal. This simulation shows that the log-normal distribution results from the sequence of two random processes acting on the signal, one being propagation over buildings of random height, and the other being random effects controlling diffraction of the rooftop fields down to street level.

Using the simulation model it is possible to investigate the dependence of the slow fading statistics on environmental and system parameters. Figure 16 shows the dependence of the standard deviations of the sector averages at 900 MHz and 1.8 GHz on the range of variation Δh_B for a uniform distribution of building heights. Random effects operating on the diffraction down to street level are seen to give a standard deviation just over 2 dB for $\Delta h_B = 0$ and $\langle h_B \rangle - h_m = 11$ m (4 story buildings). When the variation in building heights is $\Delta h_B = 6$ m ($\pm$ one story about the average), the standard deviation increases to about 4 dB at 900 MHz and 5 dB at 1.8 GHz.

Measurements of range dependence and slow fading statistics were made in Aalborg, Denmark, using four different base stations among buildings that varied from 3 to 5 stories for ranges R under 1 km. Averaging results for the four base stations gives standard deviations of 7.5 dB and 8 dB in the 900 MHz and 1.8 GHz bands, respectively [31]. The standard deviations obtained from

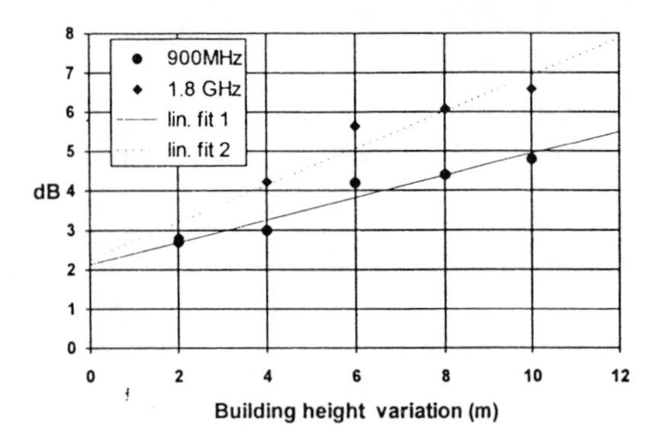

Fig. 16. Dependence of the standard deviation of the slow fading statistics on the range of building height.

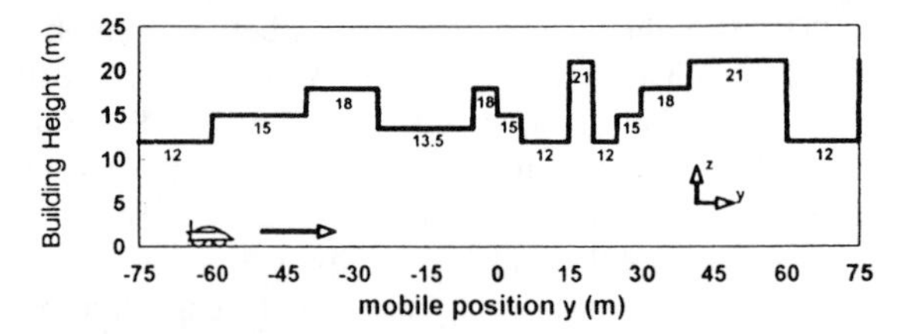

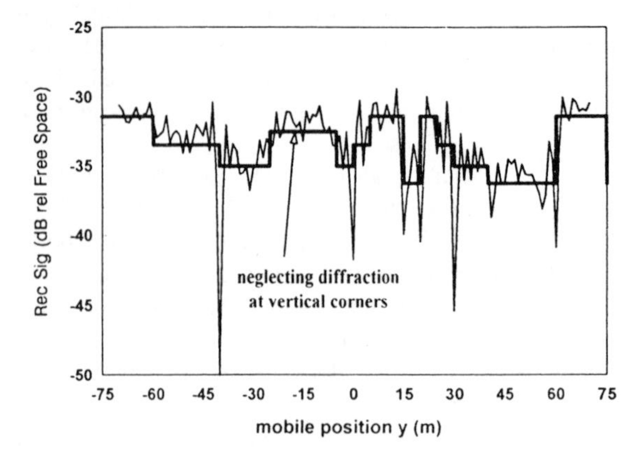

Fig. 17. Diffraction at horizontal and vertical edges of the buildings along a row gives rise to both fast fading and slow fading as the mobile moves along a street.

the measured signal are significantly higher than predicted by theory, which may be due to the inclusion of paths that are oblique to the rows of buildings, the possible inclusion of LOS paths in the measurements, and a higher percentage of the measurements for which gaps in the last row are present. Similar sets of measurements in Darmstadt, Germany, were corrected for street orientation [86]. The averages of the standard deviations for the different sets were 5.9 and 6.4 dB in the two frequency bands, which are closer to the theoretical predictions of 4 and 5 dB. Both sets of measurements give an increase of only 0.5 dB in the standard deviation as the frequency is doubled, instead of the 1 dB predicted. As predicted by the theory, slow fading patterns are highly correlated at the two frequencies, and the distribution of sector averages is log-normal [31].

Another parametric dependence that has been studied is the mobile height gain. Results for the median of the sector averages are indicated in Fig. 10 for 190-MHz propagation, and show that the height gain is the same as obtained from (7). Finally, the median sector average is found to decrease by a few decibells as Δh_B increases. This effect is partially offset by variations in building height along the last row. In any case, it would be difficult to measure this effect since it requires finding two or more urban environments that are similar in all respects except for Δh_B.

Limited studies of the effect of building height variations along the last row before the subscriber were carried out neglecting variations in building height along previous rows [92]. One such variation of building height is shown in Fig. 17, along with the sector average computed as the subscriber moves along the street. Because of lateral diffraction around the vertical edges of the buildings in this case, the sector average shows significantly greater variation than would be obtained accounting only for diffraction over the top of the building directly in front of the subscriber, which is given by the stepped plot in the lower drawing. Lateral diffraction also increases the median signal in this case by 1–2 dB.

IV. Microcellular Predictions for Generic Building Environments

The rapid development of microcellular systems employing low base station antennas and propagation over limited ranges $R < 1$ km has spawned a spectrum of approaches to the prediction problem. For microcell prediction, it is recognized that different building environments, and different propagation paths in each environment, will give rise to different characteristics. One approach to dealing with this diversity has been site-specific prediction that accounts for the geometry of individual buildings or groups of buildings, as discussed in Section V. However, this approach does not make use of the similarity of buildings in a particular environment, or other simplifying features, to achieve computational efficiency and understanding of how the building geometry influences propagation. An alternative approach is to treat individual environments and propagation paths separately by making use of building features that are generic to the environment.

In this section we investigate different propagation paths in two different building environments. We first consider LOS propagation along a street, with the buildings assumed to form a canyon with uniform walls (Subsection IV-A). For residential and commercial environments, we treat propagation over the roofs for low antennas by neglecting the cross streets and assuming the buildings to be of uniform height and organized in uniformly spaced rows (Subsection IV-B). The influence of cross streets in this environment is subsequently included to predict area coverage (Subsection IV-C). Finally, prediction in a high-rise environment is treated assuming the buildings to be so tall that propagation occurs down streets and around corners (Subsection IV-D).

A. Line-of-Sight (LOS) Propagation along Streets

If base station antennas are located on street lamps or other structures along streets, LOS propagation will

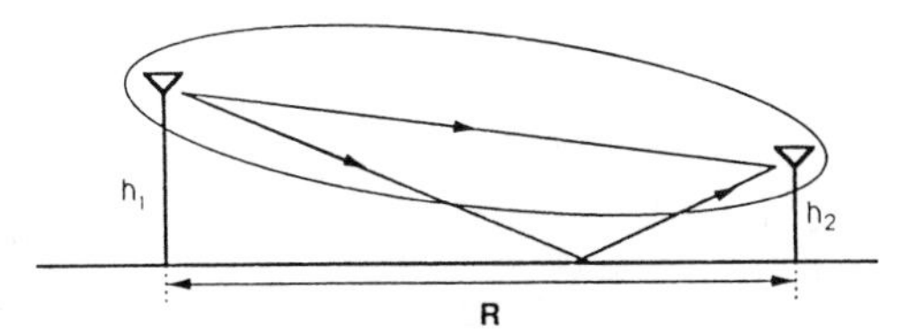

Fig. 18. Direct and ground-reflected rays, and showing the Fresnel ellipse about the direct ray.

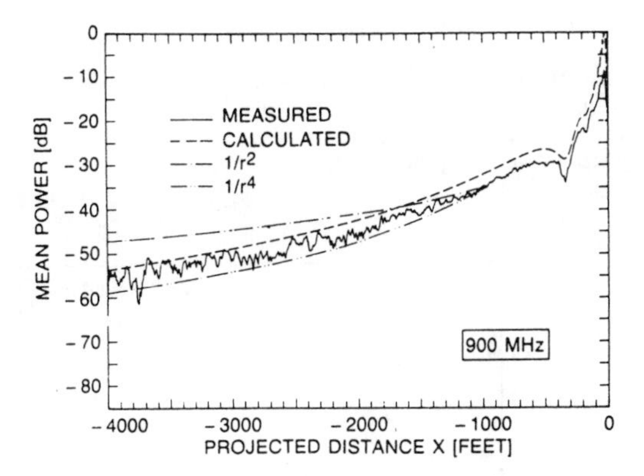

Fig. 19. Comparison of the measured signal on a LOS path with predictions made using the two-ray model—taken from [29].

be an important feature of microcellular coverage and interference. For this reason, considerable attention has been given to measuring and modeling the signal on such paths. Measurements made in rural environments confirm the applicability of a simple two-ray model consisting of the direct and reflected ray [30], [35], [63], as shown in Fig. 18. For isotropic antennas the path loss is

$$S_{\text{LOS}} = \left(\frac{\lambda}{4\pi}\right)^2 \left|\frac{1}{r_1}\exp(-jkr_1) + \Gamma\frac{1}{r_2}\exp(-jkr_2)\right|^2 \tag{13}$$

where r_1 and r_2 are the direct and reflected path lengths shown in Fig. 18. The ground reflection coefficient Γ is given by

$$\Gamma = \frac{\cos\theta - a\sqrt{\varepsilon - \sin^2\theta}}{\cos\theta + a\sqrt{\varepsilon - \sin^2\theta}} \tag{14}$$

where $a = 1/\varepsilon$ for vertical polarization and $a = 1$ for horizontal polarization. For typical ground parameters $\varepsilon = 15 - j90/F$, where F is the frequency in megahertz. The validity of the theory is seen in Fig. 19 from the close agreement with measurements made at 900 MHz with vertically polarized antennas [30].

When the signal is plotted versus R on a logarithmic scale, as in Fig. 3 for measurements at 800 MHz [35], [62], [63], it is seen that distinctly different slopes are obtained before and after a break point. The break distance R_B is that for which the Fresnel ellipse about the direct ray from base station to mobile, which is shown in Fig. 18, just touches the ground, and is approximately given by [63]

$$R_B = \frac{4h_1h_2}{\lambda}. \tag{15}$$

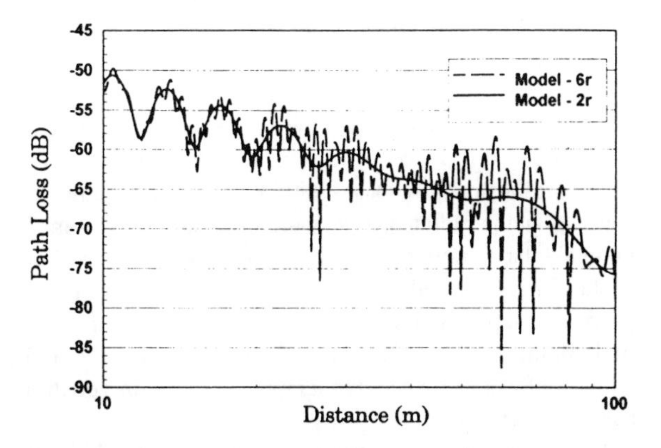

Fig. 20. Path loss computed for LOS paths using the two-ray model and a six-ray model that accounts for reflections from the buildings lining the street.

For $R < R_B$, the antenna patterns, variation of reflection coefficient with angle, and two-ray interference conspire to give a regression slope of less than 2, as see in Fig. 3, while interference between the two rays for $R > R_B$ leads to a regression slope near 4. When measurements are made over a range that includes R_B, a two-slope regression fit will give a smaller standard deviation than will a single slope fit [63].

In urban environments with buildings lining both sides of the streets, reflections can also take place at the faces of the buildings. Rays may be multiply reflected between the buildings lining the streets. In the absence of vehicular traffic, for each ray from base station to subscriber that is reflected between the buildings, there is a similar ray that is reflected once from the ground and appears to come from the image of the base station in the ground plane. Interference between the ray reflected in the ground, and the corresponding ray that is not reflected, results in a $1/R^4$ signal dependence for the pair when R is large. Building-reflected rays result in additional rapid variations about the simple two-ray model, but do not change the overall variation [30], as seen in Fig. 20, and confirmed by measurements [63].

The effect of traffic on all the ground-reflected rays remains in question. Measurements made in Manhattan, NYC, show a $1/R^2$ dependence at distances beyond R_B, suggesting that traffic might interfere with ground reflection [30].

B. Propagation over Buildings for Low Antennas

Residential/commercial environments, as the term is used here, consist of buildings that are of relatively uniform height ranging up to 5 or 6 stories. Except on line-of-sight paths down a street, propagation is primarily over the roofs of the intervening buildings. In studying propagation in this environment, the COST 231 study group in Europe [93] has employed (8) for the factor Q at small ranges, and incorporated measurements in order to extrapolate to base station antennas below the roof tops.

An alternative theoretical approach for low antennas has been developed by treating the rows of buildings as cylinders lying on the ground, as suggested in Figs. 8, 9, and 11, diffraction of the rooftop fields at the last row of buildings down to street level is as discussed in connection with (3). However, evaluation of the two-dimensional diffraction process in the plane perpendicular to the rows for low base station antennas requires the study of fields radiated by a source that is localized in the vertical plane, rather than an incident plane wave. Such a study was carried out for the special case when the horizontal separation between the base station and first row of buildings is d [83], and is further discussed in [85], which unfortunately has typographical errors in equations (13) and (15).

The reduction Q_M in the rooftop field at the Mth row past the base station due to propagation past the previous rows of buildings is found to depend on the frequency and path geometry through the dimensionless parameter g_c given by [83]

$$g_c = (H_s - h_B)\sqrt{\frac{\cos\phi}{\lambda d}} \tag{16}$$

where the factor $\cos\phi$ has been included to account for propagation oblique to the street grid [74]. Note that for these predictions Q_M replaces Q in (1). In (16) the base station height H_S may be above or below the height h_B of the surrounding buildings. Recursion relations from which Q_M can be calculated are given in [85].

The dependence of $Q_M(g_c)$ on row number M is shown in Fig. 21 for various values of the parameter g_c. In order to make the interpretation of the parameter dependence more transparent, the curves have been labeled by the relative base station height $y_0 = H_S - h_B$ in steps of 1.25 m for propagation normal to the rows ($\phi = 0$) and $d = 50$ m, $\lambda = 1/3$ m (900 MHz). For $H_S = h_B$, so that $g_c = 0$, it is seen that $Q_M = 1/M$. Since the range dependence of the signal is given by the product of Q_M^2 and the free-space path loss P_0, and since M is the integer value of R/d, it is seen that the range index n is equal to 4 when the base station is at rooftop level. For antennas below the rooftops ($y_0 < 0$), it is seen from Fig. 21 that Q_M initially decreases more rapidly than $1/M$, but quickly approaches the $1/M$ variation. Conversely, for antennas above the rooftops ($y_0 > 0$)Q_M initially decreases less rapidly than $1/M$, eventually reaching the $1/M$ dependence for large M.

To make the dependence on M more quantitative, we use the logarithmic derivative to define the negative of the slope as

$$s = -\frac{\log Q_{M+1} - \log Q_M}{\log(M+1) - \log M}. \tag{17}$$

In terms of s, the range index is given by $n = 2(1+s)$, which is plotted as a function of the relative antenna height $H_S - h_B$ in Fig. 22 for $M = 5$ and $M = 10$. For high antennas, the index is below 4, and close to 3.8, as predicted by the simple plane wave analysis, while for low antennas it is greater than 4.

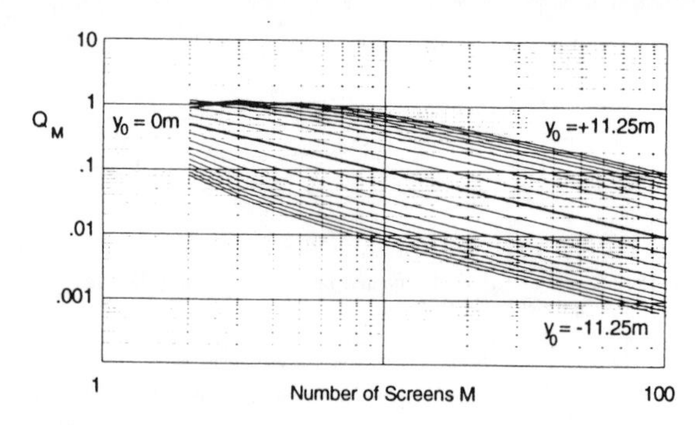

Fig. 21. Dependence on the number of intervening rows of the field-reduction factor Q_M at the top of the last row before the mobile. Base station antenna height relative to the rooftops is the parameter y_0 for a frequency of 900 MHz.

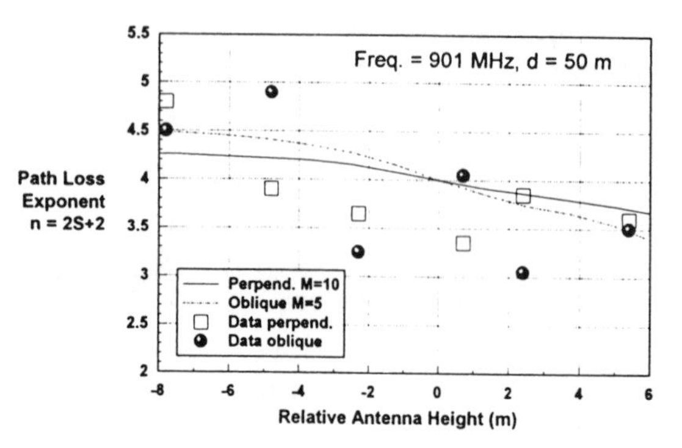

Fig. 22. Measured and computed range index n as a function of base station antenna height y_0 relative to the rooftops at 901 MHz.

Measurements at 901 and 1937 MHz have been made in the Mission and Sunset districts of San Francisco, CA, at three antenna curb heights, and for propagation perpendicular to the rows, and at an angle ϕ of approximately 60° [85], [94], [95]. In these districts the building heights are approximately 11.5 and 8 m, respectively, giving six different antenna heights relative to the buildings. The measurement range was from 100 m to 1.2 km. When expressed in terms of $\log R$, the measurement range is approximately centered on a distance of 500 m, which for the row separation $d = 50$ appropriate to these districts corresponds to $M = 10$ for propagation perpendicular to the rows, and $M = 5$ for $\phi = 60°$. The slopes taken from the regression fits to the measured data at 901 MHz are shown in Fig. 22. It is seen that the theoretical predictions for n are consistent with the trend of the measured data.

The height gain of the base station antenna is given by the variation of Q_M^2 with $H_S - h_B$. The variation of Q_M^2 with antenna height is plotted in Fig. 23 for 900-MHz propagation perpendicular to the rows of buildings, and for oblique propagation with $\phi = 60°$. In order to compare with measurements, the plots are for $M = 20$ in the perpendicular case and $M = 10$ for the oblique case, corresponding to $R \approx 1$ km. For comparison with theory, the free-space factor P_0 of (2) and the final diffraction factor P_1 of (7) were removed from the value of the 1-km intercept of the regression fit to the measured path loss.

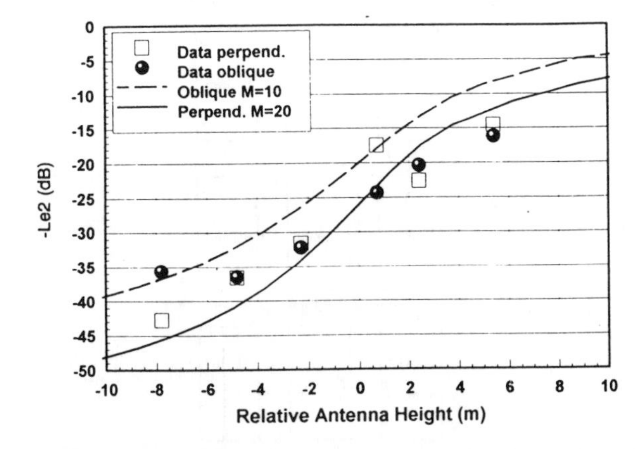

Fig. 23. Measured and computed base station height gain at 901 MHz for propagation perpendicular and oblique to the rows of buildings.

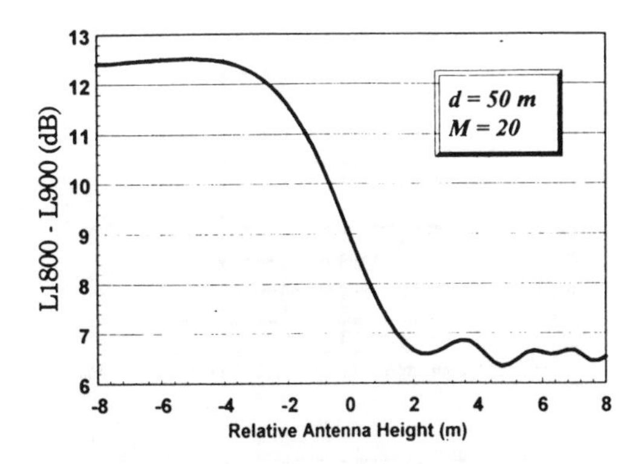

Fig. 24. Computed variation with base station height of the path loss difference between 1800- and 900-MHz signals propagating over a distance of 1 km in a residential environment.

The resulting "measured" values of Q_M^2 for perpendicular and oblique propagation are also shown in Fig. 23.

Good agreement is obtained between theory and measurements for perpendicular propagation. However, for oblique propagation there is more scatter in the reduced measurements, and for low base station antennas the theory is significantly more pessimistic than the measurements. Similar results were obtained at 1937 MHz, with the theory again giving lower values than the measurements for low base station antennas, both for perpendicular and oblique propagation.

The base station used for the measurements was a van with a telescoping mast that was parked at the curb, so that its horizontal spacing from the first row of buildings was less than the value d assumed in the theory. Moreover, the diffraction model does not account for reflections from the faces of buildings on the opposite side of the street from the base station, which may be significant for antennas below the rooftops. For the measurements, buildings opposite the base station were present at the Mission site, but not at the Sunset site. In spite of these and other possible improvements, the agreement achieved between the theory and measurements indicates that the reflection and diffraction processes addressed by the theory are fundamental in determining the received signal.

Frequency variation is contained in all three factors P_0, P_1, and Q_M^2 that comprise the received signal. Figure 24 compares the difference in path loss at 1.8 GHz and at 900 MHz as a function of the base station antenna height for propagation past 20 rows of buildings. For base stations below the height of the surrounding buildings, the difference in path loss is greater than 12 dB, while for high antennas it approaches 6 dB. Measurements indicate the same inverse relation between path loss difference and antenna height [31], [96], with differences ranging from 8 to 11 dB.

The limitation of the foregoing approach is that it does not account for cross streets, which can form a significant fraction of all paths over a small area. The next section addresses this problem.

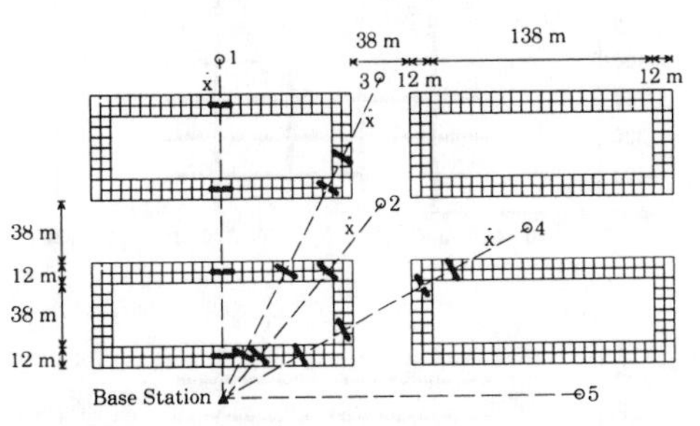

Fig. 25. Approximate footprints of two-story row houses in the Sunset district of San Francisco.

C. Coverage Area for Residential/Commercial Environments

In each city large residential/commercial areas exist that have the same generic building features. In areas with a rectangular street grid these features include average building height and width, average gap between buildings, row spacing, and block length. Propagation predictions that account for the generic building structure, and predict the sector average received signal along individual streets in the vicinity of the base station, can be expected to be useful for system design. Predictions of this type are discussed here for areas such as the Sunset and Mission districts of San Francisco that feature row houses located on a rectangular street grid. For cities that do not have a rectangular street grid, the approach discussed here can incorporate the actual street grid together with generic building information.

Figure 25 shows the approximate footprints of the houses in the Sunset district, as well as the propagation paths from a mid-block base station to several nearby subscriber locations. For simplicity, paths that involve diffraction at the vertical edges of buildings and reflections at building surfaces have been neglected. Paths 1–4 involve diffraction past rows of buildings parallel to the street on which the base station is located, and 3, 4 also involve diffraction at rows on perpendicular streets. Finally, LOS propagation occurs on path 5, which is evaluated using the two-ray model (13) for vertical polarization.

Diffraction at the rooftops is evaluated by replacing the buildings by absorbing screens located at the center of

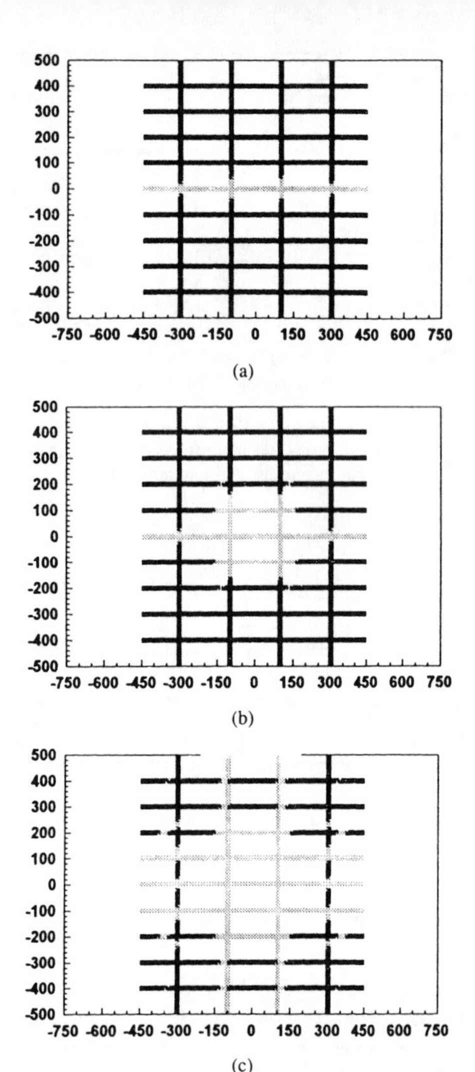

Fig. 26. Cell shape (−110-dBW contour) in the residential environment of Fig. 25 with building height $h_B = 8$ m for base station antenna heights: (a) $H_s = 4$ m; (b) $H_s = 8$ m; and (c) $H_s = 14$ m.

the building and oriented perpendicular to the propagation path [97], [98], as indicated by the heavy cross lines in Fig. 25. Because the spacing between the screens is not uniform, we have made use of Vogler's method [81], which is summarized in [97], [98]. The method is computationally slow, and Vogler offers a simplified expression, which however is not valid for more than three screens. Assuming a building height of 8 m, the −110-dB path loss contours are plotted in Fig. 26 for three base station antenna heights 4, 8, and 14 m. It is seen that the coverage area is strongly dependent on antenna height. Similar calculations for a base station located in a backyard rather than a street show that elimination of the LOS path results in a more elliptical coverage area [97], [98].

D. High-Rise Building Environments

When high buildings line the streets to form urban canyons, propagation from a low base station antenna to a subscriber at street level will take place down the streets and around corners, rather than over the buildings. Propagation on LOS paths involves reflections from the ground and building faces, as previously discussed. For non-LOS paths the signal must turn corners via multiple reflections and diffraction at the vertical edges of buildings. Studies of this process have been carried out for areas having a rectangular street grid with high-rise buildings that form an essentially continuous facade along the four sides of each block, as shown in top view in Fig. 27 [23], [36], [64], [99]–[101]. Such a generic building environment is found, for example, in sections of Manhattan.

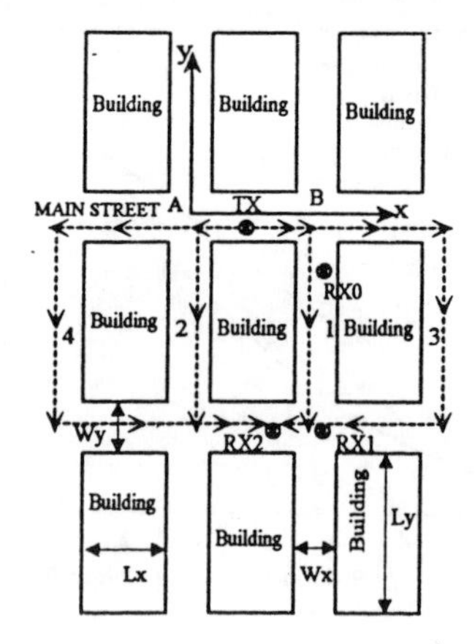

Fig. 27. Two turn routes from a transmitter to a receiver on a parallel street in a high-rise environment—taken from [101].

It is found that the signal decreases by about 20 dB when turning a corner. Thus the signal received by subscribers on streets that cross the main street on which the base station is located, such as at $RX0$ and $RX1$ in Fig. 27, will essentially be due to signals that make a single turn off the main streets, as indicated by route 1. In order to reach locations on streets parallel to the main street, such as at $RX2$ in Fig. 27, signals must make two turns, of which there are an infinite number of routes, such as 1–4.

Each one turn route is composed of an infinite number of two-dimensional (2D) ray paths that make m reflections at the buildings on the main street followed by n reflections at the buildings on the perpendicular street (RmRn rays), or followed by diffraction at one of the four building edges at the turn intersection and n reflections on the perpendicular street (RmDRn rays). Rays that are multiply diffracted at the corners of one intersection are ignored since they are significantly weaker. Note that each 2D ray is composed of two rays, one of which is reflected from the ground and appears to come from the image of the base station in the ground plane.

Using the formulation of [101], the path loss associated with an RnRm ray for vertically polarized isotropic antennas is given by

$$S_{mn} = |\Gamma(\phi_1)|^{2m} |\Gamma(\phi_2)|^{2n} S_{\text{LOS}} \tag{18}$$

where $\Gamma(\phi_1)$ and $\Gamma(\phi_2)$ are the reflection coefficients at the building faces on the main street and cross street. The factor S_{LOS} is given by (13) where r_1 and r_2 are the total unfolded path lengths of the two three-dimensional (3D)

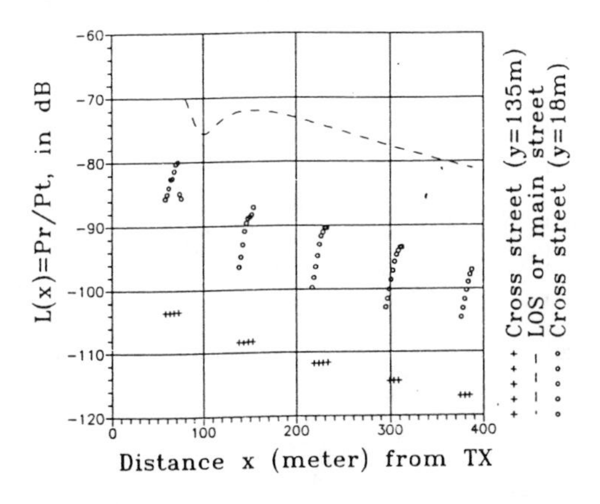

Fig. 28. Computed signal variation on the LOS street and from side-to-side across perpendicular streets at distances near to and far from the intersections with the LOS street—taken from [101].

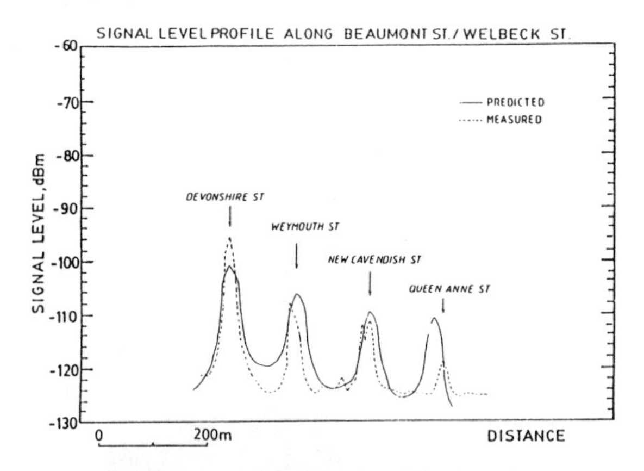

Fig. 29. Measured and computed signal variation on a street parallel to the LOS street in a high-rise environment—taken from [22].

rays. For the RmDRn rays the path loss is given by

$$S_{mDn} = |\Gamma(\phi_1)|^{2m} |\Gamma(\phi_2)|^{2n} S_{\text{LOS}} D^2(\psi) \frac{\rho_1 + \rho_2}{2\pi k \rho_1 \rho_2} \quad (19)$$

where ρ_1 and ρ_2 are the unfolded 2D ray lengths between the diffracting edge and the base station or subscriber, respectively. Assuming an absorbing boundary condition, the diffraction coefficient $D(\psi)$ is as given by (4) outside the transition region about the shadow boundary. The diffraction angle ψ is positive in the shadow region and negative in the illuminated region.

There is no single group of rays that is dominant over all locations. If both the base station and subscriber are close to the intersection where the wave turns, both RmRn and RmDRn ray groups are important in predicting the field, whereas if both are far from the intersection, the RmRn group of rays is negligible. The foregoing expressions can be generalized for two turn routes [101].

Summing the received powers of the individual rays gives the sector average. For a base station centered in the main street, the LOS signal on the main street and the signal making one turn into side streets are plotted in Fig. 28 as a function of x. The plot for $y = 0$ corresponds to LOS propagation on the main street, while those for $y = 18$ m and $y = 148$ m show the variation of the signal across the width of the cross streets at locations in Fig. 27 that are near to the main street and at the other end of the block, respectively. At a distance of 5 m into the cross street ($y =$ 18 m) after one turn, it is seen that the signal varies by 10 dB from one side of the street to the other and is 15 to 25 dB below its value on the main street. At greater distance from the main street the signal is even lower, but shows little variation from one side of the cross street to the other.

Using the diffraction coefficient for a conducting wedge, rather than absorbing, Chia *et al.* [23] have also computed signals for one- and two-turn paths for comparison with measurements made in London. Their results for propagation from a mid-block base station on the main street to locations along the first parallel street are shown in Fig. 29. As a result of turning the second corner, the signal at mid-block is seen to be about 20 dB lower than it is in the intersections. Overall, agreement between theory and measurements is quite good.

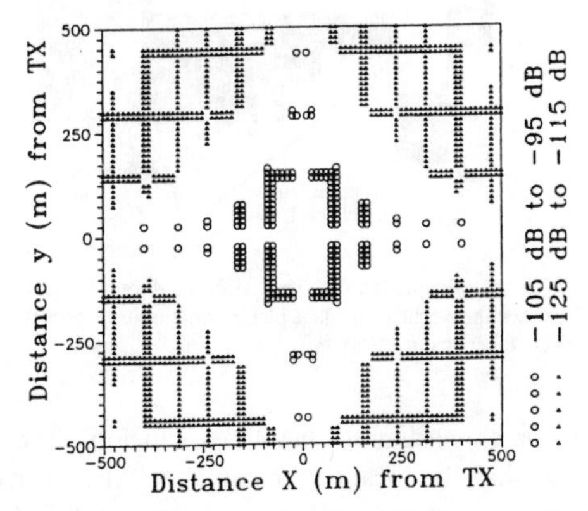

Fig. 30. Computed equi-signal contours in a high-rise environment—taken from [101].

Coverage contours are shown in Fig. 30 for a base station located at an intersection in an environment like that on Manhattan's East Side [101]. The signal has been computed for subscribers in the middle of the streets. The diamond-shape pattern of equi-signal contours is the same found in recent measurements [36]. Using the theory it is relatively simple to explore how the coverage changes when the base station is moved. For example, there is a significant asymmetry in the coverage when the base station is located near but not in the intersection, and at the curb rather than the middle of the street [101].

V. Site-Specific Outdoor Prediction

For PCS applications, various groups have been developing propagation prediction programs that incorporate the specific shape of streets and buildings at a site. The computer programs evaluate the signal at designated locations based on some form of ray tracing and diffraction theory [101]–[107], and are intended to give a deterministic prediction of the sector average signal at individual subscriber locations. No program can be expected to predict the actual fast fading pattern since even the smallest uncertainty in

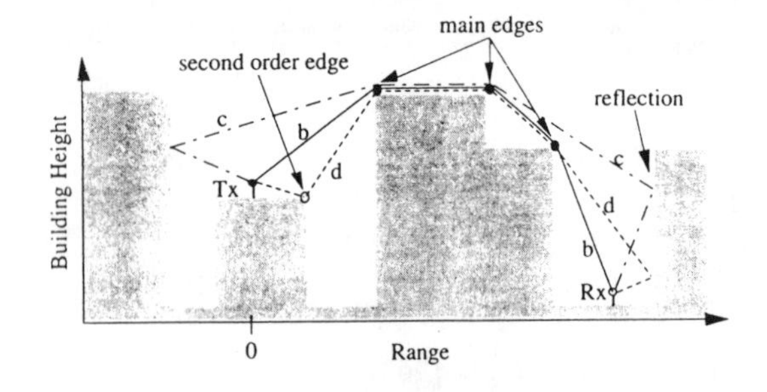

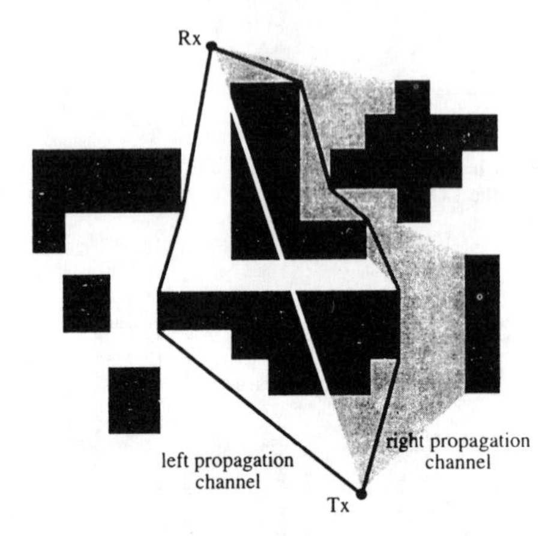

Fig. 31. Site-specific prediction using simplified ray paths restricted to either the vertical or slant plane containing transmitter and receiver—taken from [106].

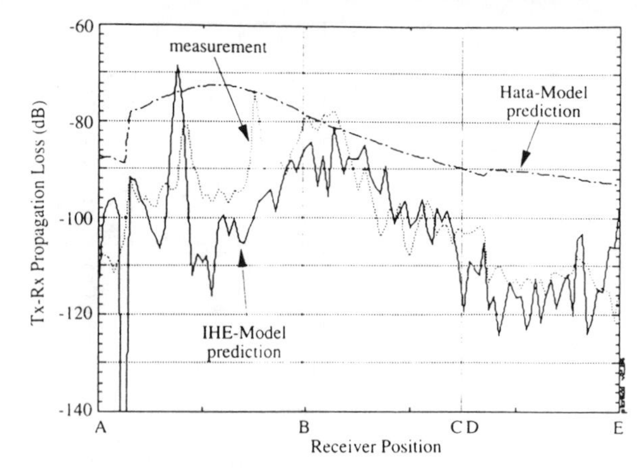

Fig. 32. Comparison of computed and measured sector average signals for test routes in Aalborg Denmark—taken from [106].

positions or dimensions of buildings will introduce phase errors that shift the locations of minima and maxima, and since moving scatterers (people, vehicles, etc.) will cause shifts in the pattern with time. However, addition of the individual ray powers will give the sector average even in the face of these uncertainties. Since this approach is based on rays, it can readily be used to predict the average power delay profile, as discussed in Section II-D.

The most difficult part of any ray approach is the determination of the ray paths. Since in all but the simplest geometries there are an infinite number of rays reaching the receiver, an essential consideration is to pre-select those rays expected to give the major contribution to the received signal. A program developed at the University of Karlsruhe simplifies the ray calculations by separating the propagation paths treated into two groups, as suggested in Fig. 31 [106]. Propagation in the vertical plane containing the base station and subscriber is treated by accounting for diffraction over the rooftops with reflections at building faces. The building faces are, of course, perpendicular to the ground, but are also assumed to be oriented perpendicular to the vertical plane when finding the reflected rays. Propagation in the slant plane, which contains the two antennas and is perpendicular to the vertical plane, takes into account reflection at building faces and diffraction at the vertical edges of buildings.

The restriction to propagation in these two planes neglects other possible ray paths, and not all rays in the two planes are included. The method can, however, incorporate terrain data. Since a full description of the program has not yet been published in a journal article, it is not entirely clear what ray paths are accounted for and what paths have been neglected, or how multiple diffraction over the rooftops is computed. It is also unclear how the program finds the paths that are included, or what reflection coefficient is used. This group and others have also developed models for predicting terrain scattering in mountainous regions [106].

Predictions made with the Karlsruhe program for a low base station antenna in a 750-m square test site at 955 MHz in Aalborg, Denmark, are compared with measurements along various segments of a drive path in Fig. 32 [106]. The program correctly predicts most features of the measure signal, except for sharp dip in the prediction just to the right of point A, and a missed peak in the measurements next to the measurement arrow. These exceptions may be due to rays that are not accounted for in the program. The Hata model is clearly inappropriate for the short measurement ranges, which never exceed 750 m.

A group at Virginia Polytechnic has also been developing a 3D ray-tracing program for treating propagation in the presence of buildings in a campus-like setting, such as that shown in Fig. 33 [105]. The program includes multiply reflected ray paths, paths that involve diffuse scattering on the last leg of the path, and diffracted rays. It is not clear how diffuse scattering and diffraction are treated, or what ray paths have been omitted. Figure 34 shows a comparison of the computed and measured impulse response at the location to the right and above the transmitter in Fig. 33. At all locations at this site the standard deviation of the error is 4.2 dB [105]. For these calculations a building reflection coefficient of 12 dB was used independent of the angle of incidence.

A group at France Telecom/PAB has developed two site-specific ray programs that make use of approximations to reduce the otherwise long computation times experienced for site-specific predictions [103], [104]. The first program uses approximate ray-tracing techniques to account for multiple reflections and diffractions and is used in conjunction with a complete database of building shape and height in

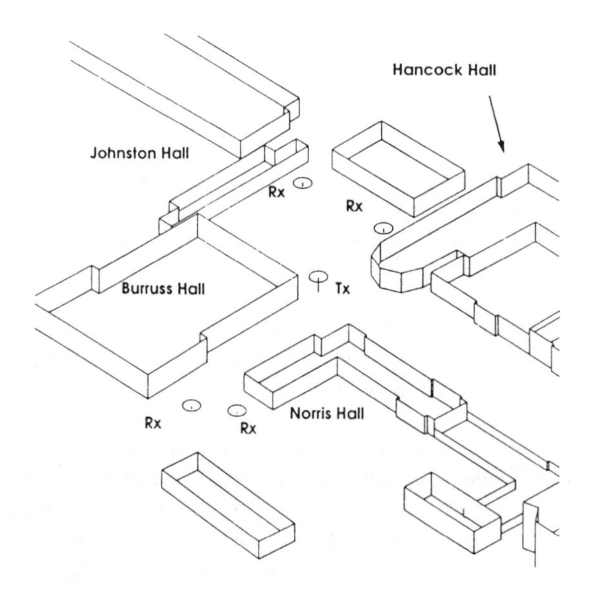

Fig. 33. Building representation in a campus environment for site-specific predictions—taken from [105].

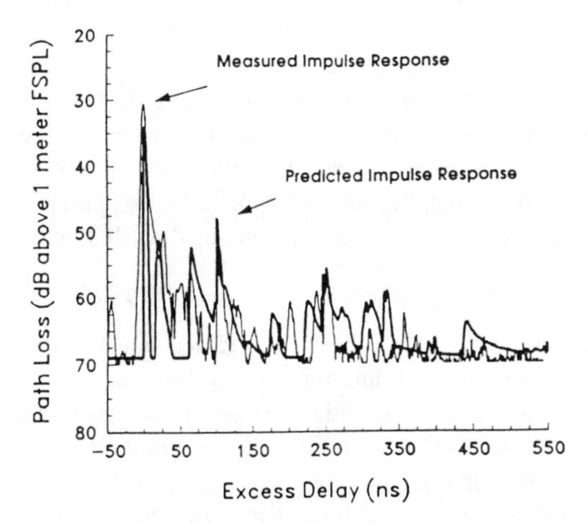

Fig. 34. Measured and computed time dependence for a pulsed source in the campus environment of Fig. 33—taken from [105].

the region covered. Best correlation with measurements was obtained using a building reflection coefficient of 7 dB [103], [104], which is substantially less than that employed at Virginia Polytechnic.

The second method used by the French group represents the environment only through the buildings lining the street on which the subscriber is located [104], and is intended for cellular applications in areas with closely spaced buildings of nearly uniform height. The method accounts for multiple reflections and one diffraction at the roof edge, and one diffraction at a vertical building edge. Good agreement is found with measurements for this limited building environment.

Knowledge of the reflection, transmission, and scattering properties of exterior and interior building walls is limited. Choosing a single coefficient independent of the angle of incidence on the basis of a best fit with measurements in a complex propagation environment can give different coefficients for different sets of measurements. Moreover, it is inadequate for treating reflection at glancing incidence where the reflection coefficient approaches unity. While some studies of the properties of realistic walls have been reported [108], [109], further work is called for.

VI. Site-Specific In-Building Prediction

The prediction of propagation inside buildings has attracted interest as a result of proposals for W-LAN and W-PBX systems. Most theoretical interest has been in site-specific predictions using either 2D ray tracing [110]–[112] for prediction on the same floor as the base station, or 3D ray tracing [113], [114] for the same and other floors. The alternatives to theoretical predictions are measurement-based schemes that assign a given path loss in decibells to major features, such as floors, walls, and rooms, between the base station and subscriber [47], [48]. Ray methods can also be used to account for signals that exit the building at exterior walls, reflect from neighboring buildings and re-enter the original building, and can be used to predict building penetration from outside sources. Because the ray approach separates out signals traveling on different paths, it is easily adapted to predict the pulse response or echo of a building.

For large open structures such as factories and atriums, and for propagation between floors, 3D ray tracing will play a role in determining the signal at different vertical elevations. However, in modern office buildings 3D ray tracing does not correctly describe the scattering processes that are associated with furnishings and ceiling features, as discussed below. In what follows we will discuss an alternative approach to dealing with furnishings and ceiling fixtures for predicting propagation over a single floor in modern office buildings, which is consistent with measurements made in large buildings. Using this approach, we then employ 2D ray tracing and show comparisons with measurements that indicate the importance of diffraction in certain cases. Finally, we will discuss propagation between floors, where diffraction plays an important role.

A. *Influence of Furniture and Ceiling Fixtures*

Typical construction of modern buildings incorporates drop ceilings of acoustical material supported by a metal frame. The space between the drop ceiling and the floor above contains light fixtures, ventilation ducts, pipes, support beams, etc. The acoustical material has a low dielectric constant so that rays incident on the ceiling will penetrate the acoustic material and be strongly scattered by the irregular structure, rather than undergo specular reflection. Similarly, furnishings placed on the floor, such as desks, cubicle partitions, filing cabinets, and work benches will scatter the rays and prevent them from reaching the floor, except in hallways. By allowing for specular reflection from floors and ceilings, 3D ray tracing fails to account for these significant phenomena. Based on the foregoing arguments, and measurements discussed below, we conclude that prop-

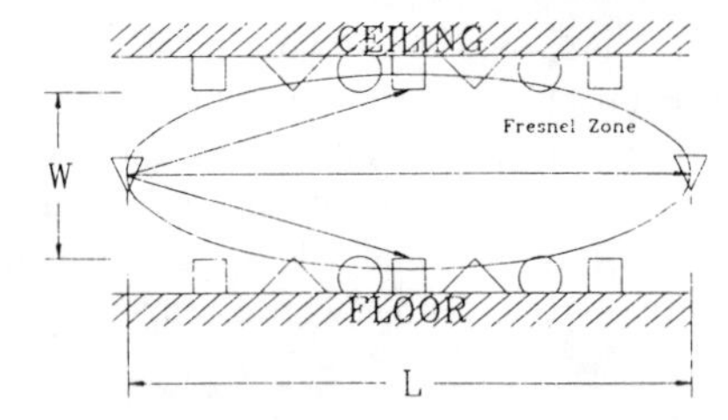

Fig. 35. Fresnel zone for propagation between transmitter and receiver in the clear space between building furnishings and ceiling fixtures—taken from [110].

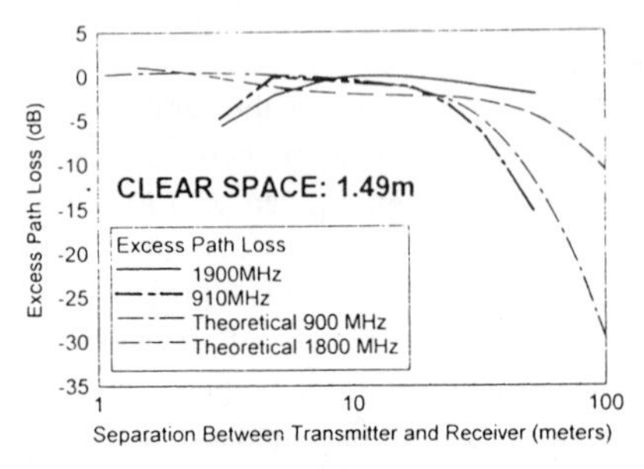

Fig. 36. Measured and computed excess path loss at 900 and 1800 MHz for a large office building having head-high cubical partitions, but no floor-to-ceiling partitions—taken from [111].

agation over one floor of such a building must take place in the clear space between the furniture and ceiling fixtures, as shown symbolically in Fig. 35.

When the transmitting and receiving antennas are located in the clear space, as in Fig. 35, the path-loss mechanism can be understood in terms of the Fresnel ellipse about them. If they are close enough so that the ellipse lies entirely within the clear space, the fields associated with the direct ray will not be affected by the presence of the scatterers [115], and the path loss will have the $1/L^2$ dependence of free space. As the separation between transmitter and receiver increases, the Fresnel ellipse will grow in size so that the scatters lie within it, as in Fig. 35, in which case the path loss is greater than that of free space. The distance at which the ellipse first encounters the scatters is W^2/λ, where W is the width of the clear space, and is seen to be proportional to frequency.

The effect of the scatters has been simulated using absorbing screens [110]. Recognizing that the exact signal pattern in some regions will be dependent on the nearby scattering objects, the path loss has been defined in terms of the average signal in a vertical cross section at a distance L from the transmitter. The path loss in excess of free space computed at 900 and 1800 MHz for $W = 1.5$ m is plotted in Fig. 36 as a function of path length L. It is seen that the excess path loss at each frequency is small out to distances of about 20 and 40 m, respectively, after which it increases dramatically.

For comparison with the theory, measurements of excess path loss were made in an office building having very large open areas furnished with 5 ft 2 in (1.57 m) high cubicle partitions, but with no floor-to-ceiling walls [111]. The clear space to the drop ceiling was $W = 1.49$ m. To achieve a true average path loss, measurements were made averaging over both ends of the path. At the receiver end, averaging was over three locations; just above the partitions, midway in the clear space, and just below the drop ceiling. For each receiver location, averaging over the transmitter location was achieved by moving it in a circle of radius 1 m. The resulting average signal was then normalized to the free-space signal (2), and smoothed to eliminate local variations.

The smoothed curves for measurements at 910 and 1900 MHz are shown in Fig. 36. Agreement between measurements and theory, both in distance and frequency dependence, show the importance of scattering from furniture and ceiling fixtures on propagation over one floor. These results also indicate that signal prediction can be made using 2D ray tracing, together with the excess path loss of Fig. 36 to account for the scattering. Additional loss will be experienced for subscribers located below the clear space, as for example when the antenna is at desk level among cubicle partitions.

In hallways with drop ceilings, the excess path loss can be simulated by imaging the ceiling scatters and base station in the floor. In this case, the width of the clear space is twice the floor-to-ceiling separation, so that the excess path loss is less than about 6 dB out to a distance of 100 m [110]. Furnished rooms whose ceilings are the concrete slab of the floor above, such as in apartment buildings and hotels, can be treated by imaging the furniture and base station in the ceiling, with the same result as obtained for hallways.

B. 2D Ray Tracing for Propagation Over a Floor

Ray procedures have been used to account for reflection and transmission at interior and exterior walls, treating the interaction as a specular process. Use of the specular approach involves two approximations, the first being that the linear extent of the wall is large enough to act as a planar reflector, and that it be electrically smooth so that the diffuse scattering does not dominate. For ray paths whose unfolded length is up to 100 m, the maximum width of the Fresnel ellipse in the horizontal plane is less than 4.1 m at 900 MHz and 2.9 m at 1.8 GHz. Since the length of walls is commonly 4 m or more, they span most of the Fresnel zone, or several zones, so that they can reasonably be considered large.

Interior walls are frequently constructed of gypsum board over steel studs, or concrete blocks. Voltage reflection coefficients Γ for a solid concrete wall, with no voids, and for a gypsum board wall, including the two layers but neglecting the studs between the layers, are shown in Fig. 37 as a function of the angle of incidence for 900-MHz radiation. Considerable variation with angle is seen, with Γ approaching unity at glancing incidence. The dielectric constant used in the calculation may be taken from handbooks [109], [112]. In some calculations, the authors have used the simplified variation shown in Fig. 37. In reading papers one must take care to understand the choice of reflection and transmission coefficients, since

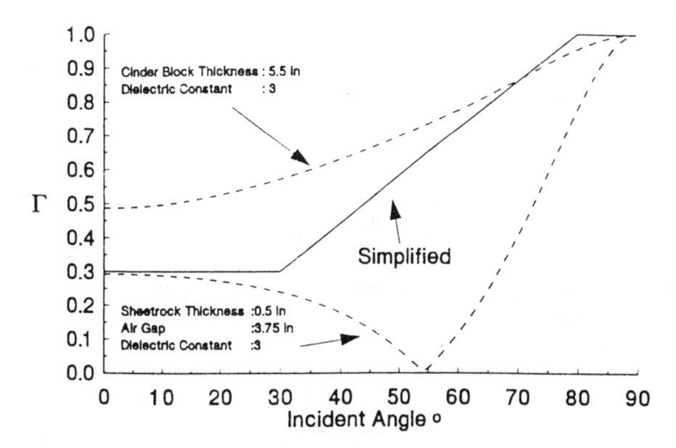

Fig. 37. Reflection coefficients for solid concrete and sheet rock walls and a simplified variation—taken from [110].

sometimes angle-independent values giving a best fit to measurements are used.

While the surfaces of walls may be smooth, the stud construction for gypsum board and interior webs and voids of concrete blocks, as well as pipes inside the walls, act to scatter waves in nonspecular directions. However, because the stud separation is larger than λ, substantial amounts of the incident energy should be carried by the specular fields.

Studies of reflection by stone exterior walls have shown that diffuse scattering resulting from surface roughness reduces the specularly reflected signal [108]. A phenomenological approach to account for diffuse scattering introduces a constant χ in the expression for the transmission coefficient as

$$T = \sqrt{\chi(1 - |\Gamma|^2)}. \tag{20}$$

For example, with $\chi = 1/2$ and using the simplified Γ of Fig. 37, at normal incidence 9% of the energy is specularly reflected, 45% is specularly transmitted, and the rest is scattered. At glancing incidence all of the energy is specularly reflected. In the case of concrete block walls, studies have shown that webbing modifies the reflection and transmission coefficients at 900 MHz, but does not introduce scattering [109]. At 1.8 GHz the webbing acts as a periodic structure that diffracts some of the incident energy into a grating order. As noted before, further studies of reflection and transmission at interior and exterior walls are called for.

The ray approach for dealing with wall reflections is shown in Fig. 38. Rays emanating from the transmitter reach the receiving point after transmission through and reflection from the walls. The signal associated with the specular component is found by adding together ray-path segments to find the complete path length L. For example, the signal associated with the ray having components r_1 and r_2 in Fig. 38 is proportional to $1/(r_1 + r_2)^2$, in addition to the excess path loss of Fig. 36. However, diffuse scattering, such as indicated by the path with segments s_1 and s_2 in Fig. 38, exhibits the multiplicative dependence $1/(s_1 s_2)^2$. Although the diffuse scattering will influence the signal in the vicinity of the scattering point, its amplitude will decrease more rapidly with distance than the specular components, and is therefore neglected.

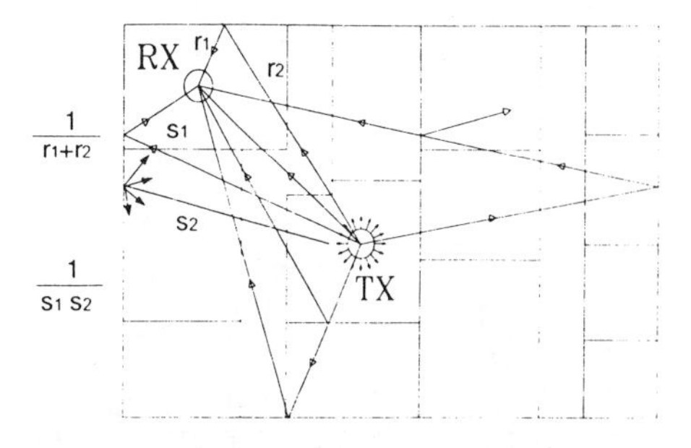

Fig. 38. Two-dimensional ray tracing accounting for specular reflection and transmission at walls—taken from [110].

Determination of all possible 2D ray paths can be carried out using a ray shooting approach, or using image theory. The ray shooting approach starts rays off from the transmitter at angular interval on the order of a degree. Each ray is traced through its interaction with the first wall, where it generates a transmitted and reflected ray. Both rays are then traced to the next interaction, and so on, building a binary tree of rays, which continues through some preset number of interactions. Since discrete rays have zero probability of intercepting a given point, a circle of finite radius proportional to the ray length is used to represent the receiver [110], [112]. The image approach is able to find the exact ray path between points by first imaging the source in the plane of each wall, one at a time, and checking that the path between the image and receiver intersects the wall in the physically existing segment. The process is repeated for double imaging in all combinations of two walls, for triple imaging in all combinations of three walls, etc., up to some predefined number. Triple images are reported to be adequate for predictions [113].

Having found the contributing rays, the path loss associated with the ith ray, assuming isotropic antennas, is given by

$$S_i = P_0 E(L_i) \prod_n |\Gamma_n(\phi_{ni})|^2 \prod_m |T_m(\phi_{mi})|^2. \tag{21}$$

Here L_i is the total unfolded path length of the ray and P_0 is the free-space factor of (2) for a distance L_i. The excess path loss of Fig. 36 is included as the factor $E(L_i)$ and $\Gamma_n(\phi_{ni})$ and $T_m(\phi_{mi})$ are the reflection and transmission coefficients at the walls encountered by the ray. Adding the path loss for the individual rays gives the sector average path loss.

Using the foregoing method, predictions have been made for a large office building, whose floor plan is shown in Fig. 39 [110]. The walls shown are floor-to-ceiling partitions of gypsum board, and the large rooms are furnished with partitioned cubicles just under 2 m high, or laboratories with work benches and shelving of the same height, leaving a clear space of 1.5 m to the drop ceiling. For 852-MHz transmission, the computed and measured signal at the locations indicated by the small circles in Fig. 39 are plotted in Fig. 40 as a function of the straight-line distance to the

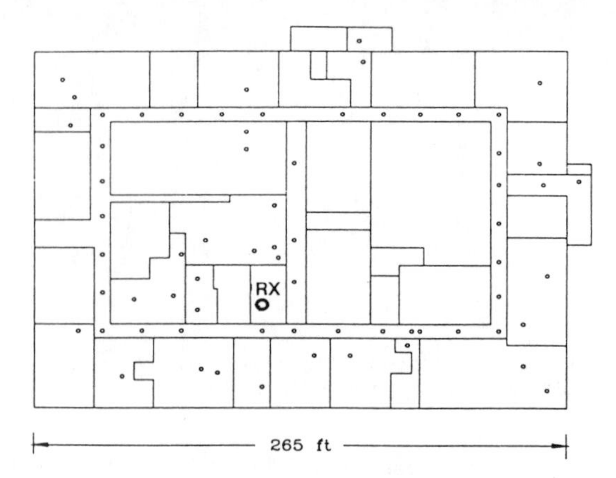

Fig. 39. Floor plan of a large office/laboratory building showing transmitter and receiver locations for signal measurements—taken from [110].

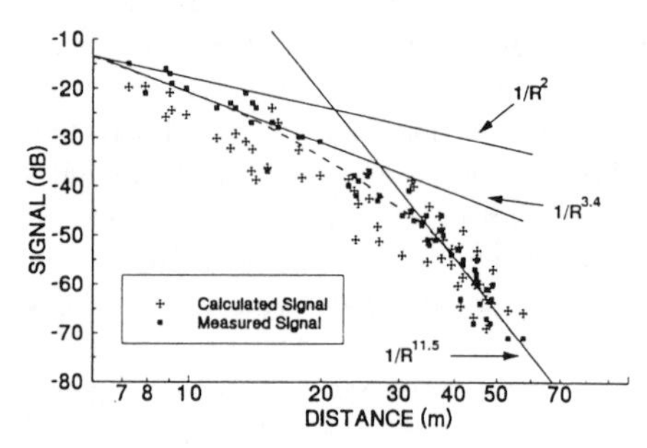

Fig. 40. Measured and computed sector averaged signal plotted versus separation distance R for the building of Fig. 39—taken from [110].

base station. A third-order polynomial curve obtained by a least squares fit to the measured data is shown in Fig. 40, together with its asymptotes for large and small distances.

For distances greater than 30 m the measurements and predictions have an average error of only 0.3 dB and a standard deviation of 4 dB, while for distances less than 30 m the predictions show a systematic error of 5.1 dB. The source of the systematic error is not known, and is the largest obtained for any of the buildings for which the program was tested. The change in slope with distance shown in Fig. 40 demonstrates the influence of the clear space on the excess path loss, as predicted by Fig. 36, and is not obtained when the excess path loss is omitted from the calculations.

A comparison has been made with measurements at 917 MHz in a university building whose floor plan is shown in Fig. 41 [47]. The walls are of masonry construction, which we have assumed to enclose classrooms or laboratories. The measured and computed signals are shown in Fig. 42. Excess attenuation for hallways has been used to compute the signal at points in the main corridor shown in Fig. 41, and the office attenuation for the points in the rooms and side halls. The signal at these two classes of points clearly follow different trends in Fig. 42 as a result of guiding by the hallway.

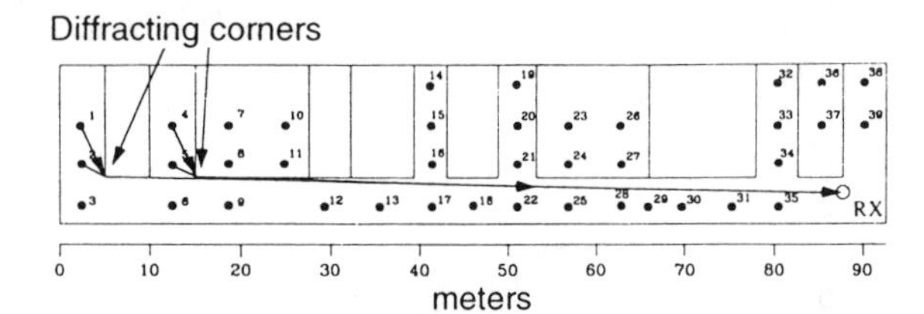

Fig. 41. Floor plan of a building at Laval University showing transmitter and receiver locations for signal measurements—taken from [46].

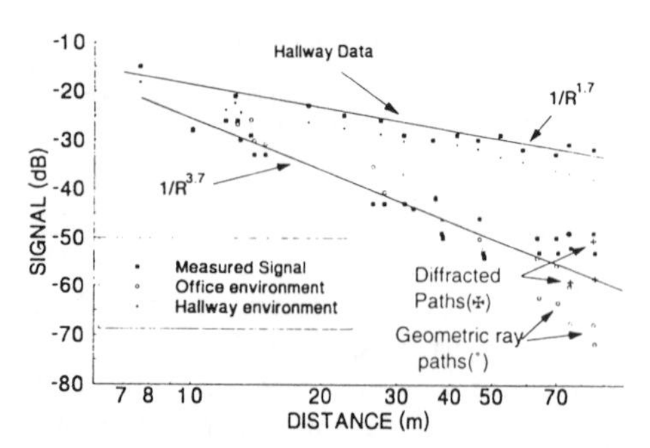

Fig. 42. Measured and computed sector averaged signal plotted versus separation distance R for the building of Fig. 41—taken from [110].

Except for the most distant points in the side halls (1, 2, 4, and 5) in Fig. 41, the predictions from Fig. 42 are in good agreement with measurements. However, at these distant points the predicted signal is substantially smaller than that measured. Since the geometric rays must undergo many transmissions and/or reflections to reach these points, the computed signal is small. However, the signal may reach these points via the simple diffraction paths shown in Fig. 31. To account for these signals, we employ the diffraction approach described in previous sections. The path loss associated with a ray path having segments $L_0, L_1, \cdots, L_M$ separated by diffraction at edges through angles $\beta_1, \cdots, \beta_M$ is given by

$$S = P_0 \frac{L}{L_0} \prod_m \frac{1}{2\pi k L_m} D^2(\beta_m) \tag{22}$$

provided each edge lies outside the shadow boundary of the previous edge. In (22), L is the total path length found by summing L_m, P_0 is the free-space path loss (2) for a distance L, and $D(\beta_m)$ is the diffraction coefficient given by (4).

The signals computed for the single diffraction paths of Fig. 41 are shown as crosses in Fig. 42. They are seen to lie close to the measured values. This result indicates that diffraction can be an important phenomenon in determining the signal at some locations. Similar results have been found in other buildings [110] for propagation on the same floor, and will be discussed in the next subsection dealing with propagation between floors. Inclusion of diffracted rays into automatic ray programs will add considerable complication to the programs.

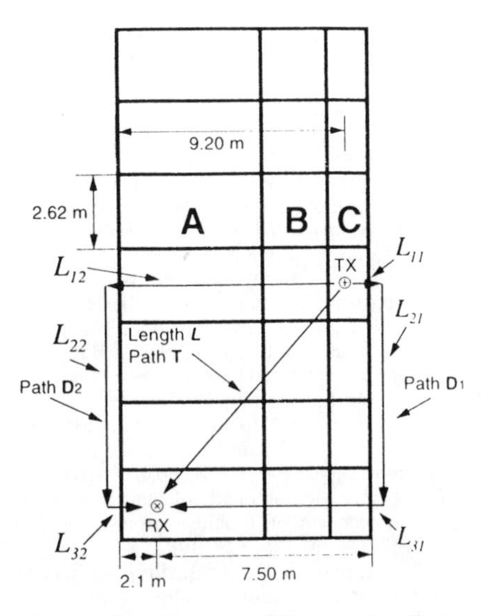

Fig. 43. Cross section of hotel building showing direct and diffracted ray paths for propagation between floors—taken from [68].

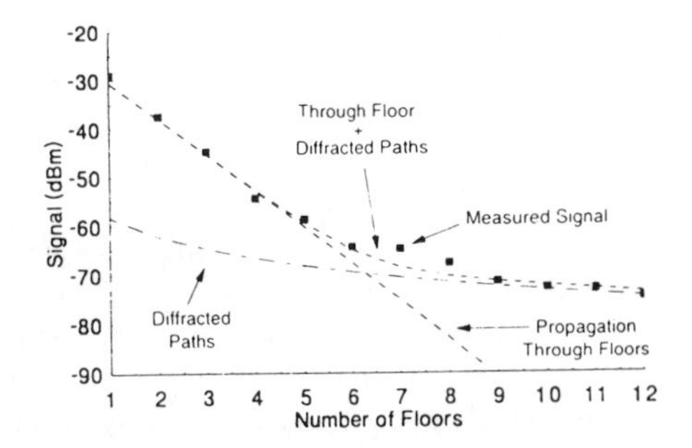

Fig. 44. Measured and computed sector averaged signal as a function of the number of floors between transmitter and receiver—taken from [68].

C. Propagation Between Floors

When the antennas are located on different floors, one possible path between them involves transmission through the intervening floors. For reinforced concrete or precast concrete floors, transmission loss has been measured at 10 dB or more [43], [48]. Floors constructed of concrete poured over corrugated steel panels show much greater loss [43]. In this case, the received signal may in fact be associated with other paths involving diffraction rather than transmission through the floor. Such paths are suggested in Fig. 43 for a very simple geometry [68]. Here a hotel is shown in cross section with the transmitter located in the hallway on one floor, and the receiver on a lower floor in the bedroom of a suite directly below the transmitter.

The direct path from a transmitter to the receiver passes through two walls and the intervening floors, which are of precast concrete. The signal associated with this path is given by the free-space factor (2) reduced by the transmission loss through the walls and floors. In addition, rays can exit the building through windows lining the hallway, diffract down along the building face, and diffract back into the lower floor through windows located there to illuminate the receiver (path D_1 in Fig. 43). A second path (D_2 in Fig. 43) involves diffraction out of and back in through the bedroom windows. The signals for these paths are given by (22) for two diffractions, which must be further reduced by the transmission loss through the windows (taken as 0.25 dB) and the walls (taken as 2.2 dB).

The measured value of the received signal, averaged over transmitter position, is plotted in Fig. 44 as a function of floor separation at 852 MHz. The signal computed for the direct path, assuming 13-dB transmission loss through the floor, and the sum of the signals computed for the two diffraction paths are also shown. It is seen that the sum of the signals for the three paths accurately predicts the measurements. The initial rapid decrease in signal arises from the transmission loss through the floor, while the later slow decrease results from the diffracted signals, which have large loss even for a separation of one floor, but vary only slowly with increased floor separation. Similar trends have been observed in other buildings [48], [116].

The example of Figs. 43 and 44 indicates the importance of diffraction paths in propagation between floors. Paths may lie outside the building or in stairwells and elevator shafts. Further work is required to gain a fuller understanding of these contributions. However, their existence limits the validity of predictions that simply add the losses assigned to major building features between antennas.

VII. Conclusion

The first and most important conclusion to be drawn from the studies reported here is that it is possible to theoretically predict the signal by using the basic wave phenomena of reflection, transmission, and diffraction, and taking into account building features. Only a few years ago such a conclusion would have been considered impossible. The accuracy that can be achieved by the predictions is related to the level of environmental detail incorporated into the predictions. Accounting for the row structure of cities is adequate to predict the range dependence in macrocellular systems, although large errors may occur for individual sites. For microcellular systems, better accuracy can be obtained by including regularly occurring generic building features. At the most sophisticated level, site-specific programs are intended to give deterministic prediction of the signal.

Reliable prediction programs can reduce the number of measurements needed for planning new systems and for developing design procedures or installation instructions. Improved predictions should also make it possible for system designers to achieve greater capacity, although the relationship between prediction accuracy and improved capacity is not clear. At another level, prediction programs can be employed during the design of a building to evaluate its impact on wireless system installation. Such programs could also be used to evaluate simple modifications to standard building designs that would reduce interference between cells on different floors. If airplanes are designed for their electromagnetic characteristics, why not buildings?

The prediction procedures discussed here are based on a few well-understood wave phenomena that reappear in different guises in the various environments. These phenomena are readily formulated in terms of ray descriptions of propagation. Purely numerical methods of solving Maxwell's equations, such as finite difference and finite element methods, are not at all applicable because of the large size of the buildings compared to wavelength. The difficulties in developing ray-based computer prediction models lie in encoding the logic for identifying the significant paths from the input data file for all structures in any type of environment, and in finding efficient ways to input building geometry. Because of these difficulties, a site-specific program written to be used universally for predictions in all environments would not be efficient for many applications. Thus while one can expect to see further development of site-specific prediction programs, it is of value to continue the development of programs that are dedicated to specific applications and environments.

REFERENCES

[1] Y. Okumura, E. Ohmori, T. Kawano, and K. Fukuda, "Field strength and its variability in VHF and UHF land-mobile radio service," *Rec. Elec. Comm. Lab.*, vol. 16, pp. 825–873, 1968.

[2] A. P. Barsis, "Determination of service area for VHF/UHF land mobile and broadcast operation over irregular terrain," *IEEE Trans. Veh. Technol.*, vol. VT-22, pp. 21–29, 1973.

[3] D. O. Reudink, "Properties of mobile radio propagation above 400 MHz," *IEEE Trans. Veh. Technol.*, vol. VT-23, pp. 143–159, 1974.

[4] K. Allsebrook and J. D. Parson, "Mobile radio propagation in British cities at frequencies in the VHF and UHF Bands," *IEEE Trans. Veh. Technol.*, vol. VT-26, pp. 313–322, 1977.

[5] V. Graziano, "Propagation correlations at 900 MHz," *IEEE Trans. Veh. Technol.*, vol. VT-27, pp. 182–189, 1978.

[6] G. D. Ott and A. Plitkins, "Urban path-loss characteristics at 820 MHz," *IEEE Trans. Veh. Technol.*, vol. VT-27, pp. 189–197, 1978.

[7] K. K. Kelly, "Flat suburban area propagation at 820 MHz," *IEEE Trans. Veh. Technol.*, vol. VT-27, pp. 198–204, 1978.

[8] N. H. Shepherd *et al.*, "Coverage prediction for mobile radio systems operating in the 800/900 MHz frequency range," *IEEE Trans. Veh. Technol.*(Special Issue), vol. 37, pp. 3–72, 1988.

[9] F. Ikegami, S. Yoshida, T. Takeuchi, and M. Umehira, "Propagation factors controlling mean field strength on urban streets," *IEEE Trans. Antennas Propagat.*, vol. AP-32, pp. 822–829, 1980.

[10] J. Walfisch and H. L. Bertoni, "A theoretical model of UHF propagation in urban environments," *IEEE Trans. Antennas Propagat.*, vol. 36, pp. 1788–1796, 1988.

[11] C. Chrysanthou and H. L. Bertoni, "Variability of sector averaged signals for UHF propagation in cities," *IEEE Trans. Veh. Technol.*, vol. 39, pp. 352–358, 1990.

[12] F. Ikegami, T. Takeuchi, and S. Yoshida, "Theoretical prediction of mean field strength for urban mobile radio," *IEEE Trans. Antennas Propagat.*, vol. 39, pp. 299–302, 1991.

[13] S. R. Saunders and F. R. Bonar, "Mobile radio propagation in built-up areas: A numerical model of slow fading," in *Proc. IEEE Vehicular Technology Conf.*, pp. 295–300, 1991.

[14] H. L. Bertoni and L. Maciel, "Theoretical prediction of slow fading statistics in urban environments," in *Proc. IEEE Int. Conf. on Universal Personal Communications* (Dallas, TX, 1992), pp. 1–4.

[15] R. Steele and V. K. Prabhu, "High-user-density digital cellular mobile radio systems," *Proc. Inst. Elec. Eng.*, pt. F, vol. 132, pp. 396–404, 1985.

[16] D. C. Cox, H. W. Arnold, and P. T. Porter, "Universal digital portable communications: A system perspective," *IEEE J. Selected Areas Commun.*, vol. SAC-5, pp. 764–773, 1987.

[17] D. C. Cox, "Universal digital portable radio communications," *Proc. IEEE*, vol. 75, pp. 436–477, 1987.

[18] R. Steele, "The cellular environment of lightweight handheld portables," *IEEE Commun. Mag.*, vol. 27, pp. 20–29, 1989.

[19] D. M. Black and D. O. Reudink, "Some characteristics of mobile radio propagation at 836 MHz in the Philadelphia area," *IEEE Trans. Veh. Technol.*, vol. VT-21, pp. 45–51, 1972.

[20] D. C. Cox, "Multipath delay spread and path loss correlation for 910-MHz urban mobile radio propagation," *IEEE Trans. Veh. Technol.*, vol. VT-26, pp. 340–344, 1977.

[21] G. L. James, *Geometrical Theory of Diffraction for Electromagnetic Waves*, 3rd ed. London, UK: Peter Peregrines, 1986.

[22] D. C. Cox, R. R. Murray, and A. W. Norris, "800 MHz attenuation measured in and around suburban houses," *AT&T Bell Lab. Tech. J.*, vol. 63, pp. 921–954, 1984.

[23] S. T. S. Chia, R. Steele, E. Green, and A. Baran, "Propagation and bit error ratio measurements for a microcellular system," *J. IRE*, vol. 57, pp. S255–S266, 1987.

[24] D. M. J. Devasirvatham, "Radio propagation studies in a small city for universal portable communications," in *Conf. Rec. IEEE VTC'88* (Philadelphia, PA, 1988), pp. 100–104.

[25] J. H. Whitteker, "Measurements of path loss at 910 MHz for proposed microcell urban mobile systems," *IEEE Trans. Veh. Technol.*, vol. 37, pp. 125–129, 1988.

[26] R. J. C. Bultitude and G. K. Bedal, "Propagation characteristics on microcellular urban mobile radio channels at 910 MHz," *IEEE J. Selected Areas Commun.*, vol. 7, pp. 31–39, 1989.

[27] P. Harley, "Short distance attenuation measurements at 900 MHz and 1.8 GHz using low antenna heights for microcells," *IEEE J. Selected Areas Commun.*, vol. 7, pp. 5–11, 1989.

[28] T. S. Rappaport, S. Y. Seidel, and R. Singh, "900 MHz multipath propagation measurements for U.S. digital cellular radiotelephone," *IEEE Trans. Veh. Technol.*, vol. 39, pp. 132–139, 1990.

[29] S. Y. Seidel, T. S. Rappaport, S. Jain, M. L. Lord, and R. Singh, "Path loss, scattering and multipath delay statistics in four european cities for digital cellular and microcellular radiotelephone," *IEEE Trans. Veh. Technol.*, vol. 40, pp. 721–730, 1991.

[30] A. J. Rustako, Jr., N. Amitay, G. J. Owens, and R. S. Roman, "Radio propagation at microwave frequencies for line-of-sight microcellular mobile and personal communications," *IEEE Trans. Veh. Technol.*, vol. 40, pp. 203–210, 1991.

[31] P. E. Mogensen, P. Eggers, C. Jensen, and J. B. Andersen, "Urban area radio propagation measurements at 955 and 1845 MHz for small and micro cells," in *Proc GLOBECOM'91* (Phoenix, AZ, 1991), pp. 1297–1302.

[32] H. H. Xia *et al.*, "Urban and suburban microcellular propagation," in *Proc. IEEE Int. Conf. on Universal Personal Communications* (Dallas, TX, 1992), pp. 5–9.

[33] S. Aguirre, K. C. Allen, and M. G. Laflin, "Signal strength measurements at 915 MHz and 1920 MHz in an outdoor microcell environment," in *Proc. IEEE Int. Conf. on Universal Personal Communications* (Dallas, TX, 1992), pp. 17–22.

[34] D. M. J. Devasirvatham, C. Banerjee, R. R. Murray, and D. A. Rappaport, "Two-frequency radiowave propagation measurements in Brooklyn," in *Proc. IEEE Int. Conf. on Universal Personal Communications* (Dallas, TX, 1992), pp. 23–27.

[35] H. H. Xia, H. L. Bertoni, L. Maciel, A. Lindsay-Stewart, R. Rowe, and L. Grindstaff, "Radio Propagation measurements and modeling for line-of-sight microcellular systems," in *IEEE Vehicular Technology Conf.* (Denver, CO, 1992).

[36] A. J. Goldsmith and L. J. Greenstein, "A measurement-based model for predicting coverage areas of urban microcells," *IEEE J. Selected areas Commun.*, vol. 11, pp. 1013–1023, 1993.

[37] D. M. J. Devasirvatham, "Time delay spread measurements of wideband radio signals within a building," *Electron. Lett.*, vol. 20, pp. 950–951, 1984.

[38] ——, "Time delay spread and signal level measurements of 850 MHz radio waves in building environments," *IEEE Trans. Antennas Propagat.*, vol. AP-34, pp. 1300–1305, 1986.

[39] J. Horikoshi, K. Tanaka, and T. Morinaga, "1.2 GHz band wave propagation measurement in concrete building for indoor radio communications," *IEEE Trans. Veh. Technol.*, vol. VT-35, no. 4, pp. 146–152, Nov. 1986.

[40] D. M. J. Devasirvatham, "Multipath time delay spread in the digital portable radio environment," *IEEE Commun. Mag.*, vol. 25, no. 6, June 1987.

[41] ——, "Multipath time delay jitter measured at 850 MHz in the portable radio environment," *IEEE J. Selected Areas Commun.*, vol. SAC-5, no. 5, June 1987.

[42] A. A. M. Saleh, and R. A. Valenzuela, "A statistical model for indoor multipath propagation," *IEEE J. Selected Areas Commun.*, vol. SAC-5, pp. 128–137, 1987.

[43] H. W. Arnold, R. R. Murray, and D. C. Cox, "815 MHz radio attenuation measured within two commercial buildings," *IEEE Trans. Antennas Propagat.*, vol. 37, pp. 1335–1339, 1989.

[44] R. J. C. Bultitude, S. Mahmoud, and W. Sullivan, "A comparison of indoor radio propagation characteristics at 910 MHz and 1.75 GHz," *IEEE J. Selected Areas Commun.*, vol. 7, pp. 20–30, 1989.

[45] T. S. Rappaport and C. D. McGillen, "UHF fading in factories," *IEEE J. Selected Areas Commun.*, vol. 7, pp 40–48, 1989.

[46] K. Pahlavan, R. Ganesh, and T. Hotaling, "Multipath propagation measurements on manufacturing floors at 910 MHz," *Electron. Lett.*, pp. 225–227, 1989.

[47] J. F. Lafortune and M. Lecours, "Measurement and modeling of propagation losses in a building at 900 MHz," *IEEE Trans. Veh. Technol.*, vol. 39, pp. 101–108, May 1990.

[48] S. Y. Seidel and T. S. Rappaport, "914 MHz path loss prediction models for indoor wireless communications in multifloored buildings," *IEEE Trans. Antennas Propagat.*, vol. 40, pp. 207–217, 1992.

[49] P. F. Driessen, M. Gimersky, and T. Rhodes, "Ray model of indoor propagation," in *Proc. of Wireless Personal Communications Conf.* (Blacksburg, VA, June 1992), pp. 17-1–17-12.

[50] R. Bultitude, R. Melangon, H. Zaghloul, G. Morrison, and M. Prokki, "The dependence of indoor radio channel multipath characteristics on transmit/receive ranges," *IEEE J. Selected Areas Commun.*, vol. 11, pp. 979–990, 1993.

[51] H. Hashemi, "The Indoor radio propagation channel," *Proc IEEE*, vol. 81, pp. 943–968, 1993.

[52] H. W. Nylund, "Characteristics of small-area signal fading on mobile circuits in the 150 MHz band," *IEEE Trans. Veh. Technol.*, vol. VT-17, pp. 24–30, 1968.

[53] G. L. Turin *et al.*, "A statistical model of urban multipath propagation," *IEEE Trans. Veh. Technol.*, vol. VT-21, pp. 1–9, 1972.

[54] W. C. Y. Lee, *Mobile Communications Engineering.* New York: McGraw-Hill, 1982, chs. 1, 6.

[55] W. C. Y. Lee and Y. S. Yeh, "Polarization diversity system for mobile radio," *IEEE Trans. Commun.*, vol. COM-20, pp. 912–913, 1972.

[56] S. A. Bergmann and H. W. Arnold, "Polarization diversity in portable communications environments," *Electron. Lett.*, vol. 22, pp. 609–610, 1986.

[57] R. G. Vaughn, "Polarization diversity in mobile communications," *IEEE Trans. Veh. Technol.*, vol. 39, pp. 177–186, 1990.

[58] D. L. Shilling, "Broadband-CDMA: A PCS wireless technology to achieve wireline quality and maximize spectral efficiency," in *Wireless Personal Communications*, M. J. Feuerstein and T. S. Rappaport, Eds. Boston, MA: Kluwer, 1993, pp. 77–91.

[59] S. Kozono and K. Watanabe, "Influence of environmental buildings on UHF land mobile radio propagation," *IEEE Trans. Commun.*, vol. COM-25, pp. 1133–1145, 1977.

[60] A. Papoulis, *Probability, Random Variables and Stochastic Processes.* New York: McGraw-Hill, pp. 266–268.

[61] M. Hata, "Empirical formula for propagation loss in land mobile radio service," *IEEE Trans. Veh. Technol.*, vol. VT-29, pp. 317–325, 1980.

[62] D. L. Shilling *et al.*, "Field test experiments using broadband code division multiple access," *IEEE Commun. Mag.*, vol. 3, no. 1, pp. 86–93, 1991.

[63] H. H. Xia, H. L. Bertoni, L. R. Maciel, A. Lindsay-Stewart, and R. Rowe, "Radio propagation characteristics for line-of-sight microcellular and personal communications," *IEEE Trans. Antennas Propagat.*, vol. 41, no. 10, pp. 1439–1447, 1993..

[64] T. Iwama *et al.*, "Investigation of propagation characteristics above 1 GHz for microcellular land mobile radio," in *Proc. IEEE Vehicular Technology Conf.*, pp. 396–400, 1990.

[65] S. J. Patsiokas, B. K. Johnson, and J. L. Dailing, "Propagation of radio signals inside buildings at 150, 450 and 850 MHz," in *Proc. IEEE Vehicular Technology Conf.* (Dallas, TX, 1986), pp. 66–72.

[66] W. Honcharenko, H. L. Bertoni, and J. Dailing, "Bi-lateral averaging over receiving and transmitting areas for accurate measurements of sector average signal strength inside buildings," *IEEE Trans. Antennas Propagat.*, submitted for publication.

[67] D. M. J. Devasirvatham, C. Banerjee, M. J. Krain, and D. A. Rappaport, "Multi-frequency radiowave propagation measurements in the portable radio environment," in *Proc. IEEE ICC'90*, pp. 1334–1340, 1990.

[68] W. Honcharenko, H. L. Bertoni, and J. Dailing, "Mechanism governing propagation between different floors in buildings," *IEEE Trans. Antennas Propagat.*, vol. 42, pp. 787–790, 1993.

[69] S. Kozono and A. Taguchi, "Mobile propagation loss and delay spread characteristics with a low base station antenna on an urban road," *IEEE Trans. Veh. Technol.*, vol. 42, pp. 103–109, 1993.

[70] E. S. Sousa, V. M. Jovanovic, and C. Daigneault, "Delay spreads measurements for the digital cellular channel in Toronto," in *Proc. Int. Symp. on Personal, Indoor and Mobile Radio Communications* (Boston, MA, 1992), pp. 80–85.

[71] W. Mohr, "Wideband propagation measurements of mobile radio channels in mountainous areas in the 1800 MHz frequency range," in *Proc. IEEE Vehicular Technology Conf.* (Secaucus, NJ, 1993), pp. 49–52.

[72] P. Withington, II and L. W. Fullerton, "An impulse radio communication system," in *Ultra-Wideband, Short-Pulse Electromagnetics*, H. L. Bertoni, L. Carin, and L. B. Felsen, Eds. New York: Plenum Press, 1993.

[73] J. C-I. Chuang, "The effects of time delay spread on portable radio communications channels with digital modulation," *IEEE Trans. Selected Areas Commun.*, vol. SAC-5, pp. 879–889, 1987.

[74] L. B. Felsen and N. Marcuvitz, *Radiation and Scattering of Waves.* Englewood Cliffs, NJ: Prentice-Hall, 1973, pp. 652–665.

[75] J. B. Keller, "Geometrical theory of diffraction," *J. Opt. Soc. Amer.*, vol. 52, pp. 116–131, 1962.

[76] R. G. Kouyoumjian and P. H. Pathak, "A uniform geometrical theory of diffraction for an edge in a perfectly conducting surface," *Proc. IEEE*, vol. 62, pp. 1448–1461, 1974.

[77] M. J. Brooking and R. Larsen, "Results of height gain measurements taken in different environments," Tech. Rep. MTR 84/42, GEC Res. Labs., Marconi Res. Cen., Chelmsford, England, 1985.

[78] G. Millington, R. Hewitt, and F. S. Immirzi, "Double knife-edge diffraction field strength predictions," *Proc.Inst. Elelc. Eng.* (Monograph 507E), pp. 419–429, 1962.

[79] S. W. Lee, "Path integrals for solving some electromagnetic edge diffraction problems," *J. Math. Phys.*, vol. 19, pp. 1434–1469, 1978.

[80] J. Boersma, "On certain multiple integrals occurring in a waveguide scattering problem," *SIAM J. Math. Anal.*, vol. 9, pp. 377–393, 1978.

[81] L. E. Vogler, "An attenuation function for multiple knife-edge diffraction," *Radio Sci.*, vol. 19, pp. 1541–1546, 1982.

[82] S. R. Saunders and F. R. Bonar, "Explicit multiple building diffraction attenuation function for mobile radio wave propagation," *Electron. Lett.*, vol. 27, pp. 1276–1277, 1991.

[83] H. H. Xia and H. L. Bertoni, "Diffraction of cylindrical and plane waves by an array of absorbing half screens," *IEEE Trans. Antennas Propagat.*, vol. 40, pp. 170–177, 1992.

[84] P. Eggers and P. Barry, "Comparison of a diffraction based radiowave propagation model with measurements," *Electron . Lett.*,, vol. 26, pp. 530–531, 1990.

[85] L. R. Maciel, H. L. Bertoni, and H. H. Xia, "Unified approach to prediction of propagation over building for all ranges of base station antenna height," *IEEE Trans. Veh. Technol.*, vol. 42, pp. 41–45, 1993.

[86] K. Low, "A comparison of CW-measurements performed in Darmstadt with the COST-231-Walfisch-Ikegami model," Rep. COST 231 TD (91) 74, Darmstadt, Germany, Sept., 1991.

[87] W. C. Y. Lee, "Studies of base-station antenna height effects on mobile radio," *IEEE Trans. Veh. Technol.*, vol. VT-29, pp. 252–260, 1980.

[88] K. Bullington, "Radio propagation for vehicular communications," *IEEE Trans. Veh. Technol.*, vol. VT-26, pp. 295–308, 1977.

[89] W. J. Vogel and J. Goldhirsh, "Tree attenuation at 869 MHz derived from remotely piloted aircraft measurements," *IEEE Trans. Antennas Propagat.*, vol. AP-34, pp. 1460–1464, 1986.

[90] J. Goldhirsh and W. J. Vogel, "Roadside tree attenuation measurements at UHF for land mobile satellite systems," *IEEE Trans. Antennas Propagat.*, vol. 35, pp. 589–596, 1987.

[91] M. F. Levy, "Diffraction studies in urban environment with

wide-angle parabolic equation method," *Electron. Lett.*, vol. 28, pp. 1491–1492, 1992.

[92] L. R. Maciel, "Signal strength prediction for pcs in microcellular environments," Ph.D. dissertation, Engineering, Polytech. Univ., New York, 1993.

[93] "Urban transmission loss models for mobile radio in the 900- and 1,800-MHz bands," Rep. COST 231 TD (90) 119 Rev. 1, Florence, Italy, Jan. 1991.

[94] L. R. Maciel, H. L. Bertoni, and H. H. Xia, "Propagation over buildings for paths oblique to the street grid," in *Proc. Int. Symp. on Personal, Indoor and Mobile Radio Communications* (Boston, MA, 1992), pp. 75–79.

[95] H. H. Xia, H. L. Bertoni, L. R. Maciel, and L. Grindstaff, "Microcellular propagation characteristics for personal communications in urban and suburban environments," *IEEE Trans. Veh. Technol.*, accepted for publication.

[96] L. Melin, M. Ronnlund, and R. Angbratt, "Radio wave propagation: A comparison between 900 and 1800 MHz," in *Proc. Vehicular Technology Conf.* (Secaucus, NJ, 1993), pp. 250–252.

[97] L. R. Maciel and H. L. Bertoni, "Cell shape for microcellular systems in residential and commercial environments," in *IEEE Conf. on Universal Personal Communications* (Ottawa, Ont., Canada, 1993), pp. 723–727.

[98] L. R. Maciel and H. L. Bertoni, "Cell shape for microcellular systems in residential and commercial environments," *IEEE Trans. Veh. Technol.*, vol. 43, no. 2, pp. 270–278, 1994..

[99] C. Berglijung and L. G. Olsson, "Rigorous diffraction theory applied to street microcell propagation," in *IEEE GLOBECOM'91* (Phoenix, AZ, 1991), pp. 1292–1296.

[100] F. G. Wagen and K. Rizk, "Simulation of radio wave propagation in urban microcellular environments," in *Proc. Conf. on Universal Personal Communications* (Ottawa, Ont., Canada, 1993), pp. 595–599.

[101] F. Niu and H. L. Bertoni, "Path loss and cell coverage of urban microcells in high-rise building environments," in *Proc. IEEE GLOBECOM'93* (Houston, TX., 1993), pp. 266–270.

[102] T. S. Rappaport, S. Y. Seidel, and K. R. Schaubach, "Site-specific propagation prediction for PCS system design," in *Wireless Personal Communications*, M. J. Feuerstein and T. S. Rappaport, Eds. Boston, MA: Kluwer, 1992, pp. 281–315.

[103] J-P. Rossi and A. J. Levy, "A ray model for decimetric radiowave propagation in urban area," *Radio Sci.*, vol. 27, pp. 971–979, 1992.

[104] ——, "Propagation analysis in cellular environment with the help of models using ray theory and GTD," in *Proc. Vehicular Technology Conf.* (Secaucus, NJ, 1993), pp. 253–256.

[105] S. Y. Seidel, K. R. Schaubach, T. T. Tran, and T. S. Rappaport, "Research in site-specific propagation modeling for PCS system design," in *Proc. Vehicular Technology Conf.* (Secaucus, NJ, 1993), pp. 261–264.

[106] T. Kurner, D. J. Cichon, and W. Wiesbeck, "Concepts and results for 3D digital terrain-based wave propagation models: An overview," *IEEE J. Selected Areas Commun.*, vol. 11, pp. 1002–1012, 1993.

[107] H. R. Anderson, "A ray-tracing propagation model for digital broadcast systems in urban areas," *IEEE Trans. Broadcast.*, vol. 39, pp. 309–317, 1993.

[108] O. Landron, M. J. Feuerstein, and T. S. Rappaport, "In situ microwave reflection coefficient measurements for smooth and rough exterior wall surfaces," in *Proc. Vehicular Technology Conf.* (Secaucus, NJ, 1993), pp. 77–80.

[109] W. Honcharenko and H. L. Bertoni, "Transmission and reflection characteristics at concrete block walls in the UHF bands proposed for future PCS," *IEEE Trans. Antennas Propagat.*, vol. 43, no. 2, pp. 232–239, 1994.

[110] W. Honcharenko, H. L. Bertoni, J. Dailing, J. Qian, and H. D. Yee, "Mechanism governing propagation on single floors in modern office buildings," *IEEE Trans. Veh. Technol.*, vol. 41, pp. 496–504, 1992.

[111] W. Honcharenko, H. H. Xia, S. Kim, and H. L. Bertoni, "Measurements of fundamental propagation characteristics inside buildings in the 900 and 1900 MHz bands," in *Proc. IEEE Vehicular Technology Conf.* (Secaucus, NJ, 1993), pp. 879–882.

[112] T. Holt, K. Pahlavan, and J. F. Lee, "A graphical indoor radio channel simulator using 2D ray tracing," in *Int. Symp. on Personal, Indoor and Mobile Radio Communications* (Boston MA, 1992), pp. 411–416.

[113] R. A. Valenzuela, "A ray tracing approach to predicting indoor wireless transmission," in *Proc. IEEE Vehicular Technology Conf.* (Secaucus, NJ, 1993), pp. 214–218.

[114] P. Kreuzgruver, P. Unterberger, and R. Gahleitner, "A ray splitting model for indoor radio propagation associated with complex geometries," in *Proc. IEEE Vehicular Technology Conf.* (Secaucus, NJ, 1993), pp. 227–230.

[115] H. L. Bertoni, A. Hessel, and L. B. Felsen, "Local properties of radiation in lossy media," *IEEE Trans. Antennas Propagat.*, vol. AP-19, pp. 226–237, 1971.

[116] J. LeBel and P. Melangon, "The development of a comprehensive indoor propagation model," in *Proc. IEEE Int. Symp. on Personal, Indoor and Mobile Radio Communications* (London, UK, 1991).

Henry L. Bertoni (Fellow, IEEE) was born in Chicago, IL, on November 15, 1938. He received the B.S. degree in electrical engineering from Northwestern University, Evanston, IL, in 1960. He received the M.S. degree in electrical engineering in 1962 and the Ph.D. degree in electrophysics in 1967, both from the Polytechnic Institute of Brooklyn (now Polytechnic University), Brooklyn, NY.

After graduation he joined the faculty of the Polytechnic. He is now Head of the Department of Electrical Engineering. His research has dealt with theoretical aspects of wave phenomena in electromagnetics, ultrasonics, acoustics, and optics. He has authored or co-authored over 90 articles on these topics. During 1982–1983 he spent a sabbatical leave at University College London as a Guest Research Fellow of the Royal Society. The research he carried out at the University College was the subject of a paper that was awarded in 1984 Best Paper Award of the IEEE Sonics and Ultrasonics Group. During the Summer of 1983 he held a Faculty Research Fellowship at USAF Rome Air Development Center, Hanscom AFB. His current research in electromagnetics deals with the theoretical prediction of UHF propagation characteristics in urban environments, and he and his students were the first to explain the mechanisms underlying characteristics observed for propagation of the Cellular Mobile Radio signals.

Dr. Bertoni is currently the Chairman of the Technical Committee on Personal Communications of the IEEE Communications Society. He is Chairman of the Hoover Medal Board of Award and has served on the ADCOM of the IEEE Ultrasonics, Ferroelectric and Frequency Control Society. He is also a member of the International Scientific Radio Union and the New York Academy of Science.

Walter Honcharenko (Member, IEEE) was born in Brooklyn, NY, on August 14, 1967. He received the B.S. and M.S. degrees in electrical engineering from the Polytechnic University, Brooklyn, in 1989 and 1991, respectively. In January 1993, he completed the Ph.D. degree in electrophysics, also at the Polytechnic University.

From 1988 to January of 1993, he was affiliated with the Center for Advanced Technology in Telecommunications at Polytechnic University. There he developed models to predict propagation characteristics and signal coverage inside buildings for future use with wireless personal communications systems. Since May 1993, he has been with the Wireless Technology Laboratory of AT&T Bell Laboratories in Whippany, NJ. He continues to work on issues related to RF propagation for telecommunication systems.

Dr. Honcharenko is a member of IEEE Antennas and Propagation and Vehicular Technology Societies.

Leandro Rocha Maciel was born in Rio de Janeiro, Brazil, on October 23, 1963. He received the B.S. and M.S. degrees in electrical engineering from the Military Institute of Engineering (IME), Rio de Janeiro, in 1986 and 1988, respectively. In 1993, he completed the Ph.D. degree in electrical engineering at the Polytechnic University of New York, Brooklyn, carrying out his research in modeling UHF propagation in urban environments, with the help of a grant from CNPq—Conselho Nacional de Desenvolvimento Cientifico e Tecnológico—of the Brazilian Govenment, and in 1992 under a grant from Telesis Technologies Laboratory.

From 1987 to 1988 he worked for the Brazilian Army (CTEx) in microwave devices measurements and rain attenuation of electromagnetic waves in the microwave band, where he developed the research for his master thesis. During the Summer of 1991, he was with Telesis Technologies Laboratory (Pac Tel), working in the FCC Experimental License Project for Personal Communication Services (PCS). Now he is a Member of the Technical Staff at AT&T Bell Laboratories, Whippany, NJ, working in the development of new wireless technologies and cellular systems worldwide.

Howard H. Xia was born in Canton, China, on August 16, 1960. He received the B.S. degree in physics from South China Normal University, Canton, in 1982. He received the M.S. degree in physics in 1986, the M.S. degree in electrical engineering in 1988, and the Ph.D. degree in electrophysics, all from Polytechnic University, Brooklyn, NY.

Since September 1990, he has been with AirTouch Communications (Formerly, PacTel Corporation) and Telesis Technologies Laboratory, Walnut Creek, CA, where he has been engaged in research and development of advanced analog and digital cellular mobile radio networks, and personal communications systems. He has published articles on the topics of outdoor/indoor radio wave propagation, spectrum sharing, and CDMA system design.

Dr. Xia serves as a member of the United States delegation to participate in IRU-R (formerly, CCIR) activities on developing Recommendations for the third-generation mobile systems—Future Public Land Mobile Telecommunication Systems (FPLMTS).

Reprinted from *IEEE Transactions on Antennas and Propagation*, Vol. 41, No. 12, Dec. 1993.

A Deterministic Approach to Predicting Microwave Diffraction by Buildings for Microcellular Systems

Thomas A. Russell, *Member, IEEE*, Charles W. Bostian, *Fellow, IEEE*, and Theodore S. Rappaport, *Senior Member, IEEE*

***Abstract*—Designers of low-power radio systems for use in urban areas would benefit from accurate computer-based predictions of signal loss due to shadowing. This paper presents a propagation prediction method that exploits a building database and considers the three-dimensional profile of the radio path. Models and algorithms are provided that allow the application of Fresnel-Kirchhoff diffraction theory to arbitrarily oriented buildings of simple shapes. Building location information used by the diffraction models is in a form compatible with a geographic information systems (GIS) database. Diffraction screens are constructed at all building edges, for both horizontal and vertical orientations, in order to consider all possible diffractions and to compute field contributions often ignored. Multiple buildings and edges of the same building that introduce multiple successive diffractions are considered with a rigorous, recursive application of the diffraction theory that requires sampling the field distribution in each aperture. Robust and computationally efficient numerical methods are applied to solve the diffraction integrals. Tests of the software implementation of these methods through example runs and comparisons with 914-MHz continuous-wave measurements taken on the Virginia Tech campus show promise for predicting radio coverage in the shadows of buildings.**

I. Introduction

RADIO signal strength prediction methods are currently evolving towards greater resolution. In an urban area the principal obstructions to free-space radio propagation are buildings, and in a typical high-power radio cell they may number in the hundreds. The complexity of their interactive effects on field strength at any given point has traditionally forced the use of gross average models for path loss as a function of distance, which may include an empirical correction factor for the degree of urbanization [1], or a theoretically derived diffraction loss based on an idealized layout of infinitely long buildings of uniform height [2]. Recently, however, several researchers have shown the feasibility of using a specific, building-by-building description of the urban environment and ray-tracing methods to determine signal strength [3], [4], [5].

In the future, wireless phone communications will serve densely populated urban areas with plentiful mini-base stations or radio ports, each of which will cover areas of several blocks at most [6], [7]. Thus the resolution of the urban area is increased and the number of included buildings reduces to a small number, increasing the significance of individual buildings (10–20 dB impact seen in downtown Ottawa [8]). An analogy may be made with terrain obstructions in a large-scale rural radio system. Methods currently employed for system design in such an area compute the diffraction introduced by each hill explicitly rather than statistically, including multiple diffraction either with geometrical construction methods [9], or with numerical sampling methods [10] to more rigorously characterize the effects.

We suggest taking a similarly careful approach to modeling the individual impacts of buildings, while recognizing the different form that these obstructions take. Of particular relevance is that hills usually fall off smoothly in height while buildings generally have sharp drop-offs. Indeed, the use of the infinite knife-edge model for terrain diffraction is based on the fact that smooth variation in the height of a hill transverse to the direction of propagation can be neglected with only small error [11]. This model, however, is inappropriate as a general approach for building diffraction, as diffraction around the corner edges is neglected. Other approaches consider corner-edge diffraction [12], but neglect rooftop diffraction paths, requiring that the transmitter and receiver are both low with respect to all of the buildings. This is appropriate for such systems but excludes hybrid or intermediate schemes, where the transmitter may be mounted on a building and may be higher than a neighboring building. It should be expected that a multitude of system designs will emerge with new personal communication systems, and a general modeling approach that allows the system designer flexibility in transmitter and receiver placement should be useful.

In this paper we propose a unified, three-dimensional approach to predicting the diffractive shadowing introduced by a collection of arbitrarily oriented buildings of simple shapes, given the locations and heights of the buildings, the transmitter, and the receiver. Finite knife-

Manuscript received November 23, 1992; revised July 14, 1993.

T. A. Russell is with Stanford Telecommunications, Inc., 1761 Business Center Drive, Reston, VA 22090.

C. W. Bostian and T. S. Rappaport are with the Bradley Department of Electrical Engineering, Virginia Polytechnic Institute and State University, Blacksburg, VA 24061.

IEEE Log Number 9214086

edge models for building edges are constructed and every possible diffraction path is simulated in order to determine the significant sources of field and to avoid missing a possibly dominant signal path. A terrain diffraction method that recursively computes the multiple diffraction introduced by successive diffracting edges through efficient numerical techniques [11] is adapted for use with finite knife edges, forming the basis of a computer program implementation. This is a deterministic site-specific approach that requires no statistical parameters or correction factors. We apply the program to a campus area and show through comparisons with measurements at 914 MHz that the transition regions are well predicted as a receiver moves into the shadow of a building, and that diffraction predictions give a worst-case prediction of signal strength.

Buildings are approximated with simple rectangular volumes identified by planar coordinates (such as Universal Transverse Mercator) of the four corners plus the height above sea level of the rooftop. More complex buildings may be composed of adjacent rectangular sections. The use of a geographic information systems (GIS) package as a software platform for input and output of the data required by the model is described in [13]. This building representation provides the information necessary to consider diffraction paths around and over the building.

II. Diffraction Prediction Method

The three-dimensional model for single diffraction by buildings introduced in [13] forms the basis of the multiple-diffraction model presented in this paper. This approach applies the Fresnel-Kirchhoff solution for diffraction through an aperture in a two-dimensional screen [14] which is approximately normal to the direction of propagation (the third dimension.) When the screen is modeled with only one dimension, this solution reduces to the infinite knife-edge model traditionally used in radio engineering.

In our case the aperture is the area surrounding a building. We must divide this complex area into simple rectangular areas in order to use the approximate form of the Kirchhoff diffraction integral consisting of the product of two one-dimensional integrals rather than a computationally intensive two-dimensional integral. The diffraction field contributions from each area are computed and summed at the observation point to give the total field, resulting in multipath diffraction.

Bachynski and Kingsmill [15] verified this multipath approach to diffraction as applied to thin, knife-edge obstacles with varying profiles transverse to the direction of propagation. They conducted scale-model measurements and saw a strong dependence on the transverse profile and substantial agreement with predictions.

A. Multiple Diffraction

Multiple diffraction effects must be considered if the wave incident on the secondary Huygens sources in the area surrounding on obstacle has been affected by a preceding edge, as in Fig. 1. The method we will use requires first evaluating the complex field at many Huygens sources, Q, in the aperture plane in order to approximate the field distribution, and then integrating over the aperture. This is the same general approach used by Whitteker [11] and Walfisch and Bertoni [2]; however, while the solutions in both [11] and [2] integrate only over the vertical dimension, our solution includes the product of integrals over both of the dimensions of the aperture plane.

When there is an intervening horizontal edge (parallel to the x axis), as in Fig. 1, the field in the aperture (at points Q) varies with y based on the degree of shadowing by the intervening edge, while the dependence on x is a function primarily of the increased path length, which can be described analytically. This important distinction allows us to characterize the field distribution with discrete function evaluations in only the y dimension. A requirement of two-dimensional sampling would lead to excessive computation time. The derivation provided in the Appendix yields the following solution:

$$E_n(P) = -ie^{ikp}\sqrt{\frac{s}{2\lambda p(s+p)}}\int_{\xi_1}^{\xi_2} e^{i\pi\xi^2/2}\,d\xi \cdot \int_{y_1}^{\infty} E_Q(y)e^{i\pi y^2/\lambda p}\,dy, \quad \text{(1a)}$$

where

$$\xi_1 = x_1\sqrt{\frac{2(s+p)}{\lambda sp}}, \quad \text{(1b)}$$

$$\xi_2 = x_2\sqrt{\frac{2(s+p)}{\lambda sp}}. \quad \text{(1c)}$$

This solution provides the component of the complex field at an observation point, P, resulting from diffraction through the rectangular aperture defined by (x_1, x_2) and (y_1, ∞) in Fig. 1, given isotropic radiation of unity amplitude at source, S. The shaded rectangles in the figure represent building faces. Lengths s and p are the distances from the origin of the aperture plane to S and P, respectively. This result reduces to Eq. (9) of [11], or Eq. (1) of [2] when the limits on x extend to $\pm\infty$. To compute the field at intermediate points Q, the points are treated as new observation points, P'.

The computational speed of the multiple diffraction procedure depends on the number of field evaluations, $E_Q(y)$, required to obtain an accurate result. This number can be minimized by performing integrations over interpolating polynomials according to the method of Whitteker, [10], [11], [16], and Stamnes et al. [17]. The coefficients of amplitude and phase interpolation functions are computed based on the field samples. The amplitude interpolator is always quadratic and the phase function is either linear or quadratic depending on the size of the computed quadratic coefficient. The aperture field is characterized with a set of up to 80 finely spaced samples appended by

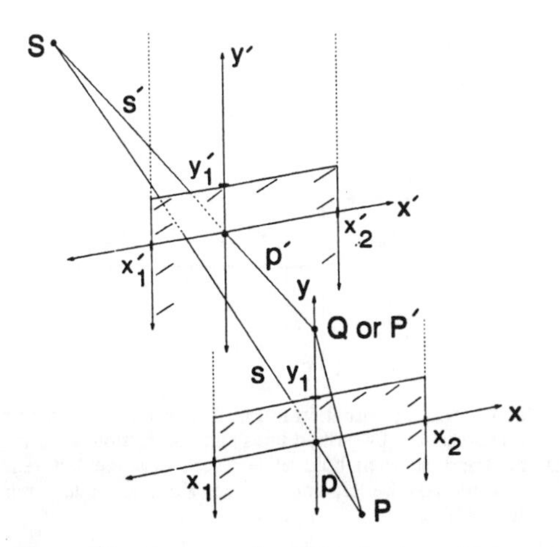

Fig. 1. Construction for computing diffraction by multiple successive building faces. The field at a Huygens source Q is found by treating it as a new observation point P'.

three more widely spaced samples. These final three samples yield a pair of amplitude and phase functions which are integrated to infinity. Spacing and number of samples are discussed in Section II.D.

The derivation of (1) is based on a diffraction field at Q caused by a horizontal edge. If, on the other hand, the edge producing the diffraction field in the aperture is a vertical edge, such as the corner edge of a building, then the opposite of the above reasoning holds: variation with x affects the degree of shadowing more than variation with y, which is assumed to be only a function of free-space expansion. We obtain for vertical edges,

$$E_n(P) = -ie^{ikp}\sqrt{\frac{s}{2\lambda p(s+p)}}\int_{x_1}^{x_2} E_Q(x)e^{i\pi x^2/\lambda p}\,dx \cdot \int_{\eta_1}^{\infty} e^{i\pi\eta^2/2}\,d\eta, \quad (2a)$$

where

$$\eta_1 = y_1\sqrt{\frac{2(s+p)}{\lambda sp}}. \quad (2b)$$

The lower limit on the range of x is $-\infty$ for a field propagating around the left side of the building, and the upper limit is ∞ for a field propagating around the right side. Note that the need for (2) instead of (1) is determined not by the nature of the diffracting edge at Q, but by the nature of the diffraction field at Q, which is a function of the previous edge.

When the field at aperture points Q is not affected significantly by prior edges (the single-diffraction case), the aperture field variation with both x and y depends only on free-space expansion, and the above results (1) and (2) reduce to

$$E_n(P) = \frac{-i}{2}E_{fs}\int_{\xi_1}^{\xi_2} e^{i\pi\xi^2/2}\,d\xi\int_{\eta_1}^{\infty} e^{i\pi\eta^2/2}\,d\eta, \quad (3a)$$

where

$$E_{fs} = \frac{e^{ik(s+p)}}{(s+p)}, \quad (3b)$$

and E_{fs} is the free-space field at P.

The total field at P is found by computing $E_n(P)$ for each of the N rectangular areas that together make up the total unobstructed portion of a diffraction screen. (The aperture area construction is described in Sections II.B and II.C.) The individual contributions are summed either in a complex field sense that displays the multipath interference between diffraction fields,

$$E(P) = \sum_{n=1}^{N} E_n(P), \quad (4)$$

or in an average power sense that provides an estimate of the local mean field strength at P,

$$E(P) = \sqrt{\sum_{n=1}^{N} E_n^2(P)}. \quad (5)$$

Both of these solutions may find application in propagation effects analysis, but the latter, (5), would be appropriate for signal strength prediction for system design.

At intermediate points, P' in Fig. 1, neither of these solutions is used. Instead, the contributing fields, $E_n(P')$, are compared and only the largest field source is used in the solution for the diffraction at the next point. This is done in order to avoid the possibly large interference effects that (4) would predict, and the consequent difficulties in integrating over the phase and amplitude variations. The alternative, the power sum method given by (5), cannot be used at intermediate points because it does not predict phase.

B. Construction of Aperture Screens

The area surrounding a building already must be separated into simple rectangular areas for evaluation reasons (see above); we take the additional step to assign an aperture area to each diffracting building edge, both vertical and horizontal, and construct a screen so as to best model that edge.

The formulation of the Kirchhoff integral in terms of Fresnel integrals requires that the aperture screen be approximately normal to the line connecting the source and observation points [14]. The faces of the buildings, however, will not in general align with such a plane. For simplicity, we make all aperture screens vertical, which requires only that the vertical component of SP is smaller than the horizontal [14]. The aperture plane is oriented normal to the horizontal projection of SP: the y axis is vertical, the x axis is horizontal, and the origin is at the point of intersection with SP. The building face is then projected onto this plane, yielding the bounds on the aperture region in terms of x and y. These bounds transform into the limits of integration according to (1b), (1c), and (2b).

1) Roof-Edge Aperture Screens: The rules for construction of aperture screens are illustrated with a simple example (Fig. 2) where the transmitter (source, S) is

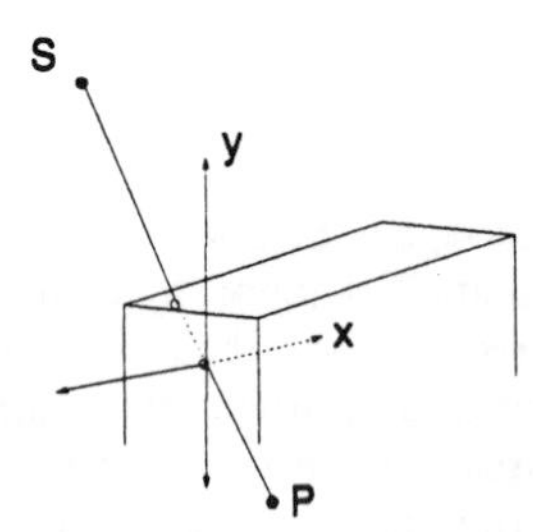

Fig. 2. Perspective view of diffraction by a rectangular building. The xy plane is constructed to model the diffraction by the roof edge of the left building side.

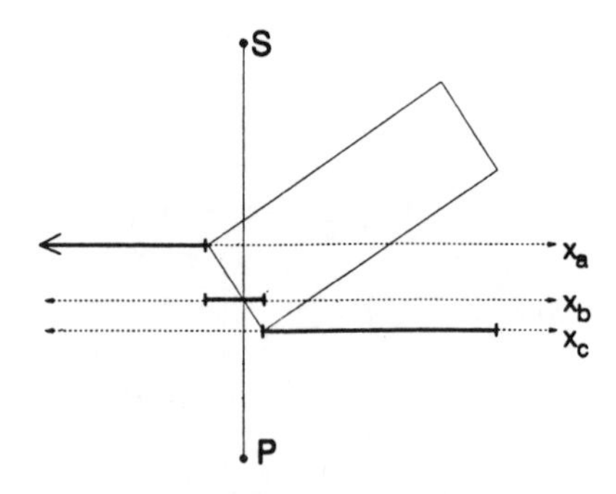

Fig. 3. Top view of the situation in Fig. 2. The x axes of three aperture planes are represented by dashed lines and the portion of each axis that defines the aperture is in bold. Plane a models the left rear corner diffraction, b models the left side rooftop diffraction, and c models the front rooftop diffraction.

higher than both the obstructing building and the receiver (observation point, P), and diffraction is introduced only by the left corner edge and the two roof edges nearest the receiver. In this example the right corner is too distant relative to the other edges to contribute relatively significant energy and the roof edges away from the receiver do not obstruct the path; therefore there is only single diffraction and no field sampling is required.

Consider first the roof edge of the building face visible on the left of the figure, through which SP passes. The aperture plane constructed for this edge is superimposed on Fig. 2. The bounds on x are illustrated in Fig. 3 with a bold section of the x_b axis. The second diffracting building face visible in Fig. 2, the front of the building, does not intersect SP though it is near enough that the roof edge will contribute diffracted field to P. The xy plane in such a case is the vertical plane that intersects the face at the vertical edge nearest to SP. The x axis of this plane is presented as x_c in Fig. 3.

2) Corner-Edge Aperture Screens: Finally, consider the diffraction field contributed by the left rear corner edge of the building. The aperture screens constructed for vertical edges always intersect the edge. The diffraction field is computed by integrating over the infinite half-plane abutting the edge. For the small geographical areas covered by microcellular systems, and fairly flat ground, there will be little obstruction by the ground. In these cases we can assume $y_1 = -\infty$ to avoid computations necessary to access terrain information. As shown in Fig. 3 (with the x_a axis), the aperture region also goes to infinity in the $-x$ direction.

3) Overall Single Diffraction Model: The three diffraction screens discussed above and shown in Fig. 3 are treated as though they are portions of the same overall aperture plane; however, we segment the infinite aperture plane into semi-infinite regions with vertical dividing lines and translate these regions along SP (a translation that has both vertical and horizontal components). When viewed along a straight line from P to S in Fig. 3, these aperture screen segments appear to form a single aperture screen. Conversely, when the apertures introduced by different building edges overlap when viewed in this way, we treat them as multiple (successive) diffraction screens. Some sample algorithms are presented in Section II.C for sorting the edges of rectangular buildings according to whether they diffract the wave multiple successive times or diffract different portions of the wave front a single time.

This construction method is designed to minimize error in estimation of diffraction by any particular edge by defining as precisely as possible the distances involved in the diffraction integral; at the wavelengths of interest for microcellular applications (less than 0.4 m) and the close proximity to large buildings, accuracy in the location of the edge is critical to accuracy in the result. In the process, some error will be incurred through the assumption of single diffraction through apertures that are offset from each other. However, it should be noted that two separated aperture segments constructed by our rules meet and share the same plane in transition regions when SP passes directly through a corner edge. (In Fig. 3, as P moves to the left x_a and x_b will meet, and as P moves to the right x_b and x_c will meet.)

4) Example: The movement of a receiver past a building is simulated for the situation pictured in Fig. 4: the transmitter is 30 m high and 150 m from the building; the receiver is 0 m high and 50 m from the building; the building is 10 m high, 50 m wide, and very long. The receiver travels a 60-m course parallel to the long edge of the building, moving from a line-of-sight situation into the shadow of the building. At the point marked in Fig. 4 the situation is identical to the one shown in Fig. 3. Fig. 5 presents the results of applying the diffraction integral (3) to each edge, showing that there are regions where each component is dominant, and demonstrating the necessity of computing each component. While the error in the power estimation would usually be less than 3 dB if only the most dominant diffraction component is computed, this component is difficult to reliably identify beforehand, particularly in the general case of multiple diffraction. The total diffraction field is represented as a power sum, or local mean field strength, computed by (5).

C. Application to a Collection of Buildings

At the receiver point the database is searched for the nearest diffracting building according to an algorithm

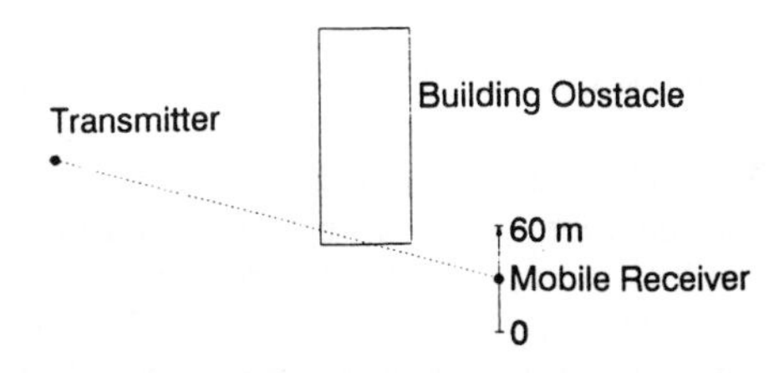

Fig. 4. Top view of the geometry used for an example receiver run. At the position indicated, this is the same situation depicted in Figs. 2 and 3.

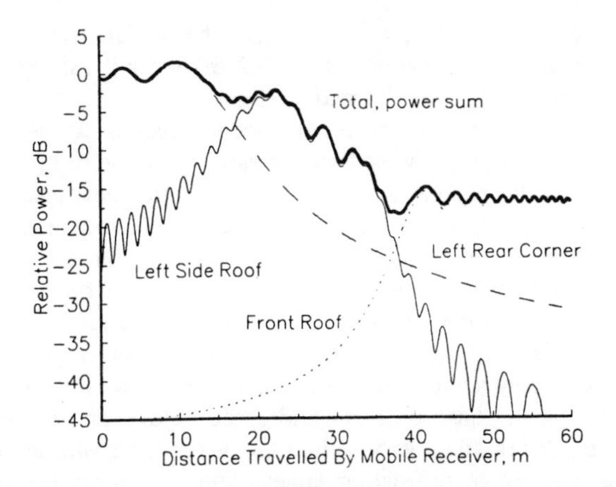

Fig. 5. Computed diffraction field contributions from the three aperture regions defined in Fig. 4. The total field represents an estimate of the local mean diffraction field strength.

described in [13], which includes a coordinate system transformation to simplify geometric manipulations. The procedure for considering each of the eight building edges as possible diffractors is demonstrated below.

1) Roof-Edge Diffraction: The first step is to construct aperture screens according to the rules in Section II.B. The results are illustrated in Figs. 6 (roof edges) and 7 (corner edges), where one of the two dimensions of each aperture is indicated with a bold line, and the other dimension is out of the page.

The four roof edges of a single building are grouped into two diffraction levels, where edges that belong to the same level define apertures that derive from a single aperture plane which has been segmented and translated. These levels are numbered according to the order in which they affect the wave; for example, in Fig. 6 the two apertures at the top of the figure are part of the first aperture plane encountered by the wave radiating from S. The two corresponding roof edges are considered level 1, and the other two edges are level 2. In this model, the level 1 edges diffract the wave once and the level 2 edges diffract the wave a second time and the edges making up a single level diffract different portions of the wave front.

The level 2 edges are considered first: both edges are checked for obstruction of more than 55% of the first Fresnel zone. This clearance threshold is traditionally used in microwave radio links [18]; however, it leads to some error in multiple diffraction, see Section III.A. If the following three conditions are true of the diffraction parameters computed for a particular aperture according to (1b), (1c), and (2b),

$$\xi_1 < -0.78, \qquad \xi_2 > 0.78, \qquad \eta_1 > -0.78,$$

then the aperture provides sufficient clearance.

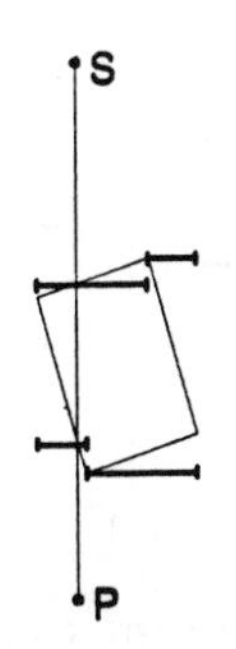

Fig. 6. Top view showing aperture constructions for all four roof edges.

If neither of the level 2 edges offers a sufficiently clear propagation path, then both edges are considered potential diffractors. However, before any time-consuming diffraction evaluations are made, each aperture should be checked to ensure that the diffraction path is capable of contributing significant fields strength. For example, if 40 dB is considered the maximum diffraction loss of interest, and any of the following are true for a particular aperture:

$$\xi_1 > 22, \qquad \xi_2 < -22, \qquad \eta_1 > 22,$$

then at least 40 dB diffraction loss will be incurred on that path and the aperture should be considered blocked.

The rooftop diffraction field at an observation point is computed by first considering the level 2 edges of the nearest diffracting building and then working towards the transmitter by proceeding on to the level 1 edges and then to prior buildings in the search for prior diffracting edges. If none are found, then the single diffraction (3) is evaluated at P; otherwise samples are taken and the multiple diffraction (1) is evaluated according to the recursive procedure described in Section II.A. An important note regarding multiple diffraction should be made. While an integration over the aperture field at a prior diffracting edge is performed based on the slightly altered geometry at each sample point in the current aperture, only at the first sample point (nearest the edge) is a search performed for prior diffractors, and samples of the field adjacent to the prior diffracting edge are only taken once.

2) Corner-Edge Diffraction: The diffraction fields contributed by the building corner edges are summed with the rooftop fields to find the total building diffraction field. Fig. 7 illustrates an example. The corners are grouped according to whether they are on the left of SP or the right, and then sorted according to the order in which they are encountered by the wave by applying a simple sorting algorithm.

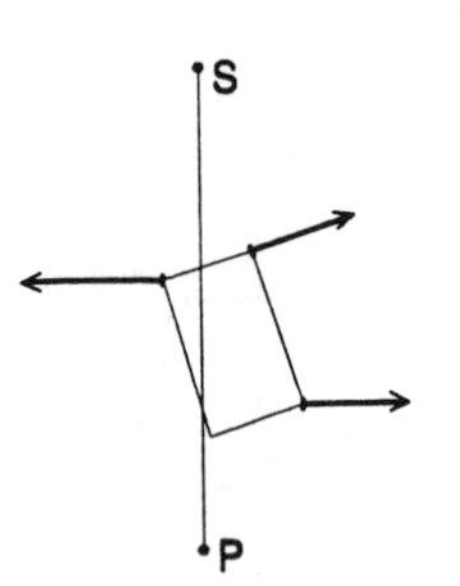

Fig. 7. Top view showing aperture constructions for the diffracting corners.

Thus, for each building we consider two corner-edge diffraction fields, one around the left of the building and one around the right. Typically only one of these will survive the test for significant diffraction field, a test which is analogous to that for roof edges. Multiple corner edges in a single group are considered as possible multiple successive diffractors, following a procedure similar to that for roof edges. Considering only a single group at a time, first the nearest corner edge to the receiver point, then each prior corner moving towards S is checked first for blockage and then for clearance. At the corner nearest S another search is conducted for prior diffracting buildings.

We note in Fig. 7 only one diffracting corner edge on the left, and two on the right (there is sufficient clearance at the corner edge on the right that is nearest P). Also, we note that the aperture plane next to the top corner in the right is at a different angle. This arises through the procedure whereby we sample the field in the aperture of the diffracting, right-side corner nearest the receiver and consider sample points as new observation points P'. The aperture plane at the preceding diffracting corner is then defined relative to the new line SP'.

3) Connected Buildings: Connected buildings are defined as those sharing an edge. This is the representation used in our model for buildings with rooftop sections of different heights; it is evidenced in the database by two corner coordinates on each of two buildings matching to within some error tolerance. Connected buildings require some exceptions to the general rules described in this section. For example, the diffraction around a corner that is connected to a taller building (and does not really exist as a diffracting edge) is not computed. Rather, the true corner of the connecting building is considered a diffraction source in the same manner as the other corners of the original building.

D. Field Sampling

The integration range over y in (1) or x in (2) is divided into two subranges: a finite region that extends at least until the ray connecting S with the sample point clears the obstructing edge, and a semi-infinite region (if the full range itself is semi-infinite) wherein the field amplitude oscillates about the free-space level, converging to this level at infinity. The first region is characterized with a large set ($\sim$ 25–80) of finely spaced samples, and the second region with three relatively widely spaced samples.

We set the sampling interval for the first integration region, Δ_1, at about $2/3\lambda$. Walfisch and Bertoni [2], using linear representations of both amplitude and phase and some different assumptions in the integral formulation, found that point spacing of less than a wavelength should result in an error in the integrand of less than 0.8%. Our error in the integrand, using quadratic polynomials, should be less. We tested the sensitivity to sample spacing by reducing it to 0.1 m and 0.165 m and increasing the number of points in order to hold approximately constant the sampling range, and saw that the variation in the computed power was less than 0.2 dB for all test cases, including up to four diffracting edges.

The sampling interval for the second region, Δ_2, is set to 3.0λ. Ref. [11] advises that the value of Δ_2 be comparable with the change in height for which the clearance of the obstructing edge changes by one Fresnel zone, corresponding to the half-period of the field oscillations, but this value varies depending on the portion of the curve sampled. Fortunately, tests that held Δ_2 to various values and varied N_1 over a wide range (thus shifting region 2) showed only small variations, except for some anomalies traced to a phase problem addressed below. These tests were performed on various building geometries with up to four successive diffracting edges. The best performance was found with $\Delta_2 = 3.0\lambda$, where variations in computed diffraction field were less than 1 dB peak-to-peak. These tests demonstrate an insensitivity to both the size of the sampled region and the portion of the curve sampled to determine the interpolating polynomials for the extrapolation to infinity. Indeed, sampling two portions of the same amplitude curve, one yielding a parabola with positive curvature and one with negative curvature, did not yield significantly different results. This was not true, however, of the phase interpolator.

The phase of the integrand is the sum of the phase of the aperture field with a $\pi y^2/\lambda p$ term that describes propagation from the obstacle to the observation point, P. Thus the phase at P due to individual Huygens sources increasing in distance from the diffracting edge shows a parabolic shape with a positive curvature and an oscillation (the aperture phase) superimposed. If the aperture phase oscillations are strong enough relative to the quadratic term, certain sets of samples will result in a phase function with a negative curvature, giving a wildly incorrect result which has been witnessed in many situations. Hence, while the result is not sensitive to small variations in the coefficients of the phase parabola, it is necessary that the function have a positive curvature so that in the extrapolation to infinity, the phase will go to positive infinity. This is ensured by requiring that the first of the three samples of region 2 is at a point of minimum phase slope. At such a point the phase curvature is zero and going positive. Fig. 8 shows the phase of the field in the aperture with the points of minimum phase slope marked. These points occur approximately where the ob-

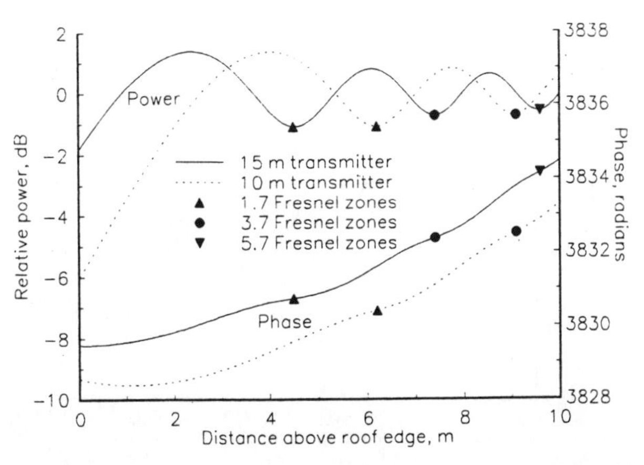

Fig. 8. Field variation in aperture with height above edge. Points of minimum phase slope and minimum amplitude are marked according to the degree of clearance over a prior diffracting edge. Results for two transmitter heights are presented. The height of both edges is 10 m.

structing edge is cleared by $n = 2k + 1.7$ Fresnel zones, where $k = 0, 1, 2, \cdots$. These are also the points of minimum amplitude.

We impose two requirements on the size of the first integration region: some minimum number of samples (we use 25) must be taken, and the sampling must continue at least until clearance is obtained. The former ensures that even where there is only some obstruction of the first Fresnel zone and the line of sight is clear, there is still sufficient characterization of the diffraction field; the latter ensures that obstructed portions of the amplitude and phase curves are fully characterized and that the samples in the second region indeed fall within the oscillatory region. These requirements are most efficiently met for a variety of shadowing depths through employing an adaptive procedure for choosing N_1.

The phase of the field in the aperture has positive curvature at the very beginning of the unobstructed region. Thus in the case of relatively deep shadowing, where many samples are taken in region 1, the minimum number of samples is exceeded by the first sample point to have clearance, and this point constitutes the boundary point. However, in the cases where the second edge is in a shallow shadow, or obstructed by a subpath obstacle, and the minimum range for region 1 extends into the oscillatory region, we must ensure positive phase curvature on region 2 by ending region 1 not at the minimum range, but extending it to the next point of minimum phase slope. The algorithm can be stated as follows: If the clearance at sample 25 is less than 0, continue sampling in region 1 until clearance is achieved; If the clearance at sample 25 is greater than or equal to 0, continue sampling in region 1 until the next integer k such that Fresnel zone clearance $n = 2k + 1.7$.

III. Results

The C-language computer program implementation of the algorithms proposed here successfully located and sorted the diffracting building edges and computed the diffraction field for the building configurations tested. The agreement of the predictions with measurements is discussed in this section. The multiple diffractions were computed in 5–10 s per successive edge on a 386SX personal computer with a math co-processor. The computation time for characterizing the field at 300 points along a one-dimensional track is on the order of a half hour, with one or two buildings obstructing the radio path at any given time.

A. Sensitivity Analyses

As a test of the multiple-diffraction procedures, consider two buildings with heights of 15 m and 10 m for buildings 1 and 2, numbered from the transmitter towards the receiver. The buildings are wide enough that the corner-edge diffraction fields are negligible; the successive edges of the buildings are at distances of 100 m, 120 m, 150 m, and 200 m from the transmitter; the receiver is at 250 m. With a receiver height of 0 m, we lower the transmitter from a height of 30 m to only 5 m and simulate the diffraction introduced at 900 MHz. The solid line in Fig. 9 plots the results of applying the algorithms described here, while the other lines represent the forced exclusion of certain edges, demonstrating the impact of neglecting certain diffractions. For example, if only the last edge before the receiver (the trailing edge of building 2) is considered, a technique often used in predicting approximate field strength, the error is on the order of 20 dB at deep levels of diffraction.

The discontinuities seen in some of the curves of Fig. 9 are the result of our assumption that edges providing clearance of at least 55% of the first Fresnel zone can be neglected. This is approximately where the single-diffraction amplitude curve first crosses 0 dB (free-space loss); as clearance increases past this point the power continues to increase before returning to oscillate about 0 dB. This clearance threshold is traditionally used in microwave radio links [18], but we found that in the multiple-diffraction situation, the field in the shadow of the second edge encountered by the wave can be near a maximum of the oscillations at the point where the first edge has a $0.55R_1$ clearance (where R_1 is the radius of the first Fresnel zone). The sudden consideration of the first edge then causes a sharp increase in power.

B. Comparisons with Measurements

The software was tested against measurements taken near two buildings on the Virginia Tech campus: Patton Hall and Davidson Hall. Only the building coordinates, the locations of the transmitter and receiver, and the frequency, 914 MHz, were input to the program; as there were no adjustments of parameters, these are "blind" tests of the models. Both of the measurement areas were chosen to isolate a single building while avoiding obvious sources of strong specular reflections. The signal radiated by a continuous-wave transmitter was received with a 6-ft-high antenna on a mobile cart and stored in the

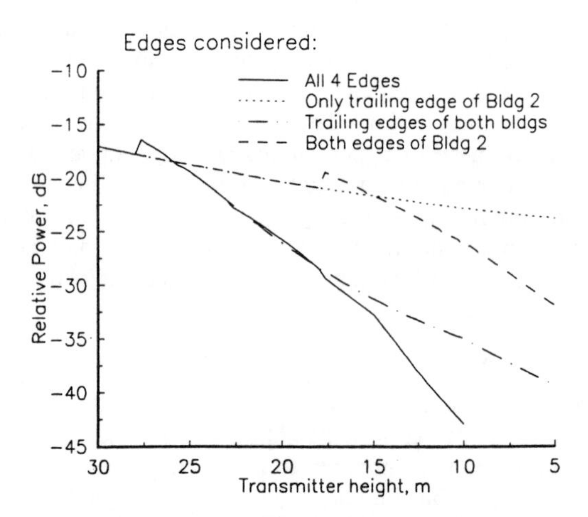

Fig. 9. Computed multiple diffraction by two buildings of heights 15 and 10 m, as a function of transmitter height. In all but the solid line, various diffracting edges are purposefully neglected to test the impact on predicted field strength.

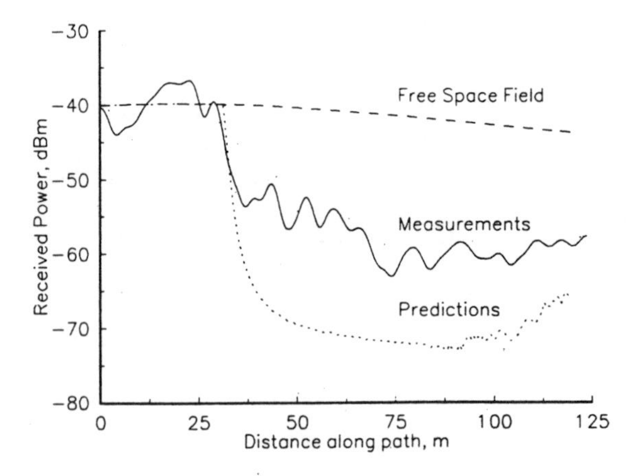

Fig. 10. Comparisons with 914-MHz measurements taken of the diffraction introduced by Patton Hall at a receiver passing in front of it. The transmitter was placed on the roof of another building.

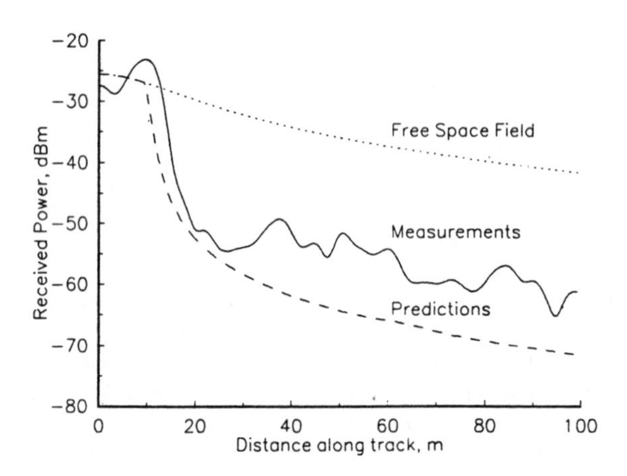

Fig. 11. Comparisons with 914-MHz measurements taken adjacent to Davidson Hall. The receiver passed down the sidewalk along one side of the building while the transmitter radiated from the top of a 15-ft-high pole around the corner of the building.

on-board computer while the cart was propelled along a sidewalk.

For the Patton Hall measurements the transmitter was placed on top of another building, providing the case of propagation from above the obstacle. The receiver began with a line-of-sight to the transmitter and passed into the shadow of Patton Hall. The measured data was filtered with a 20λ Hamming window to provide an estimate of the local mean field strength [19]. The predicted mean diffracted field strength is compared with that measured in Fig. 10. The onset of signal loss as the receiver passes behind the building is predicted, but the predictions become pessimistic when the receiver passes into a deep shadow. Note that the predicted diffraction field strength increases towards the end of the measurement track. This is due to the increasing strength of the diffraction contribution from around the far side of the building as the receiver nears that end of the building.

Davidson Hall has sections with three different heights, and provided a successful test of the algorithms for modeling a more complex building. The measurements simulate a typical situation in the proposed microcellular system architecture: the transmitter was placed 15 ft high on the sidewalk, similar to the proposed lamp-post placement of low-power base stations, and the 6-ft-high receiving antenna simulates a pedestrian with a personal communication device. The two stations were on perpendicular sidewalks that ran along two sides of the building. The transmitter was placed 50 ft from the junction of the sidewalks; the receiver began measuring at the junction, where a line-of-sight existed, and then passed into the shadow of the building. At the corner the building is 31 ft high. The comparisons with predictions, plotted in Fig. 11, show a slow drop in the signal level at 10–25 m on the track that matches well in shape but exhibits large prediction error that may be due to a 2–3-m error in locating the various elements. Again, though, as the predicted diffraction field falls below a certain threshold, the measured signal power levels off at a higher power than the prediction. A similar result was reported by Rappaport and McGillem [20] on tests in the indoor factory environment. The explanation in that case appears reasonable here too: that signal is arriving by alternate paths through scattering or reflections by other elements of the environment.

In both of our measurement runs, the mean of the measured signal never fell more than approximately 20 dB below the free-space level, whereas the predicted diffraction field fell up to 35 dB below free space. Experiments on land mobile satellite propagation [21] indicate that foliage and telephone poles typically scatter energy with about 12 to 20 dB loss, with a nearly uniform distribution with azimuth. Our measurements were taken during summer, and in the vicinity there were both leafy foliage and tall tree trunks resembling telephone poles, so the evident floor on the attenuation of the received signal is readily explained. Low-level reflections from distant buildings also are likely.

IV. Conclusion

An overall approach to prediction of diffraction by a specified assortment of buildings has been proposed as a prediction tool for a small number of buildings. Using simplified geometries for buildings, the three-dimensional profile of the radio path is surprisingly easy to access and to use as a basis of diffractive shadowing prediction, although there are a number of rules and algorithms required. The time requirement, dominated by the numerical multiple-diffraction procedures, should not be preclusive for system design requirements given a small number of shadowing buildings. In the prediction of local mean field strengths, samples can be spaced on the order of 5–10 wavelengths. A two-dimensional prediction grid for small areas thus appears feasible, particularly if a high-performance workstation is used as the computing platform. A graphics workstation would also aid in the input/output of the building information required for this site-specific approach. Unfortunately, the availability of building heights and locations may be a significant constraint at this time.

The computational precision in sensitivity analyses was about ± 1 dB. Up to 5 dB of additional error may arise from the assumption that an edge can be neglected if it presents less than 55% Fresnel zone obstruction; a different criterion is thus desirable.

It is clear that diffraction is only one process by which the signal can change direction to penetrate the shadow zone of an obstacle; reflections and other scattering mechanisms must be included in a comprehensive approach to field strength prediction. Absent these, the diffraction field is a pessimistic or worst-case estimate of the received field strength that approximates the received field best when the diffractive shadowing is not deep and strong specular reflections are not present. The measurement comparisons, though limited, appear to confirm this. The diffraction prediction models, however, are not conclusively validated in these comparisons. This is an early step in an ambitious new approach: further refinement and validation of both the approach and the models is warranted.

Appendix

The multiple-diffraction integral for perfectly absorbing finite knife edges given by (1) is derived from the following form of the Kirchhoff diffraction integral [14],

$$E(P) = \frac{-i}{\lambda} \iint_R \frac{e^{ik(r_s + r_p)}}{r_s r_p}\, dR, \tag{A1}$$

where R is the aperture region and r_s and r_p are the distances from S and P, respectively, to a Huygens source point Q in the aperture. Assume for now that the field at a Huygens source Q is a function of free-space wave expansion (no prior diffraction).

$$E(Q) = \frac{e^{ikr_s}}{r_s}, \tag{A2}$$

and can be separated out of the integrand, leaving

$$E(P) = \frac{-i}{\lambda} \iint_R E(Q) \frac{e^{ikr_p}}{r_p}\, dR. \tag{A3}$$

The first term of the integrand, $E(Q)$, describes the propagation from S to Q, and the remaining terms, e^{ikr_p} and $1/r_p$, describe the phase shift and amplitude change, respectively, introduced between Q and P. The expression for the field at Q can be broken down as follows. We first make the assumption that the normal to the aperture plane makes a small angle with SP; this allows us to make the Fresnel phase approximation [14] and expand (A2),

$$E(Q) \simeq \frac{e^{ik(s + x^2/2s + y^2/2s)}}{r_s}. \tag{A4}$$

When there is an intervening horizontal edge (parallel to the x axis), the character of the y dependence of the field at Q differs from that of the x dependence (Section II.A.). We separate them as

$$E(Q) \simeq E_Q(y) e^{ik(x^2/2s)}, \tag{A5}$$

where, given no obstacle between S and Q,

$$E_Q(y) \simeq \frac{e^{ik(s + y^2/2s)}}{r_s}, \tag{A6}$$

but in the multiple-diffraction case, we substitute explicit evaluations of the field at Q as a function of y (at $x = 0$) for the relation (A6).

The terms of the integrand in (A3) that represent the amplitude and phase change incurred between Q to P can be simplified using the standard amplitude and phase approximations [14],

$$\frac{e^{ikr_p}}{r_p} \simeq \frac{e^{ik(p + x^2/2p + y^2/2p)}}{p}. \tag{A7}$$

With the substitutions of (A5) and (A7) in (A3), and rearrangement of terms, the solution for the field at P can be written

$$E(P) = \frac{-i}{\lambda p} e^{ikp} \iint_R E_Q(y) e^{ik(x^2/2s + x^2/2p)} e^{ik(y^2/2p)}\, dR. \tag{A8}$$

The double integral can now be separated into the product of two single integrals, and the integral over x may be written in the form of a Fresnel integral through a change of variable, yielding (1),

$$E(P) = -ie^{ikp} \sqrt{\frac{s}{2\lambda p(s + p)}} \int_{\xi_1}^{\xi_2} e^{i\pi\xi^2/2}\, d\xi \cdot \int_{y_1}^{y_2} E_Q(y) e^{iky^2/2p}\, dy, \tag{A9a}$$

where

$$\xi_1 = x_1 \sqrt{\frac{2(s+p)}{\lambda sp}}, \tag{A9b}$$

$$\xi_2 = x_2 \sqrt{\frac{2(s+p)}{\lambda sp}}. \tag{A9c}$$

References

[1] Y. Okumura *et al.*, "Field strength and its variability in VHF and UHF land-mobile radio service," *Rev. Elec. Commun. Lab.*, vol. 16, nos. 9 and 10, p. 825–873, Sept.–Oct. 1968.

[2] J. Walfisch and H. L. Bertoni, "A theoretical model of UHF propagation in urban environments," *IEEE Trans. Antennas Propagat.*, vol. 36, no. 12, p. 1788–1796, Dec. 1988.

[3] F. Ikegami *et al.*, "Theoretical prediction of mean field strength for urban mobile radio," *IEEE Trans. Antennas Propagat.*, vol. 39, no. 3, pp. 299–302, Mar. 1991.

[4] A. Ranade, "Local access radio interference due to building reflections," *IEEE Trans. Commun.*, vol. 37, no. 1, pp. 70–74, Jan. 1989.

[5] K. R. Schaubach, N. J. Davis, and T. S. Rappaport, "A ray tracing method for predicting path loss and delay spread in microcellular environments," in *Proc. 42nd IEEE Veh. Technol. Conf.*, Denver, CO, May 1992, pp. 932–935.

[6] D. C. Cox, "A radio system proposal for widespread low-power tetherless communications," *IEEE Trans. Commun.*, vol. 39, no. 2, pp. 324–335, Feb. 1991.

[7] W. C. Y. Lee, "Smaller cells for greater performance," *IEEE Commun. Mag.*, vol. 29, pp. 19–23, Nov. 1991.

[8] J. H. Whitteker, "Measurements of path loss at 910 MHz for proposed microcell urban mobile systems," *IEEE Trans. Veh. Technol.*, vol. 37, no. 3, pp. 125–129, Aug. 1988.

[9] F. H. Palmer, "VHF/UHF path-loss calculations using terrain profiles deduced from a digital topographic data base," *AGARD Conf. Proc.*, no. 269, 1979, pp. 26-1–26-11.

[10] J. H. Whitteker, "Calculation by numerical integration of diffraction attenuation at VHF and UHF," in *Proc. Fifth Int. Conf. Antennas Propagat.*, IEE Conf. Publ. 274, 1987, pp. 31–34.

[11] J. H. Whitteker, "Fresnel-Kirchhoff theory applied to terrain diffraction problems," *Radio Sci.*, vol. 25, no. 5, pp. 837–851, Sept.–Oct. 1990.

[12] J.-F. Wagen, "SIP simulation of UHF propagation in urban microcells," in *Proc. 41st IEEE Veh. Technol. Conf.*, St. Louis, MO, May 1991, pp. 301–306.

[13] T. A. Russell, T. S. Rappaport, and C. W. Bostian, "Use of a building database in prediction of three-dimensional diffraction," in *Proc. 42nd IEEE Veh. Technol. Conf.*, Denver, CO, May 1992, pp. 943–946.

[14] M. Born and E. Wolf, *Principles of Optics*, 3rd ed. New York: Pergamon, 1965.

[15] M. P. Bachynski and M. G. Kingsmill, "Effect of obstacle profile on knife-edge diffraction," *IRE Trans. Antennas Propagat.*, vol. AP-10, pp. 201–205, Mar. 1962.

[16] J. H. Whitteker, "VHF/UHF propagation by diffraction—calculation by numerical integration," *AGARD Conf. Proc.*, no. 407, 1986, pp. 6-1–6-7.

[17] J. J. Stamnes, B. Spjelkavik, and H. M. Pedersen, "Evaluation of diffraction integrals using local phase and amplitude approximations," *Opt. Acta*, vol. 30, no. 2, pp. 207–222, 1983.

[18] J. D. Parsons and J. G. Gardiner, *Mobile Communication Systems*. Glasgow: Blackie, 1989.

[19] W. C. Y. Lee, "Estimate of local average power of a mobile radio signal," *IEEE Trans. Veh. Technol.*, vol. VT-34, no. 1, pp. 22–27, Feb. 1985.

[20] T. S. Rappaport and C. D. McGillem, "UHF fading in factories," *IEEE J. Select. Areas Commun.*, vol. 7, no. 1, pp. 40–48, Jan. 1989.

[21] R. M. Barts and W. L. Stutzman, "Modeling and simulation of mobile satellite propagation," *IEEE Trans. Antennas Propagat.*, vol. 40, no. 4, pp. 375–382, April 1992.

Thomas A. Russell (S'83–M'85–S'90–M'91) was born in Camden, NJ, in 1963. He received the B.S.E.E. degree from the University of Virginia, Charlottesville, in 1986, and the M.S.E.E. degree from the Virginia Polytechnic Institute and State University, Blacksburg, in 1991.

Since 1986 he has been employed with Stanford Telecommunications, Inc. in Reston, VA, where he is involved in propagation modeling, link simulation, and systems architecture studies for microwave satellite communications. Between August 1990 and December 1991 he was a Research Assistant with the Mobile and Portable Radio Research Group where his area of research was the prediction of diffraction by buildings.

Charles W. Bostian (S'67–M'67–SM'77–F'92) was born in Chambersburg, PA, on December 30, 1940. He received the B.S., M.S., and Ph.D. degrees in Electrical engineering from North Carolina State University, Raleigh, in 1963, 1964, and 1967, respectively.

After a short period as a Research Engineer with Corning Glassworks and a tour of duty in the U.S. Army, he joined the Faculty of Virginia Polytechnic Institute and State University, Blacksburg, in 1969 and is currently Clayton Ayre Professor of Electrical Engineering. His primary interests are in propagation and satellite communications.

Dr. Bostian was a 1989 IEEE Congressional Fellow with Representative Don Ritter. He is chair of the IEEE-USA Engineering R & D Policy Committee and serves on the IEEE-USA Technology Policy Council and the Congressional Fellows Committee. He is Associate Editor for Propagation of *IEEE Transactions on Antennas and Propagation*. In his off-duty hours, he is a performing folk musician, playing hammered dulcmer with the group Simple Gifts.

Theodore S. Rappaport (S'83–M'84–S'85–M'87–SM'90) was born in Brooklyn, NY on November 26, 1960. He received the B.S.E.E., M.S.E.E., and Ph.D. degrees from Purdue University in 1982, 1984, and 1987, respectively.

In 1988, he joined the Electrical Engineering faculty of Virginia Tech, Blacksburg, where he is an Associate Professor and Director of the Mobile and Portable Radio Research Group. He conducts research in mobile radio communication system design and RF propagation prediction through measurements and modeling. He guides a number of graduate and undergraduate students in mobile radio communications, and has authored or co-authored more than 70 technical papers in the areas of mobile radio communications and propagation, vehicular navigation, ionospheric propagation, and wideband communications. He holds a U.S. patent for a wide-band antenna and is co-inventor of SIRCIM, an indoor radio channel simulator that has been adopted by over 50 companies and universities.

In 1990, Dr. Rappaport received the Marconi Young Scientist Award for his contributions in indoor radio communications, and was named a National Science Foundation Presidential Faculty Fellow in 1992. He is an active member of the IEEE, and serves as senior editor of the *IEEE Journal on Selected Areas in Communications*. He is a Registered Professional Engineer in the State of Virginia and is a Fellow of the Radio Club of America. He is also president of TSR Technologies, a cellular radio and paging test equipment manufacturer.

Reprinted from *IEEE Transactions on Vehicular Technology*, Vol. 43, No. 3, Aug. 1994, pp. 487-498.

Path Loss, Delay Spread, and Outage Models as Functions of Antenna Height for Microcellular System Design

Martin J. Feuerstein, Kenneth L. Blackard, *Member, IEEE*, Theodore S. Rappaport, *Senior Member, IEEE*, Scott Y. Seidel, *Member, IEEE*, and Howard H. Xia

***Abstract*—This paper presents results of wide-band path loss and delay spread measurements for five representative microcellular environments in the San Francisco Bay area at 1900 MHz. Measurements were made with a wide-band channel sounder using a 100-ns probing pulse. Base station antenna heights of 3.7 m, 8.5 m, and 13.3 m were tested with a mobile receiver antenna height of 1.7 m to emulate a typical microcellular scenario. The results presented in this paper provide insight into the satistical distributions of measured path loss by showing the validity of a double regression model with a break point at a distance that has first Fresnel zone clearance for line-of-sight topographies. The variation of delay spread as a function of path loss is also investigated, and a simple exponential overbound model is developed. The path loss and delay spread models are then applied to communication system design allowing outage probabilities, based on path loss or delay spread, to be estimated for a given microcell size.**

I. Introduction

FOR efficient microcellular system design, it is necessary to characterize the radio channels in such a way that outage probabilities and other system performance measures can be estimated. In this paper, results of wide-band propagation path loss and multipath measurements are presented for five representative microcellular environments in the San Francisco Bay area (San Francisco and Oakland). Measurements were made using a wide-band channel sounder with a 100-ns probing pulse at 1900 MHz with base transmitter station antenna heights of 3.7 m, 8.5 m, and 13.3 m and a mobile receiver antenna height of 1.7 m. These antenna heights were used to cover the range of typical antenna heights that might be encountered in a microcellular system using lamp-post-mounted base stations at street corners. Measurement locations were chosen to coincide with places where microcellular systems will likely be deployed in urban and suburban areas.

The power delay profiles recorded at each receiver location were used to calculate path loss and delay spread. Path loss and delay spread are two important methods of characterizing channel behavior in a way that can be related to system performance measures such as bit error rate [4] and outage probability [7]. Path loss is a strong function of the propagation environment in the vicinity of the transmitter and receiver. Researchers have often used propagation models where the mean path loss decays as a function of the distance between transmitter and receiver raised to the power n, where n is called the mean path loss exponent [5]. The distance-dependent power law model (d^n) has been applied in many circumstances where there may be more than two paths between the transmitter and receiver [5], [7], [9], [10].

For the simple case of a direct path and a single ground reflection between the transmitter and receiver, the distance power law model describes the mean path loss [6], [14], [16]. Theoretically, for transmitter-receiver separation distances less than the first Fresnel zone clearance, the mean path loss exponent will be two; beyond the distance for first Fresnel zone clearance, the exponent becomes four [6], [16]. In other words, if the path loss in decibels is plotted versus the logarithm of the distance between the transmitter and receiver, a dual-slope piecewise linear curve will result from the log-log plot. The piecewise linear curve will have one slope prior to the Fresnel zone clearance break point and a different slope after the break point, where the slopes of the lines are directly related to the path loss exponents. In this paper, it is shown that this dual-slope piecewise linear model can be used accurately to characterize the measured path loss data for line-of-sight topographies, even though, in many cases, more than two rays are likely to be present between transmitter and receiver.

The rms delay spreads have been compared for the three antenna heights. In general, delay spreads were found to increase significantly with antenna height. This is an important result that has implications on the data rates that can be transmitted through these channels. In this paper, it is also

Manuscript received May 7, 1992; revised June 30, 1992 and October 5, 1992. This work was supported by Telesis Technologies Laboratory, Inc., of Walnut Creek, CA, and TSR Technologies, Inc., of Blacksburg, VA.

M. J. Feuerstein was with Mobile & Portable Radio Research Group, Virginia Polytechnic Institute & State University, Blacksburg, VA 24061 USA. He is now with US West NewVector, Bellevue, WA 98008-1329 USA.

K. L. Blackard was with Mobile & Portable Radio Research Group, Virginia Polytechnic Institute & State University, Blacksburg, VA 24061 USA. He is now with the Federal Bureau of Investigations, Quantico, VA 22135 USA.

T. S. Rappaport is with Mobile & Portable Radio Research Group, Virginia Polytechnic Institute & State University, Blacksburg, VA 24061 USA.

S. Y. Seidel was with Mobile & Portable Radio Research Group, Virginia Polytechnic Institute & State University, Blacksburg, VA 24061 USA. He is now with Bell Communications Research, Red Bank, NJ 07701-7040 USA.

H. H. Xia is with Telesis Technologies Laboratory, Walnut Creek, CA 94598 USA.

IEEE Log Number 9212490.

shown that the delay spread increases markedly as a function of path loss. An exponential overbound model has been developed so that for any of the antenna heights a worst case estimate of delay spread can be obtained for a given path loss.

The dual-slope piecewise linear model for path loss and the exponential overbound model for delay spread have been applied directly to the problem of microcellular design. By developing appropriate statistical models for the probability of a given path loss conditioned on the propagation model parameters and cell size, it is possible to determine outage probabilities [7]. If a particular path loss or delay spread criterion is used to define a system outage, then a relationship between the outage probability and maximum cell size can be obtained. The ability to predict outage probability for a specified path loss or delay spread allows microcell sizes to be optimized for a particular propagation environment, thereby providing more predictable performance compared with systems where the microcell sizes are selected arbitrarily.

This paper shows that a model that determines the Fresnel zone break point based only on antenna height and frequency can model microcellular propagation as accurately as a minimum mean square error (MMSE) fit on the data. This model is general in that we present no conditions on its usage based on street width. When reflections from surrounding buildings are considered, the location of the break point may change [16]. This paper shows that the break point predicted by the two-ray model accurately models the measured data.

II. Measurements

A. Measurement System

A time-domain channel sounder similar to the ones used in [5] and [7] was used to measure both time delay spread and path loss during summer 1991. Measurements were conducted in five microcell environments in urban and suburban areas of San Francisco and Oakland, California. A 20-MHz bandpass spectrum centered at 1900 MHz was used to transmit 100-ns-duration RF pulses with a repetition rate of 10 kHz, providing a multipath resolution of about 100 ft. The pulse train was amplified to a peak power of 10 W and fed into an omnidirectional vertically polarized antenna. The minimum measurable signal was −83 dBm for a maximum measurable path loss dynamic range of 123 dB.

The base transmitter was located inside a van that remained parked on the side of a main street during measurements. The transmitter antenna height was changed by hoisting an antenna mast from the roof of the van. Measurements were made with transmitter antenna heights of 3.7 m, 8.5 m, and 13.3 m above the ground and a mobile receiver antenna height of 1.7 m above the ground.

B. Microcellular Path Loss Measurements

The stationary van containing the transmitter was parked as close as possible to an intersection of two streets that extended radially away from the intersection. The omnidirectional transmitter antenna was raised to a height of 3.7 m above the ground, and a car with the receiver traveled along streets in the vicinity of the transmitter where the path loss was within the dynamic measurement range of the receiver. At each measurement location, the receiver car would stop as closely as possible to an intersection, and the operator recorded snapshots of the power delay profile. Power delay profiles were computed as the time average of 16 instantaneous oscilloscope snapshots of the received probing signal over a 1-s averaging interval while both the transmitter and receiver were stationary. Measurements at intersections allowed the measurement positions to be located quickly on a topographic map of the area. Many of the intersections were obstructed from the transmitter (no direct line of sight (LOS)), which is representative of shadowed microcellular environments. Results for both LOS and obstructed topographies are presented in this paper. The transmitter antenna was then raised to 8.5 and 13.3 m, and the procedure was repeated.

III. Data Processing for Path Loss and Delay Spread Analysis

A. Determining Path Loss

The time-domain channel sounder used in these tests records the magnitude squared of the complex envelope baseband pulse response of the channel [5].That is, the measurement system records

$$|r(t)|^2 = \sum_k \alpha_k^2 p^2(t - \tau_k), \tag{1}$$

where $r(t)$ is the complex envelope baseband pulse response, $p(t)$ the transmitted pulse shape, α_k the amplitudes of the particular path components, and τ_k the time delays of the individual path components. Each measurement record of this type is called a *power delay profile* because it represents the power entering the receiver as a function of the time delay. Because the transmitter used a sounding pulse with a finite temporal width, the resolution of the path time delays (τ_k) is limited to 100 ns.

A relative measure of received power, G_r, at the receiver antenna for a given power delay profile snapshot can be determined from the oscilloscope display and calibration curves

$$G_r = \sum_k \alpha_k^2. \tag{2}$$

From (2) and a calibration power delay profile recorded with a known transmitter power through a short test cable, it is possible to calculate the path loss from the measured power delay profiles. Let PL denote the channel path loss, P_T the peak transmitter power, G_T the transmitter antenna gain, G_R the receiver antenna gain, and $\Delta\tau_r$ the oscilloscope sampling time between consecutive α_k^2 points. Then, PL is given by [5]

$$\mathrm{PL(dB)} = P_T\,(\mathrm{dBm}) + G_T\,(\mathrm{dBi}) + G_R\,(\mathrm{dBi}) - P_R\,(\mathrm{dBm}), \tag{3}$$

where

$$P_R\,(\mathrm{dBm}) = P_{\mathrm{cal}}\,(\mathrm{dBm}) + 10 \cdot \log_{10}\left\{\frac{G_r \Delta\tau_r}{G_{\mathrm{cal}} \Delta\tau_{\mathrm{cal}}}\right\}. \tag{4}$$

To compute the wide-band path loss, it is necessary to compute the power received by integrating the area under the received power delay profile. The term $G_r \Delta\tau_r$ is the integrated power (total power) in a given measured profile. For a calibration run where only one 100-ns pulse is present, the total integrated power is assigned the known input power, P_{cal}. The term $G_{\mathrm{cal}} \Delta\tau_{\mathrm{cal}}$ converts the total integrated power to an equivalent peak power in a 100-ns pulse.

A frequently used model [5], [7], [9], [11] indicates that mean path loss increases exponentially with distance, that is,

$$\mathrm{PL}(d) \propto \left(\frac{d}{d_0}\right)^n \tag{5}$$

where n is the mean path loss exponent, which indicates how fast path loss increases with distance; d_0 is a reference distance; and d is the transmitter-receiver (T-R) separation distance. When plotted on a log-log scale, this power law relationship is a straight line. Absolute mean path loss in decibels is defined as the path loss in decibels from the transmitter to the reference distance d_0 plus the additional path loss described by (5) in decibels; thus,

$$\overline{\mathrm{PL}}(d)\,\mathrm{dB} = \mathrm{PL}(d_0) + 10n \log_{10}\left(\frac{d}{d_0}\right)\mathrm{dB}. \tag{6}$$

For the results presented herein, we use a close-in reference distance $d_0 = 1$ m from which to refer all measured path loss values. We assume $\mathrm{PL}(d_0)$ is due to free-space propagation from the transmitter to the 1-m reference distance. Assuming isotropic antennas, this leads to 38.0-dB path loss at 1900 MHz over a 1-m free-space path, as shown in (7)

$$\mathrm{PL}(d_0 = 1\,\mathrm{m})\,\mathrm{dB} = 20 \log_{10}\left\{\frac{4\pi\,(1\,\mathrm{m})}{\lambda\,(\mathrm{m})}\right\}\mathrm{dB}. \tag{7}$$

Because the 3-D propagation distance is larger than the 2-D ground separation when the transmitter and receiver are separated by less than 10 m, there will be some error in calculating path loss using only the 2-D ground separation. Since all measurements were made at distances greater than 10 m, we have measured separation distances using the 2-D ground separation between the transmitter and receiver, rather than the actual 3-D separation, with no loss in accuracy.

B. Two-Ray Model and Fresnel Zone Theory

In the two-ray model [6], [11], [14], [15] shown in Fig. 1, the direct ray travels from the transmitting antenna of height h_t to the receiving antenna of height h_r along the LOS path r_1. The second ray reaches the receiver antenna along the ground-reflected path r_2. This model is used commonly when the transmitting antenna is several wavelengths or more above the horizontal ground plane [6], [15]. The signal at the receiving antenna is the resultant vector sum of the direct and ground-reflected components. By using the method of images [6], [15], the received electric field can be expressed as

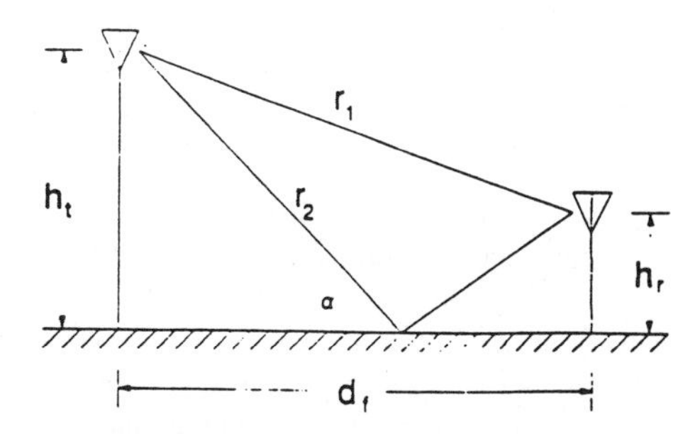

Fig. 1. Two-ray geometry for direct path and ground reflection multipath.

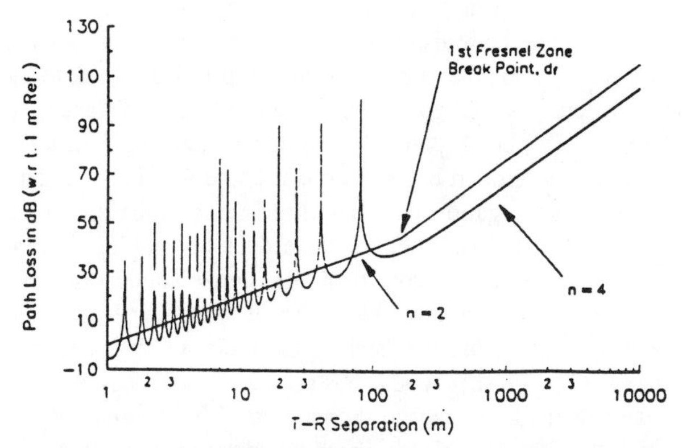

Fig. 2. Path loss versus transmitter-receiver separation for two-ray model illustrating double slope linear result for mean path loss. The example is for a transmitter height of $h_t = 3.7$ m, receiver height of $h_r = 1.7$ m, reflection coefficient of $\Gamma = -1$, and frequency of $f = 1900$ MHz.

$$|E|^2 = |E_1 - E_2|^2 = \left|\frac{1}{r_1}e^{-jkr_1} + \Gamma(\alpha)\frac{1}{r_2}e^{-jkr_2}\right|^2, \tag{8}$$

where $\Gamma(\alpha)$ is the Fresnel reflection coefficient, which is a function of the angle of incidence α and the antenna polarization. This dependence on angle and polarization is expressed as [6]

$$\Gamma(\theta) = \frac{\cos\theta - a\sqrt{\epsilon_r - \sin^2\theta}}{\cos\theta + a\sqrt{\epsilon_r - \sin^2\theta}}, \tag{9}$$

where

$$a = \begin{cases} 1/\epsilon_r, & \text{for vertical polarization} \\ 1, & \text{for horizontal polarization,} \end{cases}$$

and $\theta = 90° - \alpha$ and ϵ_r is the relative dielectric constant. Fig. 2 is a plot of the path loss in decibels relative to the received power at a 1-m reference point as a function of T-R separation for the two-ray model shown in Fig. 1, with the Fresnel reflection coefficient assumed to be $\Gamma = -1$. Fig. 2 shows two distinct regions that are separated by a "break point" there is a significant difference in the way the received signal strengths vary in these two regions. Prior to the break point, the signal strength oscillates due to the destructive and constructive interference combination of the phases of the two received rays. The mean path loss, determined by averaging out the fast fading [5], has an exponent for this region of $n = 2$. The oscillations in the signal strength cease after the break point, and the received power exhibits an $n = 4$ mean path loss exponent.

The "break point" shown in Fig. 2 can be related to Fresnel zone theory. The first Fresnel zone is defined as an ellipsoid whose foci are the transmit and receive antennas. The distance from either antenna to a point on the ellipsoid and back to the other antenna is $\lambda/2$ greater than the direct path distance, r_1, in Fig. 1. The break point will be considered as the distance for which the ground begins to obstruct the first Fresnel zone. When the propagation path has first Fresnel zone clearance, meaning that the particular obstacle does not impinge on the first Fresnel zone volume, then the signal attenuation with distance is purely due to the spherical spreading loss of the wavefront (the same mechanism as free-space propagation). Once the first Fresnel zone is obstructed, the path loss becomes greater than for free-space propagation. The distance, d_f, at which the first Fresnel zone becomes obstructed, is given by

$$d_f = \frac{1}{\lambda}\sqrt{(\Sigma^2 - \Delta^2)^2 - 2(\Sigma^2 + \Delta^2)\left(\frac{\lambda}{2}\right)^2 + \left(\frac{\lambda}{2}\right)^4}, \tag{10}$$

where $\Sigma = h_t + h_r$ and $\Delta = h_t - h_r$ [6], [15]. Equation (10) assumes a flat earth model between the transmitter and receiver. The Fresnel zone model is valid only for LOS cases where there is a direct signal path between the transmitter and receiver, and for this reason the model will be applied only to LOS topographies. For obstructed (OBS) topographies, a single regression path loss model is presented.

C. *Multiple Regression Models for Line-of-Sight Topography*

Path loss is often considered to be log-normally distributed about the mean power law described in (6) [9], [11]. The two-ray model discussed in Section III-B has been used to model propagation in many urban and suburban environments [14]–[16]. Because of the simplicity of the dual region linear curve that results from a log-log plot of path loss versus distance, the model is an excellent candidate for use in environments where a direct path and one or more reflection paths may occur. If a simple analytical model can be developed, the results can be applied directly to solving microcellular system design problems [14]. In this paper, we show that for LOS topographies a double regression model gives good results, whereas for OBS topographies a single regression model is more appropriate.

The path loss results presented in this paper for LOS topographies were obtained using two different forms of a double linear regression to compute values of the path loss exponents n_1 and n_2 for the two regions and the standard deviation σ in decibels about the best fit mean power law model in an MMSE sense for the measured data. The two multiple regression techniques divide the overall data into two subsets with a different power law exponent for each of the subsets [15]. If a logarithmic distance axis is used, then the double regression model becomes a piecewise linear model with different slopes on either side of the break point. This double regression piecewise linear model generally provides an overall MMSE best fit curve with a smaller standard deviation than a simple single linear regression model for LOS cases.

One form of the piecewise linear model forces the break point between the two linear regions to occur at a distance that has first Fresnel zone clearance, such as shown in Fig. 2. The alternate form of the model allows the break point to float so that the MMSE best fit curve on all the measured data determines the least-mean-squared error break point distance. By comparing the mean squared error for the two curve fits, it is possible to determine the difference in accuracy between the two methods of calculating the break point. If the mean squared error for the analytically derived first Fresnel break point is very close to the mean squared error for the measured MMSE break point, then there is not much advantage to using the MMSE break point. In this case, the Fresnel break point, which is based on the physical heights of the antennas, can be computed easily by (10) for any microcellular LOS system. We now show that, indeed, for LOS microcell channels, the first Fresnel zone break point is an excellent parameter for determining the break point for a piecewise linear path loss model.

In the first form of the double regression model, the break point between the two linear regions is fixed at the first Fresnel zone clearance distance determined from (10) based on the antenna heights and wavelength. The first double regression piecewise linear model for path loss PL_1 in decibels and d in meters is then

$$\mathrm{PL}_1(d) = \begin{cases} (10n_1)\log_{10}(d) + p_1, & \text{for } 1 < d < d_f \\ (10n_2)\log_{10}(d/d_f) \\ \quad +(10n_1)\log_{10}(d_f) \\ \quad +p_1, & \text{for } d > d_f, \end{cases} \tag{11}$$

where $p_1 = \mathrm{PL}(d_0)$ is the path loss in decibels at the reference distance of $d_0 = 1$ m (at 1900 MHz, $p_1 = 38.0$ dB). The unknown parameters n_1 and n_2 are the power law exponents for the two regions of the model that are optimized for MMSE with measured data [9], [5]. The power law exponents are a function of the choice of reference distance d_0 [14]. The break point between the two regions, d_f in meters, is fixed at the first Fresnel distance defined by (10). Thus, there are two unknown coefficients in this model (n_1, n_2) that completely define the slope and intercept of the line segments in the two linear regions. These two unknowns can be determined in closed form based on the MMSE criterion, where the variance or mean squared error σ^2 is the quantity to be minimized. For a total of N measurement locations, the mean squared error is

$$\sigma^2 = \frac{1}{N}\sum_{i=1}^{N}(\mathrm{PL}_i - \mathrm{PL}_1(d_i))^2, \tag{12}$$

where PL_1 is the measured path loss in decibels for the ith path loss measurement and $\mathrm{PL}_1(d_i)$ in decibels is (11) evaluated at the T-R separation distance d_i in meters for the ith measurement. The distances d_i were determined from maps and the known locations of the mobile receiver at street intersections near the base station transmitter. The two model parameters are calculated so that the mean squared error defined in (12) is minimized given the model in (11).

In the second form of the double regression model, the break point between the two linear regions is not fixed, but is instead an unknown parameter to be determined in the curve fit. Thus, the break point is the one that minimizes the mean squared error between the model and the measured data. The alternate double regression piecewise linear model for path loss PL_2 in decibels is

$$\mathrm{PL}_2(d) = \begin{cases} (10n_1^*)\log_{10}(d) + p_1, & \text{for } 1 < d < d_b \\ (10n_2^*)\log_{10}(d/d_b) \\ \quad +(10n_1^*)\log_{10}(d_b) \\ \quad +p_1, & \text{for } d > d_b, \end{cases} \tag{13}$$

where the break point d_b, n_1^*, and n_2^* are unknown parameters. Again, n_1^* and n_2^* are the path loss exponents for the model. Since the break point is allowed to float, an MMSE curve fit will produce the break point d_b, which minimizes the mean squared error. An iterative routine is used to minimize the mean squared error as defined in (12) as a function of the three unknown parameters.

D. Multiple Regression Results for Line-of-Sight Topography

Figs. 3–5 show measured path loss data as a function of T-R separation for the low, medium, and high antenna heights, respectively, for LOS topographies. Each point represents measured path loss with the mobile at a particular street intersection. Along with the experimental data points, the two forms of the double regression model that give the lowest mean squared error described in (13) with the floating MMSE break point (d_b) and (11) with the fixed Fresnel zone break point (d_f) are shown on each figure.

Table I compares the curve fit parameters and standard deviations for the two double regression models for each antenna height. From the table it can be seen that the rms errors (σ in decibels) for the two curve fits do not differ more than 0.65 dB for all three antenna heights. This indicates that the error introduced by forcing the break point at the first Fresnel zone is quite small; therefore, the double regression model with the fixed break point at the first Fresnel zone gives good results compared with the model using the MMSE optimum break point based on actual field measurements, and is based solely on operating frequency and the antenna heights. It is worth noting that, for all three antenna heights, the Fresnel break point model has n_1 consistently near two and n_2 at approximately three to four.

The results from Table I indicate that the flat earth Fresnel break point model can be used to characterize the path loss as a function of distance. If the transmit and receive antenna heights are known, along with the T-R separation, then path

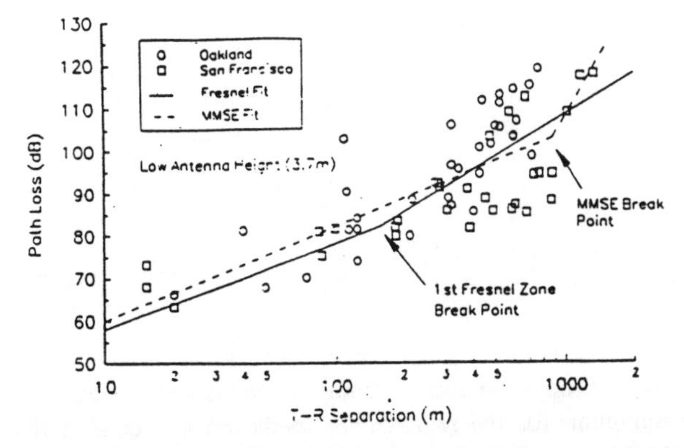

Fig. 3. Path loss versus transmitter-receiver separation for low antenna height (3.7 m) line-of-sight topography with first Fresnel zone break point and MMSE break-point curve fits.

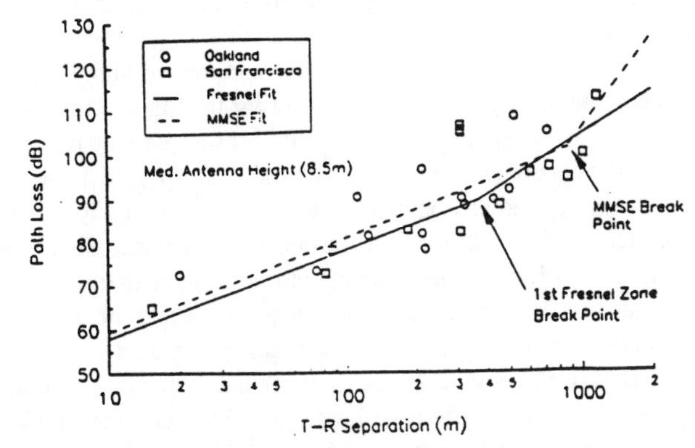

Fig. 4. Path loss versus transmitter-receiver separation for medium antenna height (8.5 m) line-of-sight topography with first Fresnel zone break point and MMSE break-point curve fits.

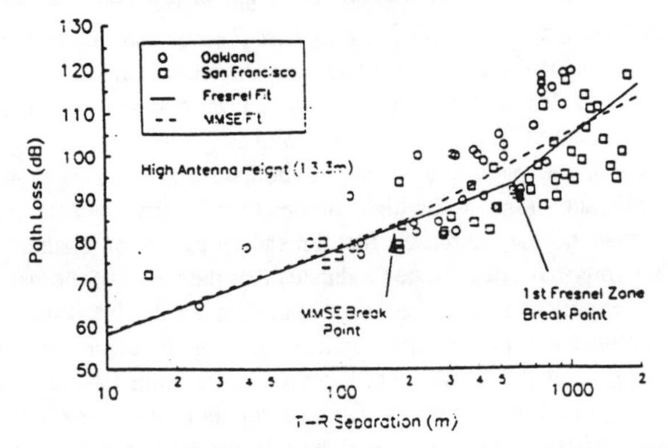

Fig. 5. Path loss versus transmitter-receiver separation for high antenna height (13.3 m) line-of-sight topography with first Fresnel zone break point and MMSE break-point curve fits.

TABLE I
POWER LAW PATH LOSS EXPONENTS, STANDARD DEVIATIONS, FIRST FRESNEL ZONE BREAK POINTS, AND MMSE BREAK-POINT CURVE FITS FOR EACH ANTENNA HEIGHT FOR LINE-OF-SIGHT TOPOGRAPHIES

	Fresnel Best-Fit				MMSE Best-Fit			
	n_1	n_2	σ (dB)	d_f (m)	n_1^*	n_2^*	σ (dB)	d_b (m)
Low (3.7 m)	2.18	3.29	8.76	159	2.20	9.36	8.64	884
Med (8.5 m)	2.17	3.36	7.88	366	2.14	6.87	7.23	884
High (13.3 m)	2.07	4.16	8.77	573	2.03	2.79	8.34	190

loss can be computed based on the two path loss exponents n_1 and n_2. From these results, letting $n_1 = 2$ is a very reasonable assumption for the region prior to the Fresnel break point. There is much more variability in the path loss exponent for the region beyond the Fresnel break point, with values of n_2 from two to seven being typical. Values for the path loss exponent beyond the break point will depend on the particular physical environment, such as building locations, terrain characteristics, and street widths.

From Table I, there is a significant trend toward higher path loss exponents beyond the break point for the higher antennas than for the lower antennas using the Fresnel best-fit model. The model parameter n_2 changes from 3.29 to 3.36 to 4.16 for the low, medium, and high antennas, respectively. It is important to note that the first Fresnel zone clearance distance (break point) for the high antenna is much larger than for the low antenna. For example, for a receiver height of 1.7 m, the high antenna (13.3 m) has a break point at $d_f = 573$ m, whereas the low antenna (3.7 m) has a break point at $d_f = 159$ m. The difference in the first Fresnel zone break-point distances is important when considering microcellular systems where the cell size may be anywhere from a few meters to about 2000 m. The 100–2000-m T-R separation range is the most critical region for microcellular applications where low-power, small-radius cells are to be implemented. For microcells, the difference in first Fresnel zone clearance distances can more than offset the difference in path loss exponents for the high versus low antennas. Although the path loss exponent beyond the break point is greater for the high antenna, the path loss values in the 100–2000-m range are significantly *less* for the high antenna than for the low antenna because of the increased first Fresnel break-point distance. This important observation indicates that the value of the path loss exponent is not nearly as important as the break-point location for microcellular applications. Fig. 6 illustrates the influence of the break-point location on the path loss curves for high and low antennas. The two curves in the figure were obtained by using the Fresnel best fit model parameters for high and low antennas from Table I in the double regression linear model of (11). From the figure, it is clear that the higher antenna provides a lower path loss over the typical microcellular coverage range, although the higher antenna has a larger path loss exponent beyond the first Fresnel zone break point.

E. Single Regression Model for Obstructed Topography

For obstructed topographies, where no direct path exists between the transmitter and receiver, a single regression power law model relating path loss to T-R separation has been used. The single regression power law model is identical to evaluating the double regression model of (11) with the break-point distance set equal to infinity ($d_f = \infty$) and the path loss exponent n_1 selected to minimize the mean squared error. By fitting the models to experimental data, results indicate that the single regression rms error is usually within 0.2 dB of the rms error for an MMSE double regression model fit using (13) for the same obstructed topography data. Because the mean squared errors are comparable for the single and double regression models, the simpler single regression can be applied without a significant loss in accuracy.

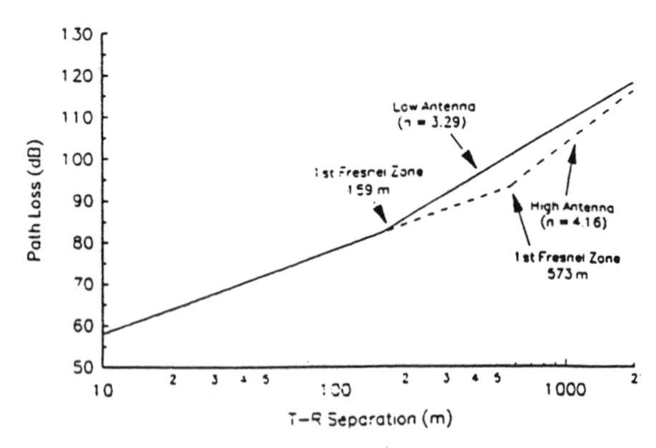

Fig. 6. Fresnel zone break-point path loss models for low and high antennas. The figure illustrates that, for microcellular applications (cell radius from 100 to 2000 m), the location of the Fresnel zone break point is more important than the path loss exponent beyond the break point in determining overall path loss.

TABLE II
POWER LAW PATH LOSS EXPONENTS, STANDARD DEVIATIONS, AND MMSE CURVE FITS FOR EACH ANTENNA HEIGHT FOR OBSTRUCTED TOPOGRAPHIES

Antenna Height	n	σ (dB)
Low (3.7 m)	2.58	9.31
Med (8.5 m)	2.56	7.67
High (13.3 m)	2.69	7.94

F. Single Regression Results for Obstructed Topography

Fig. 7 shows the measured path loss data as a function of T-R separation and the single regression curve fit for the high antenna height in obstructed topographies. Table II lists the single regression curve fit parameters for each antenna height in obstructed topographies. The rms error values (σ = standard deviation of the data about the mean power law model) for the obstructed topographies with a single regression model are generally between 7 and 10 dB. These results are similar to the values obtained for the LOS data presented in Section III-D. The rather large standard deviations indicate the need for research into more accurate LOS and OBS propagation models. The rms errors are generally larger for the obstructed cases; this observation reflects the larger degree of variability

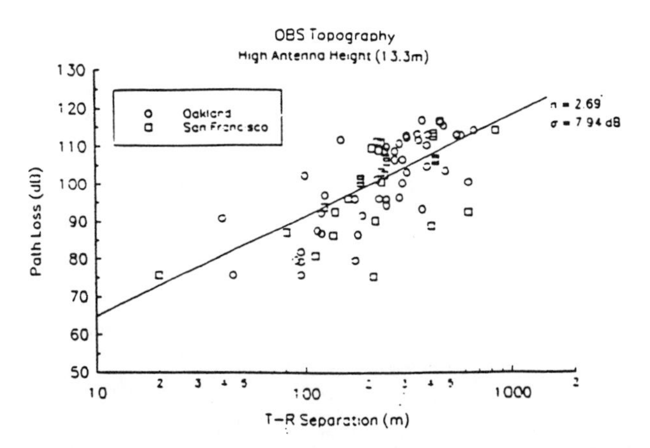

Fig. 7. Path loss versus transmitter-receiver separation for high antenna height (13.3 m) obstructed topography with single regression MMSE curve fit.

observed in non-LOS circumstances. An important result from Table II is that the path loss exponents for the low, medium, and high antennas are all $n_1 = 2.6 \pm 0.1$, indicating that in obstructed cases there is no real path loss advantage for the higher antennas. The results also indicate that path loss exponent values in the $2 < n < 3$ range accurately represent real-world measurements for microcell channels as suggested in [14]. It is also significant to note that the obstructed cases do not exhibit a pronounced break point, in contrast to the LOS cases.

G. Delay Spread Analysis for Line-of-Sight and Obstructed Topographies

One method of characterizing wide-band multipath channels is by calculating their rms delay spread (σ_d) [5]. Many researchers have shown that time delay spread can have detrimental effects on digital communications for certain portable radio channels [2], [4], although Fung and Rappaport have shown that delay spread alone does not determine the actual bit error rate at any instant [4]. The rms delay spread is the square root of the second central moment of the power delay profile and is calculated as in [5].

For this study, rms delay spread has been analyzed as a function of antenna height and path loss. Table III shows the maximum, mean, and standard deviations of the rms delay spread values as a function of the three antenna heights for all measured profiles. From the table it is clear that the mean rms delay spread increases as a function of the base station antenna height. The low antenna at 3.7 m had an average rms delay spread of 136.8 ns, whereas the high antenna at 13.3 m had an average rms delay spread of 257.9 ns. We also studied the standard deviation of the rms delay spread to determine the variability as a function of antenna height. The standard deviation of the rms delay spread increased as a function of antenna height. The delay spread increases as antenna height increases because as the antenna is raised it becomes visible to more scattering objects at greater distances. This observation brings up an interesting application to the design of microcellular systems. As discussed in Section III-C, in general a higher base station antenna provides a lower path loss for a given distance, but the higher base station antenna will also cause a larger delay spread. There will then be a tradeoff between path loss and delay spread for a given antenna height, and for a particular environment an optimum base station antenna height may be found.

TABLE III
MAXIMUM, MEAN, AND STANDARD DEVIATIONS OF THE RMS DELAY SPREADS (σ_d) OF ALL MEASUREMENT LOCATIONS FOR EACH ANTENNA HEIGHT WITH LINE-OF-SIGHT AND OBSTRUCTED TOPOGRAPHIES

Antenna Height	Max. RMS Delay (ns)	Mean RMS Delay (ns)	Std. Deviation of RMS Delay Spread (ns)
Low (3.7 m)	1,011.6	136.8	138.0
Med (8.5 m)	732.0	176.8	147.1
High (13.3 m)	1,859.5	275.9	352.0

Fig. 8 is a plot of rms delay spread versus path loss for all LOS and OBS measurement locations and for all three antenna heights. From the plot, there is clearly a trend toward increasing delay spread as path loss increases. The increasing delay spread as a function of path loss has also been observed by Devasirvatham [13] for indoor environments. As path loss increases, delay spread increases because as the LOS component becomes weaker, multipath components with larger excess delays and weaker power levels can be detected by the measurement equipment. As a result of the limited dynamic range of the measurement system display, these weak multipath components cannot be detected when a strong direct path signal is present. For system design purposes, an overbound on the path loss can be obtained by using a simple exponential model of the form $\sigma_d = e^{0.065\mathrm{PL}(d)}$, where σ_d is the rms delay spread in nanoseconds and $\mathrm{PL}(d)$ is the path loss in decibels (referred to the 1-m reference distance), as a function of the T-R separation d. The exponential model offers a simple model for delay spread as a function of path loss and distance. The exponential overbound curve is shown in Fig. 8 along with the measured data points for all measurement locations and antenna heights.

IV. MICROCELLULAR SYSTEM DESIGN

A. Calculation of Outage Probability due to Path Loss

By using the path loss models presented in Section III, along with some basic assumptions concerning user density within a microcell, approximations can be formulated for the probability of path loss for any point in the cell [7]. The microcell is assumed to be a circular region with a minimum radius of $r_{\min} = 1$ m. The path loss over a microcell is assumed to follow a log-normal distribution about the mean Fresnel break-point path loss model presented in (11). A log-normal distribution corresponds to a Gaussian distribution about the mean path loss with the values in decibels. Therefore,

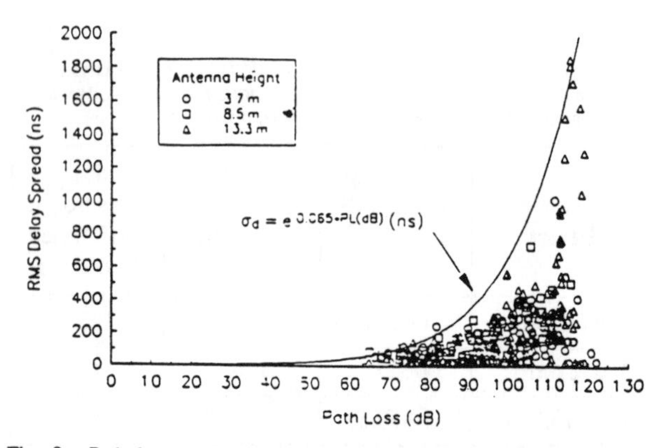

Fig. 8. Path loss versus rms delay spread for all antenna heights with line-of-sight and obstructed topographies and an exponential overbound model.

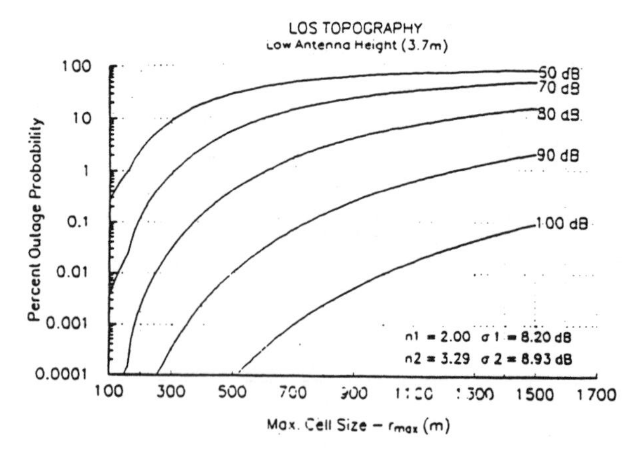

Fig. 9. Outage probability versus maximum cell size as a function of the outage path loss for line-of-sight topography with low antenna height (3.7 m).

the probability density function (pdf) of path loss conditioned on the propagation model parameters $(n_1, n_2, \sigma_1, \sigma_2)$ and the cell radius (r) can be written in terms of the Gaussian distribution as [7]

$$f(\mathrm{PL}|n_1, n_2, \sigma, r) = \frac{1}{\sqrt{2\pi}\sigma} \cdot \exp\left\{-\frac{1}{2}\left[\frac{p_1 - \mathrm{PL}_1(n_1, n_2, r)}{\sigma}\right]^2\right\}, \tag{14}$$

where p_1 is the 1-m reference path loss in decibels, σ is the standard deviation in decibels given in (12), and $\mathrm{PL}_1(n_1, n_2, r)$ is the path loss model of (11) in decibels as a function of distance $r\,(r > 1\text{ m})$ and power law exponents n_1 and n_2. For distance r less than the Fresnel break point $(r < d_f)$ then $\sigma = \sigma_1$, and beyond the break point $(r > d_f)$, then $\sigma = \sigma_2$, based on the two different regions of the path loss model. The probability can be integrated from $r_{\min}$ to $r_{\max}$ to determine the pdf conditioned solely on the model parameters $(n_1, n_2, \sigma_1, \sigma_2)$ and independent of the radial distance r [12]. The pdf of path loss over a coverage area is [7]

$$f(\mathrm{PL}|n_1, n_2, \sigma) = \int_{r_{\min}}^{r_{\max}} f(\mathrm{PL}|n_1, n_2, \sigma, r)\, f(r)\, dr, \tag{15}$$

where $f(r)$ is the pdf of users over a circular ring of radius r.

To determine the pdf of the density of users in users per meter $f(r)$, first assume that the users are distributed uniformly over the circular cell coverage area. Using the theory of functions of a single random variable, the following equation for the pdf of users with a uniform density is derived easily as a function of $r_{\max}$, the cell radius [7], [12]

$$f(r) = \frac{2r}{r_{\max}^2}. \tag{16}$$

Substituting (16) and (14) into (15), a closed-form expression for the pdf of path loss conditioned on the model parameters $f(\mathrm{PL}|n_1, n_2, \sigma)$ can be obtained. The integration over the radius range from $r_{\min}$ to $r_{\max}$ in (15) can be calculated by using a numerical integration algorithm. The resulting pdf can be integrated to obtain a cumulative distribution function (cdf), which gives the probability that path loss is within a particular decibel range throughout the microcell. From the cdf, the probability of an outage can be calculated by obtaining the probability that the path loss will be greater than the outage threshold. The outage probability curves presented in Section IV-B were generated in this manner, using values of $r_{\max}$ from 100 to 1500 m.

B. Results for Outage Probability due to Path Loss

Figs. 9–11 show the percent outage probability versus maximum cell size based on the LOS first Fresnel break-point model parameters shown in Table I for the low, medium, and high antennas. The term "outage" means that the path loss exceeds a specified level, and it is then assumed that communication is lost over the particular channel. The path loss values are with respect to the path loss at the 1-m reference distance; the actual path loss between the transmitter and receiver would be 38 dB greater than the contour values shown for a frequency of 1900 MHz. The outage probability curves are plotted for various outage thresholds in steps of 10 dB by using the method presented in Section IV-A, where each curve represents a different outage threshold. System designers can use the results obtained in Section IV-A to optimize cell sizes based on outage probabilities determined for a particular environment. Once the model parameters for a particular environment are determined or estimated, the outage probabilities due to path loss can be obtained easily.

From Figs. 9–11, several important observations can be made. The first Fresnel zone break point is clearly evident on the curves as the abrupt change in slope. Recall that, at the break point, the path loss exponent changes discontinuously from n_1 to n_2, causing the integration in (15) to change abruptly. If the standard deviation spread about the measured path loss (σ = rms error) beyond the break point is greater than

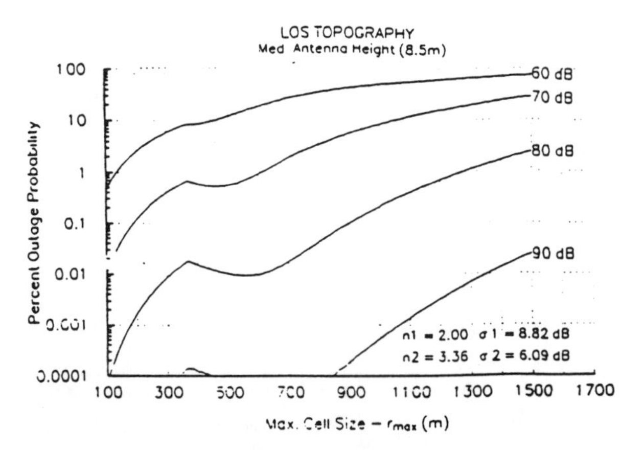

Fig. 10. Outage probability versus maximum cell size as a function of the outage path loss for line-of-sight topography with medium antenna height (8.5 m).

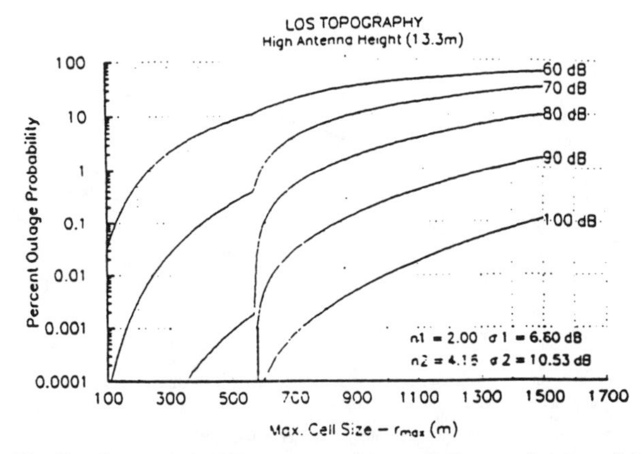

Fig. 11. Outage probability versus maximum cell size as a function of the outage path loss for line-of-sight topography with high antenna height (13.3 m).

the rms error prior to the break point, the outage probability curve will exhibit a positive increase in slope (increased outage), as seen in Figs. 9 and 11. If the rms error beyond the break point is lower than the rms error prior to the break point, then the outage probability curve will exhibit a negative change in slope (decreased outage) as seen in Fig. 10. Based on the experimental models presented in Table I, our results are inconclusive with regard to a positive or negative increase in slope beyond the break point. This is because there is no clear trend for the rms error as a function of antenna height. The low and high antenna path loss models each exhibit higher rms errors than the medium antenna path loss models.

For microcellular design, it would be advantageous to have an abrupt positive change in the slope of the outage probability curve to delineate clearly the microcell boundaries. For example, in Fig. 11, if the microcell boundary were selected at the first Fresnel zone distance of 573 m, then with a path loss outage criteria of 80 dB with respect to a 1-m reference distance, an abrupt increase in outage probability is observed at the cell boundary. In other words, the microcell boundaries are defined clearly in terms of outage probability.

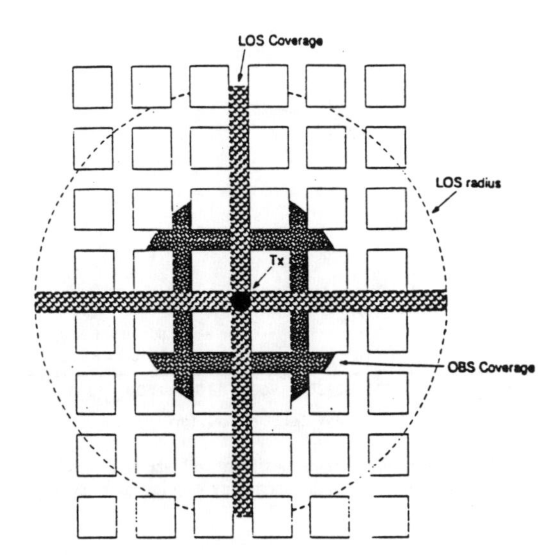

Fig. 12. Conceptual microcellular coverage area in a builtup downtown area.

This observation gives microcellular systems designers the ability to fine-tune the microcell sizes to the propagation environment. Based on estimated path loss exponents, standard deviations, and transmit/receive antenna heights, it is then possible to determine the proper choice of cell radius to achieve a certain outage probability for given path loss.

Real microcells will not be circular with LOS to all portions of the coverage area, but will have very irregular shapes [17]–[19]. A circular cell approximation for determination of the maximum cell radius based on an LOS model provides a realistic estimate of the length of a microcellular coverage region where LOS can be maintained between the transmitter and receiver. The calculation for outage probability may be repeated for obstructed locations. The resultant coverage areas may be superimposed to provide the true coverage area that incorporates both LOS and OBS propagation models, as conceptually shown in Fig. 12. In Fig. 12, the circular boundaries represent the maximum cell radii for the LOS and OBS models. The coverage area is indicated by the shaded region.

C. Calculation of Outage Probability due to Delay Spread

The relationship between delay spread and path loss, which was developed in Section III-G, can be used to estimate worst case outage probabilities based on delay spread. The exponential overbound model shown in Fig. 8 can be used to determine path loss for a specified worst case delay spread. The overbound model provides a one-to-one mapping of path loss to rms delay spread and thus represents a worst case scenario for rms delay spread. Therefore, based on a particular delay spread outage threshold, the outage probability can be computed indirectly from path loss estimates based on antenna height, T-R separation, and LOS/OBS topography. The same technique discussed in Section IV-A is used, where the path loss is obtained from the overbound model evaluated for the specified rms delay spread outage level. The system designer

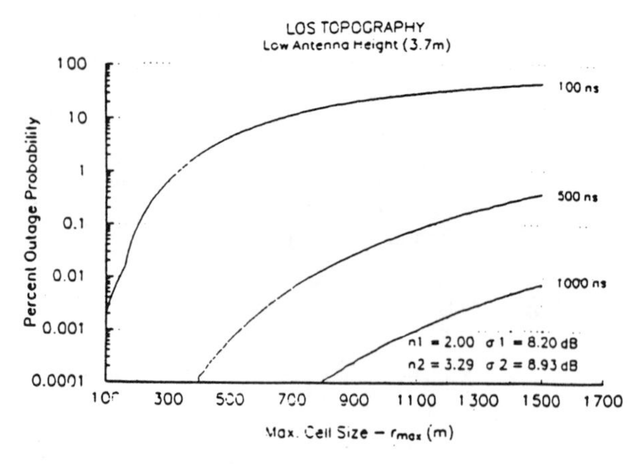

Fig. 13. Outage probability versus maximum cell size as a function of the outage delay spread for line-of-sight topography with low antenna height (3.7 m).

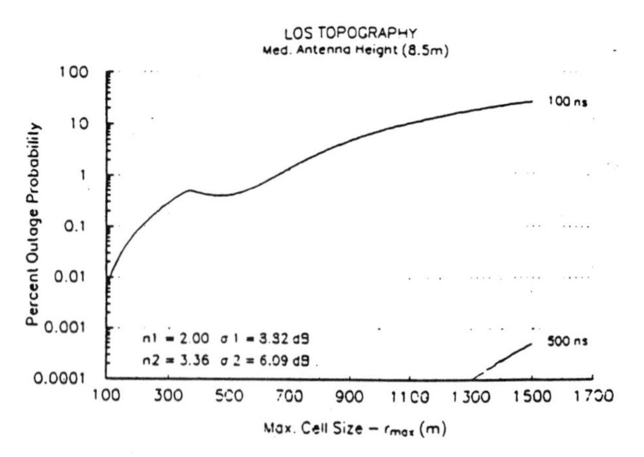

Fig. 14. Outage probability versus maximum cell size as a function of the outage delay spread for line-of-sight topography with medium antenna height (8.5 m).

is thus able to use either path loss or delay spread to define an outage.

D. Results for Outage Probability due to Delay Spread

The exponential overbound model in Fig. 8 has been used to obtain a worst case relationship between path loss and delay spread. The resulting relationship can be used in the method described in Section IV-A to calculate outage probability as a function of delay spread. Figs. 13–15 give the delay spread outage probability versus maximum cell size for LOS cases as a function of delay spread with low, medium, and high antennas, respectively. A delay spread outage occurs when the delay spread on a particular channel exceeds the threshold level. Again, because of the dual regression model, the slope changes abruptly at the first Fresnel zone boundary. The microcellular system designer can select optimum cell sizes based on delay spread as well as path loss outage criteria. For example, in Fig. 15, if the microcell size is chosen as 573 m, then the delay spread should increase abruptly beyond that point. The delay spread outage probability results represent a worst case scenario, since they were derived using the exponential overbound model.

V. Conclusions

In this paper, measurement results have been presented for five typical microcellular-type environments in urban and suburban sections of the San Francisco Bay area. These measurements were made for three different transmit antenna heights that represent typical heights for lamp-post-mounted microcell base stations. In LOS cases, path loss as a function of T-R separation has been modeled using a double regression curve fit with both a break point at the first Fresnel zone clearance distance and an MMSE optimum break point. The standard deviations of the resulting best fit models indicate that the Fresnel zone fit can be used with excellent accuracy with respect to the optimum MMSE break point, which can be found iteratively. Thus, for specified transmit and receive

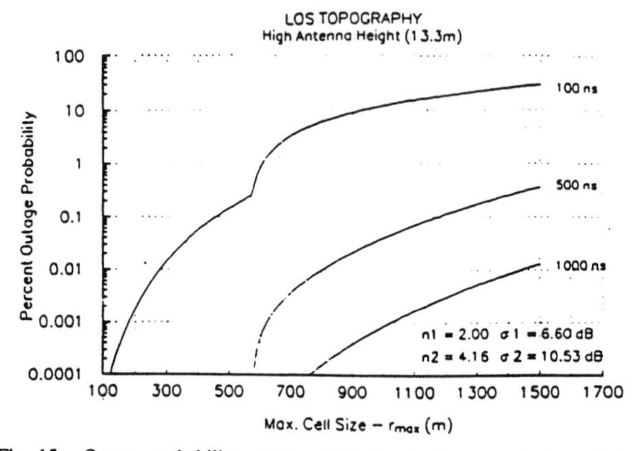

Fig. 15. Outage probability versus maximum cell size as a function of the outage delay spread for line-of-sight topography with high antenna height (13.3 m).

antenna heights and T-R separation, the mean path loss can be estimated based on power law exponents and the computed first Fresnel zone break point.

In general, there is a trend toward higher power law path loss exponents after the first Fresnel break point for the high antennas compared with the low antennas. For microcellular systems, where cell sizes are typically in the range from tens of meters to approximately 2000 m, the difference in path loss exponents is more than offset by the change in first Fresnel zone distance. As the antenna height becomes higher, the first Fresnel zone distance becomes larger. Since the path loss exponent prior to the Fresnel break point is approximately two, and the exponent beyond the break point is greater than two, the distance at which the exponents change is very important for system design. For T-R separations of 50–2000 m, the high antenna generally observed a lower path loss than the low antenna. We found that beyond the Fresnel break point, path loss exponents were typically approximately three or four.

The rms delay spread was found to be a strong function of antenna height. As the antenna height increased, the standard deviation and mean values of the rms delay spreads increased markedly. This result indicates that there is a trade off between path loss and delay spread considerations when choosing a microcellular base station antenna height. As the antenna height increases, path loss is reduced, but delay spread increases as well. An exponential overbound model has been used to give a worst case estimate of rms delay spread as a function of path loss.

The models for path loss as a function of T-R separation and for rms delay spread as a function of path loss have been incorporated into the design of microcellular-type systems. By developing an expression for the pdf of the path loss conditioned on the propagation model parameters and the cell radius, it is possible to get an expression for the outage probability as a function of cell size. If a specific path loss outage criterion is established, then an outage probability can be determined for a given cell radius. Outage probability can also be determined based on a delay spread outage criterion. Microcellular system designers can now select an optimum cell size for a particular propagation environment based on an outage probability using a path loss or delay spread outage criterion.

ACKNOWLEDGMENT

The authors would like to acknowledge the contributions of Michael D. Keitz in performing the experimental measurements and Dwayne A. Hawbaker in developing the microcellular design analysis. Helpful comments from Professor Henry Bertoni, Dr. Pete Arnold, and Dr. William C. Y. Lee, who participated in the Telesis Technology Laboratory research review board during the course of this work, are gratefully acknowledged. The comments of the reviewers that helped to improve the quality of this paper are acknowledged.

REFERENCES

[1] R. Steele, "The importance of propagation phenomena in personal communication networks," in *Proc. IEEE Conf. on Antennas and Propagation*, York, UK, Apr. 1991.

[2] J. C.-I. Chuang, "The effects of time delay spread on portable radio communications channels with digital modulation," *IEEE J. Select. Areas Commun.*, vol. SAC-5, no. 5, pp. 879–889, June 1987.

[3] R. C. Bernhardt, "The effect of path loss models on the simulated performance of portable radio systems," in *Proc. IEEE Globecom 1989*, Dallas, TX, pp. 1356–1360.

[4] T. S. Rappaport and V. Fung, "Simulation of bit error performance of FSK, BPSK, and $\pi/4$ DQPSK modulation in indoor radio flat fading channels using a measurement-based channel model," *IEEE Trans. Veh. Technol.*, vol. 40, no. 4, pp. 731–740, Nov. 1991.

[5] T. S. Rappaport, "Characterization of UHF multipath radio channels in factory buildings," *IEEE Trans. Antenn. Propagat.*, vol. 37, pp. 1058–1069, Aug. 1989.

[6] J. Griffiths, *Radio Wave Propagation and Antennas*. Englewood Cliffs, NJ: Prentice-Hall Int., 1987.

[7] D. A. Hawbaker, "Indoor wide band radio wave propagation measurements and models at 1.3 GHz and 4.0 GHz," Master's thesis, Virginia Polytechnic Inst. & State Univ., Blacksburg, VA, May 15, 1991.

[8] M. L. Skolnik, *Introduction to Radar Systems*. New York: McGraw-Hill, 1980.

[9] D. C. Cox, R. R. Murray, and A. W. Norris, "800 MHz attenuation measured in and around suburban houses," *AT&T Bell Laboratories Tech. J.*, vol. 63, pp. 921–954, July/Aug. 1984.

[10] T. S. Rappaport and C. D. McGillem, "UHF fading in factories," *IEEE J. Select. Areas Commun.*, vol. 7, no. 1, pp. 40–48, Jan. 1989.

[11] J. D. Parsons and J. G. Gardiner, *Mobile Communication Systems*. New York: Blackie/Wiley, 1989.

[12] A. Papoulis, *Probability, Random Variables, and Stochastic Processes*. New York: McGraw-Hill, 1984.

[13] D. M. J. Devasirvatham, "Time delay spread and signal level measurements of 850 MHz radio waves in building environments," *IEEE Trans. Antenn. Propagat.*, vol. AP-34, pp. 1300–1305, Nov. 1986.

[14] T. S. Rappaport and L. B. Milstein, "Effects of radio propagation path loss on DS-CDMA cellular frequency reuse efficiency for the reverse channel," *IEEE Trans. Veh. Technol.*, vol. 41, no. 3, pp. 231–242, Aug. 1992.

[15] H. H. Xia, *et al.*, "Radio propagation characteristics for line of sight microcellular and personal communications," *IEEE Trans. Antenn. Propagat.*, vol. 41, no. 10, pp. 1439–1447, Oct. 1993.

[16] A. J. Rustako, Jr., N. Amitay, G. J. Owens, and R. S. Roman, "Radio propagation at microwave frequencies for line-of-sight microcellular mobile and personal communications," *IEEE Trans. Veh. Technol.*, vol. 40, no. 1, pp. 203–210, Feb. 1991.

[17] K. Mahbobi, "Radio wave propagation in urban microcellular environment," in *Proc. 42nd IEEE Vehicular Technology Conf.*, Denver, CO, May 1992, pp. 951–955.

[18] A. J. Goldsmith and L. J. Greenstein, "An empirical model for urban microcells with applications and extensions," in *IEEE Veh. Technol. Conf.*, Denver, CO, pp. 419–422, May 1992.

[19] L. R. Maciel and H. L. Bertoni, "Cell shape for microcellular systems in residential and commercial environments," *IEEE Trans. Veh. Technol.*, vol. 43, no. 2, pp. 270–278, May 1994.

Martin J. Feuerstein was born in Memphis, TN, in 1962. He received the B.E. degree in electrical engineering and mathematics from Vanderbilt University, Nashville, TN, in 1984. In 1987 he received the M.S. degree in electrical engineering from Northwestern University in Evanston, IL. He received the Ph.D. degree in electrical engineering from the Virginia Polytechnic Institute in Blacksburg, VA, in 1990.

From 1984 to 1985, he was a Systems Engineer with the Advanced Communications Terminals Division of Northern Telecom, Inc. From 1985 to 1987, he was a Research Assistant with the Microwave Characterization Lab at Northwestern University, where he worked on microwave plasma generation and diagnostics systems. From 1987 to 1990, he was a Project Assistant with the Satellite Communications Group at Virginia Tech, where he worked on the analysis of spread spectrum position location systems. During 1990 he was a Development Engineer with Spatial Positioning Systems Inc., where he worked on 3-D laser position measurement systems. From 1991 to 1992, he was a Visiting Assistant Professor with the Mobile & Portable Radio Research Group at Virginia Tech, where he worked on radio propagation prediction techniques. Since joining US West Advanced Technologies, Boulder, CO, in 1992, he has worked on radio propagation and wireless system design.

Dr. Feuerstein is a member of Tau Beta Pi, Eta Kappa Nu, Phi Eta Sigma, and Phi Kappa Phi. He is a life member of the American Radio Relay League.

Kenneth L. Blackard (S'90–M'91) was born in Mt. Airy, NC, on July 5, 1967. He received the B.S.E.E. and M.S.E.E. degrees from Virginia Polytechnic Institute and State University in 1989 and 1991, respectively.

From 1990 to 1992, he worked with Professor T. S. Rappaport in the Mobile and Portable Radio Research Group, where he did research on radio-frequency impulsive noise and radio wave propagation. In November 1992, he joined the Federal Bureau of Investigation, where he is currently employed as a Research Engineer. His research interests are RF propagation, wireless communications, and antennas.

Mr. Blackard is a member of the Eta Kappa Nu, Phi Eta Sigma, and Phi Kappa Phi honor societies.

Theodore S. Rappaport (S'85–M'87–SM'91) was born in Brooklyn, NY, on November 26, 1960. He received B.S.E.E., M.S.E.E., and Ph.D. degrees from Purdue University in 1982, 1984, and 1987, respectively.

In 1988, he joined the Electrical Engineering Faculty of Virginia Tech, Blacksburg, where he is an Associate Professor and Director of the Mobile and Portable Radio Research Group (MPRG), a group he founded in 1990. He conducts research in mobile radio communication system design, RF propagation prediction and measurements, and digital signal processing. He guides a number of graduate and undergraduate students in mobile radio communications, and has authored or coauthored numerous papers in the areas of wireless system design and analysis, propagation, vehicular navigation, and wide-band communications. He holds several U.S. patents and is coinventor of SIRCIM and SMRCIM, indoor and microcellular radio channel software simulators that have been adopted by more than 100 companies and universities.

In 1990, Dr. Rappaport received the Marconi Young Scientist Award for his contributions in indoor radio communications, and was named a National Science Foundation Presidential Faculty Fellow in 1992. He received the 1992 IEE Electronics Letters Premiums Award for the paper "Path loss prediction in multi-floored buildings at 914 MHz," which he coauthored. He has edited two books published by Kluwer Academic Press on the subject of wireless personal communications, and has contributed chapters on the subject for the CRC Engineering Handbook series. He serves as Senior Editor of the IEEE *Journal on Selected Areas in Communications*. He also serves on the editorial boards of the IEEE Personal Communications Magazine and the International Journal for Wireless Information Networks (by Plenum). He is a Registered Professional Engineer in the State of Virginia and is a Fellow of the Radio Club of America.

Scott Y. Seidel (S'92–M'93) was born in Falls Church, VA, in 1966. He received the B.S. M.S. and Ph.D. degrees in electrical engineering from Virginia Polytechnic Institute and State University, Blacksburg, in 1988, 1989 and 1993, respectively.

From 1989 to 1993 he was a member of the Mobile and Portable Radio Research Group at Virginia Tech. His research focused on the development of both urban microcellular and indoor radio channel models using site-specific propagation prediction methods. He is coinventor of SIRCIM, an indoor radio channel simulator that has been adopted by over 100 companies and universities.

Dr. Seidel is a member of Tau Beta Pi, Eta Kappa Nu, and Phi Kappa Phi.

Howard H. Xia was born in Canton, China on August 16, 1960. He received the B.S. degree in physics from South China Normal University, Canton, China, in 1982. He received the M.S. degree in physics in 1986, the M.S. degree in electrical engineering in 1988, and the Ph.D. degree in electrophysics in 1990, all from Polytechnic University, Brooklyn, New York.

Since September 1990, he has been working with AirTouch Communications (formerly, PacTel Corporation) and Telesis Technologies Laboratory engaged in research and development of advanced analog and digital cellular mobile radio networks, and personal communications systems. He has published articles on the topics of outdoor/indoor radio wave propagation, spectrum sharing, and CDMA system design.

Dr. Xia serves as a member of United States delegations to participate ITU-R (formerly, CCIR) activities on developing recommendations for the third generation mobile systems—Future Public Land Mobile Telecommunication Systems (FPLMTS).

Reprinted from *IEEE Transactions on Vehicular Technology*, Vol. 43, No. 4, Nov. 1994, pp 879-891.

Site-Specific Propagation Prediction for Wireless In-Building Personal Communication System Design

Scott Y. Seidel, *Member, IEEE*, and Theodore S. Rappaport, *Senior Member, IEEE*

***Abstract*— This paper describes a geometrical optics based model to predict propagation within buildings for Personal Communication System (PCS) design. A ray tracing model for predicting propagation based on a building blueprint representation is presented for a transmitter and receiver located on the same floor inside a building. Measured and predicted propagation data are presented as power delay profiles that contain the amplitude and arrival time of individual multipath components. Measured and predicted power delay profiles are compared on a location-by-location basis to provide both a qualitative and a quantitative measure of the model accuracy. The concept of *effective* building material properties is developed, and the *effective* building material properties are derived for two dissimilar buildings based upon comparison of measured and predicted power delay profiles. Time delay comparison shows that the amplitudes of many significant multipath components are accurately predicted by this model. Path loss between a transmitter and receiver is predicted with a standard deviation of less than 5 dB over 45 locations in two different buildings.**

I. Introduction

MUCH WORK has been done to statistically characterize multipath propagation inside buildings in the 800 MHz to 5.8 GHz frequency range [1]–[11]. Buildings vary greatly in size, shape, and type of construction materials. The statistics of propagation measurements vary greatly from building to building and only broad conclusions related to the building type can be made.

This paper describes a geometrical optics based model to predict propagation inside buildings for personal communication system (PCS) design. A ray tracing algorithm predicts multipath impulse responses based on building blueprints. In this paper, measured and predicted propagation characteristics are compared for two different frequencies and buildings. A method for computing the *effective* building material properties for the walls in two different buildings is developed. These *effective* building material properties lead to the reflection coefficient models that give the "best fit" between measured and predicted propagation as determined from an error function that includes both multipath component amplitudes and arrival times. Knowledge of indoor propagation characteristics based on site-specific data will allow evaluation of proposed building changes or base station placement based on the physical properties of the building. The ray tracing prediction model is shown to predict path loss with an overall standard deviation of less than 5 dB throughout the two buildings. Time delay comparison shows that the amplitudes and time delays of measured power delay profiles can be predicted accurately via ray tracing.

Manuscript received March 1, 1993; revised December 13, 1993. This work was supported by NSF, MPRG Industrial Affiliates Program, and ARPA.

S. Y. Seidel is with the Radio and Personal Communications Research Department, Bell Communications Research, Red Bank, NJ 07701-5699 USA.

T. S. Rappaport is with the Mobile and Portable Radio Research Group (MPRG), Bradley Department of Electrical Engineering, Virginia Polytechnic Institute and State University, Blacksburg, VA 24061-0111 USA.

IEEE Log Number 9403804.

The field of graphical ray tracing for creating a 2-D picture of a 3-D world via computer is well developed [12], [13]. An object or group of objects called a "scene" are described in terms of their geometry and light scattering properties (color). The computer attempts to recreate a photograph of the scene for a fixed observer and one or more light sources. The graphical ray tracing is a geometrical optics model for light. To take advantage of the similarities of the graphical ray tracing and the geometrical optics models, the source code for the graphical ray tracer in [13] has been extensively modified to generate propagation impulse response data instead of pixel color. This geometrical optics based model for electromagnetic wave propagation along with a single diffraction model is used to predict multipath power delay profiles and path loss inside buildings.

Ray tracing represents the high frequency limit of the exact solution for electromagnetic fields and can give quick approximate solutions when the exact solution can not be found. In this paper, a geometrical optics based model is used to predict the propagation of radio waves in buildings. Ray tracing is a physically tractable method of predicting the delay spread and path loss of in-building radio signals, and lends itself well to rapid parallel computing. The time delays of individual multipath components can be linked to specific radio propagation paths. Ray tracing methods have been proposed for propagation prediction in microcellular environments [14]–[18] and for modeling propagation in rough terrain [19]. Ray tracing for indoor propagation has also been proposed in [20]–[26]. However, none of these applications of ray tracing for in-building propagation prediction have yet compared measured and predicted wide band power delay profiles on a location-by-location basis over a large number of measurement locations in several buildings. Yet, it is a location-by-location comparison that is required to determine the general applicability of a ray tracing propagation model.

The propagation model described in this paper uses geometrical optics to trace the propagation of direct, reflected, and transmitted fields. Singly diffracted fields are also computed. The rays, which represent a discrete local plane wave of the total field, originate from point sources and propagate in 3-D space. The lack of significant difference in propagation characteristics throughout the low microwave band [1]–[6] indicates that a ray tracing model, where surrounding objects are much greater than a wavelength at 900 MHz, and the actual frequency dependence is small, can provide accurate prediction for a variety of frequencies.

This paper presents a ray tracing method to predict the channel impulse response in the form of a power delay profile. From the power delay profile, parameters such as path loss and time delay spread of indoor radio channels may be determined. First, in Section II, the ray tracing computer code used to predict the propagation is described. In Section III, the source ray directions and the interaction of rays with objects in the building database are discussed. The path loss dependence of direct, transmitted, reflected, and diffracted rays, the algorithm for the identification of received rays, and a description of the data processing used to convert the raw ray tracing output to power delay profiles are also described in Section III. Section IV discusses the measured and predicted power delay profiles and propagation channel parameters.

II. Building Blueprint Representation

In order to implement site-specific propagation models, it is necessary to incorporate accurate site-specific building information into the propagation prediction tool. We have used AutoCAD™ to represent the significant building features, such as wall locations and building materials. This program was selected because it is considered an industry standard CAD package, and it is believed that a majority of building blueprints may be readily represented in an AutoCAD format. Since a geometrical optics based model is used, only objects that are much larger than a wavelength (large objects) at microwave frequencies are represented. Hence, it is not practical to include small-scale features within a building. For example, exit signs, door knobs, door hinges, and furniture are not included in the model. Only large objects, such as building walls or office partitions, are included in the building database. Each wall is considered to be infinitely thin. The inclusion of only large objects is justified in that the goal of this work is to predict large scale average path loss and time dispersion as influenced by major changes in the geometry surrounding the transmitter and receiver, but not the small-scale fluctuations of a narrow band signal. Fig. 1 shows the 3-D extended view of the building blueprint in AutoCAD for the second floor of Whittemore Hall, an academic building on the Virginia Tech campus.

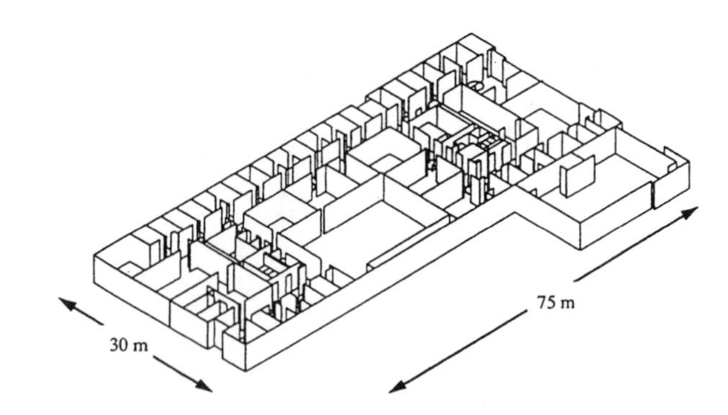

Fig. 1. 3-D AutoCAD representation of the second floor of Whittemore Hall on the Virginia Tech campus.

III. Ray Tracing for Propagation Prediction

A. Background

Our prediction technique uses "brute force" ray tracing to account for all possible propagation paths from a transmitter to multiple receivers within a specified resolution. Although graphical ray tracing programs can take advantage of sending rays in only specified directions, an electromagnetics model must account for all directions relative to both the transmitter and receiver antennas. As computation times increase, ray tracing acceleration techniques are employed to combat the computational requirements of brute force ray tracing [12].

The algorithm described here is coded in C++ for use on a UNIX based workstation. As an object-oriented language, C++ has the capability to manipulate data structures, such as vectors, objects, and the functions that match them, in a modular fashion. For example, the program can perform an intersection test for a ray and an object. Although the algorithms to determine the intersection are different for different objects, the same subroutine may be called to perform the intersection test. Due to the numerous ray-object intersection tests and extensive data arrays required for ray tracing, the program is run on a workstation. Parallel computing has been implemented to allow multiple workstations to simultaneously run the program.

B. Source Ray Directions

The transmitter and receiver are modeled as point sources in a building. In order to determine all possible rays that may leave the transmitter and arrive at the receiver, it is necessary to consider all possible angles of departure and arrival at the transmitter and receiver. Rays are launched from the transmitter at an elevation angle θ and azimuth angle ϕ relative to the standard engineering coordinate system. Antenna patterns are incorporated to include the effects of antenna beamwidth in both azimuth and elevation.

To keep all ray manipulation routines general, it is desirable that each ray tube occupy the same solid angle $d\Omega$, and each wavefront be an identical shape and size at a distance r from the transmitter. An ideal wavefront is shown in Fig. 2. Additionally, these wavefronts must be subdividable so that an increased ray resolution can be handled easily. For reference, let $r = 1$ and the total wavefront is the surface of a unit sphere. The problem then becomes one of subdividing the sphere surface into equal area "patches" that are all the same size and shape and completely cover the surface without gaps. For a 2-D (flat) surface, this problem is easy to solve. Regular triangles, squares, and hexagons can completely cover an area with equal size and shape objects without leaving gaps. Constant angular separation of the rays may be easy to visualize, but the mechanics of the problem are rather complicated. Common methods of subdividing the total wavefront, such as by using spherical coordinates, are

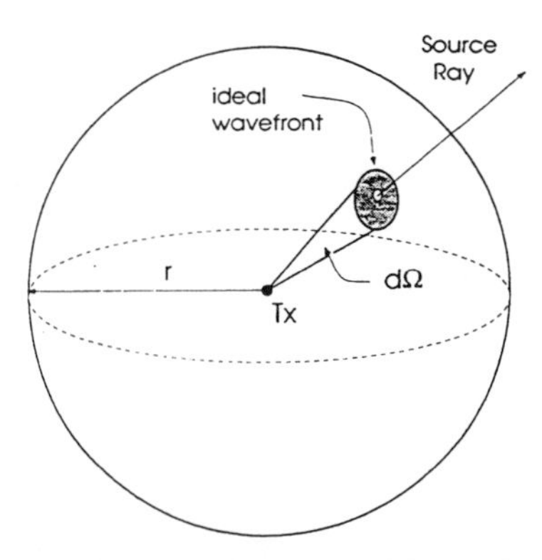

Fig. 2. Ideal wavefront represented by each source ray.

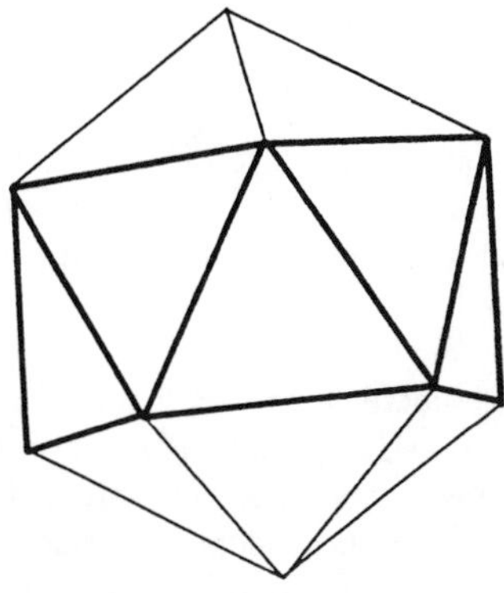

Fig. 3. A regular icosahedron that is inscribed inside a unit sphere. Rays are launched through icosahedron vertices and at intersection points of tessellated triangular faces.

insufficient due to a decrease in the angular separation between rays as they are launched in directions near the poles of the sphere.

Our solution for the source ray directions is adapted from the theory of geodesic domes [27], [28]. An icosahedron is inscribed inside the unit sphere. A regular icosahedron is a 20-sided polygon with 20 triangular faces and 12 vertices. Each vertex joins five faces. Fig. 3 shows a regular icosahedron with the "middle" faces highlighted. If rays are launched at each of the 12 vertices, each ray wavefront is an identically shaped pentagon separated by 63 degrees from each of its five nearest neighbors. A wavefront's nearest neighbors are the rays whose wavefront "patches" are adjacent to the ray wavefront.

To achieve better angular resolution in a systematic manner, each triangular face of the icosahedron is tessellated into N equal segments where N is the tessellation frequency [28]. Fig. 4 shows an example with $N = 4$. Lines parallel to one of the three sides are drawn that subdivide the triangle into smaller equilateral triangles. Rays are launched at angles that pass through the vertices of the triangles. Wavefronts are hexagonal for rays that pass through interior and edge vertices. Rays that pass through the 12 original icosahedron vertices are pentagonal. Ray wavefronts are hexagonal and pentagonal on the surface of the icosahedron. However, the surface normal of the icosahedron face is not necessarily the ray direction since the icosahedron surface must be projected onto the surface of the unit sphere to determine the true wavefront. As

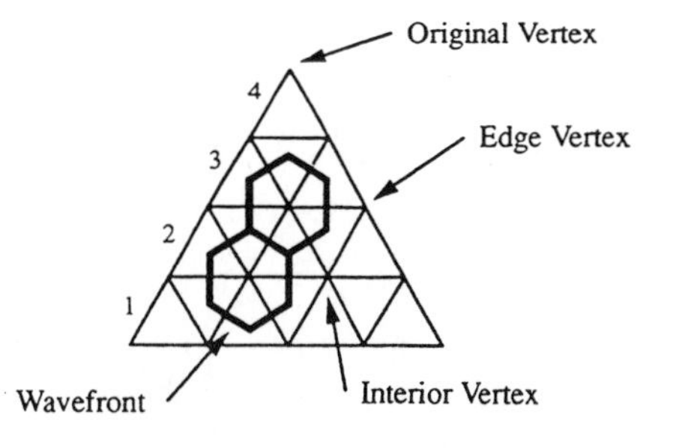

Fig. 4. Tessellation of icosahedron face. Ray wavefronts are hexagonal for edge and interior vertices.

the tessellation frequency increases, the wavefronts decrease in size, but keep their shape and relation to their nearest neighbors. The angular separation between a ray and its nearest neighbors is nearly identical for all nearest neighbors. However, the angular separations are slightly different for rays that originate from the different types of vertices. The number of source rays traced is $10N^2 + 2$ where N is the tessellation frequency [28]. The angular separation between rays decreases as the number of rays increases. This method of launching the source rays provides wavefronts that completely subdivide the surface of the unit sphere with nearly equal shape and area.

C. Tracing the Rays at a Boundary

The computer program uses ray tracing to find each ray path by which significant levels of energy radiated from the transmitting location reach the receiving point. For a given execution of the program, multiple receiving locations can be defined, so the procedure described here can be applied to each receiving point. The ray tracing is accomplished by an exhaustive search of a ray tree accounting for the decomposition of the ray at each planar intersection. First, the program determines if a line-of-sight path exists, and if so computes the received field. Next, the program traces a source ray in a previously determined direction and detects if an object intersection occurs. If no intersection is found, the process stops and a new source ray is initiated. Once the program determines that an intersection has occurred, it then checks to see if a specularly reflected or transmitted ray has an unobstructed path to the receiving location. After checking for reception, the program divides the source ray into a transmitted and reflected ray that are initiated at the intersection point on the boundary. These rays are then treated in a similar fashion to source rays. This recursion continues until a maximum number of tree levels is exceeded, the ray intensity falls below a specified threshold, or no further intersections occur. Fig. 5 shows a portion of a ray tree for one source ray.

The amplitudes of multipath reflections and transmissions are modeled by the Fresnel plane wave reflection and transmission coefficients [29]. The polarization of the wave relative to the interface determines whether the perpendicular or parallel Fresnel reflection coefficients are used. For reflections and transmissions with the floor or ceiling, the parallel coefficients are used, and the perpendicular reflection and transmission coefficients are used when the ray intersects a vertically oriented building wall. Lossy materials may be considered by

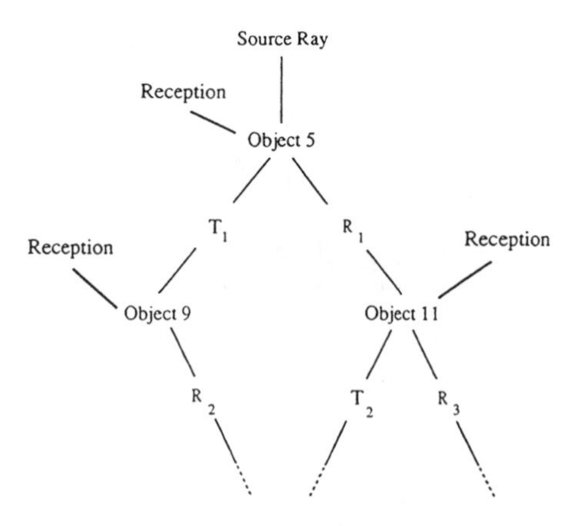

Fig. 5. Ray tree that shows how one source ray can be decomposed into many transmitted, reflected, and scattered rays from intersections with planar boundaries.

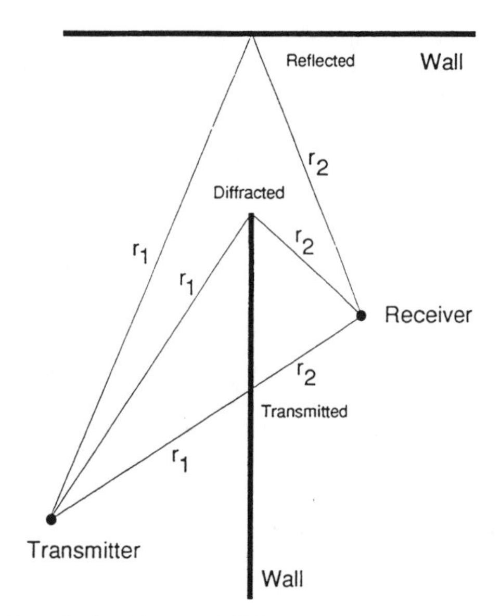

Fig. 6. Transmitted, reflected, and diffracted ray paths.

letting the relative permittivity of the dielectric constant be complex.

D. Representation of Propagation Models in the Ray Tracing Program

Direct and Specularly Reflected and Transmitted Rays: The propagation of energy from the transmitter to the receiver occurs in various modes such as by direct, reflected, transmitted, and diffracted paths. In considering the implementation of each of these in the propagation model, it is important to recognize the path loss dependence of each mode. Direct (line-of-sight) rays exhibit a $1/d^2$ power dependence according to Friis free space transmission. Specularly reflected and/or transmitted rays follow a $1/d^2$ dependence, where d represents the total ray path length. For example, the specularly reflected ray shown in Fig. 6, whose path segments are labeled r_1 and r_2, has a path loss proportional to $1/(r_1 + r_2)^2$.

Diffracted Rays: Geometrical optics (GO) fails to account for diffracted energy in the shadow regions, and the geometrical optics field is discontinuous at shadow boundaries. The Geometrical Theory of Diffraction (GTD) supplements the geometrical optics (ray tracing) by introducing a diffracted field that accounts for the nonzero fields in shadow regions and modifies the field in the GO region so that the total field is continuous. This is important since in a mobile and portable radio environment, the receiver is often shadowed from the transmitter. It is important to be able to predict the changes in the propagation as a receiver moves from an unobstructed to an obstructed location.

Single diffraction from diffracting corners in a hallway is included where the corner is modeled as a dielectric wedge. The received field is determined from the uniform geometrical theory of diffraction as modified for dielectric wedges [30], [31]. The path loss distance dependence is

$$L_i = \sqrt{\frac{r_1}{r_2(r_1 + r_2)}} \tag{1}$$

where r_1 and r_2 are as displayed in Fig. 6. In addition, the diffraction coefficient includes a distance dependence. The diffraction model accounts for all scattering in nonspecular directions. No multiple diffraction or reflection-diffraction is considered in the model.

The diffraction coefficients are the uniform geometrical theory of diffraction (UTD) coefficients developed by Kouyoumjian and Pathak [30]. The uniform diffraction coefficients are composed of an incident-diffracted and a reflected-diffracted term. The incident-diffracted term compensates for the discontinuity in the GO field at the incident shadow boundary that separates the region where the incident field can be received from the region where the incident field is blocked. The reflected-diffracted term compensates for the discontinuity in the GO field at the reflection shadow boundary that separates the region where the reflected field can be received from the region where the reflected field is blocked.

The diffraction coefficients in [30] were developed from the canonical scattering solution for perfectly conducting wedges. The canonical problem of scattering by a dielectric wedge is as yet unsolved. Hence, diffraction coefficients can not be derived directly from the solution. However, the diffraction coefficients are modified so that the continuity of the total field at the shadow boundaries is maintained. Each diffraction coefficient is made up of an incident-diffracted term and a reflected-diffracted term. These individual components of the diffraction coefficients for a perfectly conducting wedge are modified so that the total diffraction coefficient for a dielectric edge correctly compensates for the field discontinuities at the shadow boundaries in the geometrical optics solution. The diffraction coefficient becomes [31]

$$D = [(1 - T)D_i + \Gamma D_r] \tag{2}$$

where Γ and T are the reflection and transmission coefficients of the surface along the shadow boundaries given in Section IV-C. This method for computing the modified diffraction coefficients has been used to model path loss over hills or ridges [32], [33].

The implementation of diffraction in the computer program is separate from the brute force ray tracing. Since there is

TABLE I
SUMMARY OF THE VARIABLES USED TO DESCRIBE THE RAY TRACING PROPAGATION MODEL

Variable	Description [units]
f_{ti}	Field amplitude radiation pattern of the transmitter antenna
f_{ri}	Field amplitude radiation pattern of the receiver antenna
$L_i(d)$	Path loss distance dependence for the ith multipath component
d	Path length [meters]
$\Gamma(\theta_{ji})$	Reflection coefficient
$T(\theta_{ki})$	Transmission coefficient
E_i	Field strength of the ith multipath component [V/m]
E_0	Reference field strength [V/m]
e^{-jkd}	Propagation phase factor due to path length ($k=2\pi/\lambda$)

no recursion, the diffracted ray paths may be found by a straightforward search for all paths that satisfy the correct geometry for a diffracted ray. Diffracting corners are modeled as dielectric wedges and diffracted rays are found for all combinations of transmitter and receiver that each have a direct path to the diffracting wedge. The amplitude and phase of diffracted rays are determined by the uniform geometrical theory of diffraction for a dielectric wedge.

Combining Direct, Specular, and Diffracted Rays: In summary, the model implements direct, reflected, transmitted, and diffracted fields represented by the rays. Table I summarizes the variables used to describe the model. Each propagation mechanism is treated separately, and the total field is determined via coherent superposition of the individual contributions of each ray as weighted in time by a probing pulse identical to one commonly used in measurements to provide a wide band power delay profile representation of the propagation channel. The complex field amplitude of the ith ray at the receiver is given by

$$E_i = E_0 f_{ti} f_{ri} L_i(d) \prod_j \Gamma(\theta_{ji}) \prod_k T(\theta_{ki}) e^{-jkd}. \quad (3)$$

For a diffracted ray, the product of the complex reflection and transmission coefficients is replaced by the complex diffraction coefficient. Path loss is computed by referencing the result of (3) to a one-meter free space path loss.

E. Identification of Received Rays

To determine the ray traced impulse response by a brute force method, it is necessary that the estimate include only one specular ray for each actual path through the channel, regardless of how many rays are traced. Each ray represents the field in the solid angle radiating from the point transmitter. A ray is considered received if the point receiver location is included within this solid angle. The remainder of this section describes the implementation of how received rays are identified.

A perpendicular projection from the receiving location to the ray path is computed and the total (unfolded) path length, d, that the ray travels from the transmitter to the projection point is determined. A reception sphere (from [23], extended to three dimensions [18], [34], [35]) is constructed about the receiving location with a radius proportional to the unfolded path length and the angular spacing between neighboring rays at the source. If the ray intersects the reception sphere, it is received and contributes to the total received signal; otherwise, the ray is not received. Regardless of whether or not the ray is received, recursion proceeds as previously described. The reception sphere effectively accounts for the divergence of the rays from the source. If the ray from the transmitter passes through the reception sphere, then, equivalently, the point receiver intercepts the transmitted solid angle. For ray separation α sufficiently small, the ray intercepting the sphere will be an accurate measure of the ray that would pass directly through the receiving point. The physical interpretation of the reception sphere can be justified with the aid of Fig. 7. This figure is a 2-D representation of a ray being traced. Two adjacent rays, launched at $\pm\alpha$ relative to the test ray, are also shown. Note that in three dimensions, any ray will have more than two adjacent (nearest neighbor) rays and the angular separation of the adjacent rays will not necessarily coincide with the coordinate axes. As shown in the figure, a reception sphere with the correct radius can receive only one of the rays. If the radius is greater than $\alpha d/\sqrt{3}$, two of the rays could be received and would, in effect, count the same specular ray path twice. Likewise, if the radius is too small, none of the rays will intercept the sphere and the specular energy will be excluded. It is important to emphasize that the reception sphere radius is proportional to the unfolded path length from the source to receiver and is different for each ray path, and the $\sqrt{3}$ factor is due to the geometry of the circumscribed circle around the hexagonal wavefront shown in Fig. 8.

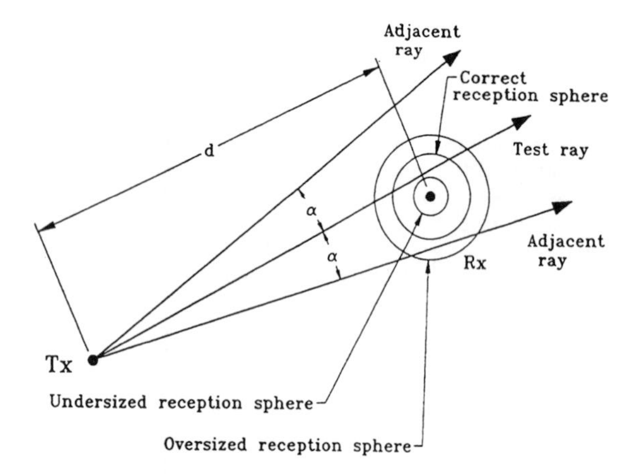

Fig. 7. 2-D view of the reception sphere. The total ray path length is d producing a reception sphere radius of $\alpha d/\sqrt{3}$ (from [18]).

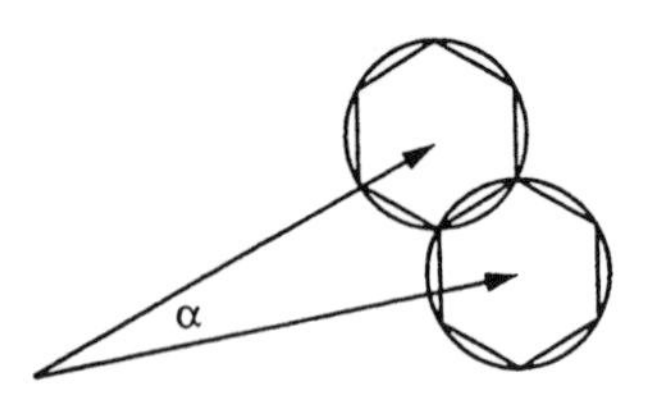

Fig. 8. Illustration of assumed wavefront for the purposes of the reception sphere. There is slight overlap between adjacent rays.

Recall Fig. 4 where α is the angular separation between rays at the source. In order to make sure that the specular point is not missed, the ray wavefront is considered as a circle circumscribed about the hexagonal (or pentagonal) wavefront shape. Fig. 8 shows two adjacent rays and their respective wavefronts. The angular radius of the circumscribed circle is $\alpha/\sqrt{3}$. Although it is possible to receive two specular points with this approach, adjacent rays that are received by overlapping reception spheres are eliminated during the processing of the raw ray tracing output. The use of a reception sphere requires that the rays must be launched such that each ray is separated from neighboring rays by a nearly constant angle, α. If nearly uniform separation is not maintained, the test ray will not be separated from adjacent rays by similar angles and the reception sphere loses its physical significance. All rays that are received are written to a raw ray tracing output file.

IV. Comparison of Measured and Predicted Propagation

A. Overview

The validation of any propagation prediction model *must* include comparison with measured data. For ray tracing methods that relate the propagation to specific objects in the physical radio channel, this comparison must include both multipath component amplitudes and arrival times in order to quantify the accuracy. This comparison must also be on a specific location-by-location basis to ensure applicability of the propagation models in a wide range of environments. The measured data are presented here as power delay profiles. The resolution is determined by the pulse width of the measurement system. Power delay profiles are computed from the predicted individual multipath components in order to compare measured and predicted propagation data.

Two different methods are used here to compare the measured and predicted propagation data. First, measured and predicted power delay profiles are compared on a location-by-location basis as a function of excess delay. An error function is defined to quantify the "difference" between the two curves. The second method of comparison is a location-by-location comparison of propagation characteristics such as path loss, COST 207 delay interval, and rms delay spread that are computed from the predicted and measured power delay profiles. Both of these methods compare measured and predicted propagation data at specific locations in a manner that incorporates both the amplitudes and arrival times of individual multipath components. Previous work in ray tracing prediction methods in buildings have not been validated by such an extensive comparison.

In [21] and [22], ray tracing was used to predict coverage of a radio system in a building, but no measured data were available for comparison. Comparison of average signal strength (path loss) as in [23] is a necessary measure of the model accuracy, but is not sufficient to determine the model's ability to determine individual multipath components, since there are many possible combinations that can predict the same average signal strength as the measured data. Only narrow band signal strength measurements were available for comparison in [23], and the work contained therein is the first to verify a ray tracing model for propagation prediction with measured data. Similarly, the multipath component arrival times must be compared for specific locations of measurements and predictions. In [26], the cumulative distribution functions of measured and predicted rms delay spreads were computed for a particular set of rooms inside a building. However, the relationship between the measured and predicted rms delay spreads was not determined, and the power delay profiles at individual locations were not compared. In addition, rms delay spread is not necessarily a good measure of a model's ability to predict multipath components accurately since two quite different power delay profiles can have the same rms delay spread. Hence, a comparison criteria that incorporates the multipath component arrival times in a power delay profile is required to determine the accuracy of a ray tracing propagation prediction model.

Here, a location-by-location power delay profile comparison and a path loss and delay interval comparison are used to quantify the prediction accuracy. Rms delay spread results are also presented, since this is a common measure of the time dispersion of a radio channel.

Because the exact building material properties are unknown, a method for minimizing the mean square error between measured and predicted power delay profiles as a function of excess delay is developed. The predicted power delay profiles are modified by implementing different amplitude and phase models for the reflection and transmission coefficients. The reflection and transmission coefficients are varied by changing the dielectric properties of the building materials used for prediction. The reflection and transmission coefficient models that give the minimum mean square error over an ensemble of measurement locations are used to determine the predicted power delay profiles presented here.

Measured and predicted power delay profiles are compared at 4.0 GHz for two buildings at 1.3 GHz in one of the buildings. This provides a qualitative comparison between measured and predicted power delay profiles for a variety of measurement locations with different surroundings. Finally, a location-by-location comparison of the propagation parameters determined from measured and predicted power delay profiles is given.

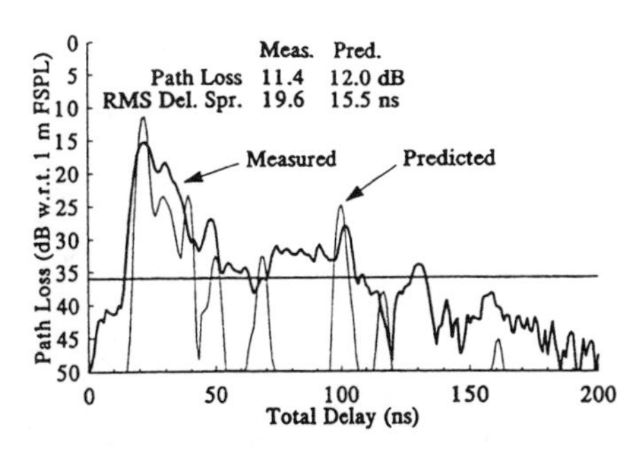

Fig. 9. Measured and predicted power delay profiles as a function of excess delay measured on the second floor of Whittemore Hall on the Virginia Tech campus.

B. Comparison Criteria and Error Function Definition

Consider a measured and a predicted power delay profile as shown in Fig. 9. Some measure of the "difference" between the two curves is desirable. A quantitative measure of the error that incorporates both the amplitudes and arrival times of the individual multipath components is a useful measure of the "difference" between measured and predicted power delay profiles. The measure used is the mean square error between the two curves on a pointwise basis at one nanosecond intervals. The error for one location is given in (4) where $M(\tau_i)$ and $P(\tau_i)$ are the amplitudes of the measured and predicted power delay profiles, respectively, sampled at time τ_i.

$$E = \sum_i (M(\tau_i) - P(\tau_i))^2. \tag{4}$$

The error between the two curves on a one nanosecond sample-by-sample basis is computed for all samples that have an amplitude above the received power threshold for either the measured or predicted power delay profile. These values are computed as the difference in decibels. Samples that are below the threshold for both measured and predicted profiles are not included in the error calculation. The threshold was determined to be the power level approximately two decibels above the noise floor as determined from the measured power delay profile on a location-by-location basis. The threshold was chosen to provide the largest possible dynamic range without allowing noise to be considered as multipath. When the sample at a constant time delay is above the threshold for one profile and below the threshold for the other, the error is considered to be the difference between the amplitude above the threshold and the threshold value. For each comparison of measured and predicted power delay profiles, an error curve that is the difference between the measured and predicted power delay profiles as a function of excess delay is computed. The best fit between measured and predicted power delay profiles minimizes the total area under all of the squared error curves. The difference between the measured and predicted power delay profiles as a function of excess delay is computed for each combination of transmitter and receiver location as indicated in (4).

C. Optimization of Effective Building Material Properties

Optimization Method: Because the exact building material properties are often unknown, a method for minimizing the mean square error between measured and predicted power delay profiles as a function of excess delay is developed. The ray tracing program determines the arrival times and path loss distance dependence of each received multipath component from the site-specific building database. The only variable that can be used to change the predicted power delay profile is the building material properties. The predicted power delay profiles are modified by implementing different amplitude and phase models for the reflection and transmission coefficients. The reflection and transmission coefficient models that give the minimum mean square error for all excess delay are then used to determine the predicted power delay profiles for 45 locations in two different buildings. The mean square error function is used for optimization since this representation incorporates the most information about the amplitudes and time delays of individual multipath components. The channel parameters, such as path loss and rms delay spread, are not used since the measured and predicted values of these parameters could be quite close even though there are major differences in the power delay profiles.

Each wall is modeled as a homogeneous infinitely thin flat planar surface infinite in extent with a constant relative dielectric constant and conductivity. In addition, for the data presented here, all walls in a building are considered to be identical. Ideally, the material properties of each individual wall represented in the database would be known in advance. However, the dielectric properties of many common building materials are, as yet, unknown. Also, the building walls are rarely (if ever) homogeneous, and are never infinite in extent or infinitely thin. Since the material properties are unknown, an alternate method is used to determine the material properties that should be used in a particular building to predict the power delay profiles. The ray tracing propagation model uses the Fresnel plane wave reflection and transmission coefficients for a single dielectric interface to determine the amplitudes of individual multipath components. Hence, the approach is to determine the *effective* material properties for the walls in the buildings under consideration. These *effective* material properties are the material properties that, when input into the Fresnel reflection and transmission coefficient

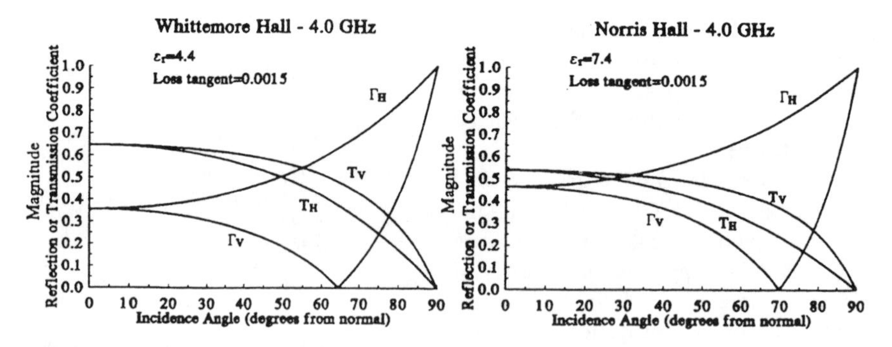

Fig. 10. Magnitude of the Fresnel transmission and reflection coefficients at 4.0 GHz for the *effective* building material properties of $\varepsilon_r = 4.4$ in Whittemore and $\varepsilon_r = 7.4$ in Norris.

equations for a single dielectric boundary, give the equivalent amplitude (in a minimum mean square error sense) of the reflections and transmissions measured in the real-world, finite-size, nonhomogeneous environment. Thus, the building material properties are optimized to minimize the area under the mean square error curve.

These *effective* material properties are computed for each combination of transmitter and receiver location in a building. The *effective* material properties are also computed for an ensemble of measurement locations at the same frequency in the same building. The mean square error is minimized for several different combinations of transmitter and receiver location at once. This provides overall "best-fit" values of the *effective* material properties for the particular ensemble when all walls of the same type are considered identical and homogeneous.

Optimization Results: Optimization was performed for three different sets of measured data. Measurements were made in Whittemore Hall at 1.3 GHz and 4.0 GHz, and in Norris Hall at 4.0 GHz. Both buildings are on the Virginia Tech campus. The measurements are described in detail in Section IV-D. Whittemore Hall is an office building with offices and classrooms that are located along a central hallway that snakes throughout the building on the second floor. Most walls are made of drywall mounted on metal studs. These walls separate individual offices from one another and the hallway. The walls that surround the stairwells at each end of the building are cinderblock and poured concrete. For the purposes of building material optimization, all walls are considered identical. Norris Hall is an older building and is constructed out of cinderblock walls. Offices are located along an L-shaped hallway. At the end of one hallway, there is a large metal door to a laboratory. This door is modeled as a perfect electric conductor with a reflection coefficient of -1 regardless of incidence angle. All of the cinderblock walls are modeled as a homogeneous dielectric.

The optimization was performed over a range of relative permittivities of 1 to 10.8. A loss tangent ($\sigma/\omega\varepsilon_0$) of 15×10^{-4} was included to incorporate a slight loss for each wall regardless of material. In Whittemore Hall, the dielectric constant that minimizes the squared error functions is $\varepsilon_r = 4.4$ at both 1.3 GHz and 4.0 GHz. These values were determined independently for the two frequencies. In Norris Hall, the best-fit *effective* dielectric constant is $\varepsilon_r = 7.4$. The resultant Fresnel reflection and transmission coefficients as a function of incidence angle at 4.0 GHz are shown in Fig. 10. These *effective* material properties were determined for each specific combination of measurement locations in the two buildings at the two frequencies. The resultant values of $\varepsilon_r = 4.4$ and 7.4 fall between the relative dielectric constant values of 2.4 for plicene cement and 8.8 for marble [36]. The values of the material properties of a concrete wall were derived from propagation measurements at 1.2 GHz in [37]. The reflection coefficients were found to vary between 0.33 and 0.4, and the relative dielectric constant was between 3.9 and 5.4 in [37]. Hence, the values determined here for the *effective* relative dielectric constant seem reasonable. The Fresnel reflection and transmission coefficients are not sensitive to small changes in the dielectric constant so that this ray tracing method may be used with confidence even when the optimal effective building materials are unknown.

D. Comparison of Measured and Predicted Power Delay Profiles

The measured and predicted power delay profiles for different measurement locations at the two different frequencies and two different buildings are now compared and discussed for both line-of-sight and obstructed measurement locations. Both the specular ray tracing and the uniform geometrical theory of diffraction were included in the model.

Whittemore Hall—1.3 GHz: The floor plan of Whittemore Hall is shown in Fig. 11. The transmitter was located at the intersection of two hallways, and is indicated by the letters "Tx" in the figure. The different receiver locations are indicated by the letters "A" through "I." These measurements were made with a direct pulse measurement system at 1.3 GHz with a 4 ns rms pulse width resolution and approximately 1.8 m high antennas. Both the transmitter and receiver were stationary during measurements and each power delay profile represents an instantaneous "snapshot" of the channel. The measurements are described in more detail in [38], [39].

The measured and predicted power delay profiles for the line-of-sight receiver at location G are shown in Fig. 12. Notice that the path loss error is less than 2 dB and the rms delay spreads are quite similar. The amplitudes and arrival times of significant individual multipath components

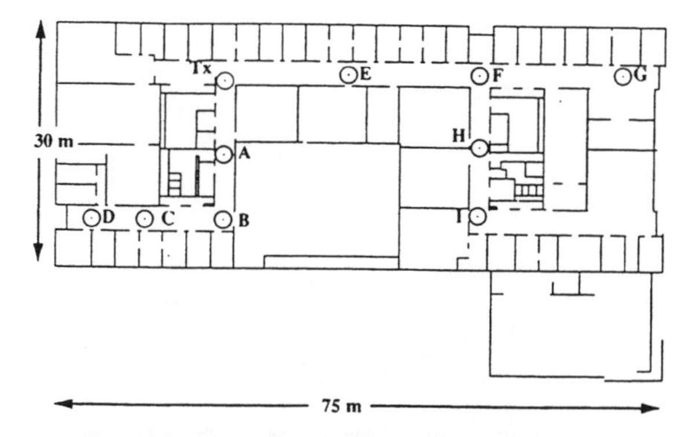

Fig. 11. Measurement locations on the second floor of Whittemore Hall at 1.3 GHz. The transmitter is indicated by a "Tx," and the receiver locations are indicated by the letters "A" through "I."

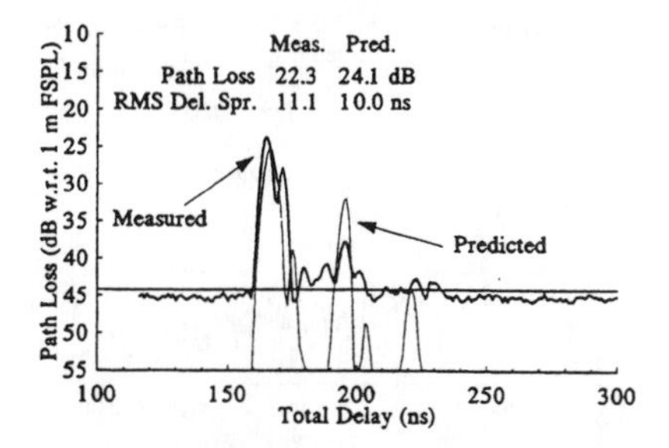

Fig. 12. Measured and predicted power delay profiles at location G in Whittemore Hall at 1.3 GHz. The path loss error is 1.8 dB and the rms delay spreads are nearly identical.

are predicted. This is a "simple" geometry as there are only a few dominant multipath components.

Whittemore Hall—4.0 GHz: Additional measurements were made on the second floor of Whittemore Hall at 4.0 GHz with a spread spectrum channel sounder. The channel sounder chip rate was 240 MHz which yielded a time resolution of about 8 ns [35]. Omnidirectional biconical antennas were used at the transmitter and receiver. The transmitter location was the same as for the 1.3 GHz measurements. The receiver was moved to different locations in the hallway throughout the same measurement area. The transmitter and receiver measurement locations are indicated in Fig. 13. For measurements at 4.0 GHz, the transmitter remained stationary, and the receiver recorded five instantaneous power delay profiles over a path length of about one meter at each location. Each instantaneous power delay profile was recorded while the receiver was stationary, and the five power delay profiles were averaged to provide a spatial average that eliminates multipath fading within one pulse period [11].

Fig. 14 shows the line-of-sight receiver measurement location W251. While the separation between the transmitter and receiver is only 7 m, there are a number of significant multipath components. The path loss error is less than one decibel and the rms delay spreads are quite similar. There is good agreement between the measured and predicted power delay profiles as the prediction includes the amplitudes and arrival times of many significant multipath components. The measured and predicted path loss at location W262 is on the order of 15 dB greater than the free space path loss at one meter for this

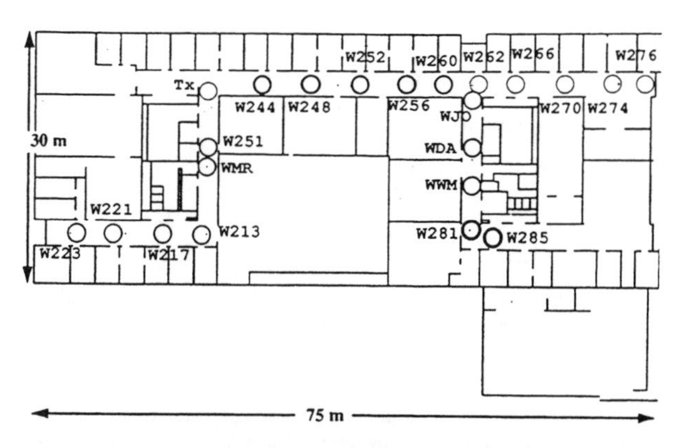

Fig. 13. Building blueprint of the second floor of Whittemore Hall that shows where power delay profile measurements were made at 4.0 GHz with a spread spectrum channel sounder with an 8 ns pulse resolution.

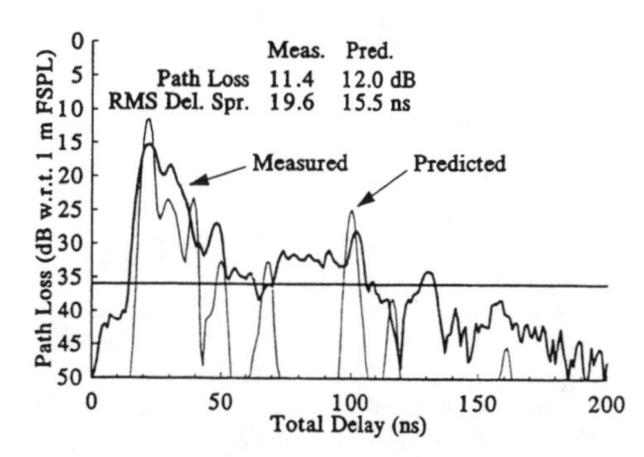

Fig. 14. Measured and predicted power delay profiles at receiver measurement location W251 in Whittemore Hall at 4.0 GHz. The path loss error is less than 1.0 dB and the rms delay spreads are quite similar.

line-of-sight location. Now consider the obstructed locations WJO and WDA. The comparison of measured and predicted power delay profiles is given in Figs. 15 and 16, respectively. The significance of these locations is that as the receiver is moved from a line-of-sight position (W262) to an obstructed position (WJO), the predicted power delay profiles accurately track the 16 dB change in path loss. The change in path loss from 31 dB to 38 dB, relative to free space path loss at one meter as the receiver moves from WJO to WDA, is also predicted. Hence, major changes in propagation characteristics are correctly tracked as the receiver undergoes significant changes in the surroundings.

Norris Hall—4.0 GHz: The locations of the transmitters and receivers for 4.0 GHz measurements in Norris Hall are shown in Fig. 17. Some measurements were made with the transmitter at location "J" where the two hallways intersect, and others were made with the transmitter at location "B" in one of the hallways. The receiver was located at the locations indicated by the letters "A" through "J." The measurement equipment and procedure were identical to the 4.0 GHz measurements made in Whittemore Hall.

Consider the transmitter at location "J" and the receiver at location "H." The measured and predicted power delay profiles at receiver location "H" are shown in Fig. 18. For this line-of-sight location, the amplitudes and arrival times of significant multipath components are predicted and the path

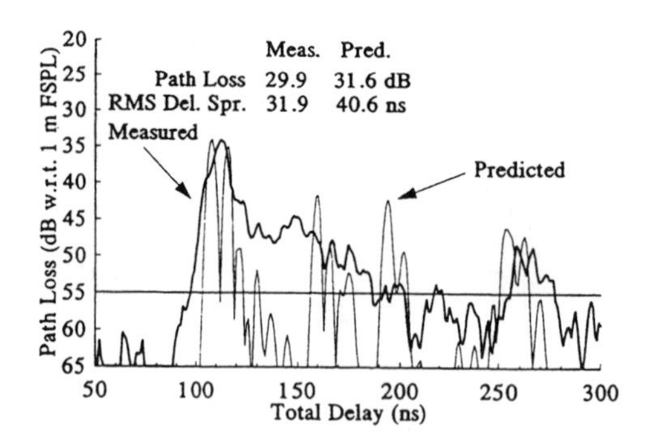

Fig. 15. Comparison of measured and predicted power delay profiles in Whittemore Hall at 4.0 GHz at receiver location WJO.

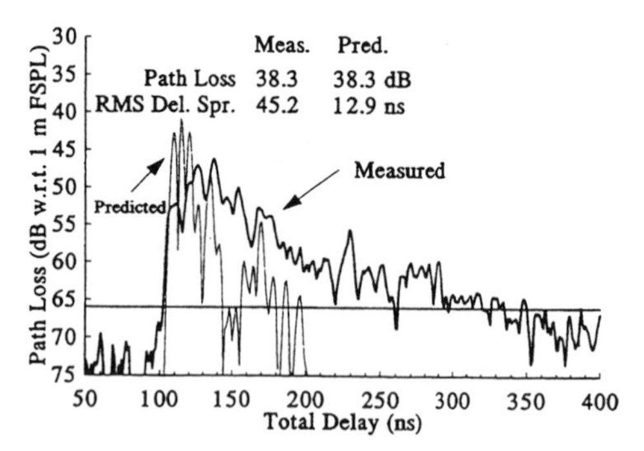

Fig. 16. Measured and predicted power delay profiles for the obstructed location WDA in Whittemore Hall at 4.0 GHz.

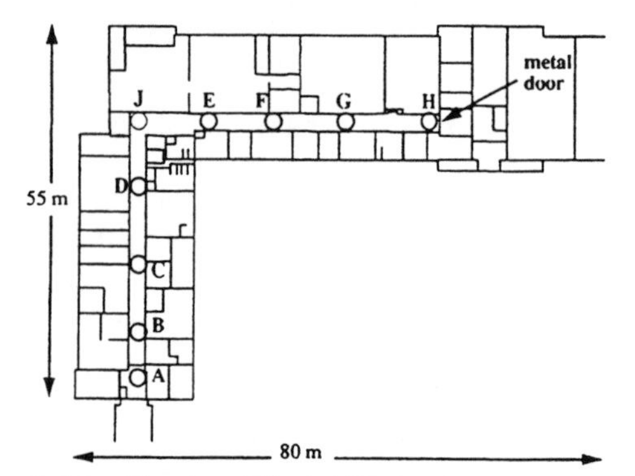

Fig. 17. Building blueprint of Norris Hall that shows the transmitter and receiver locations where power delay profiles were recorded at 4.0 GHz. At the end of the hallway near measurement location "H" was a large metal door.

loss errors are less than 4.1 dB. At location "H," the rms delay spread varies considerably between measured and predicted values, but this parameter is sensitive to small changes in the power delay profile, such as the 8 dB difference in the first arriving multipath component. The predicted power delay profile accurately predicts the arrival times of the multipath components and predicts a path loss within 4.1 dB of the measured value.

Consider now the measurement location with the transmitter moved from location "J" to location "B" with the receiver remaining at location "H" at the end of the hallway next to the metal door. The measurement topography changes from line-of-sight to heavily obstructed. In fact, the direct path

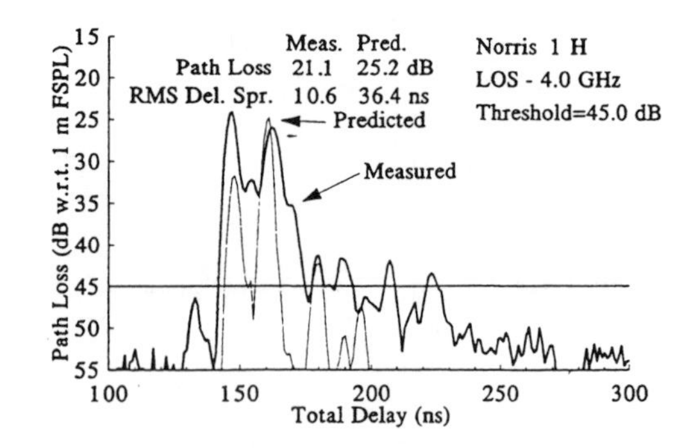

Fig. 18. Measured and predicted power delay profiles for transmitter location "J" and receiver location "H" in Norris Hall at 4.0 GHz.

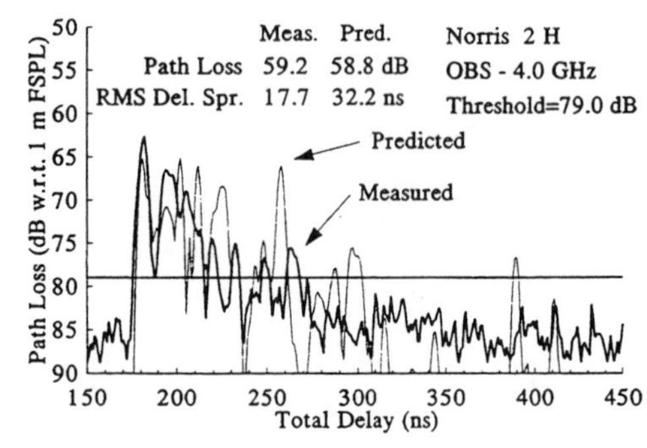

Fig. 19. Measured and predicted power delay profiles in Norris Hall at 4.0 GHz. The transmitter was at location "B," and the receiver was at location "H."

passes outside the building and reenters in the other wing of the building. The measured and predicted power delay profiles are shown in Fig. 19. Notice that the path loss error is less than one decibel and many of the amplitudes and arrival times of significant multipath components are predicted. The significance of this prediction is that as the transmitter was moved from a line-of-sight location to an obstructed location, the predicted path loss remained within several decibels of the measured path loss over a path loss change of 30 dB. Thus, changes in the propagation characteristics as a function of major changes in the surroundings of the transmitter and receiver have been accurately modeled with the propagation prediction methods described in Section III.

E. Discussion of Measured and Predicted Channel Parameters

Although path loss, rms delay spread, and delay interval are only indirectly responsible for bit error performance in wireless modems operating inside buildings, they are important channel parameters that characterize the propagation channel. Hence, a quantization of the accuracy of the model in these areas is required before applying them to wireless system design. The parameters are computed from the measured and predicted power delay profiles. In this work, the predicted power delay profiles were computed from a building model that assumes a single *effective* building material, which is determined by optimizing over the ensemble of locations in a given building at a given frequency. For each building and frequency, a loss tangent of 0.0015 was assumed and the

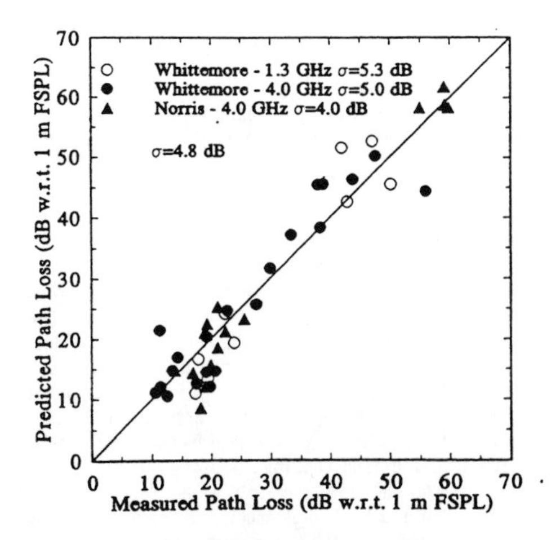

Fig. 20. Scatter plot of measured and predicted path loss for the three sets of measurements in Whittemore Hall at 1.3 and 4.0 GHz and in Norris Hall at 4.0 GHz. The standard deviation of path loss prediction error is 4.8 dB over all measurement locations in both buildings and both frequencies.

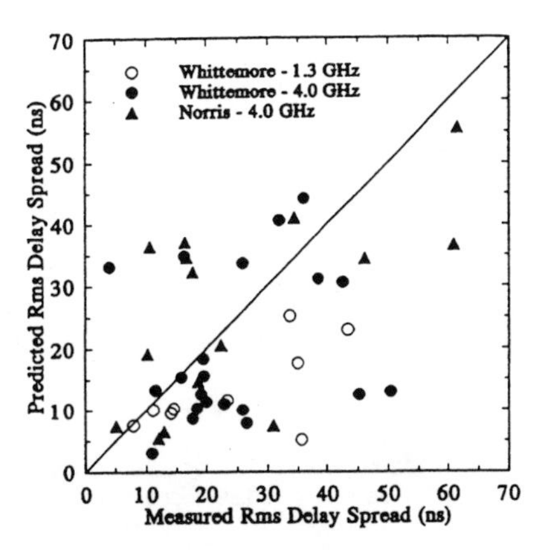

Fig. 21. Scatter plot of measured and predicted rms delay spread for all three measurement combinations.

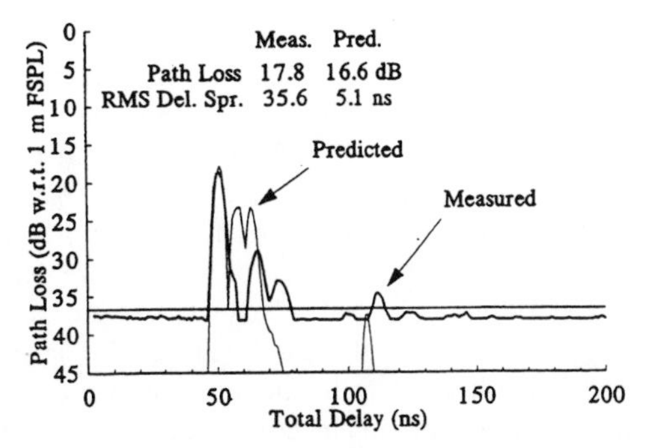

Fig. 22. Measured and predicted power delay profiles for location WHL1EVV in Whittemore Hall. Notice that the predicted component at 110 ns total delay is below the threshold leading to a large discrepancy between the measured and predicted rms delay spread. However, path loss prediction is within 1.2 dB.

relative permittivity was 4.4 for both sets of predictions in Whittemore Hall, and was 7.4 for predictions in Norris Hall. Fig. 20 shows a scatter plot of measured and predicted path loss for all measurement locations in both buildings, and at both 1.3 GHz and 4.0 GHz. The standard deviation of path loss error is 4.8 dB for all measurement locations. This standard deviation is obtained over a dynamic range of path loss that ranges from 10 dB to 60 dB above the path loss at a one meter reference distance. The individual path loss scatter points are clustered about the diagonal line that indicates where measured and predicted path loss are identical. This result shows the validity of the ray tracing propagation prediction model for path loss coverage prediction in two different buildings at two different frequencies. The model accurately tracks large changes in path loss as a function of the surroundings of the transmitter and receiver.

A scatter plot of the rms delay spread for the measured and predicted locations is shown in Fig. 21. The rms delay spread calculations show large errors at some locations. However, rms delay spread can be extremely sensitive to small changes in the power delay profile. Only samples above the threshold were used to compute the statistics. If the predicted component is below the threshold, it is not counted, even though the difference between the measured and predicted power delay profiles may be small. For example, examine the measured and predicted profiles for location WHL1EVV in Whittemore Hall shown in Fig. 22. The rms delay spread error is large, but in Fig. 22, it can be clearly seen that several multipath components are predicted with amplitudes below the threshold that must be used because of the dynamic range of the measurements. Hence, the difference between measured and predicted power delay profiles is not as large as the difference in rms delay spread indicates.

Another metric for time dispersion is the COST 207 delay interval statistic [40]. The delay interval is a measure of the span of excess delay that contains a certain portion of the total received energy. This is a parameter that considers the arrival time and amplitude of the multipath energy that is less sensitive to small changes in power delay profiles than the rms delay spread. Fig. 23 gives the scatter plot of delay interval computed for 90% of the received energy inside the delay interval window. Most of the delay intervals are less than 100 ns and are clustered about the diagonal line that indicates where measured and predicted delay interval window sizes are equal. There are several locations where the measured and predicted delay intervals differ significantly, but, in general, the 90% delay interval scatter plot is more clustered than the rms delay spread scatter plot shown in Fig. 21.

V. Conclusion

This paper has detailed the results of the ray tracing site-specific propagation model. The model incorporates both brute force ray tracing based upon geometrical optics, and diffraction using the uniform geometrical theory of diffraction. Previous work in propagation prediction has not focused on comparison of measured and predicted propagation data that includes the multipath arrival time and a specific location-by-location comparison of the measured and predicted data. Hence, a method for quantifying the difference between measured and predicted power delay profiles has been developed. The area under the square error function as a function of excess delay has been minimized to provide predicted power delay profiles

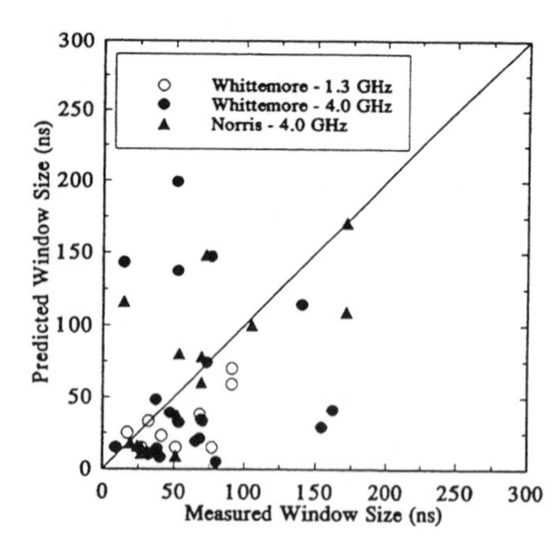

Fig. 23. Scatter plot of the measured and predicted COST 207 delay interval for 90% of the received energy.

and determine a single *effective* dielectric constant that describes all walls in a building. The *effective* building material properties were determined for each combination of building and frequency. Optimization of the error between measured and predicted power delay profiles sampled at one nanosecond intervals revealed that the *effective* relative dielectric constants are 4.4 at both 1.3 and 4.0 GHz in Whittemore Hall, and 7.4 at 4.0 GHz in Norris Hall. These values are similar to the values of other materials that could be used in building construction. Further research and more measurements are required to determine the applicability of an *effective* dielectric constant and to determine the appropriate values to be used in a site-specific propagation prediction model.

Qualitative comparison of power delay profiles showed that the amplitudes and arrival times of individual multipath components were predicted in both line-of-sight and obstructed topographies. In both buildings, as a receiver was moved from a line-of-sight location to an obstructed location, the predicted power delay profiles accurately tracked the change in path loss. This change was 16 dB in one building and 30 dB in the other. Overall, the standard deviation of the path loss error is 4.8 dB over a dynamic range of 50 dB in two buildings and two frequency bands. Comparison of the rms delay spread and COST 207 delay interval indicates that in many cases the time delay parameters of measured and predicted power delay profiles are similar.

ACKNOWLEDGMENT

The authors thank Kurt Schaubach for many stimulating discussions about ray tracing and the reviewers for their helpful comments.

REFERENCES

[1] R. J. C. Bultitude, S. A. Mahmoud and W. A. Sullivan, "A comparison of indoor radio propagation characteristics at 910 MHz and 1.75 GHz," *IEEE J. Select. Areas Commun.*, vol. 7, no. 1, pp. 20–30, Jan. 1989.

[2] D. M. J. Devasirvatham, R. R. Murray and C. Banerjee, "Time delay spread measurements at 850 MHz and 1.7 GHz inside a metropolitan office building," *Electron. Lett.*, vol. 25, no. 3, pp. 194–195, Feb. 1989.

[3] D. M. J. Devasirvatham, M. J. Krain and D. A. Rappaport, "Radio propagation measurements at 850 MHz, 1.7 GHz and 4.0 GHz inside two dissimilar office buildings," *Electron. Lett.*, vol. 26, no. 7, pp. 445–447, 1990.

[4] D. M. J. Devasirvatham, C. Banerjee, M. J. Krain and D. A. Rappaport, "Multifrequency radiowave propagation measurements in the portable environment," presented at the IEEE ICC '90, Atlanta, GA, Apr. 16–19, 1990.

[5] D. M. J. Devasirvatham, "Multifrequency propagation measurements and models in a large metropolitan commercial building for personal communications," in *IEEE Symp. Pers. Indoor Mobile Commun.*, London, Sept. 1991, pp. 98–103.

[6] D. M. J. Devasirvatham, C. Banerjee, R. R. Murray and D. A. Rappaport, "Four-frequency radiowave propagation measurements of the indoor environment in a large metropolitan commercial building," in *IEEE Globecom '91*, Phoenix, AZ, Dec. 1991, pp. 1282–1286.

[7] S. E. Alexander, "Characterising buildings for propagation at 900 MHz," *Electron. Lett.*, vol. 19, no. 29, p. 860, Sept. 29, 1983.

[8] D. C. Cox, R. R. Murray and A. W. Norris, "800 MHz attenuation measured in and around suburban houses," *AT&T Bell Labs Tech. J.*, vol. 63, pp. 921–954, July/Aug. 1984.

[9] A. A. M. Saleh and R. A. Valenzuela, "A statistical model for indoor multipath propagation," *IEEE J. Select. Areas Commun.*, vol. SAC-5, no. 2, pp. 128–137, Feb. 1987.

[10] T. S. Rappaport and C. D. McGillem, "UHF fading in factories," *IEEE J. Select. Areas Commun.*, vol. 7, no. 1, pp. 40–48, Jan. 1989.

[11] T. S. Rappaport, "Characterization of UHF multipath radio channels in factory buildings," *IEEE Trans. Antenn. Propagat.*, vol. 37, no. 8, pp. 1058–1069, Aug. 1989.

[12] A. S. Glassner, Ed., *An Introduction to Ray Tracing*. New York: Academic Press, 1989.

[13] R. T. Stevens, *Fractal Programming and Ray Tracing with C++*. Redwood City, CA: M&T Books, 1990.

[14] F. Ikegami, T. Takeuchi and S. Yoshida, "Theoretical prediction of mean field strength for urban mobile radio," *IEEE Trans. Antenn. Propagat.*, vol. 39, pp. 299–302, Mar. 1991.

[15] J. P. Rossi, J. C. Bic, A. J. Levy, Y. Gabillet and M. Rosen, "A ray launching method for radio-mobile propagation in urban area," in *IEEE Antenn. Propagat. Sympos.*, London, Ontario, Canada, June 1991, pp. 1540–1543.

[16] T. Takeuchi, M. Sako and S. Yoshida, "Multipath delay prediction on a workstation for urban mobile radio environment," in *IEEE Globecom '91*, Phoenix, AZ, Dec. 1991, pp. 1308–1312.

[17] J. P. Rossi and A. J. Levy, "A ray model for decimetric radiowave propagation in an urban area," *Radio Science*, vol. 27, no. 6, pp. 971–979, Nov.-Dec. 1992.

[18] K. R. Schaubach, N. J. Davis, IV, and T. S. Rappaport, "A ray tracing method for predicting path loss and delay spread in microcellular environments," in *42nd IEEE Veh. Technol. Conf.*, Denver, pp. 932–935, May 1992.

[19] B. Bisceglia, G. Franceschetti, G. Mazzarella, I. M. Pinto and C. Savarese, "Symbolic code approach to GTD ray tracing," *IEEE Trans. Antenn. Propagat.*, vol. AP-36, no. 10, pp. 1492–1495, Oct. 1988.

[20] P. F. Driessen, "Development of a propagation model in the 20-60 GHz band for wireless indoor communications," in *IEEE Pacific Rim Conf.*, Victoria, B.C., Canada, pp. 59–62, May 1991.

[21] J. W. McKown and R. L. Hamilton, "Ray tracing as a design tool for radio networks," *IEEE Network Mag.*, vol. 5, no. 6, pp. 27–30, Nov. 1991.

[22] P. F. Driessen *et al.*, "Ray model of indoor propagation," in *Proc. 2nd Annu. Virginia Tech Symp. on Wireless Pers. Commun.*, Blacksburg, VA, June 17–19, 1992, pp. 17-1–17-12.

[23] W. Honcharenko, H. L. Bertoni, J. Dailing, J. Qian and H. D. Yee, "Mechanisms governing UHF propagation on single floors in modern office buildings," *IEEE Trans. Veh. Technol.*, vol. 41, no. 4, pp. 496–504, Nov. 1992.

[24] J. Kiang, "Geometrical ray tracing approach for indoor wave propagation in a corridor," in *1st Int. Conf. Univ. Pers. Commun.*, Dallas, TX, Sept. 29–Oct. 2, 1992, pp. 106–111.

[25] D. I. Laurenson, A. U. H. Sheikh and S. McLaughlin, "Characterisation of the indoor mobile radio channel using a ray tracing technique," in *1992 IEEE Int. Conf. Select. Topics in Wireless Commun.*, Vancouver, B.C., Canada, June 1992, pp. 65–68.

[26] M. C. Lawton and J. P. McGeehan, "A deterministic ray launching algorithm for the prediction of radio channel characteristics in small cell environments," submitted to *IEEE Trans. Veh. Technol.*, July 1992.

[27] H. Kenner, *Geodesic Math and How to Use It*. Berkeley, CA: Univ. of California Press, 1976.

[28] M. J. Wenninger, *Spherical Models*. New York: Cambridge Univ. Press, 1979.

[29] M. Born and E. Wolf, *Principles of Optics*. New York: Pergammon Press, 1965.
[30] R. G. Kouyoumjian and P. H. Pathak, "A uniform geometrical theory of diffraction for an edge in a perfectly conducting surface," *Proc. IEEE*, vol. 62, pp. 1448–1461, Nov. 1974.
[31] W. D. Burnside, "High frequency scattering by a thin lossless dielectric slab," *IEEE Trans. Antenn. Propagat.*, vol. AP-31, no. 1, pp. 104–110, Jan. 1983.
[32] R. J. Luebbers, "Finite conductivity uniform GTD versus knife edge diffraction in prediction of propagation path loss," *IEEE Trans. Antenn. Propagat.*, vol. AP-32, no. 1, pp. 70–76, Jan. 1984.
[33] R. J. Luebbers, "Propagation prediction for hilly terrain using GTD wedge diffraction," *IEEE Trans. Antenn. Propagat.*, vol. AP-32, no. 9, pp. 951–955, Sept. 1984.
[34] S. Y. Seidel, "Site-specific propagation prediction for wireless in-building personal communication system design," Ph.D. dissertation, Dep. Elec. Eng., Virginia Polytech. Inst. and State Univ., Jan. 1993.
[35] S. Y. Seidel and T. S. Rappaport, "A ray tracing technique to predict path loss and delay spread inside buildings," in *IEEE Globecom '92*, Orlando, FL, Dec. 1992, pp. 649–653.
[36] A. R. Von Hippel, Ed., *Dielectric Materials and Applications*. Cambridge, MA: MIT Press, 1954.
[37] J. Horikoshi, K. Tanaka and T. Morinaga, "1.2 GHz band wave propagation measurements in concrete building for indoor radio communications," *IEEE Trans. Veh. Technol.*, vol. VT-35, no. 4, pp. 146–152, Nov. 1986.
[38] D. A. Hawbaker, "Indoor wide band radio wave propagation measurements and models at 1.3 GHz and 4.0 GHz," M.S. thesis, Dep. Elec. Eng., Virginia Polytech. Inst. and State Univ., May 15, 1991.
[39] T. S. Rappaport and D. A. Hawbaker, "Wide band microwave propagation parameters using circular and linear polarized antennas for indoor wireless channels," *IEEE Trans. Commun.*, vol. 40, no. 2, pp. 1–6, Feb. 1992.
[40] J. P. deWeck, P. Merki and R. Lorenz, "Power delay profiles measured in mountainous terrain," in *38th IEEE Veh. Technol. Conf.*, Philadelphia, PA, pp. 105–111, June 15, 1988.

Scott Y. Seidel (S'89–M'93), for photograph and biography, see p. 498 of the August issue of the TRANSACTIONS ON VEHICULAR TECHNOLOGY.

Theodore S. Rappaport (S'85–M'87–SM'91), for photograph and biography, see p. 498 of the August issue of this TRANSACTIONS.

Chapter 4

SPREAD SPECTRUM MULTIPLE ACCESS

This chapter provides in-depth coverage of spread spectrum multiple access, including advanced CDMA noise cancellation techniques that are the subject of intense research and development.

Reprinted from *IEEE Transactions on Information Theory*, Vol. IT-32, No. 1, Jan. 1986, pp. 85-96.

Minimum Probability of Error for Asynchronous Gaussian Multiple-Access Channels

SERGIO VERDÚ, MEMBER, IEEE

Abstract—Consider a Gaussian multiple-access channel shared by K users who transmit asynchronously independent data streams by modulating a set of assigned signal waveforms. The uncoded probability of error achievable by optimum multiuser detectors is investigated. It is shown that the K-user maximum-likelihood sequence detector consists of a bank of single-user matched filters followed by a Viterbi algorithm whose complexity per binary decision is $O(2^K)$. The upper bound analysis of this detector follows an approach based on the decomposition of error sequences. The issues of convergence and tightness of the bounds are examined, and it is shown that the minimum multiuser error probability is equivalent in the low-noise region to that of a single-user system with reduced power. These results show that the proposed multiuser detectors afford important performance gains over conventional single-user systems, in which the signal constellation carries the entire burden of complexity required to achieve a given performance level.

I. Introduction

CONSIDER a Gaussian multiple-access channel shared by K users who modulate simultaneously and independently a set of assigned signal waveforms without maintaining any type of synchronism among them. The coherent K-user receiver commonly employed in practice consists of a bank of optimum single-user detectors operating independently (Fig. 1). Since in general the input to every threshold has an additive component of multiple-access interference (because of the cross correlation with the signals of the other users), the conventional receiver is not optimum in terms of error probability. However, if the designer is allowed to choose a signal constellation with large bandwidth (e.g., in direct-sequence spread-spectrum systems), then the cross correlations between the signals can be kept to a low level for all relative delays, and acceptable performance can be achieved. Nevertheless, if data demodulation is restricted to single-user detection systems, then the cross-correlation properties of the signal constellation carry the entire burden of complexity required to achieve a given performance level, and when the power of some of the interfering users is dominant, performance degradation is too severe. For this reason and because of the availability of computing devices with increased capabilities, there is recent interest in investigating the degree of performance improvement achievable with more sophisticated receivers and, in particular, with the minimum error probability detector.

Optimum multiuser detection of asynchronous signals is inherently a problem of sequence detection, that is, observation of the whole received waveform is required to produce a sufficient statistic for any symbol decision, and hence one-shot approaches (where the demodulation of each symbol takes into account the received signal only in the interval corresponding to that symbol) are suboptimal. The reason is that the observation of the complete intervals of the overlapping symbols of the other users gives additional information about the received signal in the bit interval in question, and since this reasoning can be repeated with the overlapping bits, no restriction of the whole observation interval is optimal for any bit decision. Furthermore, since the transmitted symbols are not independent conditioned on the received realization, decisions can be made according to two different optimality criteria, namely, selection of the sequence of symbols that maximizes the joint posterior distribution (maximum-likelihood sequence detection), or selection of the symbol sequence that maximizes the sequence of marginal posterior distributions (minimum-probability-of-error detection). Moreover, the simultaneous demodulation of all the active users in the multiple-access channel can be regarded as a problem of periodically time-varying intersymbol interference, because from the viewpoint of the coherent K-user detector, the observed process is equivalent to that of a single-user-to-single-user system where the sender transmits K symbols during each signal period by modulating one out of K waveforms in a round-robin fashion.

Earlier work on multiuser detection includes, in the case of synchronous users (which reduces to an m-ary hypothesis testing problem) the receivers of Horwood and Gagliardi [9] and Schneider [16], and, in the asynchronous case, the one-shot baseband detector obtained by Poor [12] in the two-user case, and the detectors proposed by Van Etten [18] and Schneider [16] for interference-channel models with vector observations. Results on the probability of error have been obtained only for the conventional single-user receiver ([8], [14], and the references therein).

In Section II we obtain a K-user maximum-likelihood sequence detector which consists of a bank of K single-user matched filters followed by a Viterbi forward dynamic

Manuscript received May 3, 1983; revised November 5, 1984. This work was supported by the U.S. Army Research Office under Contract DAAG-81-K-0062.

The author is with the Department of Electrical Engineering, Princeton University, Princeton, NJ 08544.

IEEE Log Number 8406098.

0018-9448/86/0100-0085$01.00 ©1986 IEEE

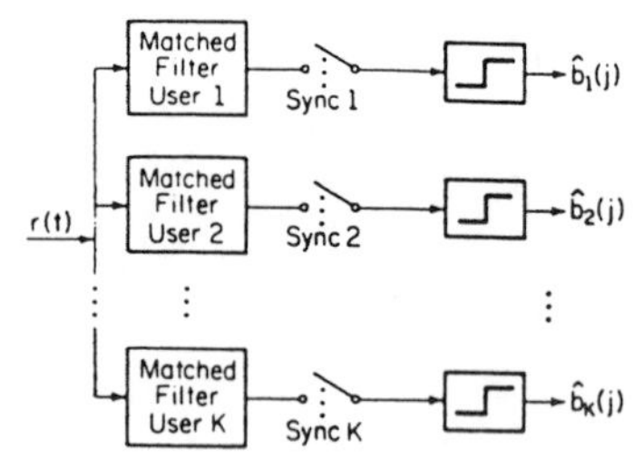

Fig. 1. Conventional multiuser detector.

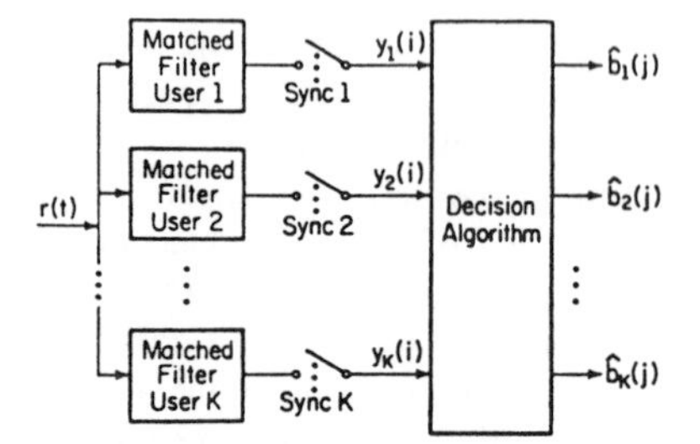

Fig. 2. Optimum K-user detector for asynchronous multiple-access Gaussian channel.

programming algorithm (Fig. 2) with 2^{K-1} states and $O(2^K)$ time complexity per bit (in the binary case). Section III is devoted to the analysis of the minimum uncoded bit error rate of multiuser detectors. This is achieved through various bounds that together provide tight approximations for all noise levels. A possible route to upper bound the error probability of the multiuser maximum-likelihood sequence detector is to generalize the approach taken by Forney in the intersymbol interference problem [3]. However, motivated by the more general structure of the multiuser problem, we introduce a different approach that, when applied to the intersymbol interference case, turns out to result in a bound that is tighter than the Forney bound. In Section IV several numerical examples illustrate the performance gains achieved by optimum multiuser detectors over conventional single-user systems.

II. Multiuser Maximum-Likelihood Sequence Detection

In this section we derive optimum decision rules for the following multiple-access model with additive linearly modulated signals in additive white Gaussian noise and scalar observations:

$$dr_t = S_t(b)\,dt + \sigma\,d\omega_t, \qquad t \in R \tag{1}$$

where

$$S_t(b) = \sum_{i=-M}^{M} \sum_{k=1}^{K} b_k(i) s_k(t - iT - \tau_k), \tag{2}$$

the symbol interval duration is equal to T (assumed to be the same for all users), $b = \{b(i) \in A_1 \times \cdots \times A_K,\ i = -M,\cdots, M\}$, and A_k, $s_k(t)$ $(= 0$ outside $[0, T])$, and $\tau_k \in [0, T)$ are the finite alphabet, the signal waveform, and the delay (modulo T with respect to an arbitrary reference), respectively, of the kth user, and ω_t is a standard Wiener process started at $t = -MT$. Without loss of generality, and for the sake of notational simplicity, we suppose that the users are numbered such that $0 \le \tau_1 \le \cdots \le \tau_K < T$. Note that even if all the transmitted symbols are assumed to be equiprobable and independent, there is not a unique optimality criterion due to the existence of several users. It is possible to select either the set of symbols that maximize the joint posterior distribution $P[b|\{r_t,\ t \in R\}]$ (globally optimum or maximum-likelihood sequence detection) or those that maximize the marginal posteriori distributions $P[b_k(i)|\{r_t,\ t \in R\}]$, $i = -M,\cdots, M$, $k = 1,\cdots, K$ (locally optimum or minimum-error-probability detection). It is shown later that the maximum-likelihood sequence detector can be implemented by a signal processing front end that produces a sequence of scalar sufficient statistics, followed by a dynamic programming decision algorithm of the forward (Viterbi) type. It can be shown [22] that the multiuser detector that minimizes the probability of error has the same structure, but it uses a backward–forward dynamic programming algorithm instead [21]. The computational complexity of the various decision algorithms will be measured and compared by their *time complexity per binary decision* (TCB), that is, the limit as $M \to \infty$ of the time required by the decision algorithm to select the optimum sequence divided by the number of transmitted bits.

Since all transmitted sequences of symbols are assumed to be equiprobable, the maximum-likelihood sequence detector selects the sequence that maximizes

$$P[\{r_t, t \in R\}|b] = C \exp(\Omega(b)/2\sigma^2) \tag{3}$$

where C is a positive scalar independent of b and

$$\Omega(b) = 2\int_{-\infty}^{\infty} S_t(b)\,dr_t - \int_{-\infty}^{\infty} S_t^2(b)\,dt. \tag{4}$$

Therefore, the maximum-likelihood sequence detector selects among the possible noise realizations the one with minimum energy. Using the definition (2), we can express the first term in the right-hand side of (4) as

$$\int_{-\infty}^{\infty} S_t(b)\,dr_t = \sum_{i=-M}^{M} b^T(i) y(i), \tag{5}$$

where $y_k(i)$ denotes the output of a matched filter for the ith symbol of the kth user, that is,

$$y_k(i) = \int_{\tau_k + iT}^{\tau_k + iT + T} s_k(t - iT - \tau_k)\,dr_t. \tag{6}$$

Hence, even though $y_k(i)$ is not a sufficient statistic for the detection of $b_k(i)$, (4) and (5) imply that the whole sequence of outputs of the bank of K matched filters y is a sufficient statistic for the selection of the most likely sequence b. This implies that the maximum-likelihood multiuser coherent detector consists of a front end of matched filters (one for each user) followed by a decision algorithm (Fig. 2), which selects the sequence b that maximizes (3) or, equivalently, (4). The efficient solution of this combinatorial optimization problem is the central issue in the derivation of the multiuser detector. The TCB of the exhaustive algorithm that computes (4) for all possible sequences has not only the inconvenient feature of being dependent on the block-size M, but it is so in an exponential way. Fortunately, $\|S(b)\|^2 = \int_{-\infty}^{\infty} S_t^2(b)\,dt$, the energy of the sequence b, has the right structure to result in decision algorithms with significantly better TCB. The key to the efficient maximization of $\Omega(b)$ lies in its sequential dependence on the symbols $b_k(i)$, which allows us to put it as a sum of terms that depend only on a few variables at a time.

Suppose that we can find a discrete-time system $x_{i+1} = f_i(x_i, u_i)$, with initial condition x_{i_0}; a transition-payoff function $\lambda_i(x_i, u_i)$; and a bijection between the set of transmitted sequences and a subset of control sequences $\{u_i, i = i_0, \cdots, i_f\}$ such that $\Omega(b) = \sum_{i=i_0}^{i_f} \lambda_i(x_i, u_i)$, subject to $x_{i+1} = f_i(x_i, u_i)$, x_{i_0}, and $b \leftrightarrow \{u_i, i = i_0, \cdots, i_f\}$. Then the maximization of $\Omega(b)$ is equivalent to a discrete-time deterministic control problem with additive cost and finite input and state spaces, and therefore it can be solved by the dynamic programming algorithm either in backward or in forward fashion. Although the decision delay is unbounded because optimum decisions cannot be made until all states share a common shortest subpath, a well-known advantage in real-time applications of the forward dynamic programming algorithm (the Viterbi algorithm) is that little degradation of performance occurs when the algorithm uses an adequately chosen fixed finite decision lag. It turns out that there is not a unique additive decomposition of the log-likelihood function $\Omega(b)$, resulting in decision algorithms with very different computational complexities. It can be shown that the Viterbi algorithm suggested by Schneider [16] has 4^K states and $O(8^K/K)$ time complexity per bit, while the decision algorithm of the multiuser detector in [20] has 2^K states and TCB = $O(4^K/K)$. By fully exploiting the sequential dependence of the log-likelihood function on the transmitted symbols, it is possible to obtain an optimum decision algorithm that exhibits a lower time complexity per bit than the foregoing. This is achieved by the additive decomposition of the log likelihood function given by the following result.

Proposition 1: Define the following matrix of signal crosscorrelations:

$$G_{ij} = \begin{cases} \int_{-\infty}^{\infty} s_{i+j}(t - \tau_{i+j}) s_j(t - \tau_j - T)\,dt, \\ \quad \text{if } i + j \le K \\ \int_{-\infty}^{\infty} s_{i+j-K}(t - \tau_{i+j-K}) s_j(t - \tau_j)\,dt, \\ \quad \text{if } i + j > K \end{cases} \tag{7}$$

for $i = 1, \cdots, K-1$ and $j = 1, \cdots, K$ (i.e., the entries of the column[1] G^j are the correlations of the signal of the jth user with the $K - 1$ preceding signals) and denote the received signal energies by $w_k = \int_0^T s_k^2(t)\,dt$, $k = 1, \cdots, K$. For any integer i, denote its modulo-K decomposition with remainder $\kappa(i) = 1, \cdots, K$, by $i = \eta(i)K + \kappa(i)$. Then we have

$$\Omega(b) = \sum_{i=i_0}^{i_f} \lambda_i(x_i, u_i) \tag{8}$$

where $i_0 = 1 - MK$, $i_f = (M+1)K$, $u_i = b_{\kappa(i)}(\eta(i)) \in A_{\kappa(i)}$, and

$$\lambda_i(x, u) = u\left[2y_{\kappa(i)}(\eta(i)) - uw_{k(i)} - 2x^T G^{\kappa(i)}\right] \tag{9}$$

with

$$x_{i+1} = \left[x_i^2 x_i^3 \cdots x_i^{K-1} u_i\right]^T, \qquad x_{i_0} = \mathbf{0}. \tag{10}$$

Proof: Utilizing the foregoing modulo-K decomposition, it is easy to check that

$$S_t(b) = \sum_{i=i_0}^{i_f} b_{\kappa(i)}(\eta(i)) s_{\kappa(i)}(t - \eta(i)T - \tau_{\kappa(i)}). \tag{11}$$

Hence we have

$$\begin{aligned} \|S(b)\|^2 = \sum_{i=i_0}^{i_f} \Bigg[& b_{\kappa(i)}^2(\eta(i)) \int_{-\infty}^{\infty} s_{\kappa(i)}^2(t - \eta(i)T - \tau_{\kappa(i)})\,dt \\ & + 2\sum_{j=i_0}^{i-1} b_{\kappa(i)}(\eta(i)) b_{\kappa(j)}(\eta(j)) \\ & \cdot \int_{-\infty}^{\infty} s_{\kappa(i)}(t - \eta(i)T - \tau_{\kappa(i)}) \\ & \cdot s_{\kappa(j)}(t - \eta(j)T - \tau_{\kappa(j)})\,dt \Bigg] \\ = \sum_{i=i_0}^{i_f} \Bigg[& b_{\kappa(i)}^2(\eta(i)) w_{\kappa(i)} \\ & + 2\sum_{l=1}^{K-1} b_{\kappa(i)}(\eta(i)) b_{\kappa(i-l)}(\eta(i-l)) \\ & \cdot \int_{-\infty}^{\infty} s_{\kappa(i)}(t - \eta(i)T - \tau_{\kappa(i)}) \\ & \cdot s_{\kappa(i-l)}(t - \eta(i-l)T - \tau_{\kappa(i-l)})\,dt \Bigg], \end{aligned} \tag{12}$$

where by agreement $b_{\kappa(i)}(\eta(i)) = 0$ for $i < i_0$. Now we show that the integral in the right-hand side of (12) is equal to $G_{K-l\,\kappa(i)}$. To that end we will examine separately the terms in which $\eta(i) = \eta(i-l)$ and $\eta(i) = \eta(i-l) + 1$. In the first case, $\kappa(i-l) = \kappa(i) - l$ and

$$\begin{aligned} & \int_{-\infty}^{\infty} s_{\kappa(i)}(t - \eta(i)T - \tau_{\kappa(i)}) \\ & \quad \cdot s_{\kappa(i-l)}(t - \eta(i-l)T - \tau_{\kappa(i-l)})\,dt \\ & = \int_{-\infty}^{\infty} s_{\kappa(i)}(t - \tau_{\kappa(i)}) s_{\kappa(i)-l}(t - \tau_{\kappa(i)-l})\,dt \\ & = G_{K-l\,\kappa(i)}. \end{aligned} \tag{13}$$

[1] The columns of the matrix G and of the row vector x_i^T are denoted by superscripts.

In the second case, $\kappa(i-l) = \kappa(i) + K - l$ and

$$\int_{-\infty}^{\infty} s_{\kappa(i)}\big(t - \eta(i)T - \tau_{\kappa(i)}\big) \cdot s_{\kappa(i-l)}\big(t - \eta(i-l)T - \tau_{\kappa(i-l)}\big)\, dt = \int_{-\infty}^{\infty} s_{\kappa(i)}\big(t - T - \tau_{\kappa(i)}\big) s_{\kappa(i)+K-l}\big(t - \tau_{\kappa(i)+K-l}\big)\, dt = G_{K-l\kappa(i)}. \tag{14}$$

From (10) and $u_i = b_{\kappa(i)}(\eta(i))$ it is clear that $b_{\kappa(i-l)}(\eta(i-l)) = x_i^{K-l}$, and therefore

$$\|S(b)\|^2 = \sum_{i=i_0}^{i_f} b_{\kappa(i)}(\eta(i)) \cdot \left[b_{\kappa(i)}(\eta(i)) w_{\kappa(i)} + 2 \sum_{j=1}^{K-1} x_i^j G_{j\kappa(i)} \right]. \tag{15}$$

Using (15) and the fact that

$$\int_{-\infty}^{\infty} S_t(b)\, dr_t = \sum_{i=i_0}^{i_f} b_{\kappa(i)}(\eta(i)) y_{\kappa(i)}(\eta(i))$$

[see (5)], the sought-after decomposition (9) follows.

The algorithm resulting from the decomposition of Proposition 1 performs K times the number of stages of that derived in [20]; however, its asymptotic complexity is considerably better since it exploits fully the separability of the log-likelihood function. In this case the dimensionality of the state space is equal to 2^{K-1}, and each state is connected to two states in the previous stage (if all the users employ binary modulation), resulting in TCB = $O(2^K)$.[2] This decomposition was introduced in [19] and shows that the part of the transition metric that is independent of the matched filter outputs is periodically time-varying with a period equal to the number of users. The nature of this behavior can be best appreciated by particularizing it to the intersymbol interference problem, in which $s_i(t) = s_j(t)$ for all $t \in [0, T]$, $i \neq j$, and $\tau_{i+1} - \tau_i = T/K$. In this case, $w_i = w_j$ and $G^i = G^j$ for $i \neq j$, and (9) reduces to the Ungerboeck metric [17, eq. (27)] when each symbol suffers the interference of $K-1$ signals. So, (9) can be viewed as the generalization of the Ungerboeck metric to a problem of periodically time-varying intersymbol interference equivalent from the receiver viewpoint, to the asynchronous multiuser model (1)–(2). The assumption that the signals of all users have the same duration can be relaxed, resulting in a decomposition similar to (9); however, the periodicity of the transition metric is lost unless the ratio between every pair of signal periods is rational.

The Viterbi algorithm resulting from the additive decomposition of Proposition 1 requires knowledge of the partial cross correlations between the signals of every pair of active users, and, unless binary antipodal modulation is employed, they also require the received signal energies. However, the signal cross correlations depend on the received relative delays, carrier phases, and amplitudes and hence cannot be determined *a priori* by the receiver. Nevertheless, the basic assumption is that the signal waveform of each user is known and that the K-user coherent receiver locks to the signaling interval and phase of each active user. Then the required parameters G_{ij} can be generated internally by cross-correlating the normalized waveform replicas stored in the receiver with the adequate delays and phases supplied by the synchronization system and by multiplying the resulting normalized cross correlations by the received amplitudes of the corresponding users. Hence the only requirement (beyond synchronization) imposed by the need for the partial cross correlations is the availability (up to a common scale factor) of the K received signal amplitudes.

The decomposition in Proposition 1 can be generalized to the case where the modulation is not necessarily linear, that is, each symbol $u \in A_k$ is mapped into a different waveform $s_k(t; u)$. It can be shown, analogously to the proof of Proposition 1, that in this case the transition metric is

$$\lambda_i(x, u) = 2 y_{\kappa(i)}(\eta(i); u) - w_{\kappa(i)}[u] - 2 \sum_{l=1}^{K-1} G_{l\kappa(i)}[x^l, u], \tag{16}$$

where $y_k(i, u)$ is the output of the matched filter of the uth waveform of the kth user; $w_k[u] = \int_0^T s_k^2(t; u)\, dt$; and

$$G_{lk}[a, u] = \begin{cases} \int s_{l+k}(t - \tau_{l+k}; a) s_k(t - \tau_k - T; u)\, dt, & \text{if } l + k \leq K \\ \int s_{l+k-K}(t - \tau_{l+k-K}; a) s_k(t - \tau_k; u)\, dt, & \text{if } l + k > K. \end{cases} \tag{17}$$

III. Error Probability Analysis of Optimum Multiuser Detectors

A. Upper and Lower Bounds

This section is devoted to the analysis of the minimum uncoded bit error probability of multiuser detectors for antipodally modulated signals in additive white Gaussian noise, that is, attention is focused on the multiple-access model (1)–(2) in the case $A_1 = \cdots = A_K = \{-1, 1\}$. Denote the most likely transmitted ith symbol by the kth user given the observations by

$$\hat{b}_k^M(i) \in \arg \max_{b \in \{-1,1\}} P\big[b_k(i) = b \mid dr_t = S_t^M(b) + \sigma\, dw_t,\ t \in R\big] \tag{18}$$

where $b = \{b(i) \in \{-1,1\}^K,\ i = -M, \cdots, M\}$ is the transmitted sequence. The goal is to obtain the finite and

[2] If the alphabet sizes are arbitrary, then (see [22]) TCB = $O(K\prod_{l=1}^K |A_l| / \sum_{l=1}^K \log |A_l|)$. If there are S clusters of synchronized users, then it is possible to reduce this complexity by a factor of S/K by taking into account that some of the cross correlations are equal to zero.

infinite horizon error probabilities,

$$P_k^M(i) = P[b_k(i) \neq \hat{b}_k^M(i)], \qquad i = -M, \cdots, M \quad (19)$$

$$P_k = \lim_{M \to \infty} P_k^M(i), \quad (20)$$

for arbitrary signal waveforms and relative delays. The existence of the limit for any given i in the right-hand side of (20) follows because $P_k^M(i)$ is equal to probability of error of an optimum detector for a multiuser signal with horizon equal to $M+1$ and complete information about $b(-M-1)$ and $b(M+1)$; hence $P_k^{M+1}(i) \geq P_k^M(i)$ for any positive integer M, $i \in \{-M, \cdots, M\}$ and $k \in \{1, \cdots, K\}$. The independence of the limit on i readily follows from the assumed stationarity of the noise and of the priors.

Our approach is to derive lower and upper bounds on the error rate for each user, which are tight in the low and high SNR regions and give a close approximation over the whole SNR range. The upper bounds are based on the analysis of two detectors that are suboptimum in terms of error probability, namely, the conventional single-user coherent detector and the K-user maximum-likelihood sequence detector. The lower bounds are the error probabilities of two optimum binary tests derived from the original problem by allowing certain side information.

The normalized difference between any pair of distinct transmitted sequences will be referred to as an error sequence, that is, the set of nonzero error sequences is[3]

$$E = \{\epsilon = \{\epsilon(i) \in \{-1, 0, 1\}^K, i = -M, \cdots, M; \; \epsilon(j) \neq \mathbf{0} \text{ for some } j\}\}.$$

The set of error sequences that are admissible conditioned on $\boldsymbol{b}$ being transmitted, that is, those that correspond to the difference between $\boldsymbol{b}$ and the sequence selected by the detector, is denoted by

$$A(\boldsymbol{b}) = \{\epsilon \in E, 2\epsilon - \boldsymbol{b} \in D\},$$

where

$$D = \{\boldsymbol{b} = \{b(i) \in \{-1, 1\}^K, i = -M, \cdots, M\}\}.$$

The admissible error sequences that affect the ith bit of the kth user of $\boldsymbol{b}$ are

$$A_k(\boldsymbol{b}, i) = \{\epsilon \in A(\boldsymbol{b}), \epsilon_k(i) = b_k(i)\}.$$

The number of nonzero components of an error sequence and the energy of a hypothetical multiuser signal modulated by the sequence ϵ are denoted, respectively, by

$$w(\epsilon) = \sum_{i=-M}^{M} \sum_{k=1}^{K} |\epsilon_k(i)|$$

and

$$\|S(\epsilon)\|^2 = \int_{-\infty}^{\infty} \left(\sum_{i=-M}^{M} \sum_{k=1}^{K} \epsilon_k(i) s_k(t - iT - \tau_k) \right)^2 dt.$$

[3] For the sake of notational simplicity, the explicit dependence of the sets of sequences on M is dropped when this causes no ambiguity.

Proposition 2: Define the following minimum distance parameters:

$$d_k^M(\boldsymbol{b}, i) = \inf_{\epsilon \in A_k^M(\boldsymbol{b}, i)} \|S(\epsilon)\|$$

and

$$d_{k,\min}^M(i) = \inf_{\boldsymbol{b} \in D} d_k^M(\boldsymbol{b}, i).$$

Then, the minimum error probability of the ith bit of the kth user is lower-bounded by

$$P_k \geq P_k^M(i)$$
$$\geq P[d_k^M(\boldsymbol{b}, i) = d_{k,\min}^M(i)] Q(d_{k,\min}^M(i)/\sigma) \quad (21)$$

and by

$$P_k \geq P_k^M(i) \geq Q(w_k/\sigma), \quad (22)$$

where $P[d_k(\boldsymbol{b}, i) = d_{k,\min}(i)]$ is the *a priori* probability that the transmitted sequence is such that one of its congruent error sequences affects the ith bit of the kth user and has the minimum possible energy.

Proof: The basic technique for obtaining lower bounds on the minimum error probability is to analyze the performance of an optimum receiver that, in addition to observing $\{r_t, t \in R\}$, has certain side information.

To obtain the first lower bound (21), the following reasoning, analogous to that of Forney [2], can be employed. Suppose that if the transmitted sequence $\boldsymbol{b}$ is such that $d_k(\boldsymbol{b}, i) = d_{k,\min}(i)$, the detector is told by a genie that the true sequence is either $\boldsymbol{b}$ or $\boldsymbol{b} - 2\boldsymbol{\delta}$, where $\boldsymbol{\delta}$ is arbitrarily chosen by the genie (independently of the noise realization) from the set

$$\arg \min_{\epsilon \in A_k(\boldsymbol{b}, i)} \|S(\epsilon)\|.$$

Under these conditions, the minimum error probability detector and the optimum sequence detector coincide and both reduce to a binary hypothesis test $\Omega(\boldsymbol{b}) \gtrless \Omega(\boldsymbol{b} - 2\boldsymbol{\delta})$. The conditional probability of error given that $\boldsymbol{b}$ is transmitted is given by

$$P[\Omega(\boldsymbol{b} - 2\boldsymbol{\delta}) > \Omega(\boldsymbol{b}) | \boldsymbol{b} \text{ transmitted}]$$
$$= P\left[\sigma \int S_t(\boldsymbol{\delta})\, d\omega_t + \|S(\boldsymbol{\delta})\|^2 < 0\right]$$
$$= Q(\|S(\boldsymbol{\delta})\|/\sigma) \quad (23)$$

where the foregoing equalities follow from (4) and the fact that $\int S_t(\boldsymbol{\delta})/\|S(\boldsymbol{\delta})\|\, d\omega_t$ is a zero-mean unit-variance Gaussian random variable. If the transmitted sequence $\boldsymbol{b}$ is such that $d_k(\boldsymbol{b}, i) > d_{k,\min}(i)$, then the error probability of the receiver with side information is trivially bounded by zero, and the lower bound (21) follows. The lower bound (22) is the kth user minimum error probability if no other user was active or, equivalently, if the receiver knew the transmitted bits of the other users.

For any error sequence ϵ such that $\epsilon(j) = \epsilon(n) \neq \mathbf{0}$, for $j \neq n$, the sequence

$$\epsilon'(m) = \begin{cases} \epsilon(m), & m \leq j \\ \epsilon(m + n - j), & m > j \end{cases}$$

satisfies $\|S(\epsilon')\| \le \|S(\epsilon)\|$. (Otherwise, one could construct a sequence with negative energy.) This implies that the infinite-horizon minimum distances, $d_k(b)$ and $d_{k\,\min}$ (i.e., one-half of the minimum rms of the difference between the signals of any pair of transmitted sequences that differ in any bit of the kth user), are achieved by finite-length error sequences, and the error rate of the kth user can be lower-bounded by

$$P_k \ge P\left[d_k(b) = d_{k,\min}\right] Q(d_{k,\min}/\sigma). \quad (24)$$

Note that since $d_{k,\min} \le d^M_{k,\min}(i)$, the bound (24) is at least as tight as (21) in the low-noise region.

The following upper bound, the error probability of the conventional single-user coherent receiver, is mainly useful in the low SNR region.

Proposition 3: Let $R_{ij} = G_{K-i\kappa(k+i)}$, for $i = 1,\cdots, K-1$ and $j = 1,\cdots, K$, that is, the entries of the column R^j are the correlations of the signal of the jth user with the $K-1$ posterior signals, and denote $I_k(\alpha, \beta) = \sum_{j=1}^{K-1}(\alpha_j G_{jk} + \beta_j R_{jk})/w_k$. The minimum error probability of the kth user is upper-bounded by

$$P_k^M(i) \le P_k \le E\left[Q\left(\sqrt{w_k}\left[1 + I_k(\alpha, \beta)\right]/\sigma\right)\right], \quad (25)$$

where the expectation is over the ensemble of independent uniformly distributed $\alpha \in \{-1,1\}^{K-1}$, $\beta \in \{-1,1\}^{K-1}$.

Proof: Suppose that $|i| < M$ and that the transmitted sequence is such that $b_k(i) = 1$. The kth matched filter output corresponding to the ith bit is a conditionally Gaussian random variable with variance equal to $w_k\sigma^2$ and mean given by

$$w_k + \sum_{l=k+1}^{K} b_l(i-1)G_{l-k\,k} + \sum_{l=1}^{k-1} b_l(i)G_{l-k+K\,k} + \sum_{l=k+1}^{K} b_l(i)R_{l-k\,k} + \sum_{l=1}^{k-1} b_l(i+1)R_{l-k+K\,k}.$$

Since the transmitted bits are assumed to be equiprobable and independent, we have

$$P_k^M(i) \le E\left[Q\left(\sqrt{w_k}\left[1 + I_k(\alpha, \beta)\right]/\sigma\right)\right], \quad (26)$$

and because the right-hand side does not depend on M, (25) follows.

We turn to the derivation of an upper bound on the error probability of the K-user maximum-likelihood sequence detector. Our approach hinges on the following definition. An error sequence $\epsilon \in E$ is *decomposable* into $\epsilon' \in E$ and $\epsilon'' \in E$ if

1) $\epsilon = \epsilon' + \epsilon''$,
2)[4] $\epsilon' < \epsilon$, $\epsilon'' < \epsilon$,
3) $< S(\epsilon'), S(\epsilon'') > \ge 0$.

As an illustration consider the following simple two-user example: $s_1(t) = s_2(t) = 1$, $0 \le t \le T/2$; $s_1(t) = -s_2(t) = -1$, $T/2 < t \le T$; and $\tau_2 - \tau_1 = T/2$. Then it is easy to check that the error sequence $\epsilon = \cdots, \mathbf{0}, [1\ 1]^T$ T, $[1\ 0]^T, \mathbf{0}, \cdots$ is decomposable into $\epsilon' = \cdots, \mathbf{0}, [1\ 0]^T$, $[1\ 0]^T, \mathbf{0}, \cdots$, and $\epsilon'' = \cdots, \mathbf{0}, [0\ 1]^T, [0\ 0]^T, \cdots$.

Proposition 4: Denote the set of error sequences that affect the ith bit of the kth user by $Z_k(i) = \{\epsilon \in E, \epsilon_k(i) \ne 0\}$. Let $F_k(i)$ be the subset of *indecomposable* sequences in $Z_k(i)$. Then the minimum-error probability of the ith bit of the kth user is upper-bounded by

$$P_k^M(i) \le \sum_{\epsilon \in F_k(i)} 2^{-w(\epsilon)} Q(\|S(\epsilon)\|/\sigma). \quad (27)$$

Proof: Formula (27) is an upper bound on the probability of error of the K-user maximum-likelihood sequence detector, that is, the receiver whose output is the sequence that maximizes $\Omega(b)$. This is the only property of the detector that we use for the proof of (27); it is not necessary to assume any specific decision algorithm, in particular that of Section II. Define the following sets of error sequences:

$$L = \left\{\epsilon \in E,\ \sigma \int S_t(\epsilon)\, dw_t \le -\|S(\epsilon)\|^2\right\} \quad (28)$$

and

$$ML(b) = \{\epsilon \in A(b),\ \Omega(b - 2\epsilon) \ge \Omega(d), \text{ for all } d \in D\} \quad (29)$$

If b is the transmitted sequence and $\epsilon \in A(b)$, then it can be shown that $\Omega(b - 2\epsilon) \ge \Omega(b)$ if and only if $\epsilon \in L$. Hence it follows that $ML(b) \subset L$ and that if $A(b) \cap L \ne \varnothing$, then the sequence detector outputs an erroneous sequence $b - 2\epsilon$, where $\epsilon \in ML(b)$. Consider the following inclusions between events in the probability space on which the transmitted sequence b, and the Wiener process $\{\omega_t, t \in R\}$ are defined:

$$\{b_k(i) \ne b_k^*(i)\} \subset \bigcup_{\epsilon \in E} \{\epsilon \in A_k(b, i) \cap ML(b)\} \subset \bigcup_{\epsilon \in F_k(i)} \{\epsilon \in A_k(b, i) \cap L\}, \quad (30)$$

where b^* is the sequence selected by the detector. The first inclusion follows from the definitions of $A_k(b, i)$ and $ML(b)$ (the converse holds if there are no ties in the maximization of $\Omega(\cdot)$). The key to the tightness of the bound in (27) is the second inclusion in (30); to verify it, we show that for every $\epsilon \in Z_k(i)$ there exists $\epsilon' \in F_k(i)$ such that

1) $\{\epsilon \in A_k(b, i)\} \subset \{\epsilon' \in A_k(b, i)\}$
2) $\{\epsilon \in ML(b)\} \subset \{\epsilon' \in L\}$.

If $\epsilon \in F_k(i)$, then $\epsilon' = \epsilon$ satisfies 1) and 2). Otherwise, $\epsilon \in Z_k(i) - F_k(i)$. We now show by induction on $w(\epsilon)$, the weight of the sequence ϵ, that there exists $\epsilon^* \in F_k(i)$ such that ϵ is decomposable into $\epsilon^* + (\epsilon - \epsilon^*)$. If a sequence of weight two is decomposable, then it is so into its two components, both of which are indecomposable since they have unit weight. Now suppose that the claim is true for any sequence whose weight is strictly less than $w(\epsilon)$.

[4] We denote $\epsilon' < \epsilon$ if $|\epsilon'_j(i)| \le |\epsilon_j(i)|$ for all $j = 1,\cdots, K$ and $i = -M,\cdots, M$.

Find the (not necessarily unique) decomposition $\epsilon = \epsilon^1 + \epsilon^2$, $\epsilon^1 \in Z_k(i)$, with largest inner product, that is,

$$\langle S(\epsilon^1), S(\epsilon^2)\rangle \geq \langle S(\epsilon^a), S(\epsilon^b)\rangle \geq 0$$

for any decomposition $\epsilon = \epsilon^a + \epsilon^b$. If $\epsilon^1 \in F_k(i)$, we have found the sought-after decomposition of ϵ. Otherwise, we can decompose ϵ^1 into $\epsilon^3 + \epsilon^4$ such that $\epsilon^3 \in F_k(i)$ because of the induction hypothesis. However, ϵ is indeed decomposable into ϵ^3 and $(\epsilon^2 + \epsilon^4)$, for otherwise the right-hand side of the equation

$$\begin{aligned}\langle S(\epsilon^2) + S(\epsilon^3), S(\epsilon^4)\rangle - \langle S(\epsilon^1), S(\epsilon^2)\rangle \\ = 2\langle S(\epsilon^3), S(\epsilon^4)\rangle - \langle S(\epsilon^3), S(\epsilon^4) + S(\epsilon^2)\rangle \quad (31)\end{aligned}$$

is strictly positive, contradicting the choice of $\epsilon^1 + \epsilon^2$ as the largest-inner-product decomposition of ϵ.

Since $\epsilon^* < \epsilon$, it follows that $\epsilon' = \epsilon^*$ satisfies property 1); to see that 2) is also fulfilled, let $\epsilon'' = \epsilon - \epsilon'$ and consider

$$\begin{aligned}\Omega(b - 2\epsilon) - \Omega(b - 2\epsilon'') \\ = 2\sigma\int [S_t(b - 2\epsilon) - S_t(b - 2\epsilon'')]\, dw_t \\ + \|S(b - 2\epsilon'') - S(b)\|^2 - \|S(b - 2\epsilon) - S(b)\|^2 \\ = -4\left(\sigma\int S_t(\epsilon')\, dw_t + \|S(\epsilon')\|^2\right) - 8\langle S(\epsilon''), S(\epsilon')\rangle \quad (32)\end{aligned}$$

where (32) follows from the fact that b is the transmitted sequence. If $\epsilon \in ML(b)$, then it is necessary that $\Omega(b - 2\epsilon) \geq \Omega(b - 2\epsilon'')$; moreover, the decomposition of ϵ into $\epsilon' + \epsilon''$ implies that $\langle S(\epsilon'), S(\epsilon'')\rangle \geq 0$. Therefore, (32) indicates that $\epsilon' \in L$, and property 2) is satisfied (see Fig. 3).

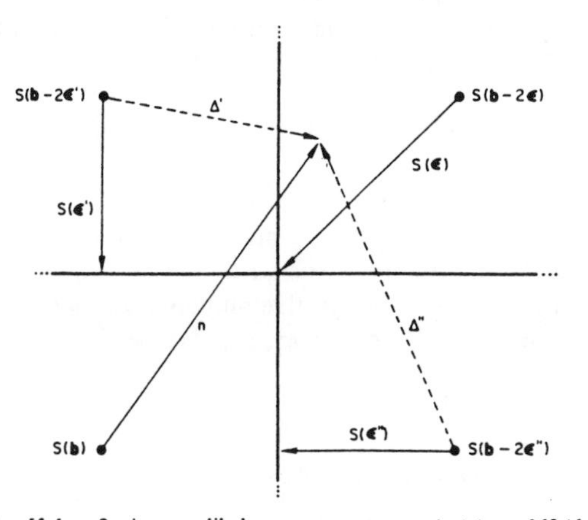

Fig. 3. If $b - 2\epsilon$ is most likely sequence *a posteriori* ($\epsilon \in ML(b)$) and $\langle S(\epsilon'), S(\epsilon'')\rangle \geq 0$, then $\|\Delta'\| \leq \|n\|$ and $\|\Delta''\| \leq \|n\|$, i.e., $\epsilon' \in L$ and $\epsilon'' \in L$. $S(b)$ is transmitted signal. n is projection of noise realization.

In order to take probabilities of the events in the right-hand side of (30), note that $\{\epsilon \in A_k(b, i)\}$ depends on the transmitted sequence but not on the noise realization; hence it is independent of the event $\{\epsilon \in L\}$. Finally,

$$P[\epsilon \in A_k(b, i)] = 2^{-w(\epsilon)} \quad (33)$$

and

$$P[\epsilon \in L] = Q(\|S(\epsilon)\|/\sigma) \quad (34)$$

are immediate from the respective definitions of $A_k(b, i)$ and L, and since the probability of the union of the events in the right-hand side of (3) is not greater than the sum of their probabilities, the upper bound of (27) follows.

One of the features of the foregoing proof is that it remains valid in the general case where the modulation is not necessarily linear, that is,

$$S_t(b) = \sum_{i=-M}^{M}\sum_{k=1}^{K} s_k(t - iT - \tau_k; b_k(i)), \qquad b \in D. \quad (35)$$

The only modification needed to state and prove Proposition 4 is to substitute $S_t(\epsilon)$ by

$$\hat{S}_t(\epsilon) = \frac{1}{2}\sum_{i=-M}^{M}\sum_{k=1}^{K} s_k(t - iT - \tau_k; \epsilon_k(i)) - s_k(t - iT - \tau_k; -\epsilon_k(i)). \quad (36)$$

It is interesting to particularize the foregoing result to the intersymbol interference problem. The Forney bound [3] corresponds to the sum of $2^{-w(\epsilon)}Q(\|S(\epsilon)\|/\sigma)$ over all *simple* sequences, that is, those containing no more than $L - 1$ consecutive zeros amid nonzero components (where L is the number of overlapping symbols). It turns out that a great proportion of simple sequences are, in fact, decomposable; however, all indecomposable sequences are simple. Hence the analysis of maximum-likelihood sequence detection of signals subject to intersymbol interference via decomposition of error sequences results in a bound that is tighter than Forney's result. This issue and the question of how to compute (27) up to any prespecified degree of accuracy are examined in [24].

In the case of bit-synchronous users ($\tau_1 = \cdots = \tau_K$), the derivation of optimum decision rules is, in contrast to Section II, a simple exercise. However, the analysis of the optimum synchronous receiver has basically the same complexity as the general case presented in this section. In particular, even though the one-shot model ($M = 0$) is sufficient, the approach of decomposition of error sequences (K-vectors in this case) is the most effective one.

B. Convergence of the Bounding Series

The limit of the right-hand side of (27) as $M \to \infty$ is an infinite series that bounds the multiuser infinite-horizon error probability. Since this upper bound is monotonic in the noise level, we can distinguish three situations depending on the actual energies, crosscorrelations, and relative delays of the signals—namely, divergence of the series for all noise levels, convergence for sufficiently low noise levels, and global convergence. Although most cases fall into the second category, we will show examples of the other two. We begin by proving a sufficient condition for local convergence of a series that overbounds (27).

Proposition 5: Partition the real line with the nontrivial semi-open intervals $R = \bigcup_{i \in Z} \Lambda_i$ defined by the points $\tau_k + iT$, $k = 1, \cdots, K$, $i \in Z$. Define $N(\epsilon)$ as the union of all intervals Λ_i for which $\int_{\Lambda_i} S_t^2(\epsilon)\, dt = 0$. Let G be a set of simple sequences such that for every pair of distinct finite sequences $\epsilon^1, \epsilon^2 \in G$ that satisfy $N(\epsilon^1) = N(\epsilon^2)$, there exist $j \in \{1, \cdots, K\}$ and $i \in Z$ such that $\epsilon_j^1(i) \neq \epsilon_j^2(i)$ and $(iT + \tau_j, (i+1)T + \tau_j) \not\subset N(\epsilon^1)$. Then there exists $\sigma_0 > 0$ such that

$$\sum_{\epsilon \in G} 2^{-w(\epsilon)} \exp\left(-\|S(\epsilon)\|^2 / 2\sigma^2\right) < \infty \tag{37}$$

for $0 < \sigma < \sigma_0$.

Proof: Define $G_{n,l} = \{\epsilon \in G$, for some $j \in Z$ the point of arrival of the first nonzero component of ϵ and the point of departure of the last nonzero component of ϵ define the interval $\bigcup_{i=j+1}^{j+n} \Lambda_i$ and exactly l of these intervals are such that $\int_{\Lambda_i} S_t^2(\epsilon)\, dt > 0\}$. For every $\epsilon \in G_{n,l}$, we have

$$(K-1) w(\epsilon) \geq n \tag{38}$$

because ϵ is simple and we cannot have more than $K - 2$ zeros surrounded by nonzeros. Define

$$\alpha^2 = \min\left\{ r > 0 \mid \text{there exists } i \in Z \text{ and } \epsilon \in G \text{ with } r = \int_{\Lambda_i} S_t^2(\epsilon)\, dt \right\}. \tag{39}$$

Notice that the existence of α is guaranteed because the signal energies are nonzero and there is a finite number of distinct signal waveforms in each interval Λ_i. From the definition of $G_{n,l}$, (38), and (39), we have

$$\sum_{\epsilon \in G_{n,l}} 2^{-w(\epsilon)} \exp\left(-\|S(\epsilon)\|^2/2\sigma^2\right) \leq 2^{(n/1-K)} |G_{n,l}| \exp\left(-l\alpha^2/2\sigma^2\right). \tag{40}$$

Now we use the assumption in the theorem to overbound the cardinality of $G_{n,l}$ by

$$|G_{n,l}| \leq n \binom{n}{l} 3^{Kl}. \tag{41}$$

To show (41) let us see how many different sequences are congruent with every choice of j in the definition of $G_{n,l}$ (for which there are n possibilities, since $\epsilon_k(0) \neq 0$ for all $\epsilon \in G$) and with every distribution of the l nonzero-energy intervals (at most $\binom{n}{l}$ possibilities). The assumption of the result states that any pair of sequences whose nonzero-energy intervals coincide cannot differ only in symbols whose intervals have zero energy. Since $S_t(\epsilon)$, $t \in \Lambda_i$ depends on ϵ through K elements at most, the number of error sequences congruent with the aforementioned choice is bounded by 3^{Kl}, and (41) follows. Substituting (41) into (40), we obtain

$$\begin{aligned} &\sum_{n=1}^{\infty} \sum_{l=0}^{n} \sum_{\epsilon \in G_{n,l}} 2^{-w(\epsilon)} \exp\left(-\|S(\epsilon)\|^2/2\sigma^2\right) \\ &\leq \sum_{n=1}^{\infty} n 2^{n/1-K} \sum_{l=0}^{n} \binom{n}{l} \left(3^{Kl} \exp\left(-l\alpha^2/2\sigma^2\right)\right) \\ &= \sum_{n=1}^{\infty} n 2^{n/1-K} \left(1 + 3^K \exp\left(-\alpha^2/2\sigma^2\right)\right)^n, \end{aligned} \tag{42}$$

which converges for

$$0 < \sigma^2 < \frac{\alpha^2}{2} \left[K \cdot \ln 3 - \ln\left(2^{1/K-1} - 1\right)\right]^{-1} \tag{43}$$

Local convergence of the Forney bound for any intersymbol interference problem was proved by Foschini [5]. Not every multiuser problem, however, results in a locally convergent bounding series. Admittedly, the implications of the sufficient condition of Proposition 5 on the waveforms and delays of the signal set are not readily apparent. As is shown by the next result, the sufficient condition of Proposition 5 is satisfied except in certain pathological cases where some degree of synchronism exists along with heavy correlation between the signals.

Proposition 6: Define the times of effective arrival and departure of the ith signal of the kth user as

$$\lambda_{iK+k}^a = \tau_k + iT + \sup\left\{\tau \in [0, T),\ \int_0^\tau s_k^2(t)\, dt = 0\right\}$$

and

$$\lambda_{iK+k}^d = \tau_k + iT + \inf\left\{\tau \in (0, T],\ \int_\tau^T s_k^2(t)\, dt = 0\right\},$$

respectively.

Suppose that a pair of distinct finite sequences $\epsilon^1, \epsilon^2 \in E$ exists such that $N(\epsilon^1) = N(\epsilon^2)$ and such that $\epsilon_k^1(i) \neq \epsilon_k^2(i)$ implies $(iT + \tau_k, (i+1)T + \tau_k) \subset N(\epsilon^1)$. Then the following two statements are true:

1) $S_t(\epsilon^1) = S_t(\epsilon^2)$, a.e.;
2) a pair of reals $\rho < \xi$ exists such that
   - $\rho = \lambda_i^a = \lambda_j^a$ for $i \neq j$,
   - $\xi = \lambda_i^d = \lambda_j^d$ for $i \neq j$, and
   - if $\epsilon_k^1(i) \neq \epsilon_k^2(i)$ then $\rho \leq \lambda_{iK+k}^a < \lambda_{iK+k}^d \leq \xi$.

Proof: 1) If $t \notin N(\epsilon^1) = N(\epsilon^2)$, then $S_t(\epsilon^1)$, $S_t(\epsilon^2)$ depend on their arguments only through those symbols that coincide; on the other hand,

$$\int_{N(\epsilon^1)} S_t^2(\epsilon^1)\, dt = 0 = \int_{N(\epsilon^2)} S_t^2(\epsilon^2)\, dt.$$

2) We show first that the effective arrival of the first symbol that differs, say $\epsilon_q(j)$, must be a point of effective multiarrival, that is, there exists $i > i_1 = jK + q$, $\lambda_i^a = \lambda_{i_1}^a$. If $\lambda_{i_1}^a$ is not a point of effective multiarrival, then we can select λ such that $\lambda_{i_1}^a < \lambda < \lambda_{iK+k}^a$, $\epsilon_k^1(i) \neq \epsilon_k^2(i)$ and

$$\int_{\lambda_{i_1}^a}^{\lambda} s_q^2(t - \tau_q - jT)\, dt > 0. \tag{44}$$

On the other hand, if $t \in (\lambda_{i_1}^a, \lambda)$, then

$$S_t(\epsilon^1) = \epsilon_q^1(j) s_q(t - \tau_q - jT) + \delta^1(t) \tag{45a}$$

and

$$S_t(\epsilon^2) = \epsilon_q^2(j) s_q(t - \tau_q - jT) + \delta^2(t), \tag{45b}$$

where $\delta^1(t) = \delta^2(t)$, a.e. in $(\lambda_{i_1}^a, \lambda)$ because the effective arrival of the rest of the unequal symbols is posterior to λ. Using 1) and (45), we obtain that $\epsilon_q^1(j) s_q(t - \tau_q - jT) = \epsilon_q^2(j) s_q(t - \tau_q - jT)$ a.e. in $(\lambda_{i_1}^a, \lambda)$, which contradicts (44) since $\epsilon_q^1(j) \neq \epsilon_q^2(j)$. Similarly, it can be shown that the effective departure of the last symbol that differs between

ϵ' and ϵ'' must be a point of effective multideparture, and 2) follows.

Proposition 6 implies that for asynchronous models, where the delays are independent and uniformly distributed, an interval of convergence exists for the bounding series with probability one. On the other hand, it is easy to see that for bit-synchronous models, the set $F_k(i)$ is finite so that (27) is finite for all noise levels in that case. Hence, the necessary ingredients for the everywhere divergence of (27) are the partial effective synchronism and the heavy cross correlation of the signal constellation. To illustrate this, consider the following example of divergence of the bounding series (27) for all noise levels.

Let $K = 6$, $s_k(t) = 1$, $t \in [0,1]$, $k = 1,\cdots,6$; $\tau_k - \tau_1 = 1/2$, $k = 2,\cdots,6$; $\tau_i - \tau_j = 0$, $i, j = 2,\cdots,6$. Consider the set of error sequences: $A_n = \{\epsilon \in E,\ \epsilon(i) = 0,\ i < 0$ and $i > n;\ \epsilon(i) \in \{[1\ -1\ 0\ 0\ 0\ 0]^T, [1\ 0\ -1\ 0\ 0\ 0]^T, [1\ 0\ 0\ -1\ 0\ 0]^T, [1\ 0\ 0\ 0\ -1\ 0]^T, [1\ 0\ 0\ 0\ 0\ -1]^T\},\ i = 0,\cdots,n-1;\ \epsilon(n) = [1\ 0\ 0\ 0\ 0\ 0]^T\}$. Notice that $\|S(\epsilon)\| = 1$, $w(\epsilon) = (2n+1)$, for all $\epsilon \in A_n$, and $|A_n| = 5^n$. It is straightforward to show that every sequence $\epsilon \in A_n$ is indecomposable; thus $A_n \in F_1$, for all $n > 1$, and the following inequality is true:

$$\begin{aligned}\sum_{\epsilon \in F_1} 2^{-w(\epsilon)} Q(\|S(\epsilon)\|/\sigma) &> \sum_{n=1}^{\infty} \sum_{\epsilon \in A_n} 2^{-w(\epsilon)} Q(\|S(\epsilon)\|/\sigma)\\ &= Q(1/\sigma) \sum_{n=1}^{\infty} \sum_{\epsilon \in A_n} 2^{-w(\epsilon)}\\ &= \frac{1}{2} Q(1/\sigma) \sum_{n=1}^{\infty} (5/4)^n.\end{aligned}$$

It follows that (27) diverges for any noise level.

C. Asymptotic Probability of Error

In this section we show that whenever the error probability upper bound (27) converges for sufficiently low noise, both bounds (24) and (27) are asymptotically tight as $\sigma \to 0$. In particular, we prove that for any $\delta > 0$ there exists $\sigma_0 > 0$ such that for all $\sigma < \sigma_0$,

$$C_k^L Q(d_{k,\min}/\sigma) \le P_k \le C_k^U (1+\delta) Q(d_{k,\min}/\sigma), \quad (46)$$

where

$$C_k^L = P[d_k(b) = d_{k,\min}] = P\left[\bigcup_{\substack{\epsilon \in F_k \text{ s.t.}\\ \|S(\epsilon)\| = d_{k,\min}}} \{\epsilon \in A(b)\}\right] \quad (47)$$

and

$$C_k^U = \sum_{\substack{\epsilon \in F_k \text{ s.t.}\\ \|S(\epsilon)\| = d_{k,\min}}} 2^{-w(\epsilon)} = \sum_{\substack{\epsilon \in F_k \text{ s.t.}\\ \|S(\epsilon)\| = d_{k,\min}}} P[\epsilon \in A(b)]. \quad (48)$$

The left-hand inequality of (46) was obtained in (24). Expression (47) follows because if $\epsilon \notin F_k$ has $\|S(\epsilon)\| = d_{k,\min}$, then it is decomposable into $\epsilon' + \epsilon''$ such that $\epsilon' \in F_k$, $\|S(\epsilon')\| = d_{k,\min}$ and $\{\epsilon \in A(b)\} \subset \{\epsilon' \in A(b)\}$. C_k^U defined in (48) is the sum of the coefficients in the bounding series (27) that correspond to the sequences with minimum energy.[5] The right-hand inequality of (46) follows from the following proposition.

Proposition 7: If $\sigma_0 > 0$ exists such that for all $0 < \sigma \le \sigma_0$

$$\sum_{\epsilon \in F_k} 2^{-w(\epsilon)} \exp(-\|S(\epsilon)\|^2/2\sigma^2) < \infty, \quad (49)$$

then

$$\lim_{\sigma \to 0} \sum_{\epsilon \in F_k} 2^{-w(\epsilon)} Q(\|S(\epsilon)\|/\sigma)/Q(d_{k,\min}/\sigma) = C_k^U. \quad (50)$$

Proof: First, we show that for any set $G \subset E$ and any constant $r \ge 0$ that satisfy

1) $\inf_{\epsilon \in G} \|S(\epsilon)\| > r$,
2) there exists σ_0 such that for all $0 < \sigma \le \sigma_0$, $\sum_{\epsilon \in G} 2^{-w(\epsilon)} \exp(-\|S(\epsilon)\|^2/2\sigma^2) < \infty$,

we have

$$\lim_{\sigma \to 0} \sum_{\epsilon \in G} 2^{-w(\epsilon)} Q(\|S(\epsilon)\|/\sigma)/Q(r/\sigma) = 0. \quad (51)$$

Consider the following inequalities:

$$\begin{aligned}\sum_{\epsilon \in G} 2^{-w(\epsilon)} Q(\|S(\epsilon)\|/\sigma)/Q(r/\sigma) &\le \sum_{\epsilon \in G} 2^{-w(\epsilon)} \exp\left([r^2 - \|S(\epsilon)\|^2]/2\sigma^2\right)\\ &\le \exp\left(\left[\inf_{\epsilon \in G} \|S(\epsilon)\|^2 - r^2\right]\left[1/2\sigma_0^2 - 1/2\sigma^2\right]\right)\\ &\quad \cdot \sum_{\epsilon \in G} 2^{-w(\epsilon)} \exp\left([r^2 - \|S(\epsilon)\|^2]/2\sigma_0^2\right) \quad (52)\end{aligned}$$

where the inequalities follow from $Q(\sqrt{x+y}) \le \exp(-x/2) Q(\sqrt{y})$ if $x \ge 0$ and $y \ge 0$, and 1) respectively. The left-hand side of (52) vanishes as $\sigma \to 0$ because as a result of 2) the series therein converges.

Particularizing this result to $r = \inf_{\epsilon \in F_k} \|S(\epsilon)\|$ and $G = \{\epsilon \in F_k$, such that $\|S(\epsilon)\| > r\}$, we see that (50) follows. Note that the infimum of $\|S(\epsilon)\|$ over both F_k and the subset $\{\epsilon \in F_k$, such that $\|S(\epsilon)\| > r\}$ are achieved because both sets have a finite subset whose elements have no greater energy than the rest of the elements in F_k and G. This implies that $r = d_{k,\min}$, and condition 1) is satisfied.

The high SNR-upper and lower bounds (46) to the kth user error probability differ by a multiplicative constant independent of the noise level. From (47) and (48) it can be seen that this constant is related to the degree of overlapping of the events $\{\{\epsilon \in A(b)\},\ \epsilon \in F_k$ and $\|S(\epsilon)\| = d_{k,\min}\}$. Typically, there exists only a pair of elements in F_k, $\{\epsilon, -\epsilon\}$, that achieve the minimum energy. Since $\{\epsilon \in A(b)\} \cap \{-\epsilon \in A(b)\} = \emptyset$ for all $\epsilon \in E$, it follows that in such case $C_k^L = C_k^U$.

An important performance measure for multiuser detectors in high SNR situations is the SNR degradation due to the existence of other active users in the channel, i.e., the

[5] It can be shown that the error probability of the multiuser maximum-likelihood sequence detector can be lower bounded by $C_k^U Q(d_{k,\min}/\sigma)$.

limit as $\sigma \to 0$ of the ratio between the effective SNR (that required by a single-user system to achieve the same asymptotic error probability) and the actual SNR. We denote this parameter as η_k, the kth user asymptotic efficiency[6] defined formally as

$$\eta_k = \sup\left\{0 \le r \le 1;\ P_k(1/\sigma) = O\left(Q\left(\sqrt{rw_k}/\sigma\right)\right)\right\}. \quad (53)$$

This parameter depends both on the signal constellation and on the multiuser detector employed. In the case of the minimum-error probability detector, (46) indicates that in the high SNR region the behavior of the kth user error probability coincides with that of an antipodal single-user system with bit-energy equal to $d^2_{k,\min}$. Therefore, the maximum achievable asymptotic efficiency is given by

$$\eta_k = d^2_{k,\min}/w_k. \quad (54)$$

Since the upper bound (46) is actually an upper bound on the error probability of the multiuser maximum-likelihood sequence detector, this detector achieves the maximum asymptotic efficiency, although it is not optimum in terms of error probability. The set of K-user asymptotic efficiencies emerge as the parameters that determine optimum performance for all practical purposes in the SNR region of usual interest. In a sequel to this paper, we derive analytical expressions, bounds, and numerical methods for the computation of these parameters which play a central role in the analysis and comparison of multiuser detectors.

IV. Numerical Examples

Two pairs of lower and upper bounds to the kth user minimum-error probability have been presented in Section III, and they have been shown to be tight asymptotically; nonetheless, it remains to ascertain the SNR level for which such asymptotic approximation is sufficiently accurate. In the sequel this question is illustrated by several examples of the computation of averages and extreme cases of the foregoing bounds with respect to the relative delays of asynchronous users. The first example is a baseband asynchronous system with two equal-energy users that employ a simple set of signal waveforms (Fig. 4). In this figure the upper bounds on the best and worst cases of the optimum detector are indistinguishable from each other, and for SNR higher than about 6 dB, from the single-user lower bound (which is also the minimum energy lower bound since $\eta_1 = 1$). Note also that the maximum interference coefficient, $\bar{I}_k = \max_{\alpha,\beta} I_k(\alpha,\beta)$ (recall Proposition 3) is one-third for all delays, and the performance of the conventional receiver varies only slightly with the relative delay.

In the next examples, we employ a set of spread-spectrum signals: three maximal-length signature sequences of length 31 generated to maximize a signal-to-multiple-access interference functional [7, table 5]. The average probability of error of the conventional receiver for equal-energy users

[6] This is not to be confused with Pitman's asymptotic measure of detection efficiency.

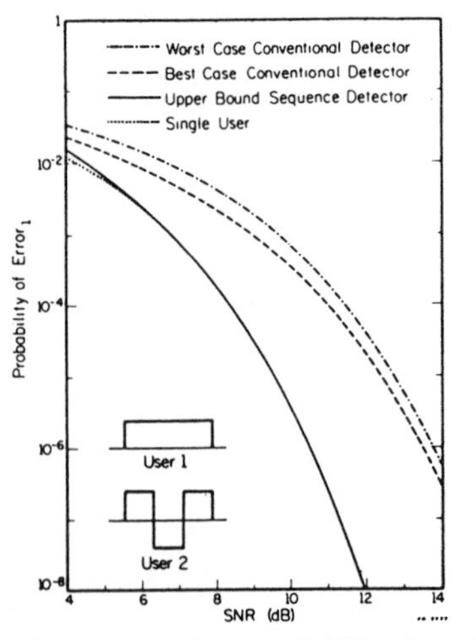

Fig. 4. Best and worst cases of error probability of user 1 achieved by conventional and optimum detectors.

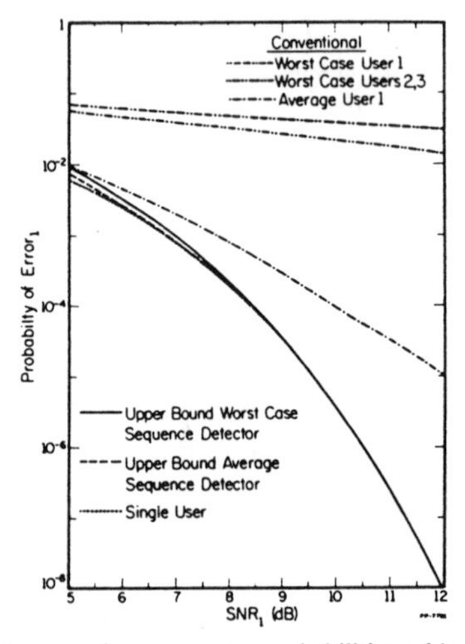

Fig. 5. Worst-case and average error probabilities achieved by conventional and optimum multiuser detectors with three active users employing m-sequences of length 31.

employing this signal set has been thoroughly studied previously [5], [14], and in Fig. 5 we reproduce (from [8, fig. 2]) the average error probability of user 1 achieved by the coherent conventional detector. Also shown in Fig. 5 are the worst cases of the conventional detector and upper bounds to the baseband worst-case and average minimum error probabilities for user 1. From the observation of Fig. 5, we can conclude that for error probabilities of 10^{-2} the average performance of the conventional detector is fairly close to the single-user lower bound, but the worst-case error probability is notably poor for the whole SNR range considered in the figure. Note, however, that since the signal set has good cross correlations for most of the relative delays, error probabilities close to the worst-case curve will occur with low probability. The worst-case and, especially, the average upper bounds on the optimum sequence detector performance are remarkably close to the single-user lower bound and show that the minimum-error

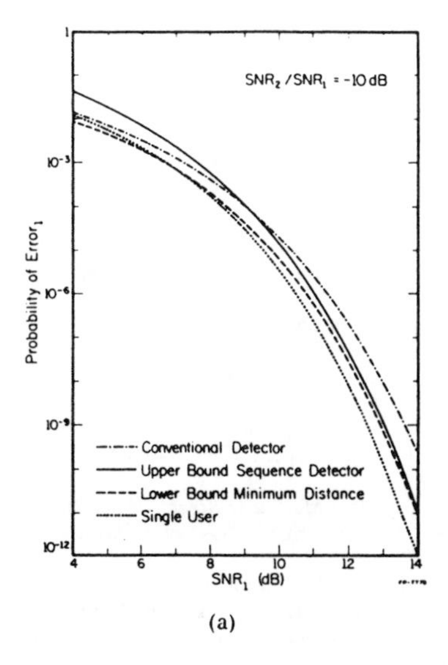

(a)

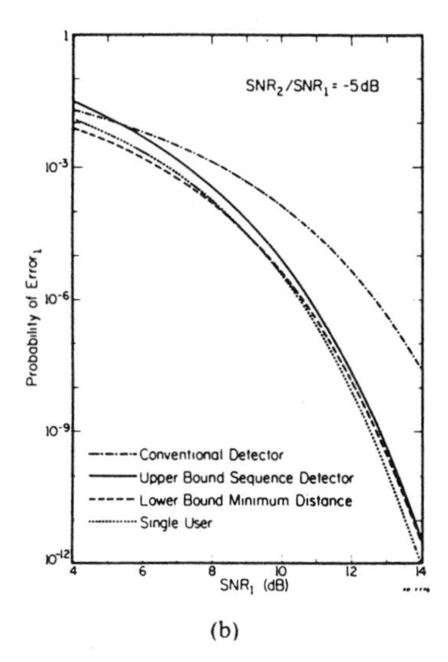

(b)

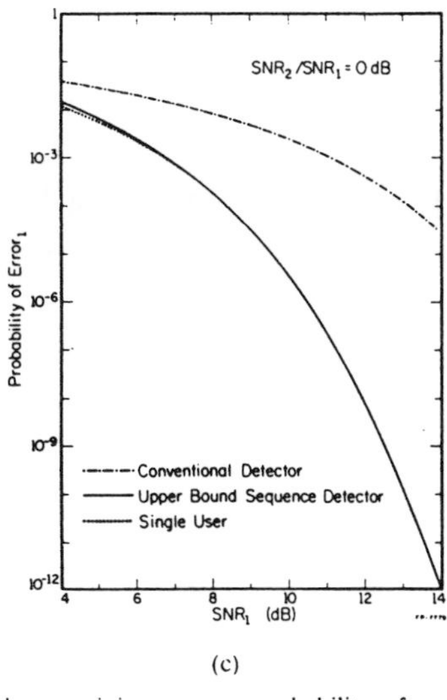

(c)

Fig. 6. Bounds on minimum error probability of user 1. Worst-case delays and two active users. (a) $w_2/w_1 = -10$ dB. (b) -5 dB. (c) 0 dB.

probability not only has a low average (around one order of magnitude better than that of the conventional detector, at 9 dB), but its dependence on the delays is negligible.

The next example investigates the *near–far* problem (i.e., the effects of unequal received energies) for two users that employ a subset of the previous set of maximal-length signature sequences. Bounds on the error probabilities corresponding to this example (worst-case relative delay between users 1 and 2) are calculated in Fig. 6 for three relative energies, namely, $SNR_2/SNR_1 = -10$ dB, -5 dB, and 0 dB. It is interesting to observe in the graphs corresponding to $SNR_2/SNR_1 = -10$ dB, -5 dB that all four bounds derived in this chapter play a role in some SNR interval; in particular, the error probability of the conventional detector is lower than the upper bound on the optimum sequence detector for small SNR. The opposite effect of an increase in the energy of the interfering users on the minimum and conventional probabilities of error is apparent: while the optimum sequence detector bounds become tighter and closer to the single-user lower bound, the conventional error probability grows rapidly until it becomes multiple-access limited (for $SNR_2/SNR_1 =$ 6.3 dB).

The results presented here open the possibility of a trade-off between the complexities of the receiver and the signal constellation in order to achieve a fixed level of performance; the actual compromise being dictated by the relative power of each user at the receiver. In multipoint-to-multipoint problems, when some active users need not be demodulated at a particular location, such a trade-off is likely to favor a multiuser detector that takes into account only those unwanted users that are not comparatively weak. If the signal constellation has moderate cross-correlation properties and the energy of the kth user is not dominant, then its minimum distance is achieved by an error sequence with only one nonzero component, and the minimum error probability approaches asymptotically the single-user bit error rate. This implies that contrary to what is sometimes conjectured, the performance of the conventional receiver is not close to the minimum error even if signals with low cross correlations are employed.

References

[1] P. R. Chevillat, "*N*-User trellis coding for a class of multiple-access channels," *IEEE Trans. Inform. Theory*, vol. IT-27, no. 1, pp. 114–120.

[2] G. D. Forney, "Lower bounds on error probability in the presence of large intersymbol interference," *IEEE Trans. Commun.*, vol. COM-20, pp. 76–77, Feb. 1972.

[3] ——, "Maximum likelihood sequence estimation of digital sequences in the presence of intersymbol interference," *IEEE Trans. Inform. Theory*, vol. IT-18, pp. 363–378, May 1972.

[4] ——, "The Viterbi algorithm," *Proc. IEEE*, vol. 61, pp. 268–278, Mar. 1973.

[5] G. J. Foschini, "Performance bound for maximum-likelihood reception of digital data," *IEEE Trans. Inform. Theory*, vol. IT-21, pp. 47–50, Jan. 1975.

[6] R. Gagliardi, "*M*-link multiplexing over the quadrature communications channel," *IEEE Commun. Mag.*, vol. 22, pp. 22–30, Sep. 1984.

[7] F. D. Garber and M. B. Pursley, "Optimal phases of maximal

sequences for asynchronous spread-spectrum multiplexing," *Electron. Lett.*, vol. 16, pp. 756–757, Sep. 11, 1980.

[8] E. A. Geraniotis and M. B. Pursley, "Error probability for direct-sequence spread-spectrum multiple-access communications-Part II: Approximations," *IEEE Trans. Commun.*, vol. COM-30, pp. 985–995, May 1982.

[9] D. Horwood and R. Gagliardi, "Signal design for digital multiple access communications," *IEEE Trans. Commun.*, vol. COM-23, pp. 378–383, Mar. 1975.

[10] J. K. Omura, "Performance bounds for Viterbi algorithms," in *1981 IEEE Int. Commun. Conf. Rec.*, pp. 2.21–2.25.

[11] C. H. Papadimitrou and K. Steiglitz, *Combinatorial Optimization: Algorithms of Complexity*. Englewood Cliffs, NJ: Prentice-Hall, 1982.

[12] H. V. Poor, "Signal detection in multiple access channels," U.S. Army Research Office Proposal (Contract DAAG29-81-K-0062 to Coordinated Science Laboratory, Univ. of Illinois), 1980.

[13] J. G. Proakis, *Digital Communications*. New York: McGraw-Hill, 1983.

[14] M. B. Pursley, D. V. Sarwate, and W. E. Stark, "Error probability for direct-sequence spread-spectrum multiple-access communications—part I: Upper and lower bounds," *IEEE Trans. Commun.*, vol. COM-30, pp. 975–984, May 1982.

[15] J. E. Savage, "Signal detection in the presence of multiple-access noise," *IEEE Trans. Inform. Theory*, vol. IT-20, pp. 42–49, Jan. 1974.

[16] K. S. Schneider, "Optimum detection of code division multiplexed signals," *IEEE Trans. Aerosp. Electron. Syst.*, vol. AES-15, pp. 181–185, Jan. 1979.

[17] G. Ungerboeck, "Adaptive maximum likelihood receiver for carrier-modulated data transmission systems," *IEEE Trans. Commun.*, vol. COM-22, pp. 624–636, May 1974.

[18] W. Van Etten, "Maximum likelihood receiver for multiple channel transmission systems," *IEEE Trans. Commun.*, vol. COM-24, pp. 276–283, Feb. 1976.

[19] S. Verdú, "Optimum sequence detection of asynchronous multiple-access communications," in *Abstracts of Papers: IEEE 1983 Int. Symp. Inform. Theory*, Sep. 1983, p. 80.

[20] ——, "Minimum probability of error for asynchronous multiple-access communication systems," in *Proc. 1983 IEEE Mil. Commun. Conf.*, vol. 1, Nov. 1983, p. 80, pp. 213–219.

[21] S. Verdú and H. V. Poor, "Backward, forward and backward-forward dynamic programming models under commutativity conditions," in *Proc. 23rd IEEE Conf. Decision Contr.*, Dec. 1984, pp. 1081–1086.

[22] S. Verdú, *Optimum Multi-user Signal Detection*, Ph.D. dissertation, Dep. Elec. Comput. Eng., Univ. of Illinois, Urbana-Champaign, Aug. 1984.

[23] A. J. Viterbi and J. K. Omura, *Principles of Digital Communication and Coding*. New York: McGraw-Hill, 1979.

[24] S. Verdú, "New bound on the error probability of maximum likelihood sequence detection of signals subject to intersymbol interference," in *Proc. 1985 Conf. Inform. Sci. Syst.*, Mar. 1985, pp. 413–418.

Reprinted from *IEEE Transactions on Information Theory*, Vol. 35, No. 1, Jan. 1989, pp. 123-136.

Linear Multiuser Detectors for Synchronous Code-Division Multiple-Access Channels

RUXANDRA LUPAS, STUDENT MEMBER, IEEE, AND SERGIO VERDÚ, SENIOR MEMBER, IEEE

Abstract —In code-division multiple-access systems, simultaneous multiuser accessing of a common channel is made possible by assigning a signature waveform to each user. Knowledge of these waveforms enables the receiver to demodulate the data streams of each user, upon observation of the sum of the transmitted signals, perturbed by additive noise. Under the assumptions of symbol-synchronous transmissions and white Gaussian noise, we analyze the detection mechanism at the receiver, comparing different detectors by their bit error rate in the low background noise region, and by their worst-case behavior in a near–far environment where the received energies of the users are not necessarily similar. Optimum multiuser detection achieves important performance gains over conventional single-user detection at the expense of computational complexity that grows exponentially with the number of users. It is shown that in the synchronous case the performance achieved by *linear* multiuser detectors (whose complexity per demodulated bit is only linear in the number of users) is similar to that of optimum multiuser detection. Attention is focused on detectors whose linear memoryless transformation is a generalized inverse of the matrix of signature waveform crosscorrelations, and on the optimum linear detector. It is shown that the generalized inverse detectors exhibit the same degree of near–far resistance as the optimum multiuser detector; the optimum linear detector is obtained subsequently, along with sufficient conditions on the signal energies and crosscorrelations to ensure that its performance is equal to that of the optimum multiuser detector.

I. Introduction

CODE-DIVISION multiple-access is a multiplexing technique where several independent users access simultaneously a multipoint-to-point channel by modulating preassigned signature waveforms. These waveforms are known to the receiver, which observes the sum of the modulated signals embedded in additive white Gaussian noise. If the assigned signals were orthogonal, then a bank of decoupled single-user detectors (matched filters followed by thresholds) would achieve optimum demodulation. In practice, however, orthogonal signal constellations are more the exception than the rule because of bandwidth or complexity limitations (the number of potential users can be very large), lack of synchronism, or other design constraints. Therefore the question of interest is how to demodulate the transmitted messages when the assigned signals are not orthogonal. In practice, demodulation strategies have been restricted to single-user detection, thereby placing the whole burden of complexity on the cross correlation properties of the signal constellation. Recently, the optimum multiuser detector for general asynchronous Gaussian channels was derived and analyzed in [1]. The optimum detector significantly outperforms the conventional single-user detector at the expense of a marked increase in computational complexity—it grows exponentially with the number of users.

The purpose of this paper is to investigate new low-complexity multiuser detection strategies that approach the performance of the optimum detector and to gain further insight into the performance of the optimum multiuser detector. Our attention in focused on symbol-synchronous channels, where the symbol epochs of all users coincide at the receiver. Although in practice this assumption rules out the important class of completely asynchronous code-division multiple-access systems, it holds in slotted channels, and its study is a necessary prerequisite for tackling the general asynchronous channel by allowing us to gain some appreciation of the main issues in the simplest possible setting.

The performance measure of interest is the probability of error of each user. In multiuser problems it is often more convenient and intuitively sound to give information concerning the error probability by means of the *efficiency*, or ratio between the *effective* signal-to-noise ratio (SNR) and the actual SNR, where the effective SNR is the one required to achieve the same probability of error in the absence of interfering users, and the actual SNR is the received energy of the user divided by the power spectral density level of the background thermal white Gaussian noise (not including interference from other users). Note that since the single-user error probability is a one-to-one function of the SNR, the efficiency gives the same information as the error probability. Its limit as the background Gaussian noise level goes to zero, the *asymptotic efficiency*, characterizes the underlying performance loss when the dominant impairment is the existence of other users rather than the additive channel noise. Denoting the power spectral density level of the background white noise by σ^2, the kth user asymptotic efficiency of a detector whose kth user error probability and energy are equal to P_k and w_k,

Manuscript received February 27, 1987; revised October 21, 1987. The material in this paper was partially presented at the 25th IEEE Conference on Decision and Control, Athens, Greece, December 1986. This work was supported in part by the U.S. Army Research Office under Contract DAAL03-87-K-0062.

The authors are with the Department of Electrical Engineering, Princeton University, Princeton, NJ 08544.

IEEE Log Number 8825697.

0018-9448/89/0100-0123$01.00 ©1989 IEEE

respectively, can be written as [1][1]

$$\eta_k = \sup\left\{0 \le r \le 1;\ \lim_{\sigma \to 0} P_k(\sigma)/Q\left(\frac{\sqrt{rw_k}}{\sigma}\right) < +\infty\right\}, \tag{1.1}$$

i.e., the logarithm of the error probability goes to zero with the same slope as the single-user bit error rate with energy $\eta_k w_k$. In this paper we compare the performance of the various multiuser detectors by means of the asymptotic efficiency. In the high[2] SNR region, the advantage of this measure over the probability of error is twofold: it quantifies the performance degradation due to the existence of other users in a simple, intuitive way, and in contrast to multiuser error probability for which only (asymptotically tight) upper and lower bounds are known [1], exact expressions for the asymptotic efficiency are feasible.

The main shortcoming of currently operational networks employing code-division multiple-access is the *near–far* problem. This refers to the situation wherein the received powers of the users are dissimilar (e.g., in mobile radio networks). Since the output of the matched filter of each user contains a spurious component which is linear in the amplitude of each of the interfering users, the error probability increases to 1/2 as the multiuser interference grows, the asymptotic efficiency becomes zero, and the conventional single-user detector is unable to recover reliably the messages transmitted by the weaker users even if signals with very low crosscorrelations are assigned to the users. However, the near–far problem is not an inherent characteristic of code-division multiple-access systems. Rather, it is the inability of the conventional single-user receiver to exploit the structure of the multiple-access interference that accounts for the ubiquity of the near–far problem in practice. We show that the optimum multiuser detector and other multiuser detectors with much lower computational complexity are *near–far resistant* under mild conditions on the signal constellation. By near–far resistance we mean the asymptotic efficiency minimized over the energies of all the interfering users. If this minimum is nonzero, and, as a consequence, the performance level is guaranteed no matter how powerful the multiuser interference, then we say that the detector is near–far resistant.

The organization of the rest of the paper is as follows. The asymptotic efficiency and the near–far resistance of both the conventional and the optimum detectors are given in Section II. In Section III, we introduce the *decorrelating* multiuser detector. This detector linearly transforms each vector of matched filter outputs with a generalized inverse of the signal crosscorrelation matrix. It is shown that, somewhat unexpectedly, the near–far resistance of the optimum multiuser detector coincides with that of the decorrelating detector whose complexity per demodulated bit is only linear in the number of users. Finally, Section IV investigates the performance of the optimum linear transformation and gives sufficient conditions on the signal energies and crosscorrelations to ensure that the asymptotic efficiency of the optimum linear transformation is equal to that of the optimum multiuser detector.

II. Single-User Detection and Optimum Multiuser Detection

Suppose that the kth user is assigned a finite energy signature waveform, $\{s_k(t),\ t \in [0,T]\}$, and that it transmits a string of bits by modulating that waveform antipodally. If the users maintain symbol synchronization and share a white Gaussian multiple-access channel, then the receiver observes

$$r(t) = \sum_{k=1}^{K} b_k(j) s_k(t - jT) + \sigma n(t), \qquad t \in [jT, jT+T] \tag{2.1}$$

where $n(t)$ is a realization of a unit spectral density white Gaussian process and $\{b_k(j) \in \{-1,1\}\}_j$ is the kth user information sequence. Assuming that all possible information sequences are equally likely, it suffices to restrict attention to a specific symbol interval in (2.1), e.g., $j = 0$.

It is easy to check that the likelihood function depends on the observations only through the outputs of a bank of matched filters:

$$y_k = \int_0^T r(t) s_k(t)\, dt, \qquad k = 1, \cdots, K \tag{2.2}$$

and therefore $y = (y_1, \cdots, y_K)$ are sufficient statistics for demodulating $b = (b_1, \cdots, b_K)$. We investigate ways of processing these sufficient statistics, which according to (2.1) and (2.2) depend on the transmitted bits in the following way:

$$y = Hb + n \tag{2.3}$$

where H is the nonnegative definite matrix of crosscorrelations between the assigned waveforms:

$$H_{ij} = \int_0^T s_i(t) s_j(t)\, dt \tag{2.4}$$

and its diagonal entries are the energies-per-bit, $H_{ii} = w_i > 0$, of each user; and n is a zero-mean Gaussian K-vector with covariance matrix equal to $\sigma^2 H$.

Conventional single-user detection is the simplest way to make decisions based on y_k; demodulation is decoupled and the multiuser interference is ignored, yielding the following decisions for the kth user:

$$\hat{b}_k^c = \operatorname{sgn} y_k.$$

On the other hand, the optimum multiuser detector selects the most likely hypothesis $\hat{b}^* = (\hat{b}_1^*, \cdots, \hat{b}_K^*)$ given the observations, which corresponds to selecting the noise

[1] $Q(x) = \int_x^\infty (1/\sqrt{2\pi}) e^{-v^2/2}\, dv$.

[2] In the numerical results of [1] and [2], the efficiency is indistinguishable from the asymptotic efficiency for SNR's higher than 7 dB.

realization with minimum energy, i.e.,

$$\hat{b}^* \in \arg \min_{b \in \{-1,1\}^K} \int_0^T \left[r(t) - \sum_{k=1}^{K} b_k s_k(t) \right]^2 dt$$
$$= \arg \max_{b \in \{-1,1\}^K} 2y^T b - b^T H b. \tag{2.5}$$

The computational complexities of the single-user detector and the optimum multiuser detector are radically different. While the time-complexity per bit (TCB) of the single-user detector is independent of the number of users, no algorithm that solves (2.5) in polynomial time in K is known. The reason for this is the nondeterministic polynomial (NP)-completeness of optimum multiuser detection (Appendix I).

The performances of the detectors are also quite different. It is straightforward to find the kth user probability of error of the conventional single-user detector:

$$P_k^c = P[y_k > 0 | b_k = -1]$$
$$= \sum_{\substack{b \in \{-1,1\}^K \\ b_k = -1}} P[y_k > 0 | b] P[b | b_k = -1]$$
$$= 2^{1-K} \sum_{\substack{b \in \{-1,1\}^K \\ b_k = -1}} Q\left(\frac{w_k - \sum_{i \neq k} b_i H_{ik}}{\sigma \sqrt{w_k}} \right). \tag{2.6}$$

In the low background noise region, the foregoing summation is dominated by the term corresponding to the least favorable bits of the interfering users, i.e., $b_i = \operatorname{sgn}(H_{ik})$. Thus the asymptotic efficiency of the conventional detector is equal to

$$\eta_k^c = \sup\left\{ 0 \le r \le 1;\ \lim_{\sigma \to 0} P_k^c / Q\left(\frac{\sqrt{r w_k}}{\sigma} \right) < +\infty \right\}$$
$$= \max{}^2 \left\{ 0, 1 - \sum_{i \neq k} \frac{|H_{ik}|}{w_k} \right\}$$
$$= \max{}^2 \left\{ 0, 1 - \sum_{i \neq k} |R_{ik}| \frac{\sqrt{w_i}}{\sqrt{w_k}} \right\} \tag{2.7}$$

where R is the matrix of normalized (unit-energy) cross correlations, i.e.,

$$H = W^{1/2} R W^{1/2} \tag{2.8}$$

where $W = \operatorname{diag}\{w_1, \cdots, w_K\}$. It follows from (2.7) that the conventional kth user detector is near–far resistant (i.e., its asymptotic efficiency is bounded away from zero as a function of the interfering users' energies) only if $R_{ik} = 0$ for all $i \neq k$, i.e., only if the kth user's signal is orthogonal to the subspace spanned by the other signals. Otherwise,

$$\bar{\eta}_k^c = \inf_{\substack{w_i \ge 0 \\ i \neq k}} \eta_k^c = 0. \tag{2.9}$$

The kth user error probability of the optimum multiuser receiver is asymptotically (as $\sigma \to 0$) equivalent to that of a binary test between the two closest hypotheses that differ in the kth bit (see [1]). The square of the Euclidean distance between the signals corresponding to these two hypotheses is equal to

$$\min_{b \in \{-1,1\}^K} \min_{\substack{d \in \{-1,1\}^K \\ d_k \neq b_k}} \left| \sum_{i=1}^{K} b_i s_i(t) - \sum_{i=1}^{K} d_i s_i(t) \right|^2$$
$$= 2 \min_{\substack{\epsilon \in \{-1,0,1\}^K \\ \epsilon_k = 1}} \epsilon^T H \epsilon. \tag{2.10}$$

Hence the asymptotic efficiency of the optimum multiuser detector is equal to

$$\eta_k = \frac{1}{w_k} \min_{\substack{\epsilon \in \{-1,0,1\}^K \\ \epsilon_k = 1}} \epsilon^T H \epsilon. \tag{2.11}$$

This is the highest efficiency attainable by any detector because as $\sigma \to 0$ the optimum multiuser detector achieves minimum probability of error for each user. In the two-user case, denoting $\rho = R_{12}$, (2.11) reduces to

$$\eta_1 = \min\left\{ 1, 1 + \frac{w_2}{w_1} - 2|\rho| \frac{\sqrt{w_2}}{\sqrt{w_1}} \right\}, \tag{2.12}$$

and similarly for user 2. Unfortunately, no explicit expressions are known for (2.11) in general. In fact, the combinatorial optimization problem in (2.11) is also NP-complete (Appendix I).

Nevertheless, it is indeed possible to obtain a closed-form expression for the near–far resistance of the optimum multiuser detector, because the minimization of the asymptotic efficiencies with respect to the energies of the interfering user waveforms reduces the combinatorial optimization problem in (2.11) to a continuous optimization problem whose solution is given by the following result.

Proposition 1: Denote by R^+ the Moore–Penrose generalized inverse[3] of the normalized crosscorrelation matrix R. If the signal of the kth user is linearly independent, i.e., it does not belong to the subspace spanned by the other signals, then

$$\bar{\eta}_k = \inf_{\substack{w_i \ge 0 \\ i \neq k}} \eta_k = \frac{1}{R_{kk}^+} \tag{2.13}$$

Otherwise, $\bar{\eta}_k = 0$.

Proof: Using (2.11) for the maximum asymptotic efficiency of the kth user, we obtain

$$\bar{\eta}_k = \min_{\substack{w_i \ge 0 \\ i \neq k}} \min_{\substack{\epsilon \in \{-1,0,1\}^K \\ \epsilon_k = 1}} \frac{1}{w_k} \epsilon^T H \epsilon$$
$$= \min_{\substack{w_i \ge 0 \\ i \neq k}} \min_{\substack{\epsilon \in \{-1,0,1\}^K \\ \epsilon_k = 1}} \frac{1}{w_k} \epsilon^T W^{1/2} R W^{1/2} \epsilon$$
$$= \min_{\substack{x \in R^K \\ x_k = 1}} x^T R x$$
$$= \min_{z \in R^{K-1}} \left(1 + 2 z^T a_k + z^T R_k z \right) \tag{2.14}$$

[3]A generalized inverse A of a matrix B is any matrix that satisfies 1. $ABA = A$ and 2. $BAB = B$. The Moore–Penrose generalized inverse is the unique generalized inverse that satisfies 3. AB and BA are Hermitian.

where R_k is obtained from R by deleting the kth row and column and a_k is the kth column of R with the kth entry removed. Henceforth, we denote such a partitioning of a symmetric matrix with respect to the kth row and column by $R=[R_k, a_k, 1]$, where the rightmost element in the square brackets is the kth diagonal entry. The minimum in the right side of (2.14) is achieved by any element z^* such that

$$R_k z^* = -a_k. \tag{2.15}$$

Because of the Fredholm theorem [11, p. 115], the solvability of (2.15) is equivalent to a_k being orthogonal to the null space of R_k. However, for all $z \in R^{K-1}$ the parabola $q(v) = v^2 + 2vz^T a_k + z^T R_k z$ has at most one zero because it is equal to the quadratic form of the nonnegative definite matrix R with a vector whose kth coordinate is v and whose other components are equal to z. Therefore, the discriminant of the parabola satisfies $(z^T a_k)^2 - z^T R_k z \le 0$; in particular, if z belongs to the null space of R_k, then $z^T a_k = 0$. So a_k is indeed orthogonal to the null space of R_k. Substituting (2.15) into (2.14) we obtain

$$\begin{aligned}\bar{\eta}_k &= 1 - z^{*T} R_k z^* \\ &= 1 - z^{*T} R_k R_k^+ R_k z^* \\ &= 1 - a_k^T R_k^+ a_k. \end{aligned} \tag{2.16}$$

Notice that the kth user is linearly dependent if and only if there exists a linear combination of the columns of R that includes the kth column and is equal to the zero vector. Therefore, if a user is linearly dependent then we can find x such that $Rx=0$ and $x_k=1$, in which case the penultimate equation in (2.14) indicates that $\bar{\eta}_k=0$. To obtain the near–far resistance of a linearly independent user, we employ the following property, which is invoked again later on.

Lemma 1: If the kth user is linearly independent, then every generalized inverse R^I of R satisfies: $(R^I R)_{kj} = \delta_{kj}$, $(RR^I)_{jk} = \delta_{jk}$ and $R^I_{kk} = R^+_{kk}$. (Analogous formulas hold for the unnormalized crosscorrelation matrix H.)

Proof of Lemma 1: Let $S = R^I R - I$. By the definition of generalized inverse, it follows that $RS = 0$, i.e., every column of S is in the null space of R. However, if the kth user is linearly independent, it is necessary that the kth element of each such column be zero. Hence $(R^I R - I)_{kj} = 0$ for all $j = 1, \cdots, K$.

Similarly, with $S = RR^I - I$ and $SR = 0$, we obtain $(RR^I)_{jk} = \delta_{jk}$. Equivalently, $RR^I u_k = u_k$, using the kth unit vector u_k. Hence, for any generalized inverses $R_1^I, R_2^I, R(R_1^I - R_2^I)u_k = 0$. However, since the kth user is linearly independent, it is necessary that the kth element of each vector in the null space of R be zero. Hence $(R_1^I - R_2^I)_{kk} = 0$.

Now we continue with the proof of Proposition 1. Partitioning R^+ with respect to the kth row and column, we have, say, $R^+ = [C, c, \gamma]$. Now, computing the submatrices of the partitioned matrix R^+R and using Lemma 1, it follows that

$$R_k c + \gamma a_k = 0 \tag{2.17}$$

and

$$c^T a_k + \gamma = 1. \tag{2.18}$$

Notice that $\gamma \neq 0$, for otherwise c would belong to the null space of R_k and would not be orthogonal to a_k, which, as we saw, is not possible. Finally, substituting (2.17) into (2.16) we obtain

$$\begin{aligned}\bar{\eta}_k &= 1 - \frac{1}{\gamma^2} c^T R_k R_k^+ R_k c \\ &= 1 - \frac{1}{\gamma^2} c^T R_k c \\ &= 1 + \frac{1}{\gamma} c^T a_k \\ &= \frac{1}{\gamma} = \frac{1}{R^+_{kk}} \end{aligned} \tag{2.19}$$

where the second, third, and fourth equations follow from the definition of generalized inverse, (2.17) and (2.18), respectively.

III. The Decorrelating Detector

In the absence of noise, the matched filter output vector is $y = Hx$. Thus if the signal set is linearly independent (i.e., H invertible), the natural strategy to follow in this hypothetical situation is to premultiply y by the inverse crosscorrelation matrix H^{-1}. The detector $\hat{x} = \operatorname{sgn} H^{-1} y$ was analyzed in [8], where its performance was quantified in the presence of noise. In [6] it was erroneously shown (cf. [3]) that this detector is optimum in terms of bit-error rate. Note that the noise components in $H^{-1}y$ are correlated, and therefore $\operatorname{sgn} H^{-1}y$ does not result in optimum decisions. It is interesting to point out that this detector does not require knowledge of the energies of any of the active users. To see this, let $\tilde{y}_k = y_k / \sqrt{w_k}$, i.e., $\tilde{y}_k$ is the result of correlating the received process with the normalized (unit-energy) signal of the kth user. Then

$$\begin{aligned}\operatorname{sgn} H^{-1} y &= \operatorname{sgn} W^{-1/2} R^{-1} W^{-1/2} y \\ &= \operatorname{sgn} W^{-1/2} R^{-1} \tilde{y} \\ &= \operatorname{sgn} R^{-1} \tilde{y},\end{aligned}$$

and therefore, the same decisions are obtained by multiplying the vector of normalized matched filter outputs by the inverse of the normalized crosscorrelation matrix. Apart from the attractive asymptotic efficiency properties shown below for the decorrelating detector, further justification for its study is provided by the fact that it is the solution to the generalized likelihood ratio test or maximum likelihood detector (e.g., [12, ch. 2], [13, p. 291]) when the energies are not known by the receiver. This approach selects the decisions that maximize the maximum of the likelihood function over the unknown parameters, i.e., (cf. (2.5))

$$\hat{b}^g \in \arg \min_{b \in \{-1,1\}^K} \min_{\substack{w_i > 0 \\ i=1,\cdots,K}} \int_0^T \left[r(t) - \sum_{k=1}^K b_k s_k(t) \right]^2 dt$$

$$= \arg \min_{b \in \{-1,1\}^K} \min_{\substack{w_i > 0 \\ i=1,\cdots,K}} y^T H^{-1} y + b^T H b - 2 b^T y$$

$$= \arg \min_{b \in \{-1,1\}^K} \min_{\substack{w_i > 0 \\ i=1,\cdots,K}} \cdot \tilde{y}^T R^{-1} \tilde{y} + b^T W^{1/2} R W^{1/2} b - 2(W^{1/2} b)^T \tilde{y}$$

$$= \operatorname{sgn}\left(\arg \min_{x \in R^K} x^T R x - 2 x^T \tilde{y}\right) = \operatorname{sgn} R^{-1} \tilde{y}.$$

Since in this paper the signal set is not constrained to be linearly independent, the above detector need not exist. In general, we consider the set $I(H)$ of generalized inverses[2] of the crosscorrelation matrix H and analyze the properties of the detector

$$\hat{x} = \operatorname{sgn} H^I y, \tag{3.1}$$

which we refer to as a *decorrelating* detector.

The kth user asymptotic efficiency achieved by a general linear transformation T can be obtained in a way similar to that of the efficiency of the conventional single-user detector $T = I$ (Section II). The first step is to find the bit error probability of the kth user:

$$P_k = P[\hat{x}_k = 1 | x_k = -1] = P[(THx + Tn)_k > 0 | x_k = -1]$$

$$= P\left[(Tn)_k > (TH)_{kk} - \sum_{j \neq k} (TH)_{kj} x_j \right]$$

$$= 2^{1-K} \sum_{\substack{x \in \{-1,1\}^K \\ x_k = -1}} P\left[(Tn)_k > (TH)_{kk} - \sum_{j \neq k} (TH)_{kj} x_j \right]. \tag{3.2}$$

Since the random variable $(Tn)_k$ is Gaussian with zero mean and variance equal to $(THT^T)_{kk}\sigma^2$, the sum in (3.2) is dominated as $\sigma \to 0$ by the term

$$2^{1-K} Q\left(\left((TH)_{kk} - \sum_{j \neq k} |(TH)_{kj}| \right) \Big/ \sigma \sqrt{(THT^T)_{kk}} \right). \tag{3.3}$$

Hence, according to definition (1.1), the kth user asymptotic efficiency achieved by the linear mapping T is

$$\eta_k(T) = \max{}^2 \left\{ 0, \frac{1}{\sqrt{w_k}} \frac{(TH)_{kk} - \sum_{j \neq k} |(TH)_{kj}|}{\sqrt{(THT^T)_{kk}}} \right\}. \tag{3.4}$$

Thus the kth user asymptotic efficiency of a decorrelating detector with matrix H^I is given by

$$\eta_k(H^I) = \max{}^2 \left\{ 0, \frac{1}{\sqrt{w_k}} \frac{(H^I H)_{kk} - \sum_{j \neq k} |(H^I H)_{kj}|}{\sqrt{(H^I H H^{I^T})_{kk}}} \right\}. \tag{3.5}$$

Proposition 2: If user k is linearly independent every $H^I \in I(H)$ satisfies

$$\eta_k(H^I) = 1/R^+_{kk}. \tag{3.6}$$

Thus for independent users the asymptotic efficiency of the decorrelating detector is independent of the energy of other users and of the specific generalized inverse selected.

Proof: If user k is linearly independent, we established in Lemma 1 that $(H^I H)_{kj} = \delta_{kj}$. Hence it follows from (3.5) that

$$\eta_k(H^I) = \frac{1}{w_k H^I_{kk}}. \tag{3.7}$$

Using the defining properties of generalized inverses (see footnote 2) it is easy to check that if $A \in I(R)$, then $W^{-1/2} A W^{-1/2} \in I(H)$, and if $B \in I(H)$, then $W^{1/2} B W^{1/2} \in I(R)$. Hence there is an obvious bijection between $I(R)$ and $I(H)$. Note that H^+ need not be the image of R^+ in this bijection. However, the inverse image of H^+, say $R^* \in I(R)$, satisfies

$$w_k H^+_{kk} = w_k (W^{-1/2} R^* W^{-1/2})_{kk} = R^*_{kk}. \tag{3.8}$$

Moreover, since user k is linearly independent, Lemma 1 implies that the denominator of (3.7) is equal to the left side of (3.8) and that the right side of (3.8) is equal to R^+_{kk}. Proposition 2 follows.

In Section IV it is shown that if user k is linearly dependent, then

$$\eta^d_k = \sup_{H^I \in I(H)} \eta_k(H^I) = \sup_{T \in R^{K \times K}} \eta_k(T) = \eta^l_k,$$

i.e., the best decorrelating detector and the best linear detector achieve the same kth user asymptotic efficiency.

Proposition 3: The near–far resistance of the decorrelating detector equals that of the optimum multiuser detector, i.e., for all $H^I \in I(H)$,

$$\inf_{\substack{w_j \geq 0 \\ j \neq k}} \eta_k(H^I) = \inf_{\substack{w_j \geq 0 \\ j \neq k}} \eta_k \equiv \bar{\eta}_k. \tag{3.9}$$

Proof: If user k is linearly independent, then according to Proposition 1 the near–far resistance of the optimum detector is equal to the asymptotic efficiency of the decorrelating detector (Proposition 2), which is independent of the energy of the other users. If user k is linearly dependent, Proposition 1 states that the near–far resistance of the optimum detector is zero, and hence the same is true for any detector.

The result of Proposition 3 is of special importance in a near–far environment, where the received signals have different energies and where the energy ratios may vary continuously over a broad scale if the positions of the users evolve dynamically. In this environment any decorrelating detector, with its linear time-complexity per bit, offers the same near–far resistance as the optimum multiuser detector, whose time-complexity per bit is exponential.

For the case where the signal set is independent, i.e., H is nonsingular (and $\eta_k^d = \eta_k(H^{-1})$ is energy-independent for all users), a geometric explanation for the equality of $\bar{\eta}_k$ and η_k^d can be given in the two-user case. Recall that the received signal y satisfies: $y = Hx + n$ and the noise autocovariance matrix is H. To have spherically symmetric noise, it is convenient to work in the $H^{-1/2}y$ domain. Here the hypotheses, denoted by A, B, C, D in Fig. 1, are at the points $H^{1/2}x$, with $x \in \{-1,1\}^2$. Since in this domain the matched filter output noise is spherically symmetric and Gaussian, the decision regions of the maximum likelihood detector, determined by the minimum Euclidean distance rule, are given by the perpendicular bisectors of the segments between the different hypotheses, and the kth user asymptotic efficiency corresponds to the square of half the minimum distance between distinct hypotheses differing in the kth bit.

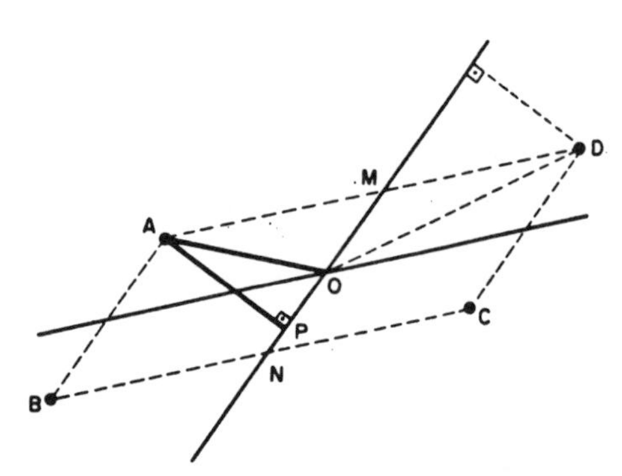

Fig. 1. Hypotheses and decision regions in two-user case.

The decision regions of the decorrelating detector are cones with a vertex at the origin, such that application of H^{-1} maps them to the coordinate axes. Thus in the $H^{-1/2}y$-domain the decision cones pass through the points $H^{1/2}e$, with e the unit vectors in R^2. These points are at the center of the sides of the parallelogram formed by the hypotheses, because the unit vectors can be represented as half the sum of adjacent hypotheses. So, the decorrelating detector decision boundaries are parallel to the parallelogram sides and intersect it at the centers of its sides. The kth bit-error probability (by symmetry we can assume that the transmitted bit was -1) is the sum of two integrals, one for each possibility for the remaining bit, of the noise density function over the region in which the kth bit is decoded as 1. In this case the kth bit-error probability can be easily computed by taking advantage of the aforementioned properties. To this end we rotate the coordinate system to let the y axis coincide with the kth-bit decision boundary and use the equal distance property of the decision boundary to the hypotheses, to observe that the two integrals are equal. We then use the spherical symmetry of the noise to identify each integral as a Q-function of the distance of the hypothesis to the decision boundary. Hence the kth user asymptotic efficiency of the decorrelating detector is equal to the square of the distance of any hypothesis to the kth bit decision boundary. Thus, in Fig. 1, $\sqrt{\eta_1}$ is the length of the shortest of the segments AM, AO and BO, and $\sqrt{\eta_1^d}$ is the length of AP. The result of Proposition 3 can now be interpreted as follows. Since η appears as the hypotenuse and η^d as the leg of a right-angled triangle, η is lower-bounded by the energy independent η^d. However, since the triangle angles vary with increasing energy of the interfering user waveform, there is a particular energy ratio for which the triangle degenerates into a line segment. This is the point when η reaches its minimum $\bar{\eta}$, which is geometrically identical with η^d. For the parallelogram formed by the hypotheses, this is the case where a diagonal is perpendicular to a side (e.g., AO perpendicular to CD).

IV. The Optimum Linear Multiuser Detector

We now turn to the question of finding the optimum linear detector. We have seen that this is a fruitful approach, since a particular type of linear detector, the decorrelating detector, offered a substantial improvement in asymptotic efficiency compared to the single-user detector, while its near–far resistance equaled that of the optimum multiuser detector. While we now know that no detector, linear or nonlinear, can outperform the decorrelating detector with respect to near–far resistance, for fixed energies it is indeed possible to obtain linear detectors that have a higher asymptotic efficiency than the one achieved by the decorrelating detector.

We find the linear detector which maximizes the asymptotic efficiency (or equivalently minimizes the probability of bit error in the low-noise region) and compare the achieved asymptotic efficiency to the ones achieved by the conventional and optimal detectors. Thus we ask which mapping T: $R^K \to R^K$ maximizes the asymptotic efficiency of the decision scheme

$$\hat{x} = \operatorname{sgn}(Ty) = \operatorname{sgn}(THx + Tn). \tag{4.1}$$

The interpretation of this optimization problem in terms of decision regions is to find the optimal partition of the K-dimension hypotheses space into K decision cones with vertices at the origin. The surfaces of these cones determine the columns of the inverse T^{-1} of the mapping sought. Application of T on the cone configuration will map the cones on quadrants, after which a sign detector is used.

The kth user asymptotic efficiency of a general linear detector, as given by (4.1) was derived in (3.4):

$$\eta_k(T) = \max{}^2\left\{0, \frac{1}{\sqrt{w_k}} \frac{(TH)_{kk} - \sum_{j\neq k} |(TH)_{kj}|}{\sqrt{(THT^T)_{kk}}}\right\}. \tag{4.2}$$

The best linear detector has the asymptotic efficiency

$$\eta_k^l = \sup_{T \in R^{K\times K}} \eta_k(T). \tag{4.3}$$

Hence the asymptotic efficiency of the best linear detector

is equal to

$$\eta_k' = \sup_{v \in R^K} \max{}^2 \left\{ 0, \frac{1}{\sqrt{w_k}} \frac{h_k^T v - \sum_{j \neq k} |h_j^T v|}{\sqrt{v^T H v}} \right\}$$

$$= \max{}^2 \left\{ 0, \sup_{v \in R^K} \eta_k(v) \right\} \tag{4.4}$$

with

$$\eta_k(v) = \frac{1}{\sqrt{w_k}} \frac{h_k^T v - \sum_{j \neq k} |h_j^T v|}{\sqrt{v^T H v}} \tag{4.5}$$

where v denotes the kth row of T. To minimize the probability of P_k, we have to maximize the argument of the Q-function, and equivalently maximize the asymptotic efficiency $\eta_k(v)$, with respect to the components of the vector v. Since the map applied on the matched filter outputs is linear, the asymptotic efficiencies of all the users can be simultaneously maximized, each such maximization yielding the corresponding row of the map to be applied. For the sake of clarity, we first consider the two-user case, for which explicit expressions for the maximum linear asymptotic efficiency can be obtained.

A. The Two-User Case

Throughout this subsection we denote the normalized crosscorrelation between the signals by $\rho = R_{12}$. We first give an explicit expression for the optimum linear detector.

Proposition 4: The kth user optimal linear transformation $T_k(y) = v^T y$ on the matched filter outputs prior to threshold detection is given by

$$v^T = \left[1; -\operatorname{sgn} \rho \min\left\{1, |\rho| (w_k/w_i)^{1/2}\right\}\right] \tag{4.6}$$

$$= \begin{cases} [1; -\operatorname{sgn} \rho], & \text{if } (w_i/w_k)^{1/2} \leq |\rho| \\ b_k^T, & \text{otherwise} \end{cases} \tag{4.7}$$

where b_k^T is the kth row of the decorrelating detector and $(i, k) \in \{(1,2),(2,1)\}$.

Proof: Without loss of generality, let $k = 1$. We have

$$H = \begin{bmatrix} w_1 & \rho\sqrt{w_1 w_2} \\ \rho\sqrt{w_1 w_2} & w_2 \end{bmatrix}, \qquad v^T = [1; v_2] \tag{4.8}$$

$$\eta_1(v) = \frac{1}{\sqrt{w_1}} \frac{h_1^T v - |h_2^T v|}{\sqrt{v^T H v}}$$

$$= \frac{1 + \rho(w_2/w_1)^{1/2} v_2 - |\rho(w_2/w_1)^{1/2} + (w_2/w_1) v_2|}{\sqrt{1 + 2\rho(w_2/w_1)^{1/2} v_2 + (w_2/w_1) v_2^2}} \tag{4.9}$$

and the objective is to maximize the right side of (4.9) with respect to v_2. We consider the case $|\rho| = 1$ separately.

a) Case $|\rho| \neq 1$: Introduce an indicator function for the absolute value term:

$$I = \begin{cases} 1, & \rho + (w_2/w_1)^{1/2} v_2 > 0 \\ -1, & \rho + (w_2/w_1)^{1/2} v_2 < 0 \\ 0, & \text{else} \end{cases} \tag{4.10}$$

Then

$$\frac{d\eta_1'}{dv_2} = -\frac{(1-\rho^2)(w_2/w_1)}{\left(1 + 2\rho(w_2/w_1)^{1/2} v_2 + (w_2/w_1) v_2^2\right)^{3/2}} (I + v_2). \tag{4.11}$$

Therefore, we should take $v_2 = -I$ when this is consistent with the definition of I as a function of v_2. Thus

$$v_2 = \begin{cases} 1, & \text{if } I = -1 \Leftrightarrow 0 < (w_2/w_1)^{1/2} < -\rho \\ -1, & \text{if } I = 1 \Leftrightarrow 0 < (w_2/w_1)^{1/2} < \rho \end{cases}. \tag{4.12}$$

As can easily be seen, both values correspond to maxima. If neither of these conditions is met, the derivative does not have a zero. The optimal value for v_2 can be determined by taking a closer look at the behavior of $d\eta_1/dv_2$, in Fig. 2.

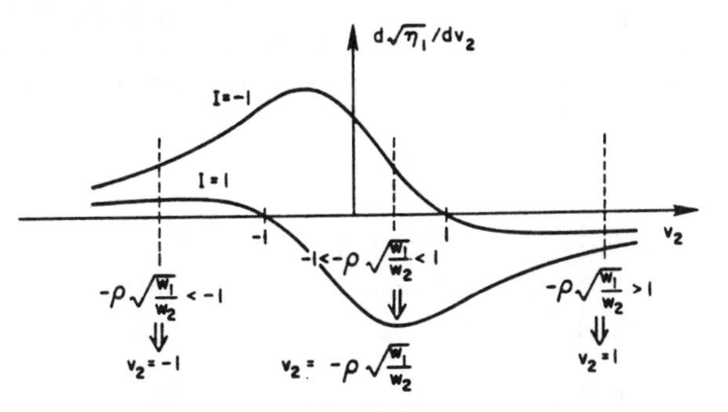

Fig. 2. Behavior of derivative in (4.11).

For both $I = 1$ and $I = -1$, the derivative of η_1 is positive for v_2 smaller than the abscissa of the zero of the derivative (which is equal to $-I$), and negative afterwards. Due to the nonlinearity of η_1 the derivative has the form corresponding to $I = -1$ for $v_2 < -\rho(w_2/w_1)^{1/2}$ and the form corresponding to $I = 1$ afterwards. Since the second branch (for $I = 1$) turns negative before the first one, we have to take the largest value of v_2 yielding a positive derivative on the first branch. It can easily be seen that in the "no-zero" case, $-1 < -\rho(w_2/w_1)^{1/2} < 1$, this is the point of discontinuity, i.e., $v_2 = -\rho(w_2/w_1)^{1/2}$. Note that for $\rho = 0$ we get $v^T = [1; 0]$, the identity transformation, as expected, since the users are then decoupled and a single-user detector is optimal. By taking the inverse of R we also see that in the no-zero case the optimal transformation vector is exactly the corresponding row of the inverse correlation matrix.

b) Case $|\rho| = 1$: Equation (4.9) becomes

$$\eta_1(v_2) = \operatorname{sgn}\left(1 + \operatorname{sgn}\rho (w_2/w_1)^{1/2} v_2\right) - (w_2/w_1)^{1/2} \tag{4.13}$$

We see that, for $(w_2/w_1)^{1/2} < 1$, any v_2 satisfying $v_2 \operatorname{sgn}\rho > -(w_1/w_2)^{1/2}$ is optimal, in particular the one given in (4.7). Otherwise, the asymptotic efficiency of the best linear transformation is zero, hence all linear transformations are equivalent.

Substituting the result of Proposition 4 into the asymptotic efficiency of (4.9), we obtain the following.

Proposition 5: The kth user asymptotic efficiency of the optimal linear two-user detector equals

$$\eta_k^l = \begin{cases} 1 - 2|\rho|(w_i/w_k)^{1/2} + w_i/w_k, & \text{if } (w_i/w_k)^{1/2} \le |\rho| \\ 1 - \rho^2, & \text{otherwise} \end{cases} \tag{4.14}$$

for $(i,k) \in \{(1,2),(2,1)\}$.

The kth user asymptotic efficiency obtained in the range $(w_i/w_k)^{1/2} < |\rho|$ equals the optimum asymptotic efficiency, obtained in (2.12). Even outside the region of optimality, the best linear detector shows a far better performance than the conventional single-user detector (see Fig. 3), since if $w_i/w_k > \rho^2$, then η_1^l is independent of w_i/w_k, whereas according to (2.7) the asymptotic efficiency of the conventional detector is equal to zero for $w_i/w_k \ge 1/\rho^2$.

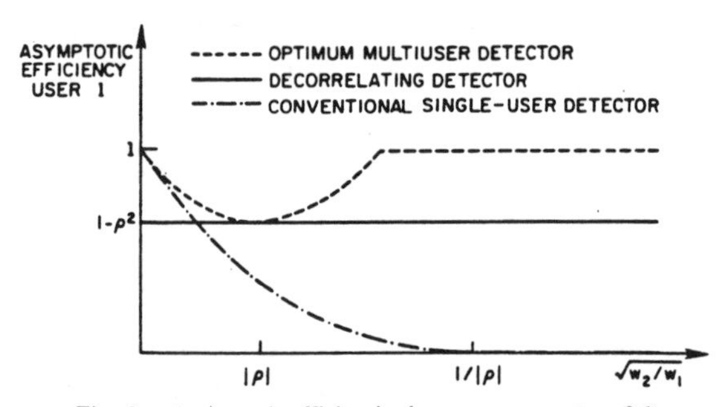

Fig. 3. Asymptotic efficiencies in two-user case ($\rho = 0.6$).

There is an intuitive interpretation of the dual behavior of the best linear detector and of the boundary point $(w_i/w_k)^{1/2} = |\rho|$. Let $k = 1$. The input to the threshold device corresponding to the first user, $z_1 = v^T y$, has three components:

$$z_1 = w_1\left[(1-\rho^2) + \rho\left(\rho + v_2 (w_2/w_1)^{1/2}\right)\right] x_1 + w_1\left[(w_2/w_1)^{1/2}\left(\rho + v_2(w_2/w_1)^{1/2}\right)\right] x_2 + \tilde{n} \tag{4.15}$$

where $\tilde{n}$ is a Gaussian random variable of variance $w_1\sigma^2[(1-\rho^2) + (\rho + v_2(w_2/w_1)^{1/2})^2]$. For $(w_2/w_1)^{1/2} > |\rho|$, the second term outweighs the second part of the first term, so the best one can do is to eliminate it by choosing $v_2 = -\rho(w_1/w_2)^{1/2}$ (the decorrelating detector). Since this minimizes the noise variance at the same time, it is the best strategy in this region. If, however, $(w_2/w_1)^{1/2} < |\rho|$, and v_2 is such that the term $\rho(\rho + v_2(w_2/w_1)^{1/2})$ is positive, it is a better policy to allow interference from user 2, which is compensated by the second part in the first term, and use the residual positive contribution in the first term to increase the SNR as compared to the decorrelating case. We have seen that this strategy leads to the same performance as the more complex maximum likelihood detector.

Note that in the two-user case the signal energies and cross correlations cannot be picked so as to allow *both* users optimal performance at the same time: for user 1 we need $(w_2/w_1)^{1/2} < |\rho| < 1$, whereas for user 2 we need $(w_2/w_1)^{1/2} > 1/|\rho| > 1$.

B. The K-User Case

Unlike Propositions 2 and 5, in the general K-user case it is not feasible to obtain an explicit expression for the asymptotic efficiency achieved by the best linear detector.

Proposition 6: The kth user asymptotic efficiency of the best linear detector equals:

$$\eta_k^l = \frac{1}{w_k}\max{}^2\left\{0, \max_{\substack{e_j \in \{-1,1\} \\ j \ne k}} \eta(e)\right\} \tag{4.16a}$$

with

$$\eta(e) = \max_{\substack{v \in R^K \\ v^T H v = 1 \\ e_j h_j^T v \ge 0 \\ j \ne k}} v_o^T H v \tag{4.16b}$$

where the ith component of v_o is equal to

$$(v_o)_i = \begin{cases} -e_i, & i \ne k \\ 1, & i = k \end{cases}.$$

Then the maximum $\eta(e)$ is achieved for $\tilde{v}$ such that

$$\tilde{v} = \frac{v_o + \sum_{j \ne k} \lambda_j e_j u_j}{\left(v_o^T H v_o + v_o^T H \sum_{j \ne k} \lambda_j e_j u_j\right)^{1/2}},$$

$$(u_j)_i = \begin{cases} 0, & i \ne j \\ 1, & i = j \end{cases} \tag{4.17}$$

$$e_j h_j^T \tilde{v} \ge 0 \text{ for } j \ne k \tag{4.18}$$

$$h_j^T \tilde{v} \ne 0 \Rightarrow \lambda_j = 0 \tag{4.19}$$

$$\lambda_j \ge 0, \qquad j \ne k. \tag{4.20}$$

Proof: Let

$$S_j^+ = \left\{x \in R^K : h_j^T x \ge 0\right\}$$

$$S_j^- = \left\{x \in R^K : h_j^T x \le 0\right\}. \tag{4.21}$$

From (4.5) we seek

$$\sup_{v \in R^K} \frac{1}{\sqrt{v^T H v}} \left(h_k^T v - \sum_{j \neq k} |h_j^T v| \right)$$

$$= \max_{\substack{e_j \in \{-1,1\} \\ j \neq k}} \sup_{v \in \cap_j S_j^{e_j}} \frac{1}{\sqrt{v^T H v}} \left(h_k^T v - \sum_{j \neq k} |h_j^T v| \right) \quad (4.22)$$

$$= \max_{\substack{e_j \in \{-1,1\} \\ j \neq k}} \eta(e), \text{ with } \eta(e)$$

$$= \sup_{v \in \cap_j S_j^{e_j}} \frac{1}{\sqrt{v^T H v}} \left(h_k^T v - \sum_{j \neq k} |h_j^T v| \right). \quad (4.23)$$

From the definition of v_o we see that the term in parentheses equals $v_0^T H v$. Now $v \in \cap_j S_j^{e_j} \Leftrightarrow e_j h_j^T v \geq 0,\ j \neq k$, and since η_k^l is invariant to scaling of v, maximization of the given functional over R^K is equivalent to maximization over the ellipsoid $v^T H v = 1$.

This proves the first part of Proposition 6. We now have to perform two maximizations where the second one has the explicit form of an exhaustive search. We turn our attention to the inner maximization in (4.16). We first show that it is possible to replace the feasible set therein by an equivalent convex set, i.e., the asymptotic efficiency is unchanged if we replace

$$\eta(e) = \sup_{\substack{v \in R^K \\ v^T H v = 1 \\ e_j h_j^T v \geq 0 \\ j \neq k}} v_o^T H v \text{ by } \eta(e) = \sup_{\substack{v \in R^K \\ v^T H v \leq 1 \\ e_j h_j^T v \geq 0 \\ j \neq k}} v_o^T H v. \quad (4.24)$$

To show (4.24), let $y = H^{1/2} v$, $z_j^T = j$th row of $H^{1/2}$. It then follows that $h_j^T v = z_j^T y$, $v_o^T H^{1/2} = y_o^T$, $v^T H v = y^T y = |y|^2$, and

$$\eta(e) = \sup_{\substack{y \in R^K \\ |y| = 1 \\ e_j z_j^T y \geq 0 \\ j \neq k}} y_o^T y = \sup_{\substack{y \in R^K \\ |y| = 1 \\ e_j z_j^T y \geq 0 \\ j \neq k}} |y_o| \, |y| \cos\alpha \quad (4.25)$$

where α is the angle between the vectors y_o and y. Since the inequality constraints are linear and partition the space into convex cones with vertex at the origin, the optimal angle α is independent of $|y|$. Either the optimal $\cos\alpha$ is nonnegative, in which case $\eta(e)$ is maximized for $|y|$ maximal in both versions, or it is negative, in which case $\eta(e) < 0$. In either case, the value of η_k^l, which involves comparison with zero, is unchanged if the maximization is performed over the interior of the ellipsoid, which completes the proof of the claim.

We now have to consider the following problem:

$$\eta(e) = \inf_{\substack{v \in R^K \\ v^T H v - 1 \leq 0 \\ -e_j h_j^T v \leq 0 \\ j \neq k}} -v_o^T H v. \quad (4.26)$$

Since this is a minimization problem of a continuous convex function on a compact convex set, it achieves a unique minimum on the set. Since all the functions are differentiable, we can apply the Kuhn–Tucker conditions (e.g., [4]), to get from condition (1),

$$-Hv_o + \lambda_o 2H\tilde{v} - \sum_{j \neq k} \lambda_j e_j h_j = 0,$$

hence

$$\tilde{v} = \frac{1}{2\lambda_o} \left(v_o + \sum_{j \neq k} \lambda_j e_j u_j \right) \quad (4.27)$$

with u_j the jth unit vector, as defined before. Equations (4.18) and (4.19) result from the Kuhn–Tucker conditions, condition (4.20) expresses the nonnegativity requirement for the λ_i. There is one more constraint to satisfy, which is $\tilde{v}^T H \tilde{v} = 1$:

$$1 = \tilde{v}^T H \tilde{v} = \frac{1}{2\lambda_o} \left(v_o^T H \tilde{v} + \sum_{j \neq k} \lambda_j e_j h_j^T \tilde{v} \right) = \frac{v_o^T H \tilde{v}}{2\lambda_o}.$$

We used condition (4.19) to get the last equality, so

$$2\lambda_o = v_o^T H \tilde{v} = \eta(e), \quad (4.28)$$

and since

$$v_o^T H \tilde{v} = \frac{1}{2\lambda_o} \left(v_o^T H v_o + \sum_{j \neq k} \lambda_j e_j h_j^T v_o \right),$$

we get

$$2\lambda_o = \left(v_o^T H v_o + v_o^T H \sum_{j \neq k} \lambda_j e_j u_j \right)^{1/2}.$$

This together with (4.27) completes the proof of Proposition 6.

In Appendix II we show an explicit procedure for finding the best linear detector characterized in Proposition 6. Its asymptotic efficiency is trivially upper- and lower-bounded by that of the optimum and decorrelating detectors, respectively. For certain values of energies and crosscorrelations these bounds are attained; sufficient conditions for this to occur are given in Propositions 7 and 8.

Proposition 7: The following are sufficient conditions on the signal energies and crosscorrelations for the best linear detector to achieve optimal kth user asymptotic efficiency:

$$\sqrt{w_k} > \max_{j=1,\cdots,K} \left(\frac{1}{|R_{kj}|} \sum_{i \neq k} \sqrt{w_i} |R_{ij}| \right). \quad (4.29)$$

Proof: In the optimality case, we show in Appendix II that $e_j h_j^T v_o > 0$ for all $j \neq k$. If we introduce $e_k = 1$ this has to hold also for $j = k$, otherwise we get negative asymptotic efficiency. We can rewrite these conditions as

$$DHD(-1-1\cdots 1\cdots -1)^T = \begin{bmatrix} H_{11} & e_1e_2H_{12} & \cdots & e_1H_{1k} & \cdots & e_1e_KH_{1K} \\ e_1e_2H_{21} & H_{22} & \cdots & e_2H_{2k} & \cdots & e_2e_KH_{2K} \\ \vdots & \vdots & \cdots & \vdots & \cdots & \vdots \\ e_1e_KH_{K1} & e_Ke_2H_{K2} & \cdots & e_KH_{Kk} & \cdots & H_{KK} \end{bmatrix} \begin{bmatrix} -1 \\ -1 \\ \cdot \\ 1 \\ \cdot \\ -1 \end{bmatrix} > 0 \quad (4.30)$$

where D is the diagonal matrix with ith diagonal element equal to e_i. We now see that a sufficient condition for the above inequality to hold for some $e_1,\cdots,e_K$ is

$$|H_{jk}| > \sum_{i \neq k} |H_{ji}|, \qquad j=1,\cdots,K.$$

The corresponding e_j are $e_j = \operatorname{sgn} H_{jk}$. Hence (4.29) follows by replacing H_{ij} by $\sqrt{w_iw_j}R_{ij}$.

Note that the above condition can be satisfied by only one user because

$$\sqrt{w_k} > \sqrt{w_j}/|R_{kj}| > \sqrt{w_j}, \qquad \text{for all } j.$$

Proposition 8: If user k is linearly independent, the following condition is sufficient for the kth row of the decorrelating detector $H^I \in I(H)$ to be the best kth user linear detector for a given set of signal energies and crosscorrelations:

$$|H^I_{jk}| \le H^I_{kk}, \qquad \text{for all } j \neq k. \quad (4.31)$$

Proof: We showed that in the terminal case $v = h^I_k/\sqrt{H^I_{kk}}$ is a maximizing vector for v_o^THv, hence there are nonnegative Kuhn–Tucker multipliers λ_k, such that, with (4.27) and (4.28),

$$v = \frac{h^I_k}{\sqrt{H^I_{kk}}} = \sqrt{H^I_{kk}}\left(v_o + \sum_{j\neq k}\lambda_j e_j u_j\right)$$

or

$$\frac{1}{H^I_{kk}}h^I_k = [(\lambda_1-1)e_1,\cdots,1,\cdots,(\lambda_K-1)e_K]^T$$

so

$$\lambda_j = 1 + e_jH^I_{jk}/H^I_{kk}, \qquad j \neq k. \quad (4.32)$$

Hence (4.31) is sufficient to ensure $\lambda_j \ge 0$ regardless of $\{e_i,\ i \neq k\}$.

Note that in the two-user case, Proposition 5 implies that the sufficient conditions found in Propositions 7 and 8 are also necessary.

Proposition 9: If user k is linearly dependent, then

$$\eta^d_k = \sup_{H^I \in I(H)} \eta_k(H^I) = \sup_{T \in R^{K\times K}} \eta_k(T) = \eta^l_k \quad (4.33)$$

i.e., for a dependent user the best decorrelating detector has the same asymptotic efficiency as the best linear detector.

Proof: Recall the bijection between $I(R)$ and $I(H)$ established in Section III. Let R^I denote the image of H^I under this bijection. Then, using (3.5) for the last equality, we can write

$$\eta^d_k = \sup_{H^I \in I(H)} \eta_k(H^I) = \sup_{R^I \in I(R)} \eta_k(W^{-1/2}R^IW^{-1/2})$$

$$= \max{}^2\left\{0,\ \sup_{R^I \in I(R)} \frac{(R^IR)_{kk} - \sum_{j\neq k}|(R^IR)_{kj}|\frac{\sqrt{w_j}}{\sqrt{w_k}}}{\sqrt{(R^IRR^{I^T})_{kk}}}\right\}. \quad (4.34)$$

Since R is nonnegative definite of rank r, it can be written using its orthonormal eigenvector matrix T and the $r\times r$ diagonal matrix Λ of nonzero eigenvalues of R, as

$$R = T\begin{bmatrix}\Lambda & 0\\ 0 & 0\end{bmatrix}T^T. \quad (4.35)$$

Then (cf. [10]), R^I is a generalized inverse of R if and only if, for some matrices U and V of appropriate dimensions, it can be written as

$$R^I = T\begin{bmatrix}\Lambda^{-1} & V\\ U & U\Lambda V\end{bmatrix}T^T. \quad (4.36)$$

Hence, using the corresponding partition of T, we can write

$$(R^IR)_{kj} = u_k^T[T_1\ T_2]\begin{bmatrix} I & 0\\ U\Lambda & 0\end{bmatrix}\begin{bmatrix}T_1^T\\ T_2^T\end{bmatrix}u_j$$
$$= u_k^T(T_1T_1^T + T_2U\Lambda T_1^T)u_j \quad (4.37)$$

$$(R^IRR^{I^T})_{kk} = u_k^T[T_1\ T_2]\begin{bmatrix}\Lambda^{-1} & U^T\\ U & U\Lambda U^T\end{bmatrix}\begin{bmatrix}T_1^T\\ T_2^T\end{bmatrix}u_k$$
$$= u_k^T(T_1\Lambda^{-1}T_1^T + T_2UT_1^T + T_1UT_2^T + T_2U\Lambda U^TT_2^T)u_k \quad (4.38)$$

and

$$\eta^d_k = \max{}^2\left\{0,\ \sup_{U\in R^{(K-r)\times r}} \frac{u_k^T(T_1+T_2U\Lambda)T_1^Tu_k - \sum_{j\neq k}|u_k^T(T_1+T_2U\Lambda)T_1^Tu_j|\frac{\sqrt{w_j}}{\sqrt{w_k}}}{\sqrt{u_k^T(T_1+T_2U\Lambda)\Lambda^{-1}(T_1+T_2U\Lambda)^Tu_k}}\right\}. \quad (4.39)$$

Since user k is dependent, $u_k^T T_2$, whose components are the kth components of the eigenvectors to eigenvalue zero, is nonzero. (Otherwise, for all x with $Rx = 0$, x_k would be zero, which implies that the kth user is linearly independent of the other users.) Hence since Λ is invertible, we can make the change of variables

$$x = (T_1 + T_2 U \Lambda)^T u_k \tag{4.40}$$

to get

$$\eta_k^d = \max{}^2 \left\{ 0, \sup_x \frac{x^T T_1^T u_k - \sum_{j \neq k} |x^T T_1^T u_j| \frac{\sqrt{w_j}}{\sqrt{w_k}}}{\sqrt{x^T \Lambda^{-1} x}} \right\}. \tag{4.41}$$

Using the same reasoning as in the proof of Proposition 6 for the best linear detector, we can write

$$\eta_k^d = \max{}^2 \left\{ 0, \max_{\substack{e_j \in \{-1,1\} \\ j \neq k}} \sup_{\substack{x \in R^r \\ x^T \Lambda^{-1} x = 1 \\ e_j x^T T_1^T u_j \geq 0 \\ j \neq k}} \frac{1}{w_k} v_o^T W^{1/2} T_1 x \right\} \tag{4.42}$$

where the ith component of v_o is equal to

$$(v_o)_i = \begin{cases} -e_i, & i \neq k \\ 1, & i = k \end{cases},$$

$$\eta_k^d = \frac{1}{w_k} \max{}^2 \left\{ 0, \max_{\substack{e_j \in \{-1,1\} \\ j \neq k}} \eta_k^d(e) \right\} \tag{4.43a}$$

with

$$\eta_k^d(e) = \sup_{\substack{x \in R^r \\ x^T \Lambda^{-1} x = 1 \\ e_j x^T T_1^T u_j \geq 0 \\ j \neq k}} v_o^T W^{1/2} T_1 x \tag{4.43b}$$

whereas the kth user asymptotic efficiency of the best linear detector equals (cf. (4.16)),

$$\eta_k^l = \frac{1}{w_k} \max{}^2 \left\{ 0, \max_{\substack{e_j \in \{-1,1\} \\ j \neq k}} \eta_k^l(e) \right\}$$

with

$$\eta_k^l(e) = \sup_{\substack{v \in R^K \\ v^T H v = 1 \\ e_j h_j^T v \geq 0 \\ j \neq k}} v_o^T H v.$$

Let

$$v^* \in \arg \eta_k^l(e) = \arg \max_{\substack{v \in R^K \\ v^T H v = 1 \\ e_j h_j v \geq 0 \\ j \neq k}} v_o^T H v. \tag{4.44}$$

We show that $x^* = \Lambda T_1^T W^{1/2} v^*$ is feasible, and that $v_o^T W^{1/2} T_1 x^* = \eta_k^l(e)$:

$$\begin{aligned} e_j x^{*T} T_1^T u_j &= e_j v^{*T} W^{1/2} T_1 \Lambda T_1^T u_j \\ &= e_j v^{*T} W^{1/2} R W^{1/2} W^{-1/2} u_j \\ &= \frac{1}{\sqrt{w_j}} e_j v^{*T} H u_j \\ &= \frac{1}{\sqrt{w_j}} e_j v^{*T} h_j \geq 0 \end{aligned} \tag{4.45}$$

since v^* is feasible,

$$\begin{aligned} x^{*T} \Lambda^{-1} x^* &= v^{*T} W^{1/2} T_1 \Lambda \Lambda^{-1} \Lambda T_1^T W^{1/2} v^* \\ &= v^{*T} H v^* = 1. \end{aligned} \tag{4.46}$$

Hence x^* is feasible, and

$$\begin{aligned} v_o^T W^{1/2} T_1 x^* &= v_o^T W^{1/2} T_1 \Lambda T_1^T W^{1/2} v^* \\ &= v_o^T H v^* = \eta_k^l(e). \end{aligned} \tag{4.47}$$

We know that $\eta_k^d \leq \eta_k^l$, since the decorrelating detector belongs to the class of linear detectors. We exhibited for each e a feasible vector x^*, which satisfied $v_o^T W^{1/2} T_1 x^* = \eta_k^l(e)$. Since from (4.43), $\eta_k^d(e) \geq v_o^T W^{1/2} T_1 x$ for all feasible x, we have, for all e, $\eta_k^d(e) \geq \eta_k^l(e)$. Hence $\eta_k^d \geq \eta_k^l$, which establishes (4.33).

Since the kth user asymptotic efficiency depends only on the kth row of the applied linear transformation, optimization of $\eta_k(H^I)$ over the class of generalized inverses for each dependent user k, yields different rows, each belonging to a different generalized inverse. Consequently, the collection of the K optimal rows need not be a generalized inverse.

Finally, notice that the near–far resistance of the optimum linear detector is equal to that of the optimum detector, since it is shown in Proposition 3 that a particular type of linear detector, namely, the decorrelating detector, achieves optimum near–far resistance.

V. Conclusion

The main contribution of this paper is the establishment of the fact that a set of appropriately chosen memoryless linear transformations on the outputs of a matched filter bank exhibits a substantially higher performance than the conventional single-user detector, while maintaining a comparable ease of computation. Moreover, the near–far resistance of all proposed detectors is shown to equal that of the optimum multiuser detector.

Even though the worst-case complexity of the algorithm used to find the best linear detector is exponential in the number of users, in a fixed-energy environment this computation needs to be carried out only once; hence the real-time time-complexity per bit is linear, in contrast to the optimum multiuser detector. Moreover, a region of signal energies and crosscorrelations exists in which the optimal linear detector achieves optimum asymptotic efficiency.

The decorrelating detector is easier to compute than the optimum linear detector, and it exhibits either the same or quite similar performance, depending on the energies and correlations. Since the decorrelating detector does not require knowledge of the transmitters' energies and it achieves the highest possible degree of near–far resistance, it is an attractive alternative to the optimum detector in situations where the received energies are not fixed. The only requirement for the signal of a user to be detected reliably by the decorrelating detector regardless of the level of multiple-access interference, is that it does not belong to the subspace spanned by the other signals—a mild constraint that should be compared to the condition necessary for reliable detection by the conventional single-user detector, i.e., that the signal is *orthogonal* to all the other signals.

The most interesting generalization of the results of this paper is the *asynchronous* code-division multiple-access channel.[4] Due to the fact that in the asynchronous case the channel has memory, a K-input K-output linear discrete-time filter will replace the memoryless linear transformation studied in this paper.

Appendix I

This appendix gives a summary of the results in [18]. We show that the problems of optimum multiuser demodulation and solving for the maximum asymptotic efficiency are nondeterministic polynomial time hard (NP-hard) in the number of users and therefore do not admit polynomial time algorithms unless such algorithms are found for a large class of well-known combinatorial problems including the traveling salesman and integer linear programming. According to (2.5), the selection of the most likely hypothesis given the observations is the following combinatorial optimization problem.

MULTIUSER DETECTION—

Instance: Given $K \in Z^+$, $y \in Q^K$ and a nonnegative definite matrix $H \in Q^{K\times K}$;

Find $\{b^* \in \{-1,1\}^K\}$ that maximizes $2b^Ty - b^THb$.

Proposition 10: MULTIUSER DETECTION is NP-hard.

Proof: The proof of NP-hardness of MULTIUSER DETECTION can be carried out by direct transformation from the following NP-complete problem [15].

PARTITION—

Instance: Given $L \in Z^+$, $\{l_i \in Z^+, i=1,\cdots,L\}$;

Question: Is there a subset $I \subset \{1,\cdots,L\}$ such that $\Sigma_{i\in I} l_i = \Sigma_{i \notin I} l_i$?

Given $l_1,\cdots,l_L$, we choose the following instance of MULTIUSER DETECTION:

$$K = L$$
$$h_{ij} = l_i l_j$$
$$y_k = 0, \qquad k = 1,\cdots,K.$$

With this choice, $\{l_1,\cdots,l_L\}$ is a "yes" instance of PARTITION if and only if

$$\max_{b\in\{-1,1\}^K} 2b^Ty - b^THb = 0. \tag{A.1}$$

Proposition 10 can be generalized [2], [18] to deal with arbitrary finite alphabets which are not part of the instance (and hence are fixed) of MULTIUSER DETECTION, i.e., the problem is inherently difficult when the number of users is large, regardless of the alphabet size. It is an open problem whether MULTIUSER DETECTION remains NP-hard when H is restricted to be Toeplitz. If this is the case, then it can be shown [18] that the problem of single-user maximum likelihood detection for intersymbol interference channels [17] is NP-hard in the length of the interference

The usefulness and relevance of Proposition 10 stem from the fact that when the users are asynchronous, the cross correlations between their signals are unknown *a priori* and the worst-case computational complexity over all possible mutual offsets is the complexity measure of interest since it determines the maximum achievable data rate in the absence of synchronism among the users. Actually, no family of signature signals is known to result in optimum demodulation with polynomial-in-K complexity for all possible signal offsets. Thus even if the designer of the signal constellation were to include as a design criterion the complexity of the optimum demodulator in addition to the bit-error-rate performance (which dictates signals with low crosscorrelations), he would not be able to endow the signal set with any structure that would overcome the inherent intractability of the optimum asynchronous demodulation problem for all possible offsets.

The performance analysis of the optimum receiver for arbitrary energies and crosscorrelations is also inherently hard. According to (2.11) the maximum achievable asymptotic efficiency is obtained as the solution to multiuser asymptotic efficiency.

MULTIUSER ASYMPTOTIC EFFICIENCY—

Instance: Given $K \in Z^+$, $k \in \{1,\cdots,K\}$, and a nonnegative definite matrix $H \in Q^{K\times K}$;

Find: the kth user maximum asymptotic efficiency,

$$\eta_k = \frac{1}{w_k} \min_{\substack{\epsilon\in\{-1,0,1\}^K \\ \epsilon_k \neq 0}} \epsilon^T H \epsilon.$$

Proposition 11: MULTIUSER ASYMPTOTIC EFFICIENCY is NP-hard.

Proof: The proof is divided in two steps. First, $-1/0/1$ KNAPSACK is polynomially transformed to MULTIUSER ASYMPTOTIC EFFICIENCY. Then, $-1/0/1$ KNAPSACK is shown to be NP-complete. In analogy to the 0/1 KNAPSACK problem (e.g., [16]) we define

$-1/0/1$ KNAPSACK—

Instance: Given $L \in Z^+$, $G \in Z^+$ and a family of not necessarily distinct positive integers

$$\{l_i \in Z^+, i=1,\cdots,L\};$$

Question: Are there integers $\epsilon_i \in \{-1,0,1\}$, $i=1,\cdots,L$ such that $\Sigma_{i=1}^{L} \epsilon_i l_i = G$?

We transform $-1/0/1$ KNAPSACK to MULTIUSER ASYMPTOTIC EFFICIENCY by adding a user. Given $\{G, l_1,\cdots,l_L\}$, denote $l_{L+1} = G$ and construct the following instance: $K = L+1$, $k = L+1$, $h_{ij} = l_i l_j$, $1 \le i, j \le K$.

The Kth user asymptotic efficiency is equal to zero if and only if $\{G, l_1,\cdots,l_L\}$ is a "yes" instance of $-1/0/1$ KNAPSACK. To see this, note that we can fix $\epsilon_k = -1$ in the right side of (2.11)

[4] *Note added in proof:* This has now been accomplished in the companion paper [19], using a different approach.

without loss of generality. Then,

$$\eta_K = \frac{1}{G^2} \min_{\substack{\epsilon_i \in \{-1,0,1\} \\ 1 \le i \le K-1}} \left\{ h_{KK} + \sum_{n=1}^{K-1} \epsilon_n \left[-2h_{nK} + \sum_{m=1}^{K-1} \epsilon_m h_{nm} \right] \right\}$$

$$= \frac{1}{G^2} \min_{\substack{\epsilon_i \in \{-1,0,1\} \\ 1 \le i \le K-1}} \left(G - \sum_{n=1}^{K-1} \epsilon_n l_n \right)^2 \quad \text{(A.2)}$$

The proof that $-1/0/1$ KNAPSACK is NP-complete can be found in [18].

Appendix II

We give here an explicit procedure for finding the maximizing vector $\tilde{v}$ given implicitly by Proposition 6. The idea is the following: condition (4.19) states that if the maximizing vector $\tilde{v}$ lies in the intersection of a subset of the delimiting hyperplanes with equations $h_j^T\tilde{v} = 0$, $j \in S$, with S the index set of the specific hyperplanes, only the λ_j, $j \in S$ are possibly nonzero and enter into the expression defining $\tilde{v}$. Thus we have $|S|$ equations with $|S|$ unknowns, which we can solve to get the λ_i and then $\tilde{v}$. To state (and prove the correctness of) an algorithm that finds the optimum linear transformation, the following terminology is used.

Definition 1: Let S be an index set $\{j_1, j_2, \cdots, j_n\}$, $0 \le n \le K-1$, with $j_1, \cdots, j_n \in \{1, \cdots, K\} - \{k\}$, labeled in increasing order. Define

$$D_S(j) = \det \begin{vmatrix} h_j^T v_o & H_{jj_1} & \cdots & H_{jj_n} \\ h_{j_1}^T v_o & H_{j_1 j_1} & \cdots & H_{j_1 j_n} \\ \cdot & \cdot & \cdot & \cdot \\ h_{j_n}^T v_o & H_{j_n j_1} & \cdots & H_{j_n j_n} \end{vmatrix}. \quad \text{(A.3)}$$

Definition 2: We introduce an indicator for the second Kuhn–Tucker condition:

$$\text{if } e_j D_S(j) > 0, \text{ then } C_S(j) = \text{yes}, \qquad \text{else } C_S(j) = \text{no}. \quad \text{(A.4)}$$

Definition 3: An n-tuple S of $\{1, \cdots, K\} - \{k\}$ is *matched* if for all $i \in S: C_{S-\{i\}}(i) = \text{no}$.

Definition 4: An n-tuple S contains a basis B if $\{h_j | j \in B\}$ is a basis for $\{h_j | j \in S\}$.

Proposition 12: The following algorithm finds a vector $\tilde{v}$ satisfying (4.17)–(4.20).

A. Search for the index set with least cardinality $S \subseteq \{1, \cdots, K\} - \{k\}$, for which λ_i, $i \in S$, are possibly nonzero

$n := 0$
all n-tuples := untried; S_o := matched
WHILE $n \le K - 2$
  WHILE there is still an untried n-tuple containing a matched basis B
    select untried matched n-tuple := S_n, contained matched basis := B
    IF for all $j \notin S_n$, $j \neq k$, $C_B(j) = \text{yes}$, RETURN S_n, B, STOP
    ELSE S_n := tried
  RETURN
$n := n + 1$
RETURN
"decorrelating detector is optimal," output $\{2, \cdots, K\} - \{k\}$, STOP.

B. Computation of the λ_i:

$i \notin B$: $\lambda_i = 0$
$i \in B$: λ_i are the solutions of the $|B|$ equations $|B|$ unknowns $h_i^T v = 0$, $i \in B$, where

$$v = v_o + \sum_{i \in B} \lambda_i e_i u_i.$$

C.

$$\tilde{v} = \frac{v_o + \sum_{i \in B} \lambda_i e_i u_i}{\left(v_o^T H v_o + v_o^T H \sum_{i \in B} \lambda_i e_i u_i \right)^{1/2}}.$$

Comment: Recall that this procedure has to be repeated for all the different $\{e_j\}$ in search of the maximal $\eta(e)$ value, until either the efficiency $\eta(e)$ reaches the upper bound given by the optimal detector, or all 2^K possibilities have been exhausted. Prior to running the algorithm, the sufficient conditions given in Propositions 8 and 9 should be checked.

Proof: Conditions (4.17) and (4.19) are obviously satisfied by construction of $\tilde{v}$ in C, and the requirement $h_j^T\tilde{v} = 0$ for the possibly nonzero λ_i in B. To prove conditions (4.18) and (4.20), consider the system of $|B|$ linear equations in $|B|$ unknowns of B. From A the set B is matched, and satisfies $C_B(j) = \text{yes}$ for all $j \neq k$, $j \notin S_n$. We have to show a) $\lambda_i \ge 0$, for all $i = 1, 2, \cdots, K$; and b) $C_B(j) = \text{yes}$ for all $j \neq k$, $j \notin S_n$ is equivalent to condition (4.18).

a) $\lambda_i = 0$, $i \notin B$, by construction of the index set S_n and B. For $i \in B$, in step B we solve $h_{j_i}^T\tilde{v} = 0$, all $i = 1, 2, \cdots, |B|$. Let $|B| = n$. Then

$$h_j^T v_o + \lambda_{j_1} e_{j_1} H_{j_1 j_1} + \cdots + \lambda_{j_n} e_{j_n} H_{j_1 j_n} = 0$$
$$h_{j_2}^T v_o + \lambda_{j_1} e_{j_1} H_{j_2 j_1} + \cdots + \lambda_{j_n} e_{j_n} H_{j_2 j_n} = 0$$
$$h_{j_n}^T v_o + \lambda_{j_1} e_{j_1} H_{j_n j_1} + \cdots + \lambda_{j_n} e_{j_n} H_{j_n j_n} = 0. \quad \text{(A.5)}$$

Denote by D_B the determinant of the coefficient matrix of the $\lambda_{j_i} e_{j_i}$. Since B is a basis and the corresponding matrix is nonnegative definite, D_B is strictly positive. Then, by Cramer's rule,

$$\lambda_{j_i} = \frac{-e_{j_i} D_{B-\{j_i\}}(j_i)}{D_B}. \quad \text{(A.6)}$$

The numerator is obtained by i row flips and i column flips to get j_i into position (1,1). Since the set B is matched, the numerator is nonnegative. As obtained above, the denominator is positive, hence $\lambda_i \ge 0$ for all $i \in B$. This completes the proof of a).

b) $h_j^T v = 0$, $j \in S_n$. For $j \notin S_n$, $j \neq k$, with the obtained values for λ compute the feasibility expressions:

$$e_j h_j^T \tilde{v} = e_j h_j^T \left(v_o + \sum_{i \in B} \lambda_i e_i u_i \right)$$
$$= \frac{e_j}{D_B} \left(D_B h_j^T v_o + \sum_{i \in B} - D_{B-\{i\}}(i) H_{ji} \right)$$
$$= \frac{1}{D_B} e_j D_B(j) > 0, \quad \text{(A.7)}$$

since $C_B(j) = \text{yes}$. The last equality is obtained by expanding along the first row of $D_B(j)$. This completes the proof of b). By construction the algorithm terminates after at most $K-2$ steps.

In part A of the algorithm notice that $n = 0$ corresponds to a solution in the interior of the feasible cone, with all λ equal to zero, and $\tilde{v} = v_o / \sqrt{v_o^T H v_o}$. The corresponding asymptotic efficiency $\eta^2(e)/w_k = v_o^T H v_o / w_k = \eta$, which is equal to the asymptotic efficiency of the maximum likelihood detector as given by (2.11). On the other hand, $n=1$ corresponds to a solution on exactly one of the delimiting hyperplanes, with exactly one λ nonzero (call it λ_j), and

$$\tilde{v}=\frac{1}{\eta(e)}\left(v_o-\frac{h_j^T v_o}{H_{jj}}u_j\right) \tag{A.8}$$

and

$$\eta^2(e)=v_o^T H v_o-\frac{\left(h_j^T v_o\right)^2}{H_{jj}}. \tag{A.9}$$

The asymptotic efficiency achieved in this case is bounded above by the one for $n=0$, since the second term is nonnegative. If the matrix H does not have a lot of structure, which is to be expected in practical applications, this is the most probable case. For increasing n the computational effort grows fast, but in most cases the algorithm will terminate for very small n.

We also have an explicit solution for the "terminal case," $n=K-1$, which corresponds to the decorrelating detector case. Then, without loss of generality, $\tilde{v}=h_k^I/\sqrt{H_{kk}^I}$, a scaled version of the kth column of any generalized inverse matrix of H (in particular of H^+) and $\eta(e)=1/H_{kk}^+$, which is equal to the kth user asymptotic efficiency of the decorrelating detector, when the scaling factor $1/w_k$ of (4.16) is taken into account. This can be shown as follows. In the terminal case $h_j^T\tilde{v}=0$, for all $j\neq k$. Hence

$$\eta(e)=\max_{\substack{v\in R^K\\ vHv=1\\ h_j^Tv=0\\ j\neq k}} v_o^T H v=\max_{\substack{v\in R^K\\ h_j^Tv=0\\ j\neq k\\ v_k h_k^T v=1}} h_k^T v=\max_{\substack{v\in R^K\\ Hv=(1/v_k)u_k}} \frac{1}{v_k}. \tag{A.10}$$

If user k is dependent, $Hv=0$ and $\eta(e)=0$. Since this was the best choice of v, we can without loss of generality replace $\tilde{v}$ by the kth row of any generalized inverse, because the resulting asymptotic efficiency cannot become negative. If user k is independent, Lemma 1 implies $HH^I u_k=u_k$, and for all v in the feasible set,

$$H^I H v=\frac{1}{v_k}H^I u_k.$$

Hence using Lemma 1 for both equations, we obtain

$$v_k=\sqrt{H_{kk}^I}=\sqrt{H_{kk}^+}.$$

The feasible set in (A.10), $F=\{v|Hv=(1/v_k)u_k\}$, is nonempty (e.g., it contains the set $\{(1/v_k)h_k^I, H^I\in I(H)\}$), and for all $v\in F$, $v_k=\sqrt{H_{kk}^+}$. Hence $\eta(e)=1/\sqrt{H_{kk}^+}$, and with (3.8), $(1/w_k)\eta^2(e)=1/R_{kk}^+$, which is the energy independent asymptotic efficiency of the decorrelating detector for independent users.

References

[1] S. Verdú, "Minimum probability of error for asynchronous Gaussian multiple-access channels," *IEEE Trans. Inform. Theory*, vol. IT-32, pp. 85–96, Jan. 1986.

[2] ——, "Optimum multi-user signal detection," Ph.D. dissertation, Dept. Elec. Comput. Eng., Univ. of Illinois, Urbana-Champaign, Coordinated Sci. Lab., Urbana, IL, Rep. T-151, Aug. 1984.

[3] ——, "Optimum multi-user asymptotic efficiency," *IEEE Trans. Commun.*, vol. COM-34, no. 9, pp. 890–897, Sept. 1986.

[4] I. N. Bronshtein and K. A. Semendyayev, *Handbook of Mathematics*. Verlag Harri Deutsch, 1985.

[5] J. Wozencraft and I. Jacobs, *Principles of Communication Engineering*. New York: Wiley, 1965.

[6] K. S. Schneider, "Optimum detection of code division multiplexed signals," *IEEE Trans. Aerosp. Electron. Syst.*, vol. AES-15, pp. 181–185, Jan. 1979.

[7] T. Kailath, *Linear Systems*. Englewood Cliffs, NJ: Prentice-Hall, 1980.

[8] R. Lupas-Golaszewski and S. Verdú, "Asymptotic efficiency of linear multiuser detectors," in *Proc. 25th IEEE Conf. Decision and Control*, Athens, Greece, Dec. 1986, pp. 2094–2100.

[9] D. Luenberger, *Optimization by Vector Space Methods*. New York: Wiley, 1969.

[10] T. L. Boullion and P. L. Odell, *Generalized Inverse Matrices*. New York: Wiley-Interscience, 1971.

[11] P. Lancaster and M. Tismenetsky, *The Theory of Matrices*. New York: Academic, 1985.

[12] H. V. Poor, *An Introduction to Signal Detection and Estimation*. New York: Springer-Verlag, 1988.

[13] C. W. Helstrom, *Statistical Theory of Signal Detection*, 2nd ed. London: Pergamon, 1968.

[14] S. Verdú, "Optimum multi-user detection is NP-hard," presented at the 1985 Int. Symp. Information Theory, Brighton, UK, June 1985.

[15] M. R. Garey and D. S. Johnson, *Computers and Intractability: A Guide to the Theory of NP-Completeness*. San Francisco, CA: Freeman, 1979.

[16] C. H. Papadimitriou and K. Steiglitz, *Combinatorial Optimization: Algorithms and Complexity*. Englewood Cliffs, NJ: Prentice-Hall, 1982.

[17] G. D. Forney, "Maximum likelihood sequence estimation of digital sequences in the presence of intersymbol interference," *IEEE Trans. Inform. Theory*, vol. IT-18, no. 3, pp. 363–378, May 1972.

[18] S. Verdú, "Computational complexity of optimum multiuser detection," *Algorithmica*, vol. 4, 1989, to appear.

[19] R. Lupas and S. Verdú, "Near-far resistance of multiuser detectors in asynchronous channels," *IEEE Trans. Commun.*, vol. 37, 1989, to appear.

Reprinted from *IEEE Transactions on Communication*, Vol. 38, No. 4, Apr. 1990, pp. 509-519

Multistage Detection in Asynchronous Code-Division Multiple-Access Communications

MAHESH K. VARANASI, MEMBER, IEEE, AND BEHNAAM AAZHANG, MEMBER, IEEE

***Abstract*—A multiuser detection strategy for coherent demodulation in an asynchronous code-division multiple-access system is proposed and analyzed. The resulting detectors process the sufficient statistics via a multistage algorithm. This algorithm is based on a successive multiple-access interference annihilation scheme. An efficient real-time implementation of the multistage algorithm with a fixed decoding delay is obtained and it is shown to require a computational complexity/symbol which is *linear* in the number of users K. Hence, the multistage detector contrasts the optimum demodulator, which is based on a dynamic programming algorithm; has a variable decoding delay; and a software complexity per symbol that is *exponential* in K [18]. Further, an exact expression for the probability of error is obtained for the two-stage detector. The probability of error computations show that the two-stage receiver is particularly well suited for "near-far" situations. In fact, performance approaches that of single-user communications as the interfering signals become *stronger*. The near-far problem is therefore alleviated. Further, significant performance gains over the conventional receiver are obtained even for relatively high bandwidth efficiency situations.**

I. Introduction

In a code-division multiple-access (CDMA) system, several users simultaneously transmit information over a common channel using preassigned code waveforms. The receiver is equipped with a knowledge of the codes of some or all of the users. It is then required to demodulate the information symbol sequences of these users, upon reception of the sum of the transmitted signals of all the users in the presence of additive noise. Examples of such situations arise in a variety of communication systems such as satellite communications, radio networks and other multipoint-to-multipoint multiple-access networks.

In a majority of the CDMA systems of practical importance, the users transmit information independently. Therefore the transmitted signals of different users arrive asynchronously at the receiver. Since their relative time delays are arbitrary, it is inevitable that the cross-correlations between signals of different users are nonzero. Low cross-correlations among code waveforms for all relative time delays are obtained by the design of a set of complex code waveforms at the expense of an increased bandwidth [15]. Since bandwidth is a valuable resource, the problem of interest is to be able to accommodate as many users as can be reliably demodulated for a given bandwidth.

The conventional approach to multiuser demodulation is to demodulate each user's signal as if it were the only one present. This receiver consists of a bank of filters matched to each user's signal

Paper approved by the Editor for Spread Spectrum of the IEEE Communications Society. Manuscript received July 13, 1988; revised December 9, 1988. This work was supported by the National Science Foundation under Grant NCR-8710844. This paper was presented at the 22nd Conference on Information Sciences and Systems, Princeton University, Princeton, NJ, March 16-18, 1988.

The authors are with the Department of Electrical and Computer Engineering, Rice University, Houston, TX, 77251.

IEEE Log Number 9034845.

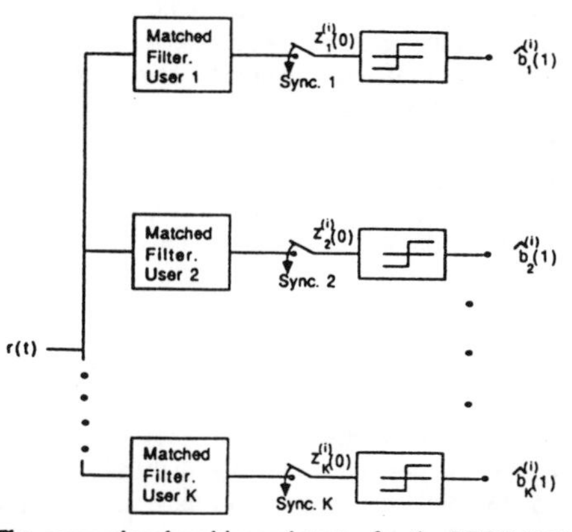

Fig. 1. The conventional multiuser detector for the PBSK-CDMA system.

waveform in corresponding time and phase synchronism as in Fig. 1. The kth user's decision statistic consists of the desired signal component, additive noise, and multiple-access interference due to the cross-correlations of the kth user's signal with the signals from the other users. This receiver makes decisions by comparing the statistic to a threshold and has the advantage of simplicity. It also lends itself to expedient modification for decentralized reception. Extensive analysis of the conventional detector for spread-spectrum systems has been undertaken over the past few years. Different methods for computation of bounds and approximations of the probability of error can be found in [2], [5], [7], [9], [14] and the references therein. These studies have revealed the usefulness of a careful design of the signal constellation wherein acceptable performance from the conventional receiver is possible due to low cross-correlations between signals of different users. However, as the number of users in a system of fixed bandwidth grows or as the relative powers of the interfering signals become large (the "near-far" problem), severe performance degradation of the suboptimum systems is observed even for relatively low bandwidth efficiencies [1], [2]. This effect has been succinctly characterized by Verdu in [19], in terms of asymptotic efficiency, a measure that quantifies the multiple-access limitation of a detector.

The optimum multiuser demodulator can be obtained based on a maximum likelihood sequence detection formulation. If all the information matrices are equiprobable, the maximum likelihood sequence detector is obtained by maximizing the joint *a posteriori* probability

$$p[B \mid \{r(t); t \in R\}] \tag{1.1}$$

where $r(t)$ denotes the received signal, B represents the information bit matrix, the (k, i)th element of which is the ith bit of the kth user denoted as $b_k^{(i)}$. If the packet length of the information bit sequence of each user is $(2P+1)$ and the number of users in the system is K, then an exhaustive maximization of the joint *a posteriori* probability needs its computation for each of the $2^{(2P+1)K}$ possible values of B. Such a brute-force maximization is practically useless. However, a significant computational gain over this exhaustive scheme was obtained by Verdu in [18] by exploiting the additive decomposability of the log-likelihood function, paving the way for its maximization by a forward dynamic programming algorithm of the Viterbi type. Although the computational and storage complexity of this efficient algorithm was independent of the packet length, it was however shown to depend *exponentially* on the number of users. It was also shown that the multiuser optimum detection problem is *NP*-hard. In addition to requiring intensive software complexity, the maximum likelihood solution also has a variable decoding delay.

From earlier discussion it is clear that despite its simplicity, the conventional receiver has limited use for CDMA systems in the desirable range of bandwidth efficiencies. In addition, its vulnerability to near-far effects is considerable. On the other hand, the optimum receiver is computationally too intensive, and has a variable decoding delay which is unacceptable in many applications. There is hence a need for suboptimum receivers which perform reliably in higher bandwidth efficiency situations and are robust to near-far effects with a reasonable computational complexity to ensure their practical implementation.

In this paper, we propose and analyze detectors based on a multiuser detection strategy of successive multiple-access interference rejection. The resulting detectors process the sufficient statistics via a multistage algorithm. An efficient real-time implementation of the multistage algorithm with a fixed decoding delay is obtained and it is shown to require a computational complexity/symbol which is *linear* in the number of users K, in obvious contrast to the optimum demodulator, which has a software complexity per symbol that is *exponential* in K. It is also shown that in a system with well-designed code waveforms, the performance of the multistage detector closely tracks that of the optimum receiver. In most cases of practical interest, the proposed receiver achieves significant performance gains over the conventional receiver. As in the case of the optimum detector, the proposed multistage receiver requires a knowledge of the signal strengths for its implementation. In some practical situations where exact values of the energies are not available, estimates can be used instead [17]. However, for the analysis presented in this paper, an exact knowledge of energies is assumed.

The rest of the paper is organized as follows. A general multiuser CDMA communication model is described in Section II. While the proposed multistage multiple-access interference rejection scheme is valid for the general CDMA system, we will for most part of the paper be concerned with the analysis of the more specific BPSK-CDMA system which we shall also describe in this section. Section III deals with the development of the multistage algorithm and some insight into the resulting detector is also presented. In Section IV, an exact expression for the probability of error is obtained for the two-stage detector. In Section V, we will present numerical results comparing the two-stage, the conventional and the single-user demodulators (i.e., matched filter operating in the absence of interfering signals) and thereby illustrate achievable gains in bandwidth efficiency and near-far immunity. In that same section, we also analyze the proposed receiver for a direct-sequence spread-spectrum multiple-access system.

II. System Description

Let us assume that there are K transmitters in the system, and each transmitter employs an S-ary signal set derived from the S' code waveforms assigned to the corresponding user. We will now introduce some notation to describe the received signal waveform. Suppose $S_k = \{s_{k,0}, s_{k,1}, \cdots, s_{k,S-1}\}$ is the S-ary signal set of the kth user where each signal is time-limited to $[0, T]$, T being the symbol duration assumed to be equal for all users. Denote the diagonal energy matrix E so that diag $\{E\} = \{E_1, E_2, \cdots, E_K\}$ where E_k is the received energy of the kth user's signal which is assumed to be independent of the transmitted symbol. Let I be the set of time indices $\{-P, -P+1, \cdots, -1, 0, 1, \cdots, P-1, P\}$, $2P+1$ being the packet length, and let I_S be a set of S indexes $\{0, 1, \cdots, S-1\}$. Assume that the kth user transmits its jth symbol by transmitting $s_{k,j}$. Representing the sequence of information symbols of the kth user by a map $b_k: I \rightarrow I_S$, we have that if $b_k(i) = j$, the ith symbol of the kth user is j. Define B to be a $K \times (2P+1)$ information matrix, representing the maps $b_1, b_2, \cdots, b_K$, so that the (k, i)th element of B is equal to $b_k(i)$. Hence, the received signal is modeled as being the sum of K signals denoted by $s(t, B)$ and additive white Gaussian noise n_t with spectral density $N_0/2$. The received process is therefore given as

$$r(t) = s(t, B) + n_t, \; t \in R$$

where

$$s(t, B) = \sum_i \sum_{k=1}^{K} s_{k, b_k(i)}(t - iT - \tau_k), \; (t \in R),$$

so that $\tau_k \in [0, T)$ represents the time delay of the kth user. Each value of the matrix B corresponds to the specification of packets transmitted by all the users and so there are $S^{K(2P+1)}$ possible maps B. The centralized demodulator that we propose in this paper is general enough to be valid for the CDMA situation just described, which we will call the general CDMA system.

However, for the sake of simplicity, we will often restrict ourselves to a more specific system wherein each user is assigned one code waveform. Denote the modulating signal of the kth user as $s_k(t)$. The kth user then antipodally modulates this signal to transmit information bits. In the general CDMA system, this corresponds to $S' = 1$, $S = 2$, $b_k(i) \in \{0, 1\}$, $s_{k,0}(t) = s_k(t)$ and $s_{k,1}(t) = -s_k(t)$, so that the transmitted signal is

$$\begin{aligned} s(t, B) &= \sum_i \sum_{k=1}^{K} (-1)^{b_k(i)} s_k(t - iT - \tau_k) \\ &= \sum_i \sum_{k=1}^{K} b_k^{(i)} s_k(t - iT - \tau_k) \end{aligned} \tag{2.2}$$

where $b_k^{(i)} \triangleq (-1)^{b_k(i)} \in \{-1, +1\}$. This system will be referred to as the BPSK-CDMA system.

The essential system parameters are the signal cross-correlations and their relative energies. Following the notation in [18] let us define the $K \times K$ cross-correlation matrices $H(i)$ such that the (k, l)th element is

$$h_{kl}(i) = \int_{-\infty}^{\infty} s_k(t - \tau_k) s_l(t + iT - \tau_l)\, dt. \tag{2.3}$$

For simplicity and without loss of generality, let us assume an ordering on the time delays τ_k such that $0 \le \tau_1 \le \tau_2 \le \cdots \le \tau_K < T$. Since the modulating signals are time-limited, $H(i) = 0 \; \forall |i| > 1$ and $H(-i) = H^T(i)$. Note that $H(1)$ is an upper triangular matrix with a zero diagonal. In the next section, the multiple-access communication system described above is considered for multistage demodulation.

III. The Multistage Detector

In this section, we present a development of the multistage scheme for the multiuser demodulation problem specified in the previous section. First the sufficient statistics for optimum demodulation are described. In Section III-A, we will derive the multistage multiple-access rejecting detector for the BPSK-CDMA system first and then extend it for the general CDMA system and present the multistage detector in algorithmic form. In Section III-B, we will address some important implementation issues.

Consider the maximum likelihood sequence detection formulation for the problem. The most likely information matrix is chosen as that which maximizes the probability in (1.1) or equivalently, the log-likelihood function [16]

$$L[B, r(t)] = 2 \int s(t, B) r(t)\, dt - \int s^2(t, B)\, dt. \tag{3.4}$$

The above expression shows that the sufficient statistics are the outputs of a bank of K matched filters matched to the modulating signal of each user, sampled in corresponding time synchronism. The output of the kth matched filter sampled at the end of the ith time interval is given as

$$z_k^{(i)}(0) = \int_{-\infty}^{\infty} r(t)s_k(t+iT-\tau_k)\,dt \quad (3.5)$$

$$= \eta_k^{(i)} + \sum_{l=k+1}^{K} h_{kl}(1)b_l^{(i-1)} + \sum_{l=1}^{K} h_{kl}(0)b_l^{(i)} + \sum_{l=1}^{k-1} h_{kl}(-1)b_l^{(i+1)} \quad (3.6)$$

where $\eta_k^{(i)}$ is the component of the statistic due to the additive channel noise. In vector notation, letting $z^{(i)}(0)$ denote $[z_1^{(i)}(0), z_2^{(i)}(0), \cdots, z_K^{(k)}(0)]^T$, we have

$$z^{(i)}(0) = \eta^{(i)} + H(1)b^{(i-1)} + H(0)b^{(i)} + H(-1)b^{(i+1)} \quad (3.7)$$

where $b^{(i)}$ is the ith column of B. The set of sufficient statistics for the maximum likelihood sequence detection of the sequence $\{b^{(i)}; \forall i\}$ is therefore $\{z^{(i)}(0), \forall i\}$. In fact, this set of sufficient statistics is minimally sufficient for the demodulation of any one bit [18].

A. Derivation

The multistage suboptimum solution to the maximization of $L[B, r(t)]$ is now developed. Consider the demodulation of the ith bit of the kth user. Assume that we are in the $(m+1)$st stage. Let the mth stage estimates of bits $b_l^{(j)}$ be denoted as $\hat{b}_l^{(j)}(m)$ for all l and for all j. We propose the $(m+1)$st stage estimate of $b_k^{(i)}$ as being

$$\hat{b}_k^{(i)}(m+1) = \arg\left\{ \max_{\substack{b_k^{(i)} \in \{+1,-1\} \\ b_l^{(j)} = \hat{b}_l^{(j)}(m)}} [L(B, r(t))] \right\}. \quad (3.8)$$

Note that the maximization is performed by setting $b_l^{(j)}$ to be equal to its mth stage estimate for all l and for all j but not $l = k$ and $j = i$ simultaneously. From (2.2), (3.5), and (3.4), an additive decomposition of the log-likelihood function can be obtained as

$$L[B, r(t)] = \sum_p \langle b^{(p)}, 2z^{(p)}(0) - H(0)b^{(p)} - 2H(1)b^{(p-1)} \rangle \quad (3.9)$$

where $\langle x, y \rangle$ denotes the inner product between the vectors x and y. Substituting from (3.9) into (3.8) and retaining only terms in the summation that involve $b_k^{(i)}$, we have

$$\hat{b}_k^{(i)}(m+1) = \arg\left\{ \max_{\substack{b_k^{(i)} \in \{+1,-1\} \\ b_l^{(j)} = \hat{b}_l^{(j)}(m)}} \left[\sum_{p=i}^{i+1} \left\langle b^{(p)}, 2z^{(p)}(0) - H(0)b^{(p)} - 2H(1)b^{(p-1)} \right\rangle \right] \right\}.$$

First each inner product is expanded, the terms that depend on $b_k^{(i)}$ are then retained. Substituting the mth stage estimates for all bits except $b_k^{(i)}$ and noting the properties of $H(\cdot)$, we have the following expression for the estimate:

$$\hat{b}_k^{(i)}(m+1) = \mathrm{sgn}\,[z_k^{(i)}(m)] \quad (3.10)$$

where

$$z_k^{(i)}(m) = z_k^{(i)}(0) - \sum_{l=k+1}^{K} h_{kl}(1)\hat{b}_l^{(i-1)}(m) - \sum_{l\neq k} h_{kl}(0)\hat{b}_l^{(i)}(m) - \sum_{l=1}^{k-1} h_{kl}(-1)\hat{b}_l^{(i+1)}(m). \quad (3.11)$$

The above result has a simple interpretation.[1] Note from (3.6), (3.10), and (3.11) that the $(m+1)$st stage decision statistic is obtained by *subtracting* the *estimate* of the multiple-access interference (denoted by $\hat{I}_k^{(i)}(m)$ in Fig. 3) which is reconstructed using the mth stage estimates of the information bits.[2] Substituting for $z_k^{(i)}(0)$ from (3.6) into (3.11), we have

$$z_k^{(i)}(m) = \eta_k^{(i)} + E_k b_k^{(i)} + \sum_{l=k+1}^{K} h_{kl}(1)d_l^{(i-1)}(m) + \sum_{l\neq k} h_{kl}(0)d_l^{(i)}(m) + \sum_{l=1}^{k-1} h_{kl}(-1)d_l^{(i+1)}(m) \quad (3.12)$$

where $d_l^{(j)}(m) \triangleq [b_l^{(j)} - \hat{b}_l^{(j)}(m)]$ is equal to $2b_l^{(j)}$ or zero depending on whether $\hat{b}_l^{(j)}(m)$ is in error or not.

We now seek to express the $(m+1)$st-stage update of the estimate of $b^{(i)}$ using matrix notation. Performing the maximization of (3.8) and letting $\hat{b}^{(i)}(m)$ denote the mth stage estimate of the vector $b^{(i)}$, we have the $(m+1)$st-stage vector decision statistic written in matrix notation as

$$z^{(i)}(m) = \eta^{(i)} + Eb^{(i)} + H(1)d^{(i-1)}(m) + [H(0) - E]d^{(i)}(m) + H(-1)d^{(i+1)}(m) \quad (3.13)$$

where $d^{(i)}(m) \triangleq b^{(i)} - \hat{b}^{(i)}(m)$. The $(m+1)$st stage update of the iteration is given as

$$\hat{b}^{(i)}(m+1) = \mathrm{sgn}\,[z^{(i)}(m)], \quad m \geq 1. \quad (3.14)$$

Having described the multistage solution, we now need to choose an initial estimate of the information bits denoted by $\hat{b}^{(i)}(1)$, $\forall i$. We choose the conventional detector for the first stage for reasons of conceptual simplicity in implementation and analysis, i.e.,

$$\hat{b}^{(i)}(1) = \mathrm{sgn}\,[z^{(i)}(0)], \quad \forall i. \quad (3.15)$$

This initial assignment implies that the proposed multistage detector aims at improving the conventional decision on the information bits. Other suboptimum initial conditions can be substituted depending on how much computation one is willing to do. It would seem important, as it would for any iterative solution, that a reasonably good initial estimate be selected. An exact quantification of this statement will be provided by the probability of error analysis in Section IV.

For the general CDMA problem, the multistage detector is similar to the one described above. It consists of a bank of SK matched filters followed by an multistage algorithm. The detector in this case can be described in the following algorithmic form.

I) Obtain sufficient statistics (stage 0) from the received signal waveform.

II) Perform M stages of processing the sufficient statistics where the mth ($m \geq 1$) stage processor acts on the statistics produced by the $(m-1)$st stage. The mth stage consists of the following procedure: i) *estimation* of the unknown symbols from the $(m-1)$st stage statistics.[3] If $m = M$, stop; else, proceed with ii).
ii) *reconstruction* of the MA interference using the estimates obtained in step i) and *subtraction* of the reconstructed MA interference from the sufficient statistics to obtain the mth stage statistics.

[1]The same result is obtained if one starts from the minimum probability of bit-error formulation [18] and proceeds in a manner analogous to the one followed here.

[2]In [4] and [3], the theoretical implication of a somewhat similar idea has been studied, albeit without reference to any particular modulation or multiple-access technique. The reader is also referred to [20] and [11] for interference cancellation techniques for demodulation in a spread-spectrum multiple-access system.

[3]The unknown symbols are estimated in a simple suboptimum fashion making use of the low cross-correlation properties among signals of different users.

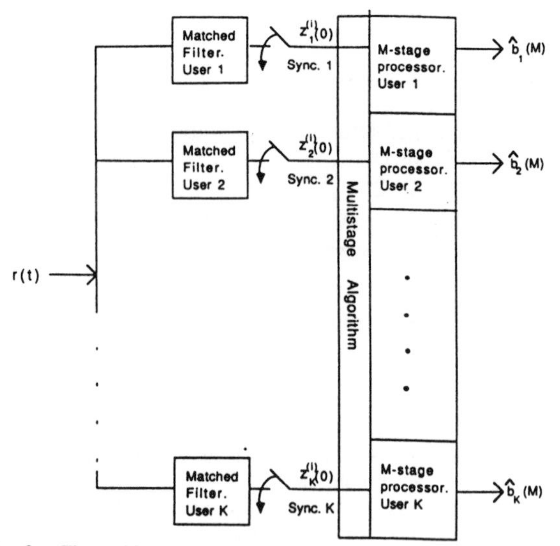

Fig. 2. The multistage multiuser detector for the BPSK-CDMA system.

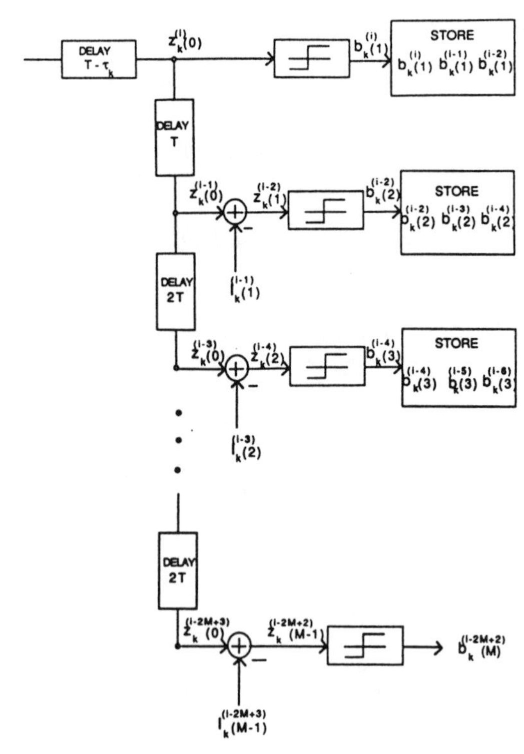

Fig. 3. A detailed implementation of the kth M-stage processor where for each $m = 1, 2, \cdots, M-1$, $\hat{I}_k^{(i-2m+1)}(m)$ denotes the estimate of the multiple-access interference reconstructed in the mth stage based on bit estimates $\hat{b}_j^{(i-2m)}(m-1)$, $\hat{b}_j^{(i-2m+1)}(m-1)$ and $\hat{b}_j^{(i-2m+2)}(m-1) \forall j \neq k$ obtained from the other $K-1$ processors.

The optimum demodulation of each symbol requires the observation of the received signal for its entire duration. Hence, it is necessary that, for the M-stage detector performance to approach near-optimality as M increases, the time interval over which the demodulator would have to process the received signal for the demodulation of a certain symbol should increase. It is shown in the Appendix that this is indeed the case and further the effective time interval over which the M-stage receiver processes the observed signal for demodulation of a symbol is quantified. As the number of stages increases by one, the time interval over which the signal is effectively processed increases so as to accommodate $2(K-1)$ more information symbols. In the Mth stage, the interference due to the information symbols that directly overlap with the symbol under consideration are rejected based on the $(M-1)$st stage receiver's estimates of these symbols. Each of the $2(K-1)$ interfering symbols is estimated by the $(M-1)$st stage receiver by rejecting the corresponding interference based on the symbol estimates from the $(M-2)$nd stage of the M-stage algorithm and so on. It is clear that the multiple-access interference is rejected with a weighted emphasis. If the probability of error decreases as the number of stages increases, the maximum weight is assigned to the symbols that directly overlap with the symbol under consideration for demodulation.

B. Implementation

A realizable implementation of the multistage algorithm should never need a symbol estimate that has either not yet been made or was obtained but not stored long enough. An efficient implementation should store symbol estimates at various stages only so long as they are needed and avoid repetition of any processing.

A detector which efficiently incorporates the multistage algorithm for the BPSK-CDMA system is shown in Figs. 2 and 3. Fig. 2 shows a bank of K matched filters followed by a set of K M-stage processors. The detailed implementation of the kth such processor is shown in Fig. 3. The signal flow is depicted by a snap-shot of the receiver at time $(i+2)T$. In order to have all elements of $z^{(i)}(0)$ available simultaneously at $(i+2)T$, the kth processor stores the sufficient statistic from kth matched filter for $T-\tau_k$ seconds. Recall from (3.13) that to obtain the mth stage estimate of $b^{(i)}$, one needs at most the estimates $\hat{b}^{(i-1)}(m-1)$, $\hat{b}^{(i)}(m-1)$, $\hat{b}^{(i+1)}(m-1)$. Repeating the argument, the one-stage estimates $\hat{b}^{(i-(m-1))}(1), \cdots, \hat{b}^{(i+(m-1))}(1)$ are required to obtain $\hat{b}^{(i)}(m)$. This accounts for a time delay of $(m-1)T$. Further, each stage of multiple-access interference rejection takes a bit duration by assumption.[4] Define decoding delay $T_d(m)$ for the mth stage estimate of a bit to be the time that elapses from the availability of its first-stage estimate to the availability of its mth stage estimate. Accounting for the normalizing delay $T-\tau_k$ for the kth user, $T_d(m) = 2(m-1)T$. For instance, at time $(i+2)T$ when $\hat{b}^{(i)}(1)$ is obtained, the M-stage receiver makes available $\hat{b}^{(i-2(m-1))}(m)$, for each $m = 1, 2, \cdots, M$, as shown in Fig. 3. We therefore have a fixed decoding delay implementation.

Consider the storage requirements of this implementation. The vector of decision statistics $z^{(i-2m+3)}(m)$ for each $m = 1, 2, \cdots, M-1$ require the estimates $\hat{b}^{(i-2(m-1))}(m-1)$, $\hat{b}^{(i-2(m-1)+1)}(m-1)$ and $\hat{b}^{(i-2(m-1)+2)}(m-1)$ which are obtained at iT, $(i+1)T$ and at the present time $(i+2)T$, respectively. It is clear that at any given time, we need to store the current bit estimate and the previous two bit estimates of each user for each $m = 1, 2, \cdots, M-1$. Therefore, the storage required is $3(M-1)K$. Moreover, the same storage is required for the multistage detector for the general CDMA system.

Let us now consider the computational requirements. Reconstruction of the MA interference in each stage requires no more than $2(k-1)$ additions/bit. Therefore, for an M-stage detector, $2(M-1)(K-1)$ additions/bit are needed. In the general CDMA system since each sufficient statistic corresponding to $z_k^{(i)}(0)$ is at most an S-vector, and since a different MA interference has to be subtracted from each component, no more than $2S(M-1)(K-1)$ additions/symbol are required. In conclusion, we have shown that the implementation of the multistage detector considered here has a fixed decoding delay and a storage complexity/symbol given by $O(MK)$ and a computational complexity/symbol given as $O(SMK)$.

Finally, in contrast to the conventional receiver, the multistage receiver, like the optimum receiver, requires a knowledge of the signal energies in addition to the time and phase synchronization with signals of all the users. The problem of estimating signal energies during a training period has been dealt with in [17]. It remains to be seen what performance gains can be achieved with the multistage receiver over the conventional demodulator. This is the topic of the next section.

[4] For convenience, construction and subtraction of multiple-access interference is assumed to require a time duration T and sign determination is assumed to be instantaneous.

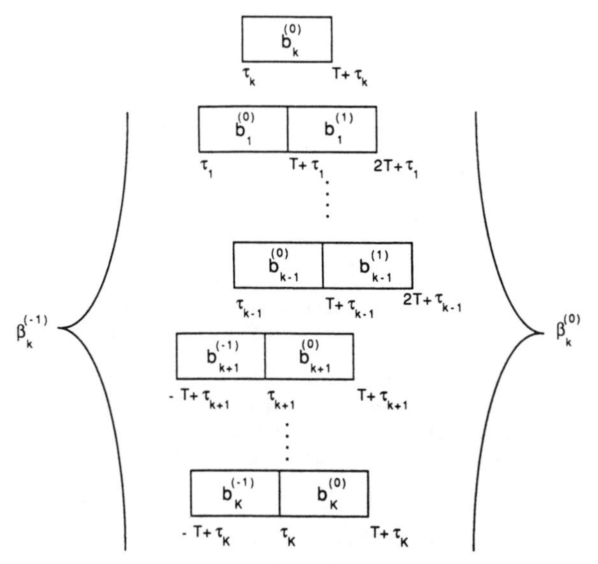

Fig. 4. The left and right interfering bit vectors $\beta_k^{(-1)}$ and $\beta_k^{(0)}$, which overlap with $b_k^{(0)}$.

IV. Probability of Error Analysis

The use of a CDMA system for multiple-access communications is motivated in part by its ability to reject disturbances due to multipath, jamming—intentional or otherwise, etc. The decoding operation (despreading in spread-spectrum systems), spreads the spectrum of these external disturbances. Even when the ambient noise is low, the addition of the spreaded disturbance will produce the effect of a higher level of ambient noise. Hence, the performance of CDMA receivers is of interest for high as well as low signal-to-noise ratios. Bit-error probability as a function of signal-to-noise ratio is therefore the relevant performance measure.[5]

An exact expression for the probability of error is obtained for the two-stage receiver in a K-user BPSK-CDMA system. Without loss of generality, consider the demodulation of the zeroth bit of the first user. Denoting the bit error probability for the 2-stage receiver as $P_e(2)$, we have that

$$P_e(2) = \frac{1}{2}[\Pr[z_1^{(0)}(1) \geq 0 | b_1^{(0)} = -1] + \Pr[z_1^{(0)}(1) < 0 | b_1^{(0)} = +1]]$$

where $z_1^{(0)}(1)$ is obtained from (3.12). From now on, we will drop the subscript 1 and the superscript 0 from $\eta_1^{(0)}$ and $b_1^{(0)}$ to replace them by η and b, respectively. Let $\beta_k^{(i-1)}$ and $\beta_k^{(i)}$ be the $(K-1)$-dimensional column vectors that represent the left and right bits that interfere with the bit $b_k^{(i)}$. Each of these vectors for $i = 0$ is illustrated in Fig. 4. Notice that $\beta_k^{(i)}$ can be written as $(b_1^{(i+1)}, \cdots, b_{k-1}^{(i+1)}, b_{k+1}^{(i)}, \cdots, b_K^{(i)})^T$. Further, define the matrix $\beta_1 = [\beta_1^{(-1)}\beta_1^{(0)}]$ to represent the bits that directly overlap with b and $\beta_2 = [\beta_2^{(-2)}\beta_k^{(0)}]$ to denote the bits that occur in the decision statistic under consideration but are not and let the error vector $\beta_1^{(i)} - \hat{\beta}_1^{(i)}(1)$ be defined as $\delta_1^{(i)}$. Analogous to the definition of β_1, define the error vector that occurs in the decision statistic under consideration as $\delta_1 = [\delta_1^{(-1)}\delta_1^{(0)}]$. Similar to the definition of $\beta_k^{(i)}$, let us define a $(K-1)$-dimensional Gaussian random vector as $\xi_k^{(i)} = (\eta_1^{(i+1)}, \cdots, \eta_{k-1}^{(i+1)}, \eta_{k+1}^{(i)}, \cdots, \eta_K^{(i)})^T$. Also let ξ_1 be a Gaussian random vector as $[\xi_1^{(-1)T}\xi_1^{(0)T}]^T$.

It is clear from (3.12) that the decision statistic $z_1^{(0)}(1)$ is a function not only of η, b, and β_1 but also of β_2 and ξ_1 because of its dependence on δ_1. The decision statistic can be written to show the dependence on these parameters as

$$z_1^{(0)}(1) = E_1 b + \eta + I(\beta_1, \beta_2, \xi_1, \delta_1). \tag{4.16}$$

The first, second, and the third term will be referred to as the *desired signal*, *additive noise*, and *residual interference*, respectively.

The residual interference is a nonlinear function of the Gaussian random vector ξ_1. In fact, it is a discrete valued function taking on a specific value for each value of the error vector δ_1. The random variables forming ξ_1 are correlated with η since they are obtained by integrating the product of the corresponding signal and a common white noise process over overlapping intervals. The $(2K-1)$-dimensional zero-mean Gaussian random vector $\xi = [\xi_1^{(-1)T}\xi_1^{(0)T}\eta]^T$ has a covariance matrix given as

$$E[\xi\xi^T] = \frac{N_0}{2}\begin{bmatrix} \tilde{H}(0) & \tilde{H}(-1) & \tilde{h}_{c1}(-1) \\ \tilde{H}(1) & \tilde{H}(0) & \tilde{h}_{c1}(0) \\ \tilde{h}_{r1}(1) & \tilde{h}_{r1}(0) & h_{1,1}(0) \end{bmatrix} \tag{4.17}$$

where these blocks are derived from the following decomposition of $H(i)$ defined in (2.3):

$$H(i) = \begin{bmatrix} h_{11}(i) & \tilde{h}_{r1}(i) \\ \tilde{h}_{c1}(i) & \tilde{H}(i) \end{bmatrix} \tag{4.18}$$

where $\tilde{h}_{c1}(i)$ is a $(K-1) \times 1$ column vector and $\tilde{h}_{r1}(i)$ is a $1 \times (K-1)$ row vector and $\tilde{H}(i)$ is a $(K-1) \times (K-1)$ matrix. Therefore, the additive noise and the residual interference are statistically dependent.

Let us now define the parameters that will be needed in describing the error probability derivation. Define the external values of the sum of the desired signal component and the residual interference for a given b and β_1 as

$$L_b(\beta_1) \triangleq \min_{\delta_1}[E_1 b - I(\cdot, \delta_1)] \text{ and}$$

$$U_b(\beta_1) \triangleq \max_{\delta_1}[-E_1 b - I(\cdot, \delta_1)]. \tag{4.19}$$

Using the fact that an element of the error vector δ_1 is such that $[\delta_1]_j \in \{2[\beta_1]_j, 0\}$, it can be verified that for $b \in \{+1, -1\}$, we can write

$$L_b(\beta_1) = -E_1 b - 2\sum_{j=2}^{K}([b_j^{(-1)}h_{1j}(1)]^+ + [b_j^{(0)}h_{1j}(0)]^+),$$

and

$$U_b(\beta_1) = -E_1 b - 2\sum_{j=2}^{K}([b_j^{(-1)}h_{1j}(1)]^- + [b_j^{(0)}h_{1j}(0)]^-)$$

where $[x]^+$ is equal to x if $x > 0$ and is equal to zero otherwise. Similarly, $[x]^-$ is equal to x if $x < 0$ and is equal to zero otherwise.

The general strategy for the evaluation of bit error probability will now be described. Let us first condition on b, β_1 and η. The probability of error can be written as

$$P_e(2) = E_{b,\beta_1}[E_\eta\{\Pr(\text{error } | b, \beta_1, \eta)\}] \tag{4.20}$$

where E_{b,β_1} denotes expectation over the ensemble of independent, uniformly distributed $b, \beta_1 \in \{-1, +1\}^{2K-1}$ and similarly, E_η denotes expectation over the Gaussian random variable η. Partitioning the range of η into three intervals defined as $A_1 \triangleq [-\infty, L_b(\beta_1)]$, $A_2 \triangleq [L_b(\beta_1), U_b(\beta_1)]$, and $A_3 \triangleq [U_b(\beta_1), \infty]$, the term in the square bracket in (4.20) can be expressed as

$$\begin{aligned} E_\eta[\Pr(\text{error } | b, \beta_1, \eta)] &= E_\eta[I_{A_1}(\eta)\Pr(\text{error } | b, \beta_1, \eta)] \\ &+ E_\eta[I_{A_2}(\eta)\Pr(\text{error } | b, \beta_1, \eta)] \\ &+ E_\eta[I_{A_3}(\eta)\Pr(\text{error } | b, \beta_1, \eta)] \end{aligned} \tag{4.21}$$

[5] Although asymptotic efficiency [19] lacks the detail of bit-error probability, it results in significant simplification in the analysis of certain detectors [10], [18]. However, no such simplicity seems to result in the case of the multistage detectors.

where $I_{A_i}(\eta)$ denotes the indicator function of the interval A_i. The rest of the derivation deals with the evaluation of the three terms in the above expression.

Let us first begin with the first and the third terms in (4.21) because they are relatively simple to obtain. Recall the definitions of $L_b(\boldsymbol{\beta}_1)$ and $U_b(\boldsymbol{\beta}_1)$ in (4.19). Observe from (4.16), that for a given $\boldsymbol{\beta}_1$, if $b = +1$ and $\eta > U_{+1}(\boldsymbol{\beta}_1)$ or if $b = -1$ and $\eta < L_{-1}(\boldsymbol{\beta}_1)$, then $\mathrm{sgn}(z_1^{(0)}) = b$. Hence, irrespective of what the error vector $\boldsymbol{\delta}_1$ may be, no error occurs. Conversely, if $b = +1$ and $\eta < L_{+1}(\boldsymbol{\beta}_1)$ or if $b = -1$ and $\eta > U_{-1}(\boldsymbol{\beta}_1)$, then $\mathrm{sgn}(z_1^{(0)}(1)) = -b$. In this case, irrespective of what $\boldsymbol{\delta}_1$, an error occurs. Summarizing,

$$\Pr[\text{error } |b, \boldsymbol{\beta}_1, \eta] = \begin{cases} 1 \text{ if } b = +1 \text{ and } \eta < L_{+1}(\boldsymbol{\beta}_1) \\ 0 \text{ if } b = -1 \text{ and } \eta < L_{-1}(\boldsymbol{\beta}_1) \end{cases}, \quad (4.22)$$

and

$$\Pr[\text{error } |b, \boldsymbol{\beta}_1, \eta] = \begin{cases} 0 \text{ if } b = +1 \text{ and } \eta > U_{+1}(\boldsymbol{\beta}_1) \\ 1 \text{ if } b = -1 \text{ and } \eta > U_{-1}(\boldsymbol{\beta}_1) \end{cases}. \quad (4.23)$$

Substituting (4.23) and (4.22) into the first and the third terms, respectively, of the expression in (4.21), and then taking the expectation of each term over η and then over $\boldsymbol{\beta}_1$ and b, it can be shown that

$$E_{b,\boldsymbol{\beta}_1}[E_\eta\{I_{A_1}(\eta)\Pr(\text{error } |b, \boldsymbol{\beta}_1, \eta)\}] = \frac{1}{2}E_{\boldsymbol{\beta}_1}\left[Q\left(-\sqrt{\frac{2}{N_0E_1}}L_{+1}(\boldsymbol{\beta}_1)\right)\right], \quad (4.24)$$

and

$$E_{b,\boldsymbol{\beta}_1}[E_\eta\{I_{A_3}(\eta)\Pr(\text{error } |b, \boldsymbol{\beta}_1, \eta)\}] = \frac{1}{2}E_{\boldsymbol{\beta}_1}\left[Q\left(\sqrt{\frac{2}{N_0E_1}}U_{-1}(\boldsymbol{\beta}_1)\right)\right] \quad (4.25)$$

where $Q(\cdot)$ is the complementary error function so that $Q(x) = 1/\sqrt{2\pi}\int_x^\infty e^{-t^2/2}\,dt$.

Next, consider the evaluation of the second term in (4.21). For a given b and $\boldsymbol{\beta}_1$ and for a realization of η in A_2 denoted by χ, the set of error vectors that give rise to an error can be written as

$$\Delta_b(\chi, \boldsymbol{\beta}_1) = \{\boldsymbol{\delta}_1 \in D(\boldsymbol{\beta}_1) \text{ such that } \mathrm{sgn}[\chi + E_1 b + I(\boldsymbol{\beta}_1, \boldsymbol{\beta}_2, \boldsymbol{\xi}_1, \boldsymbol{\delta}_1)] = -b\}$$

where $D(\boldsymbol{\beta}_1)$ is the set of all $2^{2(K-1)}$ admissible error vectors $\boldsymbol{\delta}_1$. Since every $\boldsymbol{\delta}_1 \in \Delta_b(\chi, \boldsymbol{\beta}_1)$ yields a disjoint hyperquadrant in $(z_2^{(-1)}(0), \cdots, z_K^{(-1)}(0), z_2^{(0)}(0), \cdots, z_K^{(0)}(0))$ space, the probability of this event can be expressed as

$$\Pr(\text{error } |b, \boldsymbol{\beta}_1, \eta = \chi) = q_b(\chi, \boldsymbol{\beta}_1) = \sum_{\boldsymbol{\delta}_1 \in \Delta_b(\chi, \boldsymbol{\beta}_1)} [\Pr(\hat{\boldsymbol{\beta}}_1(1) = \boldsymbol{\beta}_1 - \boldsymbol{\delta}_1 | \boldsymbol{\beta}_1, \eta = \chi, b)]$$

$$= \sum_{\boldsymbol{\delta}_1 \in \Delta_b(\chi, \boldsymbol{\beta}_1)} E_{\boldsymbol{\beta}_2}[\Pr(\hat{\boldsymbol{\beta}}_1(1) = \boldsymbol{\beta}_1 - \boldsymbol{\delta}_1 | \boldsymbol{\beta}_1, \boldsymbol{\beta}_2, \eta = \chi, b)] \quad (4.26)$$

or alternatively,

$$q_b(\chi, \boldsymbol{\beta}_1) = E_{\boldsymbol{\beta}_2}[\Pr(\boldsymbol{\delta}_1 \in \Delta_b(\chi, \boldsymbol{\beta}_1) | \boldsymbol{\beta}_1, \boldsymbol{\beta}_2, \eta = \chi, b)] \quad (4.27)$$

where each element in $\hat{\boldsymbol{\beta}}_1(1) = [\hat{\boldsymbol{\beta}}_1^{(-1)}(1)\hat{\boldsymbol{\beta}}_1^{(0)}(1)]$ is equal to the sign of the corresponding zeroth stage statistic that is Gaussian under the conditioning specified in the above equation. Therefore, the probability inside the expectation in (4.26) is equivalent to the evaluation of a $2(K-1)$-dimensional normal distribution function. For the evaluation of this probability, the conditional density of $\begin{bmatrix}\boldsymbol{\xi}_1^{(-1)} \\ \boldsymbol{\xi}_1^{(0)}\end{bmatrix}$ given $\eta = \chi$ has to be used, which can be deduced from (4.17) and easily shown to be normal with mean vector $\chi/E_1 \begin{bmatrix}\tilde{\boldsymbol{h}}_{c1}(-1) \\ \tilde{\boldsymbol{h}}_{c1}(0)\end{bmatrix}$ and covariance matrix given as

$$\frac{N_0}{2}\begin{bmatrix}\tilde{\boldsymbol{H}}(0) & \tilde{\boldsymbol{H}}(-1) \\ \tilde{\boldsymbol{H}}(1) & \tilde{\boldsymbol{H}}(0)\end{bmatrix} - \frac{N_0}{2E_1}\begin{bmatrix}\tilde{\boldsymbol{h}}_{c1}(-1) \\ \tilde{\boldsymbol{h}}_{c1}(0)\end{bmatrix}[\tilde{\boldsymbol{h}}_{r1}(1)\tilde{\boldsymbol{h}}_{r1}(0)]. \quad (4.28)$$

For details on the efficient computation of the multivariate normal distribution function see [12]. Alternatively, the evaluation of the probability in the expression for $q_b(\chi, \boldsymbol{\beta}_1)$ in (4.27) can be simplified by expressing it as the sum of multivariate normal distribution functions of dimension less than or equal to $2(K-1)$ by exploiting the nature of the set $\Delta_b(\chi, \boldsymbol{\beta}_1)$. Significant savings in computation result from such a direct evaluation of the probability $q_b(\chi, \boldsymbol{\beta}_1)$. The conditional density of $\begin{bmatrix}\boldsymbol{\xi}_1^{(-1)} \\ \boldsymbol{\xi}_1^{(0)}\end{bmatrix}$ given $\eta = \chi$, depends on χ amd hence so does the probability in (4.27). Substituting (4.27), or equivalently (4.26) in the second term of the expression in (4.21), and taking the expectation over η, b, and $\boldsymbol{\beta}_1$, we have

$$E_{b,\boldsymbol{\beta}_1}\{E_\eta[I_{A_2}(\eta)\Pr(\text{error } |b, \boldsymbol{\beta}_1, \eta)]\} = E_{b,\boldsymbol{\beta}_1}\left[\int_{L_b(\boldsymbol{\beta}_1)}^{U_b(\boldsymbol{\beta}_1)} q_b(\chi, \boldsymbol{\beta}_1) f_\eta(\chi)\,d\chi\right] \quad (4.29)$$

where $f_\eta(\cdot)$ is the marginal probability density function of η.

Finally, observe from (4.20) and (4.21) that the error probability is simply obtained as the sum of the contributions from the error events corresponding to η belonging to the intervals A_1, A_2, and A_3 from (4.24), (4.25), and (4.29), respectively. Therefore,

$$P_e(2) = \frac{1}{2}E_{\boldsymbol{\beta}_1}\left[Q\left(-\sqrt{\frac{2}{N_0E_1}}L_{+1}(\boldsymbol{\beta}_1)\right) + Q\left(\sqrt{\frac{2}{N_0E_1}}U_{-1}(\boldsymbol{\beta}_1)\right)\right] + E_b\left[E_{\boldsymbol{\beta}_1}\left\{\int_{L_b(\boldsymbol{\beta}_1)}^{U_b(\boldsymbol{\beta}_1)} q_b(\chi, \boldsymbol{\beta}_1) f_\eta(\chi)\,d\chi\right\}\right]. \quad (4.30)$$

Note, however, that for a given $\boldsymbol{\beta}_1$, we need to compute the probabilities in (4.26) or equivalently in (4.27) for each evaluation of the function $q_b(\chi, \boldsymbol{\beta}_1)$, before it can be integrated in the interval $[L_b(\boldsymbol{\beta}_1), U_b(\boldsymbol{\beta}_1)]$. There are $2^{(2K-1)}$ such integrals to compute. Needless to say, the evaluation of the exact probability of error is computationally intensive and depends exponentially on the number of users. The intensive computational requirement for analyzing this receiver is partly due to the fact that the various Gaussian noise variables that appear in the decision statistic for the two-stage receiver are all correlated with each other. Furthermore, the decision statistic is a nonlinear function of these variables.

Let us now consider the bit-error probability of the conventional detector in a K-user (Gaussian) channel. Again, without loss of generality, consider the demodulation of the zeroth bit of the first user. It is easily shown from (3.15), (3.6) and using the notation from (4.18) that this error probability can be expressed as

$$P_e(1) = E_{\boldsymbol{\beta}_1}\left\{Q\left[\sqrt{\frac{2}{N_0E_1}}(E_1 + \tilde{\boldsymbol{h}}_{r1}(1)\boldsymbol{\beta}_1^{(-1)} + \tilde{\boldsymbol{h}}_{r1}(0)\boldsymbol{\beta}_1^{(0)})\right]\right\}. \quad (4.31)$$

The expression in (4.31) will be evaluated in the numerical examples in the following section to compare the performance of the conventional detector to the two-stage receiver.

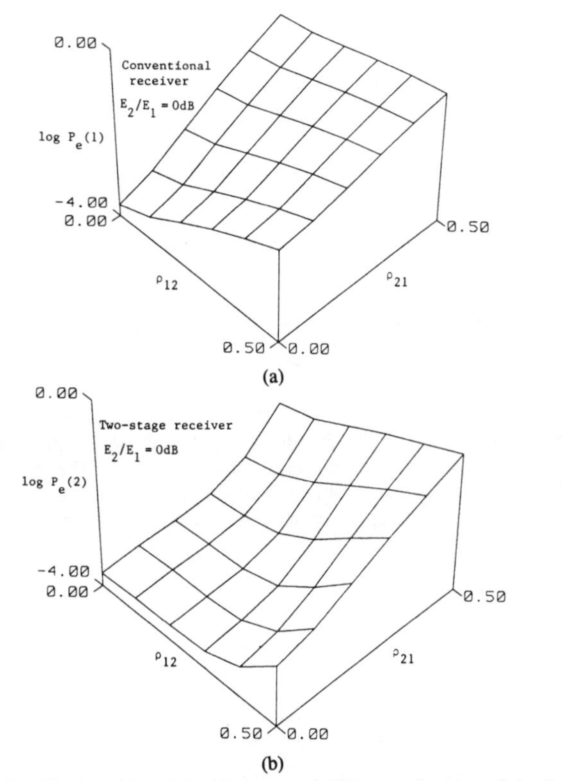

(a)

(b)

Fig. 5. The logarithm of the bit-error probability as a function of the signal cross-correlations ρ_{12} and ρ_{21} when $E_2/E_1 = 0$ dB in the two-user case; (a) conventional receiver; (b) two-stage receiver.

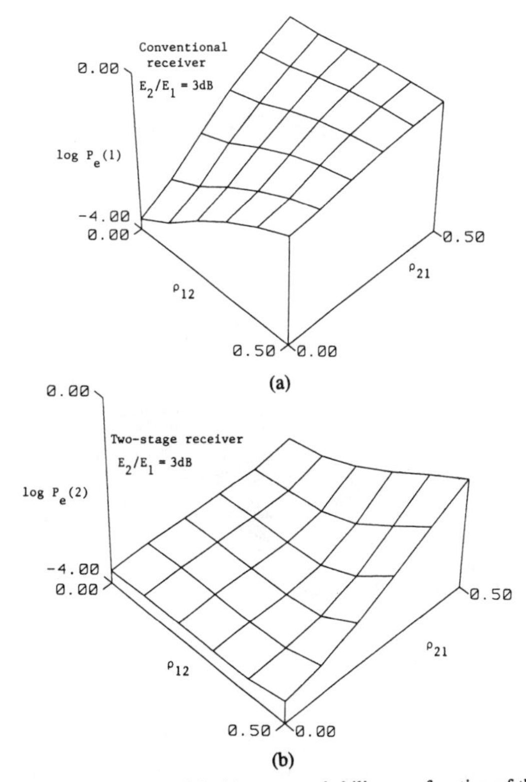

(a)

(b)

Fig. 6. The logarithm of the bit-error probability as a function of the signal cross-correlations ρ_{12} and ρ_{21} when $E_2/E_1 = 3$ dB in the two-user case; (a) conventional receiver; (b) two-stage receiver.

V. Numerical Results

In this section, we present several numerical examples to illustrate the comparison of three demodulators; the conventional detector, the two-stage detector and the matched filter operating in the absence of interfering signals. The first set of examples depict the performances as the signal cross-correlations and their relative signal strengths are varied. Then, we evaluate the performances of these detectors for an important class of BPSK-CDMA systems which employ direct-sequence spread-spectrum (DS-SS) signaling for their multiple-access capability. Several numerical examples for a two-user channel are presented for both low and high bandwidth efficiencies and also the "near–far" problems.

A. Enhanced Bandwidth Efficiency and Near-Far Immunity

For a two-user channel, the performance of any detector would depend on four variables; the normalized signal cross-correlations $\rho_{12} = (E_1E_2)^{-1/2}h_{12}(0)$ and $\rho_{21} = (E_1E_2)^{-1/2}h_{21}(1)$; the signal-to-noise ratios of the two users denoted SNR_1 and SNR_2, respectively. Figs. 5 and 6 illustrate the logarithm of the error probability of the first user as a function of ρ_{12} and ρ_{21} with $\mathrm{SNR}_1 = 8$ dB. In Fig. 5, the signal strength of the second user is the same as that of the first user. In Fig. 6, a near–far situation is considered with $E_2/E_1 = 3$ dB. That is, the interfering user is about twice as strong as the desired user. Fig. 7 depicts the logarithm of the probability of error of the first user as a function of the correlations ρ_{12}, ρ_{21} and the energy ratio E_2/E_1 in dB when ρ_{12} is set equal to ρ_{21}.

Bandwidth efficiency has a direct bearing on the signal correlations. In general, higher bandwidth efficiencies result in larger correlations. From Figs. 5 and 6 it can be observed that the region of reliable demodulation[6] in the cross-correlation space has been extended considerably by the two-stage detector. Higher bandwidth efficiencies are therefore achieved. Also, notice that the extension is greater when the interfering signal is stronger. This fact suggests that the probability of error comparison would be of interest as a function of the relative signal strengths in addition to the signal cross-correlations. Fig. 7 illustrates the fact that the region of effective detection becomes larger as the interfering signal strength increases. The improvement over the conventional receiver in near-far situations is therefore two-fold since the conventional receiver degrades with the increase of the interfering signal strength. The less dramatic improvement for large correlations suggests the need for employing alternative initial estimates and/or multiple stages. Finally, the degradation of the two-stage detector as compared to the conventional detector for a small enough interfering signal strength points to the need for a higher stage detector. For example, in a three-stage detector a good estimate of the second user's signal is made in the second stage (since the first user's signal strength is relatively higher) which is then used in the third stage to reject the MA interference more reliably than was possible in the second stage.

[6]The term "reliable demodulation" may be interpreted as demodulation with an error probability close to the single-user error probability, although closeness to the minimum possible error probability is more meaningful. Bounds for the latter have been obtained in [18].

B. Direct-Sequence Spread-Spectrum Systems

Next, we consider some examples with direct-sequence spread-spectrum signaling. The details of the system description can be found in a number of references (see, for example, [13]). However, the essentials will be repeated here for convenience. For DS-SS signals, the time-limited bandpass signal $s_k(t)$ in (2.2) can be written as

$$s_k(t) = \sqrt{2E_kT^{-1}}a_k(t)\cos(\omega_c t + \theta_k) \tag{5.32}$$

where ω_c is the carrier frequency and θ_k is the phase angle. In this format, the direct-sequence spreading waveform $a_k(t)$ assigned to the kth user is given as

$$a_k(t) = \sum_{j=0}^{N-1} a_k^{(j)}c(t - jT_c); a_k^{(j)} \in \{+1, -1\}$$

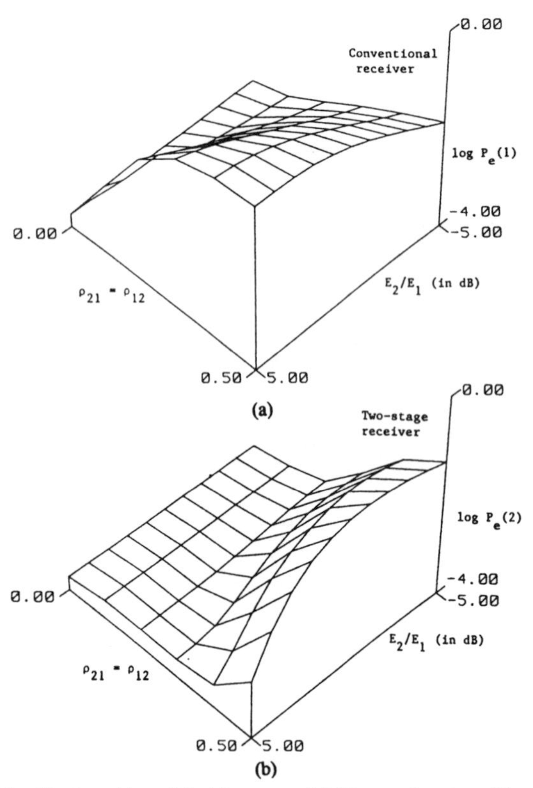

Fig. 7. The logarithm of the bit-error probability as a function of the signal cross-correlations ρ_{12} and ρ_{21} and E_2/E_1 (in dB) in the two-user case when $\rho_{12} = \rho_{21}$; (a) conventional receiver; (b) two-stage receiver.

where for each k, $\{a_k^{(j)}\}_{j=0}^{N-1}$ is the code sequence assigned to the kth transmitter. The chip waveform $c(\cdot)$ is assumed to be a unit rectangular signal restricted to be zero outside the chip interval $[0, T_c]$ and the number of chips per bit is N. If $\omega_c T$ is an integer multiple of 2π and $\gg 1$, the matrices $H(i)$ defined in Section II can be related to the normalized baseband correlation functions as in [17] so that

$$h_{kl}(0) = \begin{cases} (E_k E_l)^{1/2} \hat{R}_{kl}(\tau_k - \tau_l)\cos(\phi_{kl}) & \text{if } l \le k \\ (E_k E_l)^{1/2} \hat{R}_{lk}(\tau_l - \tau_k)\cos(\phi_{kl}) & \text{if } l > k \end{cases} \quad (5.33a)$$

$$h_{kl}(1) = \begin{cases} 0 & \text{if } l \le k \\ (E_k E_l)^{1/2} \hat{R}_{kl}(T + \tau_k - \tau_l)\cos(\phi_{kl}) & \text{if } l > k \end{cases} \quad (5.33b)$$

where $\hat{R}_{kl}(\tau) \triangleq T^{-1}\int_\tau^T a_k(t-\tau)a_l(t)\,dt$ for $0 \le \tau \le T$, is the partial continuous-time correlation function [15]. The relative phase delay is defined as $\phi_{kl} = |(\tau_k - \tau_l)\omega_c + \theta_l - \theta_k|$. The data bits $\{b_k^{(i)}; \forall i; \forall k\}$ are independent and equally likely to be +1 or −1.

The probability of bit-error of the first user for the one-stage receiver given in (4.31) is now denoted as $P_e(1)(\tau, \phi)$ where $\tau \triangleq (\tau_1, \tau_2, \cdots, \tau_K)^T$ and $\phi \triangleq (\phi_1, \phi_2, \cdots, \phi_K)^T$, to show its explicit dependence on the time and phase delays of all the users in the system. It can be shown that for each τ,

$$P_e(1)(\tau, \phi) \le P_e(1)(\tau, \phi)|_{\phi=0},$$

using the convexity of the Q-function and assuming the eye-open condition [8]. That is, the base-band case is the worst case for each set of relative time delays. Further, under the same condition, we can show that the error probability, averaged over independent and uniformly distributed time and phase delays with $\tau \in [0, T]^K$ and $\phi \in [0, 2\pi]^K$, is upper bounded by its average for the base-band case on a discrete grid of time delays corresponding to the chip boundaries, i.e.,

$$\frac{1}{(2\pi T)^{K-1}} \iint P_e(1)(\tau, \phi)\,d\tau\,d\phi \le \frac{1}{N^{K-1}} \sum_{\tau \in \{0, T_c, \cdots, (N-1)T_c\}^{K-1},\ \phi=0} P_e(1)(\tau, \phi) \triangleq \tilde{P}_e(1). \quad (5.34)$$

Moreover, the worst case error probability occurs on the worst-case grid of time delays. Therefore, we have

$$\max_{\tau,\phi} P_e(1)(\tau, \phi) = \max_{\substack{\tau \in \{0, T_c, \cdots, (N-1)T_c\}^{K-1} \\ \phi=0}} P_e(1)(\tau, \phi) \triangleq \bar{P}_e(1). \quad (5.35)$$

These and similar results for the upper bound of the maximum-likelihood sequence detector were obtained in [17]. Therefore, in order to conduct a meaningful comparison between the conventional, the two-stage and the optimum receiver, we will compute

$$\tilde{P}_e(2) \triangleq \frac{1}{N^{K-1}} \sum_{\substack{\tau \in \{0, T_c, \cdots, (N-1)T_c\}^{K-1} \\ \phi=0}} P_e(2)(\tau, \phi) \quad (5.36)$$

$$\bar{P}_e(2) \triangleq \max_{\substack{\tau \in \{0, T_c, \cdots, (N-1)T_c\}^{K-1} \\ \phi=0}} P_e(2)(\tau, \phi). \quad (5.37)$$

for the two-stage receiver for comparison to the corresponding probabilities of the conventional and the optimum receivers.

Two sets of examples are considered corresponding to both low and high bandwidth efficiencies (defined by the K/N ratios). For each set, the near-far situation is also illustrated. These examples were considered for presenting numerical results in [18] and therefore comparison to the performance of the optimum receiver can be made.

In the first set of examples in Fig. 8, we consider two spread-spectrum signals which are maximal-length signature sequences of length 31 generated to maximize a signal-to-multiple-access interference functional [6]. Average error probability of the conventional detector for these signature waveforms in the equal-energy situation has been extensively studied [7], [14]. Figs. 8(a)–(c) depict the worst case (1- and 2-STAGE U.B. in legend) the upper bound on the average error probability (1- and 2-STAGE AV.U. in legend) for the conventional and the two-stage receiver in comparison to the single user error probability for E_2/E_1 (in dB) equal to −3 dB, 0 dB, and +3 dB, respectively, corresponding to the second user being approximately half, equal and twice as strong as the first user. Notice the improvement in the performance of the two-stage receiver as the second users' signal strength increases. This behavior was also noted in [3] where it was shown that the interference channel capacity coincides with that of a single-user channel and in [19] where it was shown that the asymptotic efficiency of the optimum detector is equal to one for a sufficiently high interference. The simple explanation for this is that, if the second user has a high energy, it is estimated better by the first stage and hence is rejected more successfully.

The second set of examples [see Fig. 9(a)–(d)] correspond to high bandwidth efficiency situations. The code waveforms of the two users are depicted in Fig. 9(a). Fig. 9(a)–(c) depict the worst-case error probabilities for the relative signal strengths shown in each of the graphs. Here, the worst case error probability and the upper bound on the average error probability are the same for the one-stage and also for the two-stage receivers. Notice a similar trend again in this high bandwidth efficiency case. Fig. 9(d) illustrates the near-invariance of the error probability of the two-stage receiver as a function of the signal strength of the second user. The near–far problem is therefore alleviated. The conventional receiver on the other hand, degrades due to a strong interfering signal.

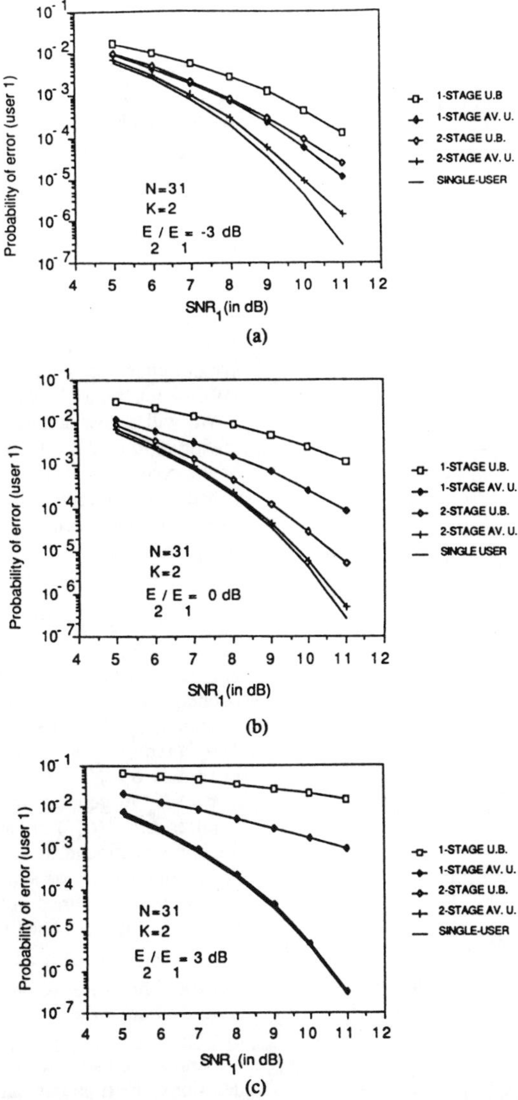

Fig. 8. A comparison between the worst case and upper bound of the average error probability of a two-user direct-sequence spread-spectrum system with $N = 31$ for the conventional receiver and the two-stage receiver and the signal-user bit-error probability; (a) $E_2/E_1 = -3$ dB, (b) $E_2/E_1 = 0$ dB, (c) $E_2/E_1 = 3$ dB.

VI. Conclusions

Optimum multiuser demodulation for coherent communications in CDMA channels is inherently a difficult problem requiring intensive computation and a variable decoding delay that is unacceptable in most practical applications. On the other hand, the conventional multiuser demodulator suffers severe degradation in the useful ranges of bandwidth efficiencies and in near–far situations. There is hence a need for suboptimum detectors which are robust to near–far effects with a reasonable computational complexity to ensure their practical implementation. In this paper, a multistage multiuser detector is proposed which is based on a successive multiple-access interference rejection scheme. This algorithm has a computational complexity that is linear in the number of users. Also, an efficient implementation of the proposed detector is demonstrated.

The probability of error of the two-stage receiver was obtained. Its computation is complicated because the decision statistic depends in a nonlinear form on some normal random variables which are dependent. From the evaluation of the probability of error, it was shown that the region of reliable demodulation in the cross-correlation space was extended considerably even by the two-stage detector. This extension was larger as the interfering signal strength increased. It is this feature that makes the multistage detector eminently suitable for demodulating relatively weak signals in the presence of strong interfering signals. The near–far problem was therefore alleviated. However, the improvement for large signal cross-correlation was less significant, thereby motivating the need for employing better initial estimates and/or higher-stage detectors for this region. Similar conclusions were obtained from two sets of direct-sequence spread-spectrum examples corresponding to low and high bandwidth efficiencies.

In conclusion, the proposed multistage demodulators would prove to be valuable in situations where there is need for partial/full centralized demodulation as well as situations where there are partial/no security restrictions within the network. For these situations, some or all of the code waveforms are available at the receiver. The corresponding time and phase synchronization capabilities are required (centralized demodulation) or can be made available. The modification of the multistage receiver to take into account only a subset of all the users can be made easily.

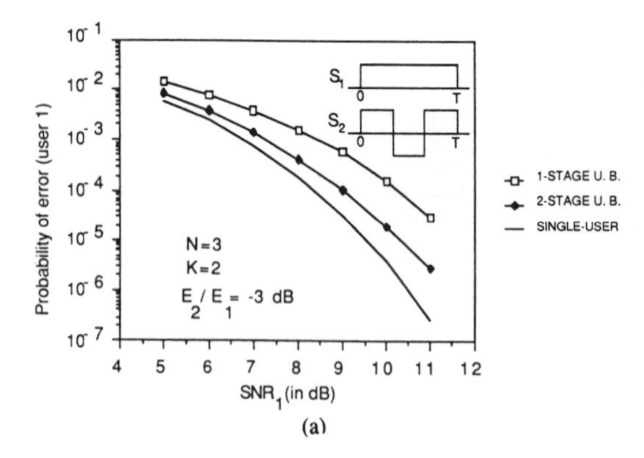

(a)

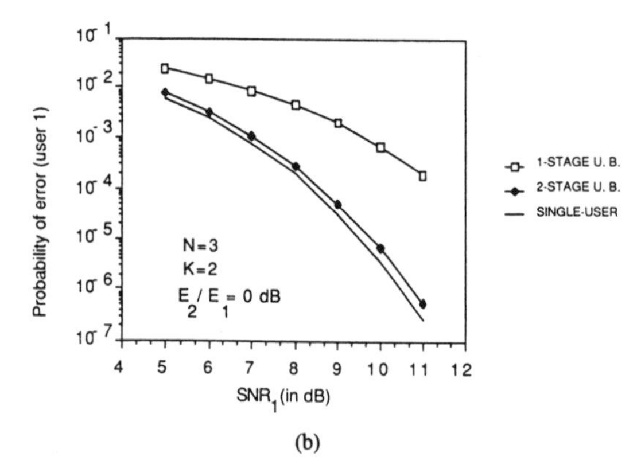

(b)

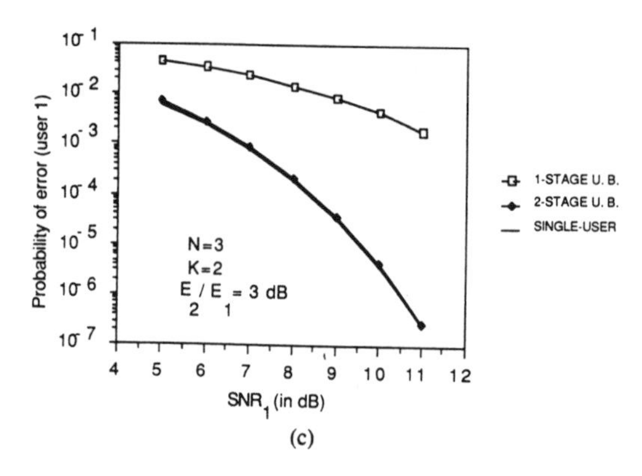

(c)

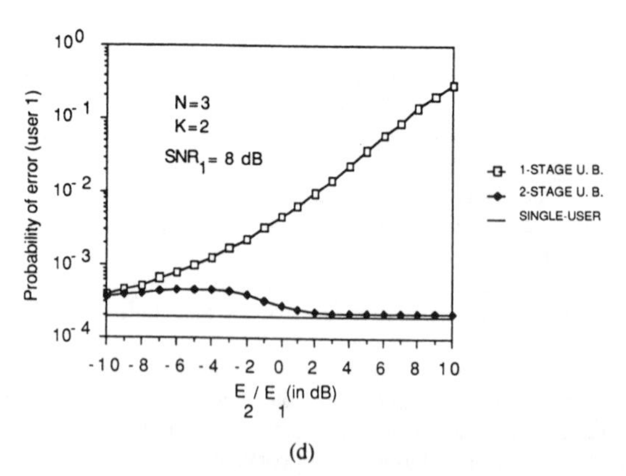

(d)

Fig. 9. A comparison between the worst case and upper bound of the average error probability of a two-user direct-sequence spread-spectrum system with $N = 3$ for the conventional receiver and the two-stage receiver and the single-user bit-error probability; (a) $E_2/E_1 = -3$ dB, (b) $E_2/E_1 = 0$ dB, (c) $E_2/E_1 = 3$ dB.

Appendix

Processing Time Interval for M-Stage Detectors

In this Appendix, the effective time interval over which the M-stage multistage detector processes the received signal in demodulating any symbol is obtained. Consider the ith symbol of the kth user, the M-stage receiver using the conventional detector's decision as the first stage, effectively processes the received signal waveform in the interval given as

$$[-(M-1)T + \gamma_k T + \tau_{\sigma_k(k+M-1)} + iT,$$
$$MT + \alpha_k T + \tau_{\sigma_k(k-M+1)} + iT]$$

where $K\gamma_k = (k+M-1) - \sigma_K(k+M-1)$, $K\alpha_k = (k-M+1) - \sigma_K(k-M+1)$, and $\sigma_K(j) = 1 + (j-1) \bmod K$. Excluding the symbol under consideration, in the general case there are $2M(K-1)$ information symbols associated with this interval.

We will arrive at this processing interval by considering the following argument on the processing time of each stage. The 1st-stage or the conventional decision for $b_k^{(i)}$ is based on $z_k^{(i)}(0)$ which is obtained from the received signal in the interval $[it + \tau_k, \tau_k + (i+1)T]$. The 2nd-stage decision is based on $z_k^{(i)}(1)$ which is obtained by rejecting the MA interference in $z_k^{(i)}(0)$ based on the 1st-stage estimates of bits that overlap with $b_k^{(i)}$. Since it is assumed that $0 \le \tau_1 \le \tau_2 \le \cdots \le \tau_K < T$, the left-most bit that overlaps with $b_k^{(i)}$ denoted as $L(b_k^{(i)})$ is $b_{k+1}^{(i-1)}$ if $k+1 \le K$ or $b_1^{(i)}$ if $k = K$ (lower end of the processing interval at this stage is $\tau_k + (i-1)T$ or $\tau_1 + iT$, respectively). Again, the 3rd-stage decision statistic $z_k^{(i)}(2)$ is obtained by rejecting the MA interference based on 2nd-stage decisions of symbols that overlap with $b_k^{(i)}$. The 2nd-stage decision on $L(b_k^{(i)})$ requires its 1st-stage decision on $L(L(b_k^{(i)})) = L^2(b_k^{(i)})$ which in turn is $b_{k+2}^{(i-2)}$ if $k+2 \le K$ or $b_1^{(i-1)}$ if $k+1 = K$ or $b_2^{(i-1)}$ if $k = K$. In general, the Mth-stage decision statistic $z_k^{(i)}(M-1)$, requires $L^{M-1}(b_k^{(i)})$ in the form of the 1st-stage decision. Following the argument above, it can be easily verified that the left-most bit that occurs in the decision statistic $z_k^{(i)}(m)$ is $L^m(b_k^{(i)}) = b_{\sigma_K(k+m)}^{(i-m+\gamma_k)}$ where σ_K and γ_k are defined in the statement of the remark. Similarly, the right-most bit that appears in $z_k^{(i)}(m)$ is denoted as $R^m(b_k^{(i)})$ and can be verified to be $b_{\sigma_K(k-m)}^{(i+m+\alpha_k)}$ where α_k is defined in the statement of the remark. Therefore, to obtain the Mth stage estimate of $b_k^{(i)}$, the first stage estimate of the $(K-1)$ left-most bits and the $(K-1)$ right-most bits that occur in the decision statistic $z_k^{(i)}(M-1)$ are needed. Since the first-stage estimates are assumed to be the conventional one-shot decisions, the time interval over which $r(t)$ is effectively processed is given as the left time epoch of $L^{M-1}(b_k^{(i)})$ and the right time epoch of $R^{M-1}(b_k^{(i)})$.

Acknowledgment

The authors are indebted to the reviewers for their thorough reviews and useful suggestions for improving the readability of this paper.

References

[1] B. Aazhang and H. V. Poor, "Performance of DS/SSMA communications in impulsive channels—Part II: Hard limiting correlation receivers," *IEEE Trans. Commun.*, vol. COM-36, pp. 88–97, Jan. 1988.

[2] ——, "Performance of DS/SSMA communications in impulsive channels—Part I: Linear correlation receivers," *IEEE Trans. Commun.*, vol. COM-35, pp. 1179–1188, Nov. 1987.

[3] A. B. Carleial, "A case where interference does not reduce capacity," *IEEE Trans. Inform. Theory*, vol. IT-21, pp. 569–570, Sept. 1975.

[4] T. M. Cover, "Some advances in broadcast channels," in *Advances in Communication Systems*, A. J. Viterbi, Ed. New York: Academic, 1975, pp. 229–260.

[5] P. K. Enge and D. V. Sarwate, "Spread-spectrum multiple-access performance of orthogonal codes: Linear receivers," *IEEE Trans. Commun.*, vol. COM-35, Dec. 1987.

[6] F. D. Garber and M. B. Pursley, "Optimal phases of maximal sequences for asynchronous spread-spectrum multiplexing," *IEE Electron. Lett.*, vol. 16, pp. 756–757, Sept. 1980.

[7] E. A. Geraniotis and M. B. Pursley, "Error probability for direct-sequence spread-spectrum multiple-access communications—Part II:

Approximations," *IEEE Trans. Commun.*, vol. COM-30, pp. 985-995, May 1982.

[8] S. Haykin, *Communication Systems*. New York: Wiley, 1983.

[9] J. S. Lehnert and M. B. Pursley, "Error probabilities for binary direct-sequence spread-spectrum communications with random signature sequences," *IEEE Trans. Commun.*, vol. COM-35, pp. 87-98, Jan. 1987.

[10] R. Lupas-Golaszewski and S. Verdu, "Asymptotic efficiency of linear multiuser detectors," Proc. 25th Conf. Decision Contr., Athens, Greece, Dec. 1986.

[11] T. Masamura, "SSMA system with intrasystem interference cancellation," *Trans. IEICE, Japan*, vol. E71, Mar. 1988.

[12] R. L. Plackett, "A reduction formula for normal multivariate integrals," *Biometrika*, 1954.

[13] M. B. Pursley, "Spread-spectrum multiple access communications," in *Multi User Communications*, G. Longo, Ed. New York: Springer-Verlag, 1981, pp. 139-199.

[14] M. B. Pursley, D. V. Sarwate, and W. E. Stark, "Error probability for direct-sequence spread-spectrum multiple-access communications—Part I: Upper and lower bounds," *IEEE Trans. Commun.*, vol. COM-30, pp. 975-984, May 1982.

[15] D. V. Sarwate and M. B. Pursley, "Crosscorrelation properties of pseudorandom and related sequences," *Proc. IEEE*, vol. 68, pp. 593-619, May 1980.

[16] H. L. van Trees, *Detection, Estimation, and Modulation Theory, Part I*. New York: Wiley, 1968.

[17] S. Verdu, "Optimum multi-user signal detection," Ph.D. dissertation, Dep. Elec. Comp. Eng., Univ. Illinois, Urbana-Champaign, 1984.

[18] —, "Minimum probability of error for asynchronous Gaussian multiple-access channels," *IEEE Trans. Inform. Theory*, vol. IT-32, pp. 85-96, Jan. 1986.

[19] —, "Optimum multiuser asymptotic efficiency," *IEEE Trans. Commun.*, vol. COM-34, Sept. 1986.

[20] A. J. Viterbi, "Spread-spectrum multiple access with binary PSK modulation can approach Shannon capacity for the aggregate Gaussian channel," Preprint, 1986.

Mahesh K. Varanasi (S'87-M'89) was born in Hyderabad, India, in 1963. He received the Bachelor's degree in electronics and communication engineering in 1984 from Osmania University, and the M.S. and Ph.D. degrees in electrical engineering in 1987 and 1989 from Rice University, Houston, TX.

He joined the faculty at the University of Colorado, Boulder, in the Fall of 1989, as an Assistant Professor of Electrical and Computer Engineering. His current research interests are in the areas of multiuser communication theory and statistical signal processing.

Dr. Varanasi was a recipient of the K. K. Nair Gold Medal, awarded to the best graduating student by the ECE department, Osmania University. He received a Rice University Fellowship for the academic year 1984-1985 and a Research Fellowship for the years 1985-1989 and is a member of Eta Kappa Nu.

Behnaam Aazhang (S'81-M'85) was born in Bandar Anzali, Iran, on December 7, 1957. He received the B.S. (with highest honors), M.S., and Ph.D. degrees in electrical and computer engineering from the University of Illinois, Urbana-Champaign, in 1981, 1983, and 1986, respectively.

From 1981 to 1985, he was a Research Assistant in the Coordinated Science Laboratory, University of Illinois. In August 1985, he joined the faculty of Rice University, Houston, TX, where he is an Assistant Professor of Electrical and Computer Engineering. His current research interests are in communication theory, information theory, and their applications with emphasis on multiuser communications, packet radio networks, and spread-spectrum systems.

Dr. Aazhang is a recipient of the NSF Engineering Initiation Award 1987-1989 and the IBM Graduate Fellowship 1984-1985, and is a member of Tau Beta Pi and Eta Kappa Nu.

Reprinted from *Proceedings of the IEEE*, Vol. 76, No. 6, June 1988, pp. 657-671.

Interference Rejection Techniques in Spread Spectrum Communications

LAURENCE B. MILSTEIN, FELLOW, IEEE

Invited Paper

Spread spectrum communication systems have many applications, including interference rejection, multiple accessing, multipath suppression, low probability of intercept transmission, and accurate ranging. Of all the potential applications, the ability of a spread spectrum system to withstand interference, both intentional and unintentional, is probably its greatest asset. Of course, any spread spectrum receiver can only suppress a given amount of interference; if the level of interference becomes too great, the system will not function properly.

Even under these latter circumstances, however, other techniques, which enhance the performance of the system over and above the performance improvement that comes automatically to systems simply by employing spread spectrum, are available for use. These techniques typically involve some type of additional signal processing and are the subject of this paper. In particular, two general types of narrow-band interference suppression schemes are discussed in depth, and a short overview is presented for several other techniques as well. The two classes of rejection schemes emphasized in the paper are 1) those based upon least-mean square estimation techniques, and 2) those based upon transform domain processing structures.

I. Introduction

The most important use of a spread spectrum communication system is that of interference suppression. As is well known [41], [42], [49], [53], the inherent processing gain of a spread spectrum system will, in many cases, provide the system with a sufficient degree of interference rejection capability. However, at times the interfering signal is powerful enough so that even with the advantage that the system obtains by spreading the spectrum, communication becomes effectively impossible. In some of these cases, the interference immunity can be improved significantly by using signal processing techniques which complement the spread spectrum modulation.

From basic detection theory [16], [42], [54], the optimal receiver for detecting a known signal in additive white gaussian noise (AWGN) consists of a parallel bank of matched filters, where the number of such filters is determined by the dimensionality of the signal set. For binary antipodal signaling (i.e., a system wherein two signals are used, say, $s_1(t)$ and $s_2(t)$, such that $s_1(t) = -s_2(t)$), only a single matched filter is required. Since a correlation receiver is equivalent to a matched filter receiver, a receiver that performs a single coherent correlation is optimum for detecting one of two known antipodal waveforms.

If the noise is gaussian but not white, the receiver is much more complex, and indeed requires the solution of an integral equation to fully specify it. However, under the special condition of an infinite observation interval, the form of the optimal receiver reduces to the cascade of a prewhitening filter and a matched filter, where the transfer function of the prewhitening filter is the inverse of the power spectral density of the noise, and the matched filter is again matched to the signal structure.

When the signal is being received in nongaussian noise, the situation becomes even more difficult. For the special case of a signal embedded in noise plus sine-wave interference, a nonlinear receiver was shown in [37] to result from the maximum-likelihood formulation of receiver design. However, a receiver of this type is not easy to implement, and the receiver of [37] is not even the most general receiver structure for this problem, since it is based upon a set of discrete input samples, not upon the actual continuous-time waveform. In this paper, we are going to describe various receiver designs which, while not optimal, demonstrate very good performance when used to detect spread spectrum waveforms received in the presence of narrow-band interference, and which are all practical structures in the sense of being implementable with state-of-the-art technologies.

If, indeed, the interference is relatively narrowband compared with the bandwidth of the spread spectrum waveform, then the technique of interference cancellation by the use of notch filters often results in a large improvement in system performance, and the purpose of this paper is to illustrate several such spectral filtering techniques. In particular, the use of tapped delay line-type structures to implement notch filters is discussed below in a good deal

Manuscript received July 28, 1987; revised February 10, 1988. This work was partially supported by the U.S. Army Research Office under Grant DAA L03-86-K-0092, and the Office of Naval Research under Contract ONR N00014-82-K-0376.

The author is with the Department of Electrical and Computer Engineering, University of California-San Diego, La Jolla, CA 92093, USA.

IEEE Log Number 8821644.

0018-9219/88/0600-0657$01.00 © 1988 IEEE

of depth. These notch-filters are used to further enhance the performance of a spread spectrum system over and above what the inherent processing gain of the system provides, and so, in this sense, they complement the spreading technique.

Consider the following: If, after spreading the spectrum of the underlying information over the maximum bandwidth available to the system, the resulting interference rejection capability is still not large enough to sufficiently attenuate any undesired signal, some additional means of interference removal must be used. With respect to the use of notch filters for this purpose, there appear to be two techniques that have received the most interest. The first technique, described in references such as [17], [20], [22], [18], uses a tapped delay line to implement either a one-sided prediction-error filter (Weiner filter [39]), or a two-sided filter. The basic rationale for the use of, say, the Weiner prediction filter for narrow-band interference suppression can be easily seen. The incoming waveform to the spread spectrum receiver consists of the desired spread spectrum signal (taken to be a binary phase-shift keyed (BPSK) direct sequence (DS) signal), thermal noise, and the narrow-band interference. Since both the DS signal and the thermal noise are wide-band processes, their future values cannot be readily predicted from their past values. On the other hand, the interference, being a narrow-band process, can indeed have its future values (and, in particular, its current value) predicted from past values. Hence, the current value, once predicted, can be subtracted from the incoming signal, leaving a waveform comprised primarily of the DS signal and the thermal noise. The same general philosophy holds for the two-sided transversal filter, except now the estimate of the present value of the interference is based upon both past and future values, and the improvement in system performance alluded to above is due to the use of both the past and the future to estimate the present.

The second technique is that of transform domain processing as described, for example, in [31]–[33]. In this technique, a tapped delay line, typically implemented with a surface acoustic wave (SAW) device, with a chirp impulse response built into the taps, is used as a real-time Fourier transformer. As described fully below, a notch filter is implemented by Fourier transforming the received waveform, using an on-off switch to perform the notching operation, and then inverse transforming.

In considering these techniques, both the similarities and the differences become evident. Both techniques can use tapped delay-line implementations, and both can be made adaptive. In the former scheme, the system can be made adaptive by using a tapped delay line with variable tap weights. These tap weights can be adapted, for example, by using the well-known least-mean-square (LMS) algorithm (see below). In the latter technique, it will be shown that an envelope detector in cascade with a threshold crossing indicator can be used to determine the location (in frequency) of the narrow-band interference and hence adjust the position of the notch (or notches) to suppress the interference.

Because both schemes use tapped delay-line implementations, both systems can be built with either SAW technology [30] or CCD [9] technology. Which technology should be used typically depends upon the required bandwidth, with SAW devices being the obvious choice for very wide-band communication (e.g., bandwidths on the order of 100 MHz or more).

One general problem associated with the actual implementation of either of these two systems is that of dynamic range. Since, by definition, these systems are intended to operate in large interference environments, the range of input levels that SAW devices and CCDs can handle is crucial. For example, a typical dynamic range for a SAW device when used as a real-time Fourier transformer is about 40 dB. Yet if the SAW device is part of a system designed for anti-jam (A-J) protection, variations of input level could be 80 dB or more.

II. Estimation-Type Filters

As we have noted, in a spread spectrum communication system employing a direct-sequence pseudonoise spreading signal, the effect of narrow-band interference on system performance is reduced due to the inherent processing gain of the system. However, when the processing gain is insufficient due to bandwidth restriction to allow satisfactory communication to take place, one technique which can at times improve the performance of the system is the method of interference rejection to be described in this section.

Fig. 1 shows the essential parts of a receiver using a suppression filter, and Figs. 2 and 3 show the two-sided and one-sided filter structures, respectively. The one-sided filter is often referred to as a prediction-error filter.

If we assume the spread spectrum signal samples taken at different taps are not correlated (see below), and if there

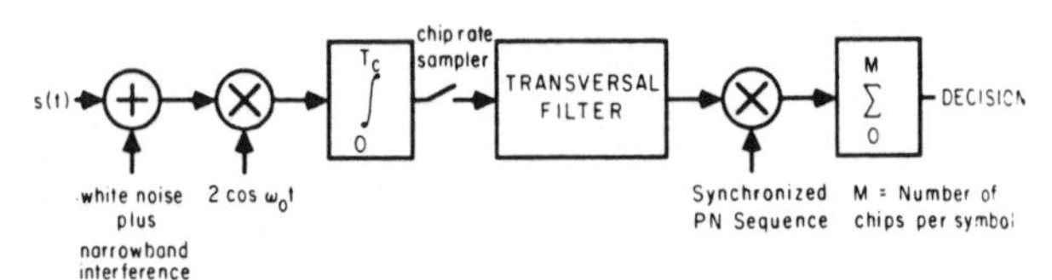

Fig. 1. Receiver block diagram.

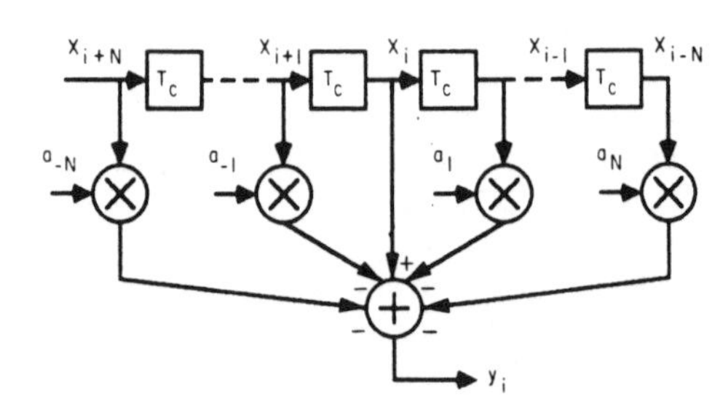

Fig. 2. Two-sided transversal filter.

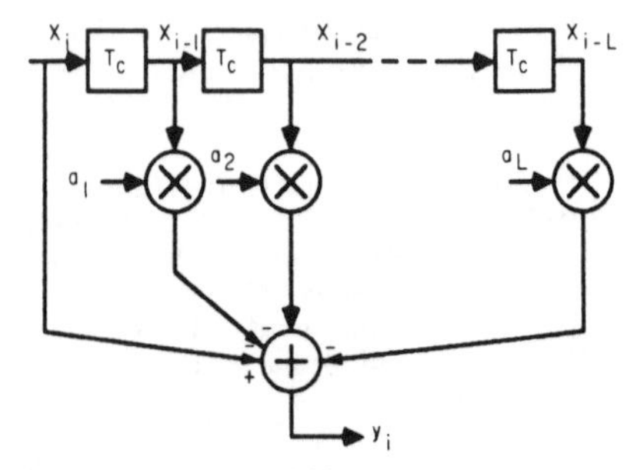

Fig. 3. Single-sided transversal filter.

is only white noise interference, the tap weights will be zero to maintain minimum output error power. If there is additional narrow-band interference, the tap weights will be adjusted to predict the input signal so that the resulting mean-squared error is minimized. The level of the interference is reduced at the expense of introducing some distortion on the desired signal.

There are many references available on this type of suppression filter, and they fall into several general categories. The first group of references emphasize the analytic enhancement in system performance achievable by the use of such filters as determined by the signal-to-noise ratio (SNR) improvement factor of the system. This improvement factor is typically defined as the ratio of the SNR with the suppression filter in the system to the SNR of the system operating without the rejection filter [17], [20], [22], [26], [27], [50], [24].

As is well known in digital communications, while SNR provides a good qualitative indication of system performance, it often does not provide a good quantitative description. To obtain the latter it is necessary to use average probability of error as the criterion-of-goodness, and both analytical [18], [19], [28] and experimental [44]–[47] investigations have been conducted.

A. Signal-to-Noise Ratio Improvement

To understand the operation of this type of suppression filter, consider the two-sided filter of Fig. 2. The received waveform consists of a binary phase-shift-keyed direct sequence spread spectrum waveform, a single tone interfering signal of known amplitude and frequency but with a random phase, and thermal noise. That is, the input to the receiver, $r(t)$, is given by

$$r(t) = s(t) + I(t) + n_w(t), \tag{1}$$

where

$$s(t) = Ac(t)\, d(t) \cos \omega_0 t \tag{2a}$$

$$I(t) = \alpha \cos [(\omega_0 + \Omega)t + \theta] \tag{2b}$$

and $n_w(t)$ is AWGN of two-sided spectral density $\eta_0/2$. In (2), A and α are constant amplitudes, θ is a random phase uniformly distributed in $[0, 2\pi]$, $d(t)$ is a random binary sequence of data symbols taking on values ± 1 with equal probability which last for T seconds, ω_0 is the carrier frequency of the transmitted signal, Ω is the frequency offset of the interference, and $c(t)$ is the spreading sequence taking on values ± 1 which last for T_c seconds, where $T_c \ll T$. The most common type of spreading sequence is a pseudonoise (PN) sequence [41], [42], [49], [54].

For simplicity, assume that we can coherently demodulate the received waveform with the reference $2 \cos \omega_0 t$, as shown in Fig. 1. In reality, to accomplish this operation before despreading is not realistic [41], but there are straightforward techniques to circumvent this problem, and in any event, the essence of the interference rejection mechanism is not dependent upon this assumption (see [46]). Hence, we assume that the sample on the central tap of the rejection filter shown in Fig. 2 at time iT_c is given by

$$x_i = d_i + V \cos (\Omega i T_c + \phi) + n_i \tag{3}$$

where, from Fig. 1 and (1) and (2), it is straightforward to show that

$$d_i = \pm A T_c \tag{4a}$$

$$V = 2\alpha \frac{\sin \frac{\Omega T_c}{2}}{\Omega} \tag{4b}$$

and

$$\phi = \theta - \frac{\Omega T_c}{2}. \tag{4c}$$

Also, for ease of notation, we let $AT_c = \sqrt{S}$, so that $d_i = \pm \sqrt{S}$, where the $\pm$ is determined by the combination of algebraic signs of the data symbol and the symbol of the spreading sequence (typically referred to as a "chip") at the current instant of time.

Let us now define two $2N$-dimensional vectors X_i and W as

$$\boldsymbol{X}_i \triangleq [x_{i+N}, x_{i+N-1}, \cdots, x_{i+1}, x_{i-1}, \cdots, x_{i-N}]^t$$

and

$$\boldsymbol{W} \triangleq [a_{-N}, a_{-N+1}, \cdots, a_{-1}, a_1, \cdots, a_N]^t$$

respectively, where t denotes transpose, $\boldsymbol{X}_i$ is the sample vector of the off-center taps at time iT_c, and $\boldsymbol{W}$ is the adjustable tap weight vector. Hence, the output sample of the filter is

$$y_i = x_i - \boldsymbol{W}^t \boldsymbol{X}_i. \tag{5}$$

Upon squaring (5), we obtain

$$y_i^2 = x_i^2 - 2x_i \boldsymbol{X}_i^t \boldsymbol{W} + \boldsymbol{W}^t \boldsymbol{X}_i \boldsymbol{X}_i^t \boldsymbol{W} \tag{6}$$

and thus, the expected value of y_i^2 (or the output power) is given by

$$\begin{aligned} E[y_i^2] &= E[x_i^2] - 2E[x_i \boldsymbol{X}_i^t]\boldsymbol{W} + \boldsymbol{W}^t E[\boldsymbol{X}_i \boldsymbol{X}_i^t]\boldsymbol{W} \\ &\triangleq E[x_i^2] - 2\boldsymbol{P}^t \boldsymbol{W} + \boldsymbol{W}^t \boldsymbol{R} \boldsymbol{W} \end{aligned} \tag{7}$$

where P and R are defined by (8) and (9), respectively, below.

Since the signal and the noise are independent, and assuming that the period of the PN sequence is sufficiently long so that the PN signal samples at different taps are approximately uncorrelated (see [41], [42], [49], [54] for the autocorrelation function of a PN sequence), then

$$\begin{aligned} \boldsymbol{P}^t &\triangleq E[x_i \boldsymbol{X}_i^t] \\ &= [J \cos N\Omega T_c, J \cos (N-1)\Omega T_c, \cdots, \\ &\qquad J \cos \Omega T_c, J \cos \Omega T_c, \cdots, J \cos N\Omega T_c] \end{aligned} \tag{8}$$

and

$$\boldsymbol{R} \triangleq E[\boldsymbol{X}_i \boldsymbol{X}_i^t] = \begin{bmatrix} S + J + \sigma_n^2 & J \cos \Omega T_c & \cdots & J \cos 2N\Omega T_c \\ J \cos \Omega T_c & S + J + \sigma_n^2 & \cdots & J \cos (2N-1)\Omega T_c \\ \cdots & \cdots & \cdots & \\ J \cos 2N\Omega T_c & J \cos (2N-1)\Omega T_c & \cdots & S + J + \sigma_n^2 \end{bmatrix} \tag{9}$$

where $J = V^2/2$ is the power of the interfering tone at the output of the integrator, σ_n^2 is the power due to the thermal noise, and S has been defined above. Equations (8) and (9) follow because, from (3), the autocorrelation function of x_i is given by

$$E[x_i x_{i+m}] = (S + \sigma_n^2)\delta(m) + J \cos m\Omega T_c$$

where $\delta(m)$ is the Kronecker delta function.

The tap weights $a_{-N}, \cdots, a_1, \cdots, a_N$ are adjusted to obtain minimum $E[y_i^2]$.[1] From (7), letting

$$\frac{\partial E[y_i^2]}{\partial a_k} = 0, \qquad k = -N, \cdots, -1, 1, \cdots, N \tag{10}$$

we obtain

$$-2\boldsymbol{P} + 2\boldsymbol{R}\boldsymbol{W}opt = 0 \tag{11}$$

or

$$\boldsymbol{W}opt = \boldsymbol{R}^{-1}\boldsymbol{P} \tag{12}$$

where $\boldsymbol{W}_{opt}$ is the optimum tap weight vector. This, of course, is the well known Wiener-Hopf equation.

In [22], it is shown that the solution to (12) is

$$a_{k_{opt}} = 2\gamma \cos k\Omega t \tag{13}$$

where

$$\gamma = \frac{J}{2(S + \sigma_n^2) + J\left[2N - 1 + \dfrac{\sin (2N + 1)\Omega T_c}{\sin \Omega T_c}\right]}. \tag{14}$$

It is also shown in [22] that the minimum output noise power is given by

$$E[e^2]_{min} = \frac{J}{1 + \dfrac{J}{2(S + \sigma_n^2)}\left[2N - 1 + \dfrac{\sin (2N + 1)\Omega T_c}{\sin \Omega T_c}\right]} + \sigma_n^2 \tag{15}$$

If the signal-to-noise ratio improvement factor G is defined as the ratio of the output SNR to the input SNR, then G_2, the improvement factor for the transversal filter of Fig. 2, is given by

$$G_2 = \frac{(S/N)_{out}}{(S/N)_{in}} = \frac{J + \sigma_n^2}{\dfrac{J}{1 + \dfrac{J}{2(S + \sigma_n^2)}\left[2N - 1 + \dfrac{\sin (2N + 1)\Omega T_c}{\sin \Omega T_c}\right]} + \sigma_n^2}. \tag{16}$$

Note that this ratio of SNR's is consistent with the definition of the SNR improvement factor given at the beginning of this section. In particular, with no suppression filter, the SNR is

$$SNR_{nf} = \frac{S}{J + \sigma_n^2}$$

while with the suppression filter in place, the SNR is given by

$$SNR_f = \frac{S}{E(e^2)_{min}}$$

where $E(e^2)_{min}$ is given by (15). Taking the ratio of SNR_{nf} to SNR_f then yields (16). Notice also that, for this problem, minimizing the output noise power corresponds to maximizing the SNR.

In the extreme case, if $\sigma_n^2 = 0$

$$G_2 = 1 + \frac{J}{2S}\left[2N - 1 + \frac{\sin (2N + 1)\Omega T_c}{\sin \Omega T_c}\right]. \tag{17}$$

From (17), we see that G_2 increases either as the number of taps increases or as the input interference-to-signal power ratio J/S increases. Fig. 4 shows an example where $2N = 10$, $\sigma_n^2 = 0$, and $J/S = 100$. Also shown on Fig. 4 are comparable

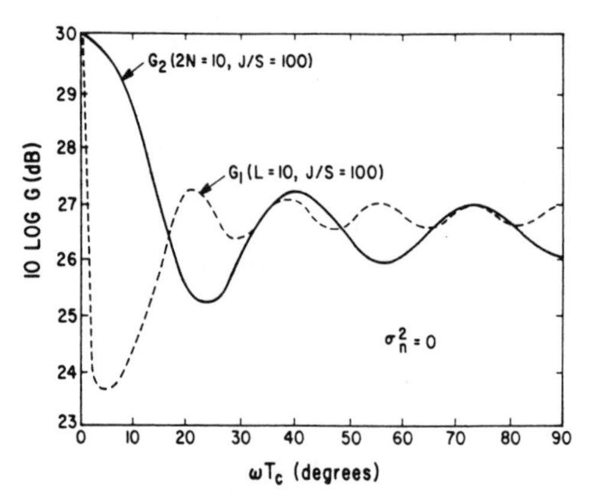

Fig. 4. Output signal-to-noise ratio improvement versus ωT.

results for G_1, the improvement factor for the prediction-error filter of Fig. 3. Analytically, this latter expression is given by

$$G_1 = \frac{J + \sigma_n^2}{2(S + \sigma_n^2)\left[\left(L + \dfrac{2(S + \sigma_n^2)}{J}\right) - \dfrac{\sin L\Omega T}{\sin \Omega T} \cos (L + 1)\Omega T\right] \Big/ \left[\left(L + \dfrac{2(S + \sigma_n^2)}{J}\right)^2 - \dfrac{\sin^2 L\Omega T}{\sin^2 \Omega T}\right] + \sigma_n^2}.$$

Finally, if one computes the transfer function of the two-sided filter, it can be shown that [22]

$$H(w) = 1 - \frac{J\left[\left(\sin \dfrac{(2N + 1)(\omega + \Omega)T_c}{2} \Big/ \sin \dfrac{(\omega + \Omega)T_c}{2}\right) + \left(\sin \dfrac{(2N + 1)(\omega - \Omega)T_c}{2} \Big/ \sin \dfrac{(\omega - \Omega)T_c}{2}\right) - 2\right]}{2(S + \sigma_n^2) + J\left[2N - 1 + \dfrac{\sin (2N + 1)\Omega T_c}{\sin \Omega T_c}\right]}.$$

Fig. 5 shows an example of a transfer function where $N = 5$, $\sigma_n^2 = 0$, $J/S = 100$, and $\Omega T_c = \pi/3$. It can be seen that $H(\omega)$ behaves as a notch filter.

[1]Note that since y_i is the difference between x_i and the weighted sum of the remaining x_{i+j}, $|j| = 1, 2, \cdots, N$, minimizing $E\{y_i^2\}$ corresponds to minimizing the mean-square error between x_i and its estimate.

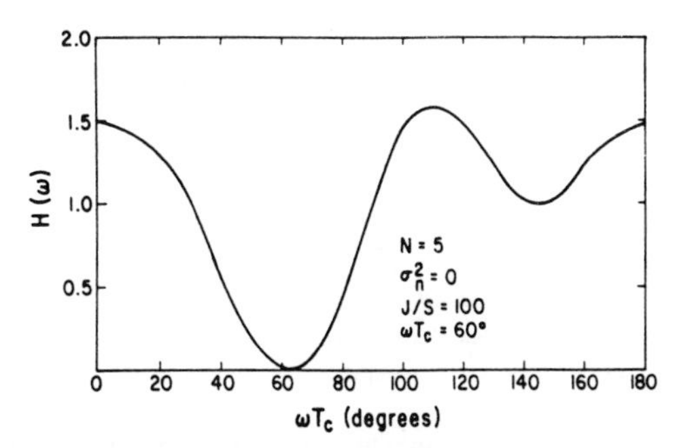

Fig. 5. Frequency response of the transversal filter.

The above analysis is especially straightforward because it corresponds to SNR at the output of the suppression filter rather than SNR at the output of the final detection filter. For analyses that incorporate the final despreading and low-pass filtering, the reader is referred to [17], [20], [26], [27]. Also, for analyses dealing with multiple narrow-band interferers rather than just a single source, references [20], [51], [24] are appropriate. Finally, while the results presented here are for BPSK systems, analogous results exist for quadrature phase-shift keyed (QPSK) receivers [23], [25]. Although the suppression filter for a QPSK signal is more complex than it is for a BPSK waveform, the results are qualitatively very similar.

B. Average Probability of Error

Let us now consider the block diagram of Fig. 6. This system is analyzed in depth in [18], from which the following results are taken. The expression for probability of error is quite lengthy and is not presented here. However, typical performance results are shown in Figs. 7–10. In Fig. 7, prob-

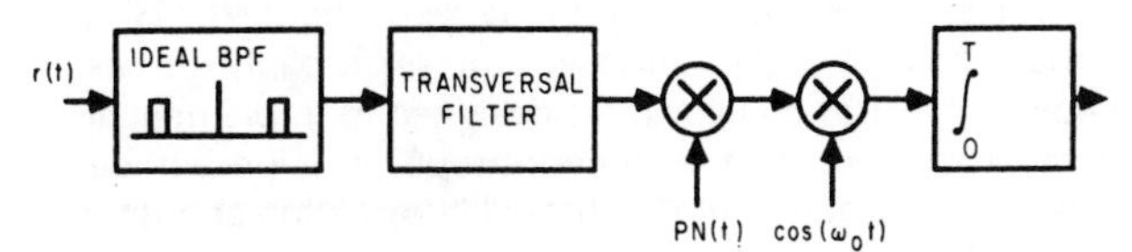

Fig. 6. Receiver block diagram.

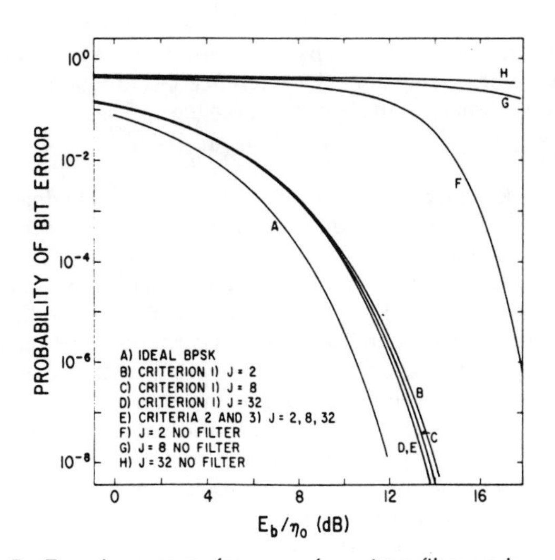

Fig. 7. Tone jammer performance for a 4-tap filter under Criteria 1, 2, and 3.

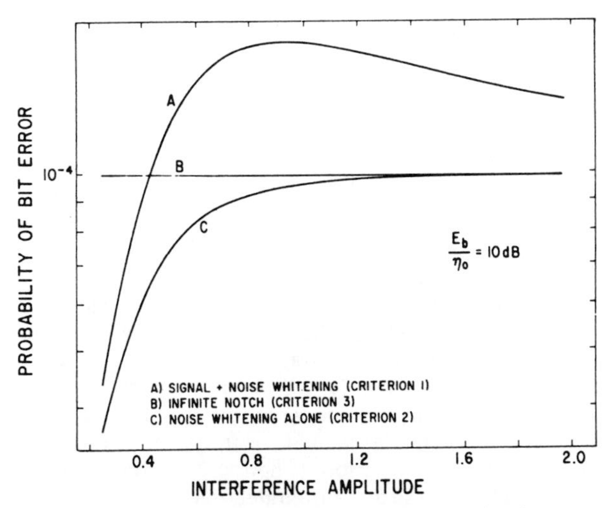

Fig. 8. Tone jammer performance with varying amplitude.

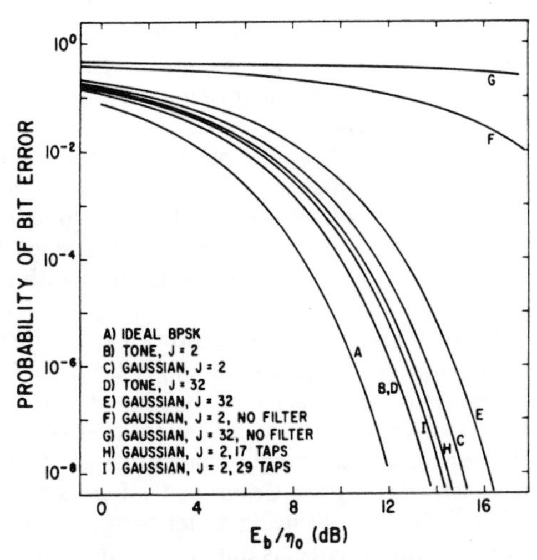

Fig. 9. Comparison between tone and narrow-band gaussian interference. Bandwidth of gaussian interference equals 10 percent of front-end bandwidth.

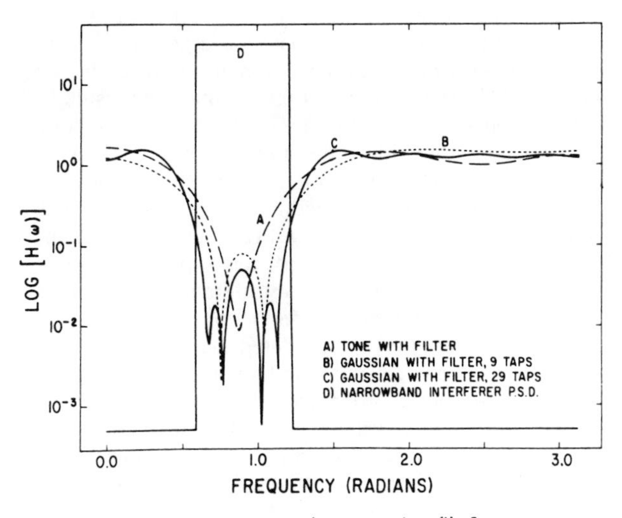

Fig. 10. Frequency responses of suppression filters.

ability of error curves are presented for a simple DS system employing a processing gain of seven (i.e., there are seven chips/bit) when a tone interferer is located at a frequency which is offset from the carrier frequency of the DS waveform by an amount equal to the symbol rate (i.e., Ω in (2b) equals $2\pi/T$).

There are three different design criteria used to set the tap weights for the system of Fig. 6. Criterion 1 corresponds to whitening (i.e., making the output samples uncorrelated) the entire received signal. It can be shown that this is equivalent to the intuitive procedure described at the beginning of this section, namely, predicting the current value of the received waveform and subtracting that predicted value from the received value [34]. Criterion 2 corresponds to whitening the noise and interference only (see the discussion on decision feedback filters near the end of this section), and the last criterion, Criterion 3, corresponds to designing the filter so that an infinitely deep notch is placed at the frequency location of the interfering tone.

In Fig. 7, the probability of error is plotted versus the ratio of energy-per-bit-to-noise spectral density, E_b/η_0, for interference powers of 2, 8, and 32 for each of the three filter design criteria. The noise spectral density η_0 is fixed at 1/8, so that when E_b equals 1/2, the resulting E_b/η_0 is 6 dB. Note that under Criterion 1, performance improves as the interference power increases from 2 to 32 (see below for the explanation). In Fig. 8, E_b/η_0 is fixed at 10 dB and the sensitivity of system performance to the amplitude of the interference is illustrated. Note that the performance under Criterion 3 is invariant to changes in interference amplitude since the tone is always completely rejected by the infinitely deep notch filter.

From Fig. 8, it is seen that the Criterion 2 leads to the best system performance. It appears that under Criterion 3, the notch deepens too rapidly (its depth is infinite for any finite J), while for Criterion 1, the notch does not deepen rapidly enough. Indeed, the seemingly strange behavior of the receiver designed using the Criterion 1 suppression filter referred to above, namely giving better performance for a higher level of interference, is due to the suboptimal balance achieved by the filter in terms of minimizing the degradation to the DS signal while maximizing the interference rejection. That is, over a fairly wide range of input signal levels, the receiver designed by this criterion is too conservative in the sense of not forcing the notch to be deep enough, and hence resulting in insufficient interference suppression. Note, however, that for a large enough interferer, the performance of the system designed under any one of the three criteria converges to the same result.

It is seen that when the interference is a pure tone, increasing the tone power has no effect on system performance when the notch is infinite. However, when the interference has a finite spectral width, it cannot be completely rejected by a suppression filter with a finite number of taps, since only a finite number of zeros can be placed in the frequency band spanned by the interference. This is illustrated in Fig. 9, where the interfering tone of (2b) has been replaced with an interferer modeled as a stationary, zero-mean, narrow-band gaussian random process. The remaining parameters of the system are fixed at those used for the tone interference, and the bandwidth of the narow-band process ω_i is set at 10 percent of the receiver front-end bandwidth. It is easily seen that the suppression filter is less effective against a source of interference with a finite (i.e., nonzero) spectrum, than it is against a tone of equivalent power. It is also seen that performance against the narrow-band gaussian process can be improved by increasing the number of taps in the suppression filter. Finally, in Fig. 10, the magnitudes of the frequency responses of the suppression filters designed under Criterion 2 are plotted for both narrow-band gaussian and tone interference. It is seen that the change in the filter notch for the gaussian interferer as the number of taps is increased from 9 to 29 becomes quite noticeable.

An alternate block diagram for a system of this type is shown in Fig. 11. This system is amenable to implementation with the type of processing inherent with either charge-coupled devices or digital signal processing, and in [18], results similar to those shown in Figs. 7–10 for the system of Fig. 6 are presented for the system of Fig. 11.

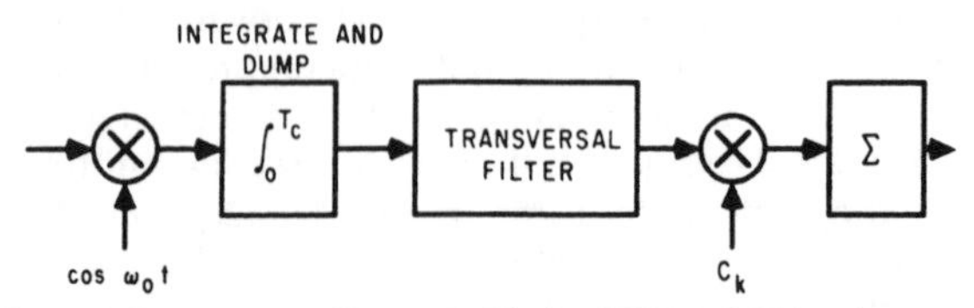

Fig. 11. Suppression filter suitable for CCD or DSP implementation.

Note that all results presented up to this point have assumed precise statistical knowledge of the interference. In reality, such information is rarely available, hence one must envision implementing an adaptive version of the rejection filter. There are a variety of adaptive algorithms that can be used, as well as a variety of receiver structures [20], [19], [14], [21], [50].

For the results presented here, the adaptive version of the system uses the Widrow–Hoff LMS algorithm [52], [53] to update the tap weights. This algorithm is probably the best known of a class of algorithms designed to implement an iterative solution to the Wiener–Hopf equation without making use of any *a priori* statistical information about the received signal. The LMS algorithm can be expressed as

$$\mathbf{W}^{(k+1)} = \mathbf{W}^{(k)} + \mu y^{(k)}\mathbf{X}^{(k)}$$

where $\mathbf{W}^{(k)}$ is the vector of tap weights, $\mathbf{X}^{(k)}$ is the vector of waveform samples on the taps, $y^{(k)}$ is the difference between the waveform sample on the reference tap (denoted $x_0^{(k)}$) and the estimate of that same sample (i.e., $y^{(k)}$ is given by

$$y^{(k)} = x_0^{(k)} - \mathbf{W}^{(k)'}\mathbf{X}^{(k)}),$$

all at the kth adaptation, and μ is a parameter which determines the rate of convergence of the algorithm. In other words, the parameters $\mathbf{W}^{(k)}$, $\mathbf{X}^{(k)}$ and $y^{(k)}$ are the same as those defined in (3), (2), and (5), respectively, except now the notation has been slightly changed to indicate the explicit dependence on the iteration value k. It is interesting to note that in most applications of the LMS algorithm [52], an external reference waveform is needed in order to correctly adjust the tap weights. However, in this particular application, the signal on the reference tap (e.g., the center tap of a two-sided symmetrical tapped delay line) serves the role of the external reference.

The analysis of the performance of an adaptive receiver

employing the LMS algorithm analysis is very difficult. An approximate analysis is presented in [19], and typical results are shown in Fig. 12. Curve *B* is the exact probability of error

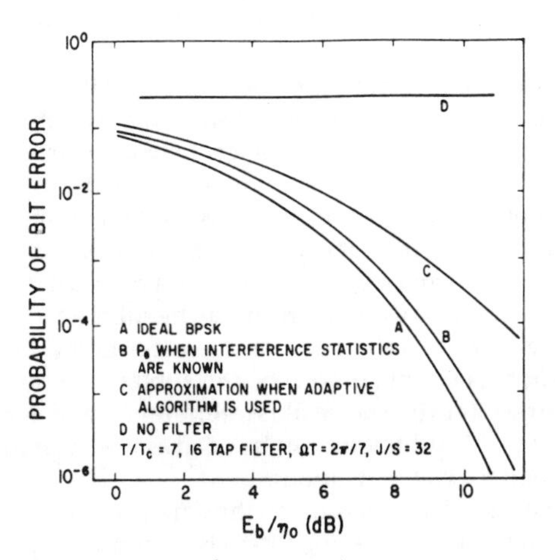

Fig. 12. Performance of adaptive receiver.

of the system when the statistics of the interference are precisely known (and hence the LMS algorithm is not needed), while curve *C* corresponds to the approximate results derived in [19]. That is, they correspond to a receiver using the LMS algorithm to adjust the tap weights. Upon comparing curves *B* and *C* of that figure, the degradation incurred by the lack of knowledge of the statistics of the interference is easily seen.

In order to explore this effect further, the system shown in Fig. 11 was implemented using charge-coupled devices, and, independently, using digital logic. The former system is described in [44], and the latter system is described in [45]. In what follows, a brief overview of the digital system is presented.

As noted in [45], direct implementation of the LMS algorithm requires two multipliers per filter tap, one to perform the update operation and a second to do the actual signal sample weighting. A block diagram showing a conventional implementation of the LMS algorithm is shown in Fig. 13.

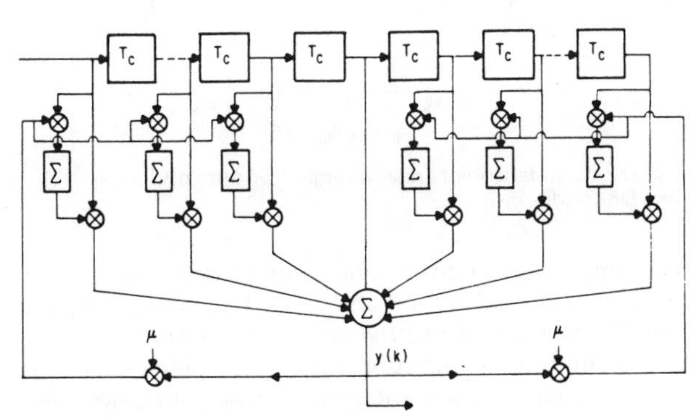

Fig. 13. Conventional implementation of LMS algorithm for a two-sided filter.

However, it is also possible to implement the algorithm using a so-called burst processing technique [44], [45], which allows the construction of a filter of arbitrary order using only two multipliers. However, the price for obtaining the lower multiplier count while still updating all weights each sample period is a loss of bandwidth.

To illustrate this technique, a test configuration was set up and used to obtain probability of error data for the system. A 7-chip PN sequence is modulated by random data and added to a tone interferer and white gaussian noise. The composite signal is then adaptively filtered and correlated. A decision based on the correlator output is compared to the actual data sent and the number of errors that are made is counted. Fig. 14 shows a series of curves obtained using this test arrangement. Curve *A* is the the-

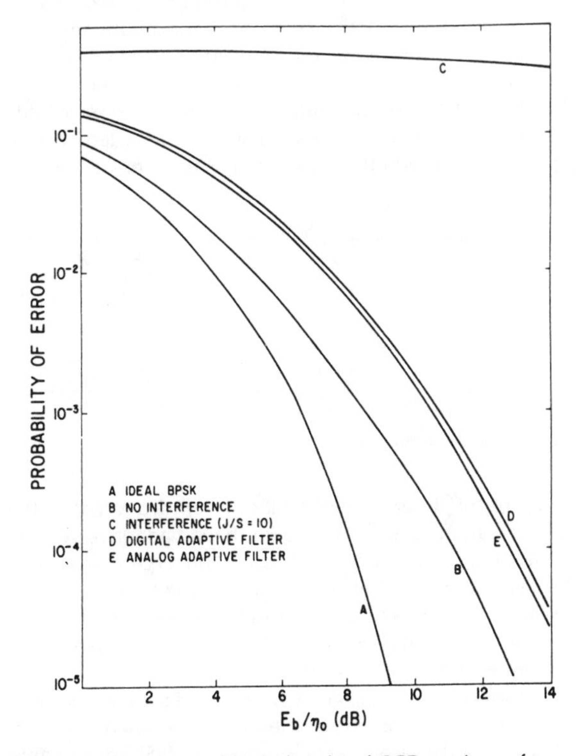

Fig. 14. Performance of both digital and CCD versions of adaptive filter.

oretical BPSK result while curve *B* is the system performance in the absence of both the interference and the adaptive filter. The degradation in performance from that of curve *A* is due to implementation losses. Curve *C* shows the effect of a tone interferer with $J/S = 10$ dB, again in the absence of the suppression filter. As expected, this results in an essentially useless system. Curve *D* shows the system performance with a 16-tap adaptive filter in the receiver. The interference suppression causes a significant improvement in the probability of error performance, although it does not completely remove the interference. The performance of the analog system of [44] for an interferer with $J/S = 10$ is given by curve *E*. The interference frequency for this result, as well as for the digital system, is $f_c/7$, where f_c is the chip rate (i.e., $f_c = 1/T_c$).

The curves of Fig. 14 demonstrate that the adaptive filter is providing a significant improvement in performance. As shown by curves *D* and *E* of Fig. 14, the results obtained with the digital system are nearly identical to those found using the analog version, although the sources of degradation differ for the two receivers. For the digital system, quantization noise was the limiting factor, whereas for the analog system, charge transfer inefficiency [9] limited the overall receiver performance.

Up to this point, the transversal filter structure has been emphasized. However, other suppression filter structures do exist, and a couple of these are discussed below. Similarly, while the LMS algorithm has been discussed here as a means of making the system adaptive, there are other algorithms that are available for the same purpose, and which can, in addition, overcome certain drawbacks in the LMS algorithm. In particular, it is well known [53] that the convergence rate of the LMS algorithm is relatively slow, it being a function of the ratio of the maximum and minimum eigenvalues of the autocorrelation matrix $\boldsymbol{R}$ of (9). A good discussion of this can be found in [52]. Because of this drawback to the LMS algorithm, other structures have been investigated and one of the most popular is the lattice filter [1], [42]. A typical lattice structure is shown in Fig. 15. The

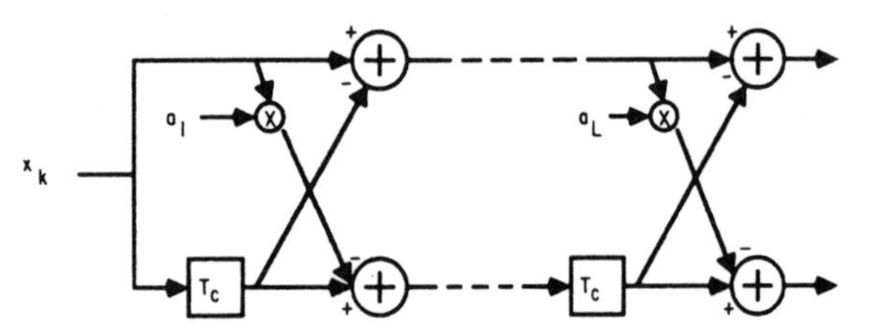

Fig. 15. Lattice filter.

layout of this filter is somewhat different from that of the transversal filter, and it is known that an adaptive version of this filter can result in much faster convergence than can the LMS algorithm, because each section of the lattice can be shown to converge individually, independent of the remaining sections (i.e., the various stages of the lattice are decoupled from one another).

In both [15] and [47], lattice filters used for narrow-band interference suppression are described. Simulation results for a two-stage lattice are described in [15], and experimental results for both three- and ten-stage lattice filters are presented in [47]. In all cases, significant improvement in performance over that of a DS system operating in the absence of a suppression filter is shown to be possible.

Another alternative to a transversal filter is a decision feedback (DF) filter [42], [21], [14], [23], [25]. One version of such a filter is shown in Fig. 16 and analyzed in [50]. The

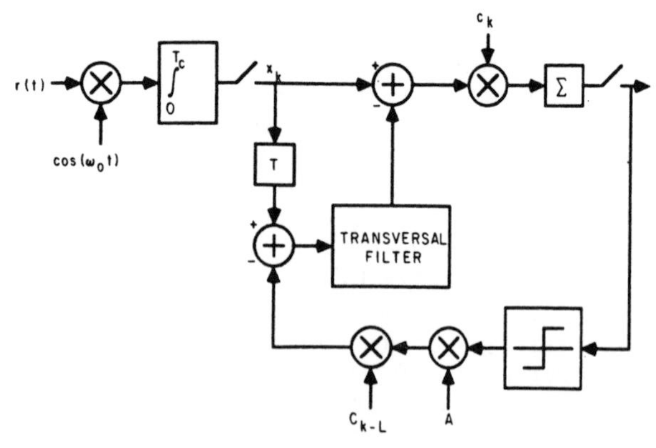

Fig. 16. Decision feedback receiver.

rationale for this scheme is that if one could whiten just the noise and interference (i.e., without the desired signal being present), the performance of the system might improve (in fact, this idea was presented briefly at the beginning of this section as the second of the three criteria described for the filter design).

The principle behind the operation of a DF filter is quite simple. Since the received waveform consists of the desired signal plus noise and interference, to whiten just the noise and interference, some means of removing the desired signal is necessary. However, since the output of the receiver is an estimate of the data symbol that has been transmitted, that estimate can be used to generate a replica of the transmitted waveform which, in turn, can be subtracted from the received signal. If the decision the receiver makes on the current data symbol is correct, the subtraction referred to above results in just noise plus interference, and hence, the output of this subtractor can be used as the input to a filter designed to whiten its input.

Of course, if the decision on the data symbol is incorrect, the input to the whitening filter consists not only of noise plus interference, but also twice the desired signal component. Hence, the possibility exists for error propagation. If one considers first the idealized case of perfect (i.e., error-free) decision feedback, results such as those obtained in [50] and presented in Fig. 17 are available. The interference

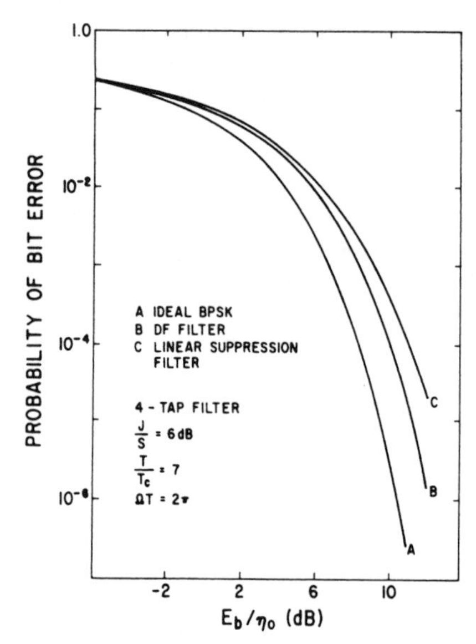

Fig. 17. Comparison of linear filter and DF filter under perfect DF conditions.

is a single tone, and the results correspond to a four tap filter, $T/T_c = 7$, $J/S = 6$ dB and $\Omega T = 2\pi$. Upon comparing the DF performance of curve B with that of curve C, which corresponds to a linear suppression filter of the same size, the potential improvement in using the DF structure is evident.

This potential improvement could, of course, be negated by the effect of decision feedback errors. Interestingly however, for this system such an effect appears to be negligible. Specifically, simulation results generated in [50] and illustrated in Fig. 18 compare the performance of the system operating under the assumption of perfect decision feed-

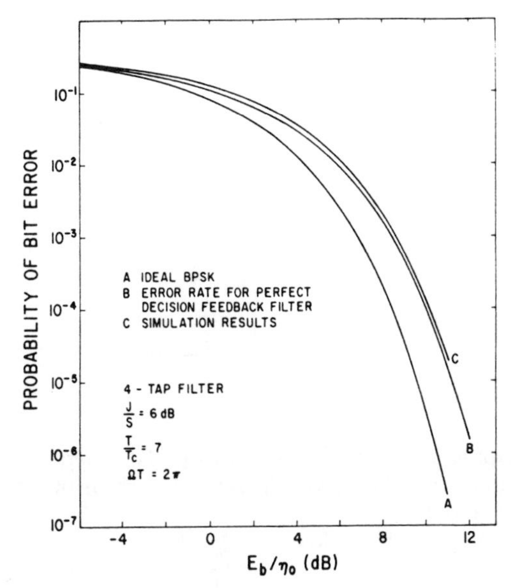

Fig. 18. BER for tone jammer for DF filter.

back to that of the system operating in the presence of error propagation, and, as can be seen, show the effect of error propagation to be minimal.

As a final point of interest, since these techniques are most appropriate for narrow-band interference suppression, a natural question that arises is "What is the definition of 'narrowband'?" While no precise answer to that question appears to be available, some perspective on the answer can be obtained by considering some of the results presented in [28]. In [28], the "worst-case" spectral density of a gaussian interferer was found. The interferer's spectral content was nonzero over only a prespecified fraction of the spread bandwidth, and the spectrum of the interference was optimized to maximize the mean-square error at the output of an infinitely-long prediction filter. The resulting interferer was then used in a system with a finite length filter, and the average probability of error of the receiver was derived.

The resulting performance can be seen in Figs. 19 and 20, corresponding to $\lambda = 0.1$ and 0.5, respectively, where λ represents the percent of the spread bandwidth occupied by the interference. In both cases, the center frequency of the interference coincides with the carrier frequency of the transmitted signal. It is seen that when $\lambda = 0.1$, the rejection filter is very effective, yet when $\lambda = 0.5$, the filter is almost worthless. Figs. 21 and 22 show the magnitudes of the transfer functions of the filters corresponding to $\lambda = 0.1$ and 0.5, respectively. Both the filters are notch-filters, but the notch corresponding to $\lambda = 0.5$ is so wide that it results in significant distortion to the desired signal.

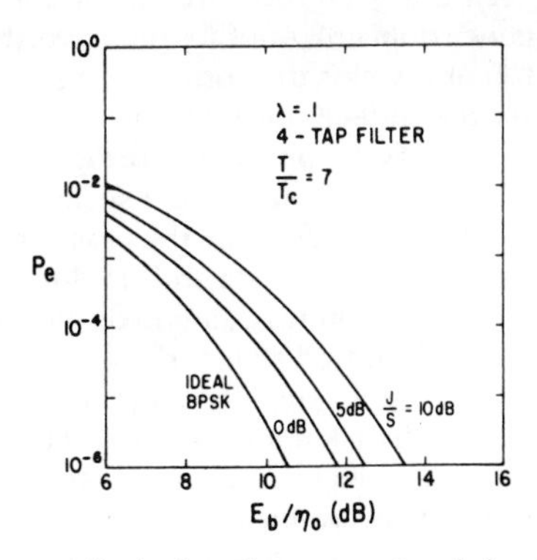

Fig. 19. P_e versus E/η_0 for interference centered about carrier frequency.

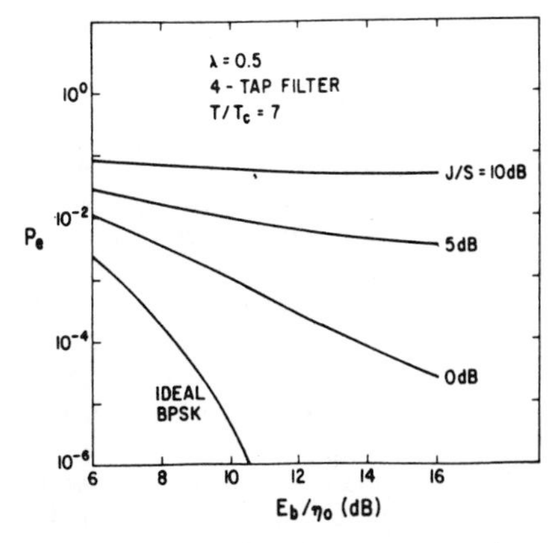

Fig. 20. P_e versus E_b/η_0 for interference centered about carrier frequency.

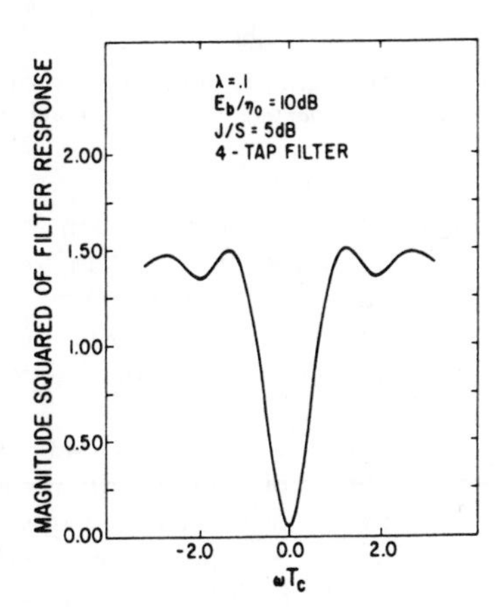

Fig. 21. Filter amplitude response.

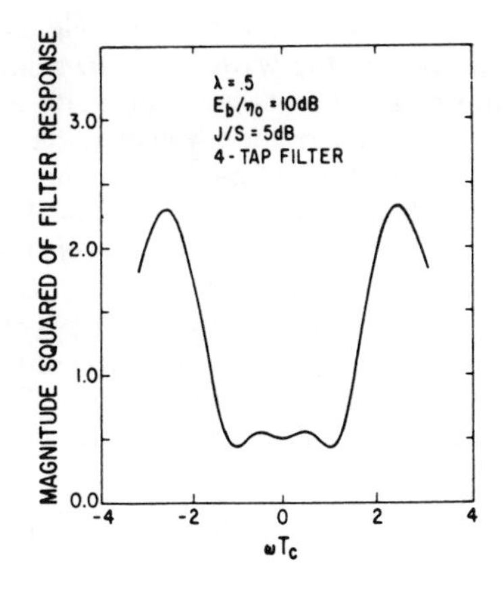

Fig. 22. Filter amplitude response.

III. Transform Domain Processing

A receiver which performs the notch filtering operation in a completely different manner from the systems described in Section II is the so-called transform domain processing system [38], [4], [36], [7], [8]. The basic building block of such a system is a device which performs a real-time Fourier transform. For spread spectrum applications, this device is typically a SAW device. In what follows in this section, we briefly review the technique of real-time Fourier transformation, and then describe and analyze the transform domain processing system. Most of the material in this section is taken from [29], [31], [32], [48].

The receiver to be analyzed is that described in [31] and shown in Fig. 23. The input consists of the sum of the transmitted signal $\pm s(t)$, AWGN $n_w(t)$, and the interference $I(t)$. The Fourier transform of the input is taken, the transform is multiplied by the transfer function of some appropriate filter $H_c(w)$, the inverse transform of the product is taken, and the resulting waveform is put through a detection filter matched to $s(t)$.

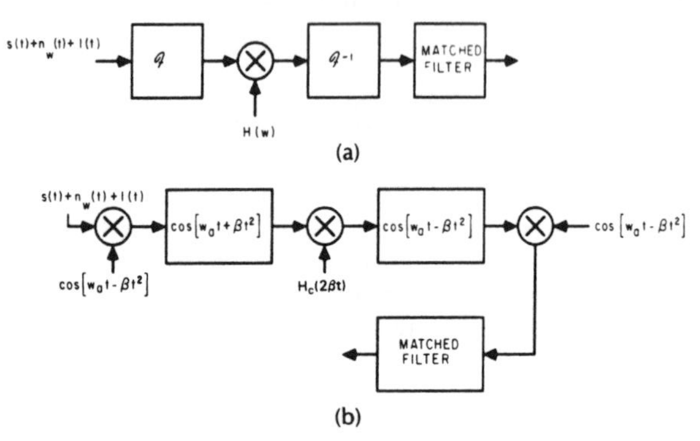

Fig. 23. Receiver block diagram for transform domain processing system. (a) Functional operations. (b) Actual implementation.

Intuitively, if one considers the spectra of the signal and interference components of the input $r(t)$, it can be seen why the receiver shown in Fig. 23 can be expected to provide interference suppression. On the one hand, we have a low level, broad-band DS spectrum; on the other hand, we have added to it a high level but narrow-band interference waveform. Since the output of the Fourier transformer shown in Fig. 23 is a waveform evolving in real-time which looks qualitatively like the one shown in Fig. 24(a), multiplying that output by the waveform shown in Fig. 24(b) should suppress a significant amount of interference power while only slightly reducing the power of the desired signal. This heuristic explanation will be shown to indeed be accurate. Note that while the abscissas in Fig. 24(a) and (b) are labeled ω, the variable ω is actually a linear function of time.

Let us now consider analyzing the performance of this receiver. Since the system is linear, the three components of the input can be treated separately. Denoting any one of them by $f(t)$, assumed nonzero for $t \in [0, T]$, the signal at the output of the first SAW device is given by

$$f_1(t) = \int_0^T f(\tau) \cos (w_a\tau - \beta\tau^2) \cos (\omega_a(t - \tau) + \beta(t - \tau)^2)\, d\tau \tag{18}$$

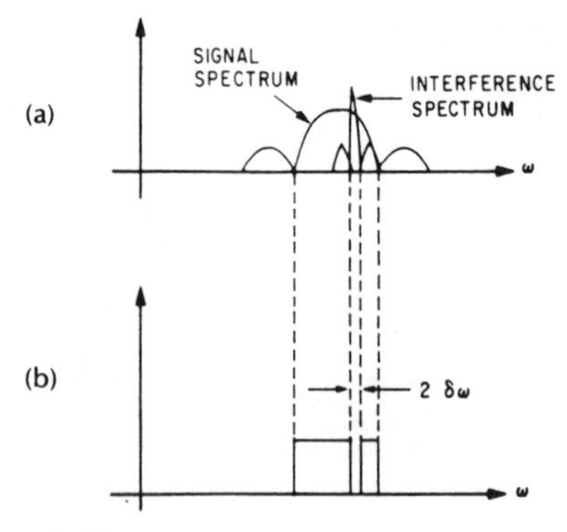

Fig. 24. Notch filter.

an expression which is valid for $t \in [T, T_1]$, where T_1 is the length of the SAW device in seconds (see below for the explanation).

Simplifying, this yields

$$\begin{aligned} f_1(t) &= (1/2) \cos (\omega_a t + \beta t^2) \int_0^T f(\tau) \cos 2\beta t\tau\, d\tau \\ &\quad + (1/2) \sin (\omega_a t + \beta t^2) \int_0^T f(\tau) \sin 2\beta t\tau\, d\tau \\ &\quad + 1/2 \int_0^T f(\tau) \cos (2\omega_a\tau - 2\beta\tau^2 \\ &\quad + 2\beta t\tau - w_a t - \beta t^2)\, d\tau \end{aligned} \tag{19}$$

$$\simeq (1/2) F_R(2\beta t) \cos (\omega_a t + \beta t^2) - (1/2) F_I(2\beta t) \cdot \sin (\omega_a t + \beta t^2) \tag{20}$$

where $F_R(\omega)$ and $F_I(\omega)$ are the real and imaginary parts, respectively, of the transform of $f(t)$, and the approximation used in going from (19) to (20) is to ignore the third term of (19), which is a double frequency term. Alternately, if it is desired to exactly cancel the third term, one can implement a system described in [31]. Note that this latter system requires twice as much equipment and, as a practical matter, is usually not needed, since the double frequency term is almost always filtered out to a sufficient degree by the receiver of Fig. 23.

The important thing to observe about (20) is that the real and imaginary components of the transform of $f(t)$ are modulating quadrature carriers, meaning both components have been individually recovered and thus the Fourier transform itself has been recovered (over a finite interval in the frequency domain). Another point worth emphasizing is that (20) only yields the correct values of $F_R(2\beta t)$ and $F_I(2\beta t)$ when $f(t)$ is indeed nonzero only for $t \in [0, T]$. Such would be the case, say, for one pulse in a digital pulse stream where the duration of each pulse is T seconds, and where $T < T_1$. However, it would not be the case for a waveform which may be greater than T seconds in duration, such as the noise. For such waveforms, (20) yields the Fourier transform of the time-truncated signal, not of the signal itself.

Note that (20) does not yield the Fourier transform of $f(t)$ for all $2\beta t$ (i.e., for all ω). Rather, (20) yields $F(\omega)$ only during that interval of time when $f(t)$ is fully contained in the tapped delay line. Since the delay line is T_1 seconds long and since the duration of $f(t)$ is T seconds, the frequency range over which (20) yields a true Fourier transform is $w \in [2\beta T, 2\beta T_1]$.

Having transformed $f(t)$, it is now desired to filter it with a filter whose transfer function is $H(w)$. In Fig. 23, it can be seen that the output of the first chirp filter is multiplied by $H_c(2\beta t)$. This function is related to the desired transfer function $H(w) = H_R(w) + jH_I(w)$ by

$$H_c(2\beta t) = 4H_R(2\beta t) \cos 2\beta tT_1 + 4H_I(2\beta t) \sin 2\beta tT_1. \quad (21)$$

The terms $\cos 2\beta tT_1$ and $\sin 2\beta tT_1$ shift the region the input signal is located in from $[0, T]$ to $[T, T_1 + T]$. This is necessary because the inverse transform filter can be shown to yield an accurate inverse transform only in the range $t \in [T_1, T_1 + T]$. This follows from the same argument used to define the region that the forward transform is valid in, namely, that the inverse transform is only valid when the entire forward transform is contained in the tapped delay line used to perform the inverse. In the remainder of this section, it is assumed that the filter is purely real, so that $H_I(w) \equiv 0$.

Proceeding further with the analysis, it is shown in [31] that the component of the final output of the system due to $f(t)$, when sampled at $t = T_1 + T$, is given by

$$f_0(T_1 + T) = \int_0^T f(t)[h_R(t) * s(t)]\, dt \quad (22)$$

where $h_R(t)$ is the inverse Fourier transform of $H_R(w)$ and $*$ denotes convolution. Finally, it is possible to implement an adaptive version of this system as described in [48] and illustrated in Fig. 25. Its operation can be seen as follows: The

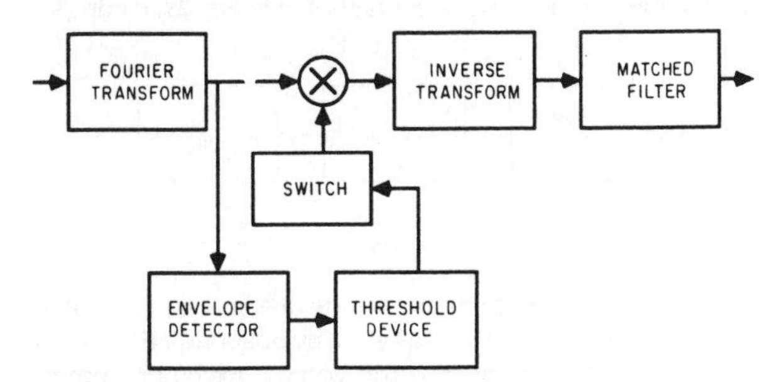

Fig. 25. Block diagram of adaptive transform domain processing receiver.

lower branch envelope detects the Fourier transformed input, and the output of the envelope detector is fed into a switch controlled by a threshold device. The upper branch passes the Fourier transformed input directly to the multiplier. The switch in the lower branch is set so that any time the output of the envelope detector exceeds a predetermined level, the output of the switch is forced to zero (and hence the lower input to the multiplier is also zero). In this manner, the adaptive notch switch is implemented.

To determine the amount of improvement in average probability of error that a technique such as transform domain processing can provide, results derived in [32] are used. The received waveform to the system shown in Fig. 23 is again given by (1) and the actual expressions needed to determine the average probability of error are given in [31] and are not repeated here. Rather, results obtained from evaluating those expressions are presented below, along with system comparisons and perspectives. Further results can be found in [32].

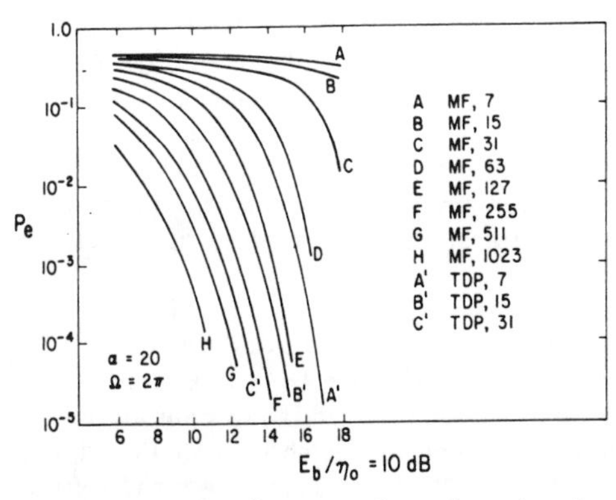

Fig. 26. Comparison of performance of transform domain processing receiver and conventional DS receiver.

Fig. 26 shows curves of average probability of error for both the transform domain processing system and a conventional system. By a "conventional system," we mean a DS receiver that is not employing the suppression filter. The curves are labeled with a number (e.g., 31) indicating the processing gain of the system (in all cases a full period of the spreading sequence is superimposed upon each data symbol), as well as either the abbreviation of MF, which stands for "matched filter," or the abbreviation TDP, which stands for "transform domain processing."

This figure indicates the processing gain needed in the conventional system (i.e., the MF system) to yield comparable performance to that of the transform domain processing system for a given interference level α. For example, from Fig. 26, corresponding to $\alpha = 20$ and $\Omega T = 2\pi$, a conventional DS spread spectrum system would have to operate with a processing gain somewhere between 255 and 511 in order to yield the same probability of error performance that the transform domain processing system yields with a processing gain of 31. Hence the improvement is a factor of about 10 dB.

As another perspective on the performance of such a system, an experimental version was implemented and various results are documented in [43] and [10]. In Fig. 27, some of these experimental results are reproduced. The spread spectrum code used for the experiment is a 63 chip PN

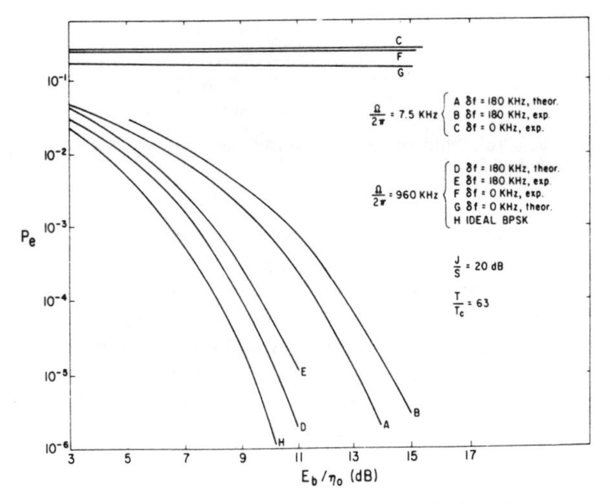

Fig. 27. Comparison of measured and theoretical results.

sequence with a chip rate of 1.875 MHz. The Fourier transform of this signal is obtained at the output of a real-time Fourier transformer which uses chirp devices with center frequencies of 15 MHz, bandwidths of 7 MHz, interaction times of 117 μs and chirp rates of 3×10^{10} Hz/s. Thus, in the frequency domain, 1 μs corresponds to 60 kHz and the main lobe of the transform of the desired signal has a width of 3.75 MHz.

Fig. 27 presents curves of probability of error versus energy-per-bit to noise spectral density ratio for the case of single tone interference. The curves are parameterized by the offset frequency Ω, and by whether-or-not the notch is employed. When the notch is indeed used, the notch width is fixed at 180 kHz. For the frequency offset, either 7.5 or 960 kHz is used. Theoretical curves are also presented in the same figure and the agreement is within a fraction of a decibel. To achieve adequate phase averaging, the single tone interferer is phase modulated with a phase excursion of $\pm\pi$ radians at 100 Hz. The signal power-to-interference power ratio is -20 dB for all the measurements shown.

Another consideration in the overall system design concerns the shape of the window used to "view" the received waveform. Rectangular windows were the ones used most often in the experiments, but it is well known that rectangular windows produce large sidelobes which can be reduced by proper weighting functions; however, these weighting functions distort the input signal itself.

In [10], an initial attempt was made to resolve this question, with typical results shown in Fig. 28. There are two curves shown on this figure, one corresponding to an unweighted system (i.e., a rectangular window) and the other corresponding to the use of a raised cosine window function. It is seen from Fig. 28 that at large values of J/S, the use of weighting provides the potential for a significant enhancement of system performance.

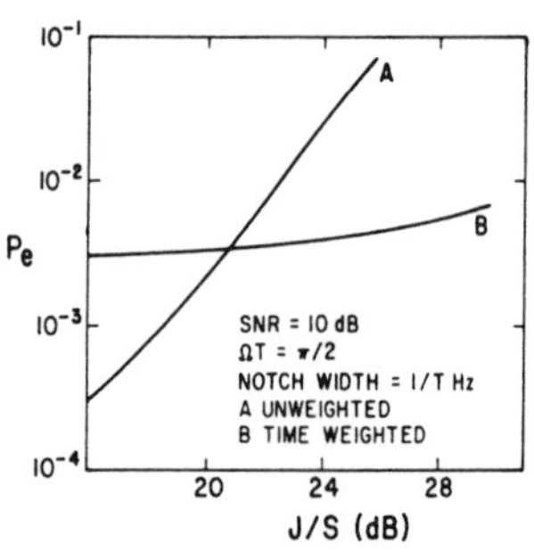

Fig. 28. Probability of error versus the interference power-to-signal power ratio.

It should be cautioned, however, that the above results were obtained under the assumption of perfect bit synchronization. Therefore, aside from the need to examine the effect of various other window shapes, it is necessary to investigate whether or not any weighting functions (other than the rectangular one) can be successfully used before synchronization has taken place. In addition, the overall sensitivity of a weighted system to timing errors must be determined.

The above results indicate the degree to which the technique of transform domain processing can reject either a constant tone or a slowly varying tone. When the interference has a larger spectral width, the technique can still be used, and, as an example, the results presented in [33] correspond to the interference being a colored gaussian random process. For this situation, the multiplying transfer function is just the inverse of the power spectral density of the noise and interference.

IV. Detection of Spread Signals

In the previous two sections, the use of interference suppression techniques was discussed from the point of view of enhancing the performance of a spread spectrum receiver for which the intended signal is embedded in interference. That is, the receiver in question was the receiver to which the message was originally transmitted.

Consider now the opposite situation, namely one whereby the receiver of interest is not the intended receiver, but one which nevertheless is attempting to determine the presence or absence of the spread spectrum waveform. The use of a spread signal by a transmitter to "hide" its waveform from unintended receivers is another well-known application of spread spectrum techniques and is referred to as low probability of intercept, or LPI for short. In turn, a receiver whose goal is to learn whether or not such a signal is indeed being transmitted is often referred to as an intercept receiver. Reference [49] provides an introductory treatment of intercept receivers.

While there are many types of intercept receivers, only the most classical one is discussed here. That one is called a total power radiometer, and is shown in Fig. 29. It consists of a bandpass filter, a square-law device and an integrator. Essentially, it looks for energy at dc by observing the output voltage of the integrator. If the voltage exceeds a predetermined threshold, signal-plus-noise is declared; if the voltage falls below the threshold, noise only is declared.

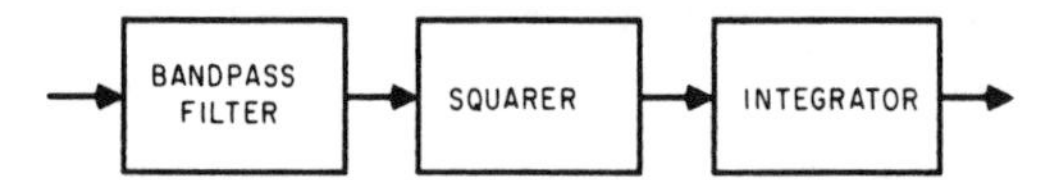

Fig. 29. Radiometer.

Suppose, however, that in addition to signal and noise, interference is also present at the input to the radiometer. This interference is not necessarily intentional interference, but could just be a conventional narrow-band waveform (i.e., not another spread spectrum signal) that happens to be present somewhere in the same frequency band as the signal the intercept receiver is attempting to detect. Then, upon squaring the composite received waveform, energy from each of the components is generated and contributes to the output voltage of the integrator. When signal is indeed present, the presence of the interference might actually aid in the detection of the signal since the interference is usually just adding to the total energy of the received waveform. However, when signal is absent, the radiometer might be deceived into believing the signal is actually present, because the presence of the interference is going to make it more likely that the integrator output voltage exceeds the threshold. In other words, the radiometer, being a device that bases its decision on the total received energy, cannot distinguish between energy due to signal, energy due to noise, and energy due to interference.

Hence, the main effect of the interference is to increase the probability of false alarm. To combat this effect, an interference suppression filter can be used to reject the interference before the composite received waveform is squared, and initial results of using TDP to accomplish this goal are described in [11]–[13]. Fig. 30 shows a block diagram of the system, and Table 1 shows some measured results of probability of detection (P_d) and probability of false alarm (P_{fa}) taken from [12]. When the results are referred to as corresponding to a "weighted signal," it simply means that the input signal is multiplied by a nonrectangular window function prior to Fourier transformation. For the data presented in Table 1, the weighting corresponds to a four-term Blackman–Harris window. Also, the ratio of interference power-to-signal power is 28 dB, while the ratio of signal power-to-thermal noise power is 0 dB.

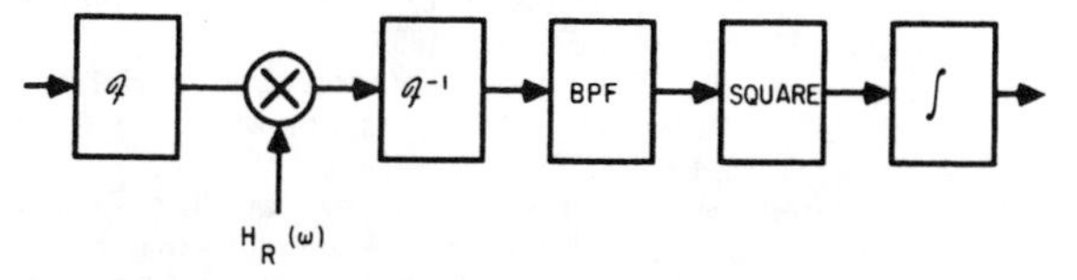

Fig. 30. Radiometer with interference suppression.

Table 1 Probability of Detection Versus False Alarm for Weighted and Un-Weighted Signal, with Excision Filter On and Off

Excision On				Excision Off	
Unweighted Signal		Weighted Signal		Weighted or Unweighted Signal	
P_{fa}	P_d	P_{fa}	P_d	P_{fa}	P_d
0.0025	0.59	0.0024	0.72	0.91	0.91
0.0042	0.73	0.0046	0.86	0.97	0.98
0.02	0.85	0.0075	0.9	0.98	0.98
		0.0094	0.92		

It is immediately seen from the entries in Table 1 that, indeed, the false alarm probability in the presence of interference, but in the absence of the suppression filter, can be so large as to make the system useless. However, when the interference suppression filter is inserted into the system, the false alarm probability is reduced to an acceptable level.

V. Other Interference Rejection Structures

The previous sections have emphasized two specific techniques for interference rejection, one employing the estimation-type filter and the other making use of transform domain processing ideas. These two interference suppression techniques were chosen in part because of the familiarity of the author with them and in part because they happen to be generating interest outside of just the academic/research community. However, a variety of other rejection schemes have been proposed and some of these are briefly described below.

If the interference is both sufficiently narrowband and sufficiently strong to allow a phase-locked loop (PLL) to achieve phase-lock on it in the presence of the desired signal and thermal noise, it is possible to form a composite estimate (i.e., one accounting for both phase and amplitude) of the interference and subtract it from the received waveform. While the idea of subtracting an estimate from the received signal sounds similar to the method described in Section II, the manner in which the estimate is obtained is quite different.

One system designed to reject interference as just described is presented in [6], and a block diagram of the receiver is shown in Fig. 31. The ratio of interference power to signal power is assumed large enough so that the PLL locks onto the frequency and phase of the interference. Again, because of the relatively large interference level, the amplitude of the output of the low pass filter shown in the lower arm of the receiver of Fig. 31 is dominated by the interference. Hence the locally generated reference to the subtractor in the upper arm becomes the desired estimate of the interference. In [35], a technique similar to that of [6] is presented, but with the addition of various circuitry designed to result in a more accurate estimate of the interference than is achieved in [6].

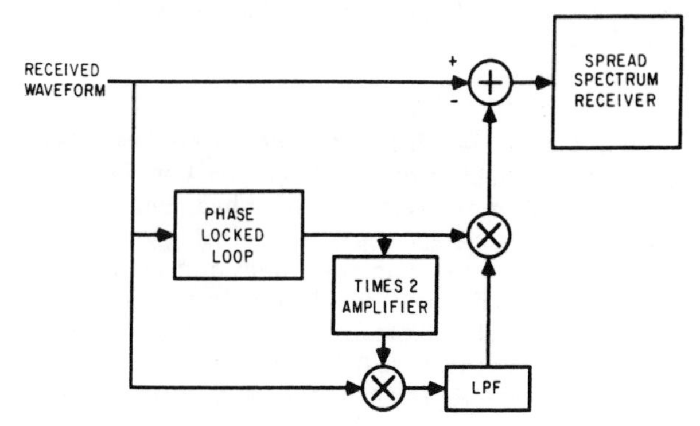

Fig. 31. Rejection scheme of [6].

A technique for canceling wide-band interference is described in [5]. Referring to Fig. 32, it is seen that if the ratio of interference power-to-signal power is sufficiently large, the output of the limiter is essentially the interference. However, the output of amplifier 1 consists of both signal plus interference, meaning that if the gain can be appropriately adjusted, the difference circuit can be used to suppress the interference. This gain adjustment is accomplished by an AGC operation.

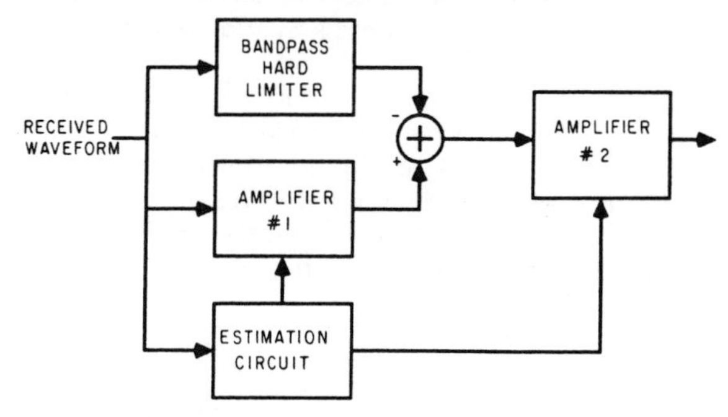

Fig. 32. Rejection scheme of [5].

The analyses presented in [6], [35], [5] correspond to noise-free conditions. Hence, it appears to be an open question as to how well these techniques perform in the very low signal-to-(thermal) noise ratio environment typical of most spread spectrum receivers.

A completely different technique for making a DS receiver more robust with respect to interference is described in [2], [3], [40]. This scheme uses an A/D converter, in conjunction with a variable threshold, to retain those chips of the spreading sequence which, when added to a strong interfering signal, are still received with their correct polarity. For this system to operate properly, it is necessary to have both a large *J/S* and a large ratio of interference power-to-noise power.

Conclusions

In this paper, a variety of interference suppression schemes designed to operate in conjunction with a DS spread spectrum receiver were described. Emphasis was placed on two general techniques, one using Wiener-type filters and the other employing transform domain processing. Both techniques were shown to have the potential of yielding a sizable improvement in system performance relative to that achievable by a conventional DS receiver, but that the improvement was subject to certain constraints.

Most notably was the constraint that the interference be relatively narrowband with respect to the DS waveform. Also, since high-speed signal processing is inherent in virtually any DS system, implementation of these schemes is limited to certain wide-band technologies such as SAW and CCD, and these technologies, in turn, have certain limitations such as dynamic range.

Regarding which scheme to use in a given scenario, it appears that each one has its own set of advantages and disadvantages. To see this, consider the simple example of a sinewave interferer at a known frequency. Whereas the estimation-type filter can, if desired, put a zero at the frequency of the sinewave and hence null it out completely, this complete nulling is typically not possible for the TDP receiver. Because the input to the TDP receiver is windowed in time before it is Fourier transformed, sidelobes are immediately put on the interference spectrum and, hence, even an infinitely deep notch over some appropriate fraction of the bandwidth of the system will not completely eliminate the interference. Alternately, because the TDP system can be made adaptive on an essentially "instantaneous" basis, without the need for an adaptive algorithm with its attendant convergence problems, the TDP receiver would seem to have an advantage over the receiver employing an estimation filter in those scenarios whereby rapid adaptivity is required.

There is still much to be learned in the area of interference rejection and this remains an active research field. In addition to the specific topics mentioned in this paper, other current areas of research include the use of interference suppression schemes to aid in the initial acquisition of the DS signal and the use of such techniques in a spread spectrum network.

Acknowledgment

The author wishes to thank Prof. R. Rao for carefully reading the manuscript and making some useful suggestions on improving the presentation of the material.

References

[1] S. T. Alexander, *Adaptive Signal Processing*. New York, NY: Springer-Verlag, 1986.

[2] F. Amoroso, "Adaptive A/D converter to suppress CW interference in DSPN spread-spectrum communications," *IEEE Trans. Commun.*, vol. COM-31, pp. 1117–1123, Oct. 1983.

[3] F. Amoroso and J. L. Bricker, "Performance of the adaptive A/D converter in combined CW and gaussian interference," *IEEE Trans. Commun.*, vol. COM-34, pp. 209–213, Mar. 1986.

[4] C. Atzeni, G. Manes, and L. Masotti, "Programmable signal processing by analog chirp transformation using SAW devices," in *Proc. 1975 Ultrasonics Symp.*, p. 572, 1973.

[5] P. W. Baier and K. J. Friederichs, "A nonlinear device to suppress strong interfering signals with arbitrary angle modulation in spread-spectrum receivers," *IEEE Trans. Commun.*, vol. COM-33, pp. 300–302, Mar. 1985.

[6] M. J. Bouvier, Jr., "The rejection of large CW interferers in spread spectrum systems," *IEEE Trans. Commun.*, vol. COM-28, pp. 254–256, Feb. 1978.

[7] J. H. Cafarella, W. M. Brown, Jr., E. Stern, and J. A. Alusow, "Acoustoelectric convolvers for programmable matched filtering in spread-spectrum system," *Proc. IEEE*, vol. 64, p. 756, 1976.

[8] P. Das, L. B. Milstein, and R. T. Webster, "Application of SAW chirp transform filter in spread spectrum communication systems," in *6th European Microwave Conf.*, pp. 261–266, Sept. 1976.

[9] A. Gersho, "Charge coupled devices: The analog shift register comes of age," *IEEE Commun. Mag.*, pp. 27–32, Nov. 1975.

[10] J. Gevargiz, M. Rosenmann, P. Das, and L. B. Milstein, "A comparison of weighted and nonweighted transform domain processing systems for narrowband interference excision," in *IEEE Military Communications Conf.*, pp. 32.3.1–32.3.4, Oct. 1984.

[11] J. Gevargiz, P. Das, and L. B. Milstein, "Implementation of a transform domain processing radiometer for DS spread spectrum signals with adaptive narrowband interference exciser," presented at the IEEE International Conf. on Communications, June 1985.

[12] J. Gervargiz, P. Das, L. B. Milstein, J. Moran, and O. McKee, "Implementation of DS-SS intercept receiver with an adaptive narrowband interference exerciser using transform domain processing and time weighting," in *IEEE Military Communications Conf.*, pp. 20.1.1–20.1.5, Oct. 1986.

[13] J. Gevargiz, P. Das, and L. B. Milstein, "Performance of a transform domain processing DS intercept receiver in the presence of finite bandwidth interference," in *IEEE Global Telecommunications Conf.*, pp. 21.5.1–21.5.5, Dec. 1986.

[14] A. A. Giordano and F. M. Hsu, *Least Square Estimation with Applications to Digital Signal Processing*. New York, NY: Wiley-Interscience, 1985.

[15] J. Guilford and P. Das, "The use of the adaptive lattice filter for narrowband jammer rejection in DS spread spectrum systems," in *Proc. IEEE International Conf. on Communications*, pp. 822–826, June 22–26, 1985.

[16] C. W. Helstrom, *Statistical Theory of Signal Detection*. New York, NY: Pergamon, 1960.

[17] F. M. Hsu and A. A. Giordano, "Digital whitening techniques for improving spread-spectrum communications performance in the presence of narrow-band jamming and interference," *IEEE Trans. Commun.*, vol. COM-26, pp. 209–216, Feb. 1978.

[18] R. A. Iltis and L. B. Milstein, "Performance analysis of narrowband interference rejection techniques in DS spread-spectrum systems," *IEEE Trans. Commun.*, vol. COM-26, pp. 209–216, Feb. 1978.

[19] ——, "An approximate statistical analysis of the Widrow LMS algorithm with application to narrow-band interference rejection," *IEEE Trans. Commun.*, vol. COM-33, pp. 121–130, Feb. 1985.

[20] J. W. Ketchum and J. G. Proakis, "Adaptive algorithms for estimating and suppressing narrow-band interference in PN spread-spectrum systems," *IEEE Trans. Commun.*, vol. COM-30, pp. 913–924, May 1982.

[21] J. W. Ketchum, "Decision feedback techniques for interference cancellation in PN spread-spectrum communication systems," in *IEEE Military Communications Conf.*, pp. 39.5.1–39.5.5, Oct. 1984.

[22] L. Li and L. B. Milstein, "Rejection of narrow-band interference in PN spread-spectrum systems using transversal filters," *IEEE Trans. Commun.*, vol. COM-30, pp. 925-928, May 1982.

[23] —, "Rejection of CW interference in QPSK systems using decision-feedback filters," *IEEE Trans. Commun.*, vol. COM-31, pp. 473-483, Apr. 1983.

[24] Z. Li, H. Yuan, and G. Bi, "Rejection of multi-tone interference in PN spread spectrum systems using adaptive filters," in *IEEE International Conf. on Communications*, pp. 24.5.1-24.5.5, June 1987.

[25] F. Lin and L. M. Li, "Rejection of finite-bandwidth interference in QPSK systems using decision-feedback filters," in *IEEE International Conf. on Communications*, pp. 24.6.1-24.6.5, June 1987.

[26] E. Masry, "Closed-form analytical results for the rejection of narrow-band interference in PN spread-spectrum systems—Part I: Linear prediction filters," *IEEE Trans. Commun.*, vol. COM-32, pp. 888-896, Aug. 1984.

[27] —, "Closed-form analytical results for the rejection of narrow-band interference in PN spread-spectrum systems—Part II: Linear interpolation filters," *IEEE Trans. Commun.*, vol. COM-33, pp. 10-19, Jan. 1985.

[28] E. Masry and L. B. Milstein, "Performance of DS spread-spectrum receivers employing interference-suppression filter under a worst-case jamming condition," *IEEE Trans. Comm.*, vol. COM-34, pp. 13-21, Jan. 1986.

[29] L. B. Milstein and P. Das, "Spread spectrum receiver using acoustic surface wave technology," *IEEE Trans. Commun.*, vol. COM-25, no. 8, pp. 841-847, Aug. 1977.

[30] —, "Surface acoustic wave devices," *IEEE Commun. Mag.*, pp. 25-33, Sept. 1979.

[31] —, "An analysis of a real-time transform domain filtering digital communication system, Part I: Narrowband interference rejection," *IEEE Trans. Commun.*, vol. COM-28, pp. 816-824, June 1980.

[32] L. B. Milstein, P. K. Das, and J. Gevargiz, "Processing gain advantage of transform domain filtering DS spread spectrum systems," *Military Communications Conf.*, pp. 21.2.1-21.2.4, Oct. 1982.

[33] L. B. Milstein and P. K. Das, "An analysis of a real-time transform domain filtering digital communication system—Part II: Wideband interference rejection," *IEEE Trans. Commun.*, vol. COM-31, pp. 21-27, Jan. 1983.

[34] L. B. Milstein and R. A. Iltis, "Signal processing for interference rejection in spread-spectrum communications," *IEEE ASSP Mag.*, pp. 18-31, Apr. 1986.

[35] A. E. S. Mostafa, M. Abdel-Kader, and A. El-Osmany, "Improvements of anti-jam performance of spread-spectrum systems," *IEEE Trans. Commun.*, vol. COM-31, pp. 803-808, Jan. 1983.

[36] G. R. Nudd and O. W. Otto, "Chirp signal processing using acoustic surface wave filters," in *1975 Ultrasonics Symp. Proceedings*, p. 350, 1975.

[37] J. Ogawa, S. J. Cho, N. Morinaga, and T. Namekawa, "Optimum detection of *M*-ary PSK signal in the presence of CW interference," *Trans. IECE Japan*, vol. E64, pp. 800-806, Dec. 1981.

[38] O. W. Otto, "Real-time Fourier transform with a surface wave convolver," *Electron Lett.*, vol. 8, p. 623, 1972.

[39] A. Papoulis, *Probability, Random Variables and Stochastic Processes.* New York, NY: McGraw-Hill, 1965, pp. 218-220.

[40] F. J. Pergal, "Adaptive threshold A/D conversion techniques for interference rejection in DSPN receiver applications," in *IEEE Military Communications Conf.*, pp. 4.7.1-4.7.7, Oct. 1987.

[41] R. L. Pickholtz, D. L. Schilling, and L. B. Milstein, "Theory of spread-spectrum communications—a tutorial," *IEEE Trans. Commun.*, vol. COM-30, pp. 855-884, May 1982.

[42] J. G. Proakis, *Digital Communications.* New York, NY: McGraw-Hill, 1983.

[43] M. Rosenmann, M. J. Gevargiz, P. K. Das, and L. B. Milstein, "Probability of error measurement for an interference resistant transform domain processing receiver," in *IEEE Military Communications Conf.*, pp. 638-640, Oct. 1983.

[44] G. I. Saulnier, P. Das, and L. B. Milstein, "Suppression of narrow-band interference in a PN spread-spectrum receiver using a CTD-based adaptive filter," *IEEE Trans. Commun.*, vol. COM-32, pp. 1227-1232, Nov. 1984.

[45] —, "An adaptive digital suppression filter for direct-sequence spread-spectrum communications," *IEEE J. Selected Areas Commun.*, vol. SAC-3, no. 5, pp. 676-686, Sept. 1985.

[46] —, "Suppression of narrow-band interference on a direct sequence spread spectrum receiver in the absence of carrier synchronization," in *IEEE Military Communications Conf.*, pp. 13-17, Oct. 1985.

[47] G. J. Saulnier, K. Yum, and P. Das, "The suppression of tone jammers using adaptive lattice filtering," in *IEEE International Conf. on Communications*," pp. 24.4.1-24.4.5, June 1987.

[48] D. Shklarsky, P. K. Das, and L. B. Milstein, "Adaptive narrowband interference suppression," in *1979 National Telecommunications Conf.*, pp. 15.2.1-15.2.4, Nov. 1979.

[49] M. K. Simon, J. Omura, R. A. Scholtz, and B. K. Levitt, *Spread-Spectrum Communications*, vols. I-III. Rockville, MD: Computer Science Press, 1985.

[50] F. Takawira and L. B. Milstein, "Narrowband interference rejection in PN spread spectrum systems using decision feedback filters," in *IEEE Military Communications Conf.*, pp. 20.4.1-20.4.5, Oct. 1986.

[51] Y.-C. Wang and L. B. Milstein, "Rejection of multiple narrowband interference in both BPSK and QPSK DS spread-spectrum systems," *IEEE Trans. Commun.*, vol. COM-36, pp. 195-204, Feb. 1988.

[52] B. Widrow *et al.*, "Adaptive noise canceling: Principles and applications," *Proc. IEEE*, vol. 63, pp. 1692-1716, Dec. 1975.

[53] B. Widrow and S. D. Stearns, *Adaptive Signal Processing.* Englewood Cliffs, NJ: Prentice-Hall, 1985.

[54] R. E. Ziemer and R. L. Peterson, *Digital Communications and Spread Spectrum.* New York, NY: MacMillan, 1985.

Laurence B. Milstein (Fellow, IEEE) received the B.E.E. degree from the City College of New York, New York, NY, in 1964, and the M.S. and Ph.D. degrees in electrical engineering from the Polytechnic Institute of Brooklyn, Brooklyn, NY, in 1966 and 1968, respectively.

From 1968 to 1974 he was employed by the Space and Communications Group of Hughes Aircraft Company, and from 1974 to 1976 he was a member of the Department of Electrical and Systems Engineering, Rensselaer Polytechnic Institute, Troy, NY. Since 1976 he has been with the Department of Electrical and Computer Engineering, University of California at San Diego, La Jolla, where he is a Professor and Department Chairman, working in the area of digital communication theory with special emphasis on spread-spectrum communication systems. He has also been a consultant to both government and industry in the areas of radar and communications.

Dr. Milstein was an Associate Editor for Communication Theory for the IEEE TRANSACTIONS ON COMMUNICATIONS and an Associate Technical Editor for the IEEE COMMUNICATIONS MAGAZINE. He was a member of the Board of Governors of the IEEE Communications Society, and is a member of Eta Kappa Nu, Tau Beta Pi, and Sigma Xi.

Reprinted from *IEEE Journal on Selected Areas in Commun.*, Vol. 12, No. 5, June 1994, pp. 796-807.

Analysis of a Simple Successive Interference Cancellation Scheme in a DS/CDMA System

Pulin Patel and Jack Holtzman

***Abstract*— Compensating for near/far effects is critical for satisfactory performance of DS/CDMA systems. So far, practical systems have used power control to overcome fading and near/far effects. Another approach, which has a fundamental potential in not only eliminating near/far effects but also in substantially raising the capacity, is multiuser detection and interference cancellation. Various optimal and suboptimal schemes have been investigated. Most of these schemes, however, get too complex even for relatively simple systems and rely on good channel estimates. For interference cancellation, estimation of channel parameters (viz. received amplitude and phase) is important. We analyze a simple successive interference cancellation scheme for coherent BPSK modulation, where the parameter estimation is done using the output of a linear correlator. We then extend the analysis for a noncoherent modulation scheme, namely *M*-ary orthogonal modulation. For the noncoherent case, the needed information on both the amplitude and phase is obtained from the correlator output. Performance of the IC scheme along with multipath diversity combining is studied.**

I. Introduction

RECENTLY, there has been considerable interest in application of DS/CDMA (Direct Sequence Code Division Multiple Access) in cellular and personal communications. The choice of CDMA is attractive because of its potential capacity increases and other technical factors such as antimultipath fading capabilities.

Compensating for the near/far effect is critical for satisfactory performance of DS/CDMA systems. Commercial digital cellular system based on CDMA uses stringent power control, described in [1], to combat near/far effects and fading. Another approach, still in the research stage, is multiuser detection. In addition to mitigating the near/far effect, multiuser detection has the more fundamental potential of significantly raising capacity by cancelling multiple access interference. Valuable fundamental investigations (e.g., [2], [3]) have demonstrated huge potential capacity and performance improvements and have also shown the complexity of optimal structures. This has motivated the search for practical schemes achieving part of this potential. More information on multiuser detectors and interference cancellers can be found in a number of references, e.g., [4]–[8].

Manuscript received June 4, 1993; revised December 11, 1993. This paper was presented in part at the Fourth WINLAB Workshop on Third Generation Wireless Information Networks, October 19–10, 1993, East Brunswick, NJ USA.

The authors are with the Wireless Information Networks Laboratory (WINLAB), Rutgers, The State University of New Jersey, Piscataway, NJ 08855-0909 USA.

IEEE Log Number 9215588.

In general, a major problem with multiuser detectors and interference cancellers is the maintenance of simplicity. Even the suboptimal linear detectors have considerably complex processing, especially in an asynchronous channel. Certain schemes, where the users' signals are detected collectively, turn out to have a complex parallel structure. An alternative to parallel cancellation is to perform successive cancellation. The serial (successive) structure is not only more simple (requires less hardware), but is also more robust in doing the cancellations (see [7]). Comparison of the performance of the parallel and successive IC schemes is done in [9]. Successive cancellation was discussed in [10]. In [11] and [12], the CDMA-IC scheme involves successive cancellation and for this it relies on a gain list of users with nonincreasing strength. Our approach also successively cancels strongest users but assumes no knowledge of the users' powers. It uses the outputs of conventional correlation receiver to rank the users instead of separate channel estimates as in [11], [12][1]. Our approach [13], [9] is closer to that of [6], which uses the despread signals to select the strongest one for cancellation (but we cancel more than just the strongest). The approach in [14] also uses the information in the despread signal for cancellation.

We analyze the IC scheme operating by successively cancelling user interferences, ranked in order of received powers with the ranking obtained from correlations of the received signal with each user's chip sequence. In [5], the *bit decisions* are successively fed back in order of decreasing signal strength to improve performance. This occurs after matched filters and a decorrelating filter (matrix inversion). We do not use a decorrelator but rather successively feed back chip sequences based on decreasing signal strengths.

To perform interference cancellation, estimation of various parameter (viz. amplitude and phase) of the signals is important. For the coherent case, we shall assume the perfect knowledge of the phase is available at the receiver; however, amplitude estimation is required, which is accomplished by using the output of the linear correlator. In the noncoherent case, the phase knowledge is not required for the demodulation, but interference cancellation requires the knowledge of both the random phase and the amplitude. This knowledge of the amplitude and phase is also extracted from the output of the linear correlator.

The paper is organized as follows: Analysis of the IC scheme for BPSK modulation is done in Section II, followed by some results in Section III. In Section III, we also analyze

[1] The cancellation in [11], [12] is actually done in the spectral domain.

the performance of the IC scheme, under flat Rayleigh fading, using order statistics and the effect of averaging the correlations on the performance of the IC scheme. In Section IV, we extend the analysis of the IC scheme to a more robust modulation scheme, namely, M-ary orthogonal modulation along with multipath combining. Concluding comments are presented in Section V.

II. Successive IC Scheme for Coherent BPSK Modulation

We consider here a simple system to obtain basic results, where the received signal is

$$r(t) = \sum_{k=1}^{K} A_k \cdot a_k(t-\tau_k) \cdot b_k(t-\tau_k) \cdot \cos(\omega_c t + \phi_k) + n(t) \quad (1)$$

$r(t)$ = received signal
K = total number of active users
A_k = amplitude of kth user
$b_k(t)$ = bit sequence of kth user at bit-rate R_b
$a_k(t)$ = spreading chip sequence of kth user at chip-rate R_c
$n(t)$ = additive white Gaussian noise (two sided power spectral density $= N_0/2$)
N = T/T_c where T = bit period and T_c = chip period

τ_k and ϕ_k are the time delay and phase of the kth user, which are assumed to be known, i.e., tracked accurately.

The bits and chips are rectangular. Their values are all i.i.d. random values with probability 0.5 of ± 1. The τ_k and ϕ_k are i.i.d. uniform random variables in $[0, T]$ and $[0, 2\pi]$, respectively, for the asynchronous case.

We assume knowledge of the spread sequences of all the users, but no knowledge of the energies of the individual users is needed. As shown in the block diagram in Fig. 1, the basic idea is decoding the strongest users (whichever they may be), and then cancelling effects from the received signal. Here, the strongest user is not known beforehand, but is detected from the strength of the correlations of each of the users' chip sequence with the received signal. The correlation values (obtained from the conventional bank of correlators) are passed on to a selector which determines the strongest correlation value and selects the corresponding user for decoding and cancellation. These correlation values (as opposed to separate power estimates) form the basis for not only estimating the amplitude but also for maintaining the order of cancellation[2]. The process is repeated until the weakest user is decoded[3]. The flow chart of the process is shown in Fig. 2. Detailed analysis of the IC scheme for coherent BPSK system (after lowpass filtering) can be found in [13]. The results of the analysis are as follows.

After j cancellations, the decision variable for the $(j+1)^{\text{st}}$ user is given by

$$\widehat{Z_{j+1}} = \frac{1}{2} A_{j+1} b_{j+1} + \frac{1}{2} C_{j+1} \quad (2)$$

[2]Using the output of the correlator also helps in obtaining the knowledge of the phase during noncoherent demodulation (discussed later in Section IV),

[3]An alternative is to limit the number of cancellations. The impact of limiting the number of cancellations on the performance of the IC scheme is demonstrated in Section IV.

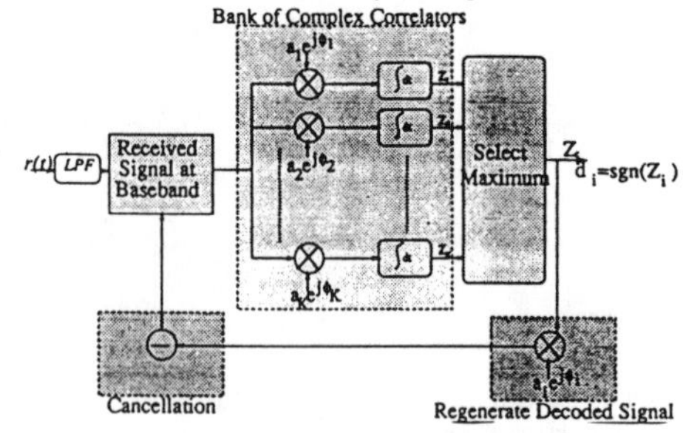

Fig. 1. Successive IC scheme with coherent BPSK modulation.

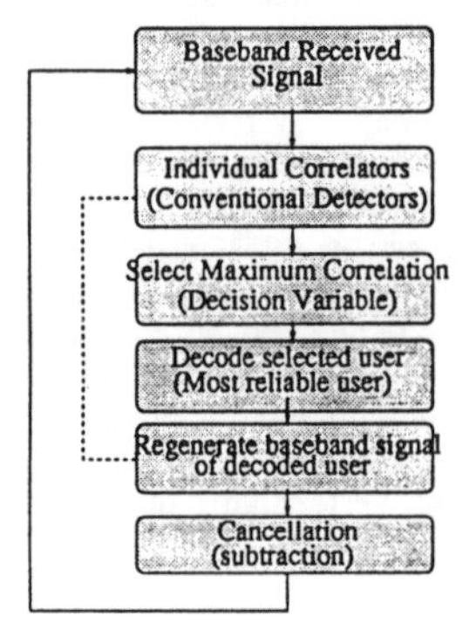

Fig. 2. Flow diagram of interference cancellation schemes.

and C_{j+1} is given by

$$C_{j+1} = \sum_{k=j+2}^{K} A_k I_{k,j+1}(\tau_{k,j+1}, \phi_{k,j+1}) + \left(n_{j+1}^I + n_{j+1}^Q\right) - \sum_{i=1}^{j} C_i I_{i,i+1}(\tau_{i,i+1}, \phi_{i,i+1}) \quad (3)$$

In the above expression, the first term is the multiple access interference of the uncancelled users, the second term is due to the Gaussian noise, and the third term is the cumulative noise due to imperfect cancellation. The cross correlation term is given by

$$I_{k,i}(\tau_{k,i}, \phi_{k,i}) = \frac{1}{T}\left[\int_0^T a_k(t-\tau_{k,i}) \cdot a_i(t) dt\right] \times \cos(\phi_k - \phi_i) \quad (4)$$

The variance of C_{j+1} conditioned on A_k as follows

$$\begin{aligned}\eta_{j+1} &= \text{Var}[C_{j+1} \mid A_k] \\ &= \sum_{k=j+2}^{K} A_k^2 \cdot \text{Var}[I_{k,j+1}(\tau_{k,j+1}, \phi_{k,j+1})] \\ &\quad + \text{Var}\left[\left(n_{j+1}^I + n_{j+1}^Q\right)\right] \\ &\quad + \sum_{i=1}^{j} \eta_i \cdot \text{Var}[I_{i,i+1}(\tau_{i,i+1}, \phi_{i,i+1})] \quad (5)\end{aligned}$$

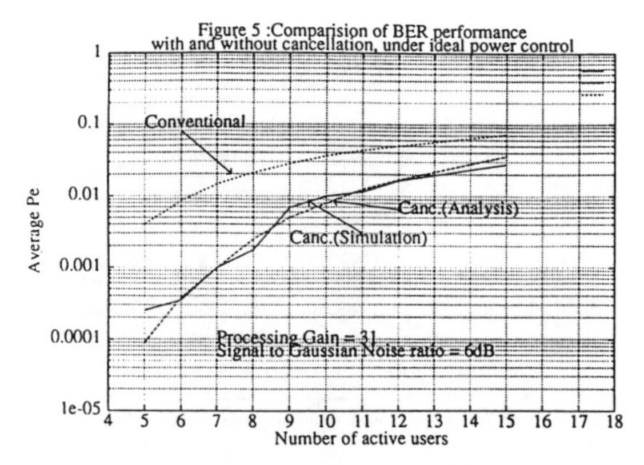

Fig. 3. Comparison of BER performance under ideal power control (synchronous case).

Further analysis yields

$$\eta_{j+1} = \begin{cases} \frac{1}{N}\sum_{k=j+2}^{K} A_k^2 + \frac{N_0}{T} + \frac{1}{N}\sum_{i=1}^{j}\eta_i & ; \text{ for Synchronous} \\ \frac{1}{3N}\sum_{k=j+2}^{K} A_k^2 + \frac{N_0}{T} + \frac{1}{3N}\sum_{i=1}^{j}\eta_i; & \text{for Asynchronous} \end{cases} \quad (6)$$

The signal-to-noise ratio, conditioned on A_k, is then given by

$$\gamma_{j+1} = \frac{\frac{1}{4}A_{j+1}^2}{\frac{1}{4}\eta_{j+1}} = \frac{A_{j+1}^2}{\frac{1}{3N}\sum_{k=j+2}^{K} A_k^2 + \frac{N_0}{T} + \frac{1}{3N}\sum_{i=1}^{j}\eta_i}; \text{ for Asynchronous Case} \quad (7)$$

To calculate the bit error rate, we shall use the Gaussian approximation [15], [16], i.e., we shall assume that the noise C_{j+1} is Gaussian with zero mean and variance η_{j+1}. The probability of bit error after the jth cancellation, conditioned on the amplitudes A_k, is then given by

$$P_e^{j+1} = P\left\{\widehat{Z_{j+1}} < 0 | b_{j+1} = +1\right\} = P\{C_{j+1} < -A_{j+1}\} = Q\left(\frac{A_{j+1}}{\sqrt{\eta_{j+1}}}\right) = Q(\sqrt{\gamma_{j=1}}) \quad (8)$$

Remark: Throughout the paper, approximations will be validated with simulations. The analysis assumes the users are cancelled in order of powers. The actual algorithm, using sample correlations to order the powers, may actually have this order changed. As will be seen, however, comparisons with simulation show good accuracy.

III. Results for Coherent BPSK

A. Performance of the IC Scheme Under Ideal Power Control

In Fig. 3, we compare the analysis and simulation results for users under ideal power control. The P_e's from analysis obtained using the Gaussian approximation agree well with the simulation. It is clear that the cancellation scheme works better than the conventional scheme even under ideal power control. Also, note that for similar P_e performance (0.01), six users in the conventional scheme can be increased to 11 users with cancellation, giving a substantial increase in capacity.

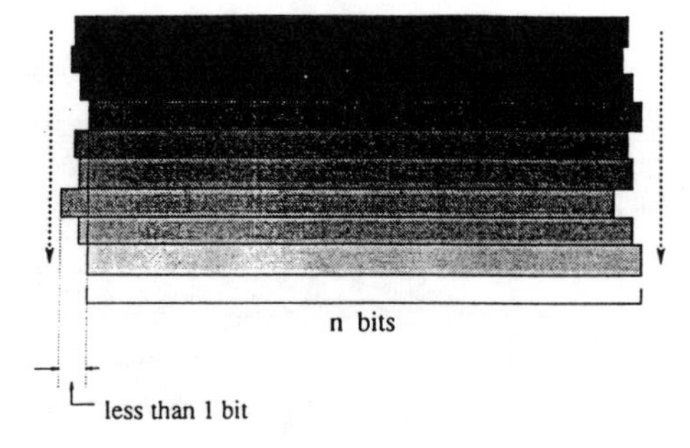

Fig. 4. Successive decoding in asynchronous channel.

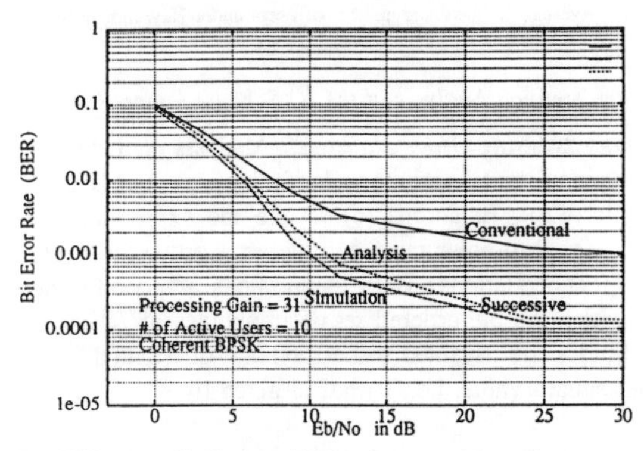

Fig. 5. BER versus E_b/N_o under ideal power control (asynchronous case).

Note on Asynchronous Channel: Introduction of asynchronism does not change the cancellation algorithm where the decoding and cancellation is done in decreasing order of the received signal strength. The only requirements (for coherent BPSK) of the cancellation algorithm are the timing and the phase knowledge (which are the same requirements for the conventional receiver). For asynchronous systems however, we must define what bits are compared with what other bits, and we propose the following. Group n bits of each user into a cancellation frame, where the maximum time between the first bit start and the last bit end is $(n + 1)$ bit times; see Fig. 4. After an entire frame is received, the correlations of the n bits of each user are averaged and the ranking of the users is obtained from these averages of correlations over n bits. This is what we used in our asynchronous simulations.[4] Issues in practical implementation are discussed in Section IV.

Fig. 5 is the BER versus the bit energy to noise ratio plots for the asynchronous case and under ideal power control. Total of 10 active users are present and the processing gain is 31.

Analysis of the cancellation scheme under fading is done next, where the order of cancellation changes as fast as the power level of the users change. More information on multiuser detectors in fading channels can be found in [17]–[19].

[4]For the results presented in this section, the ranking was obtained from averaging but the amplitude of the bit was estimated using correlation over a single bit. This was done for the purpose of controlling the simulation run times. In the sequel, we shall realize the improvement in performance by using the average of correlations over n bits for estimating the amplitude.

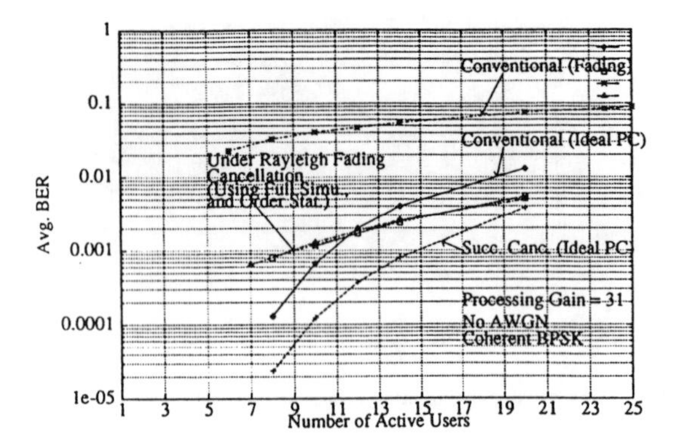

Fig. 6. Average BER versus number of users under Rayleigh fading.

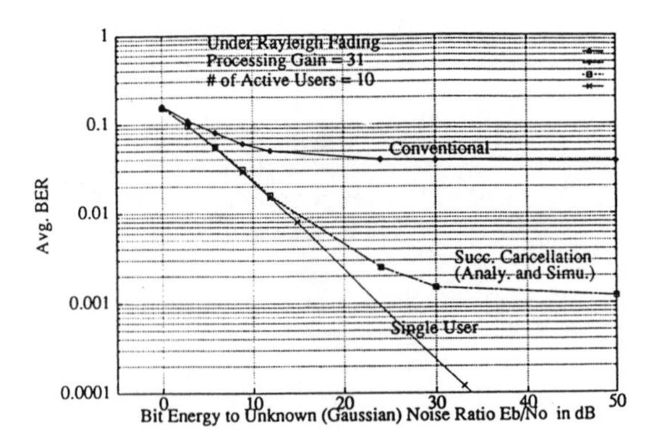

Fig. 7. Average BER versus E_b/N_o under Rayleigh fading.

B. Performance Analysis of the IC Scheme under Fading

Analysis Using Order Statistics: Analysis of the BER performance of the IC scheme under fading was done using order statistics [20]. Equations (6)-(8) are the expressions of the noise, signal-to-noise ratio, and the error probability resulting after j cancellations. These expressions are conditioned on A_k, which are the ordered set of amplitudes of K users. The amplitudes are assumed to be Rayleigh distributed with unit mean square value, i.e., its pdf is given by

$$f(x) = 2xe^{-x^2} \tag{9}$$

and its cdf (cumulative density function) is given by:

$$F(x) = 1 - e^{-x^2} \tag{10}$$

The pdf's of the ordered A_k (where A_1 is the strongest and A_K is the weakest) is denoted by $f_{A_k}(x)$ and is obtained as follows

$$f_{A_k}(x) = \frac{K!}{(K-k)!(k-1)!} F^{K-k}(x)[1-F(x)]^{k-1} f(x) \tag{11}$$

The expected values of the ordered A_k are then obtained as $E[A_k^2] = \int_0^\infty x_2 f_{A_k}(x)dx$. The denominator of (7) is approximated as Gaussian because the dominating terms for both small and large j are sums of random variables (all approximations will be validated by simulation). We have that (12) below. The error probability expression after the jth cancellation (8) is then unconditioned using the pdf of the $j+1^{\text{th}}$ strongest amplitude as follows:

$$\widehat{P_e^{j+1}} = \int_0^\infty Q\left(\frac{A_{j+1}}{\sqrt{E_{A_k}[\eta_{j+1}]}}\right) f_{A_{j+1}}(x)dx \tag{13}$$

The average probability of error is then obtained as the average of the BER resulting from all stages of cancellation and is plotted in Fig. 6.

Comparison with Simulation Results: Monte-Carlo type simulations were run for the asynchronous case. The agreement of the actual computer simulation results with the analysis using order statistics is evident from Fig. 6.

From Fig. 6, it can also be seen that the performance of the successive IC scheme is much better compared to the conventional receiver under Rayleigh fading. Note that no coding is taken into consideration. So, the BER could be further lowered by using efficient coding. It should also be noted that Rayleigh fading is a worst case for two reasons. First of all, wideband CDMA transmission mitigates the Rayleigh fading effect [21]. Second, with CDMA, a Rake receiver can resolve the multipaths and improve performance. Interference cancellation with multipath resolution is analyzed later. The performance of the successive IC scheme under Rayleigh fading is comparable to that of the conventional receiver under ideal power control. This makes the successive IC scheme fading resistant.

Fig. 7 is the plot of the BER versus the bit energy to Gaussian noise (not including multiple access interference) ratio. Though the successive IC is much better than the conventional receiver under fading, its performance is not comparable to the single user detector (optimal detector). Further improvement in the IC scheme (by using averaging, described next) would, however, make it possible to achieve the single user bound.

C. Performance as a Function of Power Unbalance

The question as to the performance of the IC scheme as a function of power unbalance actually addresses two questions:

1) How does the IC scheme perform as the power unbalance increases?
2) How does the IC improvement over the conventional receiver change as the power unbalance increases?

Both of these questions are addressed in Fig. 8[5]. Since the IC scheme is a nonlinear scheme, it is difficult to draw a gen-

[5] Although Fig. 8 could be obtained from order statistic analysis like that of Section III, it was actually generated from a simulation. This explains the lack of smoothness.

$$E_{A_k}[\eta_{j+1}] = \begin{cases} \frac{1}{N}\sum_{k=j+2}^{K} E[A_k^2] + \frac{N_0}{T} + \frac{1}{N}\sum_{i=1}^{j}\eta_i; & \text{for Synchronous} \\ \frac{1}{3N}\sum_{k=j+2}^{K} E[A_k^2] + \frac{N_0}{T} + \frac{1}{3N}\sum_{i=1}^{j}\eta_i; & \text{for Asynchronous} \end{cases} \tag{12}$$

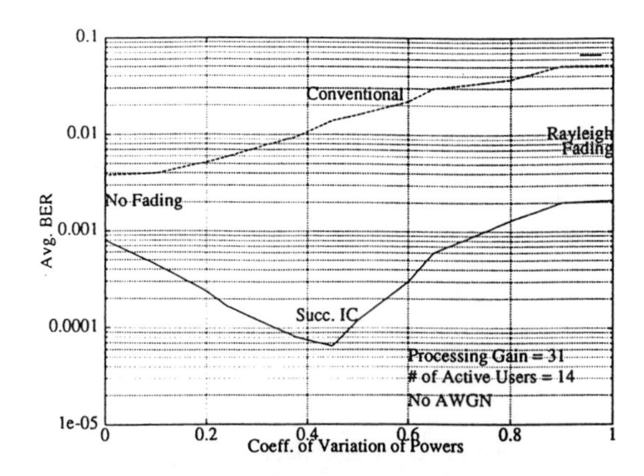

Fig. 8. Average BER versus coefficient of powers.

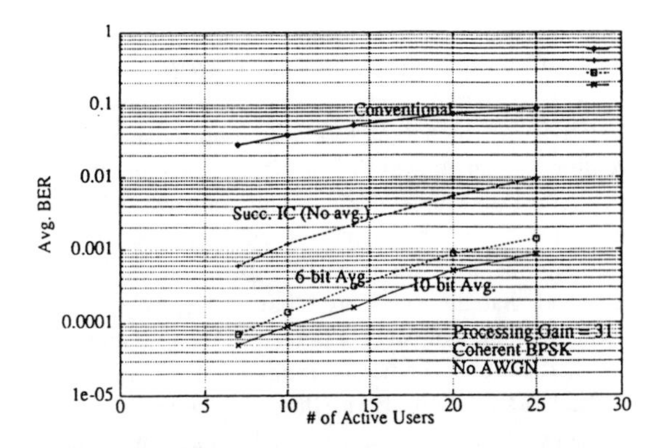

Fig. 9. Average BER versus number of active users, with no averaging, 6–bit averaging, and 10–bit averaging.

eral statement about its performance under fading with varying degrees. So to observe the effect of power unbalance on the BER performance of the IC scheme, we consider cases where the power levels of users vary with different coefficient of variations (CV). From Fig. 8, it is seen that with perfect power control (CV = 0), there is an improvement in going from conventional to IC but not as big an improvement as when CV is bigger (around 0.4). As variations increase, the performance of the IC scheme improves initially, and then starts degrading, whereas the performance of the conventional receiver starts degrading as soon as power unbalance is introduced.

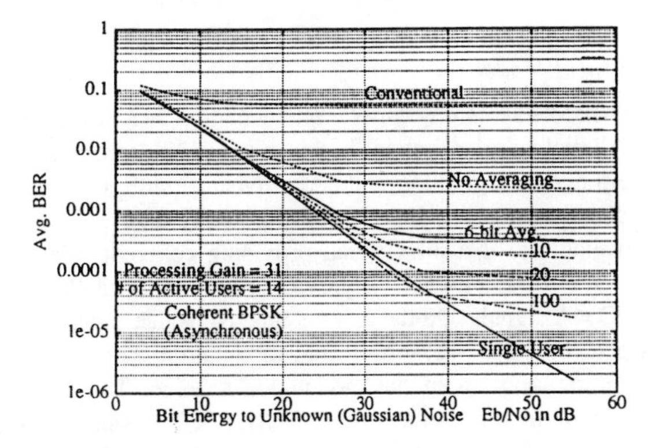

Fig. 10. Average BER versus E_b/N_o.

D. Effect of Averaging over n Bits on the Performance of IC Scheme

As described before, average correlations over n bits was used to rank the users but the estimated averaging did not use averaging. In this section, we shall study the performance improvement achieved by using average correlations to estimate the amplitude. As a matter of fact, by using the average correlations over n bits, the variance of the noise in the estimate of the amplitude decreases by a factor of $1/n$ (assuming that the correlations are independent from bit to bit). Using (6) and the above fact, we obtain the variance of the noise in the decision variable after jth cancellation as (14), see below. where n is the number of bits used for averaging. We shall now study the effect of averaging over n bits on the performance of IC Scheme under Rayleigh fading.

Fig. 9 is the plot of the average BER versus the number of active users. As the estimation of the amplitude of the bit is improved by averaging over several bits, the BER performance improves significantly in a fading environment. The improvement in the average BER (and hence the capacity) is evident from Fig. 9 as 6-bit or 10-bit averaging is used instead of no averaging. It is also clear from Fig. 10 that as the estimate of the amplitude of the user being cancelled is improved (by increasing the number of bits being averaged), the single user bound can be approached in a fading channel. This, however, assumes that the amplitude is fixed over the averaging interval which does not account for fast Rayleigh fading[6].

Note on Using Power Estimates Instead of Correlations An alternative to using correlation values to determine the order of cancellation and to perform cancellation is to use separate channel estimates. First-order analysis of the IC scheme using power estimates instead was done in [13], which was also verified by simulations. It was noted that for comparable BER performance the power estimate must be accurate within 3 dB of the actual power value, if correlation of just one bit is used. However, if correlations over 10 bits are averaged, the required accuracy of the power estimate becomes more strict (around 1 dB).

[6]For example, the Rayleigh fading amplitude correlation between a received symbol and the tenth succeeding symbol is around 0.94 for frequency of 1.8 GHz and symbol rate of 30 k symbols/s and 100 km/h velocity.

$$E_{A_k}[\eta_{j+1}] = \begin{cases} \frac{1}{N}\sum_{k=j+2}^{K} E[A_k^2] + \frac{N_o}{T} + \frac{1}{Nn}\sum_{i=1}^{j}\eta_i; & \text{for Synchronous} \\ \frac{1}{3N}\sum_{k=j+2}^{K} E[A_k^2] + \frac{N_o}{T} + \frac{1}{3Nn}\sum_{i=1}^{j}\eta_i; & \text{for Asynchronous} \end{cases} \quad (14)$$

Until now, we derived a method for analyzing a successive interference cancellation scheme in a DS/CDMA system using coherent BPSK modulation. The analysis was verified through simulations. Significant performance improvement over the conventional receiver was demonstrated, for both under ideal power control and under fading. This translates to increased capacity and fading resistance. It was also shown that averaging significantly improves the performance of the IC scheme and that the single-user bound can be achieved. In the next section, we shall extend our analysis to a more robust modulation scheme for mobile radio, namely M-ary orthogonal modulation. Here, the IC scheme is applied to the noncoherent receiver, and its performance with multipath diversity combining is studied.

IV. IC Scheme with M-ary Orthogonal Modulation

A. System Model

A CDMA system is under implementation as a next generation digital cellular system. In this system [22], a combination of orthogonal signalling and code division multiple access is used on the reverse link (mobile to base) to overcome the unavailability of a pilot signal providing coherent reference. This scheme is efficient in providing non coherent detection of CDMA signals. On the reverse link, 64 Walsh functions are used to obtain 64-ary orthogonal modulation. Fig. 11 shows the block diagram of the modulation scheme. The received version of the transmitted Offset-QPSK (OQPSK) signals from all the mobiles in a single cell during one symbol interval T_w is given by

$$r(t) = \sum_{k=1}^{K} S_{k,j}(t - \tau_k) + n(t) \tag{15}$$

Assuming that there are N multipaths per user that are arriving at the base station, the received signal is explicitly given by

$$\begin{aligned} r(t) = \sum_{k=1}^{K}\sum_{i=1}^{N}&[\alpha_k^i\sqrt{P_k}\cdot W_k^j(t-\tau_k^i)\cdot a_k((t-\tau_k^i) \\ &\times p_I(t-\tau_k^i)\cos(\omega_c t+\phi_k^i) \\ &+\alpha_k^i\sqrt{P_k}\cdot W_k^j(t-T_d-\tau_k^i)\cdot a_k(t-T_d-\tau_k^i) \\ &\times p_Q(t-T_d-\tau_k^i)\sin(\omega_c t+\phi_k^i)]+n(t) \end{aligned} \tag{16}$$

$r(t)$ = received signal
K = total number of active users
P_k = power of each chip of kth user
α_k^i = varying amplitude of the ith multipath of the kth user
$W_k^j(t)$ = j^{th} M-ary symbol of kth user ($j = 1, \ldots, M$) (Walsh symbol)
$a_k(t)$ = spreading chip sequence of kth user
$pI(t)$ = I channel short PN sequence
$p_Q(t)$ = Q channel short PN sequence
$n(t)$ = additive white Gaussian noise (two-sided power spectral density $= N_0/2$)

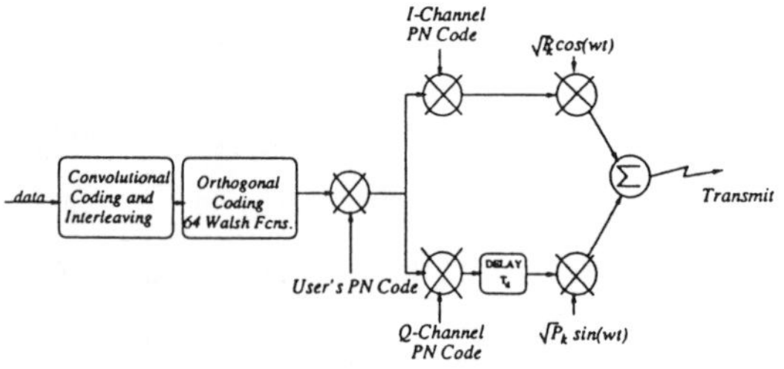

Fig. 11. DS/CDMA transmitter.

N = total number of multipaths
τ_k^i and ϕ_k^i are the time delay and phase of the ith multipath of the kth user.
τ_k are assumed to be known, i.e., tracked accurately.
T_c = chip interval
T_d = offset time ($T_c/2$)

The τ_k^i and ϕ_k^i are i.i.d. uniform random variables in $[0, T]$ and $[0,\ 2\pi]$, respectively, for the asynchronous case. No knowledge of ϕ_k^i is assumed.

For simplicity of notation, we shall denote the product of the M-ary symbol, the user PN code, and the I or Q channel PN code as follows

$$\begin{aligned} a_{k,j}^I(t) &= W_k^j(t)\cdot a_k(t)\cdot p_I(t) \\ a_{k,j}^Q(t) &= W_k^j(t-T_d)\cdot a_k(t-T_d)\cdot p_Q(t-T_d) \end{aligned} \tag{17}$$

Rewriting the received signal, we get

$$\begin{aligned} r(t) = \sum_{k=1}^{K}\sum_{i=1}^{N}&[\alpha_k^i\sqrt{P_k}\cdot a_{k,j}^I(t-\tau_k^i)\cdot\cos(\omega_c t+\phi_k^i) \\ &+\alpha_k^i\sqrt{P_k}\cdot a_{k,j}^Q(t-\tau_k^i).\sin(\omega_c t+\phi_k^i)]+n(t) \end{aligned} \tag{18}$$

We assume knowledge of the spread sequences of all the users, but no knowledge of the powers and phases of the individual users is needed. As shown in Fig. 12, the receiver correlates the received signal with the respective I & Q channel PN code and with the respective user's spreading code and with all possible 64-ary symbols for all the tractable multipaths. Each multipath that is tracked generates 64 coefficients representing the 64 possible Walsh symbols. These 64 coefficients from all the multipaths are then combined to produce a single set of 64 coefficients. A decision on the symbol is then made by selecting the largest of these 64 coefficients. This is the conventional CDMA receiver for noncoherent reception. Analysis of this receiver in the AWGN channel was done in [23]. Analysis of M-ary modulation in multipath channel was done in [24] for Rayleigh distributed multipaths and for general multipaths in [25]. We shall follow the analysis of [25] with multipath combining and apply Interference Cancellation to it. Note again that we are not considering convolutional coding and interleaving here. At the output of the lowpass filter (LPF) of the I channel, we get

$$\begin{aligned} d^I(t) &= \text{LPF}\{r(t)\cos(\omega_c t)\} \\ &= \sum_{k=1}^{K}\sum_{i=1}^{N}\left\{\left[\alpha_k^i\sqrt{P_k}\cdot a_{k,j}^I(t-\tau_k^i)\cdot\frac{\cos(\phi_k^i)}{2}\right]\right. \\ &\quad \left.+\left[\alpha_k^i\sqrt{P_k}\cdot a_{k,j}^Q(t-\tau_k^i)\cdot\frac{\sin(\phi_k^i)}{2}\right]\right\}+\frac{n_c(t)}{2} \end{aligned} \tag{19}$$

where $n_c(t)$ is the in-phase component of the Gaussian noise ($n(t)$ after lowpass filtering can be represented by $n_c(t) + j \cdot n_s(t)$). Similarly, $d^Q(t)$ is obtained. As seen in Fig. 12, the decision variable obtained from the lth multipath of the kth is given by $S_{k,m}(l)$; $m = 1, \ldots M$. Combining all L multipaths, we obtain $S_{k,m} = \sum_{l=1}^{L} S_{k,m}(l)$ and the decision on the symbol of the kth user is then obtained as $\hat{m} = \max_m [S_{k,m}]; m = 1, \ldots M$. For a particular user (say user 1), we have the following

$$S_{1,m}(l) = \left(Z_{1,m}^{II}(l) + Z_{1,m}^{QQ}(l)\right)^2 + \left(Z_{1,m}^{IQ}(l) - Z_{1,m}^{QI}(l)\right)^2 \tag{20}$$

where $Z_{1,m}^{II}(l), Z_{1,m}^{IQ}(l), Z_{1,m}^{QI}(l)$ and $Z_{1,m}^{QQ}(l)$ are the respective correlators outputs for the mth symbol of the 1st user from the lth multipath and are given by (22)–(24), see below. Similarly, we get Here, $I_1^{II}, I_1^{IQ}(l), I_1^{QQ}(l)$, and $I_1^{QI}(l)$ are the respective interferences due to all multipaths of other (K–1) users. $I_{1,1}^{II}(l), I_{1,1}^{IQ}(l), I_{1,1}^{QQ}(l)$, and $I_{1,1}^{QI}(l)$ are the self-interferences due to a user's own multipaths. $N_1^{II}(l), N_1^{IQ}(l), N_1^{QQ}(l)$, and $N_1^{QI}(l)$ are the respective thermal noises which are considered to be uncorrelated Gaussian random variables. In the next subsection, we shall apply the IC scheme.

B. Interference Cancellation

The IC scheme is a simple successive scheme (as described before), where the strongest user (hence the most reliable one) is decoded first and cancelled from the composite signal. The process is repeated until all the users are decoded. Fig. 13 is the schematic of the DS/CDMA receiver with interference cancellation and Rake receiver for multipath diversity combining, where for the strongest user each multipath that is being tracked by the rake is cancelled from the received signal (at baseband) by using the appropriate correlator outputs. The maximum decision variable output is selected as the strongest user. Say user 1 was the strongest user (selected as $\max_k S_{k,j}$). After decoding the jth symbol (picked as $\max_m (S_{1,m}); m = 1, \ldots M$ of user 1, its signal (all the multipaths of user 1) is regenerated and cancelled from the respective $I\&Q$ components at the baseband level. After making the decision on its symbol, all the L multipaths of user 1 are cancelled from the received signal (at baseband) as follows

$$\begin{aligned} d_1^I(t) &= d^I(t) - \sum_{l=1}^{L} Z_{1,j}^{II}(l)\cdot a_{1,j}^I(t-\tau_1^l) \\ &\quad - \sum_{l=1}^{L} Z_{1,j}^{IQ}(l)\cdot a_{1,j}^Q(t-\tau_1^l) \end{aligned} \tag{25}$$

Note that, out of the N multipaths that are reaching the base station, only L multipaths which are being tracked by the Rake receivers are being cancelled.

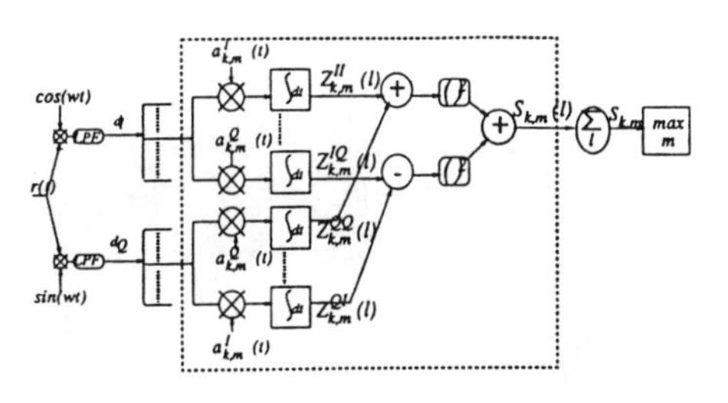

Fig. 12. Noncoherent DS/CDMA receiver.

$$\begin{aligned} Z_{1,m}^{II}(l) &= \frac{1}{T_w}\int_0^{T_w} d^I(t)a_{1,m}^I(t-\tau_1^l)dt \\ &= \begin{cases} \alpha_1^l\sqrt{P_1}\frac{\cos(\phi_1^l)}{2} + I_{1,1}^{II}(l) + I_1^{II}(l) + N_1^{II}(l)\ ; \text{if } m=j \\ I_{1,1}^{II}(l) + I_1^{II}(l) + N_1^{II}(l)\ ; \text{else} \end{cases} \end{aligned} \tag{21}$$

$$Z_{1,m}^{IQ}(l) = \begin{cases} \alpha_1^l\sqrt{P_1}\frac{\sin(\phi_1^l)}{2} + I_{1,1}^{IQ}(l) + I_1^{IQ}(l) + N_1^{IQ}(l)\ ; \text{if } m=j \\ I_{1,1}^{IQ}(l) + I_1^{IQ}(l) + N_1^{IQ}(l)\ ; \text{else} \end{cases} \tag{22}$$

$$Z_{1,m}^{QQ}(l) = \begin{cases} \alpha_1^l\sqrt{P_1}\frac{\cos(\phi_1^l)}{2} + I_{1,1}^{QQ}(l) + I_1^{QQ}(l) + N_1^{QQ}(l)\ ; \text{if } m=j \\ I_{1,1}^{QQ}(l) + I_1^{QQ}(l) + N_1^{QQ}(l)\ ; \text{else} \end{cases} \tag{23}$$

$$Z_{1,m}^{QI}(l) = \begin{cases} -\alpha_1^l\sqrt{P_1}\frac{\sin(\phi_1^l)}{2} + I_{1,1}^{QI}(l) + I_1^{QI}(l) + N_1^{QI}(l)\ ; \text{if } m=j \\ I_{1,1}^{QI}(l) + I_1^{QI}(l) + N_1^{QI}(l)\ ; \text{else} \end{cases} \tag{24}$$

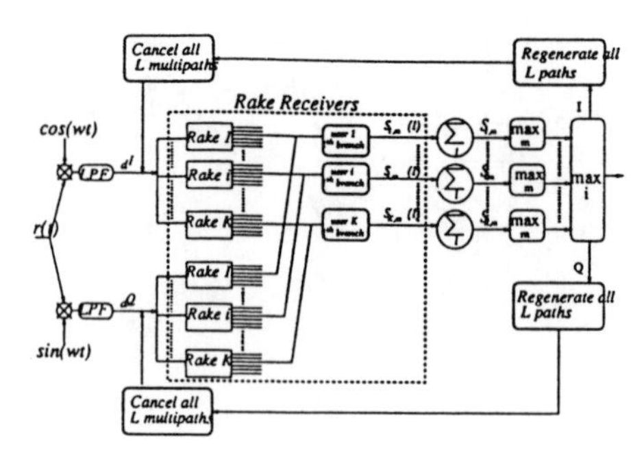

Fig. 13. Noncoherent DS/CDMA receiver with interference cancellation and multipath combining.

Simplifying this expression, we get

$$d_1^I(t) = \sum_{k=2}^{K}\sum_{l=1}^{N} s_{k,j,l}(t-\tau_k) + \frac{n_c(t)}{2} + \text{noise due to cancellation} \quad (26)$$

where $s_{k,j,l}(t-\tau_k)$ is the signal from l^{th} multipath of the kth user which is not yet decoded and cancelled.

Explicitly,

$$\begin{aligned} d_1^I(t) = &\sum_{k=2}^{K}\sum_{l=1}^{N} s_{k,j,l}(t-\tau_k^l) + \frac{n_c(t)}{2} \\ &+ \sum_{l=L+1}^{N} s_{1,j,l}(t-\tau_1^l) \\ &- C_1^I a_{1,j}^I(t-\tau_1^l) - C_1^Q a_{1,j}^Q(t-\tau_1^l) \end{aligned} \quad (27)$$

In this expression, the first term is the signals from all the multipaths of all other remaining $K-1$ users, the second term is the in-phase component of Gaussian noise, the third term is the contribution of the uncancelled multipaths of user # 1, and the last two terms are the noise due to imperfect cancellation, where C_1^I and C_1^Q are as follows:

$$C_1^I = \sum_{l=1}^{L}\{I_{1,1}^{II}(l) + I_1^{II}(l) + N_1^{II}(l)\}$$

and

$$C_1^Q = \sum_{l=1}^{L}\{I_{1,1}^{IQ}(l) + I_1^{IQ}(l) + N_1^{IQ}(l)\} \quad (28)$$

Similarly, $d_1^Q(t)$ is obtained.

Now for the next strongest user, there are only $K-2$ interferers (since we effectively cancelled out the strongest user, though not perfectly). The correlations for the next strong user (user #2) are given by

$$\begin{aligned} Z_{2,m}^{II}(l) &= \frac{1}{T_w}\int_{\tau_2^l}^{\tau_2^l+T_w} d_1^I(t)a_{2,m}^I(t-\tau_2^l)\,dt \\ &= \begin{cases} \alpha_2^l\sqrt{P_2}\frac{\cos(\phi_2^l)}{2} + C_2^I(l); & \text{if } m=j \\ C_2^I(l); & \text{else} \end{cases} \end{aligned} \quad (29)$$

where $C_2^I(l)$ is as follows

$$\begin{aligned} C_2^I(l) = &I_{2,2}^{II}(l) + I_2^{II}(l) + N_2^{II} - C_1^I I_{1,2} \\ &- C_1^Q I_{1,2} + R_1^I + R_1^Q \end{aligned} \quad (30)$$

where $R_1^{I/Q}$ is the noise due to the uncancelled multipaths of user 1 and are given by

$$R_1^{I/Q} = \sum_{l=L+1}^{N} \alpha_1^l\sqrt{P_1}\cdot I_{1,2}\cdot\frac{\cos(\phi_1^l)}{2} \quad (31)$$

$I_{1,2}$ is defined as follows

$$I_{1,2} = \frac{1}{T_w}\int_{\tau_1}^{\tau_1+T_w} a_{1,m}^{I/Q}(t-\tau_1)\cdot a_{2,j}^{I/Q}(t-\tau_2)dt \quad (32)$$

Similarly, $Z_{2,m}^{IQ}(l)$, $Z_{2,m}^{QI}(l)$, and $Z_{2,m}^{QQ}(l)$ are obtained. The decision variable for the second user is then obtained as

$$S_{2,m} = \sum_{l=1}^{L} S_{2,m}(l) \quad (33)$$

where

$$S_{2,m}(l) = \left(Z_{2,m}^{II}(l)+Z_{2,m}^{QQ}(l)\right)^2 + \left(Z_{2,m}^{IQ}(l)-Z_{2,m}^{QI}(l)\right)^2 \quad (34)$$

Now the second cancellation is given by

$$\begin{aligned} d_2^I(t) = &d_1^I(t) - \sum_{l=1}^{L} Z_{2,j}^{II}(l).a_{2,j}^I(t-\tau_2^l) \\ &- \sum_{l=1}^{L} Z_{2,j}^{IQ}(l)\cdot a_{2,j}^Q(t-\tau_2^l) \end{aligned} \quad (35)$$

Hence, following the same procedure, where for the hth cancellation we have

$$\begin{aligned} d_h^I(t) = &d_{h-1}^I(t) - \sum_{l=1}^{L} Z_{h,j}^{II}(l)\cdot a_{h,j}^I(t-\tau_h^l) \\ &- \sum_{l=1}^{L} Z_{h,j}^{IQ}(l)\cdot a_{h,j}^Q(t-\tau_h^l) \end{aligned} \quad (36)$$

Simplifying, we get

$$\begin{aligned} d_h^I(t) = &\sum_{k=h+1}^{K}\sum_{l=1}^{N} s_{k,j,l}(t-\tau_k^l) + \frac{n_c(t)}{2} \\ &+ \sum_{k=1}^{h}\sum_{l=L+1}^{N} s_{k,j,l}(t-\tau_k^l) - \sum_{i=1}^{h}\{C_i^I a_{i,j}^I(t-\tau_i^l) \\ &+ C_i^Q a_{i,j}^Q(t-\tau_i^l)\} \end{aligned} \quad (37)$$

where $C_i^I = \sum_{l=1}^{L} C_i^I(l)$ and $C_i^Q = \sum_{l=1}^{L} C_i^Q(l)$. Similarly, $d_h^Q(t)$ is obtained.

Hence, after h cancellations only $K-h$ users are remaining. The correlations for the $h+1^{\text{st}}$ user are obtained as follows:

$$\begin{aligned} Z_{h+1,m}^{II}(l) &= \frac{1}{T_w}\int_{\tau_{h+1}^l}^{\tau_{h+1}^l+T_w} d_h^I(t)a_{h+1,m}^I(t-\tau_{h+1}^l)dt \\ &= \begin{cases} \alpha_{h+1}^l\sqrt{P_{h+1}}\frac{\cos(\phi_{h+1}^l)}{2} + C_{h+1}^I(l)\ ; & \text{if } m=j \\ C_{h+1}^I(l)\ ; & \text{else} \end{cases} \end{aligned} \quad (38)$$

where $C_{h+1}^I(l)$ is as follows

$$C_{h+1}^I(l) = I_{h+1,h+1}^{II}(l) + I_{h+1}^{II}(l) + N_{h+1}^{II} - \sum_{i=1}^{h}[C_i^I I_{1,2} + C_i^Q I_{1,2}] + \sum_{i=1}^{h}\left[R_i^I + R_i^Q\right] \quad (39)$$

In this expression, the first term is the self-interference of the $h+1^{\text{st}}$ user (which is negligible), the second term is the multiple access interference from the remaining K–h–1 users, the third term is the thermal (Gaussian) noise, and the last two terms are the noise due to imperfect cancellations in the previous stages.

Further analysis of the terms in (39) yields the following (a detailed analysis can be found in the Appendix). Denoting the variance of the noise in the correlator output as follows, we get

$$\begin{aligned}\eta_{h+1}^I(l) &= \text{Var}\left[Z_{h+1\cdot m}^{II}(l)\right] \\ &= \frac{N_o}{4T_w} + \frac{1}{6N_c}\sum_{\substack{n=1 \\ n\langle\rangle l}}^{N} P_{h+1}\left(\alpha_{h+1}^l\right)^2 \\ &+ \frac{1}{6N_c}\sum_{k=h+2}^{K}\sum_{n=1}^{N} P_k(\alpha_k^n)^2 + \frac{4}{3N_c}\sum_{i=1}^{h} L\cdot\eta_i^I(l) \\ &+ \frac{1}{6N_c}\sum_{k=1}^{h}\sum_{n=L+1}^{N} P_k(\alpha_k^n)^2 \end{aligned} \quad (40)$$

where N_c is the chips/symbol. Conditioned on α_k^n, $\left(Z_{h+1,m}^{II}(l) + Z_{h+1,m}^{QQ}(l)\right)$ is approximated as a Gaussian random variable with variance η_{h+1}, given by

$$\begin{aligned}\eta_{h+1} &= \eta_{h+1}^I(l) + \eta_{h+1}^Q(l) = 2\eta_{h+1}^I(l) \\ &= \frac{N_o}{2T_w} + \frac{1}{3N_c}\sum_{\substack{n=1 \\ n\neq l}}^{N} P_{h+1}\left(\alpha_{h+1}^n\right)^2 \\ &+ \frac{1}{3N_c}\sum_{k=h+2}^{K}\sum_{n=1}^{N} P_k(\alpha_k^n)^2 + \frac{4}{3N_c}\sum_{i=1}^{h} L\eta_i \\ &+ \frac{1}{3N_c}\sum_{k=1}^{h}\sum_{n=L+1}^{N} P_k(\alpha_k^n)^2 \end{aligned} \quad (41)$$

Similarly, $\left(Z_{h+1,m}^{IQ}(l) - Z_{h+1,m}^{QI}(l)\right)$ is also a Gaussian random variable with variance η_{h+1}. Hence, $S_{h+1,m}(l)$ becomes a Chi-square distributed random variable with two degrees of freedom. Assuming that the α_n^k are i.i.d. random variables, then the variance $\eta_h + 1$ is independent of l (i.e., same for all multipaths). Therefore, the decision variable of the $h+1^{\text{st}}$ user becomes a Chi-square distributed random variable with $2L$ degrees of freedom. The recovered power of the $h+1^{\text{st}}$ user by tracking L multipaths is given by

$$c_{h+1}^2 = \sum_{l=1}^{L} P_{h+1}\left(\alpha_{h+1}^l\right)^2 \quad (42)$$

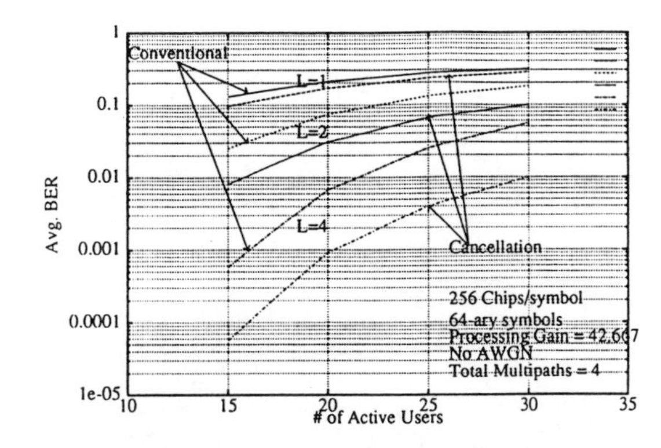

Fig. 14. Average bit error rate versus number of active users under ideal power control.

The probability of correct decision of the $h+1^{\text{st}}$ user is given by [26], [25]

$$P_c^{h+1} \int_0^\infty \left[1 - e^{-\frac{s_1}{2\eta_{h+1}}}\sum_{l=0}^{L-1}\frac{s_1^l}{(2\eta_{h+1})^l l!}\right]^{M-1} \cdot \frac{1}{(2\eta_{h+1})} \cdot \left(\frac{s_1}{c_{h+1}^2}\right)^{\left(\frac{L-1}{2}\right)} \cdot e^{-\frac{c_{h+1}^2+s_1}{2\eta_{h+1}}} \cdot I_{L-1}\left(\frac{\sqrt{c_{h+1}^2 s_1}}{\eta_{h+1}}\right) . ds_1 \quad (43)$$

where $I_L(.)$ is the modified Bessel function of the Lth order. The bit error probability, *conditioned on* c_{h+1}^2 is then given by

$$P_b^{h+1} = \frac{2^{d-1}}{M-1}(1 - P_c^{h+1}) \quad (44)$$

where $d = \log_2 M$.

C. Performance with Multipath Combining

Fig. 14 is obtained by numerically evaluating (43), (44) for ideal power control (i.e., c^2 are equal for all users). As seen in the figure, the performance of both the conventional receiver and the IC scheme improves as the number of multipaths tracked (L) increases. Also note that the performance improvement of the IC scheme over the conventional receiver also increases as the number of multipaths tracked increases.

Note that the expression for bit error, after the h cancellations, is conditioned on the ordered set of c^2. We shall use order statistics (as described before) to obtain the average bit error rate under fading. As c^2's are the combined mean squares of the amplitudes of all multipaths that are being tracked, its distribution depends on the distribution of the α_k's. Assuming α_k's to be lognormal, we can make the approximation that c^2 (which is a sum of lognormal r.v's) is also lognormal (this was justified in [27]). We, thus, consider c^2 to be lognormally distributed with mean μ and variance σ^2. The pdf and cdf of c^2 is, hence, $f(x) = \frac{1}{\sqrt{2\pi}\xi x}e^{-\frac{(\log(x)-\zeta)^2}{2\sigma^2}}$ and $F(x) = \Phi\left(\frac{\log(x)-\zeta}{\xi}\right)$, respectively. ζ and ξ are the mean and standard deviation of the normal distribution from which the lognormal distribution is obtained. The pdf of ordered c_k^2, $f_{c_k^2}(x)$, is then obtained as described earlier [using (11)].

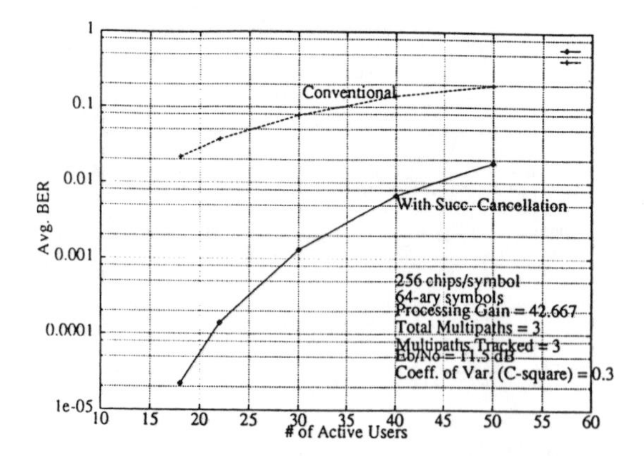

Fig. 15. Average bit error rate versus number of active users under fading.

Denoting the mean and standard deviation of the kth strongest c^2 as μ_k and σ_k, we can obtain the average BER at the kth cancellation stage, using the following approximation

$$\widehat{P_b^k} \approx \frac{2}{3}P_b^k(\mu_k) + \frac{1}{6}P_b^k(\mu_k + \sqrt{3}\sigma_k) + \frac{1}{6}P_b^k(\mu_k - \sqrt{3}\sigma_k) \quad (45)$$

This approximation has shown good accuracy in other applications (see [28]) and, more importantly, for a closely related problem with lognormal random variables [25]. For simplifying the analysis, we shall assume that there are three resolvable multipaths and all three paths are being tracked. The combined energy c^2 is assumed to be lognormally distributed with the coefficient of variation $\left(cv = \left(\frac{\sigma}{\mu}\right)\right)$ equal to 0.3. In Fig. 15, we have plotted the average BER versus the number of active users. The performance improvement of the IC scheme over the conventional scheme is evident.

D. Issues in Practical Implementations

There are two different ways of implementing the successive IC scheme, as seen in Fig. 16. One way is to successively regenerate the strong user's signal using the output of the correlator and its chip sequence, and to cancel it from the received signal and proceed likewise. An alternative way to perform this cancellation is to obtain the cross correlation (by using separate integrators which are correlating only the chip sequences of the respective users at their respective timings) and then update the output of the correlator by using the cross correlation information. Thus, this method operates at the bit level. Both these methods are equivalent analytically. The number of operations required by both methods are the same. The difference is that, in the former method, the received signal itself is being updated (as signals are being cancelled) while in the latter method the output of the correlators are being updated. The latter method has the flexibility of performing the added integrations (cross correlations) separately, whereas the former method performs the additional integrations as the users' are decoded and cancelled successively. The advantages and disadvantages of each method are being studied.

We shall assume for now that the received signal is being successively updated as the strongest user is being cancelled. We shall consider that each Rake receiver has correlators

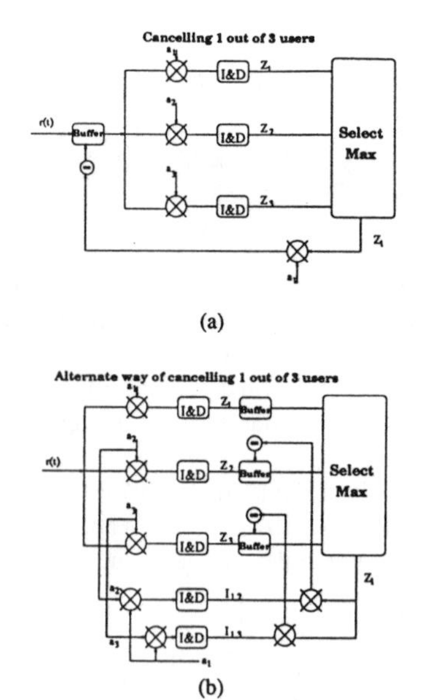

Fig. 16. Two methods of implementing successive IC schemes.

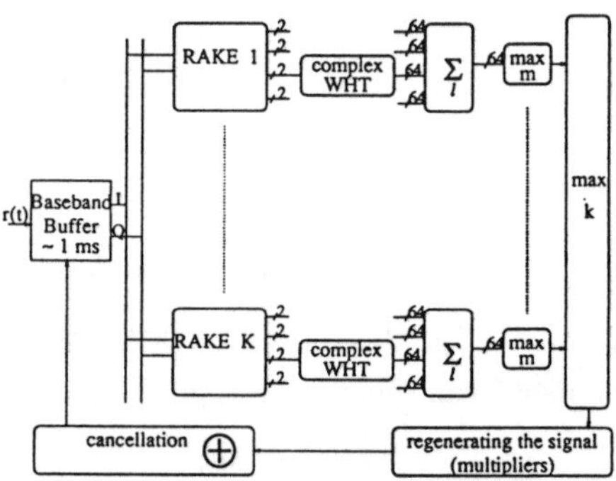

Fig. 17. Interference cancellation and multipath combining.

in parallel for tracking each multipath. A complex Walsh-Hadamard Transform (WHT) generates 64 coefficients for each multipath, corresponding to the 64 possible symbols. As can be seen from Fig. 17, for each user, each multipath that is being tracked has a Walsh decoder. The output from the WHT's from each multipath are then combined coherently, resulting in a single set of 64 coefficients. The maximum of these 64 coefficients determines the symbol sent by that particular user. Until this point, the receiver is the conventional DS/CDMA receiver.

The signal of the strongest user is then decoded and regenerated for cancellation. Note that the process of regenerating the signal and cancellation requires simple multipliers and adders. The processing delay is mainly limited by the speed of performing one Walsh-Hadamard Transform (WHT). Since successive cancellation is involved, the possible number of cancellations is limited by the speed of performing one WHT. In order to ensure a regular flow of symbols at the symbol rate R_s, the speed of the WHT must be at least $K \cdot R_s$ (where K is the possible number of cancellations). For example, in order to have at least 100 cancellations, the speed of the WHT must be at least 0.16 MHz (assuming a bit rate of 9.6

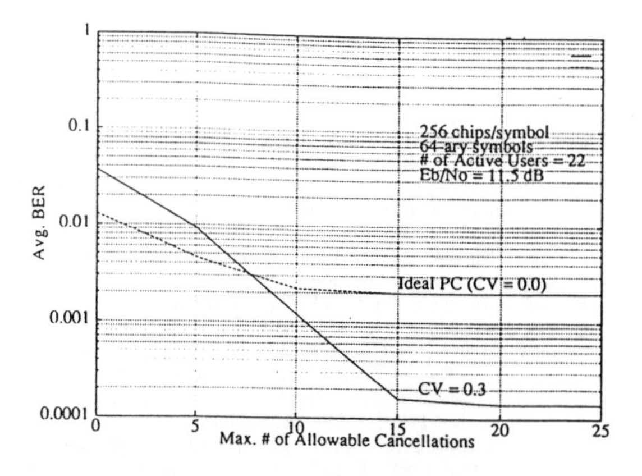

Fig. 18. Average bit error rate versus number of active users under fading.

kb/s), i.e., each WHT should take less than 6.25 μs. Thus, the processing speed of the hardware may limit the number of possible cancellations. The impact of limiting the number of cancellation on the BER performance of the IC scheme is depicted in Fig. 18. As the number of cancellations increases, the average BER decreases. Hence, there is a tradeoff of complexity versus performance. Faster processing implies a higher number of allowable cancellations which implies better BER performance.

V. Conclusion

Through this work, we have shown that by using a simple successive IC scheme, one can effectively estimate and cancel a CDMA signal and thus substantially reduce near/far effects from a CDMA system and increase the system capacity. The performance of the IC scheme with multipath diversity combining was also studied. Further work needs to be done in analyzing the processing delay involved and the practical implementation of the interference canceller as well as sensitivities to errors (such as delay tracking and quantization).

Appendix

A. Noise Analysis

In (39), the thermal noise $N_{h+1}^{II}(l)$ is expressed as follows

$$N_{h+1}^{II}(l) = \frac{1}{2T_w}\int_{\tau_{h+1}^2}^{T_w+\tau_{h+1}^2} n_c(t)a_{h+1,j}^I(t-\tau_{h+1}^l)\,dt$$
$$= \frac{1}{2T_w}\sum_{u=0}^{N_c-1}\int_{uT_c+\tau_{h+1}^2}^{(u+1)T_c+\tau_{h+1}^2} n_c(t)a_{h+1,j}^I(t-\tau_{h+1}^l)\,dt$$
$$= \frac{1}{2T_w}\sum_{u=0}^{N_c-1}\int_{uT_c+\tau_{h+1}^2}^{(u+1)T_c+\tau_{h+1}^2} \pm n_c(t)\,dt \quad (46)$$

where $N_c = \frac{T_w}{T_c}$. It can be shown that $N_{h+1}^{II}(l)$ is a Gaussian random variable with zero mean and variance $\frac{N_o}{4T_w}$. Similarly, $N_{h+1}^{IQ}(l), N_{h+1}^{QQ}(l)$, and $N_{h+1}^{QI}(l)$ are also uncorrelated Gaussian random variables with zero mean and variance $\frac{N_o}{4T_w}$.

B. Interference Analysis

In (39), the first term is the self-interference due to a user's own multipaths. The second term in (39) is the multiple access interference after h cancellations and is expressed as

$$I_{h+1}^{II}(l) = \sum_{k=h+1}^{K}\sum_{n=1}^{N}\frac{\sqrt{P_k}\alpha_k^n}{2T_w} \times \int^{T_w+\tau_{h+1}^2} \{[a_{h+1,m}^I(t-\tau_{h+1}^l)a_{k,j}^I(t-\tau_k^n)\cos(\phi_k^n)] + [a_{h+1,m}^I(t-\tau_{h+1}^l)a_{k,j}^Q(t-\tau_k^n)\sin(\phi_k^n)]\}\,dt. \quad (47)$$

The integral is the asynchronous cross correlation of random binary sequences which can be characterized as Gaussian random variable with zero mean and variance $\frac{1}{3N_c}$. Therefore, conditioned on the α_k^n, this multiple access interference is approximated as Gaussian random variable with zero mean and variance

$$\mathrm{Var}\,[I_{h+1}^{II}(l)|\alpha_k^n] = \sum_{k=h+2}^{K}\sum_{n=1}^{N}\frac{P_k(\alpha_k^n)^2}{4}\left[\frac{1}{3N_c}+\frac{1}{3N_c}\right] = \frac{1}{6N_c}\sum_{k=h+1}^{K}\sum_{n=1}^{N}P_k(\alpha_k^n)^2. \quad (48)$$

The self-interference [first term in (39)] is given by (49), see below. Following similar steps, we get

$$\mathrm{Var}\,[I_{h+1,h+1}^{II}(l)|\alpha_{h+1}^n] = \sum_{\substack{n=1\\ n\neq l}}^{N}\frac{P_{h+1}(\alpha_{h+1}^n)^2}{4}\left[\frac{1}{3N_c}+\frac{1}{3N_c}\right] = \frac{1}{6N_c}\sum_{\substack{n=1\\ n\neq l}}^{N}P_{h+1}(\alpha_{h+1}^n)^2. \quad (50)$$

The term $I_{1,2}$, as defined in (32), is also a random variable with zero mean and variance

$$\mathrm{Var}[I_{1,2}] = \mathrm{Var}\left[\frac{1}{T_w}\int_{\tau_1^2}^{T_w+\tau_1^2} a_{2,m}^I(t-\tau_2^l)a_{1,j}^I(t-\tau_1^l)\,dt\right] = \frac{2}{3N_c} \quad (51)$$

The fourth term in (39) is

$$\sum_{i=1}^{h}(C_i^I+C_i^Q)I_{1,2} = \sum_{i=1}^{h}\sum_{l=1}^{L}(C_i^I(l)+C_i^Q(l))I_{1,2} \quad (52)$$

$$I_{h+1,h+1}^{II}(l) = \sum_{\substack{n=1\\ n\neq l}}^{N}\frac{\sqrt{P_{h+1}}\alpha_{h+1}^n}{2T_w}\times\int_{\tau_{h+1}^2}^{T_w+\tau_{h+1}^2}\{[a_{h+1,m}^I(t-\tau_{h+1}^l)a_{h+1,j}^I(t-\tau_{h+1}^n)\cos(\phi_{h+1}^n)] + [a_{h+1,m}^I(t-\tau_{h+1}^l)a_{h+1,j}^Q(t-\tau_{h+1}^n)\sin(\phi_{h+1}^n)]\}\,dt. \quad (49)$$

Denoting $\eta_i^I(l) = \text{Var}[C_i^I(l)|\alpha_k^n]$, and since each multipath that is being tracked is independent of each other, $C_i^I(l)$ and $C_i^Q(l)$ are i.i.d random variables with zero mean and variance $\eta_i^I(l)$. Therefore,

$$\text{Var}\left[\sum_{i=1}^{h}(C_i^I + C_i^Q)I_{1,2}\right] = \frac{4}{3N_c}\sum_{i=1}^{h} L\eta_i^I(l) \quad (53)$$

The last term in (39), which is the uncancelled multipath of the users already decoded, is given by

$$\sum_{i=1}^{h}(R_i^I + R_i^Q) = \sum_{i=1}^{h}\sum_{l=L+1}^{N}\left\{\sqrt{P_i}\alpha_i^l I_{1,2}\frac{\cos(\phi_i^l)}{2} + \sqrt{P_i}\alpha_i^l I_{1,2}\frac{\sin(\phi_i^l)}{2}\right\} \quad (54)$$

This term is also a random variable with zero mean and variance

$$\text{Var}\left[\sum_{i=1}^{h}(R_i^I + R_i^Q)\right] = \frac{1}{6N_c}\sum_{i=1}^{h}\sum_{l=L+1}^{N} P_i(\alpha_i^l)^2 \quad (55)$$

REFERENCES

[1] K. S. Gilhousen, I. M. Jacobs, R. Padovani, A. J. Viterbi, L. A. Weaver, and C. W. III, "On the capacity of a cellular CDMA system," *IEEE Trans. Vehic. Technol.*, vol. 40, no. 2, pp. 303–311, May 1991.

[2] R. Lupas and S. Verdu, "Near-far resistance of multiuser detectors in asynchronous channels," *IEEE Trans Commun.*, vol. 38, pp. 497–507, Apr. 1990.

[3] R. Lupas and S. Verdu, "Linear multiuser detectors for synchronous code-division multiple access channel," *IEEE Trans. Inform. Theory*, pp. 123–136, Jan. 1989.

[4] M. Varanasi and B. Aazhang, "Multistage detection in asynchronous code-division multiple access communications," *IEEE Trans. Commun.*, vol. 38, pp. 509–519, Apr. 1990.

[5] A. Duel-Hallen, "Decorrelating decision-feedback multiuser detector for synchronous code-division multiple access channels," *IEEE Trans. Commun.*, Feb. 1993.

[6] S. Kubota, S. Kato, and K. Feher, "Inter-channel interference cancellation technique for CDMA mobile/personal communications base stations," in *Proc. Int. Symp. Spread Spectrum Tech. and Applicat. (ISSSTA)*, Yokohama, Japan, Dec. 1992, pp. 91–94.

[7] Y. C. Yoon, R. Kohno, and H. Imai, "Combination of an adaptive array antenna and a canceller of interference for direct-sequence spread-spectrum multiple access system," *IEEE J. Select. Areas Commun.*, vol. 8, no. 4, May 1990.

[8] A. Kajiwara and M. Nakagawa, "Spread spectrum block demodulator with high capacity crosscorrelation canceller," in *Proc. GLOBECOM*, 1991.

[9] P. Patel and J. Holtzman, "Performance comparison of a DS/CDMA system using a successive interference cancellation (IC) scheme and a parallel IC scheme under fading," to appear.

[10] A. J. Viterbi, "Very low rate convolutional codes for maximum theoretical performance of spread-spectrum multiple-access channels," *IEEE J. Select. Areas Commun.*, vol. 8, pp. 641–649, May 1990.

[11] P. Dent, B. Gudmundson, and M. Ewerbring, "CDMA-IC: a novel code division multiple access scheme based on interference cancellation," in *Proc. PIMRC*, Boston, MA, Oct. 1992, pp. 4.1.1.–4.1.5.

[12] P. Teder, G. Larsson, B. Gudmundson, and M. Ewerbring, "CDMA with interference cancellation: A technique for high capacity wireless systems," in *Proc. IEEE Int. Conf. Commun.* Geneva, Switzerland, 1993.

[13] P. Patel and J. Holtzman, "Analysis of a simple successive interference cancellation scheme in DS/CDMA system using correlations," in *Proc. GLOBECOM* Houston, TX, 1993.

[14] M. Kawabe, T. Kato, T. Sato, A. Kawahashi, and A. Kukasawa, "Advanced CDMA scheme for PCS based on interference cancellation," in *ICUPC Rec.*, pp. 1000–1003, 1993.

[15] M. B. Pursley, "Performance evaluation for phase-coded spread-spectrum multiple access communication—Part I: System analysis," *IEEE Trans. Commun.*, vol. COM-25, 1977.

[16] J. Holtzman, "On calculating DS/SSMA error probabilities," in *Proc. Int. Symp. Spread-Spectrum Tech. and Applicat.*, Yokohama, Japan, Dec. 1992.

[17] S. Vasudevan and M. K. Varanasi, "Multiuser detectors for asynchronous CDMA communication over Rician fading channels," in *Proc. Commun. Theory Mini Conf.*, Orlando, FL, Dec. 1992.

[18] Y. C. Yoon, R. Kohno, and H. Imai, "Cascaded co-channel interference cancelling and diversity combining for spread spectrum multi-access over multipath fading channels," *IEICE Trans. Commun.*, no. 2, pp. 163–168, Feb. 1993.

[19] Z. Zvonar and D. Brady, "Optimum detection in asynchronous multiple access multipath Rayleigh fading channels," to appear.

[20] H. A. David, *Order Statistics*. New York: Wiley, 1981.

[21] J. M. Holtzman and L. M. A. Jalloul, "Rayleigh fading effect reduction with wideband DS/CDMA signals," in *Proc. IEEE GLOBECOM Conf.*, Phoenix, AZ, Dec. 1991, pp. 16.7.1–16.7.5. To appear in *IEEE Trans. Commun.*.

[22] Qualcomm Inc., "Wideband spread spectrum digital cellular system," proposed EIA/TIA Interim Standard, Apr. 21, 1992.

[23] K. Kim, "On the error probability of a DS/SSMA system with a non-coherent m-ary orthogonal modulation," in *Proc. IEEE Vehic. Technol. Conf.*, Denver, CO, May 1992, pp. 482–485.

[24] Q. Bi, "Performance analysis of a CDMA system in the multipath fading environment," in *Proc. IEEE Int. Conf. on Personal, Indoor, and Mobile Radio Commun.*, Boston, MA, Oct. 1992, pp. 108–111.

[25] L. Jalloul and J. Holtzman, "Performance analysis of DS/CDMA system with noncoherent m-ary orthogonal modulation in multipath fading channels," to appear in *J. Select. Areas Commun.*, 1994.

[26] J. G. Proakis, *Digital Communications*. 2nd edn. New York: McGraw-Hill, 1989.

[27] S. C. Schwartz and Y. S. Yeh, "On the distribution function and moments of power sums with log-normal components," *Bell Syst. Tech. J.*, vol. 61, no. 7, Sept. 1982.

[28] J. Holtzman, "A simple, accurate method to calculate spread-spectrum multiple access error probabilities," *IEEE Trans. Commun.*, vol. 40, no. 3, Mar. 1992.

Pulin Patel was born in Mombasa, Kenya, on August 8, 1970. He received the B.E. degree from Stevens Institute of Technology in 1992, and the M.S. degree in electrical engineering from Rutgers University in 1994.

Since May 1992, he has been a Research Assistant at the Wireless Information Network Laboratory (WINLAB) at Rutgers University. He has been working on system-level studies of the cellular CDMA system, and interference cancellation schemes for spread-spectrum systems. His current research interests are in efficient radio resource allocation in integrated systems, multimedia wireless systems, wireless networks, and advanced spread-spectrum techniques.

Jack M. Holtzman received the B.E.E. degree from City College of New York in 1958, the M.S. degree from U.C.L.A. in 1960, and the Ph.D. degrees from the Polytechnic Institute of Brooklyn in 1967.

He worked for AT&T Bell Laboratories for 26 years on control theory, teletraffic theory, telecommunications, and performance analysis. At Bell Labs, he was supervisor of the Applied Mathematics Group and then Head of the Teletraffic Theory and System Performance Department. In 1990, he joined Rutgers University where he is Professor of Electrical and Computer Engineering and Associate Director of the Wireless Information Network Laboratory (WINLAB). He is also the Director of the Wireless Communications Certificate Program. His current areas of work are on spread spectrum, handoffs, mobility management, propagation, and system performance.

Reprinted from *IEEE Personal Communications*, Vol. 2, No. 2, Apr. 1995, pp. 45-58.

Modifying present systems could yield significant capacity increases

Multiuser Detection for CDMA Systems

ALEXANDRA DUEL-HALLEN, JACK HOLTZMAN, AND ZORAN ZVONAR

Spread spectrum has been very successfully used by the military for decades. Recently, spread-spectrum-based code division multiple access (CDMA), has taken on a significant role in cellular and personal communications. Multiple access allows multiple users to share limited resources such as frequency (bandwidth) and time. There are a number of multiple access schemes including more than one type of CDMA. We shall concentrate on one type, direct sequence CDMA (DS/CDMA). CDMA has been found to be attractive because of such characteristics as potential capacity increases over competing multiple access methods, anti-multipath capabilities, soft capacity, and soft handoff.

We shall not cover all the background of DS/CDMA, since that has been well explained in recent literature (e.g., [1]). In fact, this article may be viewed as a supplement to that literature with an update on some potential enhancements to the versions of DS/CDMA currently being developed [2-4]).[1] We will show that there is a natural modification of the present systems that is potentially capable of significant capacity increases. By "natural modification" we mean a modification that can be made conceptually clear, not that it is easy to implement. Indeed, the optimal multiuser detector is much too complex and most of the present research addresses the problem of simplifying multiuser detection for implementation. The objective of this article is to make the basic idea intuitive and then show how investigators are trying to reduce the idea to practice. We also indicate multiuser receiver structures with potentially acceptable levels of complexity and address potential obstacles for achieving theoretically predicted performance in practice. As a result of these investigations, an answer to the following question is expected: Is there a suboptimal multiuser detector that is cost effective to build with significant enough performance advantage over present day systems? A definitive answer is not yet available.

We will first review some salient features of CDMA systems needed for the discussion to follow.

Limitations of a Conventional CDMA System

A conventional DS/CDMA system treats each user separately as a signal, with the other users considered as either interference, e.g.,Multiple Access Interference (MAI), or noise. The detection of the desired signal is protected against the interference due to the other users by the inherent interference suppression capability of CDMA, measured by the processing gain. The interference suppression capability is, however, not unlimited and as the number of interfering users increases, the equivalent noise results in degradation of performance, i.e., increasing bit error rate (BER) or frame error rate. Even if the number of users is not too large, some users may be received at such high signal levels that a lower power user may be swamped out. This is the *near/far effect*: users near the receiver are received at higher powers than those far away, and those further away suffer a degradation in performance. Even if users are at the same distance, there can be an effective near/far effect because some users may be received during a deep fade. DS/CDMA systems are very sensitive to the near/far effect and the recent success of DS/CDMA has, in large part, been due to the successful implementation of relatively tight power control, with attendant added complexity. There are thus two key limits to present DS/CDMA systems:

- All users interfere with all other users and the interferences add to cause performance degradation.
- The near/far problem is serious and tight power control, with attendant complexity, is needed to combat it.

Multipath Propagation

One other aspect of CDMA that we need to review is the ability to combat multipath reception of signals [6]. Due to multiple reflections, the received signal contains delayed, distorted replicas of the original transmitted signal. First, consider what happens in a non-spread-spectrum system. When the multiple reflections, called multipath signals or simply multipaths, from one transmitted bit are received within the time duration of one bit, the received signal consists of the superposition of several signal replicas, each with its own amplitude and phase. It is important to recognize that this superposition is the addition of complex quantities. Due to the motion of the mobile (or, even of the base station in some systems), the relative phases of the received signals are continually changing. This results in successive reinforcement and interference of the superposed multipath signals, resulting in very large time variations in the received signal. Such variations are referred to as Rayleigh fading (or Rician fading, if there is a direct component in addition to the reflections). The variations due to Rayleigh fading are a serious

The work of the first two authors has been supported by an NSF TIE Project Award, No. EEC 9416209.

[1] *A comprehensive reference set on spread spectrum until 1985 is [5].*

 1070-9916/95/$04.00 © 1995 IEEE

cause of performance degradation and a communication system must be designed carefully, taking that into account.

We shall refer to systems where all of the multipath signals arrive within one bit interval as "narrowband." On the other hand, the bit rate may be so high that multipath signals from one bit arrive over a duration longer than that of one bit. Such systems will be called "wideband." The Rayleigh fading effect is less pronounced, because there are fewer multipath signals from one transmitted bit arriving during the bit duration.

CDMA systems are inherently wideband when the chip duration, as opposed to the longer bit duration, is compared to the time between multipath receptions. One can then combat multipath interference by multipath reception, whereby the different multipath arrivals are considered as independent receptions of the signal and are used to give a beneficial time diversity. This is usually done with a RAKE receiver, the name apparently taken from the action of a rake with a number of teeth pulling in a number of items simultaneously. So, instead of multipath being just a source of performance degradation, the multipaths are used to provide the benefit of diversity [7].

Interference Cancellation and Multiuser Detection

In a conventional CDMA system, all users interfere with each other. Potentially significant capacity increases and near/far resistance can *theoretically* be achieved if the negative effect that each user has on others can be canceled. A more fundamental view of this is multiuser detection, in which all users are considered as signals for each other. Then, instead of users interfering with each other, they are all being used for their mutual benefit by joint detection. The drawback of optimal multiuser detection is one of complexity so that suboptimal approaches are being sought. There is a wide range of possible performance/complexity combinations possible. Much of the present research is aimed at finding an appropriate tradeoff between complexity and performance.

Multiuser Detection in Cellular Systems

In a cellular system, a number of mobiles communicate with one base station (BS). Each mobile is concerned only with its own signal while the BS must detect all the signals. Thus, the mobile has information only about its own chip sequence while the base station has the knowledge of all the chip sequences. For this reason, as well as less complexity being tolerated at the mobile (where size and weight are critical), multiuser detection is currently being envisioned mainly for the BS, or in the reverse link (mobile to BS). It is important to realize, however, that the BS maintains information only on those mobiles in its own cell. This plays a role in the limitations on improvements to be expected in a multiuser detection system, to be discussed next.

Limitations to Improvements

Before we discuss multiuser improvements to the conventional DS/CDMA detector, it is important to define factors that limit such improvement [1, 8]. One factor is intercell interference in a system that cancels only the intracell interference[2] I. For intercell interference which is a fraction f of the intracell interference, the bound of capacity increase (all of the intracell interference is canceled) is $(1 + f)/f$. For $f = 0.55$, this factor is 2.8 [1]. Observe that with a sectorized antenna, it is conceivable to cancel users from another sector and thus improve the bound.

Another limiting factor is the fraction f_c of energy captured by a RAKE receiver. That is, a RAKE receiver with L branches or "fingers" will try to capture the power in the L strongest multipath rays, but there will be additional received power in additional rays. For the conventional detector, this is self-interference. Reference [10] gives examples of the fraction of captured power. The fraction of captured power is a function of chip rate and delay spread as well as the number of RAKE branches. So, combining the two effects (measured by f and f_c), the total interference before cancellation is $(1 + f)I$ (neglecting the smaller self-interference due to uncaptured multipath power of the desired user). Cancellation removes at most $f_c I$ so the bound on improvement is $(1 + f)/(1 + f - f_c)$. For $f_c \approx 1$, the above bound of 2.8 on capacity improvement remains. For $f_c = 0.5$, the bound is reduced to 1.5.

We will show that there is a natural modification of the present systems that is potentially capable of significant capacity increases. As a result of these investigations, an answer to the following question is expected: Is there a suboptimal multiuser detector that is cost effective to build with significant enough performance advantage over present day systems?

It should be recognized that multiuser detection is used not only to increase capacity but also to alleviate the near/far problem, and the preceding bound does not account for that benefit. Relaxing the power control requirement actually translates into a capacity benefit which is, however, more difficult to quantify than by the above simple signal/interference argument. A multiuser detector could recapture part of this reduction by reducing variability (or relax the requirements on power control).

To put these constraints on improvements into further perspective, we are assuming here that multiuser detection is a candidate primarily for the reverse link for reasons given earlier. Since the reverse link is usually more limiting than the forward link,[3] increasing the reverse link capacity will improve the overall system capacity. But increasing it beyond the forward link capacity will not further increase the overall system capacity. Thus,

- The potential capacity improvements in cellular systems are not enormous (order of magnitude) but certainly nontrivial.
- Enormous capacity improvements only on the reverse link (the candidate for multiuser detection) would only be partly used anyway in determining overall system capacity.
- Hence, the cost of doing multiuser detection must be as low as possible so that there is a per-

[2] Intracell interference is from interferers in the same cell as the desired user while intercell interference is from interferers outside the cell. It has been proposed that intercell interference be canceled by explicitly communicating this information, or by adaptive or blind methods (see [9]). This research is at an earlier stage.

[3] Some cases in which the forward link appear to be limiting are given in [62].

formance/cost tradeoff advantage to multiuser detection.

The bottom line is that there are significant advantages to multiuser detection which are, however, bounded and a simple implementation is needed.

Historical Background

The idea of interference cancellation arises in many contexts, e.g., noise cancellation in speech [11] and adaptive interference canceling as in Chapter 12 of [12]. There are thus a number of non-CDMA references with ideas similar to those being currently studied for CDMA. We should distinguish between canceling noise which has no useful purpose (as in [11] and Chapter 12 of [12]) from canceling interference which is due to other signals which are themselves to be detected. A couple of non-CDMA examples in the latter category are [13-15]. The CDMA case considered here is of the second type, where the signals being canceled are of interest also. It should be remarked, however, that the first type of cancellation also is of importance in CDMA systems, e.g., in suppressing narrowband interference (this is not discussed in this article, but is discussed in Section 5 of [3]). Both types of interference cancellation have in common the goal of removing from a desired signal a noise-like interference. But in the second type (the type considered here), the fact that the signals being removed are themselves information carrying leads to a new viewpoint, that of *simultaneously* detecting all the information carrying signals.

In a conventional CDMA system, all users interfere with each other. In multiuser detection, all users are considered as signals for each other. The drawback of optimal multiuser detection is one of complexity so that suboptimal approaches are being sought.

The first CDMA interference cancellation references we are aware of are [16, 17]. Both of these papers delineate a number of ideas that are present in much of the ongoing research. Estimates based on mean square error and maximum likelihood are discussed in [16]. Reference [17] shows how cancellation is implemented by solving simultaneous equations, in essence, by inverting a key matrix. There were subsequently a number of papers with variants of the ideas of [16, 17]. Significant theoretical steps forward were taken in [18, 19] (with earlier references), in analyzing the structure and complexity of optimal receivers. This work triggered a new research effort on suboptimal algorithms. The strong connection between MAI and intersymbol interference (ISI) was also made in [18]. There are aspects of MAI, however, that are not shared by ISI:

- The near/far problem.
- MAI is affected by the relationship among user chip sequences (codes) as well as by the imperfections of the radio channel, while ISI is due only to the channel.

Thus, while equalization (used to combat ISI) will play a role in the multiuser detectors to be discussed, it should not be expected that it can be used without modification.

Recent survey papers include [8, 9] with many further references. The rest of the references here are cited as needed in the discussion.

Multiuser Detection: Concept and Techniques

The CDMA Channel Model and Approaches to Detection

A CDMA channel with K users sharing the same bandwidth is shown in Fig. 1. The signaling interval of each user is T seconds, and the input alphabet is antipodal binary: $\{+1, -1\}$. The objective is to detect those polarities, which contain the transmitted information. During the n-th signaling interval, the input vector is $x_n = (x_n^1, \dots, x_n^K)^T$, where x_n^k is the input symbol of the k-th user. User k ($k = 1, \dots, K$) is assigned a signature waveform (or code, or spreading chip sequence) $s_k(t)$ which is zero outside $[0, T]$ and is normalized

$$\int_0^T s_k(t)^2 \, dt = 1$$

Pulse amplitude modulation is employed at the transmitter. The baseband signal of the k-th user is

$$u_k(t) = \sum_{i=0}^{\infty} x_i^k c_i^k s_k(t - iT - \tau_k), \qquad (1)$$

where τ_k is the transmission delay, and c_i^k is the complex channel attenuation. According to (1), each user's signal travels along a single path, so this model does not illustrate multipath propagation. The effect of multipath is discussed in the section on noncoherent multiuser detection. For synchronous CDMA, the delay $\tau_k = 0$ for all users. For asynchronous CDMA, the delays can be different. The channel attenuation is a complex number

$$c_i^k = \sqrt{w_i^k} \exp(j\theta_i^k),$$

where w_i^k and θ_i^k are the received power and phase of the k-th user, respectively. The received signal (at baseband) is the noisy sum of all the users' signals:

$$y(t) = \sum_{k=1}^{K} u_k(t) + z(t), \qquad (2)$$

where $z(t)$ is the complex additive white Gaussian noise (AWGN). The first step in the detection process is to pass the received signal $y(t)$ through a matched filter bank (or a set of correlators). It consists of K filters matched to individual signature waveforms followed by samplers at instances $nT + \tau_k$, $k = 1, \dots, K$, $n = 1, 2, \dots$. The outputs of the matched filter bank form a set of sufficient statistics about the input sequence x_n given $y(t)$ [18]. Thus, we will consider the equivalent discrete-time channel model which arises at the output of the matched filter bank.

For the rest of this section, we will concentrate on a very simplified DS/CDMA system. (There are a number of simplifications which will be exposed in the rest of the article. In fact, each relaxation of simplification will represent another factor to consider for the multiuser detection system.) The *simplifying assumptions* are as follows:

We Consider *Real* Channel Attenuations – The real model is convenient for analyzing coherent methods, and can be easily generalized to the complex case. In the following section, we extend our treatment to multiuser detectors for fading channels, where complex attenuations need to be considered.

Derivation of Multiuser Detectors is Presented for *Synchronous* CDMA System – The synchronous assumption considerably simplifies exposition and analysis and often permits the derivation of closed-form expressions for the desired performance measures. These are useful since similar trends are found in the analysis of the more complex asynchronous case. Furthermore, every asynchronous system can be viewed as an equivalent synchronous system with larger effective user population [20], which is often explored in burst CDMA communications. Moreover, synchronous systems are becoming more of practical interest since quasi-synchronous approach has been proposed for satellite [21] and microcell applications [22]. It should be recognized, however, that the transition from synchronous to asynchronous can considerably increase the complexity of multiuser detection. Throughout the paper, we address implementation issues and complexity increase for various detectors for the asynchronous model.

Certain Parameters are *Known* Exactly – Although multiuser detectors presented in this section take advantage of completely known amplitudes, and phases and delays do not appear in the treatment at all, the next section addresses non-coherent detectors which do not require the knowledge of amplitudes and phases. Sensitivity and robustness are discussed in a subsection titled "Issues in Practical Implementations."

For synchronous CDMA, the output signal $y(t)$ (2) for $nT \leq t < (n+1)T$ does not depend on the inputs of other users sent during past or future time intervals. Consequently, it is sufficient to consider a one-shot system with input vector $x_n = (x^1, \ldots, x^K)^T$, real positive channel attenuations (amplitudes) $c^1 = \sqrt{w^1}, \ldots, c^K = \sqrt{w^K}$ and real additive white Gaussian noise $z(t)$ with power spectral density N_0. The sampled output of the k-th matched filter (matched to the signature waveform of user k) is

$$y_k = \int_0^T y(t)s_k(t)dt = \int_0^T s_k(t)\left[\sum_{j=1}^{K} s_j(t)c^jx^j + z(t)\right]dt$$
$$= c^kx^k + \sum_{j\neq k}^{K} x^jc^j \int_0^T s_k(t)s_j(t)dt + \int_0^T s_k(t)z(t)dt \quad (3)$$

Note that y_k consists of three terms. The first is the desired information which gives the sign of the information bit x_k (which is exactly what is sought). The second term is the result of the multiple access interference (MAI), and the last is due to the noise. The second term typically dominates the noise so that one would like to remove its influence. Its influence is felt through the cross-

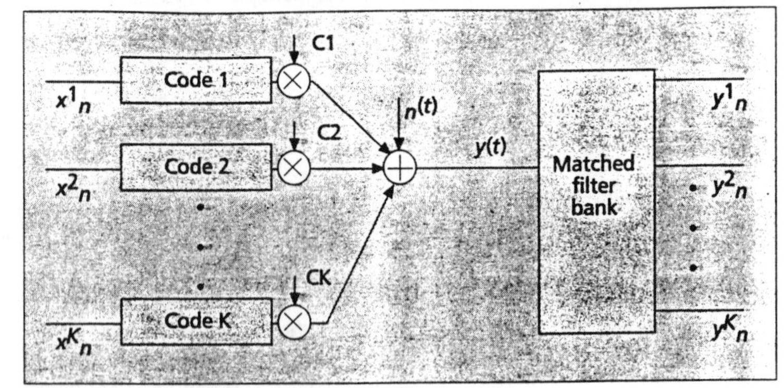

■ **Figure 1.** *The CDMA channel model.*

correlations between the chip sequences and the powers of users. If one knew the cross-correlations and the powers, then one could attempt to cancel the effect of one user upon another. This is, in fact, the intuitive motivation for interference cancellation schemes.

Suppose there are only two users in the system. Let r be the cross-correlation between the signature waveforms of the two users

$$r = \int_0^T s_1(t)s_2(t)dt.$$

In this case, the outputs of the matched filters are

$$y^1 = c^1x^1 + rc^2x^2 + z^1 \text{ and } y^2 = c^2x^2 + rc^1x^1 + z^2. \quad (4)$$

The MAI terms for users 1 and 2 are rc^2x^2 and rc^1x^1, respectively. If these terms were not present, the single user system would result. The bit error rate of the optimal detector for the single user system serves as a lower bound on the performance of any other detector. This single user bound is

$$P_k(E) = Q\left(\sqrt{\frac{w^k}{N_0}}\right), \quad (5)$$

where the Q-function

$$Q(x) = \frac{1}{\sqrt{2\pi}}\int_x^\infty \exp\left(\frac{-y^2}{2}\right)dy.$$

The conventional DS/CDMA uses the same approach as the optimal receiver for the single user system. It detects the bit from user k by correlating the received signal with the chip sequence of user k. Thus, the conventional detector makes its decision at the output of the matched filter bank:

$$\hat{x}^k = \text{sgn}(y^k). \quad (6)$$

When MAI terms are significant, as shown in (3), the bit error rate of this detector is high. Note that MAI depends both on the cross-correlations and the powers of users. In the 2-user example above, if user 1 is much stronger than user 2 (the near/far problem), the MAI term rc^1x^1 present in the signal of the second user is very large, and can significantly degrade performance of the conventional detector for that user. A multiuser detector called a successive interference canceller (decision-directed) can remedy this problem as follows. First, a decision $\hat{x}^1$ is made for the stronger user 1

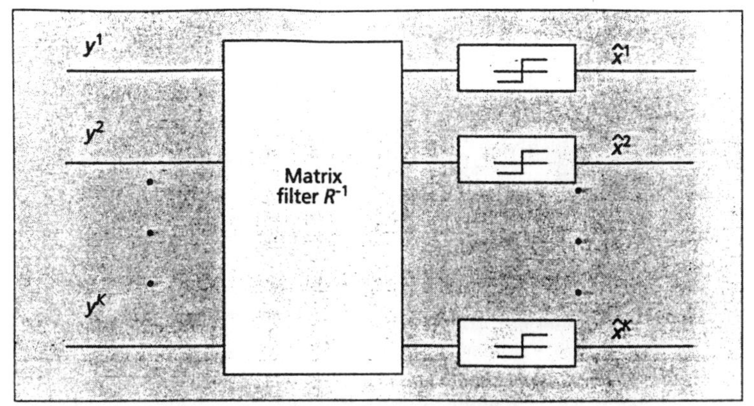

■ **Figure 2.** *The decorrelator for synchronous CDMA.*

using the conventional detector. Since user 2 is much weaker then user 1, this decision is reliable from the point of view of user 2. So, this decision can be used to subtract the estimate of MAI from the signal of the weaker user. The decision for user 2 is given by

$$\begin{aligned}\hat{x}^2 &= \text{sgn}(y^2 - rc^1x^1) \\ &= \text{sgn}(c^2x^2 + rc^1(x^1 - \hat{x}^1) + z^2)\end{aligned} \quad (7)$$

Provided the decision of the first user is correct, all MAI can be subtracted from the signal of user 2.[4] If we fix the energy of the second user, and let the energy of the first user grow, the error rate of the successive interference canceller for the second user will approach the single-user bound. Thus, this detector is successful in combating the near/far problem. This simple example motivates the use of multiuser detectors for CDMA channels. Below, we will discuss several previously proposed multiuser detectors.

The Decorrelating Detector

As a step towards the most general formulation, consider the matrix version of the equivalent discrete time model (3). The output vector $y = [y^1, y^2, \ldots, y^K]^T$ can be expressed as

$$y = RWx + z, \quad (8)$$

where R and W are $K \times K$ matrices, and z is a colored Gaussian noise vector. The components of the matrix R are given by cross-correlations between signature waveforms

$$R_{k,j} = \int_0^T s_k(t) s_j(t) dt. \quad (9)$$

The second matrix W is diagonal with $W_{k,k}$ given by the channel attenuation c_k of the k-th user. For example, in a two-user system, the matrix

$$R = \begin{pmatrix} 1 & r \\ r & 1 \end{pmatrix},$$

where r is the cross-correlation between the signature waveforms of the users (9).

Inspection of (8) immediately suggests a method to solve for x, whose components x_k contain the bit information sought. If z was identically zero, we have a linear system of equations, $y = RWx$, the solution of which can be obtained by inverting R (it is invertible in most cases of interest [23]). With a non-zero noise vector z, inverting R is still an effective procedure and actually optimal in certain circumstances, to be discussed later. This results in

$$\tilde{y} = R^{-1}y = Wx + \tilde{z} \quad (10)$$

where it is seen that the information vector x is recovered but contaminated by a new noise term (Fig. 2). From (10), the signal of the k-th user is

$$\tilde{y}^k = c^k x^k + \tilde{z}^k. \quad (11)$$

The decision is $\hat{x}^k = \text{sgn}(\tilde{y}^k)$.

Note that the decorrelating detector completely eliminates MAI. However, the power of the noise $\tilde{z}^k$ is $N_0 (R^{-1})_{k,k}$ which is greater than the noise power N_0 at the output of the matched filter (8). For example, for the two-user system with the cross-correlation r, the noise power at the output of the decorrelating filter is $N_0/(1-r^2)$. The error rate of the decorrelator is given by

$$P_k(E) = Q\left(\sqrt{\frac{w_k}{N_0 R^{-1}_{k,k}}}\right) \quad (12)$$

The performance of the decorrelating detector degrades as the cross-correlations between users increase. In the asynchronous case the decorrelating detector also reduces to matrix inversion in a burst type communications, or is given by linear, time-invariant K-input K-output filter for the infinite length transmitted data [20]. In both cases the complexity of the detector grows and several approaches have been proposed to reduce the complexity, as addressed later.

The decorrelator has several desirable features. It does not require the knowledge of the users' powers, and its performance is independent of the powers of the interfering users. This can be seen from (11). The only requirement is the knowledge of timing which is anyway necessary for the code despreading at the centralized receiver. Observe that neither signal nor noise terms depend on the powers of interferers. In addition, when users' energies are not known, and the objective is to optimize performance for the worst case MAI scenario, the decorrelator is the optimal approach [23]. In addition, the noncoherent version of the decorrelator has been developed (see the following section). These properties of the decorrelator make it very well suited for the near/far environment.

Multiuser detection is closely related to equalization for intersymbol interference (ISI) channels [7]. For example, the decorrelating detector is analogous to the zero-forcing equalizer. Similarly, the MMSE linear multiuser detector [24] (also given by a matrix inverse) is the multidimensional version of the MMSE linear equalizer for the single-user ISI channel. The linear structure of these detectors often limits their performance. In the following section, we will describe several non-linear approaches to multiuser detection.

The Optimal Detector

The objective of maximum-likelihood sequence estimation (MLSE) is to find the input sequence which maximizes the conditional probability, or likelihood of the given output sequence [7]. For the simplified synchronous CDMA problem discussed above, the maximum likelihood decision

[4] *Note, also, that r and c^1 are assumed to be known exactly for this example. We shall return to this issue.*

for the vector of bits x is given by

$$\hat{x} = \arg\left\{ \max_{x \in \{-1,+1\}^K} \left[2y^T Wx - b^T WRWb \right] \right\} \quad (13)$$

This equation dictates a search over the 2^K possible combinations of the components of the bit vector x. For asynchronous CDMA, the MLSE detector can be implemented using the Viterbi algorithm [18]. The path metrics of this algorithm were derived by identifying the asynchronous CDMA channel with a single-user channel with periodically time-varying ISI. The memory of this equivalent channel is K-1 (the number of interferers), and therefore the resulting Viterbi algorithm has 2^{K-1} states and requires K storage updates per transmission interval. Although the optimal detector has excellent performance, it is too complex for practical implementation, and we will not discuss it in greater detail. A suboptimal detector which uses a sequential decoder instead of the Viterbi algorithm was presented in [25].

Non-Linear Suboptimal Multiuser Detectors

In this section, we will consider several interference cancellation methods which utilize feedback to reduce MAI in the received signal. These algorithms can be broken into three classes:

- Multistage detectors, e.g., [24, 26-28, 33]
- Decision-feedback detectors [29-31].
- Successive interference cancellers (this idea is explicit or implicit in a number of papers).

Note: this classification is to facilitate exposition. The three categories are not actually disjoint and particular realizations of suboptimal detectors may use combinations of the three classes.

The first two classes of algorithms are decision-directed. They utilize previously made decisions of other users to cancel interference present in the signal of the desired user. These algorithms require estimation of channel parameters and coherent detection. The algorithms in the third class can use soft decisions (e.g., outputs of the correlation receivers as in [32]) rather than hard decisions to remove MAI components. They lend themselves to noncoherent implementation. The algorithms of the second and third classes employ successive interference cancellation (also proposed in [33]), which requires ordering of users according to their powers. The signals of stronger users are demodulated first and canceled from the signals of weaker users. This technique provides an efficient and practical solution to the near/far problem.

Several representatives from the three classes of non-linear detectors are described below.

Multistage Detectors – A multistage detector (Fig. 3) proposed in [26] uses (14) instead of (13):

$$\hat{x}_k(n) = \arg\left\{ \max_{\substack{x_k \in \{-1,+1\} \\ x_l = \hat{x}_l(n-1), l \neq k}} \left[2y^T Wx - b^T WRWb \right] \right\} \quad (14)$$

The n-th stage of this detector uses decisions of the (n-1)-st stage to cancel MAI present in the

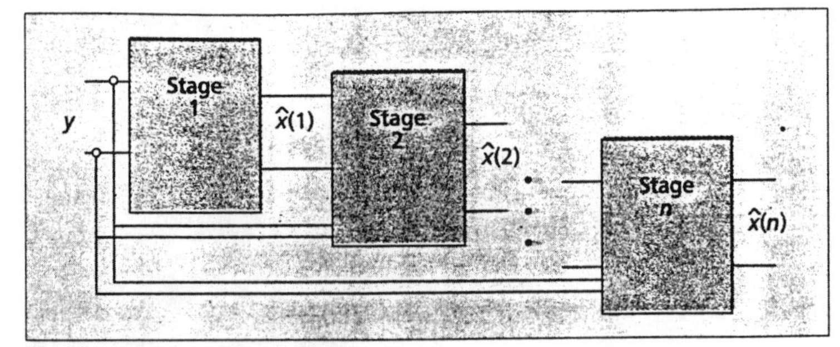

■ **Figure 3.** *The multistage detector.*

received signal. Thus, maximization is over one bit at a time, instead of over k bits, as in (13). Due to delay constraints, it is desirable to limit the number of stages to two. For example, consider a two-stage detector with the conventional first stage for the synchronous two-user system with the cross-correlation r(4). The decisions produced by the first stage (conventional) detector are $\hat{x}^1(1)$ and $\hat{x}^2(1)$ computed as in (6). The decisions of the second stage are $\hat{x}^1(2) = \text{sgn}[y^1 - rc^2\hat{x}^2(1)]$ and $\hat{x}^2(2) = \text{sgn}[y^2 - rc^1\hat{x}^1(1)]$. The performance of this two-stage detector depends on the relative energies of the users. Clearly, if the first user is stronger than the second, the decisions of the second stage for user 2 agree with those of the decision-directed successive interference canceller, described in the last paragraph of the section on the CDMA channel model. Thus, for the weaker user, the second stage produces more reliable decisions than the first stage. However, for the stronger user, feedback might not be beneficial since the decision produced by the conventional detector for the weaker user is poor. More reliable two-stage detector results if the conventional detector in the first stage is replaced by the decorrelator [28]. This example illustrates the issues which play a role in the design of multistage detectors. In summary, the two important questions are:

- How to choose the initial stage.
- How to choose the subsequent stages of processing.

A discussion of different options for the initial and subsequent stages is given, along with further references, in [27].

Decision-Feedback Detectors – The detectors proposed in [29-31] are multiuser decision-feedback equalizers, characterized by two matrix transformations: a forward filter and a feedback filter. These detectors are analogous to the decision-feedback equalizers employed in single user ISI channels [7]. However, in addition to equalization, the decision-feedback multiuser detectors employ successive cancellation. In each time frame, decisions are made in the order of decreasing user's strength, i.e., the stronger users make decisions first, allowing the weaker users to utilize these decisions. The sorting is performed by any multiuser detector with successive MAI cancellation. We will explain the rationale for using this particular order in the next section.

A diagram of the decorrelating (zero-forcing) decision-feedback detector for synchronous CDMA [30] is shown in Fig. 4. At the output of the sorter, users are ranked according to their

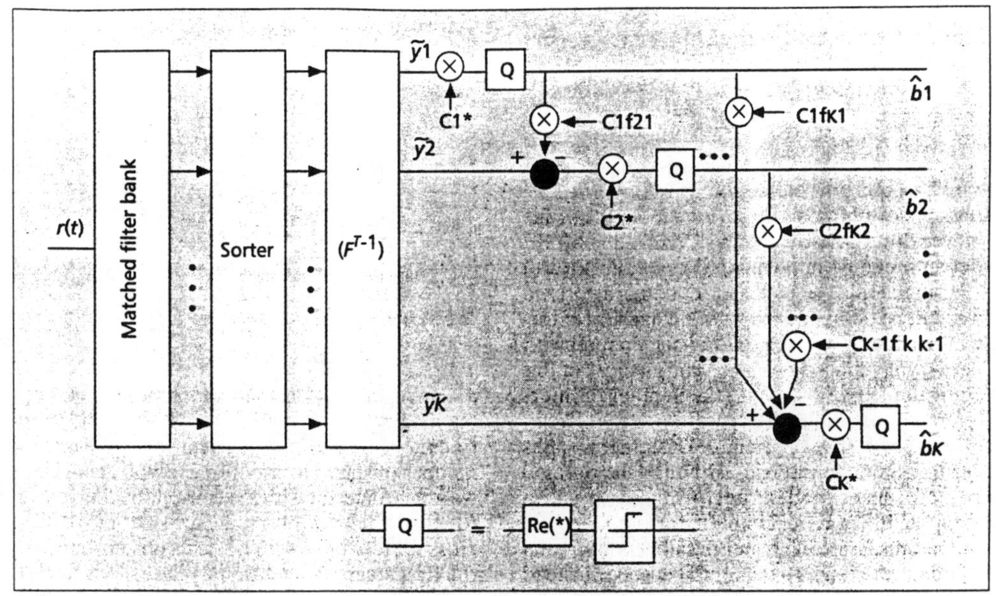

■ **Figure 4.** *The decorrelating decision-feedback detector for Synchronous CDMA.*

powers, so that the strongest user is ranked first, and the weakest is ranked last. Following the sorter, a noise whitening filter is applied. This filter is obtained by Cholesky factorization of the correlation matrix, which yields a resulting MAI matrix that is lower triangular. Consequently, at the output of the whitening filter, the signal of the k-th strongest user $\tilde{y}_k$ is given by:

$\tilde{y}_k$ = desired signal
+ MAI due to stronger users $(1,\ldots, k - 1)$ + noise.

In particular, the signal of the strongest user $\tilde{y}$ is not corrupted by MAI, and can be demodulated first. This decision is then used to subtract MAI from the signal of the second user, and so on. For the asynchronous CDMA, several decision-feedback detectors were derived in [29, 31].

The performance of the decision-feedback detector is similar to that of the decorrelator for the strongest user, and gradually approaches the single user bound as the user's power decreases relative to powers of interferers. Thus, for the decision-feedback detector, performance advantages with respect to the conventional or the decorrelating detectors are greater for relatively weaker users. This is also the case for multistage detectors with the decorrelating first stage. Figure 5 depicts typical performance of several detectors for the weakest user in a bandwidth efficient system. The signature waveforms for this asynchronous four-user CDMA system were derived from Gold sequences of length 7 [31]. (see also [28, 30].) In Fig. 5, the conventional, decorrelating, decision-feedback and multistage detectors are compared for the weakest user (user 4). The powers of all users grow, but the differences between the powers remains the same. Note that the two-stage detector with the conventional first stage is interference-limited. Both the decision-feedback and the two-stage detector with the decorrelating first stage have excellent performance in this near/far scenario.

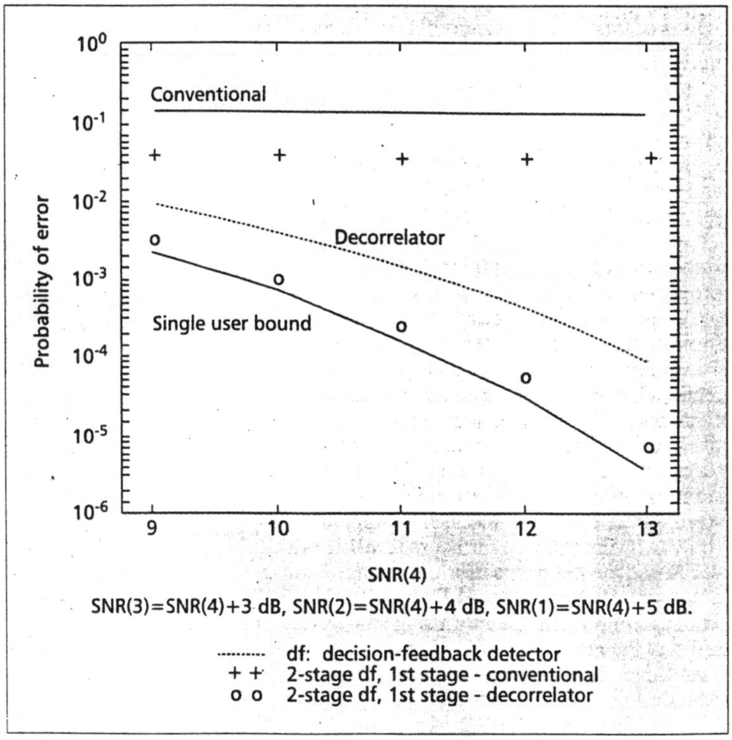

■ **Figure 5.** *Error rates for user 4 in the four-user system of [31].*

Successive Interference Cancellers – One approach to successive interference cancellation is to consider what would be the simplest augmentation to the conventional detector which would achieve some of the benefits of multiuser detection. This can be explained most simply by referring back to (3). In order to cancel the MAI, the factors $x^j c^j$ are needed, in addition to the cross-correlations. These can be obtained either with estimates of each of the factors x^j and c^j separately, i.e., separation of the bit estimate and power estimates. Alternately, one can estimate the product $x^j c^j$ directly by using the correlator output. We shall focus on the latter method because that requires the simplest augmentation to the conventional detector. It is found that using the correlator output to estimate $x^j c^j$ is sufficiently accurate to obtain improvement over the conventional detector.

As mentioned previously, it is important to cancel the strongest signal before detection of the other signals because it has the most negative effect. Also, the best estimate of signal strength is from the strongest signal for the same reason that the best bit decision is made on that signal: the strongest signal has the minimum MAI, since the strongest signal is excluded from its own MAI. This is the twofold rationale for doing successive cancellation in order of signal strength:

- Canceling the strongest signal has the most benefit.
- Canceling the strongest signal is the most reliable cancellation.

In a number of studies (see references in [8], it has been shown that this method of cancellation yields significant improvements over the conventional detector (but substantially less than the optimum multiuser detector).

Successive cancellation works by successively subtracting off the strongest remaining signal. An alternative (the parallel method) is to simultaneously subtract off all of the users' signals from all of the others. It is found [34] that when all of the users are received with equal strength, the parallel method outperforms the successive scheme (Fig. 6). When the received signals are of distinctly different strengths (the more important case), the successive method is superior in performance (Fig. 7). The important thing to note is that in both cases, both the successive and parallel interference cancellers outperform the conventional detector and the unequal power case is the more important case.

The successive cancellation must operate fast enough to keep up with the bit rate and not introduce intolerable delay. For this reason, it will presumably be necessary to limit the number of cancellations. The ability to limit the number of cancellations is consistent with the objective of controlling complexity by choosing an appropriate performance/complexity tradeoff. For more information and references on successive interference cancellation, see [8].

Multiuser Detectors for Encoded Data

Error-control codes are essential for reliable performance of cellular systems. There has not been much work so far on performance of multiuser detectors for encoded signals. When convolutional codes are employed by all users, the MLSE detector is given by the Viterbi algorithm which is more complex than the optimal detector discussed previously for the uncoded case (due to the additional memory associated with each user) [35]. Since the MLSE detector is too complex to implement, several suboptimal methods were addressed in [33, 36-40]. In [33], successive cancellation technique was presented for a CDMA system with orthogonal convolutional coding. [37] discussed multistage detection for convolutionally encoded signals. The authors divided various approaches to multiuser detection for encoded signals into two classes. The first class contains the partitioned approaches, in which a multiuser detector precedes the decoder and does not utilize the decoded data. In the algorithms of the second class, the integrated approaches, the decoded symbols of the interferers are used for MAI cancellation in the signal of the desired user. Reduced complexity receivers, combined with decoders which incorporate reliability information,

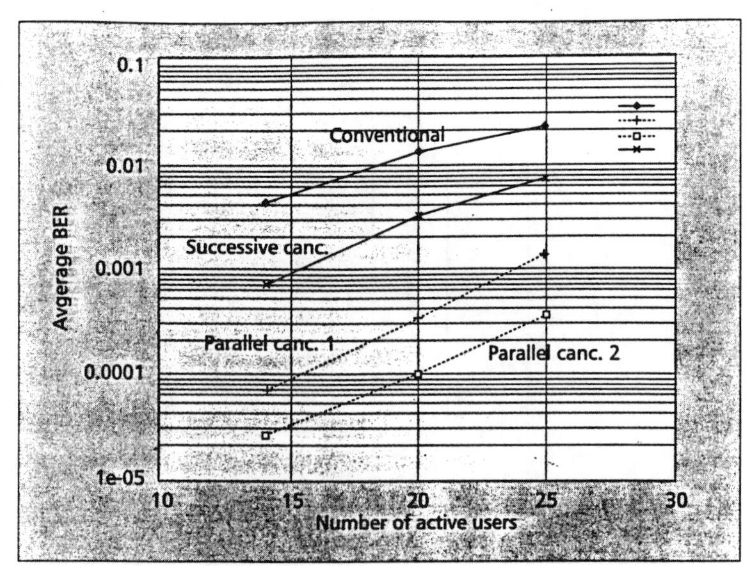

■ **Figure 6.** *BER vs. no. of active users under ideal power control (asynchronous).*

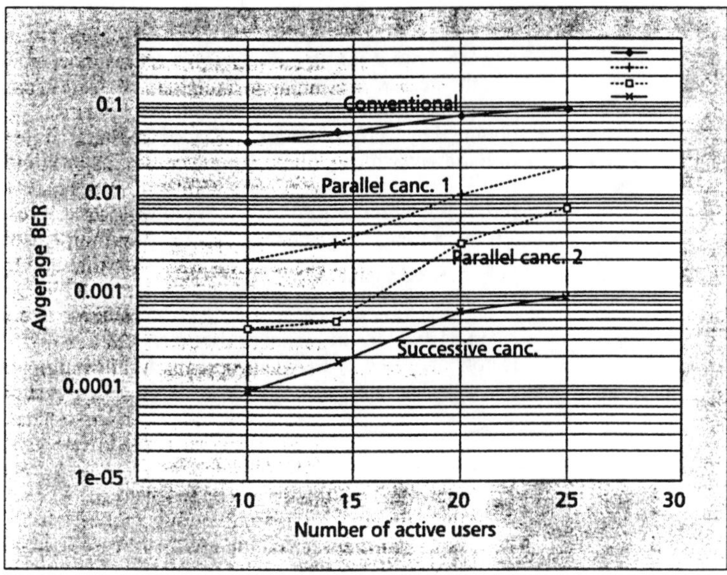

■ **Figure 7.** *BER vs. no. of active users under Rayleigh fading (asynchronous).*

were presented in [36, 38]. Applicability of turbo-codes to CDMA mobile radio system using joint detection was demonstrated in [40]. Reference [39] addressed combined multistage detection and trellis-coded modulation.

One issue to be cognizant of is that error control allows operation at a lower SNR, which can reduce the improvements available with some multiuser detectors.

Noncoherent Multiuser Detection and Multipath Fading Channels

For the sake of simplicity of exposition, most of the discussion in the second section concentrated on coherent multiuser reception. The underlying assumption is that the multiuser receiver is able to estimate and track the phase of each active user in CDMA scenario. However, as stated

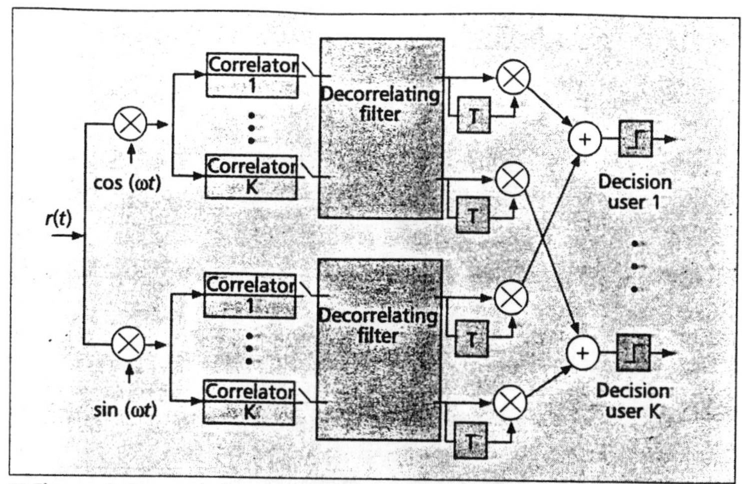

■ **Figure 8.** *Decorrelating multiuser detector for DPSK signals.*

earlier, the reverse link of a cellular CDMA system employs noncoherent reception since the pilot signal which provides a coherent reference is not available.[5] Therefore, the concept of noncoherent multiuser detection and multiuser receiver performance in multipath fading channels are of particular interest for practical CDMA systems.

As discussed in the section on multipath propagation, multipath fading presents a major limitation to the performance of wireless CDMA systems such as cellular mobile radio, indoor wireless communications, and personal communication services. In these systems MAI is enhanced by multipath propagation, and near/far effects are produced not only by the difference in the distance between the transmitter and receiver, but also by the fading on the propagation paths. While multipath propagation, usually encountered in an urban scenario, offers inherent diversity, certain scenarios in suburban areas may result in a single-path propagation [6, 41]. This depends on whether the individual multipaths can be resolved i.e., whether the chip duration is small enough relative to the separation between multipaths. In the case of single-path propagation, small but nonzero cross-correlations among chip sequences can cause a severe near/far problem in the presence of fading. When there is only a single fading path for each active user and interference is relatively strong, there are no means of diversity to overcome fading of the desired signal below the level of MAI, unless the explicit diversity using distributed antennas is introduced. Since multiuser receivers alleviate the near/far problem, they significantly improve the CDMA system performance in the single-path scenario [42-45, 61].

When the chip duration is small enough to resolve the different multipath receptions, multipath diversity is exploited to improve the performance in the presence of MAI. The conventional receiver in the case of multipath fading channel consists of a bank of RAKE receivers, one for each active user, at the base station and one for the desired user in a mobile. A summary of the research efforts on conventional reception techniques is given in [46]. Consequently, having in mind the multipath combining property of a RAKE receiver, multiuser techniques in multipath fading channels utilize some form of RAKE structure at the receiver front-end. With multipath resolution available, multiuser detection and multipath diversity reception are combined to provide the reliable receiver performance. An optimal MLSE receiver for multipath fading CDMA channel presented in [47] consists of the same front-end of coherent RAKE filters, followed by a dynamic programming algorithm of the Viterbi type. This optimal structure provides the same order of diversity and asymptotically has the same error probability as a RAKE receiver in the single-user case at the expense of high complexity which is again exponential in the number of active users.

All noncoherent multiuser techniques perform in-phase and quadrature (I & Q) demodulation before the detection. The noncoherent multiuser detection was first considered for differentially phase-shift keying (DPSK) systems [48]. The concept can be easily described as presented in Fig. 8. The decorrelating operation is performed both on in-phase and quadrature signal branches (after I & Q demodulation), so the phase information of the signal is preserved, although the phase is not being explicitly tracked. Since the decorrelating filter eliminates MAI, the same decision logic as in the single-user receiver can be applied for DPSK demodulation. Moreover, this type of detector was shown to be optimally near/far resistant.

The resulting expression for the error probability indicates that the performance loss compared to coherent reception is the same as in a single-user channel [48]. The realization and performance of this noncoherent decorrelating receiver do not depend on signal amplitudes and phases.

Since the RAKE receiver can be interpreted as a combiner of the correlators outputs and the combining method depends on the modulation type, the concept of noncoherent linear multiuser detection can be extended to the multipath fading scenario. To eliminate the effects of MAI prior to the combining process, linear multipath decorrelating receivers [49, 50] perform the decorrelating operation on KL correlator outputs, where L is the number of resolvable fading paths and K is the number of active users as depicted in Fig. 9. Consequently, the equal-gain diversity combining for DPSK signaling is performed on signals which suffer from less interference. The performance loss due to the noise enhancement in the presence of other active users is modest, for the typical mobile radio scenarios it is on the order of 3 dB compared to single-user RAKE performance over the whole range of SNRs [51]. Similar performance degradations are observed for the coherent linear multiuser receivers [52]. When a fraction of the multipath power is captured due to limited number of RAKE correlators, only a portion of MAI is eliminated by decorrelating operation and residual MAI may cause the performance degradation (recall the section on limitations to improvements). In that case additional antenna diversity was shown to be effective in reducing the effects of the residual MAI [51]. Another combination of interference cancellation and antenna diversity was analyzed in [53].

Interference cancellation techniques for multipath fading channels inherently employ the regeneration of the interfering signals. The major difference among numerous interference cancellation techniques is in the methods for the channel parame-

[5] *This is the case in [2]. Other approaches are currently being investigated.*

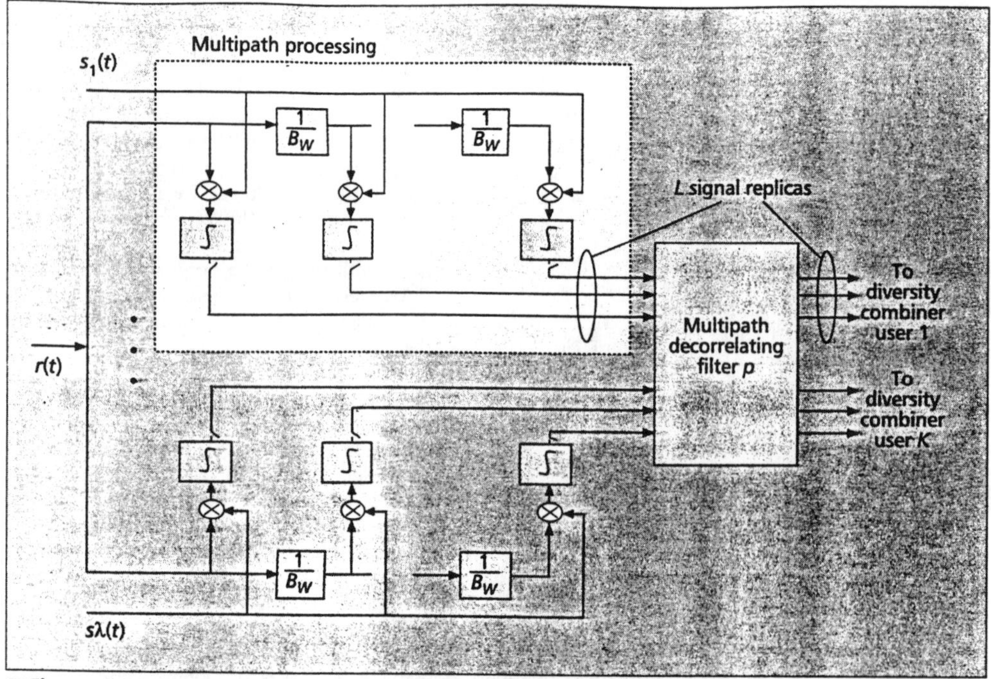

■ **Figure 9.** *Multipath decorrelating/multipath combining linear multiuser receiver for DPSK signals.*

ters estimation and interfering signal reconstruction. A successive interference cancellation receiver for noncoherent M-ary orthogonal modulation, which uses the outputs of the correlators to estimate the signal amplitudes and hence does not require any separate channel estimates, is given in [32]. A noncoherent version of successive interference cancellation employs a combination of M-ary orthogonal signaling and CDMA on the reverse link. After the *I & Q* demodulation at the receiver front-end, the received signal is correlated with respective *I & Q* chip sequences, the user's spreading code, and with all M-ary symbols obtained from Walsh functions. The interference cancellation algorithm starts by decoding the strongest user first. The amplitude of the decoded user is estimated from the correlator output and the strongest user's signal is regenerated using this estimate and the corresponding chip sequence, and canceled from the received signal. The cancellation is repeated until all users are decoded or until a limited number of cancellations are done. Since RAKE receivers are used in the front-end in CDMA multipath fading channels, each multipath arrival tracked by RAKE is canceled from the received signal using the appropriate correlator output as shown in Fig. 10.

Again, for any of the multipath multiuser detectors, the residual MAI problem could arise when the number of canceled paths is smaller than the total number of multipaths in the channel.

Issues in Practical Implementations

Complexity

Major obstacles to the application of the multiuser detectors in practical wireless systems are processing complexity and possible processing delay. The optimal multiuser MLSE receiver is clearly too complex for any application in a system with a large user population, and most of the present efforts are focused on implementation of suboptimal structures. However, even the suboptimal structures can lead to unacceptable levels of complexity. For example, the decorrelating detector in the asynchronous case results in a *K*-input *K*-output filter implementation which is stable but noncausal, so the appropriate delay has to be inserted [20]. In addition, changes in timing and addition/removal of new users results in time-varying coefficients of the detector.

The bottom line is that there are significant advantages to multiuser detection which are, however, bounded and a simple implementation is needed.

To overcome these difficulties, the sliding window decorrelating algorithm has been proposed as a practical alternative, both for infinite and finite data block lengths in asynchronous CDMA systems. Rather than having the total length of the received signal available for the construction of cross-correlation parameters, only a finite-length window of the signal is used for decorrelating operation with the correction of the edge effects. The resulting algorithm has to solve the linear system of equations described in [50]. The major cost in each iteration is due to recomputation of the linear system due to the change in relative delays among users, dynamic selection of multipath, and voice activity exploration. The computationally efficient algorithm for updating the coefficients of the decorrelating filter is proposed by exploiting the parallelism in the linear system solution [54]. The number of operations for the correction of the matrix filter coefficients is on the order of *KN*, where *K* is the number of active users in CDMA system, and *N* is the length of the sliding window. The readily available technology for high-speed zero-forcing equalizers makes the decorrelator

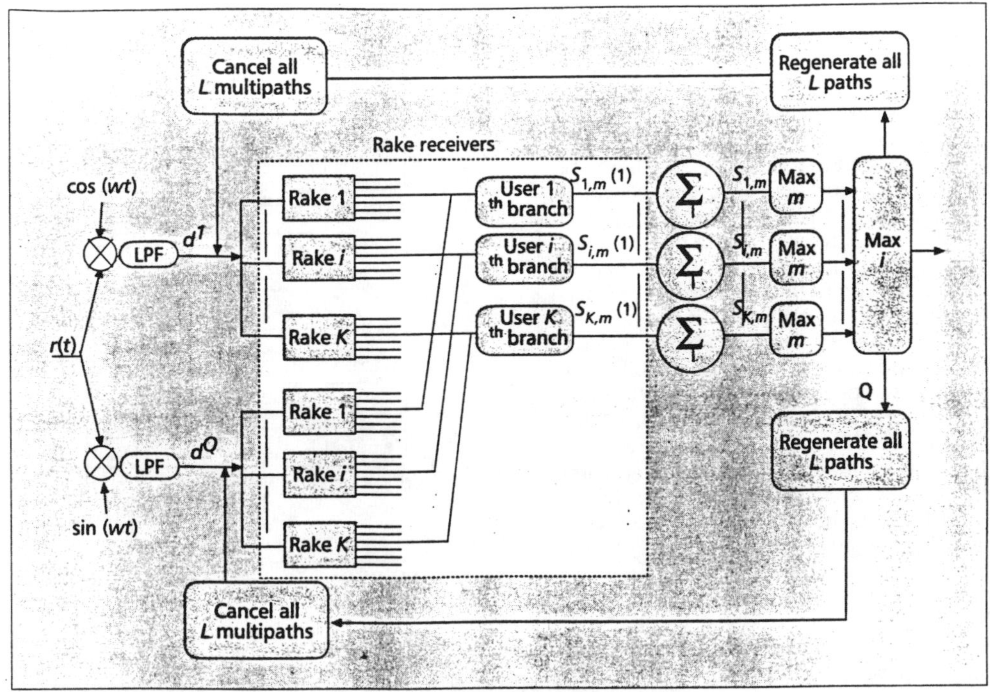

■ **Figure 10.** *Noncoherent receiver with interference cancellation and multipath combining.*

one of the simplest interference cancellation techniques to implement [54].

The successive interference cancellation scheme of the section on successive interference cancellers may be the simplest augmentation to the conventional detector. The major complexity in the multipath environment comes from tracking each multipath at the RAKE receiver, which involves, for each path M, coefficients corresponding to the possible M orthogonal symbols [32]. These coefficients are generated by a Walsh-Hadamard Transform (WHT) and the speed of performing the transform determines the processing delay of the canceller. To ensure real-time implementation, the cancellations, including the WHTs, must be done fast enough to keep up with the symbol rate in the CDMA system. The complexity of the successive interference cancellation scheme can be controlled by limiting the number of cancellations, i.e., one can achieve a compromise between performance and complexity. Observe that there is decreasing improvement with subsequent cancellations because cancellations are done in order of strength of the different users' signals.

Sensitivity and Robustness

Almost all of the discussions and analyses of multiuser detection have assumed a number of idealizations. While multiuser detection has been analyzed thoroughly for both AWGN and fading channels, less work has been reported on the underlying parameter estimation issues in CDMA channels, and consequently, on the impact of imperfect parameter estimates on the performance of multiuser receivers. Clearly, for any type of tracking error (frequency, amplitude, phase, or timing), the chip sequences being canceled will be offset and imperfect cancellation will be performed. For fading channel applications and noncoherent modulation on the reverse link, timing synchronization plays a vital role. It should also be emphasized that the chip tracking is essential for any type of DS/CDMA system and that tight code synchronization, within a fraction of a chip duration, is required for the reliable operation of the conventional detector. The pertinent question is whether the tracking error tolerable for the conventional detector is tolerable for the cancellation receiver, or how much tighter it must be for multiuser detection.

The problem of the impact of the synchronization errors on the multiuser receiver performance is inherently complex due to the nonlinear nature of the solution and can be analyzed only to a certain degree. Afterwards, one has to rely on simulations which brings another difficulty due to the coupling effect from various parameters of the CDMA system to the receiver performance. We will briefly summarize the current status in this area, which can be described more as the effort of establishing credential methodology, rather than trying to give a definite answer.

In the case of decorrelating detector, tracking errors result in the mismatch of the cross-correlation coefficients among different users. Analysis was performed in [55], assuming the Gaussian distribution of the tracking error and the performance of decorrelating detector was assessed by simulation using the standard deviation as the parameter. In this case, Gold codes of length 15 were employed, with three active CDMA users. The authors have shown that the performance degradation due to tracking errors is quite sensitive to the inequality of received powers.

The impact of tracking errors on the performance of successive interference cancellation receiver is analyzed in [56]. For numerical results presented here, the interference cancellation scheme was subject to pessimistic conditions:

- Did not use averaging of the correlator outputs for amplitude estimates which significantly improves cancellation performance.
- Assumed equal received powers (perfect power

control). The improvement over the conventional detector is much greater in the more realistic case of unequal received powers.

Figure 11 shows results for a processing gain (number of chips/bit) $N = 31$, and the total number of users is varied from five to 20. There are three curves each for the interference cancellation scheme and the conventional detector. Each curve represents different standard deviation e of tracking error, normalized with respect to chip duration ($e = 0$ is zero tracking error). The interference cancellation scheme retains superiority over the conventional detector. Similar types of results were found in [57] from simulation of the scheme of [58]. For Rayleigh fading on one path and mobile speed of 100 km/h it was reported that interference cancellation technique is less sensitive to imperfect power control and chip synchronization. For example, for error in power control with standard deviation of 1 dB with respect to nominal received power, an error of 5 percent in chip synchronization does not cause significant degradation in error probability.

While it is premature to draw any general conclusions about robustness of the interference cancellation receivers at this point, there are some promising results.

Several authors also addressed sensitivity of coherent multiuser detectors to channel mismatch (see references of [9] and [59-61]. While these investigations are still preliminary, they bring researchers closer to understanding performance advantages of multiuser detectors.

Concluding Remarks

The theoretical bases of optimal multiuser detection are well understood. Given the prohibitive complexity of optimum multiuser detectors, attention has been focused on suboptimal detectors, and the properties of these detectors are well understood by now. The next stages of investigation, involving implementation and robustness issues, are accelerating now and will lead to determination of the practical and economic feasibility of the multiuser detector. Initial studies of robustness show that robustness need not be a fatal flaw. Further investigations into practicality will include actual hardware implementations. This is the critical issue in answering the question posed at the end of the introduction to this article.

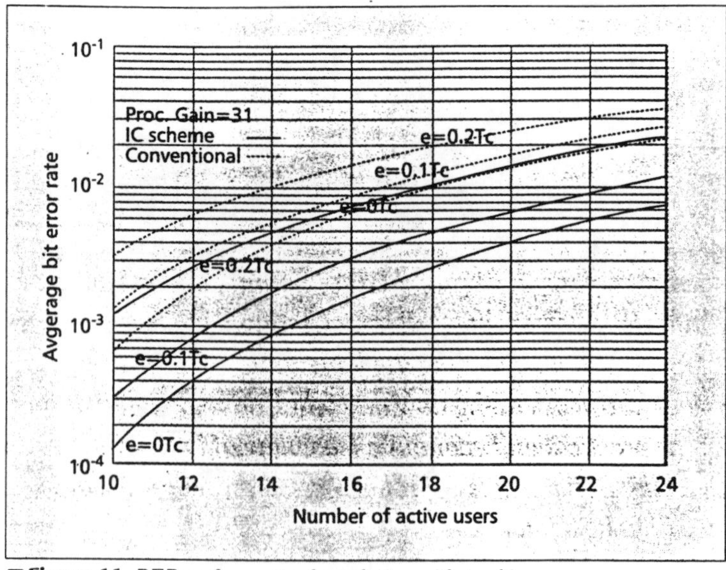

Figure 11. *BER performance degradation with tracking errors.*

References

[1] A. J. Viterbi, "The Orthogonal-Random Wave form Dichotomy for Digital Mobile Personal Communications," *IEEE Personal Commun.*, First Quarter 1994, pp. 18-24.

[2] K. S. Gilhousen *et al.*, "On the capacity of a Cellular CDMA System," *IEEE Trans. on Vehicular Tech.*, vol. VT-40, no. 2, May 1991, pp. 303-312.

[3] R. L. Pickholtz, L. B. Milstein and D. L.Schilling, "Spread Spectrum for Mobile Communications" *IEEE Trans. on Vehicular Tech.*, vol. VT-40, no. 2, May 1991, pp. 313-322.

[4] R. Kohno, R. Meidan, and L.B. Milstein, "Spread Spectrum Access Methods for Wireless Communications," *IEEE Commun. Magazine*, Jan. 1995, pp. 58-67.

[5] M. K. Simon *et al.*, "Spread Spectrum Communications, Vols. I-III, (Computer Science Press, 1985).

[6] G. Turin, "The Effects of Multipath and Fading on the Performance of Direct Sequence CDMA Systems," *IEEE JSAC*, vol. SAC-2, no. 4, July 1984, pp. 597-603.

[7] J. G. Proakis, "Digital Communications," 2nd Ed., (McGraw-Hill, 1989).

[8] J. M. Holtzman, "DS/CDMA Successive Interference Cancellation," Proc. of ISSSTA '94, Oulu, Finland, July 1994. pp. 69-78.

[9] S. Verdu, "Adaptive Multiuser Detection," Proc. of ISSSTA '94, Oulu, Finland, July 1994, pp. 43-50.

[10] L. F. Chang, "Dispersive Fading Effects in CDMA Radio Systems," Proc. of ICUPC '92 Dallas, TX, Sep. 1992, pp. 185-189.

[11] J. S. Lim, ed., Speech Enhancement, (Prentice-Hall, 1983).

[12] B. Widrow and S. D. Stearns, Adaptive Signal Processing, (Prentice-Hall, 1985).

[13] H. Nicolas, A. Giordano and J. Proakis, "MLD and MSE Algorithms for Adaptive Detection of Digital Signals in the Presence of Interchannel Interference," *IEEE Trans. on Info. Theory*, vol. IT-23, no. 5, Sep. 1977, pp. 563-575.

[14] J. Salz, "Digital Transmission Over Cross-Coupled Linear Channels," *AT&T Tech. J.*, vol. 64, no. 6, July-Aug. 1985, pp. 1147-1158.

[15] J. W. Carlin *et al.*, "An IF Cross-Pol Canceller for Microwave Radio Systems," *IEEE JSAC*, vol. SAC-5, no. 3, April 1987, pp. 502-514.

[16] K. S. Schneider, "Optimum Detection of Code Division Signals," *IEEE Trans. on Aerospace and Electronic Sys.*, vol. AES-15, no. 1, Jan. 1979, p. 181-185.

[17] R. Kohno, M. Hatori, and H. Imai, "Cancellation Techniques of Co-Channel Interference in Asynchronous Spread Spectrum Multiple Access Systems," *Electronics and Commun.*, vol. 66-A, no. 5, 1983, pp. 20-29.

[18] S. Verdu, "Minimum Probability of Error for Asynchronous Gaussian Multiple Access Channels," *IEEE Trans. on Info. Theory*, vol. IT-32, no. 1, Jan. 1986, pp. 85-96.

[19] S. Verdu, "Optimum Multiuser Asymptotic Efficiency," *IEEE Trans. on Commun.*, Vol. Com-34, No. 9, Sept. 1986, pp. 890-897.

[20] R. Lupas and S. Verdu, "Near-Far Resistance of Multiuser Detectors in Asynchronous Channel," *IEEE Trans. on Commun.*, vol. COM-38, no. 4, April 1990, pp. 496-508.

[21] R. De Gaudenzi, C.Elia, and R.Viola, "Bandlimited Quasi-Synchronous CDMA: A Novel Satellite Access Technique for Mobile and Personal Communication Systems," *IEEE JSAC*, vol. SAC-10, no. 2, Feb. 1992, pp. 328-343.

[22] A. Kajiwara and M. Nakagawa, "Microcellular CDMA System with a Linear Multiuser Interference Canceller," *IEEE JSAC*, vol. 12, no. 4, May 1994, pp. 605-611.

[23] R. Lupas and S. Verdu, "Linear Multiuser Detectors for Synchronous Code-Division Multiple-Access Channel," *IEEE Trans. on Info. Theory*, vol. IT-35, no. 1, Jan. 1989, pp. 123-136.

[24] Z. Xie, R. T. Short, and C. K. Rushforth, "A Family of Suboptimum Detectors for Coherent Multiuser Communications," *IEEE JSAC*, vol. SAC-8, no. 4, May 1990 pp. 683-690.

[25] Z. Xie, C. K. Rushforth, and R. T. Short, "Multiuser Signal Detection Using Sequential Decoding," *IEEE Trans. on Commun.*, vol. COM-38, no. 5, May 1990, pp. 578-583.

[26] M. K. Varanasi and B. Aazhang, "Multistage Detection in Asynchronous Code Division Multiple-Access Communications," *IEEE Trans. on Commun.*, vol. COM-38, no. 4, April 1990, pp. 509-519.

[27] T. R. Giallorenzi and S. G. Wilson, "Decision Feedback Multiuser Receivers for Asynchronous CDMA Systems," Proc. of GLOBECOM '93, Houston, TX, Nov.- Dec. 1993, pp. 1677-1681.

[28] M. K. Varanasi and B. Aazhang, "Near-Optimum Detection in Synchronous Code-Division Multiple Access Systems," *IEEE Trans. on Commun.*, vol. COM-39, May 1991, pp. 725-736.

[29] A. Duel-Hallen, "On Suboptimal Detection for Asynchronous Code-Division Multiple Access Channels," Proc. of the 26th Annual Conference on Information Sciences and Systems, Princeton University, Princeton, NJ, March 1992, pp. 838-843.

[30] A. Duel-Hallen, "Decorrelating Decision-Feedback Multiuser Detector for Synchronous Code-Division Multiple Access Channel," *IEEE Trans. on Commun.*, vol. COM-41, no.2, Feb. 1993 pp. 285-290.

[31] A. Duel-Hallen, "A Family of Multiuser Decision-Feedback Detectors for Asynchronous Code-Division Multiple Access Channels," To appear in *IEEE Trans. on Commun.*, Feb. 1995.

[32] P. Patel and J. Holtzman, "Analysis of a Simple Successive Interference Cancellation Scheme in DS/CDMA System," *IEEE JSAC* - Special Issue on CDMA, vol. 12, no. 5, June 1994, pp. 796-807.

[33] A. Viterbi, "Very Low Rate Convolutional Codes for Maximum Theoretical Performance of Spread-Spectrum Multiple-Access Channels," *IEEE JSAC*, vol. SAC-8, no. 4, May 1990, pp. 641-649.
[34] P. Patel and J. Holtzman, "Performance Comparison of a DS/CDMA System using a Successive Interference Cancellation (IC) Scheme and a Parallel IC Scheme under Fading," Proc. of ICC'94, New Orleans, LA , May 1994, pp. 510-515.
[35] T. R. Giallorenzi and S. G. Wilson, "Trellis-Based Multiuser Receivers for Convolutionally Coded CDMA Systems," Proc. of the 31st Allerton Conference on Comm., Control and Computing, Oct. 1993.
[36] P. Hoeher, "On Channel Coding and Multiuser Detection for DS-CDMA,"Proc. of the 2nd International Conference on Universal

The properties of optimal and suboptimal multiuser detectors are well understood now. The next stages of investigation, involving implementation and robustness issues, are accelerating now and will lead to determination of the practical and economic feasibility of the multiuser detector. Initial studies of robustness show that robustness need not be a fatal flaw.

Personal Communications, Ottawa, Canada, Oct. 1993, pp. 641-646.
[37] T. R. Giallorenzi and S. G. Wilson, "Multistage Decision Feedback and Trellis-Based Multiuser Receivers for Convolutionally Coded CDMA Systems," Technical Report UVA/538341/EE93/102, Comm. Systems Lab., Dept of Electr. Eng., Univ. Of Virginia, May 1993.
[38] M. Nasiri-Kenari and C. K Rushforth, " An Efficient Soft-Decision Decoding Algorithm for Synchronous CDMA Communications with Error-Control Coding," Proceedings of the IEEE International Symposium on Information Theory, Trondheim, Norway, June 27-July 1, 1994, p. 227.
[39] U. Fawer and B. Aazhang, "Multiuser Reception for Trellis-Based Code Division Multiple Access Communications," Proceedings of MILCOM'94, Eatontown, NJ, Oct. 1994, pp. 977-981.
[40] P. Jung, M. Naßhan and J. Blanz, "Application of Turbo-Codes to CDMA Mobile Radio System Using Joint Detection and Antenna Diversity," Proceedings of VTC'94, Stockholm, Sweden, June 1994, pp. 770-774.
[41] W. Lee, "Overview of Cellular CDMA," *IEEE Trans. on Vehicular Technology*, vol. VT-40, no. 2, May 1991, pp. 291-302.
[42] Z. Zvonar and D. Brady, "Multiuser on in Single-Path Rayleigh Fading Channels," *IEEE Trans. on Commun.*, vol. COM-42, no.4, April 1994, pp.1729-1739.
[43] A. Duel-Hallen, "Performance of Multiuser Zero-Forcing and MMSE Decision Feedback Detectors for CDMA Channels," Conference Record of the Second Communication Theory Mini-Conference in conjunction with Globecom '93, Houston, TX, Dec. 1993, pp. 82-86.
[44] P. R. Patel and J. M. Holtzman, "Analysis of Successive Interference Cancellation in M-ary Orthogonal DS-CDMA System with Single Path Rayleigh Fading," Proc. of 1994 International Zurich Seminar on Digital Communications, Zurich, Switzerland, March 1994, pp. 150-161.
[45] H. Y. Wu, A. Duel-Hallen, "Performance of Multiuser Decision-Feedback Detectors for Flat Fading Synchronous CDMA Channels," Proc. of the 28-th Annual Conference on Information Sciences and Systems, Princeton University, Princeton, NJ, March 16-18, 1994, pp. 133-138.
[46] C. Kchao and G. Stuber, "Performance Analysis of a Single Cell Direct Sequence Mobile Radio System," *IEEE Trans. on Commun.*, vol. COM-41, no. 10, Oct. 1993, pp. 1507-1516.
[47] Z. Zvonar and D. Brady, "Optimum Detection in Asynchronous Multiple-Access Multipath Rayleigh Fading," Proc. of the 26th Annual Conference on Information Sciences and Systems , Princeton University, March 1992, pp. 826-831.
[48] M. Varanasi, "Noncoherent Detection in Asynchronous Multiuser Channels," *IEEE Trans. on Info. Theory*, vol. IT-39, no. 1, Jan. 1993, pp. 157-176.
[49] Z. Zvonar and D. Brady, "Coherent and Differentially Coherent Multiuser Detectors for Asynchronous CDMA Frequency-Selective Channels," Proc. MILCOM '92, San Diego, CA, Oct. 1992, pp. 442-446.
[50] S. Wijayasuriya, J. McGeehan and G. Norton, "RAKE Decorrelating Receiver for DS-CDMA Mobile Radio Networks," *Electronic Letters*, vol. 29, no. 4, Feb. 1993, pp. 395-396.
[51] Z. Zvonar, "Multiuser Detection and Diversity Combining for Wireless CDMA Systems," Wireless and Mobile Communications, J. Holtzman and D. Goodman, eds., (Kluwer Academic Publishers, 1994), pp. 51-65.
[52] A. Klein and P. Baier, "Linear Unbiased Data Estimation in Mobile Radio System Applying CDMA," 1993. *IEEE JSAC*, vol. SAC-11, no. 7, Sep. 1993, pp. 1058-1066.
[53] R. Kohno *et al.*, "Combination of an Adaptive Array Antenna and a Canceller of Interference for Direct-Sequence Spread-Spectrum Multiple-Access System," *IEEE JSAC*, vol. SAC-8, no.4, May 1990, pp. 675-682.
[54] S. Wijayasuriya, G. Norton, and J. McGeehan, "A Novel Algorithm for Dynamic Updating of Decorrelator Coefficients in Mobile DS-CDMA," Proc. of the 4th International Symposium on Personal, Indoor and Mobile Radio Communications, Yokohama, Japan, Oct. 1993, pp. 292-296.
[55] E. Storm *et al.*, "Sensitivity Analysis of Near-Far Resistant DS-CDMA Receivers to Propagation Delay Estimation Errors," Proc. of VTC'94, Stockholm, Sweden, July 1994, pp. 757-761.
[56] F. C. Cheng and J. M. Holtzman, "Effect of Tracking Errors on DS/CDMA Successive Interference Cancellation," Proc. of Third Communication Theory Mini Conference in conjunction with Globecom '94 , San Francisco, CA, Nov. 1994, pp. 166-170.
[57] L. Levi, F. Muratore and G. Romano, "Simulation Results for a CDMA Interference Cancellation Technique in a Rayleigh Fading Channel," Proc. of 1994 International Zurich Seminar on Digital Communications, Zurich, Switzerland, March 1994, pp 162-171.
[58] P. Dent, B. Gudmundson, and M. Ewerbring, "CDMA-IC: A Novel Code Division Multiple Access Scheme Based on Interference Cancellation," Proc. of PIMRC '92, Boston, MA, Oct. 1992, pp. 98-102.
[59] S. Gray, M. Kocic and D. Brady, "Multiuser Detection in Mismatched Multiple-Access Channels," To appear in *IEEE Trans. on Commun.*.
[60] Z. Zvonar, M. Stojanovic, "Performance of Multiuser Diversity Reception in Nonselective Rayleigh Fading CDMA Channels," Proc. of the Third Communication Theory Mini-Conference (CTMC '94), San Francisco, CA, Nov. 1994, pp. 171-175.
[61] H. Y. Wu, A. Duel-Hallen, "Channel Estimation and Multiuser Detection for Frequency-Nonselective Fading Synchronous CDMA Channels," Proceedings of the 32nd Annual Allerton Conference on Communication, Control and Computing, Monticello, IL, Sep. 1994.
[62] M. Wallace and R. Walton, "CDMA Radio Network Planning," Proc. of ICUPC '94, San Diego, CA, Sept. 27 - Oct. 4, 1994, pp. 62-67.

Biographies

ALEXANDRA DUEL-HALLEN received a B.S. in mathematics from Case Western Reserve University in 1982, an M.S. in computer, information, and control engineering from the University of Michigan in 1983, and a Ph.D. in electrical engineering from Cornell University in 1987. She worked for AT&T Bell Laboratories in Columbus, Ohio during the summer of 1982, and participated in the AT&T One Year on Campus Program at the University of Michigan during the 1982-1983 academic year. She received an AT&T Ph.D. Fellowship during 1985-1987. In 1987-1990 she was a visiting assistant professor at the School of Electrical Engineering, Cornell University. In 1990-1992, she was with the Mathematical Sciences Research Center, AT&T Bell Laboratories, Murray Hill, New Jersey. She joined the Department of Electrical and Computer Engineering at North Carolina State University, Raleigh, North Carolina, in January 1993 as an assistant professor. Her current research interests are in channel equalization and spread spectrum communications. She has served as an editor for Communication Theory for the *IEEE Transactions on Communications* since 1989. During 1994, she was a secretary of the Information Theory Society Board of Governors.

JACK M. HOLTZMAN [F '95] received a B.E.E. from City College of New York, an M.S. from U.C.L.A., and a Ph.D. from the Polytechnic Institute of Brooklyn. He worked for AT&T Bell Laboratories for 26 years on control theory, teletraffic theory, telecommunications, and performance analysis. At Bell Labs, he was supervisor of the Mathematical Analysis and Consulting Group and then Head of the Teletraffic Theory and System Performance Department. In 1990 he joined Rutgers University, where he is Professor of Electrical and Computer Engineering and Associate Director of the Wireless Information Network Laboratory (WINLAB), Piscataway, New Jersey. He is also the director of the Wireless Communications Certificate Program. His current areas of work are on spread spectrum, handoffs, resource management, propagation, and wireless system performance.

ZORAN ZVONAR received a Dipl. Ing. in 1986 and an M.S. in 1989, both from the Department of Electrical Engineering, University of Belgrade, Yugoslavia, and a Ph.D. in electrical engineering from Northeastern University, Boston, in 1993. From 1986 to 1989 he was with the Department of Electrical Engineering, University of Belgrade, Belgrade, Yugoslavia, where he conducted research in the area of telecommunications. From 1993 to 1994 he was a post-doctoral investigator at the Woods Hole Oceanographic Institution, where he worked on multiple-access communications for underwater acoustic local area networks. Since 1994 he has been with the Analog Devices, Communications Division, Wilmington, Massachusetts, where he is working on the design of wireless communications systems.

Reprinted from *IEEE Personal Communications*, Vol. 1, No. 1, First Quarter, 1994, pp. 18-24.

A discussion of two alternative and rival approaches

The Orthogonal-Random Waveform Dichotomy for Digital Mobile Personal Communication

ANDREW J. VITERBI

The conversion of terrestrial wireless telephony to digital transmission technology is just beginning. However, with more than four years of experimental laboratory and field testing, we have already learned numerous practical lessons, both positive and negative, relative to the art and science of multiple-access communication by large user populations. Europe, Japan, and North America have each developed digital cellular standards. The North American experience has been the most contentious and diverse. Only here have two alternative and rival approaches been carried through to the development of detailed standards leading to imminent large-scale commercial deployments. In this article, the two alternatives are denoted as orthogonal and random waveform multiple access, and are described and discussed successively.

Orthogonal Waveform Multiple Access

The conventional approach is to give each user in a given cell, communicating with a given base station, its own disjoint frequency or time slot. Thus, in frequency-division multiple access (FDMA), users are said to be frequency-orthogonal, meaning that one user's signal (ideally) has no effect on another user's receiver tuned to a different frequency band; in time-division multiple access (TDMA), users are time-orthogonal, since transmission by different users in disjoint time intervals achieves the same result.[1] Were there only a single cell and line-of-sight propagation such that thermal noise was the only interference, each user's transmission could be viewed as a single point-to-point communication channel, which ideally could transmit error-free at rates up to the Shannon limit, a linear function of the bandwidth and the time duty factor (for TDMA) and a logarithmic function of the signal-to-(thermal) noise ratio. Unfortunately, these assumptions are practically never valid for terrestrial communication, and rarely even for satellites.[2]

The multicellular terrestrial environment is characterized instead by multipath propagation, which can cause deep fades, particularly in narrowband channels, and by other-cell user interference. The latter is caused by the so-called frequency reuse provision whereby users in different cells sufficiently separated from the given cell are also assigned the same frequency band as users in the given cell. Figure 1 shows this reuse pattern for a frequency reuse factor of 1/7, wherein all cells in the ring surrounding cell A are assigned different bands, but cell A's band is reassigned to one cell in each seven-cell cluster. The foregoing assumes that each cell's base station employs an omnidirectional antenna. Although cellular base stations are normally equipped with three-sectored antennas, the reuse factor remains the same, because each sector is assigned a different set of frequencies. The resulting other-cell interference is reduced by sectoring, which for the current analog cellular system is needed to maintain the required signal-to-interference ratio of 18 dB necessary for adequate performance with frequency modulation (FM) occupying a 30-Khz frequency slot [3].

The early digital cellular standards adopted on all three continents, as noted above, employ TDMA with some version of quaternary phase-shift keyed (QPSK) modulation.[3] Although QPSK is capable of a transmission efficiency of up to 2 b/s/Hz, all current systems require overheads for signaling, error correcting codes, and equalizer training such that the effective bit rate in every case is less than 1 b/s/Hz.[4] Coupled with a reuse factor of 1/7, it follows that for a total bandwidth

[1] *In fact, TDMA users are given both a frequency and a time assignment, but no two users occupy the same time-frequency slot simultaneously.*

[2] *One such example, involving a single isolated geosynchronous satellite supporting multiple accesses by fixed very small aperture terminals (VSATs), was considered by the author [1] and is sometimes cited to question his views [2].*

[3] *As discussed in [4], the modulation and multiple access adopted by most terrestrial TDMA systems imitate those previously implemented in early digital satellites for trunking applications with a few large users.*

[4] *Actually, the effective bit rate of the European Global System for Mobile Communications (GSM) standard is only 13/25 b/s/Hz [5], but its goal is to achieve a reuse factor of 1/4 so that the net effect would be equivalent to a reuse factor of 1/7 with 0.91 b/s/Hz.*

W Hz and data rate R b/s, the number of orthogonal slots available limits the number of simultaneous users per cell to

$$N_u \leq \frac{W/R}{7} \text{ users /cell.} \quad (1)$$

For example, with North American TDMA (the IS-54 Standard),[5] the total allocated bandwidth in each direction, W = 12.5 Mhz, and the vocoder data rate (with overhead) R = 10 kb/s, so that the number of users per cell is approximately 180.

Another serious drawback of narrowband slotted multiple access is the self-interference caused by multipath. Whenever transmissions, propagating over two or more paths, arrive at the receiver separated in time by less than the inverse bandwidth of the receiver, there will be frequent instances of severe fading caused by phase cancellation between multipath components. Two means of abating this effect are to use an equalizer for the channel propagation characteristics and to introduce a sizable margin for transmitted power. The effectiveness of the first is limited, particularly for narrowband signals and for channels that vary rapidly, as in a fast-moving vehicle. Increasing power is always an option, but not without negative consequences such as increased interference to the users of the same frequency band in other cells and electromagnetic interference to other devices.[6]

In addition to capacity limitations and multipath-fading degradation, slotted or orthogonal systems suffer from degraded service at the edges of cells and during periods immediately preceding handoff between cells. In the next section, we shall describe how wideband but nonorthogonal multiple access greatly mitigates all these effects.

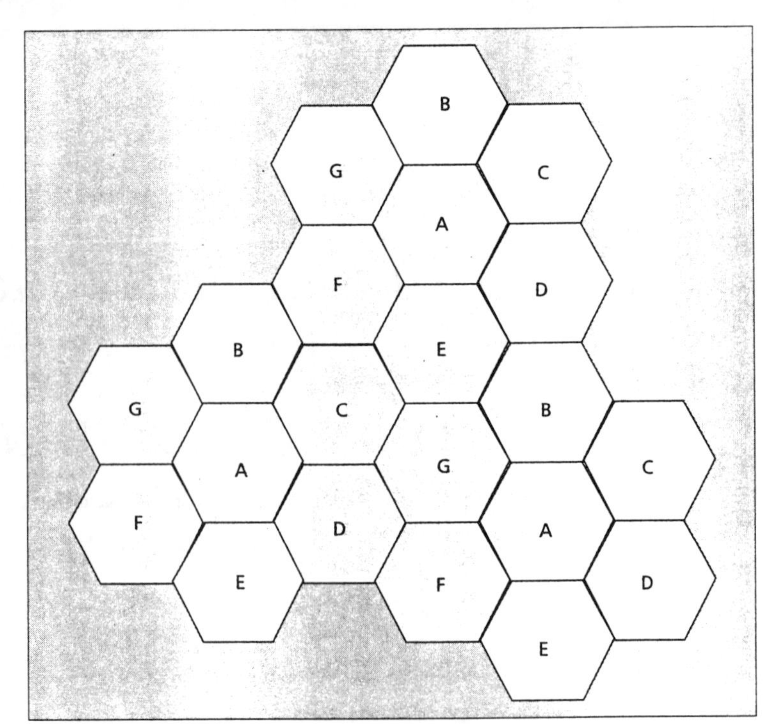

■ **Figure 1.** *Seven-cell frequency reuse clusters.*

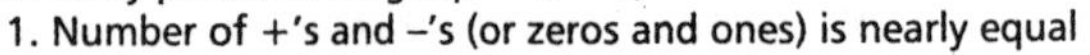

Maximum length linear feedback shift register of r stages staisfies three randomness properties:

In every period of length $p = 2^r - 1$,

1. Number of +'s and –'s (or zeros and ones) is nearly equal
2. Half the runs (of +'s and –'s have length 1

One quarter have length 2, 1/8 length 3, . . . $1/2^k$ length k, $k < r$

3. Sequence correlation $C(\tau) = \frac{1}{p}\sum_{n=1}^{p} a_n a_{n+\tau} \approx \delta(\tau)$, Kronecker delta,

thus indistinghuishable from pure random when p is large.

Note, however, when correlation is over much shorter span than p, $C(\tau)$ is a random variable with distribution similar to sum of binary i.i.d. variables just as for random sequences.

■ **Figure 2.** *(Pseudo) random sequence generator.*

Random Waveform Multiple Access

In contrast with the multiple-access methods just considered, techniques generally referred to as spread spectrum multiple access (SSMA) or code-division multiple access (CDMA) employ a waveform that for all purposes appears random to anyone but the intended receiver of the transmitted waveform. Actually, for ease of both generation and synchronization by the receiver, the waveform is pseudorandom, meaning that it can be generated by mathematically precise rules; but statistically, it nearly satisfies the properties of a truly random sequence. Figure 2 shows both the pseudorandom sequence generator, a shift register clocked at a rate proportional to the total (spread) bandwidth of the waveform, and the randomness properties it satisfies. Figure 3 shows the modulator and demodulator employing this pseudorandom waveform.[7] Two pseudorandom sequences, generated as in Fig. 2, multiply the input bit sequence (as provided from a digital source or digitized coded speech or video), the bit rate of which is much lower than the shift register's clock. The ratio of the shift register clock rate to the input bit rate (usually an integer and a power of 2) is called the spreading factor or processing gain. After conversion to analog by means of waveshaping filters, the two waveforms modulate (by binary PSK) the in-phase and quadrature components of the carrier, which are then summed and transmitted. The receiver reverses the process by removing the spreading pseudorandom sequences prior to baseband processing.

The resulting waveform is wideband, noise-like, and balanced in phase, and has a flexible timing structure, all attributes that follow from the multiplication of the baseband digital sequence by a much-higher-rate pseudorandom sequence. Most important, these features of the waveform ensure its robustness to all forms of interference, from other users and other sources, as well as to destructive multipath of the desired user's transmission. To explain this, the major attributes of DS/CDMA are listed below, roughly in order of their importance, and will subsequently be discussed in more detail:

- Universal frequency reuse.
- Fast and accurate power control.
- Constructive combining of multipath propagated components by means of a Rake Receiver.
- Soft handoff (handover) between contiguous-cell base stations using the same Rake Receiver.

[5] A reuse factor of 1/4 with 1 b/s/Hz has even been claimed by some IS-54 TDMA proponents, but thus far even a reuse factor of 1/7 appears to not be universally achievable.

[6] See, for example, S. Fist- Will GSM and D-Amps Give Way to the CDMA Push?" Australian Communications, July 1993.

[7] This describes direct sequence code-division multiple access (DS/CDMA). An alternative technique called frequency hopping (FH/CDMA) may use the same pseudorandom sequence generator but a different modulation system.

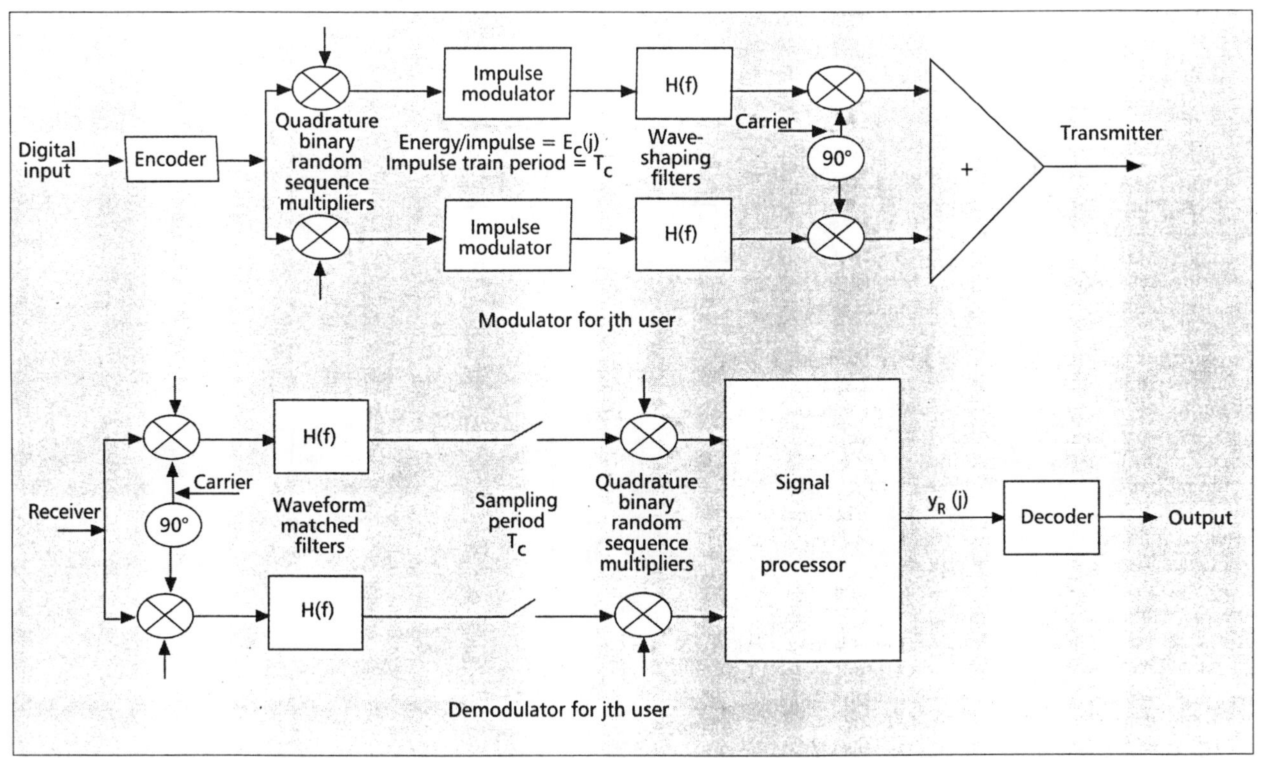

Figure 3. *Generic spread spectrum modem.*

- Autonomous capacity increases for variable-rate (speech or data) transmission.
- Natural and seamless exploitation of sectored antennas and adaptive beamforming.
- Capacity increase with forward error-correcting (FEC) coding without overhead penalty.

Universal frequency reuse means that the spectrum allocated to multicellular mobile/personal communication service may be used simultaneously by numerous users and by numerous other users in every other cell as well. The "reuse factor" previously defined for slotted orthogonal waveforms is therefore exactly 1 for CDMA. This is possible because the waveform is robust to interference, and, by the spreading process described above, each user's waveform appears as (bandlimited) thermal noise, which is the most benign form of interference.

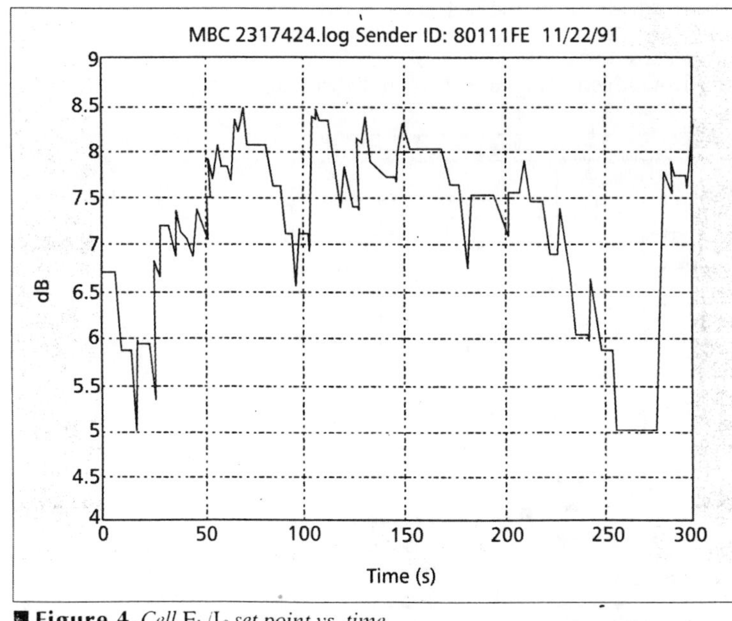

Figure 4. *Cell* E_b/I_0 *set point vs. time.*

Power control is a necessity in a system in which many users simultaneously transmit to a base station with the same frequency spectral allocation; otherwise, users transmitting from locations nearest to the base station will be received at much higher power levels, and hence swamp out users transmitting from locations at the cell's edge. Dynamic ranges on the order of 80 to 100 dB are possible, but most of this inequity can be corrected by causing users to transmit at power levels inversely proportional to their received power level (from the base station), with the latter easily determined by the automatic gain control (AGC) circuit in the mobile user's receiver. However, to achieve accurate power levels, it is necessary to send commands from the base station to the mobile user, based on the base station's received power, indicating an increase or decrease of the mobile's transmitted power level. More precisely, the purpose of power control is to guarantee a particular level of performance (e.g., one frame error in 100 or 1000), which is a function of the E_b/I_0, bit energy-to-interference density, and which varies according to channel propagation (multipath) characteristics. Figure 4 shows the required and desired E_b/I_0 levels for a typical field test experiment while a mobile unit traverses both flat and hilly areas over a 5-min interval. The accuracy of power control is shown by the experimental data of Fig. 5, a histogram of the deviation (error) in E_b/I_0 relative to the desired level. The combination of the variation in desired levels and the difference between desired and achieved levels results in a somewhat larger variation than that due to either fac-

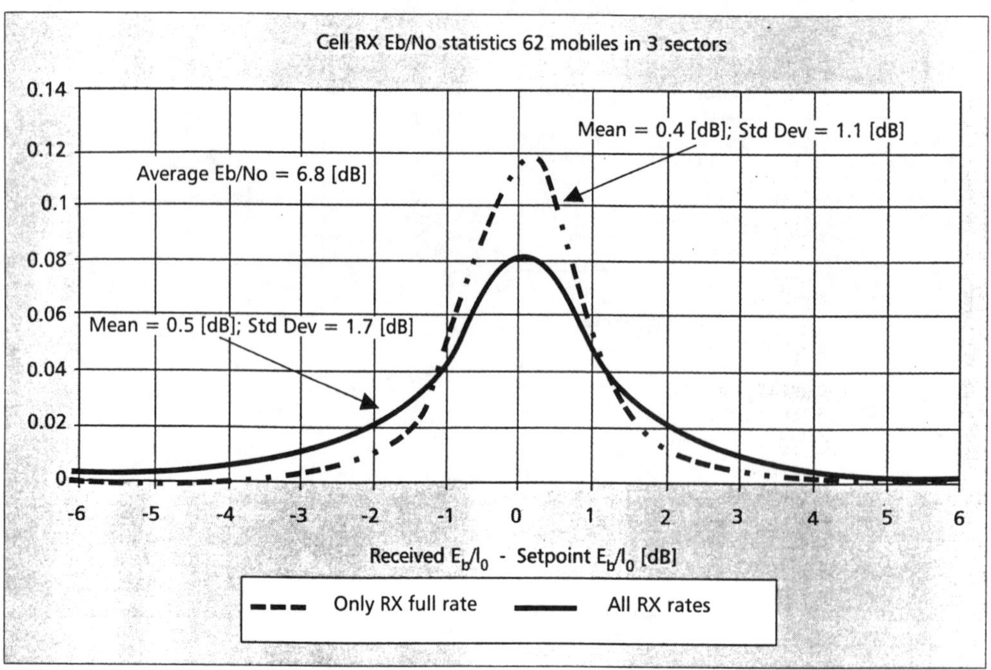

■ **Figure 5.** *Cell differential* E_b/I_0 *histograms: 62 mobiles in three sectors.*

tor. This is shown in Fig. 6, which demonstrates that the actual E_b/I_0 has a nearly log-normal distribution (normal or Gaussian in dB), with a standard deviation of less than 2.5 dB.

Perhaps the most significant result of fast and accurate power control of the wideband signal is the much-reduced transmitted power level required. Figure 7 shows the actual transmitted power levels of a CDMA mobile transmitter and an analog (AMPS) FM transmitter communicating with the same base stations at the same time, while the mobile user is traveling for about 6 min on a California freeway. The CDMA user requires between 20 dB to 30 dB less power than the AMPS user; this relative performance has been reproduced consistently in urban and suburban environments. The consequences are obvious for reduced interference to other sources and reduced health hazards, as well as for increased battery life for portables.

Turning to multipath mitigation, it is well known that the multipath components of a wideband waveform are discernible and separable provided their differential path delays exceed the inverse bandwidth of the waveform. Hence, frequency-slotted (FDMA) waveforms, which are inherently narrowband, cannot be so separated, and time- and frequency-slotted (TDMA) waveforms are only moderately better, but still too narrowband for most delays. On the other hand, if the spread waveform bandwidth exceeds 1 MHz, paths separated by 1 µs or more in delay are separable. If such paths can be acquired rapidly and tracked individually, they can be combined constructively and even coherently, thus utilizing all the received energy, rather than allowing components to combine destructively as for narrowband waveforms. The constructive combining is performed by a Rake Receiver, first proposed nearly 40 years ago [6]; present-day very large-scale integration (VLSI) application-specific integrated circuit (ASIC) technology makes it possible to implement the entire multicomponent Rake Receiver, along with all

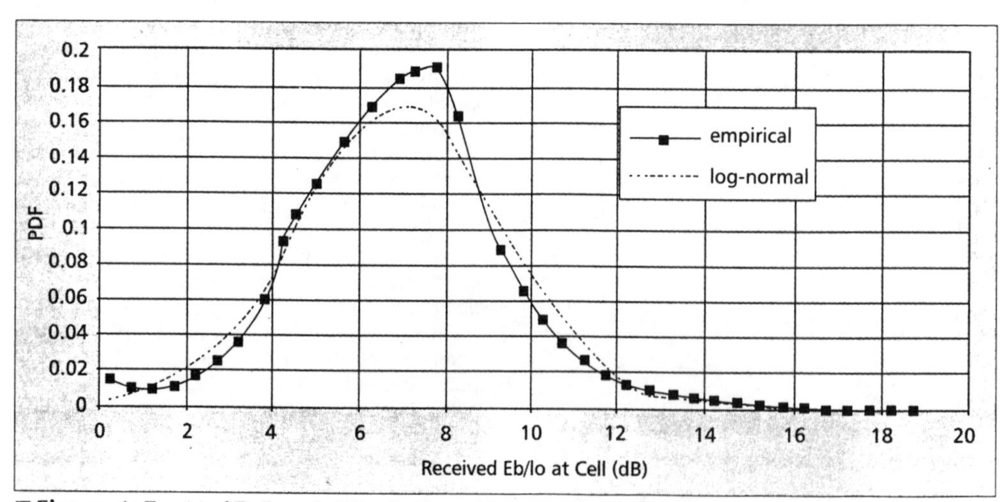

■ **Figure 6.** *Empirical* E_b/I_0 *probability density and log-normal approximation (m = 7.0 dB; sigma = 2.4 dB).*

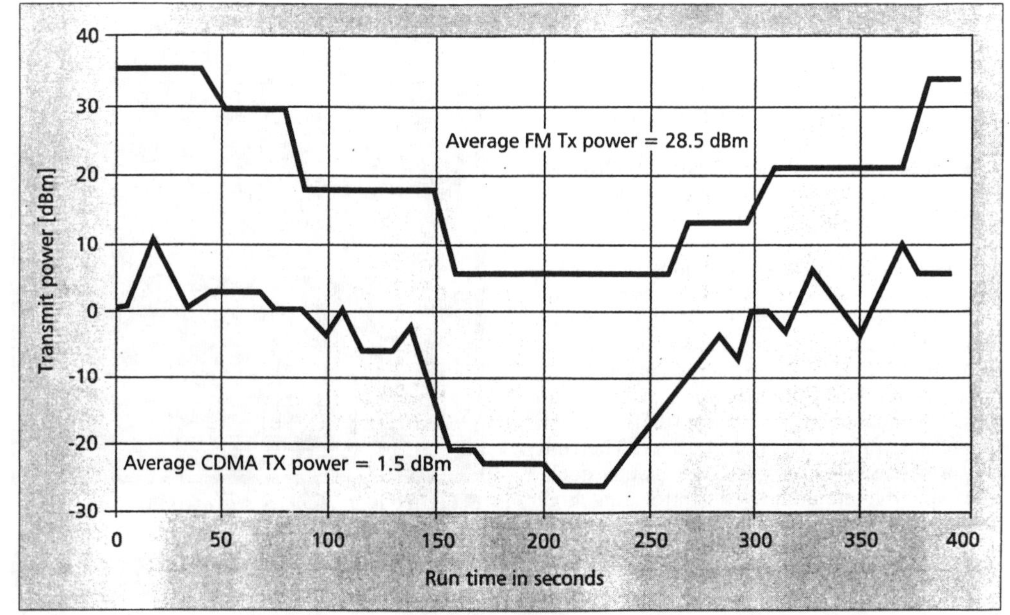

■ **Figure 7.** *CDMA vs. FM mobile transmit power, route 14.*

other baseband processing functions, on a single chip (Fig. 8).

The Rake Receiver also provides the mobile user with the capability to receive the same signal transmitted from two base stations. This enables the so-called soft handoff feature mentioned above, whereby two contiguous-cell base stations simultaneously transmit to and receive from the same mobile user.[8] This ensures reliable handoff as the mobile user moves away from the vicinity of the first cell toward the second cell, with dual communication starting when the second cell's signal begins to approach the level of the first and ending when the first cell's signal drops to well below the second. Not only does this drastically reduce the probability of a dropped call and of poor reception at cell's edge, but it also increases cell coverage (to better than double the area when the cell is not fully loaded) and cell capacity to about double of what would be possible with conventional hard handoff [7]. Note that soft handoff is possible only with universal frequency reuse.

As mentioned previously and throughout, the main thrust of spread spectrum CDMA is to render the interference from all users and all cells, sharing the same spectrum, as benign as possible. This is partly achieved by making all (interfering) signals appear random to one another. However, even better performance and greater capacity is achieved if the interference is not present. For voice traffic as well as for low-duty-cycle data traffic, considerable gains are obtained by reducing the transmitted power when the average data rate is reduced. This feature can be easily incorporated into many speech coding algorithms, particularly those based on linear predictive coding (LPC) techniques. As part of the speech analysis process, typically performed on 20-ms successive speech segments, a decision can be made that voice activity is low or absent; thus, the frame may be coded with fewer bits, and hence with a lower average power. With less power thus transmitted, there will be less interference to other users, who in turn will interfere less by similarly reducing their power during lower voice (or data) activity frames. This effect allows us to increase capacity by about a factor of 2.5.[9] Similarly, using sectored antennas, the more sectors and the narrower each beam, the less interference from users outside the sector. Thus, the interference reduction is at least proportional to the antenna gain (in the horizontal plane); actually, it is somewhat greater in the reverse (mobile-to-base station) direction because of soft handoff. This results in a proportional capacity gain.[10]

Finally, a gain is achieved by reducing the required E_b/I_0, and hence the required transmitted power, for each user by employing powerful FEC coding. While coding reduces E_b/I_0 for any multiple-access system, in slotted systems, coding redundancy causes bandwidth expansion and hence capacity reduction. In spread spectrum systems, ample bandwidth is available because of the universal reuse feature; therefore, the processing gain or spreading factor is in no way reduced by the coding redundancy. Thus, capacity is not reduced, but rather increased by the reduction in required E_b/I_0 and transmitted power.

A general approximate formula for capacity in terms of the number of users supportable per cell in the reverse (mobile-to-base-station) direction[11] is given [8] by

$$N_u = \frac{(W/R)G_v G_A}{(E_b/I_0)(1+f)}\left[(1-\eta)F\right] \quad \text{Erlangs/cell} \quad (2)$$

where W/R is the spectrum spreading factor, as before; E_b/I_0 is the required bit-energy-to-interference density (including both other-user interference and thermal or background noise); f is the ratio of total average other-cell-user interference to average same-cell other-user interference; G_v is the gain due to variable-rate voice (or data) activity; G_A is the sectored antenna gain; and the last bracketed term is a reduction due to power control limits and variability, and traffic intensity statistical variation. Erlangs are roughly equivalent to the average number of users supported continually. Note that

[8] *While the mobile user can receive, and constructively and coherently combine, both transmissions to achieve optimal diversity, the base stations, being geographically separated, must decode their receptions individually, letting the switching office arbitrate between the two if they differ, resulting in less optimal "selection diversity."*

[9] *Some slotted systems also employ voice activity detection with intermittent transmission to reduce interference. However, capacity gain is achieved only if unused time segments are reassigned to other users. The required signaling and backhaul interconnection capacity, as well as assuring a sufficient number of users to achieve trunking efficiency, make the approach much more difficult for slotted systems than for random (spread) systems, where capacity increase is achieved automatically by individual users reducing their interference to others.*

[10] *With orthogonal or slotted systems, as noted above, sectorization reduces interference, but it does not increase capacity since additional slots must be provided for each sector.*

[11] *For the forward (base-station-to-mobile) direction the formula is slightly different, although many of the features described for the reverse link are similar.*

while voice activity and sectored antennas account for the numerator gains, the effect of all other properties listed at the outset (universal frequency reuse, power control, the Rake Receiver, and FEC) is to reduce the denominator factors E_b/I_0 and f. Typical values for all these parameters currently are

$G_v = 2.5, G_A = 3, E_b/I_0 = 5$ (i.e., 7 dB), and $f = .55$.

The reduction from power control only (bracketed term assuming constant traffic intensity) is about 0.75. This results in $N_u \approx 0.73\ W/R$. Comparing this with the slotted orthogonal waveform capacity $N_u \approx 0.14\ W/R$, this represents a capacity gain greater than 5. A number of improvements are possible and currently being considered, in both reduction of E_b/I_0 (by about 1 dB or 20 percent) and in increased sectoring (six sectors would double the capacity, and adaptive beamforming might increase this even further).

However, the advantage of the random (DS/CDMA) waveform over the slotted orthogonal (FDMA or TDMA) waveform is as much qualitative, in terms of improved quality at cell edge through soft handoff and consistently lower error rates, and reduced transmitted power through power control and FEC coding, as it is quantitative, in terms of capacity and cell coverage. While the above descriptions and approximate formula may help to explain these advantages, they can only ultimately be verified by experimental field tests with cells loaded nearly to capacity. Such testing programs have been in progress for nearly four years [9, 10]. From 1989 through Summer 1993, a consortium of CDMA cellular manufacturers and service providers, listed in Table 1, have accumulated the testing statistics listed in Table 2. The CDMA technology described here has been standardized by the Telecommunication Industry Association (TIA) as document number IS-95 [11]. Additionally, several official documents, submitted to the FCC [12, 13] and international bodies, have quantitatively and qualitatively substantiated these conclusions.

Hybrid and Other Techniques

While it might appear from the foregoing that orthogonal or random multiple-access techniques are strictly disjoint, the two can sometimes be combined to advantage. In particular, it is possible to provide waveform orthogonality among all users of the same cell while maintaining mutual randomness only between users of different cells. This is rendered possible by the wide bandwidth of a spread spectrum DS/CDMA system, which, as noted above, provides considerable waveform flexibility. Orthogonality can then be achieved using just the binary (spread) waveform by first multiplying each user's binary input by a (spread) binary sequence which is orthogonal to that of every other user of the same cell, followed by multiplication by a (pseudo) random sequence which is cell-specific but common to all users of that cell. One class of such binary orthogonal sequences is the Walsh-Hadamard orthogonal set.

Note that same-cell orthogonality requires that all users be synchronized in time (to the accuracy of a small fraction of one chip). This is readily feasible, and currently implemented in IS-95, for the forward (base-station-to-mobile) link because transmissions to all mobiles originate at the same transmitter. For the reverse link it becomes much more difficult, since the mobiles are geographically distributed and closed-loop timing control (similar in nature to closed loop power control) would have to be implemented to achieve this.

At best, orthogonality of same-cell users would eliminate that part of the interference, but not other-cell interference. In equation (2), this would ideally have the effect of reducing the denominator factor $(1 + f)$ to f, which would increase capacity by a factor of 2.8 for $f = 0.55$. However, multipath may greatly reduce this advantage. If, for example, there were one primary

■ **Figure 8.** *ASIC mask plot of MSM chip.*

Service providers	Equipment suppliers
Ameritech	Alps Electric (Alpine)
Bell Atlantic Mobile	AT&T
Bell Mobility	Goldstar
GTE	Hyundai
NYNEX	Maxon
PacTelCellular	Mitsubishi
U S WEST NewVector	Motorola
	Nokia Mobile
	Northern Telecom
	OKI Electric
	Panasonic
	QUALCOMM
	Samsung
	Sony

■ **Table 1.** *CDMA test program participants.*

- 350,000+ miles of drive testing
- 450,000+ calls placed
- 1,765,000+ minutes (30 K Hrs.)
- 10 million hand-offs

■ **Table 2.** *Parameters of CDMA testing program prior to issuance of IS-95.*

path, with received power 3 dB above the sum of all other paths, the denominator interference factor would only be reduced to $(1/3 + f)$; while if the primary path power were merely equal to the sum of all others, it would be reduced to $(1/2 + f)$. In these two cases, capacity would be increased only by a factor of 1.76 and 1.48, respectively, for $f = 0.55$. The added complexity of closed-loop timing control to achieve such partial orthogonality on the reverse link may be questioned.

One other technique, called "interference cancellation," has lately gained considerable appeal in the theoretical literature [14, 15]. Here, strictly random (DS/CDMA) multiple access is employed, and interference is reduced by successively, or concurrently, demodulating each user and hence estimating each user's input bits, then remodulating (including the pseudorandom sequence multiplication) and subtracting its effect, with correct amplitude and phase, from the composite system. The process is repeated for

This technology is certain to play a major role in bringing about universal and ubiquitous personal communication services in the early part of the second wireless century.

all users in successive cancellation, or for numerous iterations in concurrent cancellation. Some authors claim, optimistically and unrealistically, that such an approach precludes or reduces the need for power control. At best, interference cancellation would have a similar effect to same-cell-user orthogonality, reducing the denominator interference factor $(1 + f)$ to something approaching f. At worst, it may lack robustness, and thus may, at times, make matters worse. In general, its processing complexity and possible processing delay make its application questionable.

Concluding Remarks

Universal frequency reuse with wideband spread spectrum (pseudorandom) waveforms is a powerful method for achieving high user capacity with low transmitter powers, as well as high-quality performance, through constructive multipath combining and soft handoff, coupled with powerful FEC coding, variable-rate source coding, and high-gain antennas (at the base station). This technology is certain to play a major role in bringing about universal and ubiquitous personal communication services in the early part of the second wireless century.

References

[1] A. J. Viterbi, "When Not to Spread Spectrum — A Sequel," *IEEE Commun. Mag.*, vol. 23, no. 4, Apr. 1985, pp. 12-17.

[2] G. Calhoun, *Wireless Access and the Local Telephone Network*, (Artech House, 1992), p. 377.

[3] W. C .Y. Lee, *Mobile Cellular Telecommunications Systems*, McGraw-Hill, 1989.

[4] A. J. Viterbi, "A Perspective on the Evolution of Multiple Access Satellite Communication," *IEEE JSAC*, vol. 10, no. 6, Aug. 1992, pp. 980-984.

[5] R. Steele, *Mobile Radio Communication*, (London: Pentech Press, 1992).

[6] R. Price and P. E. Green, Jr., "A Communication Technique for Multipath Channels," *Proc. IRE*, vol. 46, Mar. 1958, pp. 555-570.

[7] A. J. Viterbi, A.M. Viterbi, E. Zehavi, and K.S. Gilhousen, "Soft Handoff Extends CDMA Cell Coverage and Increases Reverse Link Capacity," to be published in *IEEE JSAC*.

[8] A. M. Viterbi and A. J. Viterbi, "Erlang Capacity of a Power Controlled CDMA System," *IEEE JSAC*, vol. 11, no. 6, Aug. 1993, pp. 892-900.

[9] "Next Generation Cellular: Results of the Field Trials — NAMPS, TDMA, CDMA and Microcells," presentations by industry to the Cellular Telecommunications Industry Association (CTIA), Washington, D.C., Dec. 4-5, 1991.

[10] "CDMA Digital Cellular Technology Forum," presentation by Bell Atlantic Mobile, Ameritech Mobile, GTE Mobilnet, U S WEST NewVector, and QUALCOMM, San Diego, Calif., Feb. 23-24, 1993.

[11] TIA/EIA/IS-95 Interim Standard, "Mobile Station-Base Station Compatibility Standard for Dual-Mode Wideband Spread Spectrum Cellular System," TIA, July 1993.

[12] Bell Atlantic Mobile Systems, "CDMA Field Test Results for Washington, DC," Rep. to FCC, Washington, DC, Nov. 19, 1992.

[13] Ameritech Mobile Communication, Inc. "Developmental Report on CDMA and TDMA Digital Cellular Techniques," Rep. submitted to FCC, Jan. 1993.

[14] P. W. Dent, "CDMA Subtractive Demodulation," U.S. Patent #5,218,619, issued June 8, 1993.

[15] Y. C. Yoon, R. Kohno, and H. Imai, "A Spread-Spectrum Multi-Access System with a Cascade of Co-Channel Interference Cancellers for Multipath Fading Channels," *IEEE 2nd Int'l. Symp. on Spread Spectrum Techniques and Applications (ISSSTA '92)*, Yokohama, Japan, Nov. 29, 1992, pp. 87-90.

Biography

ANDREW J. VITERBI [F '73] received S.B. and S.M. degrees from MIT in 1957 and a Ph.D. from the University of Southern California in 1962. He was a member of the project team at C.I.T. Jet Propulsion Laboratory that designed and implemented the telemetry equipment on the first successful U.S. satellite, and was one of the first communication engineers to propose digital transmission techniques for space and satellite telecommunication systems. As a professor in the UCLA School of Engineering and Applied Science, he wrote two books on digital communication theory, for which he received the 1968 IEEE Information Theory Group Outstanding Paper Award. He received the 1975 Christopher Columbus International Award, the Alexander Graham Bell Medal, the 1986 USC Annual Outstanding Engineering Graduate Award, and the 1990 Marconi International Fellowship Award. He has been associated with UC San Diego since 1975. He co-founded LINKABIT Corporation in 1968, and served as executive vice president from 1974 to 1982 and president from 1982 to 1984. In 1985, he co-founded and became vice chairman and chief technical officer of QUALCOMM, Inc., San Diego, California, which specializes in mobile satellite and terrestrial communication and signal processing technology. He is a member of the U.S. National Academy of Engineering.

Chapter 5

MODERN CODING AND MODULATION

This chapter provides seminal works on coding and modulation, including fundamental contributions in the area of trellis coding, convolutional coding, and the newly discovered turbo codes.

Reprinted from *Proceedings of the IEEE*, Vol. 61, No. 3, Mar. 1973, pp. 268-278.

The Viterbi Algorithm

G. DAVID FORNEY, JR.

Invited Paper

***Abstract*—The Viterbi algorithm (VA) is a recursive optimal solution to the problem of estimating the state sequence of a discrete-time finite-state Markov process observed in memoryless noise. Many problems in areas such as digital communications can be cast in this form. This paper gives a tutorial exposition of the algorithm and of how it is implemented and analyzed. Applications to date are reviewed. Increasing use of the algorithm in a widening variety of areas is foreseen.**

I. Introduction

THE VITERBI algorithm (VA) was proposed in 1967 [1] as a method of decoding convolutional codes. Since that time, it has been recognized as an attractive solution to a variety of digital estimation problems, somewhat as the Kalman filter has been adapted to a variety of analog estimation problems. Like the Kalman filter, the VA tracks the state of a stochastic process with a recursive method that is optimum in a certain sense, and that lends itself readily to implementation and analysis. However, the underlying process is assumed to be finite-state Markov rather than Gaussian, which leads to marked differences in structure.

This paper is intended to be principally a tutorial introduction to the VA, its structure, and its analysis. It also purports to review more or less exhaustively all work inspired by or related to the algorithm up to the time of writing (summer 1972). Our belief is that the algorithm will find application in an increasing diversity of areas. Our hope is that we can accelerate this process for the readers of this paper.

This invited paper is one of a series planned on topics of general interest—The Editor.

Manuscript received September 20, 1972; revised November 27, 1972.

The author is with Codex Corporation, Newton, Mass. 02195.

II. Statement of the Problem

In its most general form, the VA may be viewed as a solution to the problem of maximum *a posteriori* probability (MAP) estimation of the state sequence of a finite-state discrete-time Markov process observed in memoryless noise. In this section we set up the problem in this generality, and then illustrate by example the different sorts of problems that can be made to fit such a model. The general approach also has the virtue of tutorial simplicity.

The underlying Markov process is characterized as follows. Time is discrete. The state x_k at time k is one of a finite number M of states m, $1 \leq m \leq M$; i.e., the state space X is simply $\{1, 2, \cdots, M\}$. Initially we shall assume that the process runs only from time 0 to time K and that the initial and final states x_0 and x_K are known; the state sequence is then represented by a finite vector $\boldsymbol{x} = (x_0, \cdots, x_K)$. We see later that extension to infinite sequences is trivial.

The process is Markov, in the sense that the probability $P(x_{k+1} | x_0, x_1, \cdots, x_k)$ of being in state x_{k+1} at time $k+1$, given all states up to time k, depends only on the state x_k at time k:

$$P(x_{k+1} \mid x_0, x_1, \cdots, x_k) = P(x_{k+1} \mid x_k).$$

The transition probabilities $P(x_{k+1} | x_k)$ may be time varying, but we do not explicitly indicate this in the notation.

It is convenient to define the *transition* ξ_k at time k as the pair of states (x_{k+1}, x_k):

$$\xi_k \triangleq (x_{k+1}, x_k).$$

We let Ξ be the (possibly time-varying) set of transitions

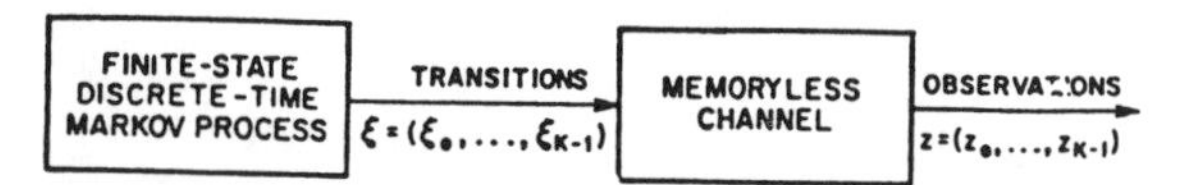

Fig. 1. Most general model.

$\xi_k=(x_{k+1}, x_k)$ for which $P(x_{k+1}|x_k)\neq 0$, and $|\Xi|$ their number. Clearly $|\Xi|\leq M^2$. There is evidently a one-to-one correspondence between state sequences x and transition sequences $\xi=(\xi_0, \cdots, \xi_{K-1})$. (We write $x \overset{1-1}{\leftrightarrow} \xi$.)

The process is assumed to be observed in memoryless noise; that is, there is a sequence z of observations z_k in which z_k depends probabilistically only on the transition ξ_k at time k:[1]

$$P(z \mid x) = P(z \mid \xi) = \prod_{k=0}^{K-1} P(z_k \mid \xi_k).$$

In the parlance of information theory, z can be described as the output of some memoryless channel whose input sequence is ξ (see Fig. 1). Again, though we shall not indicate it explicitly, the channel may be time varying in the sense that $P(z_k|\xi_k)$ may be a function of k. This formulation subsumes the following as special cases.

1) The case in which z_k depends only on the state x_k:

$$P(z \mid x) = \prod_k P(z_k \mid x_k).$$

2) The case in which z_k depends probabilistically on an output y_k of the process at time k, where y_k is in turn a deterministic function of the transition ξ_k or the state x_k.

Example: The following model frequently arises in digital communications. There is an input sequence $u=(u_0, u_1, \cdots)$, where each u_k is generated independently according to some probability distribution $P(u_k)$ and can take on one of a finite number of values, say m. There is a noise-free signal sequence y, not observable, in which each y_k is some deterministic function of the present and the ν previous inputs:

$$y_k = f(u_k, \cdots, u_{k-\nu}).$$

The observed sequence z is the output of a memoryless channel whose input is y. We call such a process a *shift-register process*, since (as illustrated in Fig. 2) it can be modeled by a shift register of length ν with inputs u_k. (Alternately, it is a νth-order m-ary Markov process.) To complete the correspondence to our general model we define:

1) the state

$$x_k \triangleq (u_{k-1}, \cdots, u_{k-\nu})$$

2) the transition

$$\xi_k \triangleq (u_k, \cdots, u_{k-\nu}).$$

The number of states is thus $|X|=m^\nu$, and of transitions, $|\Xi|=m^{\nu+1}$. If the input sequence "starts" at time 0 and "stops" at time $K-\nu$, i.e.,

$$u = (\cdots, 0, u_0, u_1, \cdots, u_{K-\nu}, 0, 0, \cdots)$$

then the shift-register process effectively starts at time 0 and ends at time K with $x_0=x_K=(0, 0, \cdots, 0)$.

[1] The notation is appropriate when observations are discrete-valued; for continuous-valued z_k, simply substitute a density $p(z_k|\xi_k)$ for the distribution $P(z_k|\xi_k)$.

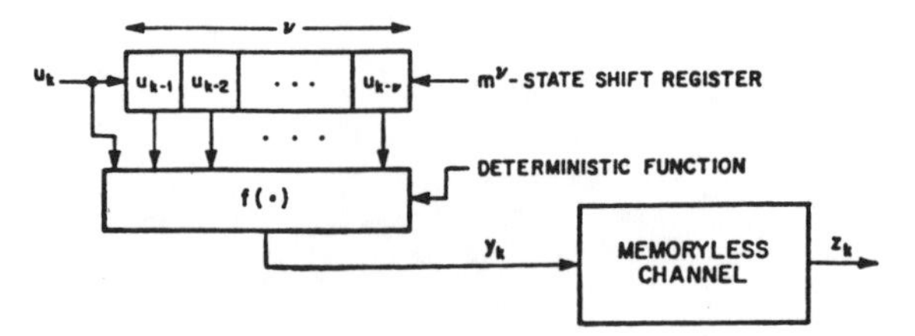

Fig. 2. Shift-register model.

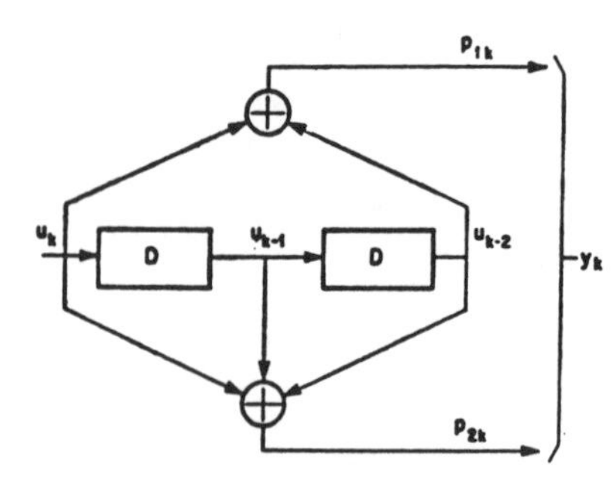

Fig. 3. A convolutional encoder.

Finally, we state the problem to which the VA is a solution. Given a sequence z of observations of a discrete-time finite-state Markov process in memoryless noise, find the state sequence x for which the *a posteriori* probability $P(x|z)$ is maximum. Alternately, find the transition sequence ξ for which $P(\xi|z)$ is maximum (since $x \overset{1-1}{\leftrightarrow} \xi$). In the shift-register model this is also the same as finding the most probable input sequence u, since $u \overset{1-1}{\leftrightarrow} x$; or also the most probable signal sequence y, if $y \overset{1-1}{\leftrightarrow} x$. It is well known that this MAP rule minimizes the error probability in detecting the whole sequence (the block-, message-, or word-error probability), and thus is optimum in this sense. We shall see that in many applications it is effectively optimum in any sense.

Application Examples: We now give examples showing that the problem statement above applies to a number of diverse fields, including convolutional coding, intersymbol interference, continuous-phase frequency-shift keying (FSK), and text recognition. The adequately motivated reader may skip immediately to the next section.

A. Convolutional Codes

A rate-$1/n$ binary convolutional encoder is a shift-register circuit exactly like that of Fig. 2, where the inputs u_k are information bits and the outputs y_k are blocks of n bits, $y_k=(p_{1k}, \cdots, p_{nk})$, each of which is a parity check on (modulo-2 sum of) some subset of the $\nu+1$ information bits $(u_k, u_{k-1}, \cdots, u_{k-\nu})$. When the encoded sequence (codeword) y is sent through a memoryless channel, we have precisely the model of Fig. 2. Fig. 3 shows a particular rate-$\frac{1}{2}$ code with $\nu=2$. (This code is the only one ever used for illustration in the VA coding literature, but the reader must not infer that it is the only one the VA can handle.)

More general convolutional encoders exist: the rate may be k/n, the inputs may be nonbinary, and the encoder may even contain feedback. In every case, however, the code may be taken to be generated by a shift-register process [2].

We might also note that other types of transmission codes (e.g., dc-free codes, run-length-limited codes, and others) can be modeled as outputs of a finite-state machine and hence fall into our general setup [49].

B. Intersymbol Interference

In digital transmission through analog channels, we frequently encounter the following situation. The input sequence u, discrete-time and discrete-valued as in the shift-register model, is used to modulate some continuous waveform which is transmitted through a channel and then sampled. Ideally, samples z_k would equal the corresponding u_k, or some simple function thereof; in fact, however, the samples z_k are perturbed both by noise and by neighboring inputs $u_{k'}$. The latter effect is called intersymbol interference. Sometimes intersymbol interference is introduced deliberately for purposes of spectral shaping, in so-called partial-response systems.

In such cases the output samples can often be modeled as

$$z_k = y_k + n_k$$

where y_k is a deterministic function of a finite number of inputs, say, $y_k = f(u_k, \cdots, u_{k-\nu})$, and n_k is a white Gaussian noise sequence. This is precisely Fig. 2.

To be still more specific, in pulse-amplitude modulation (PAM) the signal sequence y may be taken as the convolution of the input sequence u with some discrete-time channel impulse-response sequence $(h_0, h_1, \cdots)$:

$$y_k = \sum_i h_i u_{k-i}.$$

If $h_i = 0$ for $i > \nu$ (finite impulse response), then we obtain our shift-register model. An illustration of such a model in which intersymbol interference spans three time units ($\nu = 2$) appears in Fig. 4.

It was shown in [29] that even problems where time is actually continuous—i.e., the received signal $r(t)$ has the form

$$r(t) = \sum_{k=0}^{K} u_k h(t - kT) + n(t)$$

for some impulse response $h(t)$, signaling interval T, and realization $n(t)$ of a white Gaussian noise process—can be reduced without loss of optimality to the aforementioned discrete-time form (via a "whitened matched filter").

C. Continuous-Phase FSK

This example is cited not for its practical importance, but because, first, it leads to a simple model we shall later use in an example, and, second, it shows how the VA may lead to fresh insight even in the most traditional situations.

In FSK, a digital input sequence u selects one of m frequencies (if u_k is m-ary) in each signaling interval of length T; that is, the transmitted signal $\eta(t)$ is

$$\eta(t) = \cos\left[\omega(u_k)t + \theta_k\right], \qquad kT \leq t < (k+1)T$$

where $\omega(u_k)$ is the frequency selected by u_k, and θ_k is some phase angle. It is desirable for reasons both of spectral shaping and of modulator simplicity that the phase be continuous at the transition interval; that is, that

$$\omega(u_{k-1})kT + \theta_{k-1} \equiv \omega(u_k)kT + \theta_k \text{ modulo } 2\pi.$$

This is called continuous-phase FSK.

The continuity of the phase introduces memory into the modulation process; i.e., it makes the signal actually transmitted in the kth interval dependent on previous signals. To take the simplest possible case ("deviation ratio" $= \frac{1}{2}$), let the

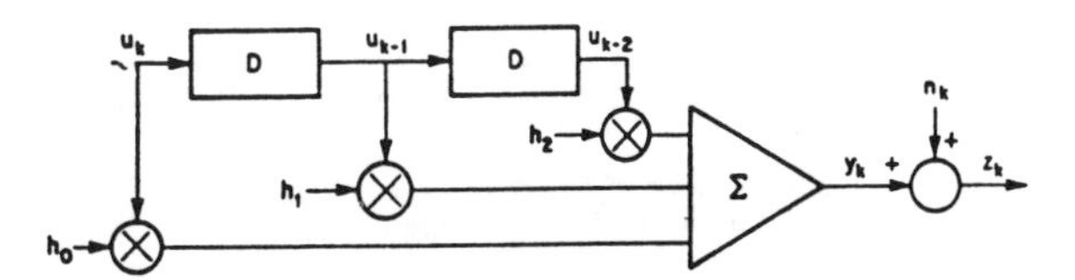

Fig. 4. Model of PAM system subject to intersymbol interference and white Gaussian noise.

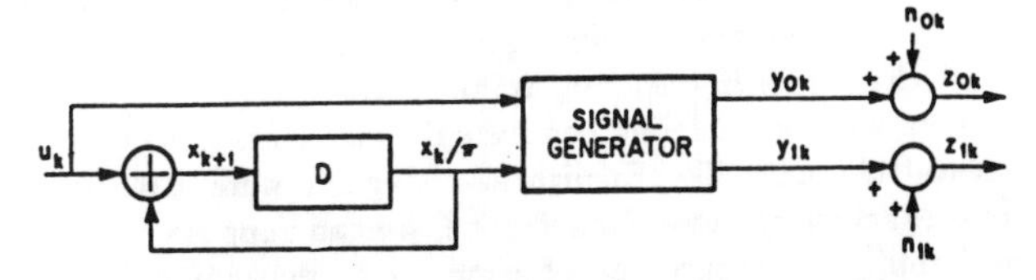

Fig. 5. Model for binary continuous-phase FSK with deviation ratio $\frac{1}{2}$ and coherent detection in white Gaussian noise.

input sequence u be binary and let $\omega(0)$ and $\omega(1)$ be chosen so that $\omega(0)$ goes through an integer number of cycles in T seconds and $\omega(1)$ through an odd half-integer number; i.e., $\omega(0)T \equiv 0$ and $\omega(1)T = \pi$ modulo 2π. Then if $\theta_0 = 0$, $\theta_1 = 0$ or π, according to whether u_0 equals zero or one, and similarly $\theta_k = 0$ or π, according to whether an even or odd number of ones has been transmitted.

Here we have a two-state process, with $X = \{0, \pi\}$. The transmitted signal y_k is a function of both the current input u_k and the state x_k:

$$y_k = \cos\left[\omega(u_k)t + x_k\right] = \cos x_k \cos \omega(u_k)t, \qquad kT \leq t < (k+1)T.$$

Since transitions $\xi_k = (x_{k+1}, x_k)$ are one-to-one functions of the current state x_k and input u_k, we may alternately regard y_k as being determined by ξ_k.

If we take $\eta_0(t) \triangleq \cos \omega(0)t$ and $\eta_1(t) \triangleq \cos \omega(1)t$ as bases of the signal space, we may write

$$y_k = y_{0k}\eta_0(t) + y_{1k}\eta_1(t)$$

where the coordinates (y_{0k}, y_{1k}) are given by

$$(y_{0k}, y_{1k}) = \begin{cases} (1, 0), & \text{if } u_k = 0, x_k = 0 \\ (-1, 0), & \text{if } u_k = 0, x_k = \pi \\ (0, 1), & \text{if } u_k = 1, x_k = 0 \\ (0, -1), & \text{if } u_k = 1, x_k = \pi. \end{cases}$$

Finally, if the received signal $\xi(t)$ is $\eta(t)$ plus white Gaussian noise $\nu(t)$, then by correlating the received signal against both $\eta_0(t)$ and $\eta_1(t)$ in each signal interval (coherent detection), we may arrive without loss of information at a discrete-time output signal

$$z_k = (z_{0k}, z_{1k}) = (y_{0k}, y_{1k}) + (n_{0k}, n_{1k})$$

where n_0 and n_1 are independent equal-variance white Gaussian noise sequences. This model appears in Fig. 5, where the signal generator generates (y_{0k}, y_{1k}) according to the aforementioned rules.

D. Text Recognition

We include this example to show that the VA is not limited to digital communication. In optical-character-recognition (OCR) readers, individual characters are scanned, salient

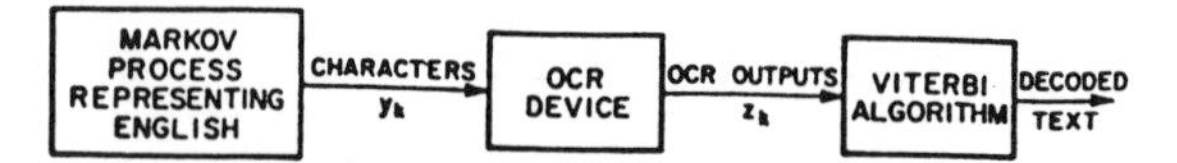

Fig. 6. Use of VA to improve character recognition by exploiting context.

features isolated, and some decision made as to what letter or other character lies below the reader. When the characters actually occur as part of natural-language text, it has long been recognized that contextual information can be used to assist the reader in resolving ambiguities.

One way of modeling contextual constraints is to treat a natural language like English as though it were a discrete-time Markov process. For instance, we can suppose that the probability of occurrence of each letter depends on the ν previous letters, and estimate these probabilities from the frequencies of $(\nu+1)$-letter combinations [$(\nu+1)$-grams]. While such models do not fully describe the generation of natural language (for examples of digram and trigram English, see Shannon [45], [46]), they account for a large part of the statistical dependencies in the language and are easily handled.

With such a model, English letters are viewed as the outputs of an m^ν-state Markov process, where m is the number of distinguishable characters, such as 27 (the 26 letters and a space). If it is further assumed that the OCR output z_k is dependent only on the corresponding input character y_k, then the OCR reader is a memoryless channel to whose output sequence we may apply the VA to exploit contextual constraints; see Fig. 6. Here the "OCR output" may be anything from the raw sensor data, possibly a grid of zeros and ones, to the actual decision which would be made by the reader in the absence of contextual clues. Generally, the more raw the data, the more useful the VA will be.

E. *Other*

Recognizing certain similarities between magnetic recording media and intersymbol interference channels, Kobayashi [50] has proposed applying the VA to digital magnetic recording systems. Timor [51] has applied the algorithm to a sequential ranging system. Use of the algorithm in source coding has been proposed [52]. Finally, Preparata and Ray [53] have suggested using the algorithm to search "semantic maps" in syntactic pattern recognition. These exhaust the applications known to the author.

III. The Algorithm

We now show that the MAP sequence estimation problem previously stated is formally identical to the problem of finding the shortest route through a certain graph. The VA then arises as a natural recursive solution.

We are accustomed to associating with a discrete-time finite-state Markov process a state diagram of the type shown in Fig. 7(a), for a four-state shift-register process like that of Fig. 3 or Fig. 4 (in this case, a de Bruijn diagram [54]). Here nodes represent states, branches represent transitions, and over the course of time the process traces some path from state to state through the state diagram.

In Fig. 7(b) we introduce a more redundant description of the same process, which we have called a *trellis* [3]. Here each node corresponds to a distinct state at a given time, and each branch represents a transition to some new state at the next

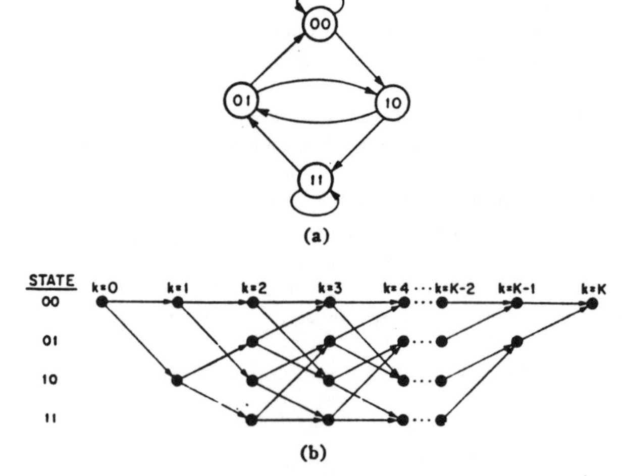

Fig. 7. (a) State diagram of a four-state shift-register process. (b) Trellis for a four-state shift-register process.

instant of time. The trellis begins and ends at the known states x_0 and x_K. Its most important property is that to every possible state sequence x there corresponds a unique path through the trellis, and vice versa.

Now we show how, given a sequence of observations z, every path may be assigned a "length" proportional to $-\ln P(x, z)$, where x is the state sequence associated with that path. This will allow us to solve the problem of finding the state sequence for which $P(x|z)$ is maximum, or equivalently for which $P(x, z)=P(x|z)P(z)$ is maximum, by finding the path whose length $-\ln P(x, z)$ is minimum, since $\ln P(x, z)$ is a monotonic function of $P(x, z)$ and there is a one-to-one correspondence between paths and sequences. We simply observe that due to the Markov and memoryless properties, $P(x, z)$ factors as follows:

$$P(x, z) = P(x)P(z \mid x)$$
$$= \prod_{k=0}^{K-1} P(x_{k+1} \mid x_k) \prod_{k=0}^{K-1} P(z_k \mid x_{k+1}, x_k).$$

Hence if we assign each branch (transition) the "length"

$$\lambda(\xi_k) \triangleq -\ln P(x_{k+1} \mid x_k) - \ln P(z_k \mid \xi_k)$$

then the total length of the path corresponding to some x is

$$-\ln P(x, z) = \sum_{k=0}^{K-1} \lambda(\xi_k)$$

as claimed.

Finding the shortest route through a graph is an old problem in operations research. The most succinct solution was given by Minty in a quarter-page correspondence in 1957 [55], which we quote almost in its entirety:

> The shortest-route problem . . . can be solved very simply . . . as follows: Build a string model of the travel network, where knots represent cities and string lengths represent distances (or costs). Seize the knot "Los Angeles" in your left hand and the knot "Boston" in your right and pull them apart. If the model becomes entangled, have an assistant untie and re-tie knots until the entanglement is resolved. Eventually one or more paths will stretch tight—they then are alternative shortest routes. . . . It is well to label the knots since after one or two uses of the model their identities are easily confused.

Unfortunately, the Minty algorithm is not well adapted to modern methods of machine computation, nor are assistants as pliable as formerly. It therefore becomes necessary to move on to the VA, which is also well known in operations research [56]. It requires one additional observation.

We denote by $\mathbf{x}_0^k$ a segment $(x_0, x_1, \cdots, x_k)$ consisting of the states to time k of the state sequence $\mathbf{x} = (x_0, x_1, \cdots, x_k)$. In the trellis $\mathbf{x}_0^k$ corresponds to a path segment starting at the node x_0 and terminating at x_k. For any particular time-k node x_k, there will in general be several such path segments, each with some length

$$\lambda(\mathbf{x}_0^k) = \sum_{i=0}^{k-1} \lambda(\xi_i).$$

The shortest such path segment is called the *survivor* corresponding to the node x_k, and is denoted $\hat{\mathbf{x}}(x_k)$. For any time $k>0$, there are M survivors in all, one for each x_k. The observation is this: the shortest complete path $\hat{\mathbf{x}}$ must begin with one of these survivors. (If it did not, but went through state x_k at time k, then we could replace its initial segment by $\hat{\mathbf{x}}(x_k)$ to get a shorter path—contradiction.)

Thus at any time k we need remember only the M survivors $\hat{\mathbf{x}}(x_k)$ and their lengths $\Gamma(x_k) \triangleq \lambda[\hat{\mathbf{x}}(x_k)]$. To get to time $k+1$, we need only extend all time-k survivors by one time unit, compute the lengths of the extended path segments, and for each node x_{k+1} select the shortest extended path segment terminating in x_{k+1} as the corresponding time-$(k+1)$ survivor. Recursion proceeds indefinitely without the number of survivors ever exceeding M.

Many readers will recognize this algorithm as a simple version of forward dynamic programming [57], [58]. By this or any other name, the algorithm is elementary once the problem has been cast in the shortest route form.

We illustrate the algorithm for a simple four-state trellis covering 5 time units in Fig. 8. Fig. 8(a) shows the complete trellis, with each branch labeled with a length. (In a real application, the lengths would be functions of the received data.) Fig. 8(b) shows the 5 recursive steps by which the algorithm determines the shortest path from the initial to the final node. At each stage only the 4 (or fewer) survivors are shown, along with their lengths.

A formal statement of the algorithm follows:

Viterbi Algorithm

Storage:

k	(time index);
$\hat{\mathbf{x}}(x_k),\ 1 \le x_k \le M$	(survivor terminating in x_k);
$\Gamma(x_k),\ 1 \le x_k \le M$	(survivor length).

Initialization:

$$k = 0;$$

$$\hat{\mathbf{x}}(x_0) = x_0; \qquad \hat{\mathbf{x}}(m) \text{ arbitrary}, \qquad m \ne x_0;$$

$$\Gamma(x_0) = 0; \qquad \Gamma(m) = \infty, \qquad m \ne x_0.$$

Recursion: Compute

$$\Gamma(x_{k+1}, x_k) \triangleq \Gamma(x_k) + \lambda[\xi_k = (x_{k+1}, x_k)]$$

for all $\xi_k = (x_{k+1}, x_k)$.

Find

$$\Gamma(x_{k+1}) = \min_{x_k} \Gamma(x_{k+1}, x_k)$$

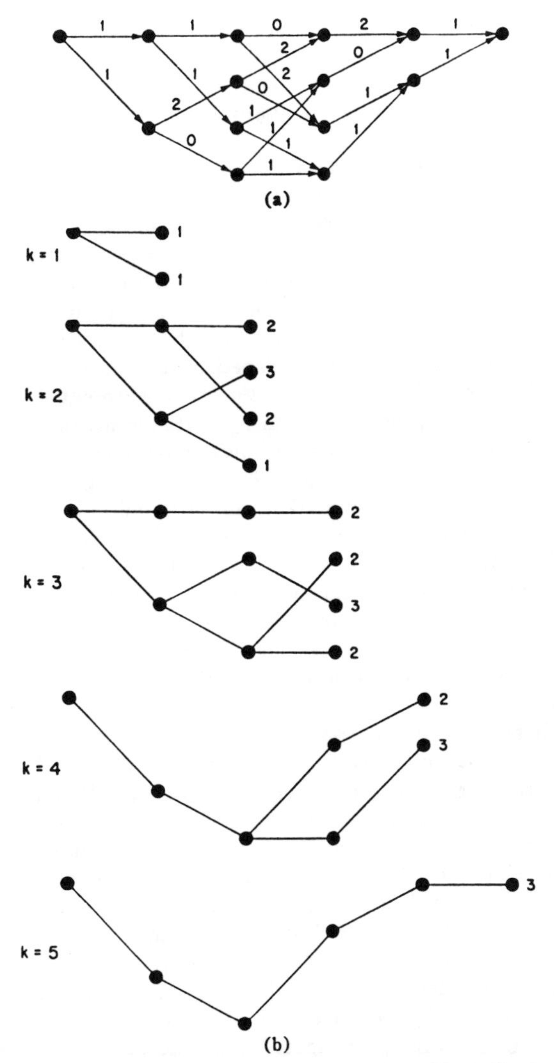

Fig. 8. (a) Trellis labeled with branch lengths; $M=4$, $K=5$. (b) Recursive determination of the shortest path via the VA.

for each x_{k+1}; store $\Gamma(x_{k+1})$ and the corresponding survivor $\hat{\mathbf{x}}(x_{k+1})$.

Set k to $k+1$ and repeat until $k=K$.

With finite state sequences $\mathbf{x}$ the algorithm terminates at time K with the shortest complete path stored as the survivor $\hat{\mathbf{x}}(x_K)$.

Certain trivial modifications are necessary in practice. When state sequences are very long or infinite, it is necessary to truncate survivors to some manageable length δ. In other words, the algorithm must come to a definite decision on nodes up to time $k-\delta$ at time k. Note that in Fig. 8(b) all time-4 survivors go through the same nodes up to time 2. In general, if the truncation depth δ is chosen large enough, there is a high probability that all time-k survivors will go through the same nodes up to time $k-\delta$, so that the initial segment of the maximum-likelihood path is known up to time $k-\delta$ and can be put out as the algorithm's firm decision; in this case, truncation costs nothing. In the rare cases when survivors disagree, any reasonable strategy for determining the algorithm's time-$(k-\delta)$ decision will work [20], [21]: choose an arbitrary time-$(k-\delta)$ node, or the node associated with the shortest survivor, or a node chosen by majority vote, etc. If δ is large enough, the effect on performance is negligible.

Also, if k becomes large, it is necessary to renormalize the lengths $\Gamma(m)$ from time to time by subtracting a constant from all of them.

Finally, the algorithm may be required to get started without knowledge of the initial state x_0. In this case it may be initialized with any reasonable assignment of initial node lengths, such as $\Gamma(m)=0$, all m, or else $\Gamma(m)=-\ln \pi_m$ if the states are known to have *a priori* probabilities π_m. Usually, after an initial transient, there is a high probability that all survivors will merge with the correct path. Thus the algorithm synchronizes itself without any special procedures.

The complexity of the algorithm is easily estimated. First, memory: the algorithm requires M storage locations, one for each state, where each location must be capable of storing a "length" $\Gamma(m)$ and a truncated survivor listing $\hat{x}(m)$ of δ symbols. Second, computation: in each unit of time the algorithm must make $|\Xi|$ additions, one for each transition, and M comparisons among the $|\Xi|$ results. Thus the amount of storage is proportional to the number of states, and the amount of computation to the number of transitions. With a shift-register process, $M=m^\nu$ and $|\Xi|=m^{\nu+1}$, so that the complexity increases exponentially with the length ν of the shift register.

In the previous paragraph, we have ignored the complexity involved in generating the incremental lengths $\lambda(\xi_k)$. In a shift-register process, it is typically true that $P(x_{k+1}|x_k)$ is either $1/m$ or 0, depending on whether x_{k+1} is an allowable successor to x_k or not; then all allowable transitions have the same value of $-\ln P(x_{k+1}|x_k)$ and this component of $\lambda(\xi_k)$ may be ignored. Note that in more general cases $P(x_{k+1}|x_k)$ is known in advance; hence this component can be precomputed and "wired in." The component $-\ln P(z_k|\xi_k)$ is the only component that depends on the data; again, it is typical that many ξ_k lead to the same output y_k, and hence the value $-\ln P(z_k|y_k)$ need be computed or looked up for all these ξ_k only once, given z_k. (When the noise is Gaussian, $-\ln P(z_k|y_k)$ is proportional simply to $(z_k-y_k)^2$.) Finally, all of this can be done outside the central recursion ("pipelined"). Hence the complexity of this computation tends not to be significant.

Furthermore, note that once the $\lambda(\xi_k)$ are computed, the observation z_k need not be stored further, an attractive feature in real-time applications.

A closer look at the trellis of a shift-register process reveals additional detail that can be exploited in implementation. For a binary shift register, the transitions in any unit of time can be segregated into $2^{\nu-1}$ disjoint groups of four, each originating in a common pair of states and terminating in another common pair. A typical such cell is illustrated in Fig. 9, with the time-k states labeled $x'0$ and $x'1$ and the time-$(k+1)$ states labelled $0x'$ and $1x'$, where x' stands for a sequence of $\nu-1$ bits that is constant within any one cell. For example, each time unit in the trellis of Fig. 8 is made up of two such cells. We note that only quantities within the same cell interact in any one recursion. Fig. 10 shows a basic logic unit that implements the computation within any one cell. A high-speed parallel implementation of the algorithm can be built around $2^{\nu-1}$ such identical logic units; a low-speed implementation, around one such unit, time-shared; or a software version, around the corresponding subroutine.

Many readers will have noticed that the trellis of Fig. 7(b) reminds them of the computational flow diagram of the fast

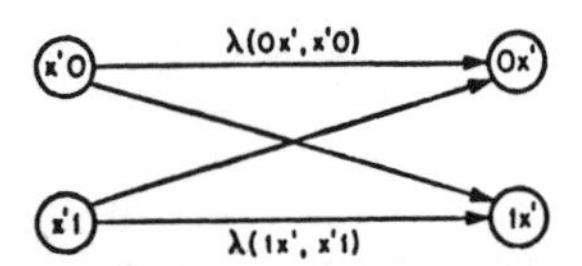

Fig. 9. Typical cell of a shift-register process trellis.

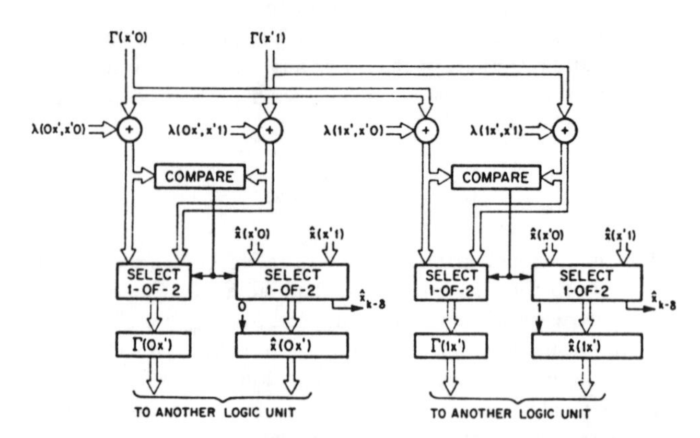

Fig. 10. Basic VA logic unit for binary shift-register process.

Fourier transform (FFT). In fact, it is identical, except for length, and indeed the FFT is also ordinarily organized cellwise. While the add-and-compare computations of the VA are unlike those involved in the FFT, some of the memory-organization tricks developed for the FFT may be expected to be equally useful here.

Because of its highly parallel structure and need for only add, compare, and select operations, the VA is well suited to high-speed applications. A convolutional decoder for a $\nu=6$ code ($M=64$, $|\Xi|=128$) that is built out of 356 transistor–transistor logic circuits and that can operate at up to 2 Mbits/s perhaps represents the current state of the art [22]. Even a software decoder for a similar code can be run at a rate the order of 1000 bits/s on a minicomputer. Such moderate complexity qualifies the VA for inclusion in many signal-processing systems.

For further details of implementation, see [22], [23], and the references therein.

IV. Analysis of Performance

Just as important as the straightforwardness of implementation of the VA is the straightforwardness with which its performance can be analyzed. In many cases, tight upper and lower bounds for error probability can be derived. Even when the VA is not actually implemented, calculation of its performance shows how far the performance of less complex schemes is from ideal, and often suggests simple suboptimum schemes that attain nearly optimal performance.

The key concept in performance analysis is that of an error event. Let x be the actual state sequence, and $\hat{x}$ the state sequence actually chosen by the VA. Over a long time x and $\hat{x}$ will typically diverge and remerge a number of times, as illustrated in Fig. 11. Each distinct separation is called an error event. Error events may in general be of unbounded length if x is infinite, but the probability of an infinite error event will usually be zero.

The importance of error events is that they are probabilistically independent of one another; in the language of probability theory they are *recurrent*. Furthermore, they allow us to calculate error probability per unit time, which is neces-

Fig. 11. Typical correct path x (heavy line) and estimated path $\hat{x}$ (lighter line) in the trellis, showing three error events.

sary since usually the probability of *any* error in MAP estimation of a block of length K goes to 1 as K goes to infinity. Instead we calculate the probability of an error event starting at some given time, given that the starting state is correct, i.e., that an error event is not already in progress at that time.

Given the correct path x, the set $\mathcal{E}_k$ of all possible error events starting at some time k is a treelike trellis which starts at x_k and each of whose branches ends on the correct path, as illustrated in Fig. 12, for the trellis of Fig. 7(b). In coding theory this is called the incorrect subset (at time k).

The probability of any particular error event is easily calculated; it is simply the probability that the observations will be such that over the time span during which $\hat{x}$ is different from x, $\hat{x}$ is more likely than x. If the error event has length τ, this is simply a two-hypothesis decision problem between two sequences of length τ, and typically has a standard solution.

The probability $P(\mathcal{E}_k)$ that *any* error event in $\mathcal{E}_k$ occurs can then be upper-bounded, usually tightly, by a union bound, i.e., by the sum of the probabilities of all error events in $\mathcal{E}_k$. While this sum may well be infinite, it is typically dominated by one or a few large leading terms representing particularly likely error events, whose sum then forms a good approximation to $P(\mathcal{E}_k)$. True upper bounds can often be obtained by flow-graph techniques [4], [5], [29].

On the other hand, a lower bound to error-event probability, again frequently tight, can be obtained by a genie argument. Take the particular error event that has the greatest probability of all those in $\mathcal{E}_k$. Suppose that a friendly genie tells you that the true state sequence is one of two possibilities: the actual correct path, or the incorrect path corresponding to that error event. Even with this side information, you will still make an error if the incorrect path is more likely given z, so your probability of error is still no better than the probability of this particular error event. In the absence of the genie, your error probability must be worse still, since one of the strategies you have, given the genie's information, is to ignore it. In summary, the probability of any particular error event is a lower bound to $P(\mathcal{E}_k)$.

An important side observation is that this lower bound applies to any decision scheme, not just the VA. Simple extensions give lower bounds to related quantities like bit probability of error. If the VA gives performance approaching these bounds, then it may be claimed that it is effectively optimum with respect to these related quantities as well [30].

In conclusion, the probability of any error event starting at time k may be upper- and lower-bounded as follows:

$$\max P(\text{error event}) \le P(\mathcal{E}_k) \le \max P(\text{error event}) + \text{other terms}.$$

With luck these bounds will be close.

Example

For concreteness, and because the result is instructive, we carry through the calculation for continuous-phase FSK of the particularly simple type defined earlier. The two-state trellis for this process is shown in Fig. 13, and the first part of

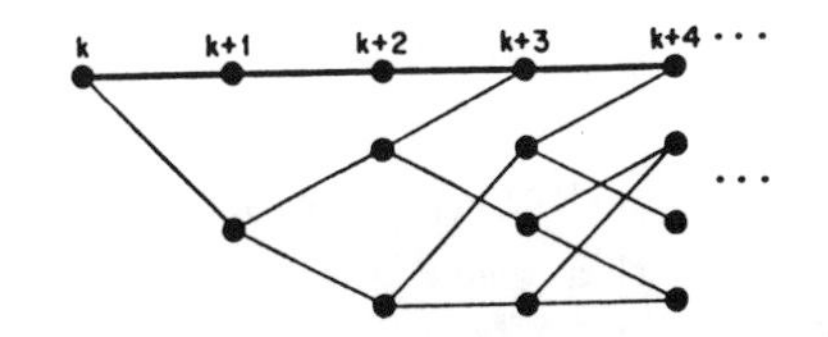

Fig. 12. Typical correct path x (heavy line) and time-k incorrect subset for trellis of Fig. 7(b).

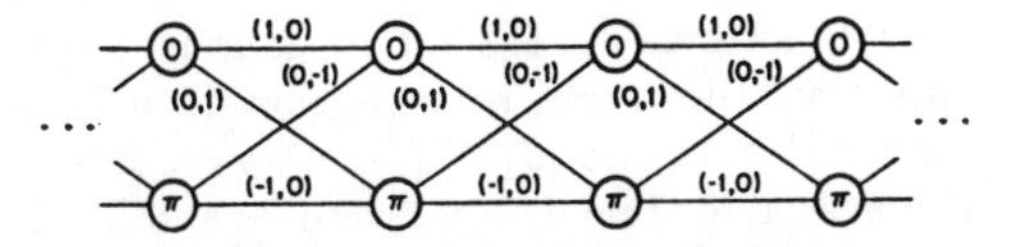

Fig. 13. Trellis for continuous-phase FSK.

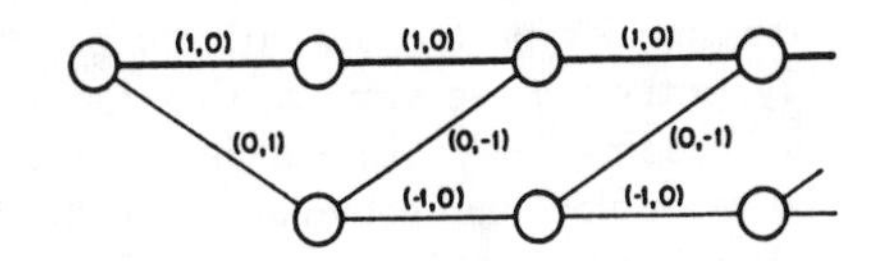

Fig. 14. Typical incorrect subset. Heavy line: correct path. Lighter lines: incorrect subset.

a typical incorrect subset in Fig. 14. The shortest possible error event is of length 2 and consists of a decision that the signal was $\{\cos \omega(1)t, -\cos \omega(1)t\}$ rather than $\{\cos \omega(0)t, \cos \omega(0)t\}$, or in our coordinate notation that $\{(0, 1), (0, -1)\}$ is chosen over $\{(1, 0), (1, 0)\}$. This is a two-hypothesis decision problem in a four-dimensional signal space [59]. In Gaussian noise of variance σ^2 per dimension, only the Euclidean distance d between the two signals matters; in this case $d=\sqrt{2}$, and therefore the probability of this particular error event is $Q(d/2\sigma)=Q(1/\sigma\sqrt{2})$, where $Q(x)$ is the Gaussian error probability function defined by

$$Q(x) \triangleq \int_x^\infty \frac{1}{\sqrt{2\pi}} e^{-y^2/2} dy.$$

By examination of Fig. 14 we see that the error events of lengths 3, 4, $\cdots$ lie at distances $\sqrt{6}$, $\sqrt{10}$, $\cdots$ from the correct path; hence we arrive at upper and lower bounds on $P(\mathcal{E}_k)$ of

$$Q(\sqrt{2}/2\sigma) \le P(\mathcal{E}_k) \le Q(\sqrt{2}/2\sigma) + Q(\sqrt{6}/2\sigma) + Q(\sqrt{10}/2\sigma) + \cdots.$$

In view of the rapid decrease of $Q(x)$ with x, this implies that $P(\mathcal{E}_k)$ is accurately estimated as $Q(\sqrt{2}/2\sigma)$ for any reasonable noise variance σ^2. It is easily verified from Fig. 13 that this result is independent of the correct path x or the time k.

This result is interesting in that the best one can do with coherent detection in a single symbol interval is $Q(1/2\sigma)$ for orthogonal signals. Thus exploiting the memory doubles the effective signal energy, or improves the signal-to-noise ratio by 3 dB. It may therefore be claimed that continuous-phase FSK is inherently 3 dB better than noncontinuous for deviation ratio $\frac{1}{2}$, or as good as antipodal phase-shift keying. (While we have proved this only for a deviation ratio of $\frac{1}{2}$, it holds for nearly any deviation ratio.) Even though the VA is quite simple for a two-state trellis, the fact that only one type of error event has any significant probability permits still simpler suboptimum schemes to achieve effectively the same performance [43], [44].[2]

V. Applications

We conclude with a review of the results, both theoretical and practical, obtained with the VA to date.

A. Convolutional Codes

It was for convolutional codes that the algorithm was first developed, and naturally it has had its greatest impact here.

The principal theoretical result is contained in Viterbi's original paper [1]; see also [5]–[7]. It shows that for a suitably defined ensemble of random trellis codes and MAP decoding, the error probability can be made to decrease exponentially with the constraint length ν at all code rates R less than channel capacity. Furthermore, the rate of decrease is considerably faster than for block codes with comparable decoding complexity (although the same as that for block codes with the same decoding delay). In our view, the most telling comparison between block and convolutional codes is that an effectively optimum block code of any specified length and rate can be created by suitably terminating a convolutional (trellis) code, but with a lower rate for the block code of course [7]. In fact, if convolutional codes were any better, they could be terminated to yield better block codes than are theoretically possible—an observation which shows that the bounds on convolutional code performance must be tight.

For fixed binary convolutional codes of nonasymptotic length on symmetric memoryless channels, the principal result [6] is that $P(\mathcal{E}_k)$ is approximately given by

$$P(\mathcal{E}_k) \approx N_d 2^{-dD}$$

where d is the free distance, i.e., the minimum Hamming distance of any path in the incorrect subset $\mathcal{E}_k$ from the correct path; N_d is the number of such paths; and D is the Bhattacharyya distance

$$D = \log_2 \sum_{Z} P(z \mid 0)^{1/2} P(z \mid 1)^{1/2}$$

where the sum is over all outputs z in the channel output space Z.

On Gaussian channels

$$P(\mathcal{E}_k) \approx N_d \exp(-dRE_b/N_0)$$

where E_b/N_0 is the signal-to-noise ratio per information bit. The tightness of this bound is confirmed by simulations [8], [9], [22]. For a $\nu = 6$, $d = 10$, $R = \frac{1}{2}$ code, for example, error probabilities of 10^{-3}, 10^{-5}, and 10^{-7} are achieved at $E_b/N_0 = 3.0$, 4.3, and 5.5 dB, respectively, which is within a few tenths of a dceibel of this bound [22].

Channels available for space communications are frequently accurately modeled as white Gaussian channels. The VA is attractive for such channels because it gives performance superior to all other coding schemes save sequential decoding, and does this at high speeds, with modest complexity, and with considerable robustness against varying channel parameters. A number of prototype systems have been implemented and tested [21]–[26], some quite original, and it seems likely that Viterbi decoders will become common in space communication systems.

Finally, Viterbi decoders have been used as elements in very-high-performance concatenated coding schemes [10]–[12], [27] and in decoding of convolutional codes in the presence of intersymbol interference [35], [60].

B. Intersymbol Interference

Application of the VA to intersymbol interference problems is more recent, and the main achievements have been theoretical. The principal result, for PAM in white Gaussian noise, is that $P(\mathcal{E}_k)$ can be tightly bounded as follows:

$$K_L Q(d_{\min}/2\sigma) \leq P(\mathcal{E}_k) \leq K_U Q(d_{\min}/2\sigma)$$

where K_L and K_U are small constants, $Q(x)$ is the Gaussian error probability function defined earlier, σ^2 is the noise variance, and $d_{\min}$ is the minimum Euclidean distance between any two distinct signals [29]. This result implies that on most channels intersymbol interference need not lead to any significant degradation in performance, which comes as rather a surprise.

For example, with the most common partial response systems, the VA recovers the 3-dB loss sustained by conventional detectors relative to full-response systems [29], [32]. Simple suboptimum processors [29], [33] can do nearly as well.

Several workers [34], [35], [42] have proposed adaptive versions of the algorithm for unknown and time-varying channels. Ungerboeck [41], [42] and Mackechnie [35] have shown that only a matched filter rather than a whitened matched filter is needed in PAM.

It seems most likely that the greatest effect of the VA on digital modulation systems will be to reveal those instances in which conventional detection techniques fall significantly short of optimum, and to suggest effective suboptimum methods of closing the gap. PAM channels that cannot be linearly equalized without excessive noise enhancement due to nulls or near nulls in the transmission band are the likeliest candidates for nonlinear techniques of this kind.

C. Text Recognition

An experiment was run in which digram statistics were used to correct garbled text produced by a simulated noisy character recognizer [47]. Results were similar to those of Raviv [48] (using digram statistics), although the algorithm was simpler and only had hard decisions rather than confidence levels to work with. It appears that the algorithm may be a useful adjunct to sophisticated character-recognition systems for resolving ambiguities when confidence levels for different characters are available.

VI. Conclusion

The VA has already had a significant impact on our understanding of certain problems, notably in the theories of convolutional codes and of intersymbol interference. It is beginning to have a substantial practical impact as well in the engineering of space-communication links. The amount of work it has

[2] De Buda [44] actually proves that an optimum decision on the phase x_k at time k can be made by examining the received waveform only at times $k-1$ and k; i.e., $(z_{0,k-1}, z_{1,k-1})$, (z_{0k}, z_{1k}). The proof is that the log likelihood ratio

$$-\ln \frac{P(z_{0,k-1}, z_{1,k-1}, z_{0k}, z_{1k} \mid x_{k-1}, x_k = 0, x_{k+1})}{P(z_{0,k-1}, z_{1,k-1}, z_{0k}, z_{1k} \mid x_{k-1}, x_k = \pi, x_{k+1})}$$

is proportional to $-z_{0,k-1} + z_{1,k-1} - z_{0k} - z_{1k}$ for any values of the pair of states (x_{k-1}, x_{k+1}). For this phase decision (which differs slightly from our sequence decision) the error probability is exactly $Q(\sqrt{2}/2\sigma)$.

inspired in the intersymbol interference area suggests that here too practical applications are not far off. The generality of the model to which it applies and the straightforwardness with which it can be analyzed and implemented lead one to believe that in both theory and practice it will find increasing application in the years ahead.

Appendix

Related Algorithms

In this Appendix we mention some processing structures that are closely related to the VA, chiefly sequential decoding and minimum-bit-error-probability algorithms. We also mention extensions of the algorithm to generate reliability information, erasures, and lists.

When the trellis becomes large, it is natural to abandon the exhaustive search of the VA in favor of a sequential trial-and-error search that selectively examines only those paths likely to be the shortest. In the coding literature, such algorithms are collectively known as sequential decoding [13]–[16]. The simplest to explain is the "stack" algorithm [15], [16], in which a list is maintained of the shortest partial paths found to date, the path on the top of the list is extended, and its successors reordered in the list until some path is found that reaches the terminal node, or else decreases without limit. (That some path will eventually do so is ensured in coding applications by the subtraction of a bias term such that the length of the correct path tends to decrease while that of all incorrect paths tends to increase.) Searches that start from either end of a finite trellis [17] are also useful.

In coding applications, sequential decoding has many of the same properties as Viterbi decoding, including the same error probability. It allows the decoding of longer and therefore more powerful codes, at the cost of a variable amount of computation necessitating buffer storage for the incoming data z. It is probably less useful outside of coding, since it depends on the decoder's ability to recognize when the best path has been found without examining other paths, and therefore requires either a finite trellis or a very large distance between the correct path and possible error events.

In the intersymbol interference literature, many of the early attempts to find optimum nonlinear algorithms used bit-error probability as the optimality criterion. The Markov property of the process leads to algorithms that are manageable but less attractive than the Viterbi [36]–[42].

The general principle of several of these algorithms is as follows.[3] First, we calculate the joint probability $P(x_k, z)$ for every state x_k in the trellis, or alternately $P(\xi_k, z)$ for every transition ξ_k. This is done by observing that

$$\begin{aligned} P(x_k, z) &= P(x_k, z_0^{k-1})P(z_k^K \mid x_k, z_0^{k-1}) \\ &= P(x_k, z_0^{k-1})P(z_k^K \mid x_k) \end{aligned}$$

since, given x_k, the outputs z_k^K from time k to K are independent of the outputs z_0^{k-1} from time 0 to $k-1$. Similarly

$$\begin{aligned} P(\xi_k, z) &= P(x_k, x_{k+1}, z) \\ &= P(x_k, z_0^{k-1})P(x_{k+1}, z_k \mid x_k, z_0^{k-1}) \\ &\quad \cdot P(z_{k+1}^K \mid x_{k+1}, x_k, z_0^k) \\ &= P(x_k, z_0^{k-1})P(x_{k+1}, z_k \mid x_k)P(z_{k+1}^K \mid x_{k+1}). \end{aligned}$$

[3] The author is indebted to Bahl *et al.* [18] for a particularly lucid exposition of this type of algorithm.

Now we note the recursive formula

$$\begin{aligned} P(x_k, z_0^{k-1}) &= \sum_{x_{k-1}} P(x_k, x_{k-1}, z_0^{k-1}) \\ &= \sum_{x_{k-1}} P(x_{k-1}, z_0^{k-2})P(x_k, z_{k-1} \mid x_{k-1}) \end{aligned}$$

which allows us to calculate the M quantities $P(x_k, z_0^{k-1})$ from the M quantities $P(x_{k-1}, z_0^{k-2})$ with $|\Xi|$ multiplications and additions using the exponentiated lengths

$$e^{-\lambda(\xi_{k-1})} = P(x_k \mid x_{k-1})P(z_{k-1} \mid \xi_{k-1}).$$

Similarly we have the backward recursion

$$\begin{aligned} P(z_k^K \mid x_k) &= \sum_{x_{k+1}} P(z_k^K, x_{k+1} \mid x_k) \\ &= \sum_{x_{k+1}} P(z_k, x_{k+1} \mid x_k)P(z_{k+1}^K \mid x_{k+1}) \end{aligned}$$

which has a similar complexity. Completion of these forward and backward recursions for all nodes allows $P(x_k, z)$ and/or $P(\xi_k, z)$ to be calculated for all nodes.

Now, to be specific, let us consider a shift-register process and let $S(u_k)$ be the set of all states x_{k+1} whose first component is u_k. Then

$$P(u_k, z) = \sum_{x_{k+1} \in S(u_k)} P(x_{k+1}, z).$$

Since $P(u_k, z) = P(u_k \mid z)P(z)$, MAP estimation of u_k reduces to finding the maximum of this quantity. Similarly, if we wish to find the MAP estimate of an output y_k, say, then let $S(y_k)$ be the set of all ξ_k that lead to y_k and compute

$$P(y_k, z) = \sum_{\xi_k \in S(y_k)} P(\xi_k, z).$$

A similar procedure can be used to estimate any quantity which is a deterministic function of states or transitions.

Besides requiring multiplications, this algorithm is less attractive than the VA in requiring a backward as well as a forward recursion and consequently storage of all data. The following amended algorithm [39], [48] eliminates the latter ugly feature at the cost of suboptimal performance and additional computation. Let us restrict ourselves to a shift-register process with input sequence $\mathbf{u}$ and agree to use only observations up to time $k+\delta$ in estimating u_k, say, where $\delta \geq \nu - 1$. We then have

$$P(u_k, z_0^{k+\delta}) = \sum_{u_{k+1}} \cdots \sum_{u_{k+\delta}} P(\mathbf{u}_k^{k+\delta}, z_0^{k+\delta}).$$

The $m^{\delta+1}$ quantities in the sum can be determined recursively by

$$\begin{aligned} P(\mathbf{u}_k^{k+\delta}, z_0^{k+\delta}) &= \sum_{u_{k-1}} P(\mathbf{u}_{k-1}^{k+\delta}, z_0^{k+\delta}) \\ &= \sum_{u_{k-1}} P(\mathbf{u}_{k-1}^{k+\delta-1}, z_0^{k+\delta-1}) \\ &\quad \cdot P(u_{k+\delta}, z_{k+\delta} \mid u_{k+\delta-1}, \cdots, u_{k+\delta-\nu}). \end{aligned}$$

While the recursion is now forward only, we now must store $m^{\delta+1}$ quantities rather than m^ν. If δ is large, this is most unattractive; if δ is close to $\nu-1$, the estimate may well be decidedly suboptimum.

These variations thus seem considerably less attractive than the VA. Nonetheless, something like this may need to be hybridized with the VA in certain situations such as tracking a finite-state source over a finite-state channel, where only the state sequence of the source is of interest.

Finally, we can consider augmented outputs from the VA. A good general indication of how well the algorithm is doing is the depth at which all paths are merged; this can be used to establish whether or not a communications channel is on the air, in synchronism, etc. [11], [28]. A more selective indicator of how reliable particular segments are is the difference in lengths between the best and the next-best paths at the point of merging; this reliability indicator can be quantized into an erasure output. Lastly, the algorithm can be altered to store the L best paths, rather than the single best path, as the survivors in each recursion, thus eventually generating a list of the L most likely path sequences.

References

Convolutional Codes

[1] A. J. Viterbi, "Error bounds for convolutional codes and an asymptotically optimum decoding algorithm," *IEEE Trans. Inform. Theory*, vol. IT-13, pp. 260–269, Apr. 1967.

[2] G. D. Forney, Jr., "Convolutional codes I: Algebraic structure," *IEEE Trans. Inform. Theory*, vol. IT-16, pp. 720–738, Nov. 1970.

[3] ——, "Review of random tree codes," NASA Ames Res. Cen., Moffett Field, Calif., Contract NAS2-3637, NASA CR 73176, Final Rep., Dec. 1967, appendix A.

[4] G. C. Clark, R. C. Davis, J. C. Herndon, and D. D. McRae, "Interim report on convolution coding research," Advanced System Operation, Radiation Inc., Melbourne, Fla., Memo Rep. 38, Sept. 1969.

[5] A. J. Viterbi and J. P. Odenwalder, "Further results on optimal decoding of convolutional codes," *IEEE Trans. Inform. Theory* (Corresp.), vol. IT-15, pp. 732–734, Nov. 1969.

[6] A. J. Viterbi, "Convolutional codes and their performance in communication systems," *IEEE Trans. Commun. Technol.*, vol. COM-19, pp. 751–772, Oct. 1971.

[7] G. D. Forney, Jr., "Convolutional codes II: Maximum likelihood decoding," Stanford Electronics Labs., Stanford, Calif., Tech. Rep. 7004-1, June 1972.

[8] J. A. Heller, "Short constraint length convolutional codes," in *Space Program Summary 37-54*, vol. III. Jet Propulsion Lab., Calif. Inst. Technol., pp. 171–177, Oct.–Nov. 1968.

[9] ——, "Improved performance of short constraint length convolutional codes," in *Space Program Summary 37-56*, vol. III. Jet Propulsion Lab., Calif. Inst. Technol., pp. 83–84, Feb.–Mar. 1969.

[10] J. P. Odenwalder, "Optimal decoding of convolutional codes," Ph.D. dissertation, Dep. Syst. Sci., Sch. Eng. Appl. Sci., Univ. of California, Los Angeles, 1970.

[11] J. L. Ramsey, "Cascaded tree codes," MIT Res. Lab. Electron., Cambridge, Mass., Tech. Rep. 478, Sept. 1970.

[12] G. W. Zeoli, "Coupled decoding of block-convolutional concatenated codes," Ph.D. dissertation, Dep. Elec. Eng., Univ. of California, Los Angeles, 1971.

[13] J. M. Wozencraft, "Sequential decoding for reliable communication," in *1957 IRE Nat. Conv. Rec.*, vol. 5, pt. 2, pp. 11–25.

[14] R. M. Fano, "A heuristic discussion of probabilistic decoding," *IEEE Trans. Inform. Theory*, vol. IT-9, pp. 64–74, Apr. 1963.

[15] K. S. Zigangirov, "Some sequential decoding procedures," *Probl. Pered. Inform.*, vol. 2, pp. 13–25, 1966.

[16] F. Jelinek, "Fast sequential decoding algorithm using a stack," *IBM J. Res. Develop.*, vol. 13, pp. 675–685, Nov. 1969.

[17] L. R. Bahl, C. D. Cullum, W. D. Frazer, and F. Jelinek, "An efficient algorithm for computing free distance," *IEEE Trans. Inform. Theory* (Corresp.), vol. IT-18, pp. 437–439, May 1972.

[18] L. R. Bahl, J. Cocke, F. Jelinek, and J. Raviv, "Optimal decoding of linear codes for minimizing symbol error rate" (Abstract), in *1972 Int. Symp. Information Theory* (Pacific Grove, Calif., Jan. 1972), p. 90.

[19] P. L. McAdam, L. R. Welch, and C. L. Weber, "M.A.P. bit decoding of convolutional codes," in *1972 Int. Symp. Information Theory* (Pacific Grove, Calif., Jan. 1972), p. 91.

Space Applications

[20] Linkabit Corp., "Coding systems study for high data rate telemetry links," Final Rep. on NASA Ames Res. Cen., Moffett Field, Calif., Contract NAS2-6024, Rep. CR-114278, 1970.

[21] G. C. Clark, "Implementation of maximum likelihood decoders for convolutional codes," in *Proc. Int. Telemetering Conf.* (Washington, D. C., 1971).

[22] J. A. Heller and I. M. Jacobs, "Viterbi decoding for satellite and space communication," *IEEE Trans. Commun. Technol.*, vol. COM-19, pp. 835–847, Oct. 1971.

[23] G. C. Clark, Jr., and R. C. Davis, "Two recent applications of error-correction coding to communications system design," *IEEE Trans. Commun. Technol.*, vol. COM-19, pp. 856–863, Oct. 1971.

[24] I. M. Jacobs and R. J. Sims, "Configuring a TDMA satellite communication system with coding, in *Proc. 5th Hawaii Int. Conf. Systems Science.* Honolulu, Hawaii: Western Periodicals, 1972, pp. 443–446.

[25] A. R. Cohen, J. A. Heller, and A. J. Viterbi, "A new coding technique for asynchronous multiple access communication," *IEEE Trans. Commun. Technol.*, vol. COM-19, pp. 849–855, Oct. 1971.

[26] D. Quagliato, "Error correcting codes applied to satellite channels," in *1972 IEEE Int. Conf. Communications* (Philadelphia, Pa.), pp. 15/13–18.

[27] Linkabit Corp., "Hybrid coding system study," Final Rep. on Contract NAS2-6722, NASA Ames Res. Cen., Moffett Field, Calif., NASA Rep. CR114486, Sept. 1972.

[28] G. C. Clark, Jr., and R. C. Davis, "Reliability-of-decoding indicators for maximum likelihood decoders," in *Proc. 5th Hawaii Int. Conf. Systems Science.* Honolulu, Hawaii: Western Periodicals, 1927, pp. 447–450.

Intersymbol Interference

[29] G. D. Forney, Jr., "Maximum-likelihood sequence estimation of digital sequences in the presence of intersymbol interference," *IEEE Trans. Inform. Theory*, vol. IT-18, pp. 363–378, May 1972.

[30] ——, "Lower bounds on error probability in the presence of large intersymbol interference," *IEEE Trans. Commun. Technol.* (Corresp.), vol. COM-20, pp. 76–77, Feb. 1972.

[31] J. K. Omura, "On optimum receivers for channels with intersymbol interference," abstract presented at the IEEE Int. Symp. Information Theory, Noordwijk, The Netherlands, June 1970.

[32] H. Kobayashi, "Correlative level coding and maximum-likelihood decoding," *IEEE Trans. Inform. Theory*, vol. IT-17, pp. 586–594, Sept. 1971.

[33] M. J. Ferguson, "Optimal reception for binary partial response channels," *Bell Syst. Tech. J.*, vol. 51, pp. 493–505, Feb. 1972.

[34] F. R. Magee, Jr., and J. G. Proakis, "Adaptive maximum-likelihood sequence estimation for digital signaling in the presence of intersymbol interference," *IEEE Trans. Inform. Theory* (Corresp.), vol. IT-19, pp. 120–124, Jan. 1973.

[35] L. K. Mackechnie, "Maximum likelihood receivers for channels having memory," Ph.D. dissertation, Dep. Elec. Eng., Univ. of Notre Dame, Notre Dame, Ind., Jan. 1973.

[36] R. W. Chang and J. C. Hancock, "On receiver structures for channels having memory," *IEEE Trans. Inform. Theory*, vol. IT-12, pp. 463–468, Oct. 1966.

[37] K. Abend, T. J. Harley, Jr., B. D. Fritchman, and C. Gumacos, "On optimum receivers for channels having memory," *IEEE Trans. Inform. Theory* (Corresp.), vol. IT-14, pp. 819–820, Nov. 1968.

[38] R. R. Bowen, "Bayesian decision procedures for interfering digital signals," *IEEE Trans. Inform. Theory* (Corresp.), vol. IT-15, pp. 506–507, July 1969.

[39] K. Abend and B. D. Fritchman, "Statistical detection for communication channels with intersymbol interference," *Proc. IEEE*, vol. 58, pp. 779–785, May 1970.

[40] C. G. Hilborn, Jr., "Applications of unsupervised learning to problems of digital communication," in *Proc. 9th IEEE Symp. Adaptive Processes, Decision, and Control* (Dec. 7–9, 1970).
C. G. Hilborn, Jr., and D. G. Lainiotis, "Optimal unsupervised learning multicategory dependent hypotheses pattern recognition," *IEEE Trans. Inform. Theory*, vol. IT-14, pp. 468–470, May 1968.

[41] G. Ungerboeck, "Nonlinear equalization of binary signals in Gaussian noise," *IEEE Trans. Commun. Technol.*, vol. COM-19, pp. 1128–1137, Dec. 1971.

[42] ——, "Adaptive maximum-likelihood receiver for carrier-modulated data transmission systems," in preparation.

Continuous-Phase FSK

[43] M. G. Pelchat, R. C. Davis, and M. B. Luntz, "Coherent demodulation of continuous-phase binary FSK signals," in *Proc. Int. Telemetry Conf.* (Washington, D. C., 1971).

[44] R. de Buda, "Coherent demodulation of frequency-shift keying with low deviation ratio," *IEEE Trans. Commun. Technol.*, vol. COM-20, pp. 429–435, June 1972.

Text Recognition

[45] C. E. Shannon and W. Weaver, *The Mathematical Theory of Communication.* Urbana, Ill.: Univ. of Ill. Press, 1949.

[46] C. E. Shannon, "Prediction and entropy of printed English," *Bell Syst. Tech. J.*, vol. 30, pp. 50–64, Jan. 1951.

[47] D. L. Neuhoff, "The Viterbi algorithm as an aid in text recognition," Stanford Electronic Labs., Stanford, Calif., unpublished.

[48] J. Raviv, "Decision making in Markov chains applied to the problem of pattern recognition," *IEEE Trans. Inform. Theory*, vol. IT-13, pp. 536–551, Oct. 1967.

Miscellaneous

[49] H. Kobayashi, "A survey of coding schemes for transmission or recording of digital data," *IEEE Trans. Commun. Technol.*, vol. COM-19, pp. 1087–1100, Dec. 1971.
[50] ——, "Application of probabilistic decoding to digital magnetic recording systems," *IBM J. Res. Develop.*, vol. 15, pp. 64–74, Jan. 1971.
[51] U. Timor, "Sequential ranging with the Viterbi algorithm," Jet Propulsion Lab., Pasadena, Calif., JPL Tech. Rep. 32-1526, vol. II, pp. 75–79, Jan. 1971.
[52] J. K. Omura, "On the Viterbi algorithm for source coding" (Abstract), in *1972 IEEE Int. Symp. Information Theory* (Pacific Grove, Calif., Jan. 1972), p. 21.
[53] F. P. Preparata and S. R. Ray, "An approach to artificial nonsymbolic cognition," *Inform. Sci.*, vol. 4, pp. 65–86, Jan. 1972.
[54] S. W. Golomb, *Shift Register Sequences.* San Francisco, Calif.: Holden-Day, 1967, pp. 13–17.
[55] G. J. Minty, "A comment on the shortest-route problem," *Oper. Res.*, vol. 5, p. 724, Oct. 1957.
[56] M. Pollack and W. Wiebenson, "Solutions of the shortest-route problem—A review," *Oper. Res.*, vol. 8, pp. 224–230, Mar. 1960.
[57] R. Busacker and T. Saaty, *Finite Graphs and Networks: An Introduction with Applications.* New York: McGraw-Hill, 1965.
[58] J. K. Omura, "On the Viterbi decoding algorithm," *IEEE Trans. Inform. Theory*, vol. IT-15, pp. 177–179, Jan. 1969.
[59] J. M. Wozencraft and I. M. Jacobs, *Principles of Communication Engineering.* New York: Wiley, 1965, ch. 4.
[60] J. K. Omura, "Optimal receiver design for convolutional codes and channels with memory via control theoretical concepts," *Inform. Sci.*, vol. 3, pp. 243–266, July 1971.

Trellis-Coded Modulation with Redundant Signal Sets Part I: Introduction

Gottfried Ungerboeck

Simple four-state trellis-coded modulation (TCM) schemes improve the robustness of digital transmission against additive noise by 3 dB without reducing data rate or requiring more bandwidth than conventional uncoded modulation schemes. With more complex schemes, coding gains up to 6 dB can be achieved. This article describes how TCM works

Trellis-Coded Modulation (TCM) has evolved over the past decade as a combined coding and modulation technique for digital transmission over band-limited channels. Its main attraction comes from the fact that it allows the achievement of significant coding gains over conventional uncoded multilevel modulation without compromising bandwidth efficiency. The first TCM schemes were proposed in 1976 [1]. Following a more detailed publication [2] in 1982, an explosion of research and actual implementations of TCM took place, to the point where today there is a good understanding of the theory and capabilities of TCM methods. In Part 1 of this two-part article, an introduction into TCM is given. The reasons for the development of TCM are reviewed, and examples of simple TCM schemes are discussed. Part II [15] provides further insight into code design and performance, and addresses recent advances in TCM.

TCM schemes employ redundant nonbinary modulation in combination with a finite-state encoder which governs the selection of modulation signals to generate coded signal sequences. In the receiver, the noisy signals are decoded by a soft-decision maximum-likelihood sequence decoder. Simple four-state TCM schemes can improve the robustness of digital transmission against additive noise by 3 dB, compared to conventional uncoded modulation. With more complex TCM schemes, the coding gain can reach 6 dB or more. These gains are obtained without bandwidth expansion or reduction of the effective information rate as required by traditional error-correction schemes. Shannon's information theory predicted the existence of coded modulation schemes with these characteristics more than three decades ago. The development of effective TCM techniques and today's signal-processing technology now allow these gains to be obtained in practice.

Signal waveforms representing information sequences are most impervious to noise-induced detection errors if they are very different from each other. Mathematically, this translates into the requirement that signal sequences should have large distance in Euclidean signal space. The essential new concept of TCM that led to the aforementioned gains was to use signal-set expansion to provide redundancy for coding, and to design coding and signal-mapping functions jointly so as to maximize directly the "free distance" (minimum Euclidean distance) between coded signal sequences. This allowed the construction of modulation codes whose free distance significantly exceeded the minimum distance between uncoded modulation signals, at the same information rate, bandwidth, and signal power. The term "trellis" is used because these schemes can be described by a state-transition (trellis) diagram similar to the trellis diagrams of binary convolutional codes. The difference is that in TCM schemes, the trellis branches are labeled with redundant nonbinary modulation signals rather than with binary code symbols.

The basic principles of TCM were published in 1982 [2]. Further descriptions followed in 1984 [3-6], and coincided with a rapid transition of TCM from the research stage to practical use. In 1984, a TCM scheme with a coding gain of 4 dB was adopted by the International Telegraph and Telephone Consultative Commit-

Reprinted from *IEEE Communications Magazine*, Vol. 25, No. 2, Feb. 1987, pp. 5-11.

0163-6804/87/0002-0005 $01.00 © 1987 IEEE

tee (CCITT) for use in new high-speed voiceband modems [5,7,8]. Prior to TCM, uncoded transmission at 9.6 kbit/s over voiceband channels was often considered as a practical limit for data modems. Since 1984, data modems have appeared on the market which employ TCM along with other improvements in equalization, synchronization, and so forth, to transmit data reliably over voiceband channels at rates of 14.4 kbit/s and higher. Similar advances are being achieved in transmission over other bandwidth-constrained channels. The common use of TCM techniques in such applications, as satellite [9-11], terrestrial microwave, and mobile communications, in order to increase throughput rate or to permit satisfactory operation at lower signal-to-noise ratios, can be safely predicted for the near future.

Classical Error-Correction Coding

In classical digital communication systems, the functions of modulation and error-correction coding are separated. Modulators and demodulators convert an analog waveform channel into a discrete channel, whereas encoders and decoders correct errors that occur on the discrete channel.

In conventional multilevel (amplitude and/or phase) modulation systems, during each modulation interval the modulator maps m binary symbols (bits) into one of $M = 2^m$ possible transmit signals, and the demodulator recovers the m bits by making an independent M-ary nearest-neighbor decision on each signal received. Figure 1 depicts constellations of real- or complex-valued modulation amplitudes, henceforth called signal sets, which are commonly employed for one- or two-dimensional M-ary linear modulation. Two-dimensional carrier modulation requires a bandwidth of 1/T Hz around the carrier frequency to transmit signals at a modulation rate of 1/T signals/sec (baud) without intersymbol interference. Hence, two-dimensional 2^m-ary modulation systems can achieve a spectral efficiency of about m bit/sec/Hz. (The same spectral efficiency is obtained with one-dimensional $2^{m/2}$-ary baseband modulation.)

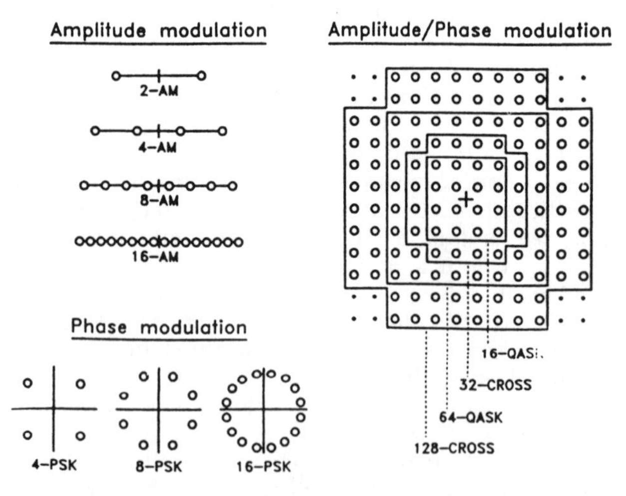

Fig. 1. Signal sets for one-dimensional amplitude modulation, and two-dimensional phase and amplitude phase modulation.

Conventional encoders and decoders for error correction operate on binary, or more generally Q-ary, code symbols transmitted over a discrete channel. With a code of rate $k/n < 1$, $n - k$ redundant check symbols are appended to every k information symbols. Since the decoder receives only discrete code symbols, Hamming distance (the number of symbols in which two code sequences or blocks differ, regardless of how these symbols differ) is the appropriate measure of distance for decoding and hence for code design. A minimum Hamming distance d^H_{min}, also called "free Hamming distance" in the case of convolutional codes, guarantees that the decoder can correct at least $[(d^H_{min} - 1)/2]$ code-symbol errors. If low signal-to-noise ratios or nonstationary signal disturbance limit the performance of the modulation system, the ability to correct errors can justify the rate loss caused by sending redundant check symbols. Similarly, long delays in error-recovery procedures can be a good reason for trading transmission rate for forward error-correction capability.

Generally, there exist two possibilities to compensate for the rate loss: increasing the modulation rate if the channel permits bandwidth expansion, or enlarging the signal set of the modulation system if the channel is band-limited. The latter necessarily leads to the use of nonbinary modulation ($M > 2$). However, when modulation and error-correction coding are performed in the classical independent manner, disappointing results are obtained.

As an illustration, consider four-phase modulation (4-PSK) without coding, and eight-phase modulation (8-PSK) used with a binary error-correction code of rate 2/3. Both systems transmit two information bits per modulation interval (2 bit/sec/Hz). If the 4-PSK system operates at an error rate of 10^{-5}, at the same signal-to-noise ratio the "raw" error rate at the 8-PSK demodulator exceeds 10^{-2} because of the smaller spacing between the 8-PSK signals. Patterns of at least three bit errors must be corrected to reduce the error rate to that of the uncoded 4-PSK system. A rate-2/3 binary convolutional code with constraint length $\nu = 6$ has the required value of $d^H_{min} = 7$ [12]. For decoding, a fairly complex 64-state binary Viterbi decoder is needed. However, after all this effort, error performance only breaks even with that of uncoded 4-PSK.

Two problems contribute to this unsatisfactory situation.

Soft-Decision Decoding and Motivation for New Code Design

One problem in the coded 8-PSK system just described arises from the independent "hard" signal decisions made prior to decoding which cause an irreversible loss of information in the receiver. The remedy for this problem is soft-decision decoding, which means that the decoder operates directly on unquantized "soft" output samples of the channel. Let the samples be $r_n = a_n + w_n$ (real- or complex-valued, for one- or two-dimensional modulation, respectively), where the a_n are the discrete signals sent by the modulator, and the w_n represent samples of an additive white Gaussian noise process. The decision rule of the optimum sequence decoder is to determine, among the set C of all coded signal sequences which a cascaded encoder and modulator can produce, the sequence $\{\hat{a}_n\}$ with minimum squared Euclidean distance (sum of squared errors) from $\{r_n\}$, that is, the sequence $\{\hat{a}_n\}$ which satisfies

$$|r_n - \hat{a}_n|^2 = \underset{\{\hat{a}_n\}\epsilon C}{Min} \sum |r_n - a_n|^2.$$

The Viterbi algorithm, originally proposed in 1967 [13] as an "asymptotically optimum" decoding technique for convolutional codes, can be used to determine the coded signal sequence $\{\hat{a}_n\}$ closest to the received unquantized signal sequence $\{r_n\}$ [12,14], provided that the generation of coded signal sequences $\{a_n\}\epsilon C$ follows the rules of a finite-state machine. However, the notion of "error-correction" is then no longer appropriate, since there are no hard-demodulator decisions to be corrected. The decoder determines the most likely coded signal sequence directly from the unquantized channel outputs.

The most probable errors made by the optimum soft-decision decoder occur between signals or signal sequences $\{a_n\}$ and $\{b_n\}$, one transmitted and the other decoded, that are closest together in terms of squared Euclidean distance. The minimum squared such distance is called the squared "free distance:"

$$d^2_{free} = \underset{\{a_n\}\neq\{b_n\}}{Min} \sum |a_n - b_n| \ ; \ \{a_n\}, \{b_n\} \epsilon C.$$

When optimum sequence decisions are made directly in terms of Euclidean distance, a second problem becomes apparent. Mapping of code symbols of a code optimized for Hamming distance into nonbinary modulation signals does not guarantee that a good Euclidean distance structure is obtained. In fact, generally one cannot even find a monotonic relationship between Hamming and Euclidean distances, no matter how code symbols are mapped.

For a long time, this has been the main reason for the lack of good codes for multilevel modulation. Squared Euclidean and Hamming distances are equivalent only in the case of binary modulation or four-phase modulation, which merely corresponds to two orthogonal binary modulations of a carrier. In contrast to coded multilevel systems, binary modulation systems with codes optimized for Hamming distance and soft-decision decoding have been well established since the late 1960s for power-efficient transmission at spectral efficiencies of less than 2 bit/sec/Hz.

The motivation of this author for developing TCM initially came from work on multilevel systems that employ the Viterbi algorithm to improve signal detection in the presence of intersymbol interference. This work provided him with ample evidence of the importance of Euclidean distance between signal sequences. Since improvements over the established technique of adaptive equalization to eliminate intersymbol interference and then making independent signal decisions in most cases did not turn out to be very significant, he turned his attention to using coding to improve performance. In this connection, it was clear to him that codes should be designed for maximum free Euclidean distance rather than Hamming distance, and that the redundancy necessary for coding would have to come from expanding the signal set to avoid bandwidth expansion.

To understand the potential improvements to be expected by this approach, he computed the channel capacity of channels with additive Gaussian noise for the case of discrete multilevel modulation at the channel input and unquantized signal observation at the channel output. The results of these calculations [2] allowed making two observations: firstly, that in principle coding gains of about 7-8 dB over conventional uncoded multilevel modulation should be achievable, and secondly, that most of the achievable coding gain could be obtained by expanding the signal sets used for uncoded modulation only by the factor of two. The author then concentrated his efforts on finding trellis-based signaling schemes that use signal sets of size 2^{m+1} for transmission of m bits per modulation interval. This direction turned out to be succesful and today's TCM schemes still follow this approach.

The next two sections illustrate with two examples how TCM schemes work. Whenever distances are discussed, Euclidean distances are meant.

Four-State Trellis Code for 8-PSK Modulation

The coded 8-PSK scheme described in this section was the first TCM scheme found by the author in 1975 with a significant coding gain over uncoded modulation. It was designed in a heuristic manner, like other simple TCM systems shortly thereafter. Figure 2 depicts signal sets and state-transition (trellis) diagrams for a) uncoded 4-PSK modulation and b) coded 8-PSK modulation with four trellis states. A trivial one-state trellis diagram is shown in Fig. 2a only to illustrate uncoded 4-PSK from the viewpoint of TCM. Every connected path through a trellis in Fig. 2 represents an allowed signal sequence. In

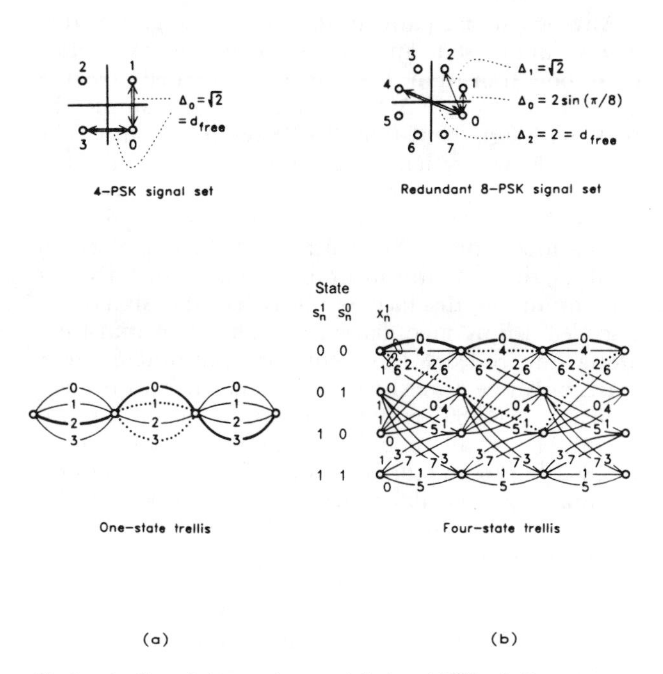

Fig. 2. (a) Uncoded four-phase modulation (4-PSK), (b) Four-state trellis-coded eight-phase modulation (8-PSK).

both systems, starting from any state, four transitions can occur, as required to encode two information bits per modulation interval (2 bit/sec/Hz). For the following discussion, the specific encoding of information bits into signals is not important.

The four "parallel" transitions in the one-state trellis diagram of Fig. 2a for uncoded 4-PSK do not restrict the sequences of 4-PSK signals that can be transmitted, that is, there is no sequence coding. Hence, the optimum decoder can make independent nearest-signal decisions for each noisy 4-PSK signal received. The smallest distance between the 4-PSK signals is $\sqrt{2}$, denoted as Δ_0. We call it the "free distance" of uncoded 4-PSK modulation to use common terminology with sequence-coded systems. Each 4-PSK signal has two nearest-neighbor signals at this distance.

In the four-state trellis of Fig. 2b for the coded 8-PSK scheme, the transitions occur in pairs of two parallel transitions. (A four-state code with four distinct transitions from each state to all successor states was also considered; however, the trellis as shown with parallel transitions permitted the achievement of a larger free distance.) Fig. 2b shows the numbering of the 8-PSK signals and relevant distances between these signals: $\Delta_0 = 2\sin(\pi/8)$, $\Delta_1 = \sqrt{2}$, and $\Delta_2 = 2$. The 8-PSK signals are assigned to the transitions in the four-state trellis in accordance with the following rules:

a) Parallel transitions are associated with signals with maximum distance Δ_2(8-PSK) = 2 between them, the signals in the subsets (0,4), (1,5), (2,6), or (3,7).
b) Four transitions originating from or merging in one state are labeled with signals with at least distance Δ_1(8-PSK) = $\sqrt{2}$ between them, that is, the signals in the subsets (0,4,2,6) or (1,5,3,7).
c) All 8-PSK signals are used in the trellis diagram with equal frequency.

Any two signal paths in the trellis of Fig. 2(b) that diverge in one state and remerge in another after more than one transition have at least squared distance $\Delta_1^2 + \Delta_0^2 + \Delta_1^2 = \Delta_2^2 + \Delta_0^2$ between them. For example, the paths with signals 0-0-0 and 2-1-2 have this distance. The distance between such paths is greater than the distance between the signals assigned to parallel transitions, Δ_2(8-PSK) = 2, which thus is found as the free distance in the four-state 8-PSK code: $d_{free} = 2$. Expressed in decibels, this amounts to an improvement of 3 dB over the minimum distance $\sqrt{2}$ between the signals of uncoded 4-PSK modulation. For any state transition along any coded 8-PSK sequence transmitted, there exists only one nearest-neighbor signal at free distance, which is the 180° rotated version of the transmitted signal. Hence, the code is invariant to a signal rotation by 180°, but to no other rotations (cf., Part II). Figure 3 illustrates one possible realization of an encoder-modulator for the four-state coded 8-PSK scheme.

Soft-decision decoding is accomplished in two steps: In the first step, called "subset decoding", within each subset of signals assigned to parallel transitions, the signal closest to the received channel output is determined. These signals are stored together with their squared distances from the channel output. In the second step, the Viterbi algorithm is used to find the signal path through the code trellis with the minimum sum of squared distances from the sequence of noisy channel outputs received. Only the signals already chosen by subset decoding are considered.

Tutorial descriptions of the Viterbi algorithm can be found in several textbooks, for example, [12]. The essential points are summarized here as follows: assume that the optimum signal paths from the infinite past to all trellis states at time n are known; the algorithm extends these paths iteratively from the states at time n to the states at time n + 1 by choosing one best path to each new state as a "survivor" and "forgetting" all other paths that cannot be extended as the best paths to the new states; looking backwards in time, the "surviving" paths tend to merge into the same "history path" at some time n − d; with a sufficient decoding delay D (so that the randomly changing value of d is highly likely to be smaller than D), the information associated with a transition on the common history path at time n − D can be selected for output.

Let the received signals be disturbed by uncorrelated Gaussian noise samples with variance σ^2 in each signal dimension. The probability that at any given time the decoder makes a wrong decision among the signals associated with parallel transitions, or starts to make a sequence of wrong decisions along some path diverging for more than one transition from the correct path, is called the error-event probability. At high signal-to-noise ratios, this probability is generally well approximated by

$$Pr(e) \simeq N_{free} \bullet Q[d_{free}/(2\sigma)],$$

where Q(.) represents the Gaussian error integral

$$Q(x) = \frac{1}{\sqrt{2\pi}} \int_x^{\infty} exp(-y^2/2)dy,$$

and N_{free} denotes the (average) number of nearest-neighbor signal sequences with distance d_{free} that diverge at any state from a transmitted signal sequence, and remerge with it after one or more transitions. The above approximate formula expresses the fact that at high

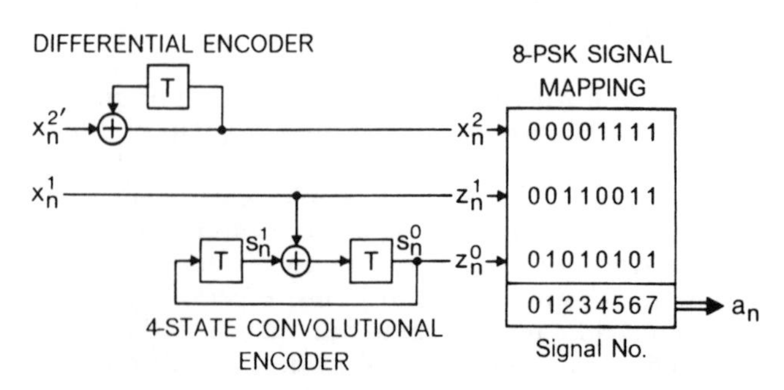

Fig. 3. Illustrates an encoder for the four-state 8-PSK code.

signal-to-noise ratios the probability of error events associated with a distance larger then d_{free} becomes negligible.

For uncoded 4-PSK, we have $d_{free} = \sqrt{2}$ and $N_{free} = 2$, and for four-state coded 8-PSK we found $d_{free} = 2$ and $N_{free} = 1$. Since in both systems free distance is found between parallel transitions, single signal-decision errors are the dominating error events. In the special case of these simple systems, the numbers of nearest neighbors do not depend on which particular signal sequence is transmitted.

Figure 4 shows the error-event probability of the two systems as a function of signal-to-noise ratio. For uncoded 4-PSK, the error-event probability is extremely well approximated by the last two equations above. For four-state coded 8-PSK, these equations provide a lower bound that is asymptotically achieved at high signal-to-noise ratios. Simulation results are included in Fig. 4 for the coded 8-PSK system to illustrate the effect of error events with distance larger than free distance, whose probability of occurrence is not negligible at low signal-to-noise ratios.

Figure 5 illustrates a noisy four-state coded 8-PSK signal as observed at complex baseband before sampling in the receiver of an experimental 64 kbit/s satellite modem [9]. At a signal-to-noise ratio of $E_s/N_0 = 12.6$ dB (E_s: signal energy, N_0: one-sided spectral noise density), the signal is decoded essentially error-free. At the same signal-to-noise ratio, the error rate with uncoded 4-PSK modulation would be around 10^{-5}.

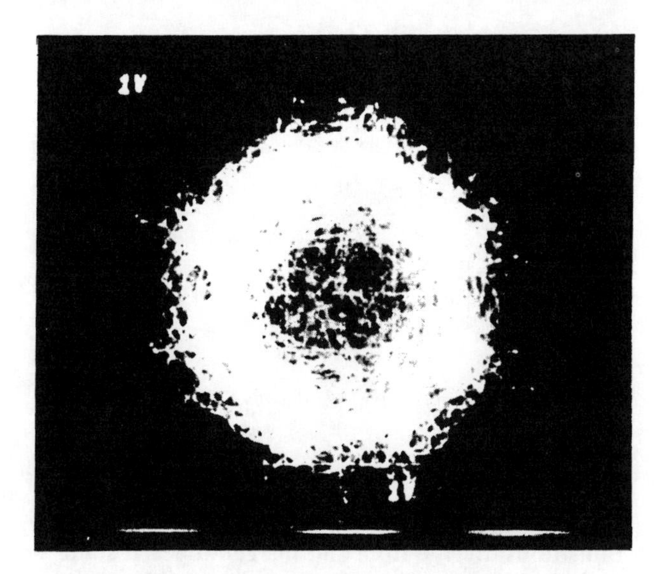

Fig. 5. Noisy four-state coded 8-PSK signal at complex baseband with a signal-to-noise ratio of $E_sN_0 = 12.6$ dB.

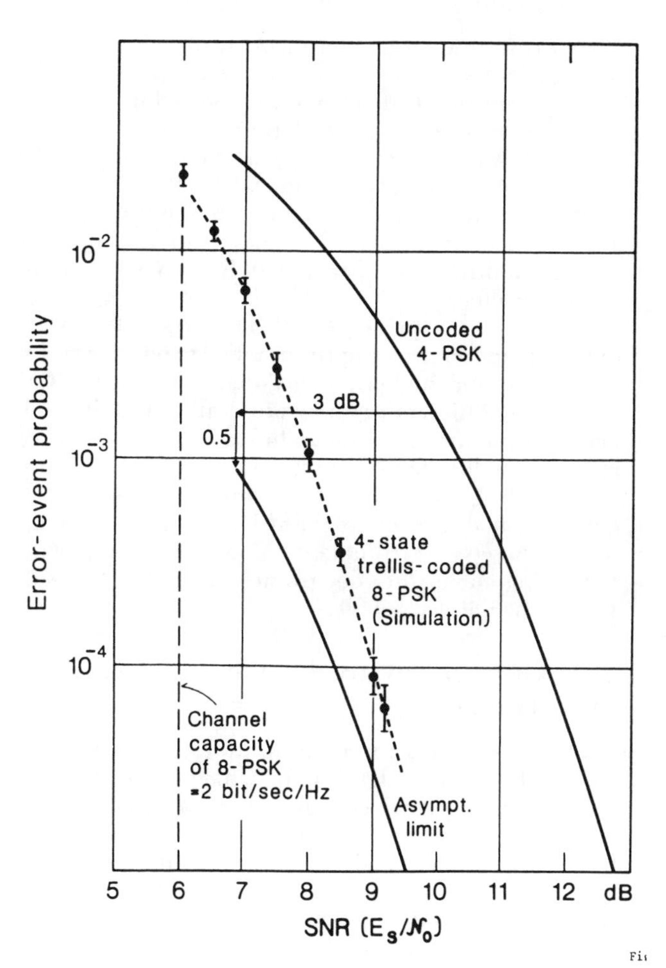

Fig. 4. Error-event probability versus signal-to-noise ratio for uncoded 4-PSK and four-state coded 8-PSK.

In TCM schemes with more trellis states and other signal sets, d_{free} is not necessarily found between parallel transitions, and N_{free} will generally be an average number larger than one, as will be shown by the second example.

Eight-State Trellis Code for Amplitude/Phase Modulation

The eight-state trellis code discussed in this section was designed for two-dimensional signal sets whose signals are located on a quadratic grid, also known as a lattice of type "Z_2". The code can be used with all of the signal sets depicted in Fig. 1 for amplitude/phase modulation. To transmit m information bits per modulation interval, a signal set with 2^{m+1} signals is needed. Hence, for m = 3 the 16-QASK signal set is used, for m = 4 the 32-CROSS signal set, and soforth. For any m, a coding gain of approximately 4 dB is achieved over uncoded modulation.

Figure 6 illustrates a "set partitioning" of the 16-QASK and 32-CROSS signal sets into eight subsets. The partitioning of larger signal sets is done in the same way. The signal set chosen is denoted by A0, and its subsets by D0, D1, . . . D7. If the smallest distance among the signals in A0 is Δ_0, then among the signals in the union of the subsets D0,D4,D2,D6 or D1,D5,D3,D7 the minimum distance is $\sqrt{2}\ \Delta_0$, in the union of the subsets D0,D4; D2,D6; D1,D5; or D3,D7 it is $\sqrt{4}\ \Delta_0$, and within the individual subsets it is $\sqrt{8}\ \Delta_0$. (A conceptually similar partitioning of the 8-PSK signal set into smaller signal sets with increasing intra-set distances was implied in the example of coded 8-PSK. The fundamental importance

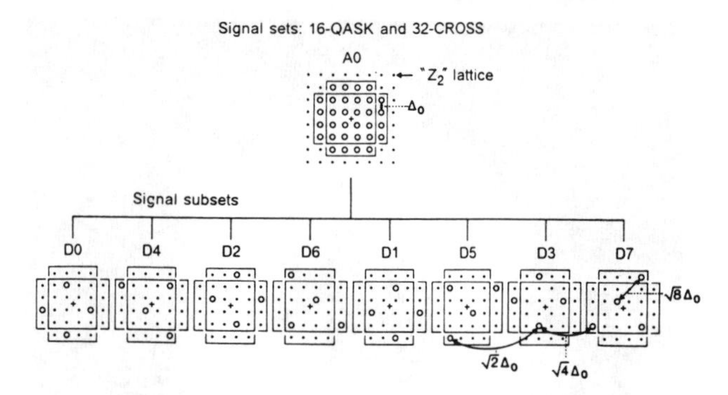

Fig. 6. Set partitioning of the 16-QASK and 32-CROSS signal sets.

of this partitioning for TCM codes will be explained in Part II.)

In the eight-state trellis depicted in Fig. 7, four transitions diverge from and merge into each state. To each transition, one of the subsets D0, . . . D7 is assigned. If A0 contains 2^{m+1} signals, each of its subsets will comprise 2^{m-2} signals. This means that the transitions shown in Fig. 7 in fact represent 2^{m-2} parallel transitions in the same sense as there were two parallel transitions in the coded 8-PSK scheme. Hence, 2^m signals can be sent from each state, as required to encode m bits per modulation interval.

The assignment of signal subsets to transitions satisfies the same three rules as discussed for coded 8-PSK, appropriately adapted to the present situation. The four transitions from or to the same state are always assigned either the subsets D0,D4,D2,D6 or D1,D5,D3,D7. This guarantees a squared signal distance of at least $2\Delta_0^2$ when sequences diverge and when they remerge. If paths remerge after two transitions, the squared signal distance is at least $4\Delta_0^2$ between the diverging transitions, and hence the total squared distance between such paths will be at least $6\Delta_0^2$. If paths remerge after three or more transitions, at least one intermediate transition contributes an additional squared signal distance Δ_0^2, so the squared distance between sequences is at least $\sqrt{5}\ \Delta_0$.

Hence, the free distance of this code is $\sqrt{5}\ \Delta_0$. This is smaller than the minimum signal distance within in the subsets D0, . . . D7, which is $\sqrt{8}\ \Delta_0$. For one particular code sequence D0-D0-D3-D6, Fig. 6 illustrates four error paths at distance $\sqrt{5}\ \Delta_0$ from that code sequence; all starting at the same state and remerging after three or four transitions. It can be shown that for any code

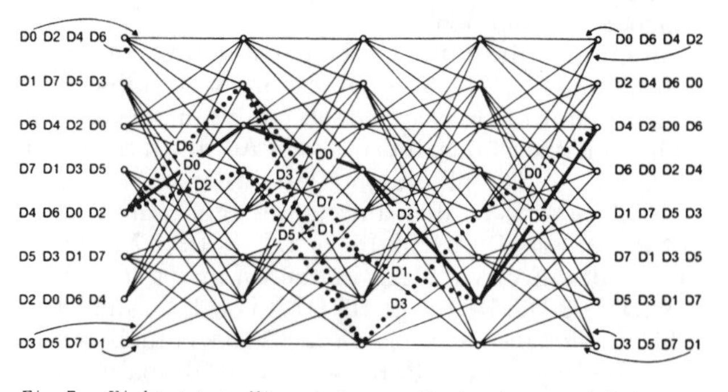

Fig. 7. Eight-state trellis code for amplitude phase modulation with "Z_2"-type signal sets; $d_{free} = \sqrt{5}\ \Delta_0$.

sequence and from any state along this sequence, there are four such paths, two of length three and two of length four. The most likely error events will correspond to these error paths, and will result in bursts of decision errors of length three or four.

The coding gains asymptotically achieved at high signal-to-noise ratios are calculated in decibels by

$$G_{c/u} = 10\ log_{10}\left[(d^2_{free,c}/d^2_{free,u})/E_{s,c}/E_{s,u})\right],$$

where $d^2_{free,c}$ and $d^2_{free,u}$ are the squared free distances, and $E_{s,c}$ and $E_{s,u}$ denote the average signal energies of the coded and uncoded schemes, respectively. When the signal sets have the same minimum signal spacing Δ_0, $d^2_{free,c}/d^2_{free,u} = 5$, and $E_{s,c}/E_{s,u} \simeq 2$ for all relevant values of m. Hence, the coding gain is 10 $\log_{10}(5/2) \simeq 4$ dB.

The number of nearest neighbors depends on the sequence of signals transmitted, that is N_{free} represents an average number. This is easy to see for uncoded modulation, where signals in the center of a signal set have more nearest neighbors than the outer ones. For uncoded 16-QASK, N_{free} equals 3. For eight-state coded 16-QASK, N_{free} is around 3.75. In the limit of large "Z_2"-type signal sets, these values increase toward 4 and 16 for uncoded and eight-state coded systems, respectively.

Trellis Codes of Higher Complexity

Heuristic code design and checking of code properties by hand, as was done during the early phases of the development of TCM schemes, becomes infeasible for codes with many trellis states. Optimum codes must then be found by computer search, using knowledge of the general structure of TCM codes and an efficient method to determine free distance. The search technique should also include rules to reject codes with improper or equivalent distance properties without having to evaluate free distance.

In Part II, the principles of TCM code design are outlined, and tables of optimum TCM codes given for one-, two-, and higher-dimensional signal sets. TCM encoder/modulators are shown to exhibit the following general structure: (a) of the m bits to be transmitted per encoder/modulator operation, $\tilde{m} \leq m$ bits are expanded into $\tilde{m}+1$ coded bits by a binary rate-$\tilde{m}$ ($\tilde{m}$+1) convolutional encoder; (b) the $\tilde{m}+1$ coded bits select one of $2^{\tilde{m}+1}$ subsets of a redundant 2^{m+1}-ary signal set; (c) the remaining m−$\tilde{m}$ bits determine one of $2^{m-\tilde{m}}$ signals within the selected subset.

New Ground Covered by Trellis-Coded Modulation

TCM schemes achieve significant coding gains at values of spectral efficiency for which efficient coded-modulation schemes were not previously known, that is, above and including 2 bit/sec/Hz. Figure 8 shows the free distances obtained by binary convolutional coding with 4-PSK modulation for spectral efficiencies smaller than 2 bit/sec/Hz, and by TCM schemes with two-dimensional signal sets for spectral efficiencies equal to or larger than 2 bit/sec/Hz. The free distances of uncoded modulation at the respective spectral effi-

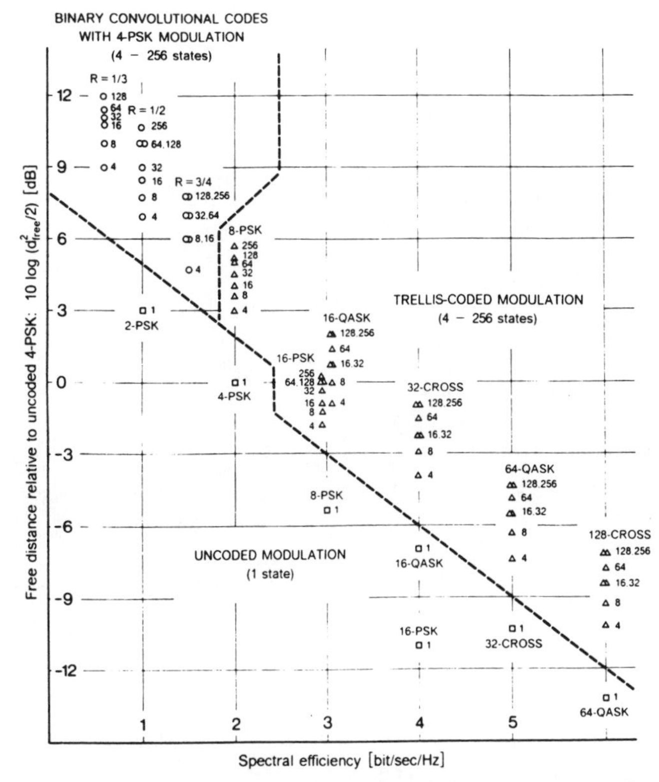

Fig. 8. Free distance of binary convolutional codes with 4-PSK modulation, and TCM with a variety of two-dimensional modulation schemes, for spectral efficiencies from 2/3 to 6 bit/sec/Hz.

ciencies are also depicted. The average signal energy of all signal sets is normalized to unity. Free distances are expressed in decibels relative to the value $d^2_{free} = 2$ of uncoded 4-PSK modulation. The binary convolutional codes of rates 1/3, 1/2, and 3/4 with optimum Hamming distances are taken from textbooks, such as, [12]. The TCM codes and their properties are found in the code tables presented in Part II (largely reproduced from [2]).

All coded systems achieve significant distance gains with as few as 4, 8, and 16 code states. Roughly speaking, it is possible to gain 3 dB with 4 states, 4 dB with 8 states, nearly 5 dB with 16 states, and up to 6 dB with 128 or more states. The gains obtained with two-state codes usually are very modest. With higher numbers of states, the incremental gains become smaller. Doubling the number of states does not always yield a code with larger free distance. Generally, limited distance growth and increasing numbers of nearest neighbors, and neighbors with next-larger distances, are the two mechanisms that prevent real coding gains from exceeding the ultimate limit set by channel capacity. This limit can be characterized by the signal-to-noise ratio at which the channel capacity of a modulation system with a 2^{m+1}-ary signal set equals m bit/sec/Hz [2] (see also Fig. 4).

Conclusion

Trellis-coded modulation was invented as a method to improve the noise immunity of digital transmission systems without bandwidth expansion or reduction of data rate. TCM extended the principles of convolutional coding to nonbinary modulation with signal sets of arbitrary size. It allows the achievement of coding gains of 3-6 dB at spectral efficiencies equal to or larger than 2 bit/sec/Hz. These are the values at which one wants to operate on many band-limited channels. Thus, a gap in the theory and practice of channel coding has been closed.

References

[1] G. Ungerboeck and I. Csajka, "On improving data-link performance by increasing the channel alphabet and introducing sequence coding," 1976 Int. Symp. Inform. Theory, Ronneby, Sweden, June 1976.

[2] G. Ungerboeck, "Channel coding with multilevel phase signals," *IEEE Trans. Information Theory*, vol. IT-28, pp. 55-67, Jan. 1982.

[3] G. D. Forney, Jr., R. G. Gallager, G. R. Lang, F. M. Longstaff, and S. U. Qureshi, "Efficient modulation for band-limited channels," *IEEE Trans. Selected Areas in Comm.*, vol. SAC-2, pp. 632-647, Sept. 1984.

[4] L. F. Wei, "Rotationally invariant convolutional channel coding with expanded signal space—Part I: 180 degrees," *IEEE Trans. Selected Areas in Comm.*, vol. SAC-2, pp. 659-672, Sept. 1984.

[5] L. F. Wei, "Rotationally invariant convolutional channel coding with expanded signal space—Part II: nonlinear codes," *IEEE Trans. Selected Areas in Comm.*, vol. SAC-2, pp. 672-686, Sept. 1984.

[6] A. R. Calderbank and J. E. Mazo, "A new description of trellis codes," *IEEE Trans. Information Theory*, vol. IT-30, pp. 784-791, Nov. 1984.

[7] CCITT Study Group XVII, "Recommendation V.32 for a family of 2-wire, duplex modems operating on the general switched telephone network and on leased telephone-type circuits," Document AP VIII-43-E, May 1984.

[8] CCITT Study Group XVII, "Draft recommendation V.33 for 14400 bits per second modem standardized for use on point-to-point 4-wire leased telephone-type circuits," Circular No. 12, COM XVII/YS, Geneva, May 17, 1985.

[9] G. Ungerboeck, J. Hagenauer, and T. Abdel Nabi, "Coded 8-PSK experimental modem for the INTELSAT SCPC system," Proc. 7th Int. Conf. on Digital Satellite Communications (ICDS-7), pp. 299-304, Munich, May 12-16, 1986.

[10] R. J. F. Fang, "A coded 8-PSK system for 140-Mbit/s information rate transmission over 80-MHz nonlinear transponders," Proc. 7th Int. Conf. on Digital Satellite Communications (ICDS-7), pp. 305-313, Munich, May 12-16, 1986.

[11] T. Fujino, Y. Moritani, M. Miyake, K. Murakami, Y. Sakato, and H. Shiino, "A 120 Mbit/s 8PSK modem with soft-Viterbi decoding," Proc. 7th Int. Conf. on Digital Satellite Communications (ICDS-7), pp. 315-321, Munich, May 12-16, 1986.

[12] G. C. Clark and J. B. Cain, *Error-Correction Coding for Digital Communications*, Plenum Press, New York and London, 1981.

[13] A. J. Viterbi, "Error bounds for convolutional codes and an asymptotically optimum decoding algorithm," *IEEE Trans. Information Theory*, vol. IT-13, pp. 260-269, April 1967.

[14] G. D. Forney, Jr., "The Viterbi algorithm," *Proc. of the IEEE*, vol. 61, pp. 268-278, March 1973.

[15] G. Ungerboeck, "Trellis-coded modulation with redundant signal sets, Part II: State of the art," *IEEE Communications Magazine*, vol. 25, no. 2, Feb. 1987.

Trellis-Coded Modulation with Redundant Signal Sets
Part II: State of the Art

Gottfried Ungerboeck

This article is intended to bring the reader up to the state of the art in trellis-coded modulation. The general principles that have proven useful in code design are explained. The important effects of carrier-phase offset and phase invariance are discussed. Finally, recent work in trellis-coded modulation with multi-dimensional signal sets is described

In this second part [1], a synopsis of the present state of the art in trellis-coded modulation (TCM) is given for the more interested reader. First, the general structure of TCM schemes and the principles of code construction are reviewed. Next, the effects of carrier-phase offset in carrier-modulated TCM systems are discussed. The topic is important, since TCM schemes turn out to be more sensitive to phase offset than uncoded modulation systems. Also, TCM schemes are generally not phase invariant to the same extent as their signal sets. Finally, recent advances in TCM schemes that use signal sets defined in more than two dimensions are described, and other work related to trellis-coded modulation is mentioned. The best codes currently known for one-, two-, four-, and eight-dimensional signal sets are given in an Appendix.

Design of Trellis-Coded Modulation Schemes

The trellis structure of the early hand-designed TCM schemes and the heuristic rules used to assign signals to trellis transitions suggested that TCM schemes should have an interpretation in terms of convolutional codes with a special signal mapping. This mapping should be based on grouping signals into subsets with large distance between the subset signals. Attempts to explain TCM schemes in this manner led to the general structure of TCM encoders/modulators depicted in Fig. 1. According to this figure, TCM signals are generated as follows: When m bits are to be transmitted per encoder/modulator operation, $\tilde{m} \leq m$ bits are expanded by a rate-$\tilde{m}/(\tilde{m}+1)$ binary convolutional encoder into $\tilde{m}+1$ coded bits. These bits are used to select one of $2^{\tilde{m}+1}$ subsets of a redundant 2^{m+1}-ary signal set. The remaining $m - \tilde{m}$ uncoded bits determine which of the $2^{m-\tilde{m}}$ signals in this subset is to be transmitted.

Set Partitioning

The concept of set partitioning is of central significance for TCM schemes. Figure 2 shows this concept for a 32-CROSS signal set [1], a signal set of lattice type "Z_2". Generally, the notation "Z_k" is used to denote an infinite "lattice" of points in k-dimensional space with integer coordinates. Lattice-type signal sets are finite subsets of lattice points, which are centered around the origin and have a minimum spacing of Δ_0.

Set partitioning divides a signal set successively into smaller subsets with maximally increasing smallest intra-set distances Δ_i, $i = 0,1, \ldots$. Each partition is two-way. The partitioning is repeated $\tilde{m}+1$ times until $\Delta_{\tilde{m}+1}$ is equal to or greater than the desired free distance of the TCM scheme to be designed. The finally obtained subsets, labeled D0, D1, ... D7 in the case of Fig. 2, will henceforth be referred to as the "subsets." The labeling of branches in the partition tree by the $\tilde{m}+1$ coded bits $z_n^{\tilde{m}}, \ldots, z_n^0$, in the order as shown in Fig. 2, results in a label $\underline{z}_n = [z_n^{\tilde{m}}, \ldots z_n^0]$ for each subset. The label reflects the position of the subset in the tree.

This labeling leads to an important property. If the labels of two subsets agree in the last q positions, but not in the bit z_n^q, then the signals of the two subsets are

Reprinted from *IEEE Communications Magazine*, Vol. 25, No. 2, Feb. 1987, pp. 12-21.

0163-6804/87/0002-0012 $01.00

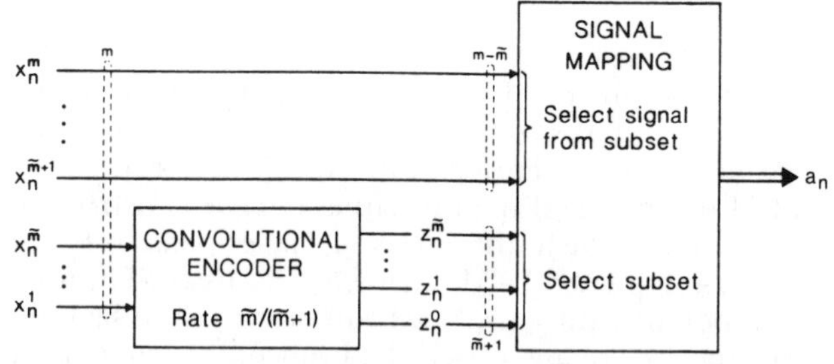

Fig. 1. General structure of encoder/modulator for trellis-coded modulation.

elements of the same subset at level q in the partition tree; thus they have at least distance Δ_q. This distance bound can be stated in a "set-partitioning lemma" and will be used in the next subsection.

The $m - \tilde{m}$ uncoded bits $x_n^m, \ldots, x_n^{\tilde{m}+1}$ are used to choose a signal from the selected subset. The specific labeling of subset signals by these bits is not particularly important at this point of the discussion. In the code trellis, the signals of the subsets become associated with $2^{m-\tilde{m}}$ parallel transitions.

The free Euclidean distance of a TCM code can now be expressed as

$$d_{free} = Min[\Delta_{\tilde{m}+1}, d_{free}(\tilde{m})],$$

where $\Delta_{\tilde{m}+1}$ is the minimum distance between parallel transitions and $d_{free}(\tilde{m})$ denotes the minimum distance between nonparallel paths in the TCM trellis diagram. In the special case of $\tilde{m} = m$, the subsets contain only one signal, and hence there are no parallel transitions.

Convolutional Codes for Trellis-Coded Modulation

At every time n, the rate-$\tilde{m}/(\tilde{m}+1)$ convolutional encoder depicted in Fig. 1 receives $\tilde{m}$ input bits, and generates $\tilde{m}+1$ coded bits which serve as the subset labels $\underline{z}_n = [z_n^{\tilde{m}}, \ldots z_n^0]$. The set of all possible sequences $\{\underline{z}_n\}$, which the encoder can generate, forms a convolutional code. A linear convolutional code of rate $\tilde{m}/(\tilde{m}+1)$ is most compactly defined by a parity-check equation which puts a constraint on the code bits in a sliding time window of length $\nu + 1$:

$$\sum_{i=0}^{\tilde{m}} (h_\nu^i z_{n-\nu}^i \oplus h_{\nu-1}^i z_{n-\nu+1}^i \oplus \ldots h_0^i z_n^i) = 0.$$

In this equation, $\oplus$ denotes modulo-2 addition. The quantity ν is called the constraint length. The quantities h_ℓ^i, $\nu \geq \ell \geq 0$; $0 \leq i \leq \tilde{m}$, are the binary parity-check coefficients of the code. Valid code sequences satisfy this equation at all times n. The equation defines only the code sequences, not the input/output relation of an encoder. A later subsection deals with minimal encoder realizations with ν binary storage elements, which is equivalent to saying that the code has 2^ν trellis states.

From the parity-check equation, one can observe that code sequences $\{\underline{z}_n\}$ can have arbitrary values for each $\tilde{m}$-tuple $[z_n^{\tilde{m}}, \ldots z_n^1]$ with an appropriate choice of the sequence $\{z_n^0\}$ so that the parity-check equation is satisfied. This property can be expressed in a "rate-$\tilde{m}/(\tilde{m}+1)$ code lemma."

Let now $\{\underline{z}_n\}$ and $\{\underline{z}'_n\} = \{\underline{z}_n \oplus \underline{e}_n\}$ be two code sequences, where $\{\underline{e}_n\}$ denotes the error sequence by which these sequences differ. Since the convolutional code is linear, $\{\underline{e}_n\}$ is also a code sequence. It follows from the "set-partitioning lemma" mentioned in the preceding subsection and the "rate-$\tilde{m}/(\tilde{m}+1)$ code lemma" that the squared free distance between non-parallel paths in the TCM trellis is bounded by [2]

$$d_{free}^2(\tilde{m}) \geq \underset{\{\underline{e}_n\} \neq \{\underline{0}\}}{Min} \sum_n \Delta_{q(\underline{e}_n)}^2.$$

Here $q(\underline{e}_n)$ is the number of trailing zeros in $\underline{e}_n$, that is, the number of trailing positions in which two subset labels $\underline{z}_n$ and $\underline{z}'_n = \underline{z}_n \oplus \underline{e}_n$ agree. For example, $q(\underline{e}_n) = 2$, if $\underline{e}_n = [e_n^m, \ldots, e_n^3, 1, 0, 0]$. The "set-partitioning lemma" states that the distance between signals in the subsets selected by $\underline{z}_n$ and $\underline{z}'_n$ is lower-bounded by $\Delta_{q(\underline{e}_n)}$. One must take $\Delta_{q(\underline{0})} = 0$, not $\Delta_{\tilde{m}+1}$. Minimization has to be carried out over all non-zero code (error) sequences $\{\underline{e}_n\}$ that deviate at, say, time 0 from the all-zero sequence $\{\underline{0}\}$ and remerge with it at a later time. The "rate-$\tilde{m}/(\tilde{m}+1)$ code lemma" assures that for any given sequence $\{\underline{e}_n\}$ there exist two coded signal sequences whose signals have at any time n the smallest possible distance between the signals of subsets whose labels differ by $\underline{e}_n$. Usually, this smallest distance equals $\Delta_{q(\underline{e}_n)}$ for all $\underline{e}_n$. If this is the case, the above bound on $d_{free}(\tilde{m})$ becomes an equation. (Only when the signal subsets contain very few signals may the bound not be satisfied with equality. A similar always true equation can then be used to compute $d_{free}(\tilde{m})$ [2].)

This equation is of key importance in the search for optimum TCM codes. It states that free Euclidean distance can be determined in much the same way as free Hamming distance is found in linear binary codes, even though linearity does not hold for TCM signal sequences. It is only necessary to replace the Hamming weights of the e_n (number of 1's in $\underline{e}_n$) by the Euclidean weights $\Delta_{q(\underline{e}_n)}^2$. It is not necessary (as some authors seem to think) to compute distance between every pair of TCM signal sequences.

Search for Optimum TCM Codes

For the one- and two-dimensional signal sets depicted in Fig. 1 of Part I [1], the minimum intra-set distances are

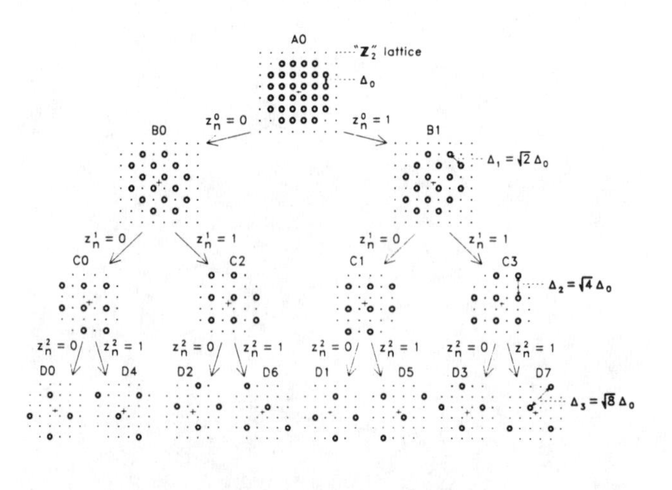

Fig. 2. Set partitioning of the 32-CROSS signal set (of lattice type "Z_2").

as follows. For 4-AM, 8-AM, ... (signal sets of type "Z_1"), $\Delta_{i+1} = 2\Delta_i$, i = 0.1, ... For 16-QASK, 32-CROSS, ... (signal sets of lattice type "Z_2"), $\Delta_{i+1} = \sqrt{2}\,\Delta_i$, i = 0,1, ... The non-lattice type signal sets 8-PSK and 16-PSK have special sequences of intra-set distances. The intra-set distances for higher-dimensional signal sets will be given when multi-dimensional TCM schemes are discussed later in this article.

For a given sequence of minimum intra-set distances $\Delta_0 \leq \Delta_1 \leq ... \Delta_{\tilde{m}}$, and a chosen value of ν, a convolutional code with the largest possible value of $d_{free}(\tilde{m})$ can be found by a code-search program described in [2]. The program performs the search for the $(\nu + 1) \cdot (\tilde{m} + 1)$ binary parity-check coefficients in a particular order and with a set of code-rejection rules such that explicit checks on the value of $d_{free}(\tilde{m})$ are very frequently avoided.

Tables of optimum codes for one-, two-, four-, and eight-dimensional TCM schemes are shown in the Appendix. Parity-check coefficients are specified in octal form, for example, $[h_6^0, ..., h_0^0] = [1,0,0,0,1,0,1]$ is written as $\underline{h}^0 = 105_8$. Equivalent codes in terms of free distance will be obtained if the parity-check coefficients of $\underline{h}^i$ are added modulo-2 to the coefficients of $\underline{h}^k$, for $i > k$ [2]. If $\Delta_i = \Delta_k$, $\underline{h}^i$ and $\underline{h}^k$ may also be interchanged. When in the code tables the free distance of a code is marked by an asterisk (*), $d_{free}(\tilde{m})$ exceeds $\Delta_{\tilde{m}+1}$, and hence the free distance occurs only between the subset signals assigned to parallel transitions. These schemes have the smallest numbers of nearest neighbors. For example, the 256-state code for "Z_1"-type signals has this property. For large values of m, this code attains a full 6 dB coding gain with only two nearest neighbors.

Two Encoder Realizations

The parity-check equation specifies only the convolutional code. Encoders for the same code can differ in the input/output relation which they realize. Figure 3 illustrates two encoders for the 8-state linear code specified in Tables II and III ($\nu = 3$) in the Appendix. One is called a systematic encoder with feedback, the other a feedback-free encoder. Both encoders are minimal, that is, they are realized with ν binary storage elements. The transformation of one minimal encoder into the other follows from the structural properties of convolutional codes described in [3]. With a systematic encoder, the input bits appear unchanged at the output. Therefore, a systematic encoder cannot generate a catastrophic code, i.e., a code with no distance increase between two trellis paths that remain distinct for an unbounded length. This is also true, although far from being obvious, for an equivalent minimal feedback-free encoder [3].

The forward and backward connections in the systematic encoder are specified by the parity-check coefficients of the code. All codes presented in the Appendix have $h_\nu^0 = h_0^0 = 1$. This guarantees the realizability of an encoder in the form shown in Fig. 3a. The reader familiar with recursive digital filters will see that the parity-check equation is used (almost directly) to compute the bit z_n^0 from the other uncoded bits. Furthermore, all codes have $h_\nu^i = h_0^i = 0$, for $i > 0$. This

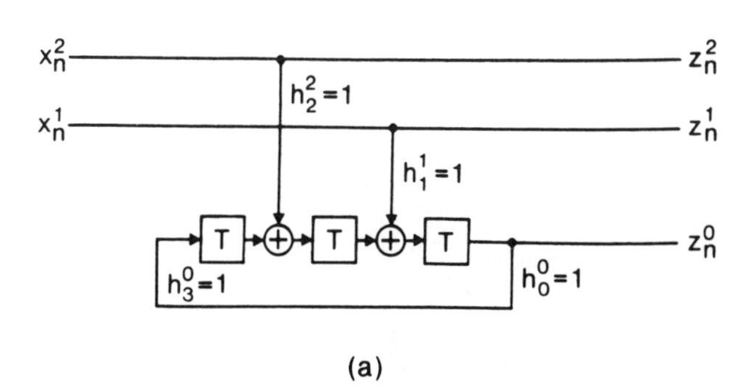

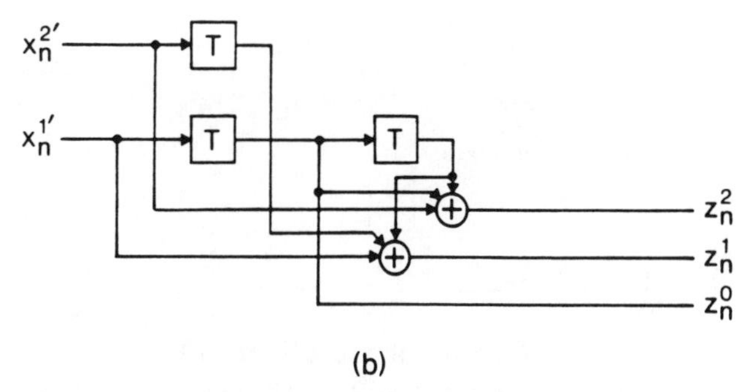

Fig. 3. Two encoders for a linear 8-state convolutional code with parity-check coefficients $\underline{h}^2 = [0,1,0,0]$, $\underline{h}^1 = [0,0,1,0]$, $\underline{h}^0 = [1,0,0,1]$ (cf. Tables II and III in the Appendix). (a) Minimal systematic encoder with feedback. (b) Minimal feedback-free encoder.

ensures that at time n the uncoded bits have no influence on the bit z_n^0, nor on the input to the first binary storage element in the encoder. Hence, whenever in the code trellis two paths diverge from or merge into a common state, the bit z_n^0 must be the same for these transitions, whereas the other bits differ in at least one bit. Signals associated with diverging and merging transitions therefore have at least distance Δ_1 between them, which reflects the second heuristic rule for good TCM codes mentioned in Part I [1].

TCM schemes for two-dimensional carrier modulation (with 8-PSK signal sets and "Z_2"-type signal sets) have up to the present time attracted the most attention. Practical realizations of these systems indicated that the effects of transmission impairments other than additive Gaussian noise on their performance need to be studied, in particular those of carrier offset.

Effects of Carrier-Phase Offset

This section addresses the problems that arise when a carrier-modulated two-dimensional TCM signal is demodulated with a phase offset $\Delta\phi$. The soft-decision decoder then operates on a sequence of complex-valued signals $\{r_n\} = \{a_n \cdot \exp(j\Delta\phi) + w_n\}$, where the a_n are transmitted TCM signals and the w_n denote additive Gaussian noise. The phase offset $\Delta\phi$ could be caused, for instance, by disturbances of the carrier phase of the received signal which the phase-tracking scheme of the receiver cannot track instantly.

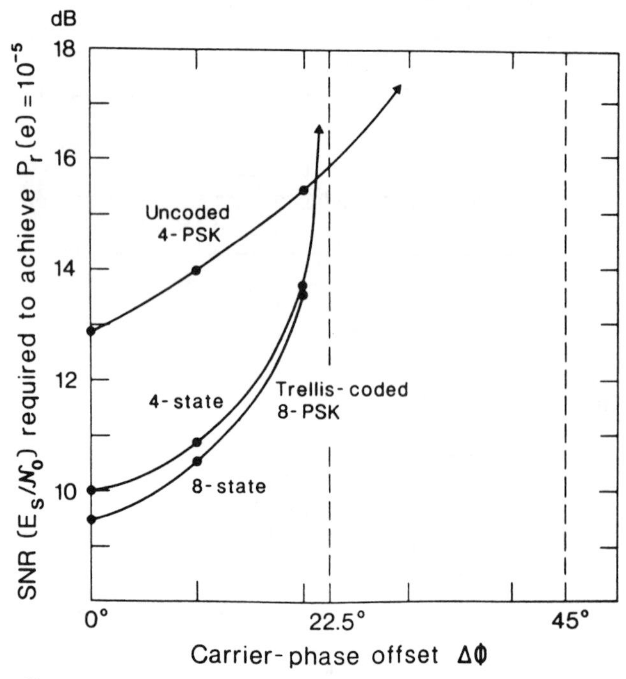

Fig. 4. Error performance of coded 8-PSK and uncoded 4-PSK in the presence of carrier-phase offset $\Delta\phi$.

Performance Degradation

The error performance of 4-state and 8-state coded 8-PSK systems in the presence of phase offset (based on unpublished work) is illustrated in Fig. 4. The figure shows the signal-to-noise ratio needed to sustain an error-event probability of 10^{-5} as a function of $\Delta\phi$. For the coded 8-PSK systems, the required signal-to-noise ratio increases with increasing values of $\Delta\phi$ until both systems fail at $\Delta\phi = 22.5°$, even in the absence of noise. In contrast, uncoded 4-PSK requires a higher signal-to-noise ratio at small phase offsets, but has an operating range up to $\Delta\phi = 45°$ in the absence of noise. These results are typical for TCM schemes.

The greater susceptibility of TCM schemes to phase offset can be explained as follows. In the trellis diagrams of TCM schemes, there exist long distinct paths with low growth of signal distance between them, that is, paths which have either the same signals or signals with smallest distance Δ_0 assigned to concurrent transitions. In the absence of phase offset, the non-zero squared distances Δ_0^2 and the squared larger distances of diverging or merging transitions add up to at least the squared free distance. However, if phase offset rotates the received signals such that received signals become located halfway between the signals of the original signal set, the difference in distance between received signals and the signals on distinct transitions that are Δ_0 apart may be reduced to zero. There may then be no difference in distance between a long segment of received signals and two distinct trellis paths, just as though the code were catastrophic. At this point, the decoder begins to fail.

Behavior of Carrier-Phase Tracking Loops

Nowadays, in most digital carrier-modulation systems, decision-directed loops are employed for carrier-phase tracking. In these loops, the phase offset is estimated from the received signal and the decoder decisions. The estimated phase offset controls the demodulating carrier phase. In a TCM receiver, if the phase offset exceeds a critical value, for example, 22.5° in the case of coded 8-PSK, the decoder decisions become essentially uncorrelated with the received signal and the mean value of the phase estimate drops to zero. Figure 5 illustrates, for 4-state coded 8-PSK [2], the mean estimate of $\Delta\phi$ ("S-curve") and its variance as a function of the actual value of the phase offset. A vanishing mean estimate, as occurs for $\Delta\phi$ between 22.5° and 157.5°, leaves the carrier-phase tracking loop in an undriven random-walk situation which can last for long periods. Eventually, the system resynchronizes when the randomly-fluctuating demodulating carrier phase approaches a value for which the received signal again resembles a valid TCM sequence. This behavior is in significant contrast to the short phase skips and rapid recovery observed in uncoded 4-PSK or 8-PSK systems. It suggests that in some cases TCM systems may require special methods to force rapid resynchronization.

Invariance of Two-Dimensional TCM Codes under Phase Rotation

TCM codes are not usually invariant to all phase rotations under which the signal set is phase invariant. Figure 5 indicates a phase symmetry of 4-state coded 8-PSK only at $\Delta\phi = 180°$, but not at other multiples of 45°. This symmetry can be verified by inspection of the code trellis presented in Fig. 2b of Part I [1]. Coded 8-PSK schemes which are invariant to phase shifts of all multiples of 45° have been found [4], but these schemes require more than four states to achieve a coding gain of 3 dB.

In general, it is desirable that TCM codes have as many phase symmetries as possible to ensure rapid carrier-phase resynchronization after temporary loss of synchronization. On the other hand, such phase invariances must be made transparent to the transmitted user

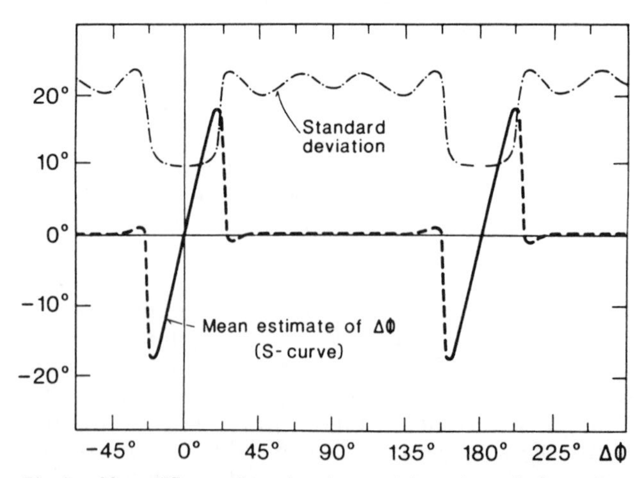

Fig. 5. Mean ("S-curve") and variance of the estimated phase offset $\Delta\hat{\phi}$ in a decision-directed carrier-phase tracking loop for 4-state coded 8-PSK versus the actual phase offset $\Delta\phi$, at a signal-to-noise ratio of 13 dB (tentative decisions used with zero delay).

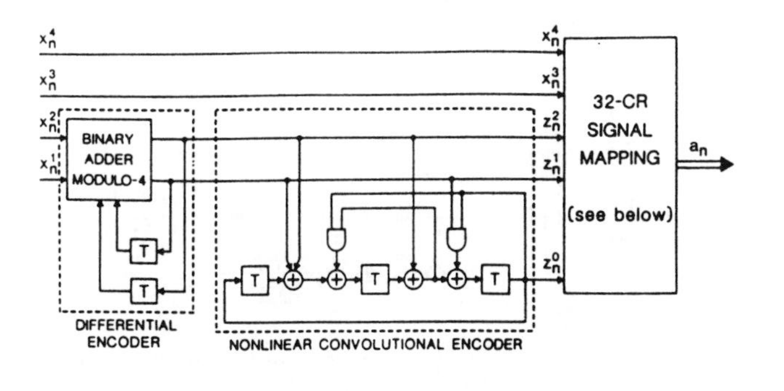

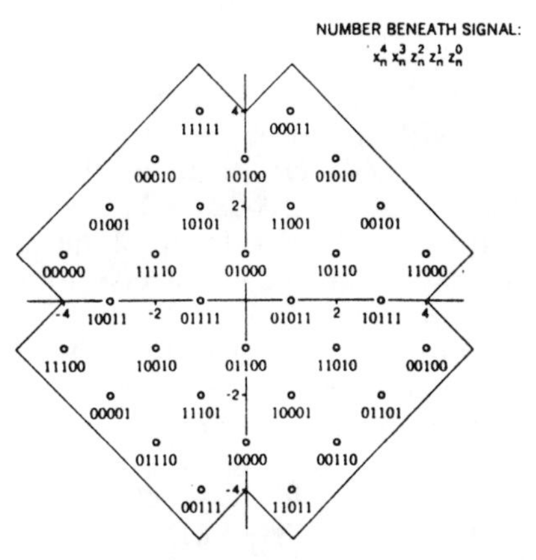

Fig. 6. Nonlinear 8-state encoder modulator with 32-CROSS signal set and differential encoding, as in CCITT Recommendation V.32.

information by some form of differential encoding and decoding. If loss of phase synchronization is very unlikely, one may argue that TCM codes without phase invariances may have the advantage that the receiver can establish absolute phase from the received signal, so that no differential encoding/decoding is required.

The problems of phase invariance and differential encoding/decoding attracted considerable attention in work toward a TCM code for use in CCITT Recommendations for voice-band modems operating full-duplex at up to 9.6 kbit/s over two-wire telephone circuits, and at up to 14.4 kbit/s over four-wire circuits. There was considerable interest in a two-dimensional 8-state code that can achieve, with 90°-symmetric QASK and CROSS signal sets, a coding gain of about 4 dB over uncoded modulation. With the known linear code (cf. Table III in the Appendix, $\nu = 3$), it was only possible (by adding parity-check coefficients in a way which does not change free distance, as mentioned in the subsection on optimum-code search) to have either no phase symmetry or a symmetry at 180° [5], [4, Part I]. A breakthrough was finally accomplished by L.F. Wei, who introduced nonlinear elements into the convolutional encoder of the 8-state code. This made the code invariant to 90° rotations while maintaining its coding gain of 4 dB [6], [5, Part I]. Figure 6 shows the resulting encoder/modulator with its differential encoder, nonlinear convolutional encoder, and signal mapping for a 32-CROSS signal set ($m = 4$), as finally adopted in the CCITT V.32 Recommendation [7, Part I]. The labeling of subsets differs slightly from that indicated in Fig. 2, but the subsets are the same. The same code was also chosen for the CCITT V.33 Draft Recommendation [8, Part I], but with 64-QASK and 128-CROSS signal sets ($m = 5,6$). In the limit of large signal sets, the number of nearest neighbors in the 8-state linear and the CCITT nonlinear code is 16.

In a late contribution to the CCITT [7], illustrated in Fig. 7, an alternative 8-state nonlinear encoder with the differential-encoding function integrated into the encoder was proposed. The coding gain and the number of nearest neighbors are identical to those of the other 8-state schemes. The trellis diagram of the alternative nonlinear code was shown in Fig. 6 of Part I [1]. Differential decoding requires that the receiver compute $x_n^1 = z_n^0 \oplus z_{n+1}^0$. Subsets are labeled as indicated in Fig. 2. The selection of signals within the subsets by the uncoded bits x_n^1, x_n^3 is worth mentioning. If $x_n^1 = 0$, only signals of the inner 16-QASK set are transmitted ($m = 3$). With non-zero values of x_n^1, outer signals of the larger 32-CROSS set are also selected ($m = 4$). Extension of this concept to larger signal sets resulted in one general signal mapping for all data rates, e.g., for $3 \leq m \leq 7$ [7]. The mapping has the additional property that it can just as well be used for uncoded modulation with modulo-4 differential encoding of the bits z_n^1, z_n^0.

The nonlinear 8-state TCM codes appear to be special cases. Similar nonlinear phase-invariant codes with 16 and more states can be constructed. However, at least for 16 states, it does not seem possible to find a code with the same 4.8 dB coding gain as can be obtained with a linear code.

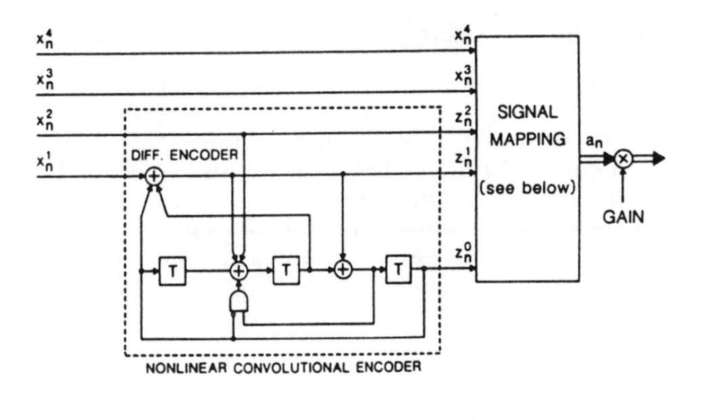

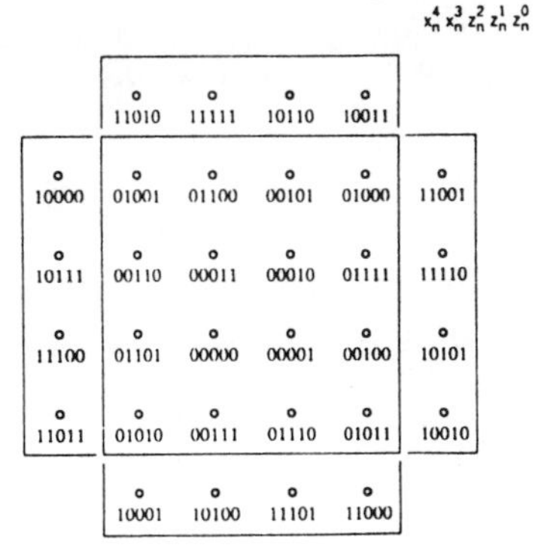

Fig. 7. Alternative nonlinear 8-state encoder modulator with integrated differential encoding and general signal mapping for 16-QASK, 32-CROSS, etc., signal sets.

Multi-Dimensional Trellis Codes

Recently, there have been a number of investigations into trellis coding with signal sets defined in more than two dimensions [3, Part I], [8-11]. In practical systems, multi-dimensional signals can be transmitted as sequences of constituent one- or two-dimensional (1-D or 2-D) signals. In this section, 2K-D TCM schemes are considered which transmit m bits per constituent 2-D signal, and hence mK bits per 2K-D signal. The principle of using a redundant signal set of twice the size needed for uncoded modulation is maintained. Thus, 2K-D TCM schemes use 2^{Km+1}-ary sets of 2K-D signals. Compared to 2-D TCM schemes, this results in less signal redundancy in the constituent 2-D signal sets.

For 2-D TCM schemes with "Z_2"-type signal sets, the minimum signal spacing Δ_0 must be reduced by approximately the factor $\sqrt{2}$ (− 3 dB) to have the same average signal power as for uncoded modulation. This loss in signal spacing needs to be more than compensated for by coding to obtain an overall improvement in free distance. The lower signal redundancy of multi-dimensional TCM schemes with "Z_{2K}"-type signal sets results only in a reduction of the minimum signal spacing by the 2K-th root of 2 (−1.5 dB for K = 2; and −0.75 dB for K = 4), so coding has to contribute less than in the case of 2-D TCM to obtain the same gain in free distance. The larger signal spacing should also make multi-dimensional TCM systems less sensitive to phase offset. Finally, it has been found that multi-dimensional TCM schemes with 90° phase invariance can be obtained with linear codes.

Four-Dimensional Trellis-Coded Modulation

The 4-D TCM schemes (K = 2) described in this subsection employ compact sets of 2^{2m+1} signals chosen from a lattice of type "Z_4" with minimum signal spacing Δ_0. Figure 8 illustrates the set partitioning of a signal set A_4^0 of type "Z_4". The general idea is to derive the set partitioning of a higher-dimensional signal set from the set partitioning of constituent lower-dimensional signal sets. In the present case, A_4^0 and its subsets are characterized by two constituent "Z_2"-type signal sets A0 and their

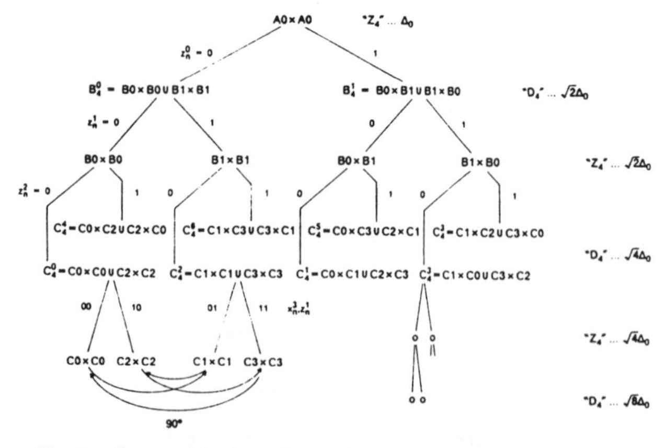

Fig. 8. Set partitioning of four-dimensional signal sets of lattice type "Z_4", also showing the effect of a 90° rotation.

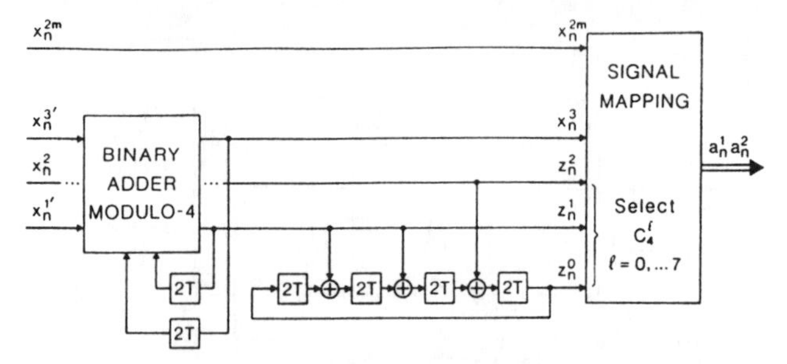

Fig. 9. Sixteen-state encoder/demodulator for four-dimensional "Z_4"-type trellis-coded modulation with differential encoding.

subsets, such as introduced in Fig. 2. This leads to a partition tree with signal sets of types "Z_4" → "D_4" → "Z_4" → "D_4" → "Z_4", etc., with minimum intra-set distances Δ_0, $\Delta_1 = \Delta_2 = \sqrt{2}\,\Delta_0$, $\Delta_3 = \Delta_4 = \sqrt{4}\,\Delta_0$, etc. The next paragraph describes the details of the partitioning process (and may be skipped by readers without specific interest in this process).

Set partitioning begins by writing $A_4^0 = A0 \times A0$ ($\times$ denotes set-product operation: the product set consists of all concatenations of elements of the first set with the elements of the second set). Substitution of $A0 = B0 \cup B1$ ($\cup$ denotes set union) yields $A_4^0 = (B0 \cup B1) \times (B0 \cup B1) = (B0 \times B0) \cup (B0 \times B1) \cup (B1 \times B0) \cup (B1 \times B1)$. The first partition divides A_4^0 into the two subsets $B_4^0 = (B0 \times B0) \cup (B1 \times B1)$ and $B_4^1 = (B0 \times B1) \cup (B1 \times B0)$. These subsets are of type "D_4", where "D_4" denotes the densest lattice known in 4-D space [12]. The minimum intra-set distance in B_4^0 and B_4^1 is $\sqrt{2}\,\Delta_0$, which is the minimum distance between constituent 2-D signals in B0 or B1, and also between one 4-D signal in $B0 \times B0$ and another in $B1 \times B1$. On the next binary partition, e.g., when B_4^0 is partitioned into subsets $B0 \times B0$ and $B1 \times B1$, no distance increase is obtained. These subsets are of type "Z_4", like A_4^0, from which they differ only in their orientation, position with respect to the origin, and scaling. Hence, their partitioning is conceptually similar to that of A_4^0. The minimum intra-set distance increases to $\sqrt{4}\,\Delta_0$ when, e.g., $B0 \times B0$ is split into subsets $C_4^0 = (C0 \times C0) \cup (C2 \times C2)$ and $C_4^4 = (C0 \times C2) \cup (C2 \times C0)$, which are now again of type "D_4".

Optimum convolutional codes are found by using the obtained sequence of minimum intra-set distances in the code-search program mentioned earlier. The codes and their asymptotic coding gains over uncoded modulation with "Z_2"-type signals are given in Table IV in the Appendix. The gains are valid for large signal sets which fill the same volume in signal space as the signal sets used for uncoded modulation. Thus, the comparison is made for the same average signal power and the same peak power of 2-D signals.

It may be helpful to discuss the 16-state code of Table IV, which achieves an asymptotic coding gain of 4.52 dB, in more detail. The code uses the eight 4-D subsets C_4^0, ... C_4^7 shown in Fig. 8, and has 64 distinct transitions in its trellis diagram. The only nearest-neighbor signals are those associated with parallel transitions, and their number at any transition is 24 (the number of nearest neighbors in a "D_4" lattice). Figure 9 depicts one possible realization of an encoder/modulator with differential

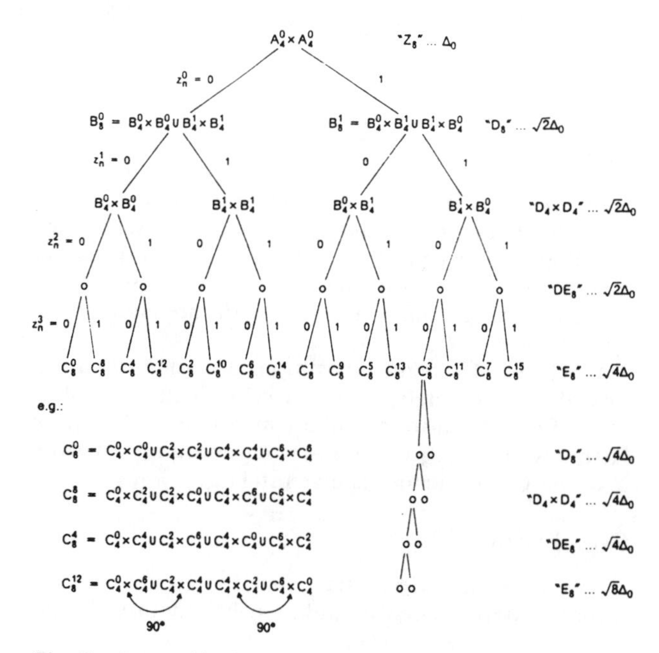

Fig. 10. Set partitioning of eight-dimensional signal sets of lattice type "Z_8", also showing the effect of a 90° rotation.

encoding. The code from Table IV was first made invariant to inversion of the bit z_n^1 by interchanging the parity-check coefficient vectors $\underline{h}^1$ and $\underline{h}^2$. Invariance to 90° rotations and the required differential encoding follow from the 90° symmetries indicated in Fig. 8, which in turn are based on the 90° symmetries in the constituent 2-D signal subsets. The subsets $C_4^0, \ldots C_4^7$, each composed of two subsets CiXCk, must be chosen individually for each value of m. The subset C0XC0 contains 2^{2m-3} signals, and may be constructed first. The other subsets CiXCk are then obtained by 90° rotations of the two constituent subsets C0 in C0XC0. For the specific case of m = 4.5, C0XC0 contains 8X8 signals, and hence the 8-ary subset C0 of Fig. 2 can be used. This construction of the 4-D subsets also suggests an efficient subset-decoding method that begins with signal decisions within the constituent 2-D subsets C0, ... C3. In general, the design of signal sets can be more complicated. References [3, Part I] and [11] discuss mapping techniques for cases where signal-set sizes are not powers of 2.

Eight-Dimensional Trellis-Coded Modulation

The technique of set partitioning of a higher-dimensional signal set based on the known partitioning of lower-dimensional sets is now applied to 8-D signal sets (K=4) of type "Z_8" = "$Z_1 \times Z_1$". Figure 10 illustrates the details. The sequence of minimum intra-set distances Δ_0, $\Delta_1 = \Delta_2 = \Delta_3 = \sqrt{2}\,\Delta_0$, $\Delta_4 = \Delta_5 = \Delta_6 = \Delta_7 = \sqrt{4}\,\Delta_0$, etc., is obtained, corresponding to a chain of lattice types "Z_8" → "D_8" → "$D_4 \times D_4$" → "DE_8" → "E_8" → "D_8", etc., where "E_8" denotes the famous Gosset lattice, the densest lattice known in 8-D space [12]. (The nomenclature "DE_8" was introduced in [9]; [11] uses "$D_8^{\perp}$".)

Codes obtained by the code-search program are given in Table V in the Appendix. The codes use 2^{4m-1} 8-D signals partitioned into 16 subsets $C_8^0, \ldots C_8^{15}$ of type "E_8". In the limit of large signal sets, the codes achieve an asymptotic coding gain of 5.27 dB over uncoded "Z_2"-type modulation. If code complexity is increased to 64 states, the only nearest neighbors are those associated with parallel transitions, and their number is 240, which is the number of nearest neighbors in an "E_8" lattice. The "E_8"-type subsets can be further partitioned into two subsets with 90° symmetries as indicated in Fig. 10. This property can be verified by observing the 90° symmetries among the constituent 4-D signals as shown in Fig. 8. Hence, 8-D codes are inherently 90° phase invariant, because their subsets have this property. Differential encoding/decoding can be performed entirely within the subsets, decoupled from the convolutional encoding function.

Other 8-D TCM schemes are obtained by choosing the 2^{4m-1} signals from another lattice type than "Z_8" in the chain of types encountered in Fig. 10, and performing the code search for the sequence of minimum intra-set distances that originates from this type. Codes with signals from "DE_8" or "E_8" are of some interest [9]-[11], although it does not seem that these codes exhibit significant advantages over the "Z_8"-type codes, if code complexities, asymptotic coding gains, and numbers of nearest neighbors are compared. This is also true for 4-D codes with "D_4" signals, as compared to codes with "Z_4" signals.

Discussion

The number of distinct transitions in the trellis diagrams of TCM codes is $2^{\nu+\tilde{m}}$. This so-called "trellis complexity" represents a measure of code (decoding) complexity. A fair comparison of TCM schemes with different signal dimensionalities requires normalization of trellis complexities and numbers of nearest neighbors to the same number of signal dimensions. In the following, normalization to two dimensions is assumed. Hence, normalized trellis complexity specifies the number of distinct trellis transitions to be dealt with by the decoder per 2-D signal or two 1-D signals received. Similarly, a normalized number of nearest neighbors indicates the number of error events with free distance that could start (on average) during the same time interval.

In Fig. 11, asymptotic coding gains of TCM schemes with large 1-D (K = 0.5) to 8-D (K = 4) signal sets are plotted versus normalized trellis complexity, $2^{\nu+\tilde{m}}/K$. Normalized numbers of nearest neighbors, N_{free}/K, are given in parentheses. At a normalized trellis complexity of 8, the "Z_2"-type 4-state code is without competition. The "Z_4"-type 16-state code, whose encoder/modulator was illustrated in Fig. 9, shows a 0.5 dB advantage over a "Z_2"-type 8-state code, e.g., the nonlinear CCITT code, and also a slightly reduced number of nearest neighbors, at the same normalized complexity of 32. Next in the order of increasing complexities, the "Z_2"-type 16-state code may be of interest, but it cannot be made invariant to 90° rotations. At a normalized complexity of 128, i.e., four times the complexity of the CCITT code, the "Z_8"- and "E_8"-type 64-state codes are found as attractive 90° phase-invariant codes. Finally, at a 32 times higher complexity than the CCITT code, the "Z_4"-type 256-state code stands out for its asymptotic coding gain and low number of nearest neighbors.

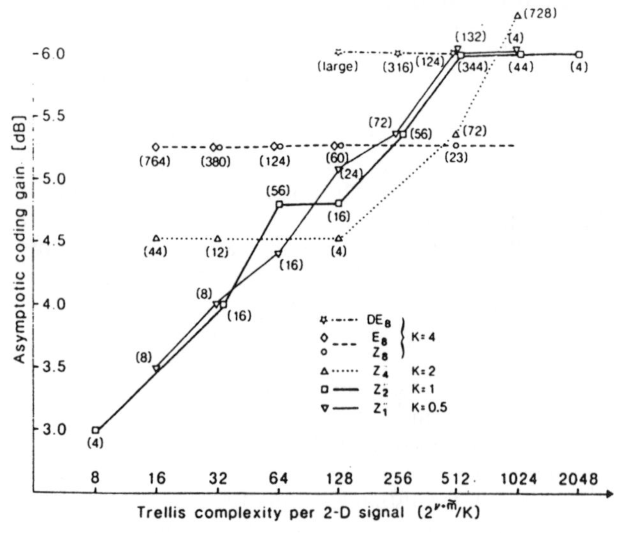

Fig. 11. Asymptotic coding gains versus trellis complexity per 2-D signal ($2^{\nu+\tilde{m}}/K$) for large 2K-D signal sets of type "Z_{2K}" for K = 0.5, 1, 2, 4; "E_8"; and "DE_8". Numbers of nearest neighbors per 2-D signal (N_{free}/K) are given in parentheses.

The asymptotic coding gain of the "DE_8" codes exceeds that of the "Z_8" and "E_8" codes by 0.75 dB, but the "DE_8" codes also have many more nearest neighbors. Hence, one may question their usefulness. Similarly, the "Z_1"-type 128-state code with the highest asymptotic coding gain of 6.28 dB shown in Fig. 11 may not be of practical interest, because of its large number of nearest neighbors.

Figure 11 gives important information about the ranking of TCM codes. However, the picture also remains somewhat incomplete. Real coding gains at given error probabilities, considering nearest and next-nearest neighbors and the boundary effect of finite signal sets, are not included. In first approximation, one may use the rule that for error rates around 10^{-5} the real coding gain is reduced by 0.2 dB for every increase in the number of nearest neighbors by the factor of 2. There is also very little published information about the carrier-phase sensitivity (a possible advantage of the multi-dimensional TCM schemes) of the TCM schemes under discussion. The complexity of subset decoding and decoder-memory requirements are further important aspects that need to be considered.

In general, one can make the following observations. At low complexity, higher-dimensional TCM schemes exhibit larger asymptotic coding gains than the lower-dimensional schemes, however, these coding gains are compromised by large numbers of nearest neighbors. In the mid-range, 4-D and 8-D TCM schemes achieve slightly larger real coding gains than the 1-D and 2-D schemes. Finally, at high trellis complexities lower-dimensional TCM schemes will eventually prevail in performance. This can be explained by the fact that these schemes have more signal redundancy available for coding than higher-dimensional TCM schemes. Overall, the differences in real coding gains are not very large, that is, they are smaller than 1 dB for the range of complexities considered.

Other Recent Work

Trellis codes have also been designed for 1-D and 2-D signal sets with nonequally-spaced ("asymmetric") signals [6, Part I], [13]. Some modest coding gains compared to schemes with equally-spaced signals are achieved when the codes have few states and small signal sets. These gains disappear for larger signal sets and higher code complexity. There are open questions about the number of nearest neighbors and sensitivity to carrier phase offset when signals are nonequally spaced.

While TCM schemes have been designed for linear modulation channels, similar developments took place in the field of continuous phase modulation (CPM) for channels requiring constant envelope signals. A summary on CPM schemes is given in [14].

Conclusion

It is probably fair to state that in recent years the theory of trellis-coded modulation has matured to the point where the achievement of further major gains seem less likely. However, there are still open questions concerning real coding gains, performance under channel impairments other than Gaussian noise, and actual implementation complexities.

The 8-state CCITT scheme was established only two years ago (1984). In the meanwhile, many manufacturers of voice-band modems and other transmission equipment have adopted the new combined coding and modulation technique. At least one manufacturer has already realized the sophisticated "Z_8"-type 64-state TCM scheme in a commercial product. In the struggle toward higher coding gains, application of more complexity is met with diminishing returns. For channels with Gaussian noise, the so-called "cut-off rate" R_0, which is smaller than channel capacity by the equivalent of about 3 dB, has been suggested as a more realistic limit [15]. TCM schemes have reached this barrier.

Acknowledgments

The author expresses his sincere appreciation for the many helpful comments and suggestions he obtained from colleagues and reviewers while writing this two-part article. Dr. G.D. Forney deserves special thanks for his many excellent technical comments and many detailed suggestions about the presentation of the material. By sending early versions of [9] and [11] to the author, Dr. L.F. Wei and Dr. D.G. Forney contributed significantly to the discussion of multi-dimensional TCM, as presented this part of the paper. Dr. V.M. Eyuboglu helped generously by verifying the correctness of the codes presented and providing still missing numbers of nearest neighbors.

Appendix: Code Tables

Tables I-III are largely reproduced from [2]. Tables IV and V have not been published previously; however, similar codes with up to 64 states were found by L.F. Wei [9]. In the tables, an asterisk (*) indicates that free distance occurs only among parallel transitions, i.e., $d_{free}(\tilde{m}) > \Delta_{\tilde{m}+1}$.

TABLE I

CODES FOR AMPLITUDE MODULATION WITH "Z_1" SIGNALS,

$\{\Delta_i, 0 \le i \le 2\} = \Delta_0, 2\Delta_0, 4\Delta_0$.

No. of states 2^ν	$\tilde{m}$	Parity check coefficients $\underline{h}^1$	$\underline{h}^0$	d^2_{free}/Δ^2_0	Asympt. coding gain [dB] $G_{4AM/2AM}$ (m = 1)	$G_{8AM/4AM}$ (m = 2)	$G_{c/u}$ ($m\to\infty$)	N_{free} ($m\to\infty$)
4	1	2	5	9.0	2.55	3.31	3.52	4
8	1	04	13	10.0	3.01	3.77	3.97	4
16	1	04	23	11.0	3.42	4.18	4.39	8
32	1	10	45	13.0	4.15	4.91	5.11	12
64	1	024	103	14.0	4.47	5.23	5.44	36
128	1	126	235	16.0	5.05	5.81	6.02	66
256	1	362	515	16.0*	—	5.81	6.02	2
256	1	362	515	17.0	5.30	—	—	

TABLE II

CODES FOR PHASE MODULATION

8-PSK: $\{\Delta_i, 0 \le i \le 2\} = 2\sin(\pi/8), \sqrt{2}, 2$;

16-PSK: $\{\Delta_i, 0 \le i \le 3\} = 2\sin(\pi/16), 2\sin(\pi/8), \sqrt{2}, 2$.

No. of states 2^ν	$\tilde{m}$	Parity-check coefficients $\underline{h}^2$	$\underline{h}^1$	$\underline{h}^0$	d^2_{free}/Δ^2_0	Asympt. coding gain [dB] $G_{8PSK/4PSK}$ (m=2)	$G_{16PSK/8PSK}$ (m=3)	N_{free} ($m\to\infty$)
4	1	—	2	5	4.000*	3.01	—	1
8	2	04	02	11	4.586	3.60	—	2
16	2	16	04	23	5.172	4.13	—	≈2.3
32	2	34	16	45	5.758	4.59	—	4
64	2	066	030	103	6.343	5.01	—	≈5.3
128	2	122	054	277	6.586	5.17	—	≈0.5
256	2	130	072	435	7.515	5.75	—	≈1.5
4	1	—	2	5	1.324	—	3.54	4
8	1	—	04	13	1.476	—	4.01	4
16	1	—	04	23	1.628	—	4.44	8
32	1	—	10	45	1.910	—	5.13	8
64	1	—	024	103	2,000*	—	5.33	2
128	1	—	024	203	2,000*	—	5.33	2
256	2	374	176	427	2.085	—	5.51	≈8.0

TABLE III

CODES FOR TWO-DIMENSIONAL MODULATION WITH "Z_2" SIGNALS,

$\{\Delta_i, 0 \le i \le 3\} = \Delta_0, \sqrt{2}\,\Delta_0, \sqrt{4}\,\Delta_0, \sqrt{8}\,\Delta_0$.

No. of states 2^ν	$\tilde{m}$	Parity-check coefficients $\underline{h}^2$	$\underline{h}^1$	$\underline{h}^0$	d^2_{free}/Δ^2_0	Asympt. coding gain [dB] $G_{16QA/8PSK}$ (m=3)	$G_{32CR/16QA}$ (m=4)	$G_{64QA/32CR}$ (m=5)	$G_{C/U}$ ($m\to\infty$)	N_{free} ($m\to\infty$)
4	1	—	2	5	4.0*	4.36	3.01	2.80	3.01	4
8	2	04	02	11	5.0	5.33	3.98	3.77	3.98	16
16	2	16	04	23	6.0	6.12	4.77	4.56	4.77	56
32	2	10	06	41	6.0	6.12	4.77	4.56	4.77	16
64	2	064	016	101	7.0	6.79	5.44	5.23	5.44	56
128	2	042	014	203	8.0	7.37	6.02	5.81	6.02	344
256	2	304	056	401	8.0	7.37	6.02	5.81	6.02	44
512	2	0510	0346	1001	8.0*	7.37	6.02	5.81	6.02	4

TABLE IV
CODES FOR FOUR-DIMENSIONAL MODULATION WITH "Z_4" SIGNALS,
$\{\Delta_i, 0 \leq i \leq 5\} = \Delta_0, \sqrt{2}\,\Delta_0, \sqrt{2}\,\Delta_0, \sqrt{4}\,\Delta_0, \sqrt{4}\,\Delta_0, \sqrt{8}\,\Delta_0$.

No. of states 2^ν	$\tilde{m}$	Parity-check coefficients $\underline{h}^4$	$\underline{h}^3$	$\underline{h}^2$	$\underline{h}^1$	$\underline{h}^0$	d^2_{free}/Δ^2_0	Asympt. coding gain [dB] $(m\rightarrow\infty)$	N_{free} $(m\rightarrow\infty)$
8	2	—	—	04	02	11	4.0	4.52	88
16	2	—	—	14	02	21	4.0*	4.52	24
32	3	—	30	14	02	41	4.0*	4.52	8
64	4	050	030	014	002	101	5.0	5.48	144
128	4	120	050	022	006	203	6.0	6.28	

TABLE V
CODES FOR EIGHT-DIMENSIONAL MODULATION WITH "Z_8" SIGNALS,
$\{\Delta_i, 0 \leq i \leq 5\} = \Delta_0, \sqrt{2}\,\Delta_0, \sqrt{2}\,\Delta_0, \sqrt{2}\,\Delta_0, \sqrt{4}\,\Delta_0, \sqrt{4}\,\Delta_0$.

No. of states 2^ν	$\tilde{m}$	Parity-check coefficients $\underline{h}^4$	$\underline{h}^3$	$\underline{h}^2$	$\underline{h}^1$	$\underline{h}^0$	d^2_{free}/Δ^2_0	Asympt. coding gain [dB] $(m\rightarrow\infty)$	N_{free} $(m\rightarrow\infty)$
16	3	—	10	04	02	21	4.0	5.27	
32	3	—	10	04	02	41	4.0	5.27	496
64	3	—	044	014	002	101	4.0*	5.27	240
128	4	120	044	014	002	201	4.0*	5.27	112

V.M. Eyuboglu and G.D. Forney [16] discovered typographical errors in the earlier published "Z_1"- and "Z_2"-type 256-state codes [2], which have now been corrected in Tables I and III.

Some of the 8-PSK codes of Table II were improved, compared to those published in [2], by using the exact expression for $d_{free}(\tilde{m})$ in the code search. The 16-PSK codes of Table II are new.

The exact numbers of nearest neighbors, N_{free}, given in the tables were taken from various sources, in particular [11] and [17]. The approximate values of N_{free}, given for some codes in Table II, are average values recently determined by the author.

References

[1] G. Ungerboeck, "Trellis-coded modulation with redundant signal sets—Part I: Introduction," *IEEE Communications Magazine*, vol. 25, no. 2, Feb. 1987.

[2] G. Ungerboeck, "Channel coding with multilevel phase signals," *IEEE Trans. Information Theory*, vol. IT-28, pp. 55-67, Jan. 1982.

[3] G. D. Forney, Jr., "Convolutional codes I: Algebraic structure," *IEEE Trans. Information Theory*, vol. IT-16, pp. 720-738, Nov. 1970.

[4] M. Oerder, "Rotationally invariant trellis codes for mPSK modulation," *1985 Internat. Commun. Conf. Record*, pp. 552-556, Chicago, June 23-26, 1985.

[5] IBM Europe, "Trellis-coded modulation schemes for use in data modems transmitting 3-7 bits per modulation interval," CCITT SG XVII Contribution COM XVII, No. D114, April 1983.

[6] AT&T Information Systems, "A trellis coded modulation scheme that includes differential encoding for 9600 bit/sec, full-duplex, two-wire modems," CCITT SG XVII Contribution COM XVII, No. D159, August 1983.

[7] IBM Europe, "Trellis-coded modulation schemes with 8-state systematic encoder and 90° symmetry for use in data modems transmitting 3-7 bits per modulation interval," CCITT SG XVII Contribution COM XVII, No. D180, October 1983.

[8] A. R. Calderbank and N. J. A. Sloane, "Four-dimensional modulation with an eight-state trellis code," *AT&T Tech. Jour.*, vol. 64, pp. 1005-1017, May-June 1985.

[9] L. F. Wei, "Trellis-coded modulation with multidimensional constellations," submitted to *IEEE Trans. Information Theory*, Aug. 1985.

[10] A. R. Calderbank and N. J. A. Sloane, "An eight-dimensional trellis code," *Proc. of the IEEE*, vol. 74, pp. 757-759, May 1986.

[11] G. D. Forney, Jr., *Coset Codes I: Geometry and Classification*, Aug. 25, 1986.

[12] N. J. A. Sloane, "The packing of spheres," *Scientific American*, vol. 250, pp. 116-125, Jan. 1984.

[13] M. K. Simon and D. Divsalar, "Combined trellis coding with asymmetric MPSK modulation," *JPL Publication* 85-24, May 1, 1985.

[14] C. E. Sundberg, "Continuous phase modulation," *IEEE Communications Magazine*, vol. 24, no. 4, pp. 25-38, April 1986.

[15] J. L. Massey, "Coding and modulation in digital communications," *Proc. 1974 Int. Zurich Seminar on Digital Communications*, Zurich, Switzerland, pp. E2(1)-(4), March 1974.

[16] V. M. Eyuboglu and G. D. Forney, Jr., private communications, Sept. 1984 and Sept. 1986.

Reprinted form *IEEE Transactions on Communications Tech.*, Vol. Com-19, No. 5, Oct. 1971, pp. 751-772.

Convolutional Codes and Their Performance in Communication Systems

ANDREW J. VITERBI, SENIOR MEMBER, IEEE

***Abstract*—This tutorial paper begins with an elementary presentation of the fundamental properties and structure of convolutional codes and proceeds with the development of the maximum likelihood decoder. The powerful tool of generating function analysis is demonstrated to yield for arbitrary codes both the distance properties and upper bounds on the bit error probability for communication over any memoryless channel. Previous results on code ensemble average error probabilities are also derived and extended by these techniques. Finally, practical considerations concerning finite decoding memory, metric representation, and synchronization are discussed.**

I. Introduction

ALTHOUGH convolutional codes, first introduced by Elias [1], have been applied over the past decade to increase the efficiency of numerous communication systems, where they invariably outperform block codes of the same order of complexity, there remains to date a lack of acceptance of convolutional coding and decoding techniques on the part of many communication technologists. In most cases, this is due to an incomplete understanding of convolutional codes, whose cause can be traced primarily to the sizable literature in this field, composed largely of papers which emphasize details of the decoding algorithms rather than the more fundamental unifying concepts, and which, until recently, have been divided into two nearly disjoint subsets. This malady is shared by the block-coding literature, wherein the algebraic decoders and probabilistic decoders have been at odds for a considerably longer period.

The convolutional code dichotomy owes its origins to the development of sequential (probabilistic) decoding by Wozencraft [2] and of threshold (feedback, algebraic) decoding by Massey [3]. Until recently the two disciplines flourished almost independently, each with its own literature, applications, and enthusiasts. The Fano sequential decoding algorithm [4] was soon found to

Paper approved by the Communication Theory Committee of the IEEE Communication Technology Group for publication without oral presentation. Manuscript received January 7, 1971; revised June 11, 1971.

The author is with the School of Engineering and Applied Science, University of California, Los Angeles, Calif. 90024, and the Linkabit Corporation, San Diego, Calif.

greatly outperform earlier versions of sequential decoders both in theory and practice. Meanwhile the feedback decoding advocates were encouraged by the burst-error correcting capabilities of the codes which render them quite useful for channels with memory.

To add to the confusion, yet a third decoding technique emerged with the Viterbi decoding algorithm [9], which was soon thereafter shown to yield maximum likelihood decisions (Forney [12], Omura [17]). Although this approach is probabilistic and emerged primarily from the sequential-decoding oriented discipline, it leads naturally to a more fundamental approach to convolutional code representation and performance analysis. Furthermore, by emphasizing the decoding-invariant properties of convolutional codes, one arrives directly to the maximum likelihood decoding algorithm and from it to the alternate approaches which lead to sequential decoding on the one hand and feedback decoding on the other. This decoding algorithm has recently found numerous applications in communication systems, two of which are covered in this issue (Heller and Jacobs [24], Cohen *et al.* [25]). It is particularly desirable for efficient communication at very high data rates, where very low error rates are not required, or where large decoding delays are intolerable.

Foremost among the recent works which seek to unify these various branches of convolutional coding theory is that of Forney [12], [21], [22], *et seq.*, which includes a three-part contribution devoted, respectively, to algebraic structure, maximum likelihood decoding, and sequential decoding. This paper, which began as an attempt to present the author's original paper [9] to a broader audience,[1] is another such effort at consolidating this discipline.

It begins with an elementary presentation of the fundamental properties and structure of convolutional codes and proceeds to a natural development of the maximum likelihood decoder. The relative distances among codewords are then determined by means of the generating function (or transfer function) of the code state diagram. This in turn leads to the evaluation of coded communication system performance on any memoryless channel. Performance is first evaluated for the specific cases of the binary symmetric channel (BSC) and the additive white Gaussian noise (AWGN) channel with biphase (or quadriphase) modulation, and finally generalized to other memoryless channels. New results are obtained for the evaluation of specific codes (by the generating function technique), rather than the ensemble average of a class of codes, as had been done previously, and for bit error probability, as distinguished from event error probability.

The previous ensemble average results are then extended to bit error probability bounds for the class of time-varying convolutional codes by means of a generalized generating function approach; explicit results are obtained for the limiting case of a very noisy channel and compared with the corresponding results for block codes. Finally, practical considerations concerning finite memory, metric representation, and synchronization are discussed. Further and more explicit details on these problems and detailed results of performance analysis and simulation are given in the paper by Heller and Jacobs [24].

While sequential decoding is not treated explicitly in this paper, the fundamentals and techniques presented here lead naturally to an elegant tutorial presentation of this subject, particularly if, following Jelinek [18], one begins with the recently proposed stack sequential decoding algorithm proposed independently by Jelinek and Zigangirov [7], which is far simpler to describe and understand then the original sequential algorithms. Such a development, which proceeds from maximum likelihood decoding to sequential decoding, exploiting the similarities in performance and analysis has been undertaken by Forney [22]. Similarly, the potentials and limitations of feedback decoders can be better understood with the background of the fundamental decoding-invariant convolutional code properties previously mentioned, as demonstrated, for example, by the recent work of Morrissey [15].

II. Code Representation

A convolutional encoder is a linear finite-state machine consisting of a K-stage shift register and n linear algebraic function generators. The input data, which is usually, though not necessarily, binary, is shifted along the register b bits at a time. An example with $K = 3$, $n = 2, b = 1$ is shown in Fig. 1.

The binary input data and output code sequences are indicated on Fig. 1. The first three input bits, 0, 1, and 1, generate the code outputs 00, 11, and 01, respectively. We shall pursue this example to develop various representations of convolutional codes and their properties. The techniques thus developed will then be shown to generalize directly to any convolutional code.

It is traditional and instructive to exhibit a convolutional code by means of a tree diagram as shown in Fig. 2.

If the first input bit is a zero, the code symbols are those shown on the first upper branch, while if it is a one, the output code symbols are those shown on the first lower branch. Similarly, if the second input bit is a zero, we trace the tree diagram to the next upper branch, while if it is a one, we trace the diagram downward. In this manner all 32 possible outputs for the first five inputs may be traced.

From the diagram it also becomes clear that after the first three branches the structure becomes repetitive. In fact, we readily recognize that beyond the third branch the code symbols on branches emanating from the two nodes labeled a are identical, and similarly for all the

[1] This material first appeared in unpublished form as the notes for the Linkabit Corp., "Seminar on convolutional codes," Jan. 1970.

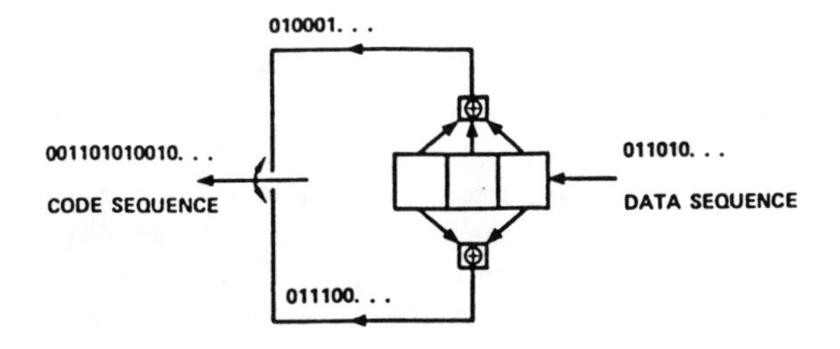

Fig. 1. Convolutional coder for $K = 3, n = 2, b = 1$.

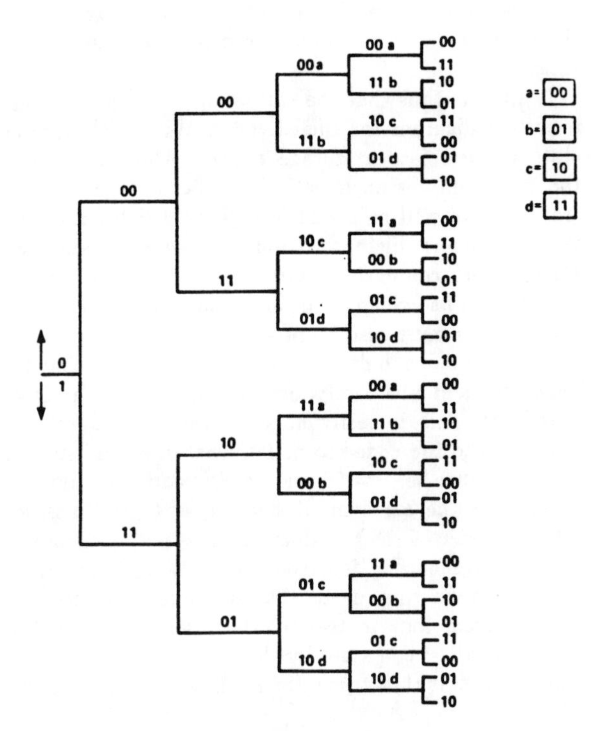

Fig. 2. Tree-code representation for coder of Fig. 1.

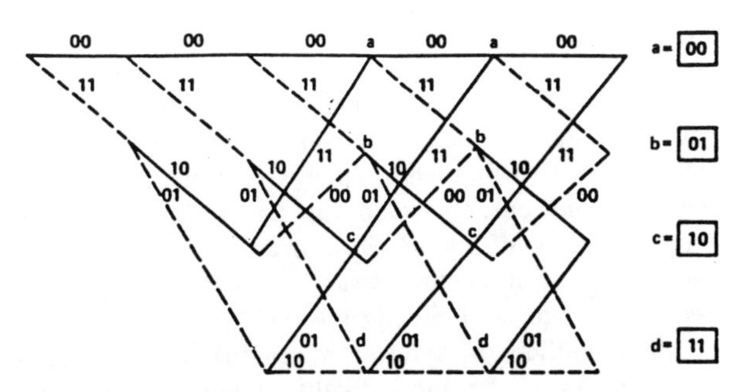

Fig. 3. Trellis-code representation for coder of Fig. 1.

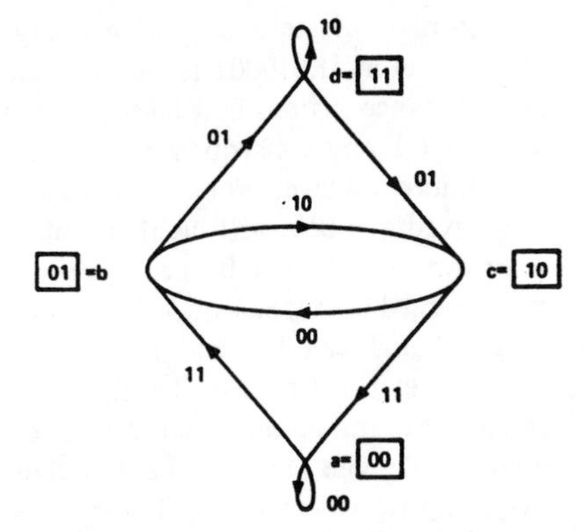

Fig 4. State-diagram representation for coder of Fig. 1.

identically labeled pairs of nodes. The reason for this is obvious from examination of the encoder. As the fourth input bit enters the coder at the right, the first data bit falls off on the left end and no longer influences the output code symbols. Consequently, the data sequences $100xy\cdots$ and $000xy\cdots$ generate the same code symbols after the third branch and, as is shown in the tree diagram, both nodes labeled a can be joined together.

This leads to redrawing the tree diagram as shown in Fig. 3. This has been called a trellis diagram [12], since a trellis is a tree-like structure with remerging branches. We adopt the convention here that code branches produced by a "zero" input bit are shown as solid lines and code branches produced by a "one" input bit are shown dashed.

The completely repetitive structure of the trellis diagram suggests a further reduction in the representation of the code to the state diagram of Fig. 4. The "states" of the state diagram are labeled according to the nodes of the trellis diagram. However, since the states correspond merely to the last two input bits to the coder we may use these bits to denote the nodes or states of this diagram.

We observe finally that the state diagram can be drawn directly by observing the finite-state machine properties of the encoder and particularly the fact that a four-state directed graph can be used to represent uniquely the input–output relation of the eight-state machine. For the nodes represent the previous two bits while the present bit is indicated by the transition branch; for example, if the encoder (machine) contains 011, this is represented in the diagram by the transition from state $b = 01$ to state $d = 11$ and the corresponding branch indicates the code symbol outputs 01.

III. Minimum Distance Decoder for Binary Symmetric Channel

On a BSC, errors which transform a channel code symbol 0 to 1 or 1 to 0 are assumed to occur independently from symbol to symbol with probability p. If all input (message) sequences are equally likely, the decoder which minimizes the overall error probability for any code, block or convolutional, is one which examines the error-corrupted received sequence $y_1y_2\cdots y_j\cdots$ and chooses the data sequence corresponding to the transmitted code sequence $x_1x_2\cdots x_j\cdots$, which is closest to the received sequence in the sense of Hamming distance; that is, the transmitted sequence which differs from the received sequence in the minimum number of symbols.

Referring first to the tree diagram, this implies that we should choose that path in the tree whose code sequence differs in the minimum number of symbols from the received sequence. However, recognizing that the transmitted code branches remerge continually, we may equally limit our choice to the possible paths in the trellis diagram of Fig. 3. Examination of this diagram indicates that it is unnecessary to consider the entire received sequence (which conceivably could be thousands or millions of symbols in length) at one time in deciding upon the most likely (minimum distance) transmitted sequence. In particular, immediately after the third branch we may determine which of the two paths leading to node or state a is more likely to have been sent. For example, if 010001 is received, it is clear that this is at distance 2 from 000000 while it is at distance 3 from 111011 and consequently we may exclude the lower path into node a. For, no matter what the subsequent received symbols will be, they will effect the distances only over subsequent branches after these two paths have remerged and consequently in exactly the same way. The same can be said for pairs of paths merging at the other three nodes after the third branch. We shall refer to the minimum distance path of the two paths merging at a given node as the "survivor." Thus it is necessary only to remember which was the minimum distance path from the received sequence (or survivor) at each node, as well as the value of that minimum distance. This is necessary because at the next node level we must compare the two branches merging at each node level, which were survivors at the previous level for different nodes; e.g., the comparison at node a after the fourth branch is among the survivors of comparisons at nodes a and c after the third branch. For example, if the received sequence over the first four branches is 01000111, the survivor at the third node level for node a is 000000 with distance 2 and at node c it is 110101, also with distance 2. In going from the third node level to the fourth the received sequence agrees precisely with the survivor from c but has distance 2 from the survivor from a. Hence the survivor at node a of the fourth level is the data sequence 1100 which produced the code sequence 11010111 which is at (minimum) distance 2 from the received sequence.

In this way we may proceed through the received sequence and at each step for each state preserve one surviving path and its distance from the received sequence, which is more generally called *metric*. The only difficulty which may arise is the possibility that in a given comparison between merging paths, the distances or metrics are identical. Then we may simply flip a coin as is done for block codewords at equal distances from the received sequence. For even if we preserved both of the equally valid contenders, further received symbols would affect both metrics in exactly the same way and thus not further influence our choice.

This decoding algorithm was first proposed by Viterbi [9] in the more general context of arbitrary memoryless channels. Another description of the algorithm can be obtained from the state-diagram representation of Fig. 4. Suppose we sought that path around the directed state diagram, arriving at node a after the kth transition, whose code symbols are at a minimum distance from the received sequence. But clearly this minimum distance path to node a at time k can be only one of two candidates: the miminum distance path to node a at time $k - 1$ and the minimum distance path to node c at time $k - 1$. The comparison is performed by adding the new distance accumulated in the kth transition by each of these paths to their minimum distances (metrics) at time $k - 1$.

It appears thus that the state diagram also represents a system diagram for this decoder. With each node or state we associate a storage register which remembers the minimum distance path into the state after each transition as well as a metric register which remembers its (minimum) distance from the received sequence. Furthermore, comparisons are made at each step between the two paths which lead into each node. Thus four comparators must also be provided.

There remains only the question of truncating the algorithm and ultimately deciding on one path rather than four. This is easily done by forcing the last two input bits to the coder to be 00. Then the final state of the code must be $a = 00$ and consequently the ultimate survivor is the survivor at node a, after the insertion into the coder of the two dummy zeros and transmission of the corresponding four code symbols. In terms of the trellis diagram this means that the number of states is reduced from four to two by the insertion of the first zero and to a single state by the insertion of the second. The diagram is thus truncated in the same way as it was begun.

We shall proceed to generalize these code representations and optimal decoding algorithm to general convolutional codes and arbitrary memoryless channels, including the Gaussian channel, in Sections V and VI. However, first we shall exploit the state diagram further to determine the relative distance properties of binary convolutional codes.

IV. Distance Properties of Convolutional Codes

We continue to pursue the example of Fig. 1 for the sake of clarity; in the next section we shall easily generalize results. It is well known that convolutional codes are group codes. Thus there is no loss in generality in computing the distance from the all zeros codeword to all the other codewords, for this set of distances is the same as the set of distances from any specific codeword to all the others.

For this purpose we may again use either the trellis diagram or the state diagram. We first of all redraw the trellis diagram in Fig. 5 labeling the branches according to their distances from the all zeros path. Now consider all the paths that merge with the all zeros for the first time at some arbitrary node j.

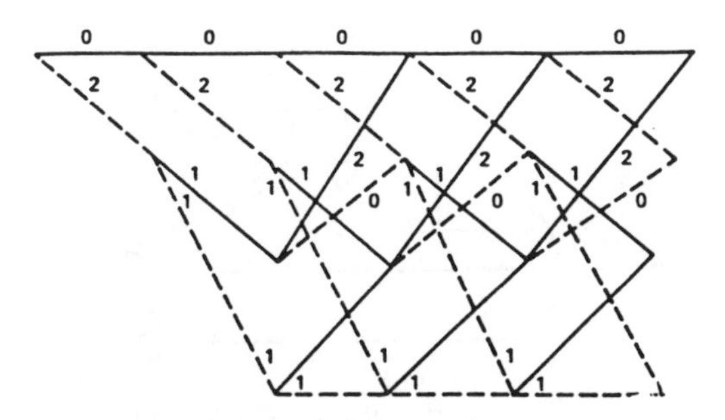

Fig. 5. Trellis diagram labeled with distances from all zeros path.

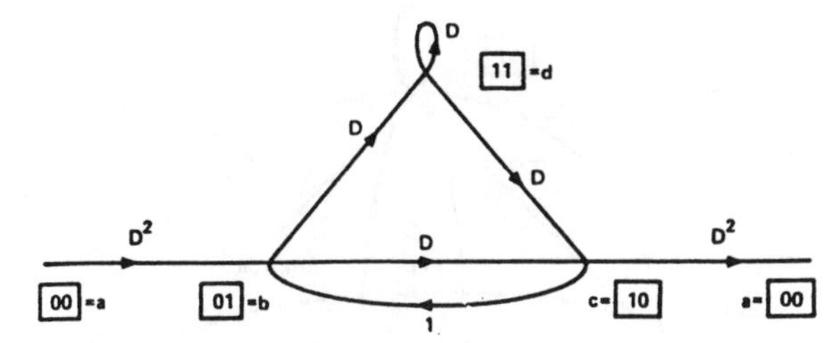

Fig. 6. State diagram labeled according to distance from all zeros path.

It is seen from the diagram that of these paths there will be just one path at distance 5 from the all zeros path and this diverged from it three branches back. Similarly there are two at distance 6 from it, one which diverged 4 branches back and the other which diverged 5 branches back, and so forth. We note also that the input bits for distance 5 path are 00 ··· 0100 and thus differ in only one input bit from the all zeros, while the distance 6 paths are 00 ··· 01100 and 00 ··· 010100 and thus each differs in 2 input bits from the all zeros path. The minimum distance, sometimes called the minimum "free" distance, among all paths is thus seen to be 5. This implies that any pair of channel errors can be corrected, for two errors will cause the received sequence to be at distance 2 from the transmitted (correct) sequence but it will be at least at distance 3 from any other possible code sequence. It appears that with enough patience the distance of all paths from the all zeros (or any arbitrary) path can be so determined from the trellis diagram.

However, by examining instead the state diagram we can readily obtain a closed form expression whose expansion yields directly and effortlessly all the distance information. We begin by labeling the branches of the state diagram of Fig. 4 either D^2, D, or $D^0 = 1$, where the exponent corresponds to the distance of the particular branch from the corresponding branch of the all zeros path. Also we split open the node $a = 00$, since circulation around this self-loop simply corresponds to branches of the all zeros path whose distance from itself is obviously zero. The result is Fig. 6. Now as is clear from examination of the trellis diagram, every path which arrives at state $a = 00$ at node level j, must have at some previous node level (possibly the first) originated at this same state $a = 00$. All such paths can be traced on the modified state diagram. Adding branch exponents we see that path $a\ b\ c\ a$ is at distance 5 from the correct path, paths $a\ b\ d\ c\ a$ and $a\ b\ c\ b\ c\ a$ are both at distance 6, and so forth, for the *generating functions* of the output sequence weights of these paths are D^5 and D^6, respectively

Now we may evaluate the generating function of all paths merging with the all zeros at the jth node level simply by evaluating the generating function of all the weights of the output sequences of the finite-state machine.[2] The result in this case is

$$T(D) = \frac{D^5}{1 - 2D}$$
$$= D^5 + 2D^6 + 4D^7 + \cdots + 2^k D^{k+5} + \cdots. \quad (1)$$

This verifies our previous observation and in fact shows that among the paths which merge with the all zeros at a given node there are 2^k paths at distance $k + 5$ from the all zeros.

Of course, (1) holds for an infinitely long code sequence; if we are dealing with the jth node level, we must truncate the series at some point. This is most easily done by considering the additional information indicated in the modified state diagram of Fig. 7.

The L terms will be used to determine the length of a given path; since each branch has an L, the exponent of the L factor will be augmented by one every time a branch is passed through. The N term is included only if that branch transition was caused by an input data "one," corresponding to a dotted branch in the trellis diagram. The generating function of this augmented state diagram is then

$$T(D, L, N)$$
$$= \frac{D^5L^3N}{1 - DL(1 + L)N}$$
$$= D^5L^3N + D^6L^4(1 + L)N^2 + D^7L^5(1 + L)^2N^3$$
$$+ \cdots + D^{5+k}L^{3+k}(1 + L)^kN^{1+k} + \cdots. \quad (2)$$

Thus we have verified that of the two distance 6 paths one is of length 4 and the other is of length 5 and both differ in 2 input bits from the all zeros.[3] Also, of the distance 7 paths, one is of length 5, two are of length 6, and one is of length 7; all four paths correspond to input sequences with three ones. If we are interested in the jth node level, clearly we should truncate the series such that no terms of power greater than L^j are included.

We have thus fully determined the properties of all paths in the convolutional code. This will be useful later in evaluating error probability performance of codes used over arbitrary memoryless channels.

[2] Alternatively, this can be regarded as the transfer function of the diagram regarded as a signal flow graph.

[3] Thus if the all zeros was the correct path and the noise causes us to choose one of the incorrect paths, two bit errors will be made.

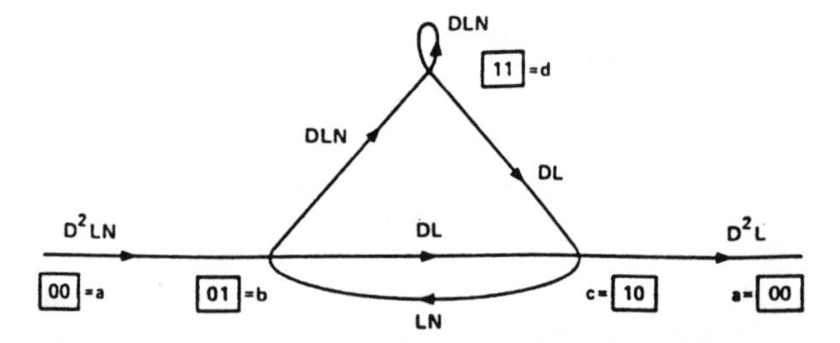

Fig. 7. State diagram labeled according to distance, length, and number of input ones.

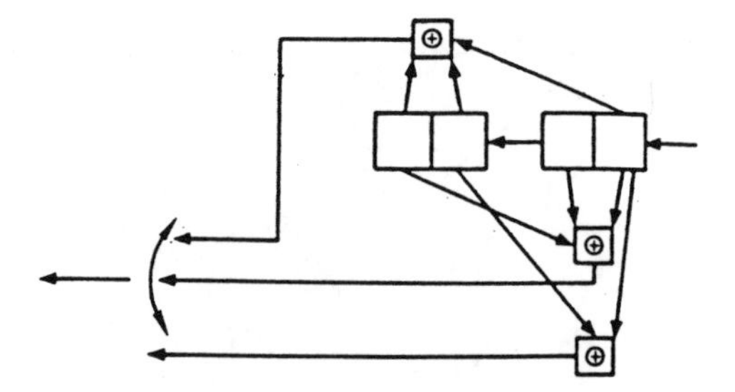

Fig. 8. Coder for $K = 2, b = 2, n = 3$, and $R = 2/3$.

V. Generalization to Arbitrary Convolutional Codes

The generalization of these techniques to arbitrary binary-tree ($b = 1$) convolutional codes is immediate. That is, a coder with a K-stage shift register and n mod-2 adders will produce a trellis or state diagram with 2^{K-1} nodes or states and each branch will contain n code symbols. The rate of this code is then

$$R = \frac{1}{n} \text{ bits/code symbol.}$$

The example pursued in the previous sections had rate $R = 1/2$. The primary characteristic of the binary-tree codes is that only two branches exit from and enter each node.

If rates other than $1/n$ are desired we must make $b > 1$, where b is the number of bits shifted into the register at one time. An example for $K = 2, b = 2, n = 3$, and consequently rate $R = 2/3$ is shown in Fig. 8 and its state diagram is shown in Fig. 9. It differs from the binary-tree codes only in that each node is connected to four other nodes, and for general b it will be connected to 2^b nodes. Still all the preceding techniques including the trellis and state-diagram generating function analysis are still applicable. It must be noted, however, that the minimum distance decoder must make comparisons among all the paths entering each node at each level of the trellis and select one survivor out of four (or out of 2^b in general).

VI. Generalization of Optimal Decoder to Arbitrary Memoryless Channels

Fig. 10 exhibits a communication system employing a convolutional code. The convolutional encoder is precisely the device studied in the preceding sections. The data sequence is generally binary ($a_j = 0$ or 1) and the code sequence is divided into subsequences where $\mathbf{x}_j$ represents the n code symbols generated just after the input bit a_j enters the coder: that is, the symbols of the jth branch. In terms of the example of Fig. 1, $a_3 = 1$ and $\mathbf{x}_3 = 01$. The channel output or received sequence is similarly denoted. $\mathbf{y}_j$ represents the n symbols received when the n code symbols of $\mathbf{x}_j$ were transmitted. This model includes the BSC wherein the $\mathbf{y}_j$ are binary n vectors each of whose symbols differs from the corresponding symbol of $\mathbf{x}_j$ with probability p and is identical to it with probability $1 - p$.

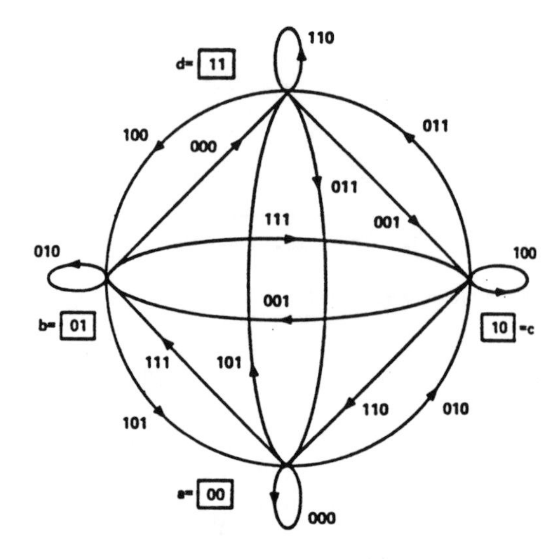

Fig. 9. State diagram for code of Fig. 8.

For completely general channels it is readily shown [6], [14] that if all input data sequences are equally likely, the decoder which minimizes the error probability is one which compares the conditional probabilities, also called likelihood functions, $P(\mathbf{y} \mid \mathbf{x}^{(m)})$, where $\mathbf{y}$ is the overall received sequence and $\mathbf{x}^{(m)}$ is one of the possible transmitted sequences, and decides in favor of the maximum. This is called a maximum likelihood decoder. The likelihood functions are given or computed from the specifications of the channel. Generally it is more convenient to compare the quantities $\log P(\mathbf{y} \mid \mathbf{x}^{(m)})$ called the log-likelihood functions and the result is unaltered since the logarithm is a monotonic function of its (always positive) argument.

To illustrate, let us consider again the BSC. Here each transmitted symbol is altered with probability $p < 1/2$. Now suppose we have received a particular N-dimensional binary sequence $\mathbf{y}$ and are considering a possible transmitted N-dimensional code sequence $\mathbf{x}^{(m)}$ which differs in d_m symbols from $\mathbf{y}$ (that is, the Hamming distance between $\mathbf{x}^{(m)}$ and $\mathbf{y}$ is d_m). Then since the channel is memoryless (i.e., it affects each symbol independently of all the others), the probability

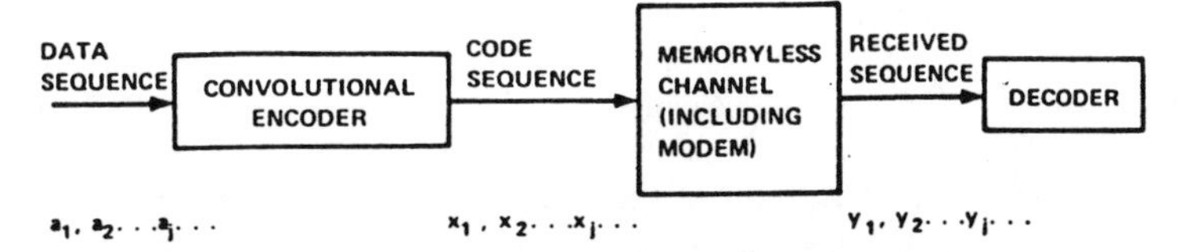

Fig. 10. Communication system employing convolutional codes.

that this $\mathbf{x}^{(m)}$ was transformed to the specific received $\mathbf{y}$ at distance d_m from it is

$$P(\mathbf{y} \mid \mathbf{x}^{(m)}) = p^{d_m}(1-p)^{N-d_m}$$

and the log-likelihood function is thus

$$\log P(\mathbf{y} \mid \mathbf{x}^{(m)}) = -d_m \log (1-p/p) + N \log (1-p)$$

Now if we compute this quantity for each possible transmitted sequence, it is clear that the second term is constant in each case. Furthermore, since we may assume $p < 1/2$ (otherwise the role of 0 and 1 is simply interchanged at the receiver), we may express this as

$$\log P(\mathbf{y} \mid \mathbf{x}^{(m)}) = -\alpha d_m - \beta \tag{3}$$

where α and β are positive constants and d_m is the (positive) distance. Consequently, it is clear that maximizing the log-likelihood function is equivalent to minimizing the Hamming distance d_m. Thus for the BSC to minimize the error probability we should choose that code sequence at minimum distance from the received sequence, as we have indicated and done in preceding sections.

We now consider a more physical practical channel: the AWGN channel with biphase[4] phase-shift keying (PSK) modulation. The modulator and optimum demodulator (correlator or integrate-and dump filter) for this channel are shown in Fig. 11.

We use the notation that x_{jk} is the kth code symbol for the jth branch. Each binary symbol (which we take here for convenience to be ± 1) modulates the carrier by $\pm \Pi/2$ radians for T seconds. The transmission rate is, therefore, $1/T$ symbols/second or $b/nT = R/T$ bit/s. The function ϵ_s is the energy transmitted for each symbol. The energy per bit is, therefore $\epsilon_b = \epsilon_s/R$. The white Gaussian noise is a zero-mean random process of one-sided spectral density N_0 W/Hz, which affects each symbol independently. It then follows directly that the channel output symbol y_{jk} is a Gaussian random variable whose mean is $\sqrt{\epsilon_s} x_{jk}$ (i.e., $+\sqrt{\epsilon_s}$ if $x_{jk} = 1$ and $-\sqrt{\epsilon_s}$ if $x_{jk} = -1$) and whose variance is $N_0/2$. Thus the conditional probability density (or likelihood) function of y_{jk} given x_{jk} is

$$p(y_{jk} \mid x_{jk}) = \frac{\exp\left[-(y_{jk} - \sqrt{\epsilon_s}\, x_{jk})^2/N_0\right]}{\sqrt{\Pi N_0}}. \tag{4}$$

The likelihood function for the jth branch of a particular

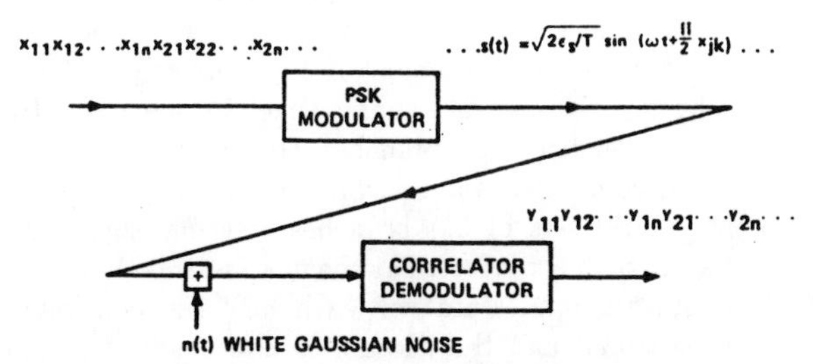

Fig. 11. Modem for additive white Gaussian noise PSK modulated memoryless channel.

code path $\mathbf{x}_j^{(m)}$

$$p(\mathbf{y}_j \mid \mathbf{x}_j^{(m)}) = \prod_{k=1}^{n} p(y_{jk} \mid x_{jk}^{(m)})$$

since each symbol is affected independently by the white Gaussian noise, and thus the log-likelihood function for the jth branch is

$$\begin{aligned}
\ln p(\mathbf{y}_j \mid \mathbf{x}_j^{(m)}) &= \sum_{k=1}^{n} \ln p(y_{jk} \mid x_{jk}^{(m)}) \\
&= -\frac{1}{N_0} \sum_{k=1}^{n} [y_{jk} - \sqrt{\epsilon_s}\, x_{jk}^{(m)}]^2 - \tfrac{1}{2} \ln \frac{\Pi}{N_0} \\
&= \frac{2\sqrt{\epsilon_s}}{N_0} \sum_{k=1}^{n} y_{jk} x_{jk}^{(m)} - \frac{\epsilon_s}{N_0} \sum_{k=1}^{n} [x_{jk}^{(m)}]^2 \\
&\quad - \frac{1}{N_0} \sum_{k=1}^{n} y_{jk}^2 - \tfrac{1}{2} \ln \frac{\Pi}{N_0} \\
&= C \sum_{k=1}^{n} y_{jk} x_{jk}^{(m)} - D
\end{aligned} \tag{5}$$

where C and D are independent of m, and we have used the fact that $[x_{jk}^{(m)}]^2 = 1$. Similarly, the log-likelihood[5] function for any path is the sum of the log-likelihood functions for each of its branches.

We have thus shown that the maximum likelihood decoder for the memoryless AWGN biphase (or quadriphase) modulated channel is one which forms the inner product between the received (real number) sequence and the code sequence (consisting of ± 1) and chooses the path corresponding to the greatest. Thus the *metric* for this channel is the inner product (5) as contrasted with the distance[6] metric used for the BSC.

[4] The results are the same for quadriphase PSK with coherent reception. The analysis proceeds in the same way, if we treat quadriphase PSK as two parallel independent biphase PSK channels.

[5] We have used the natural logarithm here, but obviously a change of base results merely in a scale factor.

[6] Actually it is easily shown that maximizing an inner product is equivalent to minimizing the Euclidean distance between the corresponding vectors.

For convolutional codes the structure of the code paths was described in Sections II–V. In Section III the optimum decoder was derived for the BSC. It now becomes clear that if we substitute the inner product metric $\Sigma y_{jk}x_{jk}^{(m)}$ for the distance metric $\Sigma d_{jk}^{(m)}$, used for the BSC, all the arguments used in Section III for the latter apply equally to this Gaussian channel. In particular the optimum decoder has a block diagram represented by the code state diagram. At step j the stored metric for each state (which is the maximum of the metrics of all the paths leading to this state at this time) is augmented by the branch metrics for branches emanating from this state. The comparisons are performed among all pairs of (or in general sets of 2^b) branches entering each state and the *maxima* are selected as the new most likely paths. The history (input data) of each new survivor must again be stored and the decoder is now ready for step $j+1$.

Clearly, this argument generalizes to any memoryless channel and we must simply use the appropriate metric $\ln P(\mathbf{y} \mid \mathbf{x}^{(m)})$, which may always be determined from the statistical description of the channel. This includes, among others, AWGN channels employing other forms of modulation.[7]

In the next section, we apply the analysis of convolutional code distance properties of Section IV to determine the error probabilities of specific codes on more general memoryless channels.

VII. Performance of Convolutional Codes on Memoryless Channels

In Section IV we analyzed the distance properties of convolutional codes employing a state-diagram generating function technique. We now extend this approach to obtain tight upper bounds on the error probability of such codes. We shall consider the BSC, the AWGN channel and more general memoryless channels, in that order. We shall obtain both the first-event error probability, which is the probability that the correct path is excluded (not a survivor) for the first time at the jth step, and the bit error probability which is the expected ratio of bit errors to total number of bits transmitted.

A. Binary Symmetric Channel

The first-event error probability is readily obtained from the generating function $T(D)$ [(5) for the code of Fig. 1, which we shall again pursue for demonstrative purposes]. We may assume, without loss of generality, since we are dealing with group codes, that the all zeros path was transmitted. Then a first-event error is made at the jth step if this path is excluded by selecting another path merging with the all zeros at node a at the jth level.

Now suppose that the previous-level survivors were such that the path compared with the all zeros at step j is the path whose data sequence is $00 \cdots 0100$ corresponding to nodes $a \cdots a\ a\ b\ c\ a$ (see Fig. 4.). This differs from the correct (all zeros) path in five symbols. Consequently an error will be made in this comparison if the BSC caused three or more errors in these particular five symbols. Hence the probability of an error in this specific comparison is

$$P_5 = \sum_{e=3}^{5} \binom{5}{e} p^e(1-p)^{5-e}. \tag{6}$$

On the other hand, there is no assurance that this particular distance five path will have previously survived so as to be compared with the correct path at the jth step. If either of the distance 6 paths were compared instead, then four or more errors in the six different symbols will definitely cause an error in the survivor decision, while three errors will cause a tie which, if resolved by coin flipping, will result in an error only half the time. Then the probability if this comparison is made is

$$P_6 = \frac{1}{2}\binom{6}{3} p^3(1-p)^3 + \sum_{e=4}^{6} \binom{6}{e} p^e(1-p)^{6-e}. \tag{7}$$

Similarly, if the previously surviving paths were such that a distance d path is compared with the correct path at the jth step, the resulting error probability is

$$P_k = \begin{cases} \displaystyle\sum_{e=(k+1)/2}^{k} \binom{k}{e} p^e(1-p)^{k-e}, & k \text{ odd} \\ \displaystyle\frac{1}{2}\binom{k}{k/2} p^{k/2}(1-p)^{k/2} + \sum_{e=k/2+1}^{k} \binom{k}{e} p^e(1-p)^{k-e}, & k \text{ even.} \end{cases} \tag{8}$$

Now at step j, since there is no simple way of determining previous survivors, we may overbound the probability of a first-event error by the sum of the error probabilities for all possible paths which merge with the correct path at this point. Note this *union bound* is indeed an upper bound because two or more such paths may both have distance closer to the received sequence than the correct path (even though only one has survived to this point) and thus the events are not disjoint. For the example with generating function (1) it follows that the first-event error probability[8] is bounded by

$$P_E < P_5 + 2P_6 + 4P_7 + \cdots + 2^k P_{k+5} + \cdots \tag{9}$$

where P_k is given by (8).

In Section VII-C it will be shown that (8) can be upper bounded by (see (39)).

$$P_k < 2^k p(1-p)^{k/2}. \tag{10}$$

Using this, the first-event error probability bound (9)

[7] Although more elaborate modulators, such as multiple FSK or multiphase modulators, might be employed, Jacobs [11] has shown that the most effective as well as the simplest system for wide-band space and satellite channels is the binary PSK modulator considered in the example of this section. We note again that the performance of quadriphase modulation is the same as for biphase modulation, when both are coherently demodulated.

[8] We are ignoring the finite length of the path, but the expression is still valid since it is an upper bound.

can be more loosely bounded by

$$P_E < \sum_{k=5}^{\infty} 2^{k-5} 2^k p(1-p)^{k/2}$$

$$= \frac{[2\sqrt{p(1-p)}]^5}{1 - 4\sqrt{p(1-p)}} = T(D) \big|_{D=2\sqrt{p(1-p)}} \quad (11)$$

where $T(D)$ is just the generating function of (1)

It follows easily that for a general binary-tree ($b = 1$) convolutional code with generating function

$$T(D) = \sum_{k=d}^{\infty} a_k D^k \quad (12)$$

the first-event error probability is bounded by the generalization of (9).

$$P_E < \sum_{k=d}^{\infty} a_k P_k \quad (13)$$

where P_k is given by (8) and more loosely upper bounded by the generalization of (11)

$$P_E < T(D) \big|_{D=2\sqrt{p(1-p)}}. \quad (14)$$

Whenever a decision error occurs, one or more bits will be incorrectly decoded. Specifically, those bits in which the path selected differs from the correct path will be incorrect. If only one error were ever made in decoding an arbitrary long code path, the number of bits in error in this incorrect path could easily be obtained from the augmented generating function $T(D, N)$ (such as given by (2) with factors in L deleted). For the exponents of the N factors indicate the number of bit errors for the given incorrect path arriving at node a at the jth level.

After the first error has been made, the incorrect paths no longer will be compared with a path which is overall correct, but rather with a path which has diverged from the correct path over some span of branches (see Fig. 12). If the correct path $\mathbf{x}$ has been excluded by a decision error at step j in favor of path $\mathbf{x}'$, the decision at step $j + 1$ will be between $\mathbf{x}'$ and $\mathbf{x}''$. Now the (first-event) error probability of (13) or (14) is for a comparison, at any step, between path $\mathbf{x}$ and any other path merging with it at that step, including path $\mathbf{x}''$ in this case. However, since the metric[9] for path $\mathbf{x}'$ is greater than the metric for $\mathbf{x}$, for on this basis the correct path was excluded at step j, the probability that path $\mathbf{x}''$ metric exceeds path $\mathbf{x}'$ metric at step $j + 1$ is less than the probability that path $\mathbf{x}''$ exceeds the (correct) path $\mathbf{x}$ metric at this point. Consequently, the probability of a new incorrect path being selected after a previous error has occurred is upper bounded by the first-event error probability at that step.

Moreover, when a second error follows closely after a first error, it often occurs (as in Fig. 12) that the erroneous bit(s) of path $\mathbf{x}''$ overlap the erroneous bit(s) of path $\mathbf{x}'$. With this in mind, we now show that for a

[9] Negative distance from the received sequence for the BSC, but clearly this argument generalizes to any memoryless channel.

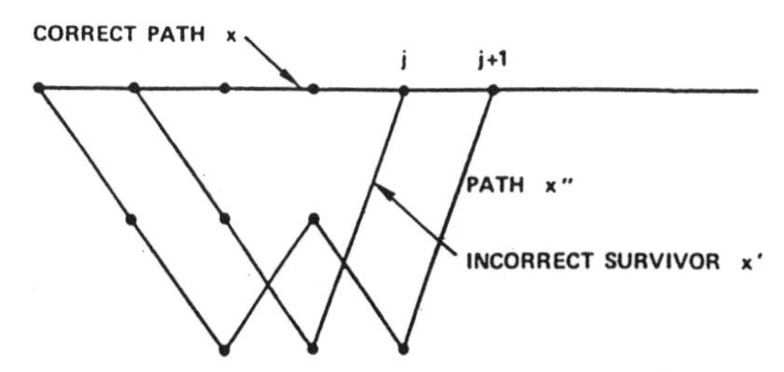

Fig. 12. Example of decoding decision after initial error has occurred.

binary-tree code if we weight each term of the first-event error probability bound at any step by the number of erroneous bits for each possible erroneous path merging with the correct path at that node level, we upper bound the bit error probability. For, a given step decision corresponds to decoder action on one more bit of the transmitted data sequence; the first-event error probability union bound with each term weighted by the corresponding number of bit errors is an upper bound on the expected number of bit errors caused by this action. Summing the expected number of bit errors over L steps, which as was just shown may result in overestimating through double counting, gives an upper bound on the expected number of bit errors in L branches for arbitrary L. But since the upper bound on expected number of bit errors is the same at each step, it follows, upon dividing the sum of L equal terms by L, that this expected number of bit errors per step is just the bit error probability P_B, for a binary-tree code ($b = 1$). If $b > 1$, then we must divide this expression by b, the number of bits encoded and decoded per step.

To illustrate the calculation of P_B for a convolutional code, let us consider again the example of Fig. 1. Its transfer function in D and N is obtained from (2), letting $L = 1$, since we are not now interested in the lengths of incorrect paths, to be

$$T(D, N) = \frac{D^5 N}{1 - 2DN}$$

$$= D^5 N + 2D^6 N^2 = \cdots + 2^k D^{k+5} N^{k+1} + \cdots. \quad (15)$$

The exponents of the factors in N in each term determine the number of bit errors for the path(s) corresponding to that term. Since $T(D) = T(D, N)\big|_{N=1}$ yields the first-event error probability P_E, each of whose terms must be weighted by the exponent of N to obtain P_B, it follows that we should first differentiate $T(D, N)$ at $N = 1$ to obtain

$$\left.\frac{dT(D, N)}{dN}\right|_{N=1}$$

$$= D^5 + 2 \cdot 2D^6 + 3 \cdot 4D^7 + \cdots + (k+1)2^k D^{k+5} + \cdots$$

$$= \frac{D^5}{(1 - 2D)^2}. \quad (16)$$

Then from this we obtain, as in (9), that for the BSC

$$P_B < P_5 + 2\cdot 2P_6 + 3\cdot 4P_7 + \cdots + (k+1)2^k P_{k+5} + \cdots \quad (17)$$

where P_k is given by (8).

If for P_k we use the upper bound (10) we obtain the weaker but simpler bound

$$P_B < \sum_{k=5}^{\infty} (k-4)2^{k-5}[4p(1-p)]^{k/2} = \left.\frac{dT(D,N)}{dN}\right|_{N=1, D=2\sqrt{p(1-p)}} = \frac{|2\sqrt{p(1-p)}|^5}{[1-4\sqrt{p(1-p)}]^2}. \quad (18)$$

More generally for any binary-tree ($b = 1$) code used on the BSC if

$$\left.\frac{dT(D,N)}{dN}\right|_{N=1} = \sum_{k=d}^{\infty} c_k D^k \quad (19)$$

then corresponding to (17)

$$P_B < \sum_{k=d}^{\infty} c_k P_k \quad (20)$$

and corresponding to (18) we have the weaker bound

$$P_B < \left.\frac{dT(D,N)}{dN}\right|_{N=1, D=2\sqrt{p(1-p)}}. \quad (21)$$

For a nonbinary-tree code ($b \neq 1$), all these expressions must be divided by b.

The results of (14) and (18) will be extended to more general memoryless channels, but first we shall consider one more specific channel of particular interest.

B. AWGN Biphase-Modulated Channel

As was shown in Section VI the decoder for this channel operates in exactly the same way as for the BSC, except that instead of Hamming distance it uses the metric

$$\sum_i \sum_{j=1}^{n} x_{ij} y_{ij}$$

where $x_{ij} = \pm 1$ are the transmitted code symbols, y_{ij} the corresponding received (demodulated) symbols, and j runs over the n symbols of each branch while i runs over all the branches in a particular path. Hence, to analyze its performance we may proceed exactly as in Section VII-A except that the appropriate pairwise-decision errors P_k must be substituted for those of (6) to (8).

As before we assume, without loss of generality, that the correct (transmitted) path $\mathbf{x}$ has $x_{ij} = +1$ for all i and j (corresponding to the all zeros if the input symbols were 0 and 1). Let us consider an incorrect path $\mathbf{x}'$ merging with the correct path at a particular step, which has k negative symbols ($x_{ij}' = -1$) and the remainder positive. Such a path may be incorrectly chosen only if it has a higher metric than the correct path, i.e.,

$$\sum_i \sum_{j=1}^{n} x_{ij}' y_{ij} \geq \sum_i \sum_{j=1}^{n} x_{ij} y_{ij}$$

or

$$\sum_i \sum_{j=1}^{n} (x_{ij}' - x_{ij}) y_{ij} \geq 0$$

where i runs over all branches in the two paths. But since, as we have assumed, the paths $\mathbf{x}$ and $\mathbf{x}'$ differ in exactly k symbols, wherein $x_{ij} = 1$ and $x_{ij}' = -1$, the pairwise error probability is just

$$\begin{aligned} P_k &= \Pr\left\{\sum_i \sum_{j=1}^{n} (x_{ij}' - x_{ij}) y_{ij} \geq 0\right\} \\ &= \Pr\left\{\sum_{r=1}^{k} (x_r' - x_r) y_r \geq 0\right\} \\ &= \Pr\left\{-2 \sum_{r=1}^{k} y_r \geq 0\right\} \\ &= \Pr\left\{\sum_{r=1}^{k} y_r \leq 0\right\} \end{aligned} \quad (22)$$

where r runs over the k symbols wherein the two paths differ. Now it was shown in Section VI that the y_{ij} are independent Gaussian random variables of variance $N_0/2$ and mean $\sqrt{\epsilon_s} x_{ij}$, where x_{ij} is the actually transmitted code symbol. Since we are assuming that the (correct) transmitted path has $x_{ij} = +1$ for all i and j, it follows that y_{ij} or y_r has mean $\sqrt{\epsilon_s}$ and variance $N_0/2$. Therefore, since the k variables y_r are independent and Gaussian, the sum $Z = \sum_{r=1}^{k} y_r$ is also Gaussian with mean $k\sqrt{\epsilon_s}$ and variance $kN_0/2$.

Consequently,

$$P_k = \Pr(Z < 0) = \int_{-\infty}^{0} \frac{\exp(-Z - k\sqrt{\epsilon_s})^2/kN_0}{\sqrt{\Pi k N_0}}\, dZ = \int_{\sqrt{2k\epsilon_s/N_0}}^{\infty} \left[\frac{\exp(-x^2/2)}{\sqrt{2\Pi}}\right] dx \triangleq \operatorname{erfc}\sqrt{\frac{2k\epsilon_s}{N_0}}. \quad (23)$$

We recall from Section VI that ϵ_s is the symbol energy, which is related to the bit energy by $\epsilon_s = R\epsilon_b$, where $R = b/n$. The bound on P_E then follows exactly as in Section VII-A and we obtain the same general bound as (13)

$$P_E < \sum_{k=d}^{\infty} a_k P_k \quad (24)$$

where a_k are the coefficients of

$$T(D) = \sum_{k=d}^{\infty} a_k D^k \quad (25)$$

and where d is the minimum distance between any two paths in the code. We may simplify this procedure considerably while loosening the bound only slightly for this channel by observing that for $x \geq 0$, $y \geq 0$,

$$\operatorname{erfc}\sqrt{x+y} \leq \exp\left(\frac{-y}{2}\right) \operatorname{erfc}\sqrt{x}. \quad (26)$$

Consequently, for $k \geq d$, letting $l = k - d$, we have from (23)

$$P_k = \operatorname{erfc}\sqrt{\frac{2k\epsilon_s}{N_0}} = \operatorname{erfc}\sqrt{\frac{2(d+l)\epsilon_s}{N_0}} \leq \exp\left(\frac{-l\epsilon_s}{N_0}\right)\operatorname{erfc}\sqrt{\frac{2d\epsilon_s}{N_0}} \tag{27}$$

whence the bound of (24), using (27), becomes

$$P_E < \sum_{k=d}^{\infty} a_k P_k \leq \operatorname{erfc}\sqrt{\frac{2d\epsilon_s}{N_0}} \sum_{k=d}^{\infty} a_k \exp\left[\frac{-(k-d)\epsilon_s}{N_0}\right]$$

or

$$P_E < \operatorname{erfc}\sqrt{\frac{2d\epsilon_s}{N_0}} \exp\left(\frac{d\epsilon_s}{N_0}\right) T(D) \Big|_{D=\exp(-\epsilon_s/N_0)}. \tag{28}$$

The bit error probability can be obtained in exactly the same way. Just as for the BSC [(19) and (20)] we have that for a binary-tree code

$$P_B < \sum_{k=d}^{\infty} c_k P_k \tag{29}$$

where c_k are the coefficients of

$$\left.\frac{dT(D, N)}{dN}\right|_{N=1} = \sum_{k=d}^{\infty} c_k D^k. \tag{30}$$

Thus following the came arguments which led from (24) to (28) we have for a binary-tree code

$$P_B < \operatorname{erfc}\sqrt{\frac{2d\epsilon_s}{N_0}} \exp\left(\frac{d\epsilon_s}{N_0}\right) \left.\frac{dT(D, N)}{dN}\right|_{N=1, D=\exp(-\epsilon_s/N_0)} \tag{31}$$

For $b > 1$, this expression must be divided by b.

To illustrate the application of this result we consider the code of Fig. 1 with parameters $K = 3$, $R = 1/2$, whose transfer function is given by (15). For this case since $R = 1/2$ and $\epsilon_s = 1/2\, \epsilon_b$, we obtain

$$P_B < \frac{\operatorname{erfc}\sqrt{5\epsilon_b/N_0}}{(1 - 2e^{-\epsilon_b/2N_0})}. \tag{32}$$

Since the number of states in the state diagram grows exponentially with K, direct calculation of the generating function becomes unmanageable for $K > 4$. On the other hand, a generating function calculation is basically just a matrix inversion (see Appendix I), which can be performed numerically for a given value of D. The derivative at $N = 1$ can be upper bounded by evaluating the first difference $[T(D, 1 + \epsilon) - T(D, 1)]/\epsilon$, for small ϵ. A computer program has been written to evaluate (31) for any constraint length up to $K = 10$ and all rates $R = 1/n$ as well as $R = 2/3$ and $R = 3/4$. Extensive results of these calculations are given in the paper by Heller and Jacobs [24], along with the results of simulations of the corresponding codes and channels. The simulations verify the tightness of the bounds.

In the next section, these bounding techniques will be extended to more general memoryless channels, from which (28) and (31) can be obtained directly, but without the first two factors. Since the product of the first two factors is always less than one, the more general bound is somewhat weaker.

C. General Memoryless Channels

As was indicated in Section VI, for equally likely input data sequences, the minimum error probability decoder chooses the path which maximizes the log-likelihood function (metric)

$$\ln P(\mathbf{y} \mid \mathbf{x}^{(m)})$$

over all possible paths $\mathbf{x}^{(m)}$. If each symbol is transmitted (or modulates the transmitter) independent of all preceding and succeeding symbols, and the interference corrupts each symbol independently of all the others, then the channel, which includes the modem, is said to be memoryless[10] and the log-likelihood function

$$\ln P(\mathbf{y} \mid \mathbf{x}^{(m)}) = \sum_i \sum_{j=1}^{n} \ln P(y_{ij} \mid x_{ij}^{(m)})$$

where $x_{ij}^{(m)}$ is a code symbol of the mth path, y_{ij} is the corresponding received (demodulated) symbol, j runs over the n symbols of each branch, and i runs over the branches in the given path. This includes the special cases considered in Sections VII-A and -B.

The decoder is the same as for the BSC except for using this more general metric. Decisions are made after each set of new branch metrics have been added to the previously stored metrics. To analyze performance, we must merely evaluate P_k, the pairwise error probability for an incorrect path which differs in k symbols from the correct path, as was done for the special channels of Sections VII-A and -B. Proceeding as in (22), letting x_{ij} and x_{ij}' denote symbols of the correct and incorrect paths, respectively, we obtain

$$\begin{aligned} P_k(\mathbf{x}, \mathbf{x}') &= \Pr\left[\sum_i \sum_{j=1}^{n} \ln P(y_{ij} \mid x_{ij}') > \sum_i \sum_{j=1}^{n} \ln P(y_{ij} \mid x_{ij})\right] \\ &= \Pr\left\{\sum_{r=1}^{k} \ln \frac{P(y_r \mid x_r')}{P(y_r \mid x_r)} > 0\right\} \\ &= \Pr\left\{\prod_{r=1}^{k} \frac{P(y_r \mid x_r')}{P(y_r \mid x_r)} > 1\right\} \end{aligned} \tag{33}$$

where r runs over the k code symbols in which the paths differ. This probability can be rewritten as

$$P_k(\mathbf{x}, \mathbf{x}') = \sum_{\mathbf{y} \in Y_k} \prod_{r=1}^{k} P(y_r \mid x_r) \tag{34}$$

where Y_k is the set of all vectors $\mathbf{y} = (y_1, y_2, \cdots, y_r, \cdots, y_k)$ for which

[10] Often more than one code symbol in a given branch is used to modulate the transmitter at one time. In this case, provided the interference still affects succeeding branches independently, the channel can still be treated as memoryless but now the symbol likelihood functions are replaced by branch likelihood functions and (33) is replaced by a single sum over i.

$$\prod_{r=1}^{k} \frac{P(y_r \mid x_r')}{P(y_r \mid x_r)} > 1. \tag{35}$$

But if this is the case, then

$$P_k(\mathbf{x}, \mathbf{x}') < \sum_{y \in Y_k} \prod_{r=1}^{k} P(y_r \mid x_r) \left[\frac{P(y_r \mid x_r')}{P(y_r \mid x_r)}\right]^{1/2}$$

$$< \sum_{\text{all } y \in Y} \prod_{r=1}^{k} P(y_r \mid x_r)^{1/2} P(y_r \mid x_r')^{1/2} \tag{36}$$

where Y is the entire space of received vectors.[11] The first inequality is valid because we are multiplying the summand by a quantity greater than unity,[12] and the second because we are merely extending the sum of positive terms over a larger set. Finally we may break up the k-dimensional sum over $\mathbf{y}$ into k one-dimensional summations over $y_1, y_2, \cdots, y_k$, respectively, and this yields

$$P_k(\mathbf{x}, \mathbf{x}') \leq \sum_{y_1} \sum_{y_2} \cdots \sum_{y_k} \prod_{r=1}^{k} P(y_r \mid x_r)^{1/2} P(y_r \mid x_r')^{1/2}$$

$$= \prod_{r=1}^{k} \sum_{y_r} P(y_r \mid x_r)^{1/2} P(y_r \mid x_r')^{1/2} \tag{37}$$

To illustrate the use of this bound we consider the two specific channels treated above. For the BSC, y_r is either equal to x_r, the transmitted symbol, or to $\bar{x}_r$, its complement. Now y_r depends on x_r through the channel statistics. Thus

$$P(y_r = x_r) = 1 - p$$

$$P(y_r = \bar{x}_r) = p. \tag{38}$$

For each symbol in the set $r = 1, 2, \cdots, k$ by definition $x_r \neq x_r'$. Hence for each term in the sum if $x_r = 0$, $x_r' = 1$ or vice versa. Hence, whatever x_r and x_r' may be

$$\sum_{y_r=0}^{1} P(y_r \mid x_r)^{1/2} P(y_r \mid x_r')^{1/2} = 2p^{1/2}(1 - p)^{1/2}$$

and the product (37) of k identical factors is

$$P_k = 2^k p^{k/2}(1 - p)^{k/2} \tag{39}$$

for all pairs of correct and incorrect paths. This was used in Section VII-A to obtain the bounds (11) and (21).

For the AWGN channel of Section VII-B we showed that the likelihood functions (probability densities) were

$$p(y_r \mid x_r) = \frac{\exp[-(y_r - \sqrt{\epsilon_s}\, x_r)^2/N_0]}{\sqrt{\Pi N_0}} \tag{40}$$

where $x_r = +1$ or -1 and

$$x_r + x_r' = 0. \tag{41}$$

Since y_r is a real variable, the space of y_r is the real line and the sum in (37) becomes the integral

$$\int_{-\infty}^{\infty} p(y_r \mid x_r)^{1/2} p(y_r \mid x_r')^{1/2}\, dy_r = \frac{1}{\sqrt{\Pi N_0}} \int_{-\infty}^{\infty} \exp\left\{-\frac{[(y_r - \sqrt{\epsilon_s}\, x_r)^2 + (y_r - \sqrt{\epsilon_s}\, x_r')^2]}{2N_0}\right\} dy_r$$

$$= \frac{1}{\sqrt{\Pi N_0}} \int_{-\infty}^{\infty} \exp\left[\frac{-(y_r^2 + \epsilon_s)}{N_0}\right] dy_r = \exp\left(\frac{-\epsilon_s}{N_0}\right)$$

where we have used (41) and $x_r^2 = x_r'^2 = 1$. The product of these k identical terms is, therefore,

$$P_k < \exp\left(\frac{-k\epsilon_s}{N_0}\right) \tag{42}$$

for all pairs of correct and incorrect paths. Inserting these bounds in the general expressions (24) and (29), and using (25) and (30) yields the bound on first-event error probability and bit error probability.

$$P_E < T(D)\big|_{D=\exp(-\epsilon_s/N_0)} \tag{43}$$

$$P_B < \left.\frac{dT(D, N)}{dN}\right|_{N=1,\, D=\exp(-\epsilon_s/N_0)} \tag{44}$$

which are somewhat (though not exponentially) weaker than (28) and (31).

A characteristic feature of both the BSC and the AWGN channel is that they affect each symbol in the same way independent of its location in the sequence. Any memoryless channel has this property provided it is stationary (statistically time invariant). For a stationary memoryless channel (37) reduces to

$$P_k(\mathbf{x}, \mathbf{x}') < \left[\sum_{y_r} P(y_r \mid x_r)^{1/2} P(y_r \mid x_r')^{1/2}\right]^k \triangleq D_0^k \tag{45}$$

where[13]

$$D_0 \triangleq \sum_{y_r} P(y_r \mid x_r)^{1/2} P(y_r \mid x_r')^{1/2} < 1. \tag{46}$$

While this bound on P_k is valid for all such channels, clearly it depends on the actual values assumed by the symbols x_r and x_r', of the correct and incorrect path, and these will generally vary according to the pairs of paths $\mathbf{x}$ and $\mathbf{x}'$ in question. However, if the input symbols are binary, x and $\bar{x}$, whenever $x_r = x$, then $x_r' = \bar{x}$,

[11] This would be the set of all 2^k k-dimensional binary vectors for the BSC, and Euclidean k space for the AWGN channel. Note also that the bound of (36) may be improved for asymmetric channels by changing the two exponents of ½ to s and $1 - s$, respectively, where $0 < s < 1$.

[12] The square root of a quantity greater than one is also greater than one.

[13] For an asymmetric channel this bound may be improved by changing the two exponents 1/2 to s and $1 - s$, respectively, where $0 < s < 1$.

so that for any input-binary memoryless channel (46) becomes

$$D_0 = \sum_y P(y \mid x)^{1/2} P(y \mid \bar{x})^{1/2} \tag{47}$$

and consequently

$$P_E < T(D) \mid_{D=D_0} \tag{48}$$

$$P_B < \left. \frac{dT(D, N)}{dN} \right|_{N=1, D=D_0} \tag{49}$$

where D_0 is given by (47). Other examples of channels of this type are FSK modulation over the AWGN (both coherent and noncoherent) and Rayleigh fading channels.

VIII. Systematic and Nonsystematic Convolutional Codes

The term *systematic* convolutional code refers to a code on each of whose branches one of the code symbols is just the data bit generating that branch. Thus a systematic coder will have its stages connected to only $n-1$ adders, the nth being replaced by a direct line from the first stage to the commutator. Fig. 13 shows an $R = 1/2$ systematic coder for $K = 3$.

It is well known that for group block codes, any nonsystematic code can be transformed into a systematic code which performs exactly as well. This is not the case for convolutional codes. The reason for this is that, as was shown in Section VII, the performance of a code on any channel depends largely on the relative distances between codewords and particularly on the minimum free distance d, which is the exponent of D in the leading term of the generating function. Eliminating one of the adders results in a reduction of d. For example, the maximum free distance code for $K = 3$ is that of Fig. 13 and this has $d = 4$, while the nonsystematic $K = 3$ code of Fig. 1 has minimum free distance $d = 5$. Table I shows the maximum minimum free distance for systematic and nonsystematic codes for $K = 2$ through 5. For large constraint lengths the results are even more widely separated. In fact, Bucher and Heller [19] have shown that for asymptotically large K, the performance of a systematic code of constraint length K is approximately the same as that of a nonsystematic code of constraint length $K(1-R)$. Thus for $R = 1/2$ and very large K, systematic codes have the performance of nonsystematic codes of half the constraint length, while requiring exactly the same optimal decoder complexity. For $R = 3/4$, the constraint length is effectively divided by 4.

IX. Catastrophic Error Propagation in Convolutional Codes

Massey and Sain [13] have defined a catastrophic error as the event that a finite number of channel symbol errors causes an infinite number of data bit errors to be decoded. Furthermore, they showed that a necessary and sufficient condition for a convolutional code to produce

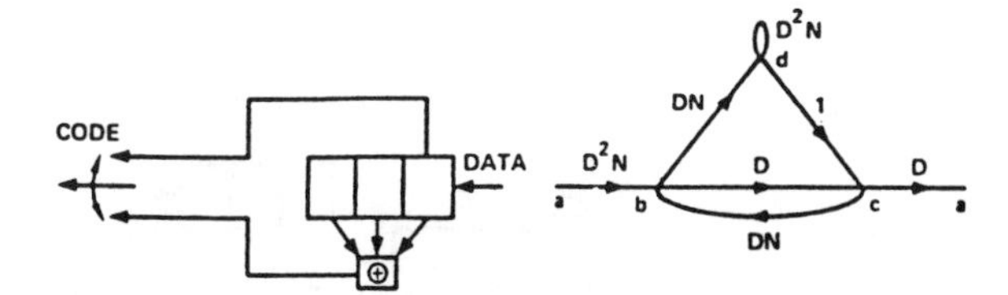

Fig. 13. Systematic convolution coder for $K = 3$ and $r = 1/2$.

TABLE I
Maximum–Minimum Free Distance

K	Systematic	Nonsystematic[a]
2	3	3
3	4	5
4	4	6
5	5	7

[a] We have excluded catastrophic codes (see Section IX); $R = \frac{1}{2}$.

catastrophic errors is that all of the adders have tap sequences, represented as polynomials, with a common factor.

In terms of the state diagram it is easily seen that catastrophic errors can occur if and only if any closed loop path in the diagram has a zero weight (i.e, the exponent of D for the loop path is zero). To illustrate this, we consider the example of Fig. 14.

Assuming that the all zeros is the correct path, the incorrect path $a\ b\ d\ d \cdots d\ c\ a$ has exactly 6 ones, no matter how many times we go around the self loop d. Thus for a BSC, for example, four-channel errors may cause us to choose this incorrect path or consequently make an arbitrarily large number of bit errors (equal to two plus the number of times the self loop is traversed). Similarly for the AWGN channel this incorrect path with arbitrarily many corresponding bit errors will be chosen with probability erfc $\sqrt{6\epsilon_b/N_0}$.

Another necessary and sufficient condition for catastrophic error propagation, recently found by Odenwalder [20] is that any nonzero data path in the trellis or state diagram produces $K-1$ consecutive branches with all zero code symbols.

We observe also that for binary-tree ($R = 1/n$) codes, if each adder of the coder has an even number of connections, then the self loop corresponding to the all ones (data) state will have zero weight and consequently the code will be catastrophic.

The main advantage of a systematic code is that it can never be catastrophic, since each closed loop must contain at least one branch generated by a nonzero data bit and thus having a nonzero code symbol. Still it can be shown [23] that only a small fraction of nonsystematic codes is catastrophic (in fact, $1/(2^n - 1)$ for binary-tree $R = 1/n$ codes. We note further that if catastrophic errors are ignored, nonsystematic codes with even larger free distance than those of Table I exist.

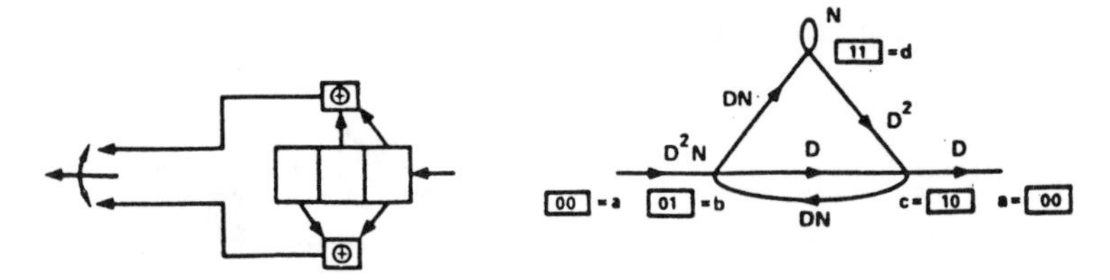

Fig. 14. Coder displaying catastrophic error propagation.

X. Performance Bounds for Best Convolutional Codes for General Memoryless Channels and Comparison with Block Codes

We begin by considering the path structure of a binary-tree[14] ($b = 1$) convolutional code of any constraint K, independent of the specific coder used. For this purpose we need only determine $T(L)$ the generating function for the state diagram with each branch labeled merely by L so that the exponent of each term of the infinite series expansion of $T(L)$ determines the length over which an incorrect path differs from the correct path before merging with it at a given node level. (See Fig. 7 and (2) with $D = N = 1$).

After some manipulation of the state-transition matrix of the state diagram of a binary-tree convolutional code of constraint length K, it is shown in Appendix I[15] that

$$T(L) = \frac{L^K(1 - L)}{1 - 2L + L^K} < \frac{L^K}{1 - 2L}$$
$$= L^K(1 + 2L + 4L^2 + \cdots + 2^k L^k + \cdots) \tag{50}$$

where the inequality indicates that more paths are being counted than actually exist. The expression (50) indicates that of the paths merging with the correct path at a given node level there is no more than one of length K, no more than two of length $K + 1$, no more than three of length $K + 2$, etc.

We have purposely avoided considering the actual code or coder configuration so that the preceding expressions are valid for all binary-tree codes of constraint length K. We now extend our class of codes to include time-varying convolutional codes. A time-varying coder is one in which the tap positions may be changed after each shift of the bits in the register. We consider the ensemble of all possible time-varying codes, which includes as a subset the ensemble of all fixed codes, for a given constraint length K. We further impose a uniform probabilistic measure on all codes in this ensemble by randomly reselecting each tap position after each shift of the register. This can be done by hypothetically flipping a coin nK times after each shift, once for each stage of the register and for each of the n adders. If the outcome is a head we connect the particular stage to the particular adder; if it is a tail we do not. Since this is repeated for each new branch, the result is that for each branch of the trellis the code sequence is a random binary n-dimensional vector. Furthermore, it can be shown that the distribution of these random code sequences is the same for each branch at each node level except for the all zeros path, which must necessarily produce the all zeros code sequence on each branch. To avoid treating the all zeros path differently, we ensure statistical uniformity by requiring further that after each shift a random binary n-dimensional vector be added to each branch[16] and that this also be reselected after each shift. (This additional artificiality is unnecessary for input-binary channels but is required to prove our result for general memoryless channels). Further details of this procedure are given in Viterbi [9].

We now seek a bound on the average error probability of this ensemble of codes relative to the measure (random-selection process) imposed. We begin by considering the probability that after transmission over a memoryless channel the metric of one of the fewer than 2^k paths merging with the correct path after differing in $K + k$ branches, is greater than the correct metric. Let $\mathbf{x}_i$ be the correct (transmitted) sequence and $\mathbf{x}_i'$ an incorrect sequence for the ith branch of the two paths. Then following the argument which led to (37) we have that the probability that the given incorrect path may cause an error is bounded by

$$P_{K+k}(\mathbf{x}, \mathbf{x}') < \prod_{i=1}^{K+k} \sum_{\mathbf{y}_i} P(\mathbf{y}_i \mid \mathbf{x}_i)^{1/2} P(\mathbf{y}_i \mid \mathbf{x}_i')^{1/2} \tag{51}$$

where the product is over all $K + k$ branches in the path. If we now average over the ensemble of codes constructed above we obtain

$$\bar{P}_{K+k} < \prod_{i=1}^{K+k} \sum_{\mathbf{x}_i} \sum_{\mathbf{x}_i'} \sum_{\mathbf{y}_i} q(\mathbf{x}_i) P(\mathbf{y}_i \mid \mathbf{x}_i)^{1/2} q(\mathbf{x}_i') P(\mathbf{y}_i \mid \mathbf{x}_i')^{1/2} \tag{52}$$

where $q(\mathbf{x})$ is the measure imposed on the code symbols of each branch by the random selection, and because of the statistical uniformity of all branches we have

$$\bar{P}_{K+k} < \{ \sum_{\mathbf{y}} [\sum_{\mathbf{x}} q(\mathbf{x}) P(\mathbf{y} \mid \mathbf{x})^{1/2}]^2 \}^{K+k} = 2^{-(K+k)nR_0} \tag{53}$$

[14] Although for clarity all results will be derived for $b = 1$, the extension to $b > 1$ is direct and the results will be indicated at the end of this Section.

[15] This generating function can also be used to obtain error bounds for orthogonal convolutional codes all of whose branches have the same weight, as is shown in Appendix I.

[16] The same vector is added to all branches at a given node level.

where

$$R_0 \triangleq -\frac{1}{n} \log_2 \{ \sum_{\mathbf{y}} [\sum_{\mathbf{x}} q(\mathbf{x}) P(\mathbf{y} \mid \mathbf{x})^{1/2}]^2 \}. \quad (54)$$

Note that the random vectors $\mathbf{x}$ and $\mathbf{y}$ are n dimensional. If each symbol is transmitted independently on a memoryless channel, such as was the case in the channels of Sections VII-A and -B, (54) is reduced further to

$$R_0 = -\log_2 \{ \sum_{y} [\sum_{x} q(x) P(y \mid x)^{1/2}]^2 \} \quad (55)$$

where x and y are now scalar random variables associated with each code symbol. Note also that because of the statistical uniformity of the code, the results are independent of which path was transmitted and which incorrect path we are considering.

Proceeding as in Section VII, it follows that a union bound on the ensemble average of the first-event error probability is obtained by substituting $\bar{P}_{K+k}$ for L^{K+k} in (50). Thus

$$\bar{P}_E < \sum_{k=0}^{\infty} 2^k \bar{P}_{K+k} < \sum_{k=0}^{\infty} 2^k 2^{-(K+k)R_0/R} = \frac{2^{-KR_0/R}}{1 - 2^{-(R_0/R-1)}} \quad (56)$$

where we have used the fact that since $b = 1$, $R = 1/n$ bits/symbol.

To bound the bit error probability we must weight each term of (56) by the number of bit errors for the corresponding incorrect path. This could be done by evaluating the transfer function $T(L, N)$ as in Section VII (see also Appendix I), but a simpler approach, which yields a simpler bound which is nearly as tight, is to recognize that an incorrectly chosen path which merges with the correct path after $K + k$ branches can produce no more $k + 1$ bit errors. For, any path which merges with the correct path at a given level must be generated by data which coincides with the correct path data over the last $K - 1$ branches prior to merging, since only in this way can the coder register be filled with the same bits as the correct path, which is the condition for merging. Hence the number of incorrect bits due to a path which differs from the correct path in $K + k$ branches can be no greater than $K + k - (K - 1) = k + 1$.

Hence we may overbound $\bar{P}_B$ by weighting the kth term of (56) by $k + 1$, which results in

$$\bar{P}_B < \sum_{k=0}^{\infty} (k + 1) 2^{-k(R_0/R-1)} 2^{-KR_0/R} = \frac{2^{-KR_0/R}}{[1 - 2^{-(R_0/R-1)}]^2}. \quad (57)$$

The bounds of (56) and (57) are finite only for rates $R < R_0$, and R_0 can be shown to be always less than the channel capacity.

To improve on these bounds when $R > R_0$, we must improve on the union bound approach by obtaining a single bound on the probability that any one of the fewer than 2^k paths which differ from the correct path in $K + k$ branches has a metric higher than the correct path at a given node level. This bound, first derived by Gallager [5] for block codes, is always less than 2^k times the bound for each individual path. Letting $Q_{K+k} \triangleq$ Pr (any one of 2^k incorrect path metrics $>$ correct path metric), Gallager [5] has shown that its ensemble average for the code ensemble is bounded by

$$\bar{Q}_{K+k} < 2^{k\rho} 2^{-(K+k)nE_0(\rho)} \quad (58)$$

where

$$E_0(\rho) = -\frac{1}{n} \log_2 \sum_{\mathbf{y}} [\sum_{\mathbf{x}} q(\mathbf{x}) p(\mathbf{y} \mid \mathbf{x})^{1/1+\rho}]^{1+\rho}, \quad 0 < \rho \leq 1 \quad (59)$$

where ρ is an arbitrary parameter which we shall choose to minimize the bound. It is easily seen that $E_0(0) = 0$, while $E_0(1) = R_0$, in which case $\bar{Q}_{K+k} = 2^k \bar{P}_{K+k}$, the ordinary union bound of (56). We bound the overall ensemble first-event error probability by the probability of the union of these composite events given by (58). Thus we find

$$\bar{P}_E < \sum_{k=0}^{\infty} \bar{Q}_{K+k} < \frac{2^{-KE_0(\rho)/R}}{1 - 2^{-(E_0(\rho)/R-\rho)}}. \quad (60)$$

Clearly (60) reduces to (56) when $\rho = 1$.

To determine the bit error probability using this approach, we must recognize that $\bar{Q}_{K+k}$ refers to 2^k different incorrect paths, each with a different number of incorrect bits. However, just as was observed in deriving (57), an incorrect path which differs from the correct path in $K + k$ branches prior to merging can produce at most $k + 1$ bit errors. Hence weighting the kth term of (60) by $k + 1$, we obtain

$$\bar{P}_B < \sum_{k=0}^{\infty} (k+1) \bar{Q}_{K+k} < \sum_{k=0}^{\infty} (k+1) 2^{-k(E_0(\rho)/R-\rho)} 2^{-KE_0(\rho)/R} = \frac{2^{-KE_0(\rho)/R}}{[1 - 2^{-(E_0(\rho)/R-\rho)}]^2}, \quad 0 < \rho \leq 1. \quad (61)$$

Clearly (61) reduces to (57) when $\rho = 1$.

Before we can interpret the results of (56), (57), (60), and (61) it is essential that we establish some of the properties of $E_0(\rho)$ $(0 < \rho \leq 1)$ defined by (59). It can be shown [5], [14] that for any memoryless channel, $E_0(\rho)$ is a concave monotonic nondecreasing function as shown in Fig. 15 with $E_0(0) = 0$ and $E_0(1) = R_0$.

Where the derivative $E_0'(\rho)$ exists, it decreases with ρ and it follows easily from the definition that

$$\lim_{\rho \to 0} E_0'(\rho) = \frac{1}{n} \sum_{\mathbf{y}} \sum_{\mathbf{x}} q(\mathbf{x}) \log_2 \frac{P(\mathbf{y} \mid \mathbf{x})}{\sum_{\mathbf{x}'} q(\mathbf{x}') P(\mathbf{y} \mid \mathbf{x}')} = \frac{1}{n} I(X^n, Y^n) \triangleq C \quad (62)$$

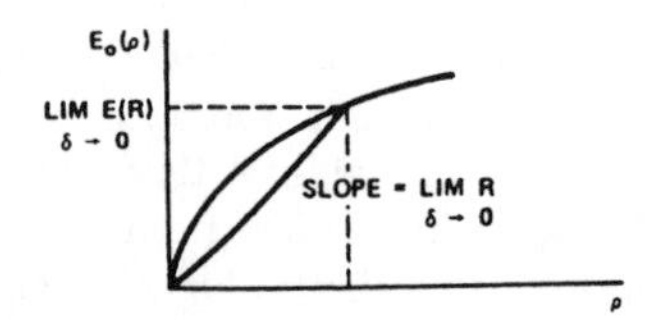

Fig. 15. Example of $E_0(\rho)$ function for general memoryless channel.

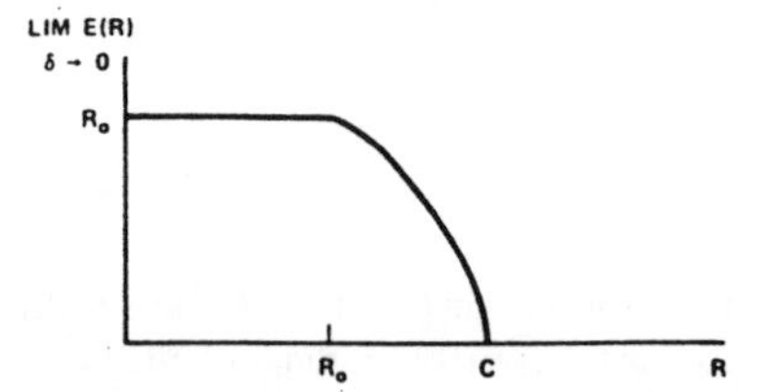

Fig. 16. Typical limiting value of exponent of (67).

the mutual information of the channel[17] where X^n and Y^n are the channel input and output spaces, respectively, for each branch sequence. Consequently, it follows that to minimize the bounds (60) and (61), we must make $\rho \leq 1$ as large as possible to maximize the exponent of the numerator, but at the same time we must ensure that

$$R < \frac{E_0(\rho)}{\rho}$$

in order to keep the denominator positive. Thus since $E_0(1) = R_0$ and $E_0(\rho) < R_0$, for $\rho < 1$, it follows that for $R < R_0$ and sufficiently large K we should choose $\rho = 1$, or equivalently use the bounds (56) and (57). We may thus combine all the above bounds into the expressions

$$\bar{P}_E < \frac{2^{-KE(R)/R}}{1 - 2^{-\delta(R)}} \tag{63}$$

$$\bar{P}_B < \frac{2^{-KE(R)/R}}{[1 - 2^{-\delta(R)}]^2} \tag{64}$$

where

$$E(R) = \begin{cases} R_0, & 0 \leq R < R_0 \\ E_0(\rho), & R_0 < R < C, \quad 0 < \rho \leq 1 \end{cases} \tag{65}$$

$$\delta(R) = \begin{cases} R_0/R - 1, & 0 < R < R_0 \\ E_0(\rho)/R - \rho, & R_0 \leq R < C, \quad 0 < \rho \leq 1. \end{cases} \tag{66}$$

To minimize the numerators of (63) and (64) for $R > R_0$ we should choose ρ as large as possible, since $E_0(\rho)$ is a nondecreasing function of ρ. However, we are limited by the necessity of making $\delta(R) > 0$ to keep the denominator from becoming zero. On the other hand, as the constraint length K becomes very large we may choose $\delta(R) = \delta$ very small. In particular, as δ approaches 0, (65) approaches

$$\lim_{\delta \to 0} E(R) = \begin{cases} R_0, & 0 < R < R_0 \\ E_0(\rho), & R_0 \leq R = E_0(\rho)/\rho < C, \\ & 0 < \rho \leq 1. \end{cases} \tag{67}$$

[17] C can be made equal to the channel capacity by properly choosing the ensemble measure $q(\mathbf{x})$. For an input-binary channel the random binary convolutional coder described above achieves this. Otherwise a further transformation of the branch sequence into a smaller set of nonbinary sequences is required [9].

Fig. 15 demonstrates the graphical determination of $\lim_{\delta \to 0} E(R)$ from $E_0(\rho)$.

It follows from the properties of $E_0(\rho)$ described, that for $R > R_0$, $\lim_{\delta \to 0} E(R)$ decreases from R_0 to 0 as R increases from R_0 to C, but that it remains positive for all rates less than C. The function is shown for a typical channel in Fig. 16.

It is particularly instructive to obtain specific bounds, in the limiting case, for the class of "very noisy" channels, which includes the BSC with $p = 1/2 - \gamma$ where $|\gamma| \ll 1$ and the biphase modulated AWGN with $\epsilon_s/N_0 \ll 1$. For this class of channels it can be shown [5] that

$$E_0(\rho) = \frac{\rho C}{1 + \rho} \tag{68}$$

and consequently $R_0 = E_0(1) = C/2$. (For the BSC, $C = \gamma^2/2 \ln 2$ while for the AWGN, $C = \epsilon_s/N_0 \ln 2$.)

For the very noisy channel, suppose we let $\rho = C/R - 1$, so that using (68) we obtain $E_0(\rho) = C - R$. Then in the limit as $\delta \to 0$ (65) becomes for a very noisy channel

$$\lim_{\delta \to 0} E(R) = \begin{cases} C/2, & 0 \leq R \leq C/2 \\ C - R, & C/2 \leq R \leq C. \end{cases} \tag{69}$$

This limiting form of $E(R)$ is shown in Fig. 17.

The bounds (63) and (64) are for the average error probabilities of the ensemble of codes relative to the measure induced by random selection of the time-varying coder tap sequences. At least one code in the ensemble must perform better than the average. Thus the bounds (63) and (64) hold for the best time-varying binary-tree convolutional coder of constraint length K. Whether there exists a fixed convolutional code with this performance is an unsolved problem. However, for small K the results of Section VII seem to indicate that these bounds are valid also for fixed codes.

To determine the tightness of the upper bounds, it is useful to have lower bounds for convolutional code error probabilities. It can be shown [9] that for all $R < C$

$$P_B \geq P_E > 2^{-K[E_L(R)/R - o(K)]} \tag{70}$$

where

$$\left.\begin{aligned} E_L(R) &= E_0(\rho) \\ R &= E_0(\rho)/\rho \end{aligned}\right\}, \quad 0 \leq \rho < \infty, \quad 0 \leq R \leq C \tag{71}$$

and $o(K) \to 0$ as $K \to \infty$. Comparison of the parametric

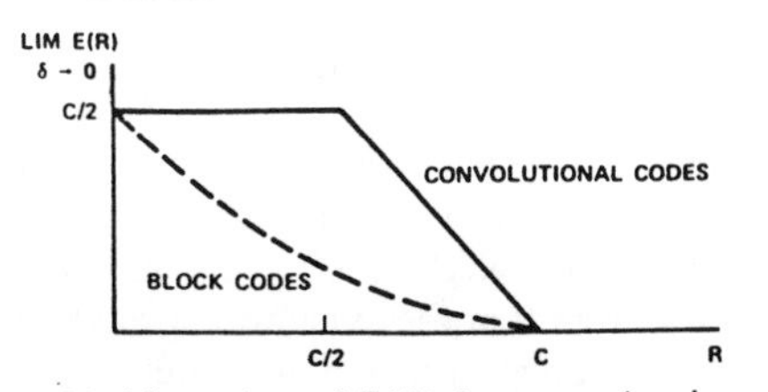

Fig. 17. Limiting values of $E(R)$ for very noisy channels.

equations (67) with (71), shows that

$$E_L(R) = \lim_{\delta \to 0} E(R)$$

for $R > R_0$ but is greater for low rates.

For very noisy channels, it follows easily from (71) and (68) that

$$E_L(R) = C - R, \qquad 0 \leq R \leq C.$$

Actually, however, tighter lower bounds for $R < C/2$ (Viterbi [9]) show that for very noisy channels

$$E_L(R) = \begin{cases} C/2, & 0 \leq R \leq C/2 \\ C - R, & C/2 \leq R < C, \end{cases} \tag{72}$$

which is precisely the result of (69) or of Fig. 17. It follows that, at least for very noisy channels, the exponential bounds are asymptotically exact.

All the results derived in this section can be extended directly to nonbinary ($b > 1$) codes. It is easily shown (Viterbi [9]) that the same results hold with $R = b/n$, R_0 and $E_0(\rho)$ multiplied by b, and all event probability upper bounds multiplied by $2^b - 1$, and bit probability upper bounds multiplied by $(2^b - 1)/b$.

Clearly, the ensemble of codes considered here is nonsystematic. However, by a modification of the arguments used here, Bucher and Heller [19] restricted the ensemble to systematic time-varying convolutional codes (i.e., codes for which b code symbols of each branch correspond to the data which generates the branch) and obtained all the above results modified only to the extent that the exponents $E(R)$ and $E_L(R)$ are multiplied by $1 - R$. (See also Section VIII.)

Finally, it is most revealing to compare the asymptotic results for the best convolutional codes of a given constraint length with the corresponding asymptotic results for the best block codes of a given block length. Suppose that K bits are coded into a block code of length N so that $R = K/N$ bits/code symbol. Then it can be shown (Gallager [5], Shannon *et al.* [8]) that for the best block code, the bit error probability is bounded above and below by

$$2^{-K[E_{Lb}(R)/R + o(K)]} < P_B < 2^{-KE_b(R)/R} \tag{73}$$

where

$$E_b(R) = \operatorname*{Max}_{0 \leq \rho \leq 1} [E_0(\rho) - \rho R]$$

$$E_{Lb}(R) \leq \operatorname*{Max}_{0 \leq \rho} [E_0(\rho) - \rho R].$$

Both $E_b(R)$ and $E_{Lb}(R)$ are functions of R which for all $R > 0$ are less than the exponents $E(R)$ and $E_L(R)$ for convolutional codes [9]. In particular, for very noisy channels they both become [5]

$$E_b(R) = E_{Lb}(R) = \begin{cases} C/2 - R \\ (\sqrt{C} - \sqrt{R})^2. \end{cases} \tag{74}$$

This is plotted as a dotted curve in Fig. 17.

Thus it is clear by comparing the magnitudes of the negative exponents of (73) and (64) that, at least for very noisy channels, a convolutional code performs much better asymptotically than the corresponding block code of the same order of complexity. In particular at $R = C/2$, the ratio of exponents is 5.8, indicating that to achieve equivalent performance asymptotically the block length must be over five times the constraint length of the convolutional code. Similar degrees of relative performance can be shown for more general memoryless channels [9].

More significant from a practical viewpoint, for short constraint lengths also, convolutional codes considerably outperform block codes of the same order of complexity.

XI. Path Memory Truncation Metric Quantization and Synchronization

A major problem which arises in the implementation of a maximum likelihood decoder is the length of the path history which must be stored. In our previous discussion we ignored this important point and therefore implicitly assumed that all past data would be stored. A final decision was made by forcing the coder into a known (all zeros) state. We now remove this impractical condition. Suppose we truncate the path memories after M bits (branches) have been accumulated, by comparing all 2^K metrics for a maximum and deciding on the bit corresponding to that path (out of 2^K) with the highest metric M branches forward. If M is several times as large as K, the additional bit errors introduced in this way are very few, as we shall now demonstrate using the asymptotic results of the last section.

An additional bit error may occur due to memory truncation after M branches, if the bit selected is from an incorrect path which differed from the correct path M branches back and which has a higher metric, but which would ultimately be eliminated by the maximum likelihood decoder. But for a binary-tree code there can be no more than 2^M distinct paths which differ from the correct path M branches back. Of these we need concern ourselves only with those which have not merged with the correct path in the intervening nodes. As was originally shown by Forney [12], using the ensemble arguments of Section X we may bound the average probability of this event by [see (58)]

$$\bar{P}_t < 2^{M\rho} 2^{-ME_0(\rho)/R}, \qquad 0 < \rho \leq 1. \tag{75}$$

To minimize this bound we should maximize the exponent $E_0(\rho)/R - \rho$ with respect to ρ on the unit interval. But this yields exactly $E_b(R)$, the upper bound exponent of (73) for block codes. Thus

$$\bar{P}_t < 2^{-ME_b(R)/R} \tag{76}$$

where $E_b(R)$ is the block coding exponent.

We conclude therefore that the memory truncation error is less than the bit error probability bound without truncation, provided the bound of (76) is less than the bound of (64). This will certainly be assured if

$$ME_b(R) > KE(R). \tag{77}$$

For very noisy channels we have from (69) and (74) or Fig. 17, that

$$\frac{M}{K} > \begin{cases} \dfrac{1}{1 - 2R/C}, & 0 \leq R \leq C/4 \\ \dfrac{1}{2(1 - \sqrt{R/C})^2}, & C/4 \leq R \leq C/2 \\ \dfrac{1 - R/C}{(1 - \sqrt{R/C})^2}, & C/2 < R < C. \end{cases}$$

For example, at $R = C/2$ this indicates that it suffices to take $M > (5.8)K$.

Another problem faced by a system designer is the amount of storage required by the metrics (or log-likelihood functions) for each of the 2^K paths. For a BSC this poses no difficulty since the metric is just the Hamming distance which is at most n, the number of code symbols, per branch. For the AWGN, on the other hand, the optimum metric is a real number, the analog output of a correlator, matched filter, or integrate-and-dump circuit. Since digital storage is generally required, it is necessary to quantize this analog metric. However, once the components y_{ik} of the optimum metric of (5), which are the correlator outputs, have been quantized to Q levels, the channel is no longer an AWGN channel. For biphase modulation, for example, it becomes a binary input Q-ary output discrete memoryless channel, whose transition probabilities are readily calculated as a function of the energy-to-noise density and the quantization levels. The optimum metric is not obtained by replacing y_{ik} by its quantized value $Q(y_{ik})$ in (5) but rather it is the log-likelihood function $\log P(\mathbf{y} \mid \mathbf{x}^{(m')})$ for the binary-input Q-ary-output channel.

Nevertheless, extensive simulation [24] indicates that for 8-level quantization even use of the suboptimal metric $\sum_k Q(y_{ik})x_{ik}^{(m)}$ results in a degradation of no more than 0.25 dB relative to the maximum likelihood decoder for the unquantized AWGN, and that use of the optimum metric is only negligibly superior to this. However, this is not the case for sequential decoding, where the difference in performance between optimal and suboptimal metrics is significant [11].

In a practical system other considerations than error performance for a given degree of decoder complexity often dictate the selection of a coding system. Chief among these are often the synchronization requirements. Convolutional codes utilizing maximum likelihood decoding are particularly advantageous in that no block synchronization is ever required. For block codes, decoding cannot begin until the initial point of each block has been located. Practical systems often require more complexity in the synchronization system than in the decoder. On the other hand, as we have by now amply illustrated, a maximum likelihood decoder for a convolutional code does not require any block synchronization because the coder is free running (i.e., it performs identical operations for each successive input bit and does not require that K bits be input before generating an output). Furthermore, the decoder does not require knowledge of past inputs to start decoding; it may as well assume that all previous bits were zeros. This is not to say that initially the decoder will operate as well, in the sense of error performance, as if the preceding bits of the correct path were known. On the other hand, consider a decoder which starts with an initially known path but makes an error at some point and excludes the correct path. Immediately thereafter it will be operating as if it had just been turned on with an unknown and incorrectly chosen previous path history. That this decoder will recover and stop making errors within a finite number of branches follows from our previous discussions in which it was shown that, other than for catastrophic codes, error sequences are always finite. Hence our initially unsynchronized decoder will operate just like a decoder which has just made an error and will thus always achieve synchronization and generally will produce correct decisions after a limited number of initial errors. Simulations have demonstrated that synchronization generally takes no more than four or five constraint lengths of received symbols.

Although, as we have just shown, branch synchronization is not required, code symbol synchronization within a branch is necessary. Thus, for example, for a binary-tree rate $R = 1/2$ code, we must resolve the two-way ambiguity as to where each two code-symbol branch begins. This is called node synchronization. Clearly if we make the wrong decisions, errors will constantly be made thereafter. However, this situation can easily be detected because the mismatch will cause *all* the path metrics to be small, since in fact there will not be any correct path in this case. We can thus detect this event and change our decision as to node synchronization (cf. Heller and Jacobs [24]). Of course, for an $R = 1/n$ code, we may have to repeat our choice n times, once for each of the symbols on a branch, but since n represents the redundancy factor or bandwidth expansion, practical systems rarely use $n > 4$.

XII. Other Decoding Algorithms for Convolutional Codes

This paper has treated primarily maximum likelihood decoding of convolutional codes. The reason for this was two-fold: 1) maximum likelihood decoding is closely related to the structure of convolutional codes and its consideration enhances our understanding of the ultimate capabilities, performance, and limitation of these codes; 2) for reasonably short constraint lengths ($K < 10$) its implementation is quite feasible[18] and worthwhile because of its optimality. Furthermore for $K \leq 6$, the complexity of maximum likelihood decoding is sufficiently limited that a completely parallel implementation (separate metric calculators) is possible. This minimizes the decoding time per bit and affords the possibility of extremely high decoding speeds [24].

Longer constraint lengths are required for extremely low error probabilities at high rates. Since the storage and computational complexity are proportional to 2^K, maximum likelihood decoders become impractical for $K > 10$. At this point *sequential decoding* [2], [4], [6] becomes attractive. This is an algorithm which sequentially searches the code tree in an attempt to find a path whose metric rises faster than some predetermined, but variable, threshold. Since the difference between the correct path metric and any incorrect path metric increases with constraint length, for large K generally the correct path will be found by this algorithm. The main drawback is that the number of incorrect path branches, and consequently the computation complexity, is a random variable depending on the channel noise. For $R < R_0$, it is shown that the average number of incorrect branches searched per decoded bit is bounded [6], while for $R > R_0$ it is not; hence R_0 is called the computational cutoff rate. To make storage requirements reasonable, it is necessary to make the decoding speed (branches/s) somewhat larger than the bit rate, thus somewhat limiting the maximum bit rate capability. Also, even though the average number of branches searched per bit is finite, it may sometimes become very large, resulting in a storage overflow and consequently relatively long sequences being erased. The stack sequential decoding algorithm [7], [18] provides a very simple and elegant presentation of the key concepts in sequential decoding, although the Fano algorithm [4] is generally preferable practically.

For a number of reasons, including buffer size requirements, computation speed, and metric sensitivity, sequential decoding of data transmitted at rates above about 100 K bits/s is practical only for hard-quantized binary received data (that is, for channels in which a hard decision -0 or $1-$ is made for each demodulated symbol). For the biphase modulated AWGN channel, of course, hard quantization (2 levels or 1 bit) results in an efficiency loss of approximately 2 dB compared with soft quantization (8 or more levels—3 or more bits). On the other hand, with maximum likelihood decoding, by employing a parallel implementation, short constraint length codes ($K \leq 6$) can be decoded at very high data rates (10 to 100 Mbits/s) even with soft quantization. In addition, the insensitivity to metric accuracy and simplicity of synchronization render maximum likelihood decoding generally preferable when moderate error probabilities are sufficient. In particular, since sequential decoding is limited by the overflow problem to operate at code rates somewhat below R_0, it appears that for the AWGN the crossover point above which maximum likelihood decoding is preferable to sequential decoding occurs at values of P_B somewhere between 10^{-3} and 10^{-5}, depending on the transmitted data rate. As the data rate increases the P_B crossover point decreases.

A third technique for decoding convolutional codes is known as *feedback decoding*, with threshold decoding [3] as a subclass. A feedback decoder basically makes a decision on a particular bit or branch in the decoding tree or trellis based on the received symbols for a limited number of branches beyond this point. Even though the decision is irrevocable, for limited constraint lengths (which are appropriate considering the limited number of branches involved in a decision) errors will propagate only for moderate lengths. When transmission is over a binary symmetric channel, by employing only codes with certain algebraic (orthogonal) properties, the decision on a given branch can be based on a linear function of the received symbols, called the *syndrome*, whose dimensionality is equal to the number of branches involved in the decision. One particularly simple decision criterion based on this syndrome, referred to as *threshold decoding*, is mechanizable in a very inexpensive manner. However, feedback decoders in general, and threshold decoders in particular, have an error-correcting capability equivalent to very short constraint length codes and consequently do not compare favorably with the performance of maximum likelihood or sequential decoding.

However, feedback decoders are particularly well suited to correcting error bursts which may occur in fading channels. Burst errors are generally best handled by using interleaved codes: that is, employing L convolutional codes so that the jth, $(L + j)$th $(2L + j)$th, etc., bits are encoded into one code for each $j = 0, 1, \cdots, L - 1$. This will cause any burst of length less than L to be broken up into random errors for the L independently operating decoders. Interleaving can be achieved by simply inserting $L - 1$ stage delay lines between stages of the convolutional encoder; the resulting single encoder then generates the L interleaved codes. The significant advantage of a feedback or threshold decoder is that the same technique can be employed in the decoder resulting in a single (time-shared) decoder rather than L decoders, providing feasible implementations for hard-quantized channels, even for protection against error bursts of thousands of bits. Details of feedback decoding

[18] Performing metric calculations and comparisons serially.

are treated extensively in Massey [3], Gallager [14], and Lucky *et al.* [16].

Appendix I

Generating Function for Structure of a Binary-Tree Convolutional Code for Arbitrary K and Error Bounds for Orthogonal Codes

We derive here the distance-invariant ($D = 1$) generating function $T(L, N)$ for any binary tree ($b = 1$) convolutional code of arbitrary constraint length K. It is most convenient in the general case to begin with the finite-state machine state-transition matrix for the linear equations among the state (node) variables. We exhibit this in terms of N and L for a $K = 4$ code as follows:

$$\begin{bmatrix} 1 & 0 & 0 & -NL & 0 & 0 & 0 \\ -L & 1 & 0 & 0 & -L & 0 & 0 \\ -NL & 0 & 1 & 0 & -NL & 0 & 0 \\ 0 & -L & 0 & 1 & 0 & -L & 0 \\ 0 & -NL & 0 & 0 & 1 & -NL & 0 \\ 0 & 0 & -L & 0 & 0 & 1 & -L \\ 0 & 0 & -NL & 0 & 0 & 0 & 1-NL \end{bmatrix} \cdot \begin{bmatrix} X_{001} \\ X_{010} \\ X_{011} \\ X_{100} \\ X_{101} \\ X_{110} \\ X_{111} \end{bmatrix} = \begin{bmatrix} NL \\ 0 \\ 0 \\ 0 \\ 0 \\ 0 \\ 0 \end{bmatrix}. \tag{78}$$

This pattern can be easily seen to generalize to a $2^{K-1} - 1$ dimensional square matrix of this form for any binary-tree code of constraint length K, and in general the generating function

$$T(L, N) = LX_{100\cdots0},$$

$$\text{where } 100 \cdots 0 \text{ contains } (K-2) \text{ zeros.} \tag{79}$$

From this general pattern it is easily shown that the matrix can be reduced to a dimension of 2^{K-2}. First combining adjacent rows, from the second to the last, pairwise, one obtains the set of $2^{K-2} - 1$ relations

$$NX_{j_1j_2\cdots j_{K-2}0} = X_{j_1j_2\cdots j_{K-2}1} \tag{80}$$

where $j_1, j_2, \cdots, j_{K-2}$ runs over all binary vectors except for the all zeros. Substitution of (80) into (78) yields a 2^{K-2}-dimensional matrix equation. The result for $K = 4$ is

$$\begin{bmatrix} 1 & 0 & -L & 0 \\ -NL & 1 & -NL & 0 \\ 0 & -L & 1 & -L \\ 0 & -NL & 0 & 1-NL \end{bmatrix} \cdot \begin{bmatrix} X_{001} \\ X_{011} \\ X_{101} \\ X_{111} \end{bmatrix} = \begin{bmatrix} NL \\ 0 \\ 0 \\ 0 \end{bmatrix}. \tag{81}$$

Defining the new variable

$$X'_{00\cdots01} = NL\,X_{00\cdots01} + X_{00\cdots11} \tag{82}$$

(which corresponds to adding the second row to NL times the first), we obtain finally a $2^{K-2} - 1$ dimensional matrix equation, which for $K = 4$ is

$$\begin{bmatrix} 1 & -N(L+L^2) & 0 \\ -L & 1 & -L \\ -NL & 0 & 1-NL \end{bmatrix} \cdot \begin{bmatrix} X'_{001} \\ X_{101} \\ X_{111} \end{bmatrix} = \begin{bmatrix} N^2L^2 \\ 0 \\ 0 \end{bmatrix}. \tag{83}$$

Note that (83) is the same as (78) for K reduced by unity, but with modifications in two places, both in the first row; namely, the first component on the right side is squared, and the middle term of the first row is reduced by an amount NL^2. Although we have given the explicit result only for $K = 4$, it is easily seen to be valid for any K.

Since in all respects, except these two, the matrix after this sequence of reductions is the same as the original but with its dimension reduced corresponding to a reduction of K by unity, we may proceed to perform this sequence of reductions again. The steps will be the same except that now in place of (80), we have

$$NX_{j_1j_2\cdots j_{K-3}01} = X_{j_1j_2\cdots j_{K-3}11} \tag{80'}$$

and in place of (82)

$$X''_{00\cdots01} = NL\,X'_{00\cdots01} + X_{00\cdots111} \tag{82'}$$

while in place of (81) the right of center term of the first row is $-(L + L^2)$ and the first component on the right side is N^2L^2. Similarly in place of (83) the center term of the first row is $-N(L + L^2 + L^3)$ and the first component on the right side is N^3L^3.

Performing this sequence of reductions $K - 2$ times in all, but omitting the last step—leading from (81) to (83)—in the last reduction, the original $2^{K-1} - 1$ equations are reduced in the general case to the two equations

$$\begin{bmatrix} 1 & -(L + L^2 + \cdots L^{K-2}) \\ -NL & 1-NL \end{bmatrix} \cdot \begin{bmatrix} X_{00-01}^{(K-3)} \\ X_{11\cdots1} \end{bmatrix} = \begin{bmatrix} (NL)^{K-2} \\ 0 \end{bmatrix} \tag{84}$$

whence it follows that

$$X_{11\cdots 1} = \frac{(NL)^{K-1}}{1 - N(L + L^2 + \cdots + L^{K-1})} \tag{85}$$

Applying (79) and the $K - 2$ extensions of (80) and (80′) we find

$$T(L, N) = LX_{100\cdots 00} = LN^{-1}X_{100\cdots 01}$$

$$= LN^{-2}X_{100\cdots 011} = \cdots = LN^{-(K-2)}X_{11\cdots 1}$$

$$= \frac{NL^K}{1 - N(L + L^2 + \cdots + L^{K-1})}$$

$$= \frac{NL^K(1 - L)}{1 - L(1 + N) + NL^K} \tag{86}$$

If we require only the path length structure, and not the number of bit errors corresponding to any incorrect path, we may set $N = 1$ in (86) and obtain

$$T(L) = \frac{L^K}{1 - (L + L^2 + \cdots + L^{K-1}} = \frac{L^K(1 - L)}{1 - 2L + L^K}. \tag{87}$$

If we denote as an upper bound an expression which is the generating function of more paths than exist in our state diagram, we have

$$T(L) < \frac{L^K}{1 - 2L}. \tag{88}$$

As an additional application of this generating function technique, we now obtain bounds on P_E and P_B for the class of orthogonal convolutional (tree) codes introduced by Viterbi [10]. For this class of codes, to each of the 2^K branches of the K-state diagram there corresponds one of 2^K orthogonal signals. Given that each signal is orthogonal to all others in $n \geq 1$ dimensions, corresponding to n channel symbols or transmission times (as, for example, if each signal consists of n different pulses out of $2^K n$ possible positions), then the weight of each branch is n. Consequently, if we replace L, the path length enumerator, by D^n in (86) we obtain for orthogonal codes

$$T(D, N) = \frac{ND^{nK}(1 - D^n)}{1 - D^n(1 + N) + ND^{nK}}. \tag{89}$$

Then using (48) and (49), the first-event error probability for orthogonal codes is bounded by

$$P_E < \frac{D_0^{nK}(1 - D_0^n)}{1 - 2D_0^n + D_0^{nK}} < \frac{D_0^{nK}(1 - D_0^n)}{1 - 2D_0^n} \tag{90}$$

and the bit error probability bound is

$$P_B < \left.\frac{dT(N, D)}{dN}\right|_{N=1, D=D_0}$$

$$= \frac{D_0^{nK}(1 - D_0^n)^2}{(1 - 2D_0^n + D_0^{nK})^2} < \frac{D_0^{nK}(1 - D_0^n)^2}{(1 - 2D_0^n)^2} \tag{91}$$

where D_0 is a function of the channel transition probabilities or energy-to-noise ratio and is given by (46).

Acknowledgment

The author gratefully acknowledges the considerable stimulation he has received over the course of writing the several versions of this paper from Dr. J. A. Heller, whose recent work strongly complements and enhances this effort, for numerous discussions and suggestions and for his assistance in its presentation at the Linkabit Corporation "Seminars on Convolutional Codes." This tutorial approach owes part of its origin to Dr. G. D. Forney, Jr., whose imaginative and perceptive reinterpretation of my original work has aided immeasurably in rendering it more comprehensible. Also, thanks are due to Dr. J. K. Omura for his careful and detailed reading and correction of the manuscript during his presentation of this material in the UCLA graduate course on information theory.

References

[1] P. Elias, "Coding for noisy channels," in *1955 IRE Nat. Conv. Rec.*, vol. 3, pt. 4, pp. 37–46.
[2] J. M. Wozencraft, "Sequential decoding for reliable communication," in *1957 IRE Nat. Conv. Record*, vol. 5, pt. 2, pp. 11–25.
[3] J. L. Massey, *Threshold Decoding*. Cambridge, Mass.: M.I.T. Press, 1963.
[4] R. M. Fano, "A heuristic discussion of probabilistic decoding," *IEEE Trans. Inform. Theory*, vol. IT-9, Apr. 1963, pp. 64–74.
[5] R. G. Gallager, "A simple derivation of the coding theorem and some applications," *IEEE Trans. Inform. Theory*, vol. IT-11, Jan. 1965, pp. 3–18.
[6] J. M. Wozencraft and I. M. Jacobs, *Principles of Communication Engineering*. New York: Wiley, 1965.
[7] K. S. Zigangirov, "Some sequential decoding procedures," *Probl. Peredach Inform.*, vol. 2, no. 4, 1966, pp. 13–25.
[8] C. E. Shannon, R. G. Gallager, and E. R. Berlekamp, "Lower bounds to error probability for coding on discrete memoryless channels," *Inform. Contr.*, vol. 10, 1967, pt. I, pp. 65–103, pt. II, pp. 522–552.
[9] A. J. Viterbi, "Error bounds for convolutional codes and an asymptotically optimum decoding algorithm," *IEEE Trans. Inform. Theory*, vol. IT-13, Apr. 1967, pp. 260–269.
[10] ——, "Orthogonal tree codes for communication in the presence of white Gaussian noise," *IEEE Trans. Commun. Technol.*, vol. COM-15, April 1967, pp. 238–242.
[11] I. M. Jacobs, "Sequential decoding for efficient communication from deep space," *IEEE Trans. Commun. Technol.*, vol. COM-15, Aug. 1968, pp. 492–501.
[12] G. D. Forney, Jr., "Coding system design for advanced solar missions," submitted to NASA Ames Res. Ctr. by Codex Corp., Watertown, Mass., Final Rep., Contract NAS2-3637, Dec. 1967.
[13] J. L. Massey and M. K. Sain, "Inverses of linear sequential circuits," *IEEE Trans. Comput.*, vol. C-17, Apr. 1968, pp. 330–337.
[14] R. G. Gallager, *Information Theory and Reliable Communication*. New York: Wiley, 1968.
[15] T. N. Morrissey, "Analysis of decoders for convolutional codes by stochastic sequential machine methods," Univ. Notre Dame, Notre Dame, Ind., Tech. Rep. EE-682, May 1968.
[16] R. W. Lucky, J. Salz, and E. J. Weldon, *Principles of Data Communication*. New York: McGraw-Hill, 1968.

[17] J. K. Omura, "On the Viterbi decoding algorithm," *IEEE Trans. Inform. Theory*, vol. IT-15, Jan. 1969, pp. 177–179.
[18] F. Jelinek, "Fast sequential decoding algorithm using a stack," *IBM J. Res. Dev.*, vol. 13, no. 6, Nov. 1969, pp. 675–685.
[19] E. A. Bucher and J. A. Heller, "Error probability bounds for systematic convolutional codes," *IEEE Trans. Inform. Theory*, vol. IT-16, Mar. 1970, pp. 219–224.
[20] J. P. Odenwalder, "Optimal decoding of convolutional codes," Ph.D. dissertation, Dep. Syst. Sci., Sch. Eng. Appl. Sci., Univ. California, Los Angeles, 1970.
[21] G. D. Forney, Jr., "Coding and its application in space communications," *IEEE Spectrum*, vol. 7, June 1970, pp. 47–58.
[22] ——, "Convolutional codes I: Algebraic structure," *IEEE Trans. Inform. Theory*, vol. IT-16, Nov. 1970, pp. 720–738; "II: Maximum likelihood decoding," and "III: Sequential decoding," *IEEE Trans. Inform. Theory*, to be published.
[23] W. J. Rosenberg, "Structural properties of convolutional codes," Ph.D. dissertation, Dep. Syst. Sci., Sch. Eng. Appl. Sci., Univ. California, Los Angeles, 1971.
[24] J. A. Heller and I. M. Jacobs, "Viterbi decoding for satellite and space communication," this issue, pp. 835–848.
[25] A. R. Cohen, J. A. Heller, and A. J. Viterbi, "A new coding technique for asynchronous multiple access communication," this issue, pp. 849–855.

Andrew J. Viterbi (S'54–M'58–SM'63) was born in Bergamo, Italy, on March 9, 1935. He received the B.S. and M.S. degrees in electrical engineering from the Massachusetts Institute of Technology, Cambridge, in 1957, and the Ph.D. degree in electrical engineering from the University of Southern California, Los Angeles, in 1962.

While attending M.I.T., he participated in the cooperative program at the Raytheon Company. In 1957 he joined the Jet Propulsion Laboratory where he became a Research Group Supervisor in the Communications Systems Research Section. In 1963 he joined the faculty of the University of California, Los Angeles, as an Assistant Professor. In 1965 he was promoted to Associate Professor and in 1969 to Professor of Engineering and Applied Science. He was a cofounder in 1968 of Linkabit Corporation of which he is presently Vice President.

Dr. Viterbi is a member of the Editorial Boards of the PROCEEDINGS OF THE IEEE and of the journal *Information and Control*. He is a member of Sigma Xi, Tau Beta Pi, and Eta Kappa Nu and has served on several governmental advisory committees and panels. He is the coauthor of a book on digital communication and author of another on coherent communication, and he has received three awards for his journal publications.

Reprinted from IEEE Journal on Selected Areas in Commun., Vol. 8, No. 4, May 1990, pp. 641-649.

Very Low Rate Convolutional Codes for Maximum Theoretical Performance of Spread-Spectrum Multiple-Access Channels

ANDREW J. VITERBI, FELLOW, IEEE

***Abstract*—This paper treats a spread-spectrum multiple-access communication system for which both spreading and error control is provided by binary PSK modulation with orthogonal convolutional codes. Performance of spread spectrum multiple access by a large number of users employing this type of coded modulation is determined in the presence of background Gaussian noise. With this approach and coordinated processing at a common receiver, it is shown that the aggregate data rate of all simultaneous users can approach the Shannon capacity of the Gaussian noise channel.**

I. Introduction

In 1967, shortly after the publication of a new decoding algorithm for convolutional codes [1], the author proposed a class of orthogonal convolutional codes [2] for which a bound on error performance was easily derivable without ensemble average arguments and which served as a further expository vehicle for the decoding algorithm which was later shown to produce maximum likelihood sequence decisions [3], [4]. This class of codes provides an improvement over orthogonal block codes in terms of complexity required to achieve a given error probability [5].

Most significantly, this class of very low rate codes is an attractive choice for spread spectrum multiple access (SSMA)- also referred to as code-division multiple access (CDMA). For such applications, each user employs a pseudorandom sequence at the symbol rate, whose period is much longer than a bit duration, which is added modulo-2 to the coded data sequence. This sequence randomizes the waveform of any user relative to all other uses, but does not spread the spectrum beyond what is provided by the orthogonal code. Although the system becomes interference limited, its total throughput is generally greater than time-division or frequency-division multiple-access (TDMA or FDMA) systems for a variety of both practical and theoretical considerations.

This paper reviews the class of very low rate orthogonal codes, its implementation and its performance in Gaussian noise. The performance in more general interference (including mutual interference among users) is then treated, with primary consideration given to its use for spread spectrum multiple access. The last section presents a new technique for achieving capacity for the SSMA channel in background Gaussian noise.

II. Orthogonal Convolutional Codes: Implementation and Performance in Very Noisy Channels

Fig. 1 shows an implementation of an *orthogonal convolutional* encoder of *constraint length K* based on a Hadamard block coder, only one of many possible implementations, but one which leads to a particularly desirable decoder implementation. The upper shift register is the convolutional coder register which is shifted at the bit rate. The lower register is shifted at the symbol rate (2^K times the bit rate). The kth stage is delayed by $2^k T$ where T is one symbol period and the kth two-position switch changes state every 2^k symbols. The code array H_K of dimension $2^K \times 2^K$ thus generated is given recursively by the matrix relation

$$H_k = \begin{bmatrix} H_{k-1} H_{k-1} \\ H_{k-1} \overline{H}_{k-1} \end{bmatrix} \qquad k = 1, 2, \cdots K \tag{1}$$

where $H_0 = 0$, and $\overline{H}_k$ is the complement to H_k.

It follows therefore that the code rate

$$r = 2^{-K} \text{ bits/code symbol} \tag{2}$$

and the code sequences corresponding to any two unmerged input bit sequences (which never coincide in $K - 1$ or more consecutive bits) will produce orthogonal code sequences.

The decoder for the Hadamard orthogonal convolutional encoder of Fig. 1 first generates[1] the inverse Hadamard transform [6] to produce the 2^K metrics corresponding to the 2^K Hadamard sequences of the signal set; these are then supplied to a conventional (serial) Viterbi decoder which performs one basic add-compare-select operation during each symbol period [5].

It is well known [5] that the bit error probability of the code operating in an additive white Gaussian (AWGN)

Manuscript received February 17, 1989; revised September 14, 1989.
The author is with QUALCOMM, Inc., San Diego, CA 92121.
IEEE Log Number 9034789.

[1]This implementation is based on that of the Hadamard block decoder first suggested by R. Green at Jet Propulsion Laboratory [6] and consequently dubbed the "Green Machine."

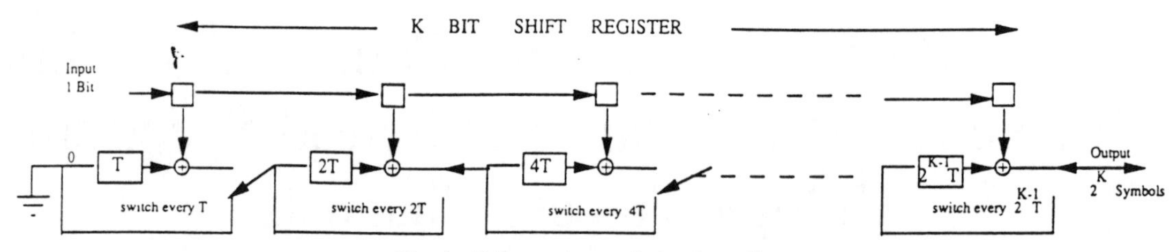

Fig. 1. Orthogonal convolutional encoder.

channel is upper bounded by

$$P_b < \frac{2^{-K E_c(r)/r}}{[1 - 2^{\delta(r)}]^2} \qquad 0 < r < c(1-\gamma) \qquad (3)$$

where

$$E_c(r) = \begin{cases} c/2, & 0 < r < c/2 \\ c - r/(1-\gamma), & c/2 \le r < c(1-\gamma) \end{cases} \qquad (4)$$

and $\delta(r) > 0$ for $\gamma > 0$.

For the AWGN channel $c = E_s/(N_o \ln 2)$ and $c/r = E_b/(N_o \ln 2)$ where E_s and E_b are, respectively, the symbol and bit energies and N_o is the noise density. It is shown in Appendix I that the same bound applies to the class of very noisy channels which we define presently.

The classical definition of very noisy (additive memoryless) channels [5] is adapted to amplitude continuous channels on the basis of each *symbol* y as

$$p(y|x = \epsilon) = p_N(y - \epsilon) \approx p_N(y) - \frac{\epsilon}{\sigma}\frac{dp_N(y)}{dy} + (1/2)\frac{\epsilon^2}{\sigma^2}\frac{d^2p_N(y)}{dy^2} \qquad (5)$$

where $p_N(y) = p_N(-y)$ is the normalized additive noise probability density[2], ϵ is the symbol mean and σ^2 is the (unnormalized) noise variance, with $\epsilon^2/\sigma^2 << 1$. The latter condition also justifies the truncation of the Taylor series of (5) beyond the second term.

The use of very low rate codes practically always implies operation over a very noisy channel in terms of signal-to-noise ratio on a symbol basis, for given a reasonably efficient *bit* signal-to-noise ratio, multiplying this by the very low code rate r yields a very low *symbol* signal-to-noise ratio. This then provides a generalization of the results for AWGN to a much wider class of channels. In particular, as shown in Appendix I, the bound of (3) and (4) also holds for asymptotically small ϵ^2/σ^2 where

$$c = \frac{\epsilon^2/\sigma^2}{2 \ln 2} \qquad (6)$$

[2]Without loss of generality, the noise density is normalized to unit variance. This is tantamount to normalizing the output of the symbol matched filter, which, however, is not required in the implementation.

III. Application to Spread Spectrum Multiple Access (SSMA)

The natural application of very low rate codes is to spread-spectrum applications wherein for other purposes the ratio of bandwidth-to-bit rate or symbol rate-to-bit rate is made very large, which virtually guarantees the very noisy character of the channel as was just noted. One such purpose is for efficient multiple access by numerous uncoordinated users. These users may all be transmitting to a single (hub) receiver or they may be transmitting among several subsets, but in all cases they share a common frequency spectrum. Spread spectrum multiple access (SSMA), sometimes also called code division multiple access (CDMA), can be implemented as shown in Fig. 2. The SSMA waveform generator for the kth user produces a quadrature-random-sequence modulated carrier which is approximately band-limited to W Hz by the quadrature low-pass filter transfer functions, $H(f)$. In each arm of the quadrature modulator for the kth user, a symbol $a_k(n)$ of the binary code sequence is multiplied by a binary random[3] sequence, $C_k^c(n)$ or $C_k^s(n)$, once for each symbol period $T = 1/W$. Note, however, that the random sequences randomize each user's signal with respect to all other users; but they do not spread the spectrum since the symbol rate W of the random sequences is the same as the code symbol rate. This is not typical of SSMA signals but is a property of the present approach in which spreading is accomplished by the encoder.

Hence, W also equals the symbol rate. The resulting binary symbols modulate impulse generators which drive the band-limited low-pass filters $H(f)$ whose outputs quadrature modulate the carrier as shown in Fig. 2(a). Note that the coded data modulation is binary phase shift keyed (BPSK) even though the carrier is quadriphase modulated by the two parallel binary random sequences. The latter is needed, as is shown in Appendix II, to ensure that other-user interference produces a random component whose variance is independent of the relative phases between users. It is further shown in Appendix II that the kth user sees an additive (non-Gaussian) noise channel

[3]In practice these are pseudorandom sequences of arbitrarily long period (equivalent to many bits). For purposes of analysis we shall take these to be Bernoulli sequences, i.e., consisting of i.i.d. binary variables, ± 1. Acquisition and synchronization of these sequences requires that either a preamble or an unmodulated pseudorandom carrier sequence be sent along with the data-modulated sequence.

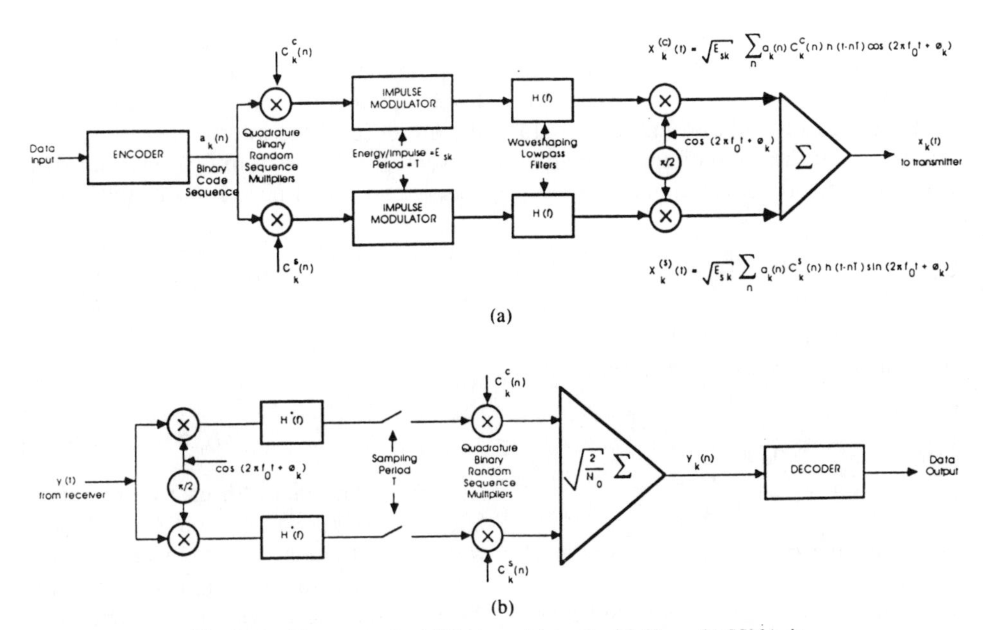

Fig. 2. (a) Binary encoder—SSMA modulator for kth User. (b) SSMA demodulator—decoder for kth user.

with signal-to-noise ratio

$$\frac{\epsilon_k^2}{\sigma_k^2} = \frac{2E_{sk}/N_o}{1 + \sum_{j \neq k} E_{sj}/N_o} = \frac{2S_k/(N_o W)}{1 + \sum_{j \neq k} S_j/(N_o W)} \tag{7}$$

where E_{sk} is the symbol energy of the kth user and $S_k = E_{sk}/T = E_{sk} W$ is the corresponding power. Note also that the (unnormalized) noise is the sum of the thermal noise $N_o W$ and the other user interference $\Sigma_{j \neq k} Sj$, which constitutes a heuristic argument in support of (7) (Ref. [7]). This is clearly a very noisy channel when the total number of users $M >> 1$. In fact, for equal user signal energies, (7) approaches $2/M$ as M becomes large. However, even for small M, $\epsilon_k^2/\sigma_k^2 << 1$ typically since the symbol energy-to-thermal noise per user is very small for a spread spectrum system.

Combining (6) and (7) implies for the kth user of a SSMA system

$$c_k = \frac{\epsilon_k^2/\sigma_k^2}{2 \ln 2} = \frac{1}{\ln 2} \frac{S_k/(N_o W)}{1 + \sum_{j \neq k} S_j/(N_o W)}$$

$$> \frac{1}{\ln 2} \frac{S_k/(N_o W)}{1 + S_T/(N_o W)} \tag{8}$$

where

$$S_T = \sum_{k=1}^{M} S_k \tag{9}$$

is the total power of all M users combined. The slight lower bound is due to not excluding the kth term of the sum in the denominator, which will be ignored henceforth.

We define also R_k, the bit rate for the kth user in bits/s and note that since W is both the bandwidth and the symbols/s, it follows that $r_K = R_k/W$.

Now from (3) and (4) it follows that to achieve an acceptable bit error rate the code rate and consequently the bit-rate-to-bandwidth ratio

$$\frac{R_k}{W} = r_k = \propto c_k = \frac{\propto}{\ln 2} \frac{S_k/(N_o W)}{1 + \sum_{j \neq k} S_j/(N_o W)}, \quad \propto < 1. \tag{10}$$

Summing over all users and letting R_T be the total combined data rate of all users, the total bit rate-to-bandwidth supportable is then according to (8) and (10),

$$\frac{R_T}{W} = \frac{\sum_{k=1}^{M} R_k}{W} = \sum_{k=1}^{M} r_k > \frac{\propto}{\ln 2} \frac{\sum_{k=1}^{M} S_k/(N_o W)}{1 + S_T/(N_o W)}$$

$$= \frac{\propto}{\ln 2} \frac{S_T/(N_o W)}{1 + S_T/(N_o W)}. \tag{11}$$

Clearly as M becomes large enough that $S_T/(N_o W) >> 1$,

$$\frac{R_T}{W} \sim \frac{\propto}{\ln 2} \frac{\text{bits/s}}{\text{Hz}}. \tag{12}$$

This is the maximum bandwidth efficiency of an SSMA system, for which each user is independently demodulated and decoded.

For SSMA applications involving messaging or digital voice, a reasonable bit rate might be 9.6 Kb/s while the spread bandwidth might be on the order of 10 MHz. Assuming equal bit and code rates for all M users, one could then assign a bandwidth-to-bit rate ratio of approximately 10^3 which implies for orthogonal convolutional codes,

$$\frac{W}{R_k} = \frac{1}{r_k} = 2^K = 2^{10}, \qquad k = 1, 2, \cdots M$$

and consequently a constraint length $K = 10$.

Still assuming a constant data rate $R_k = R$ for all users, this implies

$$\frac{R_T}{W} = \frac{MR}{W} = \frac{\propto}{\ln 2} \quad \frac{\text{b/s}}{\text{Hz}}, \propto < 1 \tag{13}$$

which implies that the number of supportable users is $M = (W/R) \propto /\ln 2$.

Thus, from this limited perspective of a dedicated channel with ample signal-to-noise ratio, the bandwidth efficiency is probably slightly below that of frequency division multiple access (FDMA) which can achieve better that 1 b/s/Hz efficiency using QPSK modulation. In a power-limited application, however, FDMA would probably achieve no higher bandwidth efficiency than (13).

On the other hand, unlike FDMA, SSMA does not require dedicated channels. Thus, an intermittent message or voice source need not transmit during quiet or inactive periods. Gilhousen *et al.* [8] have shown that exploiting this and other related practical limitations, SSMA effectively outperforms FDMA by a substantial amount. In the next section, we consider instead a promising technique which can approach the theoretical limit of multiple-access channel performance in background Gaussian noise.

It is worth noting that although by using orthogonal codes with constraint length K equal to the logorithm of the bandwidth expansion avoids any further spectrum spreading, there may be cases where other considerations lead to smaller or larger K. In the former case, the additional spreading is easily provided for bandwidth expansion in multiples of 2^K by using pseudorandom sequences whose symbol rate is higher than the code symbol rate. In the latter case, bandwidth expansions which are smaller powers of 2 can be obtained by taking $L < K$ linear combinations of the binary shift register contents, prior to selecting the orthogonal Hadamard sequence from a set of only 2^L sequences. It can be shown that as long as $L \gg E_b/N_o$, performance is only slightly degraded.

IV. Ultimate Theoretical Potential of SSMA Through Coordinated Processing and Successive Cancellation

Suppose the M transmitters implemented as shown in Fig. 2(a) now transmit, at possibly different power levels but over the same bandwidth and at the same bit and code rates, to a common (hub) receiver. (The case of different rates will be considered below and in Appendix III). Now suppose the Mth (last arriving) user transmits at a power level sufficient to achieve adequate coded performance in the presence of the interference from the preceding $M - 1$ users. It follows from (10) that this Mth user's symbol energy and signal power must satisfy the relation

$$r = r_M = \frac{\propto}{\ln 2} \frac{X_M}{1 + \sum_{j=1}^{M-1} X_j} \tag{14}$$

where

$$X_j \triangleq E_{sj}/N_o = S_j/(N_o W), \qquad j = 1, 2 \cdots M.$$

Now this additional Mth user represents additional interferences to users 1 through $M - 1$ and normally they would need to adjust their power to overcome this if they are to use the same bit and code rate. However, suppose the common receiver, upon first demodulating and decoding correctly the Mth user's signal, reencodes and remodulates at the correct phase and amplitude so it can subtract this from the composite signal thus cancelling the effect of the Mth user. This requires storage of the composite signal, and a very accurate estimate of the Mth user's amplitude and phase. The latter, of course, is also needed prior to decoding, and is obtained by conventional estimation and tracking methods but a much more accurate estimate of phase and amplitude, or equivalently, the amplitudes of in-phase and quadrature components of the signals, can be obtained by forming the inner products of the quadrature components of the received signal $y(t)$ with a local replica of the Mth transmission formed by reencoding and remodulating the data sequence for the Mth user just obtained from the demodulator-decoder. (See the processing flow diagram in Fig. 3). Once the quadrature component amplitudes have been so estimated, with a mean square accuracy which improves linearly with length of the message, these can be used to properly scale the quadrature components of the reencoded, remodulated (normalized) Mth signal and then to subtract this component from the composite signal (see Fig. 3).

Once this has been done, it follows that the power ratio X_{M-1} required by the $(M - 1)$th user need only satisfy

$$r = r_{M-1} = \frac{\propto}{\ln 2} \frac{X_{M-1}}{1 + \sum_{j=1}^{M-2} X_j}$$

and with the same estimation and subtraction performed on each successively demodulated and decoded user, the power ratio X_k required by the kth user need only satisfy

$$r = r_k = \frac{\propto}{\ln 2} \frac{X_k}{1 + \sum_{j=1}^{k-1} X_j} \qquad k = 1, 2 \cdots M. \tag{15}$$

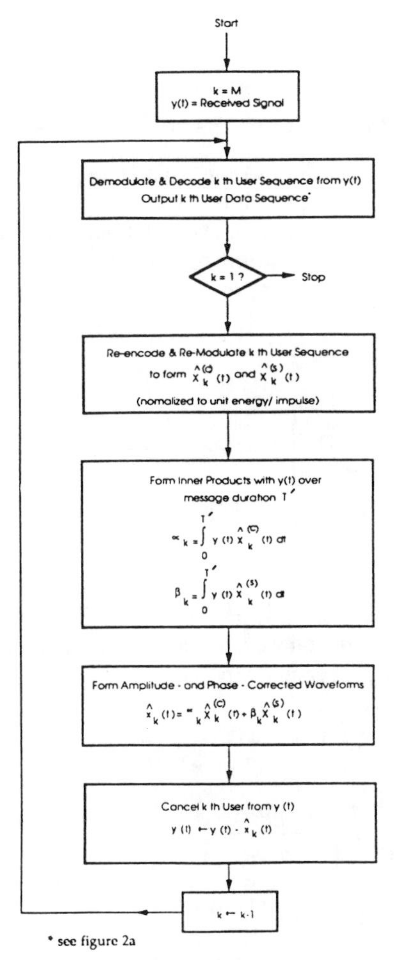

Fig. 3. Flow chart of successive cancellation.

Since all rates are equal, it follows that (15) is satisfied for all k by setting

$$X_k = X_1(1 + X_1)^{k-1} \qquad k = 1, 2, \cdots M \qquad (16)$$

as can be readily shown by substitution, and X_1 can be obtained from the condition

$$S_T/N_oW = \sum_{k=1}^{M} X_k = X_1 \sum_{k=1}^{M} (1 + X_1)^{k-1}$$

$$= (1 + X_1)^M - 1$$

whence

$$X_1 = \left(1 + S_T/(N_oW)\right)^{1/M} - 1$$

$$= e^{(1/M)[\ln(1 + S_T/N_oW)]} - 1.$$

Expanding the exponential in a Taylor series yields

$$X_1 = \frac{1}{M} \ln\left(1 + S_T/(N_oW)\right)$$

$$\cdot \left[1 + \frac{1}{2M} \ln\left(1 + S_T/(N_oW)\right) + \cdots\right]. \qquad (17)$$

Thus, it follows from (15) and (17) that for the common bit rate-to-bandwidth ratio,

$$\frac{R}{W} = r = r_1 = \frac{\propto X_1}{\ln 2}$$

$$\gtrsim \frac{\propto}{M} \log_2\left[1 + S_T/(N_oW)\right] \text{ b/s/Hz} \qquad (18)$$

where the inequality approaches equality for asymptotically large M. Hence, the total composite rate of all users,

$$R_T/W = MR/W \gtrsim \propto \log_2\left[1 + S_T/(N_oW)\right] \text{ b/s/Hz}. \qquad (19)$$

Thus, for the totality of users sharing a common bandwidth, the composite bit rate approaches within a factor $\propto$ of the Shannon limit [9]. For asymptotically large K, as shown by (3) and (4), $\propto$ can be made to approach 1.

This result, albeit predictable from information theory ([9] and [10]), is remarkable in that it can be achieved with an almost practical system. The extension to variable bit rates is also possible, as shown in Appendix III. Of course, orthogonal codes require the code rate to be an inverse power of 2. Hence, if the same bandwidth W is to be shared by all users, and multiple bit rates are used, these must differ by factors of two. Interestingly, if the higher order users employ lower bit rates and thus lower code rates 2^{-K}, they will consequently use larger constraint K and hence produce lower bit error rates, which is useful since these lower rate, higher order users are decoded before the higher rate, lower order users and any errors appear as residual interference to the successively decoded users.

Appendix III also considers the issue of residual interference, whether through imperfect subtraction because of inaccurate amplitude and phase estimation or because of bit errors[4] in previously decoded users. The conclusion is that in the limit of infinite total $S_T/(N_oW)$

$$\lim_{S_T/(N_oW)\to\infty} \frac{R_T}{W} = \frac{\log_2(1/\beta)}{1-\beta} \qquad (20)$$

where β is the residual uncanceled "error power" fraction of the previously decoded users, assuming that on the average all users have equal relative residual error power. Thus, for a sufficiently large number of users, which guarantees large $S_T/(N_oW)$ and sufficiently accurate cancellation, the composite data rate-to-bandwidth ratio can be considerably greater than 1 b/s/Hz, and consequently much greater than could be supported by FDMA or TDMA.

V. Conclusions

It is hoped that this paper finally dispels the "first myth" [11] of spread-spectrum communications: namely, that the system spreading caused by use of forward error control, with code rates less than unity, somehow reduces the "processing gain" advantage of excess spreading not

[4]Interleaving may be required to avoid burst errors, which otherwise would violate the underlying independence assumption.

dedicated to error control. Here, in fact the entire spreading is dedicated to error control and yet not only is the "processing gain" exploited fully, but with the addition of coordinated processing, the ultimate capacity of the multiple-access channel in additive white Gaussian noise is also shown to be achievable in the limit of arbitrarily long codes. This performance is, in fact, not achievable without the use of very low rate forward error control codes with consequently large bandwidth expansion, and only in this way can spread-spectrum multiple-access communication approach its ultimate potential.

Appendix I
Error Bounds for Orthogonal Codes on Very Noisy Memoryless Additive Channels

The bound for convolutional orthogonal codes is based on the bound for block orthogonal codes ([5, p. 256–257]). For the latter, for any binary codeword $x_m = \epsilon_m$ of dimension N where $\epsilon_{mn} = \pm\epsilon$.

$$P_{Em} = \sum_y (y|\epsilon_m)^{1/(1+\rho)}\left[\sum_{m'\neq m} p(y|\epsilon_{m'})^{1/(1+\rho)}\right]^\rho$$
$$= \sum_y p(y)e^{Z_m(y)/(1+\rho)}\left[\sum_{m'\neq m} e^{Z_{m'}(y)/(1+\rho)}\right]^\rho \tag{I.1}$$

where

$$Z_m(y) = \ln\frac{p(y|\epsilon_m)}{p(y)} = \sum_{n=1}^N \ln\frac{p(y_n|\epsilon_{mn})}{p(y_n)}$$
$$= \sum_{n-1}^N \ln\left[1 + \frac{\epsilon_{mn}}{\sigma}\frac{p'(y_n)}{p(y_n)} + 0\left(\frac{\epsilon^2}{\sigma^2}\right)\right]. \tag{I.2}$$

The last two steps follow from the fact that the channel is memoryless and that it is very noisy as defined in (5). (The subscript N in (5) for the normalized density function is omitted here.) Note also that all $\epsilon_{mn}^2 = \epsilon^2$ because $\epsilon_{mn} = \pm\epsilon$. Expanding $\ln(1 + X)$ in a Taylor series, it follows that

$$E_y[Z_m(y)Z_{m'}(y)]$$
$$= E\left\{\left[\sum_n \frac{\epsilon_{mn}}{\sigma}\frac{p'(y_n)}{p(y_n)} + 0\left(\frac{\epsilon^2}{\sigma^2}\right)\right]\cdot\left[\sum_k \frac{\epsilon_{m'k}}{\sigma}\frac{p'(y_k)}{p(y_k)} + 0\left(\frac{\epsilon^2}{\sigma^2}\right)\right]\right\}$$

where E_y indicates the expectation with respect to $p(y) = \Pi_n p(y_n)$.

Thus

$$E_y[Z_m(y)Z_{m'}(y)]$$
$$= \sum_{n=1}^n \epsilon_{mn}\epsilon_{m'n}E_y\left[\frac{p'(y_n)}{p(y_n)}\right]^2 + 0\left(\frac{\epsilon^4}{\sigma^4}\right) \to 0 \quad \text{for } \epsilon^2/\sigma^2 \ll 1 \tag{I.3}$$

where the last step follows from the fact that the N-dimensional binary sequences ϵ_m and $\epsilon_{m'}$ are orthogonal.

Similarly, for a and b any two positive integers,

$$\left|E_y[Z_m^a(y)Z_m^b(y)]\right| \leq 0(\epsilon^4) \to 0 \quad \text{for } \epsilon^2/\sigma^2 \ll 1.$$

Hence, $Z_m(y)$ and $Z_{m'}(y)$ are asymptotically independent for $\epsilon^2/\sigma^2 \ll 1$ (which is the very noisy condition).

Then imposing the memoryless condition, substituting definition (5) in (I.1) and utilizing the just proven independence condition, results in

$$P_{E_m} \leq E_y\left\{\prod_n\left[1 + \frac{\epsilon_{mn}}{\sigma}\frac{p'(y_n)}{p(y_n)} + \frac{\epsilon^2}{2\sigma^2}\frac{p''(y_n)}{p(y_n)}\right]^{1/(1+\rho)}\right\}$$
$$\cdot E_y\left\{\sum_{m'\neq m}\prod_n\left[1 + \frac{\epsilon_{m'n}}{\sigma}\frac{p'(y_n)}{p(y_n)} + \frac{\epsilon^2}{2\sigma^2}\frac{p''(y_n)}{p(y_n)}\right]^{1/(1+\rho)}\right\}^\rho.$$

Expanding the fractional power terms in Taylor series, deleting terms beyond quadratic and utilizing Jensen's inequality [5] on the second expectation, yields

$$P_{E_m} \leq \left\{\prod_n E_{y_n}\left[1 + \frac{\epsilon_{mn}}{\sigma(1+\rho)}\frac{p'(y_n)}{p(y_n)} + \frac{\epsilon^2}{2\sigma^2}\left(\frac{1}{1+\rho}\frac{p''(y_n)}{p(y_n)} - \frac{\rho}{(1+\rho)^2}\left[\frac{p'(y_n)}{p(y_n)}\right]^2\right)\right]\right\}$$
$$\cdot\left\{\sum_{m'\neq m}\prod_n E_{y_n}\left[1 + \frac{\epsilon_{m'n}}{\sigma(1+\rho)}\frac{p'(y_n)}{p(y_n)} + \frac{\epsilon^2}{2\sigma^2}\left(\frac{1}{1+\rho}\frac{p''(y_n)}{p(y_n)} - \frac{\rho}{(1+\rho)^2}\left[\frac{p'(y_n)}{p(y_n)}\right]^2\right)\right]\right\}^\rho$$
$$< M^\rho\left\{1 - \frac{\rho\epsilon^2}{2\sigma^2(1+\rho)^2}E\left[\frac{p'(y_n)}{p(y)}\right]^2\right\}^{(1+\rho)N}$$
$$0 < \rho < 1. \tag{I.4}$$

where the last step results from the fact that since $p(y) = p(-y)$ is an even function, $p'(y)$ is odd and $\lim_{y\to\infty} p'(y) = 0$, whence it follows that

$$E\left[\frac{p'(y_n)}{p(y_n)}\right] = \int_{-\infty}^{\infty} p'(y)\,dy = 0$$

and

$$E\left[\frac{p''(y_n)}{p(y_n)}\right] = \int_{-\infty}^{\infty} p''(y)\,dy = 2\lim_{y\to\infty} p'(y) = 0.$$

Finally, applying the Schwarz inequality to the remaining expectation in (I.4), using the fact that y is a random variable with zero mean and unit variance, we obtain

$$E(y^2)\, E\left[\frac{p'(y_n)}{p(y_n)}\right]^2 \geq E\left[y\frac{p'(y_n)}{p(y_n)}\right]^2 = \left[\int_{-\infty}^{\infty} yp'(y)\,dy\right]^2 = \left[yp(y)\Big|_{-\infty}^{\infty} - \int_{-\infty}^{\infty} p(y)\,dy\right]^2 = 1.$$

Hence,

$$E\left[\frac{p'(y)}{p(y)}\right]^2 \geq 1$$

and

$$P_{E_m} < M^{\rho} \exp\left[-\frac{N\epsilon^2}{2\sigma^2}\frac{\rho}{1+\rho}\right], \quad 0 < \rho < 1.$$

Substituting into (4.6.17) of [5], with $M = 2^k$, $N = (K + k)/r$ and E_b/N_o of the latter replaced by $\epsilon^2/(2\sigma^2 r)$ produces the final result of (4.6.23) and (4.6.24) given here as inequalities (3) and (4) with

$$c/r = 2^{K-1}\epsilon^2/(\sigma^2 \ln 2).$$

Appendix II

The transmitted signal for the kth user of the SSMA channel [Fig. 2(a)]

$$x_k(t) = \sqrt{E_{sk}} \sum_n a_k(n)\left[C_k^c(n)\, h(t - nT) \cdot \cos(2\pi f_o t + \phi_k) + C_k^s(n)\, h(t - nT) \cdot \sin(2\pi f_o t + \phi_k)\right] \tag{II.1}$$

where E_{sk} is the energy per symbol of the kth user. Without loss of generality, the filter transfer function is normalized so that for its squared magnitude

$$\int |H(f)|^2\, df = 1 \tag{II.2}$$

while its rms bandwidth

$$W = \left[\int f^2 |H(f)|^2\, df\right]^{1/2} = 1/T. \tag{II.3}$$

The received signal (with propagation losses normalized) will be

$$y(t) = \sum_{k=1}^{M} x_k(t) + n(t)$$

where $n(t)$ is background additive white Gaussian noise of one-sided density N_o. The matched-filter demodulator for the kth user, which maximizes the ratio of desired signal power-to-background noise power, is shown in Fig. 2(b). The normalization by the known noise standard deviation $(N_o/2)^{1/2}$ in the final step is for convenience in the analysis. It is straightforward to determine the mean and variance of the nth sample $y_k(n)$ for the kth user's demodulator. These are the parameters ϵ and σ^2.

The mean is simply

$$\epsilon_k = E[y_k(n)] = \sqrt{\frac{2E_{sk}}{N_o}}\, a_k(n) \int |H(f)|^2\, df = \sqrt{\frac{2E_{sk}}{N_o}}\, a_k(n) \tag{II.4}$$

where $a_k(n) = \pm 1$, while the variance is the sum of three terms: that due to background noise V_N, that due to intersymbol interference V_I, and that due to other user interference V_O,

$$\sigma_k^2 = \text{Var}[y_k(n)] = V_N + V_I + V_O \tag{II.5}$$

where

$$V_N = \frac{2}{N_o}\int \frac{N_o}{2}|H(f)|^2\, df = 1 \tag{II.6}$$

$$V_I = \sum_{\substack{i=-\infty \\ i \neq 0}}^{\infty} V_I(i),$$

$$V_I(i) = \frac{2E_{sk}}{N_o}\left[\int |H(f)|^2 \cos(2\pi i f/W)\, df\right]^2 \tag{II.7}$$

and

$$V_O = \sum_{j \neq k}^{M} V_O(j) \tag{II.8}$$

where $V_O(j)$ is the variance due to the jth interference in the kth user's demodulator.

To calculate the variance due to other-user interference, it is simplest to note that at the output of each impulse modulator of the jth user's modulator, there is a random process consisting of periodic impulses of energy E_{sj} and either sign (equiprobably). Now consider the impulse train for user j to be the limit of the periodic train of pulses of infinitesimal duration Δ and period T and area $E_{sj}^{1/2}$ and consequently amplitude $E_{sj}^{1/2}/\Delta$. Then since each pulse is equally likely to have either polarity and each user's impulse train is randomly delayed relative to every other user's, the pulse train for user j appears as a stationary random process to user k, with covariance function

$$R_j(t; \Delta) = \begin{cases} (E_{sj}/\Delta^2)(\Delta/T)(1 - |t|/\Delta) & -\Delta \leq t \leq \Delta \\ 0 & \text{otherwise.} \end{cases}$$

In the limit as $\Delta \to 0$, this approaches an impulse of area E_{sj}/T. Thus,

$$R(t) = \lim_{\Delta \to 0} R_j(t, \Delta) = (E_{sj}/T)\,\delta(t)$$

whose corresponding spectral density is uniform and of level E_{sj}/T. This applies to the impulse train out of each impulse modulator for the jth user and the two are independent provided the multiplicative random sequences are independent. Thus the jth user's signal will appear to the kth user's demodulator as the random process (see Fig. 2)

$$\nu_c^{(j)}(\tau)\cos(2\pi f_o\tau+\varphi_j) + \nu_s^{(j)}(\tau)\sin(2\pi f_o\tau+\varphi_j)$$

where $\nu_c^{(j)}(\tau)$ and $\nu_s^{(j)}(\tau)$ are independent processes, each of spectral density $(E_{sj}/T)|H(f)|^2$.

After quadrature multiplication by the carrier of user k and matched filtering, the two demodulator filter outputs due to the jth user are

$$\tfrac{1}{2}\tilde{\nu}_c(\tau)[\cos(\varphi_j-\varphi_k)-\sin(\varphi_j-\varphi_k)]$$

and

$$\tfrac{1}{2}\tilde{\nu}_s(\tau)[\sin(\varphi_j-\varphi_k)+\cos(\varphi_j-\varphi_k)]$$

where the double frequency terms are assumed eliminated by the low-pass matched filters and where $\tilde{\nu}_c(\tau)$ and $\tilde{\nu}_s(\tau)$ are matched filtered versions of $\nu_c(\tau)$ and $\nu_s(\tau)$. Hence, they are independent processes each of spectral density $(E_{sj}/T)|H(f)|^4$.

Thus the variance of two matched filter outputs at each sample time are

$$\frac{1}{4}\int\frac{E_{sj}}{T}|H(f)|^4\,df \cdot [1 \pm 2\sin(\varphi_j-\varphi_k)\cos(\varphi_j-\varphi_k)].$$

Since each of the two independent variables has an even probability density function, multiplication by the two quadrature independent binary random sequences does not change the variance. Then addition of the two and scaling by $\sqrt{2/N_o}$ results in a contribution to the variance of the kth demodulator by the jth user equal to

$$V_O(j)=\frac{E_{sj}}{N_oT}\int|H(f)|^4\,df. \tag{II.9}$$

Combining (II.5) through (II.9) it follows that Var $[y_k(n)]$ depends on $H(f)$ through the intersymbol interference (II.7) and the other-user interference (II.9). Both can be minimized, under the normalizing condition (II.2), by appropriate choice of $H(f)$ as follows from the two lower bounds.

$V_I(i)\geq 0$ for all i with equality whenever $|H(f)|^2$ = constant over a finite number of periods and equals zero otherwise, and

$$\int_{-W/2}^{W/2}|H(f)|^4\,df\int_{-W/2}^{W/2}1\,df \geq \left[\int_{-W/2}^{W/2}|H(f)|^2\cdot 1\,df\right]^2 = 1 \tag{II.10}$$

with equality iff $|H(f)|^2 = 1/W$, $|f| < W/2$. Note that in this case $\int [H(f)]^4\,df = 1/W = T$.

Thus, combining (II.4) through (II.10) leads to the conclusion that given M transmitters, each using independent quadrature-random-sequence band-limited modulators, as shown in Fig. 2, biphase modulated by a (coded) data sequence $a_k(n)$, the kth demodulator output is strictly band-limited to bandwidth W and has squared mean-to-variance given by

$$\frac{\epsilon_k^2}{\sigma_k^2}=\frac{\{E[y_k(n)]\}^2}{\text{Var}[y_k(n)]}\leq\frac{2E_{sk}/N_o}{1+\sum_{j\neq k}E_{sj}/N_o} \tag{II.11}$$

with equality iff

$$|H(f)|^2=\begin{cases}1/W, & |f|<W/2\\ 0, & \text{otherwise.}\end{cases}$$

Even though this band-limited condition is physically unrealizable without finite delay, it can be approached within an arbitrarily small approximation error by filters $H(f)$ of finite but arbitrarily increasing delay.

Appendix III
Successive Cancellation

Suppose successive cancellation is used, but with different code rates. While these cannot be arbitrary, we ignore constraints. Then the total data rate, defined as the sum of the data rates of all M users is obtained from (15) as

$$\frac{R_T}{W}=\frac{\sum_{k=1}^{M}R_k}{W}=\sum_{k=1}^{M}r_k=\frac{\propto}{\ln 2}\sum_{k=1}^{M}\frac{X_k}{1+\sum_{j=1}^{k-1}X_j} = \frac{\propto}{\ln 2}\sum_{k=1}^{M}\frac{Y_{k+1}-Y_k}{1+Y_k} \tag{III.1}$$

where

$$Y_k=\sum_{j=1}^{k-1}X_j,\quad Y_1=0,\quad Y_{M+1}=\sum_{k=1}^{M}X_k=S_T/N_oW.$$

The sum can be further lower bounded by an integral as follows:

$$\sum_{k=1}^{M}\frac{Y_{k+1}-Y_k}{1+Y_k}=\sum_{k=1}^{M}\int_{Y_k}^{Y_{k+1}}\frac{1}{1+Y_k}\,dy \geq \sum_{k=1}^{M}\int_{Y_k}^{Y_{k+1}}\frac{dy}{1+y}=\int_0^{Y_{M+1}}\frac{dy}{1+y} = \ln(1+S_T/N_oW). \tag{III.2}$$

Thus combining (III.1) and (III.2), it follows that using coordinated processing with successive cancellation, employing only binary modulation of quadrature random spread-spectrum carriers, it is possible to approach a combined capacity for the M-user multiple-access channel which is lower bounded by

$$\frac{R_T}{W} \geq \frac{\propto}{\ln 2} \ln\left(1 + S_T/N_o W\right)$$
$$= \propto \log_2\left(1 + S_T/N_o W\right). \qquad \text{(III.3)}$$

Effect of Incomplete Cancellation

Suppose that amplitude and phase estimations are only so accurate that, after cancellation, a fraction β of each cancelled user's signal power is left in the composite signal, then the interference at the kth decoding step is

$$\sum_{j=1}^{k-1} X_j + \beta \sum_{k+1}^{M} X_j.$$

In place of (III.1), we now have

$$\frac{R_T}{W} = \frac{\propto}{\ln 2} \sum_{k=1}^{M} \frac{X_k}{1 + \sum_{1}^{k-1} X_j + \beta \sum_{k+1}^{M} X_j}$$

$$= \frac{\propto}{\ln 2} \sum_{k=1}^{M} \frac{X_k}{1 + \beta \sum_{1}^{M} X_j + (1-\beta) \sum_{1}^{k-1} X_j}$$

$$\geq \frac{\propto}{\ln 2} \int_0^{S_T/N_o W} \frac{dy}{(1 + \beta S_T/N_o W) + (1-\beta) y}$$

$$= \frac{\propto}{1-\beta} \log_2 \left(\frac{1 + S_T/N_o W}{1 + \beta S_T/N_o W} \right). \qquad \text{(III.4)}$$

Note that as $\beta \to 0$, (III.4) approaches (III.3) while as $\beta \to 1$, it approaches the result for uncoordinated processing.

Note also that as $S_T/N_o W \to \infty$, the total rate approaches the limit of

$$\lim_{S_T/N_o W \to \infty} \frac{R_T}{W} = \frac{\propto \log_2 (1/\beta)}{1-\beta} \text{ b/s/Hz} \qquad \text{(III.5)}$$

which is the irreducible error due to imperfect cancellation of a fraction β of each signal. The same result is also obtained by incorporating the imperfect cancellation interference for the case of equal rate users.

Acknowledgment

The author gratefully acknowledges the contribution of R. Padovani who obtained inequality (II.10).

References

[1] A. J. Viterbi, "Error bounds for convolutional codes and an asymptotically optimum decoding algorithm," *IEEE Trans. Inform. Theory*, vol. IT-13, pp. 260-269, Apr. 1967.

[2] —, "Orthogonal tree codes for communication in the presence of white Gaussian noise," *IEEE Trans. Commun. Technol.*, vol. COM-15, pp. 238-242, Apr. 1967.

[3] G. D. Forney, Jr., "The Viterbi algorithm," *Proc. IEEE*, vol. 61, pp. 268-278, Mar. 1973.

[4] J. K. Omura, "On the Viterbi decoding algorithm," *IEEE Trans. Inform. Theory*, vol. IT-15, pp. 177-179, Jan. 1969.

[5] A. J. Viterbi and J. K. Omura, *Principles of Digital Communications and Coding*. New York: McGraw-Hill, 1979.

[6] R. R. Green, "A serial orthogonal decoder," *JPL Space Programs Summary*, vol. 37-39-IV, pp. 247-253, Jet Prop. Lab., Pasadena, CA 1966.

[7] A. J. Viterbi, "When not to spread spectrum—A sequel," *IEEE Commun. Mag.*, vol. 23, pp. 12-17, Apr. 1985.

[8] K. S. Gilhousen, I. M. Jacobs, R. Padovani, and L. A. Weaver, "Increased capacity using CDMA for mobile satellite communication," *J. IEEE Trans. Select. Areas Commun.*, see this issue, pp. 503-514.

[9] C. E. Shannon, "Communication in the presence of noise," *Proc. IRE*, vol. 37, pp. 10-21, Jan. 1949.

[10] A. D. Wyner, "Recent results in the Shannon theory," *IEEE Trans. Inform. Theory*, vol. IT-20, pp. 2-10, Jan. 1974.

[11] A. J. Viterbi, "Spread spectrum communications—Myths and realities," *IEEE Commun. Mag.*, vol. 17, pp. 11-18, May 1979.

Andrew J. Viterbi (S'54-M'58-SM'63-F'73) received the S.B. and S.M. degrees in electrical engineering from the Massachusetts Institute of Technology in 1957. He received the Ph.D. degree in electrical engineering from the University of Southern California in 1962.

He has devoted approximately equal segments of his career to academic research, industrial development, and entrepreneurial activities. On July 1, 1985, he became a founder and Vice Chairman and Chief Technical Officer of QUALCOMM, Inc., a company concentrating on mobile satellite communications for both commercial and military applications. In 1968, he cofounded LINKABIT Corporation. He was Executive Vice President of LINKABIT from 1974 to 1982. In 1982, he took over as President of M/A-COM LINKABIT, Inc. From 1984 to 1985, he was appointed Chief Scientist and Senior Vice President of M/A-COM, Inc. In his first employment after graduating from M.I.T. in 1957, he was a member of the project team at C.I.T. Jet Propulsion Laboratory which designed and implemented the telemetry equipment on the first successful U.S. satellite, Explorer I.

As a Professor in the UCLA School of Engineering and Applied Science from 1963 to 1973, Dr. Viterbi did fundamental work in digital communication theory and wrote two books on the subject, for which he received numerous professional society awards and international recognition. These include three paper awards, culminating in the 1968 IEEE Information Theory Group Outstanding Paper Award. He has also received three major society awards: the 1975 Christopher Columbus International Award (from the Italian National Research Council sponsored by the City of Genoa); the 1980 Aerospace Communications Award jointly with Dr. I. Jacobs (from AIAA); and the 1984 Alexander Graham Bell Medal (from IEEE sponsored by AT&T) "for exceptional contributions to the advancement of telecommunications." He is a member of the National Academy of Engineering and has a part-time appointment as Professor of Electrical and Computer Engineering at the University of California, San Diego.

NEAR SHANNON LIMIT ERROR - CORRECTING CODING AND DECODING : TURBO-CODES (1)

Claude Berrou, Alain Glavieux and Punya Thitimajshima

Claude Berrou, Integrated Circuits for Telecommunication Laboratory

Alain Glavieux and Punya Thitimajshima, Digital Communication Laboratory

Ecole Nationale Supérieure des Télécommunications de Bretagne, France

(1) Patents N° 9105279 (France), N° 92460011.7 (Europe), N° 07/870,483 (USA)

Reprinted from *Proceeding of the IEEE International Communications Conf. (ICC)*, Geneva Switzerland, 1993, pp. 1064-1070.

Abstract - This paper deals with a new class of convolutional codes called *Turbo-codes*, whose performances in terms of Bit Error Rate (BER) are close to the SHANNON limit. The *Turbo-Code* encoder is built using a parallel concatenation of two Recursive Systematic Convolutional codes and the associated decoder, using a feedback decoding rule, is implemented as P pipelined identical elementary decoders.

I - INTRODUCTION

Consider a binary rate R=1/2 convolutional encoder with constraint length K and memory M=K-1. The input to the encoder at time k is a bit d_k and the corresponding codeword C_k is the binary couple (X_k, Y_k) with

$$X_k = \sum_{i=0}^{K-1} g_{1i} d_{k-i} \quad mod.2 \quad g_{1i} = 0,1 \qquad (1a)$$

$$Y_k = \sum_{i=0}^{K-1} g_{2i} d_{k-i} \quad mod.2 \quad g_{2i} = 0,1 \qquad (1b)$$

where G_1: $\{g_{1i}\}$, G_2: $\{g_{2i}\}$ are the two encoder generators, generally expressed in octal form.

It is well known, that the BER of a classical Non Systematic Convolutional (NSC) code is lower than that of a classical Systematic code with the same memory M at large SNR. At low SNR, it is in general the other way round. The new class of Recursive Systematic Convolutional (RSC) codes, proposed in this paper, can be better than the best NSC code at any SNR for high code rates.

A binary rate R=1/2 RSC code is obtained from a NSC code by using a feedback loop and setting one of the two outputs X_k or Y_k equal to the input bit d_k. For an RSC code, the shift register (memory) input is no longer the bit d_k but is a new binary variable a_k. If X_k=d_k (respectively Y_k=d_k), the output Y_k (resp. X_k) is equal to equation (1b) (resp. 1a) by substituting a_k for d_k and the variable a_k is recursively calculated as

$$a_k = d_k + \sum_{i=1}^{K-1} \gamma_i a_{k-i} \qquad mod.2 \qquad (2)$$

where γ_i is respectively equal to g_{1i} if X_k=d_k and to g_{2i} if Y_k=d_k. Equation (2) can be rewritten as

$$d_k = \sum_{i=0}^{K-1} \gamma_i a_{k-i} \qquad mod.2. \qquad (3)$$

One RSC encoder with memory M=4 obtained from an NSC encoder defined by generators G_1=37, G_2=21 is depicted in Fig.1.

Generally, we assume that the input bit d_k takes values 0 or 1 with the same probability. From equation (2), we can show that variable a_k exhibits the same statistical property

$$P_r\{a_k = 0 / a_1 = \varepsilon_1, .. a_{k-1} = \varepsilon_{k-1}\} = P_r\{d_k = \varepsilon\} = 1/2 \quad (4)$$

with ε is equal to

$$\varepsilon = \sum_{i=1}^{K-1} \gamma_i \varepsilon_i \quad mod.2 \quad \varepsilon = 0,1. \quad (5)$$

Thus the trellis structure is identical for the RSC code and the NSC code and these two codes have the same free distance d_f. However, the two output sequences $\{X_k\}$ and $\{Y_k\}$ do not correspond to the same input sequence $\{d_k\}$ for RSC and NSC codes. This is the main difference between the two codes.

When punctured code is considered, some output bits X_k or Y_k are deleted according to a chosen puncturing pattern defined by a matrix $\boldsymbol{P}$. For instance, starting from a rate R=1/2 code, the matrix $\boldsymbol{P}$ of rate 2/3 punctured code is

$$P = \begin{bmatrix} 1 & 1 \\ 1 & 0 \end{bmatrix}$$

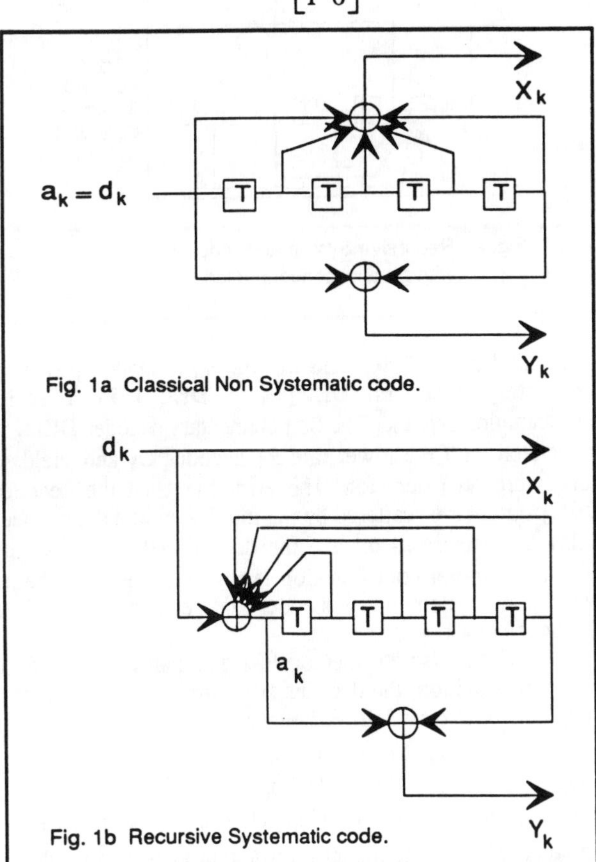

Fig. 1a Classical Non Systematic code.

Fig. 1b Recursive Systematic code.

II - PARALLEL CONCATENATION OF RSC CODES

With RSC codes, a new concatenation scheme, called parallel concatenation can be used. In Fig. 2, an example of two identical RSC codes with parallel concatenation is shown. Both elementary encoder (C_1 and C_2) inputs use the same bit d_k but according to a different sequence due to the presence of an interleaver. For an input bit sequence $\{d_k\}$, encoder outputs X_k and Y_k at time k are respectively equal to d_k (systematic encoder) and to encoder C_1 output Y_{1k}, or to encoder C_2 output Y_{2k}. If the coded outputs (Y_{1k}, Y_{2k}) of encoders C_1 and C_2 are used respectively n_1 times and n_2 times and so on, the encoder C_1 rate R_1 and encoder C_2 rate R_2 are equal to

$$R_1 = \frac{n_1 + n_2}{2n_1 + n_2} \qquad R_2 = \frac{n_1 + n_2}{2n_2 + n_1}. \qquad (6)$$

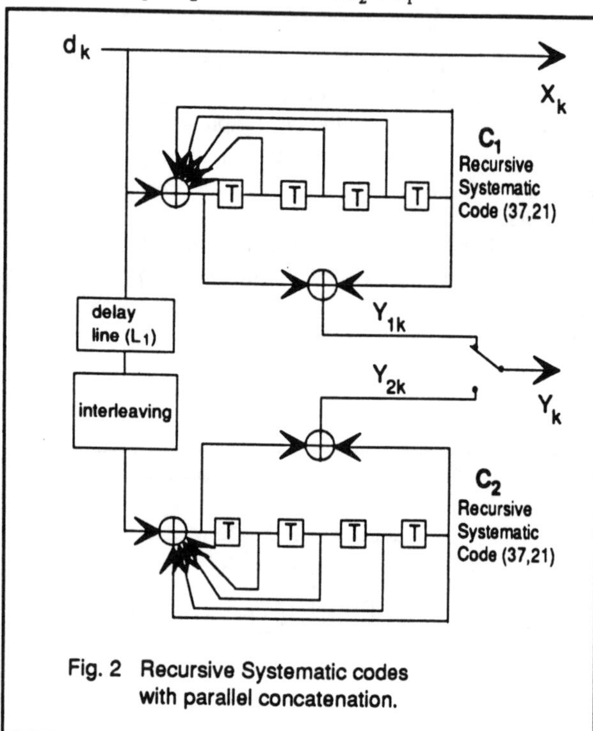

Fig. 2 Recursive Systematic codes with parallel concatenation.

The decoder **DEC** depicted in Fig. 3a, is made up of two elementary decoders (**DEC₁** and **DEC₂**) in a serial concatenation scheme. The first elementary decoder **DEC$_1$** is associated with the lower rate R_1 encoder C_1 and yields a soft (weighted) decision. The error bursts at the decoder **DEC$_1$** output are scattered by the interleaver and the encoder delay L_1 is inserted to take the decoder **DEC$_1$** delay into account. Parallel concatenation is a very attractive scheme because both elementary encoder and decoder use a single frequency clock.

For a discrete memoryless gaussian channel and a binary modulation, the decoder **DEC** input is made up of a couple R_k of two random variables x_k and y_k, at time k

$$x_k = (2d_k - 1) + i_k \qquad (7a)$$

$$y_k = (2Y_k - 1) + q_k, \qquad (7b)$$

where i_k and q_k are two independent noises with the same variance σ^2. The redundant information y_k is demultiplexed and sent to decoder **DEC$_1$** when $Y_k = Y_{1k}$ and toward decoder **DEC$_2$** when $Y_k = Y_{2k}$. When the redundant information of a given encoder (C_1 or C_2) is not emitted, the corresponding decoder input is set to zero. This is performed by the DEMUX/INSERTION block.

It is well known that soft decoding is better than hard decoding, therefore the first decoder **DEC$_1$** must deliver to the second decoder **DEC$_2$** a weighted (soft) decision. The Logarithm of Likelihood Ratio (LLR), $\Lambda_1(d_k)$ associated with each decoded bit d_k by the first decoder **DEC$_1$** is a relevant piece of information for the second decoder **DEC$_2$**

$$\Lambda_1(d_k) = Log \frac{P_r\{d_k = 1/observation\}}{P_r\{d_k = 0/observation\}}. \qquad (8)$$

where $P_r\{d_k = i \ /observation\}$, $i = 0, 1$ is the *a posteriori probability* (APP) of the data bit d_k.

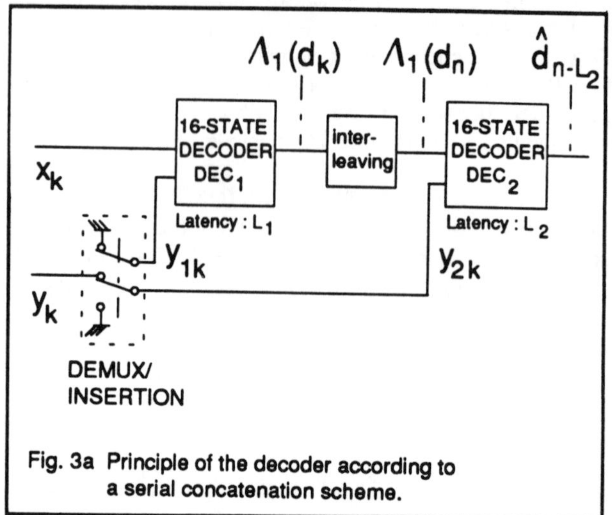

Fig. 3a Principle of the decoder according to a serial concatenation scheme.

III - OPTIMAL DECODING OF RSC CODES WITH WEIGHTED DECISION

The VITERBI algorithm is an optimal decoding method which minimizes the probability of sequence error for convolutional codes. Unfortunately this algorithm is not able to yield the APP for each decoded bit. A relevant algorithm for this purpose has been proposed by BAHL *et al.* [1]. This algorithm minimizes the bit error probability in decoding linear block and convolutional codes and yields the APP for each decoded bit. For RSC codes, the BAHL *et al.* algorithm must be modified in order to take into account their recursive character.

III - 1 Modified BAHL *et al.* algorithm for RSC codes

Consider a RSC code with constraint length K; at time k the encoder state S_k is represented by a K-uple

$$S_k = (a_k, a_{k-1} \ldots\ldots a_{k-K+1}). \qquad (9)$$

Also suppose that the information bit sequence $\{d_k\}$ is made up of N independent bits d_k, taking values 0 and 1 with equal probability and that the encoder initial state S_0 and final state S_N are both equal to zero, *i.e*

$$S_0 = S_N = (0, 0\ldots\ldots0) = \mathbf{0}. \qquad (10)$$

The encoder output codeword sequence, noted $C_1^N = \{C_1 \ldots\ldots C_k \ldots\ldots\ldots C_N\}$ is the input to a discrete gaussian memoryless channel whose output is the sequence $R_1^N = \{R_1 \ldots\ldots\ldots R_k \ldots\ldots\ldots R_N\}$ where $R_k = (x_k, y_k)$ is defined by relations (7a) and (7b).

The APP of a decoded data bit d_k can be derived from the joint probability $\lambda_k^i(m)$ defined by

$$\lambda_k^i(m) = P_r\{d_k = i, S_k = m / R_1^N\} \qquad (11)$$

and thus, the APP of a decoded data bit d_k is equal to

$$P_r\{d_k = i / R_1^N\} = \sum_m \lambda_k^i(m), \; i = 0,1. \qquad (12)$$

From relations (8) and (12), the LLR $\Lambda(d_k)$ associated with a decoded bit d_k can be written as

$$\Lambda(d_k) = Log \frac{\sum_m \lambda_k^1(m)}{\sum_m \lambda_k^0(m)}. \qquad (13)$$

Finally the decoder can make a decision by comparing $\Lambda(d_k)$ to a threshold equal to zero

$$\hat{d}_k = 1 \quad if \quad \Lambda(d_k) > 0$$

$$\hat{d}_k = 0 \quad if \quad \Lambda(d_k) < 0. \qquad (14)$$

In order to compute the probability $\lambda_k^i(m)$, let us introduce the probability functions $\alpha_k^i(m)$, $\beta_k(m)$ and $\gamma_i(R_k, m', m)$

$$\alpha_k^i(m) = \frac{P_r\{d_k = i, S_k = m, R_1^k\}}{P_r\{R_1^k\}} P_r\{d_k = i, S_k = m / R_1^k\} \quad (15)$$

$$\beta_k(m) = \frac{P_r\{R_{k+1}^N / S_k = m\}}{P_r\{R_{k+1}^N / R_1^k\}} \qquad (16)$$

$$\gamma_i(R_k, m', m) = P_r\{d_k = i, R_k, S_k = m / S_{k-1} = m'\}. \qquad (17)$$

The joint probability $\lambda_k^i(m)$ can be rewritten using BAYES rule

$$\lambda_k^i(m) = \frac{P_r\{d_k = i, S_k = m, R_1^k, R_{k+1}^N\}}{P_r\{R_1^k, R_{k+1}^N\}}. \quad (18)$$

Thus we obtain

$$\lambda_k^i(m) = \frac{P_r\{d_k = i, S_k = m, R_1^k\}}{P_r\{R_1^k\}} \frac{P_r\{R_{k+1}^N / d_k = i, S_k = m, R_1^k\}}{P_r\{R_{k+1}^N / R_1^k\}}. \qquad (19)$$

Taking into account that events after time k are not influenced by observation R_1^k and bit d_k if state S_k is known, the probability $\lambda_k^i(m)$ is equal

$$\lambda_k^i(m) = \alpha_k^i(m)\beta_k(m). \qquad (20)$$

The probabilities $\alpha_k^i(m)$ and $\beta_k(m)$ can be recursively calculated from probability $\gamma_i(R_k, m', m)$. From annex I, we obtain

$$\alpha_k^i(m) = \frac{\sum_{m'} \sum_{j=0}^{1} \gamma_i(R_k, m', m)\alpha_{k-1}^j(m')}{\sum_m \sum_{m'} \sum_{i=0}^{1} \sum_{j=0}^{1} \gamma_i(R_k, m', m)\alpha_{k-1}^j(m')} \qquad (21)$$

and

$$\beta_k(m) = \frac{\sum_{m'} \sum_{i=0}^{1} \gamma_i(R_{k+1}, m, m')\beta_{k+1}(m')}{\sum_m \sum_{m'} \sum_{i=0}^{1} \sum_{j=0}^{1} \gamma_i(R_{k+1}, m', m)\alpha_k^j(m')}. \qquad (22)$$

The probability $\gamma_i(R_k, m', m)$ can be determined from transition probabilities of the discrete gaussian memoryless channel and transition probabilities of the encoder trellis. From relation (17), $\gamma_i(R_k, m', m)$ is given by

$$\gamma_i(R_k, m', m) = p(R_k / d_k = i, S_k = m, S_{k-1} = m')$$

$$q(d_k = i / S_k = m, S_{k-1} = m')\pi(S_k = m / S_{k-1} = m') \quad (23)$$

where p(./.) is the transition probability of the discrete gaussian memoryless channel. Conditionally to ($d_k = i$, $S_k = m$, $S_{k-1} = m'$), x_k and y_k are two uncorrelated gaussian variables and thus we obtain

$$p(R_k / d_k = i, S_k = m, S_{k-1} = m') =$$

$$p(x_k / d_k = i, S_k = m, S_{k-1} = m')$$

$$p(y_k / d_k = i, S_k = m, S_{k-1} = m'). \quad (24)$$

Since the convolutional encoder is a deterministic machine, $q(d_k = i / S_k = m, S_{k-1} = m')$ is equal to 0 or 1. The transition state probabilities $\pi(S_k = m / S_{k-1} = m')$ of the trellis are defined by the encoder input statistic. Generally, $P_r\{d_k = 1\} = P_r\{d_k = 0\} = 1/2$ and since there are two possible transitions from each state, $\pi(S_k = m / S_{k-1} = m') = 1/2$ for each of these transitions.

Different steps of modified BAHL *et al.* algorithm

-Step 0 : Probabilities $\alpha_0^i(m)$ and $\beta_N(m)$ are initialized according to relation (12)

$$\alpha_0^i(0) = 1 \;\; \alpha_0^i(m) = 0 \;\; \forall m \neq 0, \; i = 0,1 \qquad (25a)$$

$$\beta_N(0) = 1 \;\; \beta_N(m) = 0 \;\; \forall m \neq 0. \qquad (25b)$$

-Step 1 : For each observation R_k, the probabilities $\alpha_k^i(m)$ and $\gamma_i(R_k, m', m)$ are computed using relations (21) and (23) respectively.

-Step 2 : When the sequence R_1^N has been completely received, probabilities $\beta_k(m)$ are computed using relation (22), and probabilities $\alpha_k^i(m)$ and $\beta_k(m)$ are multiplied in order to obtain $\lambda_k^i(m)$. Finally the LLR associated with each decoded bit d_k is computed from relation (13).

IV- THE EXTRINSIC INFORMATION OF THE RSC DECODER

In this chapter, we will show that the LLR $\Lambda(d_k)$ associated with each decoded bit d_k, is the sum of the LLR of d_k at the decoder input and of another information called extrinsic information, generated by the decoder.

Using the LLR $\Lambda(d_k)$ definition (13) and relations (20) and (21), we obtain

$$\Lambda(d_k) = Log \frac{\sum_m \sum_{m'} \sum_{j=0}^{1} \gamma_1(R_k, m', m)\alpha_{k-1}^j(m')\beta_k(m)}{\sum_m \sum_{m'} \sum_{j=0}^{1} \gamma_0(R_k, m', m)\alpha_{k-1}^j(m')\beta_k(m)}. \quad (26)$$

Since the encoder is systematic ($X_k = d_k$), the transition probability $p(x_k / d_k = i, S_k = m, S_{k-1} = m')$ in expression $\gamma_i(R_k, m', m)$ is independent of state values S_k and S_{k-1}. Therefore we can factorize this transition probability in the numerator and in the denominator of relation (26)

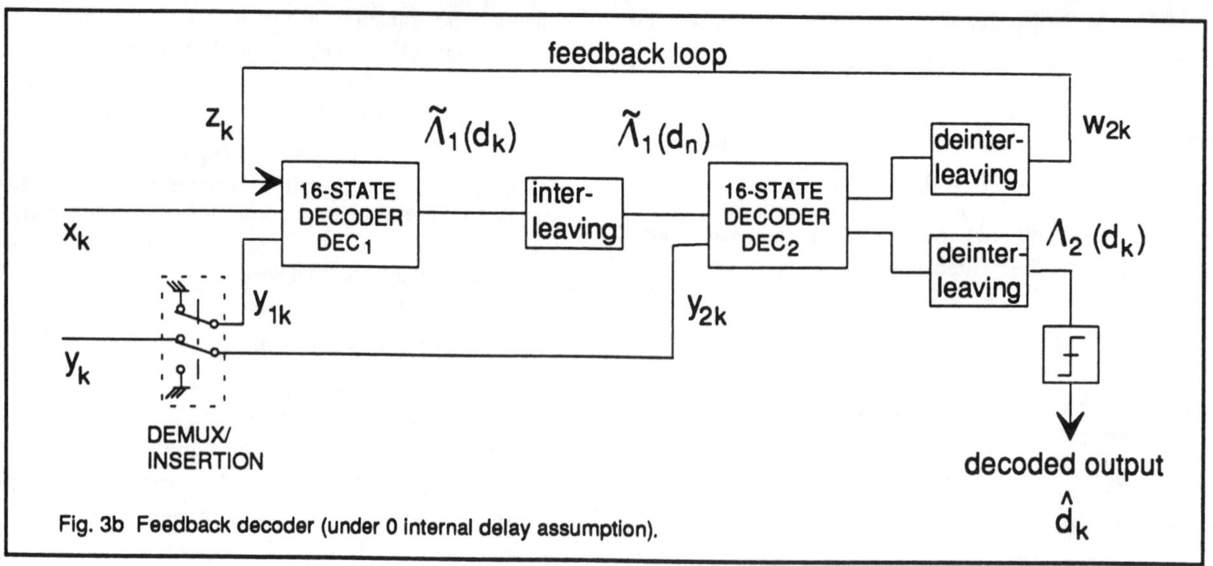

Fig. 3b Feedback decoder (under 0 internal delay assumption).

$$\Lambda(d_k) = Log\frac{p(x_k/d_k=1)}{p(x_k/d_k=0)} +$$

$$Log\frac{\sum_m\sum_{m'}\sum_{j=0}^{1}\gamma_1(y_k,m',m)\alpha_{k-1}^j(m')\beta_k(m)}{\sum_m\sum_{m'}\sum_{j=0}^{1}\gamma_0(y_k,m',m)\alpha_{k-1}^j(m')\beta_k(m)}. \quad (27)$$

Conditionally to $d_k=1$ (resp. $d_k=0$), variables x_k are gaussian with mean 1 (resp. -1) and variance σ^2, thus the LLR $\Lambda(d_k)$ is still equal to

$$\Lambda(d_k) = \frac{2}{\sigma^2}x_k + W_k \quad (28)$$

where

$$W_k = \Lambda(d_k)\,|_{x_k=0} =$$

$$Log\frac{\sum_m\sum_{m'}\sum_{j=0}^{1}\gamma_1(y_k,m',m)\alpha_{k-1}^j(m')\beta_k(m)}{\sum_m\sum_{m'}\sum_{j=0}^{1}\gamma_0(y_k,m',m)\alpha_{k-1}^j(m')\beta_k(m)}. \quad (29)$$

W_k is a function of the redundant information introduced by the encoder. In general W_k has the same sign as d_k; therefore W_k may improve the LLR associated with each decoded data bit d_k. This quantity represents the extrinsic information supplied by the decoder and does not depend on decoder input x_k. This property will be used for decoding the two parallel concatenated encoders.

V - DECODING SCHEME OF PARALLEL CONCATENATION CODES

In the decoding scheme represented in Fig. 3a, decoder $\mathbf{DEC_1}$ computes LLR $\Lambda_1(d_k)$ for each transmitted bit d_k from sequences $\{x_k\}$ and $\{y_k\}$, then the decoder $\mathbf{DEC_2}$ performs the decoding of sequence $\{d_k\}$ from sequences $\{\Lambda_1(d_k)\}$ and $\{y_k\}$. Decoder $\mathbf{DEC_1}$ uses the modified BAHL *et al.* algorithm and decoder $\mathbf{DEC_2}$ may use the VITERBI algorithm. The global decoding rule is not optimal because the first decoder uses only a fraction of the available redundant information. Therefore it is possible to improve the performance of this serial decoder by using a feedback loop.

V-1 Decoding with a feedback loop

We consider now that both decoders $\mathbf{DEC_1}$ and $\mathbf{DEC_2}$ use the modified BAHL *et al.* algorithm. We have seen in section IV that the LLR at the decoder output can be expressed as a sum of two terms if the decoder inputs were independent. Hence if the decoder $\mathbf{DEC_2}$ inputs $\Lambda_1(d_k)$ and y_{2k} are independent, the LLR $\Lambda_2(d_k)$ at the decoder $\mathbf{DEC_2}$ output can be written as

$$\Lambda_2(d_k) = f(\Lambda_1(d_k)) + W_{2k} \quad (30)$$

with

$$\Lambda_1(d_k) = \frac{2}{\sigma^2}x_k + W_{1k} \quad (31)$$

From relation (29), we can see that the decoder $\mathbf{DEC_2}$ extrinsic information W_{2k} is a function of the sequence $\{\Lambda_1(d_n)\}_{n\neq k}$. Since $\Lambda_1(d_n)$ depends on observation R_1^N, extrinsic information W_{2k} is correlated with observations x_k and y_{1k}. Nevertheless from relation (29), the greater $|n\text{-}k|$ is, the less correlated are $\Lambda_1(d_n)$ and observations x_k, y_k. Thus, due to the presence of interleaving between decoders $\mathbf{DEC_1}$ and $\mathbf{DEC_2}$, extrinsic information W_{2k} and observations x_k, y_{1k} are weakly correlated. Therefore extrinsic information W_{2k} and observations x_k, y_{1k} can be jointly used for carrying out a new decoding of bit d_k, the extrinsic information $z_k = W_{2k}$ acting as a diversity effect in an iterative process.

In Fig. 3b, we have depicted a new decoding scheme using the extrinsic information W_{2k} generated by decoder $\mathbf{DEC_2}$ in a feedback loop. This decoder does not take into account the different delays introduced by decoder $\mathbf{DEC_1}$ and $\mathbf{DEC_2}$ and a more realistic decoding structure will be presented later.

The first decoder $\mathbf{DEC_1}$ now has three data inputs, (x_k, y_{1k}, z_k) and probabilities $\alpha_{1k}^i(m)$ and $\beta_{1k}(m)$ are computed in substituting $R_k=\{x_k, y_{1k}\}$ by $R_k=(x_k, y_{1k}, z_k)$ in relations (21) and (22). Taking into account that z_k is weakly correlated with x_k and y_{1k} and supposing that z_k can be approximated by a gaussian variable with variance $\sigma_z^2 \neq \sigma^2$, the transition probability of the discrete gaussian memoryless channel can be now factored in three terms

$$p(R_k/d_k = i, S_k = m, S_{k-1} = m') = p(x_k/.)p(y_k/.)p(z_k/.) \quad (32)$$

The encoder $\mathbf{C_1}$ with initial rate R_1, through the feedback loop, is now equivalent to a rate R'_1 encoder with

$$R'_1 = \frac{R_1}{1+R_1}. \quad (33)$$

The first decoder obtains an additional redundant information with z_k that may significantly improve its performances; the term *Turbo-codes* is given for this iterative decoder scheme with reference to the turbo engine principle.

With the feedback decoder, the LLR $\Lambda_1(d_k)$ generated by decoder $\mathbf{DEC_1}$ is now equal to

$$\Lambda_1(d_k) = \frac{2}{\sigma^2}x_k + \frac{2}{\sigma_z^2}z_k + W_{1k} \quad (34)$$

where W_{1k} depends on sequence $\{z_n\}_{n \neq k}$. As indicated above, information z_k has been built by decoder $\mathbf{DEC_2}$ at the previous decoding step. Therefore z_k must not be used as input information for decoder $\mathbf{DEC_2}$. Thus decoder $\mathbf{DEC_2}$ input sequences at step p $(p \geq 2)$ will be sequences $\{\tilde{\Lambda}_1(d_n)\}$ and $\{y_{2k}\}$ with

$$\tilde{\Lambda}_1(d_n) = \Lambda_1(d_n)_{z_n=0}. \quad (35)$$

Finally from relation (30), decoder $\mathbf{DEC_2}$ extrinsic information $z_k = W_{2k}$, after deinterleaving, can be written as

$$z_k = W_{2k} = \Lambda_2(d_k)|_{\tilde{\Lambda}_1(d_k)=0} \quad (36)$$

and the decision at the decoder **DEC** output is

$$\hat{d}_k = sign[\Lambda_2(d_k)]. \quad (37)$$

The decoding delays introduced by decoder **DEC** ($\mathbf{DEC=DEC_1+DEC_2}$), the interleaver and the deinterleaver imply that the feedback information z_k must be used through an iterative process as represented in Fig. 4a, 4b. In fact, the global decoder circuit is composed of P pipelined identical elementary decoders (Fig. 4a). The pth decoder **DEC** (Fig. 4b) input, is made up of demodulator output sequences $(x)_p$ and $(y)_p$ through a delay line and of extrinsic information $(z)_p$ generated by the $(p$-1)th decoder **DEC**. Note that the variance σ_z^2 of the extrinsic information and the variance of $\tilde{\Lambda}_1(d_k)$ must be estimated at each decoding step p.

V-2 Interleaving

The interleaver uses a square matrix and bits $\{d_k\}$ are written row by row and read pseudo-randomly. This non-uniform reading rule is able to spread the residual error blocks of rectangular form, that may set up in the interleaver located behind the first decoder $\mathbf{DEC_1}$, and to give the greater free distance as possible to the concatenated (parallel) code.

VI - RESULTS

For a rate R=1/2 encoder with constraint length K=5, generators G_1=37, G_2=21 and parallel concatenation (R_1=R_2=2/3), we have computed the Bit Error Rate (BER) after each decoding step using the Monte Carlo method, as a function of signal to noise ratio E_b/N_O where E_b is the energy received per information bit d_k and N_O is the noise monolateral power spectral density. The interleaver consists of a 256x256 matrix and the modified BAHL *et al.* algorithm has been used with length data block of N=65536 bits. In

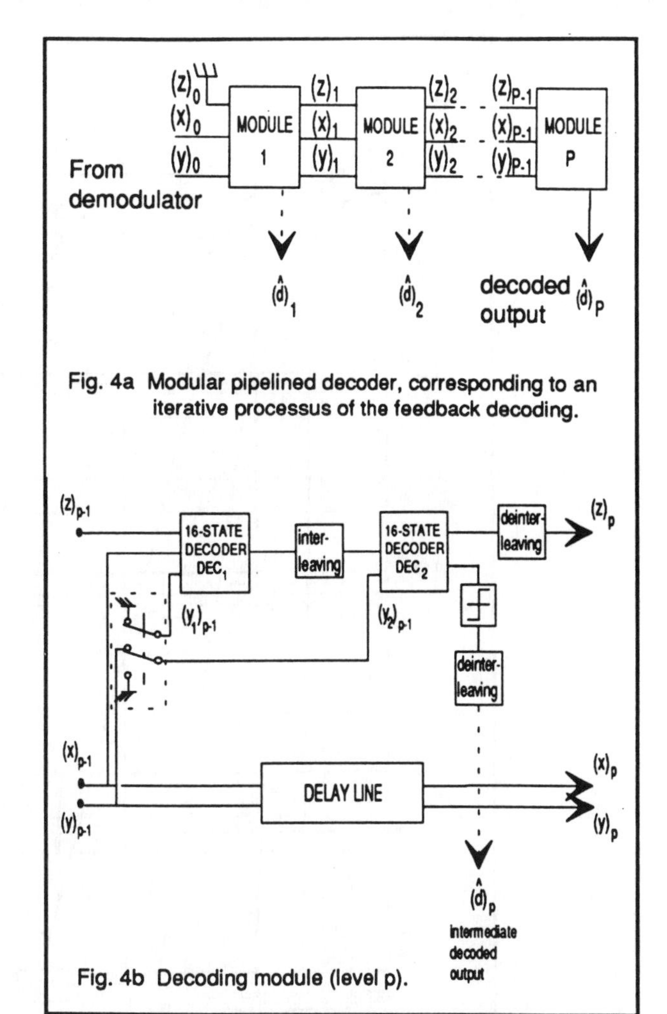

Fig. 4a Modular pipelined decoder, corresponding to an iterative processus of the feedback decoding.

Fig. 4b Decoding module (level p).

order to evaluate a BER equal to 10^{-5}, we have considered 128 data blocks i.e. approximatively 8 x10^6 bits d_k. The BER versus E_b/N_O, for different values of p is plotted in Fig. 5. For any given signal to noise ratio greater than 0 dB, the BER decreases as a function of the decoding step p. The coding gain is fairly high for the first values of p (p=1,2,3) and carries on increasing for the subsequent values of p. For p=18 for instance, the BER is lower than 10^{-5} at E_b/N_O= 0,7 dB. Remember that the Shannon limit for a binary modulation with R=1/2, is P_e = 0 (several authors take P_e =10^{-5} as a reference) for E_b/N_O= 0 dB. With parallel concatenation of RSC convolutional codes and feedback decoding, the performances are at 0,7 dB from Shannon's limit.

The influence of the constraint length on the BER has also been examined. For K greater than 5, at E_b/N_O= 0,7 dB, the BER is slightly worst at the first (p=1) decoding step and the feedback decoding is inefficient to improve the final BER. For K smaller than 5, at E_b/N_O= 0,7 dB, the BER is slightly better at the first decoding step than for K equal to 5, but the correction capacity of encoders $\mathbf{C_1}$ and $\mathbf{C_2}$ is too weak to improve the BER with feedback decoding. For K=4 (*i.e.* 8-state elementary decoders) and after iteration 18, a BER of 10^{-5} is achieved at E_b/N_O = 0,9 dB. For K equal to 5, we have tested several generators (G_1, G_2) and the best results were achieved with G_1=37, G_2=21.

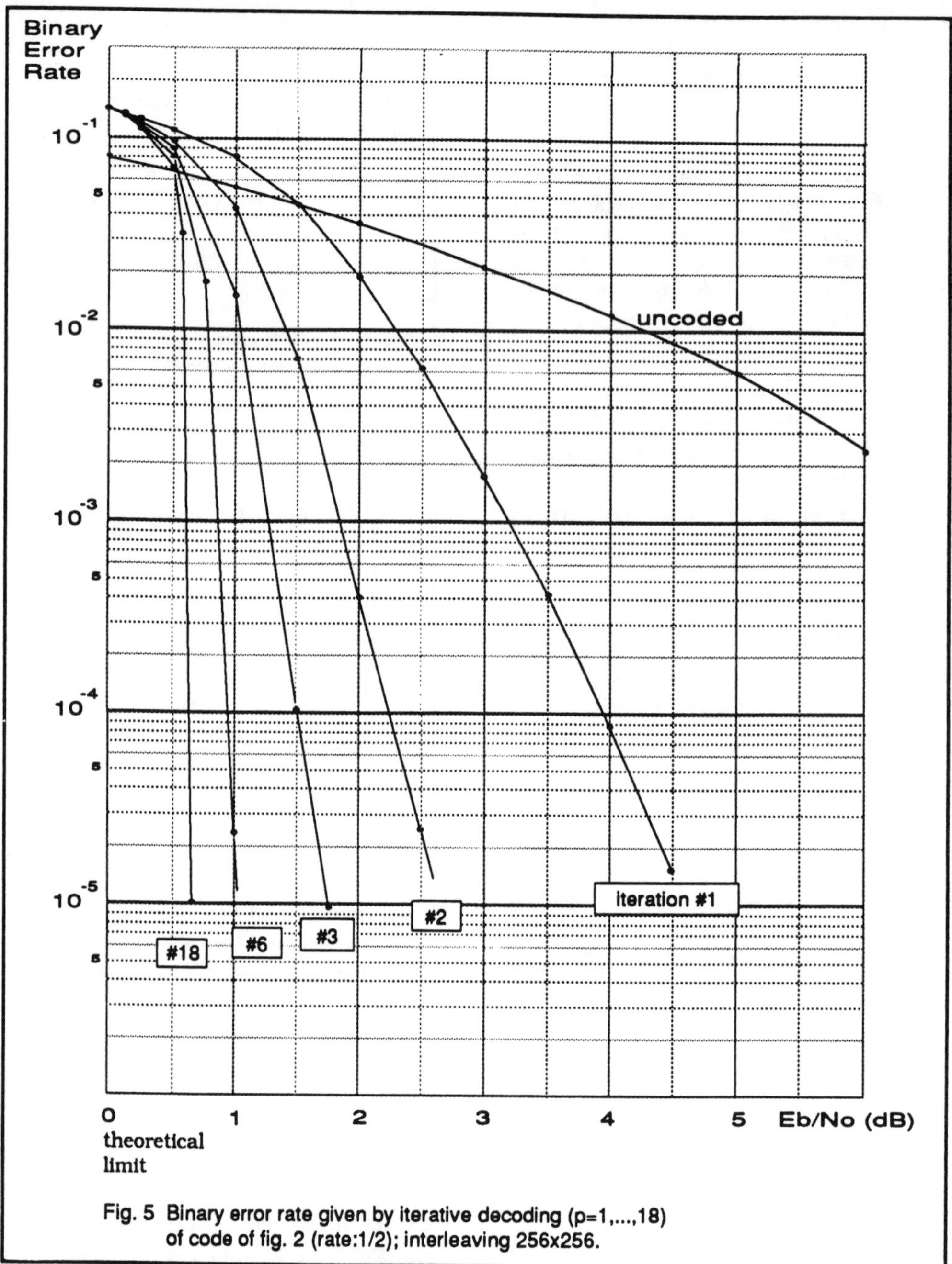

Fig. 5 Binary error rate given by iterative decoding (p=1,...,18) of code of fig. 2 (rate:1/2); interleaving 256x256.

For low signal to noise ratios, we have sometimes noticed that BER could increase during the iterative decoding process. In order to overcome this effect, we have divided the extrinsic information z_k by $[\ 1+\theta\left|\tilde{\Lambda}_1(d_k)\right|\ ]$ with $\theta = 0{,}15$.

In Fig. 6, the histogram of extrinsic information $(z)_p$ has been drawn for several values of iteration p, with all data bits equal to 1 and for a low signal to noise ratio (E_b/N_0= 0,8 dB). For p=1 (first iteration), extrinsic information $(z)_p$ is very poor about bit d_k, furthermore the gaussian hypothesis made above for extrinsic information $(z)_p$, is not satisfied! Nevertheless when iteration p increases, the histogram merges towards a gaussian law with a mean equal to 1. For instance, for p=13, extrinsic information $(z)_p$ becomes relevant information concerning data bits.

VII CONCLUSION

In this paper, we have presented a new class of convolutional codes called *Turbo-codes* whose performances in terms of BER are very close to SHANNON's limit. The decoder is made up of P pipelined identical elementary modules and rank p elementary module uses the data information coming from the demodulator and the extrinsic information generated by the rank (p-1) module. Each elementary module uses a modified BAHL *et al.* algorithm which is rather complex. A much simpler algorithm yielding weighted (soft) decisions has also been investigated for *Turbo-codes* decoding [2], whose complexity is only twice the complexity of the VITERBI algorithm, and with performances which are very close to those of the BAHL *et al.* algorithm. This new algorithm will enable encoders and decoders to be integrated in silicon with error correcting performances unmatched at the present time.

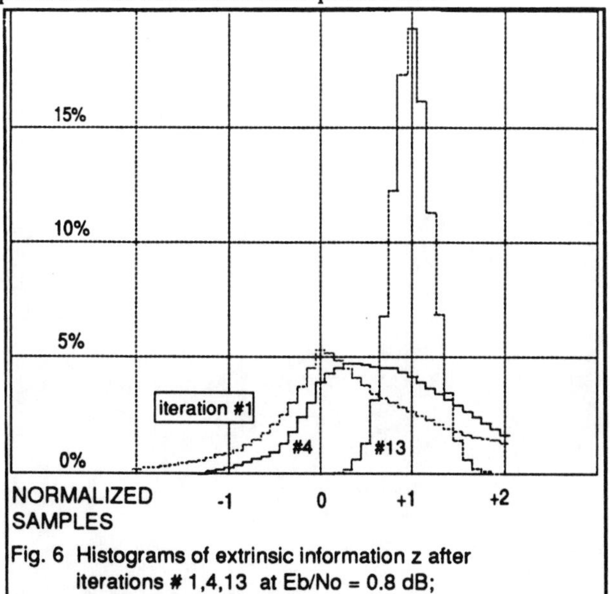

Fig. 6 Histograms of extrinsic information z after iterations # 1,4,13 at Eb/No = 0.8 dB; all information bits d=1.

ANNEX I : EVALUATION OF PROBABILITIES $\alpha_k^i(m)$ AND $\beta_k(m)$.

From relation (15) probability $\alpha_k^i(m)$ is equal to

$$\alpha_k^i(m) = \frac{Pr\{d_k = i, S_k = m, R_1^{k-1}, R_k\}}{Pr\{R_1^{k-1}, R_k\}} =$$

$$\frac{Pr\{d_k = i, S_k = m, R_k / R_1^{k-1}\}}{Pr\{R_k / R_1^{k-1}\}}. \qquad (A1)$$

The numerator of $\alpha_k^i(m)$ can be expressed from state S_{k-1} and bit d_{k-1}.

$$Pr\{d_k = i, S_k = m, R_k / R_1^{k-1}\} =$$

$$\sum_{m'} \sum_{j=0}^{1} P_r\{d_k = i, d_{k-1} = j, S_k = m, \; S_{k-1} = m', R_k / R_1^{k-1}\} \quad (A2)$$

By using BAYES rule, we can write

$$Pr\{d_k = i, S_k = m, R_k / R_1^{k-1}\} =$$

$$\sum_{m'} \sum_{j=0}^{1} \frac{Pr\{d_{k-1} = j, S_{k-1} = m', R_1^{k-1}\}}{P_r\{R_1^{k-1}\}}$$

$$P_r\{d_k = i, S_k = m, R_k / d_{k-1} = j, S_{k-1} = m', R_1^{k-1}\}. \quad (A3)$$

By taking into account that events after time (k-1) are not influenced by observation R_1^{k-1} and bit d_{k-1} if state S_{k-1} is known and from relation (17) we obtain

$$Pr\{d_k = i, S_k = m, R_k / R_1^{k-1}\} =$$

$$\sum_{m'} \sum_{j=0}^{1} \gamma_i(R_k, m' m) \alpha_{k-1}^j(m'). \quad (A4)$$

The denominator can be also expressed from bit d_k and state S_k

$$P_r\{R_k / R_1^{k-1}\} = \sum_{m} \sum_{i=0}^{1} Pr\{d_k = i, S_k = m, R_k / R_1^{k-1}\} \quad (A5)$$

and from relation (A4), we can write :

$$Pr\{R_k / R_1^{k-1}\} = \sum_{m} \sum_{m'} \sum_{i=0}^{1} \sum_{j=0}^{1} \gamma_i(R_k, m' m) \alpha_{k-1}^j(m'). \quad (A6)$$

Finally probability $\alpha_k^i(m)$ can be expressed from probability $\alpha_{k-1}^i(m)$ by the following relation

$$\alpha_k^i(m) = \frac{\sum_{m'} \sum_{j=0}^{1} \gamma_i(R_k, m' m) \alpha_{k-1}^j(m')}{\sum_{m} \sum_{m'} \sum_{i=0}^{1} \sum_{j=0}^{1} \gamma_i(R_k, m' m) \alpha_{k-1}^j(m')}. \quad (A7)$$

In the same way, probability $\beta_k(m)$ can be recursively calculated from probability $\beta_{k+1}(m)$. From relation (16), we have

$$\beta_k(m) = \frac{P_r\{R_{k+1}^N / S_k = m\}}{P_r\{R_{k+1}^N / R_1^k\}} =$$

$$\frac{\sum_{m'} \sum_{i=0}^{1} P_r\{d_{k+1} = i, S_{k+1} = m,' R_{k+2}^N, R_{k+1} / S_k = m\}}{P_r\{R_{k+1}^N / R_1^k\}}. \quad (A8)$$

By using BAYES rule, the numerator is equal to

$$P_r\{R_{k+1}^N / S_k = m\} = \sum_{m'} \sum_{i=0}^{1} P_r\{R_{k+2}^N / S_{k+1} = m'\}$$

$$P_r\{d_{k+1} = i, S_{k+1} = m,' R_{k+1} / S_k = m\}. \quad (A9)$$

By taking into account expressions of $\gamma_i(R_{k+1},m,m')$ and $\beta_{k+1}(m')$, we can write

$$\beta_k(m) = \frac{\sum_{m'} \sum_{i=0}^{1} \gamma_i(R_{k+1}, m, m') \beta_{k+1}(m')}{P_r\{R_{k+1} / R_1^k\}}. \quad (A10)$$

In substituting k by $(k+1)$ in relation (A6), the denominator of (A10) is equal to

$$Pr\{R_{k+1} / R_1^k\} = \sum_{m} \sum_{m'} \sum_{i=0}^{1} \sum_{j=0}^{1} \gamma_i(R_{k+1}, m' m) \alpha_k^j(m'). \quad (A11)$$

Finally probability $\beta_k(m)$can be expressed from probability $\beta_{k+1}(m')$, by the following relation

$$\beta_k(m) = \frac{\sum_{m'} \sum_{i=0}^{1} \gamma_i(R_{k+1}, m, m') \beta_{k+1}(m')}{\sum_{m} \sum_{m'} \sum_{i=0}^{1} \sum_{j=0}^{1} \gamma_i(R_{k+1}, m' m) \alpha_k^j(m')}. \quad (A12)$$

REFERENCES

[1] L.R. Bahl, J. Cocke, F. Jeinek and J. Raviv, "Optimal decoding of linear codes for minimizing symbol error rate", *IEEE Trans. Inform. Theory*, vol. IT-20, pp. 248-287, March 1974.

[2] C. Berrou, P. Adde, E. Angui and S. Faudeil, "A low complexity soft-output Viterbi decoder architecture", *to appear at ICC' 93.*

Chapter 6

WIRELESS LINK AND SYSTEM SIMULATION

This chapter provides easy-to-read papers that cover the fundamentals of simulation for wireless communications links and system design.

Reprinted from *IEEE Communications Magazine*, Vol. 32, No. 7, July 1994, pp. 26-35.

Simulation of Communication Systems

When both a complex system and a complex channel model are encountered, the result is typically a design or analysis problem that cannot be solved using traditional (pencil and paper) mathematical analysis. Computer-aided techniques, which usually involve some level of numerical simulation, can be a very valuable tool in these situations.

William H. Tranter and Kurt L. Kosbar

Over the past decade considerable attention has been paid to the development of computer-aided design and analysis tools that can be applied to communication systems. There are several reasons for this. Today's communication systems are much more complex than those of several decades ago. In addition, many systems operate in environments where the channel is not adequately described by a simple additive Gaussian noise model. The effects of severe bandlimiting, adjacent-channel interference, multipath, nonlinearities, and a host of other degrading effects must now be considered. When both a complex system and a complex channel model are encountered, the result is typically a design or analysis problem that cannot be solved using traditional (pencil and paper) mathematical analysis. Computer-aided techniques, which usually involve some level of numerical simulation, can be a very valuable tool in these situations. The purpose of this article is to provide a tutorial review of some of the basic techniques of communication system simulation.

Another reason for the current interest in simulation and computer-aided techniques is the widespread availability of powerful computers. These tools are currently within reach of most communication engineers and it is now possible to perform system-level simulations of complex systems at one's desk. The graphics capabilities of modern personal computers and workstations, together with laser printers, allows output to be generated in a readily usable form. These capabilities have been available for a relatively short time.

Both traditional mathematical analysis and computer simulation are based on a system model, which is typically a block diagram that describes the interconnection of the various subsystems comprising the overall system. Each functional block or subsystem is described by a signal processing operation that defines the subsystem input-output relationship. The accuracy of either the mathematical analysis or the computer simulation is dependent upon the accuracy of the system model. Thus, each and every approximation made in developing a system model must be well understood.

Computer simulation has the same goal as conventional mathematical analysis — to determine the operating characteristics and performance of a communication system. Link-level simulations typically focus on the performance measures of a communication link. Typical performance measures include the time required to initialize a link, the length of time a link can be sustained, the signal-to-noise ratio (SNR) of the recovered message in analog systems, and the symbol error rate for digital systems. Despite these similar goals, simulation often differs from mathematical analysis in a fundamental way. Simulation typically focuses on performance *estimation* while mathematical analysis nearly always involves performance *calculation*. The result of a traditional mathematical analysis is a number, while the result of a simulation is typically a random variable. This is an important distinction.

There are basically two different classes of problems that can be addressed using simulation: the transient characteristics and the steady-state characteristics of a system. The time-to-lock of a PLL used as a bit synchronizer is a typical transient characteristic. Transient characteristics are usually determined using a simulation of the specific sub-system of interest rather than using a simulation of the system as a whole. When one uses simulation to determine the performance characteristics of a system, then the entire system, including the environment in which the system operates, must be included in the simulation. Performance measures are typically steady-state characteristics. Examples are the bit error rate, mean-square error, and signal-to-noise ratios. Link-level simulation allows these problems to be addressed for arbitrarily complex systems.

Simulation should never be viewed as a substitute for mathematical analysis. Some level of analysis is necessary if one is to establish that the simulation is working correctly and that the simulation results are reasonable. This is the area of validation, which will be addressed further. Simulation, when properly used, goes hand-in-hand with traditional analysis methods. Simulation results often allow us to

WILLIAM H. TRANTER is the Schlumberger Professor of Electrical Engineering at the University of Missouri-Rolla.

KURT L. KOSBAR is an assistant professor of electrical Engineering at the University of Missouri-Rolla.

identify the most important parameters in a system and also help identify those system parameters that can be neglected. In other words, simulation results often guide analysis, since a properly developed simulation provides insights into system behavior.

While simulation is a powerful tool for both design and analysis, new problems are created when one turns to simulation. Since the continuous-time waveforms present in the system must be represented by discrete-time samples in the simulation, the waveforms must be sampled so that aliasing errors are reduced to acceptable levels. Engineering judgments are necessary for even this simple problem. The reduction of aliasing errors to negligible levels requires high sampling frequencies. High sampling frequencies in turn result in large simulation run times, which is clearly not desirable. Thus, an obvious tradeoff exists. Another problem is that the analog filters that may be present in the actual system under study must be represented by digital equivalents in the simulation. These digital equivalents always involve approximations whose nature should be understood if the simulation user is to have complete confidence in the simulation results.

A comprehensive survey of the techniques used for the simulation of communication systems would fill a rather large book [1]. In this section we will briefly consider the basic techniques used to represent signals, generate signals, and model linear systems, nonlinear systems, and time-varying systems within a simulation. We then consider the important problem of using a simulation to estimate the performance of a communication system.

Signal and System Modeling

System-level simulations can be based on time-domain techniques, frequency-domain techniques, or on a combination of these techniques. In this section we focus on the problems associated with representing time-domain signals, and modeling systems, in a digital simulation of a communication system.

Signals and Complex Envelopes

Both lowpass signals and bandpass signals are usually present in a communication system. Lowpass signals are typically information bearing signals prior to modulation and bandpass signals typically represent modulated carriers at various points in the system, such as transmitter outputs and receiver inputs. Both lowpass and bandpass signals must be represented by discrete-time sequences within the simulation. The analog signals actually present in many parts of a communications system must obviously be sampled to form the discrete-time sequences processed by the simulation. These sampled sequences must accurately specify the corresponding analog waveform if an accurate simulation is to result.

In order for sample sequences to accurately specify the analog waveforms from which the samples are formed, the sampling frequency f_s must exceed twice the highest frequency in the waveform being sampled [2]. There are a number of factors that influence the choice of the sampling frequency. Among these factors are aliasing errors, frequency warping in digital filters, and the presence of nonlinearities. Computational constraints also affect the choice of the sampling frequency. Since the simulation program must process each sample, an excessive number of samples used to represent a given waveform leads to excessive computer time requirements. We are therefore rewarded by selecting the lowest possible sampling frequency that still results in an accurate simulation. An understanding of the tradeoff between simulation accuracy and the simulation sampling frequency is important. This is usually accomplished after a simulation is developed by varying the sampling frequency and observing the changes that result in the simulation outputs.

The desire to minimize the simulation sampling frequency points us toward using signals in the simulation having lowpass-type spectra. Lowpass signals present no problem, they are sampled directly using an appropriate sampling frequency. Bandpass signals can also be directly sampled but are usually represented by equivalent lowpass signals in order to reduce the number of samples necessary to represent the signal. The complex envelope representation allows us to accomplish this.

A general modulated signal, having carrier frequency f_c is usually written in the form

$$x(t) = R(t)\cos[2\pi f_c t + \phi(t)] \tag{1}$$

where $R(t)$ represents the real envelope of $x(t)$ and $\phi(t)$ represents the phase deviation. Equation (1) can be placed in the form

$$x(t) = \mathrm{Re}\{R(t)e^{j\phi(t)}e^{j2\pi f_c t}\} \tag{2}$$

or

$$x(t) = \mathrm{Re}\{\tilde{x}(t)e^{j2\pi f_c t}\} \tag{3}$$

where the quantity $\tilde{x}(t)$ is called the complex envelope of the real signal $x(t)$. Clearly

$$\tilde{x}(t) = R(t)e^{j\phi(t)} \tag{4}$$

is a complex function of time that is independent of the carrier frequency f_c. It is important to note that the complex envelope involves signals that are usually slowly varying with respect to the carrier frequency. Since the bandwidth of a bandpass signal is usually small compared to f_c, it takes a much lower sampling frequency to represent the complex envelope, $\tilde{x}(t)$, than to represent the real-time signal $x(t)$. The result is a smaller number of samples for a given time segment of $x(t)$. The complex envelope is usually expressed in rectangular form

$$\tilde{x}(t) = x_d(t) + j\,x_q(t) \tag{5}$$

where $x_d(t)$ is the direct (or real) component of $\tilde{x}(t)$ and $x_q(t)$ is the quadrature (or imaginary) component of $\tilde{x}(t)$ [3].

Assuming that the carrier frequency is *known*, the complex envelope contains *all* of the information contained in the original signal $x(t)$. As shown by Eq. 3, $x(t)$ can be reconstructed from $\tilde{x}(t)$ by multiplying $\tilde{x}(t)$ by $e^{j2\pi f_c t}$ and taking the real part.

Signal Generation

Both deterministic and random signals exist in almost all communication systems. Models must be developed for each of these signal types that can be implemented in a digital computer simulation. Deterministic signals are usually generated using the defining equation for the signal. Equation (1), with $R(t)$ and $\phi(t)$ properly specified to represent the signal of interest, is an example. The only other concern is the choice

B*oth lowpass signals and bandpass signals must be represented by discrete-time sequences within the simulation.*

A *PN sequence generator is usually envisioned as a linear binary shift register. The characteristics of the generator are established by the feedback taps.*

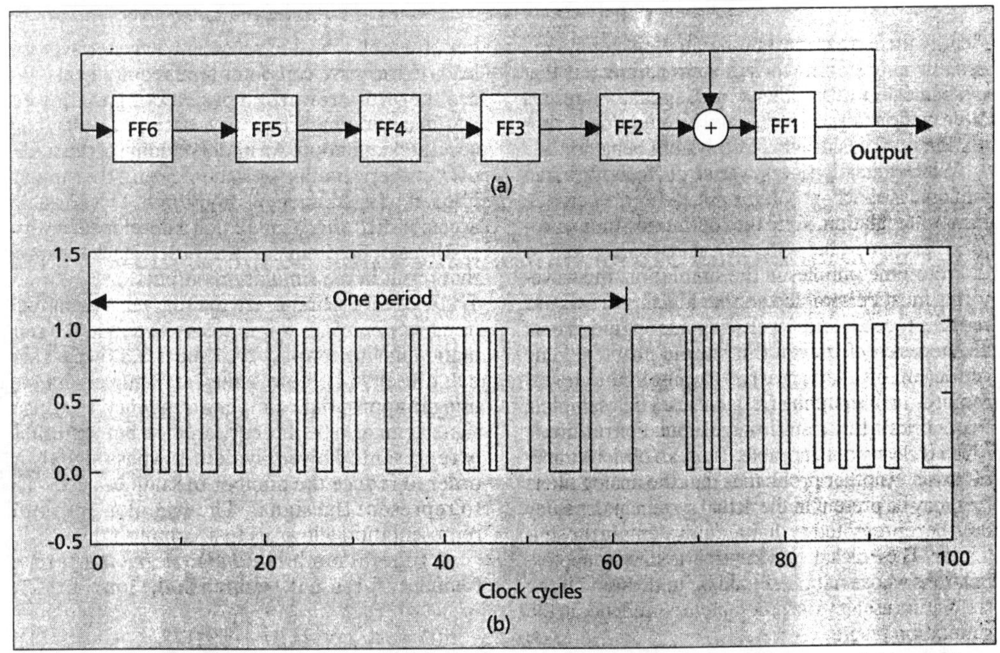

■ **Figure 1.** *PN sequence generation: a)implementator of a PN sequence generator for* m = 6; *b) resulting waveform for a given seed.*

of sampling frequency, as discussed previously.

Random signals are usually generated using either a linear congruential algorithm or a PN sequence algorithm. Although the mathematical descriptions of these two algorithms are somewhat different, they are essentially equivalent. Since a digital computer is a finite-state machine, it is not possible to generate a truly random signal on a computer and all computer-generated sequences are periodic. We are content to generate a pseudo-random sequence, which is in reality a periodic deterministic signal with a long period. Within a period the pseudo-random sequence approximates many of the properties of a random signal. We are therefore able to generate "noise-like" waveforms for use in a simulation to represent both random signals and noise, thus the term pseudo-noise (PN) sequences.

A linear congruential algorithm is defined by the expression

$$x[n+1] = (a\,x[n]+c) \bmod m \qquad (6)$$

where m is the modulus, a is the multiplier, and c is the increment. In order to improve the speed at which samples are generated, we usually set $c = 0$. The initial value of the sequence, $x[0]$, is known as the seed number of the process. Once the seed is specified, the remaining values of the sequence are specified through Eq. 6 The problem is to determine the parameters a, c, and m so that the generator defined by Eq. 6 has a sufficiently long period for the application of interest. Although the theory for accomplishing this task is well understood [4, 5] the question of what really makes a good random sequence generator, and the determination of efficient algorithms for sequence generation, still constitute an active area of research [6]. Although all simulation packages available today — suitable for communication system simulation — contain random sequence generators, the user should ensure that the operation of these generators is well understood and appropriate to the problem being investigated.

A PN sequence generator is usually envisioned as a linear binary shift register as shown in Fig. 1a. The characteristics of the generator are established by the feedback taps. The taps are defined by a polynomial and the generator achieves a maximum period of $2^m - 1$ if the polynomial is primitive [5]. For the generator shown in Figure 1a, the feedback connections are defined by

$$g(X) = 1 + X + X^6 \qquad (7)$$

which indicates feedback to the first and the sixth stages of the shift register. The register is initially placed in some state, equivalent to a seed number, and it then cycles through all possible states with period 63. This is the maximum period of $2^m - 1$ possible since the polynomial in Eq.7 is primitive and $m = 6$. The corresponding waveform is shown in Fig. 1b.

Models for Linear Systems

A model for a linear system, suitable for implementation on a digital computer is usually determined from the transfer function of the system $H(f)$, or the unit impulse response $h(t)$. If the transfer function $H(f)$, is for a lowpass type system, a computer model is easily determined directly from $H(f)$ using one of the standard digital filter synthesis techniques that map a transfer function into an equivalent digital filter. Perhaps the most popular synthesis techniques are those that yield impulse-invariant, step-invariant, and bi-linear z-transform filters [2]. All of these synthesis techniques involve approximations and it is important that the approximations be understood if the simulation user is to have confidence in the simulation result.

If, however, the starting point for a filter design is not a transfer function but an amplitude response mask, one can usually develop a linear-phase filter satisfying the requirements of the amplitude response mask. Frequency sampling filters, or finite-duration impulse response (FIR) filters based on the Parks-McClellan synthesis technique, are often useful [2]. If one is to simulate a filter with an arbitrary amplitude and phase response, it is often necessary to take frequency samples of both the desired amplitude and phase response. These samples can then be inverse transformed using the FFT to obtain the unit-pulse response $h[n]$. The input sequence can then be convolved with $h[n]$ to form the filter output. As an alternative, block FFT processing can be used. In block FFT processing the input sequence, $x[n]$, is divided into blocks of appropriate size. These blocks are then Fourier transformed using the FFT, multiplied by the filter transfer function (samples of the amplitude and phase response), and then inverse transformed to obtain the output samples. The overlap-save method [2] is typically used for these applications since $h[n]$ is a short sequence compared to $x[n]$.

We saw previously that complex envelope signal representations are generally used for bandpass signals. If the system is a bandpass system, the unit-impulse response of the system will be a bandpass signal. As such, the unit-impulse response is usually represented by the complex envelope model of the bandpass system, defined by

$$\tilde{h}(t) = h_d(t) + j\, h_q(t) \tag{8}$$

The complex envelope of the system output $\tilde{y}(t)$, is the convolution of the complex envelope of the input, represented by Eq. 5, and as given by Eq. 8. This yields

$$\tilde{y}(t) = [x_d(t) + jx_q(t)] * [h_d(t) + jh_q(t)] \tag{9a}$$

where '*' denotes convolution. The preceding expression can be written

$$\tilde{y}(t) = [x_d(t)*h_d(t) - x_q(t) * h_q(t)] + j\, [x_d(t) * h_q(t) + x_q(t) * h_d(t)] \tag{9b}$$

This yields the structure shown in Fig. 2. Since the functions $h_d(t)$ and $h_q(t)$ represent lowpass signals, computer models for these signals can be realized using the same techniques described in the preceding paragraph. Two filters will be necessary, one for $h_d(t)$ and one for $h_q(t)$.

Many of the linear systems used in a communication system involve a filtering operation. Filters, of course, have memory so that past input or output samples are used in forming the current system output. Because of this structure, filtering is computationally expensive compared to many of the other signal processing operations involved in simulation. Efficient filtering routines are therefore essential elements in any simulation program.

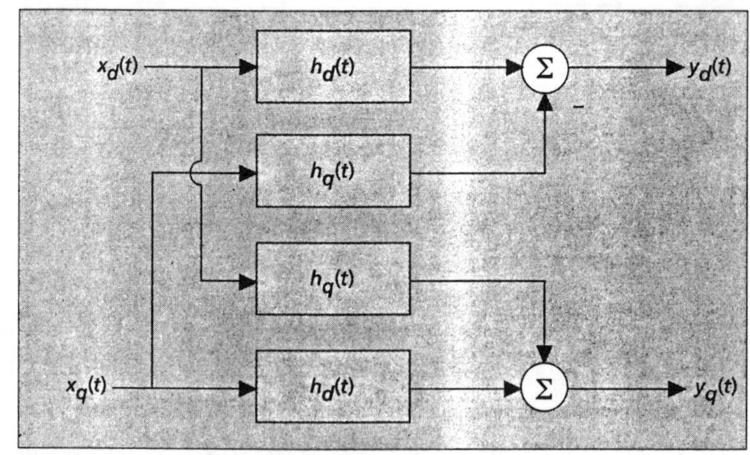

■ **Figure 2.** *Complex envelope representation of bandpass linear system.*

Models for Nonlinear and Time-Varying Systems

Nonlinear and time-varying systems present special difficulties when bandpass models for these systems are needed. While complex envelope models exist for linear, time-varying systems there is no guarantee that a complex envelope model exists for systems that are both nonlinear and time-varying. One must rely on approximation methods to model these devices.

Little can be said about the most general class of time-varying, nonlinear systems. The only method that ensures that these systems can be accurately modeled is to translate the complex envelope back to a bandpass signal and pass it though an appropriate device model. To develop more computationally efficient models, one must make assumptions about the device. In some cases one has a linear, but time-varying element. The model shown in Fig. 3 can then be used to represent the system. This is essentially a transversal filter with time-varying coefficients.

There are also a variety of models for nonlinear but time invariant systems. A well-known example is the Volterra Series expansion [7]. Unfortunately this expansion is computationally expensive and therefore rarely used. There is a special class of nonlinear devices that have very short, or no memory. In a true memoryless device, such as a square-law device, the output is only a function of the current input. If a simple sinusoid is placed into these devices, the output will have terms only at the harmonics of the input frequency and if a bandpass filter follows the memoryless nonlinearity, all but the first harmonic term can be removed. Thus a sinusoidal input produces a sinusoidal output, where the amplitude of the output may be a nonlinear function of the ampli-

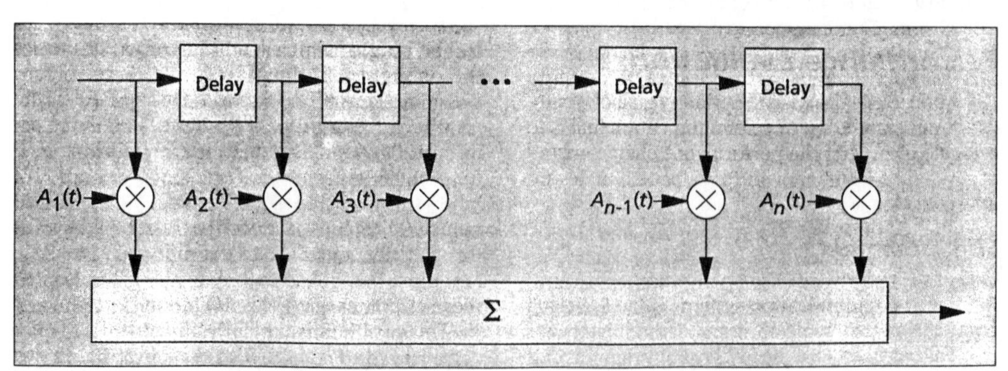

■ **Figure 3.** *Time-varying linear system model.*

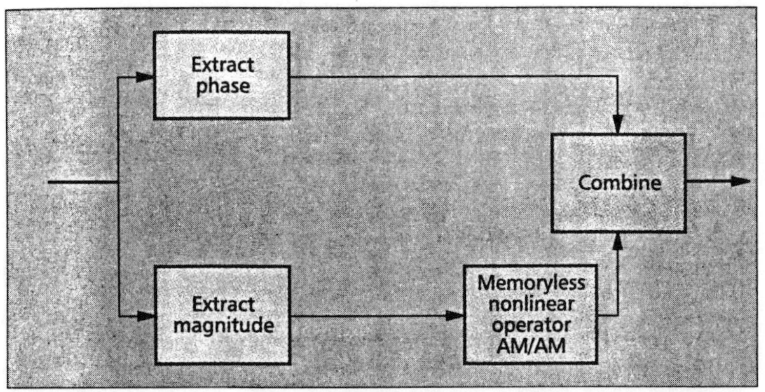

■ **Figure 4.** *AM/AM model for strictly memoryless bandpass nonlinearities.*

tude of the input. This type of device lends itself well to the complex envelope representation. As shown in Fig. 4, one merely needs to decompose the complex envelope into a magnitude and phase component, pass the magnitude through a non-linear device, and recombine it with the unaltered phase term.

Another class of interesting systems have "short" memory, i.e., the time constant of the nonlinearity is long with respect to the carrier frequency but short with respect to the message waveform. These systems can be called complex envelope memoryless systems, or envelope nonlinearities [8]. This is because the complex envelope of the output can be approximated by a memoryless, but nonlinear, function of the complex envelope of the input. Saleh [9] showed that the traveling-wave tube microwave amplifier fits this description, and the complex envelope representation of this device is shown in Fig. 5. As with the truly memoryless nonlinearity, the complex envelope of the input is decomposed into its amplitude and phase. The amplitude is both passed through a nonlinear device and used to alter the phase of the signal. If the input to the (assumed memoryless) nonlinearity is

$$x(t) = A(t)\cos[2\pi f_c t + \phi(t)] \tag{10a}$$

the output is represented by

$$y(t) = f[A(t)]\cos\{2\pi f_c t + g[A(t)] + \phi(t)\} \tag{10b}$$

The function $f[A(t)]$ is known as the AM-to-AM conversion characteristic and $g[A(t)]$ is known as the AM-to-PM conversion characteristic. For a constant envelope $x(t)$, $A(t)$ is a constant and thus $f[A(t)]$ and $g[A(t)]$ are constants. This explains the interest in constant envelope modulation techniques.

Performance Evaluation

As previously discussed, a primary goal of a computer simulation of a communication link is to evaluate or predict the performance characteristics of a system. A number of performance estimates are now considered.

SNR Estimation

One of the most widely used performance measures for analog communication systems is the SNR at a point in a system, typically at the demodulator output. The calculation of the SNR usually requires that the waveform of interest (the test waveform) be compared to a "desired" or "ideal" waveform at that point. This desired waveform is often chosen to be an amplitude-scaled and time-delayed version of the information-bearing waveform since amplitude scaling and time delay do not contribute to waveform distortion. The test waveform is then compared to the desired waveform and that portion of the test waveform that is orthogonal to the desired waveform is defined as noise. For this case the SNR estimate becomes

$$\widehat{SNR} = \frac{\rho^2}{1-\rho^2} \tag{11}$$

where ρ is the correlation coefficient between the test and desired waveforms [1].

Simulation is used to establish the test waveform for the system under study. As a simple example, if the complex envelope of the test waveform is

$$\tilde{y}(t) = Ae^{j\theta}\tilde{x}(t-\tau) + n(t) \tag{12}$$

the SNR is A^2P_x/P_n where P_x and P_n are the signal and noise powers, respectively. In most applications, the values of A, θ, τ, P_x, and P_n must be estimated before the SNR can be determined. Simulation can assist in this undertaking.

Symbol Error Rate Estimation and Monte Carlo Simulation

In digital communication systems the probability of demodulation error P_e, is typically the prime performance measure. For simplicity we will only consider binary communication systems and refer to P_e as the bit-error rate, or BER. The techniques discussed here can typically be extended to include M-ary communication systems.

The Monte Carlo (MC) method is a widely known technique for estimating the BER of a communication system [1, 10]. This method is based on the relative frequency definition of probability. A simulation is first developed that closely replicates the behavior of the system under study. The simulation will include pseudorandom data and noise sources, along with models of the devices that process the waveforms present in the system. A number of symbols are then processed by the simulation, and the experimental BER is estimated as the number of errors divided by the total number of symbols processed by the simulation. In most systems, this sample BER will be a consistent and unbiased estimate of the true BER. MC simulation is an intuitively pleasing approach that can be applied to virtually any system. It also has the side benefit of generating signals that very closely replicate the signals present in the system under study. This can be a significant advantage for validation of the simulation. If the MC estimate is consistent and unbiased, it will converge to the true BER as the number of demodulated symbols approaches infinity. Obviously simulations can only process a finite number of symbols. This raises the question of how accurate is a MC BER estimate after a finite number of symbols have been processed? To answer this question, one needs a definition of reliability, and must be able to calculate the reliability of these estimates using this definition.

This problem is typically addressed by using confidence intervals [11]. A simulation result is a sample estimate of the BER, $\hat{P}_e$, and we wish to know the true BER, P_e. To apply confidence intervals, one must be able to map $\hat{P}_e$ to an interval of the real line, $[P_L, P_H]$. This interval is a 100α percent confidence interval if

$$\Pr[P_L < P_e < P_H] > \alpha \text{ for all } P_e.$$

This mapping is in general very difficult to obtain and is not unique. Fortunately, for MC simulations with independent and identically distributed errors, there are well-known methods for finding the confidence interval. The 99 percent confidence interval for a simulation that has a $\hat{P}_e$ of 10^{-6} is shown in Fig 6. The important measure for this type of simulation is not the number of bits processed by the simulation, but the number of errors observed. A rule of thumb is that after one error has been observed, the 99 percent confidence interval covers approximately three orders of magnitude. One can also state, with 99 percent confidence, that after 10 errors, the estimated BER is within a factor of 2 of P_e, and after 100 errors the estimated BER is within a factor of 1.3 of P_e. To emphasize the drawback of MC simulation, the horizontal axis of Fig. 6 has been labeled in years of computer execution time, assuming the simulation will process one symbol per second of CPU time. For complex systems having low error rates, the Monte Carlo approach may require a considerable investment of processing time if accurate BER estimates are required.

The shortcomings of the Monte Carlo approach have been recognized for some time, and considerable research has been performed to find faster methods of BER estimation. These approaches are typically called variance-reduction techniques. Although a wide variety of techniques have been investigated, all share a common theme; by making additional assumptions about the system architecture and signal sources, one can reduce the number of symbols required to generate an estimate of a particular accuracy. Generally, the greater the number of assumptions used in the simulation, the greater the reduction in simulation execution time. While there is virtually no reward in this area for those who do not understand system behavior, there are tremendous rewards for those who can skillfully apply these techniques. There are also tremendous dangers for those who misapply the techniques.

Semi-Analytic Analysis

The semi-analytic (SA) approach places substantial demands on the analyst and system architecture, but the reward is an incredibly fast simulation [1, 10]. This approach can be described by first reviewing the simple communication system shown in Fig 7. This is obviously an analytically tractable system. The decision metric will be a Gaussian random variable, with a known mean and variance. One can calculate the probability of demodulation error, and there is no need to perform a simulation.

A more interesting, and less tractable, system is shown in Fig. 8. The transmitter now has a nonlinear power amplifier. The channel adds white Gaussian noise to the signal and passes the result through a linear filter that introduces intersymbol interference (ISI). The nonlinear amplifier and ISI cause the decision metric to be decidedly non-Gaussian, making the BER difficult to calculate. However, it is not difficult to show that the decision metric is *conditionally* Gaussian. If one specifies the transmitted data pattern, the decision metric will be Gaussian, with a mean that is a function only of the data pattern and a variance that is only a function of the noise level. The BER calculation can now be decomposed into three parts: determining the variance of the decision metric, determining the conditional mean of the decision metric, and calculating the BER by using the total probability theorem.

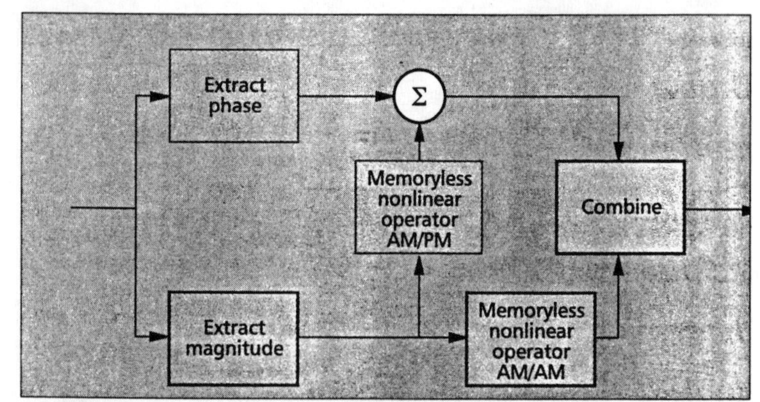

■ **Figure 5.** *AM/AM, AM/PM model for complex envelope memoryless bandpass nonlinearities.*

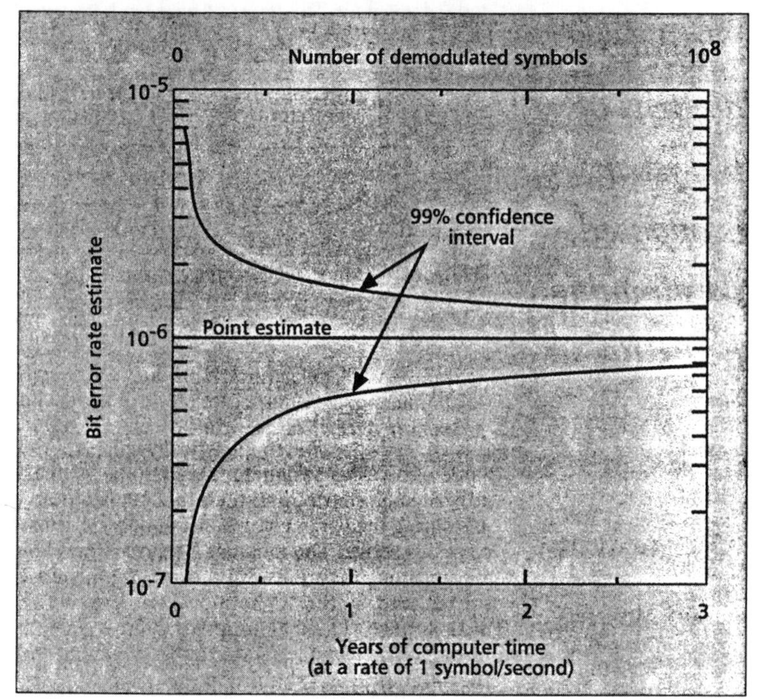

■ **Figure 6.** *Point and interval BER estimates.*

Assume the bandpass filter has a impulse response (or memory) that is n data symbols long. By total probability, the BER of this system is

$$P_e = \frac{1}{2^n}\sum_{i=1}^{2^n} Q\left(\frac{\left|E[X_i] - T\right|}{\sigma_X}\right) \qquad (12)$$

where each value of i corresponds to one of the 2^n possible data patterns, $E[X_i]$ is the mean of the decision metric of the ith data pattern, σ_X is the variance of the decision metric, T is the threshold value, and $Q[x]$ is the familiar integral

Unlike *most other simulation techniques, the semi-analytic approach calculates the BER of the system, as opposed to estimating the BER.*

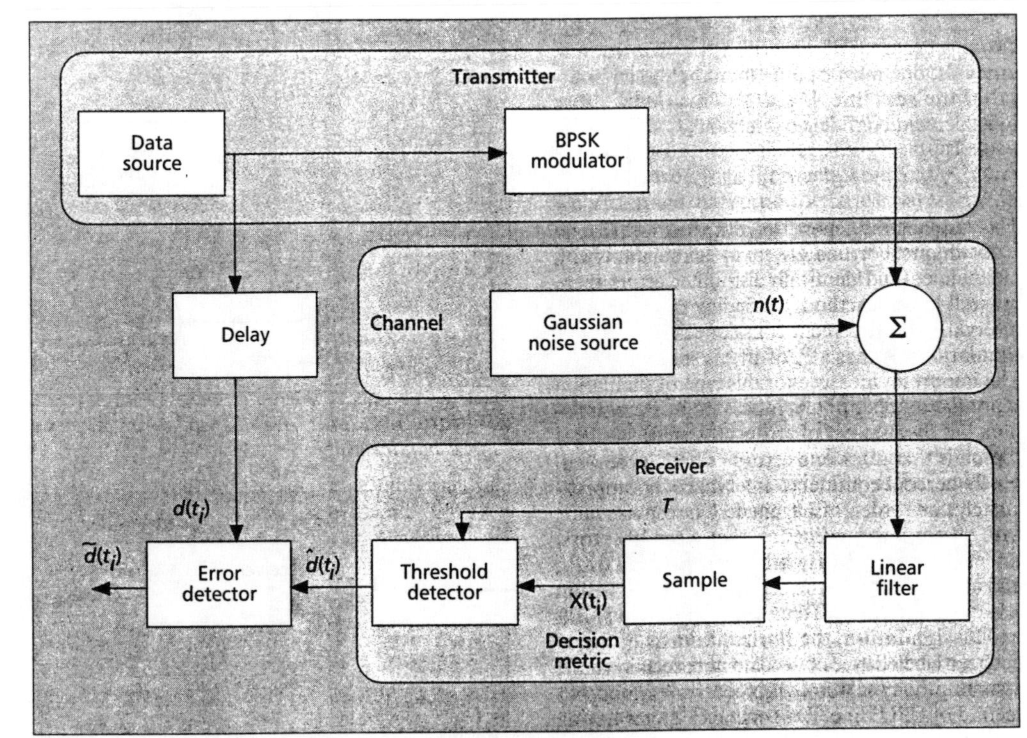

■ **Figure 7.** *Analytically tractable communication system model.*

$$Q(x)=\frac{1}{\sqrt{2\pi}}\int_x^{\infty}\exp\left(\frac{-\alpha^2}{2}\right)d\alpha . \qquad (13)$$

It is typically possible to analytically calculate the value of σ_X. This parameter may also be estimated by disabling the transmitter in the simulation, and measuring the variance of X_i when only the noise source is active. The mean of X_i can be found by reconnecting the transmitter, disabling the noise source, and using a PN sequence generator as a data source. The PN generator should cycle through all 2^n possible data patterns of length n, and the value of $E[X_i]$ should be recorded for each pattern. The BER for the system can then be calculated by inserting this data into Eq. 12. The SA approach can be used whenever one can calculate the BER of the system given the transmitted data pattern. It is most frequently used when the noise is additive and Gaussian, and the system is linear from the point of noise injection to the point where decisions are made.

Unlike most other simulation techniques, the semi-analytic approach *calculates* the BER of the system, as opposed to *estimating* the BER. It makes very efficient use of the computer resources, and once one has performed the simulation and stored the mean and variance data, they can easily calculate the BER for any SNR. The BER can therefore be determined for a range of system noise levels with a single simulation. Given all these advantages, one expects to find a significant disadvantage. The disadvantage is that one must be able to calculate the error rate of the system conditioned on the transmitted data pattern. Notice that this is typically difficult or impossible when the noise is non-Gaussian, the noise and data are correlated, the noise is not additive, the noise is non-stationary, or when there are nonlinearities after the insertion of the noise. While there is a class of systems where semi-analytic simulations are useful, there is still a need for efficient simulation techniques that place fewer demands on the system architecture.

Importance Sampling

One technique that has received considerable attention in the literature is the modified Monte Carlo, or importance sampling, (IS) technique [1, 10]. When using importance sampling, the statistics of the noise sources in the system are biased in some manner so that errors (i.e., the important events) occur with greater probability, thereby reducing the required execution time. An MC simulation is run using the biased noise source. It is possible to unbias the BER estimate of this modified simulation by applying

$$\hat{P}_{e,IS}=\frac{1}{N}\sum_{i=1}^{N}\frac{f_n(n_i)}{f_{\tilde{n}}(n_i)}I(n_i) \qquad (14)$$

Where f_n is the pdf of the original noise source, $f_{\tilde{n}}$ is the pdf of the biased noise source, n_i is a particular noise vector and $I()$ is an indicator function that is one when an error occurs and zero when the correct symbol is demodulated. This leads to the hope that after a fixed number of demodulated symbols, the IS BER estimate will be more accurate than a conventional MC BER estimate. One can show that there is virtually no limit on how much one can gain, or lose, by using IS. If an analyst is sufficiently clever to select a good IS biasing scheme for a given system, an accurate estimate of the BER can be obtained with very short computer runs. If a poor biasing scheme is selected the BER estimate may even converge at a slower rate than the MC estimate. Many different biasing methods have been suggested in the literature and, before using IS, one should ascertain if a particular biasing scheme will produce an improvement, or a degradation, over a conventional MC simulation.

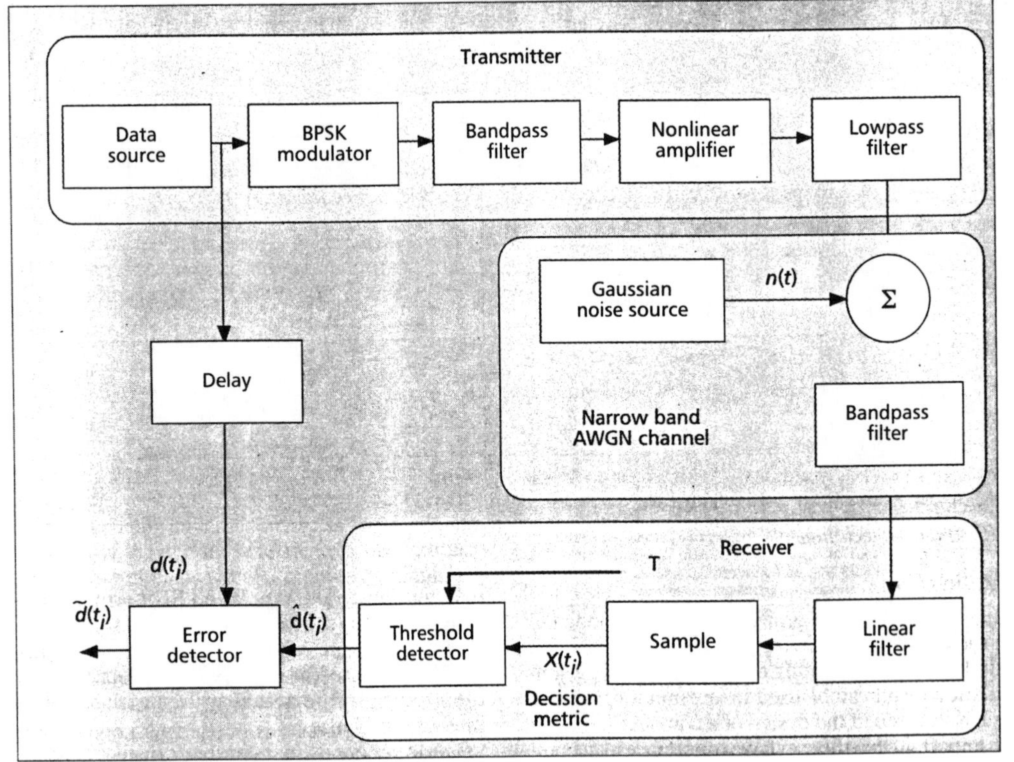

■ **Figure 8.** *Analytically tedious communication system model.*

Tail Extrapolation

The BER estimation problem is essentially a numerical integration problem. The BER of a given system is the area under the tail of an unknown probability density function (pdf). One can assume that the pdf belongs to a particular class and then perform a curve fit to the observed data. This should identify a highly likely pdf, from which one can generate a BER estimate. This is the concept behind tail extrapolation. In these simulations, one sets multiple thresholds as shown in Fig. 9. A normal MC simulation is executed, and the number of times the decision metric exceeds each threshold is recorded. A broad class of pdfs is then identified. One class that is often useful is the general exponential class

$$f_{\upsilon,\sigma,m}(x)=\frac{\upsilon}{\sqrt{8}\sigma\Gamma\left(\frac{1}{\upsilon}\right)}\exp\left(-\left|\frac{x-m}{\sqrt{2}\sigma}\right|^{\upsilon}\right) \quad (15)$$

The parameters available in this class (ν, σ and μ) are then adjusted to find the pdf that best fits the available data. The BER can be estimated by numerically evaluating the integral of the pdf for the actual threshold used in the system. It is not always clear which class of pdfs should be used for this simulation method, or how the thresholds should be chosen. As with importance sampling, in most cases it is not possible to generate a confidence interval that describes the accuracy of the BER estimate [1, 10].

Other Variance Reduction Methods

There is no shortage of techniques that can be applied to the BER estimation problem. Extreme value theory is useful for some systems and research is currently being performed on large deviation techniques. These, and many other approaches all face the same fundamental problem. One must make assumptions concerning the behavior of an analytically intractable system and then exploit the assumptions to reduce the simulation execution time. A particular technique is useful only when the design engineer can clearly identify the assumptions that were made in the analysis, and verify that the assumptions apply to the system under study. The engineer will also need to verify the accuracy of the estimate produced by the simulation. While this is often straightforward for MC simulations, it can be a much more difficult problem for advanced BER estimation techniques. Table I discusses the concerns and advantages for some of the more common simulation techniques.

Simulation of Coded Systems

Coded communication links, especially those with large coding gains, may have such low error rates that Monte Carlo techniques, and even some of the variance reduction techniques mentioned above, cannot provide accurate BER estimates with reasonable simulation execution times. Often the only feasible approach to evaluating the performance of these systems is to determine the "raw" error rate of the symbols passed through the channel. Coding theory approximations and bounds can then be used to estimate the end-to-end performance of the coded system. Caution must be exercised when using this approach since a good understanding of coding theory is necessary. In addition, a perturbation analysis should be performed to determine how small changes in the estimated uncoded error rate will influence the calculated coded probability of error. For systems with large coding gains, very small errors in the estimated BER of the uncoded system can result in unacceptably large BER estimates for the coded system.

The BER estimation problem is essentially a numerical integration problem. The BER of a given system is the area under the tail of an unknown probability density function.

Technique \ Concern	Flexibility	Analytical difficulties	Computer execution times	Accuracy of BER estimate
Monte Carlo	Applicable to any system	None	Often prohibitively long	Measurable using confidence intervals
Variance reduction techniques Importance sampling Tail extrapolation Large deviation theory Extreme value theory Others	Applicable to many systems	Must select biasing schemes, threşholds, etc. Often more of an art than a science.	Variable, from extremely short to prohibitively long	Often very difficult to access
Semi-analytic simulation	Restricted largely to systems with linear receivers and AWGN channels	Need conditional BER statistics	Very short execution times	Exact to within accuracy of DSP models
Strictly mathematical analysis	Typically restricted to highly idealized systems	Numerous and well known	Zero	Exact

■ **Table 1.** *Summary of BER estimation techniques.*

Validation

Large simulation programs are often developed to produce a reasonably simple result, such as the BER under various operating conditions. Before a simulation result can be used in any meaningful way, such as a step in the design of a complex system, it is important that the user have confidence in the simulation result. There are many reasons why a simulation result may be inaccurate, such as insufficient data to form an accurate BER estimate (discussed above), conceptual errors such as modeling inaccuracies, and software bugs. Validation of a large and complex simulation program is an important, although sometimes difficult, undertaking.

Validation of a large program does not mean a line-by-line review of the source code. This type of evaluation is time consuming, error prone, and in most cases impractical. Such a review is also unlikely to reveal conceptual errors that were made when the software models were developed. When functions are supplied by outside vendors, source code may not be supplied, making line-by-line evaluation impossible. However there are a number of techniques that can be used with reasonable success.

As with any large system, the performance of each subsystem should be evaluated before it is integrated into the simulation. While this is a necessary step, it is not sufficient to guarantee the correct operation of the final simulation. System level tests are needed to validate the overall design. When MC simulations are used, one can compare signals at selected "test-points" in the simulation with the corresponding test-points in the hardware design. This may involve plotting the time domain waveform from the simulation and comparing it to an oscilloscope trace, or may involve calculating statistics of a signal such as a histogram, mean, variance, or power spectral density. Unfortunately, some of the more advanced BER estimation techniques do not produce these intermediate signals. Perturbation analysis can be helpful when validating a simulation. One can sometimes make a few changes to the simulation, and reduce the system to one that is analytically tractable. These changes are often minor from a software standpoint, such as temporarily replacing nonlinear amplifiers with linear amplifiers, or eliminating synchronization errors by passing allowing the transmitter and receiver to share a common time base. The simulation BER estimate can then be compared to a theoretical result. When these values agree, one can gain confidence that a significant portion of the simulation is correct. One can then return to the actual system with a higher level of confidence.

Analytic bounds on system performance and simulation results complement each other. The bounds can give assurance that the simulation results are reasonable, and the simulation results can guide analysis, indicating the tightness of various bounds. Since simulation packages are becoming increasingly common, powerful, and easy to use, it may now be reasonable to have redundant simulation efforts. Two separate development teams working with different simulation packages are unlikely to make the same coding and implementation errors. Even if a single team and package is used, it is helpful to use more than one simulation approach. For example, MC simulations have long execution times, but do not suffer from some of the problems that more advanced techniques face. It is helpful to write an MC simulation, and occasionally check the performance of an advanced technique with the results of an MC simulation.

Summary: Developing A Simulation

Simulation is a useful tool for the design and analysis of communication links. Indeed, for complex systems, such as are common today, some level of simulation is often essential if insights into system behavior and performance predictions are to be made. The usual steps in developing and using such a simulation are as follows.

The first step is to develop a model of the system under study. This model often takes the form of a block diagram that defines the individual subsystems that make up the overall communication system. It is important to identify the approximations made in forming the system model. The important parameters of each subsystem must be identified so that they are carried through to the simulation.

The second step is to identify the signal processing operation necessary to define each of the subsystems in the overall communication system. At this point mathematical models for each subsystem are introduced. Thus, a choice is made concerning which signals are to be represented using complex envelope techniques. The strategy to use for representing analog filters by digital equivalents is also selected. One must appreciate the additional approximations incurred in this step.

The next step is to define the simulation products, which is the set of outputs required from the simulation. Examples are displays of the time-domain waveforms or the power-spectral density at a point in the system. If a performance prediction is to be made, such as the bit error rate for the overall communication system, the method to be used for estimating the performance must be selected. We have seen that a number of techniques may be applied to this important problem and that these techniques range from the Monte-Carlo method, which weights all errors equally and makes no assumption about the form of the decision metric, to more complex estimation schemes which do make assumptions about the decision metric. Recall that this decision allows one to expect a tradeoff between prior knowledge and computer execution time.

At this point the structure of the simulation is known and we can move to software. If a dedicated simulation language is to be used, one now selects models from the model library to implement the various subsystems in the overall communication system. One also selects a strategy for performance evaluation and this determines the estimation routines to be used in the simulation. Other simulation products, such as time-domain waveforms, spectra, and histograms are directed to a postprocessor that provides the tools for processing and displaying the data generated by a simulation. If one is developing code for a custom simulation, the previously selected signal processing and estimation strategies determine the code to be developed. After the simulation code has been developed and executed, one must ensure that the simulation results are reasonable. As previously discussed, this is the important area of validation.

In conclusion, it should be pointed out that for extremely complex systems, it is usually desirable to start out with the simplest model that incorporates only the essential features of the system under study. Simulations based on simple models are easier to verify and errors are more easily identified. The simulation can then be enhanced to include other interesting and important features of the communication system under study.

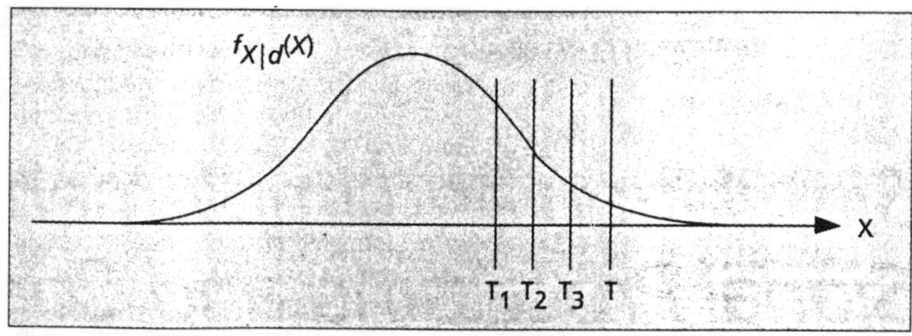

■ **Figure 9.** *Setting multiple thresholds for trail extrapolation.*

References

[1] M. Jeruchim, P. Balaban and K. Shanmugan, Simulation of Communication Systems, (New York, NY: Plenum Press, 1992).

[2] A. Oppenheim, and R. Schafer, Discrete-Time Signal Processing, (Englewood Cliffs, NY, Prentice-Hall, 1989).

[3] S. Haykin, Communication Systems, (New York, NY: Wiley, 1983).

[4] R. Rubinstein, Simulation and the Monte Carlo Method, (New York, NY: Wiley, 1981).

[5] P. Bratley, B. L. Fox and L. E. Schrage, *A Guide to Simulation* Second Edition, (Springer-Verlag, 1987).

[6] K. Park and K. W. Miller, "Random Number Generators: Good Ones are Hard to Find," *Comm. of the ACM,* vol. 31, no. 10, Oct. 1988.

[7] V. Volterra, Theory of Functionals and of Integral and Integro-Differential Equations, (Dover, New York, NY: 1959).

[8] T. T. Ha, Digital Satellite Communications, (Macmillan, New York, NY: 1986).

[9] A. A. M. Saleh, "Frequency-Independent and Frequency-Dependent Nonlinear Models of TWT Amplifiers," *IEEE Trans. on Comm.*, vol. COM-29, no. 11, Nov. 1981, pp. 1715-1720.

[10] M. Jeruchim, "Techniques for Estimating the Bit Error Rate in the Simulation of Digital Communication Systems, *IEEE JSAC*, Vol SAC-2, Jan. 1984, pp. 153-171.

[11] L. Bain and M Engelhardt, Introduction to Probability and Mathematical Statistics, (Boston, MA: Kent Pub. Co., 1987).

Biographies

WILLIAM H. TRANTER [F '85] received B.S.E.E., M.S.E.E., and Ph.D. degrees from the University of Alabama in 1964, 1965, and 1970, respectively. He joined the faculty of the University of Missouri-Rolla in 1969. His research interests are digital signal processing and computer-aided design of communication systems. His research efforts have been funded by NASA, the National Science Foundation, and G.E. Aerospace. He has recently consulted for the Education Development Corporation, Battelle Memorial Institute, and McDonnell-Douglas. He has authored numerous technical papers and is a co-author of two textbooks: *Principles of Communications* (Boston, MA: Houghton-Mifflin) and *Signals and Systems* (New York: Macmillan). From 1980 to 1985 he served as associate dean of engineering and was named Schlumberger Professor of Electrical Engineering in 1985. He is a member of Sigma Xi, Eta Kappa Nu, Tau Beta Pi, and ASEE. He currently serves as Editor-in-Chief of the *IEEE Journal on Selected Areas in Communications*.

KURT L. KOSBAR [M '88] received a B.S. in electrical and computer engineering from Oakland University, Michigan in 1982, M.S. and Ph.D. degrees in electrical engineering from the University of Southern California in 1984 and 1988 respectively. From 1982 to 1988 he was a member of the technical staff at Hughes Aircraft Company in the Space and Communications Group, where he spent the majority of this time in the Advanced Digital Signal Processing Department, designing and analyzing spacecraft-based digital demodulation systems for a variety of commercial and military applications. Since 1988, he has been an assistant professor of electrical engineering at the University of Missouri-Rolla. His primary research interests are synchronization, link level simulation of communication systems, spread spectrum communications, and adaptive equalization. His research has been supported by General Electric Aerospace, IEEE, the U.S. Air Force, the Engineering Foundation, and the National Science Foundation. He is a member of Tau Beta, Pi, Sigma Xi, Eta Kappa Nu and Phi Kappa Phi.

Reprinted from *IEEE Transactions on Vehicular Technology*, Vol. 40, No. 4, Nov. 1991, pp. 731-740.

Simulation of Bit Error Performance of FSK, BPSK, and $\pi/4$ DQPSK in Flat Fading Indoor Radio Channels Using a Measurement-Based Channel Model

Theodore S. Rappaport, *Senior Member, IEEE,* and Victor Fung, *Student Member, IEEE*

***Abstract*—As demand grows for digital wireless communication systems, the accurate prediction of average and instantaneous bit error rates (BER) for different modulation schemes will become increasingly important in system design. BER predictions will not only provide an understanding of the performance of each modulation method in the operating environment, but will also reveal the limits of data rate and channel capacity. This research uses a statistical multipath channel simulator, Simulation of Indoor Radio Channel Impulse Response Models (SIRCIM) [1], to generate realistic channel responses over local areas for both wide-band and narrow-band (CW) radio transmissions inside open plan buildings, and predicts BER for FSK, BPSK, and $\pi/4$ DQPSK modulation techniques in flat fading channels through computer simulation. The small-scale channel model, the communication system models used in the analysis, and the methods used to predict BER, are described in this paper. In fact, one of the goals of this work is to present a straightforward simulation methodology that can be readily used. The channel simulator and the systems models have been thoroughly tested, and results from average and instantaneous BER simulations are shown. The BER performances of various modulation techniques in indoor flat fading channels are presented, and it is found that BPSK offers between a 2.8 and 3.0 dB improvement over $\pi/4$ DQPSK, although the latter offers a 3 dB increase in capacity for a given spectrum allocation.**

I. Introduction

FOR low data rate systems, the error bursts due to reception during the signal nulls of a flat fading channel are a primary concern, and understanding the burstiness of the channel is necessary to implement successful antenna diversity or coding techniques. With accurate bit error rate (BER) computer simulations in flat fading environments, it becomes possible to test digital radio communication systems by using a simple baseband digital hardware BER simulator between the data source and sink as shown in Fig. 1. Digital hardware BER simulators, which operate on an applied digital data stream, are commercially available and can be programmed in real time to alter the BER. The techniques presented in this paper can be used to simulate the modulation, filtering, propagation, and detection, and lend themselves directly to the control of hardware digital simulators. The computer simulation can be run off-line to produce the instantaneous bit error rates that are applied to the hardware BER simulator in real time [3].

This paper presents the results of a simulation study that provides insight into the simulation methodology and BER performance of various modulation techniques in flat fading channels inside open plan buildings. This study uses a detailed measurement-based propagation channel model, SIRCIM (simulation of indoor radio channel impulse response models) which generates over 1000 closely spaced baseband equivalent complex impulse responses for a mobile radio operating at 1.3 GHz and traveling over a small area (a 1 m path). Each multipath component amplitude of the impulse response is correlated with multipath component amplitudes at constant excess delays and small receiver separations, and at small time delay differences for the same receiver location as described in [1], [4]. From the wideband impulse responses, a CW signal level for each impulse response is found by performing the phasor addition of individual multipath components which arrive at a receiver antenna. Thus, unlike most channel modeling techniques (i.e., [2], [5]) and simulation techniques (i.e., [6], [7]) which assume various time varying impulse responses which have no implicit correlation on the exact spatial location of a receiver within a small area, our study uses a flat fading channel model which is derived from complex impulse response models which produce multipath fading that is highly representative of indoor mobile channels within a local area. Also, unlike the multichannel hardware simulator developed by Arnold and Bodtmann [8], and the propagation simulators [9], [10] used in the European GSM group's field tests which require the users to define the propagation multipath profiles, our simulator generates the propagation multipath profiles statistically using the measurement-based model.

Manuscript received August 16, 1990; revised December 11, 1990 and May 9, 1991. This work was supported by a grant from the Computer Integrated Design, Manufacturing and Automation Center, Purdue University, West Lafayette, IN, by the Mobile and Portable Radio Research Group (MPRG) Industrial Affiliates Program of Virginia Polytechnic Institute and State University, and by Virginia's Center for Innovative Technology.

T. S. Rappaport is with the Mobile and Portable Radio Research Group, Bradley Department of Electrical Engineering, Virginia Polytechnic Institute and State University, Blacksburg, VA 24061.

V. Fung was with the Mobile and Portable Radio Research Group, Bradley Department of Electrical Engineering, Virginia Polytechnic Institute and State University, Blacksburg, VA. He is now with BNR, 1150 East Arapaho Road, Richardson, TX 75031-2328.

IEEE Log Number 9103231.

0018-9545/91$01.00

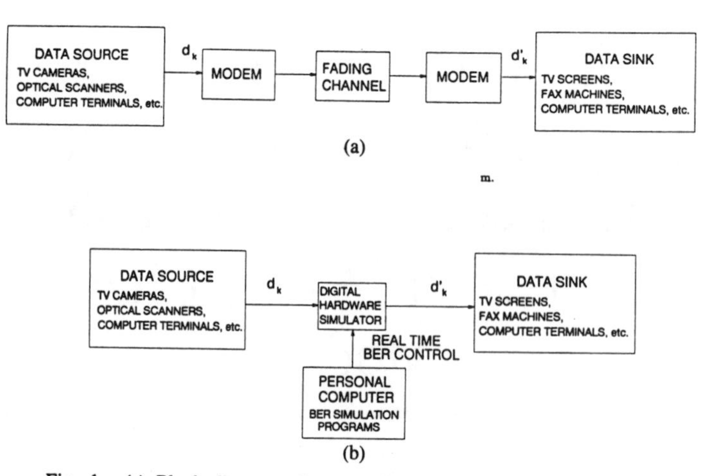

Fig. 1. (a) Block diagram of actual digital communication system. (b) Block diagram of a system using a baseband digital hardware simulator with software simulation as a driver for real time BER control.

As described subsequently, this study focuses on the BER prediction, through computer simulation, of FSK, BPSK, and $\pi/4$ DQPSK modulation in small-scale flat fading channels. We use a technique given in [6] to normalize the large scale fading effects so that only small-scale fading is considered. Our study, unlike [6], only considers flat fading channels and thus is limited in application to cases where data rates are small (on the order of 25 kb/s or less for most indoor channels) and the irreducible error rate due to intersymbol interference (ISI) is considered to be approximately known. Nevertheless, the techniques which are presented here may be extended to frequency selective fading channels since the channel characteristics are derived from wide-band impulse responses.

First, SIRCIM is described and shown to be a realistic channel modeling tool. Section III gives an overview of the BER simulation technique. Section IV describes FSK, BPSK, and $\pi/4$ DQPSK modulation techniques. Section V describes the bounded binomial sampling method that is used to ensure accuracy of simulation results. The simulation results for average BER are given in Section VI, and instantaneous BER simulation is discussed in Section VII. Section VIII summarizes the results of the research.

II. Testing of the Propagation Simulator: SIRCIM

SIRCIM is an indoor UHF multipath radio channel software simulator designed from over 50 000 wide-band and narrow-band measurements taken at five different open plan building environments. Statistics from other measurements in different buildings have recently been included. SIRCIM recreates the statistics of measured wide-band impulse responses in both line-of-sight (LOS) and obstructed (OBS) topographies, and synthesizes the phases of individual multipath components by assuming that a limited number of scatterers (determined from statistics of measured data) are randomly placed along an aisle within a building. Also, large scale path loss is generated from statistical models which are based on measurements reported in [1] and [11]. Statistical models for the number of multipath components, the variation of the multipath components, the amplitudes, phases and fading statistics of individual multipath components, and the excess delays are all based on measurements reported in [1], [11]. SIRCIM recreates the narrow-band flat fading characteristics seen by a moving receiver by vectorially summing the received signals due to the arriving multipath components, and by deterministically tracking the phase change of each multipath component as the receiver moves. Doppler shifts are included in the phase of each multipath component.

In this paper, we first study SIRCIM's reliability in producing realistic flat fading characteristics over local areas. Then, the small-scale fading distributions from SIRCIM are compared to measured data, both in terms of statistical distributions and in terms of visual similarity. Also, the signal level-crossing rate of simulated CW envelopes are compared with theoretical Rayleigh fading channels.

The average BER results for various modulation schemes presented in this paper are based on the channel characteristics of 50 simulated measurement locations and the narrow-band flat fading characteristics seen by a mobile at each location. At each location, the mobile is assumed to move along a 1 m track at a specified velocity. Along each track, a CW fading envelope is generated by SIRCIM. There are 1279 equally spaced impulse responses, and thus 1279 equally spaced CW signal strength values generated in each of the 50 CW envelopes. Twenty five of those 50 simulated channels are in LOS topography and 25 of them are in OBS topography. The angle between the line of motion of the receiver and the line drawn between transmitter and receiver is assumed to be randomly distributed from 0° to 360°. The transmitter receiver (T-R) separation is assumed to be randomly distributed for each measurement location, ranging from 2.3 to 65 m for LOS and 2.3 to 55 m for OBS. Although the effects of large scale path loss are not considered here, they may easily be incorporated for outage or co-channel interference prediction. For this study, the speed of the moving receiver is assumed to be 0.25 m/s for all simulated channels, although this may be varied in simulation. SIRCIM assumes that the scatters are aligned along aisles located to each side of the transmitter, and the width between these aisles in our analysis is assumed to be randomly distributed between 3 to 10 m for each simulation [1], [4].

As an example of SIRCIM's output, two typical spatially varying LOS channel multipath delay profiles (square magnitude impulse responses–phase is not shown for brevity) and their corresponding CW envelopes (labeled as LOS 1 and LOS 2) displayed about the median are shown in Figs. 2 and 3, respectively. Only 19 equally spaced impulse responses are shown in the figures for presentation. The two cumulative distributions of the CW envelopes about the median resulting from the impulse responses are shown in Figs. 4 and 5. Figs. 6 and 7 show two typical spatially varying OBS channel power delay profiles and their corresponding CW envelopes (labeled as OBS 1 and OBS 2). The cumulative distributions of the CW envelopes in the corresponding OBS fading channels are shown in Figs. 8 and 9. These are very realistic representations of field measured CW envelopes reported in [12].

In order to determine the overall CW fading characteristics of the 50 simulated channels, the cumulative distributions (about the local medians) of the 25 LOS and 25 OBS simulated channels were concatenated. The resulting distribution is shown in Fig. 10. It can be seen that the distribution is primarily Rayleigh above the median and tends to log-normal below the median, which is in excellent agreement with the field measured data reported in [12].

Another test of the accuracy of SIRCIM was to compare the level-crossing rates of the simulated CW envelopes with theoretical results. The level-crossing rate is a second order statistic which characterizes the frequency of a fade falling below a certain level and is important in the estimation of BER because it shows the percentage of time the signal is below the median. The resulting normalized level-crossing rate as a function of ρ (i.e., signal level/rms signal level) is shown in Fig. 11. A theoretical normalized level-crossing rate for Rayleigh fading channels is also shown in the figure for comparison. It can be seen that our result follows closely with the theoretical result and both maximum rates occur at the same level, $\rho = -3$ dB. Note that for a particular measurement location, SIRCIM may provide fading statistics far different from Rayleigh (see Figs. 2 and 4). Measurements in [12] show that particular locations have CW fading statistics which are not strictly Rayleigh.

Both the first-order cumulative distribution test and the second-order level-crossing rate test showed good comparison to measurement and theory, which indicate SIRCIM is a useful tool for BER prediction in flat fading channels.

III. Communication System Model

A general block diagram of the communication system model is shown in Fig. 12. All bandpass signals and channels described in this paper are represented by low pass complex envelope forms. Referring to Fig. 12, a pseudorandom binary bit stream, $d(t)$, is sent through the simulator program for BER simulation. The binary bit stream $d(t)$ is stored in computer memory and compared with the received bits $d'(t)$ at the output of the simulated receiver for determination of bit errors.

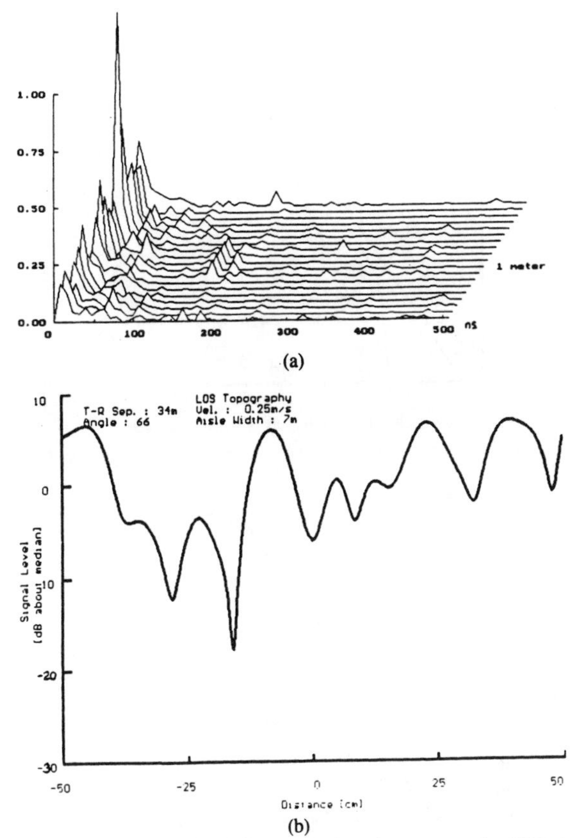

Fig. 2. (a) Example of spatially varying impulse response for LOS channel (LOS 1). (b) Simulated CW fading based on wide-band model of LOS 1.

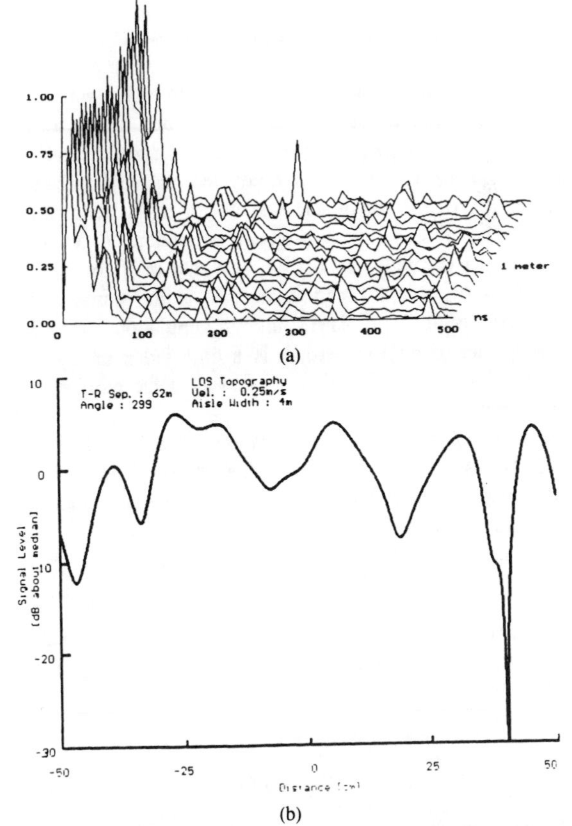

Fig. 3. (a) Example of spatially varying impulse response for LOS channel (LOS 2). (b) Simulated CW fading based on wide-band model of LOS 2.

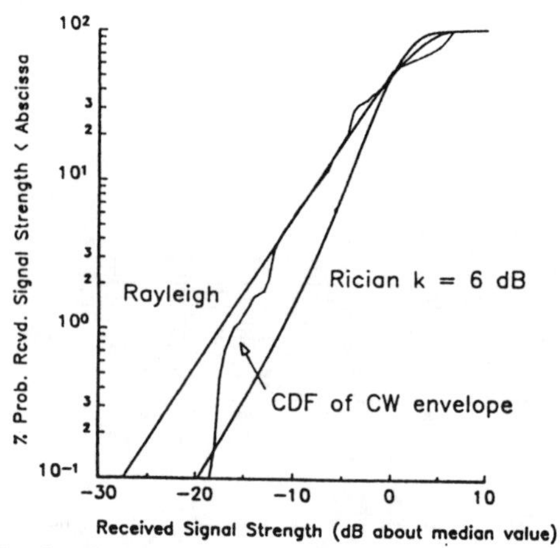

Fig. 4. Cumulative distribution of CW envelope about median of LOS 1.

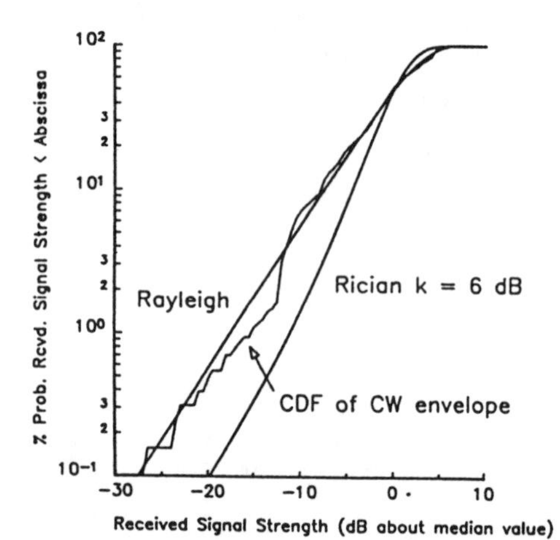

Fig. 5. Cumulative distribution of CW envelope about median of LOS 2.

Let R_l be the lth generated pseudorandom binary number which has equal probability of being 0 or 1, then $d(t)$ is a bit stream of $R_0 \cdots R_l, R_{l+1} \cdots$. In the computer simulations, the binary bit stream is converted to the baseband representation of the modulated waveform, $x(t)$, according to each modulation technique discussed in Section IV. $x(t)$ is discretized for computer simulation at a rate of $T_s = T/N$, where T is the symbol period and N is the number of samples per symbol. N is 13 in our simulation. $x(t)$ is represented as $x(kT_s)$, where k is a time index and $k = 0, 1, \cdots, N-1, \cdots, (L \cdot N) - 1$, and L is the number of symbols sent.

The 50 simulated CW fading envelopes described in Section II are used as the channels. The baseband equivalent impulse response of a flat fading channel is a delta function which has a fading amplitude

$$h_b(kT_s) = \alpha\delta(kT_s) \tag{1}$$

where α is the multiplicative factor. Assume that the fade remains constant for one symbol interval T (this is a valid assumption for flat fading channels), then the output signal of the channel is

$$\begin{aligned} y(kT_s) &= x(kT_s) * h_b(kT_s) + n(kT_s) \\ &= \alpha x(kT_s) + n(kT_s). \end{aligned} \tag{2}$$

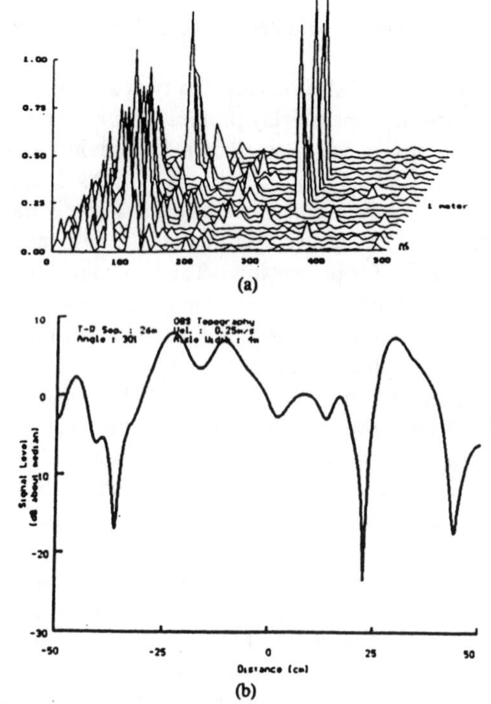

Fig. 6. (a) Example of spatially varying impulse response for OBS channel (OBS 1). (b) Simulated CW fading based on wide-band model of OBS 1.

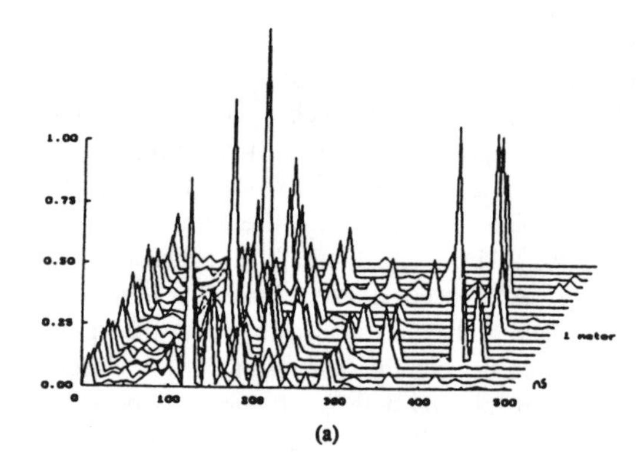

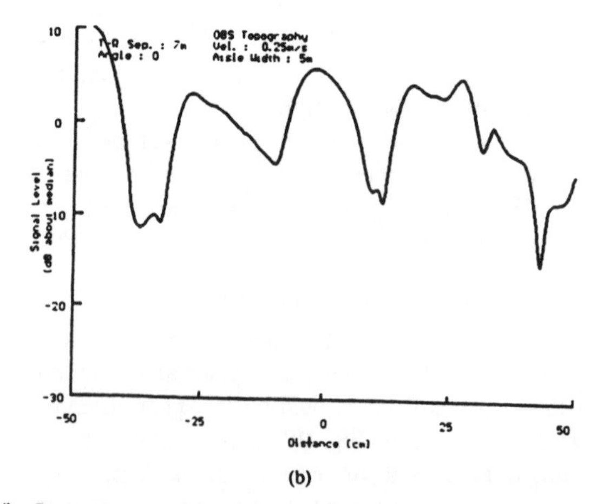

Fig. 7. (a) Example of spatially varying impulse response for OBS channel (OBS 2). (b) Simulated CW fading based on wide-band model of OBS 2.

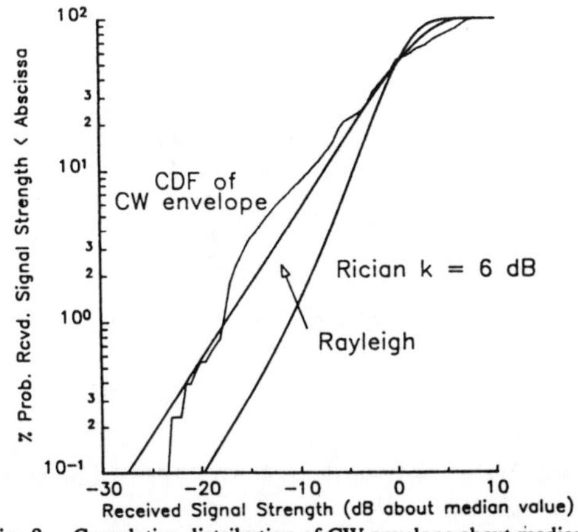

Fig. 8. Cumulative distribution of CW envelope about median of OBS 1.

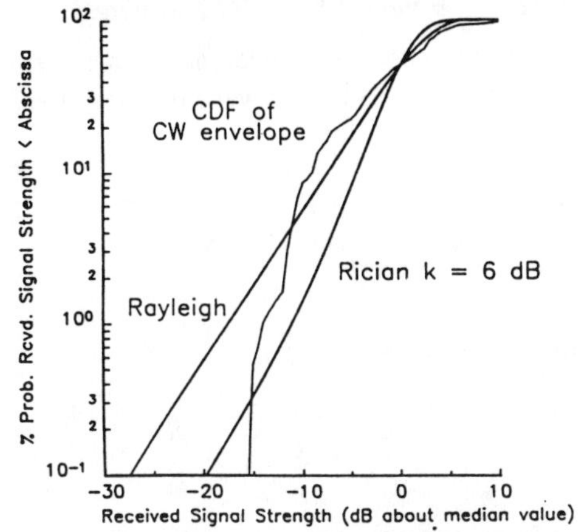

Fig. 9. Cumulative distribution of CW envelope about median of OBS 2.

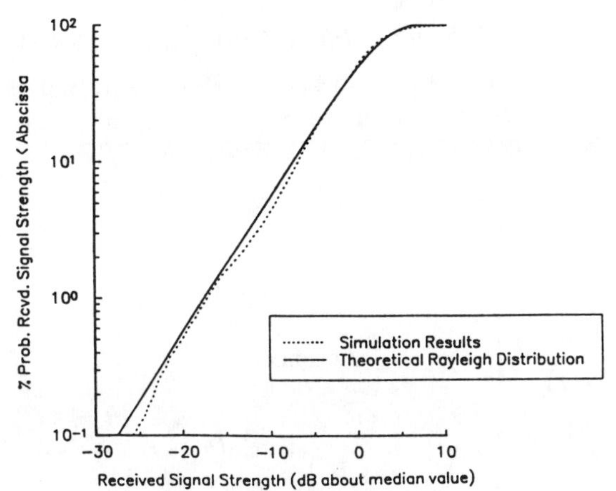

Fig. 10. Cumulative distribution of CW envelope about the median for the 50 simulated CW fading channels.

In this work, perfect phase recovery (for coherent matched filter) and time synchronization are assumed.

A. Implementation of Envelope Detection

An envelope detector is used for noncoherent detection in our systems. The input signal into a noncoherent detector is expressed as

$$y'(t) = \begin{cases} x_c(t) + n_c(t) & \text{in-phase channel} \\ x_s(t) + n_s(t) & \text{quadrature channel} \end{cases} \tag{3}$$

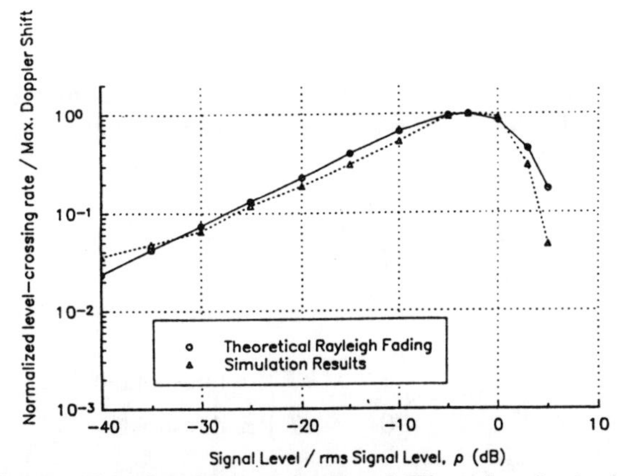

Fig. 11. Normalized level crossing rate of CW envelope for the fifty simulated CW fading along with theoretical normalized level crossing rate for Rayleigh fading channels.

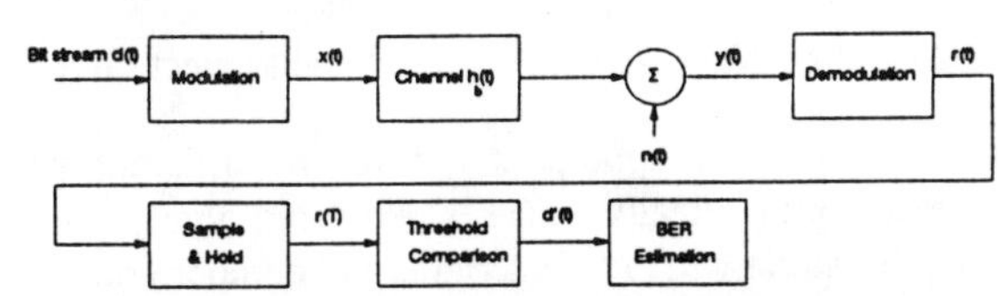

Fig. 12. Block diagram of a communication system model.

where $n_c(t)$ and $n_s(t)$ are jointly Gaussian zero mean wide-sense stationary processes [15], $x_c(t)$ and $x_s(t)$ are the in-phase (I) and quadrature (Q) components of the complex envelope $x(t)$, respectively, and $n_c(t)$ and $n_s(t)$ are the I and Q components of the bandpass noise complex envelope $n(t)$, respectively. The output of the envelope detector is given by

$$r(kT_s) = \sqrt{\left[x_c(kT_s) + n_c(kT_s)\right]^2 + \left[x_s(kT_s) + n_s(kT_s)\right]^2}$$
$$k = 0, 1, \cdots, N-1, \cdots, (L \cdot N) - 1. \tag{4}$$

B. Implementation of Matched Filter

A matched filter is used for coherent detection and is implemented by an integrate and dump filter in our analysis. The signal at the output of the matched filter during reception of the lth symbol is represented by

$$r(kT_s) = \sum_{m=lN}^{k} \left[x(mT_s) + n(mT_s)\right] \cdot T_s,$$
$$k = lN, lN+1, \cdots, lN+N-1. \tag{5}$$

Note that the lower limit on the summation denotes the start of the lth symbol in the receiver. Our convention is that the first symbol is denoted by $l = 0$.

C. Sample and Hold

A sample and hold unit is used to sample the signal coming out of the envelope detector or matched filter. For the case of envelope detection, the output signal is represented by (4) and each symbol decision is made in the middle of each symbol. For the lth symbol, the decision is made using the value of $r(kT_s)$ at time index $k = l \cdot N + (N - 1/2)$. The output of the matched filter is sampled at the end of each symbol period. For the lth symbol, the decision is made using the value of $r(kT_s)$ in (5) at time index $k = l \cdot N + N - 1$. The state of the received symbol $d'(t)$ is then determined by thresholding the detected signals as follows:

$$d'(t) = \sum_{l=0}^{L-1} R'_l \text{rect}\left(\frac{t - lT - \frac{T}{2}}{T}\right) \tag{6}$$

where

$$R'_l = \begin{cases} 1 & r(kT_s) > \text{threshold} \\ 0 & r(kT_s) \le \text{threshold} \end{cases} \tag{7}$$

and

$$k = \begin{cases} l \cdot N + \dfrac{N-1}{2} & \text{for noncoherent detection} \\ l \cdot N + N - 1 & \text{for coherent detection.} \end{cases} \tag{8}$$

The threshold is set appropriately levels for different modulation techniques. Finally, $d'(t)$ is compared with the original binary bit stream $d(t)$ to determine the number and position of bit errors. This technique enables one to determine the exact location and time correlation of bit errors. Details of the system model are given in [14].

IV. Modulation Schemes

A. Noncoherent Frequency Shift Keying (FSK)

Fig. 13 shows a baseband model of a noncoherent FSK system [15]. The two modulated waveforms represented at baseband are as follows:

$$x_1(t) = \sum_{l=0}^{L-1} A \cdot R''_l \text{rect}\left(\frac{t - lT - \frac{T}{2}}{T}\right) \tag{9}$$

where

$$R''_l = \begin{cases} 1, & R_l = 1 \\ 0, & R_l = 0 \end{cases} \tag{10}$$

$$x_2(t) = \sum_{l=0}^{L-1} A \cdot R''_l \text{rect}\left(\frac{t - lT - \frac{T}{2}}{T}\right) \tag{11}$$

where

$$R''_l = \begin{cases} 0, & R_l = 1 \\ 1, & R_l = 0 \end{cases} \tag{12}$$

and A is set to unity.

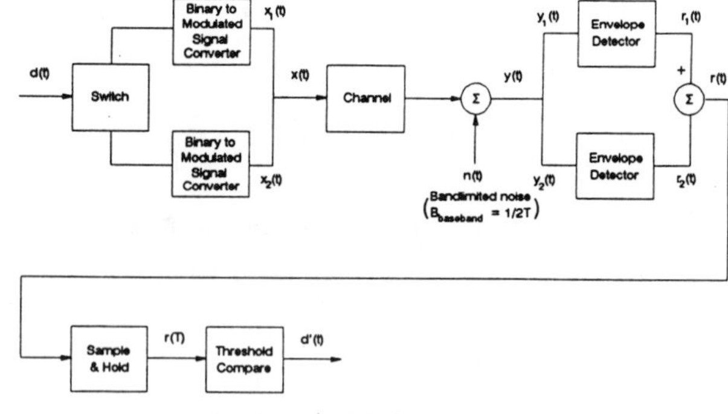

Fig. 13. A baseband model of a noncoherent FSK system.

B. Coherent Binary Phase Shift Keying (BPSK)

A baseband model of a coherent BPSK system is shown in Fig. 14 [15]. The modulated waveform is represented as

$$x_1(t) = \sum_{l=0}^{L-1} A \cdot R''_l \text{rect}\left(\frac{t - lT - \frac{T}{2}}{T}\right) \tag{13}$$

where

$$R''_l = \begin{cases} 1, & R_l = 1 \\ -1, & R_l = 0 \end{cases} \tag{14}$$

and A is set to unity.

C. $\pi/4$ Differential Phase Shift Keying ($\pi/4$ DQPSK)

Fig. 15 shows a baseband model of a $\pi/4$ DQPSK system [16]. The modulated waveform of the nth symbol represented in baseband I and Q components is given by

$$x(t) = \begin{cases} x_c(t) = \displaystyle\sum_{n=0}^{L/2-1} A \cdot \cos(\theta_n) \text{rect}\left(\frac{t - nT - \frac{T}{2}}{T}\right) \\ I \text{ channel} \\ x_s(t) = \displaystyle\sum_{n=0}^{L/2-1} A \cdot \sin(\theta_n) \text{rect}\left(\frac{t - nT - \frac{T}{2}}{T}\right) \\ Q \text{ channel} \end{cases} \tag{15}$$

where A is the amplitude and

$$\theta_n = \theta_{n-1} + \phi_n \tag{16}$$

and

$$\phi_n = \begin{cases} \pi/4, & \text{for } R_l = 1 \quad R_{l+1} = 1 \\ 3\pi/4, & \text{for } R_l = 0 \quad R_{l+1} = 1 \\ 5\pi/4, & \text{for } R_l = 0 \quad R_{l+1} = 0 \\ 7\pi/4, & \text{for } R_l = 1 \quad R_{l+1} = 0 \end{cases} \qquad \begin{matrix} n = 0, 1, \cdots, L/2 - 1 \\ l = 0, 1, \cdots, L - 1 \end{matrix}. \tag{17}$$

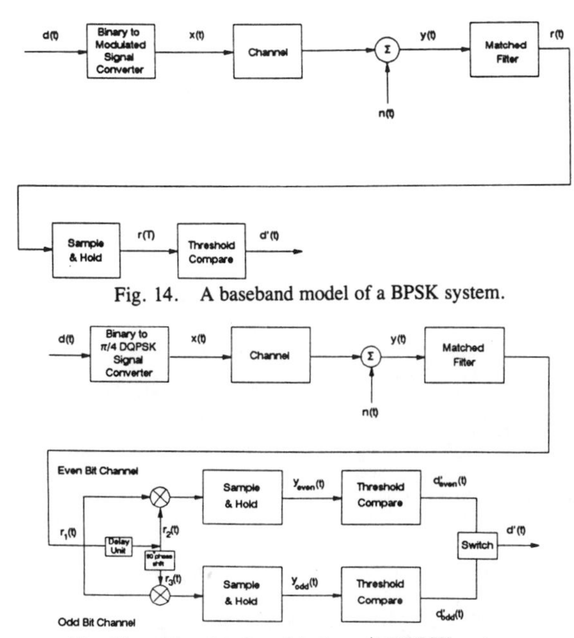

Fig. 14. A baseband model of a BPSK system.

Fig. 15. A baseband model of a $\pi/4$ DQPSK system.

θ_n and θ_{n-1} are the phases of the nth and n − 1st symbols, respectively, and A is set to unity.

Details of the simulation for all three modulation techniques are given in [14].

V. BER Calculation

We used the bounded binomial sampling method described in [13] to set confidence levels on our results. This method tells how many bits must be sent in a BER simulation from parameters provided. It also provides double boundaries in simulated BER results so that the true bit error rate lies both within certain relative precision (in terms of percentage for low bit error rates; e.g. 10^{-6}) and certain absolute precision (for high bit error rates; e.g. 10^{-2}). In our simulation, the confidence level, α, was 99%. Relative precision, δ_r, was 25% and absolution precision, δ_a', was 10^{-2}. From Fig. 1 in [13], the minimum number of bit errors (c_m) was set to 106 so that our confidence interval on BER was 25% (the exact upper and lower limit depended on the number of bit errors observed in a particular simulation run) of the median unbiased BER with 99% confidence. For an absolute lower bound on the required number of bits needed at high bit error rates, [13, fig. 2] shows the minimum number of bits (n_l) required is at least 2×10^4 for an absolute precision of 10^{-2} BER. The maximum number of bits (n_u) sent in a BER simulation depends on the BER estimated before the actual run. We assumed the BER would be on the same order of magnitude of the theoretical result in Rayleigh fading channels given in [17]. We set the maximum number of bits to approximately 10 times the bits required to produce one bit error. Our BER simulation for a particular run will terminate due to one of three possible outcomes described in [13] and the median unbiased BER estimate are calculated according to [13]. Also, a free run mode can be used to condute brute force BER estimates.

VI. BER Simulation Results

The simulation methodology was first tested in AWGN and Rayleigh fading channels, and showed excellent agreement with the theoretical results [14]. For simulation of bit errors in channels such as those shown in Figs. 2, 3, 6, and 7, the CW envelope was assumed constant over 40–80 bits at data rates near 25 kb/s. The BER results for FSK for the four typical simulated LOS and OBS channels (LOS 1, LOS 2, OBS 1 and OBS 2) described in section II are shown in Fig. 16. All four simulated channels have about the same BER performances as E_b/N_o varies from 0 to 20 dB. For E_b/N_o above 20 dB, the BER performance for LOS 2 is the worst with BER of 10^{-3} at 30 dB while OBS 2 has very small BER of 10^{-6}. In actual systems, the BER would be lower bounded by the irreducible error floor due to multipath and impulsive noise, but the comparison between four simulated channels is insightful since it shows how small-scale fading can impact BER at various locations within a building, and how the instantaneous bit error rate can differ in different channels within the same building. Such a wide variation of BER performance is realistic for mobile radio systems and cannot be derived from simple statistical channel models often used.

From inspection of CW envelopes for LOS 2 and OBS 2 in Figs. 3(b) and 7(b), it can be seen that LOS 2 has a deep fade of about 30 dB below the median while OBS 2 shows no signal with fading below 15 dB. The deep fade in LOS 2 causes bit errors even in high signal-to-noise ratios, thus explaining why LOS 2 yields a much higher BER than OBS 2 at E_b/N_o above 30 dB. The BER performances for BPSK and $\pi/4$ DQPSK in those four typical channels are shown in Figs. 17 and 18, respectively. In comparing these three modulation schemes (Figs. 16–18) on LOS 2 for E_b/N_o above 30 dB, we see that BPSK has the best BER performance and $\pi/4$ DQPSK has slightly better BER performance than FSK. However, it should be noted that $\pi/4$ DQPSK provides twice the data rate of BPSK with the given bandwidth.

Fig. 19 shows the average BER performances for the three modulation schemes (FSK, BPSK, and $\pi/4$ DQPSK) in the concatenated 50 simulated CW flat fading channels. The combined flat fading channels provide an average fading channel that might be found inside open plan buildings [12]. The BER performance for BPSK in the average channel is still the best among the three, and $\pi/4$ DQPSK performs better than FSK. Specifically, at BER = 0.01, BPSK requires an E_b/N_o = 12.1 dB, whereas for the same error performance, $\pi/4$ DQPSK requires 15.2 dB, and FSK requires 18.3 dB. The theoretical BER performance of BPSK in a Rayleigh fading channel is also shown in the figure for comparison. In the theoretical Rayleigh flat fading channel, BPSK requires an E_b/N_o of 13.8 dB at BER = 0.01. For a BER of 0.0001, BPSK in the average indoor channel requires E_b/N_o = 36.0 dB, whereas for the same error performance, $\pi/4$ DQPSK requires 39.8 dB and FSK requires in excess of 40 dB. The theoretical E_b/N_o required for BPSK in a Rayleigh fading channel is 34.1 dB. From the simulation, it appears that

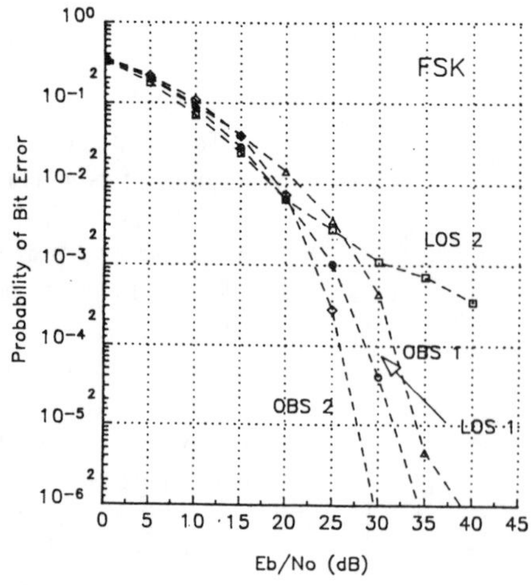

Fig. 16. Simulated BER performance for FSK system in LOS and OBS channels.

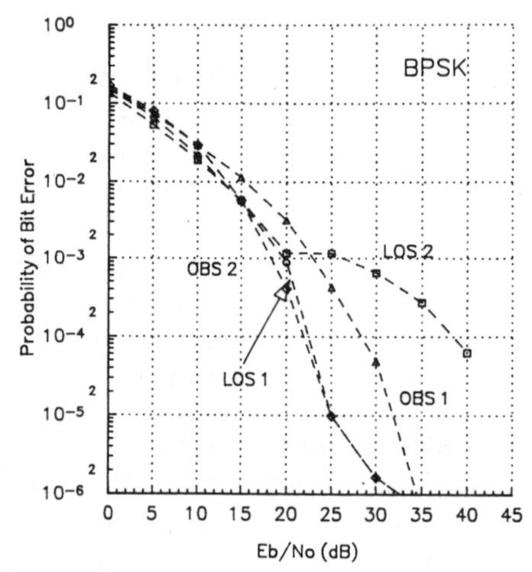

Fig. 17. Simulated BER performance for BPSK system in LOS and OBS channels.

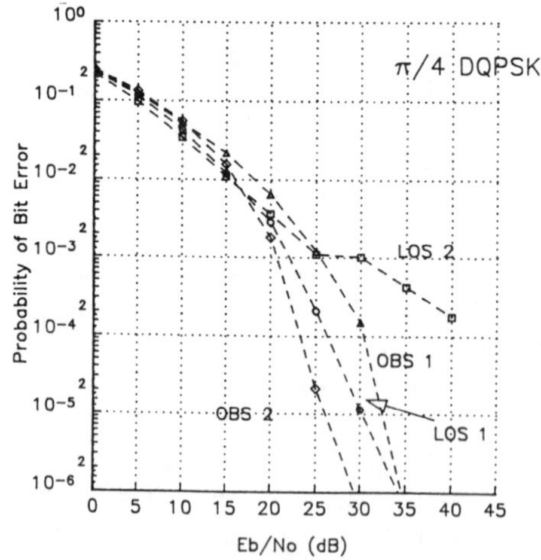

Fig. 18. Simulated BER performance for $\pi/4$ DQPSK system in LOS and OBS channels.

BPSK offers an improvement of about 3 dB over $\pi/4$ DQPSK for $E_b/N_o < 15$ dB. At $E_b/N_o > 15$ dB, BPSK only offers an improvement of about 2.8 dB. In comparing the BER performances of BPSK in the average channel and a Rayleigh fading channel, BPSK performs better in the average channel than in a Rayleigh fading channel for $E_b/N_o < 25$ dB and it performs worse in the average channel than in a Rayleigh channel for $E_b/N_o > 25$ dB. The change in slope on Fig. 19 for $\pi/4$ DQPSK at 35 dB is due to statistical uncertainty in the simulation, since each symbol error corre-

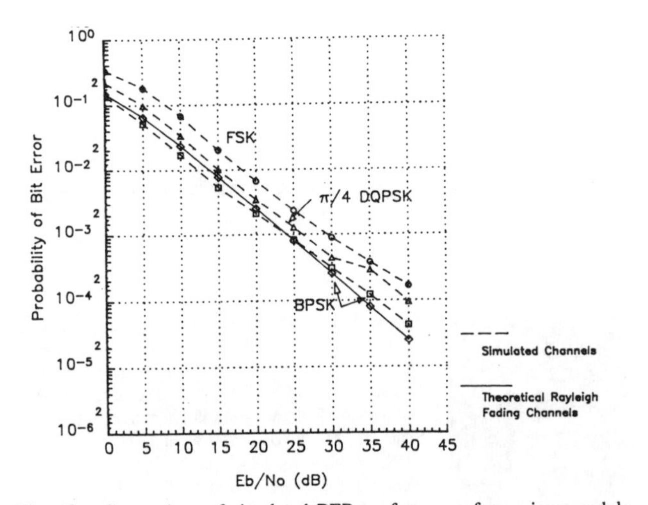

Fig. 19. Comparison of simulated BER performance for various modulation schemes in the 50 combined simulated CW fading channels.

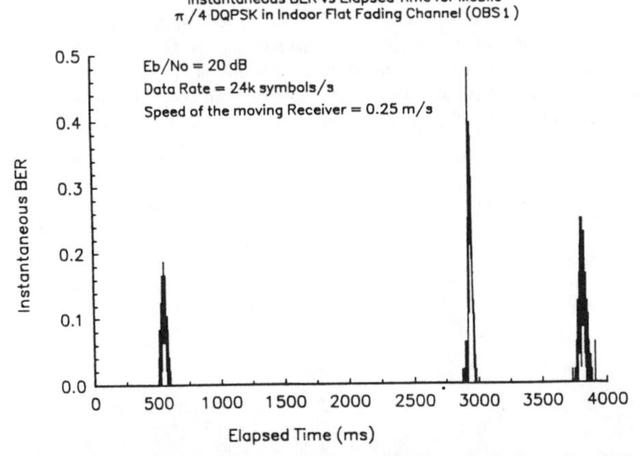

Fig. 20. Instantaneous BER for $\pi/4$ DQPSK in the indoor flat fading channel shown in Fig. 6(b) (OBS 1). Instantaneous BER values found over consecutive 24 symbol (1 ms) time intervals. Note that the instantaneous BER is directly a function of deep fades as the mobile moves.

sponds to one or two bit errors, and few errors occur over a simulation at such strong signal level. It can be seen that the BER values for all three modulation schemes are proportional to E_b/N_o, with roughly a tenfold decrease logarithmically in BER resulting for every 10 dB increase in E_b/N_o.

VII. Simulation of Instantaneous BER

The use of channel models which have accurate second order fading statistics (i.e., level crossing rate) is important for the accurate prediction of burst errors in real channels. This can be seen in Fig. 20. Fig. 20 illustrates the instantaneous BER for $\pi/4$ DQPSK operating at E_b/N_o of 20 dB in the fading channel, OBS 1, shown in Fig. 6(b). It can be seen in Fig. 20 that the times at which instantaneous BER increases is directly related to the spatial location of the receiver given in Fig. 6(b). That is, the high instantaneous BER values at the elapsed times of 500, 2850, and 3790 ms in Fig. 20 correspond exactly to the deep fades at the positions -37.5, 21.3, and 44.8 cm in Fig. 6(b). Results in Fig. 20 were generated from simulations and the instantaneous BER was computed for each block of 24 consecutive symbols. While 24 symbols provides a relatively course estimate of the instantaneous BER, it preserves the bursty nature of real world channels. The time interval over which the instantaneous BER was computed was 1 ms, although this can be adjusted in simulation. By computing a string of instantaneous BER values through simulation and storing them in a computer, it is simple to clock out the values at an approximate update rate in order to drive a digital hardware BER simulator. Such simulators are commercially available, and when used in conjunction with a computer simulation that uses a highly accurate channel model, may be advantageous in evaluating the performance of emerging digital communication services. At MPRG, a custom circuit has been designed to work with the BER simulator (BERSIM) software in a personal computer.

VIII. Summary

This paper presented the methodology and implementation of BER simulation for FSK, BPSK, and $\pi/4$ DQPSK. A novel (measurement-based) indoor radio channel simulation technique was used to generate small-scale fading for mobiles traveling in local areas. The novelty of the channel modeling is that it preserves the actual temporal and spatial correlation measured from real wide-band indoor channels, and thus can be applied to BER analysis while preserving actual second-order statistics such as level crossing rate and duration of burst errors. Thus, instantaneous BER computations are meaningful since they are very representative of real world conditions. With such a channel modeling technique, it becomes possible to drive a hardware simulator to recreate realistic bit error patterns and to investigate coding, diversity techniques, handoff, and protocols as a function of velocity.

The BER performances of FSK, BPSK, and $\pi/4$ DQPSK in a variety of indoor flat fading channels were compared, and it was shown that BPSK has a 3 dB advantage over $\pi/4$ DQPSK at low E_b/N_o values, and a 2.8 dB advantage at higher signal levels. This is offset by the 3 dB improvement on spectral efficiency offered by $\pi/4$ DQPSK, which suggests that, of the three modulation techniques compared, $\pi/4$ DQPSK is the most appropriate modulation technique for in-building flat fading mobile radio channels.

Acknowledgment

The authors wish to thank Bill Tisdale of the Contel Technology Center for his comments concerning this work, and the reviewers for their helpful comments.

References

[1] T. S. Rappaport, S. Y. Seidel, and K. Takamizawa, "Statistical channel impulse response models for factory and open plan building radio communication system design," *IEEE Trans. Commun.*, vol. 39, pp. 794–807, May 1991.

[2] W. C. Jakes, *Microwave Mobile Communications*, New York: Wiley, 1974.

[3] V. Fung and T. S. Rappaport, "A real time bit-error simulation of the performance of $\pi/4$ DQPSK in flat fading mobile radio channels," in *Proc. IEEE Int. Conf. on Commun. '91*, June 1991.

[4] S. Y. Seidel, "UHF indoor radio channel models for manufacturing environments," *Masters thesis*, Virginia Polytechnic Institute and State University, Blacksburg, Aug. 1989.

[5] A. A. M. Saleh, and R. A. Valenzuela, "A statistical model for indoor radio communications," *IEEE J. Select. Areas Commun.* vol. SAC-5, pp. 128–137, Feb. 1987.

[6] J. C-I. Chuang, "The effects of time delay spread on portable radio

communications channels with digital modulation," *IEEE J. Select. Areas Commun.*, vol. SAC-5, pp. 879-889, June 1987.
[7] R. A. Valenzuela, "Performance of quadrature amplitude modulation for indoor radio communications," *IEEE Trans. Commun.*, vol. COM-11, pp. 1236-1238, Nov. 1987.
[8] H. M. Arnold and W. F. Bodtmann, "A hybrid multichannel hardware simulator for frequency-selective mobile radio paths," *IEEE Trans. Commun.*, vol. COM-31, pp. 370-377, Mar. 1983.
[9] D. Berthoumieux and J. M. Pertoldi, "Hardware propagation simulator of the frequency fading channel at 900 MHz," *Proc. Nordic Seminar on Digital Land Mobile Radiocommun.*, Oct. 1986, pp. 214-217.
[10] H. W. Schussler *et. al.*, "A digital frequency selective fading simulator," in *Proc. Nordic Seminar on Digital Land Mobile Radiocommun.*, Oct. 1986, pp. 331-336.
[11] T. S. Rappaport, "Characterization of UHF multipath channels in factory buildings," *IEEE Trans. Antennas Propagat.*, vol. 37, pp. 1058-1069, Aug. 1989.
[12] T. S. Rappaport, and C. D. McGillem, "UHF fading in Factories," *IEEE J. Select. Areas Commun.*, vol. 7, pp. 40-48, Feb. 1989.
[13] E. L. Crow, and M. J. Miles, "A minimum cost, accurate statistical method to measure bit error rates," *Int. Conf. Comput. Commun., Rec.*, pp. 631-635, 1976.
[14] V. Fung, "Simulation of BER performance of FSK, BPSK, $\pi/4$ DQPSK in flat and frequency-selective fading channels," *Masters thesis*, Virginia Polytechnic Institute and State University, Blacksburg, VA, Aug. 1991.
[15] L. W. Couch II, *Digital & Analog Communication Systems*. New York: Macmillan, 1987.
[16] P. A. Baker, "Phase-modulation data sets for serial transmission at 2,000 and 2,400 bits per second," Part I, *AIEE Trans. Commun. Electron.*, pp. 166-171, July, 1962.
[17] J. G. Proakis, *Digital Communications*. New York: McGraw-Hill, 1983.

Theodore S. Rappaport (S'83-M'84-S'85-M'87-SM'91), for a photograph and biography please see page 730 of this issue.

Victor Fung (S'91) was born in Hong Kong in 1961. He received the B.S. degree in electrical engineering from Michigan State University, East Lansing, in 1984 and the M.S. degree in electrical engineering from Virginia Polytechnic Institute and State University, Blacksburg, in 1991.

From 1984 to 1986, he worked as a project electronics engineer in Lutron Electronics Limited, Pennsylvania. In 1986, he joined Commodore in Hong Kong, where he worked on the testing of Amiga personal computers. Between 1987 to 1989, he was a senior engineer in MiniScribe (HK) Limited, where he worked on high capacity hard disk testing. As a master's student, he worked with Dr. Rappaport in the Mobile and Portable Radio Research Group. He is now with BNR, Richardson, TX. His research interests are communication system simulation in mobile radio environments. He is currently developing software tools for simulating bit error rate performance for various modulation techniques in mobile fading channels.

Reprinted from *IEEE Journal on Selected Areas in Commun.*, Vol. 11, No. 3, Apr. 1993, pp. 393-405.

Bit Error Simulation for $\pi/4$ DQPSK Mobile Radio Communications using Two-Ray and Measurement-Based Impulse Response Models

Victor Fung, Theodore S. Rappaport, *Senior Member, IEEE,* and Berthold Thoma

***Abstract*— A combination hardware and software simulation technique that allows real-time bit-by-bit error simulation for mobile radio systems is described in this paper. The technique simulates mobile radio communication links and generates average bit error rate (BER) and bit-by-bit error patterns. The hardware simulates bit errors between a data source and sink in real time using the error patterns. Various communication system parameters (e.g., modulation scheme, data rate, signal-to-noise ratio, and receiver speed) and different channel environments (i.e., outdoor and indoor multipath fading channels) may be specified and permit performance comparison. Additive white Gaussian noise and cochannel interference effects are also simulated by the software. Using the simulation tool, we studied average BER results for $\pi/4$ DQPSK with Nyquist pulse shaping in indoor and outdoor, flat, and frequency-selective fading channels. BER results for high data rate (> 450 kb/s) transmission in channels generated by a measurement-based indoor channel model, SIRCIM [1], are compared with results in channels generated by the classic two-ray Rayleigh fading model. Simulation results show that when the ratio of rms delay spread to symbol duration is greater than about 0.04, the irreducible BER is not only a function of rms delay spread but is also a function of the temporal and spatial distribution of multipath components. In addition, an example of bit-by-bit error simulation of the transmission of a video image in a mobile radio fading channel is shown. This simulation methodology, which has been implemented in a program called BERSIM, allows subjective evaluation of link quality between a source and sink in laboratory in real time without requiring any radio frequency hardware.**

I. Introduction

THE accurate prediction of average and instantaneous BER in multipath channels will become increasingly important in system design as demand grows for digital wireless communication systems. BER predictions allow designers to determine acceptable modulation methods, coding techniques, and receiver implementations in the operating environments. However, since there are numerous system and channel parameters (e.g., signal-to-noise ratio, data rate, modulation type, cochannel interference, impulsive noise, mobile speed) that can affect BER in mobile communications, it is extremely difficult to evaluate the performance of mobile communication systems using analytical techniques alone. Also, due to the complexity and time-varying nature of mobile radio channels, it is often too complicated to design and optimize the parameters by analysis. Thus, computer simulation becomes a viable tool in performance evaluation and tradeoff analysis in the design of mobile communication systems. This paper describes a simulator methodology that can predict average BER and generate bit-by-bit error patterns for real-time bit error simulation. The bit error rate simulator (BERSIM) can simulate various popular modulation schemes (i.e., BPSK, FSK, and $\pi/4$ DQPSK), can accommodate data rates up to 15 Mb/s (for $\pi/4$ DQPSK), E_b/N_0 from 0 to 100 dB, receiver speeds up to 150 km/h, square root raised cosine pulse shaping, and cochannel interference in indoor and outdoor, flat, and frequency-selective fading channels. A two-ray Rayleigh fading channel model described by Clarke [2] and implemented by Smith [3] in software is used as one of the indoor and outdoor channel models in the simulator, where the delay between the two rays can be specified. Also, a measurement-based statistical indoor channel model, SIRCIM (Simulation of Indoor Radio Channel Impulse Response Models), developed by Rappaport and Seidel [1] is used to provide channel impulse responses for indoor channels. This paper presents BER results for both indoor and outdoor channels and illustrates how the type of channel model can dramatically impact BER. Different methods of timing recovery in dispersive fading channels impact the BER of a transmission are not considered in this study, but are implemented in BERSIM and studied in [20]. Simulations in this paper use the centroid of the spatially averaged power delay profile to determine symbol timing.

Flat and frequency-selective fading [4] are two distortion phenomena due to multipath that can cause bit errors. This paper focuses on frequency-selective fading for indoor and outdoor channels. BER results for $\pi/4$ DQPSK, BPSK, and FSK in indoor flat fading channels can be found in [5]. For high data rate indoor systems (e.g., where the ratio of rms delay spread to symbol period is greater than 10^{-2} [16]), frequency-selective fading will introduce intersymbol interference (ISI) in the data stream and cause bit errors. This research evaluates the performance of $\pi/4$ DQPSK in indoor channels at data rates of 450 kbs, the data rate proposed by Bellcore [6], 1.152 Mb/s, which is the data rate proposed for DECT (Digital European Cordless Telecommunications) [7], and a higher data rate of 2 Mb/s for comparison. For outdoor channels, the data rate of

Manuscript received February 12, 1992; revised August 1, 1992.
V. Fung is with Bell Northern Research, Richardson, TX 75081.
T. S. Rappaport and B. Thoma are with the Mobile and Portable Radio Research Group, Bradley Department of Electrical Engineering, Virginia Polytechnic Institute and State University, Blacksburg, VA 24061.
IEEE Log Number 9207827.

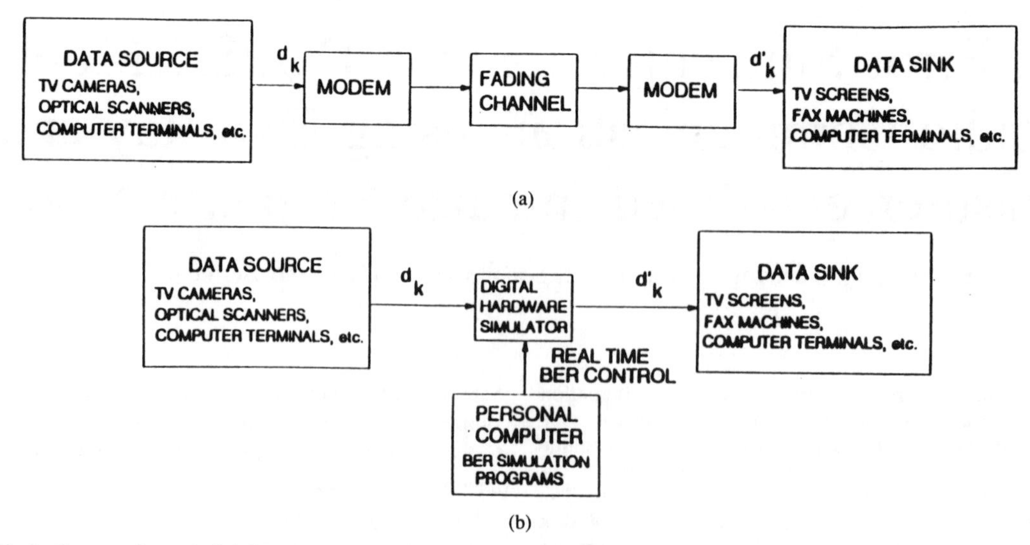

Fig. 1. (a) Block diagram of actual digital communication system. (b) Block diagram of a system using a baseband digital hardware BER simulator with software simulation as a driver for real-time BER control.

24,300 symbols/second, specified in the U.S. Digital Cellular Standard IS-54, is used [15]. $\pi/4$ DQPSK has been chosen as the modulation scheme for U.S. Digital Cellular, and it has been shown to perform well in indoor flat fading channels [5]. It is used exclusively in simulations presented in this paper. However, the simulation methods described can be applied to other modulation schemes as well.

For data transmission systems, error bursts due to signal nulls or intersymbol interference are a primary concern, and understanding the temporal distribution of errors is necessary to implement successful antenna diversity or coding techniques. With accurate BER computer simulations in fading environments, it becomes possible to test digital radio communication systems by using a simple and inexpensive baseband digital hardware BER simulator between the data source and sink as shown in Fig. 1. Digital hardware BER simulators, which operate on an applied digital data stream, can be programmed in real time to insert bit errors based on the bit-by-bit error patterns generated by the software simulator. In that case, an entire communication system operating in mobile fading environments with adjustable parameters can be simulated in baseband using the software and hardware simulator. Subjective evaluating between data source and data sink can then be performed conveniently in a laboratory.

In this paper, the communication systems used for the simulation, the two-ray channel model, the SIRCIM channel model, and the methods used to predict average BER and to generate bit-by-bit error patterns for $\pi/4$ DQPSK are described. Earlier work on these topics have already appeared in [11], [1], [5]. BER results for various system parameters in outdoor channels using a two-ray channel model, and BER results for indoor channels using SIRCIM with three different data rates are presented. The indoor channel (SIRCIM) results are then compared with the results of the classic two-ray channel model, with the same rms delay spread, to determine whether irreducible BER is dependent only on rms delay spread or if the actual distribution of multipaths also impact BER. Finally, an example of real-time digital video transmission in mobile radio channels is used to demonstrate the utility of real-time bit-by-bit error simulation and the burstiness of errors in wireless data communications.

It should be noted that the two-ray Rayleigh fading channel model has been used extensively in the literature [1], [16]–[19]. With the recent advent of channel sounders that can resolve multipath components to resolutions of a few ns, however, it has become clear that individual multipath components do not fade as a Rayleigh distribution but rather log-normal or Ricean [1]. The motivation for the present paper is to explore how different channel models and data rates affect the bit errors in wireless systems.

II. Communication System Model

A block diagram of the $\pi/4$ DQPSK system used in the simulation is shown in Fig. 2. All bandpass signals and channels described in this paper are represented and simulated using the corresponding baseband in-phase and quadrature forms [8], [5]. Referring to Fig. 2, a pseudorandom binary bit stream $d(t)$ is sent through the communication system for BER simulation. The binary bit stream $d(t)$ is stored in computer memory and is eventually compared with the received bits $d'(t)$ at the output of the simulated receiver for determination of bit errors. The number of bits sent can be determined by the bounded binomial sampling method [9], or a brute force method can be used where a large number of bits (typically several millions) are generated and the resulting errors counted.

By placing a square root raised cosine filter in both the transmitter and the receiver, both Nyquist pulse shaping and matched filtering (in flat fading channels) are achieved. Specifically, [5, eq. (15)–(17)] were used to simulate the $\pi/4$ DQPSK modem at baseband. The ideal transfer function of the square

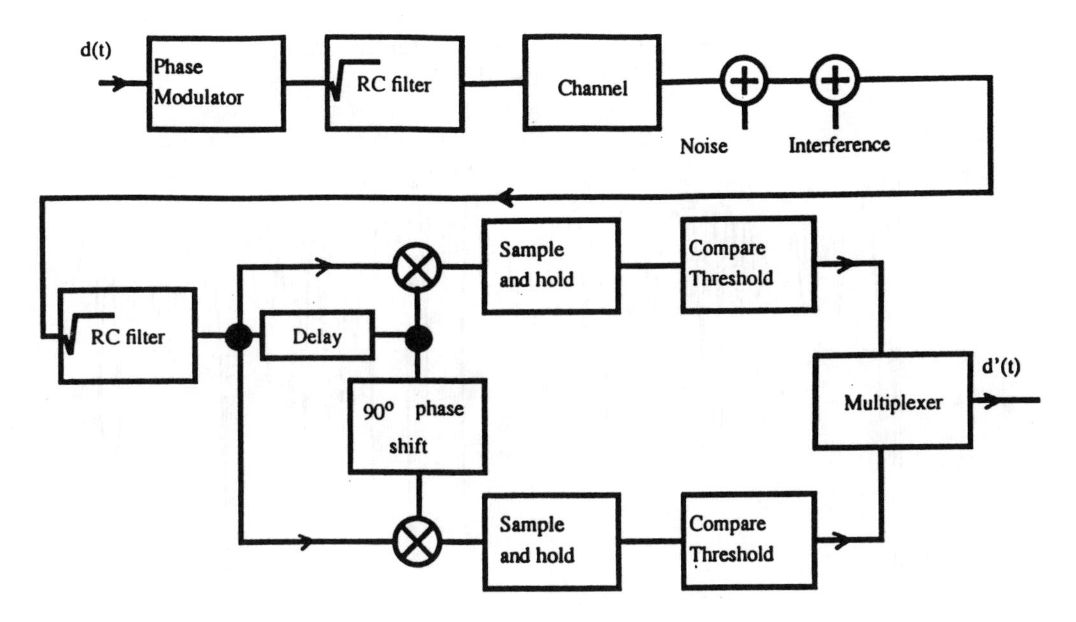

Fig. 2. Block diagram of the $\pi/4$ DQPSK system.

root raised cosine filter is defined as [10]

$$H_{\sqrt{RC}} = \begin{cases} 1 & 0 \le |f| \le \frac{r_b}{2}(1-\alpha) \\ \cos\left(\frac{\pi}{2\alpha r_b}\left[|f| - \frac{r_b}{2}(1-\alpha)\right]\right) & \frac{r_b}{2}(1-\alpha) < |f| < \frac{r_b}{2}(1+\alpha) \\ 0 & \frac{r_b}{2}(1+\alpha) < |f| \end{cases} \quad (1)$$

where α is the rolloff factor of the filter.

The impulse response of the filter is used for convolution in the time domain, and is obtained by performing an inverse Fourier transform [8]. In BERSIM, the impulse response of the square root raised cosine filter is truncated to 12 symbols in duration and is stored as a collection of amplitudes as a function of time.

The in-phase $I(t)$ and quadrature $Q(t)$ parts of the modulated signal are convolved with the impulse response of the square root raised cosine filter in the simulation [5]. The signals at the output of the filter are convolved with the time-varying complex impulse response of the simulated channel. However, the channel is considered to be stationary for an integer number of symbol durations. The generation of impulse responses for indoor and outdoor channels is described in Section III. As shown in Fig. 2, white Gaussian noise is added to the signal at the output of the channel. In the simulation, the mean signal level is held constant while the added noise power is changed for BER analysis as a function of signal-to-noise ratio. Cochannel interference is simulated by passing a different pseudorandom bit stream through an independent simulated channel and adding it to the desired signal at the input of the receiver. The cochannel interference level is set by the ratio of the desired average signal power (C) to the average interference signal power (I), which is denoted as C/I. The matched filter in the receiver is a square root raised cosine filter identical to that of the transmitter.

Different types of detectors for $\pi/4$ DQPSK modulation are known and have been shown to offer similar performance and various design advantages [11], [18], [19], [21]. In this work, IF differential detection is assumed [5], [11].

After detection, $d'(t)$ is written to computer memory in sequential order and is compared with the original bit stream $d(t)$ to determine bit errors. This technique enables one to determine the exact number and time correlation of bit errors. The errors can be recorded in their exact order of occurrence for later replay with the hardware portion of the simulator.

III. Channel Description

Channel models which have accurate first- and second-order fading statistics (i.e., level crossing rate, fading distributions of multipath components, and instantaneous delay spread) are vital for the accurate prediction of burst errors in real channels and are needed for accurate bit-by-bit error simulation. A two-ray Rayleigh fading model is one of the channel models used by BERSIM to generate both outdoor and indoor channels. Time correlation of the flat fading behavior in the two-ray model is preserved by the Doppler fading spectrum used to generate the two-ray impulse responses [8]. Also, a more accurate measurement-based statistical channel simulation model, SIRCIM [1], is used to provide the channel impulse responses for indoor channels.

A. Two-Ray Channel Model

The discrete impulse response of the flat and frequence-selective fading channels is given by the two-ray model

$$h(kT_s) = \alpha_1 e^{j\theta_1(kT_s)}\delta(kT_s) + \alpha_2 e^{j\theta_2(kT_s)}\delta(kT_s - \tau) \quad (2)$$

where T_s is the fractional symbol sampling period described in [5], α_1 and α_2 are independent and Rayleigh distributed, θ_1

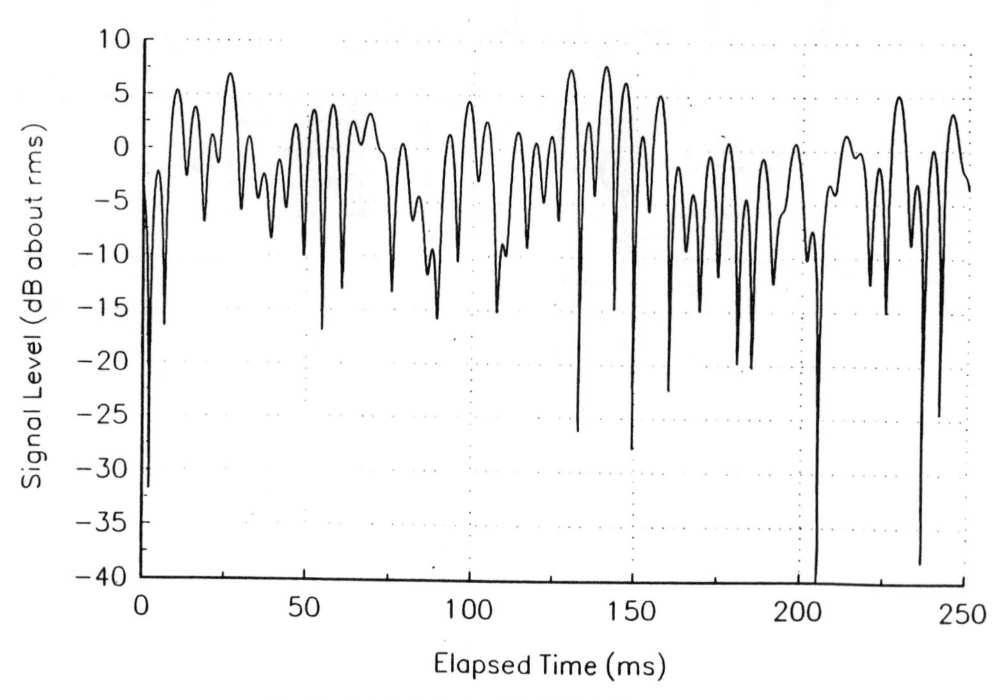

Fig. 3. A typical simulated Rayleigh fading envelope.

and θ_2 are independent and uniformly distributed over $(0, 2\pi]$, and τ is the time delay between the two rays. The power sum of $E\{\alpha_1^2\}$ and $E\{\alpha_2^2\}$ is set to unity in the simulation, so the channel has unity average gain over a simulation run. The ratio of $E\{\alpha_1^2\}$ to $E\{\alpha_2^2\}$ is the power ratio of the main ray (C) to the delayed ray (D) and is denoted as C/D. A flat fading channel is formed by setting α_2 to zero. The rate of change of phases θ_1 and θ_2 over time (or equivalently over the travel distance of the mobile) is dictated by the Doppler fading spectrum.

A software fading simulator similar to [3] has been used to generate the amplitude and phase of each ray in the two-ray model. The power spectrum of the fading envelope is given in [2]

$$s(f) = \begin{cases} \dfrac{E_0^2}{2\pi f_m \sqrt{1 - \left(\dfrac{f}{f_m}\right)^2}} & |f| < f_m \\ 0 & \text{elsewhere} \end{cases} \quad (3)$$

where E_0^2 is related to the received signal power, and $f_m = v/\lambda$ is the Doppler frequency in Hertz corresponding to a receiver speed v (m/s) and a carrier wavelength λ (in meters).

The fading spectrum is used to simulate the Rayleigh fading found in mobile radio [2], [4]. Specifically, the spectra of two independent complex Gaussian processes are shaped to the fading spectrum in (3), and then an IFFT is used to provide a complex fading envelope for a mobile travel distance of 24 wavelengths. In a manner similar to [3], 8192 samples spaced by the appropriate frequency sampling interval are used to represent each fading spectrum in the frequency domain. The value of the frequency sampling interval depends on the doppler shift, and is always chosen to provide sufficient resolution to resolve even short deep fades in the time domain. Since the fading envelope resulting from the IFFT spans a fixed travel distance, smaller sampling intervals are used for slower receiver speeds. In order to provide a real waveform from the inverse Fourier transform of each of the two complex Gaussian spectra, the real part of each complex Gaussian spectrum is made an even function by mirroring values about the carrier frequency. Similarly, the imaginary part is made an odd function about the carrier before performing an inverse Fourier transform. The real outputs of those two IFFT's are used as in-phase and quadrature baseband fading signals in the time domain [8]. The distribution of the envelope of the resulting signals in the time domain is Rayleigh, and the phase is uniformly distributed from 0 to 2π over several hundred samples. The envelope and phase generated are used as a time series to determine the amplitude and phase of the rays in (2), at a particular sample time. Since 8192 samples of interval $1/T$ in the frequency domain correspond to 8192 samples equally spaced over the duration T in the time domain, linear interpolation is used to determine the amplitude and phase of each sample point within each bit sent through the channel. The frequency-selective fading is simulated by adding a delay τ between the two rays, and is implemented in software. A typical fading envelope generated by the simulator with a Doppler frequency of 91 Hz (receiver speed = 120 km/h, λ = 0.35 m) is shown in Fig. 3. The two-ray model simulator has been thoroughly tested [8], [20], and the cumulative distribution and normalized level crossing rate agree with the theoretical results given in [1], [4], [12].

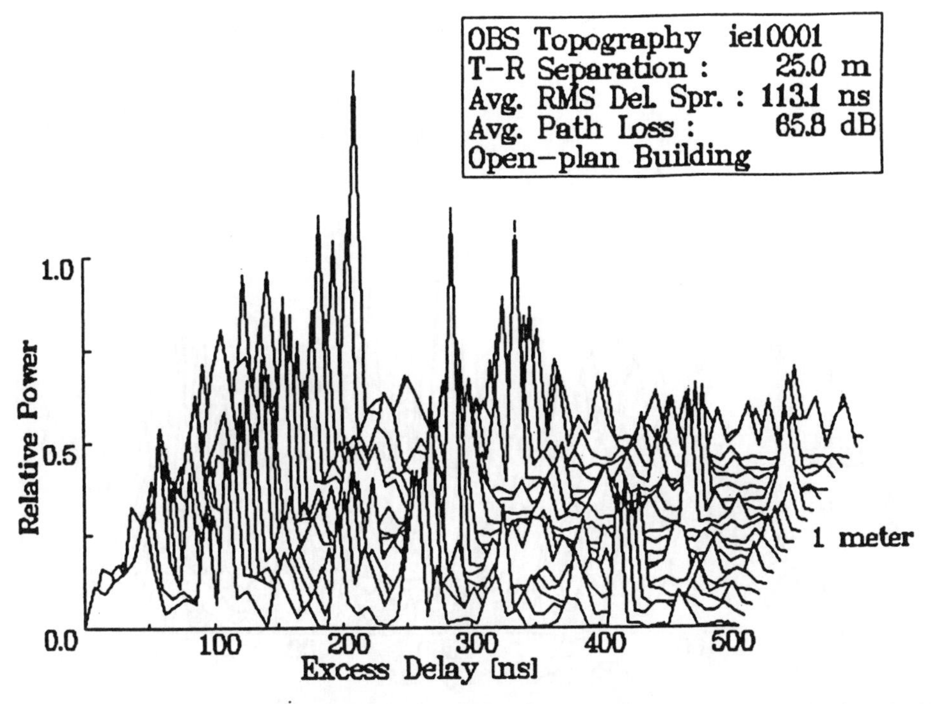

Fig. 4. Typical wideband impulse response profile for OBS channels generated by SIRCIM. Spatial averged rms delay spread is 113.1 ns.

B. SIRCIM Channel Model

SIRCIM is an indoor UHF multipath radio channel software simulator designed from over 50,000 wideband and narrowband measurements from many different buildings. SIRCIM recreates the statistics of measured wideband impulse responses in both LOS (Line of Sight) and OBS (Obstructed Sight) topographies, and generates 19 baseband complex impulse responses for indoor mobile radio channels when traveling over a small area (a 1 m path). Statistical models for the number of multipath components, the variation of the multipath components, the amplitudes, phases, and fading statistics of individual multipath components, and the excess delays are all based on measurements reported in [1], [13], and more recent data. A detailed description of the channel simulator SIRCIM can be found in [1], [13], [5]. Extensive tests have been made to verify SIRCIM before using it for the BER analysis presented here [1], [8], [20].

Examples of multipath profiles generated by SIRCIM, which were used for simulations in this paper, are shown in Figs. 4 and 5. These are typical spatially-varying OBS channel multipath delay profiles (squared magnitude impulse responses–phases are not shown for brevity) and are realistic representations of field measured multipath profiles reported in [1].

The average BER results presented in this paper are based on the channel characteristics of simulated measurement locations seen by a mobile at each location. At each location, the mobile is assumed to move along a short track at a specified velocity. The angle between the line of motion of the receiver and the line drawn between transmitter and receiver is randomly distributed from 0° to 360° for each simulated measurement track. The transmitter/receiver (T/R) separation is assumed to be 25 m, and the speed of the moving receiver is assumed to be 1 m/s for all indoor simulations, which roughly corresponds to the walking speed of a portable radio user. SIRCIM assumes that the scatterers are aligned along aisles located to each side of the transmitter, and the width between these aisles in our analysis is arbitrarily chosen to be 7 m for each simulation, although this can be varied easily [13].

SIRCIM produces 19 spatially-varying complex impulse responses spaced by quarter-wavelength increments for any carrier frequency from 900 MHz to 60 GHz. A single multipath component is assumed within a multipath delay bin of 7.8 ns. In this paper, we have used a carrier frequency of 1300 MHz for all simulations, which implies a 1 m measurement track. The simulator BERSIM can read SIRCIM channel files, and also allows the user to load in other complex channel files (either measured or computer generated) using a similar, but more general, data format as described later. Fig. 6 illustrates how time-varying impulse responses are represented as snapshots which vary over the travel distance of the mobile. BERSIM accepts complex impulse response profiles consisting of D equal-width multipath delay bins and I equally spaced profiles of separation Δs_{ch}. SIRCIM files use $D = 64$, a sampling interval $\Delta\tau_{ch}$ of 7.8125 ns (corresponding to 500 ns of excess delay for each complex impulse response snapshot), and Δs_{ch} of a quarter wavelength. However, BERSIM allows any value of $\Delta\tau_{ch}$, any Δs_{ch} less than a half wavelength of the carrier frequency, and allows D to be as large as 128.

Since mobile radio measurements always involve spatial sampling of the complex impulse response, BERSIM uses cubic spline interpolations on the amplitudes and phases of each multipath delay bin over space to generate hundreds of interpolated impulse responses between the primary profiles

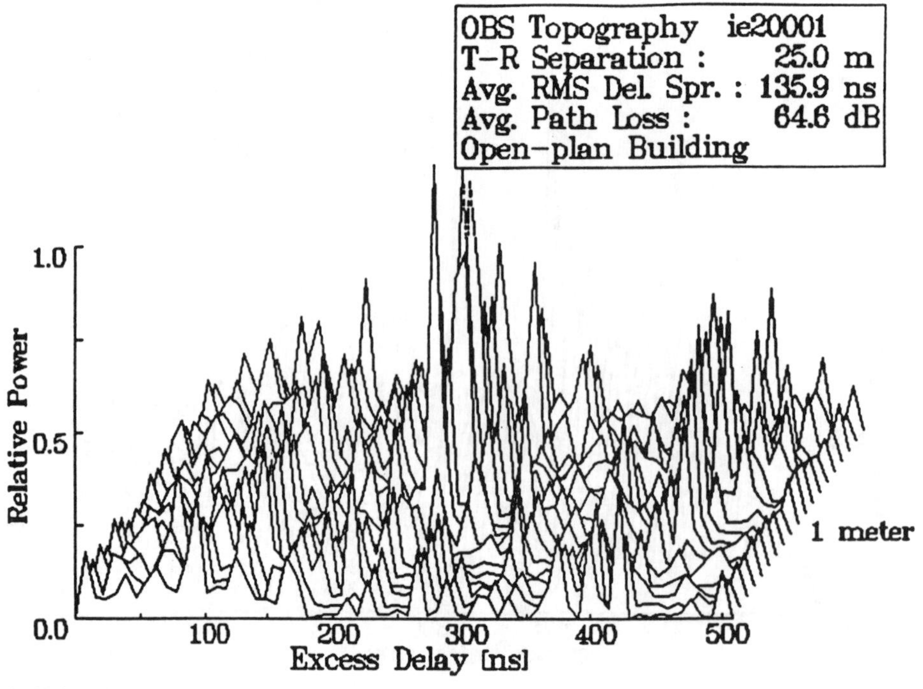

Fig. 5. Typical wideband impulse response profile for OBS channels generated by SIRCIM. Spatial averaged rms delay spread is 135.9 ns.

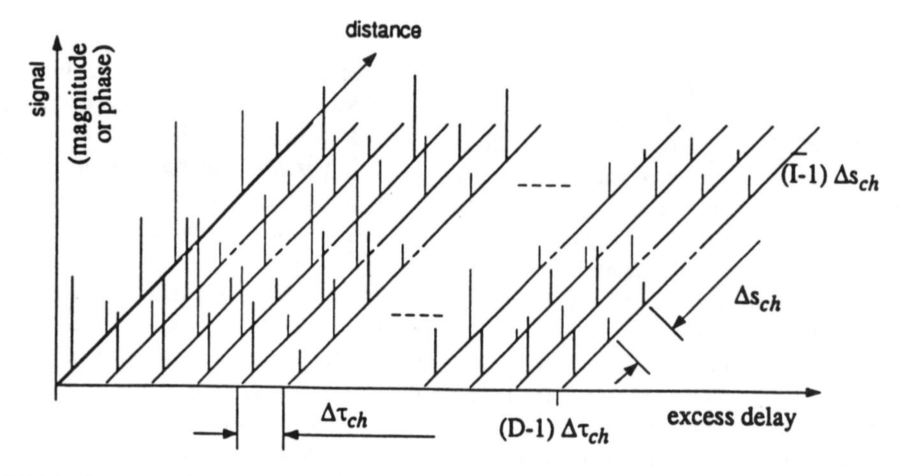

Fig. 6. Format of a BERSIM channel impulse response file. For SIRCIM files, $\Delta\tau_{ch}$ is 7.8 ns, $D = 64$, and Δs_{ch} is the spatial sampling of a quarter wavelength of the carrier frequency. For simulations in this paper, $I = 19$ for a 1 m travel distance and carrier frequency of 1300 MHz.

as shown in Fig. 7. For SIRCIM channels, this interpolation procedure can produce thousands of impulse responses from the 19 original profiles. The spatial separation Δs_{symb} of each new complex impulse response defines discrete spatial locations at which data are convolved with a static complex impulse response. The value of Δs_{symb} is determined by the velocity of the mobile, and M, the number of impulse responses after interpolation over space, is a function of the number of symbols sent during simulation and the mobile velocity [20]. Note that BERSIM interpolates each original multipath component before convolving the channel with the digital signal. The user must be sure that the spatial samples and time delay resolution of the channel data is adequate to ensure there is no aliasing over space. The high temporal resolution and quarter-wavelength spatial samples of SIRCIM ensures that each multipath component can be accurately regenerated over space.

In order to use each impulse response component for BER simulations, the time delay bins of each impulse response profile must be decimated into larger time delay bins of time period T_s, where T_s is the sampling period of a symbol. As described in [5], BERSIM uses 13 samples to represent each transmitted data symbol. Assuming that the channel is static over T_s, all multipath components within T_s are vectorially summed to form a new multipath component as shown in Fig. 8. To reduce computation time, no splining is performed over time delay at discrete spatial locations.

In our indoor simulations, the data rate is high (i.e., > 450 kb/s) and the receiver is moving slowly (i.e., 1 m/s) so that we may assume the channel remains unchanged during the

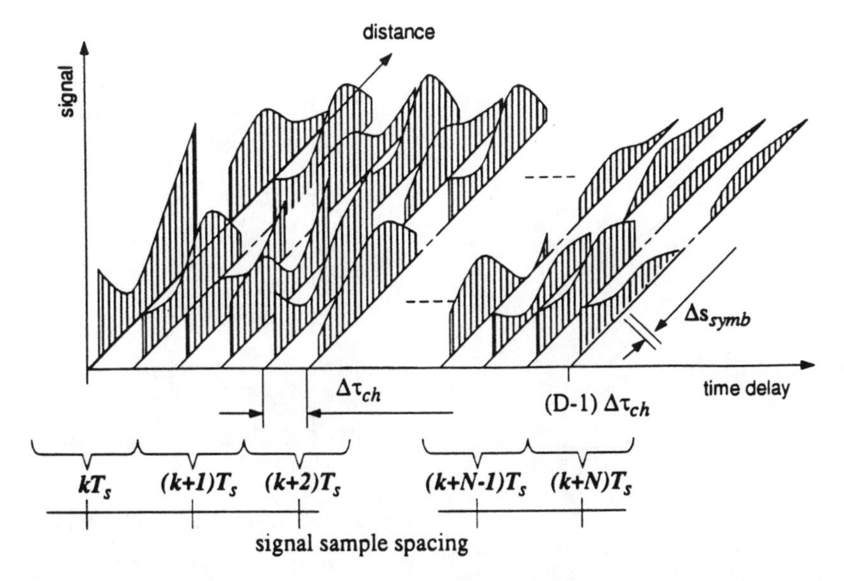

Fig. 7. Format of a BERSIM impulse response file after interpolating the amplitudes and phases of individual multipath components over space using a cubic spline technique. Note the fading behavior of individual multipath components is preserved. As shown in Fig. 8, the sample period of an applied symbol encompasses several consecutive time delay bins. The spatial separation $\Delta s_{\rm symb}$ is a function of mobile velocity.

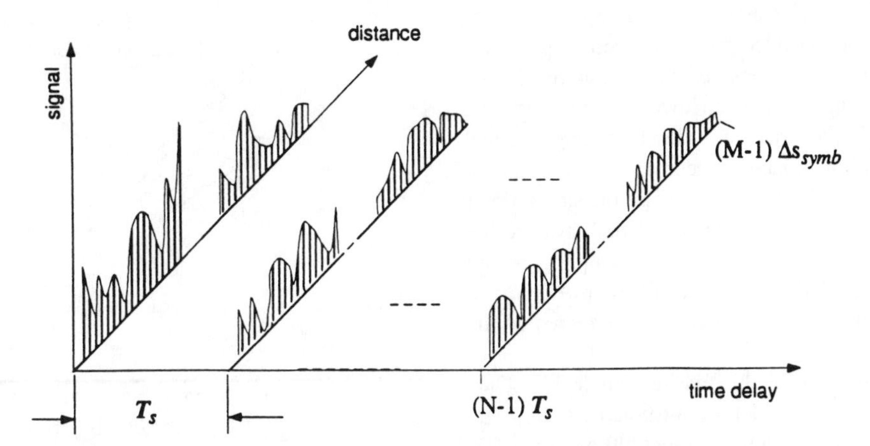

Fig. 8. Format of a BERSIM impulse response file after matching the impulse response time delay samples to the symbol samples. As described in [5], BERSIM convolves an integer number of symbols with each complex impulse response at discrete distances. M denotes the number of specific spatial locations used during a simulation run, and is a function of the velocity and the amount of data transmitted during simulation.

transmission of hundreds of symbols. For a data rate of 450 kb/s, the channel is assumed to be stationary for 38 symbols. For a data rate of 1.152 Mb/s and receiver speed of 1 m/s, the channel is assumed to be stationary during the transmission of 96 symbols. For a data rate of 2 Mb/s, the channel is assumed to be stationary for 200 symbols. These assumptions are reasonable, since the impulse response variation between two successive impulse responses over space is very small for the considered speeds.

IV. BER Calculation

Two different methods are used to determine the number of bits sent in a BER simulation and to calculate the average BER. One method is the "bounded binomial sampling" method described in [9], and the other is a brute-force "free run" method. Bounded binomial sampling tells how many bits must be sent in a BER simulation from the parameters provided. It also provides double boundaries in simulated BER results so that the three BER lies both within a certain relative precision (e.g., 50%) and a certain absolute precision (e.g., 10^{-2}). The relative precision is more appropriate to low BER, while the absolute precision is more appropriate to high BER [8]. In our simulation, the confidence level was 99%, relative precision was 25%, and absolution precision was 10^{-2}. This means that there is 99% confidence that the true bit error rate is both within 25% and 10^{-2} of the BER found by our simulation. To use the bounded binomial sampling method, the maximum number of bits sent in a BER simulation depends on the BER estimated before the actual run. We assumed the BER would be on the same order of magnitude of analytical results given in [14], or would give pessimistic estimates when analytical results were not available. We set the maximum number of bits to be approximately 10 times the number of bits required to produce one bit error using the bounded binomial sampling technique. Our BER simulation for a particular run terminates

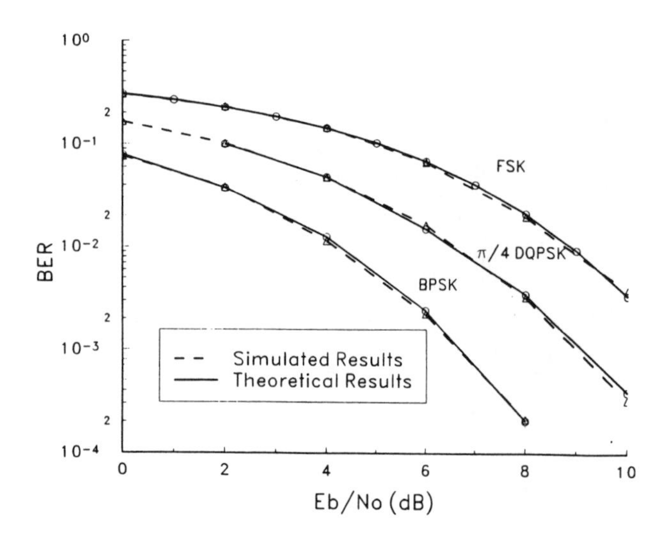

Fig. 9. BER versus E_b/N_o for an AWGN channel for $\pi/4$ DQPSK, BPSK, and FSK compared with theoretical results. The symbols are shaped using a raised cosine filter with rolloff factor $\alpha = 1$.

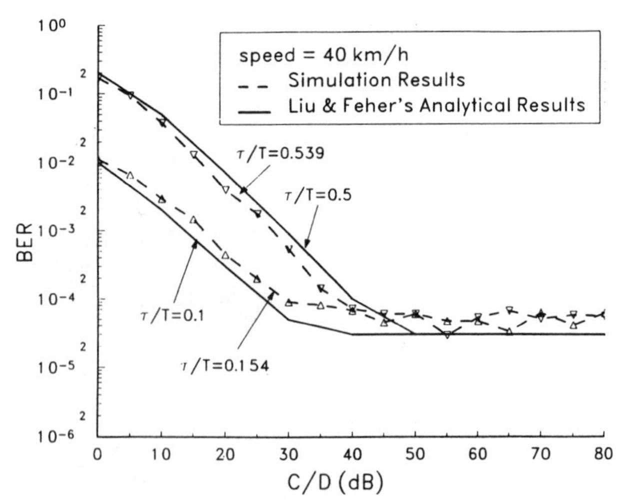

Fig. 10. BER versus C/D of $\pi/4$ DQPSK in a frequency-selective two-ray Rayleigh fading channel for two different signal delays. E_b/N_o = 100 dB, C/I = 100 dB, fc = 850 MHz, fs = 24.300 kBd, α = 0.2, v = 40 km/h.

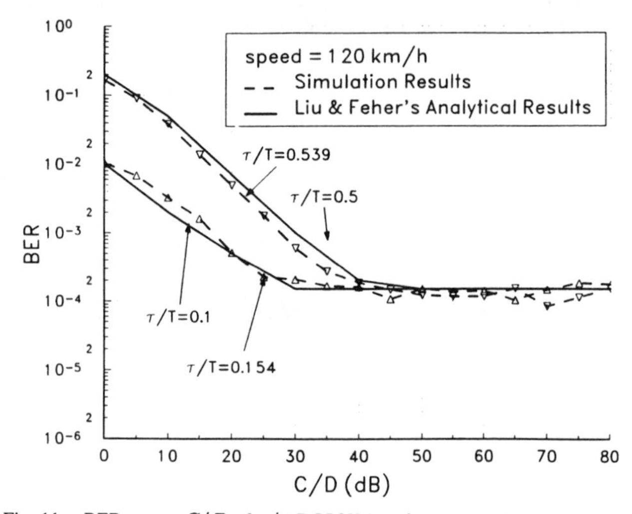

Fig. 11. BER versus C/D of $\pi/4$ DQPSK in a frequency-selective two-ray Rayleigh fading channel for two different signal delays. E_b/N_o = 100 dB, C/I = 100 dB, fc = 850 MHz, fs = 24.300 kBd, α = 0.2, v = 120 km/h.

due to one of the three possible outcomes described in [9], and the median unbiased BER are calculated according to [9]. The advantages of the bounded binomial sampling method are that it sets confidence levels on simulation results and also requires less simulation time. However, this method is not desirable in high data rate and slowly fading channels (indoor channels), where a sudden deep fade in the channel might generate numerous bit errors and cause the simulation to terminate before the mobile traveled over a small distance (i.e., a 1 m track). Thus, the bounded binomial sampling method was used in finding the BER for $\pi/4$ DQPSK in AWGN and two-ray channels with mobile velocities much greater than walking speeds.

The number of bits sent in the free-run mode is set to a very large number (on the order of millions). In this way, the BER is typical for a particular channel although the BER has no confidence bounds. The free-run method requires a longer run time; however, this ensures that the simulation will cover the channels intended and the simulation will not terminate prematurely. The free-run method was used in the BER simulations in SIRCIM channels.

V. Simulation Results

A. Simulation Results in AWGN Channels

Simulation results for various modulation techniques, including $\pi/4$ DQPSK in AWGN channels, are compared with theoretical results in Fig. 9 to test the accuracy of the simulator. It can be seen that the simulation results match well with the theoretical results presented in the literature [5], [8], [9], [18], [19], [21]. This indicates that the implementations of different modulation and detection techniques within the simulator are correct.

B. Simulation Results Using Two-Ray Channel Model

Figs. 10 and 11 show BER versus C/D (average power ratio of line-of-sight ray to delayed ray) in two-ray Rayleigh fading channels for two different receiver speeds and τ/T ratios, where τ denotes the time delay between the two rays and T is the symbol duration. As C/D increases, the channels become less frequency-selective and the random FM (fast fading) places a lower limit on the BER. By comparing the bit error floors in Figs. 10 and 11, the lower BER limit is determined by the receiver speed. The BER limits are about $2*10^{-5}$ and $2*10^{-4}$ for receiver speeds of 40 and 120 km/h, respectively. For C/D below 40 dB, the BER is dominated by frequency-selective fading (ISI). By comparing the results for C/D below 25 dB in Figs. 10 and 11, the variation of the receiver speed appears to have little effect on BER for the same C/D ratio. We have compared our simulation results with Liu and Feher's [14] analytical results in the figures for the same receiver speeds and slightly different τ/T. As shown in Figs. 10 and 11, our simulation results agree closely with results given in [14], where we have used only Monte-Carlo simulation without any analytical techniques.

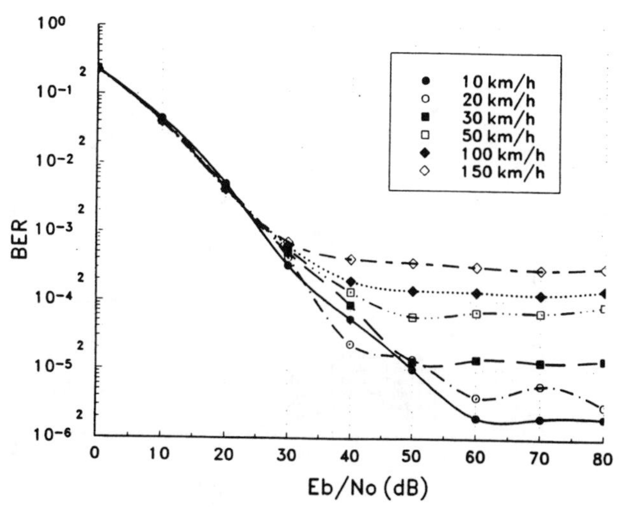

Fig. 12. BER versus E_b/N_o in a flat Rayleigh fading channel for various mobile speeds. fc = 850 MHz, fs = 24.300 kBd, α = 0.2, C/I = 100 dB.

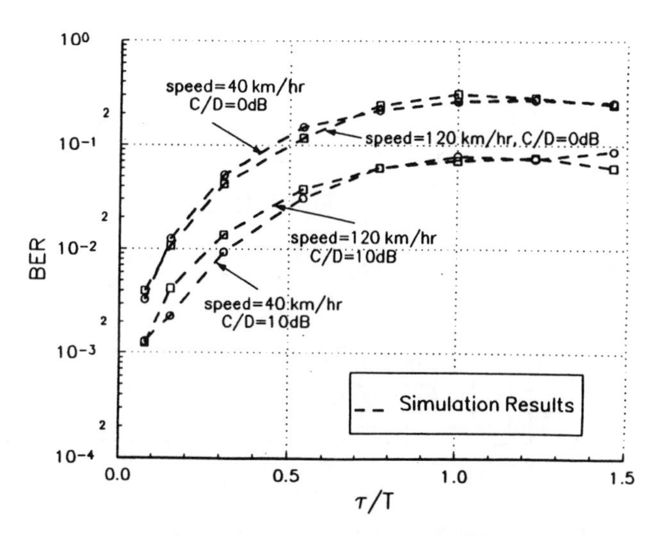

Fig. 13. BER versus t/T in frequency-selective fading channels for various receiver speeds and C/D ratios. E_b/N_o = 100 dB.

Fig. 12 shows the BER simulation results for $\pi/4$ DQPSK in flat fading channels. Results are found for various receiver speeds of 10 to 150 km/h. An irreducible bit error floor exists at E_b/N_0 above 60 dB.. The irreducible bit error floor is caused by random FM and is a function of the receiver speed. For low $E_b/N_o(E_b/N_o < 30\,\mathrm{dB})$, bit errors are caused by the additive white Gaussian noise and changing the receiver speed has minimal effect on the BER.

BER performance as a function of τ/T in frequency-selective fading channels is shown in Fig. 13. The influence of receiver speed and C/D are also shown in the figure. The results show that BER increases with the increase of τ/T from 0.077 to 0.77 for receiver speeds of 40 and 120 km/h and both C/D ratios. In our simulation, τ/T ranged from 0.077 to 1.46, and Fig. 13 shows that BER is not a strong function of the receiver speed for τ/T in that range. There is a BER ceiling of about 0.25 for τ/T greater than 1, which is determined by the C/D ratio. The BER ceiling is due to saturation of the bit errors caused by frequency-selective fading.

The effect of cochannel interference in flat fading channels is shown in Fig. 14. The BER decreases as C/I increases from 10 to 40 dB. Three different receiver speeds (40, 70 and 120 km/h) were used in our C/I simulations. Simulation results show that the variation of receiver speed makes no difference on the BER performance for a given C/I ratio. This agrees with the observation by Malupin and McNair [22]. Experimental results found in [22] from actual field measurements and analytical results for flat fading channels [14] are also shown in Fig. 14 for comparison. Results in [22] probably have a higher irreducible bit error at large C/I due to the frequency-selective fading nature of real-world channels, although it is unclear why [14] portrays better (i.e., lower BER) at E_b/N_o below 30 dB.

C. Simulation Results Using SIRCIM for Indoor Channels

Simulation results for indoor channels are based on the two channel impulse responses shown in Figs. 4 and 5. These two impulse responses are typical impulse responses for OBS

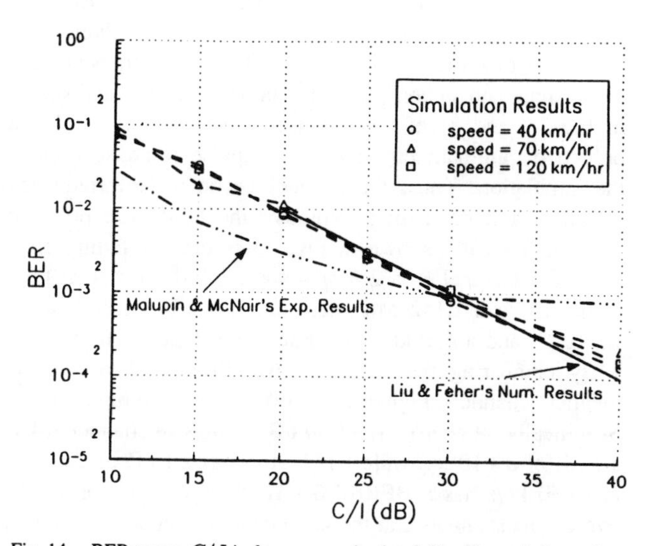

Fig. 14. BER versus C/I in frequency-selective fading channels for various τ/T. Receiver speed = 40 and 120 km/h. E_b/N_o = 100 dB, C/D = 0 dB.

channels for open-plan buildings. While we realize these two channels may not be the worst case in a particular environment, one of the goals of this work was to develop a tool that can use measurement-based models and field measurements to determine BER within a local area. We call this small-scale BER behavior, due to the small travel distances experienced by the receiver in simulation. BER results using channels in Figs. 4 and 5 should be taken as typical and are used to illustrate the difference in BER for different channels. Indeed, BERSIM may be used to produce real-time error patterns or BER results for any channel that is applied in the manner described in Section III. Since the receiver speed was chosen to be 1 m/s in simulations, one second of the time-varying channel impulse response is shown for two different channels in Figs. 4 and 5. The rms delay spreads for channels A and B are 113.1 and 135.9 ns, respectively. BER performance for $\pi/4$ DQPSK in these two channels is shown in Fig. 15. For a data rate

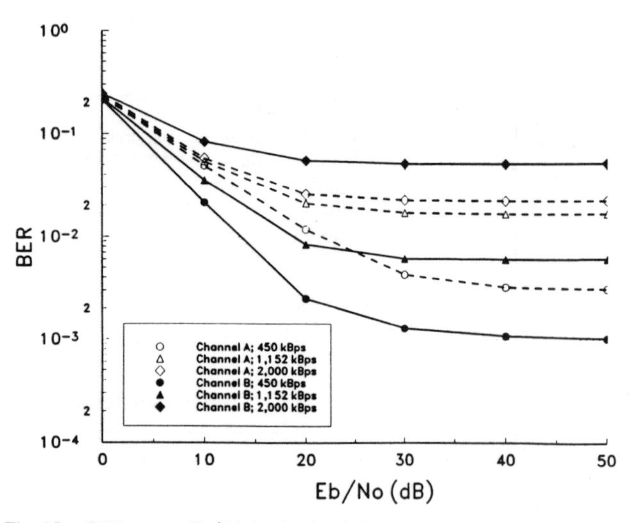

Fig. 15. BER versus E_b/N_o in simulated channel A and channel B, receiver speed 1 m/s, frequency 1.3 GHz.

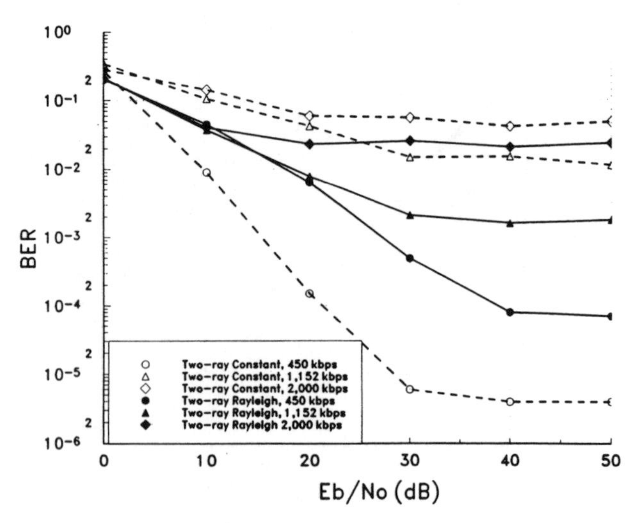

Fig. 16. BER versus E_b/N_o in Rayleigh two-ray channels and constant amplitude two-ray channels.

of 1.152 Mb/s in channel A, the irreducible bit error floor is $2 * 10^{-2}$. For a data rate of 450 kb/s, the irreducible bit error floor drops to $3 * 10^{-3}$. For 450 kb/s data transmission, the channel conversion (described in Section III-B and shown in Figs. 7 and 8) of the impulse responses [20] results in only two multipath bins for each impulse response used in the simulation. Thus, the channel becomes less frequency selective, and flat fading dominates the BER. The bit error floor of $3 * 10^{-2}$ shown in Fig. 15 is the flat fading error floor. For channel B, a 2 Mb/s data rate has a bit error floor of $6 * 10^{-2}$; a 1.152 Mb/s data rate has a bit error floor of $7 * 10^{-3}$; and a 450 kb/s data rate has a floor of 10^{-3}.

By comparing the BER for two different data rates in identical channels, higher data rates indeed have higher BER. Specifically, at an E_b/N_o of 30 dB, 2 Mb/s in channel A has a BER of $3 * 10^{-2}$, while 1.152 Mb/s has a BER of $2 * 10^{-2}$ and 450 kb/s has a BER of $5 * 10^{-3}$. A comparison of the two channels shows that the same data rate produces different small-scale BER's. The simulated 2 Mb/s transmission in channel A has a lower bit error floor than the same simulation for channel B. This is due to the fact that channel A has a lower τ/T than channel B. However, for the lower data rates and flat fading conditions ($\tau/T < 0.08$), deeper fades cause channel A to offer a higher BER floor than channel B.

Fig. 16 shows the BER performance for data rates of 450 kb/s, 1.152 Mb/s, and 2 Mb/s in an independent Rayleigh two-ray fading channel model where each ray is assumed to fade over space for a moving receiver. The purpose of this simulation was to compare the BER performance between a two-ray channel and a SIRCIM channel (channel B), where both channels have virtually identical rms delay spreads. Due to the discretization of time delays in computer simulation and the fact that BERSIM uses 13 samples per symbol and convolves an integer number of symbols with each impulse response, the rms delay spread of the two-ray channel can only be set to 170.9 ns (i.e., $1/(2 * 13(450k/2))$) for 450 kb/s, 133.5 ns (i.e., $2/(2 * 13(1.152M/2))$) for 1.152 Mb/s, and 153.8 ns (i.e., $4/(2 * 13(2M/2))$) for 2 Mb/s. The factor of two results from the fact that for a two-ray model with equal average power in each ray, the rms delay spread σ_{rms}, as computed in [1], is half of the temporal separation between the two rays. The data rate and a fixed number of samples per symbol determine the discrete values of rms delay spread that can be simulated in a two-ray model. The 170.9, 133.5, and 153.8 ns values of rms delay spread for the three two-ray channels are comparable to the 135.9 ns rms delay spread of channel B.

We also simulated two-ray multipath channels with constant (nonfading) ray amplitudes at a mobile receiver in order to compare fading channels with stationary channels. Thus, the amplitude of each ray was constant but the phase was changing based on the Doppler fading mechanism and free-space propagation. This experiment was conducted in order to determine how important the multipath amplitudes are in causing bit errors in $\pi/4$ DQPSK. The BER for 450 kb/s, 1.152 Mb/s, and 2 Mb/s in a constant amplitude two-ray model are also shown in Fig. 16 for comparison. The phase of each ray in the constant amplitude two-ray model was generated exactly the same way as for the Rayleigh two-ray channel using the method described in Section III.

By comparing Fig. 15 with Fig. 16, it can be seen that the irreducible BER in a Rayleigh two-ray fading channel is very close to the irreducible BER in channel B for Mb/s. At a data rate of 1.152 Mb/s, the Rayleigh two-ray channel has a *lower* BER than the simulated channels. The difference between errors in the Rayleigh two-ray channel and channels A and B is even more pronounced for a data rate of 450 kb/s. Since almost flat fading conditions exist for 450 kb/s, the difference in the BER performance is that the resulting narrowband fading envelope in a real-world indoor channel is not necessarily Rayleigh distributed within a local area. It can also be seen that the BER is sensitive to the shape of the profile if one considers almost equal rms delay spreads. As a result, the Rayleigh two-ray channel should *not* be used to model indoor fading channels, but rather field measurements or models based on such data should be used for modem and system design [1].

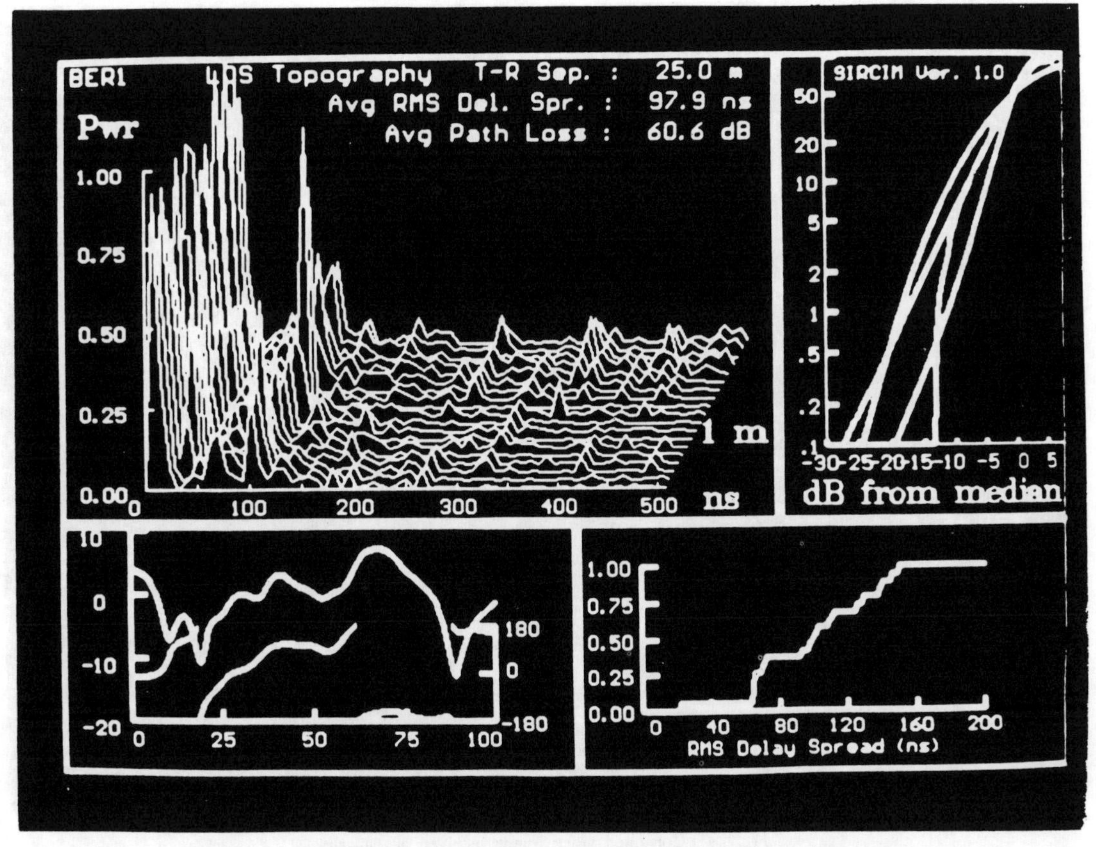

Fig. 17. An example of video image transmitted through the hardware simulator.

In summary, the irreducible BER for $\pi/4$ DQPSK is not simply a function of rms delay spread for $0.2 > \sigma_{\mathrm{rms}}/T > 0.02$ but is also a function of the temporal variations of the multipath components and their phases. Extensive work that has studied the impact of timing recovery, jitter, and multipath phase shifts on BER is available in [20].

By examining the results in the Rayleigh fading and constant amplitude two-ray models, it can be seen that the constant amplitude two-ray model has a higher bit error floor than the Rayleigh two-ray model at the data rates of 1.152 and 2 Mb/s, but has a lower BER at the data rate of 450 kb/s. This shows that the constant amplitude channel model yields worse BER results when compared with Rayleigh two-ray models for $\sigma_{\mathrm{rms}}/T > 0.08$ (a frequency-selective fading channel) because the instantaneous rms delay spread is always at its maximum (delay between the two rays/2). It also shows that the bit errors in the Rayleigh two-ray model are dominated by the delay spread, and not Rayleigh fading of multipath signals at $\sigma_{\mathrm{rms}}/T > 0.08$. Fig. 16 shows that, for σ_{rms}/T less than 0.04, the channel is less frequency selective. Thus, the Rayleigh fading in the Rayleigh two-ray channel generates more bit errors than the constant amplitude two-ray channel. As mentioned in [16], envelope fading is the most important mechanism for small delay spreads. It is interesting to note from Figs. 15 and 16 that both Rayleigh and constant envelope channels give much more optimistic BER's than the real-world-based channel models (channels A and B) produced by SIRCIM. From these results, we conclude that future computer simulations and system design should use channel models based on the real-world behavior of amplitude and phase, such as provided by SIRCIM.

VI. Real-Time Bit Error Simulation

BERSIM provides real-time BER simulation, and can play back the corrupted data at a specified rate in real time through hardware on a personal computer. Similar hardware for simulation on UNIX-based workstations is under development. The user may provide baseband data from actual hardware data sources or may read in stored data files such as video or voice data. The bit-by-bit error patterns used in the real-time simulation are generated using the method described in [5]. Specifically, a binary one is written to an external file if a bit error is encountered, and a binary zero is written if no error occurs. Once the off-line simulation is done, real-time simulation can be performed. Data streams from the data source (either externally provided or read from the computer disk) are sent to the data sink through the hardware simulator. The hardware simulator has a fast DMA controller on board that reads the bit-by-bit error pattern in real time from the source and inserts errors into the data stream according to the error pattern. The BERSIM hardware card clocks the corrupted data stream using TTL voltage levels at the specified data rate of the simulation up to 15 Mb/s. In this manner, digital communications in mobile radio channels are simulated at the baseband and a subjective evaluation of data links can be

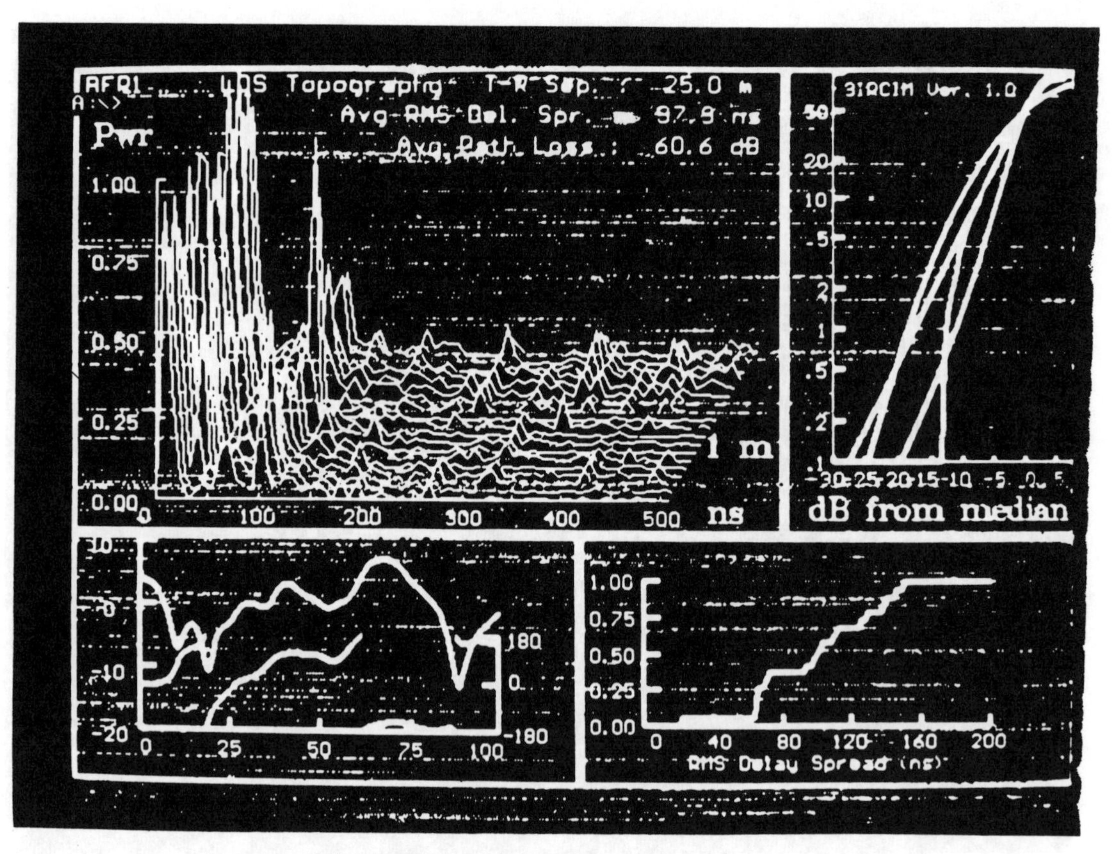

Fig. 18. Corrupted video image received at data sink.

performed easily in the laboratory. As an example, a test video image shown in Fig. 17 was transmitted through the hardware simulator. The simulation was for $\pi/4$ DQPSK operating at 48.6 Kb/s, and the video image was represented by binary pixels. The simulated digital cellular radio channel was a frequency-selective fading channel of 6 μs rms delay spread, E_b/N_0 of 20 dB, and C/I of 30 dB. The corrupted image received at the data sink is shown in Fig. 18. The bursty nature of bit errors in mobile fading channels can be seen clearly in the image. Subjective quality evaluation of wireless data links using speech and image coding in a wide range of channels, modulations, or systems can be performed easily with BERSIM.

VII. Summary

An accurate software/hardware bit-by-bit error simulator for mobile radio communications has been described. Simulation results in indoor and outdoor channels are shown and compared with theoretical results. BER results in simulated frequency-selective fading channels generated by several channel models such as two-ray, constant amplitude, and simulated indoor radio channel impulse models (SIRCIM) are presented and compared. We found that BER is not only dependent on the rms delay spread, but also on the distribution of temporal and spatial multipath components in local areas. An important result is that a two-ray Rayleigh fading model is a poor fit for indoor wireless channels and, if used, can underestimate the BER by orders of magnitude. A real-time bit error simulation of video transmission using the bit-by-bit error simulator has also been shown. The simulator, called BERSIM, has been demonstrated to be a useful tool for evaluating emerging data transmission products for digital mobile communications.

Acknowledgment

The authors wish to thank B. Tisdale of the Contel Technology Center for his comments concerning this work, and M. Keitz of Virginia Tech for designing and building the hardware simulator. BERSIM is patent-pending and is available from Virginia Tech.

References

[1] T. S. Rappaport, S. Y. Seidel, and K. Takamizawa, "Statistical channel impulse response models for factory and open plan building radio communication system design," *IEEE Trans. Commun.*, vol. 39, no. 5, pp. 794–807, May 1991.

[2] R. H. Clarke, "A statistical theory of mobile-radio reception," *Bell Syst. Tech. J.*, pp. 957–1000, July/Aug. 1968.

[3] J. I. Smith, "A computer generated multipath fading simulation for mobile radio," *IEEE Trans. Vehic. Technol.*, vol. VT-24., no. 3, pp. 39–40, Aug. 1975.

[4] W. C. Jakes, *Microwave Mobile Communications*. New York: Wiley, 1974.

[5] T. S. Rappaport and V. Fung, "Simulation of bit error performance of FSK, BPSK, and $\pi/4$ DQPSK in flat fading indoor radio channels using a measurement-based channel model," *IEEE Trans. Vechic. Technol.*, vol. VT-40, no. 4, pp. 731–741, Nov. 1991.

[6] D. C., Cox, A radio system proposal for widespread low-power tetherless communications," *IEEE Trans. Commun.*, vol. 39, no. 2, pp. 324–335, Feb. 1991.

[7] D. J. Goodman, "Trends in cellular and cordless communications," "*IEEE Commun. Mag.*, vol. 29, no. 6, June 1991.

[8] V. Fung, "Simulation of BER performance of FSK, BPSK, $\pi/4$ DQPSK in flat and frequency-selective fading channels," Masters Thesis, Virginia Polytech. Instit. and State Univ., Aug. 1991.
[9] E. L. Crow and M. J. Miles, "A minimum cost, accurate statistical method to measure bit error rates," *Int. Conf. Comput. Commun. Rec.*, pp. 631–635, 1976.
[10] K. Feher, and Engineers of Hewlett-Packard, *Telecommunications Measurements, Analysis, and Instrumentation.* Englewood Cliffs, NJ: Prentice-Hall, 1987.
[11] C. L. Liu, and K. Feher, "Noncoherent detection of $\pi/4$-DQPSK systems in a CCI–AWGN combined interference environment," in *Proc. IEEE Vehic. Technol. Conf.*, May 1989, pp. 83–94.
[12] J. D. Parsons, and J. G. Gardiner *Mobile Communication Systems.* Blackie, 1989.
[13] S. Y. Siedel, "UHF indoor radio channel models for manufacturing environments," Masters Thesis, Virginia Polytech. Instit. and State Univ., Aug. 1989.
[14] C-L. Liu, and K. Feher, "Performance of non-coherent $\pi/4$-QPSK in a frequency-selective fast Rayleigh fading channel," in *Proc. IEEE Int. Conf. Commun.*, Atlanta, GA, Apr. 1990, pp. 335.7.1–335.7.5.
[15] "Cellular system dual-mode mobile station-base station compatibility specification," Electron. Industry Assoc./Telecommun. Industry Assoc. Interim Stand. IS-54, May 1990.
[16] J. C.-I. Chuang, "The effects of time delay spread on portable radio communications channels with digital modulation," *IEEE J. Select. Areas Commun.*, vol. SAC-5, pp. 879–889, June 1987.
[17] C-L. Liu, and K. Feher, "Bit error rate performance of $\pi/4$-QPSK in a frequency-selective fast Rayleigh fading channel," *IEEE Trans. Vehic. Technol.*, vol. VT-40. no. 3, pp. 558–568, Aug. 1991.
[18] S. H. Goode, H. L. Kazecki, and Y. Shimazaki, "A comparison of limiter-discriminator, delay and coherent detection for $\pi/4$ DQPSK," in *Proc. IEEE Vehic. Technol. Conf.*, Orlando, FL, May 1990, pp. 687–694.
[19] S. Chennakeshu and G. J. Saulnier, "Differential detection of $\pi/4$-shift-DQPSK for digital cellular radio," in *Proc. IEEE Vehic. Technol. Conf.*, St. Louis., May 1991, pp. 186–191.
[20] B. Thoma, "Bit error rate simulation enhancement and outage prediction in mobile communication systems," Masters Thesis, Virginia Polytech. Instit. and State Univ., July 1992.
[21] Y. Akaiwa and Y. Nagata, "Highly efficiency digital mobile communications with a linear modulation method," *IEEE J. Select. Areas Commun.*, vol. SAC-5, pp. 890–895, June 1987.
[22] R. P. Malupin and I. M. McNair, "Bit error rate characteristics in a suburban fading environment," presented at *IEEE Vehic. Technol. Conf.*, Orlando, FL, May 1990.

Victor Fung (S'91) was born in Hong Kong in 1961. He received the B.S. degree in electrical engineering from Michigan State University, East Lansing, in 1984 and the M.S. degree in electrical engineering from Virginia Polytechnic Institute and State University, Blacksburg, in 1991.

From 1984 to 1986, he worked as a Project Electronics Engineer at Lutron Electronics Limited. In 1986, he joined Commodore in Hong Kong, where he worked on the testing of Amiga personal computers. Between 1987 to 1989, he was a Senior Engineer at MiniScribe (HK) Limited, where he worked on high-capacity hard disk testing. As a master's student, he worked with Dr. Rappaport in the Mobile and Portable Radio Research Group. He is now with BNR, Richardson, TX. His research interests are communication system simulation in mobile radio environments. He is currently developing software tools for simulating bit error rate performance for various modulation techniques in mobile fading channels.

Theodore S. Rappaport (S'83-M'84-S'85-M'87-SM'91) was born in Brooklyn, NY, on November 26, 1960. He received the B.S.E.E., M.S.E.E., and Ph.D. degrees from Purdue University in 1982, 1984, and 1987, respectively.

In 1988, he joined the Electrical Engineering faculty of Virginia Tech, Blacksburg, where he is an Associate Professor and Director of the Mobile and Portable Radio Research Group. He conducts research in mobile radio communication system design and RF propagation prediction through measurements and modeling. He guides a number of graduate and undergraduate students in mobile radio communications, and has authored or coauthored more than 70 technical papers in the areas of mobile radio communications and propagation, vehicular navigation, ionospheric propagation, and wideband communications. He holds several U.S. patents and is coinventor of SIRCIM, an indoor radio channel simulator that has been adopted by over 75 companies and universities. In 1990, he received the Marconi Young Scientist Award for his contributions in indoor radio communications, and was named a National Science Foundation Presidential Faculty Fellow in 1992. He serves as Senior Editor of the IEEE JOURNAL ON SELECTED AREAS IN COMMUNICATIONS and coedited *Wireless Personal Communications* (Kluwer Academic). He is a Registered Professional Engineer in the State of Virginia and is a Fellow of the Radio Club of America. He is also President of TSR Technologies, a cellular radio and paging test equipment manufacturer.

Berthold Thoma was born in Wenkheim, Germany, on June 29, 1963. He received the Diplom Ingenieur (FH) degree from the Fachhochschule Heilbronn, Germany in 1990.

Before entering the Fachhochschule Heilbronn in 1985, he was on a three and a half year apprenticeship with the A.W.d.H. Neckarzimmern, Germany to qualify as a trained radio engineer. During his studies in Germany, he received a Carl Duisberg scholarship to work for six months as a Research Assistant at Middlesex Polytechnic, London in the Robotics Laboratory. For his thesis at the Fachhochschule Heilbronn, he joined S-TEAM Elektronik Untereisesheim where he developed a direct-sequence spread spectrum-based LAN transceiver. In 1990, he entered Virginia Tech as a Fulbright Scholar and later joined their Mobile and Portable Radio Research Group. His graduate research focused on simulations of mobile radio communication systems to analyze bit error rate mechanisms in those systems.

Reprinted by permission from the Institute of Electronics, Information and Communication Engineers, Japan, EICE Trans. Commun., Vol. E76-B, No. 2, Feb. 1993.

INVITED PAPER *Special Issue on Land Mobile/Portable Propagation*

Performance of Decision Feedback Equalizers in Simulated Urban and Indoor Radio Channels

Theodore S. RAPPAPORT†, **Weifeng HUANG**†† *and* **Martin J. FEUERSTEIN**†††, *Nonmembers*

SUMMARY A Decision Feedback Equalizer (DFE) structure with a varying number of tap lengths was used with a recursive least squares (RLS) algorithm to determine tradeoffs between equalizer size and performance in mobile and portable digital radio systems. A mobile channel simulator, SMRCIM, was used to demonstrate how much an equalizer can improve the BER in real world urban channels. The results show that at 850 MHz, the DFE is unable to improve the BER when the mobile terminal exceeds speeds of 115 km/h for U.S. Digital Cellular systems. The performance of adaptive equalization for indoor high data rate systems was evaluated using the indoor channel simulator SIRCIM, and we found that DFEs have excellent performance for indoor radio channels. For simple structures, the BER is less than 10^{-3} at 15 dB E_b/N_o using coherent QPSK modulation. Finally, an equalizer structure for non-coherent $\pi/4$ DQPSK modulation was developed and simulation results are presented.

key words: *adaptive equalization, mobile communication channels*

1. Introduction

Most cellular telephone systems are analog systems which use spectrally inefficient frequency modulation (FM). Because of the limitation of available spectrum, more spectrally efficient methods must be adopted. With careful source encoding, some digital modulation techniques require less spectrum to send digitized voice signals, which can increase the channel capacity.[1] The Pan-European Group Speciale Mobile (GSM) and U.S. Digital Cellular telephone (USDC) standards are two examples of digital cellular telephone systems, both using time division multiple access (TDMA).

In digital transmissions, intersymbol interference (ISI) is one of the main causes of bit errors. In radio channels, ISI is caused by multipath propagation, which may be considered as the energy traveling from the transmitter to the receiver via more than one path. As a result, the channel impulse response may contain several pulses with different time delays, thereby causing distortion.

Manuscript received October 2, 1992.

† The author is with the Mobile and Portable Radio Research Group Bradley Department of Electrical Engineering Virginia Polytechnic Institute and State University Blacksburg, VA 24061-1110, USA.
†† The author is with COMSEARCH, Reston VA. USA.
††† The author is with U.S. West Advanced Technologies, Boulder, CO. USA.

Whether a channel causes significant ISI is roughly dependent on the normalized root-mean-square (rms) delay spread, σ_τ/T_s, of the channel impulse response,[2],[3] where σ_τ is the rms delay spread[4] and T_s is the symbol duration. If σ_τ/T_s is very small ($\sigma_\tau/T_s \cong 0$), the channel impulse response is close to an ideal impulse, and there will be little ISI. On the other hand, if σ_τ/T_s is large ($\sigma_\tau/T_s > 0$), the channel impulse response must be represented by several pulses with different excess delays, and this type of channel will cause severe ISI. Therefore, whether a channel will cause ISI depends not only on the channel rms delay spread, but also on the symbol rate.

In UHF mobile radio channels, there are usually many paths between the transmitter and receiver, and some of these paths have different propagation lengths. If the data rate is too high, σ_τ/T_s will not be negligible and these multipath components can cause significant ISI. Values of σ_τ determined by various measurements are given for urban cellular[5] and for indoor personal communication networks (PCN).[4]

Mobile radio channels are time-varying as a transmitter or receiver moves, so the received power changes as a function of time and location. The rate of change of the channel variation due to the movement of a transmitter or receiver can be defined by the Doppler frequency.[6] The maximum Doppler frequency f_D is

$$f_D = \frac{v}{\lambda} \tag{1}$$

where v is the velocity of the mobile and λ is the wavelength of the carrier. It is often more practical to define the channel variation by the normalized Doppler frequency f_N

$$f_N = \frac{f_D}{f_S} \tag{2}$$

where f_S is the symbol rate, because a higher symbol rate results in a smaller spatial distance that the mobile moves between two consecutive symbols for a fixed velocity, and hence a smaller change occurs between the corresponding channel impulse responses.

Fading may be divided into two categories: wideband fading and narrow band (CW) fading. The manner in which a wideband signal fades can be measured at a receiver by sending an impulse or an approximation of an impulse through the radio channel, and then summing the power of all the multipath components within the impulse response to obtain the total received power. The wideband fading describes the fluctuation of total received power in a large RF bandwidth. CW fading can be measured by sending a continuous unmodulated radio carrier and receiving the carrier at a narrow band receiver. CW fading is the propagation power loss variation at one RF frequency. CW fading has a much larger magnitude fluctuation than wideband fading.(4),(7)

Because a mobile fading channel is unknown and time-varying, an equalizer must adjust its impulse response so that the equalizer is close to the inverse filter of the channel.(8) The process of adjusting to an unknown channel is called convergence. After convergence, an adaptive equalizer can still track the channel variation; therefore, equalizers are able to improve the BER performance of mobile radio systems by adaptively canceling out the impact of multipath. There are many equalizer structures, with the decision feedback equalizer (DFE), the lattice least square decision feedback equalizer, and the maximum likelihood sequence estimator being among the good candidates for mobile communications.(9)-(13) The recursive least squares (RLS) algorithm and the least mean square (LMS) algorithm(8) are two groups of adaptive algorithms which are widely used. For mobile communications, the RLS algorithm is more appropriate because it converges faster and has better tracking ability than the LMS algorithm.(8)

In the simulations discussed in this paper, a DFE was used and an RLS algorithm was implemented to update the equalizer. A DFE is made of a feedforward transversal filter, a feedback transversal filter and a threshold detector (see Fig. 1). In Fig. 1, C_i and F_i are tap-gain values for the feedforward and feedback filters, y_i is the channel output, $\hat{d}_i$ is the output of the equalizer before the threshold detector, and d_i is the estimate of the ith symbol value. A DFE can be specified by its size (numbers of taps in the feedforward and feedback filters), space (symbol T_s spaced or fractionally spaced in the feedforward filter) and synchronization location (the location of y_k in Fig. 1).

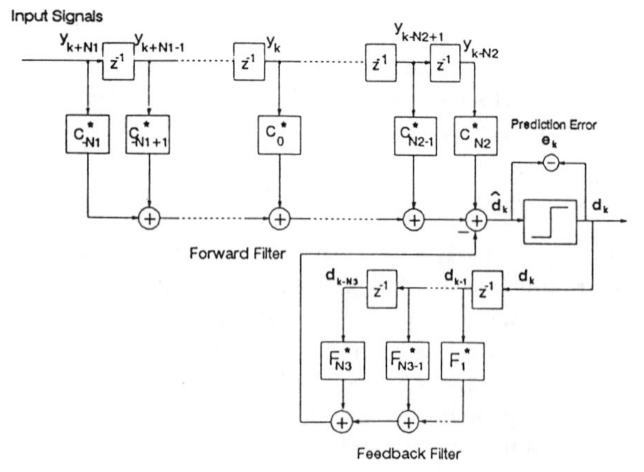

Fig. 1 The structure of a decision feedback equalizer.

In this work, a symbol-spaced equalizer was used and the DFE was always synchronized to the last tap of the feedforward filter, so N_2 was always 0 in Fig. 1. This type of arrangement can reduce the size of the DFE without sacrificing the performance.(8) For the RLS algorithm, the only variable parameter is the weighting factor $\lambda (0<\lambda<1)$,(8) where λ is used to weight the new data more heavily, so old data will not have as much effect on the adjustment of the equalizer coefficients as new data. Therefore, if the channel is time-varying, the equalizer can adjust itself relying mainly on the latest data. Usually λ is chosen between 0.8 to 1; however, if the normalized Doppler frequency increases (i.e. the mobile speeds up), λ should be decreased. However, if λ is too small, the algorithm will not be stable; therefore, once noise or other disturbances appear at the receiver, the adaptive equalizer may diverge.(8)

In the simulations of this work, perfect decision feedback was assumed. That is, instead of using detected data for training, which may contain errors, correct data are applied to the feedback filter and are used to estimate equalizer prediction error (e_k in Fig. 1). Without perfect decision feedback, if burst errors appear, the equalizer will lose track of the channel. This phenomenon is described as error propagation. The error propagation loss is about 2 dB compared to perfect decision feedback.(8)

This paper first presents the adaptive equalization simulations in two-ray urban channels, and then a real world urban radio channel impulse response model is applied to compare with the results of the two-ray channel model. Equalization in high data rate indoor systems using indoor radio channels with different rms delay spreads is also studied. Finally, an equalization structure for differential modulations is demonstrated and simulation results for this novel structure in both urban and indoor channels are presented.

2. Equalization in Two-Ray Channels

In the equalization simulation, coherent QPSK was the modulation, but no clock recovery circuit was implemented at the receiver. The receiver clock was assumed to be exactly synchronized to the transmitter at all times. Simulations were first performed for systems similar to the USDC standard.(14) The data rate was 48.6 kbps, which corresponds to a symbol duration of 41.15 μs. Each data slot contained 210 symbols, in which the first 10 symbols were used for training. Raised cosine rolloff filters with 0.35 rolloff factor were used at the transmitter. Using *square root* raised cosine filters offers slightly better performance because they implement a matched filter response.[2]

To test adaptive equalizers, one usually performs actual radio measurements or uses channel impulse response hardware simulators. However, it is difficult to control the delay spread in real-world channels, and hardware simulators are generally expensive. Even though recent research shows they are poor models of the real world,[19] for simple testing and analysis, two-ray impulse response models in which each ray has independent Rayleigh fading are widely used to evaluate the performance of adaptive equalizers and communication systems.[2],[6] The delay of the second ray is variable so that the rms delay spread of the channel can be controlled. As shown in Ref. (15), however, a two-ray model is an optimistic model for determining BER in unequalized mobile channels.

Two-ray channels were independently generated by BERSIM, a Bit Error Rate SIMulator of mobile radio channels.[3] Both rays of the channel were independently Rayleigh distributed with the equal mean power; thus, the mean rms delay spread is $\tau/2$, where τ is the delay of the second ray. The carrier frequency used was 850 MHz, and perfect decision feedback was assumed.

A DFE with an RLS algorithm was used as the equalizer with a weighting factor λ of 0.9. To limit the simulation time requirement, only 10,000 bits were used to test one BER value, so our results may lack accuracy at BER$<10^{-3}$. In the simulation, each symbol duration was divided into 12 time bins. The irreducible BER of the equalizers with different delays of the second ray (τ) was tested.

To choose the size of the DFE in a two-ray channel, a series of different DFEs were evaluated. A DFE is expressed as DFE(n_1, n_2), where n_1 is the number of taps in the feedforward filter, and n_2 is the number of taps in the feedback filter. The value of τ ranged from 0 to $2T_S$, the range most multipath components will fall in for the USDC system.[5] First, the number of feedforward taps was fixed in order to evaluate the effects of different numbers of feedback taps. The simulation results at 75 km/h are shown in Fig. 2. Although DFE(5, 1) had good performance when $\tau<T_S$, the BER for $\tau>T_S$ deteriorated. Hence, once a significant long-delay multipath component appears, the number of errors will increase significantly since only one feedback tap is available. Among the other sizes, DFE(5, 5) was the poorest, because the time span of the equalizer was much larger than the delay spread of the channel. Time span is the time range of the intersymbol interference that the equalizer can deal with. Thus, DFE(5, 5) resulted in equalizer-induced ISI because the equalizer chose finite tap-gain coefficients outside the interference range, although ideally these coefficients were supposed to be zero or very small, so the coefficients caused extra ISI that did

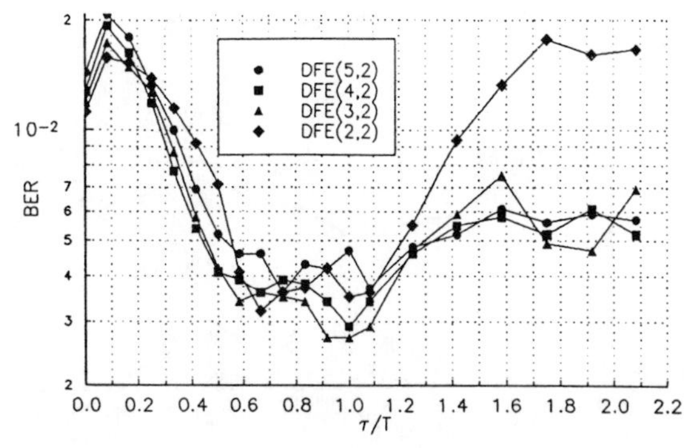

Fig. 2 BER vs. τ for a two-ray Rayleigh fading channel with 5 different numbers of feedforward taps of RLS-DFE and the number of feedback taps is 2. Carrier frequency is 850 MHz, $v=75$ km/h, $E_b/N_o=\infty$.

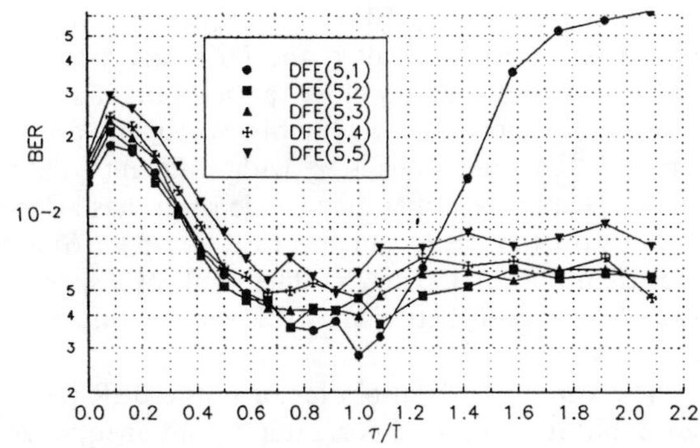

Fig. 3 BER vs. τ for a two-ray Rayleigh fading channel with 5 different numbers of feedback taps of RLS-DFE and the number of feedforward taps is 5. Carrier frequency is 850 MHz, $v=75$ km/h, $E_b/N_o=\infty$.

not exist in the original channel. The DFF(5, 2) and DFE(5, 3) were the best equalizers, but since a smaller size requires less computation, DFE(5, 2) is the proper choice. Therefore, 2 was selected as the number of feedback taps.

Once choosing the proper number of feedback taps, the number of feedforward taps was chosen in the same way, and the results are shown in Fig. 3. Although DFE(2, 2) had good performance when $\tau<T_S$, the BER increases for $\tau>T_S$. The other three choices had similar performance, so 3 was chosen as the number of feedforward taps because of its lower computational requirement. Finally, DFE(3, 2) was selected as the size of the DFE; however, DFE(2, 1) may be a better choice if all the multipath components are restricted within one symbol duration. This example shows how the worst-case multipath delay spread impacts equalizer design.

This method of finding the point of diminishing

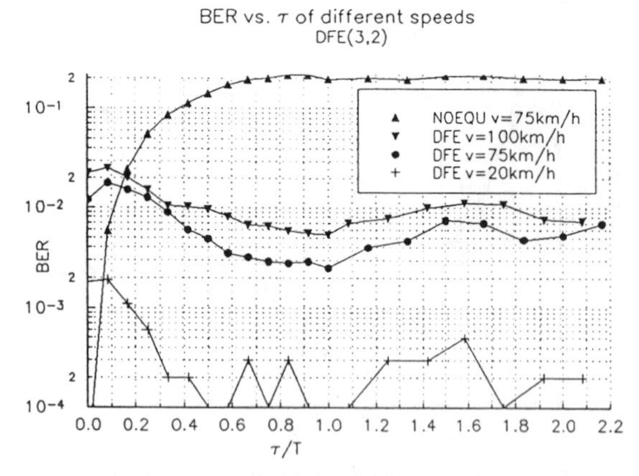

Fig. 4 BEM vs. τ of two-ray Rayleigh fading channel using DFE(3, 2) at different speeds. Carrier frequency is 850 MHz.

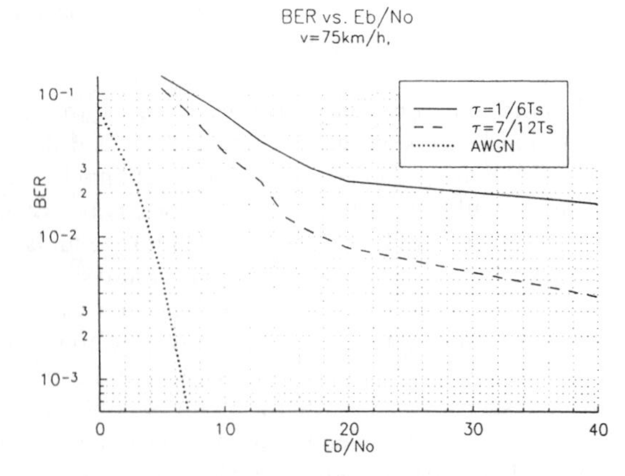

Fig. 5 BER vs. E_t/N_o using RLS-DFE(3, 2) in a two-ray Rayleigh fading channel, when $\tau/T_s=1/6$ and $\tau/T_s=7/12$. Carrier frequency is 850 MHz, $v=75$ km/h, $E_b/N_o=\infty$.

returns for the numbers of taps in the feedforward and feedback filters does not necessarily produce the optimal size of the equalizer. After some experiments, it was found that if the DFE is synchronized to the last tap of the feedforward filter, the DFE size obtained using this method is very close to the optimal size.

Figure 4 shows the irreducible BER using a DFE in mobile channels versus τ with different vehicle speeds. One can see that the vehicle speed has a large influence on the BER, and the BER of 75 km/h was better than that of 100 km/h. The result of a speed of 20 km/h was even better; the BER was less than 2×10^{-3} for all τ.

The value of τ is also a factor in the BER performance, but it is not as important as for an unequalized receiver. The DFE works better when the value of τ is large, until τ is about one symbol duration. Compared with the performance of a system without an equalizer, a DFE does not help when τ is very small. This is because when τ/T_s is small, the unequalized receiver has good performance, but a DFE still keeps a comparatively stable BER regardless the value of τ/T_s. At speeds from 75 to 100 km/h, if τ is smaller than about 1/6 to $T_s/12$, the performance *without* an equalizer is actually *better*. When the delay of the second ray is larger than $T_s/6$, the DFE can improve the BER by as much as 100 times. If the data rate is 48.6 kbps, then the critical delay of the equal mean amplitude second ray is about $\tau=4$ to 7 μs. Transforming τ to the rms delay spread σ_τ ($\sigma_\tau=\tau/2$ for the two-ray case), the critical point at which an equalizer is needed is about $\sigma_\tau=2$ to 3.5 μs. Rappaport et al.[5] indicated that 20% -30% of the locations they measured in urban areas have rms delay spreads larger than 3.5 μs.

If the vehicle speed is 20 km/h, an equalizer can improve the BER performance even when τ is about $T_s/12$ ($\sigma_\tau=T_s/24$). If the data rate is 48.6 kbps and the symbol duration is 41.15 μs, the rms delay of the critical point $T_s/24$ is lower than 2 μs. More than 50% of locations measured in Ref. (5) have an rms delay spread larger than this.

The result shows that adaptive equalization can improve the mobile system performance over a large range of rms delay spreads. The simulation shows clearly that the rate a mobile channel changes is extremely important in terms of the degree to which an adaptive equalizer can improve system performance. When a mobile channel does not change very rapidly, adaptive equalization can improve the system significantly. In this situation, only an LMS algorithm or even a non-adaptive equalizer is needed since the change within a data slot of 200 symbols is very small and the equalizer is retrained before each data slot.

The noise influence on the BER is shown in Fig. 5. Two specific τ values were chosen; $T_s/6$ and $(7/12)\,T_s$. It is shown that when the E_b/N_o is small, the noise is more important in causing errors, so the BER of the two τ's are similar. When E_b/N_o is large, the number of bit errors is determined by the value of τ. The simulated BER of a receiver without equalization in QPSK in an additive white Gaussian (AWGN) channel is also shown in Fig. 5, and these results are the same as the theoretical results presented in Ref. (3). An AWGN channel is a flat non-fading channel, so it is simulated as a stationary impulse with additive white Gaussian noise as the only source for errors.

The two-ray model simulations show that adaptive equalizer performance depends highly on the rate of channel variation. A more rapidly changing channel causes larger tracking errors. Also, an equalizer may perform worse than an unequalized system if the ISI is very small. If the delay spread is too small, a system without equalization actually performs better. Using different sized equalizers, it is possible to get drastically different results in BER. Larger sized equalizers do not necessarily have better performance because they may add their own ISI.

3. Equalization in SMRCIM Urban Channel Model

To evaluate the performance of adaptive equalization in real world channels, the Simulation of Mobile Radio Channel Impulse response Models (SMRCIM) was developed by the Mobile & Portable Radio Research Group (MPRG) at Virginia Tech. SMRCIM is a mobile channel impulse response simulator, and it is a statistical model based on various wideband measurements reported in Ref. (5). It generates impulse responses separated by a quarter wavelength ($\lambda/4$) in distance, as if the receiver moved along a fixed track in $\lambda/4$ steps. In an urban environment, the maximum excess delay of SMRCIM is set as 40 μs, which should be able to encompass most observable multipath components.[5] The 40 μs span is divided into 64 time bins, and the multipath components within each time bin are unresolvable.

According to experimental measurements, a statistical model of the probability that a multipath component appears in each time bin was set up. Multipath components in the first impulse response are generated according to the model. The mean path loss was modeled as $(d/d_o)^n$ for the excess delay of each time bin, where d is the transmitter-receiver (T-R) separation, and d_o is the reference distance. The actual path loss is log-normal distributed about the mean path loss. The multipath components are generated according to a Poisson distribution,[16] and their amplitude is correlated to the previous impulse responses.

SMRCIM generates phases of multipath components. In the first impulse response, the phases are randomly generated. After that, a deterministic model[17] is used to determine the phases of multipath components in later impulse responses. SMRCIM displays the distribution of the rms delay spread, CW fading and the amplitude distribution of the CW fading, etc.

Figure 6 is an urban channel named "URBAN5" simulated by SMRCIM. The channel has a T-R separation of 2.68 km, has a mean σ_τ of 3.80 μs, and an average path loss of 119.6 dB. There are 19 impulse responses (profiles) displayed, and they are separated by $\lambda/4$ in distance. The figure also shows the CW fading and the amplitude distribution of the CW signal.

To implement SMRCIM for adaptive equalization, cubic spline interpolation on both inphase and quadrature channel components was used to increase the original $\lambda/4$ spatial resolution of SMRCIM. The delay profiles were quantized into 12 samples per symbol duration, so the sample period was the same as the previous equalization simulations. The channel path loss was normalized so that the mean channel path loss is 0 dB.

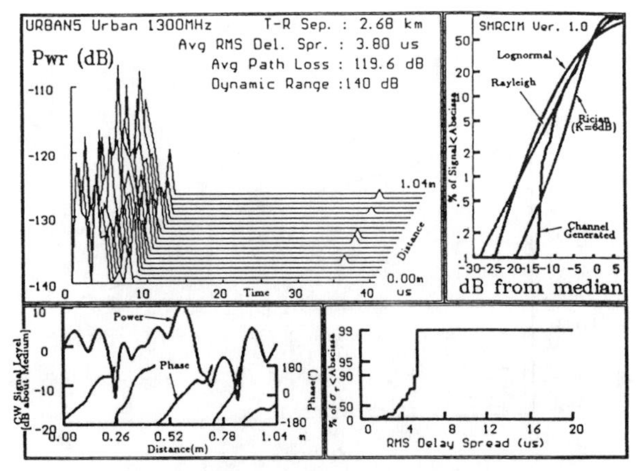

Fig. 6 An urban wide band impulse response named as URBAN5 simulated by SMRCIM at 1.3 GHz. Also shown are the distribution of the rms delay spread, narrow band fading, and narrow band signal power distribution. T-R separation is 2.68 km. The spatial mean rms delay spread is 3.80 μs.

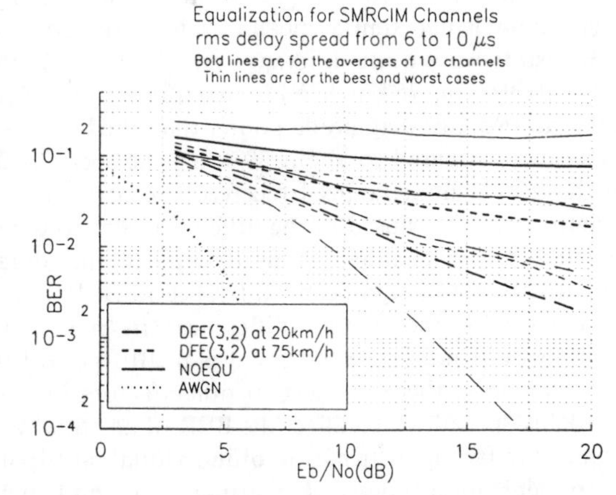

Fig. 7 BER vs. E_b/N_o in several channels generated by SMRCIM using RLS-DFE(3, 2) at 20 km/h and 75 km/h. The mean rms delay spreads of these channels are from 1 to 3 μs. Bold curves are the mean BER.

Ten SMRCIM channels at 1.3 GHz with σ_τ from 1 to 3 μs were selected for the evaluation of the DFE(3, 2) at speeds of 75 km/h and 20 km/h. The BER results of the 10 channels were averaged and plotted as bold curves in Fig. 7 and compared with the BER of a unequalized receiver. The graph also shows the average BER values were virtually the same with or without equalization at 75 km/h vehicle speeds. However, at 20 km/h, adaptive equalization shows its advantage, since the channels changed much slower. For example, Fig. 7 shows that at 75 km/h, neither a DFE(3, 2) or unequalized system could do better than about 2×10^{-2} at $E_b/N_o=20$ dB. Yet at 20 km/h velocity, a DFE(3, 2) could do better than 10^{-2} BER at 13 dB E_b/N_o.

Also, the best and the worst cases among the 10

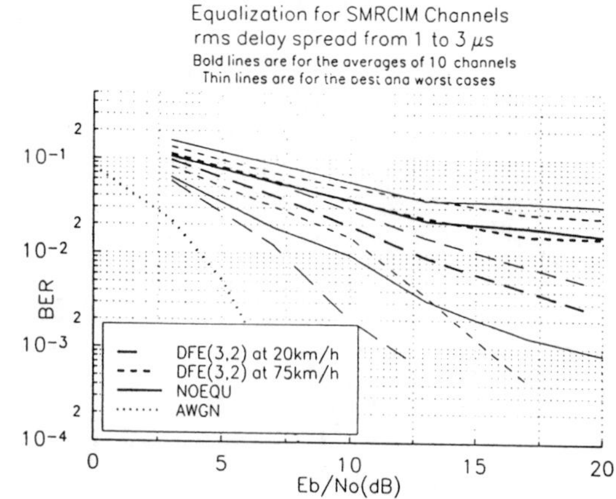

Fig. 8 BER vs. E_b/N_o in several channels generated by SMRCIM using RLS-DFE(3, 2) at 20 km/h and 75 km/h. The spatial mean rms delay spreads of these channels range from 6 to 10 μs. Bold curves are the mean BER, thin lines are the best and worst cases.

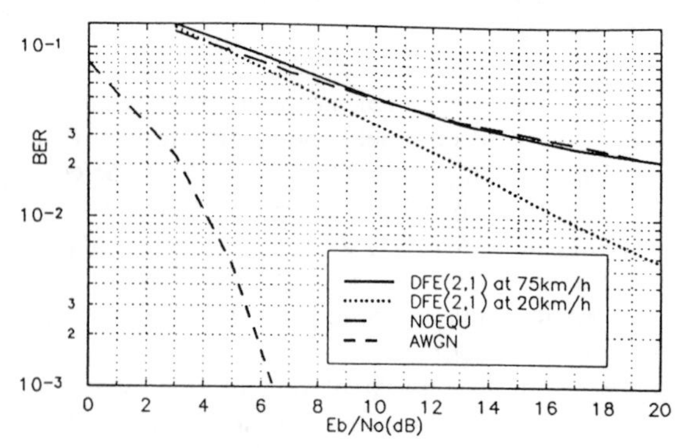

Fig. 9 Mean BER vs. E_b/N_o for 10 channels randomly generated by SMRCIM using RLS-DFE(3, 2) at 20 km/h and 75 km/h.

channels are presented in Fig. 7 to show that although these channels had similar mean rms delay spreads, the BER results were quite different. At 13 dB E_b/N_o, the worst BERs for 75 km/h and nonequalization were about an order of magnitude poorer than the best case. This ratio increased to about 30 when the speed is 20 km/h. For a nonequalized receiver, the performance variation may be caused by the differences in rms delay spread between channels (1 μs to 3 μs in rms delay spread still causes significant difference in BER). But for equalizers, these small differences in rms delay spread would not cause big changes in BER (see Fig. 4). Therefore, there are several other factors besides rms delay spread which affect the BER of an equalizer, such as the fading distribution of individual multipath component amplitudes and distribution of multipath delays.

Another 10 SMRCIM channels at 1.3 GHz with rms delay spreads from 6 to 10 μs were also generated and used in simulation runs. The results are shown in Fig. 8. In this case, using adaptive equalization provided much better results even at 75 km/h. The average BER at 20 dB E_b/N_o using equalization was 4 times lower than the results without equalization. At 20 km/h, the BER results were even more improved. It is clear that the BER performance of a system without equalization is highly dependent on σ_τ (a larger σ_τ causes higher BER). BERs of the best and the worst cases are also shown. Although these ten channels had the similar rms delay spread, the BERs changed significantly. Compared to the results in Figs. 7 and 8, the average performances of equalization in the two σ_τ regions (1~3 μs and 6~10 μs) were very close. Therefore, σ_τ does not affect the performance of a DFE significantly.

Ten random SMRCIM channels at 1.3 GHz, which had σ_τ ranging from 0.5 μs to 10 μs, were used to determine if a DFE can improve the system performance. The mean BER performance for mobile speeds of 20 km/h and 75 km/h are shown in Fig. 9. The results indicate that at 75 km/h, the adaptive equalization does not improve the BER for urban channels, and it has an overall error rate of more than 1% at 20 dB E_b/N_o, which is the same as unequalized channels. At a lower speed, however, an a DFE can significantly reduce the BER as shown in Fig. 9.

Although σ_τ is very important to the system performance without equalization, it does not affect the BER performance of an adaptive equalizer that much. On the other hand, the equalization performances are different for different channels with similar σ_τ values, so there are other factors that affect the equalization performance, such as wideband channel loss fluctuation and the distribution of multipath delays. The speed of a mobile plays a significant role in adaptive equalization in a low data rate urban system. In a system with a data rate on the same order as USDC, the BER increases significantly with increasing mobile speed. Beyond some speed, a DFE can no longer improve the BER of a mobile system. The simulation results show that at 1.3 GHz, the DFE cannot improve a USDC system BER at a receiver velocity larger than 75 km/h.

To transform the results at 1.3 GHz to 850 MHz, which is the frequency band used for U.S. cellular telephones, the Doppler frequency is kept the same for the two carrier frequencies and a new simulated velocity can be found. Let v_i, λ_i represent the velocities and wavelengths at either 1.3 GHz($i=1$) or 850 MHz($i=2$), then, for fixed f_D

$$f_D=\frac{v_1}{\lambda_1}=\frac{v_2}{\lambda_2} \tag{3}$$

Then

$$v_2 = \frac{\lambda_2 \times v_1}{\lambda_1} = \frac{f_1 \times v_1}{f_2}$$

$$= \frac{1300}{850} \times 75 = 115 \text{ km/h} \tag{4}$$

Therefore, at the 850 MHz band, our results suggest a DFE cannot improve a USDC system BER at a receiver velocity larger than 115 km/h.

4. Equalization for SIRCIM Indoor Channel Model

The Simulation of Indoor Radio Channel Impulse response Model (SIRCIM) software, developed by MPRG, is based on extensive wideband channel measurements of different indoor environments at different frequency bands.[17] A large measurement data base was collected in several different types of buildings. SIRCIM generates multipath delay profiles with the same ensemble statistics as measured data.

An indoor channel is different from an urban channel in that it usually has a very small rms delay spread and small temporal variation. Because of its small rms delay spread, indoor systems can use very high data rates without causing high BER. For instance, in an open-plan factory building, a typical σ_τ is about 100 ns and more than 90% of the profiles measured by Rappaport[4] are smaller than 150 ns. Then, if setting a criterion of $\sigma_\tau/T_s<0.1$ for 90% of total number of channels encountered, T_s can be as small as 1500 ns and the symbol rate can be as high as 0.67 million symbols-per-second (Msps). The symbol rate can be even higher for office buildings, which usually have smaller rms delay spreads. Hence, if equalizers are used, the indoor systems can use data rates up to several Mbps.

Indoor channels change very slowly, because of the slow pedestrian traffic velocities. The slow velocity causes a small Doppler frequency and a correspondingly slow variation in the channel. The normalized Doppler frequency Eq. (2) is further reduced because of the high data rates which are typical of indoor systems.

Indoor simulations in this paper used DFEs with RLS algorithms with a weighting factor of 0.99 Several SIRCIM channels with different rms delay spreads were generated for equalization with different data rates. DFEs with different sizes were tested to choose a size that has the best BER performance using the method of finding the point of diminishing return for the numbers of the taps in the feedforward and feedback filters (this technique was used in Sect. 2).

In each simulation run, a total of 18,000 bits were used. These bits were divided into 18 blocks which were evenly spread over the length the SIRCIM channel. Thus, the impulse responses at different locations with different rms delay spreads and path losses in the channel can be used to obtain a reasonably good result without an excessive amount of computation.

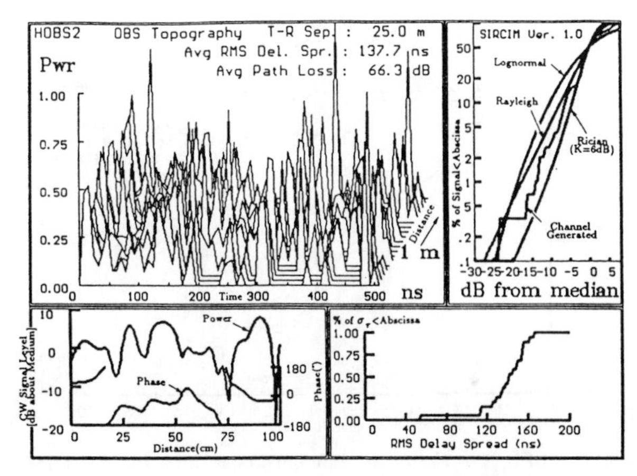

Fig. 10 An indoor wide band impulse response named as HOBS2 simulated by SIRCIM at 1.3 GHz. Also shown are the distribution of the rms delay spread, narrow band fading, and narrow band signal power distribution. The channel is simulated as 100% Obstructed in an open-plan building, T-R separation is 25 m. The spatial mean rms delay spread of HOBS2 is 138 ns.

Numerous multipath profiles for indoor channels were generated by SIRCIM. An example of a simulated channel named "HOBS2" from an open-plan building with a totally obstructed radio path is shown in Fig. 10. Since there is no line-of-sight path, the channel has a mean rms delay spread, $\sigma_\tau=138$ ns which is comparatively large, where the mean is taken as a spatial average of all the impulse responses generated across a simulated mobile travel distance of 1 meter. Figure 10 shows the channel profiles, CW fading, rms delay spread distribution, and CW envelope distribution of the simulated channel. The profiles show that there is a significant multipath component at a delay of 500 ns.

For the simulation results shown in Fig. 11, a 20 Mbps data rate was selected and the speed of the receiver was set to 10 km/h. QPSK modulation was used but no clock recovery circuit was implemented before the equalizer. Since the symbol duration T_s was 100 ns, the normalized rms delay spread, was excessively large ($\sigma_\tau/T_s=1.38$), so adaptive equalization must be used to obtain an acceptable BER. DFE(6, 5) was selected as the equalizer size for the channel HOBS2 at 20 Mbps. Compared to the result without any equalization, which has a BER of about 30%, a DFE gives a significant improvement in the BER performance. The BER is kept below 10^{-3} at an E_b/N_o of 13 dB. The BER of a non-equalization QPSK receiver in an AWGN channel was also simulated and is shown in Fig. 11, and agree with theoretical results in Ref. (19): The simulation indicates that although a DFE has good performance in a frequency-selective fading indoor channel, it is still much-worse than a receiver

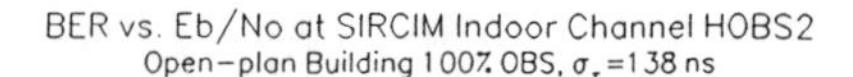

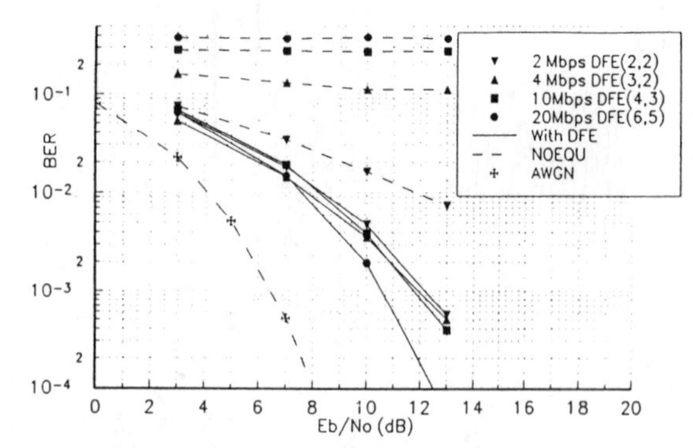

Fig. 11 BER vs. E_b/N_o at different data rates for SIRCIM indoor channel HOBS2. The sizes of DFE are obtained using the method of finding the point of diminishing return. HOBS2 has a spatial mean RMS delay spread of 138 ns. The simulation uses QPSK modulation. Carrier frequency 1.3 GHz.

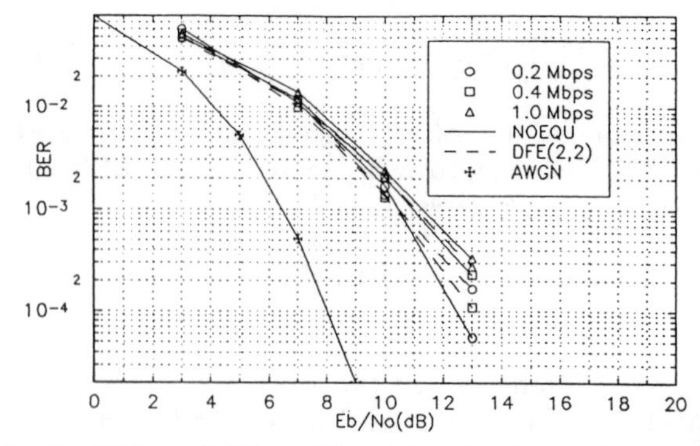

Fig. 12 BER vs. E_b/N_o at different low data rates using RLS-DFE(2,2) in SIRCIM indoor channel HOBS2, which has a spatial mean RMS delay spread of 138 ns. The simulation uses QPSK modulation.

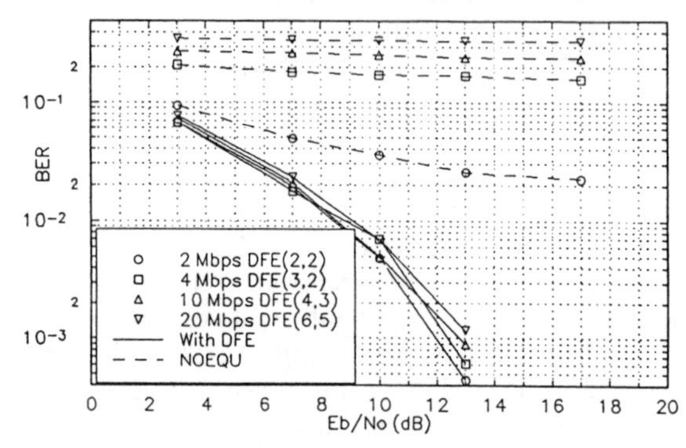

Fig. 13 BER vs. E_b/N_o at different data rates for SIRCIM indoor channel LOS4 using the RLS-DFE sizes for channel HOBS2. HARDLOS2 has a spatial mean RMS delay spread of 33.2 ns. The system uses QPSK modulation.

working in an ideal AWGN channel. The 3 to 6 dB difference is caused by the wideband fading of the indoor channel and the residual ISI in the equalizer. However, the performance of the equalizer is better than an unequalized receiver in a Rayleigh fading channel.[3]

The sizes of DFEs for 10 Mbps, 4 Mbps and 2 Mbps were found in the same way and the BER simulation results are also shown in Fig. 11. The results show that at these data rates a DFE works very well. The values of BER are below 10^{-3} at 13 dB E_b/N_o and will decrease with increasing E_b/N_o. It is clear that adaptive equalization significantly increases the indoor system performance for these data rates. The BER results for different data rates show that the data rate (or normalized Doppler frequency) does not affect the BER.

In order to determine when the DFE can no longer improve the BER, equalization at lower data rates was also tested. The results in Fig. 12 indicate when the data rate was smaller than 1 Mbps ($\sigma_\tau/T_s=0.07$), the DFE provides very limited improvement compared to non-equalization. When the data rate is 0.2 Mbps, there was very small ISI ($\sigma_\tau/T_s=0.012$). In this case, a system without equalization is better than the one with a DFE, although the difference is very small.

Other SIRCIM channels with different rms delay spreads were also generated and used to evaluate DFE performance. LOS4 is an open-plan building channel with a LOS path and has a mean σ_τ of 95.6 ns. HARDLOS2 is a channel for a hard-partitioned building with a LOS path and has a 33.2 ns mean σ_τ. A wide range of data rates were used to evaluate the performance of the adaptive equalization on these channels. The DFE sizes determined for the HOBS2 channel are also used for the equalization simulations of both LOS4 and HARDLOS2 channels.

Figure 13 shows the equalizer performance at different data rates in channel LOS4 and the results without equalization and the optimal AWGN results. When the data rate was 2 Mbps ($\sigma_\tau/T_s=0.1$), the BER without equalization remained larger than 0.01. But, if using a DFE(2, 2), the BER was below 10^{-3} at 13 dB E_b/N_o. The data rate of the system has very limited effect on the equalizer's performance as long as the equalizer is designed for the worst case delay spread. It does, however, play a significant role in the BER without equalization.

Equalization performance in the channel HARDLOS2 is shown in Fig. 14. Although the mean σ_τ for HARDLOS2 is only 33.2 ns, which is three times less than in LOS4, the DFE maintains good performance. The improvement over non-equalization was not as

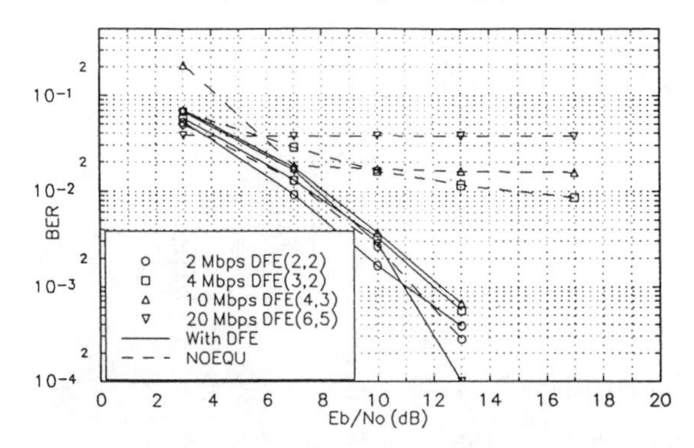

Fig. 14 BER vs. E_b/N_o at different data rates for SIRCIM indoor channel HARDLOS2 using the RLS-DFE sizes for channel HOBS2. HARDLOS2 has a spatial mean RMS delay spread of 33.2 ns. The system uses QPSK modulation.

significant as in the channel HOBS2 and LOS4. At 2 Mbps data rate (σ_τ/T_s=0.03), the results with and without equalization were very close.

Comparing the equalization results for the three indoor channels, indoor equalization systems have very consistent performance regardless of data rate and σ_τ. Since the channel changes very slowly, the channel tracking error is small.

The simulation results show that adaptive equalization can improve the BER by a large margin as long as there exists some ISI. Equalization can even improve the BER performance at σ_τ/T_s as small as 0.03. The reason for the superior performance of adaptive equalization for indoor channels is the relatively slow channel variation and the high data rate, which result in low normalized Doppler frequency.

5. Equalization for Differential Modulations

All the previous equalization simulations presented here were for coherent modulations; however, differential modulation techniques such as $\pi/4$ DQPSK are widely used because their receivers do not require a coherent detector, so the receivers are inexpensive. For coherent modulation techniques such as QPSK, information is sent through the absolute phases of the signals.

$$f(n)=\mathrm{Re}[Ae^{jwnT_s+\theta_n+\theta_0}]$$

$$\text{if } s=00, \quad \theta_n=-\frac{3\pi}{4}$$

$$\text{if } s=01, \quad \theta_n=\frac{3\pi}{4}$$

$$\text{if } s=10, \quad \theta_n=-\frac{\pi}{4}$$

$$\text{if } s=11, \quad \theta_n=\frac{\pi}{4} \tag{5}$$

where s is a symbol consisting two bits, $f(n)$ is the channel input, A is the amplitude of the signal, w is the carrier frequency, T_s is the symbol duration, θ_0 is the carrier's initial phase, and θ_n is the phase shift used to send the signal.

To detect the information, the receiver must be able to recover the exact phase of the carrier, θ_0. Also, because θ_0 changes constantly due to the movement of the transmitter, the receiver, and the medium, the receiver must be able to track the phase variations to receive the signal correctly. Therefore, coherent detection techniques such as the coherent phase locked carrier must be used.

On the other hand, the differential phase shift keying technique sends information by the difference in the phase shift between consecutive symbols. The instantaneous phase of $\pi/4$ DQPSK obeys the following rules:[18]

$$f(n)=\mathrm{Re}[Ae^{jwnT_s+\theta_n}]$$

$$\text{if } s=00, \quad \theta_n=\theta_{n-1}-\frac{3\pi}{4}$$

$$\text{if } s=01, \quad \theta_n=\theta_{n-1}+\frac{3\pi}{4}$$

$$\text{if } s=10, \quad \theta_n=\theta_{n-1}-\frac{\pi}{4}$$

$$\text{if } s=11, \quad \theta_n=\theta_{n-1}+\frac{\pi}{4} \tag{6}$$

All the variables in Eq. (6) are the same as for QPSK in Eq. (5).

By detecting the phase *difference* of the two consecutive symbols, $\theta_n-\theta_{n-1}$, the value of the second symbol can be identified using the rules in Eq. (6). Using noncoherent techniques, the original carrier phase θ_0 does not have to be recovered, and only the instantaneous phase difference is needed. Thus, the receiver is simpler to build. In the structure proposed here, a coherent detection technique is used to implement an adaptive equalizer for differential modulations. An equalizer is used as a coherent detector to recover the exact phase, although a clock recovery circuit may be used instead.

Figures 15(a) and (b) show the block diagram for the implementation of adaptive equalization for $\pi/4$ DQPSK. The equalizer is located before the differential decoder, which is between the equalizer and the decision maker. The input of the feedforward filter is the output of the channel (after RF and IF stages), which is denoted by A in the figures. The output of the equalizer after the decision maker is denoted by D, and D is recoded using the same differential coder as the transmitter. Therefore, B is

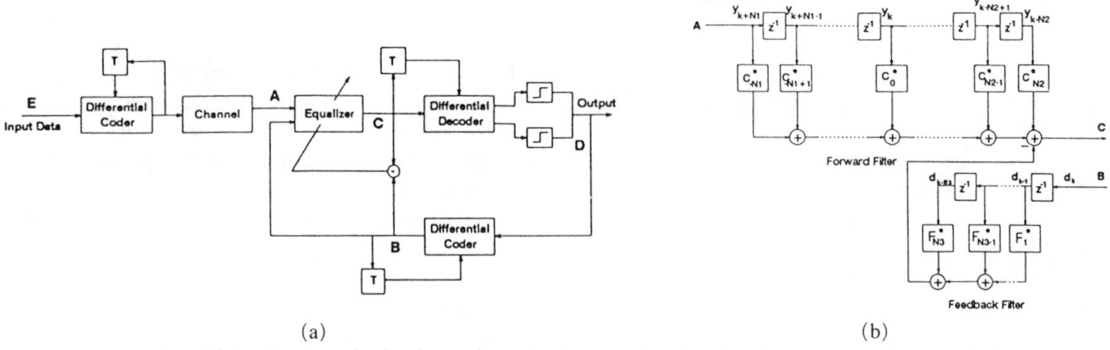

(a) (b)

Fig. 15(a) Structure for implementing adaptive equalization for differential modulations.

(b) Structure of a decision feedback equalizer for a differential modulation system.

the recoded signal, which is used as the input of the feedback filter. The error used to update the equalizer is between B and the output of the equalizer before the differential decoder, C.

The receiver detects the data by obtaining the phase difference between the equalizer's current and previous outputs. Using threshold detection, one of the four phase values in Eq. (6) is chosen, and the symbol output is detected according to the demodulation code. The phase shift is added to the accumulated data being fed into the feedback filter. The accumulated data does not necessarily have the same phase as the channel input, because only differential phase shift is important.

All adaptive equalizers, such as linear transversal equalizer, DFE, and lattice equalizers, can be used in the proposed structure and the LMS or RLS algorithms can be applied to update the equalizer.

This structure works fine when perfect decision feedback is assumed, which means using E instead of D as the input to the differential coder at the receiver (see Fig. 15). However, if detected data D is fed in, once an error appears, the equalizer will lose tracking, and error propagation appears. One of the methods to solve the error propagation problem is retraining, such as in TDMA systems, where the equalizer is retrained at the beginning of each data block and memory is used to store data states in the forward and reverse directions. Also, coding and diversity can be used to reduce the number of errors.

Simulation and Results

A DFE with an RLS algorithm using the structure in Fig. 15 was used in the simulations. To test how well this structure works, several mobile channels were applied in the evaluation of the adaptive equalization for $\pi/4$ DQPSK. The two-ray Rayleigh fading channel was the first to be implemented. Perfect decision feedback was assumed.

Figure 16 shows how a differential $\pi/4$ DQPSK DFE with an RLS algorithm works in a two-ray Rayleigh fading channel. The performance pattern

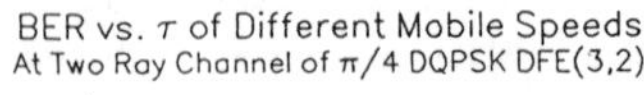

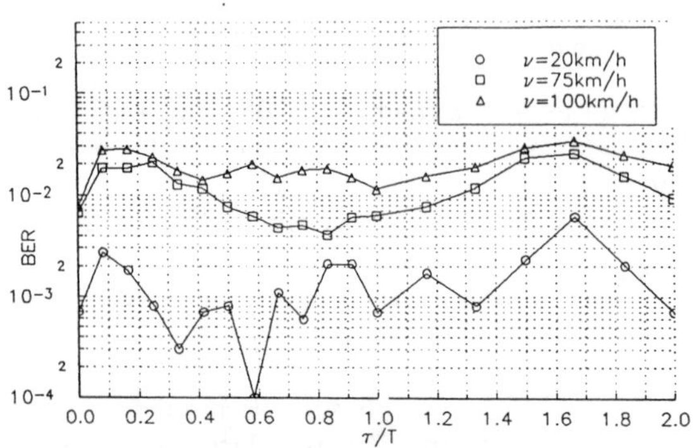

Fig. 16 BER vs. τ of two-ray Rayleigh fading channel using RLS-DFE(3, 2) and $\pi/4$-DQPSK at different speeds. Carrier frequency is 850 MHz.

was similar to that of the QPSK in Fig. 4. A DFE worked better in a slower channel, and its BER was the highest when τ ranges from $0.1T_s$ to $0.3T_s$. However, compared to equalization performance in QPSK (see Fig. 4), equalization results in $\pi/4$ DQPSK were not as good as for the QPSK except when $\tau=0$. When $v=100$ km/h, the equalizer for QPSK had a BER less than 10^{-2} at most values of τ, but for $\pi/4$ DQPSK, the BER was always larger than 10^{-2} except at $\tau=0$. When $v=$ 75 km/h, the equalizer for QPSK generally had better performance, and its lowest BER was about 2.5×10^{-3} at $\tau=T_s$, while for $\pi/4$ DQPSK, the lowest BER was more than 4×10^{-3} at $\tau=0.83T$. The performance difference at 20 km/h was larger than at 75 km/h and 100 km/h.

The indoor SIRCIM channel HOBS2 was used to evaluate the adaptive equalization for $\pi/4$ DQPSK. The DFE sizes chosen previously were used here, and the simulation results are shown in Fig. 17. Note the BERs were maintained around 10^{-3} or less at 20 dB E_b/N_o. However, compared to the results for QPSK in Fig. 11, the BER for $\pi/4$ DQPSK was still higher. For instance, at 20 Mbps, the BER of QPSK at 3 dB E_b/N_o

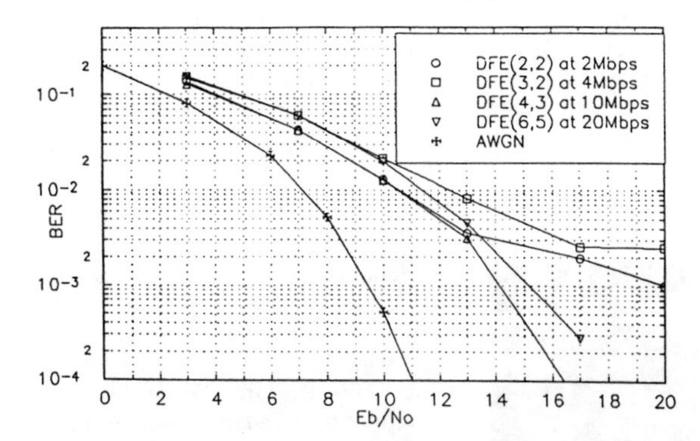

Fig. 17 BER vs. E_b/N_o of SIRCIM channel HOBS2 using RLS-DFE with $\pi/4$-DQPSK at different data rates, and the sizes of the DFE are obtained for QPSK modulation. Carrier frequency is 1.3 GHz.

was about 6.5×10^{-2}, and about 5×10^{-4} at 13 dB E_b/N_o, while for $\pi/4$ DQPSK, the BER was about 1.2×10^{-1} at 3 dB E_b/N_o, and about 3×10^{-3} at 13 dB E_b/N_o, which were significantly larger than for QPSK. The simulated BER results for an unequalized $\pi/4$ DQPSK receiver in an AWGN channel are also presented. There were several dB loss compared to the results for an AWGN channel due to the wideband fading and residual ISI. The unequalized bit-error simulation results for SIRCIM indoor channels can be found in Ref. (3).

The BER difference between the two modulations is caused by the inherent weakness of differential modulation in order to achieve a simpler receiver.[18] Without adaptive equalization, the BER for $\pi/4$ DQPSK is poorer than QPSK.[18] From the simulation results presented here, we have quantified the degradation of adaptive equalization for differential modulations using the structure proposed here.

6. Conclusion

The performance of DFEs in urban and indoor mobile channels has been evaluated. Urban channels were simulated by two-ray Rayleigh fading models. Also, a real world channel impulse response simulator, SMRCIM was developed for equalization simulations. The simulation results show that in urban mobile channels, the performance of an equalizer is highly dependent on the speed of the channel variation (or normalize Doppler frequency). More rapid channel variation results in a higher BER.

Our results show that a DFE does not necessarily improve the performance of an urban mobile radio system. The conditions for an adaptive equalizer to improve BER are that the channel has some delay spread and the channel has a slow variation. The σ_τ value at which the equalizer can improve the BER is dependent on the variation of the channel; the slower the variation, the smaller the σ_τ can be. A DFE for a USDC system will not improve the BER at 1.3 GHz, unless the receiver speed is smaller than 75 km/h, and for 850 MHz, the receiver speed should be less than 115 km/h.

The optimal size of an equalizer depends on the delay spread of the channel. A channel with a large σ_τ value needs an equalizer with more taps. For USDC channels, DFE(3, 2) is a good choice. An equalizer that is either too large or too small will affect the performance of the system.

SIRCIM indoor channel models were used to simulate indoor mobile channels. Indoor channels have small rms delay spreads, so the data rates for indoor systems can be very high compared to urban systems, and the variation of an indoor channel is much slower than an urban channel. The slow channel variation and high data rate result in good performance of the adaptive equalization.

Finally, an equalizer structure for differential modulations was proposed. The structure coherently detects the differential signals. The simulation results show that, although it has similar properties to an equalizer in a coherent modulation, the performance is not as good as a coherent modulation. The SMRCIM and SIRCIM simulation tools are available from Virginia Tech, and details of equalization programing are available in Mr. Huang's Masters thesis.

References

(1) Rappaport, T. S., "The Wireless Revolution," *IEEE Commun. Mag.*, pp. 52-71, Nov. 1991.

(2) Glance, B. and Greenstein, L. J., "Frequency Selective Fading Effects in Digital Mobile Radio with Diversity Combining," *IEEE Trans. Commun.*, vol. COM-31, no. 9, pp. 1085-1094, Sep. 1983.

(3) Rappaport, T. S. and Fung, V., "Simulation of Bit Error Performance of FSK, BPSK, and $\pi/4$ DQPSK in Flat Fading Indoor Radio Channels Using a Measurement-Based Channel Model," *IEEE Trans. Veh. Technol.*, vol. 40, no. 4, pp. 731-740, Nov. 1991.

(4) Rappaport, T. S., "Characterization of UHF Multipath Radio Channels in Factory Building," *IEEE Trans. Antennas & Propag.*, vol. 37, no. 8, pp. 1058-1069, Aug. 1989.

(5) Rappaport, T.S., Seidel, S.Y. and Singh, R., "900 MHz Multipath Propagation for U.S. Digital Cellular Radiotelephone," *IEEE Trans. Veh. Technol.* vol. 39, no. 2, pp. 132-139, May 1990.

(6) Jakes, W.C., *Microwave Mobile Communications*, Wiley, New York, 1974.

(7) Lee, W.C.Y., "Theory of Wideband Radio Propagation," *IEEE Vehicular Technology Conference 1991*, pp. 285-288, (St. Louis, MO), May 1991.

(8) Proakis, J.G., *Digital Communications*, McGraw-Hill, New York, 1989.

(9) Huang, W. and Rappaport, T.S., "A Comparison Study of Two Adaptive Equalizers," *IEEE Vehicular Technology Conference 91*, pp. 765-769, (St. Louis, MO). May

1991.
(10) D'Aria, G. and Zingarelli, V., "Results on Fast-Kalman and Viterbi Adaptive Equalizers for Mobile Radio with CEPT/GSM System Characteristics," *Proceedings of IEEE GLOBECOM*, pp. 26.3.1-26.3.5, 1988.
(11) Chennakeshy, S., et al., "Decision Feedback Equalization for Digital Cellular Radio," *Proceeding IEEE International Conference on Communications*, pp. 1492-1496, (Atlanta, GA) Apr. 1990.
(12) Lo. N.W.K., Falconer, D.D. and Sheikh, A.U.H., "Adaptive Equalization and Diversity Combining for a Mobile Radio Channel," *GLOBECOM 90*, pp. 923-927, (San Diego, CA) Dec. 1990.
(13) Sexton, T.A. and Pahlavan, K., "Channel Modeling and Adaptive Equalization of Indoor Radio Channels," *IEEE J. Sel. Areas Commun.*, vol. 7, no. 1, pp. 114-121, Jan. 1989.
(14) Electronics Industries Association, "EIA/TIA INTERIM STANDARD: Cellular System Dual-Mode Mobile Station-Base Station Compatibility IS-54," May 1990.
(15) Fung, V., Rappaport, T.S. and Thoma, B., "Bit Error Simulation for $\pi/4$ DQPSK Mobile Radio Communications Using Two-Ray and Measurement-Based Impulse Response Models," *IEEE J. Sel. Area Commun.-Special Issue on Simulation*, May 1993.
(16) Hashemi, H., "Simulation of Urban Radio Propagation," *IEEE Trans. Veh. Technol.*, vol. VT-28, pp. 213-225, Aug. 1979.
(17) Rappaport, T.S., et al., "Statistical Channel Impulse Response Models for Factory and Open Plan Building Radio Communication System Design," *IEEE Trans. Commun.*, vol. 39, no. 5, pp. 794-807, May 1991.
(18) Couch II, L.W., *Digital and Analog Communication Systems*, Macmillan, New York, 1989.
(19) Thoma, B. and Rappaport, T.S., "Simulation of Bit Error Performance and Outage Probability of $\pi/4$ DQPSK in Frequency-Selective Indoor Radio Channels Using a Measurement-Based Channel Model," *GLOBECOM '92*, (Orland, FL), Dec. 1992.

Theodore S. Rappaport was born in Brooklyn, NY on November 26, 1960. He received B.S.E.E, M.S.E.E, and Ph.D. degrees from Purdue University in 1982, 1984, and 1987, respectively. In 1988, he joined the Electrical Engineering faculty of Virginia Tech, Blacksburg, where he is an associate professor and director of the Mobile and Portable Radio Research Group. Prof. Rappaport conducts research in mobile radio communication system design and RF propagation prediction through measurements and modeling. He guides a number of graduate and undergraduate students in mobile radio communications, and has authored or coauthored more than 70 technical papers in the areas of mobile radio communications and propagation, vehicular navigation, ionospheric propagation, and wideband communications. He holds several patents and is co-inventor of SIRCIM, and SMRCIM, radio channel simulators that have been adopted by over 50 companies and universities. In 1990, he received the Marconi Young Scientist Award for his contributions in indoor radio communications, and was named a National Science Foundation Presidential Faculty Fellow in 1992. He is an active member of the IEEE, and serves as senior editor of the IEEE Journal on Selected Areas in Communications. He is a Registered Professional Engineer in the State of Virginia and is a Fellow of the Radio Club of America. Dr. Rappaport is also president of TSR Technologies, a cellular radio and paging test equipment manufacture.

Weifeng Huang received his MSEE degree in December 1991 from Virginia Polytechnic Institute and State University, working with the Mobile and Portable Radio Research Group in the area of simulation of adaptive equalization in mobile radio channels until his employment with Comsearch in May 1992.

Martin J. Feuerstein was born in Memphis, TN, in 1962. He received the B.E. degree in electrical engineering and mathematics from Vanderbilt University, Nashville, TN, in 1984. In 1987 he received the M.S. degree in electrical engineering from Northwestern University in Evanston, IL. He received the Ph.D. degree in electrical engineering from the Virginia Polytechnic Institute in Blacksburg, VA, in 1990. From 1984 to 1985, he was a systems engineer with the Advanced Communications Terminals Division of Northern Telecom, Inc. From 1985 to 1987, he was a research assistant with the Microwave Characterization Lab at Northwestern University, where he worked on microwave plasma generation and diagnostics systems. From 1987 to 1990, he was a project assistant with the Satellite Communications Group at Virginia Tech, where he worked on the analysis of spread spectrum position location systems. During 1990 he was a development engineer with Spatial Positioning Systems Inc., where he worked on 3-D laser position measurement systems. From 1991 to 1992, he was a Visiting Assistant professor with the Mobile & Portable Radio Research Group at Virginia Tech, where he worked on radio propagation prediction techniques. Since joining US WEST Advanced Technologies, Boulder, CO, in 1992, he has worked on radio propagation and wireless system design. Dr. Feuerstein is a member of Tau Beta Pi, Eta Kappa Nu, Phi Eta Sigma, and Phi Pappa Phi. He is a life member of the American Radio Relay League.

Reprinted from *IEEE Transactions on Vehicular Technology*, Vol. 41, No. 3, Aug. 1992, pp. 231-241.

Effects of Radio Propagation Path Loss on DS-CDMA Cellular Frequency Reuse Efficiency for the Reverse Channel

Theodore S. Rappaport, *Senior Member, IEEE,* and Laurence B. Milstein, *Fellow, IEEE*

Abstract—**Analysis techniques that quantitatively describe the impact of propagation path loss and user distributions on wireless direct-sequence code-division multiple-access (DS-CDMA) spread spectrum systems are presented. First, conventional terrestrial propagation models which assume a d^4 path loss law are shown to poorly describe modern cellular and personal communication system channels. Then, using both a two-ray propagation model and path loss models derived from field measurements, we analyze the impact of path loss on the frequency reuse efficiency of DS-CDMA cellular radio systems. Analysis is carried out for the reverse (subscriber-to-base) channel using a simple geometric modeling technique for the spatial location of cells, and inherent to the geometry is the ability to easily incorporate nonuniform spatial distributions of users and multiple layers of surrounding cells throughout the system. Our analysis shows the frequency reuse efficiency (F) of the reverse channel with a single ring of adjacent cells can vary between a maximum of 71% in d^4 channels with a favorable distribution of users, to a minimum of 33% in d^2 channels with a worst case user distribution. For three rings of adjacent users, F drops to 58% for the best d^4 case, and 16% for the worst d^2 case. Using the two-ray model, we show that F can vary over a wide range of values due to the fine structure of propagation path loss. The analysis techniques presented here can be extended to incorporate site-specific propagation data.**

I. Introduction

WIDESPREAD use of CDMA for personal wireless communications is likely to occur over the next several years. In 1990, more than a dozen experimental license applications were tendered by the FCC for U.S. wireless communication systems using direct-sequence spread spectrum code division multiple access (DS-CDMA). U.S. CDMA field trials are currently underway by major cellular and personal communications companies using existing cellular spectrum in the 800–900 MHz and 1.8 GHz bands. The virtues of spread spectrum in a frequency reuse mobile radio environment have been discussed by only a handful of researchers over the past two decades (see [1]-[3] for example), and in light of recent claims that CDMA could afford several times more capacity than other access techniques, analyses and simulation methods that accurately predict the capacity of CDMA systems in real world propagation environments are warranted.

Manuscript received April 1, 1991; revised October 15, 1991. This work was supported in part by DARPA/ESTO and the Mobile and Portable Radio Research Group (MPRG) Industrial Affiliates Program at Virginia Tech, and the NSF I/UCRC for Ultra-High Speed Integrated Circuits and Systems at the University of California at San Diego. Portions of this paper were presented at GLOBECOM '90, San Diego, CA.

T. S. Rappaport is with the Mobile and Portable Radio Research Group, Bradley Department of Electrical Engineering, Virginia Polytechnic Institute and State University, Blacksburg, VA 24061-0111.

L. B. Milstein is with the Department of Electrical and Computer Engineering, University of California, La Jolla, CA 92093-0407.

IEEE Log Number 9200218.

This paper studies one quantitative cellular performance measure: frequency reuse efficiency on the reverse (subscriber-to-base) channel of a CDMA cellular system that employs total frequency reuse. That is, we analyze the performance of a cellular system that allows all users in all cells to share the same reverse channel carrier frequency. In our work, we model a cellular radio system by using a concentric circle geometry that easily accommodates various distributions on the number or location of users within cells. Within each cell, ideal power control is assumed on the reverse channel, so that within a cell of interest, the same power level is received at the base from all subscribers within that cell. Subscribers in other cells are power-controlled within their own cell, and it is their radiated power that propagates as interference to the base station receiver in the cell of interest. This interference raises the noise level, and thereby reduces the number of users that can be supported at a specified average performance level.

Holtzman [9] has shown how imperfect power control in the subscriber unit degrades the frequency-reuse efficiency of a CDMA cellular system. However, except for [11], we are not aware of analyses that quantitatively describe how propagation path loss or user distributions affect the frequency reuse efficiency of such a system. As is shown subsequently, both propagation path loss and distribution of users can greatly impact frequency reuse efficiency of CDMA, even with perfect power control.

This paper is organized into five sections. Section II presents a brief overview of propagation models and demonstrates with both the classic two-ray model and measured data how modern cellular systems undergo path loss that increases with distance to the second or third power, as opposed to the commonly used d^4 model. For emerging microcellular systems, we propose a path loss law that incorporates a close-in free space reference power level to which all further users are referenced. Section III presents a frequency reuse analysis technique that uses concentric circles to model the location of adjacent cells, as well as the distribution of users within each cell. We also show how the concentric circle geometry can be equated to the conventional hexagonal geometry used in [1], [3]. In Section IV, we use this analysis technique in conjunction with path loss models developed in Section II to arrive at reverse channel frequency reuse factors for a variety of propagation channels and user distributions. Section V summarizes the results of this work.

II. Propagation

Accurate propagation modeling is vital for accurately predicting the coverage and capacity of cellular radio systems. This is particularly true in evolving CDMA systems, since capacity is interference limited instead of power limited. Unlike conventional cellular radio system design which strives for excellent base station coverage in each cell, and then relies on judicious frequency management within a service area to mitigate cochannel interference, CDMA requires no frequency planning.[1]

While more recent work has considered small scale fading and imperfections in power control [15], in this work we use a simple small scale shadow fading model, neglect frequency-selective fading effects, and consider average path loss as a function of distance between base and mobile. Thus, we consider very slow flat fading channels with no resolvable multipath components. Since we are interested in analyzing the average frequency reuse efficiency of DS-CDMA systems in different path loss environments, the use of an average distance-dependent path loss model that neglects small scale shadow fading of individual users is adopted. That is, we simplify the analysis by assuming the total received power from N independent transmitters located at a radial distance r from the base receiver is, on average, N times the power from a single subscriber at r as predicted by the path loss model. This is a valid approach for the case of any symmetric shadowing distribution about the mean path loss, and a sufficiently large number of users. However, field measurements often show that shadow fading is log-normally distributed about the mean, which is not a symmetric distribution in terms of absolute power levels. Thus, while our analysis offers insight into how different cell sizes and different path loss models impact frequency reuse, our analysis does not address second-order shadowing effects. The simulation methodology in Section IV, however, can be adopted to include shadowing effects and is described subsequently.

First, we consider the classic two-ray model (direct path and single ground reflection) to see how received signal power varies with distance between transmitter and receiver due to a simple site-specific channel model. Then, we show measured data which agrees with a path loss power law of distance to the nth power, where n is a constant that ranges between 2 and 4.

Fig. 1 illustrates the classic two-ray propagation geometry that is used to model long-haul terrestrial microwave systems. There is a π radian phase shift induced by the ground reflection [10]. This geometry does not consider other multipath or shadowing effects other than the ground reflected path. The individual LOS and ground reflected components are considered to have amplitudes which do not vary with time, and which individually undergo free space propagation as the separation distance (d) between transmitter and receiver increases. The path loss is due to the envelope produced as the RF phases of each signal component vectorially combine in different ways as the mobile travels over space. We shall show shortly that in real channels which have many multipath components that fade and are shadowed [4], [5], measured large scale path loss is comparable to those predicted from the simple geometry in Fig. 1, so long as a free space reference distance close to the transmitter is used [11], [14].

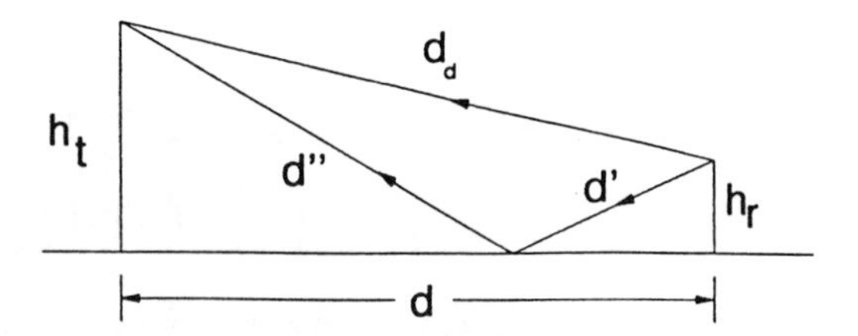

Fig. 1. Geometry for the two-ray (ground reflection) path loss model.

Historically, the geometry in Fig. 1 has been used to describe the large scale site-specific propagation phenomenon observed in terrestrial and analog cellular systems, systems which have deliberately aimed to maximize coverage distance. Traditional use of the two-ray model predicts the received signal several kilometers away, based on a received signal level at 1 km separation. It is only recently that the two-ray model of Fig. 1 has been used to model path loss in microcellular channels where small coverage areas are desired to increase frequency reuse, and thus capacity [6], [14]. Measurements in cellular and microcellular channels that provide log-distance path loss models using reference distances closer than 1 km have also been reported recently [7].

At a distance r from a transmitting antenna, if an electric field intensity given by $E = (A/r)cos(2\pi ft)$ V/m (without ground reflection) is measured in free space, then for the two-ray model of Fig. 1, the received electric field envelope at an antenna located at a distance r from the transmitting antenna is given by

$$|E(r)| = \frac{2A}{r} \sin\left(\frac{\theta}{2}\right) \tag{1}$$

where

$$\theta = \frac{2\pi}{\lambda}\left(\sqrt{(h_t+h_r)^2+r^2} - \sqrt{(h_t-h_r)^2+r^2}\right) \approx \frac{4\pi h_t h_r}{\lambda r}, \quad \text{for } r \gg h_t h_r \tag{2}$$

where h_t is the height of the transmitter antenna in meters, h_r is the height of the receiver antenna in meters, and λ is the carrier wavelength in meters. Equation (1) assumes unity gain antennas at both transmitter and receiver.[2] From (1), the

[1] CDMA can exploit resolvable multipath components using path diversity and combining (RAKE reception); however, energy in multipath components in frequency selective channels can add to the interference level of the system. Urban wide-band measurements have shown that channels can induce many delayed multipath components that have strengths within 10 dB of the strongest signal [4],[5]. We do not consider the effect of frequency-selective fading in this analysis.

[2] This analysis assumes that both the transmitter and receiver antennas offer equal gains to the direct and ground-reflected components. For very small values of r, say less than a few tens of meters, (1)–(4) are not valid and the true antenna gain patterns must be considered [14]. It can be shown that at very close range, the path loss is dominated by the direct path. The authors thank Dr. Joseph Shapira for pointing this out.

TABLE I
EXPERIMENTAL RESULTS OF WIDE-BAND PROPAGATION MEASUREMENTS IN SIX CELLULAR AND MICROCELLULAR CHANNELS; BEST-FIT EXPONENT VALUES WERE COMPUTED ASSUMING A 100 m FREE SPACE REFERENCE DISTANCE

	Antenna Height (m)	n	σ (dB)	Maximum T-R Separation (km)	Maximum rms Delay Spread (μs)	Maximum Excess Delay Spread 10 dB (μs)
Hamburgh	40	2.5	8.3	8.5	2.7	7.0
Stuttgart	23	2.8	9.6	6.5	5.4	5.8
Dusseldorf	88	2.1	10.8	8.5	4.0	15.9
Frankfurt (PA Building)	20	3.8	7.1	1.3	2.9	12.0
Frankfurt (Bank Building)	93	2.4	13.1	6.5	8.3	18.4
Kronbreg	50	2.4	8.5	10.0	19.6	51.3
All (100 m)		2.7	11.8	10.0	19.6	51.3
All (1 km)		3.0	8.9	10.0	19.6	51.3

received power is proportional to $|E(r)|^2$, and is given by

$$P(r) \propto |E(r)|^2 = \frac{4A^2}{r^2}\left(\frac{1}{2} - \frac{1}{2}\cos\theta\right). \quad (3a)$$

When the value of A is known exactly, (3a) can be used to determine the absolute electric field (and thus absolute received power) as a function of the transmitter-receiver (T-R) separation distance r. The value of A in (3a) represents the measured electric field intensity at a convenient close-in distance, which is in the far field of the transmitting antenna, i.e., the measured electric field at a distance of 1 m, 100 m, 1 km, or some other convenient distance from the transmitting antenna.

The absolute received power level as a function of r is given exactly by

$$P(r) = |E(r)|^2 * A_e/377 \quad (3b)$$

where A_e is the effective area of the receiver antenna, 377 Ω is the intrinsic impedance of free space, and A_e is proportional to the receiver antenna gain [12].

When the T-R separation, r, is much greater than $h_t h_r$, (3a) simplifies to

$$P(r) \propto \frac{A^2}{r^4}\left(\frac{4\pi h_t h_r}{\lambda}\right)^2. \quad (4)$$

The denominator of (4) suggests that power increases as distance to the fourth power, and this is sometimes quoted in the literature. This is also the propagation law assumed in [3]. It must be noted, though, that (4) is an asymptotic relationship, which holds when r is typically much larger than 1 km. As shown subsequently, (4) does not relate the actual power of distant users to close-in users, although such a model is required for meaningful CDMA system analysis and design. Rather, (4) relates the received power of very distant users to other distant users. Thus, we propose the use of a close-in free-space reference distance, α, to which the received signal strengths at all farther distances may be compared.

A. Path Loss Models

The geometry in Fig. 1 and the propagation equations (1)–(3) can be used to determine the distance-dependence of received power, and consequently the distance-dependence of an appropriate power law exponent for a two-ray flat-fading channel. If we let α denote the distance between the closest subscriber and the base station, and let α be the distance at which A in (3a) is measured, then the power of all subscribers with T-R separations greater than α can be related to the known power at α.

Consider DS-CDMA where all in-cell transmitting subscribers are power controlled to have identical signal levels P_n arriving at the base receiver. If we assume all subscribers are located at least α m away from the base station, and further assume that subscribers at α have transmitted power $P(\alpha)$, the transmitted power $P(r)$ of a subscriber at distancer$(r > \alpha)$ can be described by a log-distance model [5]

$$P(r)_{\text{dB}} = P(\alpha)_{\text{dB}} + \log_{10}\left(\frac{r}{\alpha}\right)^n. \quad (5a)$$

For this work, we assume that $P(\alpha)$ is controlled based on a free space propagation channel (direct path only) between the base and all subscribers located α m from the transmitter. The free space channel assumption for the base to the closest subscriber is not strict, and in real channels will not necessarily be valid since it depends on the antenna heights and patterns, and multipath. However, for analysis of CDMA frequency reuse, it is necessary to relate all of the power levels to some known power level $P(\alpha)$, which we consider to be caused by close-in free space propagation. For an interference limited system, one in fact may use any reasonable constant for $P(\alpha)$.[3]

The propagation path loss model in (5a) implies that if $P(\alpha)$is specified and $P(r)$ is known, either from a model (i.e., (3)) or measurement, then the path loss exponent n depends on both α and r. This can be seen be rewriting (5a)

$$n(\alpha, r) = \frac{P(r)_{\text{dB}} - P(\alpha)_{\text{dB}}}{+10\log_{10}(r/\alpha)}. \quad (5b)$$

For modern systems, α is on the order of several tens of meters, particularly in microcellular systems within heavily populated areas. Once α and $P(\alpha)$are specified, the system path loss PL in decibels, referenced to the closest user from the base

[3]The actual value of $P(\alpha)$ does have system design ramifications, since the absolute power levels will determine coverage areas. In this paper, we assume that base stations are suitably spaced so that the cellular system is not limited by link budget, but by interference.

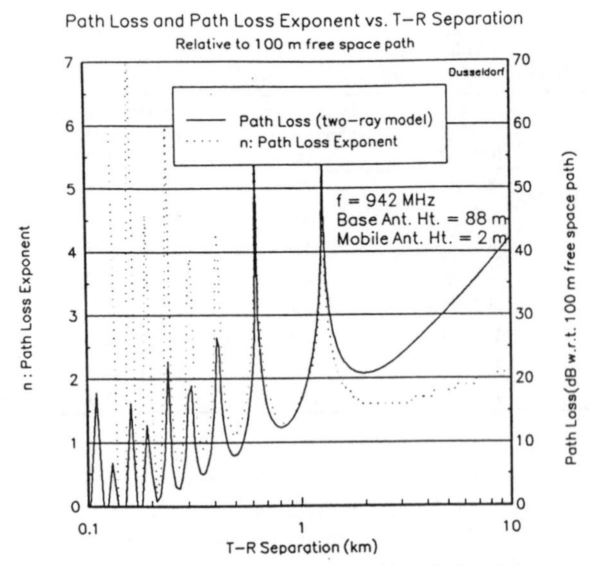

Fig. 2. Path loss and path loss exponent values for a cellular radio system as a function of T-R separation (r)and close-in reference distance α. Path loss is computed using a two-ray propagation model from (1)–(3), and path loss exponent is computed from (5b) using $f = 942$ MHz, $h_t = 88$ m, $h_r = 2$ m.

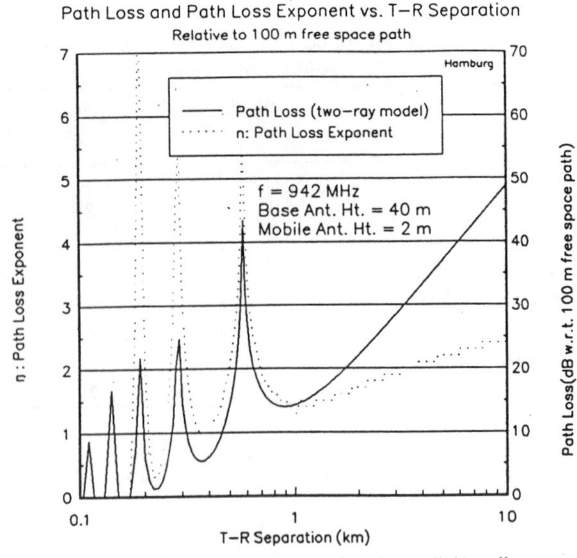

Fig. 3. Path loss and path loss exponent values for a cellular radio system as a function of T-R separation (r)and close-in reference distance α. Path loss is computed using a two-ray propagation model from (1)–(3), and path loss exponent is computed from (5b) using $f = 942$ MHz, $h_t = 40$ m, $h_r = 2$ m.

station, can be expressed as

$$PL(\alpha, r)_{dB} = P(r)_{dB} - P(\alpha)_{dB} = 10 \log_{10}\left(\frac{r}{\alpha}\right)^{n(\alpha,r)}. \quad (5c)$$

Figs. 2–4 give examples of $PL(\alpha, r)$ and $n(\alpha, r)$ for three different channels that use different antenna heights and frequencies in the 900 MHz band. Additional examples can be found in [11]. Figs. 2–4 were derived by using the two-ray model (3) to compute $P(r)$, (5b) to compute $n(\alpha, r)$, and (5c) to compute path loss, where the value of $P(\alpha)$ was assumed to be due to free space propagation ($n = 2$) over a 100 m path to the base. An important observation is that for a particular antenna configuration, the path loss exponent value, n, is certainly not constant at close-in distances, although it converges to a constant several kilometers away from the base station. It can be seen that at typical microcell fringe distances of between $d = 0.5$ km and $d = 2$ km, path loss exponents approach values between about 2.0 and 3.0 when referred to a 100 m reference. Large signal increases and deep fades caused by phase combining of the LOS and ground reflected paths occur and are predicted by the two-ray model. It should be clear that the d^4 path loss power law is a poor model at distances less than several kilometers from a terrestrial base station. This can be readily seen by observing Fig. 2 and noticing the path loss at 10 km (42 dB) is only 27 dB down from the path loss at 1 km (15 dB). In a fourth power law, this difference would be 40 dB. Between 0.1 and 1 km, the error in the fourth power model is even more striking. Even with very low antennas, Fig. 4 shows that between 0.1 and 1 km, the two-ray model predicts only 23 dB path loss isolation.

Another model assumes the power law exponent, n, in (5) is not a function of r, but is a simple constant. The proper value for n is found by minimizing the mean square error of the best fit line on a scatter plot such as shown in Fig. 5 [5]. Recent experiments have shown that the appropriate

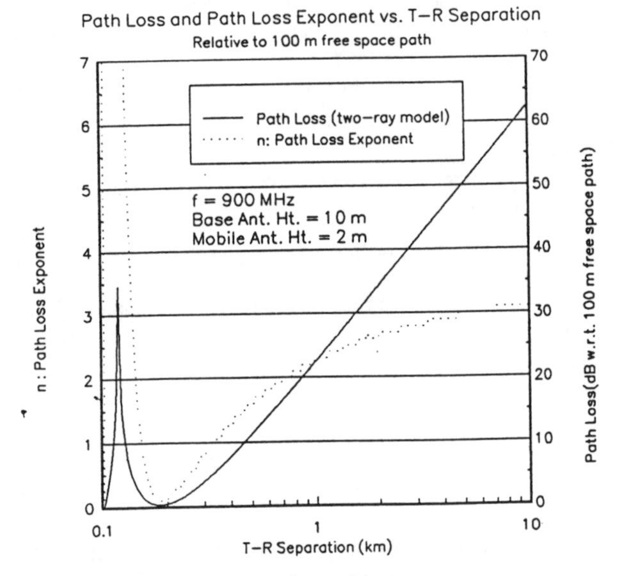

Fig. 4. Path loss and path loss exponent values for a cellular radio system as a function of T-R separation (r)and close-in reference distance α. Path loss is computed using a two-ray propagation model from (1)–(3), and path loss exponent is computed from (5b) using $f = 900$ MHz, $h_t = 10$ m, $h_r = 2$ m.

value of n is a function of the close-in reference distance used (see [5]–[8], [14]), and that d^4 predicts too much attenuation. In [5], a 500 ns probe was used to obtain an experimental database of about 6000 power delay profile measurements in six cell sites. A simple propagation model that assumes path loss is log-normally distributed about a mean path loss that falls off as a function of distance to a fixed exponent [12] was used to model the data. The measured averaged path loss provided the best (i.e., minimum mean square error) fit with a $d^{3.0}$ law when the free space reference distance was 1.0 km, but when a 0.1 km reference distance was used, the best fit was $d^{2.7}$. Fig. 5 shows the measured data and

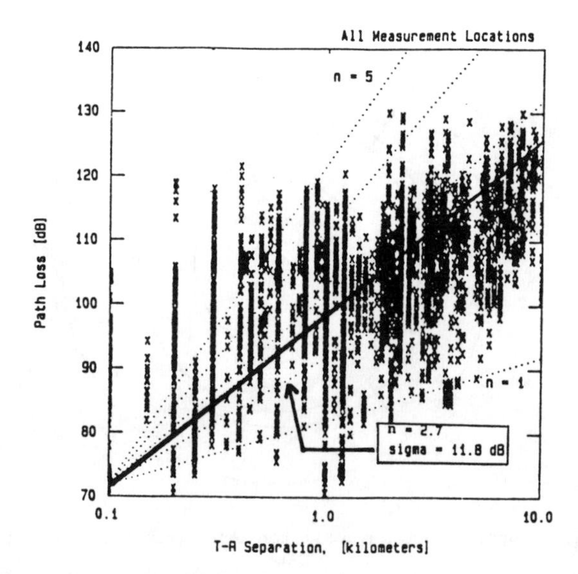

Fig. 5. Scatter plot of absolute path loss measured in six cellular and microcellular systems. The different asymptotic values for n have been computed from a 100 m reference distance using (5) in the paper. The path loss law of $d^{2.7}$ is the best linear regression fit to a log distance law of the form d with a free space leverage point at α = 100 m. Note that at distances less than 1 km, there are several locations which undergo deep fades, just as in Figs. 2–4 [5].

the minimum mean square error $d^{2.7}$ path loss law for a 100 m free space reference distance. Measurements in [5], [14] verify that measured average path loss obeys a distance power law of between d^2 to d^4 when referenced to a 100 m free space reference, and spot measurements are log-normally distributed in decibels about the average due to shadowing, with standard deviations ranging between 7 and 13 dB. Table I gives channel parameters[4] from measurements in six cellular and microcellular systems, and includes the minimum-mean-squared-error fit of n for a d^n mean path loss law [5].

When scatter plots of measured path loss data are compared with predicted path loss from a two-ray model using the correct antenna heights for base and mobile, there is reasonable agreement between the predicted location of deep fades within a kilometer from the base station [5],[6]. For example, Fig. 3 shows the two-ray prediction using a base antenna height that is approximately equal to the average base antenna height used for the measurements described in Table I and Fig. 5. By comparing path loss generated by the analytic model (i.e., Fig. 3) with measured data shown in Fig. 5, one can see both figures show large path loss values (deep fades) occurring within the first few kilometers around a base station. For distances further from the base station, the analytical model and the best fit line for the average measured path loss have comparable path loss exponents. In Section IV, both the analytical two-ray model and a simple d^n path loss model are used to compute frequency reuse efficiencies for CDMA, but the techniques shown subsequently can be used to predict performance in any distant-dependent propagation model, and may include shadow fading, as well.

[4]In Table I, maximum rms delay spread denotes the largest measured square root of the second central moment of spatially-averaged multipath power delay profiles chosen from all measurement runs in a geographic region. Maximum excess delay spread denotes the largest measured propagation delay of a multipath component, relative to the first arriving signal, having strength within 10 dB of the first arriving signal in a spatially averaged multipath power delay profile chosen from all measurement runs in a geographic region [5].

III. Propagation Effects On Frequency Reuse

We now find the cochannel interference on the reverse channel of a cellular DS-CDMA system by finding loose upper and tight lower bounds on the amount of out-of-cell (adjacent cell subscribers) interference and the in-cell interference (caused by other subscribers in the cell of interest). The adjacent cell subscribers are power-controlled by their own cell base station, and all in-cell subscribers are power controlled to provide the same signal level at the base receiver as any other in-cell subscriber.

The ratio of in-cell noise to total received noise is a figure of merit, called the frequency-reuse factor, which has values between 0 and 1. For DS-CDMA cellular, the frequency reuse factor is denoted by f, and is given by

$$f = \frac{N_0}{(N_0 + M_1 N_{a1} + M_2 N_{a2} + M_3 N_{a3} + \cdots)} \tag{6a}$$

and F is the frequency reuse efficiency, given by

$$F = f * 100\%. \tag{6b}$$

For conventional channelized cellular radio systems (including the U.S. Analog Mobile Phone standard), f is given by the number of channels assigned to a single cell divided by the total number of channels allocated to the system. This is because no cochannel interference comes from within the cell, but only from adjacent cochannel cells. The reuse factor is then dictated by the distance between cochannel cells and the path loss of the channel. For a d^4 path loss model, it can be shown that the minimum separation for a specific cochannel interference level (i.e., 18 dB) yields a seven-cell reuse geometry, or $f = 1/7$ [13]. For n values less than 4, f is less than 1/7 for channelized cellular systems.

In contrast, DS-CDMA deliberately allows for in-cell interference but does not require frequency coordination between any cell. In (6a), N_0 denotes the received noise power from all but one of the subscriber transmissions in the cell of interest (we assume one subscriber transmission is signal, not noise), and N_{ai} denotes the noise power received from all of the subscribers in one of the adjacent M_i cells located in the ith surrounding ring of the cell of interest. In general, f will be a random variable, since the noise powers from all surrounding cells will be a function of the random locations of adjacent users as well as random shadowing and voice activity, despite the fact that power control is employed in all cells. In this work, we neglect voice activity (i.e., assume all subscribers are transmitting simultaneously), assume perfect power control, and restrict the analysis to solving for *average* values of N_{ai} since we assume the shadowing of each user in each cell is symmetrically distributed about the average path loss. Note from (6a) that if only the single cell of interest is operating with a particular number of users, then a fixed signal-to-noise ratio will exist at the base station receiver for each received subscriber signal. When adjacent cells are then considered, the total noise level at the desired base station increases by a factor $1/f$. If independent users and uncorrelated spreading codes are assumed, then the noise powers add linearly after despreading at the receiver, and the increased noise will have the same effect as decreasing the number of users by a factor f to maintain the same signal-to-noise ratio as for the single cell case.

A. Assumptions

We make the following reasonable assumptions for analysis. 1) The spatial distribution of subscribers within each adjacent cell is identical, although our technique can be used to individually specify distributions in each adjacent cell (our approach accommodates various spatial distributions of users by using simple weighting factors); 2) all cells have equal area A, and there are K users per unit area. Thus, each cell has $U = KA$ users; 3) the distance of the closest subscriber to its own base is ($\alpha < d$); 4) α is the same for all cells; 5) the distance of the farthest in-cell user from the base station of interest is d; 6) the cell of interest has radius d; 7) the power received from subscribers > $7d$ (outside of the third ring) from the desired cell base is ignored; 8) ideal power control is assumed within each cell; 9) omnidirectional antennas are used at base and mobiles; 10) the received powers from all users add at the base receiver.

Fig. 6 shows the geometry used for analysis, where only the first ring of adjacent cells is shown. Our analysis technique can consider an unlimited number of surrounding cells, although here we present frequency-reuse factors for up to three surrounding rings. The center cell is the cell of interest, which contains the desired subscriber transmitting to the base in the center of the figure. Each subscriber will also interfere with adjacent base receivers, and adjacent subscribers interfere with the desired base receiver. We use wedges from an annullus made by concentric circles to represent adjacent cells in surrounding cell rings. For equal loading, the adjacent cells in each ring must each contain an area and number of users equal to that of the center cell. The geometry of the first surrounding layer is used to define the angle θ_1 over which the cell wedges of the first ring span. Because of the concentric circular geometry, all cells beyond the first ring may be related to the area and span of the first ring cells.

The geometry of Fig. 6 was chosen for analytic simplicity and differs from the traditional hex geometry in two ways. First, since the concentric circular geometry uses wedges as cells, a uniform distribution of subscribers over space means that more users will be located at the outer part of the adjacent cells (farther away from desired base), and thus will yield optimistic results for f (as compared with a uniform user distribution in a hexagonal cellular layout) if the number of users within the cell are not scaled in some way. As shown shortly, this is easily accounted for by scaling the number of users in the inner and outer parts of each adjacent cell wedge. In fact, a variety of scaling factors can be used to determine various user distributions and corresponding f values. Second, to exploit a simple geometric relationship using the law of cosines while maintaining tractability, we assume there are small concentric forbidden zones of width 2α in all of the surrounding rings where users cannot be located. Then, we locate two points on the inner and outer edge of the forbidden zone which fall on a line drawn radially from the central base station. The distance of an adjacent cell subscriber to its own base is then closely approximated by the distance from the subscriber to the point on the closest edge of the forbidden zone. Fig. 7 shows the geometry for computing distances for subscribers in both the inner and outer portion of an adjacent cell. This assumption can be made with virtually no impact on results when $\alpha < d$. While the geometry of Fig. 7 does not describe the physical shape of cells in urban streets with sharp losses around street corners, it offers significant analytical simplicity that is useful for general system design.

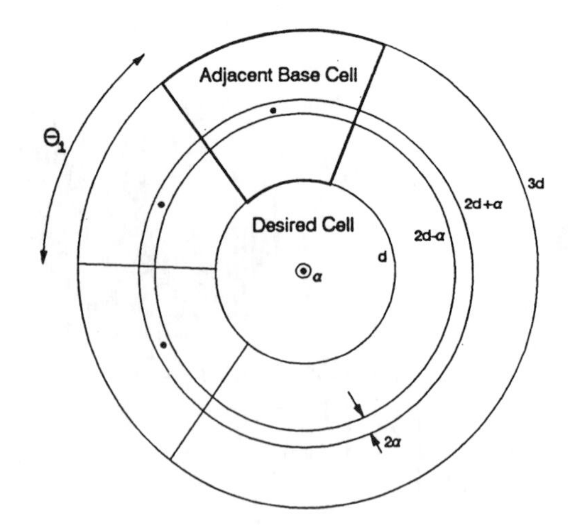

Fig. 6. Simple cell geometry used to analyze frequency reuse efficiency for CDMA.

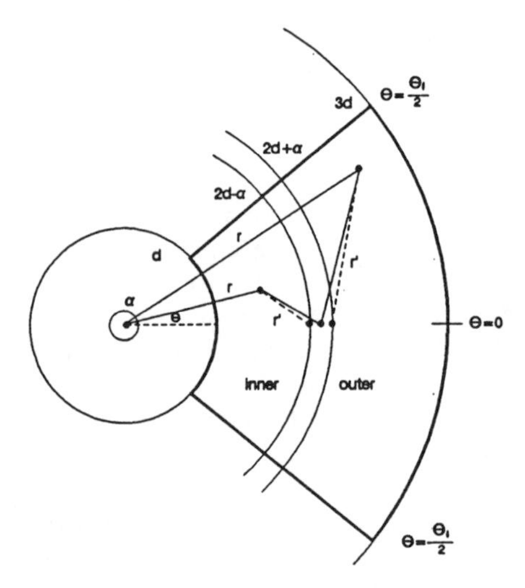

Fig. 7. By breaking the adjacent cells into two sectors (inner and outer sector), the impact of the distribution of users close to, and far from, the desired cell can be varied by using weighting coefficients. This figure shows the geometry used to relate the distances of adjacent cell subscribers to their own base cell with the distances to the base station within the center cell of interest.

B. Cell Geometries and User Distribution Weighting Factors

The area of the central cell of interest is equal to the area of all surrounding cells, and is given by

$$A = \pi d^2 - \pi \alpha^2 \approx \pi d^2. \quad (7a)$$

Let A_i denote the total area of the ith cell ring, and θ_i denote the angular span of each wedge-shaped cell in the ith ring. Neglecting α, the first surrounding ring has a total area A_1 occupied by the M_1 closest surrounding cells, and each cell wedge spans θ_1 given by (7b)

$$A_1 \approx \pi(3d)^2 - \pi(d)^2 = M_1 A$$
$$\theta_1 = 2\pi/M_1. \quad (7b)$$

From (7a) and (7b), $M_1 = 8$, thus the angular span of each closest cell is $\theta_1 = 2\pi/8 = \pi/4$ radians. For the second and all subsequent surrounding layers, it can be shown that

$$A_i = M_i A = iM_1 A = i8A, \qquad i \geq 1$$
$$\theta_i = \theta_1/i = \pi/4i. \quad (7c)$$

To account for various user distributions within the cells, it is possible to weight the number of subscribers in the inner sector and outer sector of each cell, so that the combination of users in both cell sectors sums to the desired number of users. We denote $W_{i\text{in}}$ as the weighting factor for the inner sector of a cell in the ith surrounding ring, and $W_{i\text{out}}$ as the weighting factor for the outer sector of a cell in the ith surrounding ring. The desired weighting factors are easy to obtain by first breaking the ith surrounding ring into inner and outer rings with areas that sum to A_i . Then, $W_{i\text{in}}$ and $W_{i\text{out}}$ are solved by adjusting the number of subscribers within inner and outer cell sectors to represent various spatial distributions on the users while still maintaining the desired number of users and equal area within each cell.

IV. Example Of Use Of Weighting Factors

In the first surrounding ring, the areas in the inner and outer sectors of each cell are:

$$A_{1\text{in}}/M_1 \approx \left[\pi(2d)^2 - \pi(d)^2\right] / 8 = 3A/8 \quad (7d)$$

$$A_{1\text{out}}/M_1 \approx \left[\pi(3d)^2 - \pi(2d)^2\right] / 8 = 5A/8. \quad (7e)$$

For each cell to possess equal area and $U = KA$ users, the weighting factors must satisfy

$$U = KA = KW_{1\text{in}} A_{1\text{in}}/M_1 + KW_{1\text{out}} A_{1\text{out}}/M_1$$
$$= KA[3/8W_{1\text{in}} + 5/8W_{1\text{out}}]. \quad (7f)$$

In a hexagonal (honeycomb) cellular geometry with a uniform distribution of users, half of the adjacent cell users will be closer than $r = 2d$ and half will be farther than $r = 2d$ from the central base station. To obtain a user distribution from the cellular geometry in Figs. 6 and 7, which has half of the users in an adjacent cell closer than $2d$ and half of the users farther than $2d$ from the center base station, (7d) and (7e) are solved and yield $W_{1\text{in}} = 4/3$ and $W_{1\text{out}} = 4/5$. Similarly, using (7a)–(7c) and extending (7f) to subsequent rings, one finds that $W_{2\text{in}}$ is 8/7, $W_{2\text{out}}$ is 8/9, $W_{3\text{in}}$ is 12/11, and $W_{3\text{out}}$ is 12/13. By multiplying the interference powers supplied by various cell sectors with the appropriate weighting factors, our method provides an equivalent method for solving a hexagonal cellular system with uniformly distributed users [11].

To achieve higher values of F, it is desirable to place most users far away from the center base. If no weighting is performed (i.e., if $W_{i\text{in}} = W_{i\text{out}} = 1$), this occurs naturally from the concentric circle geometry. For lower values of F, all users in every adjacent cell will be in the inner sectors, and no users will be in the outer sectors. This situation is given by $W_{i\text{out}} = 0$ for all i, and $W_{1\text{in}} = 8/3$, $W_{2\text{in}} = 16/7$, and $W_{3\text{in}} = 24/11$.

C. Comparison to Traditional Hexagonal Cell Geometry

To compare analysis results from the concentric circle geometry with traditional hexagonal cell structures considered in [1],[3],[12], one needs to relate the number of surrounding hexagonal cells to the number of surrounding wedge-shaped cells within a specified area surrounding the center cell. A hexagonal cell with major radius d occupies an area of $A_{\text{hex}} = 3\sqrt{3}\, d^2/2 = 2.598d^2$, whereas cells in Fig. 6 possess area $A = \pi d^2$. Thus, the first ring of eight wedge-shaped cells shown in Figs. 6 and 7 occupies the same area as would $8A/A_{\text{hex}}$, or 9.666, hexagonal cells. From (7c), the second ring of 16 wedge-shaped cells occupies the same area as $16A/A_{\text{hex}}$ or 19.333 hexagonal cells. The third ring of 24 wedge-shaped cells occupies the same area as 29 hexagonal cells.

It can be shown that (7g) relates the number of rings surrounding the circular center cell (Fig. 6) to the equivalent number of surrounding hexagonal cells that would surround the same center cell.

$$N_{\text{hex}} = \sum_{i=1}^{I} 8i\pi/2.598. \quad (7g)$$

In (7g), I is the total number of surrounding rings and N_{hex} is the total number of equivalent surrounding hexagonal cells. It is readily seen that the first surrounding ring of cells in the geometry proposed here will possess a greater area (i.e., 9.666 surrounding hexagonal cells) and larger number of users, and consequently will predict lower F values than the conventional "first layer" of six surrounding cells in the classic hexagonal (honeycomb) geometry. For two and three surrounding rings, the geometry of Fig. 6 can be equated to the case of 29 and 58 surrounding hexagonal cells, respectively.

V. Analysis

We presume all users obey a propagation path loss law that relates path loss to distance to the nth power relative to the reference free space level $P(\alpha)$, as given in (5a). However, we consider two distinct types of propagation models, one which uses the result of (5b) and the two-ray model to yield an exponent that is a function of distance, and another that assumes large scale path loss having the form $d^n$where the path loss exponent is assumed to be constant over distance.

All subscribers are assumed to be under power control within their own cells such that all desired mobile signals arrive with the same power P_n as the closest user within their own cell. Thus, all subscriber transmitter powers are referenced to the power transmitted by the *closest* subscriber within the cell, which is denoted as P_α .

Under perfect power control within the center cell, the received power of each of the in-cell subscribers at the base receiver is P_n , and the interfering power N_0 due to all but the single desired in-cell subscriber is given by

$$N_0 = P_n(U-1) \approx P_n U = P_n KA \tag{8a}$$

regardless of the path loss law within the cell (assuming there is sufficient dynamic range on the power transmitted by the subscriber). Without loss of generality in (8a), we replace P_n by P_α , the power received at the base due to the closest subscribers. Since we assume the same close-in propagation occurs within α m of every base station in the system, all signal levels may be referenced to the received power at a radial distance α pfrom any base station. Then, for the center cell of interest,

$$N_0 = P_\alpha KA. \tag{8b}$$

For simplicity, P_α may be assumed to be unity so that N_0 is equal to the number of users in each cell. Furthermore, K can be arbitrarily set to $1/A$, so that the noise due to in-cell users is unity, regardless of the number of users, and all subsequent cells add to the unit noise level produced by the center cell of interest. Each subscriber in the center cell has a transmitted power equal to $P_\alpha(r/\alpha)^n$, where r is the distance from mobile to base, and n is the propagation path loss exponent in (5). Note that n may be a function of r and α, as in (5b) and Figs. 2–4, or n may be assumed constant over distance, as shown in Fig. 5 and Table I.

The adjacent cell subscribers are under power control within their own cell and are a distance r' from their own base station. As shown in Fig. 7 for the first ring (i=1), the relationship between r' and r is given by (9) using the law of cosines. The true transmitter power for each subscriber in the adjacent cell can be related by P_α and r', where r' is related to r and θ by

$$\begin{aligned} r' &= \left(r^2 \sin^2\theta + (2*d*i-\alpha-r*\cos\theta)^2\right)^{1/2}, \\ &\qquad \text{for } d*i \le r \le 2*d*i-\alpha \\ r' &= \left(r^2 \sin^2\theta + (r*\cos\theta-2*d*i-\alpha)^2\right)^{1/2}, \\ &\qquad \text{for } 2*d*i+\alpha \le r \le d(2*i+1). \end{aligned} \tag{9}$$

Let $P_{ai}(r,\theta,\alpha)$ denote the power received at the center base station from a subscriber at a particular location in an adjacent ith layer cell. Then, the received power from each adjacent cell user is given by

$$P_{ai}(r,\theta,\alpha) = P_\alpha(r'/\alpha)^{n(r',\alpha)}(\alpha/r)^{n(r,\alpha)} \tag{10}$$

where r' is a function of θ as given by (9). The total interference power, N_{ai}, contributed by all subscribers in a cell in the ith surrounding ring is found by summing up the received powers from each user in the adjacent cell

$$N_{ai} = \sum_{u \epsilon A_i/M_i} P_{ai}(r,\theta,\alpha). \tag{11a}$$

It is clear from (10) that P_{ai} values in (11a) are functions of the specific locations of adjacent cell users as given by (9) and (10), and that the summation in (11a) represents a summation over a geographic cell region. To compute (11a), one can use any spatial distribution of users and any distance-dependent path loss. In real-world situations, individual values of P_{ai} will exhibit shadow fading (e.g. log-normal fading) and this can be modeled statistically in (11a) in a Monte Carlo fashion. In our work, though, we have assumed symmetric shadow fading of P_{ai} at specific T-R separations about the distance-dependent mean.

Equation (11a) is sufficiently general to solve for the power contributed from adjacent cells obeying any distance-dependent propagation path loss law. One simple way of manipulating a specified distribution of users is to employ weighting factors described in Section III-B. With weighting factors, the total noise powers contributed by users within the inner and outer sectors of adjacent cells can be easily scaled without the need to recompute the interference powers contributed by individual users. Thus, bounds on average adjacent cell noise can be obtained easily. To use the weighting factors, (11a) can be written as two separate summations as given in (11b):

$$\begin{aligned} N_{ai} = \mathbf{W}_{1\text{in}} &\sum_{u \epsilon A_{i\text{in}}/M_i} P_{ai}(r,\theta,\alpha) \\ + \mathbf{W}_{1\text{out}} &\sum_{u \epsilon A_{i\text{out}}/M_i} P_{ai}(r,\theta,\alpha). \end{aligned} \tag{11b}$$

Note that if $W_{1\text{in}}$ and $W_{1\text{out}}$ are unity, then (11b) is identical to (11a). As discussed in Section III-B, if $W_{1\text{in}} = 4/3$ and $W_{1\text{out}} = 4/5$, then a uniform spatial distribution of users in (11a) can be used in (11b) to provide the same N_{ai} as a hexagonal cellular system.

A. Results

Equation (11b) was solved numerically using a uniform spatial distribution of users for specific values of close-in reference distance and cell radius d. Then, the frequency reuse factor was computed from (6), where three different user distribution weightings in (11b) gave rise to various values for f using identical channels.

All results in this work were found using 20 000 uniformly spaced users in each adjacent cell for $d = 2$ km, and 400 000 equally spaced users within $d = 10$ km cells. Equation (6a) and symmetry were exploited to limit computations to a single cell within each ring.

Tables II and III give values of frequency reuse factor (f) as a function of path loss exponent (assumed constant over distance) for $\alpha = 0.1$ km and $\alpha = 0.05$ km, respectively, for the cases of cell boundaries at $d = 2$ km and $d = 10$ km. Only one ring of adjacent cells was considered for results given in the tables. Note that f values for $n = 4$ are comparable to those

TABLE II
FREQUENCY REUSE FACTOR FOR CDMA SYSTEMS AS A FUNCTION OF PROPAGATION EXPONENT. ONE RING OF ADJACENT CELLS CONSIDERED FOR $D = 2$ km AND $D = 10$ km, $\alpha = 0.1$ km

		Frequency Reuse Factor		
		Lower Bound	Hex Case	Upper Bound
d (km)	n	$W_1 = 3.0$ $W_2 = 0.0$	$W_1 = 1.38$ $W_2 = 0.78$	$W_1 = 1.0$ $W_2 = 1.0$
2.0	2	0.326	0.434	0.471
2.0	3	0.423	0.571	0.625
2.0	4	0.499	0.666	0.721
10.0	2	0.310	0.422	0.457
10.0	3	0.399	0.553	0.606
10.0	4	0.466	0.638	0.698

TABLE III
FREQUENCY REUSE FACTOR FOR CDMA SYSTEMS AS A FUNCTION OF PROPAGATION EXPONENT; ONE RING OF ADJACENT CELLS CONSIDERED FOR $D = 2$ km AND $D = 10$ km, $\alpha = 0.05$ km

		Frequency Reuse Factor		
		Lower Bound	Hex Case	Upper Bound
d (km)	n	$W_1 = 3.0$ $W_2 = 0.0$	$W_1 = 1.38$ $W_2 = 0.78$	$W_1 = 1.0$ $W_2 = 1.0$
2.0	2	0.316	0.425	0.462
2.0	3	0.408	0.558	0.613
2.0	4	0.479	0.646	0.707
10.0	2	0.308	0.419	0.455
10.0	3	0.396	0.550	0.603
10.0	4	0.462	0.634	0.695

given in [1],[3], where a hexagonal cellular system geometry with a comparable number of adjacent users was considered. For weighting factors that offer uniform user distributions comparable to the hexagonal geometry, we see for $d = 10$ km, f is 64% for $n = 4$, 55% for $n = 3$, and 42% for $n = 2$. A loose upper bound and tight lower bound on f are also given in Tables II and III. Weighting coefficients listed in Tables II and III differ slightly from those given in Section III-B since the term $\pi\alpha^2$ was considered in calculations which led to tabular entries, but was neglected for the general results given in (7). For a propagation model that has a constant n over distance, it is evident that f is not sensitive to α or d, but is highly sensitive to the value of the path loss exponent. Note that in free space channels ($n = 2$) with only one surrounding ring of adjacent cells, f is more than 3 dB down from the $n = 4$ result.

Although not shown in the tables, it is worth noting that when the first surrounding ring and just the inner portion of the second surrounding ring were considered in our analysis, we found $f = 0.606$ for the d uniform (hexagonal) user distribution case, which is comparable to results obtained in [3] using 18 surrounding hex cells. Using (7g) it can be shown that 18 hexagonal cells occupy the same area as the first ring

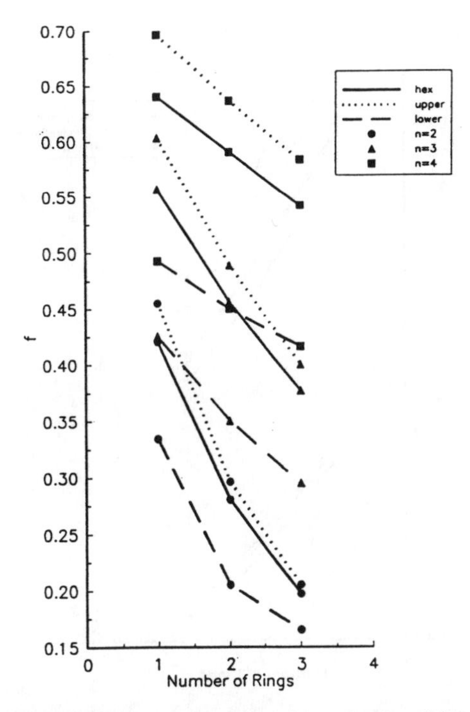

Fig. 8. Plot of the frequency reuse factor, f, as a function of the path loss exponent, n, and the number of rings surrounding the cell of interest.

and the inner portion of the second ring. Thus, our analysis yields identical results to average results in [3] which uses hexagonally shaped cells.

Fig. 8 extends the results shown in Tables II and III to include the interference effects of the second and third rings. One can see how additional rings effect the frequency reuse efficiency of CDMA. Fig. 8 shows f as a function of the number of adjacent cell rings for the case where path loss exponents remain constant with distance, with n values ranging from 2 to 4. The three weightings given in Section IV (hex, upper bound, lower bound) were used to give representative ranges on f. For the data shown in Fig. 8, = 0.1 km and $d = 10$ km were assumed. Fig. 8 shows that f drops to below 0.2 when $n = 2$ and three surrounding rings are considered. However, for $n = 4$, the hexagonal case (uniform distribution) has a frequency reuse efficiency of 54% with three layers. Note that for three surrounding rings, the $n = 3$ channel affords 3 dB more reuse than does the $n = 2$ channel, and the $n = 2$ reuse degrades by 3 dB when the surrounding layers are increased from one to three. These data indicate how the propagation environment can impact the estimate of frequency reuse.

In wireless channels, deep fading between an adjacent base and one of its subscribers will dramatically impact that subscriber's power control, and thus the power received from that unit at the cell of interest [15]. This is seen from (10), since if an adjacent user is in a deep fade while transmitting to its own base, the required transmitter power will be large and could strongly interfere with the base of interest.

We consider three different two-ray propagation channels that have a distance dependent path loss exponent. These channels were derived using (1)–(3), and possess the path loss

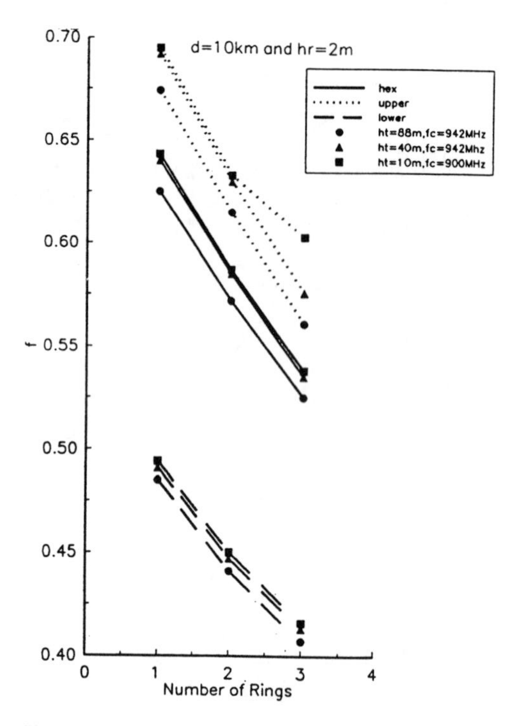

Fig. 9. Plot of the frequency reuse factor, f, as a function of three different propagation channels, for d = 10 km and α = 100 m. The propagation channels were developed from a two-ray ground reflection model. Path loss exponent values for the three channels, as a function of distance, are shown in Figs. 2–4.

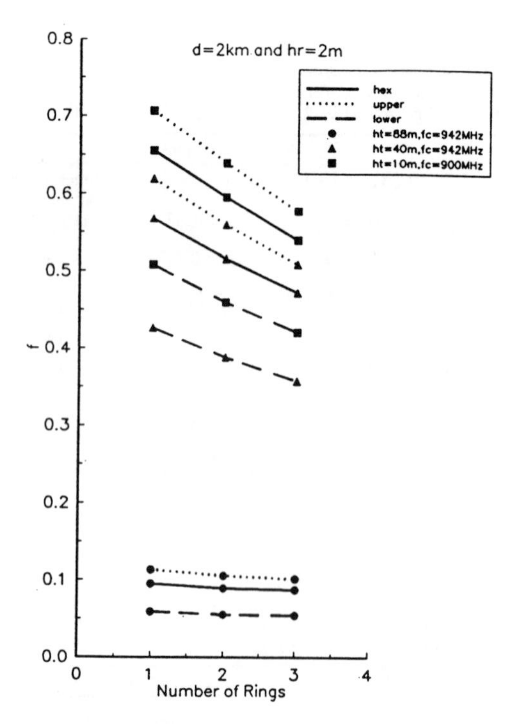

Fig. 10. Plot of the frequency reuse factor, f, as a function of three different propagation channels, for d = 2km and α= 100 m. The propagation channels were developed from a two-ray ground reflection model. Path loss exponent values for the three channels, as a function of distance, are shown in Figs. 2–4.

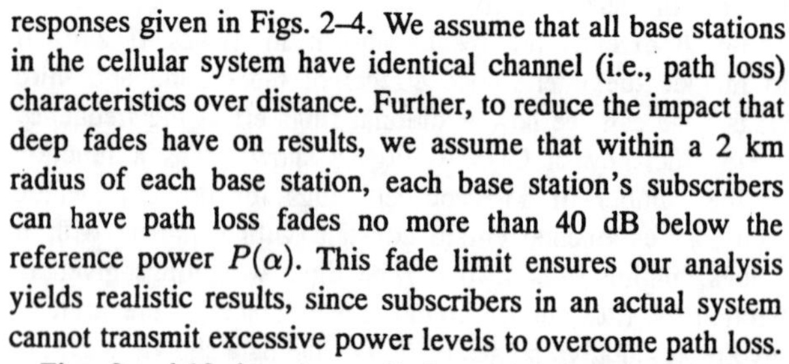

responses given in Figs. 2–4. We assume that all base stations in the cellular system have identical channel (i.e., path loss) characteristics over distance. Further, to reduce the impact that deep fades have on results, we assume that within a 2 km radius of each base station, each base station's subscribers can have path loss fades no more than 40 dB below the reference power $P(\alpha)$. This fade limit ensures our analysis yields realistic results, since subscribers in an actual system cannot transmit excessive power levels to overcome path loss.

Figs. 9 and 10 show how cell size and a distance-dependent propagation path loss exponent can impact frequency reuse. Fig. 9 shows values of f for the case of large cells (d = 10 km, α = 100 m). It is clear from Figs. 2–4 that when d = 10 km, all adjacent cells are located in the asymptotic region of d^4 path loss from the cell of interest. Fig. 9 indicates f ranges between 0.4 and 0.7, regardless of the particular two-ray channel or user distribution. By comparing Fig. 8 results with those in Fig. 9, one can see that for all user distributions, the three different distance-dependent analytical channels yield frequency reuse factors close to those offered by channels using a d^4 path loss law. This is due to the fact that the cell size is large, and natural isolation of adjacent noise power due to path loss occurs. The frequency reuse factors obtained for each of the three two-ray channels are, at most, only 5% lower than results obtained for the d^4 propagation law.

Fig. 10 shows how the same three two-ray propagation channels used to produce results in Fig. 9 can dramatically alter the frequency reuse when small cells are considered. For d = 2 km and α=100 m, path loss values shown in Figs. 2–4 were used to determine values of f. For 2 km cells, it is clear that the two-ray channels described by Figs. 2 and 3 have adjacent subscribers that provide stronger signal levels than those predicted by a d^4 path loss law. In addition, the proportion of deep faded regions within the adjacent 2 km cells is much greater than in 10 km cells. This means that a greater percentage of adjacent users in the smaller cells will need to provide greater transmitter power to overcome their own in-cell fades. While the channel represented in Fig. 4 offers frequency-reuse values virtually unchanged from those found for the same channel in a 10 km cellular system, it is readily seen that the channels portrayed by Figures 2 and 3 substantially degrade f for a 2 km system.

Fig. 10 shows the impact that close-in propagation has on frequency reuse. As seen in Fig. 2, there are many deep nulls within a 2 km radius from the base, and the first ring of subscribers will offer signal levels only 21 to 35 dB down from $P(\alpha)$. Fig. 10 shows how both of these factors lead to values of f (with three rings and uniform distribution of users in each cell) near 0.1 for the channel portrayed in Fig. 2. This is more than six times (8 dB) lower than the frequency reuse obtained using the exact same channel except with d = 10 km. Also, the channel portrayed in Fig. 2 provides f values that are 8 dB lower than results obtained for d = 2 km using the channel portrayed in Fig. 4. Fig. 10 shows the channel portrayed by Fig. 3 does not degrade f as severely, but the resulting frequency reuse factor (with three rings and uniform distribution of users in each cell) of 0.47 is 0.6 dB less than that provided by the same channel in a 10 km cell system. These results explicitly show that if CDMA is implemented in a propagation environment that has deep fades or shadowing within cells, but little isolation between adjacent cells, the frequency reuse efficiency can be severely degraded. It is worth emphasizing that the analytical channels used here are based on a particular two-ray model that uses antenna height as a parameter (see (3)). However, it should be clear that if some other physical mechanism, such as shadowing, provides the same path loss behavior as shown in Figs. 2–4, then the results in Fig. 10 will still hold, even if the physical propagation mechanisms and antenna heights are different than those which lead to (3).

From this work it becomes clear that accurate site-specific propagation models which could accurately predict the values of P_{ai} in (11a) are needed. Based on this analysis, it seems one of the major challenges that lies ahead for implementors of DS-CDMA is to strive to predict suitable base station locations that can exploit the propagation environment, so as to offer rapid path loss decay to adjacent cells while maintaining low path loss within a desired coverage area. As site-dependent propagation prediction tools are developed, it will become possible to incorporate an accurate path loss exponent that is a function of location. In addition, it should become possible to predict, *a priori*, base station placements that will exploit known shadowing properties of building structures to increase the isolation between cells. At such a time, analysis techniques such as the ones presented here will be useful for analysis, simulation, and automated installation of high-capacity CDMA systems.

VI. Conclusion

This paper has studied the frequency reuse efficiency of the reverse channel in a CDMA cellular radio system employing total frequency reuse. We considered two propagation models, both of which related path loss to the log of the T-R separation distance. One model assumed a path loss law of d^n , where n varied between 2 and 4, but was not a function of T-R separation. Experiments confirmed this range of n occurs in real channels. The other model was analytically based on the two-ray ground reflection model, and provided a path loss exponent that varied with distance between the transmitter and receiver. We showed that, even if the two ray model fails to explain the physical causes of propagation in typical urban mobile radio systems, it provides reasonable agreement to measured data, particularly within the range of 100 m to 2 km from the base antenna.

We developed an analysis technique for frequency reuse based upon the geometry of concentric circles. This geometry provides a very straightforward approach to solving for frequency reuse, and allows the use of scaler weighting factors to redistribute the location of users with any cell. In this paper, three different sets of weighting factors were used to provide average frequency reuse values that were bounded above and bounded below the result obtained for hexagonal cell sites that have a uniform distribution of users. We showed how the concentric circle geometry is related to the traditional mosaic of hexagonal cells.

General expressions for computing the frequency reuse factor for channels that possess distance-dependent path loss exponents were given and were used to solve for three specific channels. Expressions for computing frequency reuse in channels which have fixed path loss exponents were also given and used to solve for channels with exponents ranging between 2 and 4. For exponents that are not a function of location, the results showed that frequency reuse is highly dependent upon the propagation path loss exponent, particularly between $n = 2$ and $n = 3$. When three rings of adjacent users were considered (equivalent to 58 adjacent hexagonal cells), a uniform distribution of users yielded values of f that ranged from 0.197 when $n = 2$ to 0.541 when $n = 4$.

For distance-dependent path loss exponents, we found that, depending upon the isolation between cells, and the nature of fading or shadowing within the cell of interest, the frequency reuse factor can range between about 0.1 and 0.7. Such a wide range on frequency reuse suggests that accurate propagation prediction techniques are warranted for analysis, simulation, and expert system design of future CDMA personal communication systems.

Acknowledgment

The authors thank Morton Stern of Motorola, Inc. for discussions about this work.

References

[1] G. R. Cooper and R. W. Nettleton, "A spread-spectrum technique for high-capacity mobile communications," *IEEE Trans. Veh. Technol.*, vol. VT-27, pp. 264–275, Nov. 1978.

[2] K. S. Gilhousen, I. M. Jacobs, R. Padovan, and L. A. Weaver, Jr., "Increased capacity using CDMA for mobile satellite communication," *IEEE J. Select. Areas Commun.*, vol. 8, pp. 503–512, May 1990.

[3] K. S. Gilhousen *et al.* "On the capacity of a cellular CDMA system," *IEEE Trans. Veh. Technol.*, , vol. 40, pp. 303–312, May 1990.

[4] T. S Rappaport, S. Y. Seidel, and R. Singh, "900 MHz multipath propagation measurements for U. S. digital cellular radiotelephone," *IEEE Trans. Veh. Technol.*, vol. 39, pp. 132–139, May 1990.

[5] S. Y. Seidel, T. S. Rappaport, S. Jain, M.L. Lord, and R. Singh, "Path loss, scattering, and multipath delay statistics in four European cities for digital cellular and microcellular radiotelephone," *IEEE Trans. Veh. Technol.*, vol. 40, pp. 721–730, Nov. 1991.

[6] A. J. Rustako, N. Amitay, G. J. Owens, R.S Roman, "Radio propagation measurements at microwave frequencies for microcellular mobile and personal communications," in *Proc. 1989 IEEE Int. Commun. Conf.*, Boston, MA, pp. 15.5.1–15.5.5.

[7] P. Harley, "Short distance attenuation measurements at 900 MHz and 1.8 GHz using low antenna heights for microcells," *IEEE J. Select. Areas Commun.*, vol. 7, pp. 5–11, Jan. 1989.

[8] S. Mockford, A. M. D. Turkmani, and J. D. Parsons, "Local mean signal variability in rural areas at 900 MHz," in *Proc. 1990 IEEE Veh. Technol. Conf.*, Orlando, FL, pp. 610–615.

[9] J. Holtzman, "Power control and its effects on CDMA wireless systems," 1990 WINLAB Workshop," Rutgers Univ., Piscataway, NJ, Oct. 1990.

[10] K. Bullington, "Radio propagation for vehicular communications," *IEEE Trans. Veh. Technol.*, vol. VT–26, pp. 295–308, Nov. 1977.

[11] T. S. Rappaport and L. B. Milstein, "Effects of path loss and fringe user distribution on CDMA cellular frequency reuse efficiency," in *Proc. IEEE GLOBECOM '90*, San Diego, CA, pp. 500–506.

[12] W. C. Y. Lee, *Mobile Communications Engineering* New York: McGraw-Hill, 1982.

[13] ——"Spectrum efficiency in cellular," *IEEE Trans. Veh. Technol.*, vol. 38, pp. 69–75, May 1989.

[14] K. L. Blackard, M-J. Feuerstein, T. S. Rappaport, S. Y. Seidel, and H. H. Xia, "Path loss and delay spread models as functions of antenna height for microcell system design," presented at 42nd IEEE Veh. Technol. Conf., Denver, CO, May 1992.

[15] L. B. Milstein, T. S. Rappaport, and R. Barghouti, "Performance evaluation for cellular CDMA," *IEEE J. Select. Areas Commun.*, vol. 10, pp. 680–689, May 1992.

Theodore S. Rappaport (S'83–M'84–S'85–M'87–SM'91) was born in Brooklyn, NY, on November 26, 1960. He received the B.S.E.E., M.S.E.E., and Ph.D. degrees from Purdue University, West Lafayette, IN, in 1982, 1984, and 1987, respectively.

In 1988 he joined the electrical engineering faculty of Virginia Polytechnic Institute and State University, Blacksburg, where is he is an Associate Professor and Director of the Mobile and Portable Radio Research Group. He conducts research in mobile radio communication system design, simulation, and RF propagation through measurements and modeling. He has authored or coauthored technical papers in the areas of mobile radio communications and propagation, vehicular navigation, ionospheric propagation, and wide-band communications.

Dr. Rappaport holds a U.S. patent for a wide-band antenna, and is co-inventor of SIRCIM, an indoor radio channel simulator that has been adopted by more than 50 companies and universities. In 1990 he received the Marconi Young Scientist Award for his contributions in indoor radio communications and was named an NSF Presidential Faculty Fellow in 1992. He is an active member of the IEEE Communications and Vehicular Technology Societies and serves as a Senior Editor of the IEEE JOURNAL ON SELECTED AREAS IN COMMUNICATIONS. He is a Registered Professional Engineer in the State of Virginia, and is a Fellow of the Radio Club of America. He is also President of TSR Technologies, a cellular radio and paging test equipment manufacturer.

Laurence B. Milstein (S'66–M'68–SM'77–F'85) received the B.E.E. degree from the City College of New York, New York, NY, in 1964, and the M.S. and Ph.D. degrees in electrical engineering from the Polytechnic Institute of Brooklyn, Brooklyn, NY, in 1966 and 1968, respectively.

From 1968 to 1974 he was employed by the Space and Communications Group of Hughes Aircraft Company, and from 1974 to 1976 he was a member of the Department of Electrical and Systems Engineering, Rensselaer Polytechnic Institute, Troy, NY. Since 1976 he has been with the Department of Electrical and Computer Engineering, University of San Diego, La Jolla, CA, where he is a Professor and former Department Chairman, working in the area of digital communication theory with special emphasis on spread-spectrum communications systems. He has also been a consultant to both government and industry in the areas of radar and communication.

Dr. Milstein was an Associate Editor for Communication Theory for the IEEE TRANSACTIONS ON COMMMUNICATIONS, an Associate Editor for Book Reviews for the IEEE TRANSACTIONS ON INFORMATION THEORY, and an Associate Technical Editor for the *IEEE Communications Magazine*. He has been on the Board of Governors of the IEEE Communications Society, and was the Vice President of Technical Affairs in 1990 and 1991. He is currently a member of the Board of Governors of the IEEE Information Theory Society, and is a member of Eta Kappa Nu and Tau Beta Pi.

Chapter 7

DIGITAL RECEIVERS: IMPLEMENTATION & DESIGN

This chapter presents fundamental concepts used to design digital receivers and describes methods to implement digital radios and real time digital signal processing algorithms in both hardware and software.

Reprinted from *IEEE Communications Magazine*, Vol. 33, No. 1, Jan. 1995, pp. 68-78.

Advanced Digital Receiver Principles and Technologies for PCS

The synergy between digital radio communications and VLSI signal processing is revolutionizing the design of wireless terminals. Driving this synergy are certain fundamental paradigms in modern communication theory, digital signal processing, and VLSI design.

Heinrich Meyr and Ravi Subramanian

The explosive growth of wireless communications has given rise to a variety of portable voice, data, and messaging systems. In many cases, PCS is defined by some combination of these various systems. Future PCS terminals for example, can be expected to work in 900 MHz terrestrial cellular systems, in the 1.9 GHz PCS band, as well as in 1.6 GHz satellite-mobile systems. The defining feature of PCS is that of coverage. PCS terminals must enable users of PCS services to truly be accessible anywhere and anytime.

Advances in, and new applications for, each of these radio systems are being spawned at an unprecedented rate. The complexity of radio communication systems is significantly increasing with the application of more sophisticated multiple-access and digital modulation techniques in order to accommodate the tremendous growth in the number of subscribers. Employing advanced digital communications principles is the norm. Furthermore, the requirements for frequency-agile terminals are also growing as frequency-hopping techniques find their way into commercial systems.

Rapid advances in solid-state integrated circuit (IC) technology are fueling the growth in commercial wireless communications systems. New technologies are being spurred on by the desire to produce high-performance, low-power, small-size, low-cost, and high-efficiency components. Today's IC and process designers are teaming up with communications engineers to define the new directions in which to push semiconductor processes for PCS. Digital technologies have seen a breakthrough in both performance and size through the use of complementary metal-oxide semiconductor (CMOS) device scaling. These technologies can be expected to scale rapidly owing to the demand for denser and faster memory chips as well as digital microprocessors. Unlike digital logic, which seems to gain speed with increased packing density, analog performance is often compromised with higher levels of integration.

The increasing use of integrated circuits in radio has resulted in significant improvements in the reliability and performance of receivers. The number of parts that require tuning or whose characteristics could significantly change over time or temperature has dropped markedly. Manufacturing costs have also fallen due to improved procedures in assembly and testing. Rapid advances in the packaging of integrated circuits have contributed significantly to compact designs of wireless terminals. Technologies, such as multi-chip module (MCM) allow designers to integrate many individual chip die into a single substrate structure. This new structure reduces the size of the inter-chip capacitances and minimizes the power consumed to drive signals from one chip to another. This drives down the total power consumed by the system since most of the power is now consumed for processing rather than for interfacing signals from chip to chip.

These trends are giving rise to the increasing use of silicon-based signal processing techniques, which include discrete-time analog signal processing techniques (such as switched-capacitor systems and delta-sigma modulators) as well as advanced digital signal processing (DSP) techniques. Traditionally, DSP techniques were used for premodulation and post-detection functions in radio communications receivers, such as speech and channel codecs. Over the last ten years, DSP techniques have been used extensively for advanced digital communications transceiver designs, finding their way into detection, equalization, demodulation, frequency synthesis, and channel filtering. In addition, a totally new and broad range of digital algorithms are emerging for which there exists no analog counterpart.

The Communications System Design Space

Design Space and the Role of DSP

Traditionally, the communications engineer has been concerned with a two-dimensional design space comprising power and bandwidth. Employing advanced very large-scale integration (VLSI) technology, the implementation of complex digital algorithms has

HEINRICH MEYR is a professor of electrical engineering at the Aachen University of Technology (RWTH Aachen).

RAVI SUBRAMANIAN is with AT&T Bell Laboratories in Crawford Hill, New Jersey.

0163-6804/95/$04.00 1995 © IEEE

become feasible and economical. As a consequence, the communications engineer nowadays is faced with a three-dimensional design space comprising power, bandwidth, and complexity. This has enabled the design engineer to trade the physical figures of merit (bandwidth and power) against signal processing complexity. Also, the capability of realizing highly complex algorithms has made it possible to approach the theoretic performance limits imposed by Shannon theory. Advanced source and channel coding algorithms are already used in many commonplace products such as CD players and satellite receivers. The success of the personal communications revolution critically depends on the ability to efficiently transmit information over the scarce physical resource bandwidth.

When it comes to realizing digital receivers, system modeling and algorithm design closely interact with the architectural choices for implementation. As a consequence, the more traditional approach — presently pursued by many companies and research institutions — where algorithm design (found in "theory" groups) is well-separated from the implementation (found in "VLSI" or "DSP" groups) should be abandoned in favor of an interdisciplinary "center-of-gravity view" where model/algorithm, architecture, and tools are closely interacting centers of gravity, as illustrated in Fig. 1.

For the purposes of solving a given detection or estimation problem in digital radio modem design, it is not sufficient to have a good circuit architecture. It is of paramount importance to use a good algorithm that is reliable and well-conditioned. There is little point in trying to exploit a desirable architecture if the associated algorithm is not stable. A better algorithm (even the best), on the other hand, may not correspond to a suitable computational architecture. In order to avoid this type of conflict, it is essential to develop algorithms and architectures in parallel. In this way, it is often possible to achieve the best of both worlds for a given application.

The task of *interactively exploring the design space* therefore calls for the "tall thin" engineer rather than the engineer specializing in algorithms, software, hardware, or tools alone. This is not to say that specialists are unimportant. Rather, for this particular task, a specialist is not the best-prepared engineer. Moreover, in order to meet the stringent time-to-market requirements, the use of proper DSP Design Automation (DSPDA) CAD tools is of crucial importance.

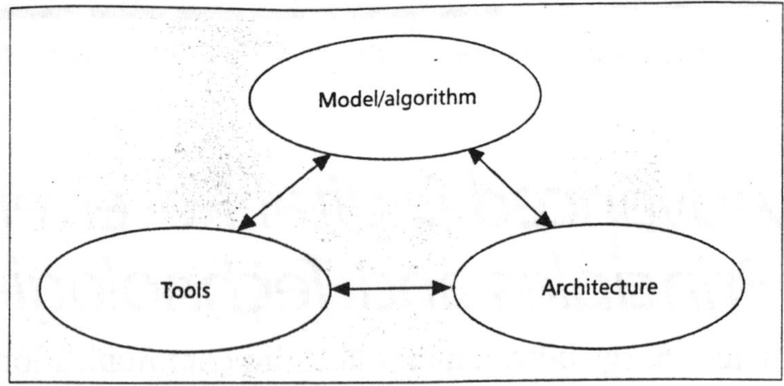

■ **Figure 1.** *"Centers of gravity" view of communication systems design.*

Communication System Model

In his seminal paper [41], Viterbi summarizes the three basic lessons learned from information theory:

- Separate source and channel encoding.
- Never discard information prematurely.
- Make the channel look like an AWGN channel.

These lessons immediately translate into important considerations for practical transmission system and receiver design. The basic structure of the information-theoretic system model according to these lessons is depicted in Fig. 2.

Lesson 3 states that the channel capacity is maximized if the decoder faces a symbol sequence a that is corrupted by Gaussian noise n only. Consequently, any receiver designed according to the three lessons above can be decomposed into two fundamental parts as shown in Fig. 3. The sole task of the inner receiver is to (ideally) produce a sequence of symbols a that is disturbed by additive Gaussian noise. Stated differently, the inner transmission system has to provide a "good" channel for the outer receiver, i.e., the source and (possibly) channel decoding system. To accomplish this, the inner transmission system must be adapted to the channel characteristics.

Principles for Deriving Receiver Structures

From the phrase "the *sole* task of the inner receiver..." one must not conclude that the design of the inner receiver is a trivial task that can be accomplished easily and with little complexity. More often than not, quite the contrary is true. Most of the design effort of a digital receiver as well as most of the hardware and software resources are usually spent on the inner receiver and not on the design of codes.

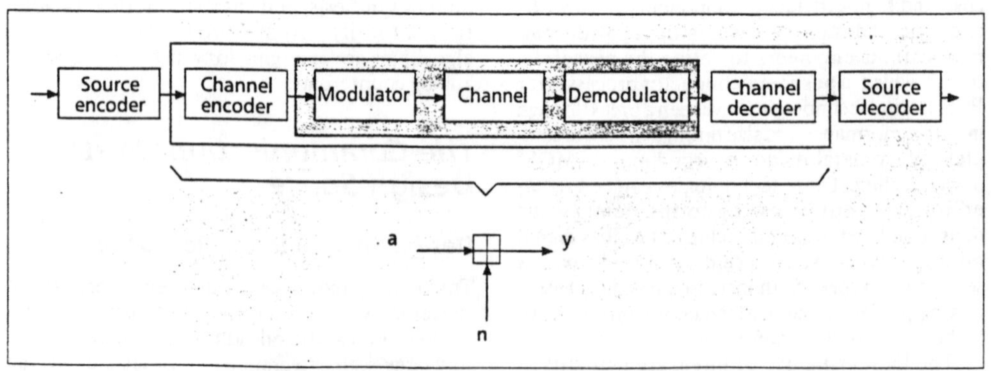

■ **Figure 2.** *Information theoretic system model.*

Most of the design effort of a digital receiver as well as most of the hardware and software resources are usually spent on the inner receiver and not on the design of codes.

The Derived Structure Approach — Detection and synchronization are the key functions to be performed by any digital receiver. For both functions, algorithms can be systematically derived. The basic idea behind this derived structure approach is that no a priori guess about the receiver structure is made. Rather, the algorithms are derived based on a mathematical model of the system and a suitably chosen performance criterion. Constructing the model of the system at hand (channel, coder, etc.) is in many instances the most difficult task. This task, however, is of tremendous value in itself since it forces the designer to organize and analyze the problem more thoroughly. It also allows for systematically incorporating a priori knowledge that may be available.

The key features of the derived structure approach are the following:

- Algorithms can be systematically derived for a given modulation/coding scheme and channel. These algorithms are optimal with respect to the chosen performance criterion.
- If, for complexity reasons, the optimal algorithms are not feasible, they form the basis from which suboptimal yet realizable solutions can be derived.
- It is possible to derive lower (not necessarily best-case) performance bounds. This is of great importance since it is often difficult to assess the performance of highly complex systems analytically.

Parameter Estimation and Detection — To illustrate the concept of systematic algorithm derivation, consider the simple transmission model

$$r(t) = \sum_{k=0}^{N-1} a_k g(t - kT - \varepsilon T)e^{j\phi} + n(t)$$

with transmitter pulse forming filter $g(t)$ and additive noise $n(t)$. From the received signal $r(t)$, the data symbol sequence $\boldsymbol{a} = \{a_0, a_1, \ldots, a_{N-1}\}$ and the synchronization parameters $\theta = (\varepsilon, \phi)$ with (relative) timing offset ε and phase ϕ have to be jointly estimated.

The performance criterion leading to minimum probability of error is the maximum a posteriori (MAP) criterion [20]

$$P\{(\boldsymbol{a}, \theta | r\} \rightarrow max$$

which maximizes the conditional probability of the desired quantities, given the observed received signal $r(t)$. If all possible data sequences $\boldsymbol{a}$ and sync parameters θ are equally likely (i.e. no prior knowledge is available), the MAP criterion becomes identical to the maximum likelihood (ML) criterion:

$$P\{r|\ (\boldsymbol{a}, \theta)\} \rightarrow max$$

which maximizes the likelihood of the observation, given the desired parameters. In other words, we look for the set of parameters $(\boldsymbol{a}, \theta)$, now interpreted as trial parameters, which has most likely generated the received signal. In general, however, an exhaustive search over all $(\boldsymbol{a}, \theta)$, is far too complex to implement so that suboptimal solutions must be derived [23].

For symbol detection, a lower bound is given by Shannon's channel capacity formula. For parameter estimation, an equally important lower bound exists, namely the Cramer-Rao bound (CRB) on estimator accuracy $var\{\hat{\theta}\,|\,\theta\}$. If the observation r is linearly dependent on the parameters θ and the noise is additive Gaussian, the ML estimate $\hat{\theta}_{ML}$ satisfies the CRB with equality, i.e. $\hat{\theta}_{ML}$ is an efficient estimate [20]. Note that in our simple transmission model, this is not the case.

PCS Examples — To illustrate the principles described above, we begin with a simple example, and then move on to more complicated scenarios. AWGN Channels. In the simplest case of an AWGN channel (e.g., satellite), the inner receiver consists of a downconverter, a sampler, and a synchronization unit where symbol-timing, carrier-phase, and possibly frequency recovery are performed. Synchronization can be interpreted as estimation and correction of the (slowly) varying channel parameters, here $\theta = (\varepsilon, \phi, \Omega)$ where Ω is the (relative) frequency shift [23]. The estimator unit produces values $\hat{\theta} = (\hat{\varepsilon}, \hat{\phi}, \hat{\Omega})$ that are subsequently used in correction units as if they were the true parameters; this is the widely applied principle of synchronized detection. In a well-designed receiver, the synchronization errors and implementation losses should manifest themselves solely in a slight degradation of the required minimum signal-to-noise ratio (SNR) to achieve a specified performance target. This is indeed achieved in practice.

While the bit error rate (BER) is the appropriate performance measure for the outer receiver, the performance measure relevant for the inner receiver (for AWGN channels) is the error variance of the estimated parameters $\hat{\theta}$. Notice that it is inappropriate to design the inner receiver for optimal BER (as is frequently done), since it violates Shannon's lesson 3: the inner receiver must provide for a "good" channel for the outer coding system, and this is achieved by minimizing the error in the synchronization parameters. As mentioned above, for both kinds of receivers (inner and outer) there exist fundamental performance bounds [20, 29]

- Channel Capacity

$$C = \frac{1}{2} \times \log_2\left(1 + \frac{E_b}{N_o}\right)$$

- Cramer-Rao Bound on Estimator Accuracy

$$var\{\hat{\theta}\,|\,\theta\} \geq J^{-1}$$

where J is the Fisher information matrix. Both bounds are logarithmic measures, but note the fundamental difference between the two: for the channel capacity the energy per bit, E_b/N_o, is involved, while the estimator variance depends on the energy per symbol, E_s/N_o. This is seen, for example, by evaluating the Fisher information matrix J that arises when the parameters $\theta = (\varepsilon, \phi, \Omega)$ are jointly estimated; for this problem, the following bounds on error variances can be established [24]:

- Relative Frequency Offset

$$var\left(\hat{\Omega}\middle|\Omega\right) = \frac{12}{N(N-1)^2} \times \frac{1}{2} \times \left(\frac{E_s}{N_o}\right)^{-1}$$

- Phase Offset

$$var(\hat{\theta}|\theta) = \frac{1}{N} \times \frac{1}{2} \times \left(\frac{E_s}{N_o}\right)^{-1}$$

- Relative Timing Offset

$$var(\hat{\varepsilon}|\varepsilon) = \frac{1}{N} \times \frac{\int |G(\omega)|^2 d\omega}{\int \omega^2 |G(\omega)|^2 d\omega} \times \frac{1}{2} \times \left(\frac{E_s}{N_o}\right)^{-1}$$

where $G(\omega)$ is the transmitter pulse shaping transfer function, and N denotes the estimation window length (in symbol intervals). The energy per symbol (relevant for synchronization) is linked with the energy per bit (relevant for detection) via the code rate R_c (number of source bits per channel symbol) via $E_s/N_o = R_c(E_b/N_o)$. The minimum E_b/N_o tolerated by the detector can be made small by using powerful codes with low code rates R_c, but then E_s/N_o may become very small. On the other hand, any practical synchronizer requires a minimum E_s/N_o to be operational which, in turn, sets a lower limit on the code rate that can be achieved in practice; e.g., see Massey's instructive papers [18, 19] for more information. The systematic design of digital synchronization algorithms by invoking principles of estimation theory for timing, phase and frequency recovery is treated in detail in [24].

Fading Channels — In the case of fading channels, the synchronization task is far more complex since, instead of the static (or slowly varying) parameters of the Gaussian channel, randomly varying channel parameters have to be continuously estimated and compensated for. Furthermore, in order to comply with Shannon's lesson 3, the transmission system should provide for and make use of various kinds of diversity, e.g., time, space, or multipath diversity. Exploiting as many diversity mechanisms as possible is effective in bridging deep fades and leveling out wide variations in the information-bearing signal. As an example, Fig. 4 shows the structure of an inner receiver for linear modulation formats and frequency-selective fading channels [8]. The sampled front-end output signals $r_d(k)$ ($d = 1,...,D$ antennas, D RF front ends) are organized in frames where the stream of useful data is interspersed with blocks of training signals.

In the synchronization path (lower part of Fig. 4), the training blocks are extracted and used for feedforward estimation of time-varying channel parameters such as the start-of-frame time instant k, frequency offset Ω and, most importantly, the (sampled) diversity channel impulse response vectors h_d. The receiver structure is therefore based on the so-called *separation principle* that makes it possible to break the two-fold randomness of random data and randomly varying channel(s) which, if channel estimation and random data detection were to be performed simultaneously, would lead to an explosion in complexity. The separation of detection and synchronization paths enables the channel(s) to be identified from known symbols alone: this process is not disturbed by random data. Conversely, the channel estimates so obtained can be assumed to be known during random data detection.

From the estimated channel(s), coefficient sets of various receiver components are computed. As

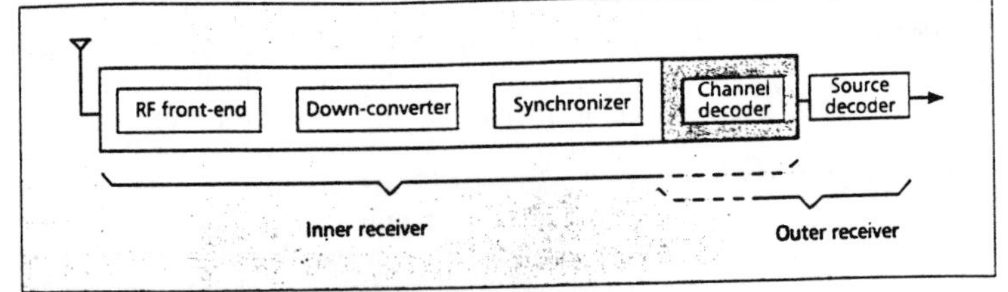

■ **Figure 3.** *Fundamental parts of a digital receiver.*

illustrated by Fig. 4, these may encompass sets of prefilter (whitening matched filter) coefficients w_d, combiner weights p, and equalizer taps f. Notice that with feedforward synchronization, this computation is just a mathematical mapping from estimated channel parameters to the coefficient sets. As opposed to recursive-type decision-feedback synchronizers, no lag error is introduced by this mapping, nor has error propagation a chance to thwart continuous receiver operation.

In the detection path (upper part of Fig. 4), the received signals are processed using the appropriate coefficient sets. In this receiver design example, antenna diversity, time diversity via interleaved channel coding, and multipath diversity via energy equalization and detection are made use of simultaneously. On fading channels, it is particularly important to avoid discarding relevant information prematurely (Shannon's lesson 2). This is achieved by transferring soft decisions between the functional blocks, e.g., equalizer, channel decoder, and source decoder.

Spread Spectrum — The design of advanced spread-spectrum multiple access systems is another example of how the key receiver design principles described earlier can be applied to produce a high-performance digital communications link. The advances in DSP techniques, A/D technology, and the speed and density of digital circuits have made these digital receivers realizable with reasonable complexity and cost.

Direct-sequence spread-spectrum systems operating in environments with multipath diversity can exploit multipath through the use of RAKE receivers. These receivers use despreading to extract the symbol energies spread over several signal paths, and use diversity combining to combine these energies to maximize the received symbol-to-noise ratio. These operations are all founded on the principle of 1) making the channel appear as an AWGN channel and 2) ensuring that the receiver uses all the information available to it at reasonable cost and complexity, including multipath versions of the transmitted signal.

Digital spread-spectrum receiver designers exploit the fact that there are actually three regimes of DSP going on in such a receiver [36]. If digitization of the received signal occurs at IF, then the first regime encountered is that of demodulation processing, where the receiver operations focus on capturing all the PN (pseudonoise) chip energy into one or two samples. This is followed by a PN-chip-rate processing regime, where code synchronization and despreading operations take place. The symbol energy is captured at the output of the despreader. Finally, the receiver operates in a symbol-rate regime, where symbol energy is maximized, estimators are provided for various parameters necessary for detection, and final data-detection takes place. Subsequent operations are performed on a symbol-by-symbol or block-of-symbols basis.

The key requirements for PCS portable terminals are performance, cost, power consumption, and size.

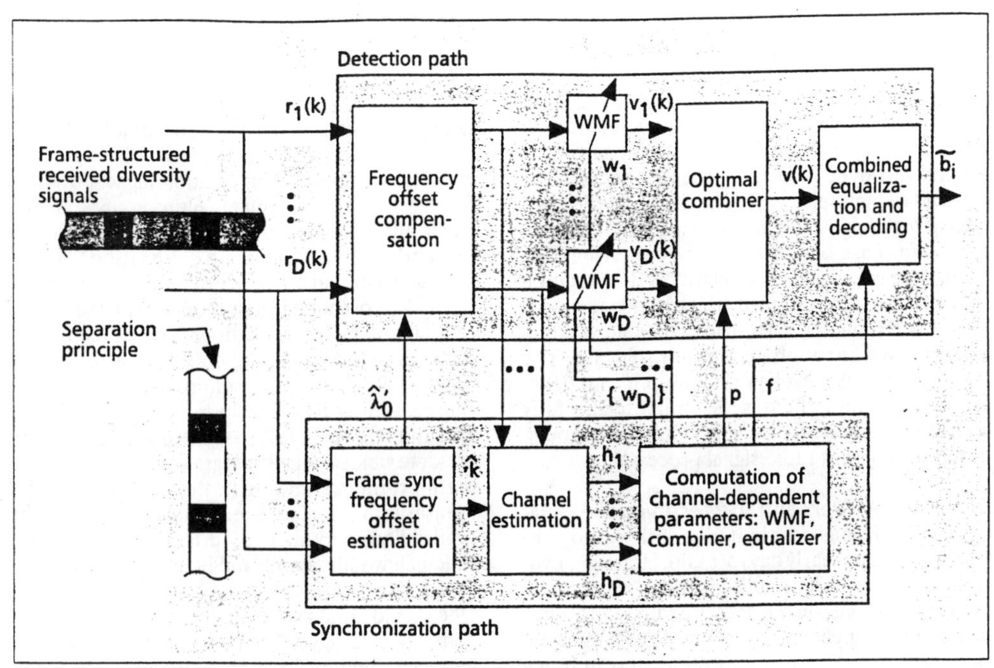

■ **Figure 4.** *The inner receiver decomposed into detection and estimation paths.*

Accurate code synchronization is essential for reliable performance of these links. Without accurate and reliable code synchronization, the receiver cannot guarantee performance to any degree.

The code synchronization operation typically consists of two stages of processing. When a receiver is first turned on, the proper code phase for the local replica of the spreading code needs to be guessed. The receiver must operate at several times the PN-chip rate in order to complete the acquisition operations. Once a reliable first guess is made, the particular code phase is tracked in a code-tracking loop. Here, variations of the classical delay-locked loop dominate most solutions. In systems with short spreading codes, however, scanning-window tracking techniques based on capturing despreader outputs in FIFOs can result in simpler implementation and control complexity [36].

PCS Radio Transceiver Architectures

Portable Radio System Requirements

The key requirements for PCS portable terminals are performance, cost, power consumption, and size. The bill of materials (BOM) of most cellular and cordless transceivers today is dominated by the radio signal processing components. These account for more than 50 percent of the BOM in most consumer phones, and they include all components related to RF/IF filtering, mixing, demodulation, and frequency synthesis. The target terminal size is typically on the order of 50 to 150 cc. This invariably requires using advanced integration techniques coupled with an intelligent partitioning of the radio subsystems.

Achieving low power consumption is done through technology and system-level trade-offs. The actual metrics are the standby-time and talk-time of the terminal between charging. Depending on which mobile communication system the terminal is designed for, the functions required to be performed by the mobile during standby widely vary. Designers need to optimize the current consumed by all the RF components, as well as the baseband DSPs, digital application-specific integrated circuits (ASICs), A/Ds, and any mixed-signal devices. At the system level, terminal designers need to understand the system operation requirements such as time-division-multiplex (TDM) or voice-activity (VA)-based operation. Optimizing power consumption here mainly involves firmware engineering for micro-controllers and DSPs as well as relegating certain firmware functions to customized hardware accelerators. Many digital processors, for example, feature power-down modes that allow for "shutting down" peripherals and certain computational units. But they do not always allow for fast ramp-ups, so that care should be exercised in their use.

Radio Receiver Principles

A radio receiver front-end is defined by an architecture chosen in order to satisfy a set of requirements. These include sensitivity, dynamic range, selectivity, signal blocking, intermodulation performance, and, of course, manufacturability, and cost. The tools of mixing, filtering, and gain control are at the designer's disposal. The typical approach used in the design of portable radio receivers considers two main desirable properties of the equipment: selectivity and sensitivity. Limitations on transmitter power and poor antenna orientation require a receiver to perform well at input levels down to 1 microvolt. The noise figure of the front-end now becomes a significant factor in defining the receiver sensitivity. Moreover, because the receiver may have to work with very low input signal levels in the presence of very large levels of unwanted signal, the linearity of the early stages of the receiver are of key importance.

The Superheterodyne Principle — The key functional requirements for a receiver are that its front-end structure must 1) accurately translate the desired RF signal to a frequency band where the processing required to extract the message is relatively cheap and 2) condition the signal (i.e., filter and gain control) to remove adjacent channel interference while maintaining receiver sensitivity. The great majority of radio receivers employ the superheterodyne principle because the fixed IFs can lead to high selectivity. That is, a first IF (high) is used for image-rejection, and a second IF (low) is used for ease of highly-selective filtering. The primary reason for this popularity is the predominance of certain standard IF frequencies, which makes components of IF strips available at a reasonable cost through economies of scale. The superheterodyne receiver offers significant benefits. Chief among them is the ability to carefully control gain, noise figure, and selectivity through different stages of radio signal processing.

Digital IF Receivers and DSP Techniques — Analog hardware has dominated mobile radio subsystems since the early days of radio. However, DSP techniques are well established today in baseband modem functions, such as speech coding and decoding, interleaving, channel coding and decoding, detection, and equalization, as well as in frequency synthesis. These techniques are now finding their way into IF processing sections of radio receivers.

The digital conversion (or Digital IF) receiver is characterized by the principle of using the inherent aliasing property of sampling to realize a demodulator. The digital IF receiver typically uses a superheterodyne architecture, where analog frequency translation is used to move the received RF spectrum to a low IF, and digitization occurs at the low IF. Thereafter, DSP techniques are used to extract the signal. The DSP techniques used include direct digital frequency synthesis (DDS), digital downconversion, high-speed digital filtering, and multirate techniques such as decimation and interpolation.

Figure 5 shows a typical configuration for a digital IF receiver. Configurations may vary depending on whether a single on double conversion superhet structure is used. The digital conversion principle exploits the fact that the lowpass I and Q components are expressible in terms of bandpass samples. The receiver is then taking samples of a passband signal, and numerically demodulating to producing I and Q samples serially. In fact, it is easy to show that to sample and reconstruct a passband signal centered at f_{IF}, the sampling rate f_s should meet the criterion

$$f_S = \frac{4f_{IF}}{2k-1}$$

where k is the image number. Recall that there are folded spectra present when the bandlimited passband signal is sampled, as shown in Fig. 5.

The simplest implementation occurs when the $k=1$ image is selected, and the IF signal is sampled at $4f_{IF}$. From an implementation point of view, a designer can explore how to trade-off a lower sampling rate for more complex DSP. Any IF sampling procedure will yield I and Q samples which are not coincident in time. As the sampling rate is dropped (i.e., k is increased) from $4f_{IF}$, these samples will be further apart in time. Thus, any detection procedure based on these samples will suffer from this time distortion. As a result, digital processing techniques to reconstruct pulse shapes will be necessary in order to compensate for the resulting distortions in I/Q samples.

There are several factors driving the use of digital IF processing. They include the advances and availability of low-cost, high-speed A/D converters, ever-increasing levels of integration, higher price/performance value, lower manufacturing costs, greater flexibility, and increased robustness and maintainability (i.e., little or no calibration). These characteristics are already making digital IF the technology of choice in non-mobile applications like cellular and PCS base-stations. One of the biggest advantages of the digital conversion receiver is the possibility it opens for multimode terminals which share a similar RF/IF strip, possibly with switchable filters, and a DSP-based programmable processor to yield savings in integration and testing. The main problem standing in the way of its implementation in portable terminals centers on the availability of very high-speed, large-dynamic range, wide-bandwidth sample-and hold circuits and high-performance A/D converters at reasonable cost. One should always note, however, that analog functions, where feasible, are often smaller, cheaper, and less power hungry. These benefits may offset the fact that the analog solution may suffer some increased manufacturing cost.

Direct-Conversion Receivers and DSP Techniques — The direct-conversion receiver structure is shown in Fig. 6. The received RF signal is filtered in a duplexer, and passed through a low-noise amplifier. After bandpass filtering, the signal is split, and fed to a quadrature mixer. The mixer is fed with the local oscillator (LO), and a 90-degree phase-shifted version of the LO, which is in fact at the RF carrier frequency. The RF passband is thus translated to baseband, and the receiver must now process the full RF spectrum at baseband. This requires extremely high dynamic range as well as extremely high selectivity, together with low noise and proper amplitude and phase balancing between the I and Q arms. Figure 6 shows the processing involved in one signal path. D_1 denotes the desired signal, and I_i denotes interferers.

The greatest advantage of the direct conversion architecture is its low component count. Its simplicity, however, is deceptive. Few commercial receivers employ this architecture. The much acclaimed benefits of a wide tuning range and high selectivity are limited by the dynamic range of the receiver. Among the challenges in its realization are the need for a very high-gain low-noise mixer (to combat $1/f$ amplifier noise at baseband) as well as rapid large-DC-offset cancellation techniques. Almost all the gain required to achieve the desired sensitivity in this architecture must come from the front-end low-noise amplifier. Furthermore, because of the LO signal path, the antenna needs to be isolated from the mixer. Building in reverse isolation of up to 80 dB on some RF IC components is no easy task. Furthermore, since all the selectivity in this receiver comes from the baseband channel filters, one requires very high-dynamic range, low-noise, highly-selective filters. It was not until the late '80s that commercial solutions were announced for simple FSK paging receivers at 450 MHz and, more recently, at 930 MHz. Recently, some 900-MHz cellular and wireless LAN products have emerged using this architecture.

Digital signal processing techniques are well established in baseband modem functions, as well as in frequency synthesis.

General-purpose programmable digital signal processor chips continue to climb up the performance curve, driven by market needs, technological advances, and architectural innovations.

Achieving a competitive direct conversion solution for commercial cellular mobile systems is a still a big challenge. Some researchers have focused on designing specific components in the RF chain to optimize performance, oftentimes going to more expensive IC processes to achieve the desired results. Others have taken the route of using more conventional IC processes, and trading-off analog performance requirements at the expense of increased DSP. In fact, the direct conversion receiver is a good example of an architecture where a successful competitive solution requires close interaction between analog high-frequency design, analog baseband design, and signal conversion technology.

Novel Low-cost Digital Architectures — Recently, some receiver structures that do not employ *I/Q* demodulation and A/D conversion when going from the IF regime to the baseband regime have appeared commercially. Many of these structures focus on distorting the IF signal (typically by limiting) and passing the output into a high-speed digital circuit, which produces digital information about the phase of the IF signal at symbol-rate intervals. This, together with the digital version of the filtered received signal strength information (RSSI), typically produced by a limiting amplifier followed by an A/D converter, produces a digital representation of the amplitude and phase of the received signal sample, allowing digital reconstruction in the form of $a_k e^{j\theta_k}$. Thus, only one A/D converter is required per received signal branch. Some receivers throw away the amplitude information, and operate solely with the phase information. thereby suffering a penalty in performance. Whether this penalty is deemed to be significant or not depends on the system under consideration and the minimum performance requirements. Techniques such as antenna diversity can offer significant performance benefits to offset the degradation of suboptimum detection schemes. Several researchers have realized such systems on dedicated low-power custom integrated circuits for cellular and cordless systems using these techniques [14, 16].

Key Digital Signal Processing Technology Issues for PCS

DSP Technology Trends

General-purpose, programmable, digital signal processor chips continue to climb up the performance curve, driven by market needs, technological advances, and architectural innovations. DSP chips have recently been the fastest growing segment of the semiconductor industry, with an annual growth rate of more than 35 percent, and will continue that growth pattern for the next several years, according to market research firm Forward Concepts Inc. This high growth for DSPs is fueled by expanding applications in modems, disk drives, digital cellular radio, digital cordless telephony, and emerging PCS applications.

Both fixed-and floating-point DSP technologies are rapidly evolving to meet the PCS market's shifting requirements. Today's DSP market breaks down to 80 percent fixed-point and 20 percent floating point (Forward Concepts). Broadly speaking, a floating point DSP has characteristically taken three to four times as much current per instruction as a fixed point processor (e.g., TI TMS320LC31). More DSP manufacturers are unveiling true 3V versions of both processors to address the low-power needs of battery-operated portables. New tiers of power-down modes allow careful control of the current consumed in the application being executed. New core architectures are taking shape slowly. And the demand for application-specific digital signal processors based on the DSP-core methodology is growing. This is bringing chip vendors, ASIC houses, and microcontroller suppliers into the DSP arena. Finally, complementing the obvious values of DSP's raw computing power, the availability of an ever-growing spectrum of silicon-level DSP-system integration and customization options has helped DSPs proliferate rapidly into vertical applications.

Today, DSP instruction sets are becoming more efficient in the execution of complex signal processing tasks. In many instances, they are implemented in a single machine cycle through the use of hardware accelerators or enhanced cores. Clocks are now distributed only to the required switching circuits within the DSP. Designs are stripped of all dynamic logic. The most advanced DSPs are manufactured today on 0.6- or 0.5-micron CMOS processes. This has significantly reduced the internal load capacitances, which in turn offer reductions in current consumption. Moreover, the reduction in geometry leads directly to increased performance from the same design as it can now sustain a faster cycle time and more functionality. DSPs are fully reaping the benefits of the advances in silicon-CMOS technologies.

The Role of DSPs, ASICs, and Microcontrollers

There are typically three digital "building blocks" available to the system designer: programmable general-purpose DSPs, microprocessors, and ASICs.

A general-purpose DSP is a software-programmable integrated circuit used for speech coding, modulation, security and integrity of information (through signal coding and scrambling), detection, equalization, and associated modem tasks such as frequency, symbol timing recovery, and gain control. Moreover, the DSP is also used for flexibility in application, and the ability to add new features with minimum redesign and re-engineering.

Microprocessors are usually used to implement protocol stacks, system software, and interface software. Most of the cellular systems require an 8- or 16-b microprocessor dedicated to these functions to address the software size and complexity. Microprocessors are better suited to perform the non-repetitive, control-oriented. input/output operations, as well as all housekeeping chores.

ASICs are used for various purposes in PCS. They are utilized for high-throughput tasks in the area of digital filtering, synchronization, equalization. and channel coding or decoding. The ASIC is often likely to also provide the glue logic to interface components. In some systems, the complete digital receiver is "poured" into silicon as an ASIC. coupled with a microcontroller, without a general-purpose DSP. ASICs have historically been

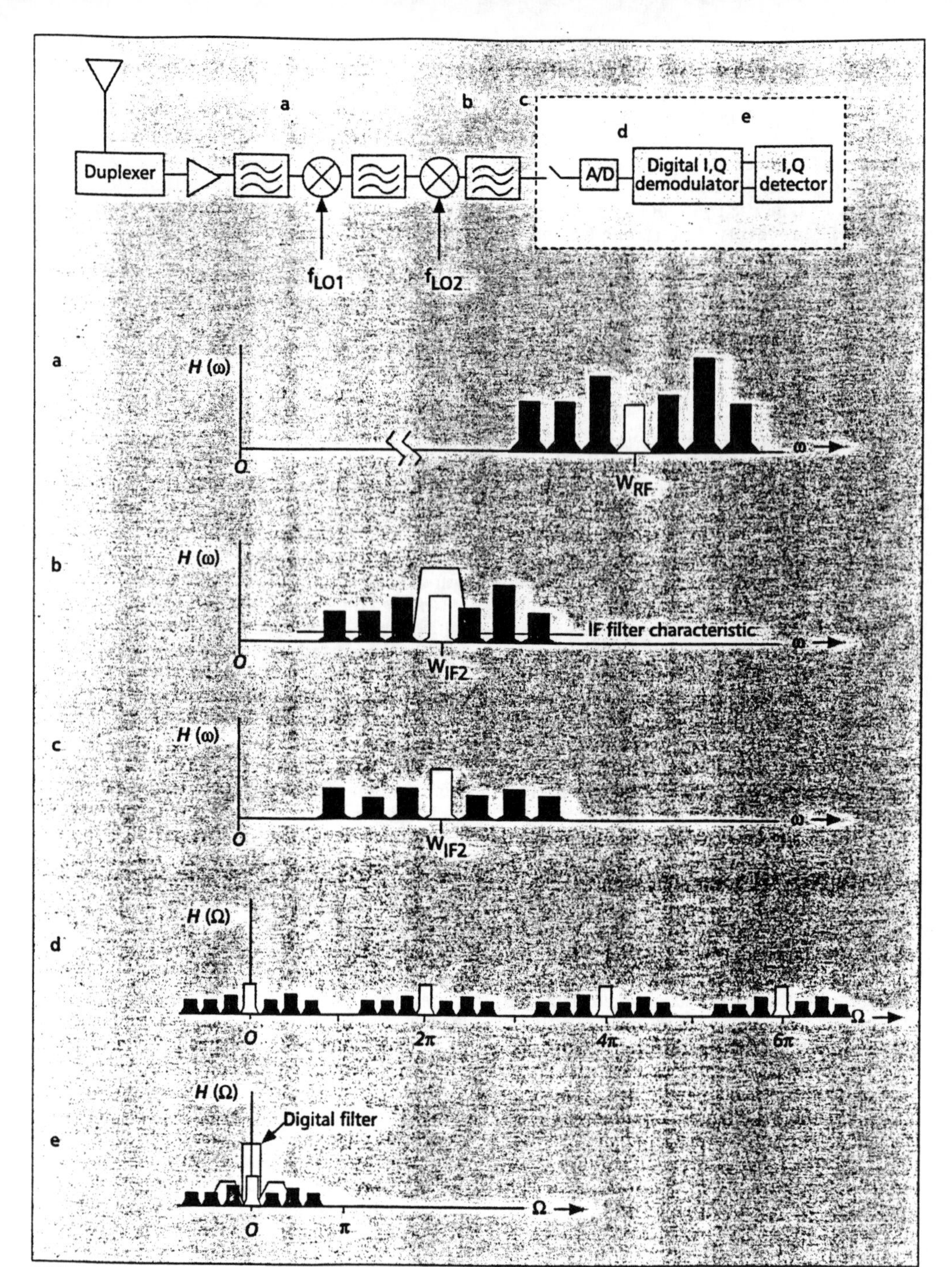

■ **Figure 5.** *Digital IF receiver principles.*

There are typically three digital "building blocks" available to the system designer: programmable general-purpose DSPs, microprocessors, and ASICs.

used because of their lower power dissipation per function. In certain applications like spread-spectrum systems, digital receiver designs require ASIC solutions in order to execute the wideband processing functions such as despreading and code synchronization. This is primarily because the chip-rate processing steps cannot be supported by current general purpose DSPs.

Over the last few years, as cellular and cordless manufacturers have brought out first-and second-generation digital cellular and cordless solutions, programmable general-purpose digital signal processors are slowly being transformed into accelerator-based DSP-microcontroller hybrids. This transformation is a result of the severe pressures being put on reducing power consumption. As firmware solutions become finalized, cycle-hungry portions of algorithms (e.g., equalizers) are being "poured into silicon," using various VLSI architectural ideas. This has given rise, for example, to new DSPs with hardware accelerators for Viterbi decoding, vectorized processing, and specialized phase-domain functions. They have been driven by applications in digital cellular telephones, such as GSM and NADC(IS-54), and cellular data (e.g., CDPD).

In the PCS arena, design costs and timescales will become more important that chip processing cost. In many markets, product life-cycles will be very short. To compete successfully, companies

In many markets, product life-cycles will be very short. To compete successfully, companies will need to be able to turn systems concepts into silicon VLSI very rapidly.

will need to be able to turn systems concepts into silicon VLSI very rapidly. This puts a premium on high-quality computer-aided design, and not least on system-level tools that enable algorithm and architecture exploration.

Tools

Design Automation Tools for analysis and modeling in the algorithm and architecture space are critical for efficient simulation and realization of DSP systems. These tools are characterized by a particular design methodology, which, for digital communications and signal processing systems, must be structured to cope with the time scale explosion problem faced by digital system designers. This time scale explosion problem in modeling can best be thought of as requiring a toolset to be able to handle time scales from 1 s (channel correlation time) to 1 ns (circuit gate delay). Allowing different levels of abstraction at different points in the time scale lead to efficient and powerful tools. This is typically done, for example, by partitioning modeling levels into functional, structural, and physical domains.

Exploring the Algorithm Space

Algorithms in DSP are comprehensively specified using block diagram descriptions. The blocks represent signal processing operations or functions, such as filters, whereas the interconnecting signals represent the data flow. Such block diagram specifications are used as input specification in state-of-the-art design environments for digital communication systems [3, 4, 15]. Preferably, these block diagrams are specified graphically in a hierarchical manner. Besides comprehensiveness, block diagram specifications allow for very structured simulation programs with a maximum reusability of components. For reusability purposes and for fast exploration of the algorithm space, blocks are parameterizable. Parameterizable simulations, in turn, allow for automatically evaluating different algorithmic variants based on the same block diagram structure.

As for any type of computation, a fundamental feature of block-diagram oriented simulation is the underlying processing paradigm. According to the nature of digital communication systems, multirate and variable-rate processing must be supported. Data-flow-oriented and time-driven approaches are applied in practice. The market is dominated by the design tools SPW (from Cadence) and COSSAP (from Synopsys), which mainly differ in the way they represent systems with multiple sampling rates and in the optimization strategies performed at the system level [3, 31]. While SPW uses a time driven approach, COSSAP is based on data flow semantics. See [33] for a lucid discussion of the merits of both tools. Recent work [4, 27] has demonstrated that heterogeneous specification languages are very powerful for that purpose. These may include various types of data flow and state diagram descriptions.

Exploring the Architecture Space

As discussed above, algorithm and architecture of digital communication systems are strongly interdependent. Therefore, joint optimization of both aspects is highly desirable. In the past, system design on the one hand and implementation on the other hand have been separate tasks. In general, different development teams have been in charge of these tasks. The system specifications had to be translated manually into implementations, which is an error prone process. To allow for a smooth interaction between the exploration of both the algorithm and implementation spaces, a tool supporting this two-fold exploration should be based on the same input specification, i.e., a block diagram description. This also offers a safe and rapid transition from algorithm design to architecture implementation.

Software-Based Architectures — As mentioned above, a wide range of functionality can be implemented on custom DSP processors. Once the choice of an appropriate target DSP processor has been made, the key optimization criteria for this type of implementations are throughput and RAM/ROM memory consumption.

Hence, design tools for software-based architectures must support the selection of a target platform and allow trade-offs between the above criteria. According to the requirement that both algorithms and architectures are based on identical input specifications, the task of such tools is to generate optimized real-time code from generic block diagram specifications [31]. The main issue is to exploit the algorithm-specific features of digital signal processors. Vectorization plays a very important role in this context. Most optimizations improve one of the implementation parameters at the cost of at least one other parameter. Therefore, optimization strategies must not always be applied. Instead, the user must be able to select the ones that lead to the prescribed implementation goals.

While memory consumption of an application can be determined exactly off-line, throughput cannot. Therefore, the generated program must be run on the actual hardware or on a hardware simulator to extract this parameter. Many signal processing algorithms, e.g., synchronization based on preamble structures, are in need of realistic input data. Consequently, the DSP hardware must be coupled to the simulation environment. This allows, for example, generation of the transmitted data and simulation of the channel distortion on that data on the simulation computer, while running the receiver implementation on the actual hardware. Exact comparisons between simulation model and DSP implementation can be carried out. If the DSP hardware is unavailable or not accessible via interprocessor communication to the simulation host, a DSP hardware simulator may also be coupled to the system simulation [43]. This coupling is powerful, as users can develop, debug, and functionally test DSP firmware within a system simulation environment.

Hardware-Based Architectures — In the case of applications requiring a very high data throughput or very large production volumes, special-purpose hardware in the form of ASICs is the preferred implementation choice. Many different implementation technologies such as full-custom layout, standard-cell-based design, and gate arrays in different feature sizes are available. Here one has to select based on production volume, required throughput, time to market, testability, power consumption, etc. The technology chosen for a cer-

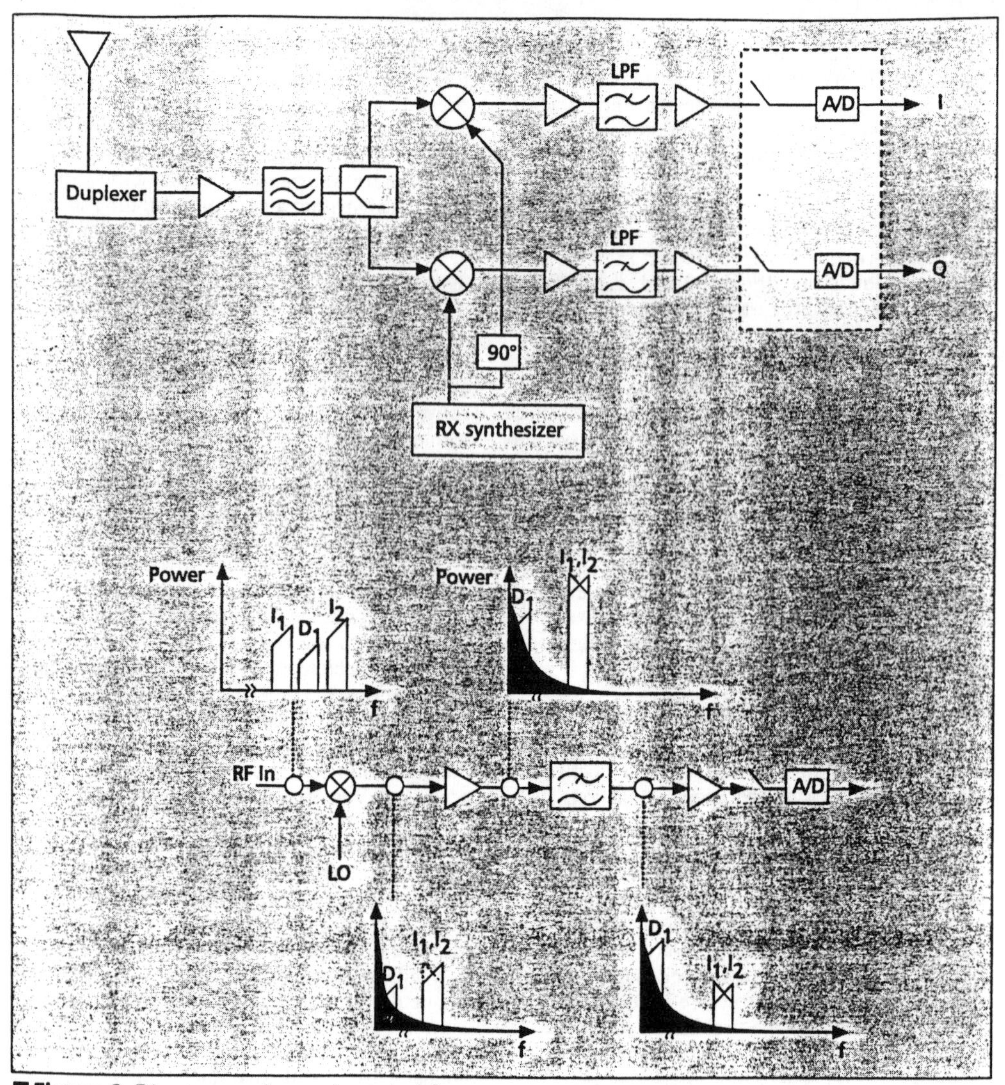

■ **Figure 6.** *Direct conversion receiver principles.*

In the case of applications requiring a very high data throughput or very large production volumes, special-purpose hardware in form of ASICs is the preferred implementation choice.

tain product may change during its life cycle (e.g., prototype realized as several field programmable gate arrays, consumer product realized as one ASIC).

To minimize nonrecurrent engineering cost and to maintain the advantages of using the high level block diagram tools, the communication system designer should use a standardized intermediate format as an interface to industry-standard hardware design tools supporting the path to hardware, such as logic synthesis [7], time-based discrete event simulation, and physical design. Hardware description languages (e.g., VHDL [11] or Verilog) serve best for this purpose. Today all major electronic design automation suppliers offer VHDL and Verilog based logic synthesis tools [2, 5, 21, 38, 40].

Today's algorithm design environments offer a fairly sophisticated way to generate synthesizable VHDL descriptions from the algorithm specification. There are many options for this transformation process. The resulting architecture may be processor-like or lowly multiplexed, consisting of several processing units and memories. A time-driven synthesis environment has a straightforward link to VHDL generation [17, 32, 39], but the algorithm design capabilities are restricted by the required configuration style. Data-flow-driven algorithm specification offers the possibility to generate hardware descriptions for different area and throughput requirements [44]. A library of reusable components which encompasses not only the behavior but also different implementation choices [12] in terms of area, throughput, and power consumption, is of great advantage in achieving a shorter time to market. Recently, design automation software suppliers have announced behavioral synthesis tools which allow a higher level of abstraction in the VHDL input than the current logic synthesis tools. Instead of register transfer level (RTL) code describing every storage cell and every operation, behavioral coding style allows an design description which is closer to that of software-based implementations. Special purpose synthesis tools for signal processing applications based on the programming language Silage [10] have been developed by IMEC, Belgium [28,26] and are in-part also available commercially [22].

Future Directions in Modeling and Tools

Whatever the route, a hierarchical approach to "pouring an algorithm into silicon" is giving rise to new

Algorithm-specific design know-how can be incorporated anywhere from the abstract algorithm description down to the actual physical architecture implementation.

tools and methodologies that give system designers the power to explore the communications, DSP, and VLSI design parameter space to make design trade-offs. These tools exploit the fundamental design concepts of hierarchy, modularity, regularity, and locality. In a nutshell, these concepts give rise to well-defined interfaces in a toolset that provide the appropriate level of abstraction to the designer at different levels of the design hierarchy. This is essential to master the ever-increasing complexity of the upcoming generations of personal communication systems. The CAD tools for system design integrating algorithm and architecture will further move toward abstraction, while offering the means for user interaction at any level. In particular, algorithm-specific design know-how can be incorporated anywhere from the abstract algorithm description down to the actual physical architecture implementation.

Conclusion

Our focus in this article has been on the new principles underlying the design and implementation of modern digital wireless terminals for PCS. We introduced the modern centers-of-gravity model, which we believe is emerging as the basis for the successful design and implementation of advanced digital communication systems. Central to this model are design principles that enable engineers to systematically derive digital receiver structures and explore algorithm and architecture trade-offs using sophisticated tools. DSP technology is critical in the implementation of these digital receiver structures efficiently. Finally, CAD tools for digital communications system design and design space exploration will be of crucial importance in the efficient execution of these designs. The tools and the technology are advancing at a phenomenal pace today, and it is this growth that will clearly be the enabler of universal low-cost wireless digital communications.

References

[1] A. Aghamohammadi, "Adaptive Phase Synchronization and Automatic Gain Control of Linearly-Modulated Signals on Frequency-Flat Fading Channels," Ph.D. Dissertation, Aachen University of Technology, 1989.

[2] "Booledozer User Manual," Altium, An IBM Company.

[3] B. Barrera and E. A. Lee, "Multirate Signal Processing in Comdisco's SPW," Proceedings of ICASSP, Toronto, April, 1991.

[4] J. Buck *et al.*, "Ptolemy: A Platform for Heterogeneous Simulation and Prototyping," Proc. 1991 European Simulation Conf., Copenhagen, Denmark, June, 1991.

[5] "Synergy User Manual," Cadence Design Systems,555 River Oaks Parkway, San Jose, CA 95134, USA.

[6] "COSSAP User's Manual," Synopsys Inc., 700 E. Middlefield Rd., Mountain View, CA 94043, USA.

[7] A. J. de Geus, "Logic synthesis speeds ASIC design," *IEEE Spectrum*, 1989, no. 8, pp. 27-31.

[8] S. A. Fechtel and H. Meyr "Combined Equalization, Decoding and Antenna Diversity Combining for Mobile / Personal Digital Radio Transmission using Feedforward Synchronization," Proc. IEEE Int. Conf. Vehicul. Technol., VTC'93, Secaucus, NJ, USA, pp. 633-636, May 1993.

[9] S. A. Fechtel "Verfahren und Algorithmen der robusten Synchronisation f\"ur die Daten\"ubertragung \"uber dispersive Schwundkan\"ale (in German)," Ph.D. Thesis, Aachen Univ. Tech. (RWTH), Wissenschaftsverlag Mainz, Aachen, 1993.

[10] D. Genin *et al.*, "DSP Specification Using the Silage Language," Proc. ICASSP'90, 1990, pp. 1057-1060.

[11] "IEEE Standard VHDL Language Reference Manual, IEEE Std 1076-1987," (IEEE, March, 1988).

[12] R. Jain *et al.*, "Computer-Aided Design of a BPSK Spread-Spectrum Chip Set," IEEE J. of Solid State Circuits, Jan. 1992, vol. 27, no. 1, pp. 44-57.

[13] R. Jain *et al.*, "Computer-Aided Design of a BPSK Spread-Spectrum Chipset," IEEE Journal on Selected Areas in Communications, Special Issue on Computer-Aided Modeling, Analysis and Design of Communication Links, vol. 11, no. 3, pp. 348-358," April 1993.

[14] S. Kato *et al.*, "A Very-Low-Power High-Quality Signal Transmission Baseband LSIC for PCS: Advanced Signal Processing and One-Chip Integration," to appear in Proc. GLOBECOM 1994.

[15] J. Kunkel, "COSSAP: A Stream Driven Simulator," Proc. IEEE International Workshop on Microelectronics in Communications, Interlaken, Switzerland, March 1991.

[16] C. P. LaRosa and M. Carney, "A Fully Digital Hardware Detector for /4-QPSK," Proc. VTC 1992, pp. 293-297.

[17] L. Lundberg, "Generating VHDL for Simulation and Synthesis from a High-Level DSP Design Tool," EURO-VHDL'91 Proceedings 1991, pp.321-327, Institute for Microelectronics, Stockholm, Sweden, Sept. 1991.

[18] J. L. Massey, "Coding and Modulation for Code-Division Multiple Accessing," Proc. 3rd Int. Workshop on DSP Techniques Applied to Space Communications, ESTEC, Noordwijk, The Netherlands, pp.. 3.1-3.17, Sep. 1992.

[19] J. L. Massey, "Deep-Space Communications and Coding: A Marriage Made in Heaven," Advanced Methods for Satellite and Deep Space Communications, Bonn, Germany, pp. 1-17, Sept. 1992.

[20] J. L. Melsa and D. L. Cohn, "Decision and Estimation Theory," (McGraw-Hill, 1978).

[21] "Autologic User Manual," Mentor Graphics Corporation, 1001 Ridder Park Drive, San Jose, CA 95131, USA.

[22] "DSP Station User's Manual," Mentor Graphics, 1001 Ridder Park Drive, San Jose, CA 95131, USA.

[23] H. Meyr and G. Ascheid, "Synchronization in Digital Communications-vol. 1," John Wiley, 1990.

[24] H. Meyr *et al.*, "Synchronization in Digital Communications-vol. 2," [John Wiley (in preparation)].

[25] M. L. Moher and J. H. Lodge, "TCMP-A Modulation and Coding Strategy for Rician Fading Channels," *IEEE JSAC*, vol. 7, no. 9, pp. 1347-1355, Dec. 1989.

[26] S. Note *et al.*, "Cathedral-III: Architecture-Driven High-Level Synthesis for High Throughput (DSP) Applications," in Proc. of the 28th ACM/IEEE Design Automation Conference, 1991, pp. 597-602.

[27] M. Pankert *et al.*, "Dynamic Data Flow and Control Flow in High Level DSP Code Synthesis," Proc. of ICASSP, Adelaide, South Australia, April, 1994.

[28] L. Philips *et al.*,"Silicon Integration of Digital User-end Mobile Communication Systems," in Proc. ICC '93, May 1993, pp. 212-216.

[29] J. G. Proakis, Digital Communications, (McGraw-Hill, 1989).

[30] D. Pucknell and K. Eshragian, "Basic VLSI Design: Systems and Circuits," (Prentice Hall, Australia, 1988).

[31] S. Ritz *et al.*, "High Level Software Synthesis for the Design of Communication Systems," *IEEE JSAC*, Special Issue on Computer-Aided Modeling, Analysis and Design of Communication Links, Vol. 11, No. 3, pp. 348-358," April, 1993.

[32] "SPW User's Manual," Cadence Design Systems, 919 E. Hillsdale Blvd., Foster City, CA 94404, USA.

[33] K. S. Shanmugam, "Simulation and Implementation Tools for Signal Processing and Communication Systems, " *IEEE Commun. Mag.*, July 1994, vol. 32, no. 7, pp. 36-40.

[34] I. Shimizu *et al.*, "New Digital Mobile Radio Technologies," *NTT Review*, Jan. 1992, pp. 64-70.

[35] R. Subramanian, "Synchronization Systems for Spread-Spectrum Receivers," Ph.D. Thesis, University of California, Berkeley, Dec. 1991.

[36] R. Subramanian, "Digital Synchronization Subsystems for Spread-Spectrum Receivers," Proc. IEEE Workshop on VLSI in Communications, Sept. 15-171, 1993, Lake Tahoe, CA, pp. 22-34.

[37] "Behavioral Compiler User Guide," Synopsys Inc., 700 E. Middlefield Rd., Mountain View, CA 94043, USA.

[38] "Design Compiler Reference Manual," Synopsys Inc., 700 E. Middlefield Rd., Mountain View, CA 94043, USA.

[39] P. B. Tjahjadi, *et. al.*, "Vanda -a CAD System for Communication Signal Processing Circuits Design," in VLSI Signal Processing IV, eds. H. Moscovitz, K. Yao, and R. Jain (IEEE Press, 1990).

[40] "SilcSyn User Manual," Viewlogic Systems, Inc., 293 Boston Post Road West, Marlboro, MA 01752-4615, USA.

[41] A. J. Viterbi, "Wireless Digital Communication: A View Based on Three Lessons Learned," *IEEE Commun. Mag.*, vol. 29, no. 9, Sept. 1991, pp. 33-36.

[42] N. Weste, and K. Eshraghian, "Principles of CMOS VLSI Design: A Systems Perspective," 2nd Ed., (Addison Wesley, 1993).

[43] P. Wilhelmsen *et al.*, "Simulating a Programmable DSP at the System Level," DSP Applications, pp. 6-18, Nov. 1993.

[44] P. Zepter and T. Grotker "Generating Synchronous Timed Descriptions of Digital Receivers from Dynamic Data Flow System Level Configurations," Proc. of European Design And Test Conference (poster session), Feb. 1994.

Biographies

HEINRICH MEYR [F '86] received M.S. and Ph.D. degrees from the ETH in Zurich, Switzerland. He is a professor of electrical engineering at the Aachen University of Technology (RWTH Aachen), Germany and has worked extensively for the last 20 years in the areas of communication theory, synchronization, and digital signal processing. His research has been applied to the design of many industrial products. At the RWTH Aachen, he heads an institute involved in analysis and design of complex signal processing systems for communications applications. His group has developed the fastest Viterbi-Decoder chip to date, a chipset DIRECS for one of the first all-digital receivers operating over 100Mb/s, and the prototype of the COSSAP simulation package.

RAVI SUBRAMANIAN received a B.S.E.E. from the California Institute of Technology and a Ph.D. in EECS from the Univ. of California at Berkeley, in 1987 and 1991, respectively. His doctoral work focused on receiver algorithms and architectures for synchronization in CDMA spread-spectrum receivers. From 1989 to 1992, he was at Teknekron Communication Systems (TCSI), where he worked on digital transceiver design for DSP-based USDC and Japanese PDC terminals. Since October 1992, he has been with AT&T Bell Laboratories, Holmdel, New Jersey, involved in research, design, and development of advanced TDMA and CDMA integrated digital receivers for cordless and cellular applications.

Reprinted from *IEEE Communications Magazine*, Vol. 33, No. 5, May 1995, pp. 26-38.

The Software Radio Architecture

As communications technology continues its rapid transition from analog to digital, more functions of contemporary radio systems are implemented in software, leading toward the software radio. What distinguishes software radio architectures? What new capabilities are more economically accessible in software radios than digital radios? What are the pitfalls? And the prognosis?

Joe Mitola

JOE MITOLA is a consulting scientist with The MITRE Corporation.

SPW is a registered trademark of the Alta Group.

TMS320 is a registered trademark of Texas Instruments, Inc.

Matlab is a registered trademark of The MathWorks Inc.

SPOX is a registered trademark of Spectrum Microsystem.

COSSAP is a registered trademark of SYNOPSIS, Inc.

Matcad is a registered trademark of MathSoft, Inc.

e are poised on the threshold of another revolution in radio systems engineering. Throughout the '70s and '80s radio systems migrated from analog to digital in almost every respect from system control to source and channel coding to hardware technology. And now the software radio revolution extends these horizons by liberating radio-based services from chronic dependency on hard-wired characteristics, including frequency band, channel bandwidth, and channel coding. This liberation is accomplished through a combination of techniques that includes multi-band antennas and RF conversion; wideband Analog to Digital (A/D) and Digital to Analog (D/A) conversion (A/D/A conversion); and the implementation of IF, baseband, and bitstream processing functions in general-purpose programmable processors. The resulting software-defined radio (or "software radio") in part extends the evolution of programmable hardware, increasing flexibility via increased programmability. And, in part, it represents an ideal that may never be fully implemented but that nevertheless simplifies and illuminates tradeoffs in radio architectures that seek to balance standards compatibility, technology insertion, and the compelling economics of today's highly competitive marketplaces.

Radio Architecture Evolution

Of the many possible definitions of architecture, the one that best relates systems, technology and economics is best suited to this discussion. We shall therefore define "architecture" as the comprehensive, consistent set of functions, components and design rules according to which systems of interest may be organized, designed, and constructed. A specific architecture entails a partitioning of functions and components such that functions are assigned to components and interfaces among components correspond to interfaces among functions.

When such functions and interfaces are defined in formal design rules via a public forum, the resulting architectures are called "open." The full economic benefits of open architectures require the existence of a large commercial base which sometimes fails to emerge in spite of openness. As system complexity increases, architecture becomes more critical because of its power to either simplify and facilitate system development (a "powerful" architecture) or to complicate development and impede progress (a "weak" architecture).

Radio architectures may be plotted in the phase space of network organization versus channel data rate, as shown in Fig. 1. These architectures have evolved from early point-to-point and relatively chaotic peer networks (e.g., citizens band and push-to-talk mobile military radio networks) toward more hierarchical structures with improved service quality. In addition, channel data rates continue to increase through multiplexing and spectrum spreading. In a multiple-hierarchy application, a single radio unit, typically a mobile terminal, participates in more than one network hierarchy. A software radio terminal, for example, could operate in a GSM network, an AMPS network, and a future satellite mobile network. Multiband multimode military radios and future Personal Communications Systems (PCS) that seamlessly integrate multimedia services across such diverse access modes represent the high end of that evolution and the focus of this discussion. The complexity of functions, components and design rules of these architectures continues to increase with each generation, as shown in Table 1. In particular, future seamless multimode networks will require radio terminals and base stations with agile RF bands, channel access modes, data rates, Bit Error Rates (BERs), power, and functionality. Software radios have emerged to increase quality of service through such agility. At the same time, software radio architectures simplify hardware component tradeoffs and provide new ways of managing the complexity of rapidly emerging standards.

The Canonical Software Radio Architecture

The components of the canonical software radio consist of a power supply, an antenna, a multiband RF converter, and a single chip containing

Generation	Illustrative Functions	Typical Components	Key Design Rules
Analog	Transmit and receive, channel select, squelch	Power, antenna, packaging (discrete analog baseband)	Channel allocations, power limits, standard modulations (AM, FM)
Early digital microwave	Transmit or receive protected modes, BER control	Analog + quadrature modems + Forward Error Control (FEC)	Analog + operations/management and bitstream multiplex interfaces, adjacent channel power envelopes
Analog Mobile Cellular Radio (MCR)	Analog + signaling and control	Analog + digital modems + embedded control processors	Analog + early digital + cell site and frequency plan + handoff protocol
Spread spectrum (CDMA/FH)	Code synchronization, code management, BER control	Analog + early digital + de-/spreading devices + embedded control processors	Analog + code design + peer network protocols + digital voice channel
TDMA MCR	Analog + early digital + analog MCR + spread spectrum + diversity + directivity	Analog + analog MCR + spread spectrum + multibeam antennas	Analog + early digital + analog MCR + digital voice channel + privacy and authentication
Future seamless multimode multimedia networks	Digital MCR + agile directional power management + data rate management + BER agility + mode handover + location reporting	Digital MCR + high programmability + agile modulators + multiband antennas + multiband RF	Digital MCR + mode handover criteria and protocols + end-to-end encryption + software defined services

Digital MCR = CDMA, TDMA, Frequency Hop (FH), and related hybrids.

■ Table 1. *Key elements of radio architecture evolution. Complexity of functions, components, and design rules increases with successive generations.*

A/D/A converters with an on-chip general purpose processor and memory that perform the radio functions and required interfaces illustrated in Fig. 2. The canonical mobile software radio terminal interfaces directly to the user (e.g., via voice, data, fax, and/or multimedia). The canonical base station interfaces to the public switched telephone network (PSTN). Fully instrumented base stations support operations and maintenance, developers and researchers via services development workstation(s). The placement of the A/D/A converters as close to the antenna as possible and the definition of radio functions in software are the hallmarks of the software radio. Thus, although software radios use digital techniques, software-controlled digital radios are generally not software radios. The key difference is the total programmability of software radios, including programmable RF bands, channel access modes, and channel modulation.

Contemporary radio designs mix analog hard-

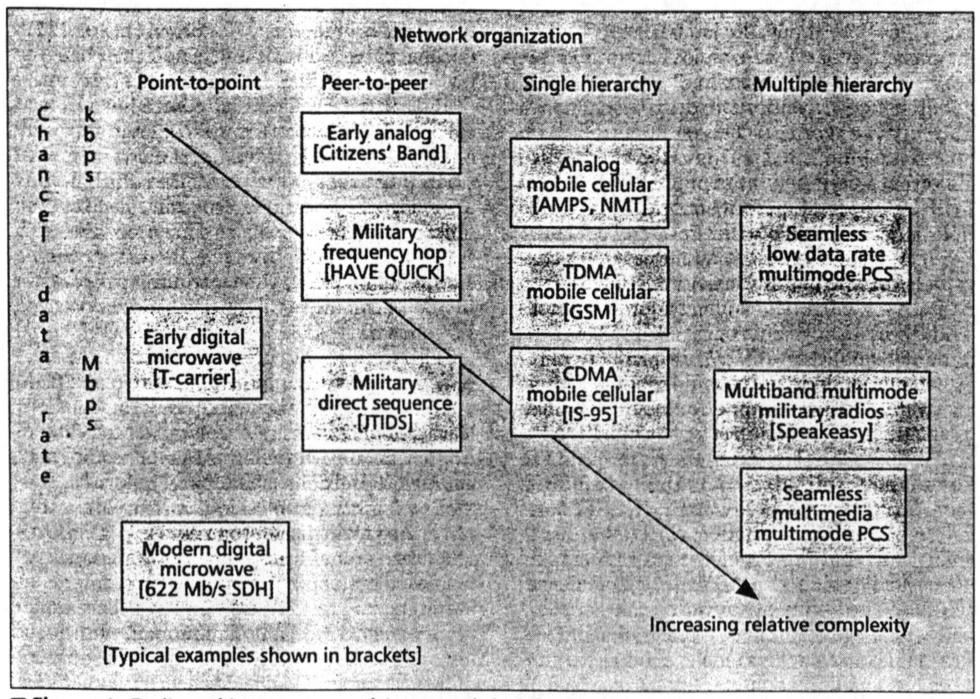

■ Figure 1. *Radio architectures are evolving towards high levels of complexity.*

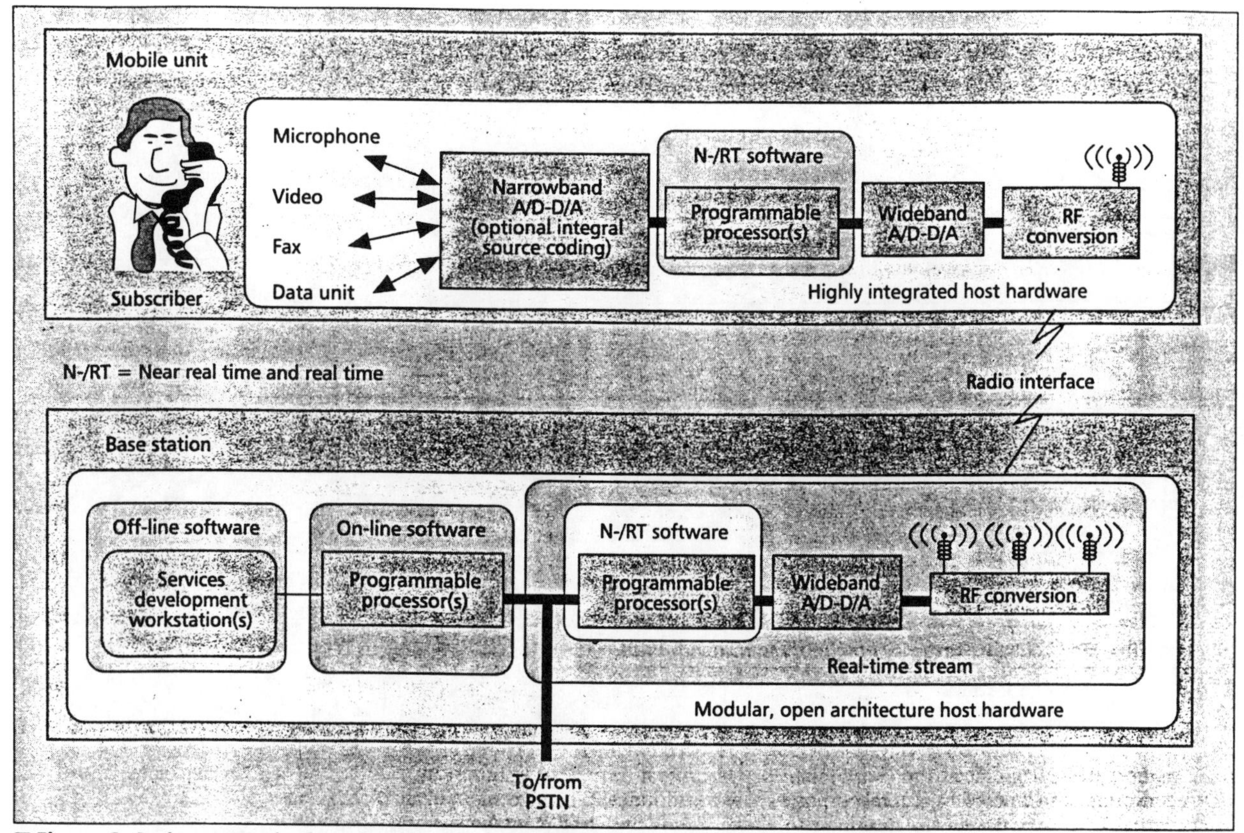

■ **Figure 2.** *In the canonical software radio, hardware is simple and functions are software-defined.*

ware, digital hardware, and software technologies. It is instructive to consider software radios *per se* to better understand benefits, pitfalls and relationships to other technologies. Software radios have become practical as costs per millions of instructions per second (MIPS) of digital signal processors (DSPs) and general purpose central processor units (CPUs) have dropped below U.S. $10 per MIPS. The economics of software radios become increasingly compelling as demands for flexibility increase while these costs continue to drop by a factor of two every few years. At the same time, absolute capacities continue to climb into the hundreds of millions of floating-point operations per second (MFLOPS) per chip, at which point software radios are compatible with commercial TDMA and CDMA applications. In addition, A/D/A converters available in affordable single-board open architecture configurations offer bandwidths of tens of MHz with the dynamic range required for software radio applications [1]. Multimedia requirements for the desktop and palmtop continue to exert downward pressure on parts count and on power consumption of such chip sets, pushing the software radio technology from the base station to the mobile terminal. Although the tradeoffs among analog devices, low-power ASICs, DSP cores and embedded microprocessors in handsets remain fluid, cutting-edge base stations employ software radio architectures. Finally, the multiband, multimode flexibility of software radios appears central to the goal of seamless integration of PCS, land mobile and satellite mobile services (including truly nomadic computing) toward which many of us aspire.

But software radio engineering is fraught with pitfalls. It is difficult to engineer wideband, low-loss antennas and RF converters. It is also difficult to accurately estimate processing demand of applications and processing capacity of reprogrammable DSP/CPU configurations. In addition, sustaining required data rates across interprocessor interfaces is problematic. There are several vendor-unique high capacity wideband signal buses, but open architecture standards for this critical element of the software radio architecture have not yet emerged. DSP function libraries continue to expand, and block diagram-based "integrated environments" exist. But we do not yet have the ability to mix and match real-time software tools and modules from different software suppliers as we can mix and match VME boards today. Most of the pitfalls can be avoided, however. And continued progress in these difficult areas will further reduce costs and time to market.

This article therefore provides a tutorial review of software radio architectures and technology, highlighting benefits, pitfalls, and lessons learned. This includes a closer look at the canonical functional partitioning of channel coding into antenna, RF, IF, baseband, and bitstream segments. A more detailed look at the estimation of demand for critical resources is key. This leads to a discussion of affordable hardware configurations, the mapping of functions to component hardware, and related software tools. This article then concludes with a brief treatment of the economics and likely future directions of software radio technology. Companion articles in this issue address the critical technologies of A/D convert-

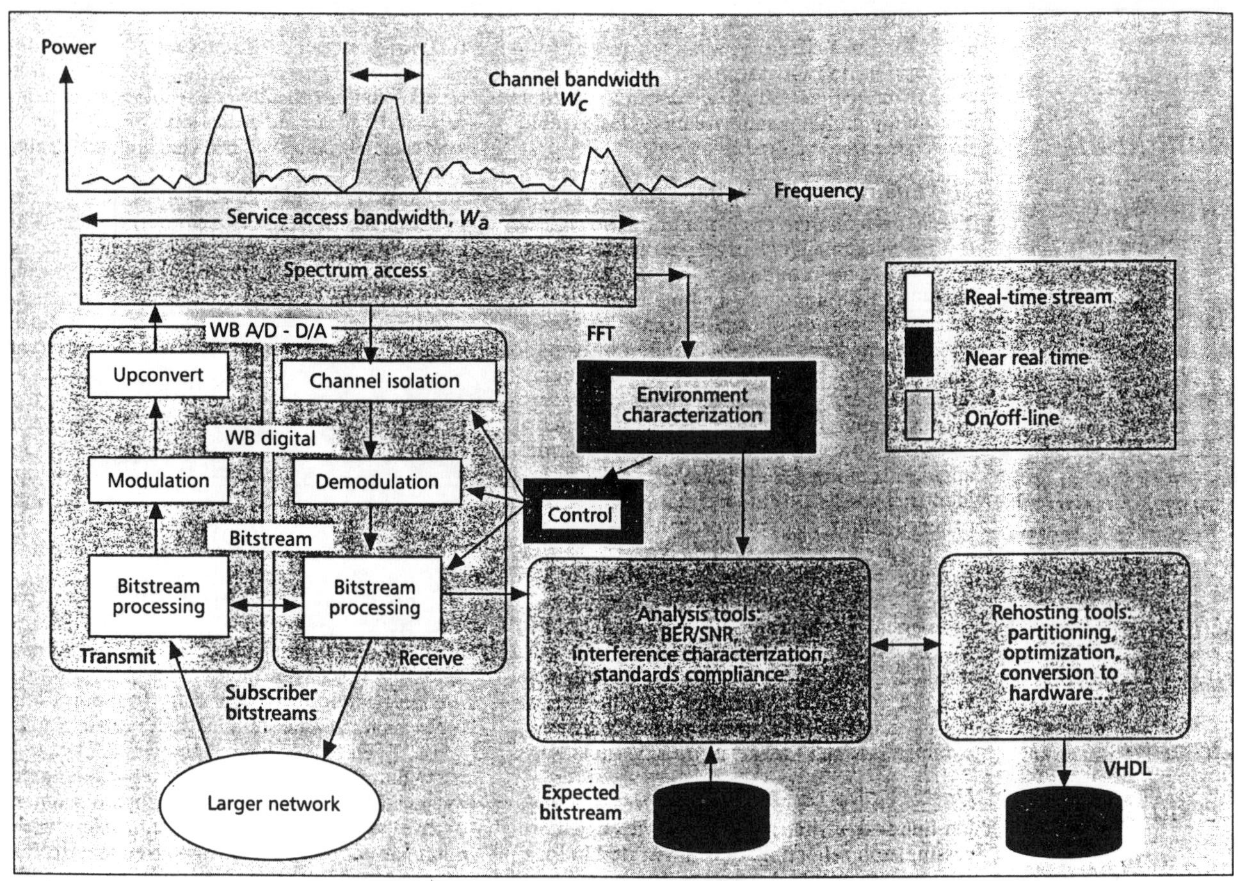

■ **Figure 3.** *Software radio functional architecture in a mobile cellular base station application.*

ers and DSP cores and cutting-edge applications in greater detail.

Software Radio Overview

The software radio architecture is widely applicable to trunk radios, peer networks, air and sea traffic management, mobile military communications, and satellite mobile systems. For simplicity, this overview describes the software radio architecture in a mobile cellular/PCS setting.

In an advanced application, a software radio does not just transmit: it characterizes the available transmission channels, probes the propagation path, constructs an appropriate channel modulation, electronically steers its transmit beam in the right direction, selects the appropriate power level, and then transmits. Again, in an advanced application, a software radio does not just receive: it characterizes the energy distribution in the channel and in adjacent channels, recognizes the mode of the incoming transmission, adaptively nulls interferers, estimates the dynamic properties of desired-signal multipath, coherently combines desired-signal multipath, adaptively equalizes this ensemble, trellis decodes the channel modulation, and then corrects residual errors via forward error control (FEC) decoding to receive the signal with lowest possible BER.

Finally, the software radio supports incremental service enhancements through a wide range of software tools. These tools assist in analyzing the radio environment, defining the required enhancements, prototyping incremental enhancements via software, testing the enhancements in the radio environment, and finally delivering the service enhancements via software and/or hardware.

The Real-Time Channel Processing Stream

The canonical software radio architecture includes the channel processing stream, the environment management stream and associated software tools illustrated in Fig. 3. The real-time channel processing stream incorporates channel coding and radio access protocols. Channel processing is characterized by discrete time point-operations such as the translation of a baseband signal to an intermediate frequency (IF) by multiplying a discrete time-domain baseband waveform by a discrete reference carrier to yield a sampled IF signal. The time between samples is on the order of tens of microseconds to hundreds of nanoseconds. Such point-operations require hundreds of MIPS and/or MFLOPS to Giga-FLOPS with strictly isochronous performance. That is, sampled data values must be computationally produced and consumed within timing windows on the order of the time between samples in order to maintain the integrity of the signals represented therein. Input/output (I/O) data rates of this stream approach a gigabit per second per A/D converter. Although these data rates are decimated through processing, it is challenging to sustain isochronism through I/O interfaces and hard real-time embedded software in this stream. Multiprocessing is

Technology limitations that require hardware-based delivery are overcome by mapping critical elements of the service enhancement to hardware via VHDL.

therefore best organized as a pipeline with sequential functions of the stream assigned to serially interconnected processors, i.e., a multiple instruction multiple data-stream (MIMD) multiprocessing architecture [2].

The Environment Management Stream

The near-real-time environment management stream continuously characterizes radio environment usage in frequency, time and space. This characterization includes channel identification and the estimation of other parameters such as channel interference levels (depending on the specific signaling and multiple access scheme) and subscriber locations. The environment management stream employs block operations such as fast Fourier transforms (FFTs), wavelet transforms, and matrix multiplies for beam forming. Channel identification results are needed in times on the order of hundreds of microseconds to hundreds of milliseconds, while power levels may be updated in milliseconds and subscriber locations may be updated less frequently. The block structure of such operations is readily accommodated by a MIMD parallel processor. The interface between this highly parallel environment management stream and the pipelined channel processing streams must synchronize the environment management parameters to the channel processing streams.

On-Line and Off-Line Software Tools

On-line and off-line systems analysis, signal processing, and rehosting tools illustrated in Fig. 3 allow one to define incremental service enhancements. For example, an enhanced beamformer, equalizer and trellis decoder may be needed to increase subscriber density. These enhancements may be prototyped and linked into the channel processing stream, allowing one to debug the algorithm(s), to experiment with parameter settings, and to determine the service value (e.g., in improved subscriber density) and resources impact (e.g., on processing resources, I/O bandwidth, and time delays).

Software-based enhancements may be organized around managed objects, collections of data and associated executable procedures that work with object resource brokers and conform to related open architecture software interface standards such as the Common Object Resource Broker (CORBA) [3]. Enhancements may then be delivered over the air to other software radio nodes, as contemplated in the future software-defined telecommunications architectures being considered by ITU-T and embraced by NTT [4] and others [5, 6]. A well integrated set of analysis and rehosting tools leads to the creation of incremental software enhancements relatively quickly, with service upgrades provided over-the-air as software-defined networks proliferate. Technology limitations that require hardware-based delivery are overcome by mapping critical elements of the service enhancement to hardware via VHDL.

Partitioning of The Channel Processing Stream

The classical canonical model of communications concatenates source encoder, channel encoder, channel (which adds noise, interference, and distortion), channel decoder, and source decoder. The channel encoder/decoder and related radio access functions constitute the real-time stream. The canonical software radio architecture partitions classical channel coding and decoding into the channel access segments of Fig. 4. These segments are: antennas, RF conversion, IF processing, baseband processing and bitstream processing. This canonical partitioning is useful because of the significant differences in functionality between segments, because of the strong cohesion among functions within a segment; because of large changes in bandwidth due to decimation within a segment; and because of the ease with which these particular segments are mapped to affordable open-architecture hardware. Further rationale for this partitioning is based on set-theoretic considerations [7]. This partitioning also structures the estimation of first-order resource requirements so that they may be combined in ways that accurately predict system performance.

The Antenna Segment

The antenna(s) of the software radio span multiple bands, up to multiple octaves per band with uniform shape and low losses to provide access to available service bands. In military applications, for example, a mobile terminal may need to employ VHF/UHF line of sight frequencies, UHF satellite communications, and HF as a backup mode. Switched access to such multiple bands requires octave bandwidth antennas and/or multiple antennas per band and an agile frequency reference in the RF Segment. In addition, multiple antenna elements may be part of a beam forming network for interference reduction or space division multiple access (SDMA).

The relationship between interference cancellation capacity and the number of antenna elements varies. A single auxiliary element, for example, can reduce interference of a large number of interferers (e.g., those in the "back lobe"). Algorithms that reduce interference through non-spatial techniques (e.g., cyclostationary algorithms) can also reduce a large number of interferers with one or with no auxiliary antenna elements. Beam forming of N antenna elements can place $N - 1$ adaptive nulls on interferers sufficiently separated in azimuth, but coherent multipath may require a distinct null for each distinct multipath direction, reducing the number of interferers accordingly. Polarization scrambling from nearby reflecting surfaces may require $2N + 1$ elements for N paths. The structure of the antenna array(s) determines the number of distinct physical and logical signal processing paths in the RF conversion and IF processing segments. As a result, the competing demands for directional selectivity, multipath compensation and interference suppression versus wideband low-loss antennas versus affordability define the tradeoffs of the antenna segment.

The RF Conversion Segment

RF conversion includes output power generation, preamplification, and conversion of RF signals to and from standard intermediate frequencies (IFs) suitable for wideband A/D/A conversion. In most radio bands, RF conversion will be analog. Certain critical RF problems are exacerbated in the software radio. These include the need for amplifier

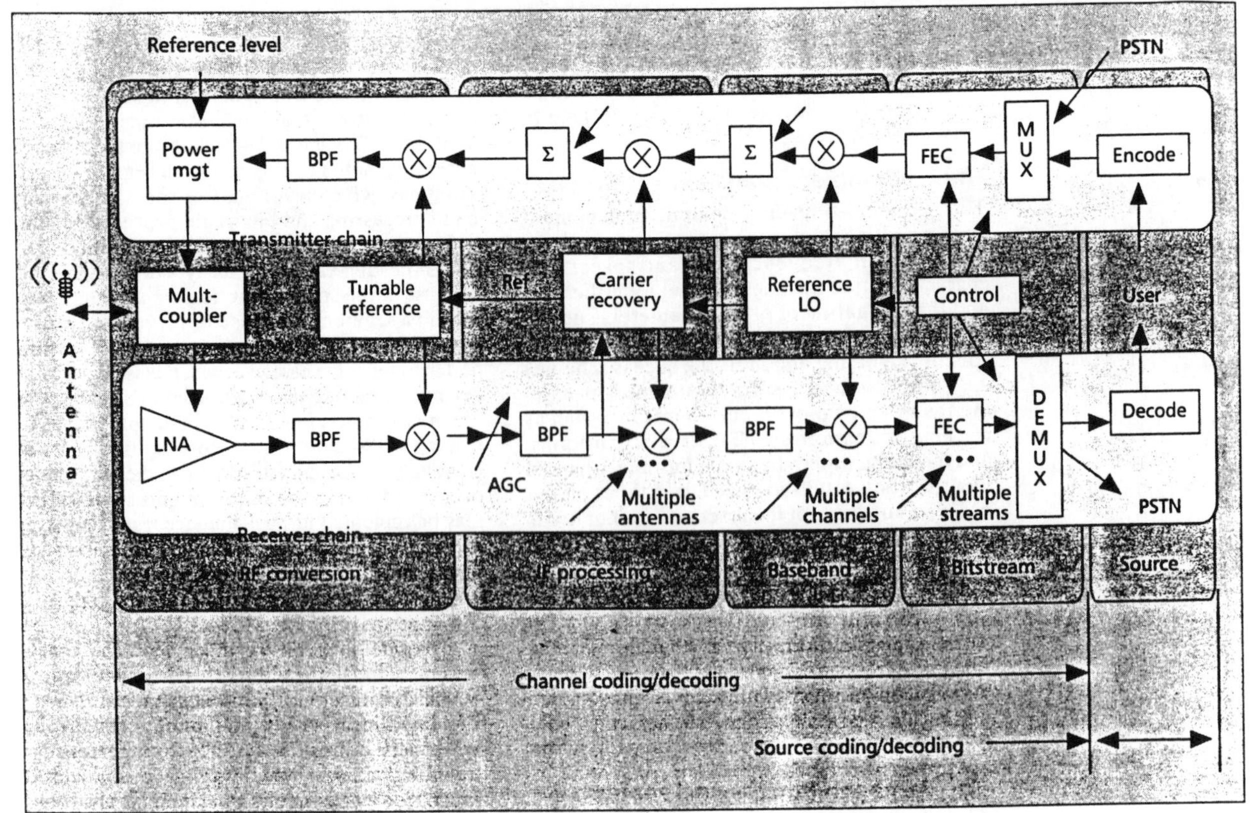

■ **Figure 4.** *The canonical software radio functional architecture maximizes cohesion and minimizes coupling.*

linearity and efficiency across the access band. RF shielding of processors may also be necessary to avoid the introduction of processor clock harmonics into the analog RF/IF circuits. Cositing of multiple transmitters also creates electromagnetic interference (EMI) problems, but these are about the same for software radios as for cosite collections of multiple discrete hardware radios.

Placement of the A/D/A Converters Is Key

W_a, the bandwidth of the IF to be digitized, determines what kinds of A/D techniques are feasible. According to the Nyquist criterion for band limited signals f_s, the sampling rate of the A/D converter, must be at least twice W_a. Practical systems typically require modest oversampling:

$$f_s > 2.5\, W_a$$

Wideband A/D/A converters access broad instantaneous segments of spectrum, typically 10 to 50 MHz. Such wide access may also be achieved in parallel subbands of more modest 1 to 10 MHz bandwidths each. The dynamic range of each parallel subband depends on the dynamic range of the A/D/A converters. Since the product of dynamic range times sampling rate is approximately constant for a given A/D/A technology, narrower subbands generally increase the useful dynamic range, albeit at the cost of increased system complexity. The many issues related to filtering and A/D conversion are deferred to the companion article on A/D converters in this issue [1].

The placement of wideband A/D/A conversion before the final IF and channel isolation filters achieves three key architectural objectives:

- It enables digital signal processing before detection and demodulation.
- It reduces the cost of mixed channel access modes by consolidating IF and baseband processing into programmable hardware.
- It focuses the component tradeoffs to a single central issue: providing the computational resources (I/O bandwidth, memory, and processing capacity) critical to each architecture segment, subject to the size, weight, power, and cost constraints of the application.

The IF Processing Segment

The IF processing segment maps the transmit and receive signals between modulated baseband and IF. The IF receiver processing segment includes wideband digital filtering to select a service band from among those available. Furthermore, IF filtering recovers medium band channels (e.g., a 200 kHz TDMA channel in GSM) and/or wideband subscriber channels (e.g., a 2 MHz CDMA channel) and converts the signal to baseband. The complexity of frequency conversion and filtering is the first order determinant of the processing demand of the IF segment. In a typical application, a 12.5 MHz mobile cellular band is sampled at 30.72 MHz (M samples per second). Frequency translation, filtering and decimation requiring 100 operations per sample equates to more than 3000 MIPS of processing demand. Although such microprocessors are on the horizon, contemporary implementations offload this computationally intensive demand to dedicated chips such as the Harris Decimating Down Converter (DDC) or Gray digital receiver chip. Spreading and de-spreading of CDMA, also an IF processing function, creates demand that is proportional to the bandwidth of

Application	Radio function	Segment	First order demand drivers
Analog Voice Channel	Companding	Source	Speech bandwidth (W_v) and sampling rate
(Receiver)	Gap suppression	Bitstream	Gap identification algorithm
	FM modulation	Baseband	Interpolation required (W_{fm}/W_v)
	Up conversion	IF	IF carrier, f_i, and FM bandwidth: $W_i = W_{fm}$
	Band selection	IF	Access bandwidth (W_a)
	Channel selection	IF	Channel bandwidth (W_c)
	FM demodulation	Baseband	f_i, W_{fm}
	DS0 reconstruction	Bitstream	Speech bandwidth
TDMA (TDM)	Voice codec	Source	Choice of voice codec
	FEC coding	Bitstream	Code rate, algorithm complexity
	Framing	Bitstream	Frame rate (R_f)
	MSK modulation	Baseband	Baud rate (R_b)
	Up conversion	IF	f_i, $R_b/2$
(Receiver)	Band selection	IF	Access bandwidth (W_a)
	Channel selection	IF	Channel bandwidth (W_c)
	Demodulation	Baseband	Baud rate (R_b) or channel bandwidth (W_c)
	Demultiplexing	Bitstream	Frame rate (R_f)
	FEC decoding	Bitstream	Code rate, algorithm complexity
	DS0 reconstruction	Source	Voice codec

■ **Table 2.** *Illustrative functions, segments, and resource demand drivers.*

the spreading waveform (typically the chip rate) times the baseband signal bandwidth. This is so computationally intensive that with current technology limitations, it is typically assigned to dedicated chips as well.

The Baseband Processing Segment

The baseband segment imparts the first level of channel modulation onto the signal (and conversely demodulates the signal in the receiver). Predistortion for nonlinear channels would be included in baseband processing. Trellis coding and soft decision parameter estimation also occur in the baseband processing segment. The complexity of this segment therefore depends on the bandwidth at baseband W_b, the complexity of the channel waveform, and the complexity of related processing (e.g., soft decision support). For typical digitally encoded baseband waveforms such as binary phase shift keying (BPSK), quadrature phase shift keying (QPSK), Gaussian minimal shift keying (GMSK), and 8-PSK with channel symbol (baud) rates of R_b:

$$R_b/3 < W_b < 2^*R_b$$

In the transmission side of the baseband segment, such waveforms are generated one sample at a time (a "point operation"). If three samples are generated for the highest frequency component, demand falls between R_b and 6^*R_b. Greater oversampling decreases the transmitted power of spectral artifacts, but also increases transmit power and processing demand. In the receiver, digital baseband modulations require timing recovery which typically includes the integration of baud intervals over time. If baud interval is measured in transitions of a high-speed clock, some timing-sensitive signal structures (e.g., TDMA) and some synchronization algorithms require up to 96 b precision integer arithmetic in the clock recovery loop(s), and such extended precision arithmetic may not be readily available, particularly on newer chips. Analog baseband modulation, such as FM voice, may also be encoded and demodulated in software in the baseband segment, with a processing demand of less than 1 MIPS per subscriber. Such software simulation of analog modulations helps achieve backwards compatibility with analog standards.

The Bitstream Segment

The bitstream segment digitally multiplexes source-coded bitstreams from multiple users (and, conversely, frames and demultiplexes them). The bitstream segment imparts forward error control (FEC) onto the bitstream, including bit interleaving and block and/or convolutional coding and/or automatic repeat request (ARQ) detection and response. Frame alignment, bit-stuffing, and radio link encryption occur in the bitstream segment. Encryption requires the isolation of encrypted bits from clear bits, resulting in the requirement to partition and isolate bitstream hardware accordingly. Final trellis-coded modulation (TCM) decisions occur in the bitstream segment. Final TCM converts soft/delayed decision parameters from the baseband segment to final bit decisions. The complexity of this segment depends on multiplexing, framing, FEC, encryption, and related bit manipulation operations.

Signaling, control and operations, administration and maintenance (OA&M) functions are also provided in the bitstream segment. The demand associated with these functions depends on the signaling, control and operations systems. Demand increases linearly with the number of simultaneously active subscribers. These functions are event-driven and typically impart an order of magnitude less computational demand than baseband processing. These functions, may, however, require access to distributed data bases, not all of which will be local to the base station. Thus, although the processing demand is relatively small, the timing requirements may be severe.

The Source Segment

The source segment differs between the mobile terminal and the base station. In the mobile terminal, the source segment consists of the user and the source encoders and decoders. Here, the relatively narrowband voice and fax A/D/A converters are typically located in the handset, palmtop, or workstation. In the base station, on the other hand, the source segment consists of the interface to the PSTN for access to remote source coding. Conversion of protocols required for interoperability with the PSTN creates processing demand in the base station's source segment. Conversion of DS0 64 kb/s PCM to RPE-LTP (GSM [11]), for example, would create 1 to 2 MIPS of demand per subscriber.

End-to-End Timing Budgets

Time delays are introduced in the IF, baseband, bitstream and source segments due to finite I/O and processing resources that empty and fill buffers in finite but sometimes random amounts of time. The end-to-end accumulation of these delays must be kept within bounds of isochronism at each segment-to-segment interface. These bounds depend on signal type and the larger network architecture. Thus, for example, end-to-end voice delay should be less than 150 ms, but the external network may consume 100 ms of this timing budget, leaving only 50 ms for the software radio. Maintaining such budgets in software radios is complicated (compared to digital radios) by queuing delays internal to and between processors.

Typical application	RF f_c	W_a	IF W_i	Channel code	Baseband W_i	Bitstream States	FEC	Multiplexing	Privacy	Source
FM mobile	VHF	30 kHz	30 kHz	FM	4 kHz	Inf	—	(PTT)	None	Compand
MCR voice	UHF	25 MHz	25 kHz	FM	4 kHz	Inf	—	(CA)	None	Compand
MCR control		"	"	MSK	10 kb/s	4	Conv	TDM	None	Data
GSM	UHF	25 MHz	200 kHz	GMSK	270.83 kb/s 13 kb/s	4	CPF	TDM	Encrypt	RPE-LTP
Mil FH	VHF	60 MHz	30 kHz Agile	FH-QPSK	16 kb/s	4	Conv ARQ	(PTT)	Encrypt	LPC/Delta
CDMA	UHF	175 MHZ	50 MHz	M-PSK	8 kb/s	M	Conv	(CD)	Privacy	LPC
JTIDS [12]	L-Band	250 MHz	3 MHz	MSK	150 kb/s	4	Multiple	TDM	Encrypt	LPC
Microwave	SHF	20 MHz	20 MHz	64 QAM	90 Mb/s	64	Multiple	TDM	Bit Stuff	PCM

(Implicit Multiplexing): (PTT) = Push to Talk; (CA) = Circuit Assigned; (CD) = Code Division
CPF: GSM specifies various modes of Convolutional, Parity and Fire codes [8].

■ **Table 3.** *Critical applications parameters bound offered demand.*

Estimating Resource Requirements

Resources critical to the software radio architecture include I/O bandwidth, memory and processing capacity. Good estimates of the demand for these resources results in a well informed mapping of the above segments to appropriate hardware. Depending on the details of the hardware, the critical resource may be memory, bus, or I/O bandwidth, or a particular embedded processor. Identification of the critical resource can be accomplished quickly using the techniques highlighted in this section. When such critical resources are identified early in the design process, order of magnitude shortfalls in performance can be avoided. By identifying first-order demand drivers; aggregating demand for each major system resource; comparing demand systematically to hardware/software capacity; and then managing the critical resources appropriately, such shortfalls can be avoided.

Standardized Measures of Demand and Capacity

Since many contemporary processors include pipelined floating point arithmetic sections or single instruction butterfly operations, MIPS and MFLOPS are *not* interchangeable. Both types of operations, however, require processor clock cycles, allowing one to express demand in a common measure of millions of operations per second (MOPS) where an operation is the amount of work that can be accomplished by a given resource (CPU, floating point unit, etc.) in a single clock cycle of a standard width (e.g., 32 b). Using this measure, required MIPS and MFLOPS may both be expressed in MOPS. In addition, I/O, direct memory access, auxiliary I/O processor throughput, memory and bus bandwidths may all be expressed in MOPS where the operand is a standard data word and the operation is store or fetch. MOPS may then be accumulated independently for each potentially critical resource (CPU, DSP unit, floating point processor, I/O bus, memory bank, etc.). Finally, software demand may be translated rigorously to equivalent MOPS. Benchmarking is the key to this last step since hand coded assembly language algorithms often out-perform high order language (HOL) code by an order of magnitude and HOL often outperforms code-generating software tools by an order of magnitude. Rigorous analysis of performance is accomplished by comparing demand and capacity in standards MOPS for each resource using queueing theory. This identifies bottlenecks and yields useful estimates of performance.

Estimate Demand In The Context of The Canonical Data Flow

Table 2 shows how functional architecture parameters drive the resource demand of the associated segment in illustrative applications. Although the associated demand may exceed the capacity available with a given generation of devices, the capacity estimates serve to identify the hardware that best supports a given segment. By determining the number of operations required per point-operation (e.g., a filtering stage) and multiplying by the critical parameter (e.g., the data rate of the stream being filtered), one can quickly arrive at demand estimates that frame the related implementation decisions.

Table 3 shows critical parameters of processing demand for illustrative applications. The demand depends to a first-order approximation on the signal bandwidths and on the complexity of key operations within IF, baseband, bitstream, and source segments as follows:

$$D = D_{if} + N^{*}(D_{bb} + D_{bs} + D_s) + D_o$$

Where D_{if}, the IF processing demand, is proportional to the acquisition bandwidth, and to the complexity of the service band isolation filter and of the subscriber channel isolation filter(s). N is the maximum number of simultaneous subscribers to be serviced by the node. The baseband demand D_{bb} is proportional to the bandwidth of a single subscriber channel and the complexity of demodulation. The bitstream demand D_{bs} is proportional to the data rate and the complexity of the forward error control algorithm, bit stuffing/interleaving, signaling and control and miscellaneous operations functions. The source segment demand D_s depends on the complexity of the interface to the PSTN. And finally, the management overhead processing demand D_o depends on the signaling system, the radio access network overhead and on the degree of instrumentation that is provided to OA&M, developers and/or researchers.

Table 4 shows how key parameters and processing demand are related for an illustrative application. This demand must be met by processors with sufficient capacity to support real-time performance, including isochronism and end-to-end delay budgets. At present, most IF processing is off-loaded to special-purpose digital receiver chips because general purpose processors with sufficient MOPS are not yet cost effective compared to specialized hardware. This tradeoff changes approximately every 18 months, with the trend in favor of the general-purpose processor. In the table, the total baseband plus bitstream processing demand of 4.7 MOPS per user is within the capabilities of nearly all DSP chips. As a result, several subscribers may share a single DSP chip. Aggregate demand (all users, including overhead) of 142.6 MOPS is nominally within the capacity of a Quad TMS320 C50 board, but when multiplexing more than one user's stream into a single processor, memory buffer sizes, bus bandwidth and fan-in/fan-out may be more critical than processing capacity.

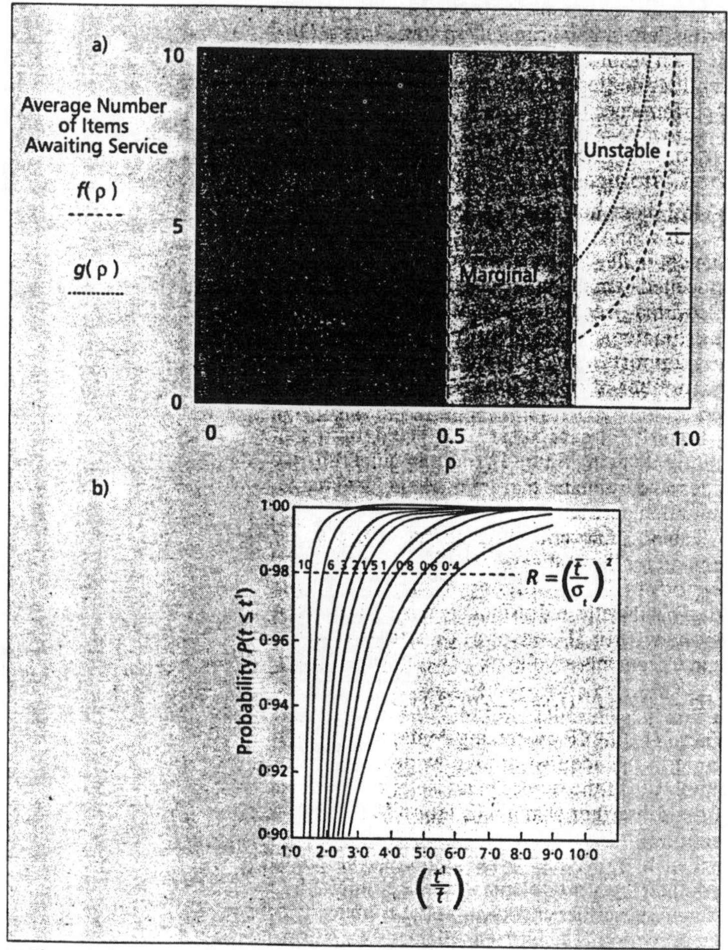

■ **Figure 5.** *a) Facility utilization determines performance, underscoring the criticality of demand estimates; b) ratio of variance to mean determines reliability, $P(t \leq t')$ for specific offered demand [Gamma distribution courtesy of Tebbs & Collins ©1977, McGraw-Hill (UK), reproduced with permissionof McGraw-Hill].*

Facility Utilization Accurately Predicts Performance

The most significant design parameter of the mapping of processing demand to processor capacity is resource utilization (of the CPU, DSP chip, memory, bus, etc.). Resource utilization, ρ, is the ratio of offered demand to available capacity. Figure 5a shows how the average number of items awaiting service at the resource varies as a function of utilization. (See [9] for an accessible treatment of material covered at a deeper level in [8] and [10].) The curve $f(\rho)$ represents exponentially distributed service times, while $g(\rho)$ represents uniform service times; system performance generally lies on or between these two curves. Rarely does usable processing capacity exceed 50 percent of stand-alone capacity as indicated by operating regions. Robust performance occurs when ρ is less than 0.5. The undesired events that result in service degradation will occur with noticeable regularity for $0.5 < \rho < 0.75$. For $\rho > 0.75$, the system is generally unstable, with queue overflows regularly destroying essential information.

An analysis of variance is required to establish the risk that the required time delays will be exceeded. The incomplete Gamma distribution of Fig. 5b characterizes the risk of exceeding a specified delay through the ratio of the specified delay to the average delay (under various relationships of the mean to the variance). Software radios work well if there is a 98 percent probability of staying within specified performance using the curve for $R = 1$ in the figure.

The simplified analysis presented above illustrates the key ideas with the processor as the critical resource. But in many cases, I/O, the backplane bus or memory access will be the critical resource. The approach is readily adapted to any critical resource. Simulation can be used to refine the estimates obtained from this simple model. Management of the system bottlenecks requires the further insights gained through benchmarking and rapid prototyping. But there is no free lunch. Software radios require up to ten times the raw processing power of ASICs and special purpose chips of the same technology. Software radios therefore lag special purpose hardware implementations by about one or two hardware generations. Thus, software radio architectures have appeared first in base stations using heterogeneous multiprocessing hardware, which is discussed briefly in the following section.

Heterogeneous Multiprocessing Hardware

Segment boundaries among antennas, RF, IF, baseband, bitstream, and source segments evolved because they make it easy to build software radios on parallel, pipelined, heterogeneous multiprocessing hardware. Such partitioning maps the segments to affordable open hardware architectures, as illustrated in Fig. 6. In this example, the VME chassis hosts the RF, IF, baseband, and bitstream segments, while the workstation houses the user interface, research tools, development

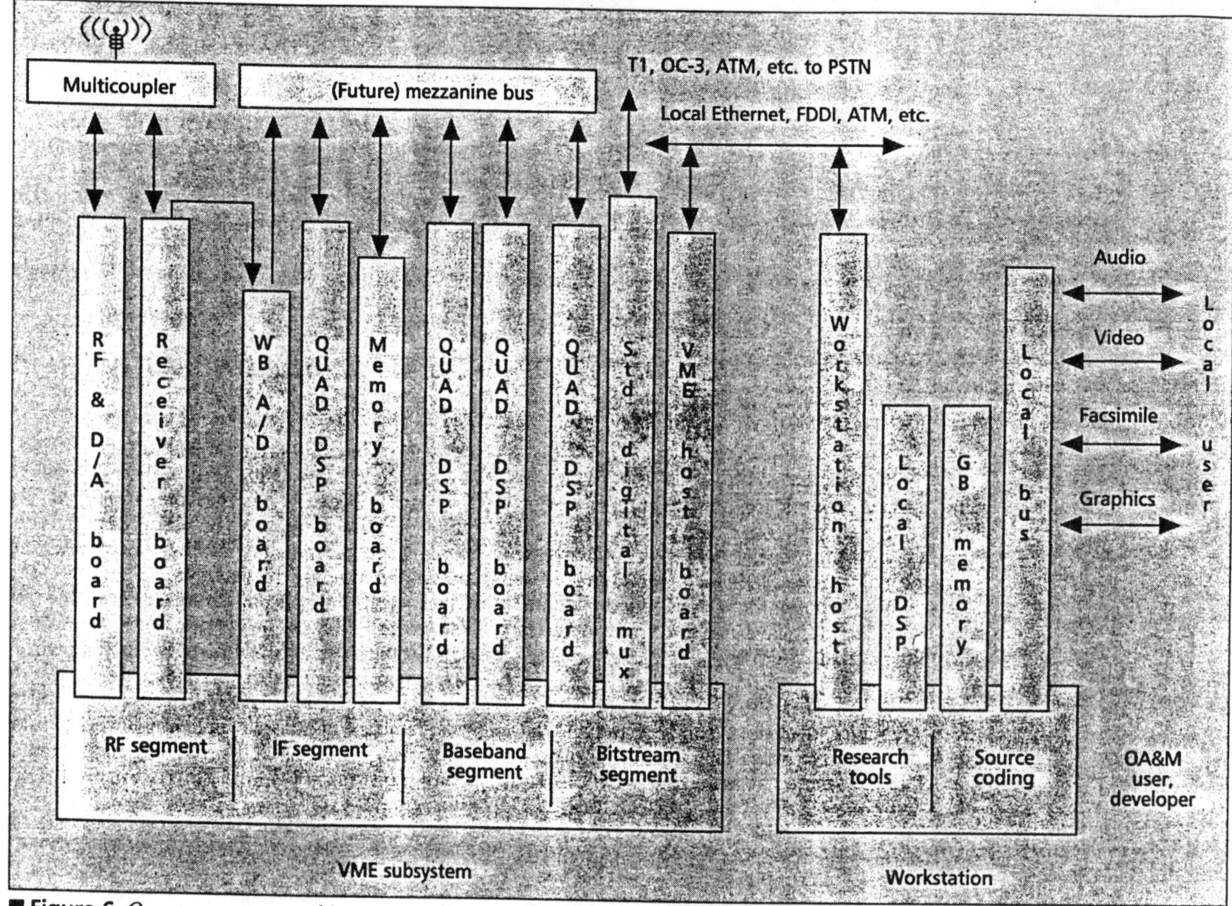

■ **Figure 6.** *Open component architecture supplies processing capacity with affordable technology insertion.*

tools and local source coding/decoding (Table 5).

In the most cost effective leveraging of open architectures, VME systems are configured by experienced developers as heterogeneous multiprocessors. The VME host serves as systems control processor . The DSP processors support the real-time channel processing stream, sometimes configured as one DSP per subscriber channel. The path from A/D to the first filtering/decimation stage typically uses a dedicated point-to-point mezzanine interconnect such as DT Connect™ (Data Translation). Timing the data transfers across this bus with the point operations of the first filtering and decimation stage introduces inefficiencies that reduce throughput. Fan-out from IF processing to multiple baseband processing DSPs is also typically accomplished via a dedicated point-to-point mezzanine path. Current implementations use customized fiber channel, TMS320/C50 or transputer links. As suggested in the figure, however, an open architecture optical mezzanine bus or local ATM switch should supplant today's vendor-unique point-to-point interconnects. An open architecture mezzanine bus with a capacity of 125 MHz x 16 bits per channel with an aggregate capacity of eight simultaneous channels would fill a critical void in contemporary open architecture standards.

Instead of configuring such heterogeneous multiprocessors at the board level, one might chose a preconfigured system. Mercury™, for example, offers a mix of SHARC 21060 (Analog Devices), PowerPC RISC, and Intel i860 chips with a unique Raceway interconnect with nominally three paths at 160 MB/sec interconnect capacity. Arrays of WE32s were used in AT&T's DSP-3 system. Arrays of i860s are available from Sky, CSPI, and others. Of particular note is the militarized TOUCHSTONE system which is also based on the i860. Such turnkey systems integrate block programming languages, automatic translation to C

Segment	Parameter	Illustrative value	Demand estimate
IF	W_a	10 MHz (2.5 oversampling)	
	IF Filter	100 OPS/ Hz	D_{if} = 2500 MOPS*
Users	N	30/ cell site	
Baseband	W_c	30 kHz	
	Demodulator	50 OPS/Hz	D_{bb}= 1.5 MOPS
Bitstream	R_b	32 kb/s	
	FEC, Signaling	100 OPS/b/s	D_{bs} = 3.2 MOPS
Source	CELP Codec	1.6 MOPS/user	D_s = 1.6 MOPS/user
Signaling	SS7	2 MOPS/site	D_o = 2 MOPS
Aggregate	DSP MOPS		D = 142.6 MOPS per cell site

* D_{if} is off-loaded to dedicated digital hardware and is therefore not included in D.

■ **Table 4.** *Illustrative processing demand: analog mobile cellular base station.*

Segment	Module	Characteristics	Illustrative manufacturers
RF	RF/IF	HF, VHF, UHF	Watkins Johnson, Steinbrecher
(A/D/A)	A/D	1 to 70 Msa/sec	Analog Devices, Pentek, DT
IF, BB	DSP	4 x 30 MFLOPS	AMD, TI, Intel, Mercury, Sky
IF	Digital Rx	30.72 MHz filters	Harris, Gray, E-Systems
IF	Memory	64 MB at 30.72 MHz	TI, Harris, TRW
BS, SC	Hosts	M680 x 0	Motorola, Force, Intel
SC	Workstation	50 to 100 Specmarks	Sun, Hewlett-Packard, DEC
(BB = Baseband; BS = Bitstream; SC = Source Coding)			

■ **Table 5.** *Mapping of segments to open architecture VME modules.*

and machine code and extensive debugging support but at relatively high cost of entry. Vendor-unique turnkey systems may be excellent choices when there is a good match between the application and the vendor's standard configured system. Vendor-unique details tend to limit hardware and software choices to the original vendor.

Architecture Tradeoffs

Software radios ideally place all IF, baseband, bitstream and source processing in a single processor. The assessment of the feasibility of the software radio centers on comparing estimated demand to the capacities of the available processors. Implementations back off from the ideal single-CPU implementation where driven to do so by processor, memory, or interconnect technology limitations, or to achieve cost advantages (e.g., of off-loading filtering to a Harris chip so that the DSP is less expensive). Software tasks are then structured into managed objects designed to run on any DSP or CPU with access to the data and sufficient processing capacity.

Open Architecture Software Tools Remain Problematic

Figure 7 shows how the time criticality of performance varies considerably with functionality. No signal processing environments in existence fully span the range from hard real-time IF processing through off-line support. Each tool and environment on the market excels in one or more aspects of the required support, although a few integrated environments approach this ideal. UNIX works well as both VME host and workstation host in laboratory systems. Portable VME systems often use a DOS/Windows laptop for user interface and display in lieu of a workstation, with a DOS/EISA or UNIX/VME signal processing configuration. SPOX and Harris Corporation's

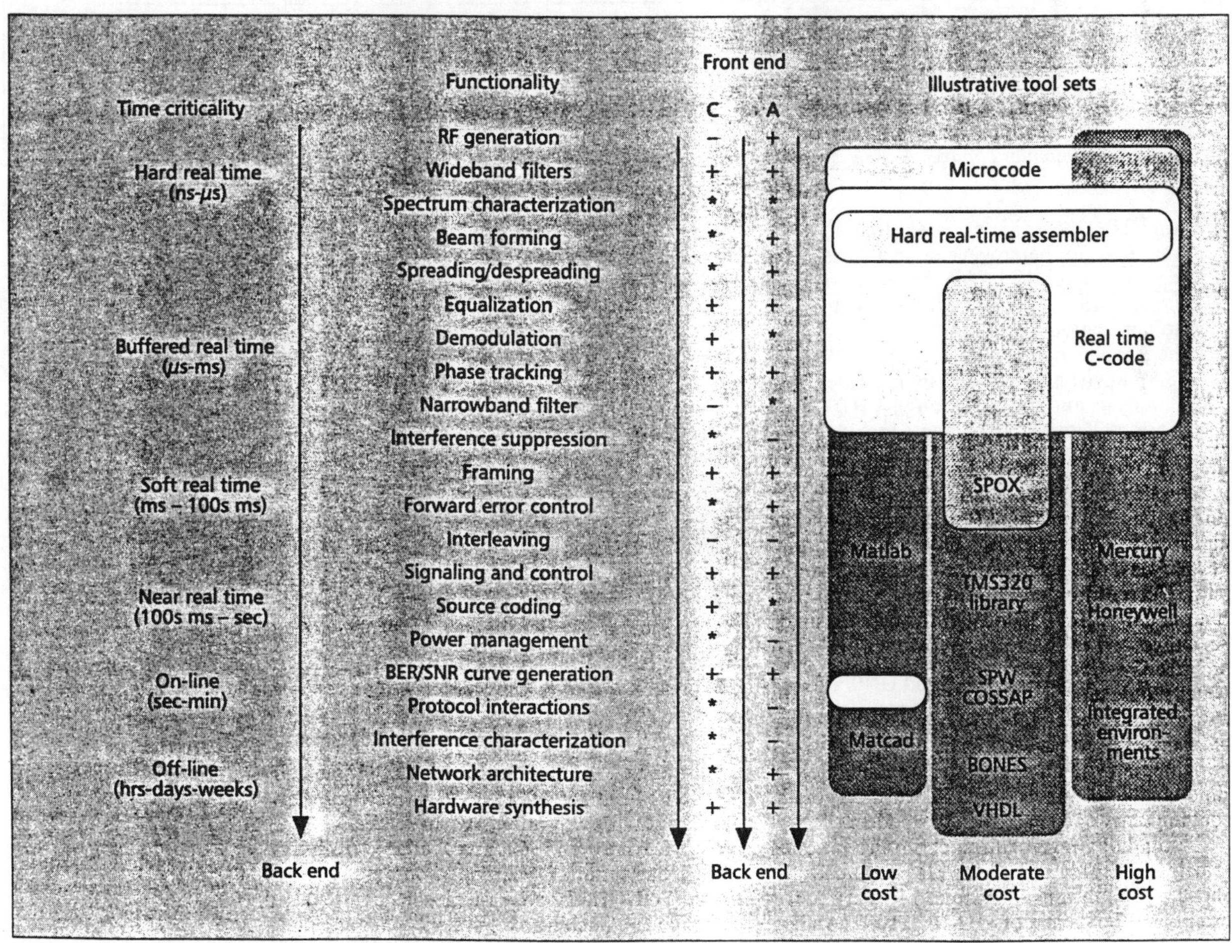

■ **Figure 7.** *Software tools span the function space, but tool integration is far from seamless.*

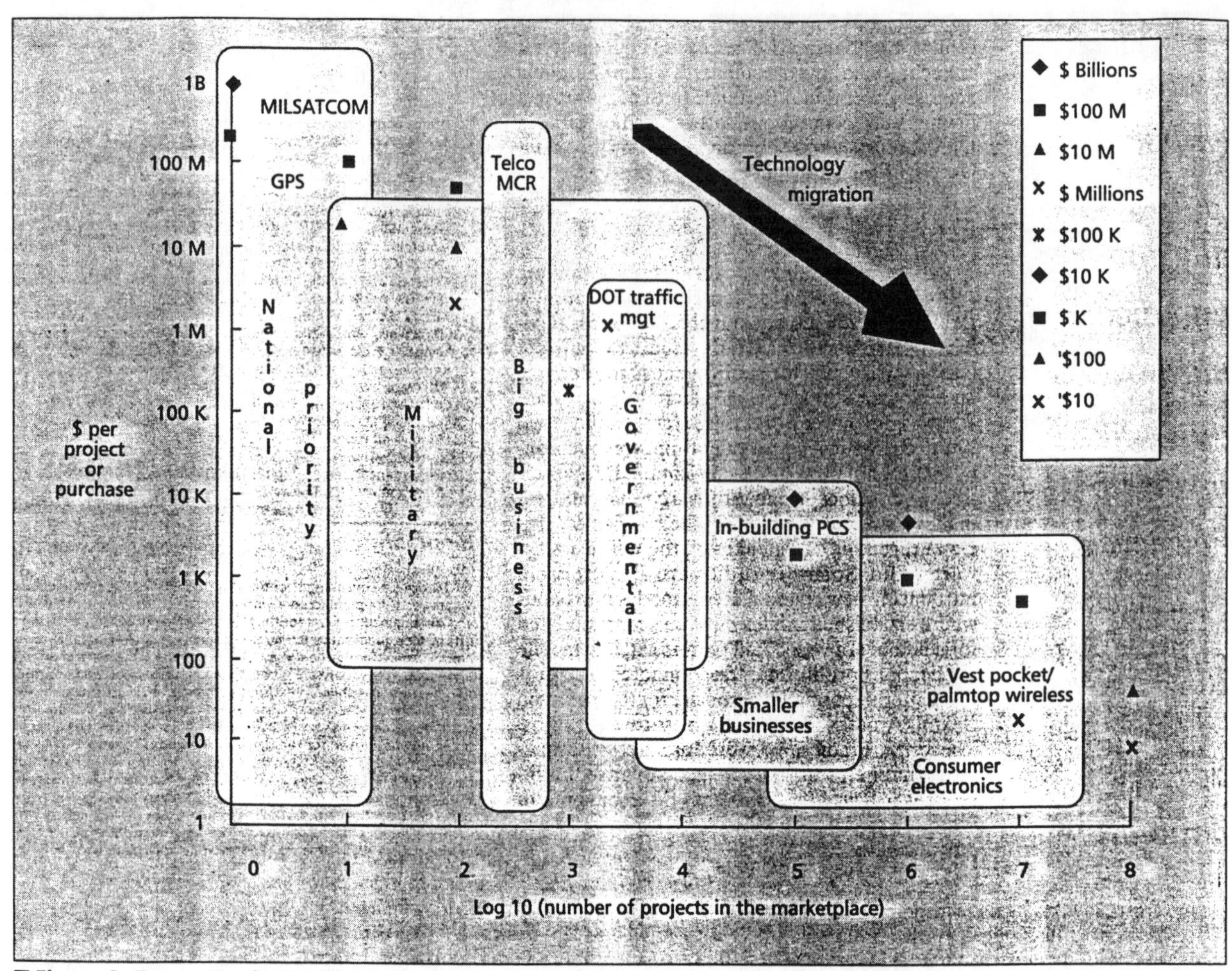

■ **Figure 8.** *Economics of project size versus number of projects drive software radio technology migration.*

real-time UNIX excel in near-real-time support. Software tools for heterogeneous distributed processing are developing rapidly, but lack extended precision arithmetic and standardized application-to-application interfaces.

The Signal Processing Workbench (SPW) and COSSAP environments each span a wide range from analysis and design through VHDL code generation. These environments include block diagram languages that allow the designer to express processing in a signal flow representation. Block diagram languages are useful for concept development. The compilers can also map these to an intermediate language (typically C) which can be compiled and downloaded to target DSP chips (with up to an order of magnitude of performance degradation compared to hand coded C). Real-time access to the DSP status and performance is relatively hard to come by, however. Mercury and UNISYS' Militarized Touchstone environments provide counter examples, offering real-time operating systems with integrated performance monitoring and debugging capabilities. Some operating systems require one to build specialized status reporting tasks to send data of interest to the workstation. This is not a big impediment on a small project, but on larger, more complex projects, the integrated environments have significant advantages in reduced development times.

Rehosting tools tend to be aligned by vendor with Vendor-X providing a tool that will download C code to Vendor-X's own DSP hardware. The Signal Processing Workbench (SPW), Alta Corp., is a notable counter-example, with a very complete set of rehosting tools including targeting for several popular DSP cores and translation to VHDL for rapid transition to product hardware. Code libraries of such tool sets will evolve to standards-based suites of managed objets as object resource broker technlogy becomes more widely implemented in DSP tools.

Economics

Figure 8 shows the rough order of magnitude relationship between size of a project (or purchase) and the number of such projects (or potential purchases) in the marketplace. Software radios are now in the segment of the market dominated by the military, big business (e.g., the telephone and wireless service providers) and governments. The companion applications articles describe projects with costs on the order of a few million dollars to more than ten million, for example. Over time, the software radio will continue to move down and to the right as the size, power and cost of general purpose DSP chips, A/D and D/A converters and related interconnect and memory allow. In addition, object management software technology applied to software radios offers powerful

Software embodiment of traditionally analog and digital hardware functions opens up new levels of service quality and channel access flexibility.

tools for managing the increased complexity of emerging radio network standards within affordable acquisition and maintenance budgets. Within a few hardware generations, software radios will be in the vest-pocket and palmtop, providing the seamless communications services we aspire to.

Conclusions

Software embodiment of traditionally analog and digital hardware functions opens up new levels of service quality and channel access flexibility. In applications where access to multiple bands with multiple radio access modes is a necessity, the software radio can reduce hardware size, weight and power through fewer radio units. The need for such improvement appeared as chronic "interoperability" problems in military applications over a decade ago. Software radios address military issues as discussed in the companion article in this issue on Speakeasy, the military software radio. Software radios are also useful in introducing new channel access modes into bands where established modes must be accommodated for a number of years. As PCS and satellite mobile applications proliferate, the commercial sector is encountering continuing pressure for the increased flexibility and seamlessness of software radios. The software radio is a powerful architecture framework that helps us deliver such advanced radio services in a way that leverages the economics of contemporary microelectronics and software technologies.

References

[1] Bensley *et al.*, Introduction to Parallel Supercomputing, The MITRE Corporation, Bedford, MA, 1988.

[2] Pentek Corporation, (Laurel, MD: SPSI, 1994).

[3] The Common Object Resource Broker: Architecture and Specification, OMG 91.12.1, (Maynard, MA: Digital Equipment Corp. *et al.*, 1992).

[4] Kondo, Matsuo and Suzuki, "A Software-Defined Architecture Concept for Telecommunication Information Systems," ICC '94, (NY: IEEE Press, 1994).

[5] Voruganti, "A Global Network Management Framework for the 90's," ICC 94, (NY: IEEE Press, 1994), p. 1094.

[6] Celestino and Claude, "Building a Platform Support Environment for TMN Applications", ICC 94, (NY: IEEE Press, 1994) p.1847.

[7] Mitola "Software Radios: Survey, Critical Evaluation and Future Directions," Proceedings of the National Telesystems Conference, (NY: IEEE Press, May 1992).

[8] Kant, *Introduction to Computer System Performance Evaluation*, (NY: McGraw Hill, 1992).

[9] Collins and Tebbs, *Real Time Systems*, (UK: McGraw Hill, 1977).

[10] Kleinrock, *Queueing Systems*, vol. II, (NY: Macmillan, 1982).

[11] Mouly and Pautet, *The GSM System for Mobile Communications*, (France: Palaiseau, 1992).

[12] Ziemer and Petersen, *Digital Communications and Spread Spectrum Systems*, (NY: Macmillan, 1985), p. 694.

Biography

JOE MITOLA is a consulting scientist at the MITRE Corporation, McLean, Virginia, a not-for-profit corporation that operates federally funded research and development centers for the U.S. Government. His research interests include quantitative methods in systems engineering, radio technology, and information theory. He advises the U.S. Government on large-scale systems architectures, telecommunications, and information processing technologies. Prior to joining MITRE in 1993, he was chief scientist of Electronic Systems, E-Systems Melpar, culminating more than a decade of positions of increasing responsibility. Earlier in his career, he worked for the Harris Corporation, Advanced Decision Systesms, and ITT. He began his career at the U.S. Department of Defense in 1967. He received a B.S.E.E. from Northeastern University in 1972 and an M.S.E. from Johns Hopkins University in 1974. In addition, he passed the Ph.D. qualifying examinations in Information Processing (Articificial Interlligence) and Systems Theory at the University of Maryland in 1985. His e-mail address is: jmitola@mitre.org.

Reprinted from *IEEE Communications Magazine*, Vol. 33, No. 5, May 1995, pp. 39-45.

Analog-to-Digital Converters and Their Applications in Radio Receivers

Rapid advances in hardware development of analog-to-digital converters (ADCs) have paved the way for development of radio receivers using digitization at the IF, and in some cases, at the RF. The constraints placed on these receivers due to hardware limitations of these devices are discussed and some examples of high-speed, state-of-the-art ADCs are given.

Jeffery A. Wepman

s advances in technology provide increasingly faster and less expensive digital hardware, more of the traditional analog functions of a radio receiver will be replaced with software or digital hardware. The ultimate goal in radio receiver design is to directly digitize the RF signal at the output of the receive antenna and hence implement all receiver functions in either digital hardware or software. Trends in receiver design have evolved toward this goal by incorporating digitization closer and closer to the receive antenna for systems at increasingly higher frequencies and wider bandwidths. Applications for these receivers are expected to increase rapidly in areas such as mobile cellular, satellite, and personal communications services (PCS) systems.

The analog-to-digital converter (ADC) is a key component in these radio receivers. ADCs most often used for wideband digitization at the RF or IF are organized as shown in Fig. 1. Key parameters of ADCs are effected by specific ADC circuit elements. For example, accuracy and linearity are primarily determined by the sample-and-hold circuitry, while jitter in the sampling clock can introduce noise in the desired output of the ADC. The quantizer establishes the resolution of the ADC. For a given burst rate, the buffers limit the sustainable throughput.

This article discusses some of the key parameters for ADCs used in radio receiver applications. The requirements, practical limitations, and potential problems for ADCs are also discussed. Conversion methods of practical ADCs such as flash, successive approximation, sigma-delta, bandpass sigma-delta, and subranging are not presented here. A discussion of these techniques can be found in [1].

Certain commercial equipment, components, instruments, or materials are identified in this article to provide some examples of current technology. In no case does such identification imply recommendation or endorsement by the National Telecommunications and Information Administration, nor does it imply that the material or equipment identified is necessarily the best available for the purpose. Furthermore, examples of technology identified in this article are not intended to be all inclusive; they represent only a sampling of what is available.

JEFFERY A. WEPMAN is an electronics engineer with the Institute for Telecommunication Sciences (ITS), National Telecommunications and Information Administration (NTIA), U.S. Department of Commerce.

Sampling Methods and Analog Filtering

The sampling process is of critical importance in radio receivers using digitization at the RF or IF. The content of the resulting sampled signal waveform is highly dependent on the relationship between the sampling rate employed and the minimum and maximum frequency components of the analog input signal. Some common sampling techniques that utilize a uniform spacing between the samples include Nyquist sampling, oversampling, quadrature sampling, and bandpass sampling (also called downsampling or direct downconversion). Sampling techniques with nonuniform spacing between the samples do exist, but they are not widely used and therefore are not considered in this article.

Any time a continuous-time analog signal is uniformly sampled, the spectrum of the original signal $F(f)$ is repeated at integer multiples of the sampling frequency (i.e., $F(f)$ becomes periodic). This is an inherent effect of sampling and cannot be avoided. This phenomenon is shown graphically in Fig. 2. Figure 2a shows the spectrum of the original analog signal $F(f)$. Figure 2b shows the spectrum of the sampled signal $F_s(f)$ using a sampling rate of $f_s = 2f_{max}$.

Nyquist Sampling

The general sampling theorem for sampling a bandlimited analog signal (a signal having no frequency components above a certain frequency f_{max}) requires that the sampling rate be at least two times the highest frequency component of the analog signal $2f_{max}$. This ensures that the original signal can be reconstructed exactly from the samples. A sampling rate of two times the highest frequency component of the analog signal is

0163-6804/95/$04.00 1995 © IEEE

called the Nyquist sampling rate. Figure 2b shows an example of sampling a bandlimited signal with a maximum frequency of f_{max} at the Nyquist rate ($f_s = 2f_{max}$). Note that the copies of the spectrum of the analog signal $F(f)$ that are present in the spectrum of the sampled signal $F_s(f)$ do not overlap. As the sampling rate is increased beyond the Nyquist rate, the copies of the spectrum of the analog signal $F(f)$ that are present in the spectrum of the sampled signal $F_s(f)$ are spread even farther apart. This is shown in Fig. 2c. Sampling a bandlimited signal at rates equal to or greater than the Nyquist rate guarantees that spectrum overlap (often called aliasing) does not occur and that the original analog signal can be reconstructed exactly [2, 3].

Out-of-Band Energy

Two practical problems arise when sampling at the Nyquist rate: defining what a bandlimited signal truly is in a practical sense and analog filtering before the ADC stage. A theoretically defined bandlimited signal is a signal with no frequency components above a certain frequency. When considering real signals such as an RF signal at the input of a radio receiver, however, signals of all frequencies are always present. It is a matter of the amplitude of these frequencies that is important. In particular, the relative amplitude of the undesired signals to the desired signal is important. When digitizing an RF or IF signal at the Nyquist rate in a radio receiver, undesired signals (above one-half the sampling rate) of a sufficient amplitude can create spectrum overlap and distort the desired signal. This phenomenon is illustrated in Fig. 3. Figure 3a shows the spectrum of the analog input signal with its desired and undesired components. If this signal is sampled at two times the highest frequency in the desired signal f_d, the resulting spectrum of the sampled signal $F_s(f)$ is shown in Fig. 3b. Note that spectrum overlap has occurred here (i.e., the spectrum of the undesired signal occurs within the spectrum of the desired signal). This causes distortion in the reconstructed desired signal.

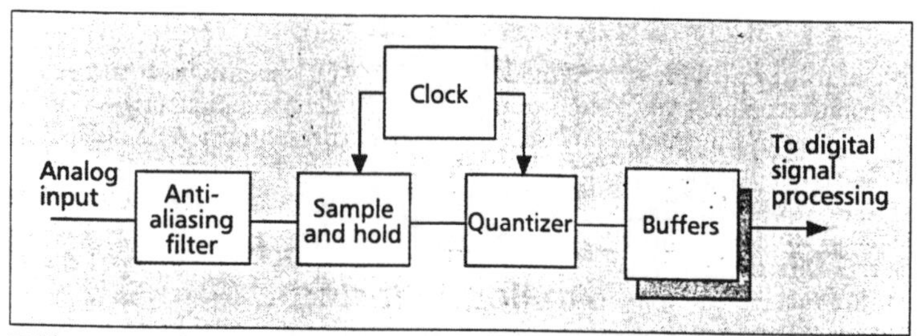

■ **Figure 1.** *Elements of ADCs used for wideband digitization at the RF or IF.*

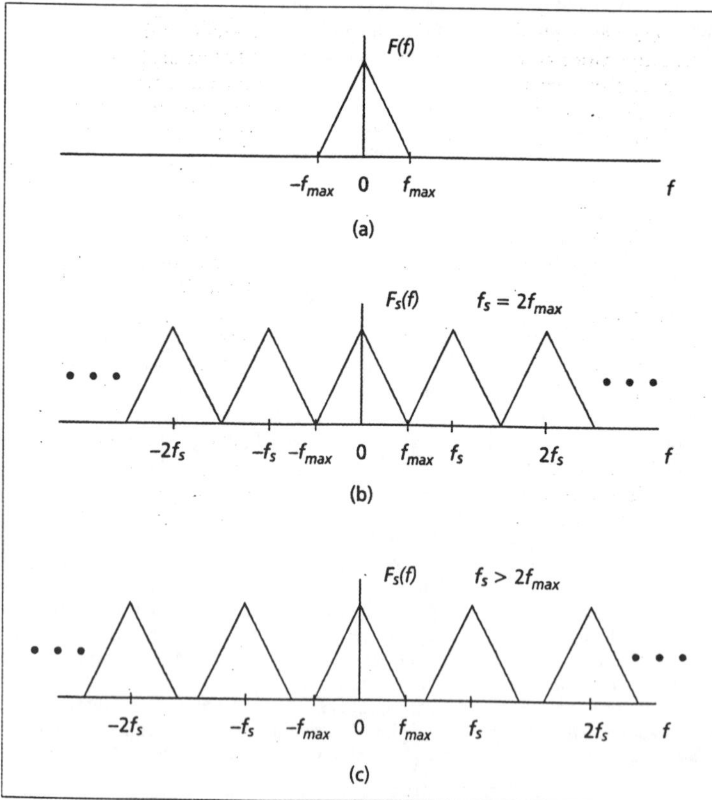

■ **Figure 2.** *Spectrum of: a) a bandlimited continuous-time analog signal, b) the signal sampled at $f_s = 2f_{max}$, and c) the signal sampled at $f_s > 2f_{max}$.*

This effect raises an important question: What is the relative amplitude of the signals occurring above one-half of the sampling rate at which distortion of the desired signal due to spectrum overlap begins to predominate the distortion due to ADC nonlinearities? Nonlinearities in the ADC cause spurious responses in the ADC output spectrum. Distortion due to spectrum overlap can be said to predominate distortion due to ADC nonlinearities when the undesired signals appearing in the Nyquist band (DC to one-half the sampling rate) due to spectrum overlap exceed the largest spurious response of the ADC due to nonlinearities. Therefore, undesired signals appearing in the Nyquist band due to spectrum overlap must be lower in power than the largest spurious response of the ADC. In other words, distortion of the desired signal is predominated by ADC nonlinearities (and not spectrum overlap) if signals higher in frequency than $f_s/2$ are lower in power than the largest spurious response of the ADC. This can be quite a stringent requirement. Depending upon the specific type of radio system, it is possible that this requirement can be relaxed.

To determine ways to "relax" this requirement, the following questions should be asked: How much distortion of the desired signal is tolerable? Does the bandwidth and frequency content of both the desired signal in the Nyquist band and the undesired signals above the Nyquist band effect the distortion of the desired signal? These questions are best answered by considering the details of the specific radio communication system, such as the type of source information (voice, data, video, etc.), desired signal bandwidth, modulation and coding techniques, undesired signal characteristics (bandwidth, power, and type of signal), and finally, the performance criterion used to evaluate the quality of the reception of the desired signal.

Realizable Anti-Aliasing Filters

Analog filtering before the ADC stage is intimately related to the definition of bandlimiting. Where the definition of bandlimiting deals with the content of the signals that may be present, analog filtering before the ADC represents a signal processing stage where certain frequencies can be attenuated. It is important to know both the signals that can be present before filtering and the amount of attenuation that the filter offers for different frequencies. With knowledge of both of these, the true spectrum of the signal to be digitized can be determined. Sampling at the Nyquist rate presents a large and often impractical demand on the filter used before digitization (anti-aliasing filter). Ideally, an anti-aliasing filter placed before an ADC would pass all of the desired frequencies up to some cutoff frequency and provide infinite attenuation for frequencies above the cutoff frequency. Then sampling at the Nyquist rate would be two times the cutoff frequency and no spectrum overlap would occur. Unfortunately, practically realizable filters cannot provide this type of "brickwall" response. The attenuation of real filters increases more gradually from the cutoff frequency to the stop band. Therefore, for a given cutoff frequency on a real filter, sampling at two times this cutoff frequency will produce some spectrum overlap. The steeper the transition from the passband to the stop band and the more attenuation in the stop band, the less the sampled signal will be distorted by spectrum overlap. In general, more complicated filters are required to achieve steeper transitions and higher attenuation in the stop band. Therefore, more complicated filters are required to reduce the distortion in the sampled signal due to spectrum overlap for a given sampling rate. Limitations on the practical implementation of analog filters make high-order, steep rolloff filters difficult to realize. Also, as the steepness of the rolloff is increased, the phase response tends to become more nonlinear. This can create distortion of the desired receive signal since different frequencies within a signal can be delayed in time by different amounts (i.e., phase shifted by amounts not proportional to frequency).

Oversampling Eases Requirements on the Anti-Aliasing Filter

Sampling at rates greater than the Nyquist sampling rate is called oversampling. One of the benefits of oversampling is that the copies of the spectrum of the analog signal $F(f)$ that are present in the spectrum of the sampled signal $F_s(f)$ become increasingly separated as the sampling rate is increased beyond the Nyquist rate. For an analog signal with a given frequency content and a given anti-aliasing filter with a cutoff frequency of f_c, sampling at the Nyquist rate (two times the cutoff frequency) produces a certain amount of distortion due to spectrum overlap. When sampling at a higher rate, a simpler anti-aliasing filter with a more gradual transition from passband to stop band and less stop band attenuation can be used without any increase in the distortion due to spectrum overlap. Therefore, oversampling can minimize the requirements of the anti-aliasing filter. The tradeoff, of course, is that increasingly faster ADCs are required to digitize relatively low frequency signals.

Quadrature Sampling Reduces Required Sampling Rate

In quadrature sampling the signal to be digitized is split into two signals. One of these signals is multiplied by a sinusoid to downconvert it to a zero center frequency and form the in-phase component of the original signal. The other signal is multiplied by a 90-degree phase-shifted sinusoid to downconvert it to a zero center frequency and form the quadrature-phase component of the original signal. Each of these components occupies only one-half of the bandwidth of the original signal and can be sampled at one-half the sampling rate required for the original signal. Therefore, quadrature sampling reduces the required sampling rate by a factor of two at the expense of using two phase-locked ADCs instead of one.

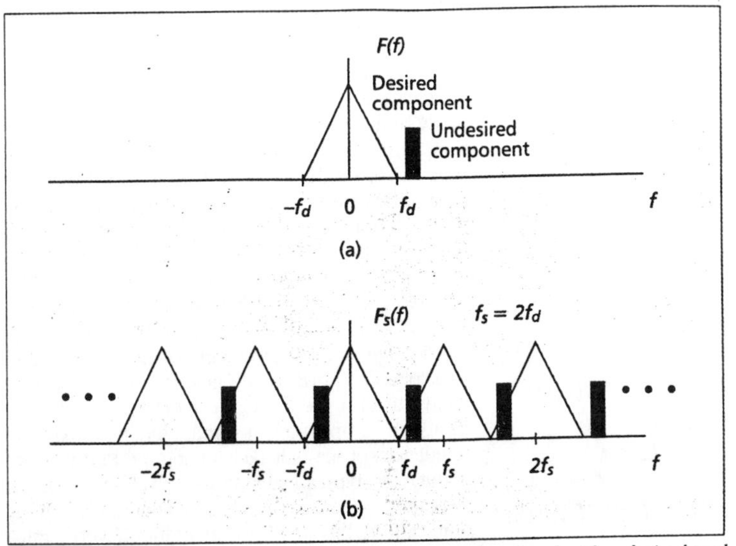

■ **Figure 3.** *Spectrum of: a) a continuous-time analog signal with a desired and undesired component and b) the signal sampled at* $f_s = 2f_d$.

Bandpass Sampling for Direct Downconversion

Sampling at rates lower than $2f_{max}$ can still allow for an exact reconstruction of the information content of the analog signal if the signal is a bandpass signal. An ideal bandpass signal has no frequency components below a certain frequency f_l and above a certain frequency f_h. For a bandpass signal, the minimum requirements on the sampling rate to allow for exact reconstruction are that the sampling rate be at least two times the *bandwidth* $f_h - f_l$ of the signal. To ensure that spectrum overlap does not occur, when sampling rates are between two times the bandwidth of the bandpass signal and two times the highest frequency in the bandpass signal, the sampling frequency f_s must satisfy

$$\frac{2f_h}{k} \le f_s \le \frac{2f_l}{(k-1)}$$

where k is restricted to integer values that satisfy

$$2 \le k \le \frac{f_h}{(f_h - f_l)}$$

and $(f_h - f_l) \le f_l$ [3].

Bandpass sampling can be used to downconvert a signal from a bandpass signal at an RF or IF to a bandpass signal at a lower IF. Since the bandpass signal is repeated at integer multiples of the sampling frequency, selecting the appropriate spectral replica of the original bandpass signal provides the downconversion function.

Bandpass sampling holds promise for radio receivers that digitize directly at the RF or IF, since the desired input signals to radio receivers are normally bandpass

Bandpass sampling holds promise for radio receivers that digitize directly at the RF or IF, since the desired input signals to radio receivers are normally bandpass signals. Theoretically, bandpass sampling allows sampling rates to be much lower than those required by sampling at two or more times the highest frequency content of the bandpass signal. This means that ADCs with slower sampling rates (and therefore potentially higher performance, lower power consumption, or lower cost) may be used. An important practical limitation, however, is that the ADC must still be able to effectively operate on the highest frequency component in the signal. This specification is usually given as the analog input bandwidth for the ADC. Conventional ADCs are designed to operate on signals having maximum frequencies of one-half the sampling rate. Performance of the ADC typically degrades with increasing input frequency. When using ADCs for bandpass sampling applications, the specifications of the converter must be examined to determine the behavior at higher frequency inputs. In addition, when bandpass sampling, stringent requirements on analog bandpass filters (steep rolloffs) are needed to prevent distortion of the desired signal from strong adjacent channel signals.

Effects of Quantization Noise, Distortion, and Receiver Noise

This section addresses the relationships among quantization noise, harmonic distortion, and receiver noise. The ADCs best suited to RF and IF processing that have widespread availability use uniform quantization. In uniform quantization, the voltage difference between each quantization level is the same. Other methods of quantization include logarithmic (A-law and μ-law), adaptive, and differential quantization. These methods are currently used in source coding. A discussion of these quantization techniques may be found in [1].

In uniform quantization, the analog signal cannot be represented exactly with only a finite number of discrete amplitude levels. Therefore, some error is introduced into the quantized signal. The error signal is the difference between the analog signal and the quantized signal. Statistically, the error signal is assumed to be uniformly distributed within a quantization level. Using this assumption, the mean squared quantization noise power P_{qn} is

$$P_{qn} = \frac{q^2}{12R}$$

where q is the quantization step size and R is the input resistance of the ADC [4]. In an ideal ADC, this representation of the quantization noise power is accurate to within a dB for input signals that are not correlated with the sampling clock.

If, on the other hand, the analog input into an ADC is periodic, the error signal is also periodic. This periodic error signal includes harmonics of the analog input signal and results in harmonic distortion. Furthermore, harmonics that fall above the Nyquist frequency appear in the Nyquist band due to aliasing. Dithering is commonly used to reduce this harmonic distortion. In one implementation of this technique, thermal noise is added at the ADC input to provide a relatively flat noise power spectrum over the Nyquist bandwidth; the thermal noise plus the quantization noise combine so that the quantization error is uniform. A simple way to implement this technique is to provide amplifier gain to boost the receiver noise to, or several dB above, the level of the quantization noise. The implications of this are examined next.

Commercially available ADCs typically have a full scale range (FSR) of 1 to 20 V. The FSR of the ADC is the difference between the maximum and the minimum analog input voltages to the ADC. Dividing the FSR by the number of quantization levels 2^B, where B is the number of bits of the ADC, provides the quantization step size q. For an 8 b ADC with a FSR of 2.5 V, the quantization step size is 9.77 mV. To compute the quantization noise power, the effective input resistance of the ADC must be known.

Components in radio receivers typically have a 50-ohm input and output impedance. The input impedance of ADCs is usually higher than this and is not well specified. Therefore, when interfacing an RF component with an ADC, as is necessary for digitization at the RF or IF, this impedance mismatch must be considered. A simple method of impedance matching is to place a 50-ohm resistive load at the input of the ADC. This forces the effective input resistance of the ADC to be close to 50 ohms. The quantization noise power can then be computed. Assuming a 50-ohm effective input resistance R to the ADC in this example, the quantization noise power equals –38 dBm. For a noise-limited receiver, the receiver noise power P_{rn} can be computed as the thermal noise power in the given receiver bandwidth (BW) plus the receiver noise figure (NF). This is given as

$$P_{rn} = -174dBm + 10\log_{10}\text{BW}(Hz) + \text{NF}(dB).$$

For a receiver with a 10-MHz BW and a 6-dB NF, the receiver noise power is –98 dBm. Therefore, to boost the receiver noise to the quantization noise power level requires a gain of 60 dB. For an ADC of higher resolution, less gain would be needed, since the quantization noise power would be smaller. Also, wider receiver bandwidths and higher receiver noise figures would require less gain since the receiver noise power would be larger. Nevertheless, for most practical receiver and ADC combinations, automatic gain control is necessary to assure that the least significant bit (LSB) represents a uniform noise input while the peak power does not exceed the ADC's FSR.

Important Specifications

In this section, theoretical signal-to-noise ratio (SNR) due to quantization noise and aperture jitter is discussed. Practical specifications for real ADCs are then presented.

Theoretical Signal-to-Noise Ratio Specifications

For radio receiver applications where the amplitude of the desired signal falls within the ADC's FSR, and the bandwidth of the desired signal is equal to the Nyquist bandwidth, the SNR of an ADC is a useful specification. The theoretical SNR of ADCs is generally thought of as $6B$ (dB), where B

is the number of bits of resolution of the ADC. A more precise expression providing the maximum possible theoretical SNR can be derived based on some assumptions about the noise and the input signal. First, it is assumed that the noise present is due to quantization error only. The amplitude of this quantization noise is assumed to be a random variable uniformly distributed over one quantization step. Assuming a sinusoidal input with an amplitude equal to the FSR of the ADC, the maximum possible theoretical SNR is given as

$$SNR = 6.02B + 1.76 + 10\log_{10}\left(\frac{f_s}{2f_{max}}\right) \quad (dB)$$

where f_s is the sampling frequency and f_{max} is the maximum frequency of the input analog signal [5, 6]. The commonly stated theoretical SNR of $6B$ (dB) is an approximation to this equation when $f_s = 2f_{max}$ and the 1.76 dB is neglected. From this equation, note that as the sampling frequency is increased beyond the Nyquist rate of $2f_{max}$, the SNR increases. This occurs because the quantization noise power, which is fixed and independent of bandwidth, is spread out over an increasingly wider band as the sampling frequency is increased. This lessens the amount of the quantization noise that falls within the Nyquist band (DC to f_{max}). Consequently, this means that oversampling increases the maximum possible SNR. Such oversampling is sometimes used to realize a greater maximum SNR than at first appears possible. An 8 b ADC, with a sampling rate of 20 Msamples/s, for example, can provide 68 dB rather than 48 dB of maximum SNR for 100 kHz signals in the passband if appropriate digital filtering is used to recover the 100 kHz signal.

Besides being limited by the quantization step size (resolution), the SNR of the ADC is also limited by aperture jitter. Aperture jitter is the variation in time of the exact sampling instant. Aperture jitter can be caused externally by jitter in the sampling clock, or internally since the sampling switch does not open at precise times. Aperture jitter causes a phase modulation of the sampled signal and thus results in an additional noise component in the sampled signal [7]. The maximum analog input frequency of the ADC is limited by this aperture jitter since the SNR due to aperture jitter (SNR_{aj}) degrades as the input frequency increases. The SNR_{aj} is given as

$$SNR_{aj} = 20\log_{10}\left(\frac{1}{2\pi f_{max}\, t_a}\right)$$

where t_a is the aperture jitter of the ADC [5]. For sampling at the Nyquist rate, where $f_s = 2f_{max}$, both the SNR due to quantization noise and the SNR due to aperture jitter can be combined to give the overall SNR [8]. Distortion of the ADC output signal can also be introduced because each sample is taken over a finite time period in real sample and hold circuits. This distortion occurs as the input signal varies over this time period.

Specification	Application	Definition
Signal-to-noise ratio (SNR)	Desired signal BW equal to Nyquist BW	MS signal power / MS power of residual error
Spurious free dynamic range (SFDR)	Desired signal BW less than Nyquist BW	MS signal Power / MS power of the largest spurious product
Noise power ratio (NPR)	Desired signal spectrum contains many narrowband channels	Power spectral density of noise outside frequency band of notch filter[1] / Power spectral density of noise inside frequency band of notch filter
Full-power analog input BW	Bandpass sampling	Range from frequency where output amplitude falls to 3 dB less than maximum[2]

[1] With an input signal having a bandlimited, flat noise spectrum and a narrow band of frequencies removed by a notch filter.

[2] For a full-scale input signal.

■ **Table 1.** *Summary of ADC specifications for radio receiver applications.*

Practical Specifications for Real ADCs

The SNR in a real ADC can be determined by measuring the residual error. Residual error is the combination of quantization noise, random noise, and nonlinear distortion (i.e., all of the undesired components of the output signal from the ADC). The residual error for an ADC is found by using a sinusoidal input into the ADC. An estimate of the input signal is subtracted from the output of the ADC and the remaining signal is the residual error. The mean squared power of the residual error is then computed. The SNR is then found by dividing the mean squared (MS) power of the input signal by the mean squared power of the residual error.[1]

A specification sometimes used for real ADCs instead of the SNR is the effective number of bits. This specification is defined as *the number of bits required in an ideal ADC so that the mean squared noise power in the ideal ADC equals the mean squared power of the residual error in the real ADC.*

One definition of the spurious free dynamic range (SFDR) assumes a sinusoidal input to the ADC. In this case, the SFDR is the ratio of the sinusoidal signal power to the peak power of the largest spurious signal in the ADC output spectrum. SFDR allows one to assess how well an ADC can simultaneously detect a very small signal in the presence of a very large signal. Hence, it is an important specification for ADCs used in radio receiver applications. A common misconception is that the SFDR of the ADC is equivalent to the SNR of the ADC. In fact, there is typically a large difference between the SFDR and the SNR of an ADC. The SNR is the ratio between the signal power and the power of the residual error. The SFDR, however, is the ratio between the signal power and the peak power of only the largest spurious product. Since the power of the residual error includes quantization noise, random noise, and nonlinear distortion within the entire Nyquist band, the power of the residual error can be much higher than the peak power of the largest spurious product. Hence, the SFDR can be much larger than the SNR [4].

The SFDR specification is useful for applications when the desired signal bandwidth is smaller than the Nyquist bandwidth. In this case, a wide band of frequencies is digitized and results in a

[1] *The SNR is often (and more accurately) called the Signal-to-Noise Plus Distortion Ratio when distortion is included with the noise, as in this case.*

Resolution (number of bits)	Sampling rate (Msamples/s)	Manufacturer
6	4000	Rockwell International
8	750	Signal Processing Technology
8	2000	Hewlett-Packard
8	3000	*
10	70	Pentek
12	50	Hughes Aircraft
12	100	*
14	24	Hughes Aircraft
18	10	Hewlett-Packard

* Device in development. Work is being sponsored by the Advanced Research Projects Agency (ARPA) of the U.S. Department of Defense.

Table 2. *Examples of current high-speed ADC technology.*

given SNR. The desired signal is then obtained by using a narrowband digital bandpass filter on this entire band of frequencies. The SNR is improved by this digital filtering process since the power of the residual error is decreased by filtering. The SFDR specification for the ADC becomes very important since a spurious component may still fall within the bandwidth of the digital filter, and hence the SFDR, unlike the SNR, does not necessarily improve by the digital filtering process. However, several techniques are available to improve the SFDR. Dithering (discussed previously) improves the SFDR of ADCs. Additionally, post-digitization processing techniques such as phase-plane compensation [9], state variable compensation [10], and projection filtering [11] have been used to improve SFDR.

For an ideal ADC, the maximum SFDR occurs at a full-scale input level. In practical ADCs, the maximum SFDR occurs at input levels at least several dB below the full-scale input level. This occurs because as the input levels approach full-scale, the response of the ADC becomes more nonlinear and more distortion is exhibited. Additionally, due to random fluctuations in the amplitude of real input signals, as the input signal level approaches the FSR of the ADC, the probability of the signal amplitude exceeding the FSR increases. This causes additional distortion from clipping. Therefore, it is extremely important to avoid input signal levels that closely approach the full-scale level in ADCs. Prediction of the SFDR for practical ADCs is difficult, therefore measurements are usually required to characterize the SFDR of the ADC.

In the preceding discussion on SFDR, a sinusoidal ADC input signal was assumed. However, intermodulation distortion (IMD) due to multi-tone inputs is important in ADCs used for wideband radio receiver applications. To characterize this IMD due to multi-tone inputs, another definition of the SFDR could be used. In this case, the SFDR is the ratio of the combined signal power of all of the multi-tone inputs to the peak power of the largest spurious signal in the ADC output spectrum. A current example of test equipment to generate multi-tone inputs produces up to 48 tones.

The noise power ratio (NPR) specification is useful in applications such as mobile cellular radio, where the spectrum of a signal to be digitized consists of many narrowband channels and where adjacent channel interference can degrade system performance. Particularly, the NPR provides information on how well an ADC can limit crosstalk between channels [9].

The NPR is measured by using a noise input signal into the ADC. This noise signal has a flat spectrum that is bandlimited to a frequency that is less than one-half the sampling frequency. Additionally, a narrow band of frequencies is removed from the noise signal using a notch filter. This noise spectrum is used as the input signal to the ADC. The frequency spectrum of the output of the ADC is then determined. The NPR is then computed by dividing the power spectral density of the noise outside the frequency band of the notch filter by the power spectral density of the noise inside the frequency band of the notch filter [4].

When using an ADC in a bandpass sampling application where the maximum input frequency into the ADC is actually higher than one-half the sampling frequency, the full-power analog input bandwidth is an important specification. A common definition of full-power analog input bandwidth (although not universal) is the range from DC to the frequency where the amplitude of the output of the ADC falls to 3 dB below the maximum output level. This assumes a full-scale input signal to the ADC. Typically, the ADC is operated at input frequencies below this bandwidth. Aside from full-power analog input bandwidth, it is important to examine the behavior of the other specifications such as SNR, SFDR, and NPR at the desired operating frequencies since these specifications may vary with frequency. Table 1 provides a summary of the important ADC specifications for radio receiver applications.

ADC Applications Tradeoffs

The performance of ADCs continues to improve at a rapid rate. For radio receiver applications using digitization at the RF or IF, ADCs with both high sampling rates and high performance are desired. Unfortunately, there is a tradeoff between these two requirements. As a general trend, although not always true, the higher the performance of the ADC, the lower its maximum sampling rate will be. Table 2 shows some examples of current high-speed ADC technology for various ADC resolutions.

When selecting an ADC for a specific radio receiver application, in addition to the sampling rate, one must consider critical specifications that characterize the ADC performance such as the SNR, SFDR, and NPR. The ADC specifications most important for various applications are listed in Table 3. In certain applications such as channelized PCS, the Universal Mobile Telecommunication System (UMTS), the Future Public Land Mobile Telecommunication System (FPLMTS), and mobile cellular systems, instead of digitizing the entire band with a single high-speed ADC, parallel ADCs used to digitize narrower bandwidths are often practical ADC architectures. In this case, ADCs with better performance can be used since the demands of a high sampling rate are relieved.

Summary

This article provides an introduction to some of the important factors that must be considered when using ADCs in radio receiver applications. These factors include the choice of sampling method, the amount and effects of out-of-band energy, the analog filtering required, the effects of quantization noise, receiver noise, and distortion, and the critical ADC specifications for radio receiver applications. The differences between Nyquist sampling, oversampling, and bandpass sampling were discussed. It was shown that sampling at the Nyquist rate presents a large and often impractical demand on the anti-aliasing filter. Oversampling eases the requirements on the anti-aliasing filter. A very steep rolloff bandpass filter is required for bandpass sampling when there are strong signals present in adjacent channels. Quadrature sampling was shown to reduce the required sampling rate by a factor of two at the expense of using two phase-locked ADCs instead of one. It was also shown that, in general, some sort of gain control is required for proper digitization of signals at the RF or IF in a receiver. ADC specifications of particular importance to radio receiver applications such as SNR, SFDR, full-power bandwidth, and NPR were examined and some examples of high-speed, state-of-the-art ADCs were given. As ADC performance continues to improve, digitization at the RF and IF in radio receivers, at increasingly higher frequencies, will be used in an increasingly broad range of applications.

Acknowledgments

The author would like to thank Daniel Moulin of The MITRE Corporation for providing the table on critical ADC specifications and performance issues for typical applications.

Typical applications	Critical ADC specifications	Performance issues
Spread spectrum	SNR SFDR NPR	SNR for quantization of small signals in an environment with strong interference SFDR for spatial filtering NPR for interchannel crosstalk
Wideband digital receivers	SFDR	SFDR for accurate detection of low-level signals in an environment with strong interference
Radar	SNR SFDR Overvoltage recovery	SNR for clutter cancellation SFDR for Doppler processing
Cellular mobile and PCS	SNR SFDR NPR	SNR and SFDR for wide bandwidth channelized receivers NPR for interchannel crosstalk
Spectrum analysis	SNR SFDR	SNR and SFDR for high fidelity measurements
Digital sampling oscilloscopes	SNR DNL*	SNR for better amplitude resolution DNL for accurate representation of waveform

* Differential nonlinearity (DNL) is the maximum amount of deviation of any quantization step in the ADC from the theoretical quantization step size of $FSR/2^B$.

■ **Table 3.** *Critical ADC specifications and performance issues for typical applications.*

References

[1] J. A. Wepman, J. R. Hoffman, and J. E. Schroeder, "An initial study of RF and IF digitization in radio receivers," NTIA Report 95-xxx (in preparation), 1995.
[2] F. G. Stremler, *Introduction to Communication Systems*, (Reading, MA: Addison-Wesley, Inc., 1977), pp. 112-120.
[3] E. O. Brigham, *The Fast Fourier Transform and its Applications*, (Englewood Cliffs, NJ: Prentice Hall, Inc., 1988), pp. 83-86 and 320-337.
[4] D. Asta, "Recent dynamic range characterization of analog-to-digital converters for spectral analysis applications," Massachusetts Institute of Technology, Lincoln Laboratory, Lexington, MA, Project Report AST-14, July, 1991.
[5] R. M. Lober, "A DSP-based approach to HF receiver design: Higher performance at a lower cost," *RF Design*, vol. 16, no. 8, 1993, pp. 92-100.
[6] M. Amarandos and S. Andrezyk, "Considerations in the development of a low cost, high performance receiver based on DSP techniques," *DSP Applications*, vol. 2, no. 12, Dec., 1993, pp. 1- 14.
[7] R. Groshong and S. Ruscak, "Undersampling techniques simplify digital radio," *Electronic Design*, vol. 39, no. 10, May, 1991, pp. 67-78.
[8] Harris Semiconductor Corporation, *Digital Signal Processing Databook*. (Melbourne, FL: Harris Semiconductor Corporation, 1994), pp. 6-3, 8-7.
[9] N. W. Spencer, "Comparison of state-of-the-art analog-to-digital converters," Massachusetts Institute of Technology, Lincoln Laboratory, Lexington, MA, Project Report AST-4, March, 1988.
[10] F. H. Irons and T. A. Rebold, "Characterization of high-frequency analog-to-digital converters for spectral analysis applications," Massachusetts Institute of Technology, Lincoln Laboratory, Lexington, MA, Project Report AST-2, Nov., 1986.
[11] N. T. Thao and M. Vetterli, "Optimal MSE signal reconstruction in oversampled A/D conversion using convexity," *Proc. ICASSP '92*, 1992, vol. 4, pp. 165-168.

Biography

JEFFERY A. WEPMAN received B.S. and M.S. degrees in electrical engineering from the University of Arizona in 1981 and 1985, respectively. He has completed all requirements except the dissertation for his Ph.D. in electrical engineering from the University of Colorado at Boulder. Since 1986, he has worked as an electronics engineer with the Institute for Telecommunication Sciences (ITS), National Telecommunications and Information Administration (NTIA), U.S. Department of Commerce, Boulder, Colorado. During this time, he worked on a wide variety of telecommunication projects, including propagation measurement system design, development, analysis, and implementation. From 1991 through 1994, he was a project leader for the Personal Communication Services (PCS) outdoor propagation measurements at ITS. He helped develop the recently patented Digital Sampling Channel Probe, ideal for making outdoor impulse response measurements to characterize wideband propagation in the radio channel. He was the recipient of two Department of Commerce Bronze Medal Awards for his work in advanced measurement system development and PCS measurements. His current work has included research on radio receivers using digitization at the RF or IF.

Reprinted from *IEEE Communications Magazine*, Vol. 33, No. 5, May 1995, pp. 46-54.

The DSP Bottleneck

The stumbling blocks in development of a commercially usable software radio have been the availability of hardware, in particular the fast DSP required. As high performance ADCs have become available commercially, with the sample rates and SFDRs required, hybrid techniques using specialized digital hardware, operating under software control, have become more common.

Rupert Baines

The attractions of the software radio — flexibility and ease of adaptation — are dependent on the manner in which not only coefficients of a filter, but virtually any aspect of a program that implements radio functions, can so easily be changed. This is in contrast with their conventional hard-wired equivalents. Of course, this software requires hardware (in the form of dedicated or programmable processors) to realize these advantages. And it has been here that until very recently, ideas have outpaced the available technology: somewhat paradoxically, the constraint on the software radio has been the hardware. But suddenly that is changing: a new generation of devices is becoming available, processing power is escalating, and implementations are fast enough to be realistically usable.

For simplicity, this discussion will concentrate on the receive side of the radio. The techniques are just as applicable on the transmitter side, but then the challenges lie in subtly different areas. Rather than the DSP-intensive task of trying to recover corrupted/distorted signals from noise, the designer has the somewhat easier synthesis task, but then has the tricky problem of providing a broadband linear power amplifier if the software radio is to be heard by anyone.

The software radio approach employing wideband A/D converters, DSP and general purpose CPUs is more expensive than the traditional hardware intensive technology. For a single channel, software radios now cost significantly more than dedicated hardware. Hard-wired receivers are cheap and perfectly adequate for commodity applications such as home tuners, cellular handsets and the vast majority of radio systems. However, the two advantages of the software approach are flexibility and concentration, and for some systems these are irresistible. Flexibility — being able to switch channels, change modulation schemes or receive different types of signals — has always been valuable to the military, and explains their involvement in much of the early development of technology. But now flexibility has an added attraction for radio designers faced with a plethora of new standards (e.g., different cellular protocols) or varying conditions. For example, PCS base stations can adapt to new protocols by downloading software, or can optimize performance by dynamically assigning channels and capacity between several sites. An optimal receiver realization does not require matched filter hardware, which would be specific to an air interface/protocol, but rather a simple analog (generic) prefilter and sufficiently high sample rate. That rate need not be specific, either, (i.e., it need not be an exact multiple of a symbol rate as long as it is sufficiently high, and sufficient processing power exists) so a single sample rate can be used for channels of differing bandwidths and symbol rates.

The second benefit comes from concentration: multiple channels share the front end radio stage, while having cheap digital processing for each. While the software radio is more expensive than a single traditional receiver, the cost per channel drops rapidly. (At present, the breakeven point lies between four and ten channels.) This is extremely attractive in cellular and PCS applications, where a typical base station will contain 20 or more radio receivers (Fig. 1a and b).

DSP Requirements for Contemporary Applications

In narrowband digital radios, baseband signals are typically digitized and processed with a channel bandwidth of tens or hundreds of kHz (30kHz for AMPS or IS54/IS136, 200kHz for GSM, 1.25 MHz for CDMA, etc.). Figure 2 shows the block diagram of such a narrowband receiver, as used in digital cellular. The radio stages are analog, tuning into a particular frequency slot and using conventional filters to exclude other signals. The software radio requirement is to digitize the entire band (perhaps 25MHz) and to perform IF processing, baseband, bitstream and other functions in software. In the United States, operators have a 12.5MHz band, suggesting a 30MSPS rate minimum (2.5x to allow for real filters, etc.). However, over-sampling this signal is useful to

RUPERT BAINES works in the Communications Division of Analog Devices.

shift aliases out of band and simplify filtering, so faster sampling rates, or narrower bandwidths, are used. For example the AirNet system operates on a 5MHz band, requiring a rate of at least 12.5MSPS, which increases to more than 25 MSPS with oversampling. Several currently commercially available U.S. systems employ 30.72MSPS x 12 bits per sample, corresponding to a convenient map of one FFT slot to a frequency bin in an FFT which the system uses to manage the environment. (See related textbox on this page.)

The DSP must be fast enough to perform isochronous operations on this rapid flow of data. As discussed in the companion article, this results in IF processing requirements on the order of 500 MIPS/MFLOPS to upwards of 10 GFLOPS, baseband processing on the order of 10 to 100 MIPS/MFLOPS per channel, and additional processing burdens for bitstream, overhead, and control that put extreme pressure on the DSP architecture.

Tackling the Front End: RF/IF Functions

A modern programmable DSP typically offers up to 200 MIPS or 50 MFLOPS. (Table 1 shows some commercially available DSPs.) As even the most rudimentary demodulation or tuning requires 10 operations per sample, this would limit it to filtering signals with a carrier of a few hundred kHz. (All modern DSPs will do a single-cycle multiply-accumulate with simultaneous internal transfers, so a FIR filter requires just one cycle per tap.) Obviously, more complex modulation, analysis, estimation, and processing require more DSP horsepower, so any practical system will be more seriously limited. In addition, in reality, data fetches and architectural constraints will also play a significant role in determining the effectiveness of those MIPS, but it is clear that they are limited to operations on fairly low bandwidths. A good FIR/IIR channel selection filter could require about 100 operations per sample at 30 Msa/s, or 3000 MIPS. Using a naive brute-force approach, we would require 15 to 60 DSPs cooperating for this section alone, repeated for every channel. As a result, even with faster devices, software on DSPs still cannot be used for the down conversion itself, but must still essentially operate at baseband (albeit a much wider baseband up to a few MHz). Does this mean we are forced to abandon our aim? Not at all; we merely abandon general purpose hardware at IF. Instead, dedicated but highly programmable filtering hardware is used at the earliest stages, doing much of the filtering and processing in fast digital logic, reducing the processing load that must be done to a level that can be handled in software. As long as this specialized hardware is versatile and is controllable from software, this hybrid architecture still meets our requirements!

One of the most common implementations of such a hybrid architecture is a time-domain approach, using a digital down-converter, such as the Harris HSP50016. This chip contains a synthesizer, a quadrature pair of digital multipliers (which act as mixers in the digital domain), and some clever filters that implement both low-pass and decimation (Haugenauer filters). The device extracts a narrow band baseband signal from a wideband digital input, and decimates it to a reduced data rate, proportional to the bandwidth of interest, that a DSP can cope with. Figure 3a contains a block diagram of the DDC, while Fig. 3b shows the fre-

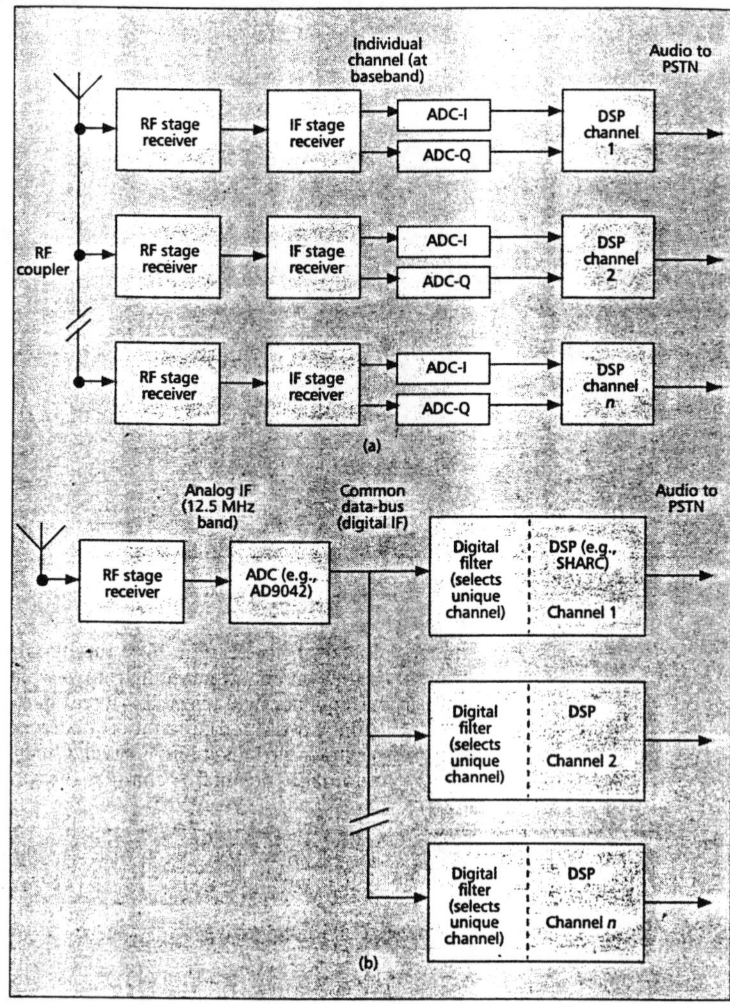

■ **Figure 1.** *Where there are many receivers in one place (e.g., cellular basestations) the wideband approach has attractions of concentration and economy.*

"Magic Numbers" Rapidly Emerging for Cellular and PCS Applications

Applications need anywhere from 5-20 MHz of spectrum with SFDR of at least 80 dB.

GSM is probably the most demanding: blocking requirements include a full signal at –13 dBm, with a simultaneous weak one at –104 dBm; or requirement to distinguish adjacent channels, both at the same strength of –43dB. This > 90 dB dynamic range is still very tough to achieve and few commercial systems are up to it.

However, the U.S. standards are easier. The European PCN system (DCS1800) and its American sibling (TAG5/PCS1900), although a GSM variant, has a 10 dB less demanding radio specification; several manufacturers have systems that should meet the requirements.

Other common (minimum) sample rates are:

- 30.72 MSPS (U.S. analog cellular — 1024 times 30 kHz channel for FFT).
- 38.88 MSPS (IS-54/IS136 digital cellular).
- 40.96 MSPS (military receivers).
- 39 MSPS (GSM, from 13 MHz system clock).
- 50 MSPS (4x over-sampling of a cellular 12.5 MHz band).

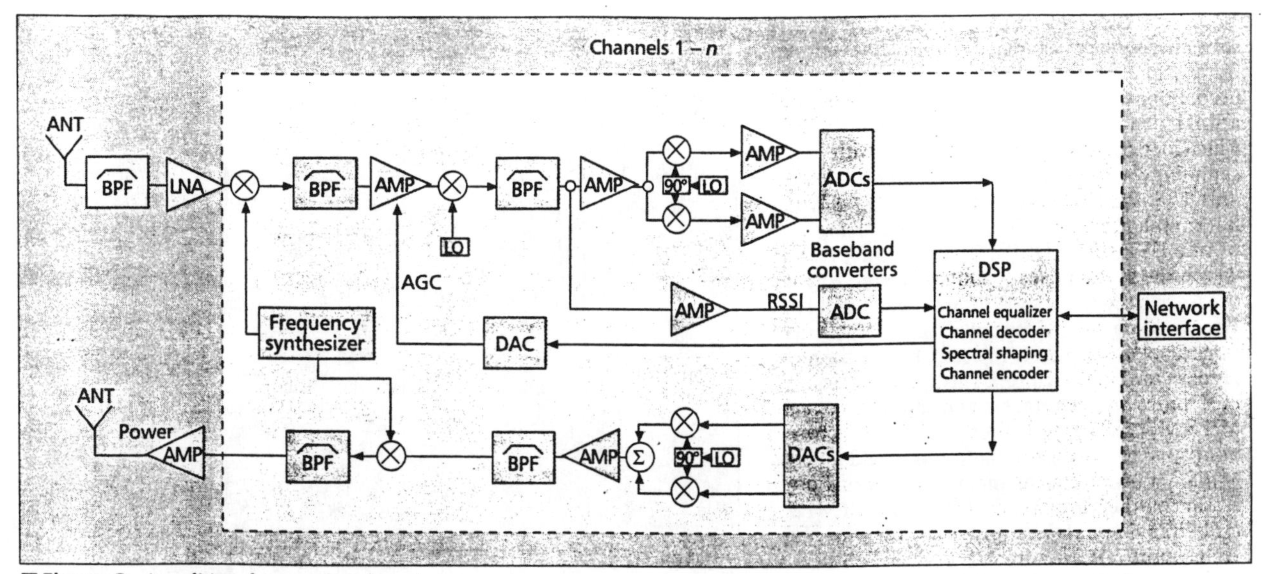

■ **Figure 2.** *A traditional narrowband receiver.*

quency plots of the operation. First the signal is sampled at the wideband sampling rate f_s, which shifts it in frequency by f_c (since it is a real signal, the spectrum is two sided); then it is quadrature filtered to extract the band of interest; and finally, the programmable decimator changes the data rate to f_n — giving a slower (more manageable) data stream containing only the desired information. The DDC has a maximum input rate of 52 MSPS, and can operate with 16 b words to give spurious free dynamic range (SFDR) of more than 100 dB. Depending on the clock rates and

Part	Manufacturer	Speed and clock	On-chip memory	Data width	Dedicated I/O	Price*	Comments
ADSP21062 SHARC	ADI	40 MFLOP 120 MIPS 40 MHZ	32 K words RAM	32 b float	Fast I/O port 240 Mbyte/s Six link ports 40 Mbyte/s each Two 40 MHz bidirectional serial ports	$120	Second-generation multiprocessing. Memory and I/O allow consistently peak performance without bottlenecks.
TMS320C32	TI	40 MFLOP 20 MIPS 40 MHz	512 words RAM	32 b float	Two 10 MHz serial ports	$25	Lowest cost; broad commercial base.
TMS 320C40	TI	50 MFLOP 25 MIPS 50 MHz	2 K words RAM	32 b fixed	Six link ports (20 Mbytes/s)	$150	First-generation multi-processing.
TMS 320C80	TI	Theoretical peak 2 GOPS	12 K words	32 b	Crossbar I/O 400 Mbytes/s	$400	On-chip MIMD multiprocessing (4 DSPS); very powerful. Integrated RISC master processor.
DSP96002	Motorola	60 MFLOP 20 MIPS 40 MHz	2 K words program 2 * 512 words data RAM	32 b float	None specific (two external bus ports, used for data and program as well as I/O)	$100	Multiprocessing interconnect limited; determind by external circuits.
DSP32C	AT&T	25 MFLOP 12.5 MIPS 50 MHz	1.5 K words RAM	32 b	Serial port, general-purpose buses		

*Approximate price in volume. This chart only discusses "conventional" DSPs, and even then is guilty of mixing apples and oranges. There are other extremely powerful processors that some would argue should be included. The Intel i860 is a fast numeric engine, while the newest DEC Alpha, the 21164, is the "fastest microprocessor on the planet" (> 450 Floating point SPECmarks with a 300 MHz clock). However, these devices are intended for the multifaceted world of general-purpose computing. While some can compete with DSPs they are not intended nor optimized for the single-minded I/O intensive operations that the devices listed above excel at.

■ **Table 1.** *Commercial DSPs.*

control parameters, the DDC can be configured in a variety of ways (i.e., real or quadrature outputs) and for a variety of frequency/bandwidth combinations. Given a programmable clock of some description, it is essentially under software control.

Figure 4 shows how the DDC reduces the gigabit per second data rate from the IF A/D converter to digital baseband data rates compatible with follow-on DSPs. The RF conversion stage contains two mixers and associated filters necessary to create a well-filtered wideband IF. The first analog filter rejects out of band components that are to be completely rejected, before the signal is passed to the first mixer. This operates to select the access band of interest and to reject out of band interference. Another mixer then moves this to the second IF. Normally, this would finally be mixed down to baseband, where it would be sampled — perhaps at 25 MSPS for a 10 MHz bandwidth. (A suitable candidate would be the AD9042 — a cost-effective 12-bit ADC that offers 80 dB SFDR and 41 MSPS rate). An attractive alternative to baseband sampling is IF sampling — digitizing the signal in the second or third Nyquist zone, and the desired signals will be aliassed inband by the sampling. This requires a wider analog input bandwidth on the ADC but greatly eases the requirements on the amplifier and filter, as the harmonics will be generated far from the pass-band.

This can be combined with oversampling, which significantly improves the SNR; every time the sampling rate is doubled, the effective noise floor drops by 3dB, as the total noise is spread over more frequencies. When it comes to channelise to narrow band the gain is dramatic — a 30 kHz channel sampled at 40.96 MSPS has a processing gain of some 28 dB!

The programmable clock input rate of the DDC selects a given channel; the filters reject other components (once more, the coefficients are controllable); and the decimator gives an output at a rate usable to follow-on DSPs. A number of DDCs can be used in parallel, with offset clocks, to select a number of channels from the wideband system.

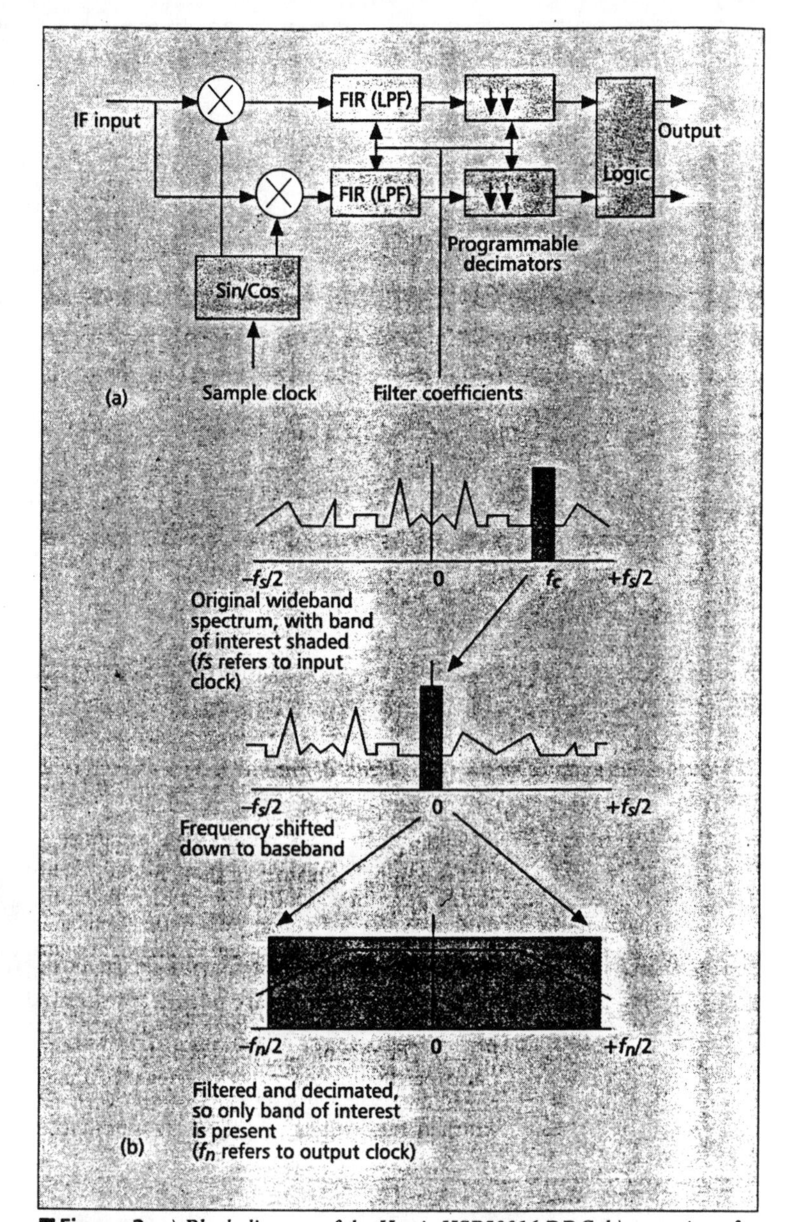

■ **Figure 3.** *a) Block diagram of the Harris HSP50016 DDC; b) operation of digital downconverter.*

Spectrum Monitoring Using DSPs Alone

Purely software-based spectrum monitoring has been implemented using the fast Fourier transform (FFT). The FFT extracts frequency domain information from a series of time domain samples, to resolve the signal into a set of frequency bins. With no oversampling, each frequency bin corresponds to a channel. The FFT can then be used to compute a power spectrum that reveals the signal level of that channel during the period of data acquisition (the FFT epoch or window). A series of such FFTs then creates a sliding window filter bank in the time domain. In practical systems, the power per channel yields key information for well managed channel assignments. Such an FFT is generated in newer PCS base stations in dense urban areas, for dynamic spectrum allocation. The base station selects which channels to use so as to optimize efficiency and minimize noise. In military and emerging PCS applications, such spectrum monitoring may be used to help the base station determine which protocols are being used, and how best to configure its channels and operations among the different systems. A completely nonstructured system is not required; for most commercial applications the receiver will know which type of signal it wants to receive (or rather, which selection of types). A given digital standard will have predictable channel bandwidth, modulation scheme, spectral density and so on, and most include training sequences and preambles to provide even more help in the equalization and recovery. However, a base station will need to perform an overview to decide which signals it is receiving, and which protocols they are using, checking bandwidth and spectral properties so that it can select the appropriate software to tune and recover that signal from all the others.

The time domain trace of a single frequency

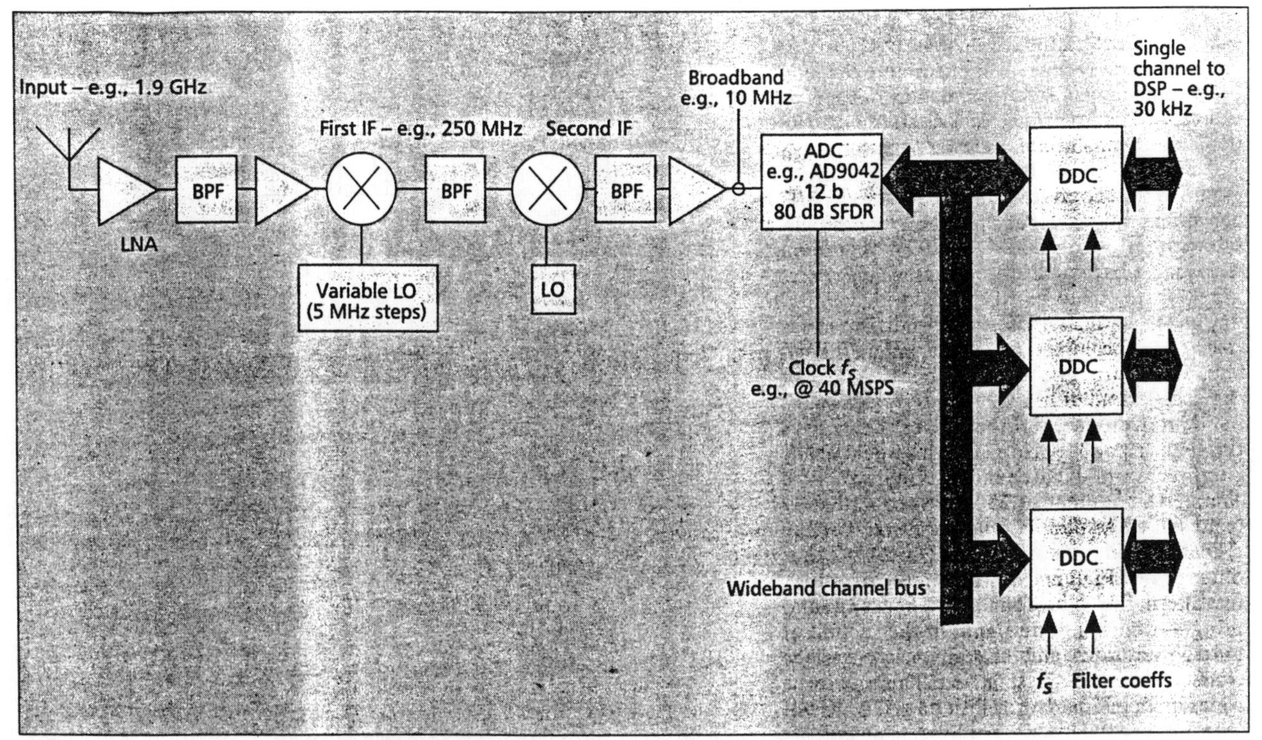

■ **Figure 4.** *An example receiver using digital downconverters for channel tuning and selection.*

bin in the power spectrum yields a series of levels that is a sampled history of that channel. (This can be described as a filter bank or transmultiplexer). With one frequency bin for each channel, the successive power spectrum results yield signal amplitude information, and this is sometimes used as a straightforward parallel AM radio receiver, operating on N channels at once (see Fig. 5). The rate of sampling each channel is the rate of performing power spectra, so each FFT must be executed quickly enough to sample the AM bandwidth at the Nyquist rate. A 4kHz modulating bandwidth would require at least 8kHz sampling —or a FFT/power spectrum at 1.25ms intervals. The SHARC performs a 1024 point complex FFT in just 0.46 ms, so that the power spectrum is computed in less than the required 1.25 ms. Rather than using a series of filters or downconverters, one per channel, the FFT performs the required operations in parallel. The parameters of the FFT can be easily changed, reconfiguring the receiver. Other modulation schemes can be recovered in a similar way. For example, an elementary FSK (frequency shift keying) receiver can be implemented

Function / Radio class	RF receiver	IF stage, tuning	Processing 1, baseband	Processing 2, source	Example
Classical (commodity)	Analog-LNA, first.	Analog, IF mixers, PLLs, mixer, etc.	Analog hardware filters.	Analog hardware (depends on modulation).	Analog cellular.
Digital	"	"	Baseband digitization. Digital equalization (carrier, timing recovery) and channel decoding.	Speech decoding, error correction done in DSP.	Current and future digital cellular, PCS handsets (e.g., GSM, IS95). Current digital cellular base stations.
Hybrid	"	IF sampling or wideband ADC.	Digital hardware downconverters, Haugenauer filters.	"	Next-generation commerical basestations (for both digital and analog systems).
Software	"	"	Software-based. Filterbank, Costas loop, integrate and dump, etc. Move to non-structured design.	"	Research. Military. Emerging commercial

■ **Table 2.** *A comparison of different radio architectures.*

directly with two frequency bins per channel, one for the low tone and one for the higher. This is a highly simplified example. In practical implementations, bit synchronization, Costas loops, integrate and dump filters, etc., increase the performance requirements of the DSP. Sophisticated algorithms to transform the signal to a more usable form often increase the DSP requirements to multiple processors. As the versatility increases, the amount of processing power required is not insignificant. And this leads us inevitably to consider multiprocessing.

The DSP Bottleneck: Multiprocessing Then and Now

Software radios invariably require the use of parallel and sequential partitioning of algorithms to get required processing power. DSPs are getting ever faster, but it will be a while before we can use a single "ultimate chip" to do everything. Instead, the idea of using multiprocessing to share the effort seems attractive. Traditionally, real-time multiprocessing with DSPs has been fiendishly tricky. Fortunately, recent products have addressed the issue, and it is becoming much simpler. One approach is to integrate a general purpose CPU and a dedicated DSP core on the same die. Motorola, for example, has released a "Communications Processor," the 68356, which includes a general purpose micro-controller with a 56000 series DSP. For those applications that required both general purpose operations and a DSP, this is an attractive approach. For IF and baseband processing, however, which require significant DSP with minimal supervisory functions, it is less relevant. A different approach has been taken by AT&T, who integrated a dedicated Viterbi decoder as co-processor on some of their 16xx series DSPs. This drastically eases the processing load for systems requiring an equalizer (e.g., GSM handsets). It is very likely that more DSPs will

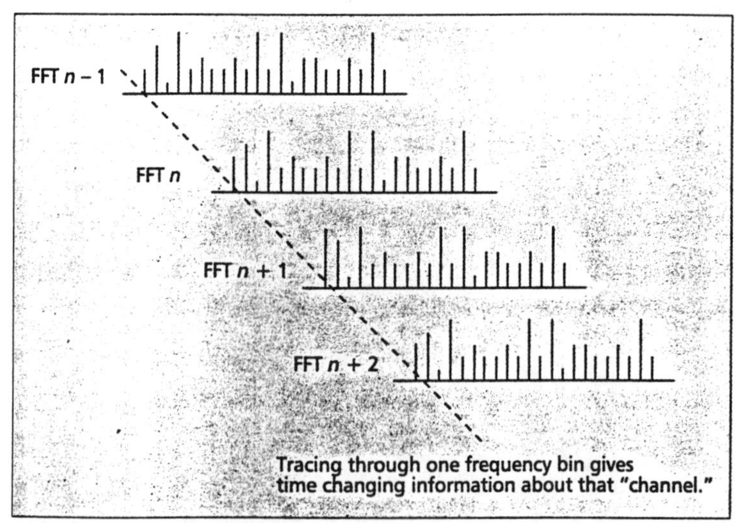

■ **Figure 5.** *A filter bank (such as FFT implementation) can be a way of tuning many channels in parallel.*

support such features in the future. (See text box on next page.)

If the system is to operate in real time (which is usually the requirement), then the data must be able to get in and out of the DSP, which can pose I/O bottleneck problems. For fast data access, fast memory is required, but this is expensive. If several processors are to be used, there must be efficient methods of linking them and coordinating their actions. Traditional DSP architectures were not well-suited to multiprocessing. It is a characteristic of a DSP (as contrasted to a conventional microprocessor) that it must operate on a continuous flow of data. A typical system might require the equivalent of four bus operation each cycle, with an instruction read, two operand fetches (e.g., latest sample of data from the ADC, and filter coefficient), and an I/O transfer (e.g., output of result). Unfortunately, most DSP implementations do not support this many data buses. As a result, the effective processing rate

Integration level	Example	Issues
Within-processor	Superscalar units within MIPS R4000 or DEC Alpha	Very design intensive, requires sophisticated compilers, only feasible for volume applications. However, may be most efficient and cost-effective. Within chip I/O is easy. Can have closely-coupled memory and cache.
Within-chip	Co-processor (e.g., 68356) Multiprocessor (e.g., TI MVP)	Smaller system. Can be cheaper. Can use specialized co-processor (e.g., Viterbi engine). Can integrate DSP and μP. Potential I/O bottleneck to off-chip parts which may be expensive. Less flexibility — processor mix is predefined.
Board	Multiple DSPs on board (e.g., several C40s or SHARCs)	Cost-effective, versatile. May be hard to program. On-chip memory very valuable. System analysis to avoid I/O bottlenecks. Need high I/O bandwidth and interprocessor support to achieve true potential.
System	Multi-VME rack	Custom or semi-custom system. Tradionally hard to program/support. Expensive. However, can give any level of performance required. Can mix different processors.

■ **Table 3.** *Degrees of parallelism.*

Second-Generation DSP Multiprocessing

On-chip bus arbitration:
- Fixed and rotating priority.
- Bus locking and timed release.
- Single cycle bus mastership latency.

These allow the processors to work together smoothly, without requiring the designer to worry about conflicts or the processors to waste time.

Unified address space:
- Direct interprocessor memory read/write/interrupt.
- Broadcast writes.
- Reflective semaphore.
- Read-Modify-Write locking.
- Global memory sharing/arbitration.

Allow DSPs to share and partition memory without conflicts. This is essential for "parallel" architectures, but does simplify pipeline or dataflow systems, too.

Transparent DMA:
- Dedicated I/O processor.
- Dual port memory allows:
  - –DMA chaining/routing.
  - –Multiple DMA channels.

Allows transfers to be coincident with computation, while dual porting allows simulataneous reads and writes — avoiding processor stalls.

On-chip memory:
- Sufficiently large ultra-fast on-chip memory.

Avoids bus bottlenecks. Memory can be faster and cheaper than if external (e.g., 4 Mb)

High-speed I/O
This is crucial.
Current devices (e.g., SHARC):
- Up to 240 Mbyte/s through external port.
- 40 Mbyte/s through each link port.
- 160 Mbyte/s max through six link ports.
- Two general purpose 40 MHz bidirectional serial ports.

decreases, as operands must be passed through a keyhole — the bus bottleneck (Fig. 6a).

On and Off Board Memory

The situation gets much worse if reasonable quantities of memory are required (as they are for any realistic radio, with demodulators or channel decoders, etc.). Commodity memories (SRAMS) are designed for the Von-Neuman processor architectures, where only one access per cycle is required. Using such a memory leaves the DSP designer with substantial bus bottleneck problems, and the need to perform heroic data-flow analysis to try to determine system feasibility. Furthermore, the need for speed means that very fast (i.e., very expensive) memories are required to reduce the impact of this bottleneck. However, the on-chip memory aspect deserves a little bit of explanation. As shown in Fig. 6b, the ideal DSP system integrates memory on-chip, which allows the program sequencer, the external I/O, and all parts of the computational unit to have simultaneous single-cycle access. (In fact, two accesses per cycle are possible under some circumstances.) Additionally, the system designer doesn't have to provide board space for memory ICs, or include cost and power budgets for fast external SRAM. Buses within the chip are faster and less power demanding than the long and capacitatively loaded burden of IC to IC connection. Similarly, the I/O peripherals, which were on-chip all along, no longer require interrupts to go off-chip, through the keyhole of a single bus, to the memory and back to the chip. Instead, a dedicated bus can be used within the IC for direct access whenever desired. Finally, there are commercial implications: the designer no longer needs to do a complex data-flow analysis to assess true performance, and time-to-market is reduced. Some estimate that the performance penalty for using off-chip memory in a real system is that single-cycle operations take an average of 2.31 cycles, equivalent to operating at 43 percent of rated speed (Table 4).

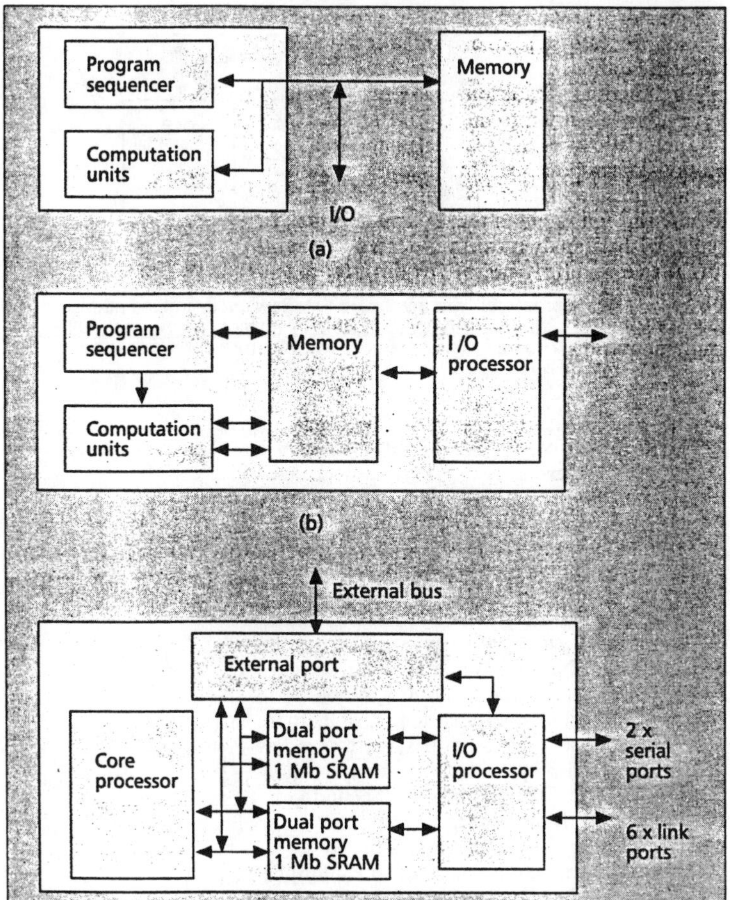

■ **Figure 6.** *a) Traditional DSP architecture. Note the bus-bottleneck; b) Multiple buses and internal memory improve DSP architecture; c) a real world implementation.*

I/O Issues in Multiprocessing

As discussed earlier, the amount of I/O data flow in such a system can be staggering. The ADC could easily produc 60 Msa/s, each 12 or more bits in resolution, requiring the processor system to be reading data at rates of at least 120 Mbytes/s. While there will be redundancy, filtering, and removal of unwanted channels, it is likely that the output data-flow will be roughly comparable, giving an upper bound of some 240 Mbytes/s, just for the

Instruction execution	Relative Frequency	Cycles
Using core registers	30%	1
Using 1 memory location	30%	2
Dual operand fetch	40%	3

Concurrent I/O occurs 10 percent of the time.
Therefore, a typical single cycle instruction requires 2.31 cycles on average:
2.31 = [(0.4*3) + (0.3*2) + (0.3*1) [all times 1.1 for the concurrent I/O penalty].
Or, to put it differently, the DSP only operates at 43 percent of rated processing power if using off-chip memory.
Since each of these off-chip fetches requires power for chip-to-chip drive circuitry, similar savings are possible in power consumption.

■ **Table 4.** *The penalty of off-chip memory access.*

data I/O. In addition, in a real system it is likely that intermediate results, program information, and the like will need to be shared between multiprocessors, necessitating dedicated high-speed interprocessor links. This emphasizes the essential importance of the processor communications architecture. Conventional designs have relatively few data paths, often funneling through bottlenecks. Instead, software radio applications require several parallel routes for information: bulk data I/O, processor to processor links, and access to both global and local memory structures. Ideally, these would be all simultaneous with dedicated DMA or I/O processors.

But most DSPs were not designed to be used as multiprocessing systems. (The TMS320C40 is the notable exception.) They did not support arbitration or prioritization, forcing the programmer to spend both development time and precious processing bandwidth in software solutions. There were no hardware systems for global memory addressing/sharing, nor for sempahores or signaling. And so on... This is somewhat surprising, given the prevalence of multiprocessor DSP systems and the fact that these techniques, and their implementation in silicon, had been seen with Inmos' Transputer (well engineered if not a staggering commercial success). This changed with the release of Texas Instrument's deservedly popular TMS320C40, which corrected several of these gaps with a DSP that was suited to multiprocessing. Indeed, it is fair to describe the C40 as the seminal DSP multiprocessor. Figure 7a shows a typical implementation of a C40 multiprocessor. The second off-chip bus makes it easier to access local memory and helps solve the bus bottleneck, while the dedicated processor-processor links allow for dedicated data-flow paths. In other words, the processor's electrical layout and the algorithms logical layout are mapped, drastically improving the transfer efficiency between processors (and, not incidentally, simplifying the design analysis). A more recent processor from TI goes even further: the C80, or MVP, includes four DSP cores with a RISC supervisor. However, while the MVP is undeniably powerful, it is also undeniably expensive, and does not seem to have been used widely, in contrast to the C40.

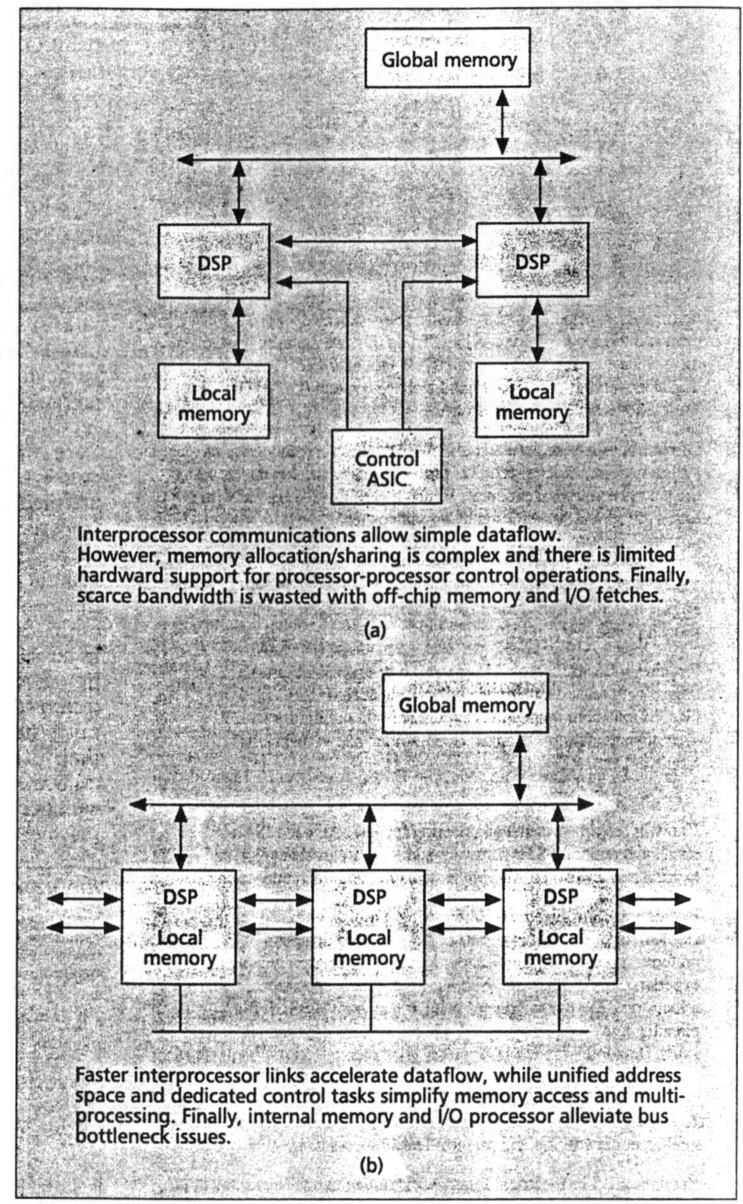

■ **Figure 7.** *a) First-generation DSP multiprocessing; b) second-generation DSP multiprocessing.*

Just as innovation in the ADC realm has suddenly made wideband receivers and IF sampling with excellent SFDR feasible, the processor market has advanced. Second generation DSP multiprocessing is now a reality, learning from the experience of the C40. While that had some of the features required, it has limits in how deep the multiprocessing ran. In particular, while there were the processor-processor links, there were few other hardware tools provided, e.g., for arbitration, semaphores, etc. (The text box on "Second-Generation DSP Multiprocessing" summarizes some of these features.) The SHARC is the first processor to meet the requirements of true multiprocessor DSP as shown in the box. Of course there will be subsequent devices that also implement these features, and no doubt will add new ones. Figure 7b shows a cluster/dataflow multiprocessor architecture based on several SHARCs. In contrast to the earlier case, this system has much greater effective I/O bandwidth and dramatically simplified coordination. The need for fast I/O and a dedicated parallel processor to manage it is hardly surprising; if a single computation core is too slow, one can add more DSPs. But if they cannot read and write fast enough, radio applications are not easy and may not be possible.

Summary

The software radio concept is critically dependent on high performance general purpose processor hardware architectures. The stumbling blocks in development of a commercially usable software radio have been the availability of hardware, in particular the fast DSP required. As high performance ADCs have become available commercially, with the sample rates and SFDRs required, hybrid techniques using specialized digital hardware, operating under software control (e.g., digital down-converters) have become more common. These techniques use digital mixers (multipliers) to select a narrowband channel from the fast, wideband data coming out of the ADC, then filter and decimate it before passing a manageable flow to the general purpose DSPs. However, there are more elegant algorithms — for example, nonstructured receivers, or parallel approaches that recover and tune an entire band at once. These require faster calculation, implying dedicated hardware or, better yet, multiprocessing DSPs. Recent developments in multiprocessing architectures (starting with the TI C40) make this much easier. However, until the SHARC, DSPs were still seriously constrained in their I/O bandwidth, level of multiprocessing support or, crucially, on-chip memory. Now that these issues have been addressed, we can expect to see many more developments in this arena reaching the marketplace.

Biography

RUPERT BAINES works in the Communications Divison of Analog Devices, Wilmington, Massachusetts. He received a B.S.E.E. and Diploma in engineering from Hull University, England, in 1986 and 1987, respectively, and an M.B.A. from IESE, Barcelona, Spain, in 1991. His e-mail address is: rupert.baines@analog.com.

Reprinted from *IEEE Communications Magazine*, Vol. 29, No. 12, Dec. 1991, pp. 87-97.

Implementation of Coded Modems

Trellis-coded modems use DSP and LSIC technologies to deliver robust operation and high reliability from miniature hardware.

by Shuzo Kato, Masahiro Morikura, and Shuji Kubota

Other articles in this Special Issue on Bandwidth and Power Efficient Modulation discuss coding, modulation, demodulation, and the problems associated with channel impairments caused by fading. This article discusses trellis-coding techniques that allow robust modem operation despite various channel impairments and to miniaturize reliable modems by employing advanced DSP and LSIC technologies. (Studies have been performed on another type of coded modulation, continuous-phase modulation, but this technology will not be discussed in this paper due to space limitations. [1-2])

Modulation can be separated into two tasks: coding and signal mapping to maximize free distance (minimum Euclidian distance) between coded signal sequences, and modulation of a carrier by the aforementioned coded signal sequences in a multilevel quadrature modulation scheme. Demodulation involves three tasks: demodulation of the received signal and synchronization of a local reference's phase to the carrier used in the modulator; extraction of symbol timing so that the received signal can be sampled and the recovered clock synchronized with the transmitter clock; and decoding of the demodulated signal according to the Viterbi algorithm.

Although the basic functions of coded modems (except coding and signal-mapping methods) are similar to those of conventional modems employing forward error correction (FEC), special attention must be paid to the carrier phase error, carrier slip, and Viterbi decoder parameters for better performance and hardware miniaturization.

This article presents modem implementation techniques that have been proposed to achieve good error probability (P_e) performance in various channel environments and to allow miniaturization of trellis-coded modems. We will discuss the application of trellis-coded modems, coding and mapping, a quadrature modulator subsystem, and quadrature demodulator subsystem, including carrier-phase offset, carrier slip, automatic frequency control (AFC), and automatic gain control (AGC). We will also discuss Viterbi decoder design techniques and the implementation by DSPs and LSICs, as well as demodulators for Rician fading channels.

Shuzo Kato is a group leader for signal processing, NTT Radio Communications System Laboratories.

Masahiro Morikura is a Senior Research Engineer with NTT.

Shuji Kubota is a Senior Research Engineer with NTT.

Applications

Coded modems have been widely used for digital data transmission over band-limited channels, such as voiceband channels. Various coded modulation schemes have been studied low to high speed applications in order to achieve higher frequency utilization efficiency and better P_e performance. Before the advent of trellis-coded modems, uncoded modems had been used and their maximum transmission rate was 9.6 kb/s for digital data transmission over voiceband channels [3].

A trellis-coded modulation scheme was adopted in 1984 by the International Telegraph and Telephone Consultative Committee (CCITT) for use in high-speed voiceband modems [4]. They are trellis-coded 32 quadrature amplitude modulation TC-32QAM, and TC-128QAM, which can achieve data transmission at 14.4 kb/s over voiceband channels [5]. These trellis-coded modems are designed for additive white Gaussian noise (AWGN) channel environments and are now in commercial use. Because their transmission rates are relatively low, digital signal processing technologies have been extensively used as a counter measure for channel impairments and to efficiently construct these modems.

On the other hand, in other communication systems such as terrestrial microwave, cellular, and satellite systems, coded modems have not yet been used commercially. Current terrestrial microwave systems employ 16, 64, or 256 QAM systems with separate forward error correction (FEC) and cellular mobile communication systems such as GSM (Special Group on Mobile Communication), TIA (Telecommunication Industry Association), or Japanese standards employ Gaussian-filtered minimum shift keying (GMSK) or $\pi/4$-quadrature phase shift keying (QPSK) systems with separate forward error correction. Current satellite communication systems employ mainly QPSK modulation with separate forward error correction.

Coded modems have not been in commercial use

0163-6804/91/$01.00 1991© IEEE

in these fields for these reasons:

- Current systems, coupled with high-coding-rate FEC, can achieve close frequency utilization efficiency, as shown in Fig. 1, and also E_b/N_o (where E_b is the energy per bit and N_o is the single-sided noise power spectral density) to that of trellis-coded modulation, provided that slight bandwidth expansion is allowed.
- Trellis-coded modems require more complicated hardware and more accurate components.
- Performance capabilities of coded modems in non-AWGN environments have not yet been fully documented.
- In satellite communication systems, transmission capacity is limited not only by the frequency band but also by satellite transmission power.

In satellite communication systems, there are three major types of applications. The first is the use of low-speed, single channel per carrier (SCPC) trunk transmission systems to achieve better P_e performance while retaining the present frequency utilization efficiency [6]. The transmission channel in this application is characterized by AWGN and the transmission bit rate is relatively low, such as 64 kb/s (see Fig. 2).

The second application is in high-speed continuous or time division multiple access (TDMA) communication systems such as 140 Mb/s transmission using a single 72 MHz bandwidth transponder to meet higher hierarchy transmission bit rates that cannot be realized by present QPSK modulation systems [7-8]. Thus, in these systems, modems are operated in continuous or burst mode in AWGN environments.

The third application is to allow mobile satellite communication systems to achieve higher frequency utilization efficiency in Rician fading channels. Most current research activities target this field, and one of them is to transmit a 4.8 kb/s digital voice data over a 5 kHz Rician fading satellite channel [9-10]. This field is suitable for trellis-coded modulation application, because transmission bit rate is up to approximately 10 kb/s which can easily be handled by DSPs. Although this application bit rate is close to that of trellis-coded modems for voice-band data transmission, special techniques to combat fading are required for robust transmission. Moreover, a trellis-coded 256QAM scheme has been studied in a multipath fading environment for high speed terrestrial microwave transmission systems [11]. The features of these trellis-coded modulation systems are summarized in Table 1.

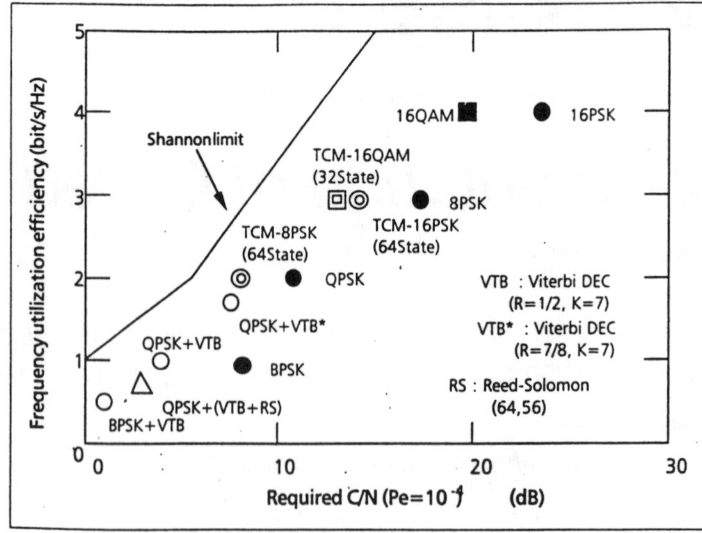

■ **Figure 1.** *Relationship between frequency utilization efficiency and required C/N of coded modulation schemes and alternatives*

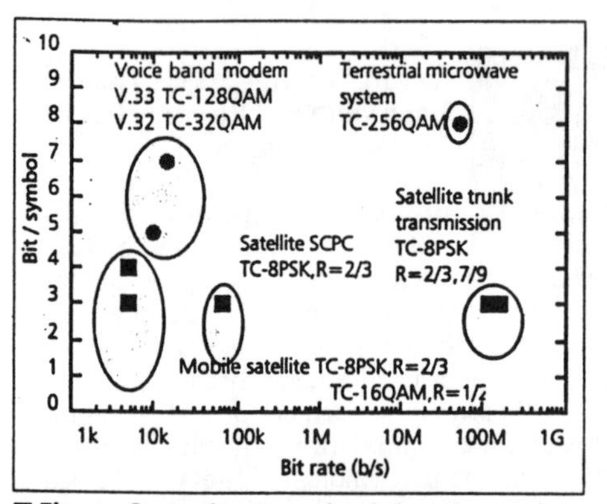

■ **Figure 2.** *Applications of coded modems*

■ **Table 1.** *Application fields of trellis coded modulation*

			Operation speed		Transmission mode		Channel impairment			Typical bandwidth
			low	high	Continuous	Burst	AWGN	Rician	Rayleigh	
Voiceband data transmission			O		O		O			3.1kHz/ch
Satellite Communication	Mobile		O		O			O	O	5kHz/ch
	Fixed	low speed	O		O		O			45kHz/ch *1
		High speed		O		O	O			72MHz/140Mbit/s *2
Cellular mobile communication			O			O			O	60kHz/3ch *3
Terrestrial microwave transmission				O	O				O	20MHz/(52Mbit/sX2) *4

*1 : SCPC (Signal channel per carrier) system (INTELSAT)
*2 : TAT-8 transmission over 72MHz transponder
*3 : TIA standard
*4 : 52Mbit/s 256QAM system (NTT)

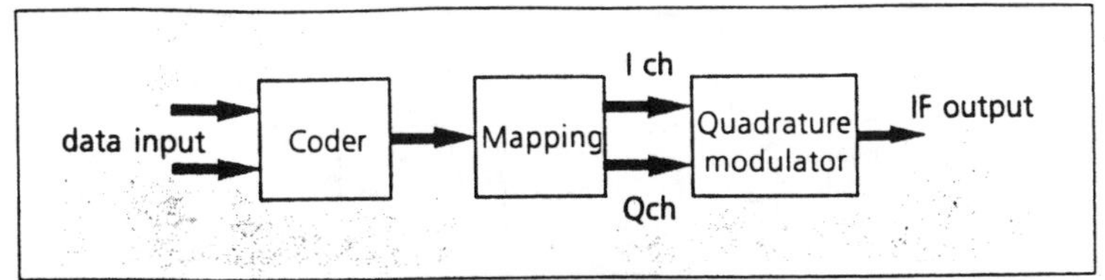

■ **Figure 3.** *Block diagram of a coded modulator*

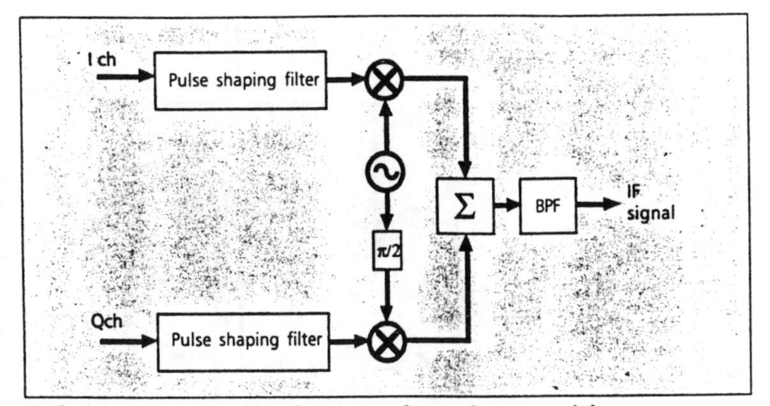

■ **Figure 6.** *Functional block diagram of a quadrature modulator*

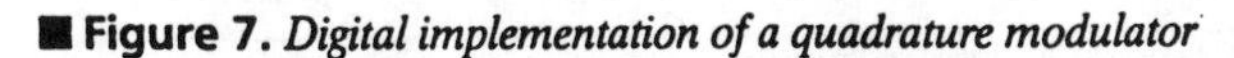

■ **Figure 7.** *Digital implementation of a quadrature modulator*

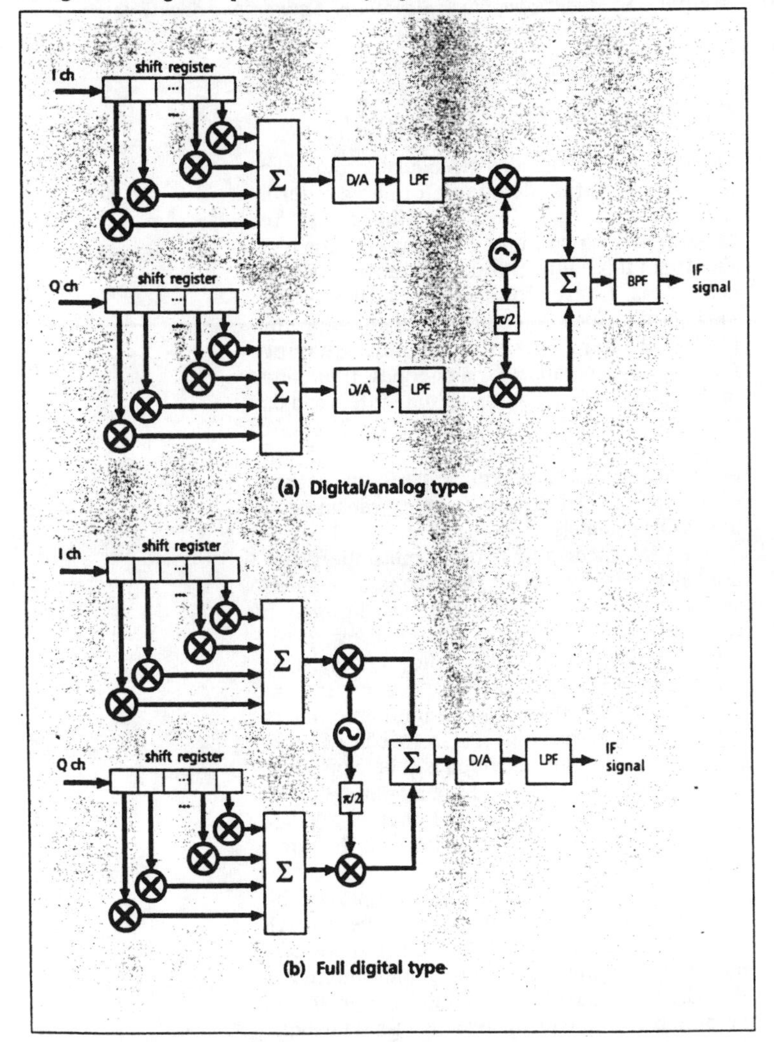

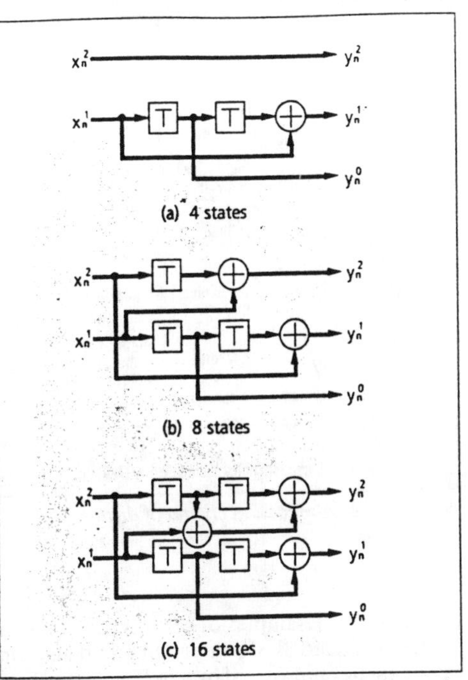

■ **Figure 4.** *Realization of an 8PSK convolutional coder*

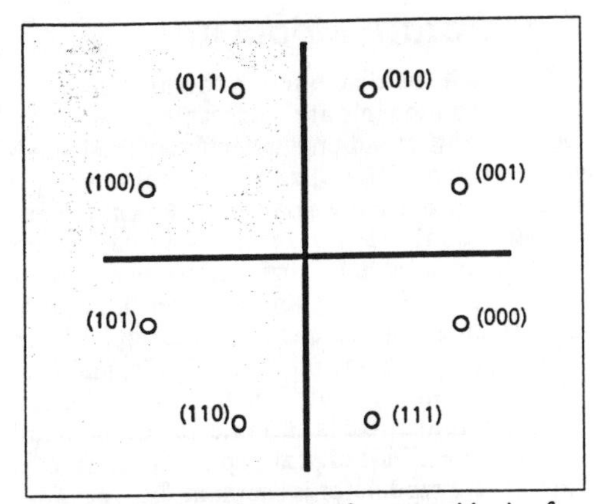

■ **Figure 5.** *Signal mapping by set partitioning for 8PSK*

Encoders

A block diagram of a coded modulator is shown in Fig. 3. Input data are trellis-coded, then mapped for transmission by a quadrature modulator. This block diagram is very similar to that of a conventional QPSK modulator with forward error correction except that in the trellis-coded modulator, the transmission phase of the carrier is chosen (set partitioning is carried out) in a "Mapping" block to maximize minimum Euclidian distance between coded signal sequences [3]. A number of encoders have been developed for various numbers of states of the convolutional encoders and for various modulation schemes [12-13]. Moreover, various mapping methods have been proposed to increase minimum Euclidian distance between coded signal sequences [14]. Because these encoders and mapping circuits are easily constructed using logic circuits, it is not difficult to implement encoders and map-

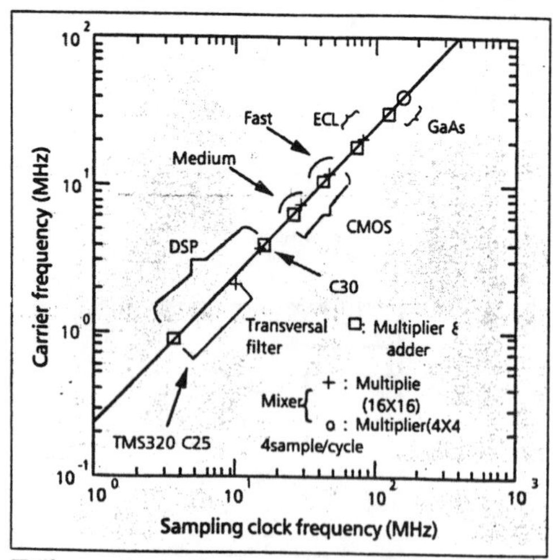

■ **Figure 8.** *Application criteria of digital multipliers*

ping circuits. Examples of convolutional coders for trellis-coded 8PSK modulation and mapping by set partitioning are shown in Fig. 4 and Fig. 5, respectively.

Quadrature Modulator

As it will be discussed in the next section, coded modems are vulnerable to both carrier-phase offset and amplitude offset of the demodulated signal. Digitalization of modulators is one of the promising ways to solve these problems. A functional block diagram of a quadrature modulator suitable for digital implementation is shown in Fig. 6. There are two levels for digital implementation of a quadrature modulator, as shown in Figures 7a and 7b. In Fig. 7a, multipliers are not implemented in digital circuits and this configuration is suitable for high speed applications. Full digital implementation of a quadrature modulator is shown in Fig. 7b.

The application criteria of digital multipliers (as of this year) is shown in Fig. 8 [15]. In this figure, four samples per cycle and full dedication of the DSP to multiplication (no other operation in the DSP except multiplication) are assumed. As seen from this figure, a multiplier for a 12 MHz carrier frequency (6 MHz baseband signal clock frequency) can be digitalized by employing CMOS masterslice LSI technologies. Digitalization of the low-pass filter is more speed-sensitive because it requires full adder operations in addition to multiplications when realized as a digital transversal filter. The maximum available clock frequency of baseband signals is also shown in Fig. 8. As seen from this figure, the maximum baseband bit rate is about 10 Mb/s with CMOS masterslice LSI technologies. Because multipliers for high-speed modems (higher than 50 Mb/s) must be operated at carrier frequencies of 70 MHz or 140 MHz, it is difficult to digitalize these modems with present technologies. For these applications, monolithic IC-implemented multipliers are a good alternative because they achieve very good performance of the small phase and amplitude error of the modulated signal [16].

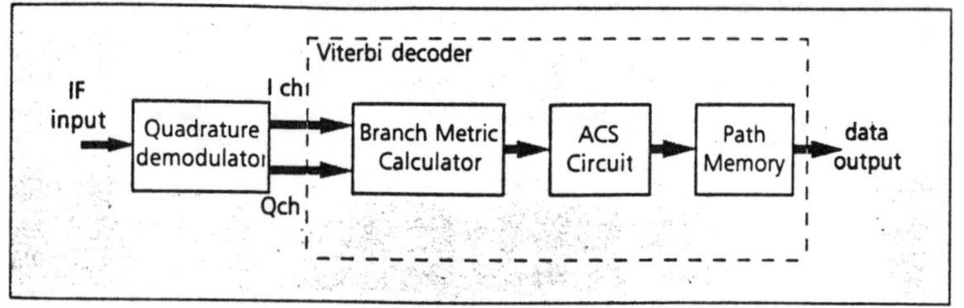

■ **Figure 9.** *Block diagram of a coded demodulator*

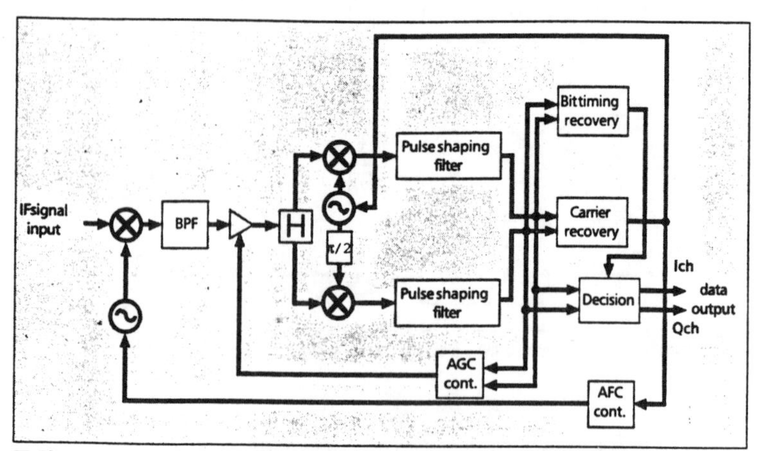

■ **Figure 10.** *Functional block diagram of a quadrature demodulator*

Quadrature Demodulator

The operation of a coded demodulator composed of a quadrature demodulator and a Viterbi decoder is shown in Fig. 9. A diagram of the quadrature demodulator functions is shown in Fig. 10. In this configuration, the received signal is bandpass filtered for noise reduction before mixing with a local reference and then shaped by baseband filters. However, the pulse-shaping function can be realized by a bandpass filter, and in this case baseband pulse-shaping filters can be replaced by simple harmonic cutting filters. A carrier recovery circuit is required for coherent demodulation to achieve better P_e performance than differential detection, especially for transmission in AWGN channels.

For high-speed applications, multipliers cannot be implemented in digital circuits, and digital signal processing technologies are usually applied for other functions such as carrier recovery, bit timing recovery, and so on, as shown in Fig. 11a. For relatively low-speed applications, multipliers and all other functions can be implemented as digital circuits, as shown in Fig. 11b, to reduce P_e performance degradation, which can be caused by the carrier-phase offset and the amplitude error of the received signal. The application criteria of digital circuit technologies to multipliers is the same as that of modulators.

Because trellis-coded modulation employs higher multilevel modulation than conventional uncoded modulation to achieve the same frequency utilization efficiency, coded modulation must face larger degradation due to modem implementation imperfections than conventional modulation. This requires more accuracy for each component,

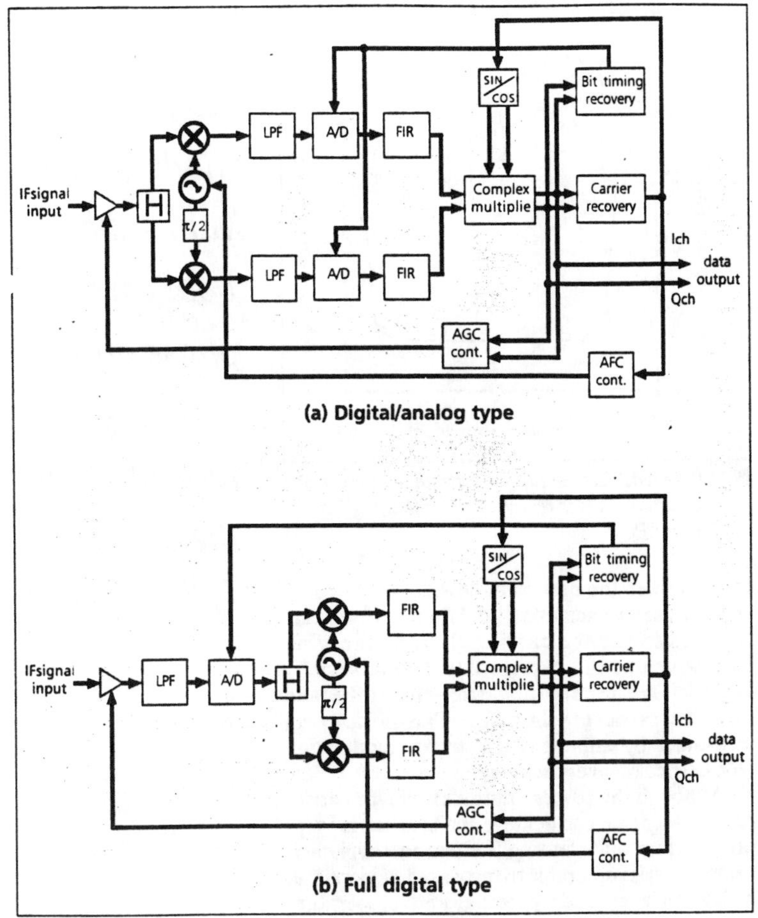

■ **Figure 11.** *Digital implementation of a quadrature demodulator*

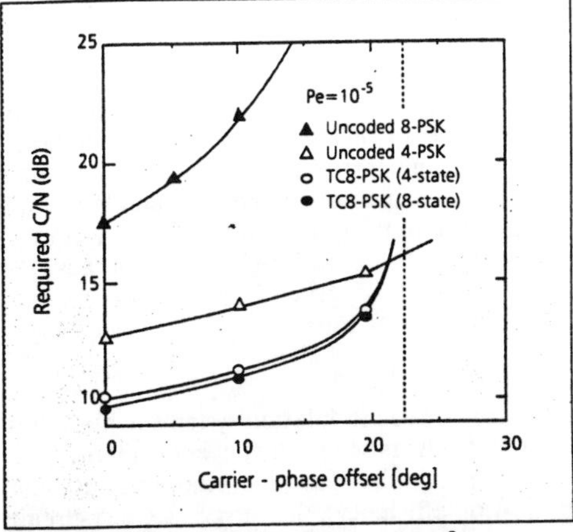

■ **Figure 12.** *Required C/N ($Pe=1x10^{-5}$) versus carrier-phase offset*

the recovered carrier phase, and the demodulated signal amplitude.

Degradation

The required carrier power to noise power ratios (C/N_s) to achieve $P_e = 1x10^{-4}$ of trellis-coded (TC) 8 PSK [13], uncoded QPSK, and uncoded 8 PSK systems versus the carrier-phase offset are shown in Fig. 12. As shown in this figure, the C/N degradation of TC8 PSK systems sharply increases when the recovered carrier-phase offset approaches 22.5 degrees. This is natural because this modulation scheme is basically 8 PSK modulation. However, the C/N degradation of uncoded 8 PSK systems is much higher than that of TC8 PSK systems and rapidly increases with increasing value of the carrier-phase offset. In other words, TC8 PSK systems have extremely large P_e performance degradation when the carrier offset is approximately 22.5 degrees, but the C/N degradation slope versus the carrier-phase offset of TC8 PSK systems is not much different from that of QPSK systems in a practical carrier offset range of less than 10 degrees. On the other hand, uncoded 8PSK systems have a larger slope of the C/N degradation versus the carrier-phase offset than that of TC8 PSK systems. Moreover, TC8 PSK systems achieve much better P_e performance against the recovered carrier-phase offset than that of uncoded 8 PSK systems.

The carrier-phase offset is mainly caused by a narrow-band carrier filter when the received carrier frequency offset is not small enough compared with the carrier filter passband. In addition, the carrier-phase offset is caused by the phase and amplitude offset of quadrature modulators and demodulators. To cope with the former problem, automatic frequency control (AFC) for the received carrier is required, while the latter can be reduced by DSP or LSIC implementation of quadrature modulators and demodulators.

Automatic frequency control.

The adjustment of the received carrier frequency to the center frequency of the carrier filter is carried out by AFC for the received carrier to restrict the recovered carrier-phase offset. For continuous transmission, AFC is relatively easy because a single demodulator demodulates continuous signals transmitted from one transmitter. It can spend a relatively large amount of time to acquire and track the received signals.

On the other hand, realization of the carrier AFC in TDMA communication systems is rather difficult, because a single demodulator must demodulate various received burst signals transmitted from various transmitters with different frequency offsets. Ideally, it is the best way to track the carrier frequency of each burst. However, in practical TDMA systems, AFC is used to track an average frequency of all received burst signals for hardware simplicity. Therefore, inter-burst frequency offsets still exist, and they cause the recovered carrier-phase offset. In addition to this recovered carrier-phase offset, the carrier-

■ **Figure 13.** *Carrier slip performance of various carrier recovery schemes*

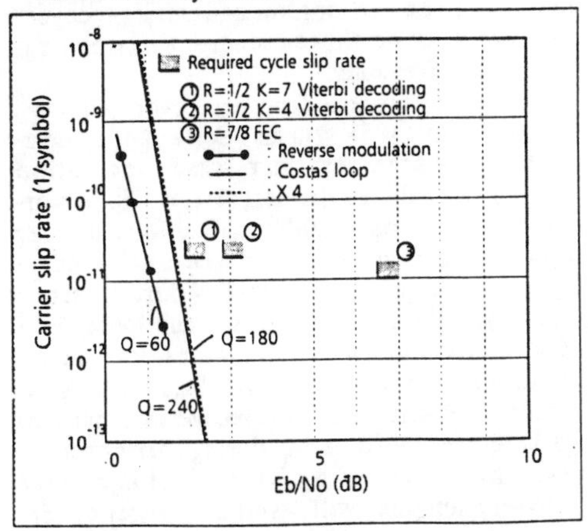

phase offset is also caused by electrical length change of the carrier recovery loop due to secular and temperature changes. When the degradation caused by the recovered carrier-phase offset is too large, a phase-compensated carrier filter is suitable for keeping the same noise bandwidth and less phase error at the output of the filter [17]. It is important to decrease the carrier-phase error because this will significantly degrade P_e performance as well as carrier slipping rate.

Carrier slip

Because coded modulation systems achieve better P_e performance than conventional modulation systems (while both achieve the same frequency utilization efficiency), the carrier recovery circuits must operate in lower C/N environments than that of conventional systems. Thus, there is the possibility of poor carrier slip rate, if the carrier filter bandwidth is not narrow enough. On the other hand, a too-narrow bandwidth carrier filter will cause a large recovered carrier-phase offset. Therefore, the carrier filter bandwidth must be chosen to meet both requirements.

How are similar problems managed in conventional systems? In typical conventional systems, such as very small aperture terminal (VSAT) or TDMA systems, QPSK modulation coupled with a rate one-half convolutional encoding and Viterbi decoding (with a constraint length (K) of 7) can achieve a P_e of 1×10^{-4} at a C/N of 4.0 dB (see Fig. 13). That is to say, a carrier recovery circuit for this QPSK modem must stably operate at a C/N of 4.0 dB.

Fortunately, threshold C/Ns of coded modems are not so low compared with that of conventional modems employing high coding gain FEC. For example, trellis-coded 8 PSK can achieve a P_e of 1×10^{-4} at a C/N of 8. 4 dB. This situation is worse by 3 dB than that for QPSK systems without forward error correction. However, its threshold C/N is still better than the above mentioned conventional system by 4.4 dB. Therefore, it is relatively easy to achieve low carrier slip rates for trellis-coded modems in continuous transmission mode by employing several techniques that have been developed for conventional systems, because a relatively long time for carrier recovery is allowed in this mode. However, this is a difficult problem for burst modem design because the carrier must be recovered in a very short time for each burst and a high S/N recovered carrier and a small amount of the carrier-phase offset must be achieved. A bandwidth (or quality factor, Q) of a carrier filter and carrier recovery schemes must be chosen to meet these contradictory requirements.

Among the proposed carrier recovery schemes, such as X4 (for QPSK systems), Costas, and reverse-modulation schemes, the reverse-modulation scheme can achieve the best carrier-slip performance with a reasonably small Q for a carrier filter [18]. This good performance is achieved by retaining the received signal S/N during carrier recovery, while other schemes typically loose 12 dB of S/N during carrier recovery for QPSK signal demodulation. Moreover, for TC8 PSK signal demodulation, the X8 carrier recovery scheme typically looses 18 dB of S/N during carrier recovery. Therefore, using X8 or Costas carrier recovery schemes will result in a larger differ-

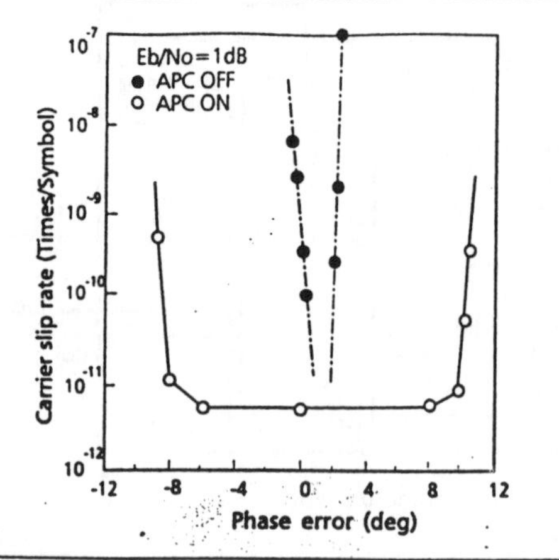

■ Figure 14. *Carrier slip performance of reverse modulation carrier recovery scheme using digital Costas APC*

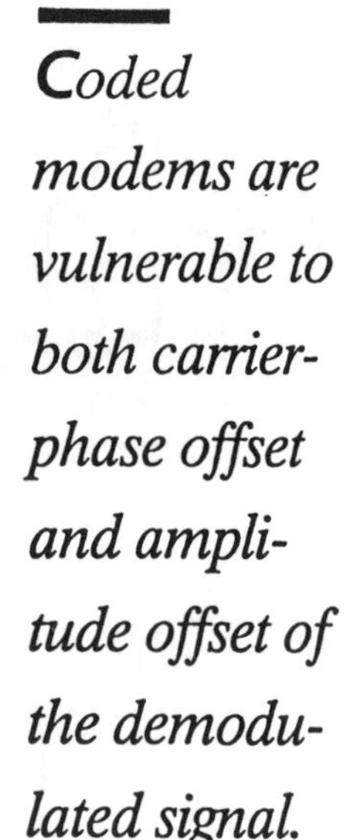

ence between the narrow bandwidth of the carrier filter used to achieve high S/N recovered carrier and to reduce carrier slip and the wide bandwidth of the carrier filter to reduce P_e degradation caused by the carrier-phase offset and to recover carrier in a short time. This difficulty can be avoided by employing the reverse-modulation type carrier recovery scheme.

Although the reverse-modulation type carrier recovery scheme can achieve the best carrier slipping performance among them in low C/N environments, it is vulnerable to recovered carrier offset which can be caused by secular and temperature changes. Therefore, in practical applications, the employment of the automatic carrier phase control (APC) technique for the recovered carrier is recommended. Carrier slip performances of the reverse modulation type carrier-recovery circuit with/without APC are shown in Fig. 14 [15].

By employing the aforementioned techniques, trellis-coded modems with low carrier slipping rates and small P_e degradation can be realized.

Automatic gain control

As we will mention in the following section, the Viterbi decoder requires five-bit soft decision, that is to say approximately ± 3% of amplitude offset is permissible. If the received signal's amplitude offset becomes 6%, approximately 0.2 dB E_b/N_0 degradation is generated at $P_e = 1\times10^{-4}$. Thus, very accurate automatic gain control (AGC) is required for coded modulation systems. This is one of the difficult problems to be overcome, especially for TDMA communication systems in which each burst may have different levels.

Viterbi Decoder

Viterbi decoders are essential components for trellis-coded demodulators and they require bulky hardware when the transmission bit rate is high. In the first part of this section, we'll take a look at the size of each component of a Viterbi decoder.

The number of states (N_s) of a Viterbi decoder is given as:

$$N_s = 2^{B\times(K-1)} \quad (1)$$

where B is the number of input bits for the encoder

In practical applications, the employment of the APC technique for recovered carrier is recommended.

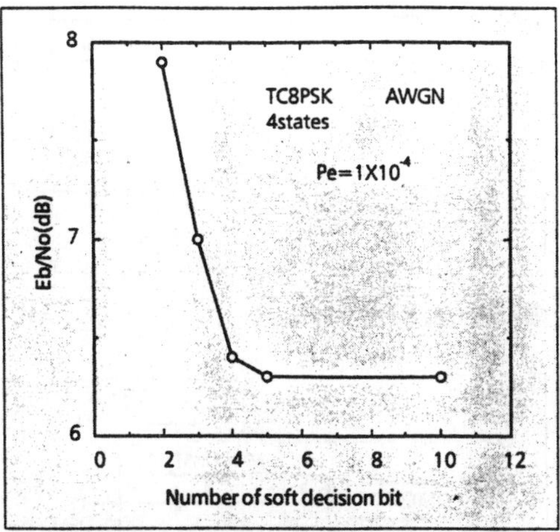

■ **Figure 15.** *Required Eb/No (Pe = 1x10⁻⁴) versus number of soft decision bits*

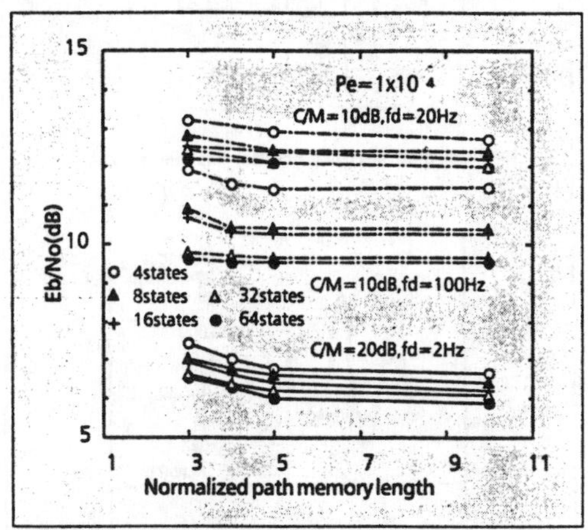

■ **Figure 16.** *Required Eb/No (Pe = 1x10⁻⁴) versus normalized path memory length by K (over Rician fading channels)*

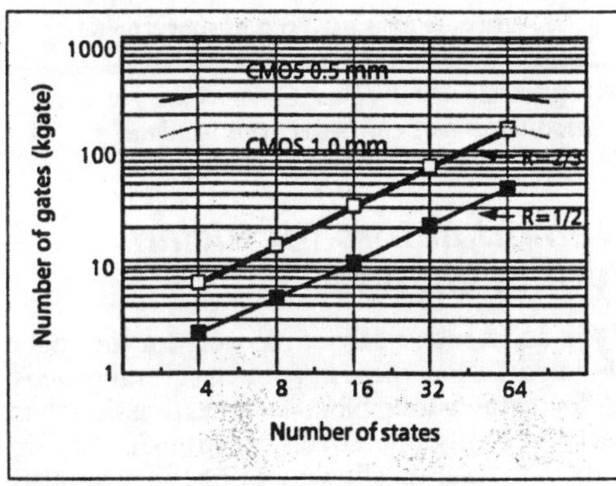

■ **Figure 17.** *Viterbi decoder hardware size (number of gates) versus number of states*

and K is the constraint length. N_s corresponds to the number of add, compare, and select (ACS) circuits which carry out addition of branch metrics, compare two branch metrics (assuming rate one-half scheme) and select one of two paths as a survivor. The number of gates in the ACS circuit, G_{acs}, is given by this equation:

$$G_{acs} = N_s x \qquad (2)$$

where, G_{add} represents the number of gates for the adder, G_{comp} represents the number of gates

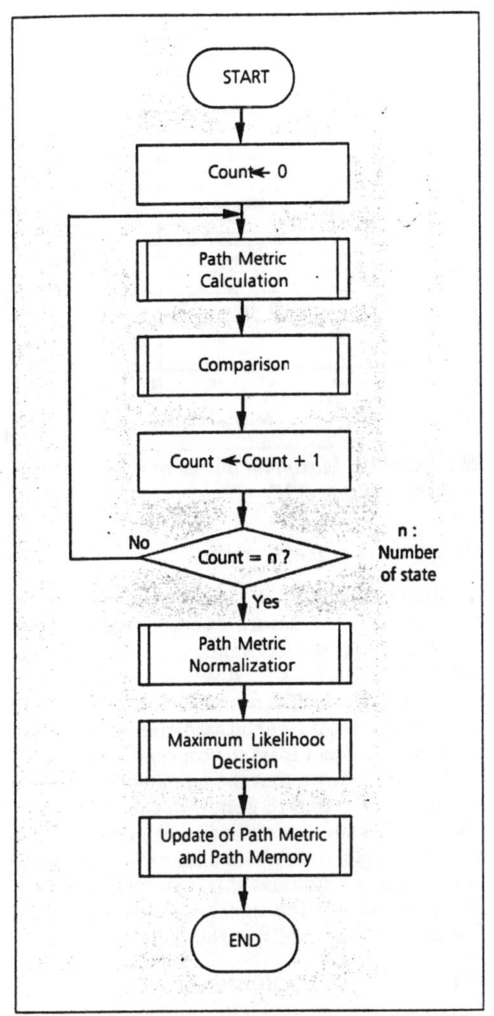

■ **Figure 18.** *Viterbi decoding algorithm for DSP implementation*

for the comparator, G_{sel} represents the number of gates for the selector and G_{reg} represents the number of gates for the path metric register.

Similarly, the number of gates of the path memory circuits, G_{pm} is:

$$G_{pm} = N_s \times PML \times G_{pmc} \qquad (3)$$

where, PML represents the path memory length and G_{pmc} represents the number of gates for each path memory cell. Therefore, the total number of gates needed for a Viterbi decoder is:

$$G_{total} = G_{acs} + G_{pm} \qquad (4)$$

If you assume a coding rate of one-half and a constraint length of seven for simplicity, then the ACS circuits and the path memory circuits each share approximately 50 percent of the total Viterbi decoder hardware.

As seen from Equation (1), the ACS circuit hardware increases in proportion to the branch metric bits, and the path memory circuit hardware size varies according to the number of states and the path memory length. Therefore, it is very important to choose optimum parameters to design minimum-hardware Viterbi decoders for coded modulation.

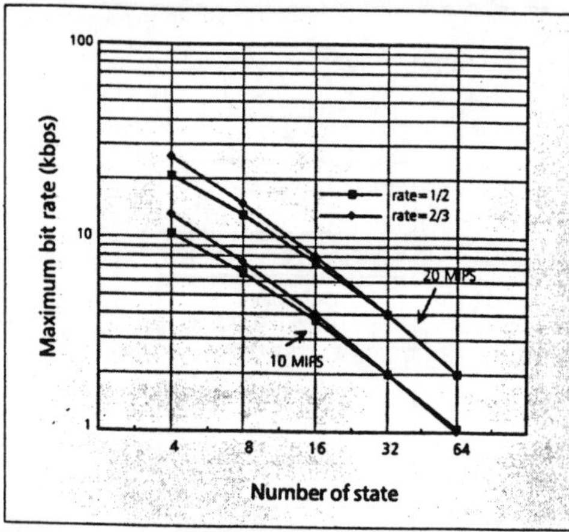

■ **Figure 19.** *Maximum bit rate of Viterbi decoders using DSP*

Parameter optimization

As shown in Fig. 15, P_e performance improvement in TC8 PSK saturates at the five soft decision bits. Because a conventional Viterbi decoder for QPSK systems in AWGN environments requires three soft decision bits, five soft decision bits increase the ACS circuit hardware size by approximately 70 percent from the conventional size.

It is well known that the minimum path memory length needed to generate the theoretical coding gain is approximately five to six times the constraint length in AWGN channels. Computer simulation results show this is also valid for trellis-coded modulation signal transmission even in Rician fading channels, as shown in Fig. 16.

DSP and LSIC implementations

Because satellite communications are both power and frequency-band limited, forward error correction has been used in various systems. A number of LSIC implementation of relatively high-speed Viterbi decoders and DSP implementation of relatively low-speed Viterbi decoders have been carried out for satellite and other communication systems.

For high-speed applications, Viterbi decoders of up to 25 MHz with a constraint length of seven and coding rates from one-half to fifteen-sixteenths are commercially available [19-20]. A new Viterbi decoder configuration for high-speed operation was proposed, and a prototype Viterbi decoder offering a 600 Mb/s per chip operation has been fabricated [21-22]. The number of gates versus number of states for trellis-coded 8 PSK modulation (coding rate = 2/3) is shown in Fig. 17, with that for one-half rate Viterbi decoders as a reference. Constructing a two-thirds rate Viterbi decoder with a constraint length of four in a single chip LSIC is possible by employing CMOS 1.0 mm-process, and CMOS 0.5 mm-process LSI technologies will be used in the near future for decoders with larger number of states.

A number of DSP implementations have been carried out for low-speed Viterbi decoders. The Viterbi decoding flow for DSP implementations is shown in Fig. 18, and the maximum bit rate of a Viterbi decoder realized as a single DSP is shown in Fig. 19. A Viterbi decoder with a single DSP chip can operate at up to 20 kb/s four states and two kb/s with 64 states.

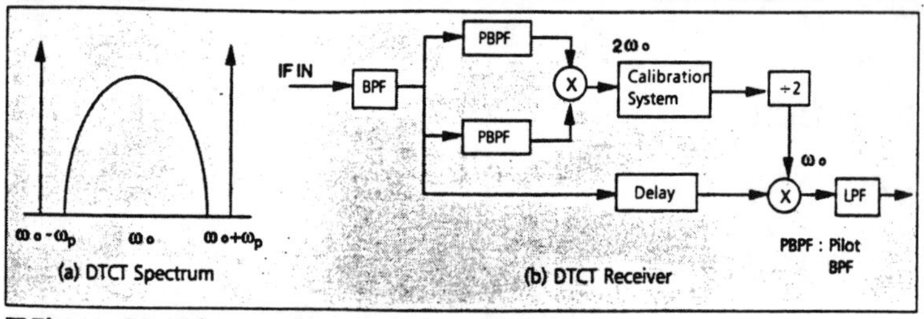

■ **Figure 20.** *Pilot signal aided coherent demodulation*

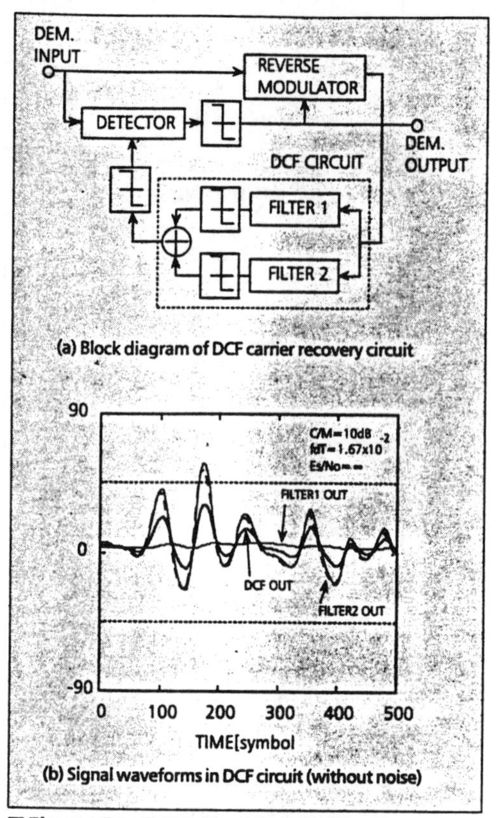

■ **Figure 21.** *DCF (Dual Carrier Filter) reverse modulatin type carrier recovery scheme*

Demodulators for Rician fading channel

To assist robust carrier recovery for the coded modulation, there have been several proposals for methods to employ pilot tones to assist robust carrier recovery for the trellis-coded modulation signals. An diagram showing pilot tones on both sides of the main signal spectrum (dual tone calibration technique (DTCT)) is shown in Fig. 20 [9]. In these systems, received unmodulated carriers are passed through a narrow-band filter to increase signal-to-noise ratio and used to detect received signals as shown in Fig. 20b. An alternative method, a dual carrier filter (DCF) reverse modulation type carrier recovery circuit, has been proposed for robust coherent detection in Rician fading environment [23]. This carrier recovery circuit achieves lower carrier slip rates and better P_e performance than conventional coherent detection. A diagram of DCF reverse modulation carrier recovery circuit and its carrier tracking performance are shown in Fig. 21.

In Rician or Rayleigh fading channels, one promis-

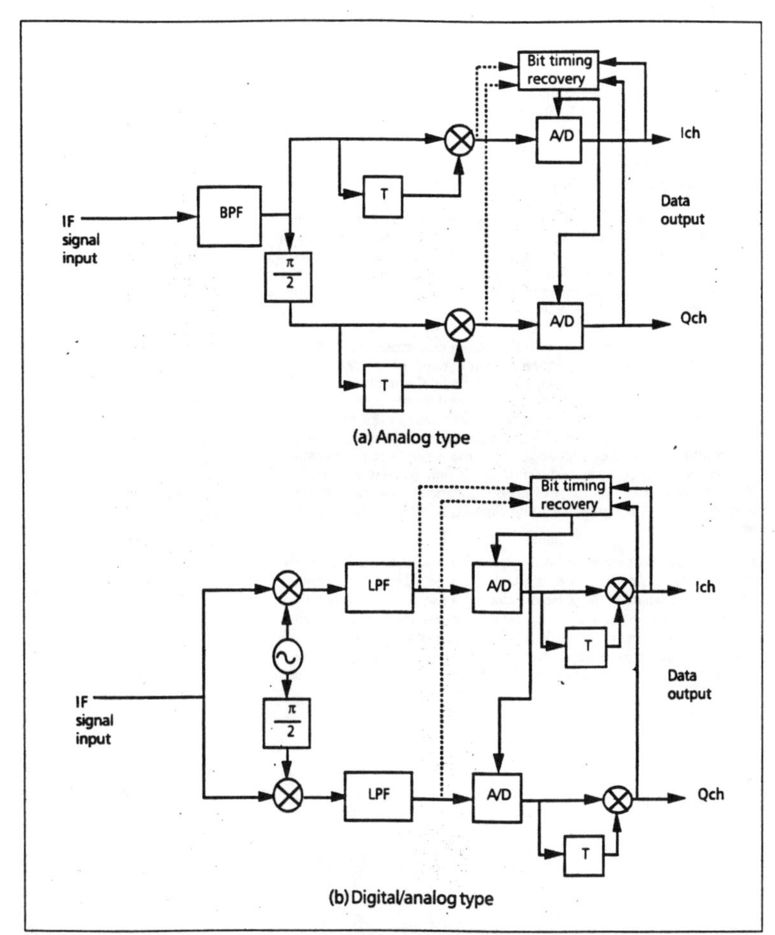

■ **Figure 22.** *Block diagram of a differential demodulator*

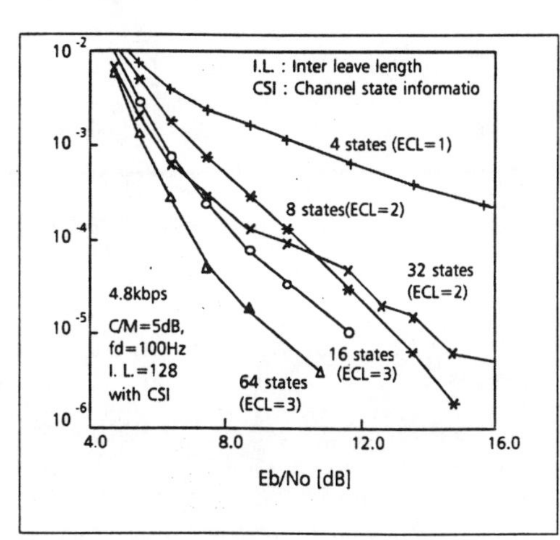

■ **Figure 23.** *Pe performance of TC8 PSK in Rician fading channels*

ing way to achieve good P_e performance is the use of differential encoding and differential detection [24]. This differential detection requires no carrier recovery circuits, and hardware implementation of demodulators is simpler, as shown in Fig. 22.

In AWGN environments, a Viterbi decoder with the larger number of states, N_S, produces higher coding gain because the larger number of states increases Euclidian distances. However, in Rician or Rayleigh fading channels, P_e performance is dominated not by the number of states but by the effective code length (ECL), as shown in Fig. 23 [25]. For example, P_e performance of TC8 PSK systems with 16 states for coding and decoding is better than that of TC8 PSK systems with 32 states in Rician fading channels(C/M = 5 dB). For minimizing hardware and achieving better P_e performance, this fact must be taken into account for designing coded modems operating in fading channels.

Conclusion

Trellis-coded modems naturally require more complicated hardware designs and more accurate components than ordinary modems. However, recent advances in digital technologies such as LSIC implementation technologies, DSP technologies, and so on, have made it possible to deploy more spectrally-efficient modulation schemes in various fields. In Rician and Rayleigh fading environments, which are typical for mobile satellite and cellular mobile systems, differential detection schemes have been studied. However, from a hardware perspective, with recent advances in digital signal-processing technologies, there are no significant difficulties in employing coherent demodulation. Application of coded modems to TDMA communication systems will be rather difficult because their vulnerability to received signal impairments, but these problems will be solved with advanced digital signal-processing technologies in the near future.

Acknowledgments

The authors would like to thank Dr. H. Yamamoto and Dr. S. Samejima of NTT Radio Communication Systems Laboratories, and to to Mr. T. Dohi and K. Kobayashi of NTT Radio Communication Systems Laboratories for their helpful discussions and guidance.

References

[1] T. Aulin and C. E. Sundberg: "Continuous phase modulation, Part I: Full response signaling," *IEEE Trans. Communications*, vol. COM-29, No. 3,pp.196-209, Mar. 1981.
[2] H. Miyakawa, H. Harashima, and Y. Tanaka: "A new digital modulation scheme, multi-code binary CPFSK," Proc. 3rd Int. Conf. on Digital satellite Communications, pp. 105-112, Nov. 1975.
[3] G. Ungerboeck: "Channel coding with multilevel/phase signals," *IEEE Trans. Information Theory*, vol. IT-28, pp. 55-67, Jan. 1982.
[4] CCITT Study Group XVII: "Recommendation V.32 for a family of 2-wire, duplex modems operating on the general switched telephone network and on leased telephone-type circuits," Document AP VII-43-E May 1984.
[5] CCITT Study Group XVII: "Draft recommendation V. 33 for 14400 bits per second modem standardized for use on point-to-point 4-wire leased telephone-type circuits," Circular No. 12, COM XVII/YS, Geneva, May 17, 1985.
[6] G. Ungerboeck, J. Hagenauer and T. Abdel Nabi: "Coded 8-PSK experimental modem for the INTELSAT SCPC system," Proc. 7th Int. Conf. on Digital Satellite Communications, pp. 299-304, May, 1986.
[7] S. A. Rhodes, R. J. Fang and P. Y. Chang: "Coded octal phase shift keying in TDMA satellite communications," *COMSAT Technical Review*, vol. 13 No. 2, Fall 1983.
[8] T. Fujino, Y. Moritani, M. Miyake, K. Murakami, Y. Sakato and H. Shiino: "A 120 Mb/s 8PSK modem with soft-Viterbi decoding," Proc. 7th Int. Conf. on Digital Satellite Communications pp.315-321, May 1986.
[9] W. Rafferty and D. Divsalar: "Modulation and coding for mobile satellite channels," in Conf. Rec., Int. Conf. Commun., 1988, pp. 1105-
[10] M. L. Moher and J. Lodge: "TCMP-A modulation and coding strategy for Rician fading channels," *IEEE Trans. Selected Areas in Comm.*, vol. SAC-7, pp.1347-355, December 1989.
[11] H. Ohtsuka, Y. Saito and S. Komaki: "Super multi-carrier trellis-coded 256 QAM digital microwave radio," in Conf. Rec., Global Commun., 1988, pp. 244-249.
[12] G. Ungerboeck: "Trellis-coded modulation with redundant signal sets Part I: Introduction," *IEEE Communications Magazine*, vol. 25, pp.5-11, February 1987.
[13] G. Ungerboeck: "Trellis-coded modulation with redundant signal sets Part II: State of the art," *IEEE Communications Magazine*, vol. 25, pp.12-21, February 1987.
[14] J. Du and M. Kasahara: "Improvement of the information bit

error rate of trellis code modulation systems," *IEICE Trans.*, vol. E 72, pp. 609-614, May 1989 (Japan).

[15] S. Kato, M. Morikura, M. Umehira, K. Enomoto and S. Kubota: "Application of advanced microelectronics to large-scale communication equipment - Compact and maintenance-free TDMA equipment," *IEEE Trans. Selected Areas in Comm.*, vol. SAC-8, pp.1551-1564, October 1990.

[16] M. Umehira, H. Kikuchi, S. Konaka and S. Kato: "High-speed precise monolithic multiplier with radiation hardness using silicon bipolar SST," *IEEE Electron. Lett.*, vol. 22, No. 14, 1986.

[17] M. Umehira, S. Kubota, K. Enomoto and S. Kato: "Compact LSI-and MMIC-implemented burst modem for low Eb/No operation," in Conf. Rec., Global Commun., 1987, pp.268-2273.

[18] S. Kato, T. Arita and K. Morita: "Onboard digital signal processing technologies for present and future TDMA and SCPC systems," *IEEE Trans. Selected Areas in Comm.*, vol. SAC-5, pp.685-700, May 1987.

[19] T. Ishitani, K. Tansho, N. Miyahara, S. Kubota and S. Kato: "A Scarce-State-Transition Viterbi-decoder VLSI for bit error correction," *IEEE Trans. Solid-State Circuit*, vol. SC-22, pp.575-582, August 1987.

[20] S. Kubota, S. Kato and T. Ishitani: "Novel Viterbi decoder VLSI implementation and its performance," *IEEE Trans. Communication*, scheduled for publication 1991.

[21] G. Fettweis and H. Meyer: "High-speed parallel Viterbi decoding: Algorithm and VLSI-architecture," *IEEE Communication Magazine*, pp. 46-55, May 1991.

[22] G. Fettweis, H. Dawid and H. Meyer: "Minimized method Viterbi decoding: 600 Mb/s per chip," in Conf. Rec., Global Commun., 1990, pp. 1712-1716.

[23] K. Kobayashi, T. Sakai, S. Kubota, M. Morikura and S. Kato: "A new carrier recovery circuit for land mobile satellite communication," Proc. of International Symposium on Personal, Indoor and Mobile Radio Communications, September, 1991.

[24] M. K. Simon, D. Divsalar: "Doppler-corrected differential detection of MPSK," *IEEE Trans. Communication*, vol. 37, pp. 99-109, February 1989.

[25] D. Divsalar and M. K. Simon: "The design of trellis-coded MPSK for fading channels: performance criteria," *IEEE Trans. Commun.*, COM-36, pp.1004-1012, Sep. 1988.

Biography

Shuzo Kato is a group leader for signal processing, NTT Radio Communication Systems Laboratories. He has been serving as a secretary of Satellite and Space Communications Committee, COMSOC, IEEE. Shuji Kubota and Masahiro Morikura are senior research engineers in the satellite communication/signal-processing group of NTT Radio Communication Systems Laboratories.

Masahiro Morikura was born in Osaka, Japan on October 26, 1956. He received the B.E. and M.E. degrees from Kyoto University, Japan in 1979 and 1981. He has been with NTT Electrical Communications Laboratories since 1981, working on research and development of TDMA equipment for satellite communications. From 1988 to 1989, he was with the Communications Research Center, Canada, as a Guest Scientist. He is currently a senior research engineer in the satellite communication signal processing group of radio communication systems laboratories, NTT. Since joining NTT in 1981, he has been engaged in research and development of TDMA equipment and synchronization for satellite communications systems.

Shuji Kubota was born in Tokyo, Japan on January 2, 1957. He received the B.E. degree from the University of Electro-Communications, Tokyo, Japan in 1980. He is currently a senior research engineer in the satellite communication signal processing group of radio communication systems laboratories, NTT. Since joining NTT in 1980, he has been engaged in research and development of forward error correction schemes and modulation schemes for satellite communication systems.

Chapter 8

ADAPTIVE ARRAYS AND SPATIAL FILTERING

This chapter presents adaptive antenna arrays and emerging spatial filtering concepts that will certainly shape the wireless industry in the years to come.

Reprinted from *IEEE Transactions on Vehicular Technology*, Vol. 42, No. 4, Nov. 1993, pp. 377-384.

Signal Acquisition and Tracking with Adaptive Arrays in the Digital Mobile Radio System IS-54 with Flat Fading

Jack H. Winters, *Senior Member*

***Abstract*— This paper considers the dynamic performance of adaptive arrays in wireless communication systems. With an adaptive array, the signals received by multiple antennas are weighted and combined to suppress interference and combat desired signal fading. In these systems, the weight adaptation algorithm must acquire and track the weights even with rapid fading. Here, we consider the performance of the Least-Mean-Square (LMS) and Direct Matrix Inversion (DMI) algorithms in the North American digital mobile radio system IS-54. We show that implementation of these algorithms permits the use of coherent detection, which improves performance by 1 dB over differential detection. Results for two base station antennas with flat Rayleigh fading show that the LMS algorithm has large tracking loss for vehicle speeds above 20 mph, but the DMI algorithm can acquire and track the weights to combat desired signal fading at vehicle speeds up to 60 mph with less than 0.2 dB degradation from ideal performance with differential detection. Similarly, interference is also suppressed with performance gains over maximal ratio combining within 0.5 dB of the predicted ideal gain.**

I. Introduction

ANTENNA arrays with optimum combining reduce the effects of multipath fading of the desired signal and suppress interfering signals, thereby increasing both the performance and capacity of wireless systems. To be practical, though, the implemented combining algorithms must be able to rapidly acquire and track the desired and interfering signals.

Most previous theoretical and computer simulation studies of the increase in performance and capacity with optimum combining, e.g., [1]–[6], assumed ideal tracking of the desired and interfering signals. In the computer simulation study where block-by-block adaptation was considered [7], the data rate was at least 5 orders of magnitude greater than the fading rate. Although this is appropriate for the indoor radio system studied in [7] which used kbps data rates at 900 MHz, digital mobile radio systems have a much lower data-to-fading-rate ratio. For example, in the North American digital cellular system IS-54 [8] with a data rate of 24.3 ksymbols/s in the 800 MHz band, at 60 mph the data-to-fading ratio is only 300, while in the Western European GSM [8] it is around 2 000. In a previous experiment [4]–[6] that demonstrated the feasibility of optimum combining with a three-fold increase in capacity (suppression of two equal-power interferers with eight antennas), the Least-Mean-Square (LMS) algorithm tracked these signals with data-to-fading-rate ratios as low as 25. However, the tracking error loss could not be measured because of A/D quantization noise. Furthermore, this experimental system had many more antennas than interferers, which is not typical of most wireless systems.

Here we consider the dynamic performance of adaptive arrays in wireless communication systems. Specifically, we consider the performance of the LMS and Direct Matrix Inversion (DMI) algorithms in tracking the desired and interfering signals in the digital mobile radio system IS-54. We show that implementation of these algorithms permits the use of coherent detection, which improves performance by 1 dB over differential detection. Results for two base station antennas and flat Rayleigh fading show that the LMS algorithm has large tracking loss at speeds above 20 mph. However, the DMI algorithm can acquire and track the weights to combat desired signal fading at vehicle speeds up to 60 mph with less than 0.2 dB degradation from the ideal (perfect tracking) performance of optimum combining with differential detection. Similarly, interference is also suppressed with performance gains over maximal ratio combining within 0.5 dB of the predicted ideal gain.

In Section II, we determine the performance of optimum combining with ideal signal tracking. In Section III we study the performance of the LMS and DMI algorithms for acquisition and tracking of the signals in IS-54. A summary and conclusions are presented in Section IV.

II. Ideal Performance

A. Weight Equation

Fig. 1 shows a block diagram of an M antenna element adaptive array. The complex baseband signal received by the ith element in the kth symbol interval $x_i(k)$ is multiplied by a controllable complex weight $w_i(k)$. The weighted signals are then summed to form the array output $s_o(k)$. The output signal is subtracted from a reference signal $r(k)$ (described in Section III) to form an error signal $\epsilon(k)$. Weight generation circuitry determines the weights from the received signals and the error signal. In this paper, we are interested in determining the weights that minimize the mean-square error, i.e., $|\epsilon^2(k)|$.

Let the weight vector $\boldsymbol{w}$ be given by

$$\boldsymbol{w} = [w_1 w_2 \cdots w_M]^T \tag{1}$$

Manuscript received September 8, 1992; revised October 26, 1992.
The author is with AT&T Bell Laboratories, Holmdel, NJ 07733.
IEEE Log Number 9211016.

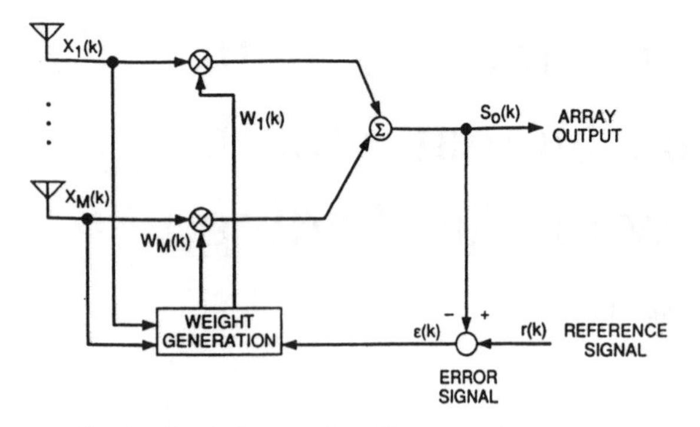

Fig. 1. Block diagram of an M element adaptive array.

where the superscript T denotes transpose, and the received signal vector $\boldsymbol{x}$ is given by

$$\boldsymbol{x} = [x_1 x_2 \cdots x_M]^T. \tag{2}$$

The received signal consists of desired signal, thermal noise, and interference and, therefore, can be expressed as

$$\boldsymbol{x} = \boldsymbol{x}_d + \boldsymbol{x}_n + \sum_{j=1}^{L} \boldsymbol{x}_j \tag{3}$$

where $\boldsymbol{x}_d$, $\boldsymbol{x}_n$, and $\boldsymbol{x}_j$ are the received desired signal, noise, and jth interfering signal vectors, respectively, and L is the number of interferers. Furthermore, let $s_d(k)$ and $s_j(k)$ be the desired and jth interfering signals, with

$$E[s_d^2(k)] = 1 \tag{4}$$

$$E[s_j^2(k)] = 1 \quad \text{for } 1 \leq j \leq L. \tag{5}$$

Then $\boldsymbol{x}$ can be expressed as

$$\boldsymbol{x} = \boldsymbol{u}_d s_d(k) + \boldsymbol{x}_n + \sum_{j=1}^{L} \boldsymbol{u}_j s_j(k) \tag{6}$$

where $\boldsymbol{u}_d$ and $\boldsymbol{u}_j$ are the desired and jth interfering signal propagation vectors, respectively.

The received signal (desired-signal-plus-interference-plus-noise) correlation matrix is given by

$$\boldsymbol{R}_{xx} = E\left[\left(\boldsymbol{x}_d + \boldsymbol{x}_n + \sum_{j=1}^{L} \boldsymbol{x}_j\right)^* \left(\boldsymbol{x}_d + \boldsymbol{x}_n + \sum_{j=1}^{L} \boldsymbol{x}_j\right)^T\right] \tag{7}$$

where the superscript * denotes complex conjugate and the expectation is taken with respect to the signals $s_d(k)$ and $s_j(k)$. Assuming the desired signal, noise, and interfering signals are uncorrelated, the expectation is evaluated to yield

$$\boldsymbol{R}_{xx} = \boldsymbol{u}_d^* \boldsymbol{u}_d^T + \sigma^2 \boldsymbol{I} + \sum_{j=1}^{L} \boldsymbol{u}_j^* \boldsymbol{u}_j^T \tag{8}$$

where σ^2 is the noise power and $\boldsymbol{I}$ is the identity matrix. Note that $\boldsymbol{R}_{xx}$ varies with the fading and that we have assumed that the fading rate is much less than the bit rate.

We define the received desired signal to noise ratio ρ as

$$\rho = \frac{E[|u_{di}|^2]}{\sigma^2} \qquad i = 1, \cdots, M \tag{9}$$

the interference-to-noise ratio (INR) as

$$\text{INR} = \frac{E[|u_{ji}|^2]}{\sigma^2} \qquad i = 1 \text{ to } M,\ j = 1 \text{ to } L \tag{10}$$

and the signal-to-noise-plus-interference ratio (SINR) as

$$\text{SINR} = \frac{\rho}{1 + \text{INR} \cdot L} \tag{11}$$

where u_{di} and u_{ji} are the ith elements of $\boldsymbol{u}_d$ and $\boldsymbol{u}_j$, respectively, and the expected value now is with respect to the propagation vectors.

The equation for the weights that minimize the mean-square error (and maximize the output SINR) is [9]

$$\boldsymbol{w} = \boldsymbol{R}_{xx}^{-1} \boldsymbol{u}_d^* \tag{12}$$

where the superscript -1 denotes the inverse of the matrix. Note that scaling of the weights by a constant does not change the output SINR. In (12), we have assumed that $\boldsymbol{R}_{xx}$ is nonsingular so that $\boldsymbol{R}_{xx}^{-1}$ exists. If not, we can use pseudoinverse techniques [10] to solve for $\boldsymbol{w}$. These optimum-combining weights are the same as those in [5], as shown in Appendix A.

B. Optimum Combiner Performance

We determine the performance of ideal optimum combining in the digital mobile radio system IS-54 in the following manner. We first determine the bit error rate (BER) with the IS-54 modulation technique, $\pi/4$-shifted differential quadrature phase shift keying (DQPSK), for ideal maximal ratio combining. With maximal ratio combining, the received signals are combined to maximize the signal-to-noise ratio at the array output, which is the optimum combining algorithm when interference is not present. Analytical results are presented for both differential detection and coherent detection, since both cases are studied in Section III. We then determine the reduction in the receive SINR required for a given BER, with optimum combining as compared to maximal ratio combining when interference is present. This gain with optimum combining is determined using analytical results with one interferer and Monte Carlo simulation with $L \geq 2$. Although these gains are generated only for coherent detection of binary PSK (BPSK), these results are also applicable to both coherent and differential detection of DQPSK, but at different BER's. This is because, for given receive SINR, the output SINR with optimum combining is independent of the modulation and detection technique, as can be seen from the equations in Section II-A. Thus the gain with optimum combining and BPSK at a given receive SINR will be similar to that of DQPSK at the same receive SINR—only the corresponding BER will be different. Note that IS-54 specifies a maximum BER of 2×10^{-2}, and, therefore, our results are generated for BER's around 10^{-2}.

With coherent detection of DQPSK and maximal ratio combining, the average BER with flat Rayleigh fading is approximately given by

$$\overline{BER} = 2P_E - P_E^2 \tag{13}$$

where (from [11])

$$P_E = 2^{-M}\left(1 - \sqrt{\frac{\rho}{2+\rho}}\right)^M \cdot \sum_{k=0}^{M} \binom{M-1+k}{k} 2^{-k}\left(1 + \sqrt{\frac{\rho}{2+\rho}}\right)^k. \tag{14}$$

With differential detection of DQPSK and maximal ratio combining, the average BER with flat Rayleigh fading can be shown to be given by

$$\overline{BER} = \int_0^\infty p(\gamma) P_E(\gamma)\, d\gamma \tag{15}$$

where $P_E(\gamma)$ is given by (from [12])

$$P_E(\gamma) = e^{-\gamma}\left[\frac{1}{2} I_0(\gamma/\sqrt{2}) + \sum_{k=1}^{\infty} (\sqrt{2}-1)^k I_k(\gamma/\sqrt{2})\right] \tag{16}$$

where I_k is the kth order modified Bessel function of the first kind, and $p(\gamma)$ is the probability density of the signal-to-noise ratio after maximal ratio combining and is given by [13]

$$p(\gamma) = \frac{\gamma^{M-1} e^{-\gamma/\rho}}{\rho^M (M-1)!}. \tag{17}$$

Note that with differential detection of differential BPSK (DBPSK), the average BER is $1/2(1+\rho)^{-M}$.

Fig. 2 shows the average bit error rate versus ρ (SINR with INR $= -\infty$ dB) for DQPSK and $M = 1$, 2 with maximal ratio combining. Results are also shown for DBPSK, which requires 3 dB lower ρ for the same BER with coherent detection. Note that for $M = 1$, with both DQPSK and DBPSK, differential detection requires a 0.4 dB higher ρ for a given BER than coherent detection. For $M = 2$, differential detection of DBPSK requires a 0.7 dB higher ρ than coherent detection, while differential detection of DQPSK requires a 1.0 dB higher ρ than coherent detection. Differential detection of DQPSK requires a 11.2 dB SINR for a 10^{-2} BER.

Now, let us consider the BER with ideal optimum combining. The BER with optimum combining and flat Rayleigh fading in the presence of noise only is given by the results above for maximal ratio combining. With one interferer that also experiences flat Rayleigh fading, the BER for coherent detection of BPSK is given by [1, eq. (25)]. For multiple interferers with flat Rayleigh fading, this BER can be determined by Monte Carlo simulation as described in [1].

Fig. 3 shows the gain in dB of ideal optimum combining over maximal ratio combining with two antennas for one to six interferers versus the interference-to-noise ratio (INR). This gain was determined from the reduction in the required SINR for a 10^{-3} BER at the receiver with coherent detection of BPSK. The gain occurs because optimum combining is suppressing interference in addition to increasing desired signal-to-noise ratio. A 10^{-3} BER was chosen because the

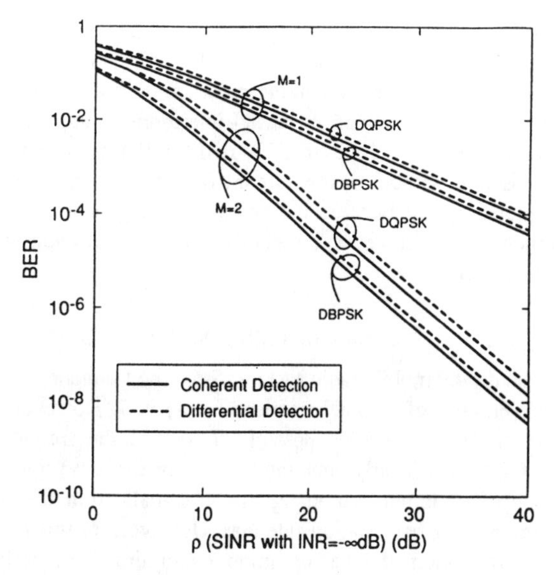

Fig. 2. Average BER versus E_b/N_0 for coherent and differential detection of DQPSK and DBPSK with $M = 1$ and 2 with flat Rayleigh fading and maximal ratio combining.

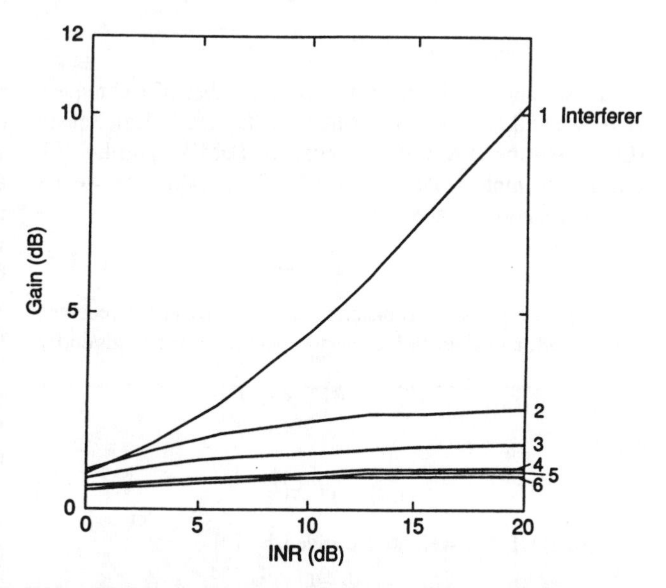

Fig. 3. Gain in dB of ideal optimum combining over maximal ratio combining with two antennas for one to six interferers versus INR at a 10^{-3} BER for coherent detection of BPSK.

11.1 dB SINR required for a 10^{-3} BER with maximal ratio combining and coherent detection of BPSK [11], [14] is close to the 11.2 dB SINR required for a 10^{-2} BER with maximal ratio combining and differential detection of DQPSK. Thus Fig. 3 also shows the gain for a 10^{-2} BER with differential detection of DQPSK. As shown in [1], the gain does not vary significantly for BER's between 10^{-2} and 10^{-3}.

With two antennas, optimum combining can completely suppress one interferer. Thus the maximum gain with optimum combining and one interferer is $10\log_{10}(10^{\mathrm{INR}/10} + 1)$ dB, which is approximately INR for large INR. However, this gain can only be achieved without desired signal fading. With fading, as shown in [4]–[6], the complete suppression of one interferer results in the loss of one order of diversity against multipath fading, which corresponds to a 12.9 dB increase in the SINR required at a 10^{-3} BER [11], [14]. Thus to achieve gain, optimum combining must trade off a partial loss in diversity improvement for partial interference suppression. The resulting gain is approximately half (in dB) the maximum gain possible without desired signal fading. With more than one interferer and two receive antennas, the gain is seen to be much lower. However, the gain is almost 1 dB even with six interferers.

III. Performance of LMS and DMI in IS-54

In the digital mobile radio system IS-54, the frequency reuse factor (number of channel frequency sets) is 7. However, as shown in [6], it may be possible to reduce the frequency reuse factor to 4 (nearly doubling the system capacity) through the use of optimum combining of the signals from the two existing receive base station antennas. However, for this result in [6], we assumed ideal optimum combining, i.e., perfect tracking of the desired and interfering signals by the combining algorithm at the base station. Below, we consider the dynamic performance of optimum combining in IS-54.

A. Weight Generation

The weights can be calculated by a number of techniques. Here, we will consider two techniques: the Least Mean Square (LMS) and the Direct Matrix Inversion (DMI) algorithm [9]. For digital implementation of the LMS algorithm, the weight update equation is given by

$$\boldsymbol{w}(k+1) = \boldsymbol{w}(k) + \mu \boldsymbol{x}^*(k)\epsilon(k) \tag{18}$$

where μ is a constant adjustment factor, $x(k)$ is the received signal vector in the kth bit interval, and the error is given by

$$\epsilon(k) = r(k) - s_0(k) \tag{19}$$

where

$$s_0(k) = \boldsymbol{w}^T \boldsymbol{x}(k). \tag{20}$$

With DMI, the weights are given by [9]

$$\boldsymbol{w} = \hat{\boldsymbol{R}}_{xx}^{-1} \boldsymbol{r}_{xd} \tag{21}$$

where the estimated receive signal correlation matrix is given by

$$\hat{\boldsymbol{R}}_{xx} = 1/K \sum_{j=1}^{K} \boldsymbol{x}^*(j)\boldsymbol{x}^T(j) \tag{22}$$

where K is the number of samples used, and the estimated reference signal correlation vector is given by

$$\hat{\boldsymbol{r}}_{xd} = 1/K \sum_{j=1}^{K} \boldsymbol{x}^*(j) r(j). \tag{23}$$

Note that, as before, we have assumed that $\hat{\boldsymbol{R}}_{xx}$ is nonsingular. If not, pseudoinverse techniques can be used [10].

The LMS algorithm is the least computationally-complex weight adaptation algorithm. However, the rate of convergence to the optimum weights depends on the eigenvalues of $\boldsymbol{R}_{xx}$, i.e., on the power of the desired and interfering signals [9]. Thus weaker interference will be acquired and tracked at a slower rate than the desired signal, and the desired signal will be tracked at a slower rate during a fade (when accurate tracking is most important).

The DMI algorithm is the most computationally-complex algorithm because it involves matrix inversion. However, DMI has the fastest convergence, and the rate of convergence is independent of the eigenvalues of $\boldsymbol{R}_{xx}$, i.e., signal power levels. One issue with the DMI algorithm is its modification for tracking time-varying signals. Here we consider calculating the weights at each symbol interval using one of two data weighting functions: 1) a sliding window (fixed K in (22) and (23)) and 2) an exponential forgetting function on $\hat{\boldsymbol{R}}_{xx}$ and $\hat{\boldsymbol{r}}_{xd}$, namely,

$$\hat{\boldsymbol{R}}_{xx}(k+1) = \beta \hat{\boldsymbol{R}}_{xx}(k) + \boldsymbol{x}^*(k)\boldsymbol{x}^T(k) \tag{24}$$

$$\hat{\boldsymbol{r}}_{xx}(k+1) = \beta \hat{\boldsymbol{r}}_{xd}(k) + \boldsymbol{x}^*(k) r(k) \tag{25}$$

where β is the forgetting factor.

For $M = 2$, the DMI algorithm has about the same computational complexity as the LMS algorithm. In particular, weight calculation from the inversion of the 2×2 correlation matrix (21) does not even require division by the determinant, since this is only a weight scale factor that does not affect the output SINR. For larger M, since the complexity of matrix inversion grows with M^3 (versus M for LMS), DMI becomes very computation intensive. However, the matrix inversion can be avoided by using recursive techniques based on least-square estimation or Kalman filtering methods [9], which greatly reduce complexity (to the order of M^2) but have performance that is similar to DMI [9]. Similarly, pseudoinverse techniques [10] can be used if $\hat{\boldsymbol{R}}_{xx}^{-1}$ does not exist. Therefore, our performance results for DMI should also apply to these recursive techniques.

Next, consider reference signal generation. Since this signal is used by the adaptive array to distinguish between the desired and interfering signals, it must be correlated with the desired signal and uncorrelated with any interference. Now, the digital mobile radio system IS-54 [8] uses time division multiple access (TDMA), with three user signals in each channel and each user transmitting two blocks of 162 symbols in each frame. For mobile to base transmission, each block includes a 14-symbol synchronization sequence starting at the 15th symbol. This sequence is common to all users in a given time slot (block), but is different for each of the six time slots per frame. Since base stations operate asynchronously, signals from other cells have a high probability of having different timing (since there are 972 symbols per frame) and being uncorrelated with the sequence in the desired signal. Thus as proposed in [6], for weight acquisition we will use the known 14-symbol synchronization sequence as the reference signal. DMI is used to determine the initial weights using this sequence, since accurate initial weights are required. Note that the weights must be reacquired for each block, because with a 24.3 k symbols/s data rate and fading rates as high as 81 Hz, the fading may change completely between blocks received from a given user. After weight acquisition, the output signal consists mainly of the desired signal, and (during proper operation) the data is detected with a bit error rate that is not more than 10^{-2} to 10^{-1}. Thus we can use the detected data as the reference signal, using either the LMS or DMI algorithm for tracking.[1] In our simulation results shown below, we did not consider the effect of data errors on the reference signal; i.e., the reference signal symbols were the same as the transmitted symbols.

Note that since the modulation technique is DQPSK, the error of interest is only the relative phase between adjacent symbols, rather than the error vector $r(k) - s_o(k)$ in (19). Indeed, the LMS algorithm can use the phase error of each symbol, i.e., $/r(k) - /s_o(k)$, where $/y$ is the phase of y, as the error signal.[2] This results in no amplitude control of $s_o(k)$, but the amplitude is not used for DQPSK detection anyway. However, we found better tracking with the error vector (19) and, therefore, used (19) for our results shown below. Note that with the DMI algorithm we do not have the option of using the phase error—we must use the error vector (19).

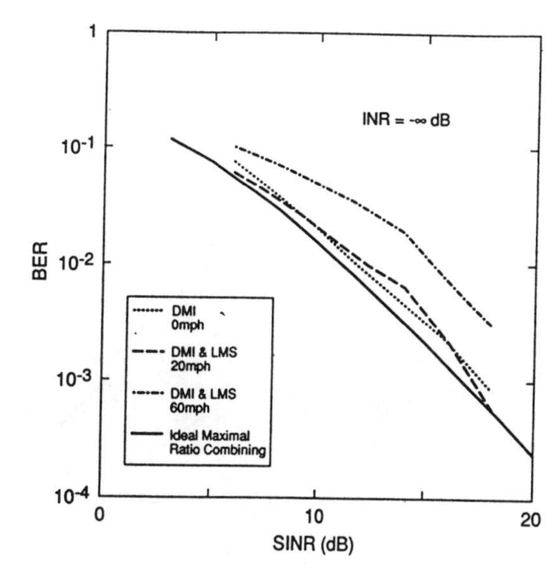

Fig. 4. BER versus SINR for vehicle speeds of 0, 20, and 60 mph with DMI for acquisition and LMS for tracking.

B. Results

To determine the performance of the acquisition and tracking algorithms in IS-54, we used IS-54 computer simulation programs written by S. R. Huszar and N. Seshadri. We modified the transmitter, fading simulator, and receiver programs for flat Rayleigh fading with one interferer and added our optimum combining algorithms with both coherent and differential detection. Specifically, the transmitted desired signal consisted of blocks of 162 symbols with $\pi/4$-shifted DQPSK modulation. The symbols in each block were randomly generated 2-bit symbols for symbols 1–14 and 29–162, and a synchronization sequence for symbols 15–28. This signal, sampled at 8 times the symbol rate, was filtered by a square root cosine rolloff filter with a rolloff factor of 0.35. For the interfering signal, randomly generated symbols, independent of the desired signal symbols, were used for the data, and a synchronization sequence that is orthogonal to that to the desired signal was used for symbols 15–28. The relative timing of the interfering and desired signals was adjustable in increments of 1/8 of the symbol duration. Independent, flat Rayleigh fading for each signal at the two receive antennas was generated by multiplying each signal by a complex Gaussian random number, which varied at the fading rate [13]. The received signals were then weighted, combined, and filtered by a square root cosine rolloff filter, followed by coherent or differential detection.

Let us first consider the performance with DMI for acquisition and LMS for tracking with differential detection without interference. Fig. 4 shows the BER versus SINR for vehicle speeds of 0, 20, and 60 mph at 900 MHz, corresponding to fading rates of 0, 27, and 81 Hz. Computer simulation results are shown for the BER over 178 blocks ($\approx$ 28 000 symbols, which should be adequate for BER $> 10^{-3}$), along with theoretical results for maximal ratio combining (15). At 0 mph, the fading channel was constant over each block, but independent between blocks. Also, LMS tracking was not used at 0 mph, and thus the results show the accuracy of weight acquisition by DMI. DMI is seen to have less than 1-dB implementation loss for BER's between 10^{-3} and 10^{-1}. At 20 and 60 mph, the tracking performance of the LMS algorithm is poor. For SINR below 14 dB, the LMS algorithm tracks so poorly that the best BER is obtained with $\mu = 0$, i.e., if LMS tracking is not used. This lack of tracking causes little degradation at 20 mph, but a several dB loss in performance at 60 mph. For SINR above 14 dB, the LMS algorithm improves performance, with the best μ equal to 0.08. At 20 mph, the performance with the LMS algorithm is about the same as at 0 mph. However, at 60 mph there is a 4.2-dB implementation loss at 10^{-2} BER. Thus the LMS algorithm is not satisfactory for optimum combining in IS-54.[3]

Next, consider DMI for both acquisition and tracking with differential detection without interference. Fig. 5 shows the average BER versus SINR with DMI and vehicle speeds of 0 and 60 mph. For these results, we used DMI with a 14-symbol sliding window ($K = 14$ in (22) and (23)), which gave us the best results for a 10^{-2} BER at 60 mph. At this BER, DMI has a negligible increase in implementation loss at 60 mph as compared to 0 mph.

Although differential detection is typically used in mobile radio because of phase tracking problems, we can also use coherent detection with optimum combining. This is because

[1] We do tracking in each block (starting from the synchronization sequence) in the forward direction for symbols 29 to 162, and in the reverse direction for symbols 14 to 1.

[2] Since DQPSK also has constant amplitude, the constant modulus algorithm can also be used to generate an error signal, i.e., $\epsilon(k) = s_o(k) - s_o(k)/|s_o(k)|$, for the LMS algorithm, as shown in [15]. A reference signal is, therefore, not needed, but this means that the receiver can acquire and track an interfering signal rather than the desired signal, and, therefore, the algorithm cannot be used for optimum combining when interference is present.

[3] In [15] it was shown that the LMS algorithm was satisfactory for diversity combining and equalization using the constant modulus algorithm for error signal generation in GSM, with data-to-fading-rate ratios as low as 1700. However, as mentioned before, this technique cannot distinguish between the desired and interfering signals.

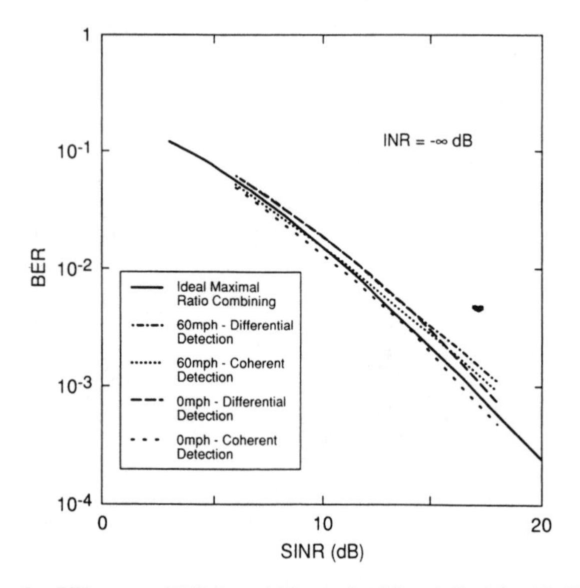

Fig. 5. BER versus SINR for vehicle speeds of 0 and 60 mph with DMI for acquisition and tracking.

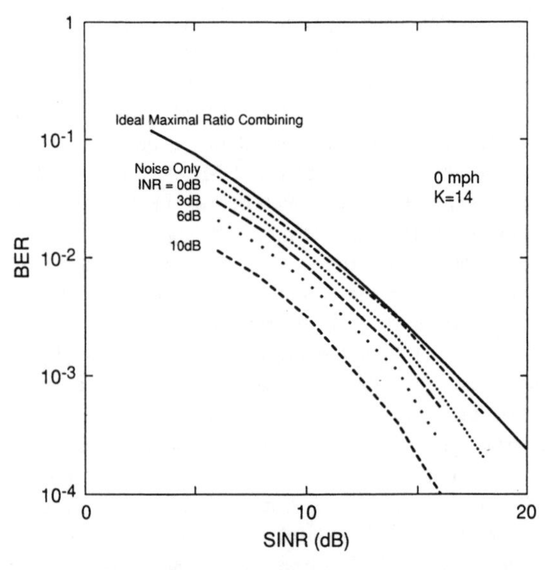

Fig. 6. BER versus SINR with one interferer for a vehicle speed of 0 mph with DMI for acquisition and tracking.

optimum combining requires coherent combining of the received signals, which means that the weights must track the received signal phase, and the array output signal phase should match the phase of the coherent reference signal. Thus coherent detection of the array output is possible, which, as shown in Section II, decreases the required SINR for a 10^{-2} BER by 1.0 dB with ideal phase tracking.[4] With the LMS algorithm, however, tracking is so poor that coherent detection is worse than differential detection. On the other hand, with the DMI algorithm, there is improvement with coherent detection. Fig. 5 shows that coherent detection decreases the required SINR for a 10^{-2} BER by 1 dB, resulting in performance that is 0.3 dB better than the theoretical performance of differential detection (but 0.7 dB worse than ideal coherent detection). At 60 mph, the performance degrades by an additional 0.5 dB; i.e., the performance is 0.2 dB worse than ideal differential detection (and 1.2 dB worse than ideal coherent detection). Thus the use of coherent rather than differential detection cancels most of the implementation loss of DMI at 60 mph.

Finally, consider the dynamic performance of optimum combining for interference suppression. For the results shown below, the symbol timing for the desired and interfering signals was the same. Our results showed that this was the worst case since there was a slight improvement in performance with timing offset between the two signals (see below).

With the LMS algorithm, even at 20 mph the performance does not improve with the INR, showing that the algorithm is not accurately tracking the interferer.

However, with DMI, the performance improvement with INR agrees with ideal tracking results. Fig. 6 shows the average BER versus SINR at 0 mph with one interferer with INR $= -\infty$, 0, 3, 6, and 10 dB. DMI with a 14-symbol sliding window and coherent detection was used as before.

[4]Note that this is significant in comparison to the 3.6 dB gain with optimum combining in IS-54 with 2 receive antennas [6]. Also, it is almost half of the 2.5 dB gain needed for a frequency reuse factor of 3 rather than 4 (and an additional 33% capacity increase).

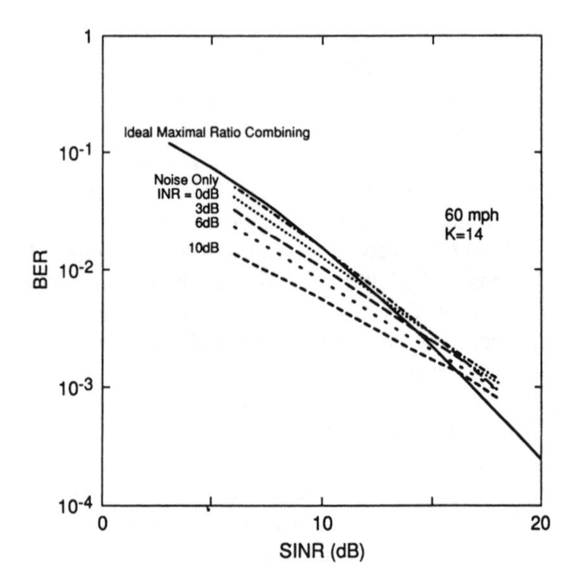

Fig. 7. BER versus SINR with one interferer for a vehicle speed of 60 mph with DMI for acquisition and tracking, and $K = 14$.

The required SINR for a 10^{-2} BER is 10.2, 9.5, 8.6, and 6.5 dB for INR = 0, 3, 6, and 10 dB, respectively, which is within 0.5 dB of the predicted gain shown in Fig. 3.

Fig. 7 shows the average BER versus SINR at 60 mph with one interferer. Again, a 14-symbol sliding window was used since this gave the best results at a 10^{-2} BER. At a 10^{-2} BER these results show a gain with INR that is within 0.5 dB of the gain shown in Fig. 3. The implementation loss increases the SINR, though, resulting in poor performance at a 10^{-3} BER with $K = 14$. However, note that the optimum window size for a given BER is determined by a tradeoff of two effects. As the window size decreases, the weights have more error due to the averaging of fewer samples, but less error caused by channel variation over the window. Our results showed that as SINR increases, the performance is improved by decreasing K.

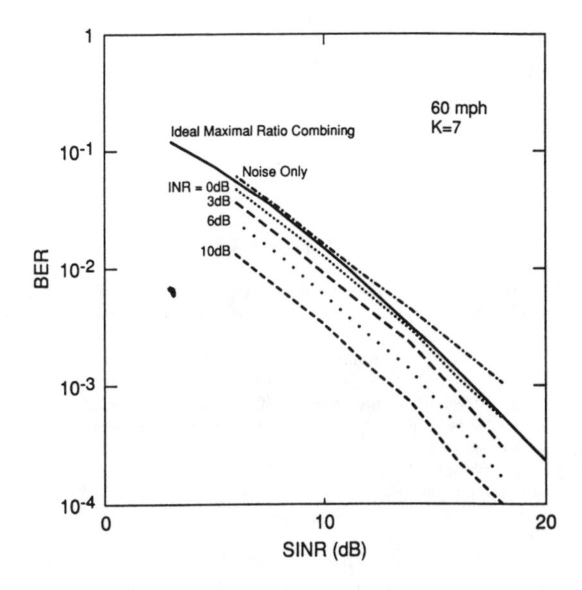

Fig. 8. BER versus SINR with one interferer for a vehicle speed of 60 mph with DMI for acquisition and tracking, and $K = 7$.

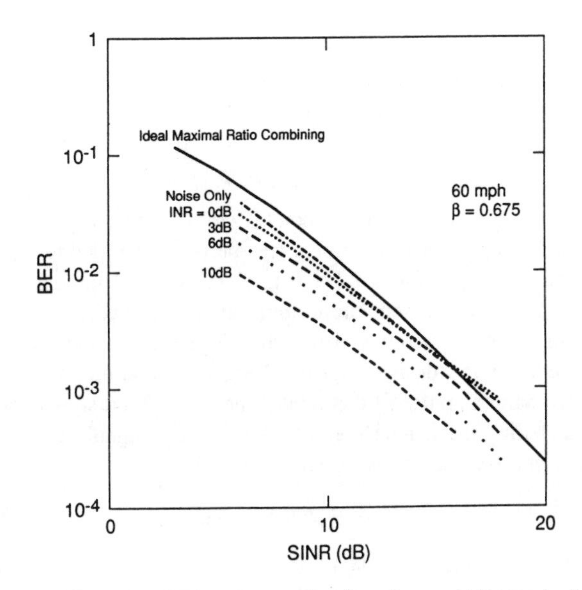

Fig. 9. BER versus SINR with one interferer for a vehicle speed of 60 mph with DMI for acquisition and tracking, and exponential weighting with $\beta = 0.675$.

Fig. 8 shows the performance with $K = 7$, which gave the best results at a 10^{-3} BER. At this BER, with interference, the improvement of optimum combining is seen to be close to that with $K = 14$ at a 10^{-2} BER (Fig. 7). Furthermore, with $K = 7$ at a 10^{-2} BER, the improvement with interference is similar to that with $K = 14$. However, with noise only, the BER for a given SINR is higher with $K = 7$ than with $K = 14$, because fewer samples are averaged to determine the weights.

Fig. 9 shows the performance of DMI with exponential weighting for $\beta = 0.675$. This β gave the best results for BER $= 10^{-2}$ and 10^{-3}. With noise only, the BER is seen to be lower than with either $K = 14$ or 7, and at a 10^{-2} BER the performance is close to that of ideal maximal ratio combining with coherent detection (i.e., 1.0 dB lower SINR than the curve shown for ideal maximal ratio combining with differential detection). With interference at a 10^{-2} BER, the gain with optimum combining is close to the predicted ideal gain; i.e., the performance is slightly better than DMI with a sliding window and $K = 14$. However, at a 10^{-3} BER with interference, the performance is slightly worse than that shown in Fig. 8 with $K = 7$. Thus either the sliding window or the exponential weighting technique can be used to generate accurately the optimum combining weights, even at 60 mph.

Finally, Fig. 10 shows the effect of timing offset between the desired and interfering signals. Results were generated for a 10 dB SINR at 60 mph with $K = 14$, as in Fig. 7. These results show that the BER varies with timing offset by less than 12% (< 0.4 dB improvement in SINR at a 10^{-2} BER), with the best performance when the interfering and desired signals are offset by approximately half the symbol duration.

IV. Summary and Conclusions

In this paper, we have studied the dynamic performance of adaptive arrays in wireless communication systems. Specifically, we studied the performance of the LMS and DMI weight adaptation algorithms in IS-54 with data to fading rates as low as 300. We showed that implementation of optimum combining allows the use of coherent detection, which improves performance by over 1 dB as compared to differential detection. Although the performance of the LMS algorithm was not satisfactory, results showed that the DMI algorithm acquired the weights in the synchronization sequence interval and tracked the desired signal for vehicle speeds up to 60 mph with less than 0.2 dB degradation from the ideal performance with differential detection at a 10^{-2} BER. Similarly, an interfering signal was also suppressed with performance gains over maximal ratio combining within 0.5 dB of the predicted ideal gain. Thus our results indicate that we can obtain close to the ideal performance improvement of optimum combining even in rapidly fading environments.

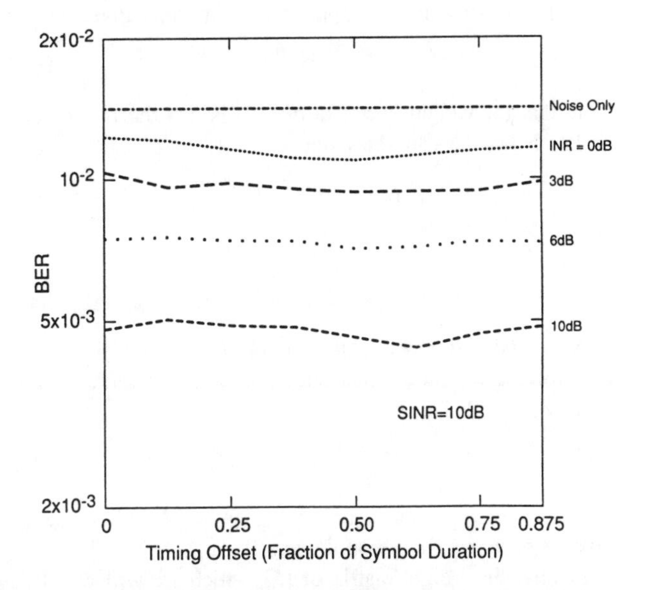

Fig. 10. Effect of timing offset on the BER for a vehicle speed of 60 mph with DMI for acquisition and tracking, $K = 14$, and SINR = 10 dB.

Acknowledgment

We gratefully acknowledge useful discussions with G. D. Golden, J. Salz, and N. Seshadri.

Appendix A

To relate the weight equation (12) to that of [5, Eq. (11)], we need to consider three differences between the analysis given here and in [5]. First, in [4]–[6], we considered the generation of $N = L + 1$ separate outputs at the receiver, each with minimum mean-square error, while here we consider only the output of the desired signal. Using the notation of [4]–[6], the channel matrix $\boldsymbol{C}$ that relates the transmitted signal vector (including the L interferers) to the received signal vector $\boldsymbol{x}$ at a given time is given in our notation by

$$\boldsymbol{C} = [\boldsymbol{u}_d \boldsymbol{u}_1 \cdots \boldsymbol{u}_L]. \tag{A-1}$$

Thus the weight matrix $\boldsymbol{W}$ for the optimum linear combiner that generates N output signals is given by (from (12))

$$\boldsymbol{W} = \alpha \boldsymbol{R}_{xx}^{-1} \boldsymbol{C}^* \tag{A-2}$$

with the vector $\boldsymbol{s}$ at the output of the combiner given by

$$\boldsymbol{s} = \boldsymbol{W}^T \boldsymbol{x}. \tag{A-3}$$

Note that the weight vector $\boldsymbol{w}$ of (12) is just the first column of $\boldsymbol{W}$. Now, we can show that

$$\boldsymbol{R}_{xx} = \sigma^2 \boldsymbol{I} + \boldsymbol{C}\boldsymbol{C}^\dagger \tag{A-4}$$

and from (A-2),

$$\boldsymbol{W} = \alpha[\sigma^2 \boldsymbol{I} + \boldsymbol{C}\boldsymbol{C}^\dagger]^{-1} \boldsymbol{C}^*. \tag{A-5}$$

A second difference is that in [4]–[6] we considered the zero-forcing weights, which can be obtained from (A-5) in the limit, $\sigma^2 \to 0$, i.e.,

$$\boldsymbol{W} = \alpha[\boldsymbol{C}\boldsymbol{C}^\dagger]^{-1} \boldsymbol{C}^*. \tag{A-6}$$

Note that $[\boldsymbol{C}\boldsymbol{C}^\dagger]^{-1}$ exists only when $N = M$. Otherwise, the inverse becomes the pseudoinverse.

Finally, the weight matrix of [5], which we will denote as $\boldsymbol{W}_{[5]}$, was defined as the transpose of the weight matrix given here, i.e.,

$$\boldsymbol{s} = \boldsymbol{W}_{[5]} \boldsymbol{x} \tag{A-7}$$

and is given in [5] as

$$\boldsymbol{W}_{[5]} = \lim_{\sigma^2 \to 0} [\sigma^2 \boldsymbol{I} + \boldsymbol{C}^\dagger \boldsymbol{C}]^{-1} \boldsymbol{C}^\dagger. \tag{A-8}$$

Although (A-6) and (A-8) look similar, note that $\boldsymbol{C}\boldsymbol{C}^\dagger$ (A-6) is an $M \times M$ matrix, while $\boldsymbol{C}^\dagger \boldsymbol{C}$ (A-8) is an $N \times N$ matrix. However, the weights can be shown to be equal (with a scalar multiple) in the limit $\sigma^2 \to 0$. The change in the weight equation was done to put it the form for DMI (21).

References

[1] J. H. Winters, "Optimum combining in digital mobile radio with cochannel interference," *IEEE J. Select. Areas Commun.*, vol. SAC-2, July 1984.

[2] J. H. Winters, "Optimum combining for indoor radio systems with multiple users," *IEEE Trans. Commun.*, vol. COM-35, Nov. 1987.

[3] J. H. Winters, "On the capacity of radio communication systems with diversity in a Rayleigh fading environment," *IEEE J. Select. Areas Commun.*, vol. SAC-5, June 1987.

[4] J. H. Winters, J. Salz, and R. D. Gitlin, "The capacity of wireless communication systems can be substantially increased by the use of antenna diversity," in *Proc. 1st Int. Conf. Universal Personal Commun.*, Sept. 1992, pp. 28–32.

[5] J. H. Winters, J. Salz, and R. D. Gitlin, "The capacity increase of wireless communication systems with antenna diversity," in *Proc. 1992 Conf. Inform. Sciences Syst.*, vol. II, Mar. 18–20, 1992, pp. 853–858.

[6] J. H. Winters, J. Salz, and R. D. Gitlin, "Adaptive antennas for digital mobile radio," *Proc. IEEE Adaptive Antenna Syst. Symposium*, Melville, NY, Nov. 1992, pp. 81–87.

[7] S. A. Hanna, M. El-Tanany, and S. A. Mahmoud, "An adaptive combiner for co-channel interference reduction in multi-user indoor radio systems," in *Proc. IEEE Veh. Technol. Conf.*, St. Louis, MO, May 19–22, 1991, pp. 222–227.

[8] D. J. Goodman, "Trends in cellular and cordless communications," *IEEE Commun. Mag.*, vol. 29, pp. 31–40, June 1991.

[9] R. A. Monzingo and T. W. Miller, *Introduction to Adaptive Arrays.* New York: Wiley, 1980.

[10] A. Dembo and J. Salz, "On the least squares tap adjustment algorithm in adaptive digital echo cancellers," *IEEE Trans. Commun.*, vol. 38, pp. 622–628, May 1990.

[11] P. Bello and B. D. Nelin, "Predetection diversity combining with selectively fading channels," *IRE Trans. Commun. Syst.*, vol. CS-10, p. 32, Mar. 1962.

[12] P. G. Proakis, *Digital Communications.* New York: McGraw-Hill, 1983, p. 175.

[13] W. C. Jakes, Jr., *et al.*, *Microwave Mobile Communications.* New York: Wiley, 1974.

[14] J. H. Winters, "Switched diversity with feedback for DPSK mobile radio systems," *IEEE Trans. Veh. Technol.*, vol. VT-32, pp. 134–150, Feb. 1983.

[15] T. Ohgane, "Characteristics of CMA adaptive array for selective fading compensation in digital land mobile radio communications," *Electron. Commun. Japan*, Part 1, vol. 74, no. 9, pp. 43–53, 1991.

Jack H. Winters (S'77–M'81–SM'88) received his B.S.E.E. degree from the University of Cincinnati, Cincinnati, OH, in 1977 and M.S. and Ph.D. degrees in electrical engineering from Ohio State University, Columbus, Ohio, in 1978 and 1981, respectively.

Since 1981, he has been with AT&T Bell Laboratories, where he is in the Wireless Communications Research Department. He has studied signal processing techniques for increasing the capacity and reducing signal distortion in fiber optic, mobile radio, and indoor radio systems and is currently studying adaptive arrays and equalization for mobile radio.

Reprinted from *IEEE Communications Magazine*, Vol. 33, No. 5, May 1995, pp. 62-68.

Direction Finding and "Smart Antennas" Using Software Radio Architectures

Software radio architectures, originally developed for military applications, are now becoming economically viable in commercial products because of the rapid advance of DSP technology.

Joseph Kennedy and Mark C. Sullivan

Operational tests and demonstrations of systems based on software radios are currently being performed in the 800 MHz mobile cellular radio band. These field trials are for mobile unit geolocation systems and adaptive phased array "smart antenna" appliqués. The geolocation system trials are in response to the demand for high confidence geolocation of mobile units for enhanced emergency 911 service and for use in the U.S. Department of Transportation's Intelligent Vehicle Highway System (IVHS) initiative. The smart antenna array appliqués address the cellular service providers' need for more user channel capacity and/or geographic coverage from existing base station installations. Software radio architectures were selected because of their ability to provide superior performance at low life cycle cost. These systems use 4 to 8 wideband coherent channels and fully characterize the arriving RF energy to either geolocate the emitter or to maximize the carrier-to-interference (C/I) ratio.

Software Radio Background

The application of software radio techniques to military communications was conceived in the 1970s. The initial goal was to provide higher fidelity in the HF band, where over-the-horizon communications via ionospheric propagation were important. The HF band is ideally suited for testing software radio principles for two reasons. First, advanced signal processing is necessary in the HF band because of the highly dynamic nature of the channel due to ionospheric propagation and also because of the high density of signals. Second, the HF band spans only a few tens of megahertz, and generally only about 10 to 20 percent of the band will propagate over the horizon. Software radios capable of processing even hundreds of kilohertz of bandwidth may thus be effective in operating over RF bands of potential interest.

Software radio architectural concepts were reduced to practice in about 1980 in the first operational military digital HF communications system. The hardware required to implement the system filled many racks and consumed a great deal of power. The system provided simultaneous coverage of only 750 kHz of the HF band with a dynamic range of less than 60 dB. Still, the ability to employ sophisticated signal processing algorithms, even in this modest hardware configuration, proved the worth of software radio techniques.

Exponential increases in DSP technology performance coupled with reductions in size and power (Moore's Law[1]) over the last 20 years have opened up applications of software radios to line-of-sight RF bands. Throughout the '80s and '90s, these systems have been upgraded through five generations.

The first few generations of software radios were developed to provide enhanced fidelity and flexibility unavailable from analog receivers. Changing needs could be met quickly and inexpensively with software upgrades. This ability to adapt to difficult problems makes the software radio attractive in commercial telecommunications applications, where infrastructure equipment might otherwise quickly become obsolete due to cumulative double digit growth in markets, rapidly evolving air standards, and requirements for new services. The action of Moore's Law has driven the cost of the software radio down at a rate greater than other competitive technologies, providing the ultimate motivation for their use in the commercial telecommunications industry.

Application to Cellular/PCS Infrastructure

The current generation of commercial wireless products employs readily available DSP technology to digitize the RF spectrum with great fidelity and then perform all of the base station functions using software radio architectures. Most of these functions are performed by executing software on a general-purpose DSP processor, but the first stage of frequency conversion and filtering is performed by a specialized decimating

JOSPEH KENNEDY is the director of the Advanced Communications Systems group at Engineering Research Associates.

MARK C. SULLIVAN is a scientist with Engineering Research Associates.

[1] *In 1979, Intel co-founder Gordon Moore noted that the achievable density of transistors on chips appeared to double every 18 months.*

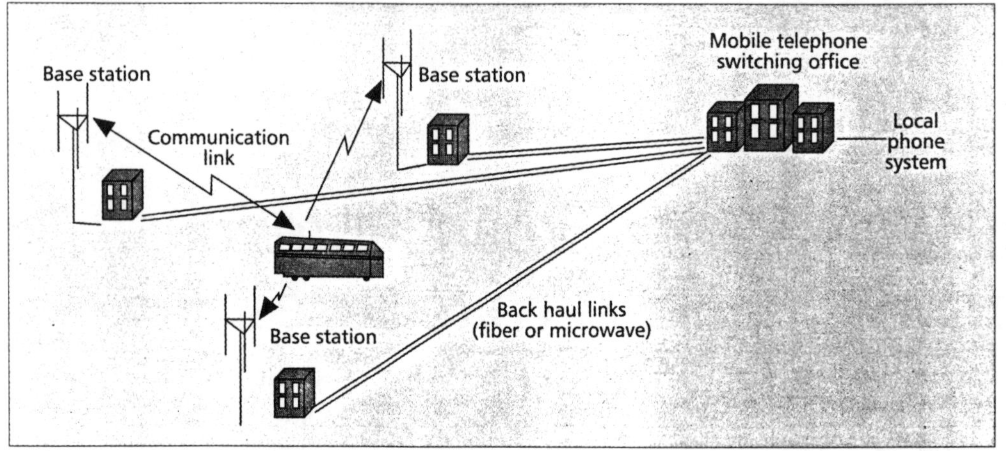

■ **Figure 1.** *Typical system layout.*

The action of Moore's Law has driven the cost of the software radio down at a rate greater than other competitive technologies.

downconverter chip produced by Harris Semiconductor and others. This circuit implements a cascaded integrator-comb filter followed by a conventional transversal filter to provide computationally efficient sample rate conversion from the tens of megahertz at the A/D converter to the tens of kilohertz occupied by the signal [2].

Using software radio techniques, a common hardware platform may provide all of the current base station functions and yet retain enough flexibility to meet future requirements through alternative software application programs. The products are configured from a set of hardware and software modules, interconnected using industry standard mechanical, electrical and software interfaces. These modules may be combined to provide user-configurable equipment suites including complete Cellular/PCS macro- and microcell base stations and "Smart Antenna" appliqués. Software selectable options include domestic and international air standards, base station controller-to-switch interfaces, fixed omni/sector or "smart antennas," and new service features such as fraud detection and mobile terminal geolocation.

CAPITAL Direction Finding Operational Demonstration

One example of commercial equipment using software radio architectures is the Cellular Applied to IVHS Tracking & Location (CAPITAL) project, currently undergoing field trials in the Washington, D.C. suburbs. The concept behind CAPITAL is that automobile traffic congestion may be monitored by tracking the progress of vehicles equipped with cellular phones. This comprehensive program demonstrates all of the capabilities of such a system, beginning with the collection and processing of cellular data and proceeding to the dissemination of traffic information to remote users and in-vehicle equipment. The operational test makes extensive use of the in-place cellular infrastructure for both wide area automobile traffic congestion monitoring and communications. Geolocation equipment has been colocated at selected towers to collect cellular phone usage statistics and to geolocate phones on designated roadways.

Cellular-based wide area surveillance techniques are being evaluated for use in automobile traffic monitoring applications because of their comparative advantage over magnetic sensors buried in the roadway, video cameras, and other conventional techniques. These advantages include:

- Area coverage based on square miles, not vehicle count or road miles.
- No disruption of road service for installation or repair.
- Orders of magnitude lower in cost than loop-based approaches.
- Very high reliability with low maintenance costs.
- Other uses include fleet management and emergency assistance.
- Geolocation equipment could be owned and operated by cellular service providers.

CAPITAL geolocates mobile users of cellular radios through the reception and processing of RF signals emitted by ordinary mobile cellular telephones. In this application the content of the signal is of no interest and the privacy of the user is protected by not demodulating the audio portion of the signal and by replacing the electronic serial number with a randomly chosen identifier. Geolocation is achieved using either the Reverse Voice Channel (RVC) or the Reverse Control Channel (RECC) transmitted by the mobile "probe" vehicle. The system geolocates the emitter to within ~100 meters using a combination of techniques. The line of bearing to the emitter is estimated at each receiving site using the Alternating Projection (AP) maximum likelihood direction-finding algorithm [4, 7]. Time of arrival for each signal is also measured with respect to a GPS time reference so that time-difference of arrival (TDOA) techniques may also be employed to fix the location. If the RVC is being processed, then the frequency of the Supervisory Audio Tone (SAT) is used to determine if the received signals are from a common mobile unit, or from interfering mobile transmitters.

RF signals are received at multiple receiving sites, characterized by angle of arrival, time stamped, and then passed to a common site where the geolocation is determined. A typical layout is shown in Fig. 1.

Geolocation equipment based on software radios is located at selected towers and interconnected

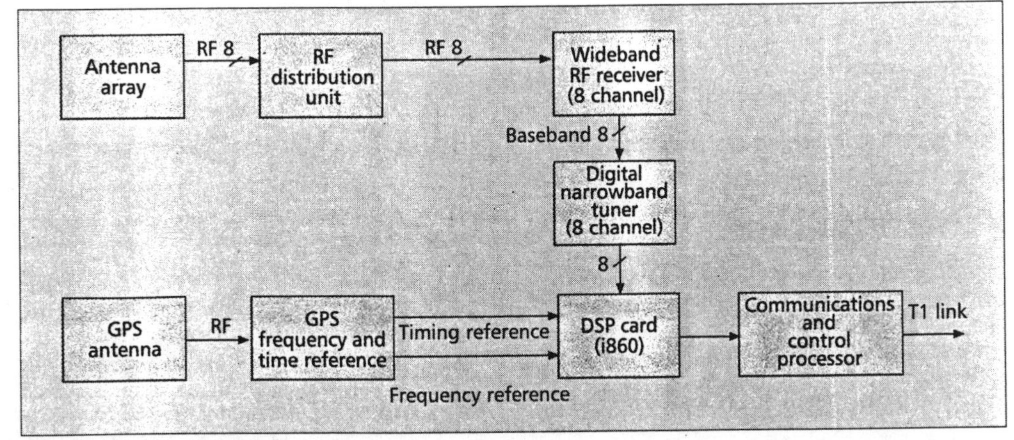

■ **Figure 2.** *Hardware block diagram.*

through existing back haul links to a control subsystem at the Mobile Telephone Switching Office (MTSO). In Fig. 1, a transit bus is shown communicating through a base station via a cellular telephone. Geolocation equipment located at one or two other sites are receiving the mobile cellular transmission and estimating a position. The algorithms employed by the system to combat multipath and interference would be all but impossible to implement without the use of software radio technology. Geolocation accuracy is enhanced by the following features:

- GPS-derived clocks provide less than 100 ns differential timing error from site to site.
- GPS-based tower siting removes any siting-based bias errors.
- Coherent multichannel time delay allows DSP-based spatial filtering to be used to combat multipath and interference effects.
- Phase is calibrated to hundredths of a degree through DSP.

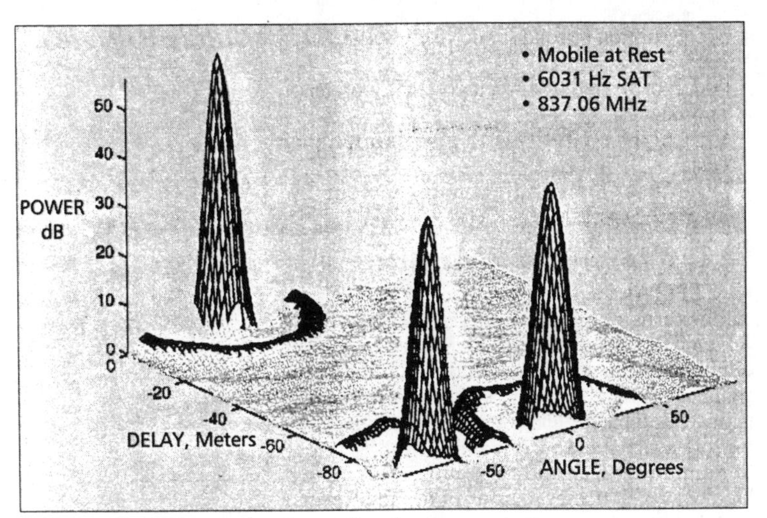

■ **Figure 3.** *PDAgram showing mulitpath.*

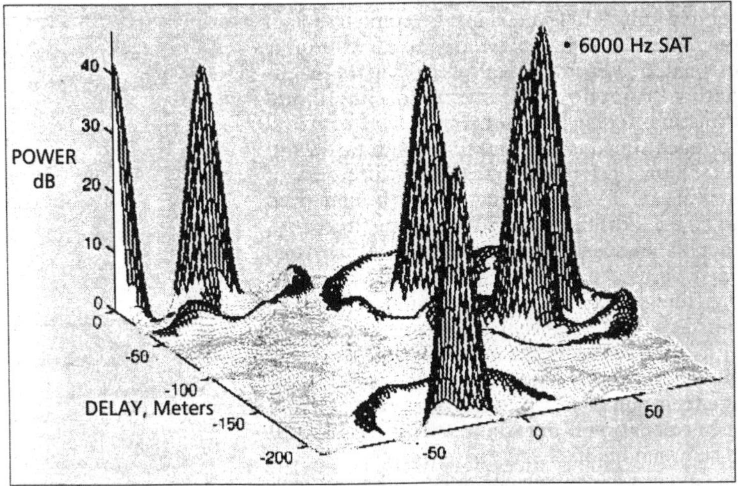

■ **Figure 4.** *PDAgram showing multipath and interference.*

A block diagram of the CAPITAL hardware is shown in Fig. 2. Signals from the eight-element antenna array are block downconverted to an IF before the entire band of interest is digitized. Eight digital tuner ASICs coherently downconvert each narrowband channel being monitored to a zero IF complex baseband format. The resulting narrowband data is then processed by Intel i860 chips in the Signal Processing Unit.

Software implemented in the "C" language enumerates and isolates the components of the received signal using the AP algorithm and spatial filtering. The SAT frequency of each component is then measured to distinguish between components of the desired signal, which will all exhibit the same frequency, and those of co-channel interference, which will generally be assigned a different frequency. The isolated common SAT signals are then cross correlated to identify the component of the desired signal that arrives earliest. The components arriving later are deemed to have resulted from multipath propagation and are discarded. The line of bearing and time of arrival of the direct-ray component are sent to a central site over a T1 link to complete the geolocation of the emitter.

The major impediment to reliable geolocation of mobile units in the cellular telephone band is severe multipath propagation. The effects of this phenomena must be mitigated in order to achieve accurate position estimates with low variance. Typical examples of the multipath encountered in the suburban Washington, D.C. operating environment were collected using instrumentation in the geolocation equipment. Specular multipath can be characterized in terms of power, delay, and angle-of-arrival relative to the direct ray.

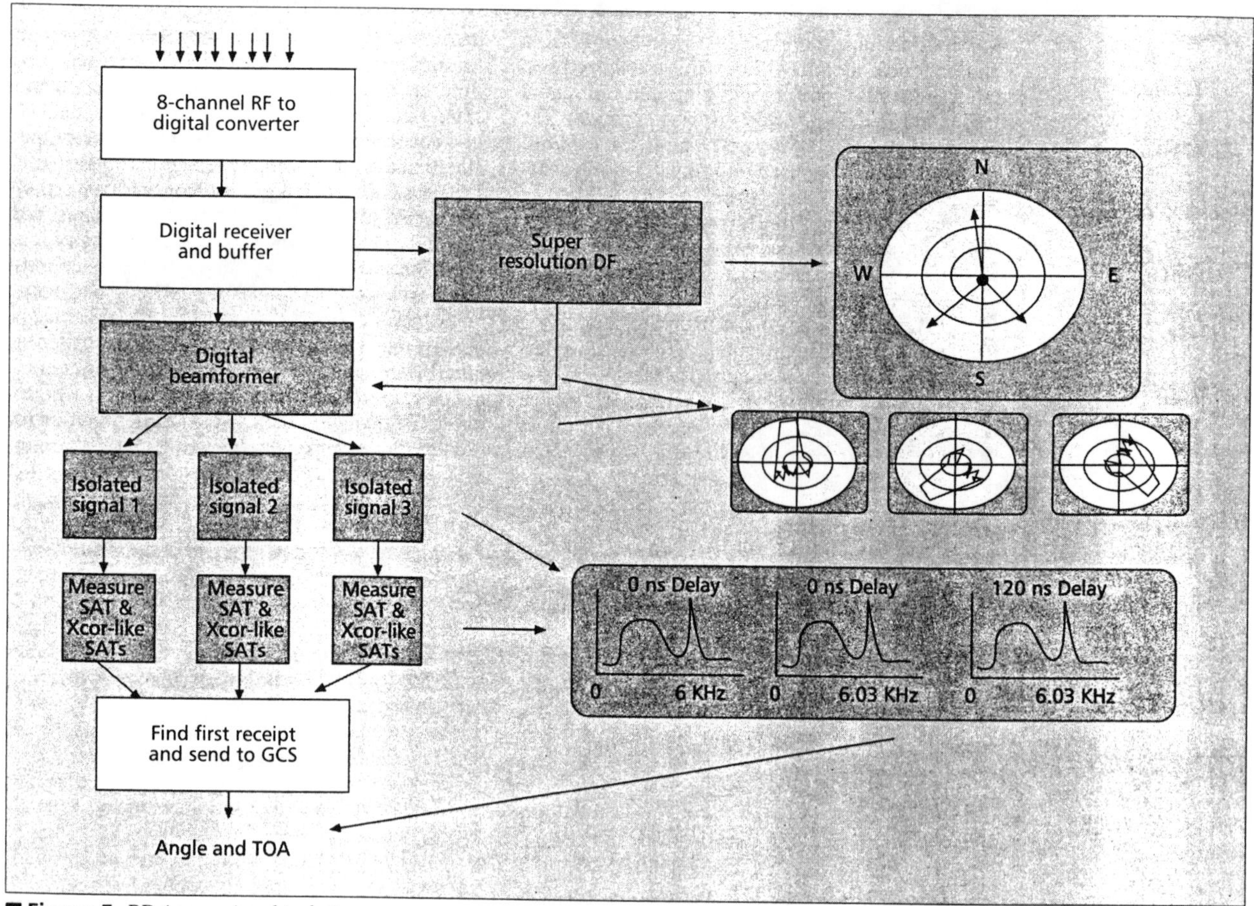

■ **Figure 5.** *PDAgram signal isolation process.*

The ideal tool to measure these parameters is the software radio, with its inherent fidelity and digital storage capability. The signal parameters may be displayed in a "PDAgram" (Power-Delay-Angle diagram) such as the example shown in Fig. 3. This PDAgram was generated from data collected by the CAPITAL system and shows three signal components, the direct ray shown on the "0" delay axis and two multipath components with differential path length delays of about 80 meters. Figure 4 shows another PDAgram obtained from a signal data record exhibiting multipath and interference. One signal exhibits a SAT frequency of 6000 Hz and the other, interfering, signal originates from a cell using a 6030 Hz SAT frequency assignment. Note that the interferer also has a multipath component present.

This record was processed to determine direction of arrival for the desired signal. The desired mobile signal multipath components, the interfering mobile (shown in a different color), and interfering mobile multipath component had to be removed to determine the correct angle and time of arrival for the "correct" direct ray. Once the signal components have been isolated, the software examines the SAT to determine which components came from interfering cellular users or from multipath reflections. The process used to isolate the signals is shown in Fig. 5. This process is employed in the DF equipment to raise the confidence in the geolocation estimates by being sure the "right" signal is being geolocated. The result is accurate geolocation needed for Emergency 911 service.

"Smart Antenna" Arrays

Another commercial application of software radio technology is the "Smart Antenna" appliqué for existing base stations. A field trial at a service provider's network this Spring will be used to evaluate the ability of the "Smart Antenna" to increase C/I for an Advanced Mobile Phone Service (AMPS) system in an area with a very high density of users.

Increased user capacity and expanded geographic coverage from existing installations are the two strong economic motivations voiced by service providers as the impetus for these developments. The need for increased area coverage is of primary concern in rural environments and for prospective PCS providers. Increases in user capacity are desired by existing service providers for high-use areas where they are faced with loss of potential revenue, customer dissatisfaction, and reduction in spectrum available for AMPS created by the introduction of digital voice services.

Focusing on the capacity challenge, the real need is to make more voice channels available in a given geographic area in the finite spectrum allocated. When translating this operational need to a technical requirement for an antenna system, the need is not for a steerable "higher gain" array, but instead an array which can increase the carrier to interference ratio (C/I) at both the base station and the mobile terminal. If this can be achieved, then more RF channels can be added to the base stations in the area without driving the C/I below an

Preliminary analyses indicate that software-radio-based "Smart Antenna" appliqués can enhance the C/I in both the forward and reverse link to allow frequency reuse of 3 to 1.

acceptable limit. Suppose that the channels in a conventional system without an array have been allocated to maximize capacity subject to a minimum acceptable C/I. If the smart array is employed, the C/I will increase, allowing the addition of more channels until the C/I is again brought to a minimum acceptable level. By increasing the C/I of the desired signal for both the forward and reverse links, the smart array provides more channel capacity.

There are three approaches which can be used when designing a smart array to maximize C/I. The first and most obvious is to create gain on the array in the intended direction using antenna aperture. This is done by using physically directive elements, or by combining the outputs of more elements to create aperture, and thus gain. It is important to note that it is not the gain that is of value, but the directivity which has value in trying to maximize C/I. The "lack of gain" in the direction of interference, ideally a null, may be more effective than gain in the direction of the desired signal.

The second approach is the mitigation of multipath fading. C/I budgets are generally designed with a significant margin for slow and fast fading. For conventional mobile radio systems, fast fading induced by multipath propagation requires an additional link margin of 8 dB [3]. If the smart array can minimize the effects of destructive specular multipath combining, most of this margin may be recovered. The multipath components present in the PDAgrams shown in Figs. 3 and 4 are typical of those found in a long-term experiment performed in the Washington, D. C. area to characterize the mobile radio propagation environment. Preventing

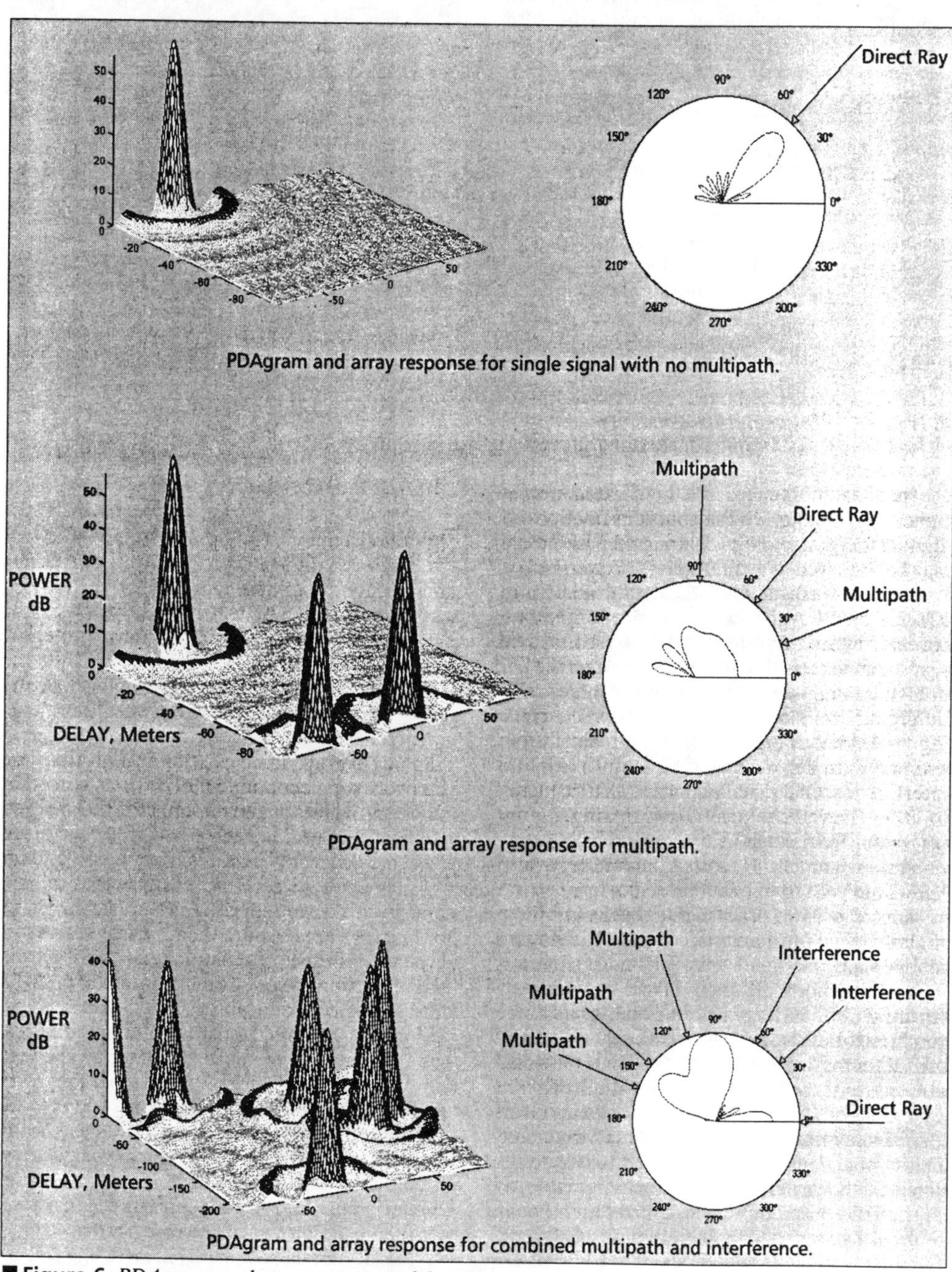

■ **Figure 6.** *PDAgrams and array responses of the "Smart Antenna."*

the significant losses in C/I induced by multipath propagation is one very effective use of a smart antenna array.

The final approach used to maximize C/I is the identification and nulling of interferers. When designing an antenna element combiner, it is generally much easier to form sharp nulls than beams for a given number of elements in the array. It is generally easier to place a 10 dB null on an interferer than to create a beam with 10 dB of gain in the direction of the desired signal. The operation of the "Smart Antenna" is illustrated in Fig. 6. In the first pair of graphs, the PDAgram shows a single component and the corresponding array response shows a single beam. In the second pair of graphs, the PDAgram shows two multipath components. The array response now includes gain with phase shifts to coherently combine these components with the direct ray. In the third pair of graphs a source of interference is present with multipath and the array response shows corresponding nulls.

Other important factors to consider in designing "Smart Antenna" appliqués are the ability to form beams on the reverse control channel access requests which arrive at unpredictable directions and times, the ability to interface to an existing base station without violating proprietary data interface agreements and disrupting hand-off management, and the ability to maintain the necessary linearity in the equipment with increased dynamic range (> 80 dB) imposed by higher antenna gains and more channels per base station.

Software radio architectures support the implementation of all three approaches to maximize C/I and also address the factors mentioned above. The architecture for the N-channel beamformer is shown in Fig. 7. Each antenna element is followed by a wideband RF to baseband converter, where the entire operating RF band is moved to baseband at very high dynamic range. The digital hardware is a combination of custom boards and commercial "off-the-shelf" boards. Digital tuner ASICs coherently downconvert each narrowband channel to a zero IF complex baseband format. The digital beamformer is implemented in assembly language software executed by a Texas Instruments TMS320C40 DSP chip. Assembly language was chosen after a "C" language implementation failed to execute in real time. Each of the resulting narrowband signals is then upconverted digitally and summed to produce a single wideband signal that is then converted back to an analog format using a high-speed D/A converter. An RF block upconverter completes the process by translating the wideband IF back to the original cellular RF band.

The signal processing algorithms provide a fully adaptive receive beam and a switched transmit beam with four azimuthal beams per 120 degree sector. The net aperture gain is in excess of 20 dBi. The adaptive receive beams are created by a specialized variant of the constant modulus algorithm (CMA) [1, 6, 7] which executes in software and maximizes C/I through a combination of spatial beam steering, interferer null steering, and maximal ratio combining of multipath.

The beamformer software stores the incoming signals in random access memory (RAM) while the beam steering coefficients are being calculated.

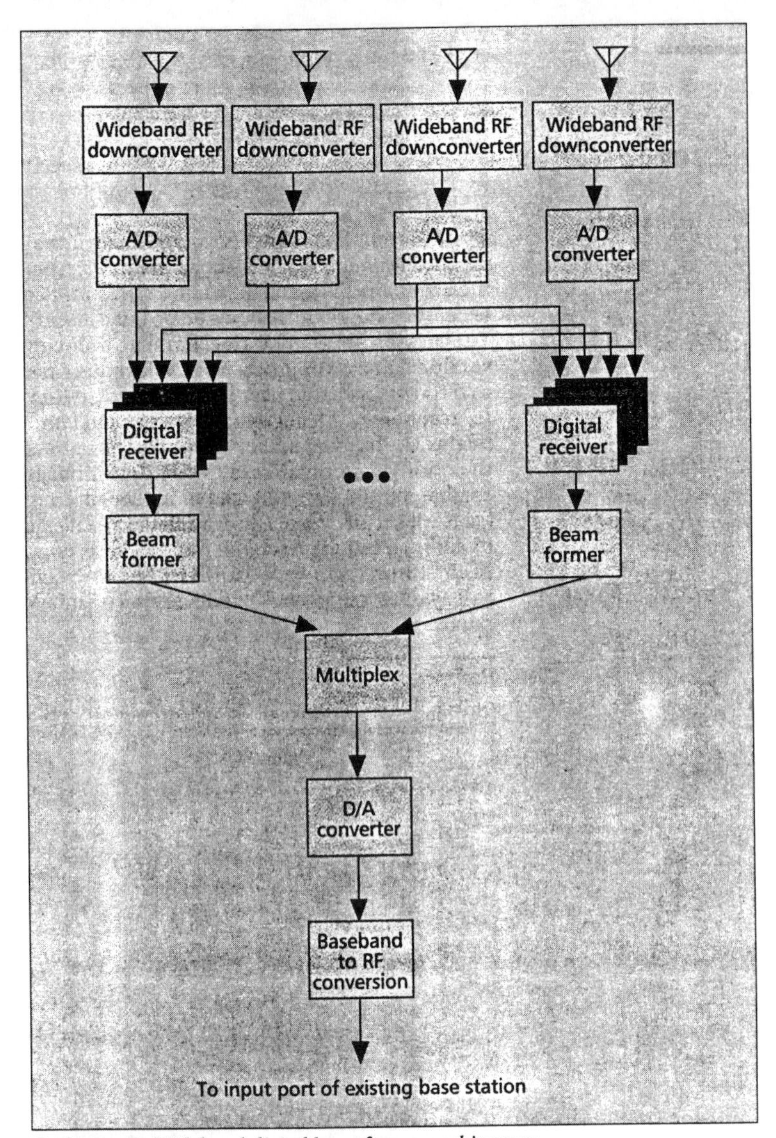

■ **Figure 7.** *Wideband digital beamformer architecture.*

Once they are calculated, the data is read out of RAM through the beamformer. Signals with unpredictable arrival times and directions of arrival may thus be accommodated using the RAM delay buffer to give the beamforming algorithm enough time to initialize the steering vector. The algorithm also separates and identifies signals by SAT frequency so that the beamformer response has gain in the directions of desired signal components and nulls in the directions of interfering components.

This architecture also uses newly developed techniques to implement "software transmission." Once the spatial filtering has been accomplished, the signal is reconstituted digitally and converted with high fidelity to RF so that the beamformed signals can be "received" by a standard base station through the omni or sector antenna port.

In current cellular systems with omnidirectional or sectored antennas, interference limits the choice of frequencies in each cell to one out of every seven. Preliminary analyses indicate that software-radio-based "Smart Antenna" appliqués can enhance the C/I in both the forward and reverse link to allow frequency reuse of 3 to 1. Use of this

In the future, software radios will be the technology of choice in an expanding array of mobile radio applications.

type of a smart array represents a greater than 2 to 1 increase in capacity and supports a 3 to 1 increase in call arrival rate for the same probability of blocked calls.

Summary

Software radio architectures, originally developed for military applications, are now becoming economically viable in commercial products because of the rapid advance of DSP technology. The advantages of digital techniques over conventional analog approaches are much the same as those of CD audio over the now-obsolete vinyl LP technology: higher dynamic range and better fidelity. As the demand for radio spectrum grows, the ability of sophisticated DSP algorithms to combat multipath fading and to reduce interference will become increasingly valuable as a means of adding capacity to cellular and PCS systems. In the future, software radios will be the technology of choice in an expanding array of mobile radio applications.

References

[1] D. N. Godard, "Self-recovering equalization and carrier tracking in two-dimensional data communication systems," *IEEE Trans. on Commun.*, vol. COM -28, no. 11, Nov. 1980, pp. 1867-1875.

[2] E. B. Hogenauer, "An economical class of digital filters for decimation and interpolation," *IEEE Trans. on Acoustics, Speech, and Signal Processing*, vol. ASSP-29, no. 2, April 1981, pp. 155-162.

[3] W. C. Y. Lee, Mobile Cellular Telecommunications, 2nd Ed., (New York: McGraw-Hill, 1995).

[4] S. K. Oh and C. K. Un, "Simple computational methods of the AP algorithm for maximum likelihood localization of multiple radiating sources," *IEEE Trans. on Signal Processing*, vol. 40, no. 11, Nov. 1992, pp. 2848-2854.

[5] J. O. Smith and J. S. Abel, "Closed-form least-squares source location estimation from range-difference measurements," *IEEE Trans. on Acoustics, Speech, and Signal Processing*, vol. ASSP-35, no. 12, Dec. 1987, pp. 1661-1669.

[6] B. J. Sublett, R. P. Gooch, and S. H. Goldberg, "Separation and bearing estimation of co-channel signals," Proceedings of the 1989 IEEE Military Communications Conference, Oct. 1989.

[7] I. Ziskind and M. Wax, "Maximum likelihood localization of multiple sources by alternating projection," *IEEE Trans. on Acoustics, Speech, and Signal Processing*, vol. 36, no. 10, pp. 1553-1560, Oct. 1988.

Biographies

JOSPEH KENNEDY is the director of the Advanced Communications Systems group at Engineering Research Associates, Vienna, Virginia, where he is currently responsible for the development of smart antenna and geolocation technology for government and commercial markets. He received an M.B.A. from Virginia Polytechnic Instutite, and M.S and B.S. degrees in electrical engineering from the George Washington University and the University of Virginia, respectively.

MARK C. SULLIVAN [M '85] is a scientist at Engineering Research Associates, Vienna, Virginia, where he is currently developing adaptive beamformers for commercial applications. He received a Ph. D. in information technology in 1994, an M.S. in statistical science in 1991, and a B.I.S. in signal processing in 1989, from George Mason University in Fairfax, Virginia. His e-mail address is sullivan@era.com.

Reprinted from *IEEE Transactions on Vehicular Technology*, Vol. 43, No. 3, Aug. 1994, pp. 680-690.

Analytical Results for Capacity Improvements in CDMA

Joseph C. Liberti, Jr., *Student Member, IEEE,* and Theodore S. Rappaport, *Senior Member, IEEE*

***Abstract*—In this paper, we examine the performance enhancements that can be achieved by employing spatial filtering in code division multiple access (CDMA) cellular radio systems. The goal is to estimate what improvements are possible using narrow-beam adaptive antenna techniques, assuming that adaptive algorithms and the associated hardware to implement these systems can be realized. Simulations and analytical results are presented which demonstrate that steerable directional antennas at the base station can dramatically improve the reverse channel performance of multicell mobile radio systems, and new analytical techniques for characterizing mobile radio systems which employ frequency reuse are described using the wedge-cell geometry of [1]. We also discuss the effects of using directional antennas at the portable unit. Throughout this paper we will use phased arrays and steerable, fixed pattern antennas to approximate the performance of adaptive antennas in multipath-free environments.**

I. Introduction

CURRENT day mobile radio systems are becoming congested due to growing competiton for spectrum. Many different approaches have been proposed to maximize data throughput while minimizing spectrum requirements for future wireless personal communications services [2], [3]. One way to increase capacity without added spectrum is to reduce cell sizes [4]. For this reason, cell sizes in emerging cellular communication systems are much smaller than cells used in land mobile cellular systems designed previously. This, however, also leads to increased infrastructure (base station) costs. Furthermore, to maximize capacity in CDMA systems, power control is required [5].

The reverse link (the link from the mobile unit to the base station) presents the most difficulty in CDMA cellular systems for several reasons. First of all, the base station has complete control over the relative power of all of the transmitted signals on the forward link; however, because of different radio propagation paths between each user and the base station, the transmitted power from each portable unit must be dynamically controlled to prevent any single user from driving the interference level too high for all other users [1]. Second, transmit power is limited by battery consumption at the portable unit, therefore there are limits on the degree to which power may be controlled. Finally, to maximize performance, all users on the forward link may be synchronized much more easily than users on the reverse link [6].

Adaptive antennas at the base station and possibly at the portable unit may mitigate these problems. In the limiting case of infinitesimal beamwidth and infinitely fast tracking ability, adaptive antennas can provide for each user a unique channel that is free from interference. All users within the system would be able to communicate at the same time using the same frequency channel, in effect providing space division multiple access (SDMA) [7]. In addition, a perfect adaptive antenna system would be able to track individual multipath components and combine them in an optimal manner to collect all of the available signal energy [8]. In this paper, we will investigate the effects of spatial filtering by simulating a phased array and by simulating antenna patterns with fixed patterns but adjustable boresight angles. Furthermore, multipath is not considered.

Clearly, the perfect adaptive antenna system described above is not feasible since it requires infinitely large antennas (or alternatively, infinitely high frequencies). This raises the question of what gains might be achieved using reasonably sized antenna arrays which operate at UHF and microwave frequencies.

While both TDMA and CDMA systems have been proposed for emerging personal communication systems, CDMA is more naturally suited to the pseudo-SDMA environment. This is because co-channel users do not have to be synchronized with each other in a CDMA system. As the advantages of SDMA are realized, the interference levels seen by each simultaneous CDMA user drop, and the bit-error performance will improve for each CDMA user. On the other hand, when no SDMA is achieved, CDMA performance is no worse than the case where omnidirectional antennas are used at both the base station and the portable unit. In a single cell TDMA system, users must be reassigned to new time slots to take any advantage of SDMA.

For interference limited asynchronous reverse channel CDMA over an additive white Gaussian noise (AWGN) channel, operating with perfect power control with no interference from adjacent cells and with omnidirectional

Manuscript received September 30, 1993; revised March 31, 1994.

The authors are with the Mobile and Portable Radio Research Group, Bradley Department of Electrical Engineering, Virginia Tech, Blacksburg, VA 24061.

IEEE Log Number 9403205.

antennas used at the basestation, the bit error rate (BER), P_b, is approximated by [6]

$$P_b \approx Q\left(\sqrt{\frac{3N}{K-1}}\right) \tag{1.1}$$

where K is the number of users in a cell and N is the spreading factor. $Q(Y)$ in (1.1) is the standard Q-function, the probability that $y > Y$ when y is a zero-mean, unit variance, Gaussian distributed random variable. Equation (1.1) assumes that the signature sequences are random and that K is sufficiently large to allow the Gaussian approximation described in [6] to be applied.

To illustrate how directive antennas can improve the reverse link in a single cell CDMA system, consider the case in which each portable unit has an omnidirectional antenna, and the base station tracks each user in the cell using a directive beam. Assume that a beam pattern, $G(\varphi)$, is formed such that the pattern has a maximum in the direction of the desired user.

Such a directive pattern can be formed at the base station using an N-element adaptive array illustrated in Fig. 1 [14], [16]. The array has N elements, each of which has K adaptive linear filters (ALF's) associated with it, if there are K users in the cell. Each ALF operates on the I and Q components of the signal from a single antenna array element. The resulting I components from all of the ALF's are summed and the Q components are summed to form the signal at the array port. Each of the ALF's may be adapted using a variety of techniques such as the use of training sequences, decision directed adaptation, and property restoral algorithms [9], [7], [10]. In the case of a narrowband array, each ALF simply takes the form of a complex tap weight. For wideband arrays, each ALF may take the form of a linear transversal filter or lattice filter [9].

When the ALF's are implemented digitally, it is possible to use a different set of ALF filter coefficients for each desired user, giving each desired user a distinct beam pattern. Each element of the array would have K ALF's associated with it, for a total of NK ALF's. The ALF coefficients for each of the K sets of N ALF's are adapted independently for each desired user.

Assume that a beam pattern, $G(\varphi)$, with no variation in the θ direction, such as that illustrated in Fig. 2, can be formed by the array. The pattern, $G(\varphi)$, can be steered through 360° in the horizontal (φ) plane such that the desired user (user 0) is always in the main beam of the pattern.

We assume that K users in the single cell CDMA system are uniformly distributed throughout a two-dimensional cell (in the horizontal plane, $\theta = \pi/2$). On the reverse link, the power received from the desired mobile signal is $P_{r;0}$. The powers of the signals incident at the base station antenna from the $K - 1$ interfering users are given by $P_{r;i}$ for $i = 1 \cdots K - 1$. Then the average total interference power, I, seen by a single desired user, measured in the received signal at the array port (as shown in

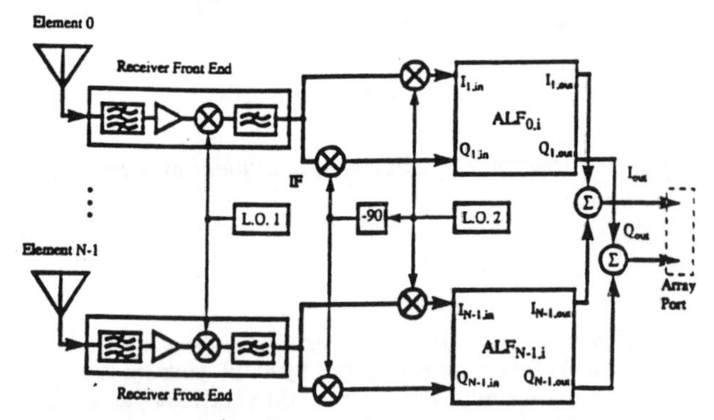

Fig. 1. A generalized adaptive antenna array with N elements. The inputs from each antenna are mixed down to an intermediate frequency and divided into I and Q components. The I and Q components from each antenna are filtered by an adaptive linear filter ($ALF_{i,j}$ is the ALF corresponding to the ith element and the jth user). The I outputs from each ALF are summed to provide I_{out}. Similarly the Q outputs from each ALF are summed to provide Q_{out}. I_{out} and Q_{out} form the signal which is available to the receiver.

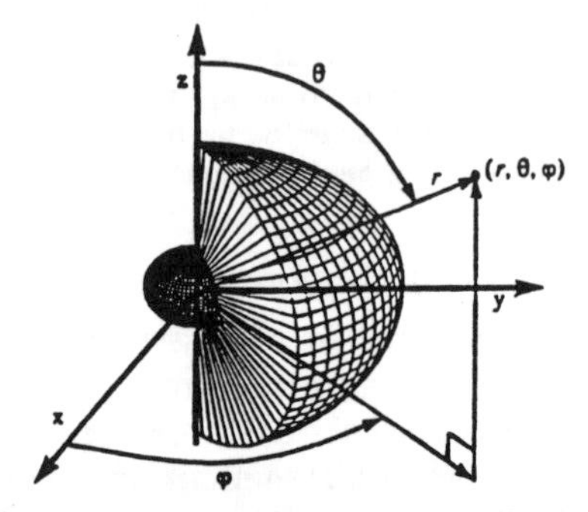

Fig. 2. An idealized flat-top power pattern with a 60° beamwidth and a −6 dB sidelobe level. This pattern has no variation in the θ direction (the elevation plane) for $0 \le \theta < \pi$. This coordinate system is used throughout this paper.

Fig. 1) of the base station antenna array, which is steered to user 0, is given by

$$I = E\left\{\sum_{i=1}^{K-1} G(\phi_i)P_{r,i}\right\} \tag{1.2}$$

where φ_i is the direction of the ith user in the horizontal plane, measured from the x-axis. No interference from outside the cell contributes to total received interference in (1.2). If perfect power control is applied such that the power incident at the base station antenna from each user is the same,[1] then $P_{r;i} = P_c$ for each of the K users, and the average interference power seen by user 0 is given by

$$I = P_c E\left\{\sum_{i=1}^{K-1} G(\phi_i)\right\}. \tag{1.3}$$

Assuming that users are independently and identically distributed throughout the cell, the average total interference power received at the central base station may be

[1]While this work considers adaptive antennas at the base station, power control could be implemented using a reference omnidirectional antenna at the base station to receive all mobile signals.

expressed as

$$I = P_c(K - 1) \int_0^R \int_0^{2\pi} f(r, \varphi)G(\varphi)\, d\varphi\, dr \quad (1.4)$$

where $f(r, \varphi)$ is the probability density function describing the geographic distribution of users throughout the cell. Assuming that users are uniformly distributed in the cell, we have

$$I = P_c \frac{(K - 1)}{2\pi} \int_0^{2\pi} G(\varphi)\, d\varphi. \quad (1.5)$$

The directivity of an antenna which has no variation in the θ direction is [11]

$$D = \frac{2\pi}{\int_0^{2\pi} G(\varphi)\, d\varphi}. \quad (1.6)$$

Therefore the average total interference seen by a user in the central cell is given by

$$I = \frac{P_c(K - 1)}{D}. \quad (1.7)$$

In order to develop simple bit error rate expressions for simultaneous asynchronous interference limited CDMA users when directive antennas are used, we assume that the bit-error-rate expression of (1.1) can be expressed as

$$P_b \approx Q(\sqrt{3N \times CIR}) \quad (1.8)$$

where N is the spreading factor, and CIR is the ratio of the power of the desired signal to the total interference. In (1.8), it is assumed that M interfering users, each with a received power level of P/M, have the same effect on bit error performance as one interfering user with a received power P. This assumption is known to be inaccurate when the powers of users are widely different and when the number of users is small [12]; however, it provides first order approximation for the case of a large number of users.

Using the fact that the power of the desired signal, weighted by the array pattern, is P_c and using (1.7), the bit error rate for user 0 is given by

$$P_b = Q\left(\sqrt{\frac{3DN}{K - 1}}\right). \quad (1.9)$$

Thus, (1.9) holds for any single cell system with perfect power control when base station antenna pattern which has no variation in the θ direction. Equation (1.9) is useful in showing that the probability of error for a CDMA system is related to the beam pattern of a receiver. If we use the idealized antenna pattern illustrated in Fig. 2 to approximate a realizable directive antenna pattern then it is immediately apparent that the gain of the antenna directly contributes to the performance of a CDMA system. For instance, if $K = 250$, and $N = 511$, with omnidirectional antennas at the base station, an average bit error rate of 6.6×10^{-3} is obtained per user. Using the flat-top beam pattern shown in Fig. 2 with a side lobe level of 0.25 and a main beamwidth of 60°, the directivity of the antenna is 2.67 or 4.3 dB. The bit error rate with the directive antenna at the base station is 2.5×10^{-5}, a BER improvement of two orders of magnitude.

This example illustrates the possible improvements that can be achieved using adaptive antennas at the base station. In the remainder of this paper, we remove the constraint that users in adjacent cells do not inerfere with the received signal, and develop a general analysis technique which is confirmed by simulation.

Section II describes analytical techniques used to determine bit error rates in cellular CDMA systems employing adaptive antennas.

Section III presents simulations in which we compare the performance of five base station antenna configurations, three of which use adaptively steerable antennas at the base station. It is assumed that the portable units use omnidirectional antennas. We also compare the simulation results with the analytical results developed in Section II.

In Section IV, the effects of adaptive antennas at the portable unit are examined using several different base station configurations. Furthermore, we demonstrate the two distinctly different effects achieved by using directive antennas at the portable unit versus using directive antennas at the base station. Finally, Section V summarizes the results of this paper.

II. Reverse Channel Performance with Adaptive Antennas at the Base Station

The use of adaptive antennas at the base station receiver is a logical first step in improving capacity for several reasons. First of all, space and power constraints are not nearly as critical at the base station as they are at the portable unit. Second, the physical size of the array does not pose as much difficulty at the base station as at the portable unit.

Note that adaptive antennas may also be used at the base station for directing energy in the forward channel, in which case the analysis is similar to the reverse channel case because of the perfect power control assumption. The only difference on the forward link is that interferers are other base stations, rather than portable users. Since the transmitter and receiver typically operate in two different frequency bands in a duplex manner, the adaptive antennas at the base station transmitter would be adjusted by performing a transformation on the tap weights adapted for the receiver, and copying the new weights to the transmitting antennas [9]. This is reasonable if an assumption of retrodirectivity on similar frequency bands is appropriate. If the multipath components arriving in the reverse channel do not have the same angles of arrival as those in the forward channel, then it is no longer appropriate to derive the transmitter tap weights from the received signal.

Equation (1.9) is only valid when a single cell is considered. To consider the effects of adaptive antennas when CDMA users are simultaneously active in several adjacent cells, we must first define the geometry of the cell region. For simplicity, we consider the geometry proposed in [1] with a single layer of surrounding cells, as illustrated in Figs. 3 and 4.

Let $d_{i,j}$ represent the distance from the ith user to base j as illustrated in Fig. 3. Let $d_{i,0}$ represent the distance from the ith user to base station 0, the center base station.

Assume that path loss in dB between user i and base j is given by a simple distance dependent path loss relationship such that the power received at base station j, from the transmitter of user i, $P_{r;i,j}$, is given by

$$P_{r;i,j} = P_{T;i}\left(\frac{\lambda}{4\pi d_{ref}}\right)^2\left(\frac{d_{ref}}{d_{i,j}}\right)^n \tag{2.1}$$

where n is the path loss exponent typically ranging between 2 and 4, and d_{ref} is a close-in reference distance [1].

If we assume that perfect power control is applied to the ith user, and all other users in cell j, by base j, such that power $P_{c;j}$ is received as base j, then the power transmitted by user i, $P_{t;i}$, is given by

$$P_{t;i} = P_{c;j}\left(\frac{4\pi d_{ref}}{\lambda}\right)^2\left(\frac{d_{i,j}}{d_{ref}}\right)^n. \tag{2.2}$$

The power received at base station 0 from user i, $P_{r;i,0}$ is given by

$$P_{r;i,0} = P_{t;i}\left(\frac{\lambda}{4\pi d_{ref}}\right)^2\left(\frac{d_{ref}}{d_{i,0}}\right)^n. \tag{2.3}$$

Substituting (2.2) into (2.3), the power received at base 0 from user i, in adjacent cell, j, is given by

$$P_{r;i,0} = P_{c;j}\left(\frac{d_{i,j}}{d_{i,0}}\right)^n \tag{2.4}$$

To analyze (2.4), we consider the geometry shown in Fig. 4.

From the law of cosines,

$$d_{i,j}^2 = (2R)^2 + (d_{i,0})^2 - 2(2Rd_{i,0})\cos\varphi_{i,0} \tag{2.5}$$

Substituting (2.5) into (2.4), the power received at base 0 from user i is given by

$$P_{r;i,0} = P_{c;j}\left(1 + \left(\frac{2R}{d_{i,0}}\right)^2 - \frac{4R}{d_{i,o}}\cos\varphi_{i,0}\right)^{n/2}. \tag{2.6}$$

To determine the average out-of-cell interference power incident on the central base station, we assume that users are uniformly distributed in a typical adjacent cell from $r = R$ to $r = 3R$ and from $\varphi = -\pi/8$ to $\pi/8$. Thus, we use a modified geometry from [1] where eight equal area cells surround the center cell. The probability density function (pdf) for the spatial distribution of users in a single adjacent cell is given by

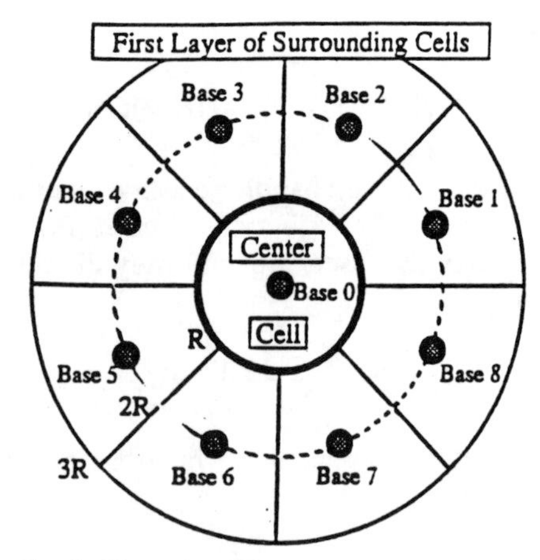

Fig. 3. The wedge cell geometry proposed in [1].

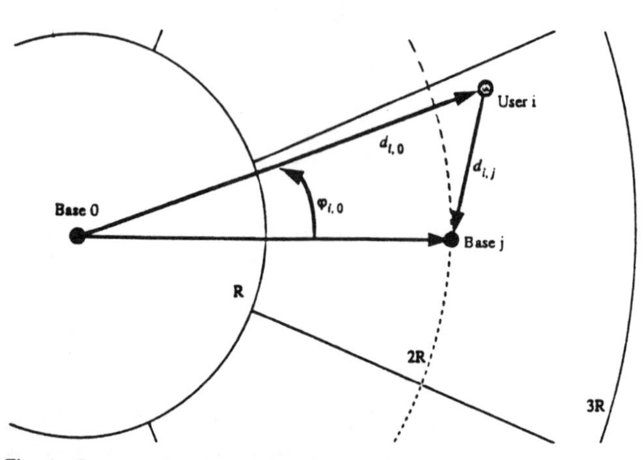

Fig. 4. Geometry for determining $d_{i,j}$ as a function of $d_{i,0}$, the distance between user i and the central base station and $\varphi_{i,0}$, the angle of the user i relative to the line between the central base station and base station j.

$$f(r, \varphi) = \frac{r}{\pi R^2} \qquad R < r < 3R: \qquad -\pi/8 < \varphi < \pi/8. \tag{2.7}$$

Let χ represent the expected value of the interference power from a single user in one of the adjacent cells when omnidirectional basestation antennas are used.

$$\chi = P_{c;j}\int_R^{3R}\int_{-\pi/8}^{\pi/8} f(r, \varphi) \cdot \left(\left(1 + \left(\frac{2R}{r}\right)^2 - \frac{4R}{r}\cos\varphi\right)\right)^{n/2} dr\, d\varphi \tag{2.8}$$

If it is assumed that all nine base stations control power such that $P_{c;j} = P_c$, then given a value of n, we can express the expected value of central cell interference power for a single adjacent cell user as

$$\chi = \beta P \tag{2.9}$$

where

$$\beta = \int_R^{3R} \int_{-\pi/8}^{\pi/8} f(r, \varphi) \left(\left(1 + \left(\frac{2R}{r}\right)^2 - \frac{4R}{r}\cos\varphi\right)\right)^{n/2} dr\, d\varphi \qquad (2.10)$$

Table I lists the values of β for several values of n.

When omnidirectional antennas are used at both the base station and the portable unit, β is related to the reuse factor, f, which is defined in [1], for a single layer of adjacent cells, as

$$f = \frac{N_0}{N_0 + N_{a1}M_1} \qquad (2.11)$$

where N_0 is the total interference, seen by a desired user in the central cell, at the central base station on the reverse link, N_{a1} is the total interference seen by the desired central cell user from all users in a single adjacent cell. M_1 is the number of cells which are immediately adjacent to the central cell, which is always eight for the geometry considered in this paper.

This reuse factor is a measure of the impact of users in adjacent cells on the performance of the link between a user in the central cell and the central base station.

When power control is performed as described in this section, such that the power received from each mobile unit in the base station controlling that unit is P_c, then (2.11) may be expressed as

$$f = \frac{(K-1)P_c}{(K-1)P_c + 8K\beta P_c} \approx \frac{1}{1 + 8\beta} \quad \text{for} \quad K \gg 1 \qquad (2.12)$$

where we have assumed that there are K users in each of the nine cells. For $n = 4$, from Table I, $\beta = 0.05513$, and, from (2.12), $f = 0.693$, implying that 31% of the interference power received at the central base station is due to users in adjacent cells. Note that, when omnidirectional antennas are used at both the base station and the portable unit, the value of the reuse factor, f, is determined by the cell geometry, the power control scheme, and the path loss exponent.

When omnidirectional antennas are used at both the base station and the portable unit, the total interference seen on the reverse link by the central base station is the sum of the interference from users within the central cell, $(K - 1)P_c$, and users in adjacent cells, $8K\beta P_c$.

$$I = (K-1)P_c + 8K\beta P_c. \qquad (2.13)$$

Let us assume that for the mth user in the central cell, an antenna beam from the base station with pattern, $G(\varphi)$, may be formed with maximum gain in the direction of user m. It is assumed that perfect power control is applied such that all base stations controlling reverse link received power to the same level, P_c. The average interference power contributed by a single user in the central cell

TABLE I
VALUES OF β AS A FUNCTION OF THE PATH LOSS EXPONENT, n AS DETERMINED BY (2.10)

n	β
2	0.14962
3	0.08238
4	0.05513

is thus given by

$$E[P_{r;i,0} | 0 < r < R] = P_c \int_0^R \int_0^{2\pi} \frac{r}{\pi R^2} G(\varphi)\, dr\, d\varphi = \frac{P_c}{D} \qquad (2.14)$$

where D is the directivity of the beam with pattern $G(\varphi)$ and the average received power at the base, $P_{r;i,0}$ from an interfering user in the central cell is directly a function of the base station directive gain. Then the average interference power at the array port of the antenna array at the base station, as shown in Fig. 1, due to a single user in an adjacent cell is given by

$$E[P_{r;i,0} | R < r < 3R] = \frac{1}{8}\sum_{p=0}^{7} \int_R^{3R} \int_{-\pi/8}^{\pi/8} G\left(\varphi + \frac{p\pi}{4}\right) \frac{r}{\pi R^2} \cdot P_c \left(1 + \left(\frac{2R}{r}\right)^2 - \frac{4R}{r}\cos(\varphi)\right)^{n/2} dr\, d\varphi. \qquad (2.15)$$

Here a special case is considered. If $G(\varphi)$ is piecewise constant over the region $(2p - 1)(\pi/8) < \varphi < (2p + 1)(\pi/8)$ for $p = 0 \cdots 7$, then the antenna pattern may be expressed as

$$G(\varphi) = \sum_{p=0}^{7} G_p U\left(\varphi - \frac{p\pi}{4}\right) \qquad (2.16)$$

where

$$U(\varphi) = \begin{cases} 1 & |\varphi| < \pi/8 \\ 0 & |\varphi| \geq \pi/8 \end{cases}. \qquad (2.17)$$

Substituting (2.16) into (2.15), we obtain,

$$E[P_{r;i,0} | R < r < 3R] = P_c \frac{1}{8}\sum_{p=0}^{7} G_p \int_R^{3R} \int_{-\pi/8}^{\pi/8} \frac{r}{\pi R^2} \left(1 + \left(\frac{2R}{r}\right)^2 - \frac{4R}{r}\cos(\varphi)\right)^{n/2} dr\, d\varphi. \qquad (2.18)$$

The directivity of the antenna pattern described by (2.16) is

$$D = \frac{8}{\sum_{p=0}^{7} G_p} \qquad (2.19)$$

Therefore, (2.18) may be rewritten, using (2.10) and (2.19), as

$$E[P_{r,i,0} \mid R < r < 3R] = \frac{P_c \beta}{D}. \tag{2.20}$$

It can be shown that (2.20) remains valid when the beam pattern, $G(\varphi)$, is rotated in the φ plane. Therefore (2.20) is appropriate when $G(\varphi)$ is piecewise constant over $(2p - 1)(\pi/8) < \varphi - \varphi_d < (2p + 1)(\pi/8)$ for any angle φ_d between $-\pi/8$ and $\pi/8$.

Using (2.20) with (1.7), the total interference power at the array port (in Fig. 1) of the center base station receiver is given by

$$I = \frac{(K - 1)P_c + 8KP_c\beta}{D}. \tag{2.21}$$

Substituting (2.21) into the (1.8), using the fact that the desired signal power at the array port is P_c, we obtain an average bit-error probability for the CDMA system employing a piecewise constant directive beam:

$$P_b \approx Q\left(\sqrt{\frac{3ND}{K(1 + 8\beta) - 1}}\right). \tag{2.22}$$

For $K >> 1$, P_b is approximated by

$$P_b \approx Q\left(\sqrt{\frac{3ND}{K(1 + 8\beta)}}\right). \tag{2.23}$$

Equation (2.23) relates the probability of error to the number of users per cell, the directivity of the base station antenna, and the propagation path loss exponent through the value of β. It is assumed that perfect power control is applied as described in Section I, with all base stations controlling reverse link received power to the same level, P_c.

III. Simulation of Adaptive Antennas at the Base Station for Reverse Channel Performance

To explore the utility for (2.23) and to verify its accuracy, we considered five base station antenna patterns which are illustrated in Fig. 5. These antenna patterns are assumed to be directed such that maximum gain is in the direction of the desired mobile users. The first-base station antenna pattern is an omnidirectional pattern which models that used in traditional cellular systems. This configuration, shown in Fig. 5(a) was used as a model for standard omnidirectional systems without adaptive antennas.

In order to make a fair comparison between the effects of various antenna types on bit error rate as a function of directivity, and given the fact that the simulations were performed in two dimensions only, antenna gains cited in this section are defined by (1.6) which is restated here:

$$D = \frac{2\pi}{\int_0^{2\pi} G(\varphi)\, d\varphi}. \tag{3.1}$$

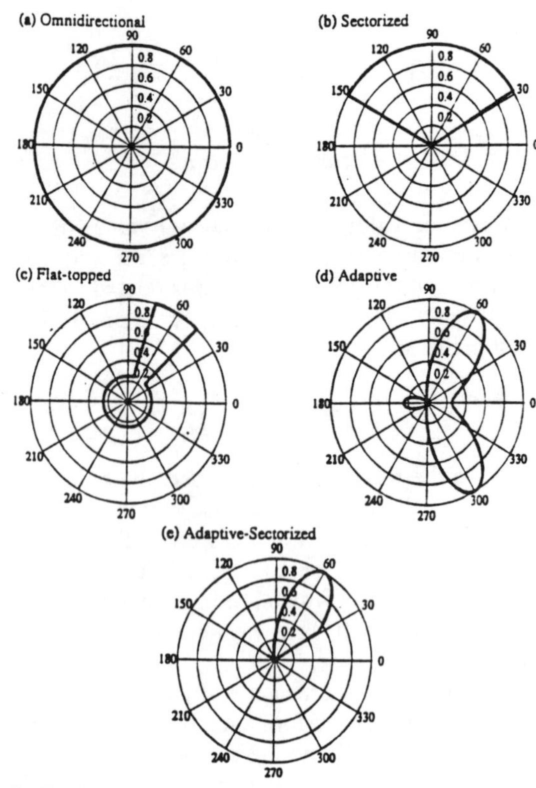

Fig. 5. The five base station antenna patterns used in this study. These patterns are shown for the case when the desired user is at an angle φ = 60° from the X axis. Shown here are (a) the omnidirectional pattern, (b) the 120° sectorized pattern, (c) the flat-topped pattern, (d) the three element binomial phased array (referred to as the "adaptive" pattern in this paper), and (e) the binomial phased array pattern overlaid with a 120° sectorization pattern (referred to as the "adaptive-sectorized" pattern).

The second configuration, illustrated in Fig. 5(b), used 120° sectorization at the base station. In our model, the base station used three sectors, one covering the region from 30° to 150°, the second covering the region from 150° to 270°, and the third covering the region from −90° to 30°. The first sector is illustrated in Fig. 5(b) since this sector would be active when the desired user is at an angle of 60°. In this system, only interfering users within view of the same sector as the desired user were included in the CIR calculation. The effective gain of this antenna is 4.8 dB.

The third simulated base station configuration, shown in Fig. 5(c), used a "flat-topped" beam pattern similar to that shown in Fig. 2. The main beam was 30° wide with uniform gain in the main lobe. Side lobes were simulated by assuming a uniform side lobe gain which was 6 dB below the main beam gain. From (1.6), the directivity of this beam is 5.1 dB.

The fourth configuration, which used a simple three element linear array, is illustrated in Fig. 5(d). This is the beam pattern formed by a binomial phased array with elements spaced a half wavelength apart. The axis of the array is in the φ = 0° direction. Like all linear arrays, this array exhibits a pattern which is symmetric about the axis of the array (the X-axis, as shown in Fig. 2), therefore a mirror image of the main beam is also present as illustrated in Fig. 5(d). This array is not capable of adaptively nulling interfering signals; therefore we expect the performance of this array to be poorer than that of a truly adaptive system. On the other hand, we did assume that the array was able to direct the one of the two main beam components in the direction of the desired user. For each desired user, the phase was computed for each element of the array and the new beam pattern was formed at the center cell base station. While the three-dimensional gain of a binomial phased array is constant at 4.3 dB regardless of scan angle, the two-dimensional gain defined by (3.1), which is more appropriate for comparison given our assumption of users in the horizontal plane only, varies between 2.6 and 6.0 dB, depending on scan angle, with the higher gain corresponding to broadside scan angles.

The pattern for the fifth simulated base station configuration, a sectorized adaptive antenna, is shown in Fig. 5(e). Beginning with the sectorizing system whose pattern is illustrated in Fig. 5(b), we added a three element linear phased array to each sector. The linear array for each sector is aligned such that the broadside direction is in the same direction as the center of the sector. This base station configuration actually uses a total of nine elements, however, only three elements are used to track any given user. For example, in Fig. 5(e), the desired user is at an angle of $\varphi = 60°$, therefore the first sector (30° to 150°) was active. The three-element linear array used for the first sector was used to further reduce beamwidth of the pattern. It was assumed that the backlobe of each of the antenna for each sector was negligible so that only users within the 120° wedge of each sector were illuminated by the beam of the phased array. This accounts for the sharp cutoff at $\varphi = 30°$. Due to the elimination of the backlobes, this pattern has a significantly higher gain than the beam pattern shown in Fig. 5(d). At broadside (for a user in the center of a sector), the gain of this pattern is approximately 10.7 dB.

To evaluate the performance of these systems, a simulation was designed using the simple wedge geometry illustrated in Fig. 3. Users were randomly placed throughout the region with an average of K users per cell. Each user was assigned to one of the nine cells based on geographical location. The path loss from each user to the base station assumed to follow (2.1). Perfect power control was applied to each user within its own cell, as described in Section II, such that the incident power from each user at the in-cell base station antenna was a constant.

The carrier to interference ratio was calculated for each user in the central cell and the bit error rate was determined for each in-cell user by assuming that all users were asynchronous and by applying the Gaussian approximation. We define $P_{i,j,k}$ as the component of the received power at the array port (shown in Fig. 1) of the base station antenna array (weighted by the array pattern) at the kth base station from the ith user associated with cell j. The CIR for the ith user in the central cell (cell $j = 0$) was calculated from

$$CIR_i = \frac{\overbrace{P_{i,0,0}}^{\text{Desired Signal}}}{\underbrace{\sum_{\substack{n=0 \\ n \neq i}}^{K-1} P_{n,0,0}}_{\text{In-cell Interference}} + \underbrace{\sum_{m=1}^{8} \sum_{n=0}^{K-1} P_{n,m,0}}_{\text{Out-of-cell Interference}}} \tag{3.2}$$

The bit error rate for the ith user in cell 0 on the reverse link was determined by first calculating the CIR for the ith user from (3.2) then using that value in (1.8), which is restated here:

$$P_{b,i} = Q(\sqrt{3N \times CIR_i}) \tag{3.3}$$

where N is the spreading factor. For each of the simulations performed in this study, a spreading factor of $N = 511$ was used. It was assumed that any portable unit in the nine-cell region (except for the desired user) contributed to the interference level of the desired user in the central cell.

This calculation was carried out for every user in the central cell and the resulting bit error rates were averaged to obtain an average bit error rate for the cell. For instance, if there were 2700 users in the nine cells and 300 users in the central cell, then the bit error rate was determined for the 300 users in the central cell, and 2699 interfering users contributed to each CIR computation. Each base station configuration was simulated for user densities ranging from 25 to 500 users per cell, in steps of 25.

Fig. 6 shows average bit error rates resulting from the simulation for the five previously described antenna patterns for several values of path loss exponent, n. The three element linear array, whose pattern is shown in Fig. 6(d), was able to achieve almost an order of magnitude improvement in BER despite the large backlobe. By eliminating the large back lobe, but still retaining significant side lobes, the flat-top pattern, shown in Fig. 6(c), achieves a BER which is better than two orders of magnitude less than the BER when omnidirectional antennas are used at the base station, with fewer than 200 users per cell.

The average bit error rate alone is not a sufficient metric of system performance. Rather, the distribution of BER's over the user population is a second-order measure which provides insight about the performance of a CDMA cellular system. Fig. 5 relates the average BER to the BER which is not exceed by 50, 90, 95, and 99% of the users. Note that for a given bit error rate, two to four times as many users many be supported using directional antennas as for omnidirectional antennas. It is useful to note that these increases in performance were made by applying relatively modest requirements to the base-station adaptive antenna. The flat top antenna was specified to have a 30° beamwidth and a side lobe level that was only 6 dB below the main lobe.

It should be noted that these bit-error-rate improve-

TABLE II

RELATIONSHIP BETWEEN THE AVERAGE BIT ERROR RATE, AND $P_{e,x}$ WHERE $P_{e,x}$ IS DEFINED SUCH THAT x% OF THE USERS IN THE CENTRAL CELL HAVE A BIT ERROR RATE WHICH IS LESS THAN $P_{e,x}$. THIS IS FOR THE CASE OF $K = 200$, AND A PATH LOSS EXPONENT OF $n = 2$. NOTE THAT THERE IS A MUCH WIDER RANGE OF BIT ERROR RATES FOR THE HIGHER GAIN ANTENNAS. FOR EXAMPLE, 2 USERS OR 1% OF THE USER POPULATION EXPERIENCED A BER WHICH WAS WORSE THAN 1.5e-3 WHEN THE SECTORIZED ANTENNA PATTERN WAS CONSIDERED

Base Station Pattern	Avg BER	$P_{e,50}$	$P_{e,90}$	$P_{e,95}$	$P_{e,99}$
Omni	3.0e-2	3.1e-2	3.2e-2	3.2e-2	3.2e-2
Sectorized	6.1e-4	5.3e-4	1.0e-3	1.1e-3	1.5e-3
Adaptive	2.9e-3	2.6e-3	6.4e-3	7.2e-3	7.7e-3
Flat-topped	4.0e-4	3.9e-4	5.2e-4	5.6e-4	6.5e-4
Adaptive-Sectorized	1.6e-7	6.5e-8	3.0e-7	4.7e-7	2.4e-6

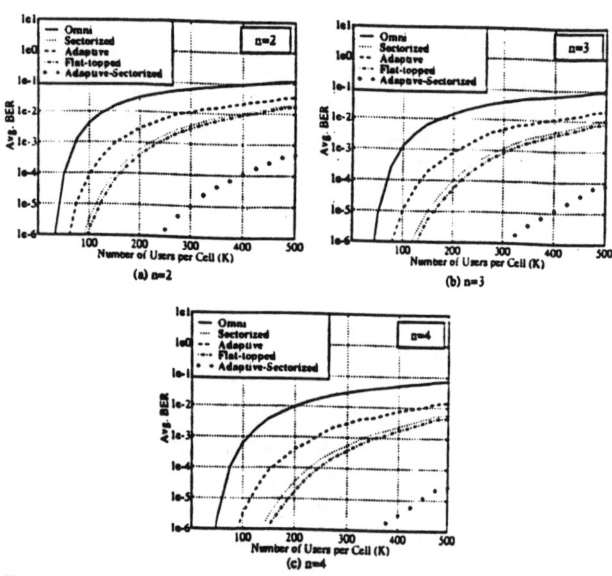

Fig. 6. BER using adaptive antennas at the base station for (a) $n = 2$, (b) $n = 3$, and (c) $n = 4$. These results were developed through simulation by averaging the BER of every user in the center cell.

ments are primarily due to the directivity of the antenna array. The improvements are also dependent on the geographical distribution of interfering users, but in the case of uniformly distributed users, as noted in Section II, the improvement is approximately equivalent to increasing the carrier-to-interference ratio by the gain of the directional antenna.

Even more drastic improvements were available when sectorization was combined with the adaptive antenna approach. Adding the three element array to the sectorized system, as shown in Fig. 5(e), provided a reduction in BER of three orders of magnitude for 200 users per cell.

Fig. 6 shows results calculated results from (2.23) for four of the antenna patterns shown in Fig. 5. By comparing Figs. 6 and 7, it can be seen that for omnidirectional antennas, 120° sectorization, and the flat-topped pattern, the calculated bit error rates from (2.23) matches the simulation results exactly, even for a relatively small number of users ($K = 25, 50$). For the case of the binomial phased

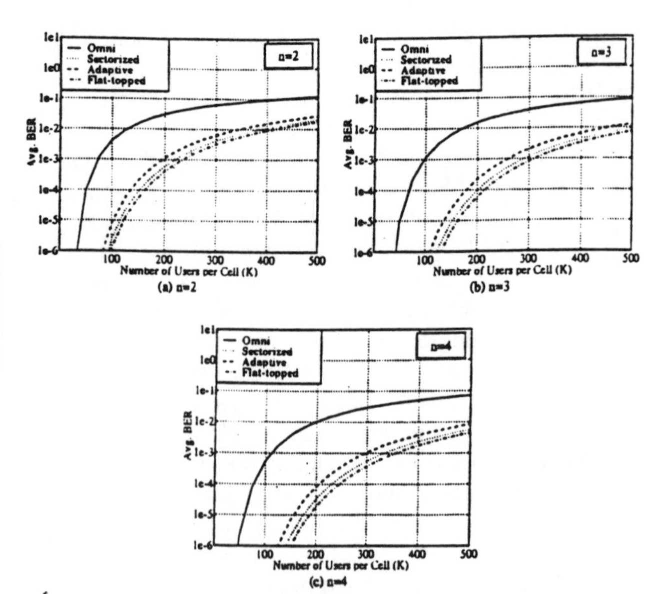

Fig. 7. Plots of analytical results using equation 2.23 with two-dimensional directivities of 1.0, 2.67, 3.0, and 3.2 for the omni, adaptive, sectorized and flat-topped patterns, respectively.

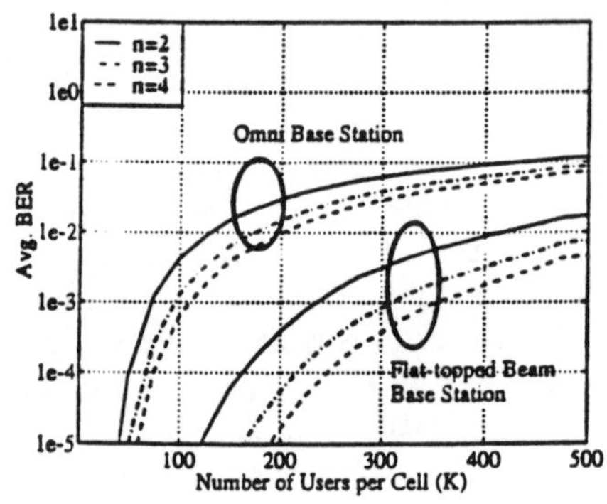

Fig. 8. BER for the omni and flat-topped beam systems as a function of n.

array, the analytical results for P_b are optimistic by almost an order of magnitude when $K < 200$ for all values of n. For $K > 350$, the analytical results for P_b are only smaller than the simulation results by a factor of 0.3 or less. Unlike the omnidirectional, sectorized, and flat-topped patterns, the binomial phased array did not exhibit constant two-dimensional gain as a function of scan angle. Therefore, the use of the three-dimensional directive gain as an "average" gain in (2.23) is an approximation. By comparing Figs. 6 and 7 it may be concluded that a smaller value of average directive gain might result in a better match between the simulated and analytical results for the binomial phased array. Nevertheless, these figures demonstrate the accuracy of (2.23) when compared with extensive simulations.

As noted in [13], use of a path loss exponent of $n = 4$ can result in overly optimistic estimates of system capacity and performance. The different base station antenna configurations demonstrate varying sensitivity to the path loss exponent, n. As illustrated in Fig. 8, the flat-topped beam system is highly sensitive to changes in the path loss exponent. This is reasonable to expect since, when the CIR is large, the bit error rate is more sensitive to relatively small changes in interference power.

IV. Simulation of Adaptive Antennas at the Portable Unit to Improve Reverse Channel Performance

In this section, we examine how the reverse channel is affected by using adaptive antennas at a portable transmitter. A flat-topped beam shape, as illustrated in Fig. 2, was used to model an adaptive antenna at the portable transmitter. Since space is extremely limited on the portable unit, the gain achievable by the portable unit antenna will be considerably less than that at the base station. For this study, it was assumed that the portable unit could achieve a beamwidth of 60° with a side lobe level that was 6 dB down from the main beam. This corresponds to an antenna with a directivity of 4.3 dB. The pattern is similar to that shown in Fig. 5(c) except that the beamwidth is wider in this case.

It was assumed that each portable unit was capable of perfectly aligning the boresight of its adaptive antenna with the base station associated with that portable unit. In this manner, portable units could radiate maximum energy to the desired base station, while reducing battery power proportional to the directivity of the portable antenna.

Portable units with adaptive antennas were simulated for each of the five base station patterns described in Section III. As in Section III, average values of P_b were found by averaging the bit error rates of each user in the central cell, subjected to interference from the central cell and all immediately adjacent cells. The resulting bit error rates for these systems are shown in Fig. 9. Note that, comparing Fig. 6 and Fig. 9, the bit error rates for the reverse channel are improved when directive antennas are used at the portable unit. For omnidirectional base stations, the BER is only decreased by a small amount (20% or less) for $K > 200$ when steerable directive antennas are used at the portable unit. However, for highly directive base station antenna patterns such as the adaptive-sectorized pattern, the BER was decreased by an order of magnitude for $K > 300$.

In Fig. 10, we have defined the BER factor as the ratio of the BER with adaptive antennas at all portable units to the BER without adaptive antennas at the portable units. A small BER factor indicates that adding adaptive antennas improved the BER significantly. For example, a BER factor of 0.5 indicates that using an adaptive antenna at the mobile unit resulted in a reduction in BER of 50% compared with the case of omnidirectional antennas at the mobile unit.

As shown in Fig. 10, the adaptive sectorized base station pattern improved greatly by adding adaptive antennas at the portable unit. The resulting BER for this base station configuration when using adaptive antennas at the portable unit was decreased by an order of magnitude compared with the BER when omnidirectional antennas were used at the portable unit. In general, the more directive base station configurations benefitted more from adding adaptive antennas at the portable unit. Using a 60° beamwidth flat-topped pattern with a −6 dB side lobe level at the portable unit, the reverse channel BER for omnidirectional base stations was only improved slightly over the case of omnidirectional antennas at the portable. For directive antennas at the base station, the improvements were more dramatic, as illustrated in Fig. 10.

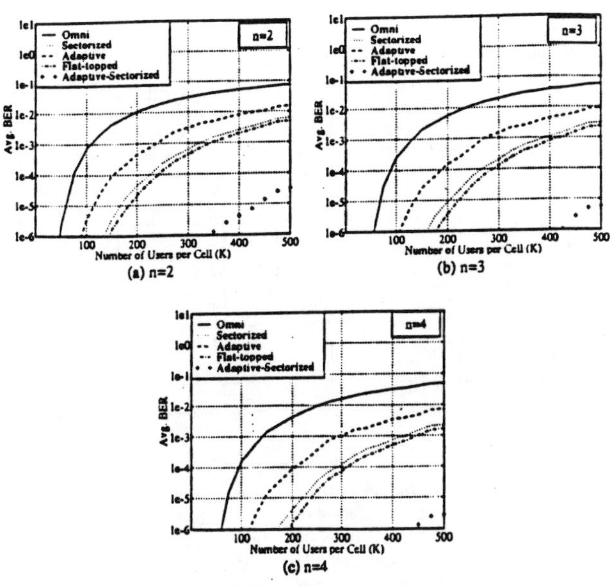

Fig. 9. BER for five different base station configurations using adaptive antennas at the portableunit for (a) $n = 2$, (b) $n = 3$, and (c) $n = 4$. These results were developed through simulation by averaging the BER of every user in the central cell.

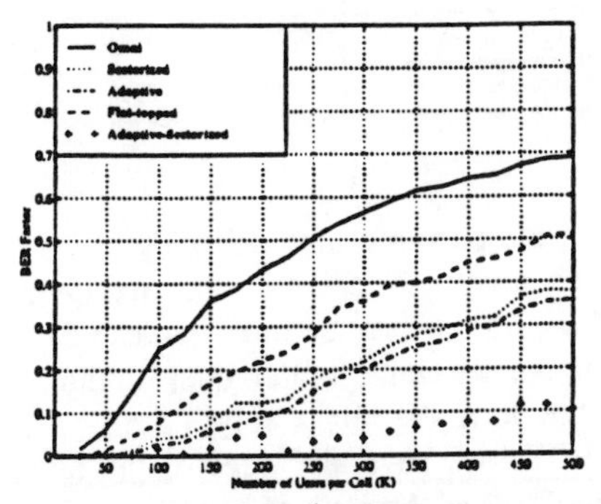

Fig. 10. BER factor, defined as the ratio of the BER with adaptive antennas at the portable unit to the BER without adaptive antennas at the portable, for five different base station configurations using when using adaptive antennas at the portable unit. This comparison is made for $n = 4$.

The relatively small improvements obtained by using adaptive antennas at the portable unit can be explained by the fact that when omnidirectional antennas are used at the mobile unit, no more than 1-0.455, or 0.545, of the total interference power is due to users in adjacent cells (see Table III where $f = 1/(1 + 8\beta)$). When using adaptive antennas at the mobile unit, all users in the central cell will appear no different to the central base station than if they had used omnidirectional antennas. Thus, adaptive

TABLE III
RATIO OF IN-CELL INTERFERENCE TO TOTAL INTERFERENCE, f, AS A FUNCTION OF PATH LOSS EXPONENT, FOR FIVE BASE STATION ANTENNA PATTERNS WITH OMNIDIRECTIONAL ANTENNAS AT THE PORTABLE UNIT

Base station antenna pattern	n=2	n=3	n=4
Omni	0.4535	0.6012	0.6927
Sectorized	0.4532	0.6008	0.6924
Adaptive	0.4524	0.6002	0.6920
Flat-topped	0.4534	0.6011	0.6926
Adaptive-sectorized	0.4531	0.6007	0.6922
$\frac{1}{1+8\beta}$ (Eq. 2.13) (values of β from Table 2.1)	0.4552	0.6028	0.6939

TABLE IV
RATIO OF IN-CELL INTERFERENCE TO TOTAL INTERFERENCE, f, AS A FUNCTION OF PATH LOSS EXPONENT, FOR FIVE BASE STATION ANTENNA PATTERNS WITH ADAPTIVE ANTENNAS AT THE PORTABLE UNIT. THIS DATA IS FROM THE SIMULATION DESCRIBED IN SECTION IV

Base station antenna pattern	n=2	n=3	n=4
Omni	0.6752	0.8155	0.8826
Sectorized	0.6749	0.8153	0.8824
Adaptive	0.6753	0.8152	0.8822
Flat-topped	0.6751	0.8154	0.8826
Adaptive-sectorized	0.6747	0.8150	0.8823

antennas at the portable unit will *only reduce out-of-cell interference levels*. Therefore, the maximum improvement in *CIR*, on the reverse link, that can be achieved by using adaptive antennas rather than omnidirectional antennas at the portable unit is only 3.5 dB.

Table III shows several values of the reuse factor, f, defined in (2.12) as the ratio of in-cell interference to total interference, for several base station patterns when omnidirectional antennas are used at the portable unit. Similarly, Table IV shows values of f when steerable, directional antennas, with directivities of 4.3 dB, are used at the portable units.

Comparing Tables III and IV, it can be concluded that the use of adaptive antennas at the base station does nothing to improve the reuse factor, f; however the use of adaptive antennas at the portable unit does allow f to be improved. When omnidirectional antennas are used at the portable unit, f is entirely determined by the cell geometry, the power control scheme, and path loss exponent, n, which is a function of propagation and not easily controlled by system designers. Using adaptive antennas at the portable unit, it is possible to tailor f to a desired value which is greater than the reuse factor obtained using omnidirectional antennas at the portable unit. Ideally, driving f to unity would allow system design to much less sensitive to the intercell propagation environment, when perfect power control is assumed.

This is an important result for CDMA cellular systems because it indicates that use of adaptive antennas at the portable unit could help to allow greater capacity through more efficient reuse, and for more frequent reuse of signature sequences throughout a large coverage area.

V. CONCLUSIONS

It was shown in this study that adaptive antennas, with relatively modest bandwidth requirements, and no interference nulling capability, both at the base station and at the portable, can provide large improvements in BER, as compared to omnidirectional systems. Analytical expressions which relate the average BER of a CDMA user to the antenna directivity and propagation environment were derived and used to determine capacity improvements offered by a number of antenna patterns. It was demonstrated in Section III that the linear phased array provided an order of magnitude of improvement over the omnidirectional base station. The low-gain (5.1 dB) flat-top pattern provided almost two orders of magnitude of improvement over the omnidirectional system. In addition, it was shown that up to three orders of magnitude of improvement can be achieved by adding a simple three element linear array to a three-sector base station. In terms of capacity, the results of Section III indicate that using adaptive antennas at the base station can allow the number of users to increase by a factor of 2 to 4, while maintaining an average BER of 10^{-3} on the reverse link.

The bit error rate on the reverse channel is further improved by adding adaptive antennas at the portable unit. Using a 4.3 dB gain antenna at the portable, the bit error rate for the directive base station configurations (but not the omnidirectional base station) was at least half of the bit error achieved without directive antennas at the portable unit. For the highly directive adaptive sectorized base station, the improvement was over an order of magnitude for user densities less than 425 users/cell when each user employed an adaptive antenna.

Since the directivity of portable unit adaptive antennas is limited by the size of a handheld device, improvements achieved on the reverse channel at the portable are not as dramatic as gains achieved by adaptive antennas at the base station. In addition, cost issues may limit the application of portable unit adaptive antennas. However, the reduction in reverse channel BER may be critical in extremely high traffic environments. In addition, the portable unit is required to track the only current base station, while adaptive antennas at the base station must track every user in the cell. It should be noted, however, most importantly, Tables III and IV showed the increase in reuse efficiency which portable adaptive antennas provide. By using modest gains at the portable unit, such antennas ameliorate the loss in capacity due to intercell propagation through interference control.

In short, adaptive antennas at the base station can have a major effect on bit-error-rate performance, but cannot impact the reuse factor, f. Conversely, it has been shown in this paper that adaptive antennas at the portable unit can provide no more than a 3.5 dB improvement in reverse channel *CIR*; however, they allow the reuse factor,

f, to be altered. It should be noted, however, that the use of directional antennas at the portable unit can only result in an increase in reuse factor of approximately 1/3.

It was assumed throughout this study that the adaptive algorithms and hardware could be designed to meet the specified requirements on beamwidth, side lobe level, and tracking ability. It should be noted that, unlike the arrays discussed in this paper, a properly designed adaptive array can null out interference. Conversely, tracking a large number of users with an adaptive array is nontrivial, and it was assumed that each of the base station arrays described here were able to track all of the portable units without error.

The multipath channel was not considered in detail in this study; however, it will be significant in developing algorithms for successful adaptive antenna steering. Rather than tracking users, the adaptive array in a multipath environment must track the angle of arrival of multipath components in order to distinguish the maximum signal. This problem is currently under investigation. Furthermore, efforts are currently underway to develop bit error rate expressions which are accurate for small numbers of simultaneous CDMA users with non-identical power levels.

REFERENCES

[1] T. S. Rappaport and L. B. Milstein, "Effects of radio propagation path loss on DS-CDMA cellular frequency reuse efficiency for the reverse channel," *IEEE Trans. Veh. Technol.*, vol. 41, no. 3, Aug. 1992.

[2] G. R. Cooper and R. W. Nettleton, "A spread-spectrum technique for high-capacity mobile communications," *IEEE Trans. Veh. Technol.*, vol. VT-27, Nov. 1978.

[3] A. Salmasi, "An overview of advanced wireless telecommunication systems employing code division multiple access," *Conf. Mobile, Portable & Personal Commun.*, Kings College, England, Sept. 1990.

[4] W. C. Y. Lee, *Mobile Cellular Telecommunications Systems.* New York: McGraw Hill, 1989.

[5] K. S. Gilhousen *et al.*, "On the capacity of a cellular CDMA system," *IEEE Trans. Veh. Technol.*, vol. 40, May 1991.

[6] M. B. Pursley, "Performance evaluation for phase-coded spread spectrum multiple-access communications with random signature sequences," *IEEE Trans. Commun.*, vol. COM-25, Aug. 1977.

[7] W. A. Gardner, S. V. Schell, and P. A. Murphy, "Multiplication of cellular radio capacity by blind adaptive spatial filtering," *IEEE Conf. Sel. Topics Wireless Commun. Mobile*, Vancouver, B.C., Canada, Jun 1992.

[8] S. C. Swales, M. A. Beach, D. J. Edwards, and J. P. McGeehan, "The performance enhancement of multibeam adaptive base-station antennas for cellular land mobile radio systems," *IEEE Trans. Veh. Technol.*, vol. 39, Feb. 1990.

[9] R. T. Compton, *Adaptive Antennas.* Englewood Cliffs, NJ: Prentice Hall, 1988.

[10] B. Agee, "Solving the near-far problem: Exploitation of spatial and spectral diversity in wireless personal communication networks," in *Proceedings Third Virigina Tech Symp. Wireless Personal Commun.*, June 1993.

[11] W. L. Stutzman and G. A. Thiele, *Antenna Theory and Design.* New York: Wiley, 1981.

[12] R. K. Morrow and J. S. Lehnert, "Bit-to-bit error dependence in slotted DS/SSMA packet systems with random signature sequences," *IEEE Trans. Commun.*, vol. 37, Oct. 1989.

[13] L. B. Milstein, T. S. Rappaport, and R. Barghouti, "Performance evaluation for cellular CDMA," *IEEE JSAC*, vol. 10, May 1992.

[14] B. Widrow, P. E. Mantey, L. J. Griffiths, and B. B. Goode, "Adaptive antenna systems," *Proc. IEEE*, vol. 55, no. 12, Dec. 1967.

[15] R. Kohno, H. Imai, M. Hatori, and S. Pasupathy, "Combination of an adaptive array antenna and a canceller of interference for direct-sequence spread-spectrum multiple-access system," *IEEE JSAC*, vol. 8, May 1990.

[16] S. Anderson, M. Millnert, Mats Viberg, and Bo Wahlberg, "An adaptive array for mobile communication systems," *IEEE Trans. Veh. Technol.*, vol. 40, Feb. 1991.

Joseph C. Liberti Jr. (S'89) was born in Ft. Belvoir, VA, Blacksburg, on October 20, 1967. He received the B.S.E.E. and M.S.E.E. degrees from Virginia Tech in 1989 and 1991 respectively.

Since 1990 he has been with the Mobile and Portable Radio Research Group (MPRG) at Virginia Tech. He is currently working towards the Ph.D. degree as a Bradley Fellow in Electrical Engineering. He has interned with Motorola, Inc., at the Applied Research group in Boynton Beach, FL, and has worked developing cellular monitoring equipment for TSR Technologies, Inc. in Blacksburg. His research interests include adaptive signal processing, PCS and cellular system design, and radiowave propagation measurement and predition. His current work involves the development of an adaptive antenna testbed for CDMA communications systems and he contributes to the development of site specific propagation prediction software at the MPRG.

Theodore S. Rappaport was born in Brooklyn, NY on November 26, 1960. He received B.S.E.E., M.S.E.E., and Ph.D. degrees from Purdue University in 1982, 1984, and 1987, respectively. In 1988, he joined the Electrical Engineering faculty of Virginia Tech, Blacksburg, where he is an associate professor in the Mobile and Portable Radio Research Group (MPRG), a group he founded in 1990. He conducts research in mobile radio communication system design, RF propagation prediction and measurements, and digital signal processing. He guides a number of graduate and undergraduate students in mobile radio communications, and has authored or co-authored numerous papers in the areas of wireless system design and analysis, propagation, vehicular navigation, and wideband communications. Dr. Rappaport holds several U.S. patents and is co-inventor of SIRCIM and SMRCIM, indoor and microcellular radio channel software simulators that has been adopted by more than 100 companies and universities. In 1990, he received the Marconi Young Scientist Award for his contributions in indoor radio communications, and was named a National Science Foundation Presidential Faculty Fellow in 1992. He received the 1992 IEE Electronics Letters Premium award for the paper "Path Loss Prediction in Multi-Floored Buildings at 914 MHz" which he co-authored. Dr. Rappaport has edited two tooks published by Kluwer Academic Press on the subject of wireless personal communications, and has contributed chapters on the subject for the CRC Engineering Handbook series. He is an active member of the IEEE, and serves as a senior editor of the IEEE Journal on Selected Areas in Communications. He also serves on the editorial boards of the IEEE Personal Communications Magazine and the International Journal for Wireless Information Networks (by Plenum). Dr. Rappaport is a Registered Professional Engineer in the State of Virginia and is a Fellow of the Radio Club of America.

Reprinted from *IEEE Transactions on Vehicular Technology*, Vol. 43, No. 3, Aug. 1994, pp. 691-698

Capacity Improvement with Base-Station Antenna Arrays in Cellular CDMA

Ayman F. Naguib, *Student Member, IEEE,* Arogyaswami Paulraj, *Fellow, IEEE,* and Thomas Kailath, *Fellow, IEEE*

***Abstract*—In this paper, the use of antenna array at base-station for cellular CDMA is studied. We present a performance analysis for a multicell CDMA network with an antenna array at the base-station for use in both base-station to mobile (downlink) and mobile to base-station (uplink) links. We model the effects of path loss, Rayleigh fading, log-normal shadowing, multiple access interference, and thermal noise, and show that by using an antenna array at the base-station, both in receive and transmit, we can increase system capacity several fold. Simulation results are presented to support our claims.**

I. Introduction

THE increasing demand for mobile communication services without a corresponding increase in RF spectrum allocation motivates the need for new techniques to improve spectrum utilization. One approach for increased spectrum efficiency in digital cellular is the use of spread spectrum code-division multiple-access (CDMA) technology [1], [2]. Despite the high capacity offered by CDMA technology, the expected demand is likely to outstrip the projected capacity with the introduction of Personal Communication Networks (PCN). One approach that shows real promise for substantial capacity enhancement is the use of spatial processing with cell site antenna array [3]–[11]. By using spatial processing at the cell site, we can estimate the array response vector and use optimum directional receive and transmit beams to improve system performance and increase capacity. Such improved antenna processing can be incoporated into the proposed CDMA transmission standards. The increase in system capacity by using antenna arrays in CDMA comes from reducing the amount of co-channel interference from other users within its own cell and neighboring cells. This reduced interference transforms to an increase in capacity. The currently proposed IS-95 CDMA standard already incorporates a degree of spatial processing through the use of simple sectored antennas at the cell site. It employs three receive and transmit beams of width 120° each to cover the azimuth. Sectoring nearly triples system capacity in CDMA. While it might appear that even narrower sectors might yield further capacity gains, simple planar wavefront assumptions used in sectoring are not valid for narrow beams that employ large apertures. Simple sectoring, therefore, suffers from significant losses and motivates the need for "smart antennas" that adapt to a dynamic spatial channel seen by the cell site antenna array.

In this paper, we study the capacity improvement of multicell CDMA cellular system with base-station antenna array for both the downlink and the uplink. As in the proposed IS-95 CDMA standard, we assume that the uplink and the downlink occupy different frequency bands. We adopt the Rayleigh fading and log-normal shadowing model in [12] to model signal level. In this model, the fast fading around the local mean has a Rayleigh distribution. Due to shadowing, the local mean fluctuates around the area mean with a log-normal distribution and standard deviation σ_s, which varies between 6 to 12 dB, depending on the degree of shadowing. We also assume that the received signal power falls off with distance according to a fourth power law. That is, the path loss between the user and the cell site is proportional to r^{-4} where r is the distance between user and cell site.

In the next section, we analyze system capacity in uplink, where each signal propagates through a distinct path and arrives at the base station with independent fading. In Section III, we also analyze system capacity in downlink where all signals received at the mobile from the same base station undergo the same fading. Next, in Section IV we present simulation results. Finally, Section V contains our concluding remarks.

II. Mobile-To-Base Link

We assume that the cell site alone uses a multielement antenna array to receive and transmit signals from and to the mobile. No antenna arrays are considered for the mobile due to practical difficulties in implementing such a concept. Consider a scenario where there are N users randomly distributed around each cell site at varying ranges. We assume that the receiver is code locked onto every user but does not know the direction of-arrival (DOA) of these users. Each user transmits a PN code modulated bit stream with a spreading factor (processing gain) of L. Let P be the received signal power at the cell site, let the system noise power (excluding interference from other in-

Manuscript received December 1, 1993; revised April 10, 1994. This research was supported in part by the SDIO/IST Program managed by the Army Research Office under grant DAAH04-93-G-0029. This paper was presented in part in the 27-th Asilomar Conference on Computer, Signals, and Systems.

The authors are with the Information Systems Laboratory, Stanford University, Stanford, CA 94305 USA.

IEEE Log Number 9403206.

0018-9545/94$04.00 © 1994 IEEE

band users) be σ^2 and, finally, let M be the number of antenna elements. Assuming *perfect instantaneous* power control, the interference from a mobile within a given mobile's cell will arrive at cell site with same power P. Since mobiles in other cells are power controlled by their cell sites, the interference power from such mobiles, when active, at the desired user's cell site is given by [1]

$$I_{i_k} = P\left(\frac{r_{i_k}^{(k)}}{r_{i_k}^{(o)}}\right)^4 \frac{\|\alpha_{i_k}^{(o)}\|^2}{\|\alpha_{i_k}^{(k)}\|^2} = P \cdot \beta_{i_k}^2 \tag{1}$$

where $r_{i_k}^{(k)}$ is the distance from the i_k-th user in the k-th cell to its cell site, $\alpha_{i_k}^{(k)}$ is a zero mean complex Gaussian random variable that represents the corresponding amplitude fade along that path and combines both the Rayleigh fading and log-normal shadowing effects (i.e., $\|\alpha_{i_k}^{(k)}\|$ has a Rayleigh distribution whose mean square value $\mathcal{E}\{\|\alpha_{i_k}^{(k)}\|^2\}$ is log-normal; i.e., $10 \log_{10} \mathcal{E}\{\|\alpha_{i_k}^{(k)}\|^2\}$ is normally distributed with zero mean and variance σ_s^2), $r_{i_k}^{(o)}$ is the distance between the same i_k-th mobile in the k-th cell and the desired user's cell site (i.e., cell site o), and finally $\alpha_{i_k}^{(o)}$ is the corresponding amplitude fade. Note that in [1], only the effects of the log-normal shadowing is considered. Note also that since the mobile will be controlled by the cell site that has minimum attenuation $\beta_{i_k} \leq 1$ [1].

Fig. 1 shows a desired signal and interference signals from mobiles within cells and outer cells for both omnidirectional beams and directional beams. Clearly, directional beams reduce the interference power and boast the signal to interference-plus-noise ratio. To be able to form such beams, we need to estimate the array response vector, or the spatial signature, of the desired user mobile. Using this estimate of the array response vector, we can form a beam towards each mobile.

Assuming a narrowband signal model, the $M \times 1$ output of an array of M sensors at the cell site can be written as

$$x(t) = \sum_{i_o=1}^{N} \psi_{i_o}\sqrt{P} b_{i_o}\left(\left\lfloor\frac{t-\tau_{i_o}}{T}\right\rfloor\right) c_{i_o}(t-\tau_{i_o}) a_{i_o} + \sum_{k=1}^{K}\sum_{i_k=1}^{N} \psi_{i_k}\sqrt{P}\beta_{i_k} b_{i_k}\left(\left\lfloor\frac{t-\tau_{i_k}}{T}\right\rfloor\right) \cdot c_{i_k}(t-\tau_{i_k}) a_{i_k} + n(t) \tag{2}$$

where K is the number of interfering cells, a_{i_k} is the $M \times 1$ array response vector for signal arriving from the i_k-th mobile in the k-th cell and we assume that $a_{i_k}^* a_{i_k} = 1$, $c_{i_k}(t)$ is the code used by that user, $b_{i_k}(\cdot)$ is the bit of duration T, τ_{i_k} is the propagation delay, ψ_{i_k} is a Bernoulli variable with probability of success v that models the *voice activity* of the same user (i.e., a user will be talking with probability v), and n is the thermal noise vector with zero-mean and covariance

$$\mathcal{E}\{n(t)n^*(\tau)\} = \frac{\sigma^2}{M} I \qquad t = \tau \tag{3}$$

$$= 0 \qquad t \neq \tau. \tag{4}$$

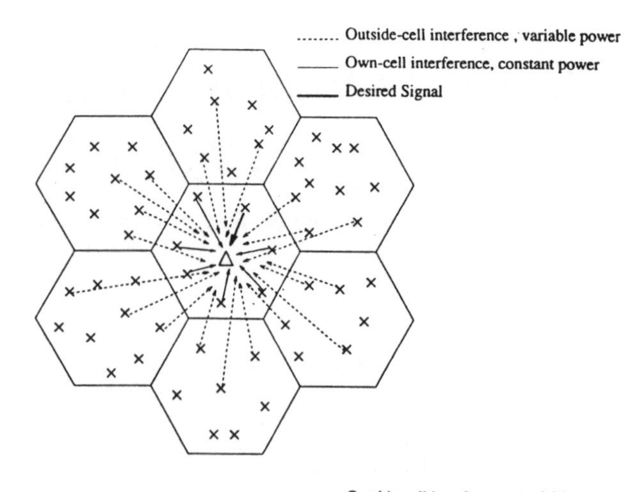

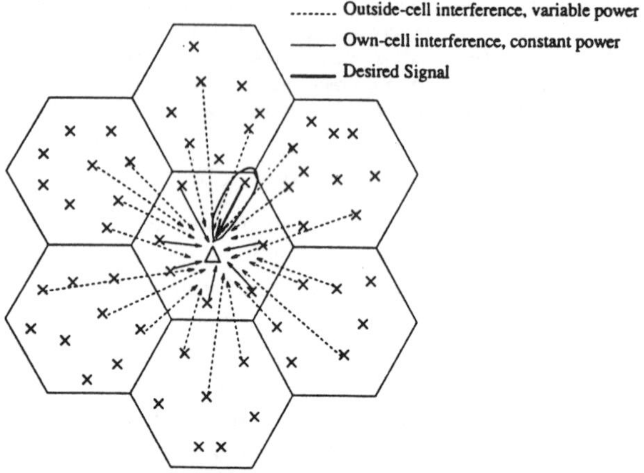

Fig. 1. Interference in uplink with and without beamforming.

Equation (3) implies that the noise is both *temporally* and *spatially* white. For the desired user, let a_o, τ_o, c_o, and $b_o(\cdot)$ be the array response vector, the time delay, the used code, and the transmitted bits, which are assumed to be i.i.d. binary random variables taking values ± 1 with equal probability, respectively. The antenna outputs are correlated with the desired user's code c_o to yield one sample vector per bit. Without loss of generality, assume that $\tau_o = 0$. The post-correlation signal vector for the desired user's l-th bit is given by

$$z_o(l) = \int_{t_1}^{t_2} x(t) c_o(t)\, dt = s_o(l) a_o + \sum_{i_0=2}^{N} \psi_{i_o} I_{i_o}(l) a_{i_o} + \sum_{k=1}^{K}\sum_{i_k=1}^{N} \psi_{i_k} I_{i_k}(l) a_{i_k} + n_T(t) \tag{5}$$

where $t_1 = (l-1)T$, $t_2 = lT$, and

$$s_o(l) = \int_{t_1}^{t_2} \sqrt{P} b_o\left(\left\lfloor\frac{t}{T}\right\rfloor\right) c_o(t) c_o(t)\, dt, \tag{6}$$

$$I_{i_o}(l) = \int_{t1}^{t2} \sqrt{P} b_{i_o}\left(\left\lfloor \frac{t - \tau_{i_o}}{T} \right\rfloor\right) c_{i_o}(t - \tau_{i_o}) c_o(t)\, dt, \quad (7)$$

$$I_{ik}(l) = \int_{t1}^{t2} \sqrt{P} \beta_{ik} b_{ik}\left(\left\lfloor \frac{t - \tau_{ik}}{T} \right\rfloor\right) c_{ik}(t - \tau_{ik}) c_o(t)\, dt, \quad (8)$$

$$n_T(l) = \int_{t1}^{t2} c_o(t) n(t)\, dt. \quad (9)$$

In order to combine the array outputs to estimate the signal from the desired mobile, we need to determine the array response vector for the wavefront arriving from this user. In general, in CDMA systems the number of users will far exceed the number of antennas. Therefore, subspace methods of direction-of-arrival estimation (e.g., MUSIC [13] and ESPRIT [14]) are not applicable. In [3], we showed that the array response vector of the desired mobile a_o can be estimated from the pre-correlation and post-correlation array covariances R_{xx} and $R_{z_o z_o}$, where

$$R_{xx} = \mathcal{E}\{xx^*\}$$

$$R_{z_o z_o} = \mathcal{E}\{z_o z_o^*\}$$

as the generalized principal eigenvector of the matrix pair $(R_{xx}, R_{z_o z_o})$. Using this estimate of a_o, the post-correlation antenna outputs are combined via beamforming to estimate the signal from the desired user. The decision variable, which is the output of the beamformer, is then given by

$$\begin{aligned} d_o(l) &= a_o^* z_o(l) \\ &= s_o(l) + n_1(l) + n_2(l) + n_T(l) \\ &= L\sqrt{P} b_o(l) + \sum_{i_o=2}^{N} \psi_{i_o} I_{i_o} a_o^* a_{i_o} \\ &\quad + \sum_{k=1}^{K} \sum_{i_k=1}^{N} \psi_{ik} I_{ik} a_o^* a_{ik} + a_o^* n_T(l). \end{aligned} \quad (10)$$

The first term $s_o(l)$ is due to signals from the desired user, the second term n_1 is due to interference from users within its own cell, the third term n_2 is due to interference from users outside the cell; both are zero mean, and n_T is due to the additive thermal noise, which is normal with zero mean and variance equal to Var $\{n_T\} = L\sigma^2/M$. Additionally, we assume that each user's code consists of a sequence of L i.i.d. binary random variables taking values ± 1 with equal probability. As noted in [15], with asynchronous transmission, random-sequence codes give approximately the same analytical results for nonrandomly chosen codes. Under this assumption and using the results in [15], we can show that the variances of n_1 and n_2 are given by

$$\text{Var}\{n_1\} = LP \sum_{i_o=2}^{N} \psi_{i_o} \|a_o^* a_{i_o}\|^2, \quad (11)$$

$$\text{Var}\{n_2\} = LP \sum_{k=1}^{K} \sum_{i_k=1}^{N} \psi_{ik} \beta_{ik}^2 \|a_o^* a_{ik}\|^2. \quad (12)$$

These variances are themselves random variables that depend on the voice activity of the users, their array response vectors, and fading and shadowing effects. The faded energy-per-bit to interference-plus-noise densities ratio can be written as

$$\frac{E_b}{N_o + I_o} = \frac{L}{\frac{\sigma^2}{MP} + I_1 + I_2} \quad (13)$$

where I_1 and I_2 are the interference-to-signal power ratios due to own cell and outer cell users respectively, and are given by

$$I_1 = \sum_{i_o=2}^{N} \psi_{i_o} \|a_o^* a_{i_o}\|^2, \quad (14)$$

$$I_2 = \sum_{k=1}^{K} \sum_{i_k=1}^{N} \psi_{ik} \beta_{ik}^2 \|a_o^* a_{ik}\|^2. \quad (15)$$

The probability of outage is defined as the probability of the bit error rate exceeding a certain threshold P_o required for acceptable performance. As noted in [1], with efficient modems and powerful convolutional codes, adequate performance (BER $< 10^{-3}$) is achieved with $E_b/(N_o + I_o) < 7$ dB. Let S be the $E_b/(N_o + I_o)$ value required to achieve the level of performance, then the outage probability is

$$\begin{aligned} P_{out} &= \Pr(\text{BER} > P_o) = \Pr\left(\frac{E_b}{N_o + I_o} < S\right) \\ &= \Pr\left(I_1 + I_2 > \frac{L}{S} - \frac{\sigma^2}{MP}\right). \end{aligned} \quad (16)$$

This expression gives the outage probability as a function of the random variables I_1 and I_2. The distribution of the random variables I_1 and I_2 depends on the number of active users, their relative distances, their array response vectors, array parameters, and fading and shadowing effects. The capacity of the system in terms of maximum cell loading can be found by finding the maximum N such that for the required BER, P_{out} will not exceed the present threshold. To obtain P_{out} as a function of N, we need to specify the array (i.e., the number of sensors, spacing between them, and their arrangement) to be able to find the distribution of $\|a_o^* a_{ik}\|^2$, and hence the distribution of I_1 and I_2.

To simplify the evaluation, we use the following first order approximation. As we pointed out earlier, the effect of forming a beam towards the desired user is to reduce the effective number of interferers to those mobiles that fall within the beam formed towards the desired mobile. Since the number of those inteferers is random, we approximate this effect by replacing the $\|a_o^* a_{ik}\|^2$ term in I_1 and I_2 by a Bernoulli random variable χ_{ik} that has a probability of success $B/2\pi$ where B is the effective bandwidth and is equal to $\mathcal{E}\{\|a_o^* a_{i_k}\|^2\}$. This random variable represents the *interference activity* of the users, i.e., a mobile will cause interference to the desired mobile if it falls within its beam. In this case, we can write I_1 and I_2 as

$$I_1 = \sum_{i_o=2}^{N} \psi_{i_o}\chi_{i_o} = \sum_{i_o=2}^{N} \phi_{i_o} \tag{17}$$

$$I_2 = \sum_{k=1}^{K}\sum_{i_k=1}^{N} \psi_{i_k}\chi_{i_k}\beta_{i_k}^2 = \sum_{k=1}^{K}\sum_{i_k=1}^{N} \phi_{i_k}\beta_{i_k}^2 \tag{18}$$

where $\phi_{i_k} = \psi_{i_k}\chi_{i_k}$ is a Bernoulli random variable with probability of success $\overline{v} = vB/2\pi$. The distribution of $f = \|\alpha_{i_k}^{(o)}\|/\|\alpha_{i_k}^{(k)}\|$ is given by [12]

$$\Pr(f < r) = \frac{1}{\sqrt{\pi}} \int_{-\infty}^{\infty} \frac{\exp(-u^2)}{1 + r^{-2}10^{-2\sigma_s u/10}} du. \tag{19}$$

For a large number of users, the random variable I_2 (interference due to $K \cdot N$ users) can be approximated by a Gaussian random variable with mean $\mu_i N$ and variance $\sigma_i^2 N$ that depend on $\overline{v}$, the degree of shadowing σ_s, and the number of interfering cells K. We have evaluated the mean and variance of I_2 using Monte Carlo integration considering only the first two tiers of interfering cells (i.e., $K = 18$) and these were found to be given by

$$\mu_i = 0.523\overline{v}, \tag{20}$$

$$\sigma_i^2 = 0.463\overline{v} - 0.274\overline{v}^2. \tag{21}$$

Also, the random variable I_1 has a binomial distribution with parameters $(N - 1, \overline{v})$. Let $L/S - \sigma^2/PM = \delta$. Since I_1, I_2, and all ϕ_{i_k} are independent, we can use the results in [1] to shows that

$$P_{out} = \sum_{k=0}^{N-1} \binom{N-1}{k} \overline{v}^k (1 - \overline{v})^{N-1-k} \cdot Q\left(\frac{\delta - k - \mu_i N}{\sqrt{\sigma_i^2 N}}\right) \tag{22}$$

where

$$Q(x) = \frac{1}{\sqrt{2\pi}} \int_x^{\infty} e^{-y^2/2} dy.$$

This equation gives the outage probability as a function of the number of mobiles per cell that can be supported. Note that this reduces to the result in [1] when no antenna arrays are used at the cell site. The results of evaluating (22) as a function of cell loading and beamwidth are shown in Fig. 3. Also, simulations to evaluate the accuracy of the above approximation are shown in Fig. 5 and discussed in Section IV.

III. Base-to-Mobile Link

Consider now the base-to-mobile link. We assume a similar scenario as in the uplink. With antenna array at the cell site, the cell site must also beamform on the downlink in order to effectively increase the system capacity. To be able to form such beams, the cell site needs to have an estimate of the transmit array response vector to each mobile. However, in the current standard, frequencies for the uplink and downlink differ by 45 MHz. In this case, the receive and transmit response vectors can be significantly different [16], [17]. Hence, reciprocity between uplink and downlink does not hold and the beamformer weights used for reception cannot be used for transmission. A method of performing transmission beamforming is the feedback method [18], [19], where training signals, or tones, are periodically transmitted from the cell site to all mobiles on the downlink. From the received signal information that the mobiles feedback to the cell site on uplink, it is possible to estimate the downlink spatial channel, and thus estimate the transmit array response vector.

All signals received at the mobile from the same base station will have propagated over the same path, hence they will experience the same fading and path loss. Therefore, we assume that cell site transmits the same power to all mobiles controlled by that cell site. With this assumption, the power of each signal arriving at the desired mobile from the k-th cell site is given by

$$P_k = P \frac{\|\alpha_k^{(o)}\|^2}{(r_k^{(o)})^4} = P \cdot \beta_k^2 \tag{23}$$

where $\alpha_k^{(o)}$ represents the fading and shadowing experience by all signals arriving at the desired mobile from the k-th cell site, and $r_k^{(o)}$ is the distance between the desired mobile from its cell site. As in [1], we assume that the power received by the mobile from its cell site is the largest among all other signals from other cell sites (otherwise the mobile would switch to the cell site whose received power is maximum). That is, we assume that

$$\beta_o < \beta_k \quad k = 1, \cdots, K. \tag{24}$$

Fig. 2 shows desired signal and interference powers seen by the desired mobile for both omni- and directional beams.

Assuming N users per cell randomly distributed around each cell site at varying ranges, we can write the received signal at the mobile of interest as

$$x_o(t) = \sum_{i_o=1}^{N} \psi_{i_o}\sqrt{P}\beta_o b_{i_o}\left(\left\lfloor\frac{t - \tau_{i_o}}{T}\right\rfloor\right) c_{i_o}(t - \tau_{i_o}) a_{i_o}^* a_o^{(o)} + \sum_{k=1}^{K}\sum_{i_k=1}^{K} \sigma_{i_k}\sqrt{P}\beta_k b_{i_k}\left(\left\lfloor\frac{t - \tau_{i_k}}{T}\right\rfloor\right) \cdot c_{i_k}(t - \tau_{i_k}) a_{i_k}^* a_o^{(k)} + n(t), \tag{25}$$

where $n(t)$ is the background noise received by the mobile, and $a_o^{(k)}$ is the transmit array response vector of the desired mobile as seen by the k-th cell site. All other notations remain the same as in the previous section. The mobile correlates the received signal by its code to yield

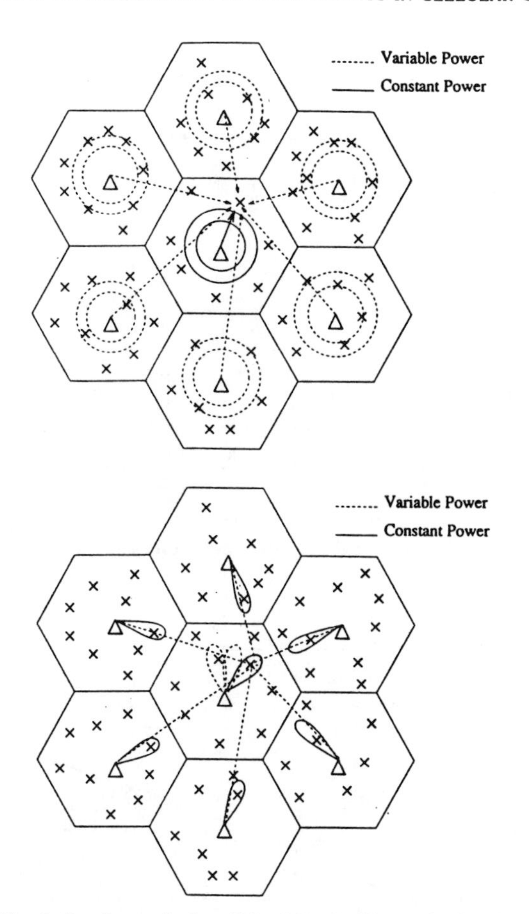

Fig. 2. Interference in downlink with and without beamforming.

the decision variable

$$d_o = s_o(l) + n_1(l) + n_2(l) + n_T(l)$$
$$= L\beta_o\sqrt{P}b_o(l) + \sum_{i_o=2}^{N} \psi_{i_o} I_{i_o} \boldsymbol{a}_{i_o}^* \boldsymbol{a}_o^{(o)}$$
$$+ \sum_{k=1}^{K} \sum_{i_k=1}^{N} \sigma_{i_k} I_{i_k} \boldsymbol{a}_{i_k}^* \boldsymbol{a}_o^{(k)} + n_T(l), \quad (26)$$

where I_{i_k} is defined as before but with β_k instead of β_{i_k} and

$$n_T(t) = \int_{t1}^{t2} c_o(t)n(t)\, dt. \quad (27)$$

As in the uplink, the first term s_o is due to the desired signal from the cell site to the desired user, the second term n_1 is due to interference from the same cell site into the desired user, which is zero mean, the third term n_2 is due to interference from other cell sites into the desired user, and n_T is due to the additive thermal noise and it is normal with zero mean and variance equal to Var $\{n_T\}$ = $L\sigma^2$

In the proposed CDMA standard IS-95, orthogonal codes (Walsh codes) are used on the downlink for all users within a cell, i.e., in the ideal case (no multipath) there is no cross-correlation between those signals and the interference due to signals from its own cell site n_1 is zero. However, we assume here that there will be cross-correlation between those signals, which represents a worst case. Hence, similar to the uplink case, we can show that the variance of n_1 and n_2 is given by

$$\text{Var}\{n_1\} = LP_o \sum_{i_o=2}^{N} \psi_{i_o} \|\boldsymbol{a}_{i_o}^* \boldsymbol{a}_o^{(o)}\|^2 \quad (28)$$

$$\text{Var}\{n_2\} = L \sum_{k=1}^{K} P_k \sum_{i_k=1}^{N} \psi_{i_k} \|\boldsymbol{a}_{i_k}^* \boldsymbol{a}_o^{(k)}\|^2, \quad (29)$$

and the energy-per-bit to interference-plus-noise densities ratio can be written as

$$\frac{E_b}{N_o + I_o} = \frac{L}{\frac{\sigma^2}{P_o} + G_1 + G_2}, \quad (30)$$

where G_1 and G_2 are the interference-to-signal power ratios due to their own cell and outer cell signals, respectively, and are given by

$$G_1 = \sum_{i_o=2}^{N} \psi_{i_o} \|\boldsymbol{a}_{i_o}^* \boldsymbol{a}_o^{(o)}\|^2 \quad (31)$$

$$G_2 = \sum_{k=1}^{K} \sum_{i_k=1}^{N} \psi_{i_k} \frac{P_k}{P_o} \|\boldsymbol{a}_{i_k}^* \boldsymbol{a}_o^{(k)}\|^2. \quad (32)$$

The corresponding outage probability is then given by

$$P_{out} = \Pr(\text{BER} > P_o) = \Pr\left(\frac{\sigma^2}{P_o} + G_1 + G_2 > \frac{L}{S}\right). \quad (33)$$

Unlike the uplink case, the distribution of G_2 does not yield itself to analysis (here we have only K independent fading variables, while in the uplink case we had $K \cdot N$ fading variables, and when N is large we were able to model I_2 as Gaussian). Therefore, we resort to simulations to estimate P_{out} as a function of cell loading and number of sensors, from which we can obtain the system capacity (maximum cell loading) as a function of cell loading and number of sensors. The results of these simulations are shown in Figs. 6-9 and are discussed in Section IV.

IV. Simulation and Numerical Results

In all of our simulations and numerical results, we consider only the first two tiers of interfering cells, which means that $K = 18$ cells. We assume that the voice activity factor v is 0.375. We assume that for adequate performance, the required BER is 10^{-3} which corresponds to $E_b/(N_o + I_o)$ of 7 dB. We also assume that the processing gain L is 128. Finally, we assume that σ_s is 8 dB.

For the uplink, the outage probability was computed using (22). The results are summarized in Fig. 3. From this figure, it is shown that by using antenna array to form narrow beams towards the desired mobiles, a many-fold increase in system capacity can be obtained. For example,

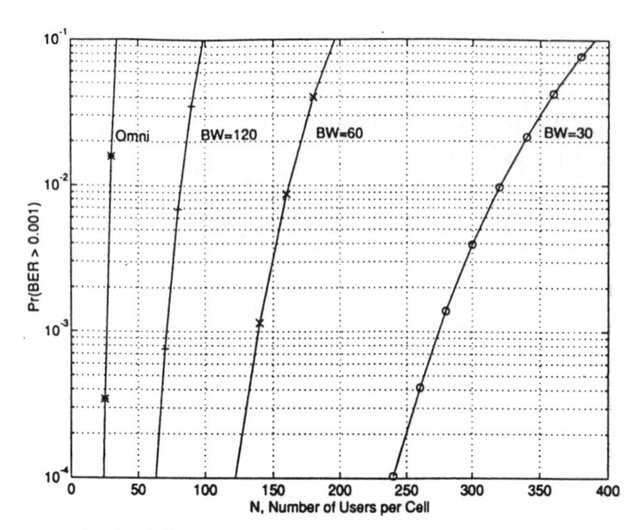

Fig. 3. Uplink outage probability as a function of beamwidth.

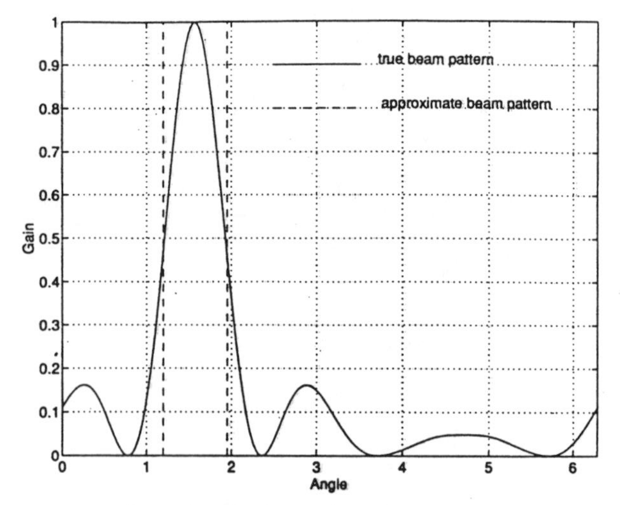

Fig. 4. Actual and approximate beam patterns.

for 0.01 outage probability, the uplink system capacity goes up from 31 users per cell for the single antenna case to about 320 users per cell if we use an array, such that we have beams with beamwidth of 30°. Also, to evaluate the accuracy of the approximation we used, we simulated the system (based on (14)–(16)) with a cell site circular antenna array with nine elements and radius equal to $\lambda/2$ corresponding to half-power beamwidth of 42°. In fact, the beamwidth was taken to be slightly more than 42° to account for the interference energy picked up through the side lobes of the array pattern. Fig. 4 shows the actual array pattern versus the approximate pattern. In Fig. 5, we plot both the outage probability computed from (22) and from simulations, which indicate good agreement between the simulation results and the approximation.

For the downlink, results for the outage probability were obtained by simulations based on (31)–(33). A circular array with one, five, and seven elements and $\lambda/2$ spacing was used in the simulation. For all other parameter values above, the histogram of $E_b/(N_o + I_o)$ is obtained for each M and N value from 20,000 runs. In each run, 19 Rayleigh random variables with mean square value that have a log-normal distribution with $\sigma_s = 8$ dB are generated, and the maximum of these is taken to be that of the desired mobile's cell site. Also, we assume that the mobile is positioned on the boundary between cells, which represents a worst case situation. Some of the generated histograms of $G_1 + G_2$ are shown in Figs. 6, 7, and 8. The generated histograms are used to estimate the probability of outage as function of cell loading and number of sensors. These results are summarized in Fig. 9, which also shows a many fold increase in capacity by using antennas to form narrow beams towards the desired user. Note that as we mentioned before, if orthogonal codes are used on the downlink and in the case of no multipath, interference will be primarily due to outside cell interference and the corresponding cell loading N at which outage will occur will be larger.

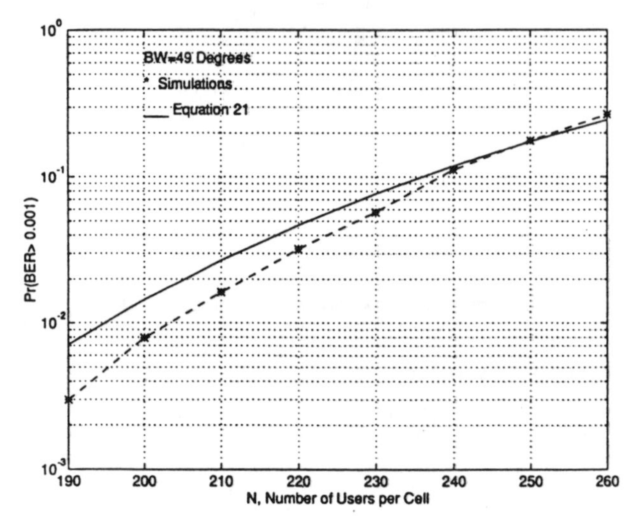

Fig. 5. Uplink outage probability: simulation vs. approximation.

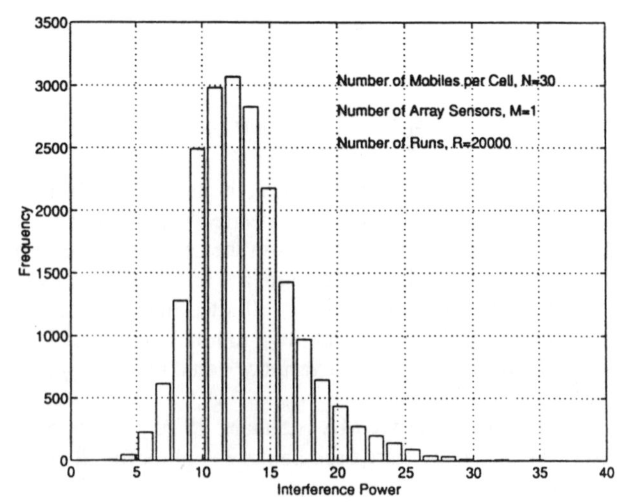

Fig. 6. Histograms of $G_1 + G_2$ for $M = 1$, $N = 30$.

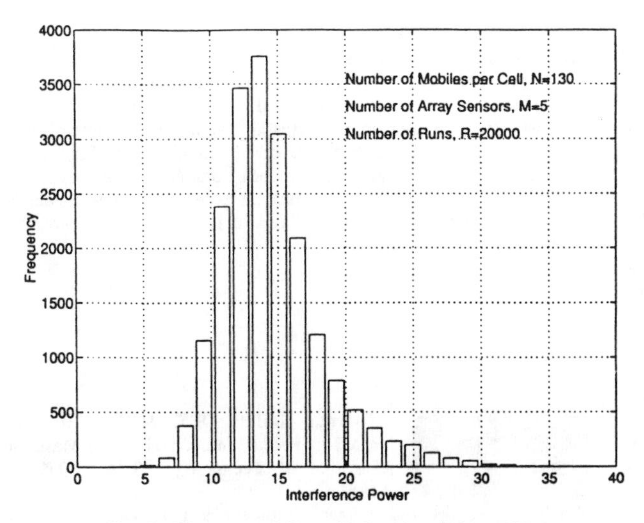

Fig. 7. Histogram of $G_1 + G_2$ for $M = 5, N = 130$.

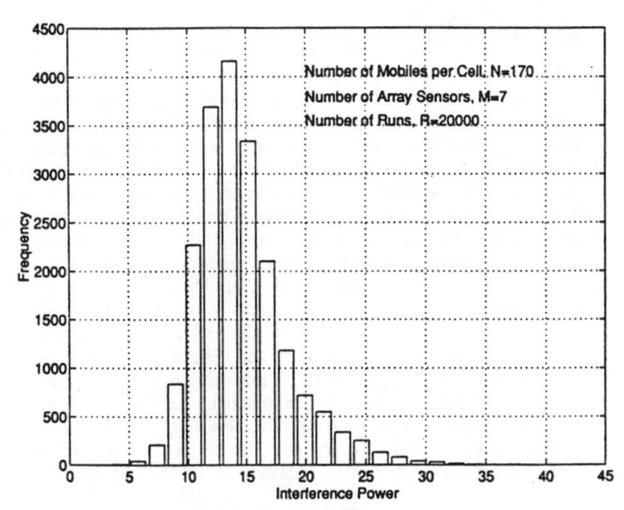

Fig. 8. Histogram of $G_1 + G_2$ for $M = 7, N = 170$

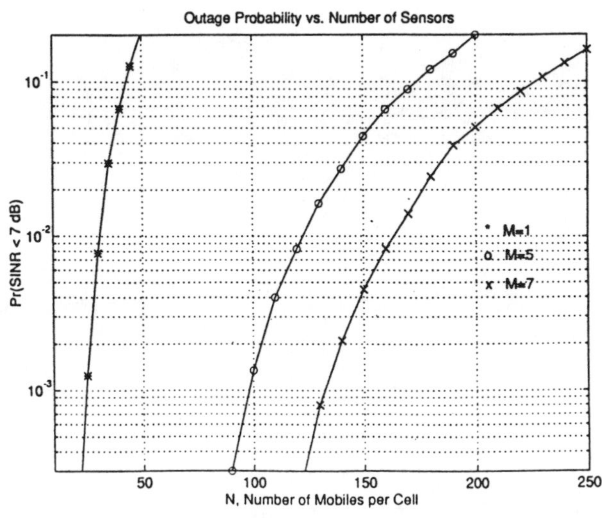

Fig. 9. Downlink outage probability vs. number of array sensors.

V. Conclusions

We have studied the capacity improvement for CDMA cellular communications systems with base-station antenna array for both uplink and downlink. The outage probability was evaluated as a function of cell loading, array parameters, fading and shadowing effects, and voice activity. Our analytical and simulation results show that there can be a substantial increase in system capacity by incorporating antenna arrays at the base-station. Our approach uses spatial processing to determine the dynamic spatial wavefront at the cell site and constructs a robust beamformer. Our model, used in this paper, does not include the effects of multipath which will be presented in a different paper.

References

[1] K. S. Gilhousen, I. M. Jacobs, R. Padovani, A. Viterbi, L. A. Weaver, and C. Wheatly, "On the capacity of a cellular CDMA system," *IEEE Trans. Veh. Technol.*, vol. 40, no. 2, pp. 303-312, May 1991.

[2] A. M. Viterbi and A. J. Viterbi, "Erlang capacity of a power controlled CDMA system," *IEEE J. Select. Areas Commun.*, vol. 11, no. 6, pp. 892-900, Aug. 1993.

[3] B. Suard, A. Naguib, G. Xu, and A. Paulraj, "Performance analysis of CDMA mobile communication systems using antenna arrays," in *Proc. ICASSP'93*, vol. VI, Minneapolis, MN, pp. 153-156, April 1993.

[4] A. F. Naguib, A. Paulraj, and T. Kailath, "Capacity improvement with base-station antenna arrays in cellular CDMA," in *Proc. 27th Asilomar Conf. on Signals, Systems and Computers*, vol. II, Pacific Grove, GA, pp. 1437-1441, Nov. 1993.

[5] J. C. Liberti and T. S. Rappaport, "Reverse channel performance improvements in CDMA cellular communication systems employing adaptive antennas," in *Proc. GLOBECOM'93*, vol. I, pp. 42-47, 1993.

[6] V. Weerackody, "Diversity for the direct-sequence spread spectrum system using multiple transmit antennas," in *Proc. ICC'93*, vol. III, Geneva, Switzerland, May 1993.

[7] P. Balaban and J. Salz, "Optimum diversity combining and equalization in data transmission with application to cellular mobile radio—Part I: Theoretical considerations," *IEEE Trans. Commun.*, vol. 40, no. 5, pp. 885-894, May 1992.

[8] S. C. Swales, M. A. Beach, D. J. Edwards, and J. P. McGeehn, "The performance enhancement of multibeam adaptive base station antennas for cellular land mobile radio systems," *IEEE Trans. Veh. Technol.*, vol. 39, no. 1, pp. 56-67, Feb. 1990.

[9] J. Winters, J. Saltz, and R. Gitlin, "The capacity of wireless communication systems can be substantially increased by the use of antenna diversity," in *Proc. Conf. on Information Science and Systems*, vol. II, Princeton, NJ, pp. 853-858, Oct. 1992.

[10] S. Anderson, M. Millnert, M. Viberg, and B. Wahlberg, "An adaptive array for mobile communication systems," *IEEE Trans. Veh. Technol.*, vol. 40, no. 1, pp. 230-236, Feb. 1991.

[11] R. Kohno, H. Imai, and S. Pasupathy, "Combination of an adaptive antenna array and a canceller of interference for direct-sequence spread spectrum multiple-access system," *IEEE J. Select. Areas Commun.*, vol. 8, no. 4, pp. 675-682, May 1990.

[12] R. C. French, "The effect of fading and shadowing on channel reuse in mobile radio," *IEEE Trans. Veh. Technol.*, vol. VT-28, no. 8, pp. 171-181, Aug. 1979.

[13] R. O. Schmidt, "A signal subspace approach to multiple-emitter location and spectral estimation," Ph.D. dissertation, Stanford Univ., Stanford, CA 94305, 1981.

[14] A. Paulraj, R. Roy, and T. Kailath, "Estimation of signal parameters by rotational invariance techniques (ESPRIT)," in *Proc. of 19th Asilomar Conf. on Circuits, Systems and Comp.*, 1985.

[15] W.-P. Yung, "Direct sequence spread-spectrum code-division-multiple access cellular systems in Rayleigh fading and log-normal shadowing channel," in *Proc. ICC'91*, vol. II, pp. 871-876, 1991.

[16] G. Xu, H. Liu, W. Vogel, H. Lin, S. Jeng, and G. Torrence, "Ex-

perimental studies of space-division-multiple-access schemes for spectral efficient wireless communications," submitted to SuperCom/ICC'94, May 1994.

[17] J. H. Winters, "Signal aquisition and tracking with adaptive arrays in wireless systems," in *Proc. 43rd Veh. Technol. Conf.*, vol. I, pp. 85–88, Nov. 1993.

[18] D. Gerlach and A. Paulraj, "Base-station transmitting antenna arrays with mobile to base feedback," in *Proc. 27th Asilomar Conf. on Signals, Systems and Computers*, Pacific Grove, CA, pp. 1432–1436, Nov. 1993.

[19] Y. Akaiwa, "Antenna selection diversity for framed digital signal transmission in mobile radio channel," in *Proc. VTC'89*, vol. I, pp. 470–473, 1989.

Ayman F. Naguib (S'94) was born on August 21, 1964 in Cairo, Egypt. He received the B.S. degree with honors and the M.S. degree in electrical engineering from Cairo University, Cairo, Egypt, in 1987 and 1990, respectively. He received the M.S. degree in statistics from Stanford University, Stanford, CA, in 1993. He is currently working towards the Ph.D. degree in electrical engineering at Stanford University, Stanford, CA.

From 1987 to 1989, he spent his military service at the Signal Processing Laboratory, Military Technical College, Cairo, Egypt. From 1989 to 1990, he was employed with Cairo University as a Research and Teaching Assistant in the Communication Theory Group. Since 1990, he has been a Research Assistant in the Information Systems Laboratory, Stanford University. His current research interests are in the areas of digital and mobile communications.

Arogyaswami Paulraj (SM'85–F'91) was educated at the Naval Engineering College, India, and at the Indian Institute of Technology, New Delhi (Ph.D. 1973).

A large part of his career to date has been spent in research laboratories in India where he supervised the development of several electronic systems. His contributions include a sonar receiver 1973–74, a surface ship sonar 1976–83, a parallel computer 1988–91, and telecommunications systems. He has held visiting appointments at several universities—Indian Institute of Technology, New Delhi, 1973–74, Loughborough University of Technology, U.K., 1974–75, Stanford University, CA, USA, 1983–86. His research has spanned several disciplines, emphasizing estimation theory, sensor signal processing, antenna array processing, parallel computer architectures/algorithms and communication systems. He is currently a Professor of Electrical Engineering at Stanford University working in the area of mobile communications.

He is the author of about 90 research papers and holds several patents and has won a number of national awards in India for his contributions to technology development.

Thomas Kailath (S'57–M'62–F'70) was educated in Poona, India, and at the Massachusetts Institute of Technology (S.M., 1959; Sc.D., 1961).

From October 1961 to December 1962, he worked at the Jet Propulsion Laboratories, Pasadena, CA, where he also taught part-time at the California Institute of Technology. He then came to Stanford University, where he served as Director of the Information Systems Laboratory from 1971 through 1980, as Associate Department Chairman from 1981 to 1987, and currently holds the Hitachi America Professorship in Engineering. He has held short-term appointments at several institutions around the world. He has worked in a number of areas including information theory, communications, computation, control, signal processing, VLSI design, statistics, linear algebra and operator theory; his recent interests include applications of signal processing, computation and control to problems in semiconductor manufacturing and wireless communications. He is the author of *Linear Systems*, Prentice Hall, 1980, and *Lectures on Wiener and Kalman Filtering*, Springer-Verlag, 1981.

Dr. Kailath has held Guggenheim, Churchill and Royal Society fellowships, among others, and received awards from the IEEE Information Theory Society and the American Control Council, in addition to the Technical Achievement and Society Awards of the IEEE Signal Processing Society. He served as President of the IEEE Information Theory Society in 1975, and has been awarded honorary doctorates by Linköping University, Sweden, and by Strathclyde University, Scotland. He is a Fellow of the Institute of Mathematical Statistics and a member of the National Academy of Engineering and the American Academy of Arts and Sciences.